CHEMICAL
ENGINEERING DESIGN

CHEMICAL ENGINEERING DESIGN

Fifth edition

RAY SINNOTT

GAVIN TOWLER

ELSEVIER

AMSTERDAM • BOSTON • HEIDELBERG • LONDON
NEW YORK • OXFORD • PARIS • SAN DIEGO
SAN FRANCISCO • SINGAPORE • SYDNEY • TOKYO

Butterworth-Heinemann is an imprint of Elsevier

Butterworth-Heinemann is an imprint of Elsevier
The Boulevard, Langford Lane, Kidlington, Oxford, OX5 1GB
30 Corporate Drive, Suite 400, Burlington, MA 01803, USA

Fifth edition 2009
Reprinted 2010, 2011

British Library Cataloguing-in-Publication Data
A catalogue record for this book is available from the British Library

Library of Congress Cataloguing-in-Publication Data
A catalog record for this book is available from the Library of Congress

ISBN: 978-0-7506-8551-1

For information on all Butterworth-Heinemann publications visit our Web site at www.elsevierdirect.com

Printed and bound in China
11 12 13 14   8 7 6 5 4 3

Working together to grow
libraries in developing countries

www.elsevier.com | www.bookaid.org | www.sabre.org

ELSEVIER   BOOK AID
            International   Sabre Foundation

Ray Sinnott began his career in design and development with several major companies, including DuPont and John Brown. He later joined the Chemical Engineering Department at the University of Wales, Swansea, UK, publishing the first edition of Chemical Engineering Design in 1983. He is a Chartered Engineer, Eur. Ing. and Fellow of the Institute of Chemical Engineers.

Gavin Towler is Director of Development at UOP LLC. Gavin is accountable for managing technology development and delivery for UOP's businesses in oil refining, petrochemicals, gas processing, adsorbents, catalysts and renewable fuels and chemicals. As adjunct professor at Northwestern University, he teaches the final year chemical engineering design classes. He is a Chartered Engineer and Fellow of the Institute of Chemical Engineers.

CONTENTS

3 FUNDAMENTALS OF ENERGY BALANCES AND ENERGY UTILIZATION

4 FLOW-SHEETING

5 PIPING AND INSTRUMENTATION

6 COSTING AND PROJECT EVALUATION

7 MATERIALS OF CONSTRUCTION

8   DESIGN INFORMATION AND DATA

9   SAFETY AND LOSS PREVENTION

10 EQUIPMENT SELECTION, SPECIFICATION AND DESIGN

11 SEPARATION COLUMNS (DISTILLATION, ABSORPTION AND EXTRACTION)

12 HEAT-TRANSFER EQUIPMENT

13 MECHANICAL DESIGN OF PROCESS EQUIPMENT

14 GENERAL SITE CONSIDERATIONS

APPENDICES

Preface

When I wrote the preface to the fourth edition I said that unless a suitable co-author could be found, the fourth was likely to be the last edition of Chemical Engineering Design.

I was surprised and pleased when Gavin Towler approached the publishers with a view to carrying on the volume and, also, preparing a new version for the American market. I did not think a co-author would be found who had the necessary combination of engineering background, academic contacts and, above all, time. Fortunately, Gavin meets all these criteria and has produced an excellent fifth edition. In particular, he has brought the treatment of flow-sheeting, safety and costing up-to-date, reflecting current design practice. He has done this whilst maintaining the basic style of the book and my aim to produce a book that would be used. Thank you Gavin.

EUR. ING. R. K. SINNOTT

The most useful book I owned as an undergraduate was "*Coulson and Richardson's Chemical Engineering, Volume 6: Chemical Engineering Design*" by Ray Sinnott. This book not only served as a manual for how to complete my design project, but was also the first place I would look for simple explanations, quick design methods and practical equipment details of every facet of chemical engineering. Throughout my career in process synthesis, design and development I have found Ray's book to be a useful source of information. I have acquired several other design textbooks and though each has its merits, none is so comprehensive or practical. As an industrial practitioner of design I came to appreciate the emphasis that Ray gave to subjects such as instrumentation, safety, materials selection and mechanical design, which are often neglected in the more theoretical books written by academics.

Shortly after I began teaching design at Northwestern University, I contacted Elsevier to ask if Ray would consider writing a modified version of his book aimed specifically at a North American audience. The alterations that I proposed were to change the references from British codes and standards to the appropriate American design codes, add a discussion of U.S. and Canadian safety and environmental legislation, include more information on computer-aided design and make some minor changes in terminology. Ray had recently decided to retire and did not want to produce another edition; however, we agreed that I would take on the project and the resulting book was published in 2008 as "*Chemical Engineering Design: Principles, Practice and Economics of Plant and Process Design*". In developing the American edition I kept to Ray's original format and style, while updating the material and drawing on my own experience to provide additional examples.

This 5th edition of Volume 6 in the Coulson and Richardson series includes some of the updated material that was introduced in the American edition, as well as other additions that I thought would enhance the book. The main changes are discussed below, and I hope that educators, students and practitioners will find the new content valuable.

Most industrial process design is now carried out using commercial design software. Extensive reference has been made to commercial process and equipment design software throughout the book and new examples with screen shots from various programs have been added. Many commercial software vendors provide licenses of their software for educational purposes at nominal fees. I strongly recommend that students should be introduced to commercial software as early and often as possible. The use of academic design and costing software should be discouraged. Academic programs usually lack the quality control and technical support required by industry, and the student is unlikely to use such software after graduation. Detailed examples of the use of computer tools in process simulation, costing and detailed design of distillation columns and heat exchangers have been added to Chapters 4, 6 and 12. All computer aided design tools must be used with some discretion and engineering judgement on the part of the designer. This judgement mainly comes from experience, but I have followed Ray's philosophy of trying to provide helpful tips on how to best use computer tools.

Chemical engineers work in a very diverse set of industries and many of these industries have their own design conventions and specialized equipment. I have attempted to broaden the range of process industries represented in the examples and problems, but where space or my lack of expertise in the subject has limited coverage of a particular topic, references to design methods available in the general literature are provided. The treatment of unit operations in Chapter 10 has been expanded to include more separation processes practised in gas processing, fine chemicals and pharmaceuticals manufacture, with new sections on adsorption, membrane separations, chromatography and ion exchange. New example design projects from a range of process industries have been added in Appendix E.

Standards and codes of practice are an essential part of engineering. There have been substantial changes in the British codes and standards since the 4th edition, as older British standards have been replaced by common European standards. The references to design codes have been updated to reflect these changes. Although this edition is written primarily for a British and European audience, the book is widely used internationally, and in some cases I have also included references to American standards where these are the most commonly used worldwide. A discussion of British and European safety and environmental legislation has been added to chapters 9 and 14; similar information for the U.S.A. and Canada is given in the American edition. The section on safety in chapter 9 has been significantly expanded. Most chemical engineers now work in an international environment and many will work in several countries during their career. The design engineer should follow corporate policy or obtain legal advice on which codes, standards and laws apply locally, and should always refer to the original source references of laws, standards and codes of practice, as they are updated frequently.

The treatment of costing and process economics in Chapter 6 has been updated and the cost correlations reflect recent price data rather than index updates of older data. Most of the costs have been given in U.S. dollars on a U.S. Gulf Coast basis, as this was the basis of the source data and most international engineering companies develop costs in U.S. dollars. Examples and problems have also been given in Euros and British Pounds Sterling and the section on converting prices from one location basis to another has been expanded. Where possible, the terminology used in the international engineering and construction industry has been used. All the examples are given in metric units, but some also use the American conventional units for illustrative purposes, as it is important for students to learn to convert data from American sources.

I have continued to follow Ray's model of describing the tools and methods that are most widely used in industrial process design and deliberately avoiding idealized conceptual methods developed by researchers that have not yet gained wide currency in industry. The reader can find good descriptions of these methods in the research literature and in more academic textbooks. A short section on optimization has been added to chapter 1, and several of the examples and problems have been modified to illustrate how experienced industrial designers optimize their designs.

In the preface to the 1st edition, Ray wrote: *"The art and practice of design cannot be learned from books. The intuition and judgement necessary to apply theory to practice will come only from practical experience. I trust that this book will give its readers a modest start on that road."* I certainly got my start in design using Ray's book and I hope that this new edition will prove as useful to future readers.

Gavin Towler

How to Use This Book

This book has been written primarily for students on undergraduate courses in Chemical Engineering and has particular relevance to their design projects. It should also be of interest to new graduates working in industry who find they need to broaden their knowledge of unit operations and design. Some of the earlier chapters of the book can also be used in introductory chemical engineering classes and by other disciplines in the chemical and process industries.

As a Design Course Textbook

Chapters 1 to 9 and 14 cover the basic material for a course on process design and include an explanation of the design method, including considerations of safety, costing, and materials selection. Chapters 2, 3 and 8 contain a lot of background material that should have been covered in earlier courses and can be quickly skimmed as a reminder. If time is short, chapters 4, 6 and 9 deserve the most emphasis. Chapters 10 to 13 cover equipment selection and design, including mechanical aspects of equipment design. These important subjects are often neglected in the chemical engineering curriculum. The equipment chapters can be used as the basis for a second course in design or as supplementary material in a process design class.

As an Introductory Chemical Engineering Textbook

The material in Chapters 1, 2, 3 and 6 does not require any prior knowledge of chemical engineering and can be used as an introductory course in chemical engineering. Much of the material in chapters 7, 9, 10 and 14 could also be used in an introductory class. There is much to be said for introducing design at an early point in the chemical engineering curriculum, as it helps the students have a better appreciation of the purpose of their other required classes, and sets the context for the rest of the syllabus. Students starting chemical engineering typically find the practical applications of the subject far more fascinating than the dry mathematics they are usually fed. An appreciation of economics, optimization and equipment design can dramatically improve a student's performance in other chemical engineering classes. If the book is used in an introductory class, then it can be referred to throughout the curriculum as a guide to design methods.

Supplementary Material

Many of the calculations described in the book can be performed using spreadsheets. Templates of spreadsheet calculations and equipment specification sheets are available in Microsoft Excel format on-line and can be downloaded from http://Elsevierdirect.com/companions.

Additional supplementary material, including Microsoft PowerPoint presentations to support most of the chapters, can be downloaded from a restricted site for instructors http://textbooks.elsevier.com.

Acknowledgements

As in my prefaces to the earlier editions of this book, I would like to acknowledge my debt to those colleagues and teachers who have assisted me in a varied career as a professional engineer. I would particularly like to thank Professor J. F. Richardson for his help and encouragement with earlier editions of this book. Also, my wife, Muriel, for her help with the typescripts of the earlier editions.

EUR. ING. R. K. SINNOTT
Coed-y-bryn, Wales

I would like to thank the many colleagues at UOP and elsewhere who have worked with me, shared their experience, and taught me all that I know about design. Particular thanks are due to Dr. Rajeev Gautam for allowing me to pursue this project and to Dick Conser, Peg Stine and Dr. Andy Zarchy for the time they spent reviewing my additions to Ray's book and approving the use of examples and figures drawn from UOP process technology. Ray has provided me with many thoughtful review comments and suggestions and I am of course very grateful that he allowed me to take on this edition and gave me complete freedom to make the changes I felt were necessary. My regular job at UOP keeps me very busy and I worked on this book in evenings and weekends, so it would not have been possible without the love, support and understanding of my wife, Caroline, and our children Miranda, Jimmy and Johnathan.

GAVIN P. TOWLER
Inverness, Illinois

Material from the ASME Boiler and Pressure Vessel Code is reproduced with permission of ASME International, Three Park Avenue, New York NY 10016. Material from the API Recommended Practices is reproduced with permission of the American Petroleum Institute, 1220 L Street, NW Washington, DC 20005. Material from British Standards is reproduced by permission of the British Standards Institution, 389 Chiswick High Road, London, W4 4AL, United Kingdom. Complete copies of the codes and standards can be obtained from these organizations.

We are grateful to Aspen Technology Inc. and Honeywell Inc. for permission to include the screen shots that were generated using their software to illustrate the process simulation and costing examples. Laurie Wang of Honeywell also provided valuable review comments. The material safety data sheet in Appendix I is reproduced with permission of Fischer Scientific Inc. Aspen Plus®, Aspen Kbase, Aspen ICARUS and all other AspenTech product names or logos are trademarks or registered trademarks of Aspen Technology Inc. or its subsidiaries in the United States and/or in other countries. All rights reserved.

The supplementary material contains images of processes and equipment from many sources. We would like to thank the following companies for permission to use these images: Alfa-Laval, ANSYS, Aspen Technology, Bete Nozzle, Bos-Hatten Inc., Chemineer, Dresser, Dresser-Rand, Enardo Inc., Honeywell, Komax Inc., Riggins Company, Tyco Flow Control Inc., United Valve Inc., UOP LLC, and The Valve Manufacturer's Association.

Jonathan Simpson of Elsevier was instrumental in launching and directing this project and provided guidance and editorial support throughout the development of the book. We would also like to thank Renata Corbani for her excellent work in assembling the book and managing the production process.

1 INTRODUCTION TO DESIGN

Chapter Contents

Key Learning Objectives

- How design projects are carried out and documented in industry
- Why engineers in industry use codes and standards and build margins into their designs
- How to improve a design using optimization methods
- Why experienced design engineers very rarely use rigorous optimization methods in industrial practice

1.1. INTRODUCTION

This chapter is an introduction to the nature and methodology of the design process, and its application to the design of chemical manufacturing processes.

1.2. NATURE OF DESIGN

This section is a general discussion of the design process. The subject of this book is chemical engineering design, but the methodology described in this section applies equally to other branches of engineering.

Chemical engineering has consistently been one of the highest-paid engineering professions. There is a demand for chemical engineers in many sectors of industry, including the traditional processing industries: chemicals, polymers, fuels, foods, pharmaceuticals and paper, as well as other sectors such as electronic materials and devices, consumer products, mining and metals extraction, biomedical implants, and power generation.

The reason that companies in such a diverse range of industries value chemical engineers so highly is the following:

Starting from a vaguely defined problem statement such as a customer need or a set of experimental results, chemical engineers can develop an understanding of the important underlying physical science relevant to the problem and use this understanding to create a plan of action and set of detailed specifications, which if implemented, will lead to a predicted financial outcome.

The creation of plans and specifications and the prediction of the financial outcome if the plans were implemented is the activity of chemical engineering design.

Design is a creative activity, and as such can be one of the most rewarding and satisfying activities undertaken by an engineer. The design does not exist at the start of the project. The designer begins with a specific objective or customer need in mind, and by developing and evaluating possible designs, arrives at the best way of achieving that objective; be it a better chair, a new bridge, or for the chemical engineer, a new chemical product or production process.

When considering possible ways of achieving the objective, the designer will be constrained by many factors, which will narrow down the number of possible designs. There will rarely be just one possible solution to the problem, just one design. Several alternative ways of meeting the objective will normally be possible, even several best designs, depending on the nature of the constraints.

These constraints on the possible solutions to a design problem arise in many ways. Some constraints will be fixed and invariable, such as those that arise from physical laws, government regulations, and standards. Others will be less rigid, and can be relaxed by the designer as part of the general strategy for seeking the best design. The constraints that are outside the designer's influence can be termed the external constraints. These set the outer boundary of possible designs, as shown in Figure 1.1. Within this boundary there will be a number of plausible designs bounded by the other constraints, the internal constraints, over which the designer has some control; such as, choice of process, choice of process conditions, materials, and equipment.

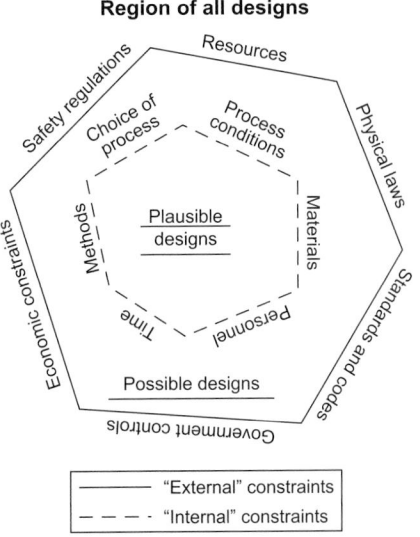

Figure 1.1.   Design constraints.

Economic considerations are obviously a major constraint on any engineering design: plants must make a profit. Process costing and economics are discussed in Chapter 6.

Time will also be a constraint. The time available for completion of a design will usually limit the number of alternative designs that can be considered.

The stages in the development of a design, from the initial identification of the objective to the final design, are shown diagrammatically in Figure 1.2. Each stage is discussed in the following sections.

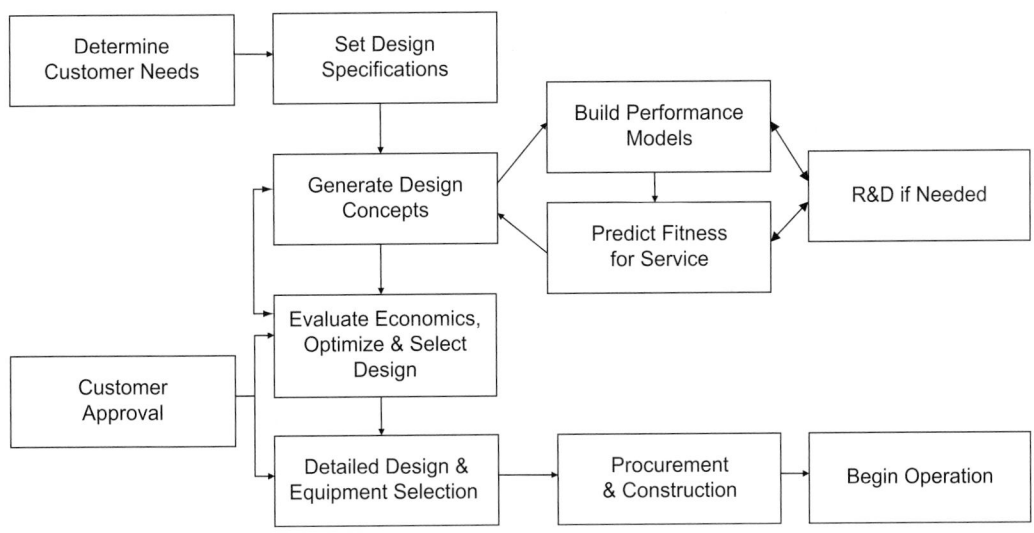

Figure 1.2.   The design process.

Figure 1.2 shows design as an iterative procedure; as the design develops the designer will be aware of more possibilities and more constraints, and will be constantly seeking new data and ideas, and evaluating possible design solutions.

1.2.1. The Design Objective (The Need)

All design starts with a perceived need. In the design of a chemical process, the need is the public need for the product, creating a commercial opportunity, as foreseen by the sales and marketing organization. Within this overall objective the designer will recognize sub-objectives; the requirements of the various units that make up the overall process.

Before starting work, the designer should obtain as complete, and as unambiguous, a statement of the requirements as possible. If the requirement (need) arises from outside the design group, from a customer or from another department, then the designer will have to elucidate the real requirements through discussion. It is important to distinguish between the needs that are 'must haves' and those that are 'should haves'. The 'should haves' are those parts of the initial specification that may be thought desirable, but that can be relaxed if required as the design develops. For example, a particular product specification may be considered desirable by the sales department, but may be difficult and costly to obtain, and some relaxation of the specification may be possible, producing a saleable but cheaper product. Whenever possible, the designer should always question the design requirements (the project and equipment specifications) and keep them under review as the design progresses. It is important for the design engineer to work closely with the sales or marketing department or with the customer directly, to have as clear as possible an understanding of the customer's needs.

When writing specifications for others, such as for the mechanical design or purchase of a piece of equipment, the design engineer should be aware of the restrictions (constraints) that are being placed on other designers. A well-thought-out, comprehensive, specification of the requirements for a piece of equipment defines the external constraints within which the other designers must work.

1.2.2. Setting the Design Basis

The most important step in starting a process design is translating the customer need into a design basis. The design basis is a more precise statement of the problem that is to be solved. It will normally include the production rate and purity specifications of the main product, together with information on constraints that will influence the design such as:

1. The system of units to be used.
2. The national, local or company design codes that must be followed.
3. Details of raw materials that are available.
4. Information on potential sites where the plant might be located, including climate data, seismic conditions and infrastructure availability. Site design is discussed in detail in Chapter 14.
5. Information on the conditions, availability and price of utility services such as fuel gas, steam, cooling water, process air, process water and electricity, that will be needed to run the process.

The design basis must be clearly defined before design can be begun. If the design is carried out for a client then the design basis should be reviewed with the client at the start of the project. Most companies use standard forms or questionnaires to capture design basis information. An example template is given in Appendix G and can be downloaded in MS Excel format from the on-line material at http://elsevierdirect.com/companions.

1.2.3. Generation of Possible Design Concepts

The creative part of the design process is the generation of possible solutions to the problem for analysis, evaluation and selection. In this activity, most designers largely rely on previous experience, their own and that of others. It is doubtful if any design is entirely novel. The antecedence of most designs can usually be easily traced. The first motor cars were clearly horse-drawn carriages without the horse; and the development of the design of the modern car can be traced step by step from these early prototypes. In the chemical industry, modern distillation processes have developed from the ancient stills used for rectification of spirits; and the packed columns used for gas absorption have developed from primitive, brushwood-packed towers. So, it is not often that a process designer is faced with the task of producing a design for a completely novel process or piece of equipment.

Experienced engineers usually prefer the tried and tested methods, rather than possibly more exciting but untried novel designs. The work that is required to develop new processes, and the cost, are usually underestimated. Commercialization of new technology is difficult and expensive and few companies are willing to make multi-million dollar investments in technology that is not well proven (known as 'me third' syndrome). Progress is made more surely in small steps; however, when innovation is wanted, previous experience, through prejudice, can inhibit the generation and acceptance of new ideas (known as 'not invented here' syndrome).

The amount of work, and the way it is tackled, will depend on the degree of novelty in a design project. Development of new processes inevitably requires much more interaction with researchers and collection of data from laboratories and pilot plants.

Chemical engineering projects can be divided into three types, depending on the novelty involved:

1. Modifications, and additions, to existing plant; usually carried out by the plant design group.
2. New production capacity to meet growing sales demand, and the sale of established processes by contractors. Repetition of existing designs, with only minor design changes, including designs of vendor's or competitor's processes carried out to understand whether they have a compellingly better cost of production.
3. New processes, developed from laboratory research, through pilot plant, to a commercial process. Even here, most of the unit operations and process equipment will use established designs.

The majority of process designs are based on designs that previously existed. The design engineer very seldom sits down with a blank sheet of paper to create a new design from scratch, an activity sometimes referred to as 'process synthesis'. Even in industries such as pharmaceuticals, where research and new product development are critically important, the types of process used are often based on previous designs for similar products, so as to make use of well-understood equipment and smooth the process of obtaining regulatory approval for the new plant.

The first step in devising a new process design will be to sketch out a rough block diagram showing the main stages in the process; and to list the primary function (objective) and the major constraints for each stage. Experience should then indicate what types of unit operations and equipment should be considered. The steps involved in determining the sequence of unit operations that constitutes a process flow sheet are described in Chapter 4.

The generation of ideas for possible solutions to a design problem cannot be separated from the selection stage of the design process; some ideas will be rejected as impractical as soon as they are conceived.

1.2.4. Fitness Testing

When design alternatives are suggested, they must be tested for fitness for purpose. In other words, the design engineer must determine how well each design concept meets the identified need. In the field of chemical engineering it is usually prohibitively expensive to build several designs to find out which one works best (a practice known as prototyping, which is common in other engineering disciplines). Instead, the design engineer builds a mathematical model of the process, usually in the form of computer simulations of the process, reactors and other key equipment. In some cases, the performance model may include a pilot plant or other facility for predicting plant performance and collecting the necessary design data. In other cases, the design data can be collected from an existing full-scale facility or can be found in the chemical engineering literature.

The design engineer must assemble all of the information needed to model the process so as to predict its performance against the identified objectives. For process design this will include information on possible processes, equipment performance, and physical property data. Sources of process information and physical properties are reviewed in Chapter 8.

Many design organizations will prepare a basic data manual, containing all the process 'know-how' on which the design is to be based. Most organizations will have design manuals covering preferred methods and data for the more frequently used design procedures. The national standards are also sources of design methods and data. They are also design constraints, as new plants must be designed in accordance with the national standards. If the necessary design data or models do not exist then research and development work is needed to collect the data and build new models.

Once the data have been collected and a working model of the process has been established, the design engineer can begin to determine equipment sizes and costs. At this stage it will become obvious that some designs are uneconomical and they can be rejected without further analysis. It is important to make sure that all of the designs that are considered are fit for the service, i.e., meet the customer's 'must have' requirements. In most chemical engineering design problems this comes down to producing products that meet the required specifications. A design that does not meet the customer's objective can usually be modified until it does so, but this always adds extra costs.

1.2.5. Economic Evaluation, Optimization and Selection

Once the designer has identified a few candidate designs that meet the customer objective, the process of design selection can begin. The primary criterion for design selection is usually economic performance, although factors such as safety and environmental impact may also play a strong role. The economic evaluation usually entails analysing the capital and operating costs of the process to determine the return on investment, as described in Chapter 6.

The economic analysis of the product or process can also be used to optimize the design. Every design will have several possible variants that make economic sense under certain conditions. For example, the extent of process heat recovery is a trade-off between the cost of energy and the cost of heat exchangers (usually expressed as a cost of heat-exchange area). In regions where energy costs are high, designs that use a lot of heat-exchange surface to maximize recovery of waste heat for re-use in the process will be attractive. In regions where energy costs are low, it may be more economical to burn more fuel and reduce the capital cost of the plant. The mathematical techniques that have been developed to assist in the optimization of plant design and operation are discussed briefly in Section 1.9.

When all of the candidate designs have been optimized, the best design can be selected. Very often, the design engineer will find that several designs have very close economic performance, in which case the safest design or that which has the best commercial track record will be chosen. At the selection stage an experienced engineer will also look carefully at the candidate designs to make sure that they are safe, operable and reliable, and to ensure that no significant costs have been overlooked.

1.2.6. Detailed Design and Equipment Selection

After the process or product concept has been selected, the project moves on to detailed design. Here the detailed specifications of equipment such as vessels, exchangers, pumps and instruments are determined. The design engineer may work with other engineering disciplines, such as civil engineers for site preparation, mechanical engineers for design of vessels and structures and electrical engineers for instrumentation and control.

Many companies engage specialist Engineering, Procurement and Construction (EPC) companies, commonly known as contractors, at the detailed design stage. The EPC companies maintain large design staffs that can quickly and competently execute projects at relatively low cost.

During the detailed design stage there may still be some changes to the design and there will certainly be ongoing optimization as a better idea of the project cost structure is developed. The detailed design decisions tend to focus mainly on equipment selection though, rather than on changes to the flow-sheet. For example, the design engineer may need to decide whether to use a U-tube or a floating-head exchanger, as discussed in Chapter 12, or whether to use trays or packing for a distillation column, as described in Chapter 11.

1.2.7. **Procurement, Construction and Operation**

When the details of the design have been finalized, the equipment can be purchased and the plant can be built. Procurement and construction are usually carried out by an EPC firm unless the project is very small. Because they work on many different projects each year, the EPC firms are able to place bulk orders for items such as piping, wire, valves, etc., and can use their purchasing power to get discounts on most equipment. The EPC companies also have a great deal of experience in field construction, inspection, testing and equipment installation. They can therefore normally contract to build a plant for a client cheaper (and usually also quicker) than the client could build it on their own.

Finally, once the plant is built and readied for start-up, it can begin operation. The design engineer will often then be called upon to help resolve any start-up issues and teething problems with the new plant.

1.3. **THE ANATOMY OF A CHEMICAL MANUFACTURING PROCESS**

The basic components of a typical chemical process are shown in Figure 1.3, in which each block represents a stage in the overall process for producing a product from the raw materials. Figure 1.3 represents a generalized process; not all the stages will be needed for any particular process and the complexity of each stage will depend on the nature of the process. Chemical engineering design is concerned with the selection and arrangement of the stages, and the selection, specification and design of the equipment required to perform the function of each stage.

Stage 1. Raw material storage: Unless the raw materials (also called feed stocks or feeds) are supplied as intermediate products (intermediates) from a neighbouring plant, some provision will have to be made to hold several days, or weeks, storage to smooth out fluctuations and interruptions in supply. Even when the materials come from an adjacent plant some provision is usually made to hold a few hours, or even days, inventory to decouple the processes. The storage required depends on the nature of the raw materials, the method of delivery, and what assurance can be placed on the continuity of supply. If materials are delivered by ship (tanker or bulk carrier) several weeks' stocks may be necessary; whereas if they are received by road or rail, in smaller lots, less storage will be needed.

Stage 2. Feed preparation: Some purification and preparation of the raw materials will usually be necessary before they are sufficiently pure, or in the right form, to

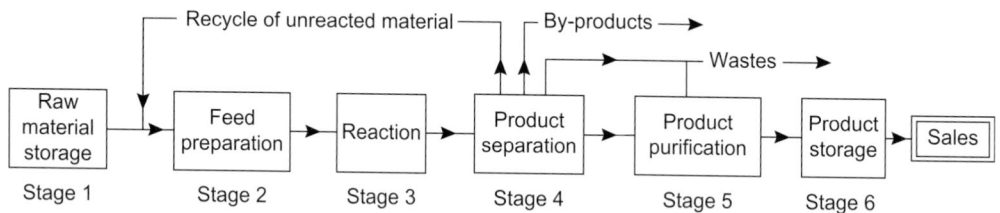

Figure 1.3.   Anatomy of a chemical process.

be fed to the reaction stage. For example, acetylene generated by the carbide process contains arsenic and sulphur compounds, and other impurities, which must be removed by scrubbing with concentrated sulphuric acid (or other processes) before it is sufficiently pure for reaction with hydrochloric acid to produce dichloroethane. Feed contaminants that can poison process catalysts, enzymes or micro-organisms must be removed. Liquid feeds need to be vaporized before being fed to gas-phase reactors and solids may need crushing, grinding and screening.

Stage 3. Reaction: The reaction stage is the heart of a chemical manufacturing process. In the reactor the raw materials are brought together under conditions that promote the production of the desired product; almost invariably, some by-products will also be formed, either through the reaction stoichiometry, by side-reactions, or from reactions of impurities present in the feed.

Stage 4. Product separation: After the reactor(s) the products and by-products are separated from any unreacted material. If in sufficient quantity, the unreacted material will be recycled to the reaction stage or to the feed purification and preparation stage. The by-products may also be separated from the products at this stage. In fine chemical processes there are often multiple reaction steps, each followed by one or more separation steps.

Stage 5. Purification: Before sale, the main product will often need purification to meet the product specifications. If produced in economic quantities, the by-products may also be purified for sale.

Stage 6. Product storage: Some inventory of finished product must be held to match production with sales. Provision for product packaging and transport is also needed, depending on the nature of the product. Liquids are normally dispatched in drums and in bulk tankers (road, rail and sea), solids in sacks, cartons or bales.

The amount of stock that is held will depend on the nature of the product and the market.

Ancillary Processes

In addition to the main process stages shown in Figure 1.3, provision must be made for the supply of the services (utilities) needed; such as, process water, cooling water, compressed air and steam. Facilities are also needed for maintenance, fire fighting, offices and other accommodation, and laboratories; see Chapter 14.

1.3.1. Continuous and Batch Processes

Continuous processes are designed to operate 24 hours a day, 7 days a week, throughout the year. Some down time will be allowed for maintenance and, for some processes, catalyst regeneration. The plant attainment or operating rate is the percentage of the available hours in a year that the plant operates, and is usually between 90% and 95%.

$$\text{Attainment \%} = \frac{\text{hours operated}}{8760} \times 100$$

Batch processes are designed to operate intermittently, with some, or all, of the process units being frequently shut down and started up. It is quite common for batch plants to use a combination of batch and continuous operations. For example, a batch reactor may be used to feed a continuous distillation column.

Continuous processes will usually be more economical for large scale production. Batch processes are used when some flexibility is wanted in production rate or product specifications.

The advantages of batch processing are:

1. Batch processing allows production of multiple different products or different product grades in the same equipment.
2. In a batch plant, the integrity of a batch is preserved as it moves from operation to operation. This can be very useful for quality control purposes.
3. The production rate of batch plants is very flexible, as there are no turn-down issues when operating at low output.
4. Batch plants are easier to clean and maintain sterile operation.
5. Batch processes are easier to scale up from chemist's recipes.
6. Batch plants have low capital for small production volumes. The same piece of equipment can often be used for several unit operations.

The drawbacks of batch processing are:

1. The scale of production is limited.
2. It is difficult to achieve economies of scale by going to high production rates.
3. Batch-to-batch quality can vary, leading to high production of waste products or off-spec product.
4. Recycle and heat recovery are harder, making batch plants less energy efficient and more likely to produce waste by-products.
5. Asset utilization is lower for batch plants as the plant almost inevitably is idle part of the time.
6. The fixed costs of production are much higher for batch plants on a $/unit mass of product basis.

Choice of Continuous versus Batch Production

Given the higher fixed costs and lower plant utilization of batch processes, batch processing usually only makes sense for products that have high value and are produced in small quantities. Batch plants are commonly used for:

- Food products
- Pharmaceutical products such as drugs, vaccines and hormones
- Personal care products
- Specialty chemicals

Even in these sectors, continuous production is favoured if the process is well understood, the production volume is large and the market is competitive.

1.4. **THE ORGANIZATION OF A CHEMICAL ENGINEERING PROJECT**

The design work required in the engineering of a chemical manufacturing process can be divided into two broad phases.

Phase 1: Process design, which covers the steps from the initial selection of the process to be used, through to the issuing of the process flow-sheets; and includes the selection, specification and chemical engineering design of equipment. In a typical organization, this phase is the responsibility of the Process Design Group, and the work is mainly done by chemical engineers. The process design group may also be responsible for the preparation of the piping and instrumentation diagrams.

Phase 2: Plant design, including the detailed mechanical design of equipment, the structural, civil and electrical design, and the specification and design of the ancillary services. These activities will be the responsibility of specialist design groups, having expertise in the whole range of engineering disciplines.

Other specialist groups will be responsible for cost estimation, and the purchase and procurement of equipment and materials.

The sequence of steps in the design, construction and start-up of a typical chemical process plant is shown diagrammatically in Figure 1.4 and the organization of a typical project group is shown in Figure 1.5. Each step in the design process will not be as neatly separated from the others as is indicated in Figure 1.4; nor will the sequence of events be as clearly defined. There will be a constant interchange of information between the various design sections as the design develops, but it is clear that some steps in a design must be largely completed before others can be started.

A project manager, often a chemical engineer by training, is usually responsible for the co-ordination of the project, as shown in Figure 1.5. In addition to co-ordinating the activities of the different specialist groups involved in the design, the project manager will ensure that intermediate deliverables identified in the project plan are completed on time and that the project is kept close to the planned budget.

As was stated in Section 1.2.1, the project design should start with a clear specification defining the product, capacity, raw materials, process and site location. If the project is based on an established process and product, a full specification can be drawn up at the start of the project. For a new product, the specification will be developed from an economic evaluation of possible processes, based on laboratory research, pilot plant tests and product market research.

Some of the larger chemical manufacturing companies have their own project design organizations and carry out the whole project design and engineering, and possibly construction, within their own organization. More usually, the design and construction are carried out under contract by one of the international Engineering, Procurement and Construction (EPC) contracting firms.

The technical 'know-how' for the process could come from the operating company or could be licensed from the contractor or a technology vendor. The operating company, technology provider and contractor will work closely together throughout all stages of the project.

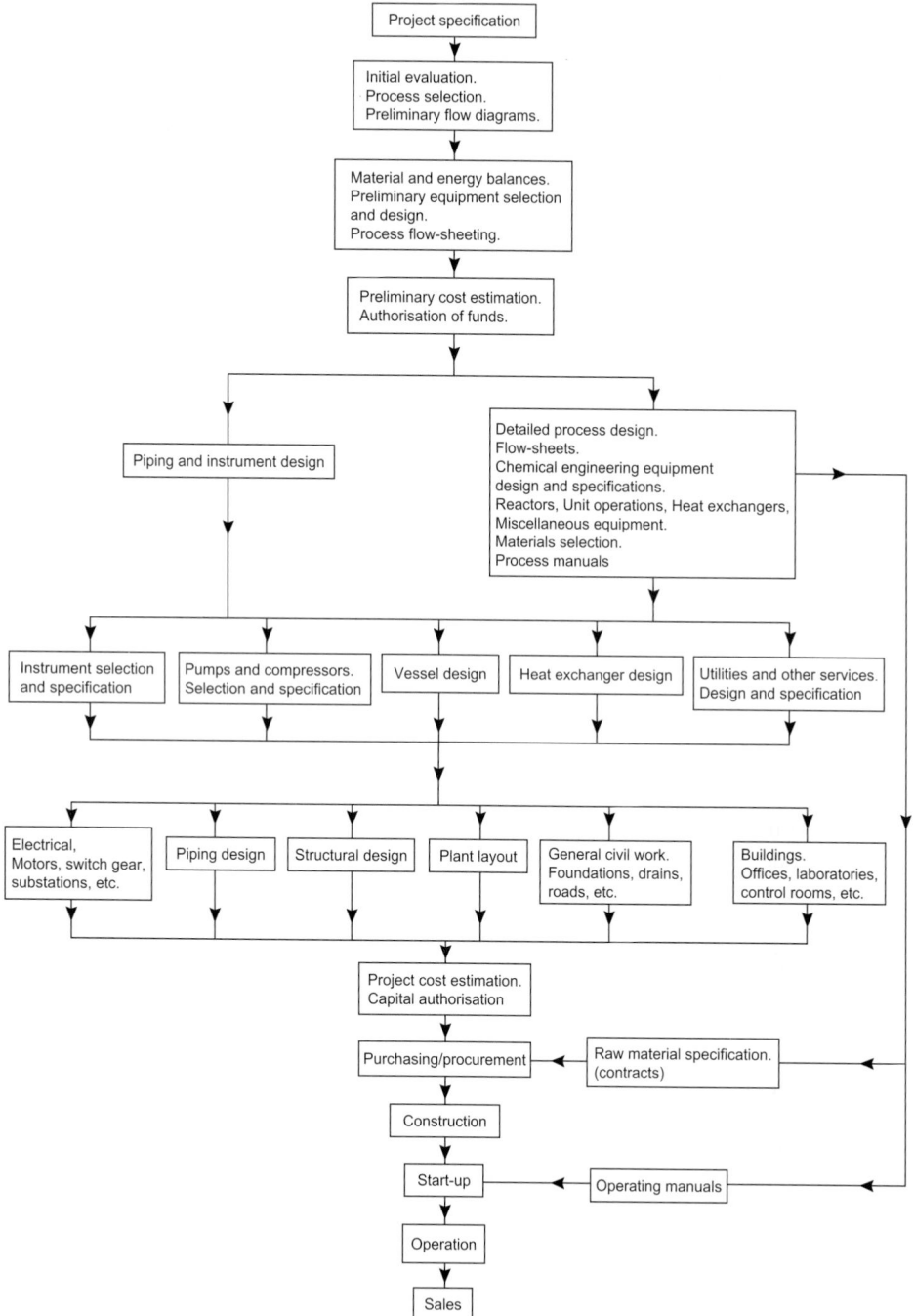

Figure 1.4.   The structure of a chemical engineering project.

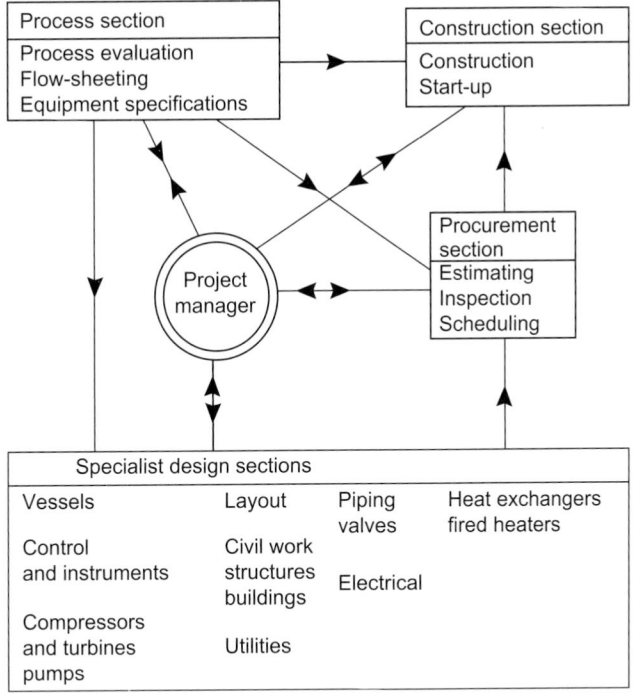

Figure 1.5. Project organization.

On many modern projects, the operating company may well be a joint venture between several companies. The project may be carried out between companies based in different parts of the world. Good teamwork, communications and project management are therefore critically important in ensuring that the project is executed successfully.

1.5. PROJECT DOCUMENTATION

As shown in Figure 1.5 and described in Section 1.4, the design and engineering of a chemical process requires the co-operation of many specialist groups. Effective cooperation depends on effective communications, and all design organizations have formal procedures for handling project information and documentation. The project documentation will include:

1. General correspondence within the design group and with:
    government departments
    equipment vendors
    site personnel
    the client
2. Calculation sheets
    design calculations
    cost estimates
    material and energy balances

3. Drawings
    flow-sheets
    piping and instrumentation diagrams
    layout diagrams
    plot/site plans
    equipment details
    piping diagrams (isometrics)
    architectural drawings
    design sketches
4. Specification sheets
    the design basis
    feed and product specifications
    an equipment list
    sheets for equipment, such as: heat exchangers, pumps, heaters, etc.
5. Health, Safety and Environmental information:
    materials safety data sheets (MSDS forms)
    HAZOP or HAZAN documentation (see Chapter 9)
    emissions assessments and permits
5. Purchase orders
    quotations
    invoices

All documents are assigned a code number for easy cross referencing, filing and retrieval.

Calculation Sheets

The design engineer should develop the habit of setting out calculations so that they can be easily understood and checked by others. It is good practice to include on calculation sheets the basis of the calculations, and any assumptions and approximations made, in sufficient detail for the methods, as well as the arithmetic, to be checked. Design calculations are normally set out on standard sheets. The heading at the top of each sheet should include: the project title and identification number, the revision number and date and, most importantly, the signature (or initials) of the person who checked the calculation. A template calculation sheet is given in Appendix G and can be downloaded in MS Excel format from the on-line material at http://elsevierdirect.com/companions.

Drawings

All project drawings are normally drawn on specially printed sheets, with the company name; project title and number; drawing title and identification number; drafter's name and person checking the drawing; clearly set out in a box in the bottom right-hand corner. Provision should also be made for noting on the drawing all modifications to the initial issue.

Drawings should conform to accepted drawing conventions, preferably those laid down by the national standards. The symbols used for flow-sheets and piping and instrument diagrams are discussed in Chapters 4 and 5. In most design offices, computer-aided design (CAD) methods are now used to produce the drawings required for all the aspects of a project: flow-sheets, piping and instrumentation, mechanical and civil work. While the released versions of drawings are usually drafted by a professional, the design engineer will often need to mark up changes to drawings or make minor modifications to flow-sheets, so it is useful to have some proficiency with the drafting software.

Specification Sheets

Standard specification sheets are normally used to transmit the information required for the detailed design, or purchase, of equipment items; such as, heat exchangers, pumps, columns, pressure vessels, etc.

As well as ensuring that the information is clearly and unambiguously presented, standard specification sheets serve as check lists to ensure that all the information required is included.

Examples of equipment specification sheets are given in Appendix G. These specification sheets are referenced and used in examples throughout the book. Blank templates of these specification sheets are available in MS Excel format in the on-line material at http://elsevierdirect.com/companions. Standard worksheets are also often used for calculations that are commonly repeated in design.

Process Manuals

Process manuals are usually prepared by the process design group to describe the process and the basis of the design. Together with the flow-sheets, they provide a complete technical description of the process.

Operating Manuals

Operating manuals give the detailed, step by step, instructions for operation of the process and equipment. They would normally be prepared by the operating company personnel, but may also be issued by a contractor or technology licensor as part of the technology transfer package for a less experienced client. The operating manuals are used for operator instruction and training, and for the preparation of the formal plant operating instructions.

Review, Checking and Sign-off

Despite the best efforts of the engineers working on a design, mistakes will almost always be made in the design calculations and drawings and in transcription of numbers between different computer programs or between computer programs and other project documentation. It is important to eliminate as many of these mistakes as possible before purchasing and construction begin, as they could potentially require costly fixes later in the project.

Design engineers are usually accountable for the quality of their own work, and it is a good idea to develop the habit of checking calculations throughout the design process. Quick mass and energy balances, of the kind introduced in Chapters 2 and 3, can often be used to confirm that answers are approximately correct. Various short-cut methods and rules of thumb are introduced throughout this book that can also be useful in checking more detailed design calculations.

In industrial projects, the design is also checked or reviewed by a senior design engineer or supervisor, who must then sign off on the calculation, drawing or specification sheet to indicate that it is satisfactory. The forms used for calculations and specification sheets usually have a space to indicate who reviewed and approved release of the design. In some cases, the reviewer must be a licensed or certified engineer; a Chartered Engineer in the UK or a Professional Engineer in the USA.

1.6. CODES AND STANDARDS

The need for standardization arose early in the evolution of the modern engineering industry; Whitworth introduced the first standard screw thread to give a measure of interchangeability between different manufacturers in 1841. Modern engineering standards cover a much wider function than the interchange of parts. In engineering practice they cover:

1. Materials, properties and compositions.
2. Testing procedures for performance, compositions and quality.
3. Preferred sizes; for example, tubes, plates, sections, etc.
4. Methods for design, inspection and fabrication.
5. Codes of practice for plant operation and safety.

The terms *standard* and *code* are used interchangeably, though *code* should really be reserved for a code of practice covering say, a recommended design or operating procedure; and *standard* for preferred sizes, compositions, etc.

All of the developed countries, and many of the developing countries, have national standards organizations, responsible for the issue and maintenance of standards for the manufacturing industries, and for the protection of consumers.

In the UK, preparation and promulgation of national standards are the responsibility of the British Standards Institution (BSI). The preparation of the standards is largely the responsibility of committees of persons from the appropriate industry, the professional engineering institutions and other interested organizations. All the published British standards are listed in the *British Standards Institute Catalogue*, which is available on-line at www.BSI-global.com. Many of the British standards are now being harmonized with the European standards.

The International Organization for Standardization (ISO) coordinates the publication of international standards. ISO is a network of the national standards institutes of 157 countries and has published over 16,500 international standards. Many of the ISO standards that are adopted are variants of national standards or, in some cases, standards that have been developed by commercial organizations. Information on ISO standards can be obtained from www.iso.org. Another important international agency is the International Electrotechnical Commission (IEC), which was founded in 1906, 41 years before ISO. The IEC sets standards for electrical and electronic equipment, and hence plays an important role in setting standards for process instrumentation and control systems; for more information see www.iec.ch. Many IEC standards are co-sponsored by ISO.

The European countries used to each maintain their own national standards, but under Article 7 of European Parliament Directive 98/34/EC, new national standards for member states of the European Union must be in conformance with European or international standards. European standards are set by the European Committee for Standardization (CEN), The European Committee for Electrotechnical Standardization (CENELEC) and the European Telecommunications Standards Institute (ETSI). European standards must be obtained from the national member organizations such as BSI; details of the process of harmonization of European standards are given on the CEN website at www.cen.eu.

In the USA, the government organization responsible for coordinating information on standards is the National Institute of Standards and Technology (NIST); standards are issued by Federal, State and various commercial organizations. The principal ones of interest to chemical engineers are those issued by the American National Standards Institute (ANSI), the American Petroleum Institute (API), the American Society for Testing Materials (ASTM), the American Society of Mechanical Engineers (ASME) (pressure vessels and pipes), the National Fire Protection Association (NFPA) (safety), the Tubular Exchanger Manufacturers Association (TEMA) (heat exchangers) and the Instrumentation, Systems and Automation Society (ISA) (process control). Most Canadian provinces apply the same standards used in the USA.

Lists of codes and standards and copies of the most current versions can be obtained from the national standards agencies or by subscription from commercial web sites such as IHS (www.ihs.com).

As well as the various national standards and codes, the larger design organizations will have their own (in-house) standards. Much of the detail in engineering design work is routine and repetitious, and it saves time and money, and ensures conformity between projects, if standard designs are used whenever practicable.

Equipment manufacturers also work to standards to produce standardized designs and size ranges for commonly used items; such as electric motors, pumps, heat exchangers, pipes and pipe fittings. They will conform to national standards, where they exist, or to those issued by trade associations. It is clearly more economic to produce a limited range of standard sizes than to have to treat each order as a special job.

For the designer, the use of a standardized component size allows for the easy integration of a piece of equipment into the rest of the plant. For example, if a standard range of centrifugal pumps is specified the pump dimensions will be known, and this facilitates the design of the foundation plates, pipe connections and the selection of the drive motors: standard electric motors would be used. For an operating company, the standardization of equipment designs and sizes increases interchangeability and reduces the stock of spares that must be held in maintenance stores.

Though there are clearly considerable advantages to be gained from the use of standards in design, there are also some disadvantages. Standards impose constraints on the designer. The nearest standard size will normally be selected on completing a design calculation (rounding up) but this will not necessarily be the optimum size; though as the standard size will be cheaper than a special size, it will usually be the best choice from the point of view of initial capital cost. The design methods given in the codes and standards are, by their nature, historical, and do not necessarily incorporate the latest techniques.

The use of standards in design is illustrated in the discussion of the pressure vessel design in Chapter 13. Relevant design codes and standards are cited throughout the book.

1.7. DESIGN FACTORS (DESIGN MARGINS)

Design is an inexact art; errors and uncertainties arise from uncertainties in the design data available and in the approximations necessary in design calculations. Experienced designers include a degree of over-design known as a design factor, design margin or safety factor, to ensure that the design that is built meets product specifications and operates safely.

In mechanical and structural design, the design factors that are used to allow for uncertainties in material properties, design methods, fabrication and operating loads are well established. For example, a factor of around 4 on the tensile strength, or about 2.5 on the 0.1% proof stress, is normally used in general structural design. The recommended design factors are set out in the codes and standards. The selection of design factors in mechanical engineering design is illustrated in the discussion of pressure vessel design in Chapter 13.

Design factors are also applied in process design to give some tolerance in the design. For example, the process stream average flows calculated from material balances are usually increased by a factor, typically 10%, to set the maximum flows for equipment, instrumentation, and piping design. This factor will give some flexibility in process operation. Where design factors are introduced to give some contingency in a process design, they should be agreed within the project organization, and clearly stated in the project documents (drawings, calculation sheets and manuals). If this is not done, there is a danger that each of the specialist design groups will add its own 'factor of safety'; resulting in gross and unnecessary over-design. Companies often specify design factors in their design manuals.

When selecting the design factor, a balance has to be made between the desire to make sure the design is adequate and the need to design to tight margins to remain competitive. Greater uncertainty in the design methods and data requires the use of bigger design factors.

1.8. SYSTEMS OF UNITS

Most of the examples and equations in this book use SI units; however, in practice the design methods, data and standards that the designer will use are often only available in the traditional scientific and engineering units. Chemical engineering has always used a diversity of units; embracing the scientific CGS and MKS systems, and both the American and British engineering systems. Those engineers in the older industries will also have had to deal with some bizarre traditional units; such as degrees Twaddle or degrees API for density and barrels for quantity. Although almost all of the engineering societies have stated support for the adoption of SI units, this is unlikely to happen world-wide for many years. Furthermore, much useful historic data will always be in the traditional units and the design engineer must know how to understand and convert this information. In a globalized economy, engineers are expected to use different systems of units even within the same company, particularly in the contracting sector where the choice of units is at the client's discretion. Design engineers must therefore have a familiarity with SI, metric and customary units, and a few of the examples and many of the exercises are presented in customary units.

It is usually the best practice to work through design calculations in the units in which the result is to be presented; but, if working in SI units is preferred, data can be converted to SI units, the calculation made, and the result converted to whatever units are required. Conversion factors to the SI system from most of the scientific and engineering units used in chemical engineering design are given in Appendix D.

Some license has been taken in the use of the SI system. Temperatures are given in degrees Celsius (°C); kelvin (K) is only used when absolute temperature is required in the calculation. Pressures are often given in bar (or atmospheres) rather than in Pascals (N/m^2), as this gives a better feel for the magnitude of the pressures. In technical calculations the bar can be taken as equivalent to an atmosphere, whatever definition is used for atmosphere. The abbreviations bara and barg are often used to denote bar absolute and bar gauge; analogous to psia and psig when the pressure is expressed in pound force per square inch. When bar is used on its own, without qualification, it is normally taken as absolute.

For stress, N/mm^2 has been used, as these units are now generally accepted by engineers, and the use of a small unit of area helps to indicate that stress is the force per unit area at a point (as is also pressure). The corresponding traditional unit for stress is the ksi or thousand pounds force per square inch. For quantity, kmol are generally used in preference to mol, and for flow, kmol/h instead of mol/s, as this gives more sensibly sized figures, which are also closer to the more familiar lb/h.

For volume and volumetric flow, m^3 and m^3/h are used in preference to m^3/s, which gives ridiculously small values in engineering calculations. Litres per second are used for small flow rates, as this is the preferred unit for pump specifications.

Plant capacities are usually stated on an annual mass flow basis in metric tons per year. Unfortunately, the literature contains a variety of abbreviations for metric tons per year, including tonnes/y, metric tons/y, MT/y (also kMTA = thousand metric tons per year), mtpy and the correct term, t/y. The non-standard abbreviations have occasionally been used, as it is important for design engineers to be familiar with all of these terms. The unit t denotes a metric ton of 1000 kg. In this book the unit ton is generally used to describe a short ton or US ton of 2000 lb (907 kg) rather than a long ton or UK ton of 2240 lb (1016 kg), although some examples use long tons. The long ton is closer to the metric ton. A thousand metric tons is usually denoted as a kiloton (kt); the correct SI unit gigagram (Gg) is very rarely used.

In the US, the prefixes M and MM are often used to denote thousand and million, which can be confusing to anyone familiar with the SI use of M as an abbreviation for *mega* ($\times 10^6$). This practice has generally been avoided, except in the widely used units MMBtu (million British thermal units) and the common way of abbreviating $1 million as $1 MM.

Most prices have been given in US dollars, denoted US$ or $, reflecting the fact that the data originated in the United States. Examples using British pounds sterling (£) and Euros (€) have been included in the chapter on process economics (Chapter 6).

Where, for convenience, other than SI units have been used on figures or diagrams, the scales are also given in SI units, or the appropriate conversion factors are given in

TABLE 1.1. Approximate Conversions between Customary Units and SI Units

Quantity	Customary Unit	SI Unit Approx.	Exact
Energy	1 Btu	1 kJ	1.05506
Specific enthalpy	1 Btu/lb	2 kJ/kg	2.326
Specific heat capacity	1 Btu/lb°F	4 kJ/kg°C	4.1868
Heat transfer coeff.	1 Btu/ft^{2}h°F	6 W/m^2 °C	5.678
Viscosity	1 centipoise	1 mNs/m^2	1.000
	1 lb$_f$/ft h	0.4 mNs/m^2	0.4134
Surface tension	1 dyne/cm	1 mN/m	1.000
Pressure	1 lbf/in^2 (psi)	7 kN/m^2	6.894
	1 atm	1 bar	1.01325
		10^5 N/m^2	
Density	1 lb/ft^3	16 kg/m^3	16.0185
	1 g/cm^3	1 kg/m^3	
Volume	1 US gal	3.8×10^{-3} m^3	3.7854×10^{-3}
Flow-rate	1 US gal/min	0.23 m^3/h	0.227

1 US gallon = 0.84 imperial gallons (UK)
1 barrel (oil) = 42 US gallons $\approx$ 0.16 m^3 (exact 0.1590)
1 kWh = 3.6 MJ

the text. Where equations are presented in customary units a metric equivalent is generally given.

Some approximate conversion factors to SI units are given in Table 1.1. These are worth committing to memory, to give some feel for the units for those more familiar with the traditional engineering units. The exact conversion factors are also shown in the table. A more comprehensive table of conversion factors is given in Appendix D.

1.9. OPTIMIZATION

Optimization is an intrinsic part of design: the designer seeks the best, or optimum, solution to a problem.

Many design decisions can be made without formally setting up and solving a mathematical optimization problem. The design engineer will often rely on a combination of experience and judgment, and in some cases the best design will be immediately obvious. Other design decisions have such a trivial impact on process costs that it makes more sense to make a close guess at the answer than to properly set up and solve the optimization problem. In every design though, there will be several problems that require rigorous optimization. This section introduces the techniques for formulating and solving optimization problems, as well as some of the pitfalls that are commonly encountered in optimization.

In this book, the discussion of optimization will, of necessity, be limited to a brief overview of the main techniques used in process and equipment design. Chemical engineers working in industry use optimization methods for process operations far more than they do for design, as discussed in Section 1.9.11. Chemical engineering students would benefit greatly from more classes in operations research methods, which are generally part of the Industrial Engineering curriculum. These methods are used in almost every industry for planning, scheduling and supply-chain management: all critical operations for plant operation and management. There is an extensive literature on operations research methods and several good books on the application of optimization methods in chemical engineering design and operations. A good overview of operations research methods is given in the classic introductory text by Hillier and Lieberman (2002). Applications of optimization methods in chemical engineering are discussed by Rudd and Watson (1968), Stoecker (1989), Biegler *et al.* (1997), Edgar and Himmelblau (2001) and Diwekar (2003).

1.9.1. The Design Objective

An optimization problem is always stated as the maximization or minimization of a quantity called the objective. For chemical engineering design projects, the objective should be a measure of how effectively the design meets the customer's needs. This will usually be a measure of economic performance. Some typical objectives are given in Table 1.2.

The overall corporate objective is usually to maximize profits, but the design engineer will often find it more convenient to use other objectives when working on sub-components of the design. The optimization of sub-systems is discussed in more detail in Section 1.9.4.

The first step in formulating the optimization problem is to state the objective as a function of a finite set of variables, sometimes referred to as the decision variables:

$$z = f(x_1, x_2, x_3, ..., x_n) \tag{1.1}$$

where

z = objective

$x_1, x_2, x_3, ..., x_n$ = decision variables

This function is called the objective function. The decision variables may be independent, but they will usually be related to each other by many constraint equations. The optimization problem can then be stated as maximization or minimization of the objective function subject to the set of constraints. Constraint equations are discussed in the next section.

Design engineers often face difficulties in formulating the objective function. Some of the economic objectives that are widely used in making investment decisions lead to intrinsically difficult optimization problems. For example, discounted cash flow

TABLE 1.2. Typical Design Optimization Objectives

Maximize	Minimize
Project net present value	Project expense
Return on investment	Cost of production
Reactor productivity per unit volume	Total annualized cost
Plant availability (time on stream)	Plant inventory (for safety reasons)
Process yield of main product	Formation of waste products

rate of return (DCFROR) is difficult to express as a simple function and is highly non-linear, while net present value (NPV) increases with project size and is unbounded unless a constraint is set on plant size or available capital. Optimization is therefore often carried out using simple objectives such as 'minimize cost of production'. Health, safety, environmental and societal impact costs and benefits are difficult to quantify and relate to economic benefit. These factors can be introduced as constraints, but few engineers would advocate building a plant in which every piece of equipment was designed for the minimum legally permissible safety and environmental performance.

An additional complication in formulating the objective function is the quantification of uncertainty. Economic objective functions are generally very sensitive to the prices used for feeds, raw materials and energy, and also to estimates of project capital cost. These costs and prices are forecasts or estimates and are usually subject to substantial error. Cost estimation and price forecasting are discussed in Sections 6.3 and 6.4. There may also be uncertainty in the decision variables, either from variation in the plant inputs, variations introduced by unsteady plant operation, or imprecision in the design data and the constraint equations. Optimization under uncertainty is a specialized subject in its own right and is beyond the scope of this book. See Chapter 5 of Diwekar (2003) for a good introduction to the subject.

1.9.2. Constraints and Degrees of Freedom

The constraints on the optimization are the set of equations that bound the decision variables and relate them to each other.

If we write $\mathbf{x}$ as a vector of n decision variables, then we can state the optimization problem as:

$$\begin{aligned} \text{Optimize (Max. or Min.)} \quad & z = f(\mathbf{x}) \\ \text{subject to (s.t.):} \quad & \mathbf{g}(\mathbf{x}) \leq 0 \\ & \mathbf{h}(\mathbf{x}) = 0 \end{aligned} \tag{1.2}$$

where

  z = the scalar objective
  $f(\mathbf{x})$ = the objective function
  $\mathbf{g}(\mathbf{x})$ = a m_i vector of inequality constraints
  $\mathbf{h}(\mathbf{x})$ = a m_e vector of equality constraints

The total number of constraints is $m = m_i + m_e$.

Equality constraints arise from conservation equations (mass, mole, energy and momentum balances) and constitutive relations (the laws of chemistry and physics, correlations of experimental data, design equations, etc.). Any equation that is introduced into the optimization model that contains an "=" sign will become an equality constraint. Many examples of such equations can be found throughout this book.

Inequality constraints generally arise from the external constraints discussed in Section 1.2: safety limits, legal limits, market and economic limits, technical limits set by design codes and standards, feed and product specifications, availability of resources, etc. Some examples of inequality constraints are as follows:

Main product purity ≥ 99.99 wt%

Feed water content ≤ 20 ppmw

NO_x emissions ≤ 50 kg/yr

Production rate $\leq 400,000$ metric tons per year

Maximum design temperature for ASME Boiler and Pressure Vessel Code Section VIII Division $2 \leq 900°F$

Investment capital $\leq \$50$ MM (50 million dollars)

The effect of constraints is to limit the parameter space. This can be illustrated using a simple two-parameter problem:

$$\text{Max.} \quad z = x_1^2 + 2x_2^2$$
$$\text{s.t.} \quad x_1 + x_2 = 5$$
$$x_2 \leq 3$$

The two constraints can be plotted on a graph of x_1 vs x_2, as in Figure 1.6.

In the case of this example, it is clear by inspection that the set of constraints does not bound the problem. In the limit $x_1 \rightarrow \infty$, the solution to the equality constraint is $x_2 \rightarrow -\infty$, and the objective function gives $z \rightarrow \infty$, so no maximum can be found. Problems of this kind are referred to as 'unbounded'. For this problem to have a solution we need an additional constraint of the form:

$$x_1 \leq a \qquad \text{(where } a > 2)$$
$$x_2 \geq b \qquad \text{(where } b < 3)$$
$$\text{or} \qquad h(x_1, x_2) = 0$$

to define a closed search space.

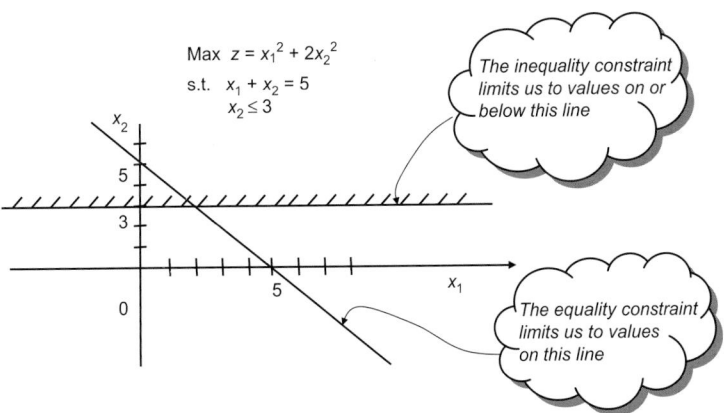

Figure 1.6. Constraints on a simple optimization problem.

It is also possible to over-constrain the problem. For example, if we set the problem:

$$\text{Max.} \quad z = x_1^2 + 2x_2^2$$
$$\text{s.t.} \quad x_1 + x_2 = 5$$
$$x_2 \leq 3$$
$$x_1 \leq 1$$

In this case, it can be seen from Figure 1.7 that the feasible region defined by the inequality constraints does not contain any solution to the equality constraint. The problem is therefore infeasible as stated.

Degrees of Freedom

If the problem has n variables and m_e equality constraints then it has $n - m_e$ degrees of freedom. If $n = m_e$ then there are no degrees of freedom and the set of m_e equations can be solved for the n variables. If $m_e > n$ then the problem is over-specified. In most cases, however, $m_e < n$ and $n - m_e$ is the number of parameters that can be independently adjusted to find the optimum.

When inequality constraints are introduced into the problem, they generally set bounds on the range over which parameters can be varied and hence reduce the space in which the search for the optimum is carried out. Very often, the optimum solution to a constrained problem is found to be at the edge of the search space, i.e., at one of the inequality constraint boundaries. In such cases, that inequality constraint becomes an equality and is said to be 'active'. It is often possible to use engineering insight and understanding of chemistry and physics to simplify the optimization problem. If the behaviour of a system is well understood, then the design engineer can decide that an inequality constraint is likely to be active. By converting the inequality constraint into an equality constraint, the number of degrees of freedom is reduced by one and the problem is made simpler.

This can be illustrated by a simple reactor optimization example. The size and cost of a reactor are proportional to residence time, which decreases as temperature is increased. The optimal temperature is usually a trade-off between reactor cost and the formation of by-products in side reactions; but if there were no side reactions, then

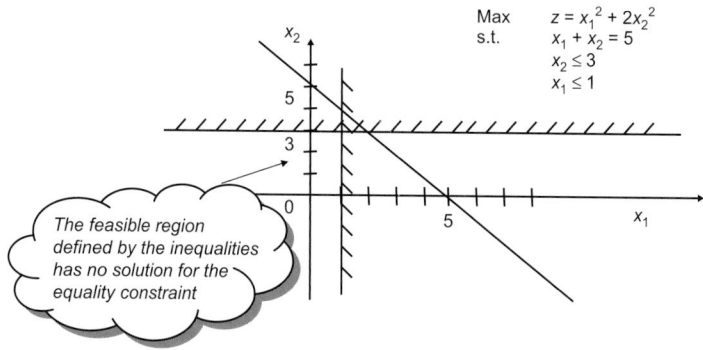

Figure 1.7.   An over-constrained problem.

the next constraint would be the maximum temperature allowed by the pressure vessel design code. More expensive alloys might allow for operation at higher temperatures. The variation of reactor cost with temperature will look something like Figure 1.8, where T_A, T_B and T_C are the maximum temperatures allowed by the vessel design code for alloys A, B and C respectively.

The design engineer could formulate this problem in several ways. It could be solved as three separate problems, one corresponding to each alloy, each with a constraint on temperature $T < T_{alloy}$. The design engineer would then pick the solution that gave the best value of the objective function. The problem could also be formulated as a mixed integer non-linear program with integer variables to determine the selection of alloy and set the appropriate constraint (see Section 1.9.10). The design engineer could also recognize that alloy A costs a lot less than alloy B, and the higher alloys only give a relatively small extension in the allowable temperature range. It is clear that cost decreases with temperature, so the optimum temperature will be T_A for alloy A and T_B for alloy B. Unless the design engineer is aware of some other effect that has an impact on cost as temperature is increased, it is safe to write $T = T_A$ as an equality constraint and solve the resulting problem. If the cost of alloy B is not excessive then it would be prudent to also solve the problem with $T = T_B$, using the cost of alloy B.

The correct formulation of constraints is the most important step in setting up an optimization problem. Inexperienced engineers are often unaware of many constraints and consequently find 'optimal' designs that are dismissed as unfeasible by more experienced designers.

1.9.3. Trade-offs

If the optimal value of the objective is not at a constraint limit then it will usually be determined by a trade-off between two or more effects. Trade-offs are very common in design, because better performance in terms of increased purity, increased recovery or reduced energy or raw materials use usually comes at the expense of higher capital expense, operating expense or both. The optimization problem must capture the trade-off between cost and benefit.

A well-known example of a trade-off is the optimization of process heat recovery. A high degree of heat recovery requires close temperature approaches in the heat exchangers (see Section 3.17), which leads to high capital cost as the exchangers require more surface area. If the minimum temperature approach is increased then the

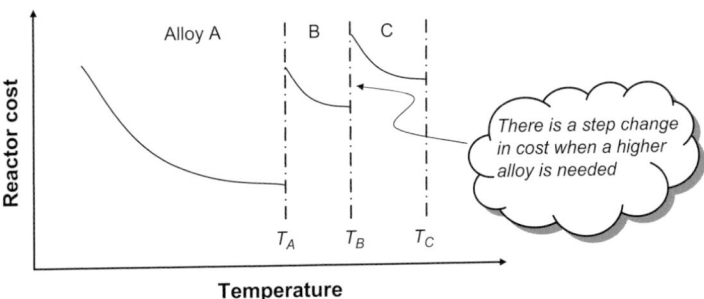

Figure 1.8. Variation of reactor cost with temperature.

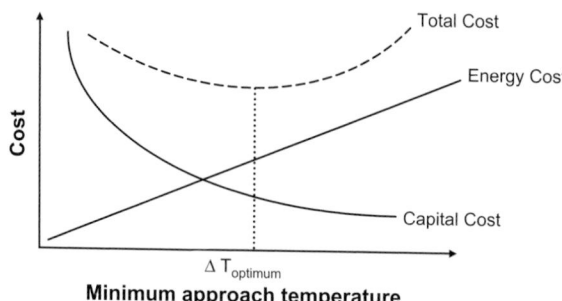

Figure 1.9.   The capital-energy trade-off in process heat recovery.

capital cost is reduced but less energy is recovered. We can plot the capital cost and energy cost against the minimum approach temperature, as shown schematically in Figure 1.9. If the capital cost is annualized (see Section 6.7) then the two costs can be added to give a total cost. The optimum value of the approach temperature, $\Delta T_{optimum}$, is then given by the minimum point in the total cost curve.

Some common trade-offs encountered in design of chemical plants include:

- More separations equipment and operating cost vs lower product purity.
- More recycle costs vs increased feed use and waste formation.
- More heat recovery vs cheaper heat-exchange network.
- Higher reactivity at high pressure vs more expensive reactors and higher compression costs.
- Fast reactions at high temperature vs product degradation.
- Marketable by-products vs more plant expense.
- Cheaper steam and electricity vs more off-site capital cost.

Stating an optimization problem as a trade-off between two effects is often useful in conceptualizing the problem and interpreting the optimal solution. For example, in the case of process heat recovery, it is usually found that the shape of the total cost curve in Figure 1.9 is relatively flat over the range $15°C < \Delta T_{optimum} < 40°C$. Knowing this, most experienced designers would not worry about finding the value of $\Delta T_{optimum}$, but would instead select a value for the minimum temperature approach within the range $15°C$ to $40°C$, based on knowledge of the customer's preference for high energy efficiency or low capital expense.

1.9.4.   Problem Decomposition

The task of formally optimizing the design of a complex processing plant involving several hundred variables, with complex interactions, is formidable, if not impossible. The task can be reduced by dividing the process into more manageable units, identifying the key variables and concentrating work where the effort involved will give the greatest benefit. Sub-division, and optimization of the sub-units rather than the whole, will not necessarily give the optimum design for the whole process. The optimization of one unit may be at the expense of another. For example, it will usually be satisfactory to optimize the reflux ratio for a fractionating column independently of the rest of the plant; but if the column is part of a separation stage following a reactor, in which the product is separated from the unreacted materials, then the design of the column will interact with, and may well determine, the optimization of the reactor design. Care must always be taken to ensure that sub-components are not optimized at the expense of other parts of the plant.

1.9.5. Optimization of a Single Decision Variable

If the objective is a function of a single variable, x, the objective function $f(x)$ can be differentiated with respect to x to give $f'(x)$. Any stationary points in $f(x)$ can then be found as the solutions of $f'(x) = 0$. If the second derivative of the objective function is greater than zero at a stationary point then the stationary point is a local minimum. If the second derivative is less than zero then the stationary point is a local maximum and if it is equal to zero then it is a saddle point. If x is bounded by constraints, then we must also check the values of the objective function at the upper and lower limiting constraints. Similarly, if $f(x)$ is discontinuous, then the value of $f(x)$ on either side of the discontinuity should also be checked.

This procedure can be summarized as the following algorithm:

$$\begin{aligned} \text{Min.} \quad z &= f(x) \\ \text{s.t.} \quad x &\geq x_L \\ x &\leq x_U \end{aligned} \qquad (1.3)$$

1. Solve $f' = \dfrac{df(x)}{dx} = 0$ to find values of x_S.
2. Evaluate $f'' = \dfrac{d^2 f(x)}{dx^2}$ for each value of x_S. If $f'' > 0$ then x_S corresponds to a local minimum in $f(x)$.
3. Evaluate $f(x_S)$, $f(x_L)$ and $f(x_U)$.
4. If the objective function is discontinuous then evaluate $f(x)$ on either side of the discontinuity, x_{D1} and x_{D2}.
5. The overall optimum is the value from the set $(x_L, x_S, x_{D1}, x_{D2}, x_U)$ that gives the lowest value of $f(x)$.

This is illustrated graphically in Figure 1.10a for a continuous objective function. In Figure 1.10a, x_L is the optimum point, even though there is a local minimum at x_{S1}. Figure 1.10b illustrates the case of a discontinuous objective function. Discontinuous functions are quite common in engineering design, arising, for example, when changes in temperature or pH cause a change in metallurgy. In Figure 1.10b the optimum is at x_{D1}, even though there is a local minimum at x_S.

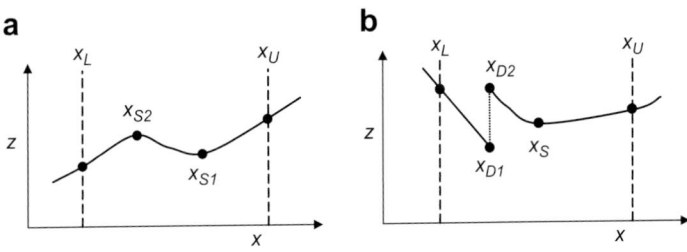

Figure 1.10. Optimization of a single variable between bounds.

If the objective function can be expressed as a differentiable equation then it is usually also easy to plot a graph like those in Figure 1.10 and quickly determine whether the optimum lies at a stationary point or a constraint.

1.9.6. Search Methods

In design problems, the objective function very often can not be written as a simple equation that is easily differentiated. This is particularly true when the objective function requires solving large computer models, possibly using several different programs and requiring several minutes, hours or days to converge a single solution. In such cases, the optimum is found using a search method. The concept of search methods is most easily explained for single variable problems, but search methods are at the core of the solution algorithms for multivariable optimization as well.

Unrestricted Search

If the decision variable is not bounded by constraints then the first step is to determine a range in which the optimum lies. In an unrestricted search we make an initial guess of x and assume a step size, h. We then calculate $z_1 = f(x)$, $z_2 = f(x + h)$, and $z_3 = f(x - h)$. From the values of z_1, z_2 and z_3, we determine the direction of search that leads to improvement in the value of the objective, depending on whether we wish to minimize or maximize z. We then continue increasing (or decreasing) x by successive steps of h until the optimum is passed.

In some cases, it may be desirable to accelerate the search procedure, in which case the step size can be doubled at each step. This gives the sequence $f(x + h)$, $f(x + 3h)$, $f(x + 7h)$, $f(x + 15h)$, etc.

Unrestricted searching is a relatively simple method of bounding the optimum for problems that are not constrained. In engineering design problems it is almost always possible to state upper and lower bounds for every parameter, so unrestricted search methods are not widely used in design.

Once a restricted range that contains the optimum has been established, restricted range search methods can be used. These can be broadly classified as direct methods that find the optimum by eliminating regions in which it does not lie, and indirect methods that find the optimum by making an approximate estimate of $f'(x)$.

Regular Search (Three-Point Interval Search)

The three-point interval search starts by evaluating $f(x)$ at the upper and lower bounds, x_L and x_U, and at the centre point $(x_L + x_U)/2$. Two new points are then added in the midpoints between the bounds and the centre point, at $(3x_L + x_U)/4$ and $(x_L + 3x_U)/4$, as shown in Figure 1.11. The three adjacent points with the lowest values of $f(x)$ (or the highest values for a maximization problem) are then used to define the next search range.

By eliminating two of the four quarters of the range at each step, this procedure reduces the range by half each cycle. To reduce the range to a fraction ε of the initial range therefore takes n cycles, where $\varepsilon = 0.5^n$. Since each cycle requires calculating $f(x)$ for two additional points, the total number of calculations is $2n = 2 \log \varepsilon / \log 0.5$.

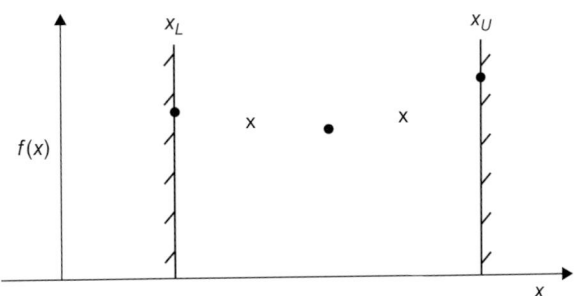

Figure 1.11. Regular search.

The procedure is terminated when the range has been reduced sufficiently to give the desired precision in the optimum. For design problems it is usually not necessary to specify the optimal value of the decision variables to high precision, so ε is usually not a very small number.

Golden-Section Search

The golden-section search, sometimes called the golden mean search, is as simple to implement as the regular search, but is more computationally efficient if $\varepsilon < 0.29$. In the golden-section search only one new point is added at each cycle.

The golden-section method is illustrated in Figure 1.12. We start by evaluating $f(x_L)$ and $f(x_U)$ corresponding to the upper and lower bounds of the range, labelled A and B in the figure. We then add two new points, labelled C and D, each located a distance ωAB from the bounds A and B, i.e., located at $x_L + \omega(x_U - x_L)$ and $x_U - \omega(x_U - x_L)$. For a minimization problem, the point that gives the highest value of $f(x)$ is eliminated. In Figure 1.12, this is point B. A single new point, E, is added, such that the new set of points $AECD$ is symmetric with the old set of points $ACDB$.

For the new set of points to be symmetric with the old set of points, $AE = CD = \omega AD$. But we know $DB = \omega AB$, so $AD = (1 - \omega)AB$ and $CD = (1 - 2\omega)AB$

so

$$(1 - 2\omega) = \omega(1 - \omega)$$

$$\omega = \frac{3 \pm \sqrt{5}}{2}$$

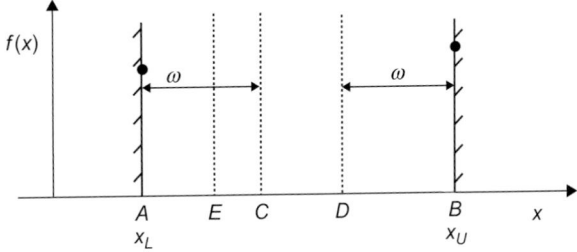

Figure 1.12. Golden-section search.

Each new point reduces the range to a fraction $(1 - \omega) = 0.618$ of the original range. To reduce the range to a fraction ε of the initial range therefore requires $n = \log \varepsilon / \log 0.618$ function evaluations.

The number $(1 - \omega)$ is known as the golden mean. The significance of this number has been known since ancient times. Livio (2002) gives a very entertaining account of its history and occurrence in art, architecture, music and nature.

Quasi-Newton Method

Newton's method is a super-linear indirect search method that seeks the optimum by solving $f'(x)$ and $f''(x)$ and searching for where $f'(x) = 0$. The value of x at step $k + 1$ is calculated from the value of x at step k using:

$$x_{k+1} = x_k - \frac{f'(x_k)}{f''(x_k)} \tag{1.4}$$

and the procedure is repeated until $(x_{k+1} - x_k)$ is less than a convergence criterion or tolerance, ε.

If we do not have explicit formulae for $f'(x)$ and $f''(x)$, then we can make a finite difference approximation about a point, in which case:

$$x_{k+1} = x_k - \frac{[f(x_k + h) - f(x_k - h)]/2h}{[f(x_k + h) - 2f(x) + f(x_k - h)]/h^2} \tag{1.5}$$

Care is needed in setting the step size, h, and the tolerance for convergence, ε. The Quasi-Newton method generally gives fast convergence unless $f''(x)$ is close to zero, in which case convergence is poor.

All of the methods discussed in this section are best suited for unimodal functions, i.e., functions with no more than one maximum or minimum within the bounded range.

1.9.7.   Optimization of Two or more Decision Variables

A two-variable optimization problem can be stated as:

$$\begin{aligned} \text{Min.} \quad & z = f(x_1, x_2) \\ \text{s.t.} \quad & h(x_1, x_2) = 0 \\ & g(x_1, x_2) \leq 0 \end{aligned} \tag{1.6}$$

For simplicity, all problems will be stated as minimization problems from here on. A maximization problem can be rewritten as Min. $z = -f(x_1, x_2)$.

With two parameters, we can plot contour lines of z on a graph of x_1 vs x_2 and hence get a visual representation of the behaviour of z. For example, Figure 1.13 shows a schematic of a contour plot for a function that exhibits a local minimum of < 30 at about (4,13) and a global minimum of < 10 at about (15,19). Contour plots are useful for understanding some of the key features of multivariable optimization that become apparent as soon as we consider more than one decision variable.

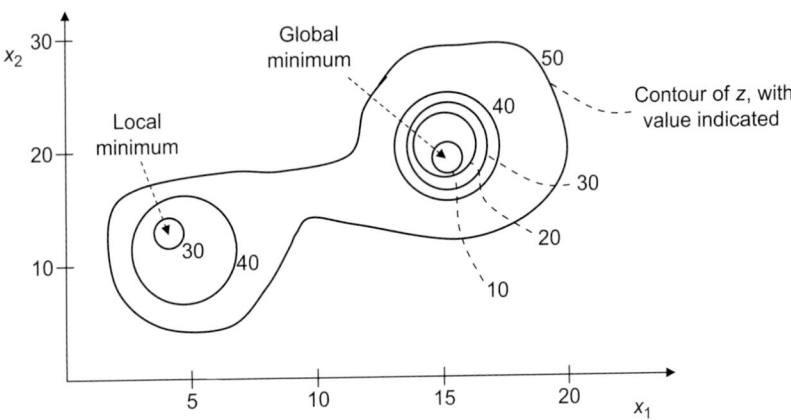

Figure 1.13. Optimization of two decision variables.

Convexity

Constraint boundaries can also be plotted in the (x_1, x_2) parameter space, as illustrated in Figure 1.14. If the constraints are not linear, then there is a possibility that the feasible region may not be convex. A convex feasible region, illustrated in Figure 1.14a is one in which any point on a straight line between any two points inside the feasible region also lies within the feasible region. This can be stated mathematically as:

$$x = \alpha x_a + (1 - \alpha)x_b \in \text{FR}$$
$$\forall x_a, x_b \in \text{FR}, 0 < \alpha < 1 \tag{1.7}$$

where
 x_a, x_b = any two points belonging to the feasible region
 FR = the set of points inside the feasible region bounded by the constraints
 α = a constant

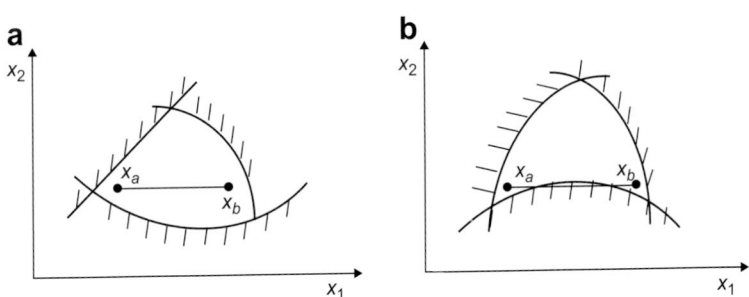

Figure 1.14. Convexity for a two-variable problem. (a) Convex feasible region. (b) Non-convex feasible region.

If any two points in the feasible region can be found such that some point on a straight line between them lies outside of the feasible region, then the feasible region is non-convex, as illustrated in Figure 1.14b.

The importance of convexity is that problems with a convex feasible region are more easily solved to a global optimum. Problems with non-convex feasible regions are prone to convergence to local minima.

Searching in Two Dimensions

The procedures for searching in two dimensions are mostly extensions of the methods used for single variable line searches:

1. Find an initial solution (x_1, x_2) inside the feasible region.
2. Determine a search direction.
3. Determine step lengths δx_1 and δx_2.
4. Evaluate $z = f(x_1 + \delta x_1, x_2 + \delta x_2)$.
5. Repeat steps 2 to 4 until convergence.

If x_1 and x_2 are varied one at a time then the method is known as a univariate search and is the same as carrying out successive line searches. If the step length is determined so as to find the minimum with respect to the variable searched then the calculation steps towards the optimum, as shown in Figure 1.15a. This method is simple to implement, but can be very slow to converge. Other direct methods include pattern searches such as the factorial designs used in statistical design of experiments (see, for example, Montgomery, 2001), the EVOP method (Box, 1957) and the sequential simplex method (Spendley *et al.*, 1962).

Indirect methods can also be applied to problems with two or more decision variables. In the steepest descent method (also known as the gradient method), the search direction is along the gradient at point (x_1, x_2), i.e., orthogonal to the contours of $f(x_1, x_2)$. A line search is then carried out to establish a new minimum point where the gradient is re-evaluated. This procedure is repeated until the convergence criterion is met, as shown in Figure 1.15b.

Problems in Multivariable Optimization

Some common problems that are encountered in multivariable optimization can be described for a two-variable problem and are illustrated in Figure 1.16. In

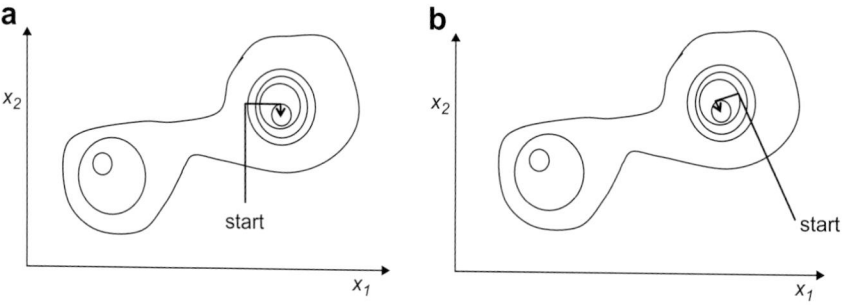

Figure 1.15.   Search methods. (a) Univariate search. (b) Steepest descent.

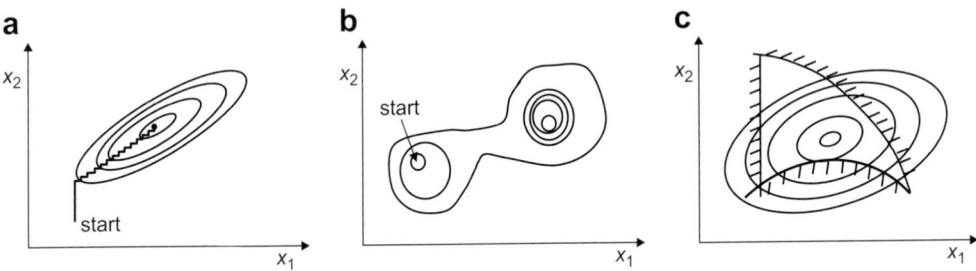

Figure 1.16. Common problems in multivariable optimization. (a) Slow convergence. (b) Convergence to local optimum. (c) Non-convex feasible region.

Figure 1.16a, the shape of the contours is such that a univariate search would be very slow to converge. Using an indirect method such as steepest descent would be more appropriate in this case. Figure 1.16b shows the problem of convergence to a local optimum. In this scenario, different answers are obtained for different initial solutions. This problem can be overcome by using pattern searches with a larger grid or by using probabilistic methods such as simulated annealing or genetic algorithms that introduce some possibility of moving away from a local optimum. An introduction to probabilistic methods is given in Diwekar (2003). Probabilistic methods are also useful when faced with a non-convex feasible region, as pictured in Figure 1.16c.

Multivariable Optimization

When there are more than two decision variables it is much harder to visualize the parameter space, but the same issues of initialization, convergence, convexity and local optima are faced. The solution of large multivariable optimization problems is at the core of the field of operations research. Operations research methods are widely used in industry, particularly in manufacturing facilities, as discussed in Section 1.9.11.

The following sections give only a cursory overview of this fascinating subject. Readers who are interested in learning more should refer to Hillier and Lieberman (2002) and the other references cited below.

1.9.8. Linear Programming

A set of continuous linear constraints always defines a convex feasible region. If the objective function is also linear and $x_i > 0$ for all x_i, then the problem can be written as a linear program (LP). A simple two-variable illustration of a linear program is given in Figure 1.17.

Linear programs always solve to a global optimum. The optimum must lie on the boundary at an intersection between constraints, which is known as a *vertex* of the feasible region. The inequality constraints that intersect at the optimum are said to be active and have $h(\mathbf{x}) = 0$, where $\mathbf{x}$ is the vector of decision variables.

Many algorithms have been developed for solution of linear programs, of which the most widely used are based on the SIMPLEX algorithm developed by Dantzig

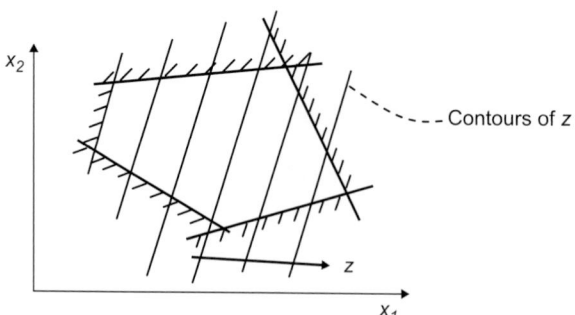

Figure 1.17.   A linear program.

(1963). The SIMPLEX method introduces slack and surplus variables to transform the inequality constraints into equalities. For example, if:

$$x_1 + x_2 - 30 \leq 0$$

we can introduce a slack variable, S_1, and write:

$$x_1 + x_2 - 30 + S_1 = 0$$

The resulting set of equalities is solved to obtain a feasible solution, in which some of the slack and surplus variables will be zero, corresponding to active constraints. The algorithm then searches the vertices of the feasible region, increasing the objective at each step until the optimum is reached. Details of the SIMPLEX method are given in most optimization or operations research textbooks. See, for example, Hillier and Lieberman (2002) or Edgar and Himmelblau (2001). There have been many improvements to the SIMPLEX algorithm over the years, but it is still the method used in most commercial solvers.

Some problems that can occur in solving linear programs are illustrated in Figure 1.18. In Figure 1.18a, the contours of the objective function are exactly parallel to one of the constraints. The problem is said to be degenerate and has an infinite number of solutions along the line of that constraint. Figure 1.18b shows a problem where the feasible region is unbounded. This situation does not usually

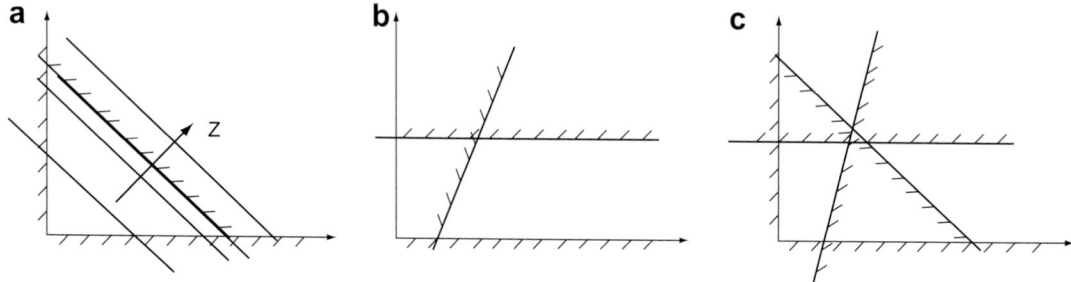

Figure 1.18.   Problems in linear programming. (a) Objective function parallel to a constraint (degenerate problem). (b) Feasible region unbounded. (c) No feasible region.

occur in engineering design unless the problem has been badly formulated. The situation in Figure 1.18c is more common, in which the problem is over-constrained and there is no feasible region.

Linear programming can be used to solve very large problems, with thousands of variables and constraints. The method is widely used in operations, particularly in optimization of oil refineries and petrochemical plants. It is used a lot less in design, as design problems almost inevitably contain many non-linear equations.

1.9.9. Non-Linear Programming

When the objective function and/or the constraints are non-linear, then the optimization must be solved as a non-linear program (NLP). Three main methods are used for solving a NLP.

Successive Linear Programming (SLP)

In successive linear programming, $f(\mathbf{x})$, $\mathbf{g}(\mathbf{x})$ and $\mathbf{h}(\mathbf{x})$ are linearized at an initial point. The resulting LP is solved to give an initial solution, and $f(\mathbf{x})$, $\mathbf{g}(\mathbf{x})$ and $\mathbf{h}(\mathbf{x})$ are linearized again at the new point. The procedure is then repeated until convergence. If the new point is outside the feasible region, then the nearest point lying inside the feasible region is used.

With SLP there is no guarantee of convergence or global optimality. The method is widely used, nonetheless, as it is a simple extension of linear programming. It should be noted that whenever discontinuous linear functions are used to approximate a non-linear function then the problem behaves like a SLP. There is no guarantee of convexity or convergence to the optimal solution.

Successive Quadratic Programming (SQP)

The SQP algorithm is similar to SLP, but instead approximates $f(\mathbf{x})$ as a quadratic function and uses quadratic programming methods that give faster convergence than SLP. SQP works well for highly non-linear problems with relatively few variables, for example, optimizing a process simulation or the design of a single piece of equipment. Biegler *et al.* (1997) suggest SQP is the best method for problems with fewer than fifty variables and where the gradients must be found numerically.

Reduced Gradient Method

Reduced gradient methods are related to the SIMPLEX algorithm. The method linearizes the constraints and introduces slack and surplus variables to transform the inequalities into equalities. The n-dimensional vector $\mathbf{x}$ is then partitioned into $n - m$ independent variables, where m is the number of constraints. A search direction is determined in the space of the independent variables and a quasi-Newton method is used to determine an improved solution of $f(\mathbf{x})$ that still satisfies the non-linear constraints. If all the equations are linear, this reduces to the SIMPLEX method (Wolfe, 1962). Various algorithms have been proposed, using different methods for carrying out the search and returning to a feasible solution, for example the generalized reduced gradient (GRG) algorithm (Abadie and Guigou, 1969) and the MINOS algorithm (Murtagh and Saunders, 1978, 1982).

Reduced gradient methods are particularly effective for sparse problems with a large number of variables. A problem is said to be sparse if each constraint involves only a few of the variables. This is a common situation in design problems, where many of the constraints are written in terms of only one or two variables. Reduced gradient methods also work better when many of the constraints are linear, as less computational time is spent linearizing constraints and returning the solution to the feasible region. Because of the decomposition of the problem, fewer calculations are required per iteration, particularly if analytical expressions for the gradients are known (which is usually not the case in design). The reduced gradient method is often used in optimizing large spreadsheet models.

All of the non-linear programming algorithms can suffer from the convergence and local optima problems described in Section 1.9.7. Probabilistic methods such as simulated annealing and genetic algorithms can be used if it is suspected that the feasible region is non-convex or multiple local optima are present.

1.9.10. Mixed Integer Programming

Many of the decisions faced in operations involve discrete variables. For example, if we need to ship 3.25 trucks of product from plant A to plant B each week, we could send three trucks for 3 weeks and then four trucks in the fourth week, or we could send four trucks each week, with the fourth truck only one-quarter filled, but we cannot send 3.25 trucks every week. Some common operational problems involving discrete variables include:

- Production scheduling: determine the production schedule and inventory to minimize the cost of meeting demand. This is particularly important for batch plants, when the plant can make different products.
- Trans-shipment problems and supply chain management: satisfy demands at different producing plants and sales destinations from different supply points, warehouses and production facilities.
- Assignment problems: schedule workers to different tasks.

Discrete variables are also sometimes used in process design, for example, the number of trays or the feed tray of a distillation column, and in process synthesis, to allow selection between flow-sheet options, as described below.

Discrete decisions are addressed in operations research by introducing integer variables. When integer variables are introduced, a linear program becomes a mixed-integer linear program (MILP) and a non-linear program becomes a mixed-integer non-linear program (MINLP). Binary integer variables are particularly useful, as they can be used to formulate rules that enable the optimization program to choose between options. For example, if we define y as a binary integer variable such that:

if $y = 1$ a feature exists in the optimal solution, and
if $y = 0$ the feature does not exsist in the optimal solution,

then we can formulate constraint equations such as:

$$\sum_{i=1}^{n} y_i = 1 \qquad \text{choose only one of } n \text{ options}$$

$$\sum_{i=1}^{n} y_i \leq m \qquad \text{choose at most } m \text{ of } n \text{ options}$$

$$\sum_{i=1}^{n} y_i \geq m \qquad \text{choose at least } m \text{ of } n \text{ options}$$

$$y_k - y_j \leq 0 \qquad \text{if item } k \text{ is selected, item } j \text{ must be selected, but not vice versa}$$

$$\left.\begin{array}{l} g_1(x) - M y \leq 0 \\ g_2(x) - M(1-y) \leq 0 \\ M \text{ is a large scalar value} \end{array}\right\} \qquad \text{either } g_1(x) \leq 0 \text{ or } g_2(x) \leq 0$$

The last rule listed above can be used to select between alternative constraints.

Mixed-Integer Programming Algorithms

Although integer variables are convenient for problem formulation, if too many integer variables are used, the number of options explodes in a combinatorial manner and solution becomes difficult. MILP problems can be solved efficiently using methods such as the 'branch and bound' algorithm. The branch and bound method starts by treating all integer variables as continuous and solving the resulting LP or NLP to give a first approximation. All integer variables are then rounded to the nearest integer to give a second approximation. The problem is then partitioned into two new integer problems for each integer variable that had a non-integral solution in the first approximation. In one branch a constraint is added that forces the integer variable to be greater than or equal to the next highest integer, while in the other branch a constraint is added that forces the variable to be equal to or less than the next lowest integer. For example, if a variable was found to have an optimal value $y = 4.4$ in the first approximation, then the new constraints would be $y \geq 5$ in one branch and $y \leq 4$ in the other. The branched problems are then solved to give new first approximations, and the branching procedure is repeated until an integer solution is found.

When an integer solution is found, it is used to set a bound on the value of the objective. For example, in a minimization problem, the optimal solution must be less than or equal to the bound set by this integral solution. Consequently, all branches with greater values of the objective can be discarded, as forcing the variables in these branches to integer values will lead to deterioration in the objective rather than improvement. The procedure then continues branching on all the non-integral integer variables from each first approximation, and setting new bounds each time an improved integer solution is found, until all of the branches have been bounded and the optimal solution has been obtained. See Hillier and Lieberman (2002) or Edgar and Himmelblau (2001) for details of the algorithm and examples of its application.

The branch and bound method can be used for MINLP problems, but it requires solving a large number of NLP problems and is, therefore, computationally intensive. Instead, methods such as the Generalized Benders' Decomposition and Outer Approximation algorithms are usually preferred. These methods solve a master MILP problem to initialize the discrete variables at each stage and then solve a NLP sub-problem to optimize the continuous variables. Details of these methods are given in Biegler *et al.* (1997) and Diwekar (2003).

Superstructure Optimization

Binary integer variables can be used to formulate optimization problems that choose between flow-sheet options. For example, consider the problem of selecting a reactor. We can set up a unit cell consisting of a well-mixed reactor, a plug-flow reactor and a bypass in parallel, each with a valve upstream, as illustrated in Figure 1.19a. If a binary variable is used to describe whether the valve is open or closed and a constraint is introduced such that only one of the valves is open, then the optimization will select the best option. A set of such unit cells can be built into a superstructure, incorporating additional features such as recycles, as shown schematically in Figure 1.19b. A more rigorous superstructure that encompasses other options such as side-stream feeds to the PFR was developed by Kokossis and Floudas (1990).

The optimization of such a superstructure can identify reactor networks or mixing arrangements that would not be intuitively obvious to the design engineer. Similar superstructure formulations have been proposed for other process synthesis problems such as distillation column sequencing, design of heat-exchange networks and design of site utility systems. Biegler *et al.* (1997) give an excellent overview of the use of superstructure-based methods in process synthesis.

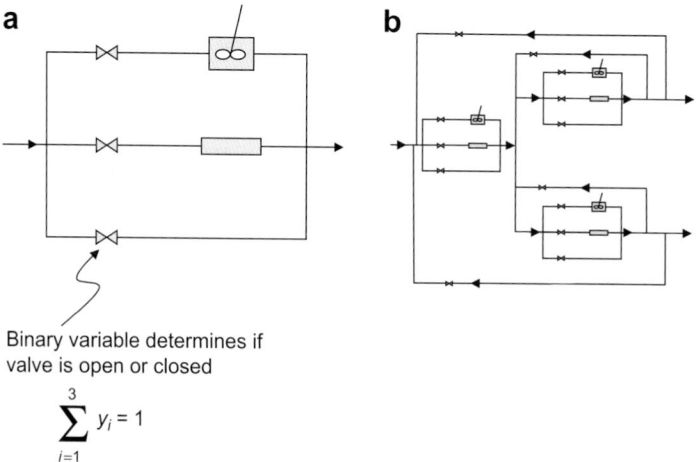

Binary variable determines if valve is open or closed

$$\sum_{i=1}^{3} y_i = 1$$

Figure 1.19.   Application of integer programming to reactor design. (a) Unit cell of reactor options. (b) Superstructure of unit cells and recycles.

1.9.11. Optimization in Industrial Practice

Optimization of Process Operations

Perhaps not surprisingly, operations research methods are widely used in process operations. Few manufacturing plants do not use LP or MILP tools for planning and scheduling. Supply chain management is very important to economic performance, and is usually carried out using large MILP models. The models used in industry for these purposes are often not very sophisticated, but proper formulation of constraints and the ability to solve robustly with a large number of variables are usually more important features of tools for these applications.

Optimization of Batch and Semi-Continuous Processes

In batch operation, there will be periods when product is being produced, followed by non-productive periods when the product is discharged and the equipment prepared for the next batch. The rate of production will be determined by the total batch time, productive plus non-productive periods.

$$\text{Batches per year} = \frac{8760 \times \text{plant attainment}}{\text{batch cycle time}} \tag{1.8}$$

where the 'plant attainment' is the fraction of the total hours in a year (8760) that the plant is in operation.

Annual production = quantity produced per batch × batches per year.

$$\text{Cost per unit of production} = \frac{\text{annual cost of production}}{\text{annual production rate}} \tag{1.9}$$

With many batch operations, the production rate decreases during the production period; for example, batch reactors and plate and frame filter presses. There is then an optimum batch size, or optimum cycle time, that gives the minimum cost per unit of production.

For some continuous processes, the period of continuous production will be limited by gradual changes in process conditions. Examples include the deactivation of catalysts or the fouling of heat-exchange surfaces. Production is lost during the periods when the plant is shut down for catalyst renewal or equipment clean-up. As with batch processes, there is an optimum cycle time to give the minimum production cost. The optimum time between shut-downs can be found by determining the relationship between cycle time and cost per unit of production (the objective function) and using one of the optimization techniques outlined in this section to find the minimum.

With discontinuous processes, the period between shut-downs will usually be a function of equipment size. Increasing the size of critical equipment will extend the production period, but at the expense of increased capital cost. The designer must strike a balance between the savings gained by reducing the non-productive period and the increased investment required.

In some batch plants, several trains of identical equipment are operated in a sequence that allows some degree of heat recovery or enables downstream equipment to operate continuously. In this type of plant the time allowed for each operation in the sequence is optimized so that an overall schedule for the plant can be developed. Scheduling of batch processes is described in Biegler *et al.* (1997).

Optimization in Process Design

Few, if any, industrial designs are rigorously optimized. This is because:

1. The cost of building rigorous models of reactor kinetics and hydraulics that give accurate prediction of by-product yields is usually not justified. The amount of time available for the project is usually insufficient for such models to be built.
2. The uncertainty in the forecasts of future prices is usually so large that it dominates most differences between design alternatives.
3. Regardless of the quality of the tools used, or the experience of the estimator, it is usually impossible to make a capital cost estimate within $\pm$ 15% without completing a substantial amount of design work (see Chapter 6). Many design decisions are thus made on the basis of sketchy cost estimates. The cost of going back and revisiting these design decisions at a later stage in the project when more design detail is available is usually not justified.
4. Criteria such as safety, operability, reliability and flexibility are of vital importance in process design. These features make the design more robust to variations in the design assumptions and operating requirements. A safe, operable and reliable plant will often require more expense above the cost of the 'optimal' design. This extra expense is difficult to trade-off against the non-financial benefits of having a process that is easier to run.
5. In most cases there are several 'near optimal' designs. The difference between the values of the objective obtained for each of these is often not statistically significant, given the errors in prices, cost estimates and yields.

In industrial process design, optimization usually involves carrying out sufficient analysis to be certain that the design is reasonably close to the optimum. The most important things for the design engineer to understand are:

1. What are the constraints on the design?
2. Which constraints are hard (inviolable) and which are soft (can be modified)?
3. Where are the discontinuities in cost? For example, what conditions cause a change to a more costly metallurgy or a different design code?
4. What are the main design trade-offs?
5. How does the objective function vary with the main process parameters?
6. What are the major cost components of the process (both capital and operating costs) and what radical changes could be made to the process to reduce these costs?

Experienced design engineers usually think through these questions carefully, to satisfy themselves that their design is 'good enough'. Only very occasionally do they formulate an optimization problem and solve it rigorously.

Example 1.1

Optimize the design of a distillation column to separate 225 metric tons per hour of an equimolar mixture of benzene, toluene, ethylbenzene, paraxylene and orthoxylene with minimum total annualized cost. The feed is a saturated liquid at 330 kPa. The recovery of toluene in the distillate should be greater than 99%, and the recovery of ethylbenzene in the bottoms should be greater than 99%.

Solution

The first step is to determine the design factor. If we assume a design factor of 10%, then the equipment should be designed for a flow rate of 248 metric tons per hour (t/h). This flow rate is used in simulating the process for the purpose of sizing equipment, but energy consumption must be based on the reboiler and condenser duties expected for a 225 t/h feed rate.

This is a single distillation column, which is easy to model in any of the commercial simulation programs. UniSim Design™ (Honeywell Inc.) was used for the purpose of this example. The simulation was set up using the component recoveries of toluene and ethylbenzene as column specifications, which gave rapid convergence. The simulation of this problem is described in more detail in Chapter 4. Tray sizing calculations were run using the UniSim Design tray sizing utility. A tray spacing of 0.61 m (2 feet) was assumed, and other tray parameters were left at the UniSim Design default values. Two metres were added to the column height to allow space for a sump and demister. Sieve trays were used and the stage efficiency was assumed to be 80%.

To optimize the design, we need to formulate an objective function. The distillation column has the following cost components:

- Capital costs: column shell, internals, condenser, receiver drum, reboiler, pumps, piping, instrumentation, structure, foundations, etc.
- Operating costs: cost of heating for the reboiler and cost of cooling for the condenser.

The purchased equipment costs can be estimated based on information from the process simulation using the cost correlations given in Section 6.3. The column shell is a pressure vessel and the design can be completed using the methods given in Chapter 13. The details of how to complete these calculations are not important here, but Example 13.2 and Example 6.2 provide detailed explanations of the method followed. Carbon steel construction was assumed. The purchased equipment costs can be converted into an installed capital cost by multiplying by an installation factor. For the purposes of this example the installation factor can be assumed to be 4.0 (see Section 6.3). The installed capital costs can be converted into an annual capital charge by dividing by 3, using a rule of thumb that is developed in Section 6.7.6.

The operating costs are simple to estimate from the condenser and reboiler duties if the cost of energy is known. For this example, the cost of heat is taken as $5.5/GJ and the cost of cooling is $0.2/GJ.

The objective function can then be written as:

$$
\text{Min.: Total annualized cost (TAC)} = \text{cost of heating} + \text{cost of cooling} \\
+ \text{ annualized capital cost} \\
= 5.5Q_r + 0.2Q_c \\
+ (4/3)(\textstyle\sum \text{purchased equipment costs})
$$

where

Q_r = annual reboiler energy consumption (GJ/yr)

Q_c = annual condenser energy consumption (GJ/yr)

The optimization problem is strictly a MINLP, as we need to consider discrete variables (number of trays, feed tray) as well as continuous variables (reflux ratio, reboiler duty, etc.). This problem is actually relatively easy to formulate and solve rigorously, but instead we will step through the calculation to illustrate how an experienced industrial designer would approach the problem.

Table 1.3 gives the results of several iterations of the optimization.

1. To begin, we need to find a feasible solution. As an initial guess we can use 40 trays with the feed on tray 20. The column converges with a reflux ratio of 3.34 and diameter 5.49 m. This is large, but not unreasonable for such a high flow rate. Looking at the components of the total annualized cost, the capital is contributing $0.8 MM/yr and energy is contributing $8.6 MM/yr, so the costs are dominated by energy cost. It is clear that adding more stages and reducing the reflux ratio will reduce the total cost. (If capital costs were dominating then we would reduce the number of stages.) There is no upper hard constraint on column height, but there is a soft constraint. At the time of writing there are only 14 cranes in the world that can lift a column taller than 80 m. There are 48 cranes that can lift a column up to 60 m. We can therefore expect that the cost of lifting a column > 60 m height will go up as it becomes more expensive to rent the necessary equipment for installation. We can start by assuming a soft constraint that the maximum height must be less than 60 m.

2. Using 90 trays with feed on tray 45 gives a reflux ratio of 2.5 and diameter 4.42 m. The column height is 56 m, which allows some space for vessel supports and clearance for piping at the column base and still comes in under the 60-m target. The capital cost increases to $0.95 MM/yr, while energy cost

TABLE 1.3.   Optimization Results

Iteration Number	1	2	3	4	5	6	7	8	9
Number of trays	40	90	120	70	80	76	84	80	80
Feed tray	20	45	60	35	40	38	42	27	53
Column height (m)	26.4	56.9	75.2	44.7	50.8	48.4	53.2	50.8	50.8
Column diameter (m)	5.49	4.42	4.42	4.42	4.42	4.42	4.42	4.42	4.57
Reflux ratio	3.34	2.50	2.48	2.57	2.52	2.54	2.51	2.48	2.78
Reboiler duty, Q_r (GJ)	34.9	28.3	28.2	28.8	28.5	28.6	28.4	28.2	30.4
Condenser duty, Q_c (GJ)	33.9	27.3	27.2	27.8	27.5	27.6	27.4	27.2	29.4
Annualized capital cost (MM$/y)	0.82	0.95	1.25	0.83	0.89	0.87	0.91	0.89	0.94
Annual energy cost (MM$/y)	8.59	6.96	6.93	7.10	7.01	7.04	6.99	6.93	7.50
Total annualized cost (MM$/y)	9.41	7.91	8.18	7.93	7.900	7.905	7.904	7.82	8.44

is reduced to $6.96 MM/yr, giving a total annualized cost of $7.91 MM/yr and savings of $1.5 MM/yr relative to the initial design.

3. We should explore whether going to an even taller column would make sense. We can investigate this by increasing the installation factor from 4 to 5 for the column shell to allow for the higher cost of using one of the larger cranes. If we increase the number of trays to 120, the column height is 75 m, which will give a total height of close to 80 m when installed. The total annualized cost increases to $8.2 MM/yr, so we can conclude that it is probably not economical to go to a total height above 60 m. We can notice though that the reflux ratio didn't change much when we added extra trays. This suggests that we are getting close to minimum reflux. It might therefore be worth backing off from the maximum column height constraint to see if there is an optimum number of trays.

4. Adding a design with 70 trays and feed on tray 35 (roughly half way between 40 and 90) gives reflux ratio 2.57 and total annualized cost $7.94 MM/yr. This is not an improvement on the 90-tray design, so the optimum must be between 70 and 90 trays.

5. A design with 80 trays and feed on tray 80 (half way between 70 and 90) gives reflux ratio 2.52 and total annualized cost $7.900 MM/yr. This is better than 70 or 90 trays. If we wanted to proceed further to establish the optimum we could continue reducing the search space using a regular search until we get to the optimum number of trays. Instead, an experienced designer would note that the difference in cost within the range examined ($0.03 MM/yr) is relatively small compared with the error in the capital cost estimate ($\pm$ 30%, or $0.29 MM/yr). Since the optimum appears to be fairly flat with respect to number of trays over the range 70 to 90, it is reasonable to take the optimum as 80 trays. (As a confirmation, iterations 6 and 7, with 76 and 84 trays indicate that the optimum indeed lies at 80 $\pm$ 2 trays).

6. Having fixed the number of trays at 80, we should now optimize the feed tray. We start by adding two new points, with the feed tray at trays 27 and 53. These give total annualized costs of $7.82 MM/yr and $8.43 MM/yr respectively. The minimum cost is given by the lower bound on feed tray location. If we try a higher feed tray (say, tray 26) the UniSim™ tray sizing utility gives a warning 'head loss under downcomers is too large'. We could overcome this warning by modifying the tray design, but once again we can notice that the annualized cost savings that we have gained by optimizing feed tray ($0.08 MM/yr) is small compared to the error in the capital cost, so the design with feed tray 27 is close enough to optimum.

The column design is thus set at 80 trays, with feed on tray 27, giving a column 50.8 m high and 4.42 m diameter.

The solution obtained is 'good enough', but is not rigorously optimal. Several possible variations in flow scheme were not considered. For example, we could have examined use of feed preheat, intermediate stage condensers or reboilers, or more efficient column internals such as high-efficiency trays or structured packing. The column cost may also be reduced if different diameters or different internals were used in the rectifying and stripping sections. In the broader context of the process, it may be possible to supply the heat required for the reboiler using heat recovered from elsewhere in the process, in which case the cost of energy will be reduced and the capital energy trade-off will be altered. In the overall process context, we could also question whether the column needs such high recoveries of toluene and ethylbenzene, since the high recoveries clearly lead to a high reflux rate and column energy cost.

1.10.   REFERENCES

Abadie, J. and Guigou, J. (1969) Gradient réduit generalisé, Électricité de France Note HI 069/02.

Biegler, L. T., Grossman, I. E., and Westerberg, A. W. (1997) *Systematic Methods of Chemical Process Design* (Prentice Hall).

Box, G. E. P. (1957) *Applied Statistics* 6, 81. Evolutionary operation: a method for increasing industrial productivity.

Dantzig, G. B. (1963) *Linear Programming and Extensions* (Princeton University Press).

Diwekar, U. (2003) *Introduction to Applied Optimization* (Kluwer Academic Publishers).

Edgar, T. E. and Himmelblau, D. M. (2001) *Optimization of Chemical Processes*, 2nd edn (McGraw-Hill).

Hillier, F. S. and Lieberman, G. J. (2002) *Introduction to Operations Research*, 7th edn (McGraw-Hill).

Kokossis, A. C. and Floudas, C. A. (1990) *Chem. Eng. Sci.* 45(3), 595. Optimization of complex reactor networks – 1. Isothermal operation.

Livio, M. (2002) The Golden Ratio (Random House).

Montgomery, D. C. (2001) *Design and Analysis of Experiments*, 5th edn (Wiley).

Murtagh, B. A. and Saunders, M. A. (1978) *Mathematical Programming* 14, 41. Large-scale linearly constrained optimization.

Murtagh, B. A. and Saunders, M. A. (1982) *Mathematical Programming Study* 16, 84. A projected lagrangian algorithm and its implementation for sparse non-linear constraints.

Rudd, D. F. and Watson, C. C. (1968) *Strategy of Process Design* (Wiley).

Spendley, W., Hext, G.R. and Himsworth, F. R. (1962) *Technometrics* 4, 44.

Stoecker, W. F. (1989) *Design of Thermal Systems* 3rd edn (McGraw-Hill)

Wolfe, P. (1962) *Notices Am. Math. Soc.* 9, 308. Methods of non-linear programming.

1.11.   NOMENCLATURE

		Dimensions in $MLT\theta$
a	A constant	—
b	A constant	—
$f(x)$	General function of x	—
$f'(x)$	First derivative of function of x with respect to x	—

$f''(x)$	Second derivative of function of x with respect to x	—
FR	The set of points contained in a feasible region	—
$\mathbf{g(x)}$	A m_i vector of inequality constraints	—
$g(x)$	General inequality constraint equation in x	—
$\mathbf{h(x)}$	A m_e vector of equality constraints	—
$h(x)$	General equality constraint equation in x	—
h	Step length in a search algorithm	—
M	A large scalar constant	—
m	Number of constraints	—
m_e	Number of equality constraints	—
m_i	Number of inequality constraints	—
n	Number of variables	—
Q_c	Condenser duty in distillation	ML^2T^{-3}
Q_r	Reboiler duty in distillation	ML^2T^{-3}
$S_1, S_2 \dots$	Slack and surplus variables	—
T	Temperature	θ
T_{alloy}	Maximum allowed temperature for an alloy	θ
T_A, T_B, T_C	Maximum allowed temperature for alloys A, B and C	θ
U	Overall heat transfer coefficient	$MT^{-3}\theta^{-1}$
$\mathbf{x}$	A vector of n decision variables	—
$x_1, x_2 \dots$	Continuous variables	—
$y_1, y_2 \dots$	Integer (discrete) variables	—
z	The objective (in optimization)	—
α	A constant between 0.0 and 1.0	—
ε	Fraction of search range or tolerance for convergence	—
ΔT	Temperature difference	θ
$\Delta T_{optimum}$	The optimal minimum temperature approach in heat recovery	θ

| $\delta x_1, \delta x_2$ | Small increments in x_1 and x_2 | — |
| ω | Ratio used in golden-section search (= 0.381966) | — |

Suffixes

D1	lower side of a discontinuity
D2	upper side of a discontinuity
i	i^{th} variable
j	j^{th} variable
k	k^{th} iteration
L	lower bound
S	stationary point
U	upper bound

1.12. PROBLEMS

1.1. Develop project plans for the design and construction of the following processes. Use Figure 1.2 as a guide to the activities that must occur. Estimate the overall time required from launching the project to the start of operation.

  i)   A petrochemical process using established technology, to be built on an existing site.

  ii) A process for full scale manufacture of a new drug, based on a process currently undergoing pilot plant trials.

  iii) A novel process for converting cellulose waste to fuel products.

  iv) A spent nuclear fuel reprocessing facility.

  v)   A solvent recovery system for an electronics production facility.

1.2. You are the project manager of a team that has been asked to complete the design of a chemical plant up to the stage of design selection. You have three engineers available (plus yourself) and the work must be completed in 10 weeks. Develop a project plan and schedule of tasks for each engineer. Be sure to allow sufficient time for equipment sizing, costing and optimization. What intermediate deliverables would you specify to ensure that the project stays on track?

1.3. A separator divides a process stream into three phases: a liquid organic stream, a liquid aqueous stream, and a gas stream. The feed stream contains three components, all of which are present to some extent in the separated steams. The composition and flow rate of the feed stream are known. All the streams will be at the same temperature and pressure. The phase equilibrium constants for the three components are available.

    i) How many design variables must be specified in order to calculate the output stream compositions and flow rates?

    ii) How would you optimize these variables if the objective of the separator was to maximize recovery of condensable components into the organic liquid stream? What constraints might limit the attainable recovery?

1.4. A rectangular tank with a square base is constructed from 5-mm steel plates. If the capacity required is 8 cubic meters determine the optimum dimensions if the tank has:

  i) a closed top

 ii) an open top.

1.5. Estimate the optimum thickness of insulation for the roof of a house given the following information. The insulation will be installed flat on the attic floor.

    Overall heat transfer coefficient for the insulation as a function of thickness, U values (see Chapter 12):

Thickness, mm	0	25	50	100	150	200	250
U, $Wm^{-2}K^{-1}$	20	0.9	0.7	0.3	0.25	0.20	0.15

    The cost of insulation, including installation, is $120/m^3$. Capital charges (see Chapter 6) are 20% per year. The cost of fuel, allowing for the efficiency of the heating system is $8/GJ. The cost of cooling is $5/GJ. Average temperatures for any region of the United States or Canada can be found online at www.weather.com (under the averages tab). Assume the house is heated or cooled to maintain an internal temperature in the range 21 to 27°C.

    Note: the rate at which heat is being lost or gained is given by $U \times \Delta T$, W/m^2, where U is the overall coefficient and ΔT the temperature difference; see Chapter 12.

1.6. What is the optimum practical shape for an above-ground dwelling, to minimize the heat losses through the building fabric? When is (or was) this optimum shape used? Why is this optimum shape seldom used in richer societies?

1.7. Hydrogen is manufactured from methane by either steam reforming (reaction with steam) or partial oxidation (reaction with oxygen). Both processes are endothermic. What reactor temperature and pressure would you expect to be optimal for these processes? What constraints might apply?

1.8. Ethylene and propylene are valuable monomers. A key step in the recovery of these materials is fractionation of the olefin from the corresponding paraffin (ethane or propane). These fractionation steps require refrigeration of the overhead condenser and very large distillation columns with many stages. Raising the pressure at which the column operates improves the performance of the refrigeration system, but increases the number of stages needed. Formulate the objective function for optimizing the recovery of ethylene from an ethylene-ethane mixture. What are the key constraints? What will be the main trade-offs?

1.9. If you had to design a plant for pasteurizing milk, what constraints would you place on the design?

1.10. A catalytic process was designed to make 150,000 metric tons per year of product with a net profit of $0.25/lb of product. The catalyst for the process costs $10/lb and it takes 2 months to shut down the process, empty the old catalyst, re-load fresh catalyst and re-start the process. The feed and product recovery and purification sections can be pushed to make as much as 120% of design basis capacity. The reactor section is sized with sufficient catalyst to make 100% of design basis when operated with fresh catalyst at 500°F. The reactor can only be operated at temperatures up to 620°F, for safety reasons. The reactor weight hourly space velocity (lb of product per hour per lb of catalyst) is given by the equation:

$$\text{WHSV} = 4.0 \times 10^6 \exp\left(\frac{-8000}{T}\right) \exp\left(-8.0 \times 10^{-5} \times t \times T\right)$$

where t = time on stream in months
          T = temperature

Find the optimal temperature versus time profile for the reactor and determine how long the process should be operated before the catalyst is changed out. (Hint: the initial temperature does not have to be 500°F.)

1.11. The portfolio of investment projects below has been proposed for a company for next year:

Project	NPV (MM$)	Cost (MM$)
A	100	61
B	60	28
C	70	33
D	65	30
E	50	25
F	50	17
G	45	25
H	40	12
I	40	16
J	30	10

i) Develop a spreadsheet optimization program to select the optimal portfolio of projects to maximize total portfolio net present value (NPV), given a total budget of $100 million. (This is a simple MILP.)
ii) How would the portfolio and NPV change if the budget was increased to $110 million?
iii) Because of corporate cost-cutting, the budget is reduced to $80 million. Which projects are now funded and what is the new NPV?

iv) Based on your answers to parts (i) to (iii), can you draw any conclusions on which projects are likely to be funded regardless of the financial situation?

v) Can you see any problems with this project selection strategy? If so, how would you recommend they should be addressed?

2 FUNDAMENTALS OF MATERIAL BALANCES

Chapter Contents

Key Learning Objectives

- How to use mass balances to understand process flows
- How to select a system boundary and design basis
- How to estimate reactor and process yields

2.1. INTRODUCTION

Material balances are the basis of process design. A material balance taken over the complete process will determine the quantities of raw materials required and products produced. Balances over individual process units set the process stream flows and compositions, and provide the basic equations for sizing equipment.

A good understanding of material balance calculations is essential in process design. In this chapter the fundamentals of the subject are covered, using simple examples to illustrate each topic. Practice is needed to develop expertise in handling what can often become very involved calculations. More examples and a more detailed discussion of the subject can be found in the numerous specialist books written on material and energy balance computations. Several suitable texts are listed under the heading of 'Further reading' at the end of this chapter.

For complex processes, material balances are usually completed using process simulation software, as described in Chapter 4. Significant time and effort can be wasted in process simulation if the fundamentals of material and energy balances are not properly understood. Careful attention must be paid to selecting the best basis and boundaries for material and energy balances, to prediction of yields, and to understanding recycle, purge and by-pass schemes. Short hand calculations, of the type illustrated in this chapter, should always be used to check process simulation results. Short calculations can also be used to accelerate convergence of flowsheet simulations by providing good initial estimates of recycle and make-up streams.

Material balances are also useful tools for the study of plant operation and trouble shooting. They can be used to check performance against design; to extend the often limited data available from the plant instrumentation; to check instrument calibrations; and to locate sources of material loss. Material balances are essential to obtaining high quality data from laboratory or pilot plants.

2.2. THE EQUIVALENCE OF MASS AND ENERGY

Einstein showed that mass and energy are equivalent. Energy can be converted into mass, and mass into energy. They are related by Einstein's equation:

$$E = mc^2 \tag{2.1}$$

where

E = energy, J
m = mass, kg
c = the speed of light *in vacuo*, 3×10^8 m/s

The loss of mass associated with the production of energy is significant only in nuclear reactions. Energy and matter are always considered to be separately conserved in chemical reactions.

2.3. **CONSERVATION OF MASS**

The general conservation equation for any process system can be written as:

$$\text{Material out} = \text{Material in} + \text{Generation} - \text{Consumption} - \text{Accumulation}$$

For a steady-state process the accumulation term is zero. Except in nuclear processes, mass is neither generated nor consumed; but if a chemical reaction takes place a particular chemical species may be formed or consumed in the process. If there is no chemical reaction the steady-state balance reduces to

$$\text{Material out} = \text{Material in}$$

A balance equation can be written for each separately identifiable species present, elements, compounds or radicals; and for the total material. Balances can be written for mass or for number of moles.

Example 2.1

2000 kg of a 5% slurry of calcium hydroxide in water is to be prepared by diluting a 20% slurry. Calculate the quantities required. The percentages are by weight.

Solution

Let the unknown quantities of the 20% slurry and water be X and Y respectively.
Material balance on $Ca(OH)_2$

$$\begin{array}{cc} In & Out \end{array}$$

$$X\frac{20}{100} = 2000 \times \frac{5}{100} \qquad \text{(a)}$$

Balance on water

$$X\frac{(100-20)}{100} + Y = 2000\frac{(100-5)}{100} \qquad \text{(b)}$$

From equation (a), $X = 500$ kg.
Substituting into equation (b) gives $Y = \underline{1500\ kg}$
Check material balance on total quantity:

$$X + Y = 2000$$

$$500 + 1500 = 2000, \text{correct}$$

2.4. **UNITS USED TO EXPRESS COMPOSITIONS**

When specifying a composition as a percentage it is important to state clearly the basis: weight, molar or volume.

The abbreviations w/w, wt% and %wt are used for mass (weight) basis. Volume basis is usually abbreviated vol%, LV% or v/v.

Example 2.2

Technical grade hydrochloric acid has a strength of 28% w/w, express this as a mole fraction.

Solution

Basis of calculation: 100 kg of 28% w/w acid.

$$\text{Molecular mass: water 18, HCl 36.5}$$

$$\text{Mass HCl} = 100 \times 0.28 = 28 \, \text{kg}$$

$$\text{Mass water} = 100 \times 0.72 = 72 \, \text{kg}$$

$$\text{kmol HCl} = \frac{28}{36.5} = 0.77$$

$$\text{kmol water} = \frac{72}{18} = \underline{4.00}$$

$$\text{Total mols} = 4.77$$

$$\text{mol fraction HCl} = \frac{0.77}{4.77} = 0.16$$

$$\text{mol fraction water} = \frac{4.00}{4.77} = \underline{0.84}$$

$$\text{Check total} \qquad 1.00$$

Within the accuracy needed for technical calculations, volume fractions can be taken as equivalent to mole fractions for gases, up to moderate pressures (say 25 bar).

Trace quantities are often expressed as parts per million (ppm). The basis, weight or volume, needs to be stated, for example, ppmw or ppmv.

$$\text{ppm} = \frac{\text{quantity of component}}{\text{total quantity}} \times 10^6$$

Note: 1 ppm = 10^{-4} per cent.

Minute quantities are sometimes quoted in ppb, parts per billion. This refers to an American billion (10^9), not a UK billion (10^{12}).

2.5. STOICHIOMETRY

The stoichiometric equation for a chemical reaction states unambiguously the number of molecules of the reactants and products that take part; from which the quantities can be calculated. The equation must balance.

With simple reactions it is usually possible to balance the stoichiometric equation by inspection, or by trial and error calculations. If difficulty is experienced in balancing complex equations, the problem can always be solved by writing a balance for each element present. The procedure is illustrated in Example 2.3.

Example 2.3

Write out and balance the overall equation for the manufacture of vinyl chloride from ethylene, chlorine and oxygen.

Solution

Method: write out the equation using letters for the unknown number of molecules of each reactant and product. Make a balance on each element. Solve the resulting set of equations.

$$A(C_2H_4) + B(Cl_2) + C(O_2) = D(C_2H_3Cl) + E(H_2O)$$

Balance on carbon

$$2A = 2D, \quad A = D$$

on hydrogen

$$4A = 3D + 2E$$
$$\text{substituting } D = A \text{ gives } E = \frac{A}{2}$$

on chlorine

$$2B = D, \text{ hence } B = \frac{A}{2}$$

on oxygen

$$2C = E, \quad C = \frac{E}{2} = \frac{A}{4}$$

putting $A = 1$, the equation becomes

$$C_2H_4 + \frac{1}{2}Cl_2 + \frac{1}{4}O_2 = C_2H_3Cl + \frac{1}{2}H_2O$$

multiplying through by the largest denominator to remove the fractions

$$4C_2H_4 + 2Cl_2 + O_2 = 4C_2H_3Cl + 2H_2O$$

2.6. CHOICE OF SYSTEM BOUNDARY

The conservation law holds for the complete process and any sub-division of the process. The system boundary defines the part of the process being considered. The flows into and out of the system are those crossing the boundary and must balance with material generated or consumed within the boundary.

Any process can be divided up in an arbitrary way to facilitate the material balance calculations. The judicious choice of the system boundaries can often greatly simplify what would otherwise be difficult and tortuous calculations.

No hard and fast rules can be given on the selection of suitable boundaries for all types of material balance problems. Selection of the best sub-division for any particular process is a matter of judgment, and depends on insight into the structure of the problem, which can only be gained by practice. The following general rules serve as a guide:

1. With complex processes, first take the boundary round the complete process and if possible calculate the flows in and out. Raw materials in, products and by-products out.
2. Select the boundaries to sub-divide the process into simple stages and make a balance over each stage separately.
3. Select the boundary round any stage so as to reduce the number of unknown streams to as few as possible.
4. As a first step, include any recycle streams within the system boundary (see Section 2.14).

Example 2.4

The diagram shows the main steps in a process for producing a polymer. From the following data, calculate the stream flows for a production rate of 10,000 kg/h.

Reactor selectivity for polymer   100%
Slurry polymerization   20 wt% monomer/water
Conversion   90% per pass
Catalyst   1 kg/1000 kg monomer
Short stopping agent   0.5 kg/1000 kg unreacted monomer

Filter wash water approx. 1 kg/1 kg polymer
Recovery column yield 98% (percentage recovered)
Dryer feed ~5% water, product specification 0.5% H_2O
Polymer losses in filter and dryer ~1%

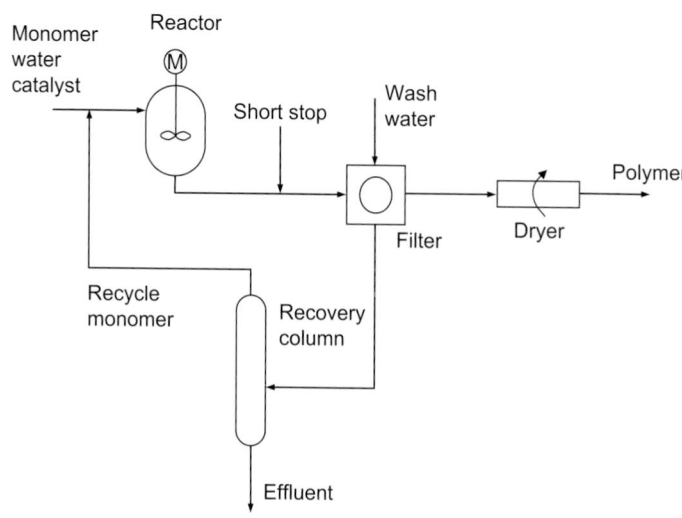

Solution

Only those flows necessary to illustrate the choice of system boundaries and method of calculation are given in the solution.

Basis: 1 h.

Take the first system boundary round the filter and dryer.

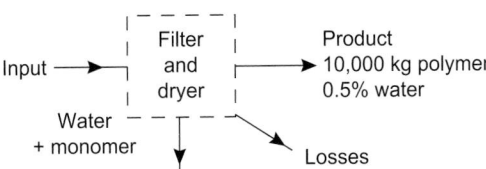

With 1% loss, polymer entering sub-system

$$= \frac{10,000}{0.99} = 10,101 \text{ kg}$$

Take the next boundary round the reactor system; the feeds to the reactor can then be calculated.

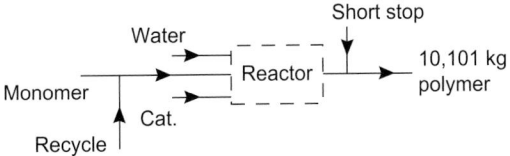

At 90% conversion per pass, monomer feed

$$= \frac{10,101}{0.9} = 11,223 \text{ kg}$$

Unreacted monomer $= 11{,}223 - 10{,}101 = 1122 \text{ kg}$

Short-stop, at 0.5 kg/1000 kg unreacted monomer

$$= 1122 \times 0.5 \times 10^{-3} = 0.6 \text{ kg}$$

Catalyst, at 1 kg/1000 kg monomer

$$= 11,223 \times 1 \times 10^{-3} = 11 \text{ kg}$$

Let water feed to reactor be F_1, then for 20% monomer

$$0.2 = \frac{11,223}{F_1 + 11,223}$$

$$F_1 = \frac{11,223(1 - 0.2)}{0.2} = 44,892 \text{ kg}$$

Now consider filter-dryer sub-system again.

Water in polymer to dryer, at 5% (neglecting polymer loss)

$$= \frac{10,101 \times 0.05}{0.95} = \underline{\underline{532\,\text{kg}}}$$

Balance over reactor–filter–dryer sub-system gives flows to recovery column.

$$\text{water}: 44,892 + 10,101 - 532 = \underline{\underline{54,461\,\text{kg}}}$$

$$\text{unreacted monomer} = \underline{\underline{1122\,\text{kg}}}$$

Now consider recovery system

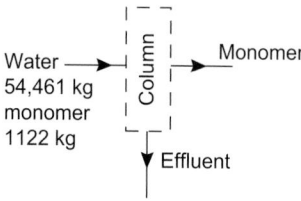

With 98% recovery, recycle to reactor

$$= 0.98 \times 1122 = \underline{\underline{1100\,\text{kg}}}$$

Composition of effluent is 23 kg monomer, 54,461 kg water.
Consider reactor monomer feed

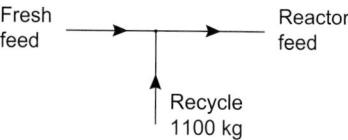

Balance round tee gives fresh monomer required

$$= 11,223 - 1100 = \underline{\underline{10,123\,\text{kg}}}$$

Note that this 12-line calculation would have required setting up a recycle and three stream-adjusts in a process simulator, illustrating that simple problems can often be solved more easily by hand or spreadsheet calculations if the boundaries for mass balances are chosen carefully.

2.7. CHOICE OF BASIS FOR CALCULATIONS

The choice of the basis for a calculation often determines whether the calculation proves to be simple or complex. As with the choice of system boundaries, no all-embracing rules or procedures can be given for the selection of the right basis for any problem. The selection depends on judgment gained by experience. Some guide rules are:

1. Time: choose the time basis in which the results are to be presented; for example kg/h, t/y, unless this leads to very large or very small numbers, where rounding errors can become problematic.
2. For batch processes use one batch.
3. Choose as the mass basis the stream flow for which most information is given.
4. It is often easier to work in moles, rather than weight, even when no reaction is involved.
5. For gases, if the compositions are given by volume, use a molar basis, remembering that volume fractions are equivalent to mole fractions up to moderate pressures.

2.8. NUMBER OF INDEPENDENT COMPONENTS

A balance equation can be written for each independent component. Not all the components in a material balance will be independent.

Physical Systems, No Reaction

If there is no chemical reaction the number of independent components is equal to the number of distinct chemical species present.

Consider the production of a nitration acid by mixing 70% nitric and 98% sulphuric acid. The number of distinct chemical species is 3; water, sulphuric acid, nitric acid.

Chemical Systems, Reaction

If the process involves chemical reaction, the number of independent components is not necessarily equal to the number of chemical species, as some may be related by the chemical equation. In this situation the number of independent components can be calculated by the following relationship:

$$\begin{aligned} \text{Number of independent components} = {} & \text{Number of chemical species} \\ & - \text{Number of independent} \\ & \quad \text{chemical equations} \end{aligned} \quad (2.2)$$

Example 2.5

If nitration acid is made up using oleum in place of the 98% sulphuric acid, there will be four distinct chemical species: sulphuric acid, sulphur trioxide, nitric acid and water. The sulphur trioxide will react with the water producing sulphuric acid so there are only three independent components.

Reaction equation $SO_3 + H_2O \rightarrow H_2SO_4$

No. of chemical species	4
No. of reactions	1
No. of independent equations	3

2.9. CONSTRAINTS ON FLOWS AND COMPOSITIONS

It is obvious, but worth emphasizing, that the sum of the individual component flows in any stream cannot exceed the total stream flow. Also, the sum of the individual molar or weight fractions must equal 1. Hence, the composition of a stream is completely defined if all but one of the component concentrations are given.

The component flows in a stream (or the quantities in a batch) are completely defined by any of the following:

1. Specifying the flow (or quantity) of each component.
2. Specifying the total flow (or quantity) and the composition.
3. Specifying the flow (or quantity) of one component and the composition.

Example 2.6

The feed stream to a reactor contains: ethylene 16%, oxygen 9%, nitrogen 31%, and hydrogen chloride. If the ethylene flow is 5000 kg/h, calculate the individual component flows and the total stream flow. All percentages are by weight.

Solution

$$\text{Percentage HCl} = 100 - (16 + 9 + 31) = \underline{44}$$

$$\text{Percentage ethylene} = \frac{5000}{\text{total}} \times 100 = \underline{16}$$

$$\text{hence total flow} = 5000 \times \frac{100}{16} = \underline{\underline{31,250 \text{ kg/h}}}$$

$$\text{so, oxygen flow} = \frac{9}{100} \times 31,250 = \underline{\underline{2813 \text{ kg/h}}}$$

$$\text{nitrogen} = 31,250 \times \frac{31}{100} = \underline{\underline{9687 \text{ kg/h}}}$$

$$\text{hydrogen chloride} = 31,250 \times \frac{44}{100} = \underline{\underline{13,750 \text{ kg/h}}}$$

General rule: the ratio of the flow of any component to the flow of any other component is the same as the ratio of the compositions of the two components.

The flow of any component in Example 2.6 could have been calculated directly from the ratio of the percentage to that of ethylene, and the ethylene flow.

$$\text{Flow of hydrogen chloride} = \frac{44}{16} \times 5000 = \underline{\underline{13,750 \text{ kg/h}}}$$

2.10. GENERAL ALGEBRAIC METHOD

Simple material-balance problems involving only a few streams and with a few unknowns can usually be solved by simple direct methods. The relationship between the unknown quantities and the information given can usually be clearly seen. For more complex problems, and for problems with several processing steps, a more formal algebraic approach can be used. The procedure is tedious if the calculations have to be done manually, but should result in a solution to even the most intractable problems, providing sufficient information is given for the problem to have a solution.

Algebraic symbols are assigned to all the unknown flows and compositions. Balance equations are then written around each sub-system for the independent components (chemical species or elements).

Material-balance problems are particular examples of the general design problem discussed in Chapter 1. The unknowns are compositions or flows, and the relating equations arise from the conservation law and the stoichiometry of the reactions. For any problem to have a unique solution it must be possible to write the same number of independent equations as there are unknowns.

Consider the general material-balance problem where there are N_s streams each containing N_c independent components. Then the number of variables, N_v, is given by:

$$N_v = N_c \times N_s \tag{2.3}$$

If N_e independent balance equations can be written, then the number of variables, N_d, that must be specified for a unique solution, is given by:

$$N_d = (N_s \times N_c) - N_e \tag{2.4}$$

Consider a simple mixing problem:

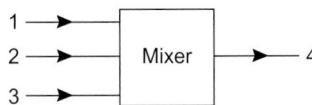

Let F_n be the total flow in stream n, and $x_{n,m}$ the concentration of component m in stream n. Then the general balance equation can be written

$$F_1 x_{1,m} + F_2 x_{2,m} + F_3 x_{3,m} = F_4 x_{4,m} \tag{2.5}$$

A balance equation can also be written for the total of each stream:

$$F_1 + F_2 + F_3 = F_4 \tag{2.6}$$

but this could be obtained by adding the individual component equations, and so is not an additional independent equation. There are m independent equations, the number of independent components.

This problem has $4m$ variables and m independent balance equations. There will be a unique solution to the problem if we specify $3m$ variables, for example, the molar flows of all m components in the three feeds or in two feeds and the product.

Consider a separation unit, such as a distillation column, that divides a process stream into two product streams. Let the feed rate be 10,000 kg/h; composition benzene 60%, toluene 30%, xylenes 10%.

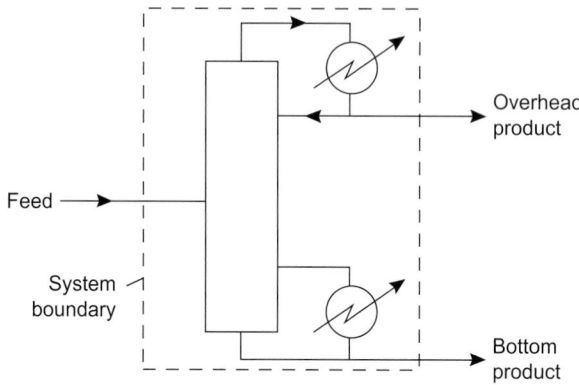

There are three streams, feed, overheads and bottoms, and three independent components in each stream.

Number of variables (component flow rates) = 9
Number of independent material balance equations = 3
Number of variables to be specified for a unique solution = 9 − 3 = 6

Three variables are specified; the feed flow and composition fixes the flow of each component in the feed.

Number of variables to be specified by designer = 6 − 3 = 3. Any three component flows can be chosen.

Normally the top composition and flow or the bottom composition and flow would be chosen.

If the primary function of the column is to separate the benzene from the other components, the maximum toluene and xylenes in the overheads would be specified; say, at 5 kg/h and 3 kg/h, and the loss of benzene in the bottoms also specified; say, at not greater than 5 kg/h. Three flows are specified, so the other flows can be calculated.

Benzene in overheads = benzene in feed − benzene in bottoms

$$0.6 \times 10,000 - 5 = \underline{\underline{5995 \text{ kg/h}}}$$

Toluene in bottoms = toluene in feed − toluene in overheads

$$0.3 \times 10,000 - 5 = \underline{\underline{2995 \text{ kg/h}}}$$

Xylenes in bottoms = xylenes in feed − xylenes in overheads

$$0.1 \times 10,000 - 3 = \underline{\underline{997 \text{ kg/h}}}$$

2.11. TIE COMPONENTS

In Section 2.9 it was shown that the flow of any component was in the same ratio to the flow of any other component, as the ratio of the concentrations of the two components. If one component passes unchanged through a process unit it can be used to tie the inlet and outlet compositions.

This technique is particularly useful in handling combustion calculations where the nitrogen in the combustion air passes through unreacted and is used as the tie component. This is illustrated in Example 2.8.

This principle is also often used in experiments to measure the flow of a process stream by introducing a measured flow of some easily analysed, compatible and inert material.

Example 2.7

Carbon dioxide is added at a rate of 10 kg/h to an air stream and the air is sampled at a sufficient distance downstream to ensure complete mixing. If the analysis shows 0.45% v/v CO_2, calculate the air-flow rate.

Solution

Normal carbon dioxide content of air is 0.03%

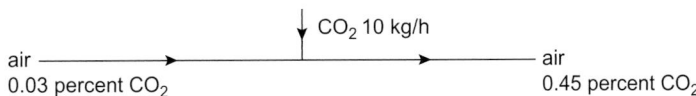

air ——————————→ air
0.03 percent CO_2 0.45 percent CO_2

Basis: kmol/h, as percentages are by volume.

$$\text{kmol/h } CO_2 \text{introduced} = \frac{10}{44} = 0.2273$$

Let X be the air flow.
Balance on CO_2, the tie component

$$CO_2 \text{ in} = 0.0003X + 0.2273$$
$$CO_2 \text{ out} = 0.0045X$$
$$X(0.0045 - 0.0003) = 0.2273$$
$$X = 0.2273/0.0042 = 54 \text{ kmol/h}$$
$$= 54 \times 29 = \underline{\underline{1560 \text{ kg/h}}}$$

Example 2.8

In a test on a furnace fired with natural gas (composition 95% methane, 5% nitrogen), the following flue gas analysis was obtained: carbon dioxide 9.1%, carbon monoxide 0.2%, oxygen 4.6%, nitrogen 86.1%, all percentages by volume.

Calculate the percentage excess air flow (percentage above stoichiometric).

Solution

$$\text{Reaction: } CH_4 + 2O_2 \rightarrow CO_2 + 2H_2O$$

Note: the flue gas analysis is reported on a dry basis, any water formed having been condensed out. Nitrogen is the tie component.

Basis: 100 mol, dry flue gas; as the analysis of the flue gas is known, the moles of each element in the flue gas (flow out) can be easily calculated and related to the flow into the system.

Let the quantity of fuel (natural gas) per 100 mol dry flue gas be X.
Balance on carbon, moles in fuel = moles in flue gas

$$0.95\,X = 9.1 + 0.2, \text{hence } X = 9.79 \text{ mol}$$

Balance on nitrogen (composition of air O_2 21%, N_2 79%):
Let Y be the flow of air per 100 mol dry flue gas.

$$N_2 \text{ in air} + N_2 \text{ in fuel} = N_2 \text{ in flue gas}$$

$$0.79\,Y + 0.05 \times 9.79 = 86.1, \text{hence } Y = 108.4 \text{ mol}$$

Stoichiometric air: from the reaction equation 1 mol methane requires 2 mol oxygen,

$$\text{so, stoichiometric air} = 9.79 \times 0.95 \times 2 \times \frac{100}{21} = \underline{\underline{88.6 \text{ mol}}}$$

$$\text{Percentage excess air} = \frac{(\text{air supplied} - \text{stoichiometric air})}{\text{stoichiometric air}} \times 100$$

$$= \frac{108.4 - 88.6}{88.6} = \underline{\underline{22 \text{ per cent}}}$$

Note that we simplified the problem by neglecting the carbon dioxide present in the ambient air. From Example 2.7, this is 0.03%, i.e., $108.4 \times 0.00003 = 0.0325$ mol per 100 mol of dry flue gas, or roughly one third of one per cent error. This does not significantly affect the calculation of the percentage excess air.

2.12. EXCESS REAGENT

In industrial reactions, the components are seldom fed to the reactor in exact stoichiometric proportions. A reagent may be supplied in excess to promote the desired reaction; to optimize the use of an expensive reagent; or to ensure complete reaction of a reagent, as in combustion.

The percentage excess reagent is defined by the following equation:

$$\text{Per cent excess} = \frac{\text{quantity supplied} - \text{stoichiometric}}{\text{stoichiometric quantity}} \times 100 \qquad (2.7)$$

It is necessary to state clearly to which reagent the excess refers.

Example 2.9

To ensure complete combustion, 20% excess air is supplied to a furnace burning natural gas. The gas composition (by volume) is methane 95%, ethane 5%.

Calculate the moles of air required per mole of fuel.

Solution

Basis: 100 mol gas, as the analysis is volume percentage.

Reactions: $CH_4 + 2O_2 \rightarrow CO_2 + 2H_2O$
$$C_2H_6 + 3.5O_2 \rightarrow 2CO_2 + 3H_2O$$

Stoichimetric moles of O_2 required $= 95 \times 2 + 5 \times 3.5 = 207.5$
With 20% excess, moles of O_2 required $= 207.5 \times 1.2 = 249$
Moles of air $(21\% \ O_2) = 249 \times 100/21 = 1185.7$
Air per mole of fuel $= 1185.7/100 = \underline{11.86 \ mol}$

2.13. CONVERSION, SELECTIVITY AND YIELD

It is important to distinguish between conversion and yield. Conversion is to do with reactants; yield with products.

Conversion

Conversion is a measure of the fraction of the reagent that reacts.

To optimize reactor design and minimize by-product formation, the conversion of a particular reagent is often less than 100%. If more than one reactant is used, the reagent on which the conversion is based must be specified.

Conversion is defined by the following expression:

$$\text{Conversion} = \frac{\text{amount of reagent consumed}}{\text{amount supplied}}$$
$$= \frac{(\text{amount in feed stream}) - (\text{amount in product stream})}{(\text{amount in feed stream})} \quad (2.8)$$

This definition gives the total conversion of the particular reagent to all products.

Example 2.10

In the manufacture of vinyl chloride (VC) by the pyrolysis of dichloroethane (DCE), the reactor conversion is limited to 55% to reduce carbon formation, which fouls the reactor tubes.

Calculate the quantity of DCE fed to the reactor to produce 5000 kg/h of VC.

Solution

Basis: 5000 kg/h VC (the required quantity).

$$\text{Reaction: } C_2H_4Cl_2 \rightarrow C_2H_3Cl + HCl$$

Molar weights: DCE 99, VC 62.5

$$\text{kmol/h VC produced} = \frac{5000}{62.5} = 80$$

From the stoichiometric equation, 1 kmol DCE produces 1 kmol VC. Let X be DCE feed in kmol/h:

$$\text{Per cent conversion} = 55 = \frac{80}{X} \times 100$$

$$X = \frac{80}{0.55} = \underline{\underline{145.5 \text{ kmol/h}}}$$

In this example, the small loss of DCE to carbon and other products has been neglected. All the DCE reacted has been assumed to be converted to VC.

Selectivity

Selectivity is a measure of the efficiency of the reactor in converting reagent to the desired product. It is the fraction of the reacted material that was converted into the desired product. If no by-products are formed, then the selectivity is 100%. If side reactions occur and by-products are formed, then the selectivity decreases. Selectivity is always expressed as the selectivity of feed A for product B, and is defined by the equation:

$$\text{Selectivity} = \frac{\text{moles of B formed}}{\text{moles of B that could have been formed if all A reacted to give B}}$$

$$= \frac{\text{moles of B formed}}{\text{moles of A consumed} \times \text{stoichiometric factor}} \tag{2.9}$$

Stoichiometric factor = moles of B produced per mole of A reacted in the reaction stoichiometric equation

Selectivity is usually improved by operating the reactor at low conversion. At high conversion, the reactor has low concentrations of at least one reagent and high concentrations of products, so reactions that form by-products are more likely to occur.

Reagents that are not converted in the reactor can be recovered and recycled. Reagents that become converted to by-products usually cannot be recovered and the by-products must be purified for sale or else disposed as waste (see Section 6.4.8). The optimum reactor conditions thus usually favour low reactor conversion to give high selectivity for the desired products when all of these costs are taken into account.

Yield

Yield is a measure of the performance of a reactor or plant. Several different definitions of yield are used, and it is important to state clearly the basis of any yield numbers. This is often not done when yields are quoted in the literature, and judgment must be used to decide what was intended.

The yield of product B from feed A is defined by:

$$\text{Yield} = \frac{\text{moles of B formed}}{\text{moles of A supplied} \times \text{stoichiometric factor}} \qquad (2.10)$$

For a reactor, the yield is the product of conversion and selectivity:

$$\text{Reaction yield} = \text{Conversion} \times \text{Selectivity}$$

$$= \frac{\text{moles A consumed}}{\text{moles A supplied}} \times \frac{\text{moles B formed}}{\text{moles A consumed} \times \text{stoichiometric factor}} \qquad (2.11)$$

With industrial reactors, it is necessary to distinguish between 'reaction yield' (chemical yield), which includes only chemical losses to side products; and the overall 'reactor yield', which also includes physical losses, such as losses by evaporation into vent gas.

If the conversion is near 100%, it may not be worth separating and recycling the unreacted material; the overall reactor yield would then include the loss of unreacted material. If the unreacted material is separated and recycled, the overall yield *taken over the reactor and separation step* would include any physical losses from the separation step.

Plant yield is a measure of the overall performance of the plant and includes all chemical and physical losses.

Plant yield (applied to the complete plant or any stage)

$$= \frac{\text{moles of product produced}}{\text{moles of reagent supplied to the process} \times \text{stoichiometric factor}} \qquad (2.12)$$

Where more than one reagent is used, or product produced, it is essential that product and reagent to which the yield refers is clearly stated.

The plant yield of B from A is the product of the reactor selectivity of feed A for product B and the separation efficiency (recovery) of each separation step that handles product B or reagent A.

Example 2.11

In the production of ethanol by the hydrolysis of ethylene, diethyl ether is produced as a by-product. A typical feed stream composition is: 55% ethylene, 5% inerts, 40% water; and product stream: 52.26% ethylene, 5.49% ethanol, 0.16% ether, 36.81% water, 5.28% inerts. Calculate the selectivity of ethylene for ethanol and for ether.

Solution

$$\text{Reactions: } C_2H_4 + H_2O \rightarrow C_2H_5OH \qquad (a)$$

$$2C_2H_5OH \rightarrow (C_2H_5)_2O + H_2O \qquad (b)$$

Basis: 100 mol feed (easier calculation than using the product stream).

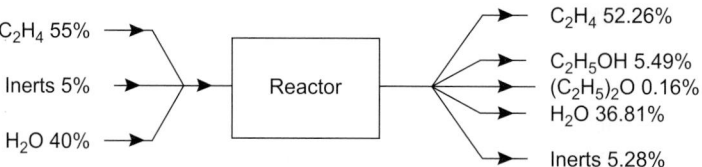

Note: the flow of inerts will be constant as they do not react, and it can thus be used to calculate the other flows from the compositions.

Feed stream   ethylene   55 mol
                     inerts   5 mol
                     water   40 mol

Product stream

$$\text{ethylene} = \frac{52.26}{5.28} \times 5 = 49.49 \ \text{mol}$$

$$\text{ethanol} = \frac{5.49}{5.28} \times 5 = 5.20 \ \text{mol}$$

$$\text{ether} = \frac{0.16}{5.28} \times 5 = 0.15 \ \text{mol}$$

Amount of ethylene reacted $= 55.0 - 49.49 = 5.51 \ \text{mol}$

$$\text{Selectivity of ethylene for ethanol} = \frac{5.20}{5.51 \times 1.0} \times 100 = \underline{\underline{94.4\%}}$$

As 1 mol of ethanol is produced per mol of ethylene the stoichiometric factor is 1.

$$\text{Selectivity of ethylene for ether} = \frac{0.15}{5.51 \times 0.5} \times 100 = \underline{\underline{5.44\%}}$$

The stoichiometric factor is 0.5, as 2 mol of ethylene produce 1 mol of ether.
   Note that the conversion of ethylene, to all products, is given by:

$$\text{Conversion} = \frac{\text{mols fed} - \text{mols out}}{\text{mols fed}} = \frac{55 - 49.49}{55} \times 100$$
$$= 10 \ \%$$

The selectivity based on water could also be calculated but is of no real interest as water is relatively inexpensive compared with ethylene. Water is clearly fed to the reactor in considerable excess.
   The yield of ethanol based on ethylene is:

$$\text{Reaction yield} = \frac{5.20}{55 \times 1.0} \times 100 = 9.45\%$$

Example 2.12

In the chlorination of ethylene to produce dichloroethane (DCE), the conversion of ethylene is reported as 99.0%. If 94 mol of DCE are produced per 100 mol of ethylene reacted, calculate the selectivity and the overall yield based on ethylene. The unreacted ethylene is not recovered.

Solution

$$\text{Reaction}: C_2H_4 + Cl_2 \rightarrow C_2H_4Cl_2$$

The stoichiometric factor is 1.

$$\text{Selectivity} = \frac{\text{moles DCE produced}}{\text{moles ethylene reacted} \times 1} \times 100$$

$$= \frac{94}{100} \times 100 = \underline{\underline{94\%}}$$

$$\text{Overall yield(including physical losses)} = \frac{\text{moles DCE produced}}{\text{moles ethylene fed} \times 1} \times 100$$

99 moles of ethylene are reacted for 100 moles fed, so

$$\text{Overall yield} = \frac{94}{100} \times \frac{99}{100} = \underline{\underline{93.1\%}}$$

Note that we get the same answer by multiplying the selectivity (0.94) and conversion (0.99).

The principal by-product of this process is trichloroethane.

Sources of Conversion, Selectivity and Yield Data

If there is minimal by-product formation, then the reactor costs (volume, catalyst, heating, etc.) can be traded off against the costs of separating and recycling unconverted reagents to determine the optimal reactor conversion. More frequently, the selectivity of the most expensive feeds for the desired product is less than 100%, and by-product costs must also be taken into account. The reactor optimization then requires a relationship between reactor conversion and selectivity, not just for the main product, but for all the by-products that are formed in sufficient quantity to have an impact on process costs.

In simple cases, when the number of by-products is small, it may be possible to develop a mechanistic model of the reaction kinetics that predicts the rate of formation of the main product and by-products. If such a model is fitted to experimental data over a suitably wide range of process conditions, then it can be used for process optimization. The development of reaction kinetics models is described in most reaction engineering textbooks. See, for example, Levenspiel (1998), Froment and Bischoff (1990) and Fogler (2005).

In cases where the reaction quickly proceeds to equilibrium, the yields are easily estimated as the equilibrium yields. Under these circumstances, the only possibilities for process optimization are to change the temperature, pressure or feed composition, so as to obtain a different equilibrium mixture. The calculation of reaction equilibrium is easily carried out using commercial process simulation programs.

When the number of components, or reactions, is too large, or the mechanism is too complex to deduce with statistical certainty, then response surface models can be used instead. Methods for the statistical design of experiments can be applied, reducing the amount of experimental data that must be collected to form a statistically meaningful correlation of selectivity and yield to the main process parameters. See Montgomery (2001) for a good introduction to the statistical design of experiments.

In the early stages of design, the design engineer will often have neither a response surface, nor a detailed mechanistic model of the reaction kinetics. Few companies are prepared to dedicate a laboratory or pilot plant and the necessary staff to collecting reaction kinetics data until management has been satisfied that the process under investigation is economically attractive. A design is thus needed before the necessary data set has been collected. Under such circumstances, the design engineer must select the optimal reactor conditions from whatever data is available. This initial estimate of reactor yield may come from a few data points collected by a chemist or taken from a patent or research paper. The use of data from patents is discussed in Section 8.2. For the purposes of completing a design, only a single estimate of reactor yield is needed. Additional yield data taken over a broader range of process conditions gives the designer greater ability to properly optimize the design.

2.14.   RECYCLE PROCESSES

Processes in which a flow stream is returned (recycled) to an earlier stage in the processing sequence are frequently used. If the conversion of a valuable reagent in a reaction process is appreciably less than 100%, the unreacted material is usually separated and recycled. The return of reflux to the top of a distillation column is an example of a recycle process in which there is no reaction.

In mass balance calculations the presence of recycle streams makes the calculations more difficult.

Without recycle, the material balances on a series of processing steps can be carried out sequentially, taking each unit in turn; the calculated flows out of one unit become the feeds to the next. If a recycle stream is present, then at the point where the recycle is returned the flow will not be known as it will depend on downstream flows not yet calculated. Without knowing the recycle flow, the sequence of calculations cannot be continued to the point where the recycle flow can be determined.

Two approaches to the solution of recycle problems are possible:

1. The cut and try ('tear') method. The recycle stream flows can be estimated and the calculations continued to the point where the recycle is calculated. The estimated flows are then compared with those calculated, and a better estimate

is made. The procedure is continued until the difference between the estimated and the calculated flows is within an acceptable tolerance.

2. The formal, algebraic, method. The presence of recycle implies that some of the mass balance equations must be solved simultaneously. The equations are set up with the recycle flows as unknowns and solved using standard methods for the solution of simultaneous equations.

With simple problems that have only one or two recycle loops, the calculation can often be simplified by the careful selection of the basis of calculation and the system boundaries. This is illustrated in Examples 2.4 and 2.13.

The solution of more complex material balance problems involving several recycle loops is discussed in Chapter 4.

Example 2.13

The block diagram shows the main steps in the balanced process for the production of vinyl chloride from ethylene. Each block represents a reactor and several other processing units. The main reactions are:

Block A, chlorination

$$C_2H_4 + Cl_2 \rightarrow C_2H_4Cl_2, \text{ yield on ethylene } 98\%$$

Block B, oxyhydrochlorination

$$C_2H_4 + 2HCl + 0.5O_2 \rightarrow C_2H_4Cl_2 + H_2O$$
$$\text{yields: on ethylene } 95\%, \text{on HCl } 90\%$$

Block C, pyrolysis

$$C_2H_4Cl_2 \rightarrow C_2H_3Cl + HCl, \text{ yields : on DCE}$$
$$99\%, \text{ on HCl } 99.5\%$$

The HCl from the pyrolysis step is recycled to the oxyhydrochlorination step. The flow of ethylene to the chlorination and oxyhydrochlorination reactors is adjusted so that the production of HCl is in balance with the requirement. The conversion in the pyrolysis reactor is limited to 55%, and the unreacted dichloroethane (DCE) is separated and recycled.

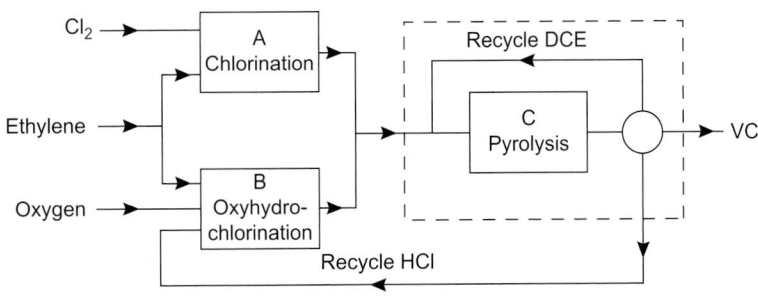

Using the yield figures given, and neglecting any other losses, calculate the flow of ethylene to each reactor and the flow of DCE to the pyrolysis reactor, for a production rate of 12,500 kg/h vinyl chloride (VC).

Solution

Molecular weights: VC 62.5, DCE 99.0, HCl 36.5.

$$VC \text{ per hour} = \frac{12,500}{62.5} = 200 \text{ kmol/h}$$

Draw a system boundary round each block, enclosing the DCE recycle within the boundary of step C.

Let the flow of ethylene to block A be X and to block B be Y, and the HCl recycle be Z.

Then the total moles of DCE produced $= 0.98X + 0.95Y$, allowing for the yields, and the moles of HCl produced in block C

$$= (0.98X + 0.95Y)0.995 = Z \qquad (a)$$

Consider the flows to and from block B

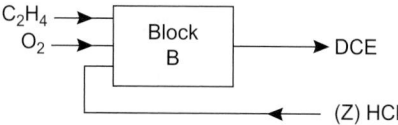

The yield of DCE based on HCl is 90%, so the moles of DCE produced

$$= 0.5 \times 0.90Z$$

Note: the stoichiometric factor is 0.5 (2 mol HCl per mol DCE).

The yield of DCE based on ethylene is 95%, so

$$0.5 \times 0.9Z = 0.95Y$$

$$Z = 0.95 \times 2Y/0.9$$

Substituting for Z into equation (a) gives

$$Y = (0.98X + 0.95Y)0.995 \times \frac{0.9}{2 \times 0.95} \qquad (b)$$

$$Y = 0.837X$$

Turning to block C, total VC produced $= 0.99 \times$ total DCE, so

$$0.99(0.98X + 0.95Y) = 200 \text{ kmol/h}$$

Substituting for Y from equation (b) gives $X = \underline{\underline{113.8 \text{ kmol/h}}}$

and $\qquad\qquad Y = 0.837 \times 113.8 = \underline{\underline{95.3 \text{ kmol/h}}}$

HCl recycle from equation (a)

$$Z = (0.98 \times 113.8 + 0.95 \times 95.3)0.995 = \underline{\underline{201.1 \text{ kmol/h}}}$$

Note: overall yield on ethylene $= \dfrac{200}{(113.8 + 95.3)} \times 100 = \underline{\underline{96\%}}$

2.15. PURGE

It is usually necessary to bleed off a portion of a recycle stream to prevent the build-up of unwanted material. For example, if a reactor feed contains inert components that are not separated from the recycle stream in the separation units, these inerts would accumulate in the recycle stream until the stream eventually consisted almost entirely of inerts. Some portion of the stream must be purged to keep the inert level within acceptable limits. A continuous purge would normally be used. Under steady-state conditions:

Loss of inert in the purge = Rate of feed of inerts into the system

The concentration of any component in the purge stream is the same as that in the recycle stream at the point where the purge is taken off. So the required purge rate can be determined from the following relationship:

[Feed stream flow rate] × [Feed stream inert concentration] =
[Purge stream flow rate] × [Specified (desired) recycle inert concentration]

Example 2.14

In the production of ammonia from hydrogen and nitrogen the conversion, based on either raw material, is limited to 15%. The ammonia produced is condensed from the reactor (converter) product stream and the unreacted material is recycled. If the feed contains 0.2% argon (from the nitrogen separation process), calculate the purge rate required to hold the argon in the recycle stream below 5.0%. Percentages are by volume.

Solution

Basis: 100 mol feed (purge rate will be expressed as moles per 100 mol feed, as the production rate is not given).

Process diagram:

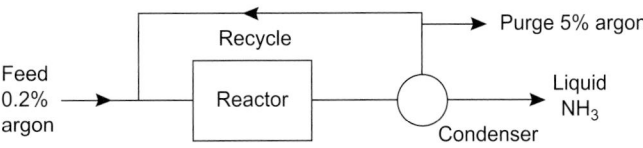

Volume percentages are taken as equivalent to mol%.

Argon entering system with feed $= 100 \times 0.2/100 = 0.2$ mol.
Let purge rate per 100 mol feed be F.
Argon leaving system in purge $= F \times 5/100 = 0.05\ F$.
At the steady state, argon leaving = argon entering

$$0.05F = 0.2$$

$$F = \frac{0.2}{0.05} = \underline{\underline{4}}$$

Purge required: 4 mol per 100 mol feed.

2.16.   BY-PASS

A flow stream may be divided and some part diverted (by-passed) around some units. This procedure is often used to control stream composition or temperature.

Material balance calculations on processes with by-pass streams are similar to those involving recycle, except that the stream is fed forward instead of backward. This usually makes the calculations easier than with recycle.

2.17.   UNSTEADY-STATE CALCULATIONS

All the previous material balance examples have been steady-state balances. The accumulation term was taken as zero, and the stream flow rates and compositions did not vary with time. If these conditions are not met, the calculations are more complex. Steady-state calculations are usually sufficient for the calculations of the process flow-sheet (Chapter 4). The unsteady-state behaviour of a process is important when considering the process start-up and shut-down, and the response to process upsets.

Batch processes are also examples of unsteady-state operation; although the total material requirements can be calculated by taking one batch as the basis for the calculation, unsteady-state balances are needed for determination of reaction and separation times.

The procedure for the solution of unsteady-state balances is to set up balances over a small increment of time, which will give a series of differential equations describing the process. For simple problems these equations can be solved analytically. For more complex problems computer methods are used.

The general approach to the solution of unsteady-state problems is illustrated in Example 2.15.

The behaviour of processes under non-steady-state conditions is a complex and specialized subject and is beyond the scope of this book. It can be important in process design when assessing the behaviour of a process from the point of view of safety and control.

The use of material balances in the modelling of complex unsteady-state processes is discussed in the books by Myers and Seider (1976) and Henley and Rosen (1969).

Example 2.15

A hold tank is installed in an aqueous effluent-treatment process to smooth out fluctuations in concentration of the effluent stream. The effluent feed to the tank normally contains no more than 100 ppm of acetone. The maximum allowable concentration of acetone in the effluent discharge is set at 200 ppm. The surge tank working capacity is 500 m³ and it can be considered to be perfectly mixed. The effluent flow is 45,000 kg/h. If the acetone concentration in the feed suddenly rises to 1000 ppm, due to a spill in the process plant, and stays at that level for half an hour, will the limit of 200 ppm in the effluent discharge be exceeded?

Solution

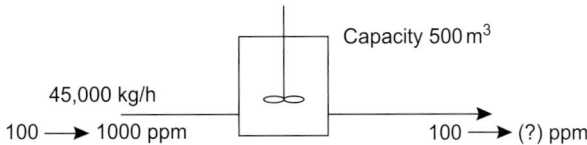

Basis: increment of time Δt.

To illustrate the general solution to this type of problem, the balance will be set up in terms of symbols for all the quantities and then actual values for this example substituted. Let:

material in the tank $= M$
flow rate $= F$
initial concentration in the tank $= C_0$
concentration at time t after the feed concentration is increased $= C$
concentration in the effluent feed $= C_1$
change in concentration over time increment $\Delta t = \Delta C$
average concentration in the tank during the time increment $= C_{av}$.

Then, as there is no generation in the system, the general material balance (Section 2.3) becomes:

$$\text{Input} - \text{Output} = \text{Accumulation}$$

Material balance on acetone.

Note: as the tank is considered to be perfectly mixed the outlet concentration will be the same as the concentration in the tank.

$$\text{Acetone in} - \text{Acetone out} = \text{Acetone accumulated in the tank}$$

$$FC_1\Delta t - FC_{av}\Delta t = M(C + \Delta C) - MC$$

$$F(C_1 - C_{av}) = M\frac{\Delta C}{\Delta t}$$

Taking the limit, as $\Delta t \to 0$

$$\frac{\Delta C}{\Delta t} = \frac{dC}{dt}, \quad C_{av} = C$$

$$F(C_1 - C) = M\frac{dC}{dt}$$

Integrating

$$\int_0^t dt = \frac{M}{F} \int_{C_0}^{C} \frac{dC}{(C_1 - C)}$$

$$t = -\frac{M}{F} \ln\left[\frac{C_1 - C}{C_1 - C_0}\right]$$

Substituting the values for the example, and noting that the maximum outlet concentration occurs at the end of the half-hour period of high inlet concentration:

$$t = 0.5 \text{ h}$$
$$C_1 = 1000 \text{ ppm}$$
$$C_0 = 100 \text{ ppm (normal value)}$$
$$M = 500 \text{ m}^3 = 500,000 \text{ kg}$$
$$F = 45,000 \text{ kg/h}$$
$$0.5 = -\frac{500,000}{45,000} \ln\left[\frac{1000 - C}{1000 - 100}\right]$$
$$0.045 = -\ln\left[\frac{1000 - C}{900}\right]$$
$$e^{-0.045} \times 900 = 1000 - C$$
$$C = \underline{\underline{140 \text{ ppm}}}$$

So the maximum allowable concentration is not exceeded.

2.18.   GENERAL PROCEDURE FOR MATERIAL-BALANCE PROBLEMS

The best way to tackle a problem will depend on the information given; the information required from the balance; and the constraints that arise from the nature of the problem. No all embracing, best method of solution can be given to cover all possible problems. The following step-by-step procedure is given as an aid to the efficient solution of material-balance problems. The same general approach can be usefully employed to organize the solution of energy balances, and other design problems.

Procedure

Step 1. Draw a block diagram of the process.
Show each significant step as a block, linked by lines and arrows to show the stream
   connections and flow direction.
Step 2. List the available data.
Show on the block diagram the known flows (or quantities) and stream compositions.
Step 3. List all the information required from the balance.
Step 4. Decide the system boundaries (see Section 2.6).

Step 5. Write out the chemical reactions involved for the main products and by-products.

Step 6. Note any other constraints, such as: specified stream compositions, azeotropes, phase or reaction equilibrium, tie substances (see Section 2.11).

The use of phase equilibrium relationships and other constraints in determining stream compositions and flows is discussed in more detail in Chapter 4.

Step 7. Note any stream compositions and flows that can be approximated.

Step 8. Check the number of conservation (and other) equations that can be written, and compare with the number of unknowns. Decide which variables are to be design variables; see Section 2.10.

This step would be used only for complex problems.

Step 9. Decide the basis of the calculation; see Section 2.7.

The order in which the steps are taken may be varied to suit the problem.

2.19. REFERENCES (FURTHER READING)

Chopey, N. P. (ed.) (1984) *Handbook of Chemical Engineering Calculations* (McGraw-Hill).

Felder, R. M. and Rousseau, R. W. (1995) *Elementary Principles of Chemical Processes*, 6th edn. (Pearson).

Fogler, H. S. (2005) *Elements of Chemical Reaction Engineering*, 3rd edn. (Prentice Hall).

Froment, G. F. and Bischoff, K. B. (1990) *Chemical Reactor Analysis and Design*, 2nd edn. (Wiley).

Henley, E. J. and Rosen, E. M. (1969) *Material and Energy Balance Computations* (Wiley).

Himmelblau, D. M. (1982) *Basic Principles and Calculations in Chemical Engineering* (Prentice-Hall).

Levenspiel, O. (1998) *Chemical Reaction Engineering*, 3rd edn. (Wiley).

Montgomery, D. C. (2001) *Design and Analysis of Experiments*, 5th edn. (Wiley).

Myers, A. L. and Seider, W. D. (1976) *Introduction to Chemical Engineering and Computer Calculations* (Prentice-Hall).

Rudd, D. F., Powers, G. J., and Siirola, J. J. (1973) *Process Synthesis* (Prentice-Hall).

Whitwell, J. C. and Toner, R. K. (1969) *Conservation of Mass and Energy* (McGraw-Hill).

Williams, E. T. and Jackson, R. C. (1958) *Stoichiometry for Chemical Engineers* (McGraw-Hill).

2.20. NOMENCLATURE

		Dimensions in **MLT**
C	Concentration after time t, Example 2.15	—
C_{av}	Average concentration, Example 2.15	—
C_0	Initial concentration, Example 2.15	—
C_1	Concentration in feed to tank, Example 2.15	—
ΔC	Incremental change in concentration, Example 2.15	—
F	Flow rate	$\mathbf{MT}^{-1}$
F_n	Total flow in stream n	$\mathbf{MT}^{-1}$

F_1	Water feed to reactor, Example 2.4	MT^{-1}
M	Quantity in hold tank, Example 2.15	M
N_c	Number of independent components	—
N_d	Number of variables to be specified	—
N_e	Number of independent balance equations	—
N_s	Number of streams	—
N_v	Number of variables	—
t	Time, Example 2.15	T
Δt	Incremental change in time, Example 2.15	T
X	Unknown flow, Examples 2.8, 2.10, 2.13	MT^{-1}
$x_{n,m}$	Concentration of component m in stream n	—
Y	Unknown flow, Examples 2.8, 2.13	MT^{-1}
Z	Unknown flow, Example 2.13	MT^{-1}

2.21. PROBLEMS

2.1. The composition of a gas derived by the gasification of coal is, volume percentage: carbon dioxide 4, carbon monoxide 16, hydrogen 50, methane 15, ethane 3, benzene 2, balance nitrogen. If the gas is burnt in a furnace with 20% excess air, calculate:
(a) the amount of air required per 100 kmol of gas,
(b) The amount of flue gas produced per 100 kmol of gas,
(c) the composition of the flue gases, on a dry basis.
Assume complete combustion.

2.2. Ammonia is removed from a stream of air by absorption in water in a packed column. The air entering the column is at 760 mmHg pressure and 20°C. The air contains 5.0% v/v ammonia. Only ammonia is absorbed in the column. If the flow rate of the ammonia–air mixture to the column is 200 m³/s and the stream leaving the column contains 0.05% v/v ammonia, calculate:
(a) The flow rate of gas leaving the column.
(b) The mass of ammonia absorbed.
(c) The flow rate of water to the column, if the exit water contains 1% w/w ammonia.

2.3. The off-gases from a gasoline stabilizer are fed to a steam reforming plant to produce hydrogen.

The composition of the off-gas, molar%, is: CH_4 77.5, C_2H_6 9.5, C_3H_8 8.5, C_4H_{10} 4.5.

The gases entering the reformer are at a pressure of 2 bara and 35°C and the feed rate is 2000 m³/h.

The reactions in the reformer are:

$$1.\ C_nH_{2n+2} + n(H_2O) \rightarrow n(CO) + (2n + 1)H_2$$
$$2.\ CO + H_2O \rightarrow CO_2 + H_2$$

The molar conversion of C_nH_{2n+2} in reaction (1) is 96% and of CO in reaction (2) 92%.

Calculate:

(a) the average molecular mass of the off-gas,
(b) the mass of gas fed to the reformer, kg/h,
(c) the mass of hydrogen produced, kg/h.

2.4. Allyl alcohol can be produced by the hydrolysis of allyl chloride. Together with the main product, allyl alcohol, di-allyl ether is produced as a by-product. The conversion of allyl chloride is typically 97% and the selectivity to alcohol is 90%, both on a molar basis. Assuming that there are no other significant side reactions, calculate masses of alcohol and ether produced, per 1000 kg of allyl chloride fed to the reactor.

2.5. Aniline is produced by the hydrogenation of nitrobenzene. A small amount of cyclo-hexylamine is produced as a by-product. The reactions are:

$$1.\ C_6H_5NO_2 + 3H_2 \rightarrow C_6H_5NH_2 + 2H_2O$$
$$2.\ C_6H_5NO_2 + 6H_2 \rightarrow C_6H_{11}NH_2 + 2H_2O$$

Nitrobenzene is fed to the reactor as a vapour, with three times the stoichiometric quantity of hydrogen. The conversion of the nitrobenzene, to all products, is 96%, and the selectivity for aniline is 95%.

The unreacted hydrogen is separated from the reactor products and recycled to the reactor. A purge is taken from the recycle stream to maintain the inerts in the recycle stream below 5%. The fresh hydrogen feed is 99.5% pure, the remainder being inerts. All percentages are molar.

For a feed rate of 100 kmol/h of nitrobenzene, calculate:

(a) the fresh hydrogen feed,
(b) the purge rate required,
(c) the composition of the reactor outlet stream.

2.6. In the manufacture of aniline by the hydrogenation of nitrobenzene, the off-gases from the reactor are cooled and the products and unreacted nitrobenzene condensed. The hydrogen and inerts, containing only traces of the condensed materials, are recycled.

Using the typical composition of the reactor off-gas given below, estimate the stream compositions leaving the condenser.

Composition, kmol/h: aniline 950, cyclo-hexylamine 10, water 1920, hydrogen 5640, nitrobenzene 40, inerts 300.

2.7. In the manufacture of aniline, the condensed reactor products are separated in a decanter. The decanter separates the feed into an organic phase and an aqueous phase. Most of the aniline in the feed is contained in the organic phase and most of the water in the aqueous phase. Using the data given below, calculate the stream compositions.

Data:

Typical feed composition, including impurities and by-products, weight%: water 23.8, aniline 72.2, nitrobenzene 3.2, cyclo-hexylamine 0.8.

Density of aqueous layer 0.995, density of organic layer 1.006. Therefore, the organic layer will be at the bottom.

Solubility of aniline in water 3.2% w/w, and water in aniline 5.15% w/w.

Partition coefficient of nitrobenzene between the aqueous and organic phases: $C_{organic}/C_{water} = 300$.

Solubility of cyclo-hexylamine in the water phase 0.12% w/w and in the organic phase 1.0% w/w.

2.8. In the manufacture of aniline from nitrobenzene the reactor products are condensed and separated into aqueous and organic phases in a decanter. The organic phase is fed to a striping column to recover the aniline. Aniline and water form an azeotrope, composition 0.96 mol fraction aniline. For the feed composition given below, make a mass balance round the column and determine the stream compositions and flow rates. Take as the basis for the balance 100 kg/h feed and 99.9% recovery of the aniline in the overhead product. Assume that nitrobenzene leaves with the water stream from the base of the column.

Feed composition, w/w basis: water 2.4, aniline 73.0, nitrobenzene 3.2, cyclo-hexylamine trace.

2.9. Guaifenesin (guaiacol glyceryl ether, 3-(2-methoxyphenoxy)-1,2-propane-diol, $C_{10}H_{14}O_4$) is an expectorant that is found in cough medicines such as Actifed™ and Robitussin™. U.S. 4,390,732 (to Degussa) describes a preparation of the active pharmaceutical ingredient (API) from guaiacol (2-methoxyphenol, $C_7H_8O_2$) and glycidol (3-hydroxy propylene oxide, $C_3H_6O_2$). When the reaction is catalysed by NaOH, the reaction yield is 93.8%. The product is purified in a thin-film evaporator giving an overall plant yield of 87%.

(a) Estimate the feed flow rates of glycidol and guaiacol that would be needed to produce 100 kg/day of the API.

(b) Estimate how much product is lost in the thin-film evaporator.

(c) How would you recover the product lost in the evaporator?

2.10. 11-[N-ethoxycarbonyl-4-piperidylidene]-8-chloro-6,11-dihydro-5H-benzo-[5,6]-cyclohepta-[1,2-b]-pyridine ($C_{22}H_{23}ClN_2O_2$) is a non-sedative anti-histamine, known as Loratadine and marketed as Claritin™. The preparation of the active pharmaceutical ingredient (API) is described in U.S. 4,282,233 (to Schering). The patent describes reacting 16.2 g of 11-[N-methyl-4-piperidylidene]-8-chloro-6,11-dihydro-5H-benzo-[5,6]-cyclohepta-[1,2-b]-pyridine ($C_{20}H_{21}ClN_2$) in 200 ml of benzene with 10.9 g of ethyl-chloroformate ($C_3H_5ClO_2$) for 18 h. The mixture is cooled, poured into ice water and separated into aqueous and organic phases. The organic layer is washed with water and evaporated to dryness. The residue is triturated (ground to a fine powder) with petroleum ether and recrystallized from isopropyl ether.

(a) What is the reaction by-product?

(b) The reaction appears to be carried out under conditions that maximize both selectivity and conversion (long time at low temperature), as might be expected given the cost of the raw material. If the conversion is 99.9% and the selectivity for the desired ethoxycarbonyl substituted compound is 100%, how much excess ethylchloroformate remains at the end of the reaction?

(c) What fraction of the ethylchloroformate feed is lost to waste products?

(d) Assuming that the volumes of water and isopropyl ether used in the quenching, washing and recrystallization steps are the same as the initial solvent volume, and that none of these materials are re-used in the process, estimate the total mass of waste material produced per kg of the API.

(e) If the recovery (plant yield) of the API from the washing and recrystallization steps is 92%, estimate the feed flow rates of 11-[N-methyl-4-piperidylidene]-8-chloro-6,11-dihydro-5H-benzo-[5,6]-cyclohepta-[1,2-b]-pyridine and ethylchloroformate required to produce a batch of 10 kg of the API.

(f) How much API could be produced per batch in a 3.8-m^3 (1000-US gal) reactor?

(g) What would be the advantages and disadvantages of carrying out the other process steps in the same vessel?

Note: Problems 2.5 to 2.8 can be taken together as an exercise in the calculation of a preliminary material balance for the manufacture of aniline by the process described in detail in Appendix F, Problem F.8. Structures for the compounds in problems 2.9 and 2.10 can be found in the Merck Index, but are not required to solve the problems.

3 FUNDAMENTALS OF ENERGY BALANCES AND ENERGY UTILIZATION

Chapter Contents

Key Learning Objectives

- How to use energy balances
- How to calculate the power consumed in compressing a gas
- Methods used for recovering process waste heat
- How to use the pinch design method to optimize process heat recovery

3.1. INTRODUCTION

As with mass, energy can be considered to be separately conserved in all but nuclear processes.

The conservation of energy differs from that of mass in that energy can be generated (or consumed) in a chemical process. Material can change form, new molecular species can be formed by chemical reaction, but the total mass flow into a process unit must be equal to the flow out at the steady state. The same is true of energy, but not of stream enthalpy. The total enthalpy of the outlet streams will not equal that of the inlet streams if energy is generated or consumed in the process; such as that due to heat of reaction. Energy can exist in several forms, including chemical energy, heat, mechanical energy and electrical energy. The total energy is conserved, but energy can be transformed from one kind of energy to another.

In process design, energy balances are made to determine the energy requirements of the process: the heating, cooling and power required. In plant operation, an energy balance (energy audit) on the plant will show the pattern of energy use, and suggest areas for conservation and savings.

Most energy balances are carried out using process simulation software, but design engineers occasionally have to incorporate energy balances into spreadsheet models or other computer programs. It is therefore important for the design engineer to have a good grasp of the basic principles of energy balances. In this chapter, the fundamentals of energy balances are reviewed briefly and examples are given to illustrate the use of energy balances in process design. The methods used for process energy recovery and conservation are also discussed.

More detailed accounts of the principles and applications of energy balances are given in the texts covering material and energy-balance calculations that are cited at the end of Chapter 2.

3.2. CONSERVATION OF ENERGY

As for material (Section 2.3), a general equation can be written for the conservation of energy:

Energy out $=$ Energy in $+$ generation $-$ consumption $-$ accumulation

This is a statement of the first law of thermodynamics.

An energy balance can be written for any process step. Chemical reaction will evolve energy (exothermic) or consume energy (endothermic). For steady-state processes the accumulation of both mass and energy will be zero.

Energy can exist in many forms and this, to some extent, makes an energy balance more complex than a material balance.

3.3. FORMS OF ENERGY (PER UNIT MASS OF MATERIAL)

3.3.1. Potential Energy

Energy due to position:

$$\text{Potential energy per unit mass} = gz \qquad (3.1)$$

where

z = height above some arbitrary datum, m
g = gravitational acceleration (9.81 m/s^2).

3.3.2. Kinetic Energy

Energy due to motion:

$$\text{Kinetic energy per unit mass} = \frac{u^2}{2} \tag{3.2}$$

where u = velocity, m/s.

3.3.3. Internal Energy

The energy associated with molecular motion. The temperature T of a material is a measure of its internal energy U:

$$U = \mathrm{f}(T) \tag{3.3}$$

3.3.4. Work

Work is done when a force acts through a distance:

$$W = \int_0^l F \, \mathrm{dx} \tag{3.4}$$

where

F = force, N
x and l = distance, m.

Work done on a system by its surroundings is conventionally taken as negative; work done by the system on the surroundings as positive.

Where the work arises from a change in pressure or volume:

$$W = \int_{v_1}^{v_2} P \, \mathrm{d}v \tag{3.5}$$

where

P = pressure, Pa (N/m^2)
v = volume per unit mass, m^3/kg.

To integrate this function, the relationship between pressure and volume must be known. In process design an estimate of the work done in compressing or expanding a gas is often required. A rough estimate can be made by assuming either reversible adiabatic (isentropic) or isothermal expansion, depending on the nature of the process.

For isothermal expansion (expansion at constant temperature):

$$Pv = \text{constant}$$

For reversible adiabatic expansion (no heat exchange with the surroundings):

$$Pv^\gamma = \text{constant}$$

where γ = ratio of the specific heats, C_p/C_v.

The compression and expansion of gases is covered more fully in Section 3.13.

3.3.5. Heat

Energy is transferred either as heat or work. A system does not contain 'heat', but the transfer of heat or work to a system changes its internal energy.

Heat taken in by a system from its surroundings is conventionally taken as positive and that given out as negative.

3.3.6. Electrical Energy

Electrical, and the mechanical forms of energy, are included in the work term in an energy balance. Electrical energy will only be significant in energy balances on electrochemical processes.

3.4. THE ENERGY BALANCE

Consider a steady-state process represented by Figure 3.1. The conservation equation can be written to include the various forms of energy.

For a unit mass of material:

$$U_1 + P_1 v_1 + u_1^2/2 + z_1 g + Q = U_2 + P_2 v_2 + u_2^2/2 + z_2 g + W \qquad (3.6)$$

The suffixes 1 and 2 represent the inlet and outlet points respectively. Q is the heat transferred across the system boundary; positive for heat entering the system, negative for heat leaving the system. W is the work done by the system; positive for work going from the system to the surroundings, and negative for work entering the system from the surroundings.

Equation 3.6 is a general equation for steady-state systems with flow.

In chemical processes, the kinetic and potential energy terms are usually small compared with the heat and work terms, and can normally be neglected.

It is convenient, and useful, to take the terms U and Pv together; defining the stream enthalpy, usually given the symbol H, as:

$$H = U + Pv$$

Enthalpy is a function of temperature and pressure. Values for the more common substances have been determined experimentally and are given in the various handbooks (see Chapter 8). Enthalpy can be calculated from specific and latent heat data; see Section 3.5.

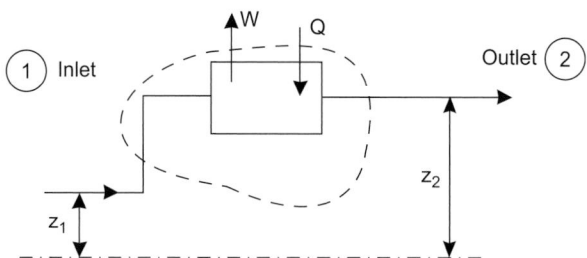

Figure 3.1. General steady-state process.

If the kinetic and potential energy terms are neglected, equation 3.6 simplifies to:

$$H_2 - H_1 = Q - W \qquad (3.7)$$

This simplified equation is usually sufficient for estimating the heating and cooling requirements of the various unit operations involved in chemical processes.

As the flow-dependent terms have been dropped, the simplified equation is applicable to both static (non-flow) systems and flow systems. It can be used to estimate the energy requirement for batch processes.

For many processes the work term will be zero, or negligibly small, and equation 3.7 reduces to the simple heat-balance equation:

$$Q = H_2 - H_1 \qquad (3.8)$$

Where heat is generated in the system; for example, in a chemical reactor:

$$Q = Q_p + Q_s \qquad (3.9)$$

Q_s = heat generated in the system. If heat is evolved (exothermic processes) Q_s is taken as *positive*, and if heat is absorbed (endothermic processes) it is taken as *negative*.

Q_p = process heat added to the system to maintain required system temperature.

Hence:

$$Q_p = H_2 - H_1 - Q_s \qquad (3.10)$$

H_1 = enthalpy of the inlet stream
H_2 = enthalpy of the outlet stream.

Example 3.1

Balance with no chemical reaction. Estimate the steam and the cooling water required for the distillation column shown in the figure.

Steam is available at 25 psig (274 kN/m^2 abs), dry saturated.
The rise in cooling water temperature is limited to 30°C.
Column operates at 1 bar.

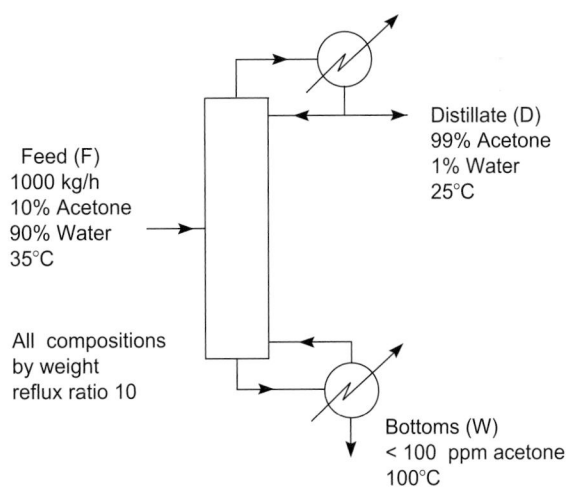

Feed (F)
1000 kg/h
10% Acetone
90% Water
35°C

Distillate (D)
99% Acetone
1% Water
25°C

All compositions
by weight
reflux ratio 10

Bottoms (W)
< 100 ppm acetone
100°C

Solution

Material Balance

It is necessary to make a material balance to determine the top and bottoms product flow rates.
    Balance on acetone, acetone loss in bottoms neglected.

$$1000 \times 0.1 = D \times 0.99$$
$$\text{Distillate, } D = 101 \text{ kg/h}$$
$$\text{Bottoms, } B = 1000 - 101 = 899 \text{ kg/h}$$

Energy Balance

The kinetic and potential energy of the process streams will be small and can be neglected.
    Take the first system boundary to include the reboiler and condenser.

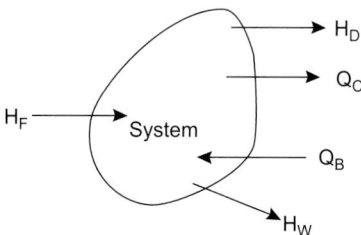

Inputs: reboiler heat input Q_B + feed sensible heat H_F.
Outputs: condenser cooling Q_C + top and bottom product sensible heats $H_D + H_B$.

    The heat losses from the system will be small if the column and exchangers are properly lagged (typically less than 5%) and will be neglected.

Basis: 25°C, 1 h.
Heat capacity data, from Coulson *et al.* (1999), average values:

Acetone:	25°C to 35°C	2.2 kJ/kg K
Water:	25°C to 100°C	4.2 kJ/kg K

Heat capacities can be taken as additive.

    Feed, 10% acetone = $0.1 \times 2.2 + 0.9 \times 4.2 = 4.00$ kJ/kg K
    Distillate, 99% acetone, taken as acetone, 2.2 kJ/kg K
    Bottoms, taken as water, 4.2 kJ/kg K.

Q_C must be determined by taking a balance round the condenser.

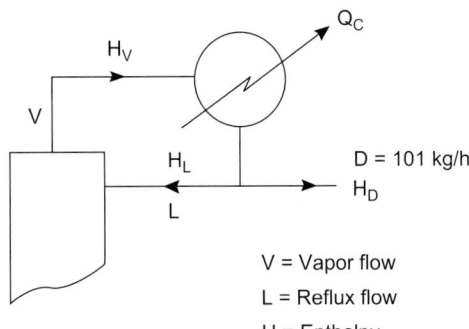

V = Vapor flow
L = Reflux flow
H = Enthalpy

Reflux ratio (see Chapter 11):

$$R = \frac{L}{D} = 10$$
$$L = 10 \times 101 = 1010 \text{ kg/h}$$
$$V = L + D = 1111 \text{ kg/h}$$

From vapour – liquid equilibrium data:

$$\text{boiling point of 99\% acetone/water} = 56.5°\text{C}$$

At steady state:

$$\text{input} = \text{output}$$

$$H_V = H_D + H_L + Q_C$$

Hence

$$Q_C = H_V - H_D - H_L$$

Assume complete condensation.

$$\text{Enthalpy of vapour } H_V = \text{latent} + \text{sensible heat}$$

There are two ways of calculating the specific enthalpy of the vapour at its boiling point.

(1) Latent heat of vaporization at the base temperature + sensible heat to heat the vapour to the boiling point.
(2) Latent heat of vaporization at the boiling point + sensible heat to raise liquid to the boiling point.

Values of the latent heat of acetone and water as functions of temperature are given in Coulson *et al.* (1999), so the second method will be used.

$$\text{Latent heat acetone at } 56.5°\text{C}(330 \text{ K}) = 620 \text{ kJ/kg}$$
$$\text{Water at } 56.5°\text{C}(330 \text{ K}) = 2500 \text{ kJ/kg}$$

Taking latent heats as additive:

$$H_V = 1111[(0.01 \times 2500 + 0.99 \times 620) + (56.5 - 25)2.2]$$
$$= 786,699 \text{ kJ/h}$$

The enthalpy of the top product and reflux are zero, as they are both at the base temperature. Both are liquid, and the reflux will be at the same temperature as the product. Hence:

$$Q_C = H_V = 786,699 \text{ kJ/h } (218.5 \text{ kW})$$

Q_B is determined from a balance over the complete system.

$$\text{Input } = \text{ Output}$$
$$Q_B + H_F = Q_C + H_D + H_W$$
$$H_F = 1000 \times 4.00(35 - 25) = 40,000 \text{ kJ/h}$$
$$H_B = 899 \times 4.2(100 - 25) = 283,185 \text{ kJ/h}$$

(boiling point of bottom product taken as 100°C).

Hence:

$$Q_B = Q_C + H_B + H_D - H_F$$
$$= 786,699 + 283,185 + 0 - 40,000$$
$$= 1,029,884 \text{ kJ/h } (286.1 \text{ kW})$$

Q_B is supplied by condensing steam.

$$\text{Latent heat of steam } = 2174 \text{ kJ/kg at } 274 \text{ kN/m}^2$$
$$\text{Steam required } = \frac{1,029,884}{2174} = \underline{473.7 \text{ kg/h}}$$

Q_C is removed by cooling water with a temperature rise of 30°C.

$$Q_c = \text{water flow} \times 30 \times 4.2$$
$$\text{Water flow } = \frac{786,699}{4.2 \times 30} = \underline{6244 \text{ kg/h}}$$

3.5.   CALCULATION OF SPECIFIC ENTHALPY

Tabulated values of enthalpy are available only for the more common materials. In the absence of published data the following expressions can be used to estimate the specific enthalpy (enthalpy per unit mass).

For pure materials, with no phase change:

$$H_T = H_{T_d} + \int_{T_d}^{T} C_p \, dT \qquad (3.11)$$

where

H_T = specific enthalpy at temperature T
C_p = specific heat capacity of the material, constant pressure
T_d = the datum temperature.

If a phase transition takes place between the specified and datum temperatures, the latent heat of the phase transition is added to the sensible-heat change calculated by equation 3.11. The sensible-heat calculation is then split into two parts:

$$H_T = H_{T_d} + \int_{T_d}^{T_p} C_{p1} \, dT + \Delta H_P + \int_{T_p}^{T} C_{p2} \, dT \qquad (3.12)$$

where

T_p = phase transition temperature
C_{p1} = specific heat capacity of first phase, below T_p

C_{p2}= specific heat capacity of second phase, above T_p
ΔH_p = the latent heat of the phase change.

The specific heat at constant pressure will vary with temperature and to use equations 3.11 and 3.12, values of C_p must be available as a function of temperature. For solids and gases C_p is usually expressed as an empirical power series equation:

$$C_p = a + bT + cT^2 + dT^3 \tag{3.13a}$$

or

$$C_p = a + bT + cT^{-1/2} \tag{3.13b}$$

Absolute (K) or relative (°C) temperature scales may be specified when the relationship is in the form given in equation 3.13a. For equation 3.13b absolute temperatures must be used.

Example 3.2

Estimate the specific enthalpy of ethyl alcohol at 1 bar and 200°C, taking the datum temperature as 0°C.

C_p liquid 0°C 24.65 cal/mol°C
　　　　100°C 37.96 cal/mol°C
C_p gas (t°C) $14.66 + 3.758 \times 10^{-2}t - 2.091 \times 10^{-5}t^2 + 4.740 \times 10^{-9}t^3$ cal/mol°C
Boiling point of ethyl alcohol at 1 bar = 78.4°C
Latent heat of vaporization = 9.22 kcal/mol.

Solution

Note: as the data taken from the literature are given in cal/mol the calculation is carried out in these units and the result converted to SI units.

As no data are given on the exact variation of the C_p of the liquid with temperature, use an equation of the form $C_p = a + bt$, calculating a and b from the data given; this will be accurate enough over the range of temperature needed.

$$a = \text{value of } C_p \text{ at } 0°C, \quad b = \frac{37.96 - 24.65}{100} = 0.133$$

$$H_{200°C} = \int_0^{78.4} (24.65 + 0.133t) \, dt + 9.22 \times 10^3 + \int_{78.4}^{200}$$

$$(14.66 + 3.758 \times 10^{-2}t - 2.091 \times 10^{-5}t^2 + 4.740 \times 10^{-9}t^3) \, dt$$

$$= \left[24.65t + 0.133t^2/2 \right]_0^{78.4} + 9.22 \times 10^3 + \left[14.66t \right._{78.4}^{200}$$

$$\left. + 3.758 \times 10^{-2}t^2/2 - 2.091 \times 10^{-5}t^3/3 + 4.740 \times 10^{-9}t^4/4 \right]$$

$$= 13.95 \times 10^3 \text{ cal/mol}$$

$$= 13.95 \times 10^3 \times 4.18 = 58.31 \times 10^3 \text{ J/mol}$$

Specific enthalpy = 58.31 kJ/mol
Molecular weight of ethyl alcohol, $C_2H_5OH = 46$
Specific enthalpy = $58.31 \times 10^3/46 = \underline{1268 \text{ kJ/kg}}$

3.6.   MEAN HEAT CAPACITIES

The use of mean heat capacities often facilitates the calculation of sensible-heat changes; mean heat capacity over the temperature range t_1 to t_2 is defined by the following equation:

$$C_{pm} \int_{t_1}^{t_2} dT = \int_{t_1}^{t_2} C_p \ dT \tag{3.14}$$

Mean specific heat values are tabulated in various handbooks. If the values are for unit mass, calculated from some standard reference temperature, t_r, then the change in enthalpy between temperatures t_1 and t_2 is given by:

$$\Delta H = C_{pm,t_2}(t_2 - t_r) - C_{pm,t_1}(t_1 - t_r) \tag{3.15}$$

where t_r is the reference temperature from which the values of C_{pm} were calculated.

If C_p is expressed as a polynomial of the form: $C_p = a + bt + ct^2 + dt^3$, then the integrated form of equation 3.14 will be:

$$C_{pm} = \frac{a(t - t_r) + \frac{b}{2}(t^2 - t_r^2) + \frac{c}{3}(t^3 - t_r^3) + \frac{d}{4}(t^4 - t_r^4)}{t - t_r} \tag{3.16}$$

where t is the temperature at which C_{pm} is required.

If the reference temperature is taken at $0°C$, equation 3.16 reduces to:

$$C_{pm} = a + \frac{bt}{2} + \frac{ct^2}{3} + \frac{dt^3}{4} \tag{3.17}$$

and the enthalpy change from t_1 to t_2 becomes

$$\Delta H = C_{pm,t2}\ t_2 - C_{pm,t1}\ t_1 \tag{3.18}$$

The use of mean heat capacities is illustrated in Example 3.3.

Example 3.3

The gas leaving a combustion chamber has the following composition: CO_2 7.8, CO 0.6, O_2 3.4, H_2O 15.6, N_2 72.6, all volume percentage. Calculate the heat removed if the gas is cooled from $800°C$ to $200°C$.

Solution

Mean heat capacities for the combustion gases are readily available in handbooks and texts on heat and material balances. The following values are taken from K. A. Kobe, *Thermochemistry of Petrochemicals*, reprint No. 44, Pet. Ref. 1958; converted to SI units, J/mol°C, reference temperature 0°C.

T, °C	N_2	O_2	CO_2	CO	H_2O
200	29.24	29.95	40.15	29.52	34.12
800	30.77	32.52	47.94	31.10	37.38

Heat extracted from the gas in cooling from 800°C to 200°C, for each component:

$$= M_c(C_{pm,800} \times 800 - C_{pm,200} \times 200)$$

where M_c = moles of that component.

Basis: 100 mol gas (as analysis is by volume), substitution gives:

$$
\begin{aligned}
CO_2 \quad & 7.8(47.94 \times 800 - 40.15 \times 200) & = 236.51 \times 10^3 \\
CO \quad & 0.6(31.10 \times 800 - 29.52 \times 200) & = 11.39 \times 10^3 \\
O_2 \quad & 3.4(32.52 \times 800 - 29.95 \times 200) & = 68.09 \times 10^3 \\
H_2O \quad & 15.6(37.38 \times 800 - 34.12 \times 200) & = 360.05 \times 10^3 \\
N_2 \quad & 72.6(30.77 \times 800 - 29.24 \times 200) & = 1362.56 \times 10^3 \\
& & = 2038.60 \text{ kJ/100 mol} \\
& & = \underline{20.39 \text{ kJ/mol}}
\end{aligned}
$$

3.7. THE EFFECT OF PRESSURE ON HEAT CAPACITY

The data on heat capacities given in the handbooks, and in Appendix C, are usually for the ideal gas state. Equation 3.13a should be written as:

$$ C_p^\circ = a + bT + cT^2 + dT^3 \tag{3.19} $$

where the superscript $^\circ$ refers to the ideal gas state.

The ideal gas values can be used for the real gases at low pressures. At high pressures the effect of pressure on the specific heat may be appreciable.

Edmister (1948) published a generalized plot showing the isothermal pressure correction for real gases as a function of the reduced pressure and temperature. His chart, converted to SI units, is shown as Figure 3.2. Edmister's chart was based on hydrocarbons, but can be used for other materials to give an indication of the likely error if the ideal gas specific heat values are used without corrections.

The method is illustrated in Example 3.4.

Example 3.4

The ideal state heat capacity of ethylene is given by the equation:

$$ C_p^\circ = 3.95 + 15.6 \times 10^{-2}T - 8.3 \times 10^{-5}T^2 + 17.6 \times 10^{-9}T^3 \text{ J/mol K} $$

Estimate the value at 10 bar and 300 K.

Solution

Ethylene:

critical pressure	50.5 bar
critical temperature	283 K

$$
\begin{aligned}
C_p^\circ &= 3.95 + 15.6 \times 10^{-2} \times 300 - 8.3 \times 10^{-5} \times 300^2 + 17.6 \times 10^{-9} \times 300^3 \\
&= 43.76 \text{ J/mol K}
\end{aligned}
$$

$$ P_r = \frac{10}{50.5} = 0.20 $$

$$ T_r = \frac{300}{283} = 1.06 $$

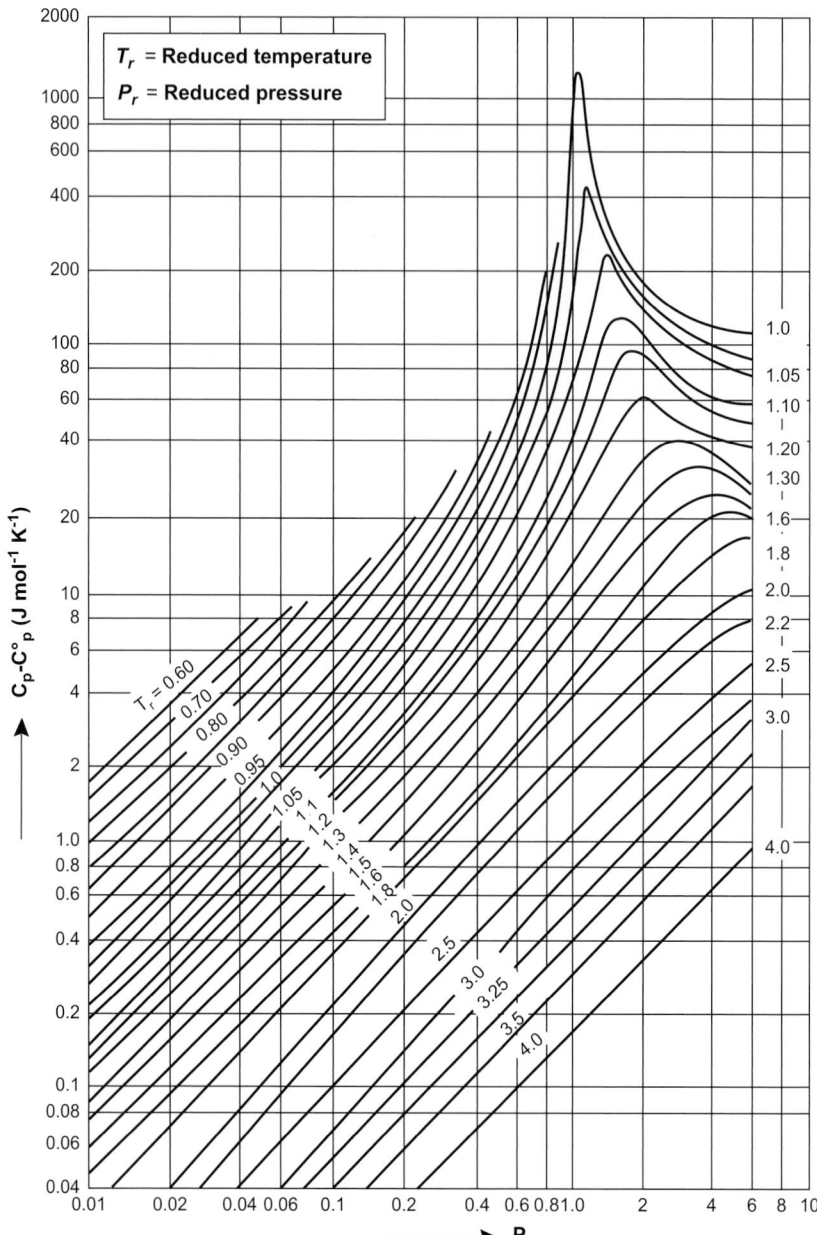

Figure 3.2.   Excess heat capacity chart (reproduced from Sterbacek *et al.* (1979), with permission).

From Figure 3.2:

$$C_p - C_p^\circ \approx 5 \text{ J/mol K}$$

So

$$C_p = 43.76 + 5 \approx 49 \text{ J/mol K}$$

The error in C_p if the ideal gas value were used uncorrected would be approximately 10%. This would clearly lead to large errors in predicting heat-exchanger duties.

3.8. ENTHALPY OF MIXTURES

For gases, the heats of mixing are usually negligible and the heat capacities and enthalpies can be taken as additive without introducing any significant error into design calculations; as was done in Example 3.3.

$$C_p(\text{mixture}) = x_a C_{pa} + x_b C_{pb} + x_c C_{pc} + \dots \tag{3.20}$$

where x_a, x_b, x_c, etc., are the mole fractions of the components a, b, c (or weight fractions if the specific heat per unit mass is used).

For mixtures of liquids and for solutions, the heat of mixing (heat of solution) may be significant, and so must be included when calculating the enthalpy of the mixture.

For binary mixtures, the specific enthalpy of the mixture at temperature t is given by:

$$H_{\text{mixture},t} = x_a H_{a,t} + x_b H_{b,t} + \Delta H_{m,t} \tag{3.21}$$

where $H_{a,t}$ and $H_{b,t}$ are the specific enthalpies of the components a and b and $-\Delta H_{m,t}$ is the heat of mixing when 1 mol of solution is formed, at temperature t.

Heats of mixing and heats of solution are determined experimentally and are available in the handbooks for the more commonly used solutions. If no values are available, judgment must be used to decide if the heat of mixing for the system is likely to be significant.

For hydrocarbon mixtures, the heat of mixing is usually small compared with the other heat quantities, and can usually be neglected when carrying out a heat balance to determine the process heating or cooling requirements. The heats of solution of organic and inorganic compounds in water can be large, particularly for the strong mineral acids and alkalis.

3.8.1. Integral Heats of Solution

Heats of solution are dependent on concentration. The integral heat of solution at any given concentration is the cumulative heat released, or absorbed, in preparing the solution from pure solvent and solute. The integral heat of solution at infinite dilution is called the *standard integral heat of solution*.

Tables of the integral heat of solution over a range of concentration, and plots of the integral heat of solution as a function of concentration, are given in the handbooks for many of the materials for which the heat of solution is likely to be significant in process design calculations.

The integral heat of solution can be used to calculate the heating or cooling required in the preparation of solutions, as illustrated in Example 3.5.

Example 3.5

A solution of NaOH in water is prepared by diluting a concentrated solution in an agitated, jacketed, vessel. The strength of the concentrated solution is 50% w/w and

2500 kg of 5% w/w solution is required per batch. Calculate the heat removed by the cooling water if the solution is to be discharged at a temperature of 25°C. The temperature of the solutions fed to the vessel can be taken to be 25°C.

Solution

Integral heat of solution of NaOH − H_2O, at 25°C:

mols H_2O/mol NaOH	$-\Delta H^{\circ}_{soln}$ kJ/mol NaOH
2	22.9
4	34.4
5	37.7
10	42.5
infinite	42.9

Conversion of weight per cent to mol/mol:

$$50\% \text{ w/w} = 50/18 \div 50/40 = 2.22 \text{ mol } H_2O/\text{mol NaOH}$$
$$5\% \text{ w/w} = 95/18 \div 5/40 = 42.2 \text{ mol } H_2O/\text{mol NaOH}$$

From a plot of the integral heats of solution versus concentration,

$$-\Delta H^{\circ}_{soln} \quad 2.22 \text{ mol/mol} = 27.0 \text{ kJ/mol NaOH}$$
$$42.2 \text{ mol/mol} = 42.9 \text{ kJ/mol NaOH}$$

Heat liberated in the dilution per mol NaOH = 42.9 − 27.0 = 15.9 kJ
Heat released per batch = mol NaOH per batch × 15.9

$$= \frac{2500 \times 10^3 \times 0.05}{40} \times 15.9$$
$$= 49.7 \times 10^3 \text{ kJ}$$

Heat transferred to cooling water, neglecting heat losses,

<u>49.7 MJ per batch</u>

In Example 3.5, the temperature of the feeds and final solution have been taken as the same as the standard temperature for the heat of solution, 25°C, to simplify the calculation. Heats of solution are analogous to heats of reaction, and examples of heat balances on processes where the temperatures are different from the standard temperature are given in the discussion of heats of reaction, Section 3.10.

3.9. ENTHALPY-CONCENTRATION DIAGRAMS

The variation of enthalpy for binary mixtures is conveniently represented on a diagram. An example is shown in Figure 3.3. The diagram shows the enthalpy of mixtures of ammonia and water versus concentration; with pressure and temperature as parameters. It covers the phase changes from solid to liquid to vapour, and the enthalpy values given include the latent heats for the phase transitions.

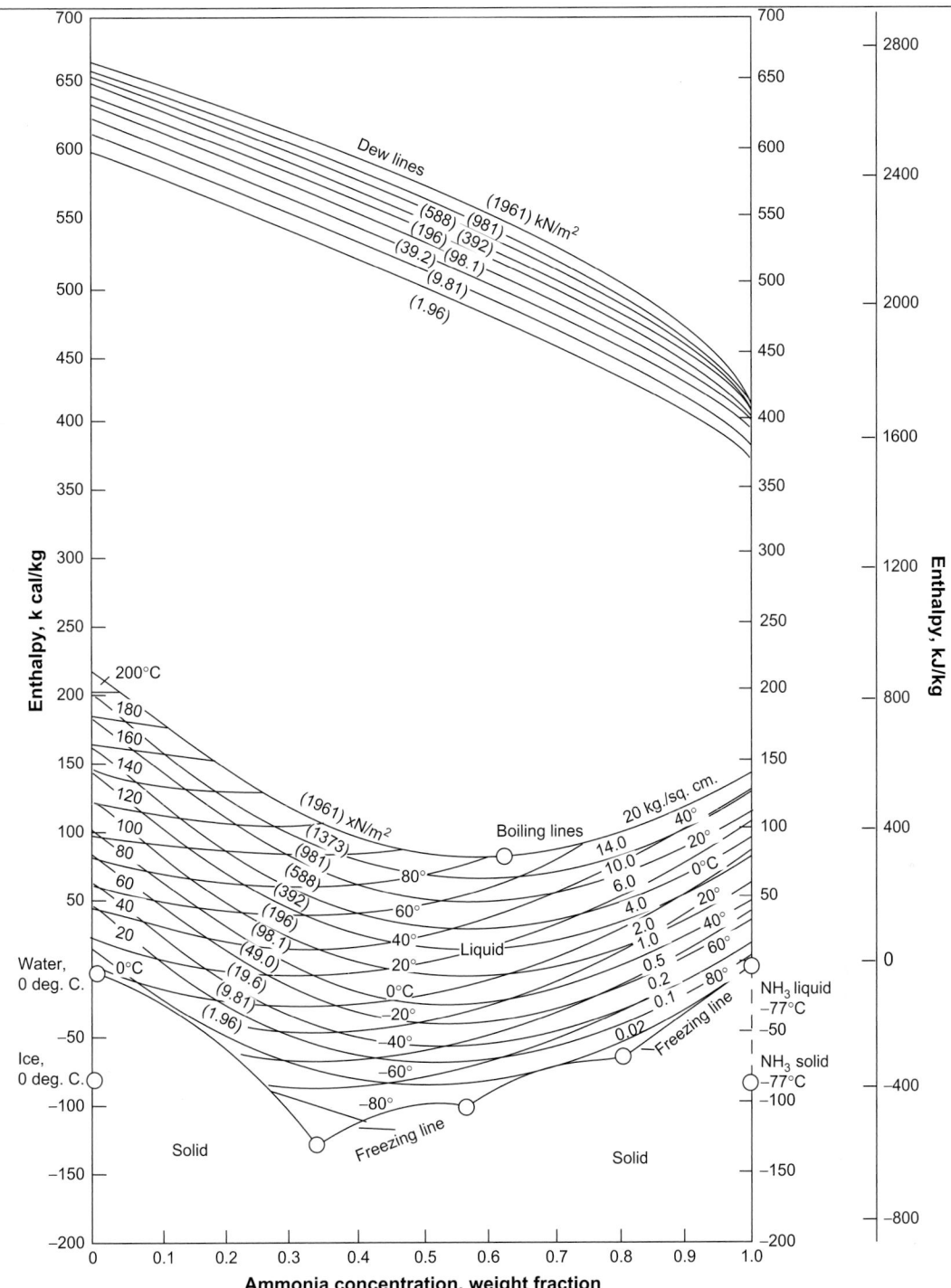

Figure 3.3. Enthalpy-concentration diagram for aqueous ammonia. Reference states: enthalpies of liquid water at 0°C and liquid ammonia at −77°C are zero (Bosnjakovic, *Technische Thermodynamik*, T. Steinkopff, Leipzig, 1935).

The enthalpy is per kg of the mixture (ammonia + water).
Reference states:

enthalpy ammonia at $-77°C$ = zero
enthalpy water at $0°C$ = zero.

Enthalpy-concentration diagrams greatly facilitate the calculation of energy balances involving concentration and phase changes; this is illustrated in Example 3.6.

Example 3.6

Calculate the maximum temperature reached when liquid ammonia at $40°C$ is dissolved in water at $20°C$ to form a 10% solution.

Solution

The maximum temperature will occur if there are no heat losses (adiabatic process). As no heat or material is removed, the problem can be solved graphically in the enthalpy concentration diagram (Figure 3.3). The mixing operation is represented on the diagram by joining the point A representing pure ammonia at $40°C$ with the point B representing pure water at $20°C$. Mixtures of this concentration lie on a vertical line at the required concentration, 0.1. The temperature of the mixture is given by the intersection of this vertical line with the line AB. This method is an application of the 'lever rule' for phase diagrams. For a more detailed explanation of the method and further examples see Himmelbau (1995) or any of the general texts on material and energy balances listed at the end of Chapter 2.

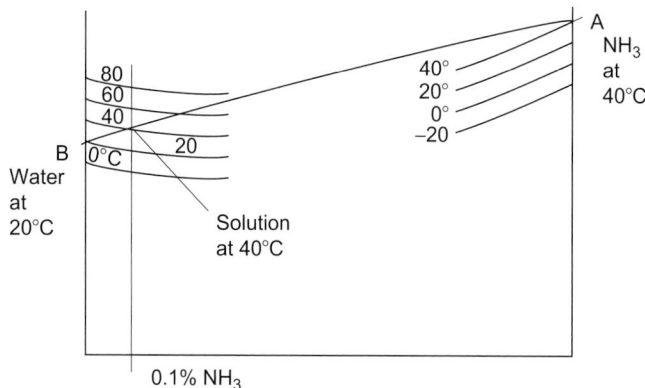

Enthalpy-concentration diagrams are convenient for binary mixtures with high heats of mixing. They are used widely in the fertilizer industry and in other industries that are concerned with making inorganic acids and bases.

3.10. HEATS OF REACTION

If a process involves chemical reaction, heat will normally have to be added or removed. The amount of heat given out in a chemical reaction depends on the conditions under which the reaction is carried out. The standard heat of reaction is the heat released when the reaction is carried out under standard conditions: pure components, pressure 1 atm (1.01325 bar) and temperature usually, but not necessarily, 25°C.

Values for the standard heats of reactions are given in the literature, or can be calculated by the methods given in Sections 3.11 and 3.12.

When quoting heats of reaction the basis should be clearly stated, either by giving the chemical equation; for example:

$$NO + \frac{1}{2}O_2 \rightarrow NO_2 \quad \Delta H_r^\circ = -56.68 \text{ kJ}$$

(the equation implies that the quantity of reactants and products are moles), or by stating to which quantity the quoted value applies:

$$\Delta H_r^\circ = -56.68 \text{kJ per mol NO}_2$$

The reaction is exothermic and the enthalpy change ΔH_r° is therefore *negative*. The heat of reaction $-\Delta H_r^\circ$ is *positive*. The superscript $^\circ$ denotes a value at *standard* conditions and the subscript r implies that a chemical reaction is involved.

The state of the reactants and products (gas, liquid or solid) should also be given, if the reaction conditions are such that they may exist in more than one state; for example:

$$H_2(g) + \tfrac{1}{2}O_2(g) \rightarrow H_2O(g), \quad \Delta H_r^\circ = -241.6 \text{ kJ}$$
$$H_2(g) + \tfrac{1}{2}O_2(g) \rightarrow H_2O \ (l), \quad \Delta H_r^\circ = -285.6 \text{ kJ}$$

The difference between the two heats of reaction is the latent heat of the water formed.

In process design calculations, it is usually more convenient to express the heat of reaction in terms of the moles of product produced, for the conditions under which the reaction is carried out, kJ/mol product.

Standard heats of reaction can be converted to other reaction temperatures by making a heat balance over a hypothetical process, in which the reactants are brought to the standard temperature, the reaction carried out, and the products then brought to the required reaction temperature; as illustrated in Figure 3.4.

$$\Delta H_{r,t} = \Delta H_r^\circ + \Delta H_{\text{prod.}} - \Delta H_{\text{react.}} \tag{3.22}$$

where

$-\Delta H_{r,t}$ = heat of reaction at temperature t,
$\Delta H_{\text{react.}}$ = enthalpy change to bring reactants to standard temperature
$\Delta H_{\text{prod.}}$ = enthalpy change to bring products to reaction temperature, t.

For practical reactors, where the reactants and products may well be at temperatures different from the reaction temperature, it is best to carry out the heat balance over the actual reactor using the standard temperature (25°C) as the datum temperature; the standard heat of reaction can then be used without correction.

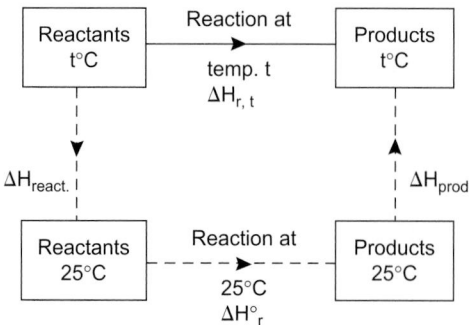

Figure 3.4. ΔH_r at temperature t.

It must be emphasized that it is unnecessary to correct a heat of reaction to the reaction temperature for use in a reactor heat-balance calculation. To do so is to carry out two heat balances, whereas with a suitable choice of datum only one need be made. For a practical reactor, the heat added (or removed) Q_p to maintain the design reactor temperature will be given by (from equation 3.10):

$$Q_p = H_{\text{products}} - H_{\text{reactants}} - Q_r \qquad (3.23)$$

where

H_{products} is the *total* enthalpy of the product streams, including unreacted materials and by-products, evaluated from a datum temperature of 25°C;

$H_{\text{reactants}}$ is the total enthalpy of the feed streams, including excess reagent and inerts, evaluated from a datum of 25°C;

Q_r is the total heat generated by the reactions taking place, evaluated from the standard heats of reaction at 25°C (298 K).

$$Q_r = \sum -\Delta H_r^{\circ} \times (\text{mol of product formed}) \qquad (3.24)$$

where $-\Delta H_r^{\circ}$ is the standard heat of reaction per mol of the particular product.

Note: A negative sign is necessary in equation 3.24 as Q_r is positive when heat is evolved by the reaction, whereas the standard enthalpy change will be negative for exothermic reactions. Q_p will be negative when cooling is required (see Section 3.4).

3.10.1. Effect of Pressure On Heats of Reaction

Equation 3.22 can be written in a more general form:

$$\Delta H_{r,P,T} = \Delta H_r^{\circ} + \int_1^P \left[\left(\frac{\partial H_{\text{prod.}}}{\partial P} \right)_T - \left(\frac{\partial H_{\text{react.}}}{\partial P} \right)_T \right] dP$$
$$+ \int_{298}^T \left[\left(\frac{\partial H_{\text{prod.}}}{\partial T} \right)_P - \left(\frac{\partial H_{\text{react.}}}{\partial T} \right)_P \right] dT \qquad (3.25)$$

If the effect of pressure is likely to be significant, the change in enthalpy of the products and reactants, from the standard conditions, can be evaluated to include both the effects of temperature and pressure (for example, by using tabulated values of enthalpy) and the correction made in a similar way to that for temperature only.

Example 3.7

This example illustrates the manual calculation of a reactor heat balance.

Vinyl chloride (VC) is manufactured by the pyrolysis of 1,2-dichloroethane (DCE). The reaction is endothermic. The flow rates to produce 5000 kg/h at 55% conversion are shown in the diagram (see Example 2.13).

The reactor is a pipe reactor heated with fuel gas, gross calorific value 33.5 MJ/m³. Estimate the quantity of fuel gas required.

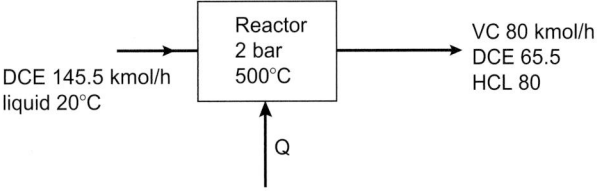

Solution

$$\text{Reaction}: C_2H_4Cl_2(g) \rightarrow C_2H_3Cl(g) + HCl(g) \quad \Delta H_r^\circ = 70,224 \text{ kJ/kmol}$$

The small quantity of impurities, less than 1%, that would be present in the feed has been neglected for the purposes of this example. Also, the selectivity for VC has been taken as 100%. It would be in the region of 99% at 55% conversion.

Heat capacity data, for vapour phase:

$$C_p^\circ = a + bT + cT^2 + dT^3 \text{ kJ/kmol K}$$

	a	$b \times 10^2$	$c \times 10^5$	$d \times 10^9$
VC	5.94	20.16	−15.34	47.65
HCl	30.28	−0.761	1.325	−4.305
DCE	20.45	23.07	−14.36	33.83

for liquid phase: DCE at 20°C, $C_p = 116$ kJ/kmol K, taken as constant over temperature rise from 20 to 25°C.

Latent heat of vaporization of DCE at 25°C = 34.3 MJ/kmol.

At 2 bar pressure the change in C_p with pressure will be small and will be neglected. Take base temperature as 25°C (298 K), the standard state for ΔH_r°.

$$\text{Enthalpy of feed} = 145.5 \times 116(293 - 298) = -84,390 \text{ J/h} = -84.4 \text{ MJ/h}$$

$$\text{Enthalpy of product stream} = \int_{298}^{773} \sum (n_i C_p) \, dT$$

Component	n_i (mol/h)	$n_i a$	$n_i b \times 10^2$	$n_i c \times 10^5$	$n_i d \times 10^9$
VC	80	475.2	1612.8	−1227.2	3812.0
HCl	80	2422.4	−60.88	106.0	−344.4
DCE	65.5	1339.5	1511.0	−940.6	2215.9
$\sum n_i C_p$		4237.1	3063.0	−2061.8	5683.5

$$\int_{298}^{773} n_i C_p \ dT = \int_{298}^{773} (4237.1 + 3063.0 \times 10^{-2}T - 2061.8 \times 10^{-5}T^2$$
$$+ 5683.5 \times 10^{-9}T^3)dT = 7307.3 \ \text{MJ/h}$$

Heat consumed in system by the endothermic reaction $= \Delta H_r^\circ \times$ moles produced

$$= 70,224 \times 80 = 5,617,920 \ \text{kJ/h} = 5617.9 \ \text{MJ/h}$$

Heat to vaporize feed (gas phase reaction)

$$= 34.3 \times 145.5 = 4990.7 \ \text{MJ/h}$$

Heat balance:

$$\text{Output} = \text{Input} + \text{Consumed} + Q$$

$$Q = H_{\text{product}} - H_{\text{feed}} + \text{Consumed}$$

$$= 7307.3 - (-84.4) + (5617.9 + 4990.7) = 18,002.3 \ \text{MJ/h}$$

Taking the overall efficiency of the furnace as 70%, the gas rate required

$$= \frac{\text{Heat input}}{(\text{calorific value} \times \text{efficiency})}$$

$$= \frac{18,002.3}{33.5 \times 0.7} = \underline{\underline{768 \ \text{m}^3/\text{h}}}$$

3.11.   STANDARD HEATS OF FORMATION

The standard enthalpy of formation ΔH_f° of a compound is defined as the enthalpy change when one mole of the compound is formed from its constituent elements in the standard state. The enthalpy of formation of the elements is taken as zero. The standard heat of any reaction can be calculated from the heats of formation $-\Delta H_f^\circ$ of the products and reactants; if these are available or can be estimated.

Conversely, the heats of formation of a compound can be calculated from the heats of reaction; for use in calculating the standard heat of reaction for other reactions.

The relationship between standard heats of reaction and formation is given by equation 3.26 and illustrated by Examples 3.8 and 3.9.

$$\Delta H_r^\circ = \sum \Delta H_f^\circ, \text{ products} - \sum \Delta H_f^\circ, \text{ reactants} \qquad (3.26)$$

A comprehensive list of enthalpies of formation is given in Appendix C and is available in MS Excel format in the on-line material at http://elsevierdirect.com/companions/.

As with heats of reaction, the state of the materials must be specified when quoting heats of formation.

Example 3.8

Calculate the standard heat of the following reaction, given the enthalpies of formation:

$$4NH_3(g) + 5O_2(g) \rightarrow 4NO(g) + 6H_2O(g)$$

Standard enthalpies of formation kJ/mol:

$NH_3(g)$	-46.2
$NO(g)$	$+90.3$
$H_2O(g)$	-241.6

Solution

Note: the enthalpy of formation of O_2 is zero.

$$\Delta H_r^\circ = \sum \Delta H_f^\circ, \text{ products } - \sum \Delta H_f^\circ, \text{ reactants}$$
$$= (4 \times 90.3 + 6 \times (-241.6)) - (4 \times (-46.2))$$
$$= \underline{\underline{-903.6 \text{ kJ/mol}}}$$

Heat of reaction $- \Delta H_r^\circ = 904$ kJ/mol

3.12. HEATS OF COMBUSTION

The heat of combustion of a compound $-\Delta H_c^\circ$ is the standard heat of reaction for complete combustion of the compound with oxygen. Heats of combustion are relatively easy to determine experimentally.

The heats of other reactions can be easily calculated from the heats of combustion of the reactants and products.

The general expression for the calculation of heats of reaction from heats of combustion is

$$\Delta H_r^\circ = \sum \Delta H_c^\circ, \text{ reactants } - \sum \Delta H_c^\circ, \text{ products} \qquad (3.27)$$

Note: the product and reactant terms are the opposite way round to that in the expression for the calculation from heats of formation (equation 3.26).

For compounds containing nitrogen, the nitrogen will not be oxidized to any significant extent in combustion and is taken to be unchanged in determining the heat of combustion.

Caution. Heats of combustion are large compared with heats of reaction. Do not round off the numbers before subtraction; round off the difference.

Two methods of calculating heats of reaction from heats of combustion are illustrated in Example 3.9.

Example 3.9

Calculate the standard heat of reaction for the following reaction: the hydrogenation of benzene to cyclohexane.

(1) $C_6H_6(g) + 3H_2(g) \rightarrow C_6H_{12}(g)$

(2) $C_6H_6(g) + 7\frac{1}{2}O_2(g) \rightarrow 6CO_2(g) + 3H_2O(l)$ $\Delta H_c^\circ = -3287.4\ kJ$
(3) $C_6H_{12}(g) + 9O_2 \rightarrow 6CO_2(g) + 6H_2O(l)$ $\Delta H_c^\circ = -3949.2\ kJ$
(4) $C(s) + O_2(g) \rightarrow CO_2(g)$ $\Delta H_c^\circ = -393.12\ kJ$
(5) $H_2(g) + \frac{1}{2}O_2(g) \rightarrow H_2O(l)$ $\Delta H_c^\circ = -285.58\ kJ$

Note: unlike heats of formation, the standard state of water for heats of combustion is liquid. Standard pressure and temperature are the same 25°C, 1 atm.

Solution
Method 1

Using the more general equation 3.26

$$\Delta H_r^\circ = \sum \Delta H_f^\circ, \text{ products } - \sum \Delta H_f^\circ, \text{ reactants}$$

the enthalpy of formation of C_6H_6 and C_6H_{12} can be calculated, and from these values the heat of reaction (1).
    From reaction (2)

$$\Delta H_c^\circ(C_6H_6) = 6 \times \Delta H_c^\circ(CO_2) + 3 \times \Delta H_c^\circ(H_2O) - \Delta H_f^\circ(C_6H_6)$$

$$3287.4 = 6(-393.12) + 3(-285.58) - \Delta H_f^\circ(C_6H_6)$$

$$\Delta H_f^\circ(C_6H_6) = -3287.4 - 3215.52 = 71.88\ kJ/mol$$

From reaction (3)
$$\Delta H_c^\circ(C_6H_{12}) = -3949.2 = 6(-393.12) + 6(-285.58) - \Delta H_f^\circ(C_6H_{12})$$
$$\Delta H_f^\circ(C_6H_{12}) = 3949.2 - 4072.28 = -123.06\ kJ/mol$$
$$\Delta H_r^\circ = \Delta H_f^\circ(C_6H_{12}) - \Delta H_f^\circ(C_6H_6)$$
$$\Delta H_r^\circ = (-123.06) - (71.88) = -195\ kJ/mol$$
Note: enthalpy of formation of H_2 is zero.

Method 2

Using equation 3.27
$$\Delta H_r^\circ = (\Delta H_c^\circ(C_6H_6) + 3 \times \Delta H_c^\circ(H_2)) - \Delta H_c^\circ(C_6H_{12})$$
$$= (-3287.4 + 3(-285.88)) - (-3949.2) = \underline{-196\ kJ/mol}$$
Heat of reaction $\Delta H_r^\circ = -196\ kJ/mol$

3.13. COMPRESSION AND EXPANSION OF GASES

The work term in an energy balance is unlikely to be significant unless a gas is expanded or compressed as part of the process. To compute the pressure work term (equation 3.5):

$$-W = \int_{v_1}^{v_2} P\ dv$$

a relationship between pressure and volume during the expansion is needed.

If the compression or expansion is isothermal (at constant temperature) then for unit mass of an ideal gas:

$$Pv = \text{constant} \tag{3.28}$$

and the work done,

$$-W = P_1 v_1 \quad \ln\frac{P_2}{P_1} = \frac{RT_1}{M_w}\ln\frac{P_2}{P_1} \tag{3.29}$$

where

P_1 = initial pressure
P_2 = final pressure
v_1 = initial volume
M_w = molecular mass (weight) of gas.

In industrial compressors or expanders, the compression or expansion path will be 'polytropic', approximated by the expression:

$$Pv^n = \text{constant} \tag{3.30}$$

The work produced (or required) is given by the general expression (see Coulson *et al.*, 1999):

$$W = P_1 v_1 \frac{n}{n-1}\left[\left(\frac{P_2}{P_1}\right)^{(n-1)/n} - 1\right] = Z\frac{RT_1}{M_w}\frac{n}{n-1}\left[\left(\frac{P_2}{P_1}\right)^{(n-1)/n} - 1\right] \tag{3.31}$$

where

Z = compressibility factor (1 for an ideal gas)
R = universal gas constant, $8.314\,\text{JK}^{-1}\,\text{mol}^{-1}$
T_1 = inlet temperature, K
W = work done, J/kg.

The value of n will depend on the design and operation of the machine.

The energy required to compress a gas, or the energy obtained from expansion, can be estimated by calculating the ideal work and applying a suitable efficiency value. For reciprocating compressors the isentropic work is normally used ($n = \gamma$) (see Figure 3.7); and for centrifugal or axial machines the polytropic work (see Figure 3.6 and Section 3.13.2).

3.13.1. Mollier Diagrams

If a Mollier diagram (enthalpy–pressure–temperature–entropy chart) is available for the working fluid, the isentropic work can be easily calculated.

$$W = H_1 - H_2 \tag{3.32}$$

where

H_1 is the specific enthalpy at the pressure and temperature corresponding to point 1, the initial gas conditions,
H_2 is the specific enthalpy corresponding to point 2, the final gas condition.

Point 2 is found from point 1 by tracing a path (line) of constant entropy on the diagram.

The method is illustrated in Example 3.10.

Example 3.10

Methane is compressed from 1 bar and 290 K to 10 bar. If the isentropic efficiency is 0.85, calculate the energy required to compress 10,000 kg/h. Estimate the exit gas temperature.

Solution

From the Mollier diagram, shown diagrammatically in Figure 3.5:

$$H_1 = 4500 \text{ cal/mol}$$
$$H_2 = 6200 \text{ cal/mol (isentropic path)}$$
$$\text{Isentropic work} = 6200 - 4500$$
$$= 1700 \text{ cal/mol}$$

For an isentropic efficiency of 0.85:

$$\text{Actual work done on gas } \frac{1700}{0.85} = 2000 \text{ cal/mol}$$

So, actual final enthalpy

$$H' = H_1 + 2000 = 6500 \text{ cal/mol}$$

From the Mollier diagram, if all the extra work is taken as irreversible work done on the gas, the exit gas temperature = 480 K

Molecular weight of methane = 16
Energy required = (moles per hour) × (specific enthalpy change)

$$= \frac{10,000}{16} \times 2000 \times 10^3$$

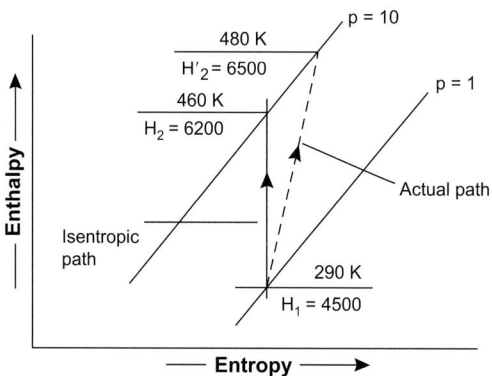

Figure 3.5.   Mollier diagram, methane.

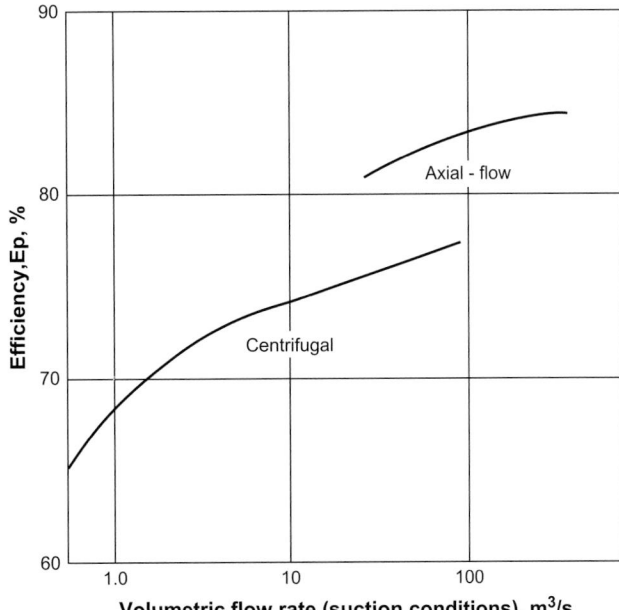

Figure 3.6. Approximate polytropic efficiencies of centrifugal and axial-flow compressors.

$$= 1.25 \times 10^9 \text{ cal/h}$$
$$= 1.25 \times 10^9 \times 4.187 \text{ J/h}$$
$$= 5.23 \times 10^9 \text{ J/h}$$

$$\text{Power} = \frac{5.23 \times 10^9}{3600} = 1.45 \text{ MW}$$

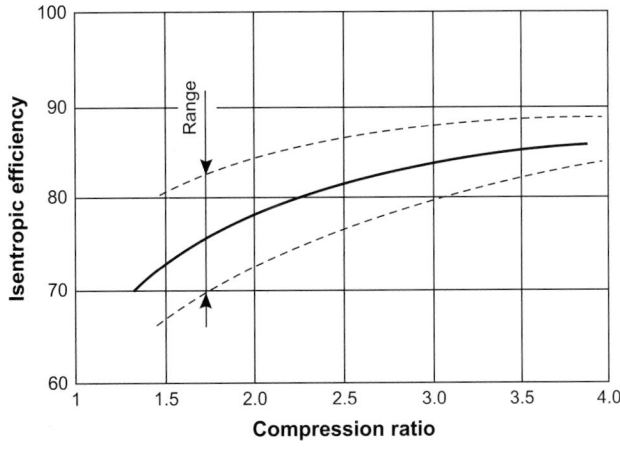

Figure 3.7. Typical efficiencies for reciprocating compressors.

3.13.2. Polytropic Compression and Expansion

If no Mollier diagram is available, it is more difficult to estimate the ideal work in compression or expansion processes.

Equation 3.31 can be used if the compressibility Z and polytropic coefficient n are known. Compressibility can be plotted against reduced temperature and pressure, as shown in Figure 3.8.

At conditions away from the critical point:

$$n = \frac{1}{1 - m} \tag{3.33}$$

where

$$m = \frac{(\gamma - 1)}{\gamma E_p} \qquad \text{for} \quad \text{compression} \tag{3.34}$$

$$m = \frac{(\gamma - 1)E_p}{\gamma} \qquad \text{for} \quad \text{expansion} \tag{3.35}$$

and E_p is the polytropic efficiency, defined by:

$$\text{for compression } E_p = \frac{\text{polytropic work}}{\text{actual work required}}$$

$$\text{for expansion } E_p = \frac{\text{actual work obtained}}{\text{polytropic work}}$$

An estimate of E_p can be obtained from Figure 3.6.

The outlet temperature can be estimated from:

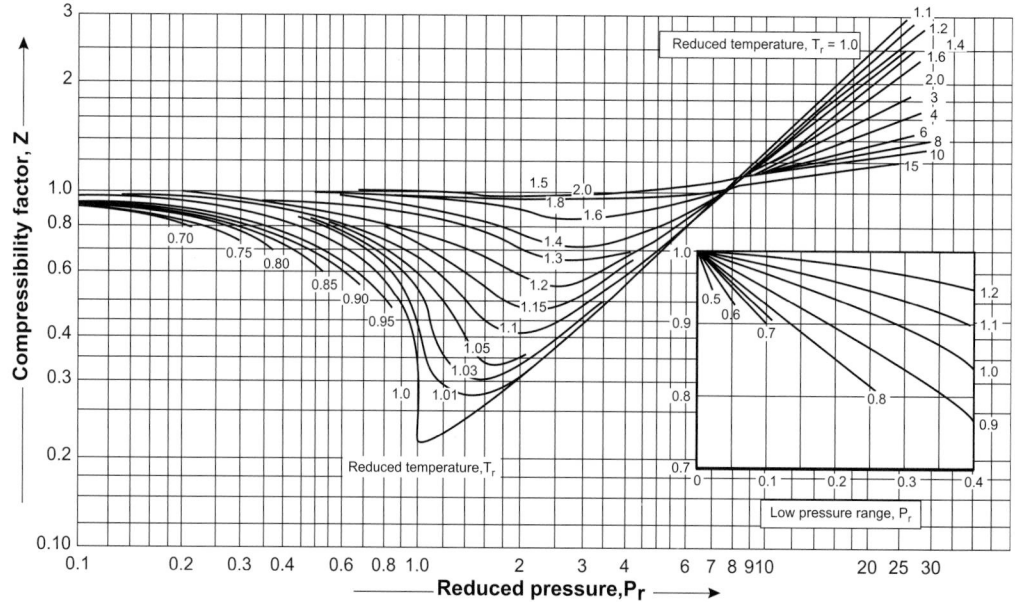

Figure 3.8. Compressibility factors of gases and vapours.

$$T_2 = T_1 \left(\frac{P_2}{P_1}\right)^m \qquad (3.36)$$

Close to the critical conditions these equations should not be used. The procedure for calculation of polytropic work of compression or expansion close to the critical point is more complex (Schultz, 1962) and it is easiest to make such calculations using process simulation programs.

Example 3.11

Estimate the power required to compress 5000 kmol/h of HCl at 5 bar, 15°C, to 15 bar.

Solution

For HCl, $P_c = 82$ bar, $T_c = 324.6$ K

$$C_p^\circ = 30.30 - 0.72 \times 10^{-2}T + 12.5 \times 10^{-6}T^2 - 3.9$$
$$\times 10^{-9}T^3 \text{ kJ/kmol K}$$

Estimate T_2 from equations 3.34 and 3.35.
For diatomic gases $\gamma \approx 1.4$.

Note: γ could be estimated from the relationship:

$$\gamma = \frac{C_p}{C_v}$$

At the inlet conditions, the flow rate in m³/s

$$= \frac{5000}{3600} \times 22.4 \times \frac{288}{273} \times \frac{1}{5} = 6.56$$

From Figure 3.6, $E_p = 0.73$.
From equations 3.34 and 3.35:

$$m = \frac{1.4 - 1}{1.4 \times 0.73} = 0.391$$

$$T_2 = 288 \left(\frac{15}{5}\right)^{0.39} = 442 \text{ K}$$

$$T_{r(\text{mean})} = \frac{442 + 288}{2 \times 324.6} = 1.12$$

$$P_{r(\text{mean})} = \frac{5 + 15}{2 \times 82} = 0.12$$

at $T_{(\text{mean})}$, $C_p^\circ = 29.14$ kJ/kmol K

Correction for pressure from Figure 3.2, 2 kJ/kmol K

$$C_p = 29.14 + 2 \approx 31 \text{ kJ/kmolK}$$

From Figure 3.8, under mean conditions:

$$Z = 0.98$$

From equation 3.33:

$$n = \frac{1}{1 - 0.391} = 1.64$$

From equation 3.31:

$$W \text{ polytropic} = 0.98 \times 288 \times 8.314 \times \frac{1.64}{1.64 - 1} \left(\left(\tfrac{15}{5}\right)^{(1.64-1)/1.64} - 1 \right)$$
$$= 3219 \text{ kJ/kmol}$$

$$\text{Actual work required} = \frac{\text{polytropic work}}{E_p} = \frac{3219}{0.73} = 4409 \text{ kJ/kmol}$$

$$\text{Power} = \frac{4409 \times 5000}{3600} = 6124 \text{ kW, say } \underline{6.1 \text{ MW}}$$

3.13.3. Multistage Compressors

Single-stage compressors can only be used for low pressure ratios. At high pressure ratios, the temperature rise is too high for efficient operation.

To cope with the need for high pressure generation, the compression is split into a number of separate stages, with intercoolers between each stage. The interstage pressures are normally selected to give equal work in each stage.

For a two-stage compressor the interstage pressure is given by:

$$P_i = \sqrt{(P_1 \times P_2)} \tag{3.37}$$

where P_i is the intermediate-stage pressure.

Example 3.12

Estimate the power required to compress 1000 m^3/h air from ambient conditions to 700 kN/m^2 gauge, using a two-stage reciprocating compressor with an intercooler.

Solution

Take the inlet pressure, P_1, as one atmosphere = 101.33 kN/m^2, absolute.
   Outlet pressure, $P_2 = 700 + 101.33 = 801.33$ kN/m^2, absolute.
   For equal work in each stage the intermediate pressure, P_i

$$= \sqrt{(1.0133 \times 10^5 \times 8.0133 \times 10^5)}$$
$$= 2.8495 \times 10^5 \text{ N/m}^2$$

For air, take ratio of the specific heats, γ, to be 1.4.
   For equal work in each stage the total work will be twice that in the first stage.
   Take the inlet temperature to be 20°C, At that temperature the specific volume is given by

$$v_1 = \frac{29}{22.4} \times \frac{293}{273} = 1.39 \text{ m}^3/\text{kg}$$

Isentropic work done, $-W = 2 \times 1.0133 \times 10^5$

$$\times 1.39 \times \frac{1.4}{1.4 - 1} \left(\left(\frac{2.8495}{1.0133} \right)^{(1.4-1)/1.4} - 1 \right)$$

$$= 338,844 \text{ J/kg} = 339 \text{ kJ/kg}$$

From Figure 3.7, for a compression ratio of 2.85 the efficiency is approximately 84%. So work required

$$= 339/0.84 = \underline{404 \text{ kJ/kg}}$$

$$\text{Mass flow rate} = \frac{1000}{1.39 \times 3600} = 0.2 \text{ kg/s}$$

$$\text{Power required} = 404 \times 0.2 = \underline{80 \text{ kW}}$$

3.13.4. Electrical Drives

The electrical power required to drive a compressor (or pump) can be calculated from a knowledge of the motor efficiency:

$$\text{Power} = \frac{-W \times \text{mass flow-rate}}{E_e} \tag{3.38}$$

where

$-W$ = work of compression per unit mass (equation 3.31)
E_e = electric motor efficiency.

TABLE 3.1. Approximate Efficiencies of Electric Motors

Size (kW)	Efficiency (%)
5	80
15	85
75	90
200	92
750	95
>4000	97

The efficiency of the drive motor will depend on the type, speed and size. The values given in Table 3.1 can be used to make a rough estimate of the power required.

3.14.   ENERGY-BALANCE CALCULATIONS

As with mass balances, energy balances for complex design problems are most easily set up and solved using commercial process simulation software, as described in Chapter 4.

Process simulation software can also be used to help build simple energy balances in spreadsheet models, for example by entering stream data to calculate mixture heat capacities, to calculate stream enthalpies or to estimate heats of reaction.

When setting up process simulation models, the design engineer needs to pay careful attention to operations that have an impact on the energy balance and heat use within the process. Some common problems to watch out for include:

1. Avoid mixing streams at very different temperatures. This usually represents a loss of heat (or cooling) that could be better used in the process.
2. Avoid mixing streams at different pressures. The mixed stream will be at the lowest pressure of the feed streams. The higher pressure streams will undergo cooling as a result of adiabatic expansion. This may lead to increased heating or cooling requirements or lost potential to recover shaft work during the expansion.
3. Segment heat exchangers to avoid internal pinches. This is particularly necessary for exchangers where there is a phase change. When a liquid is heated, boiled and superheated, the variation of stream temperature with enthalpy added looks like Figure 3.9. Liquid is heated to the boiling point (A-B), then the heat of vaporization is added (B-C) and the vapour is superheated (C-D). This is a different temperature-enthalpy profile than a straight line between the initial and final states (A-D). If the stream in Figure 3.9 were matched against a heat source that had a temperature profile like line E-F in Figure 3.10, then the exchanger would appear feasible based on the inlet and outlet temperatures, but would in fact be infeasible because of the cross-over of the temperature profiles at B. A simple way to avoid this problem is to break up the preheat, boiling and superheat into three exchangers in the simulation model, even if they will be

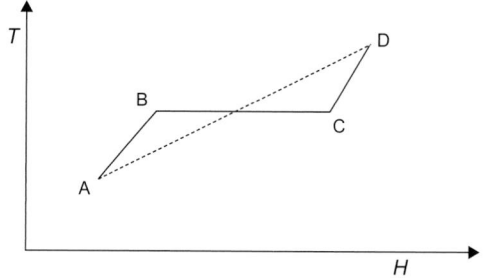

Figure 3.9.   Temperature-enthalpy profile for a stream that is vaporized and superheated.

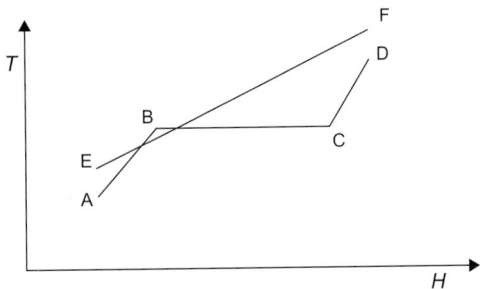

Figure 3.10. Heat transfer to a stream that is vaporized and superheated.

carried out in a single piece of equipment in the final design. The same problem also occurs with condensers that incorporate desuperheat and subcooling.

4. Check for heat of mixing. This is important whenever acids or bases are mixed with water. If the heat of mixing is large then two or more stages of mixing with intercoolers may be needed. If a large heat of mixing is expected, but is not predicted by the model, then check that the thermodynamic model includes heat of mixing effects.

5. Remember to allow for process inefficiency and design margins. For example, when sizing a fired heater, if process heating is carried out in the radiant section only, then the heating duty calculated in the simulation is only 60% of the total furnace duty (see Section 12.17). The operating duty will then be the process duty divided by 0.6. The design duty must be increased further by a suitable design factor, say 10%. The design duty of the fired heater is then $1.1/0.6 = 1.83$ times the process duty calculated in the simulation.

3.15. UNSTEADY-STATE ENERGY BALANCES

All the examples of energy balances considered previously have been for steady-state processes, where the rate of energy generation or consumption did not vary with time and the accumulation term in the general energy-balance equation was taken as zero.

If a batch process is considered, or if the rate of energy generation or removal varies with time, it is necessary to set up a differential energy balance, similar to the differential material balance considered in Chapter 2. For batch processes, the total energy requirements can usually be estimated by taking a single batch as the time basis for the calculation; but the maximum rate of heat generation must also be estimated to size any heat-transfer equipment needed.

The application of a differential energy balance is illustrated in Example 3.13.

Example 3.13
Differential Energy Balance

In the batch preparation of an aqueous solution, the water is first heated to 80°C in a jacketed, agitated vessel; 1000 Imp. gal. (4545 kg) is heated from 15°C. If the jacket area is 300 ft^2 (27.9 m^2) and the overall heat-transfer coefficient can be taken as

50 Btu ft^{-2} h^{-1} °F^{-1} (285 Wm^{-2} K^{-1}), estimate the heating time. Steam is supplied at 25 psig (2.7 bar).

Solution

The rate of heat transfer from the jacket to the water will be given by the following expression:

$$\frac{dQ}{dt} = U_j A (t_s - t) \tag{a}$$

where

dQ is the increment of heat transferred in the time interval dt, and
U_j = the overall-heat transfer coefficient
t_s = the steam-saturation temperature
t = the water temperature.

The incremental increase in the water temperature dt is related to the heat transferred dQ by the energy-balance equation:

$$dQ = MC_p \, dt \tag{b}$$

where MC_p is the heat capacity of the system.
Equating equations (a) and (b)

$$MC_p \frac{dt}{dt} = U_j A (t_s - t)$$

Integrating

$$\int_0^{t_B} dt = \frac{MC_p}{U_j A} \int_{t_1}^{t_2} \frac{dt}{(t_s - t)}$$

Batch heating time

$$t_B = -\frac{MC_p}{U_j A} \ln \frac{t_s - t_2}{t_s - t_1}$$

For this example

$$MC_p = 4.18 \times 4545 \times 10^3 \text{ JK}^{-1}$$

$$U_j A = 285 \times 27 \text{ WK}^{-1}$$

$$t_1 = 15°C, \ t_2 = 80°C, \ t_s = 130°C$$

$$t_B = -\frac{4.18 \times 4545 \times 10^3}{285 \times 27.9} \ln \frac{130 - 80}{130 - 15}$$

$$= 1990s = \underline{33.2 \text{ min}}$$

In this example, the heat capacity of the vessel and the heat losses have been neglected for simplicity. They would increase the heating time by 10 to 20%.

3.16. ENERGY RECOVERY

Process streams at high pressure or temperature, and those containing combustible material, contain energy that can be usefully recovered. Whether it is economic to recover the energy content of a particular stream depends on the value of the energy that can be usefully extracted and the cost of recovery. The value of the energy is related to the marginal cost of energy at the site, as discussed in Section 6.4.4. It may be worthwhile recovering energy from a process stream at a site where energy costs are high but not where the primary energy costs are low. The cost of recovery will be the capital and operating cost of any additional equipment required. If the savings exceed the total annualized cost, including capital charges, then the energy recovery will usually be worthwhile. Maintenance costs should be included in the annualized cost (see Chapter 6).

Some processes, such as air separation, depend on efficient energy recovery for economic operation, and in all processes the efficient use of energy-recovery techniques will reduce product cost.

Some of the techniques used for energy recovery in chemical process plants are described briefly in the following sections. The references cited give fuller details of each technique. Miller (1968) gives a comprehensive review of process energy systems; including heat exchange, and power recovery from high-pressure fluid streams.

Kenney (1984) reviews the application of thermodynamic principles to energy recovery in the process industries.

3.16.1. Heat Exchange

The most common energy-recovery technique is to use the heat in a high-temperature process stream to heat a colder stream. This saves part or all of the cost of heating the cold stream, as well as part or all of the cost of cooling the hot stream. Conventional shell and tube exchangers are normally used. The cost of the heat-exchange surface that is required may be increased, due to the reduced temperature driving forces compared to heating and cooling with utilities, or decreased, due to needing fewer exchangers. The cost of recovery will be reduced if the streams are located conveniently close.

The amount of energy that can be recovered depends on the temperature, flow, heat capacity, and temperature change possible, in each stream. A reasonable temperature driving force must be maintained to keep the exchanger area to a practical size. The most efficient exchanger will be the one in which the shell and tube flows are truly counter-current. Multiple tube pass exchangers are usually used for practical reasons. With multiple tube passes the flow is part counter-current and part co-current and temperature crosses can occur, which reduce the efficiency of heat recovery (see Section 4.5.4 and Chapter 12).

The hot process streams leaving a reactor or a distillation column are frequently used to preheat the feed streams ('feed-effluent' or 'feed-bottoms' exchangers).

In an industrial process there will be many hot and cold streams and there will be an optimum arrangement of the streams for energy recovery by heat exchange. The problem of synthesizing a network of heat exchangers has been the subject of much research and is covered in more detail is Section 3.17.

3.16.2. Waste-Heat Boilers

If the process streams are at a sufficiently high temperature and there are no attractive options for process-to-process heat transfer, then the heat recovered can be used to generate steam.

Waste-heat boilers are often used to recover heat from furnace flue gases and the process gas streams from high-temperature reactors. The pressure, and superheat temperature, of the stream generated depend on the temperature of the hot stream and the approach temperature permissible at the boiler exit (see Chapter 12). As with any heat-transfer equipment, the area required increases as the mean temperature driving force (log mean ΔT) is reduced. The permissible exit temperature may also be limited by process considerations. If the gas stream contains water vapour and soluble corrosive gases, such as HCl or SO_2, the exit gas temperature must be kept above the dew point.

Hinchley (1975) discusses the design and operation of waste heat boilers for chemical plant. Both fire-tube and water-tube boilers are used. A typical arrangement of a water-tube boiler on a reformer furnace is shown in Figure 3.11 and a fire-tube boiler in Figure 3.12.

The application of a waste-heat boiler to recover energy from the reactor exit streams in a nitric acid plant is shown in Figure 3.13.

The selection and operation of waste-heat boilers for industrial furnaces is discussed in the *Efficient Use of Energy*, Dryden (1975).

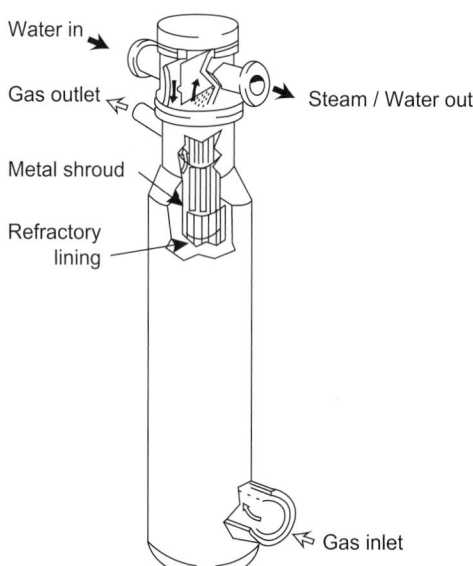

Figure 3.11.   Reformed gas waste-heat boiler arrangement of vertical U-tube water-tube boiler (Reprinted by permission of the Council of the Institution of Mechanical Engineers from the Proceedings of the Conference on Energy Recovery in the Process Industries, London, 1975.)

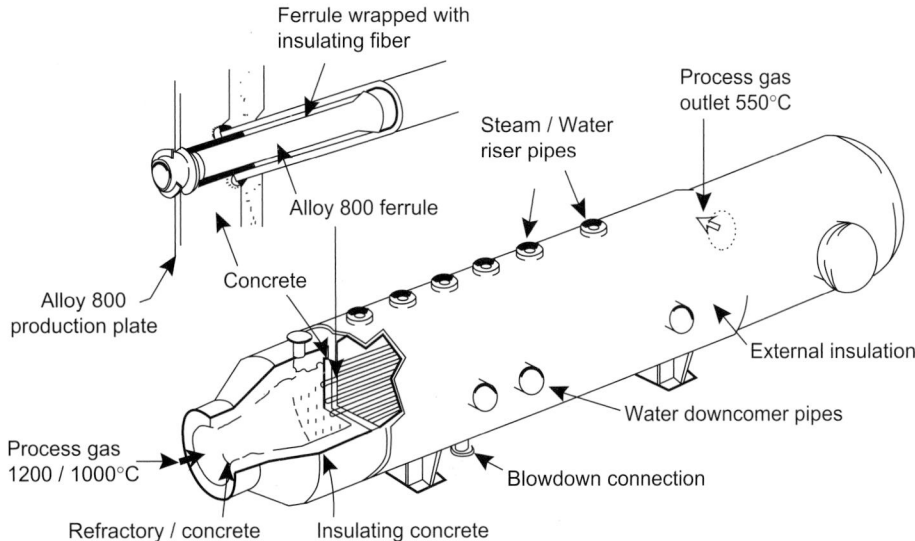

Figure 3.12. Reformed gas waste-heat boiler, principal features of typical natural circulation fire-tube boilers (Reprinted by permission of the Council of the Institution of Mechanical Engineers from the Proceedings of the Conference on Energy Recovery in the Process Industries, London, 1975.)

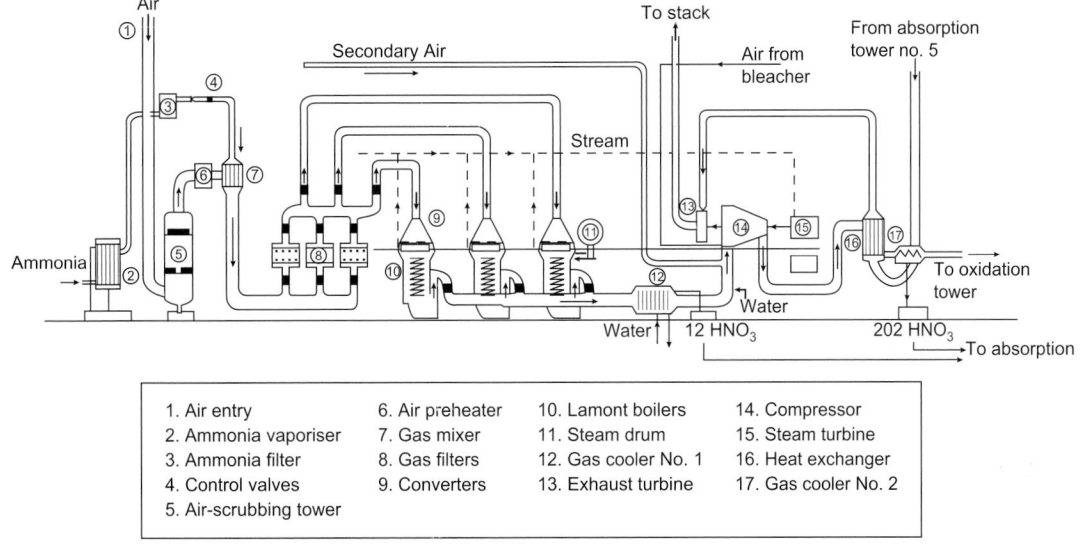

1. Air entry	6. Air preheater	10. Lamont boilers	14. Compressor
2. Ammonia vaporiser	7. Gas mixer	11. Steam drum	15. Steam turbine
3. Ammonia filter	8. Gas filters	12. Gas cooler No. 1	16. Heat exchanger
4. Control valves	9. Converters	13. Exhaust turbine	17. Gas cooler No. 2
5. Air-scrubbing tower			

(From Nitric Acid Manufacture, Miles (1961), with permission)

Figure 3.13. Connections of a nitric acid plant, intermediate pressure type.

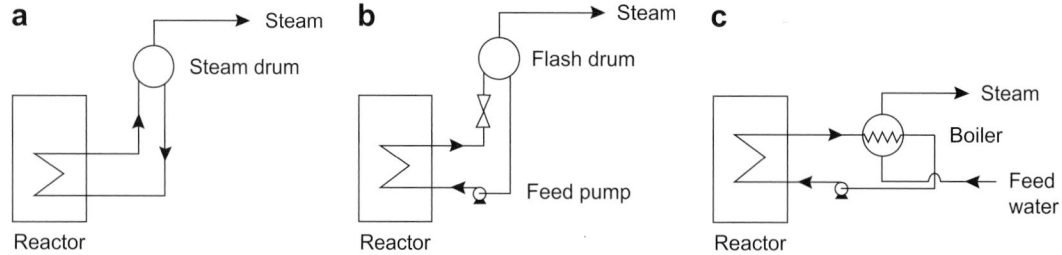

Figure 3.14.   Steam generation.

3.16.3. High-Temperature Reactors

If a reaction is highly exothermic, cooling will be needed. If the reactor temperature is high enough, the heat removed can be used to generate steam. The lowest steam pressure normally used in the process industries is 2.7 bar (25 psig) and steam is normally distributed at a header pressure of around 8 bar (100 psig); so any reactor with a temperature above 200°C is a potential steam generator.

Three systems are used:

1. Figure 3.14a. An arrangement similar to a conventional water-tube boiler. Steam is generated in cooling pipes within the reactor and separated in a steam drum.
2. Figure 3.14b. Similar to the first arrangement but with the water kept at high pressure to prevent vaporization. The high-pressure water is flashed to steam at lower pressure in a flash drum. This system would give more responsive control of the reactor temperature.
3. Figure 3.14c. In this system a heat-transfer fluid, such as Dowtherm (see Green and Perry (2007) and Singh (1985) for details of heat-transfer fluids), is used to avoid the need for high-pressure tubes. The steam is raised in an external boiler.

3.16.4. Low-Grade Fuels

Process waste products that contain significant quantities of combustible material can be used as low-grade fuels, for raising steam or direct process heating. Their use will only be economic if the intrinsic value of the fuel justifies the cost of special burners and other equipment needed to burn the waste. If the combustible content of the waste is too low to support combustion, the waste must be supplemented with higher calorific value primary fuels.

Reactor Off-Gases

Reactor off-gases (vent gas) and recycle stream purges are often of high enough calorific value to be used as fuels. The calorific value of a gas can be calculated from the heats of combustion of its constituents; the method is illustrated in Example 3.14.

Other factors which, together with the calorific value, determine the economic value of an off-gas as a fuel are the quantity available and the continuity of supply.

Waste gases are best used for steam raising, rather than for direct process heating, as this decouples the source from the use and gives greater flexibility.

Example 3.14
Calculation of Waste-Gas Calorific Value

The typical vent-gas analysis from the recycle stream in an oxyhydrochlorination process for the production of dichloroethane (DCE) (British patent BP 1,524,449) is given below, percentages on volume basis.

O_2 7.96, $CO_2 + N_2$ 87.6, CO 1.79, C_2H_4 1.99, C_2H_6 0.1, DCE 0.54

Estimate the vent gas calorific value.

Solution

Component calorific values, from Perry and Chilton (1973):

CO	67.6 kcal/mol = 283 kJ/mol
C_2H_4	372.8 = 1560.9
C_2H_6	337.2 = 1411.9

The value for DCE can be estimated from the heats of formation. Combustion reaction:

$$C_2H_4Cl_2(g) + 2\frac{1}{2}O_2(g) \rightarrow 2CO_2(g) + H_2O(g) + 2HCl(g)$$

ΔH_f° from Appendix D

$$
\begin{aligned}
CO_2 &= -393.8\,°kJ/mol \\
H_2O &= -242.0 \\
HCl &= -92.4 \\
DCE &= -130.0 \\
\Delta H_c^\circ &= \sum \Delta H_f^\circ \text{ products} - \sum \Delta H_f^\circ \text{ reactants} \\
&= [2(-393.8) - 242.0 + 2(-92.4)] - [-130.0] \\
&= -1084.4 \text{ kJ}
\end{aligned}
$$

Estimation of vent gas calorific value, basis: 100 mol.

Component	mol/100 mol		Calorific value (kJ/mol)		Heating value
CO	1.79	×	283.0	=	506.6
C_2H_4	1.99		1560.9		3106.2
C_2H_6	0.1		1411.9		141.2
DCE	0.54		1084.4		585.7
				Total	4339.7

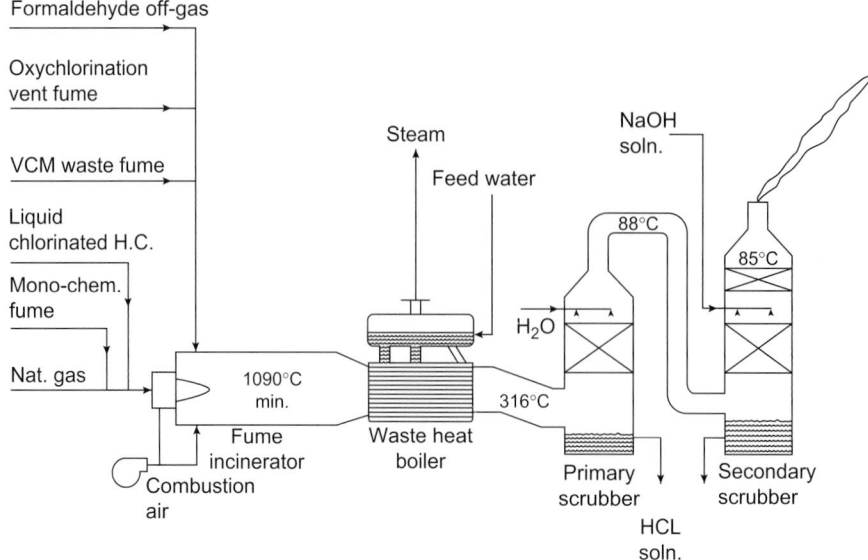

Figure 3.15.   Typical incinerator-heat recovery-scrubber system for vinyl-chloride-monomer process waste. (Courtesy of John Thurley Ltd.)

$$\text{Calorific value of vent gas} = \frac{4339.7}{100} = 43.4 \, \text{kJ/mol}$$

$$= \frac{43.4}{22.4} \times 10^3 = \underline{\underline{1938 \, \text{kJ/m}^3}} \, (52 \, \text{Btu/ft}^3) \text{ at 1 bar, } 0°\text{C}$$

This calorific value is very low compared to $37 \, \text{MJ/m}^3$ ($1000 \, \text{Btu/ft}^3$) for natural gas. The vent gas is barely worth recovery, but if the gas has to be burnt to avoid pollution it could be used in an incinerator such as that shown in Figure 3.15, giving a useful steam production to offset the cost of disposal.

Liquid and Solid Wastes

Combustible liquid and solid waste can be disposed of by burning, which is usually preferred to dumping. Incorporating a steam boiler in the incinerator design will enable an otherwise unproductive, but necessary, operation to save energy. If the combustion products are corrosive, corrosion-resistant materials will be needed, and the flue gases must be scrubbed to reduce air pollution. An incinerator designed to handle chlorinated and other liquid and solid wastes is shown in Figure 3.15. This incinerator incorporates a steam boiler and a flue-gas scrubber. The disposal of chlorinated wastes is discussed by Santoleri (1973).

Dunn and Tomkins (1975) discuss the design and operation of incinerators for process wastes. They give particular attention to the need to comply with the current clean-air legislation, and the problem of corrosion and erosion of refractories and heat-exchange surfaces.

3.16.5. High-Pressure Process Streams

Where high-pressure gas or liquid process streams are throttled to lower pressures, energy can be recovered by carrying out the expansion in a suitable turbine.

Gas Streams

The economic operation of processes that involve the compression and expansion of large quantities of gases, such as ammonia synthesis, nitric acid production and air separation, depends on the efficient recovery of the energy of compression. The energy recovered by expansion is often used to drive the compressors directly; as shown in Figure 3.13. If the gas contains condensable components, it may be advisable to consider heating the gas by heat exchange with a higher temperature process stream before expansion. The gas can then be expanded to a lower pressure without condensation and the power generated increased.

The process gases do not have to be at a particularly high pressure for expansion to be economical if the flow rate is high. For example, Luckenbach (1978) in U.S. patent 4,081,508 describes a process for recovering power from the off-gas from a fluid catalytic cracking process by expansion from about 2 to 3 bar (15 to 25 psig) down to just over atmospheric pressure (1.5 to 2 psig).

The energy recoverable from the expansion of a gas can be estimated by assuming polytropic expansion; see Section 3.13.2 and Example 3.15.

The design of turboexpanders for the process industries is discussed by Bloch *et al.* (1982).

Example 3.15

Consider the extraction of energy from the tail gases from a nitric acid adsorption tower.

Gas composition, kmol/h:

O_2	371.5
N_2	10,014.7
NO	21.9
NO_2	trace
H_2O	saturated at 25°C

If the gases leave the tower at 6 atm, 25°C, and are expanded to, say, 1.5 atm, calculate the turbine exit gas temperatures without preheat, and if the gases are preheated to 400°C with the reactor off-gas. Also, estimate the power recovered from the preheated gases.

Solution

For the purposes of this calculation it will be sufficient to consider the tail gas as all nitrogen, flow 10,410 kmol/h.

$$P_c = 33.5 \text{ atm}, \ T_c = 126.2 \text{ K}$$

Figure 3.6 can be used to estimate the turbine efficiency.

$$\text{Exit gas volumetric flow rate} = \frac{10,410}{3600} \times 22.4 \times \frac{1}{1.5} \approx 43 \text{ m}^3/\text{s}$$

From Figure 3.6, $E_P = 0.75$

$$P_r \text{ inlet} = \frac{6}{33.5} = 0.18$$

$$T_r \text{ inlet} = \frac{298}{126.2} = 2.4$$

Using equations 3.33 and 3.35, for N_2 $\gamma = 1.4$

$$m = \frac{1.4 - 1}{1.4} \times 0.75 = 0.21$$

$$n = \frac{1}{1 - m} = \frac{1}{1 - 0.21} = 1.27$$

Without preheat

$$T_2 = 298 \left(\frac{1.5}{6.0}\right)^{0.21} = 223 \text{ K}$$
$$= -50°\text{C}$$

This temperature would be problematic. Acidic water would condense out, probably damaging the turbine.

  With preheat

$$T_2 = 673 \left(\frac{1.5}{6.0}\right)^{0.21} = 503 \text{ K} = 230°\text{C}$$

From equation 3.31, the work done by the gas as a result of polytropic expansion is:

$$= -1 \times 673 \times 8.314 \times \frac{1.27}{1.27 - 1} \left\{ \left(\frac{1.5}{6.0}\right)^{(1.27-1)/1.27} - 1 \right\}$$

$$= 6718 \text{ kJ/kmol}$$

$$\text{Actual work} = \text{polytropic work} \times Ep$$
$$= 6718 \times 0.75 = \underline{\underline{5039 \text{kJ/kmol}}}$$

$$\text{Power output} = \text{work/kmol} \times \text{kmol/s} = 5039 \times \frac{10,410}{3600}$$
$$= 14,571 \text{ kJ/s} = \underline{14.6\text{MW}}$$

  This is a significant amount of power and will probably justify the cost of the expansion turbine.

Liquid Streams

As liquids are essentially incompressible, less energy is stored in a compressed liquid than a gas; however, it is often worth considering power recovery from high-pressure liquid streams (>15 bar), as the equipment required is relatively

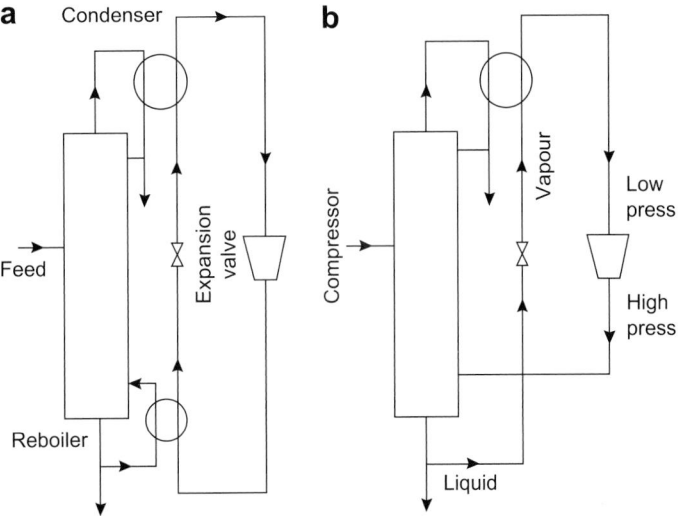

Figure 3.16. Distillation column with heat pump. (a) Separate refrigerant circuit. (b) Using column fluid as the refrigerant.

simple and inexpensive. Centrifugal pumps are used as expanders and are often coupled directly to other pumps. The design, operation and cost of energy recovery from high-pressure liquid streams is discussed by Jenett (1968), Chada (1984) and Buse (1985).

3.16.6. Heat Pumps

A heat pump is a device for raising low-grade heat to a temperature at which the heat can be used. It pumps the heat from a low temperature source to the higher temperature sink, using a small amount of energy relative to the heat energy recovered.

Heat pumps are increasingly finding applications in the process industries. A typical application is the use of the low-grade heat from the condenser of a distillation column to provide heat for the reboiler; see Barnwell and Morris (1982) and Meili (1990). Heat pumps are also used with dryers, heat being abstracted from the exhaust air and used to preheat the incoming air.

Details of the thermodynamic cycles used for heat pumps can be found in most textbooks on Engineering Thermodynamics, and in Reay and MacMichael (1988). In the process industries heat pumps operating on the mechanical vapour compression cycle are normally used. A vapour compression heat pump applied to a distillation column is shown in Figure 3.16a. The working fluid, usually a commercial refrigerant, is fed to the reboiler as a vapour at high pressure and condenses, giving up heat to vaporize the process fluid. The liquid refrigerant from the reboiler is then expanded over a throttle valve and the resulting wet vapour is fed to the column condenser. In the condenser the liquid portion of the refrigerant is evaporated, taking heat from the condensing process vapour. The refrigerant vapour is then compressed and recycled to the reboiler, completing the working cycle.

If the conditions are suitable, the process fluid can be used as the working fluid for the heat pump. This arrangement is shown in Figure 3.16b. The hot process liquid at high pressure is expanded over the throttle value and fed to the condenser, to provide cooling to condense the vapour from the column. The vapour from the condenser is compressed and returned to the base of the column. In an alternative arrangement, the process vapour is taken from the top of the column, compressed and fed to the reboiler to provide heating.

The efficiency of a heat pump is measured by the coefficient of performance:

$$COP = \frac{\text{energy delivered at higher temperature}}{\text{energy input to the compressor}} \qquad (3.39)$$

The *COP* depends principally on the working temperatures. Heat pumps are more efficient (higher *COP*) when operated over a narrow temperature range. They are thus most often encountered on distillation columns that separate close-boiling compounds. Note that the *COP* of a heat pump is not the same as the *COP* of a refrigeration cycle (Section 6.4.4).

The economics of the application of heat pumps in the process industries is discussed by Holland and Devotta (1986). Details of the application of heat pumps in a wide range of industries are given by Moser and Schnitzer (1985).

3.17. HEAT-EXCHANGER NETWORKS

The design of a heat exchanger network for a simple process with only one or two streams that need heating and cooling is usually straightforward. When there are multiple hot and cold streams, the design is more complex and there may be many possible heat exchange networks. The design engineer must determine the optimum extent of heat recovery, while ensuring that the design is flexible to changes in process conditions and can be started up and operated easily and safely.

In the 1980s, there was a great deal of research into design methods for heat-exchanger networks; see Gundersen and Naess (1988). One of the most widely applied methods that emerged was a set of techniques termed *pinch technology*, that was developed by Bodo Linnhoff and his collaborators at ICI, Union Carbide and the University of Manchester. The term derives from the fact that in a plot of the system temperatures versus the heat transferred, a *pinch* usually occurs between the hot stream and cold stream curves (see Figure 3.22). It has been shown that the pinch represents a distinct thermodynamic break in the system and that, for minimum energy requirements, heat should not be transferred across the pinch, Linnhoff *et al.* (1982).

TABLE 3.2.   Data for Heat Integration Problem

Stream Number	Type	CP (kW/°C)	Ts (°C)	Tt (°C)	Heat Load (kW)
1	hot	3.0	180	60	360
2	hot	1.0	150	30	120
3	cold	2.0	20	135	230
4	cold	4.5	80	140	270

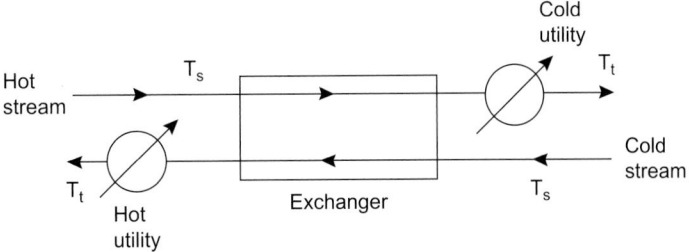

Figure 3.17. Two-stream exchanger problem.

In this section the fundamental principles of the pinch technology method for energy integration will be outlined and illustrated with reference to a simple problem. The method and its applications are described fully in a guide published by the Institution of Chemical Engineers, IChemE (1994); see also Douglas (1988), Smith (2005) and El-Halwagi (2006).

3.17.1. Pinch Technology

The development and application of the method can be illustrated by considering the problem of recovering heat between four process streams: two hot streams that require cooling, and two cold streams that must be heated. The process data for the streams is set out in Table 3.2. Each stream starts from a source temperature T_s, and is to be heated or cooled to a target temperature T_t. The heat-capacity flow rate of each stream is shown as CP. For streams where the specific heat capacity can be taken as constant, and there is no phase change:

$$CP = mC_p \tag{3.40}$$

where

m = mass flow rate, kg/s

C_p = average specific heat capacity between T_s and T_t kJ kg$^{-1\circ}$C^{-1}.

The heat load shown in Table 3.2 is the total heat required to heat, or cool, the stream from the source to the target temperature.

There is clearly scope for energy integration between these four streams. Two require heating and two cooling; and the stream temperatures are such that heat can be transferred from the hot to the cold streams. The task is to find the best arrangement of heat exchangers to achieve the target temperatures.

Simple Two-Stream Problem

Before investigating the energy integration of the four streams shown in Table 3.2, the use of a temperature-enthalpy diagram will be illustrated for a simple problem involving only two streams. The general problem of heating and cooling two streams from source to target temperatures is shown in Figure 3.17. Some heat is exchanged between the streams in the heat exchanger. Additional heat, to raise the cold stream to the target temperature, is provided by the hot utility (usually steam) in the heater; and

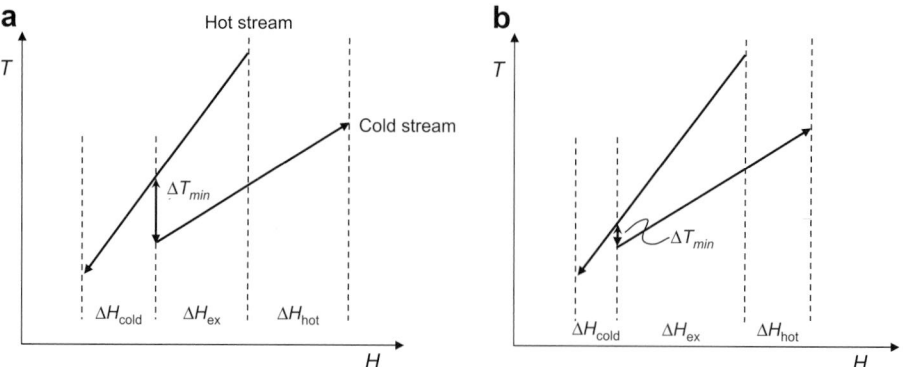

Figure 3.18.   Temperature-enthalpy (*T-H*) diagram for two-stream example.

additional cooling to bring the hot stream to its target temperature, by the cold utility (usually cooling water) in the cooler.

In Figure 3.18a the stream temperatures are plotted on the y-axis and the enthalpy change in each stream on the x-axis. This is known as a temperature-enthalpy (*T-H*) diagram. For heat to be exchanged, a minimum temperature difference must be maintained between the two streams. This is shown as ΔT_{min} on the diagram. The practical minimum temperature difference in a heat exchanger will usually be between 5°C and 30°C; see Chapter 12.

The slope of the lines in the *T-H* plot is proportional to $1/CP$, since $\Delta H = CP \times \Delta T$, so $dT/dH = 1/CP$. Streams with low heat-capacity flow rate thus have steep slopes in the *T-H* plot and streams with high heat-capacity flow rate have shallow slopes.

The heat transferred between the streams is given by the range of enthalpy over which the two curves overlap each other, and is shown on the diagram as ΔH_{ex}. The heat transferred from the hot utility, ΔH_{hot}, is given by the part of the cold stream that is not overlapped by the hot stream. The heat transferred to the cold utility, ΔH_{cold}, is similarly given by the part of the hot stream that is not overlapped by the cold stream. The heats can also be calculated as:

$$\Delta H = CP \times (\text{temperature change})$$

Since we are only concerned with changes in enthalpy, we can treat the enthalpy axis as a relative scale and slide either the hot stream or the cold stream horizontally. As we do so, we change the minimum temperature difference between the streams, ΔT_{min}, and also the amount of heat exchanged and the amounts of hot and cold utilities required.

Figure 3.18b shows the same streams plotted with a lower value of ΔT_{min}. The amount of heat exchanged is increased and the utility requirements have been reduced. The temperature driving force for heat transfer has also been reduced, so the heat exchanger has both a larger duty and a smaller log-mean temperature difference. This leads to an increase in the heat transfer area required and in the capital cost of the exchanger. The capital cost increase is partially offset by capital cost savings in the heater and cooler, which both become smaller, as well as by savings in the costs of hot

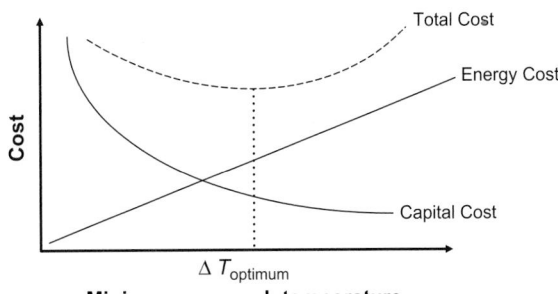

Figure 3.19.   The capital-energy trade-off in process heat recovery.

and cold utility. In general, there will be an optimum value of ΔT_{min}, as illustrated in Figure 3.19. This optimum is usually rather flat over the range 10°C to 30°C.

The maximum feasible heat recovery is reached at the point where the hot and cold curves touch each other on the *T-H* plot, as illustrated in Figure 3.20. At this point, the temperature driving force at one end of the heat exchanger is zero and an infinite heat-exchange surface is required, so the design is not practical. The exchanger is said to be *pinched* at the end where the hot and cold curves meet. In Figure 3.20, the heat exchanger is pinched at the cold end.

It is not possible for the hot and cold streams to cross each other, as this would be a violation of the 2nd law of thermodynamics.

Four-Stream Problem

In Figure 3.21a the hot streams given in Table 3.2 are shown plotted on a temperature-enthalpy diagram.

As the diagram shows changes in the enthalpy of the streams, it does not matter where a particular curve is plotted on the enthalpy axis, as long as the curve runs between the correct temperatures. This means that where more than one stream appears in a temperature interval, the stream heat capacities can be added to form a composite curve, as shown in Figure 3.21b.

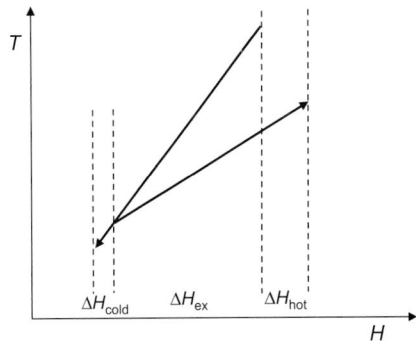

Figure 3.20.   Maximum feasible heat recovery for two-stream example.

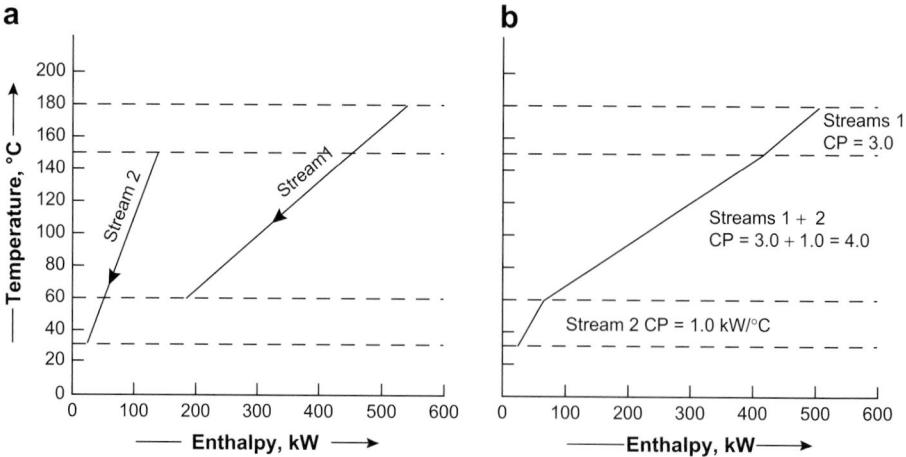

Figure 3.21.   Hot stream temperature vs enthalpy (a) Separate hot streams. (b) Composite hot streams.

In Figure 3.22, the composite curve for the hot streams and the composite curve for the cold streams are drawn with a minimum temperature difference, the displacement between the curves, of 10°C. This implies that in any of the exchangers to be used in the network, the temperature difference between the streams will not be less than 10°C.

As for the two-stream problem, the overlap of the composite curves gives a target for heat recovery and the displacements of the curves at the top and bottom of the diagram give the hot and cold utility requirements. These will be the minimum values

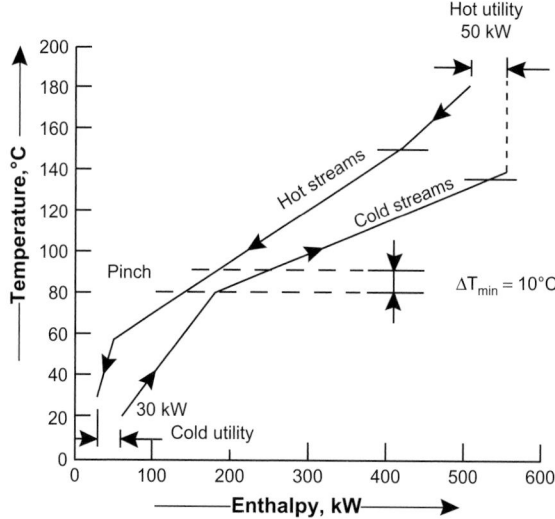

Figure 3.22.   Hot and cold stream composite curves.

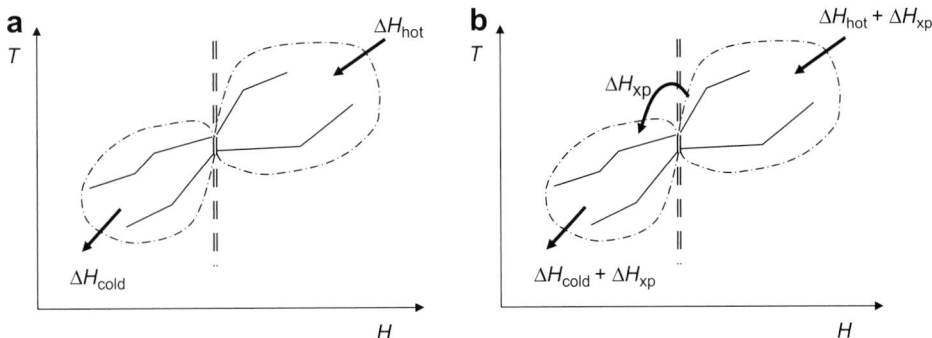

Figure 3.23. Pinch decomposition.

needed to satisfy the target temperatures. This is valuable information. It gives the designer target values for the utilities to aim for when designing the exchanger network. Any design can be compared with the minimum utility requirements to check if further improvement is possible.

In most exchanger networks the minimum temperature difference will occur at only one point. This is termed the *pinch*. In the problem being considered, the pinch occurs at between 90°C on the hot stream curve and 80°C on the cold stream curve.

For multi-stream problems, the pinch will usually occur somewhere in the middle of the composite curves, as illustrated in Figure 3.22. the case when the pinch occurs at the end of one of the composite curves is termed a threshold problem and is discussed in Section 3.17.5.

Thermodynamic Significance of the Pinch

The pinch divides the system into two distinct thermodynamic regions. The region above the pinch can be considered a heat sink, with heat flowing into it, from the hot utility, but no heat flow out of it. Below the pinch the converse is true. Heat flows out of the region to the cold utility. No heat flows across the pinch, as shown in Figure 3.23a.

If a network is designed in which heat is transferred from any hot stream at a temperature above the pinch (including hot utilities) to any cold stream at a temperature below the pinch (including cold utilities), then heat is transferred across the pinch. If the amount of heat transferred across the pinch is ΔH_{xp}, then in order to maintain energy balance the hot utility and cold utility must both be increased by ΔH_{xp}, as shown in Figure 3.23b. Cross-pinch heat transfer thus always leads to consumption of both hot and cold utilities that is greater than the minimum values that could be achieved.

The pinch decomposition is very useful in heat-exchanger network design, as it decomposes the problem into two smaller problems. It also indicates the region where heat-transfer matches are most constrained, at or near the pinch. When multiple hot or cold utilities are used there may be other pinches, termed utility pinches, that cause further problem decomposition. Problem decomposition can be exploited in algorithms for automatic heat-exchanger network synthesis.

TABLE 3.3.   Interval Temperatures for $\Delta T_{min} = 10°C$

Stream	Actual Temperature		Interval Temperature	
1	180	60	175	55
2	150	30	145	25
3	20	135	(25)	140
4	80	140	85	(145)

3.17.2.   The Problem Table Method

The problem table is a numerical method for determining the pinch temperatures and the minimum utility requirements, introduced by Linnhoff and Flower (1978). It eliminates the sketching of composite curves, which can be useful if the problem is being solved manually. It is not widely used in industrial practice any more, due to the wide availability of computer tools for pinch analysis (see Section 3.17.7).

The procedure is as follows:

1. Convert the actual stream temperatures T_{act} into interval temperatures T_{int} by subtracting half the minimum temperature difference from the hot stream temperatures, and by adding half to the cold stream temperatures:

$$\text{hot streams } T_{int} = T_{act} - \frac{\Delta T_{min}}{2}$$

$$\text{cold streams } T_{int} = T_{act} + \frac{\Delta T_{min}}{2}$$

The use of the interval temperature rather than the actual temperatures allows the minimum temperature difference to be taken into account. $\Delta T_{min} = 10°C$ for the problem being considered; see Table 3.3.

2. Note any duplicated interval temperatures. These are bracketed in Table 3.3.
3. Rank the interval temperatures in order of magnitude, showing the duplicated temperatures only once in the order; see Table 3.4.
4. Carry out a heat balance for the streams falling within each temperature interval:

TABLE 3.4.   Ranked Order of Interval Temperatures

Rank	Interval ΔT_n (°C)	Streams in interval
175°C		
145	30	−1
140	5	4 − (2 + 1)
85	55	(3 + 4) − (1 + 2)
55	30	3 − (1 + 2)
25	30	3 − 2

Note: Duplicated temperatures are omitted. The interval ΔT and streams in the intervals are included as they are needed for Table 3.5.

TABLE 3.5. Problem Table

Interval	Interval Temp. (°C)	ΔT_n (°C)	$\Sigma CP_c - \Sigma CP_h{}^*$ (kW/°C)	ΔH (kW)	Surplus or Deficit
	175				
1	145	30	−3.0	−90	s
2	140	5	0.5	2.5	d
3	85	55	2.5	137.5	d
4	55	30	−2.0	−60	s
5	25	30	1.0	30	d

* *Note*: The streams in each interval are given in Table 3.4.

For the nth interval:

$$\Delta H_n = (\Sigma CP_c - \Sigma CP_h)(\Delta T_n)$$

where

ΔH_n = net heat required in the nth interval
ΣCP_c = sum of the heat capacities of all the cold streams in the interval
ΣCP_h = sum of the heat capacities of all the hot streams in the interval
ΔT_n = interval temperature difference = $(T_{n-1} - T_n)$.

See Table 3.5.

5. 'Cascade' the heat surplus from one interval to the next down the column of interval temperatures; see Figure 3.24a.
   Cascading the heat from one interval to the next implies that the temperature difference is such that the heat can be transferred between the hot and cold streams. The presence of a negative value in the column indicates that the

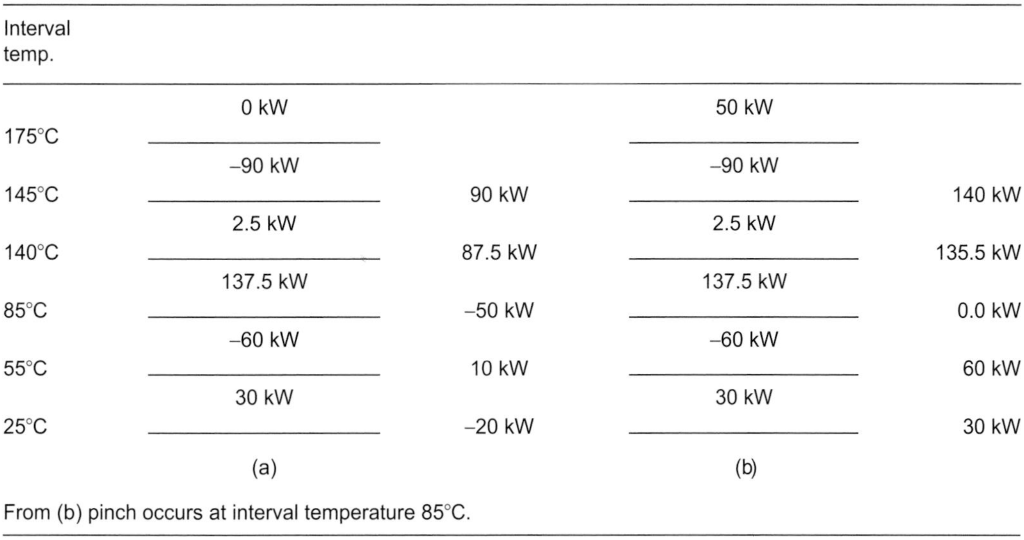

From (b) pinch occurs at interval temperature 85°C.

Figure 3.24. Heat cascade.

temperature gradient is in the wrong direction and that the exchange is not thermodynamically possible.

This difficulty can be overcome if heat is introduced into the top of the cascade:

6. Introduce just enough heat to the top of the cascade to eliminate all the negative values; see Figure 3.24b.

Comparing the composite curve, Figure 3.22, with Figure 3.24b shows that the heat introduced to the cascade is the minimum hot utility requirement and the heat removed at the bottom is the minimum cold utility required. The pinch occurs in Figure 3.24b where the heat flow in the cascade is zero. This is as would be expected from the rule that for minimum utility requirements no heat flows across the pinch. In Figure 3.24b the pinch is at an interval temperature of 85°C, corresponding to a cold stream temperature of 80°C and a hot stream temperature of 90°C, as was found using the composite curves.

It is not necessary to draw up a separate cascade diagram. This was done in Figure 3.24 to illustrate the principle. The cascaded values can be added to the problem table as two additional columns; see Example 3.16.

Summary

For maximum heat recovery and minimum use of utilities:

1. Do not transfer heat across the pinch
2. Do not use hot utilities below the pinch
3. Do not use cold utilities above the pinch.

3.17.3. Heat-Exchanger Network Design

Grid Representation

It is convenient to represent a heat-exchanger network as a grid; see Figure 3.25. The process streams are drawn as horizontal lines, with the stream numbers shown in square boxes. Hot streams are drawn at the top of the grid, and flow from left to right. The cold streams are drawn at the bottom, and flow from right to left. The stream heat capacities CP are shown in a column at the end of the stream lines.

Heat exchangers are drawn as two circles connected by a vertical line. The circles connect the two streams between which heat is being exchanged; that is, the streams that would flow through the actual exchanger. Heaters and coolers can be drawn as a single circle, connected to the appropriate utility. If multiple utilities are used, then these can also be shown as streams. Exchanger duties are usually marked

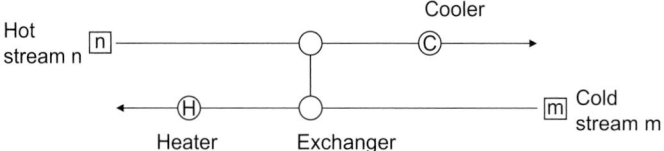

Figure 3.25.   Grid representation.

under the exchanger and temperatures are also sometimes indicated on the grid diagram.

Network Design for Maximum Energy Recovery

The analysis carried out in Figure 3.22 and Figure 3.24 has shown that the minimum utility requirements for the problem set out in Table 3.2 are 50 kW of the hot and 30 kW of the cold utility; and that the pinch occurs where the cold streams are at 80°C and the hot streams are at 90°C.

The grid representation of the streams is shown in Figure 3.26. The vertical lines represent the pinch and separate the grid into the regions above and below the pinch. Note that the hot and cold streams are off-set at the pinch, because of the difference in pinch temperature.

For maximum energy recovery (minimum utility consumption) the best performance is obtained if no cooling is used above the pinch. This means that the hot streams above the pinch should be brought to the pinch temperature solely by exchange with the cold streams. The network design is therefore started at the pinch; finding feasible matches between streams to fulfil this aim. In making a match adjacent to the pinch, the heat capacity CP of the hot stream must be equal to or less than that of the cold stream. This is to ensure that the minimum temperature difference between the curves is maintained. The slope of a line on the temperature-enthalpy diagram is equal to the reciprocal of the heat capacity. So, above the pinch the lines will converge if CP_{hot} exceeds CP_{cold} and as the streams start with a separation at the pinch equal to ΔT_{min}, the minimum temperature condition would be violated. Every hot stream must be matched with a cold stream immediately above the pinch, otherwise it will not be able to reach the pinch temperature.

Below the pinch the procedure is the same; the aim being to bring the cold streams to the pinch temperature by exchange with the hot streams. For streams adjacent to the pinch the criterion for matching streams is that the heat capacity of the cold stream must be equal to or greater than the hot stream, to avoid breaking the minimum temperature difference condition. Every cold stream must be matched with a hot stream immediately below the pinch.

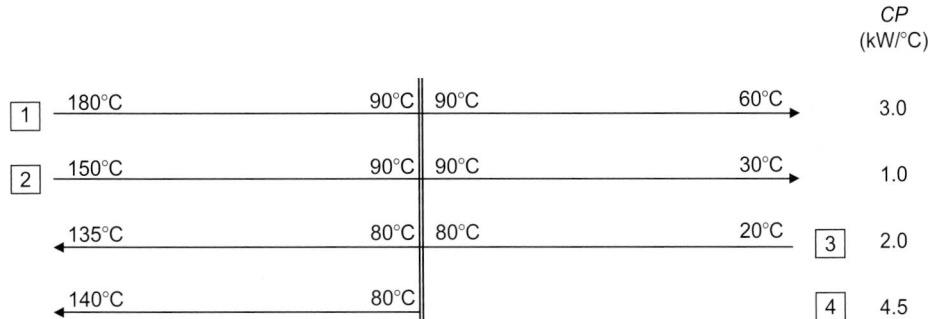

Figure 3.26. Grid for four-stream problem.

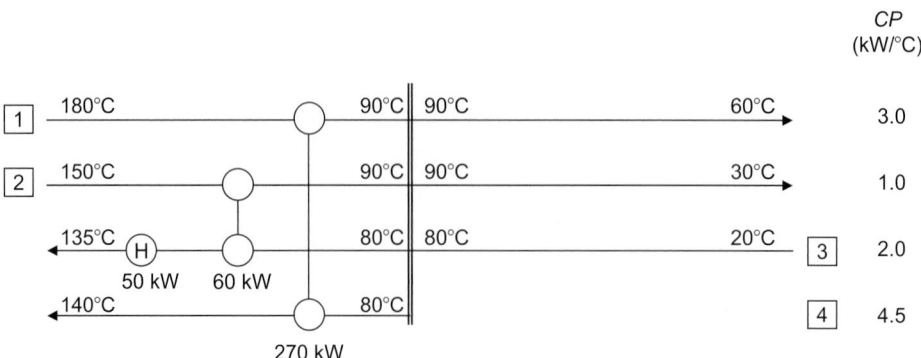

Figure 3.27.   Network design above the pinch.

Network Design Above the Pinch

$$CP_h \leq CP_c$$

1. Applying this condition at the pinch, stream 1 can be matched with stream 4, but not with 3.

   Matching streams 1 and 4 and transferring the full amount of heat required to bring stream 1 to the pinch temperature gives:

   $$\Delta H_{ex} = CP(T_s - T_{pinch})$$

   $$\Delta H_{ex} = 3.0(180 - 90) = 270 \text{ kW}$$

   This will also satisfy the heat load required to bring stream 4 to its target temperature:

   $$\Delta H_{ex} = 4.5(140 - 80) = 270 \text{ kW}$$

2. Stream 2 can be matched with stream 3, whilst satisfying the heat-capacity restriction. Transferring the full amount to bring stream 2 to the pinch temperature:

   $$\Delta H_{ex} = 1.0(150 - 90) = 60 \text{ kW}$$

3. The heat required to bring stream 3 to its target temperature, from the pinch temperature, is:

   $$\varDelta H = 2.0(135 - 80) = 110 \text{ kW}$$

   So a heater will have to be included to provide the remaining heat load:

   $$\Delta H_{hot} = 110 - 60 = 50 \text{ kW}$$

This checks with the value given by the problem table, Figure 3.24b.

The proposed network design above the pinch is shown in Figure 3.27.

Network Design Below the Pinch

$$CP_h \geq CP_c$$

1. Stream 4 begins at the pinch temperature, $T_s = 80°C$, and so is not available for any matches below the pinch.
2. A match between stream 1 and 3 adjacent to the pinch will satisfy the heat capacity restriction but not one between streams 2 and 3. So 1 is matched with 3 transferring the full amount to bring stream 1 to its target temperature:

$$\Delta H_{ex} = 3.0(90 - 60) = 90\,kW$$

3. Stream 3 requires more heat to bring it to the pinch temperature; amount needed:

$$\Delta H = 2.0(80 - 20) - 90 = 30\,kW$$

This can be provided from stream 2, as the match is now away from the pinch. The rise in temperature of stream 3 will be given by:

$$\Delta T = \Delta H/CP$$

So transferring 30 kW will raise the temperature from the source temperature to:

$$20 + 30/2.0 = 35°C$$

and this gives a stream temperature difference on the outlet side of the exchanger of:

$$90 - 35 = 55°C$$

So the minimum temperature difference condition, 10°C, will not be violated by this match.

4. Stream 2 needs further cooling to bring it to its target temperature, so a cooler must be included; cooling required:

$$\Delta H_{cold} = 1.0(90 - 30) - 30 = 30\,kW$$

Which is the amount of the cold utility predicted by the problem table.

The proposed network for maximum energy recovery is shown in Figure 3.28.

Stream Splitting

If the heat capacities of streams are such that it is not possible to make a match at the pinch without violating the minimum temperature difference condition, then the heat capacity can be altered by splitting a stream. Dividing the stream will reduce the mass flow rates in each leg and hence the heat capacities. This is illustrated in Example 3.16.

Similarly, if there are not enough streams available to make all of the required matches at the pinch, then streams with large CP can be split to increase the number of streams.

Guide rules for stream matching and splitting are given in the Institution of Chemical Engineers Guide, IChemE (1994) and by Smith (2005).

Summary

The guide rules for devising a network for maximum heat recovery are as follows.

1. Divide the problem at the pinch.
2. Design away from the pinch.
3. Above the pinch match streams adjacent to the pinch, meeting the restriction: $CP_h \leq CP_c$
4. Below the pinch match streams adjacent to the pinch, meeting the restriction: $CP_h \geq CP_c$
5. If the stream matching criteria cannot be satisfied, split a stream.
6. Maximize the exchanger heat loads.
7. Supply external heating only above the pinch, and external cooling only below the pinch.

3.17.4. Minimum Number of Exchangers

The network shown in Figure 3.28 was designed to give the maximum heat recovery, and will therefore give the minimum consumption, and cost, of the hot and cold utilities.

This will not necessarily be the optimum design for the network. The optimum design will be that which gives the lowest total annualized cost; taking into account the capital cost of the system, in addition to the utility and other operating costs. The number of exchangers in the network, and their size, will determine the capital cost.

In Figure 3.28 it is clear that there is scope for reducing the number of exchangers. The 30-kW exchanger between streams 2 and 3 can be deleted and the heat loads of the cooler and heater increased to bring streams 2 and 3 to their target temperatures.

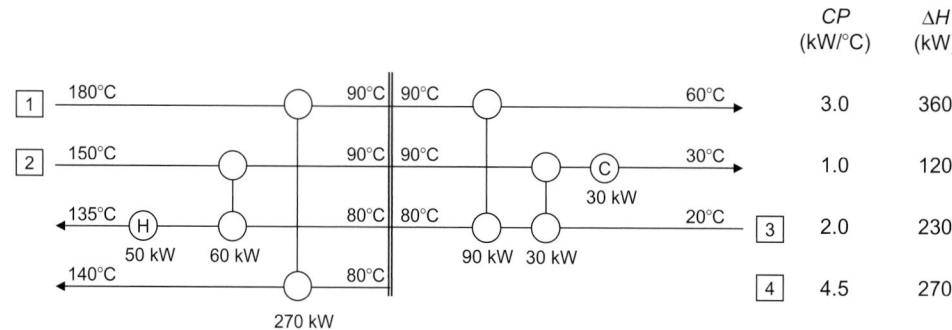

Figure 3.28.   Proposed heat-exchanger network for $\Delta T_{min} = 10°C$.

Heat would cross the pinch and the consumption of the utilities would be increased. Whether the revised network would be better, or more economic, depends on the relative cost of capital and utilities and the operability of each design. For any network, there will be an optimum design that gives the least annual cost: capital charges plus utility and other operating costs. The estimation of capital and operating costs are covered in Chapter 6.

To find the optimum design, it is necessary to cost a number of alternative designs, seeking a compromise between the capital costs, determined by the number and size of the exchangers, and the utility costs, determined by the heat recovery achieved.

For simple networks Holmann (1971) has shown that the minimum number of exchangers is given by:

$$Z_{min} = N' - 1 \qquad (3.41)$$

where

Z_{min} = minimum number of exchangers needed, including heaters and coolers
N' = the number of streams, including the utilities.

For complex networks a more general expression is needed to determine the minimum number of exchangers:

$$Z_{min} = N' + L' - S \qquad (3.42)$$

where

L' = the number of internal loops present in the network
S = the number of independent branches (subsets) that exist in the network.

A loop exists where a closed path can be traced through the network. There is a loop in the network shown in Figure 3.28. The loop is shown in Figure 3.29. The presence of a loop indicates that there is scope for reducing the number of exchangers.

For a full discussion of equation 3.42 and its applications see Linnhoff *et al.* (1979), IChemE (1994) and Smith (2005).

In summary, to seek the optimum design for a network:

1. Start with the design for maximum heat recovery. The number of exchangers needed will be equal to or less than the number for maximum energy recovery.

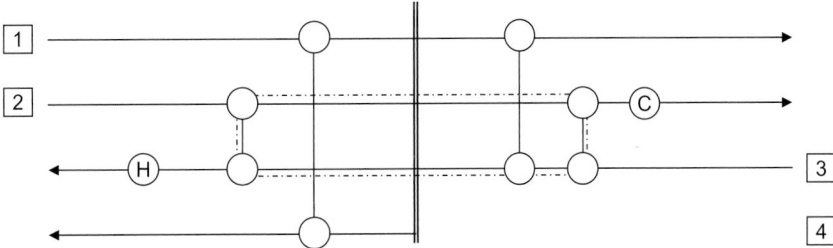

Figure 3.29. Loop in network.

2. Identify loops that cross the pinch. The design for maximum heat recovery will usually contain loops.
3. Starting with the loop with the least heat load, break the loops by adding or subtracting heat.
4. Check that the specified minimum temperature difference ΔT_{min} has not been violated. If the violation is significant, revise the design as necessary to restore ΔT_{min}. If the violation is small then it may not have much impact on the total annualized cost and can be ignored.
5. Estimate the capital and operating costs, and the total annual cost.
6. Repeat the loop breaking and network revision to find the lowest cost design.
7. Consider the safety, operability and maintenance aspects of the proposed design.

3.17.5. Threshold Problems

Problems that show the characteristic of requiring only either a hot utility or a cold utility (but not both) over a range of minimum temperature differences, from zero up to a threshold value, are known as threshold problems. A threshold problem is illustrated in Figure 3.30.

To design the heat-exchanger network for a threshold problem, it is normal to start at the most constrained point. The problem can often be treated as one half of a problem exhibiting a pinch.

Threshold problems are often encountered in the process industries. A pinch can be introduced in such problems if multiple utilities are used, as in the recovery of heat to generate steam, or if the chosen value of ΔT_{min} is greater than the threshold value.

The procedures to follow in the design of threshold problems are discussed by Smith (2005) and IChemE (1994).

3.17.6. Process Integration: Integration of Other Process Operations

The pinch technology method can give many other insights into process synthesis, beyond the design of heat-exchanger networks. The method can also be applied to the

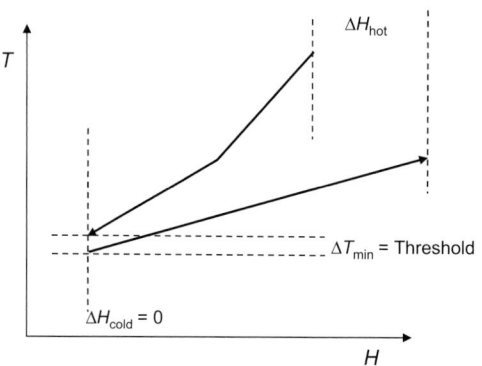

Figure 3.30.   Threshold problem.

integration of other process units; such as, separation columns, reactors, compressors and expanders, boilers and heat pumps. The wider applications of pinch technology are discussed in the Institution of Chemical Engineers Guide, IChemE (1994) and by El-Halwagi (2006) and Smith (2005).

The techniques of process integration have been expanded for use in optimizing mass transfer operations, and have been applied in waste reduction, water conservation and pollution control; see El-Halwagi (1997) and Dunn and El-Halwagi (2003).

3.17.7. Computer Tools for Heat-Exchanger Network Design

Most pinch analysis in industry is carried out using commercial pinch analysis software. Programs such as Aspen HX-Net™ (Aspen Technology Inc.), SUPER-TARGET™ (Linnhoff March Ltd.) and UniSim ExchangerNet™ (Honeywell Inc.) allow the design engineer to plot composite curves, optimize ΔT_{min}, set targets for multiple utilities and design the heat-exchanger network.

Most of these programs are able to automatically extract stream data from process simulation programs, although great care should be taken to check the extracted data. There are many possible pitfalls in data extraction; for example, not recognizing changes in the *CP* of a stream or partial vaporization or condensation of a stream, any of which could lead to a kink in the stream *T-H* profile. See Smith (2005) for more information on data extraction.

The commercial pinch technology tools also usually include automatic heat-exchanger network synthesis features. The automatic synthesis methods are based on MINLP optimization of superstructures of possible exchanger options (see Chapter 1 for discussion of MINLP methods). These tools can be used to arrive at a candidate network, but the optimization must be properly constrained so that it does not introduce a large number of stream splits and add a lot of small exchangers. Experienced designers seldom use automatic heat-exchanger network synthesis methods, as it usually requires more effort to turn the resulting network into something practical than it would take to design a practical network manually. The NLP optimization capability of the software is widely used though, for fine tuning the network temperatures by exploitation of loops and stream split ratios.

Example 3.16

Determine the pinch temperatures and the minimum utility requirements for the streams set out in the table below, for a minimum temperature difference between the streams of 20°C. Devise a heat-exchanger network to achieve the maximum energy recovery.

Stream Number	Type	Heat-Capacity Flow Rate kW/°C	Source Temp. °C	Target Temp. °C	Heat Load kW
1	hot	40.0	180	40	5600
2	hot	30.0	150	60	2700
3	cold	60.0	30	180	9000
4	cold	20.0	80	160	1600

Company Name
Address

PROBLEM TABLE ALGORITHM

Form XXXXX-YY-ZZ

Project Name								Sheet	1 of 1	
Project Number										
REV	DATE	BY	APVD	REV	DATE	APVD	BY	DATE	BY	APVD

1. Minimum temperature approach

ΔT_{min} 20 °C

2. Stream data

Stream No.	Actual temperature (°C) Source	Target	Interval temperature (°C) Source	Target	Heat capacity flow rate CP (kW/°C)	Heat load (kW)
1	180	40	170	30	40	5600
2	150	60	140	50	30	2700
3	30	180	40	190	60	9000
4	80	160	90	170	20	1600
5						0
6						0
7						0
8						0

3. Problem table

Interval	Interval temp (°C)	Interval ΔT (°C)	Sum CP_c - sum CP_h (kW/°C)	dH (kW)	Cascade (kW)	(kW)
1	190				0	2900
		20	60	1200		
2	170				-1200	1700
		0	60	0		
3	170				-1200	1700
		30	40	1200		
4	140				-2400	500
		50	10	500		
5	90				-2900	0
		40	-10	-400		
6	50				-2500	400
		10	20	200		
7	40				-2700	200
		10	-40	-400		
8	30				-2300	600

Figure 3.31. Problem table algorithm spreadsheet.

Solution

The problem table to find the minimum utility requirements and the pinch temperature can be built in a spreadsheet. The calculations in each cell are repetitive and the formula can be copied from cell to cell using the cell copy commands. A spreadsheet template for the problem table algorithm is available in MS Excel format in the online material at http://elsevierdirect.com/companions/. The use of the spreadsheet is illustrated in Figure 3.31 and described below.

First calculate the interval temperatures, for $\Delta T_{min} = 20°C$:
hot streams $T_{int} = T_{act} - 10°C$
cold streams $T_{int} = T_{act} + 10°C$.

Stream	Actual Temp. °C		Interval Temp. °C	
	Source	Target	Source	Target
1	180	40	170	30
2	150	60	140	50
3	30	180	40	190
4	80	160	90	170

In the spreadsheet this can be done by using an IF function to determine whether the source temperature is smaller than the target temperature, in which case the stream is a cold stream and should have $\Delta T_{min}/2$ added.

Next rank the interval temperatures, ignoring any duplicated values. In the spreadsheet this is done using the LARGE function. Determine which streams occur in each interval. For a stream to be present in a given interval the largest stream interval temperature must be greater than the lower end of the interval range and the lowest stream interval temperature must also be greater than or equal to the lower end of the interval range. This can be calculated in the spreadsheet using IF, AND and OR functions. Once the streams in each interval have been determined it is possible to calculate the combined stream heat capacities. These calculations are not strictly part of the problem table, so they have been hidden in the spreadsheet (in columns to the right of the table).

The sum of *CP* values for the cold streams minus that for the hot streams can then be multiplied by the interval ΔT to give the interval ΔH, and the interval ΔH values can be cascaded to give the overall heat flow. The amount of heat that must be put in to prevent the heat flow from becoming negative is the lowest value in the column, which can be found using the SMALL function. The final column then gives a cascade showing only positive values, with zero energy cascading at the pinch temperature.

In the last column 2900 kW of heat have been added to eliminate the negative values in the previous column; so the hot utility requirement is 2900 kW and the cold utility requirement, the bottom value in the column, is 600 kW.

The pinch occurs where the heat transferred is zero; that is at interval number 4, interval temperature 90°C.

So, at the pinch hot streams will be at:

$$90 + 10 = 100°C$$

and the cold streams will be at:

$$90 - 10 = 80°C$$

Note that in the table both stream 1 and stream 4 had an interval temperature of 170°C, which led to a duplicate line in the list of ranked interval temperatures. Strictly, this line could have been eliminated, but since it gave a zero value for the ΔT, it did not affect the calculation. The programming of the spreadsheet is a lot easier if duplicate temperatures are handled in this manner.

To design the network for maximum energy recovery: start at the pinch and match streams, following the rules on stream heat capacities for matches adjacent to the pinch. Where a match is made: transfer the maximum amount of heat.

The proposed network is shown in Figure 3.32.

The methodology followed in devising this network was as follows.

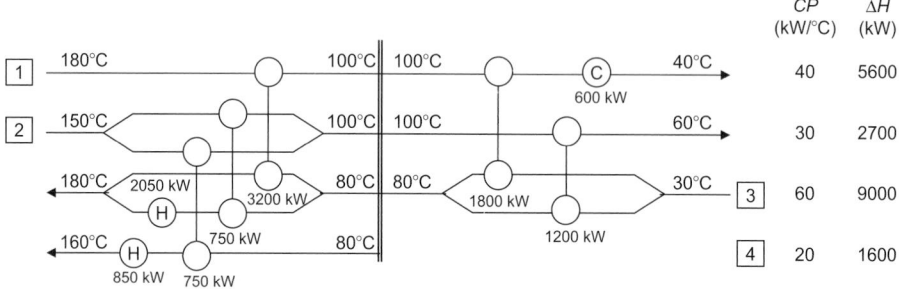

Figure 3.32.   Proposed heat-exchanger network for Example 3.16.

Above Pinch

1. $CP_h \leq CP_c$
2. Can match stream 1 or 2 with stream 3 but neither stream can match with stream 4. This creates a problem, since if we match stream 1 with 3 then stream 2 will not be able to make a match at the pinch. Likewise, if we match stream 2 with 3 then stream 1 will not be able to make a match at the pinch.
3. Check the heat available in bringing the hot streams to the pinch temperature:
   stream 1 $\Delta H = 40.0(180 - 100) = 3200\,\text{kW}$
   stream 2 $\Delta H = 30.0(150 - 100) = 1500\,\text{kW}$.
4. Check the heat required to bring the cold streams from the pinch temperature to their target temperatures:
   stream 3 $\Delta H = 60.0(180 - 80) = 6000\,\text{kW}$
   stream 4 $\Delta H = 20.0(160 - 80) = 1600\,\text{kW}$.
5. If we split stream 3 into two branches with CP of 40.0 and 20.0 then we can match the larger branch with stream 1 and transfer 3200 kW, which satisfies (ticks off) stream 1.
6. We now have two cold streams, both with CP of 20.0 and one hot stream (2) with CP of 30.0. We need to split stream 2 into two branches. As an initial guess, these can both have CP of 15.0. We can then match one branch of stream 2 with the smaller branch of 3 and transfer 750 kW, and the other branch with stream 4, also for 750 kW, which then ticks off stream 2.

7. Include a heater on the smaller branch of stream 3 to provide the balance of the heat required:

$$\Delta H_{\text{hot}} = 6000 - 3200 - 750 = 2050 \text{ kW}$$

8. Include a heater on stream 4 to provide the balance of the heat required:

$$\Delta H_{\text{hot}} = 1600 - 750 = 850 \text{ kW}$$

Check sum of heater duties $= 2050 + 850 = 2900 \text{ kW} = $ hot utility target.

Below Pinch

9. $CP_{\text{h}} \geq CP_{\text{c}}$
10. Note that stream 4 starts at the pinch temperature so cannot provide any cooling below the pinch.
11. Cannot match stream 1 or 2 with stream 3 at the pinch.
12. Split stream 3 to reduce CP.
13. Check the heat available from bringing the hot streams from the pinch temperature to their target temperatures:

    stream 1 $\Delta H = 40.0(100 - 40) = 2400 \text{ kW}$
    stream 2 $\Delta H = 30.0(100 - 60) = 1200 \text{ kW}$.

14. Check the heat required to bring the cold streams from their source temperatures to the pinch temperature:
    stream 3 $\Delta H = 60.0(80 - 30) = 3000 \text{ kW}$
    stream 4 is at the pinch temperature.

15. Note that stream 1 cannot be brought to its target temperature of 40°C by full interchange with stream 3 as the source temperature of stream 3 is 30°C, so ΔT_{min} would be violated. So transfer 1800 kW to one leg of the split stream 3.

16. Check temperature at exit of this exchanger:

$$\text{Temp out} = 100 - \frac{1800}{40} = 55°C, \text{ satisfactory}$$

17. Provide cooler on stream 1 to bring it to its target temperature, cooling needed:

$$\Delta H_{\text{cold}} = 2400 - 1800 = 600 \text{ kW}$$

18. Transfer the full heat load from stream 2 to second leg of stream 3; this satisfies both streams.

Note that the heating and cooling loads, 2900 kW and 600 kW respectively, match those predicted from the problem table.

Note also that in order to satisfy the pinch decomposition and the stream matching rules we ended up introducing a large number of stream splits. This is quite common in heat-exchanger network design. None of the three split fractions was optimized, so substantial savings as well as simplification of the network could be possible. For example, loops exist between the branches of stream 3 and stream 1 and between the branches of stream 3 and stream 2. With the current split ratios, these loops cannot be eliminated, but with other ratios it might be possible to eliminate one or two exchangers.

The introduction of multiple stream splits is often cited as a drawback of the pinch method. Stream splits can be problematic in process operation. For example, when an oil or other multicomponent stream is heated and partially vaporized then the stream is a two-phase mixture. It is difficult to control the splitting of such streams to give the required flow rate in each branch. Experienced designers usually constrain the network to avoid multiple stream splits whenever possible, even if this leads to designs that have higher than minimum utility consumption.

3.18. REFERENCES

Barnwell, J. and Morris, C. P. (1982) *Hyd. Proc.* **61** (July) 117. Heat pump cuts energy use.

Bloch, H. P., Cameron, J. A., Danowsky, F. M., James, R., Swearingen, J. S. and Weightman, M. E. (1982) *Compressors and Expanders: Selection and Applications for the Process Industries* (Dekker).

Bosnjakovic, F. (1935) *Technische Thermodynamik* (T. Steinkopff).

Buse, F. (1981) *Chem. Eng., NY* **88** (Jan 26th) 113. Using centrifugal pumps as hydraulic turbines.

Chada, N. (1984) *Chem. Eng., NY* **91** (July 23rd) 57. Use of hydraulic turbines to recover energy.

Coulson, J. M., Richardson, J. F., Backhurst, J., and Harker, J. H. (1999) *Chemical Engineering: Volume 1*, 6th edn (Butterworth-Heinemann).

Douglas, J. M. (1988) *Conceptual Design of Chemical Processes* (McGraw-Hill).

Dryden, I. (ed.) (1975) *The Efficient Use of Energy* (IPC Science and Technology Press).

Dunn, K. S. and Tomkins, A. G. (1975) *Inst. Mech. Eng. Conference on Energy Recovery in the Process Industries,* London, Waste heat recovery from the incineration of process wastes.

Dunn, R. F. and El-Halwagi, M. M. (2003) *J. Chem. Technol. Biotechol.* **78**, 1011. Process integration technology review: background and applications in the chemical process industry.

Edmister, W. C. (1948) *Pet. Ref.* **27** (Nov.) 129(609). Applications of thermodynamics to hydrocarbon processing, part XIII heat capacities.

El-Halwagi, M. M. (1997) *Pollution Prevention Through Process Integration: Systematic Design Tools* (Academic Press).

El-Halwagi, M. M. (2006) *Process Integration* (Academic Press).

Green, D. W. and Perry, R. H. (2007) *Perry's Chemical Engineers' Handbook*, 8th edn (McGraw-Hill).

Gundersen, T. and Naess, L. (1988). *Comp. and Chem. Eng.*, **12**, No. 6, 503. The synthesis of cost optimal heat-exchanger networks an industrial review of the state of the art.

Himmelblau, D. M. (1995) *Basic Principles and Calculations in Chemical Engineering*, 6th edn (Pearson).

Hinchley, P. (1975) *Inst. Mech. Eng. Conference on Energy Recovery in the Process Industries*. London Waste heat boilers in the chemical industry).

Holmann, E. C. (1971) PhD Thesis, University of South California, *Optimum networks for heat exchangers*.

Holland, F. A. and Devotta, S. (1986) *Chem. Engr, London*, No. 425 (May) 61. Prospects for heat pumps in process applications.

IChemE (1994) *User Guide on Process Integration for Efficient Use of Energy*, revised edn (Institution of Chemical Engineers, London).

Jenett, E. (1968) *Chem. Eng., NY* 75 (April 8th) 159, (June 17th) 257 (in two parts). Hydraulic power recovery systems.

Kenney, W. F. (1984) *Energy Conversion in the Process Industries* (Academic Press).

Kobe, K. A. (1958) *Pet. Ref.* **44**. Thermochemistry of petrochemicals.

Linnhoff, B. and Flower, J. R. (1978) *AIChEJ* 24, 633 (2 parts) synthesis of heat exchanger networks.

Linnhoff, B., Mason, D. R., and Wardle, I. (1979) *Comp. and Chem. Eng* 3, 295. Understanding heat exchanger networks.

Linnhoff, B., Townsend, D. W., Boland, D., Hewitt, G. F., Thomas, B. E. A., Guy, A. R., and Marsland, R. H. (1982) *User Guide on Process Integration for the Efficient Use of Energy*, 1st edn London Institution of Chemical Engineers).

Luckenbach, E.C. (1978) U.S. 4,081,508, to Exxon Research and Engineering Co. Process for reducing flue gas contaminants from fluid catalytic cracking regenerator.

Meili, A. (1990) *Chem. Eng. Prog.* 86(6), 60. Heat pumps for distillation columns.

Miller, R. (1968) *Chem. Eng., NY* 75 (May 20th) 130. Process energy systems.

Moser, F. and Schnitzer, H. (1985) *Heat Pumps in Industry* (Elsevier).

Perry, R. H., Chilton, C. H. (1973) *Chemical Engineers Handbook* 5th edn. (McGraw-Hill)

Reay, D. A. and Macmichael, D. B. A. (1988) *Heat Pumps: Design and Application*, 2nd edn (Pergamon Press).

Santoleri, J. J. (1973) *Chem. Eng. Prog.* **69** (Jan.) 69. Chlorinated hydrocarbon waste disposal and recovery systems.

Shultz, J. M. (1962) *Trans. ASME* **84** (*Journal of Engineering for Power*) (Jan.) 69, (April) 222 (in two parts). The polytropic analysis of centrifugal compressors.

Singh, J. (1985) *Heat Transfer Fluids and Systems for Process and Energy Applications* (Marcel Dekker).

Smith, R. (2005) *Chemical Process Design and Integration* (Wiley).

Sterbacek, Z., Biskup, B., and Tausk, P. (1979) *Calculation of Properties Using Corresponding-state Methods* (Elsevier).

3.19. NOMENCLATURE

		Dimensions in $MLT\theta$
A	Area	L^2
a	Constant in specific heat equation (equation 3.13)	$L^2T^{-2}\theta^{-1}$
B	Bottoms flow rate	MT^{-1}
b	Constant in specific heat equation (equation 3.13)	$L^2T^{-2}\theta^{-2}$
CP	Stream heat-capacity flow rate	$ML^2T^{-2}\theta^{-1}$

CP_c	Stream heat-capacity flow rate, cold stream	$ML^2T^{-2}\theta^{-1}$
CP_h	Stream heat-capacity flow rate, hot stream	$ML^2T^{-2}\theta^{-1}$
C_p	Specific heat at constant pressure	$L^2T^{-2}\theta^{-1}$
C_{pa}	Specific heat component a	$L^2T^{-2}\theta^{-1}$
C_{pb}	Specific heat component b	$L^2T^{-2}\theta^{-1}$
C_{pc}	Specific heat component c	$L^2T^{-2}\theta^{-1}$
C_{pm}	Mean specific heat	$L^2T^{-2}\theta^{-1}$
C_{p1}	Specific heat first phase	$L^2T^{-2}\theta^{-1}$
C_{p2}	Specific heat second phase	$L^2T^{-2}\theta^{-1}$
C_v	Specific heat at constant volume	$L^2T^{-2}\theta^{-1}$
C_p°	Ideal gas state specific heat	$L^2T^{-2}\theta^{-1}$
c	Constant in specific heat equation (equation 3.13)	$L^2T^{-2}\theta^{-3}$ or $L^2T^{-2}\theta^{-1/2}$
ΣCP_c	Sum of heat-capacity flow rates of cold streams	$ML^2T^{-2}\theta^{-1}$
ΣCP_h	Sum of heat-capacity flow rates of hot streams	$ML^2T^{-2}\theta^{-1}$
D	Distillate flow rate	MT^{-1}
E_e	Efficiency, electric motors	—
E_p	Polytropic efficiency, compressors and turbines	—
F	Force	MLT^{-2}
g	Gravitational acceleration	LT^{-2}
H	Enthalpy	ML^2T^{-2}
H_a	Specific enthalpy of component a	L^2T^{-2}
H_B	Enthalpy of bottom product stream (Example 3.1)	ML^2T^{-3}
H_b	Specific enthalpy of component b	L^2T^{-2}
H_D	Enthalpy of distillate stream (Example 3.1)	ML^2T^{-3}
H_F	Enthalpy feed stream (Example 3.1)	ML^2T^{-3}
$H_{products}$	Total enthalpy of products	ML^2T^{-3}
$H_{reactants}$	Total enthalpy of reactants	ML^2T^{-3}
H_T	Specific enthalpy at temperature T	L^2T^{-2}
H_V	Enthalpy of vapour stream (Example 3.1)	ML^2T^{-3}
ΔH	Change in enthalpy	ML^2T^{-2}
ΔH_{cold}	Heat transfer from cold utility	ML^2T^{-3}
ΔH_{ex}	Heat transfer in exchanger	ML^2T^{-3}
ΔH_{hot}	Heat transfer from hot utility	ML^2T^{-3}
ΔH_n	Net heat required in n^{th} interval	ML^2T^{-3}
$\Delta H_{react.}$	Enthalpy change to bring reactants to standard temperature	ML^2T^{-3}

ΔH_p	Latent heat of phase change	L^2T^{-2}
$\Delta H_{\text{prod.}}$	Enthalpy change to bring products to standard temperature	ML^2T^{-3}
ΔH_{xp}	Cross-pinch heat transfer	ML^2T^{-3}
$-\Delta H_{m,t}$	Heat of mixing at temperature t	L^2T^{-2}
$-\Delta H_{r,t}$	Heat of reaction at temperature t	L^2T^{-2}
$-\Delta H^{\circ}_c$	Standard heat of combustion	L^2T^{-2}
ΔH°_f	Standard enthalpy of formation	L^2T^{-2}
$-\Delta H^{\circ}_r$	Standard heat of reaction	L^2T^{-2}
$-\Delta H^{\circ}_{soln}$	Integral heat of solution	L^2T^{-2}
L	Liquid flow rate	MT^{-1}
L'	Number of internal loops in network	—
l	Distance	L
M	Mass	M
M_c	Moles of component c	M
M_w	Molecular mass (weight)	—
m	Polytropic temperature exponent	—
m	Mass flow rate	MT^{-1}
N	Number of cold streams, heat-exchanger networks	—
N'	Number of streams	—
n	Expansion or compression index (equation 3.30)	—
n_i	Molar flow of component i	MT^{-1}
P	Pressure	$ML^{-1}T^{-2}$
P_c	Critical pressure	$ML^{-1}T^{-2}$
P_i	Inter-stage pressure	$ML^{-1}T^{-2}$
P_r	Reduced pressure	—
P_1	Initial pressure	$ML^{-1}T^{-2}$
P_2	Final pressure	$ML^{-1}T^{-2}$
Q	Heat transferred across system boundary	ML^2T^{-2} or ML^2T^{-3}
Q_B	Reboiler heat load (Example 3.1)	ML^2T^{-3}
Q_C	Condenser heat load (Example 3.1)	ML^2T^{-3}
Q_p	Heat added (or subtracted) from a system	ML^2T^{-2} or ML^2T^{-3}
Q_r	Heat from reaction	ML^2T^{-2} or ML^2T^{-3}
Q_s	Heat generated in the system	ML^2T^{-2} or ML^2T^{-3}
R	Reflux ratio (Example 3.1)	—
R	Universal gas constant	$L^2T^{-2}\theta^{-1}$
S	Number of independent branches	—
T	Temperature, absolute	θ
T_1	Initial temperature	θ
T_2	Final temperature	θ

T_{act}	Actual stream temperature	θ
T_c	Critical temperature	θ
T_d	Datum temperature for enthalpy calculations	θ
T_{int}	Interval temperature	θ
T_n	Temperature in nth interval	θ
T_p	Phase-transition temperature	θ
T_{pinch}	Pinch temperature	θ
T_r	Reduced temperature	—
T_s	Source temperature	θ
T_t	Target temperature	θ
ΔT	Change in temperature	θ
ΔT_{min}	Minimum temperature difference (minimum approach) in heat exchanger	θ
ΔT_n	Internal temperature difference	θ
t	Temperature, relative scale	θ
$\mathbf{t}$	Time	$\mathbf{T}$
t_r	Reference temperature, mean specific heat	θ
t_s	Steam saturation temperature	θ
U	Internal energy per unit mass	$\mathbf{L^2T^{-2}}$
U_j	Jacket heat transfer coefficient	$\mathbf{MT^{-3}\theta^{-1}}$
u	Velocity	$\mathbf{LT^{-1}}$
V	Vapour flow rate	$\mathbf{MT^1}$
v	Volume per unit mass	$\mathbf{M^{-1}L^3}$
x	Distance	$\mathbf{L}$
x_a	Mole fraction component a in a mixture	—
x_b	Mole fraction component b in a mixture	—
x_c	Mole fraction component c in a mixture	—
W	Work per unit mass	$\mathbf{L^2T^{-2}}$
Z	Compressibility factor	—
z	Height above datum	$\mathbf{L}$
Z_{min}	Minimum number of heat exchangers in network	—
γ	Ratio of specific heats	—

3.20. PROBLEMS

3.1. A liquid stream leaves a reactor at a pressure of 100 bar. If the pressure is reduced to 3 bar in a turbine, estimate the maximum theoretical power that could be obtained from a flow rate of 1000 kg/h. The density of the liquid is 850 kg/m^3.

3.2. Calculate the specific enthalpy of water at a pressure of 1 bar and temperature of 200°C. Check your value using steam tables. The specific heat capacity of water can be calculated from the equation:

$$C_p = 4.2 - 2 \times 10^{-3}t$$

where t is in °C and C_p in kJ/kg.

Take the other data required from Appendix C.

3.3. A gas produced as a by-product from the carbonization of coal has the following composition, mole per cent: carbon dioxide 4, carbon monoxide 15, hydrogen 50, methane 12, ethane 2, ethylene 4, benzene 2, balance nitrogen. Using the data given in Appendix C, calculate the gross and net calorific values of the gas. Give your answer in MJ/m³, at standard temperature and pressure.

3.4. In the manufacture of aniline, liquid nitrobenzene at 20°C is fed to a vaporizer where it is vaporized in a stream of hydrogen. The hydrogen stream is at 30°C, and the vaporizer operates at 20 bar. For feed rates of 2500 kg/h nitrobenzene and 366 kg/h hydrogen, estimate the heat input required. The nitrobenzene vapour is not superheated.

3.5. Aniline is produced by the hydrogenation of nitrobenzene. The reaction takes place in a fluidized bed reactor operating at 270°C and 20 bar. The excess heat of reaction is removed by a heat transfer fluid passing through tubes in the fluidized bed. Nitrobenzene vapour and hydrogen enter the reactor at a temperature of 260°C. A typical reactor off-gas composition, mole per cent, is: aniline 10.73, cyclo-hexylamine 0.11, water 21.68, nitrobenzene 0.45, hydrogen 63.67, inerts (take as nitrogen) 3.66. Estimate the heat removed by the heat transfer fluid, for a feed rate of nitrobenzene to the reactor of 2500 kg/h.

The specific heat capacity of nitrobenzene can be estimated using the methods given in Chapter 8. Take the other data required from Appendix C.

3.6. Hydrogen chloride is produced by burning chlorine with an excess of hydrogen. The reaction is highly exothermic and reaches equilibrium very rapidly. The equilibrium mixture contains approximately 4% free chlorine but this is rapidly combined with the excess hydrogen as the mixture is cooled. Below 200°C the conversion of chlorine is essentially complete.

The burner is fitted with a cooling jacket, which cools the exit gases to 200°C. The gases are further cooled, to 50°C, in an external heat exchanger.

For a production rate of 10,000 tonnes per year of hydrogen chloride, calculate the heat removed by the burner jacket and the heat removed in the external cooler. Take the excess hydrogen as 1% over stoichiometric. The hydrogen supply contains 5% inerts (take as nitrogen) and is fed to the burner at 25°C. The chlorine is essentially pure and is fed to the burner as a saturated vapour. The burner operates at 1.5 bar.

3.7. A supply of nitrogen is required as an inert gas for blanketing and purging vessels. After generation, the nitrogen is compressed and stored in a bank of cylinders, at a pressure of 5 barg. The inlet pressure to the compressor is 0.5 barg, and the temperature is 20°C. Calculate the maximum power required to compress 100 m³/h. A single-stage reciprocating compressor will be used.

3.8. Hydrogen chloride gas, produced by burning chlorine with hydrogen, is required at a supply pressure of 600 kN/m², gauge. The pressure can be achieved by either operating the burner under pressure or by compressing the hydrogen chloride gas. For a production rate of hydrogen chloride of 10,000 kg/h, compare the power requirement of compressing the hydrogen supply to the burner with that to compress the product hydrogen chloride. The chlorine feed is supplied at the required pressure from a vaporizer. Both the hydrogen and chlorine feeds are essentially pure. Hydrogen is supplied to the burner one percent excess of the stoichiometric requirement.

A two-stage centrifugal compressor will be used for both duties. Take the polytropic efficiency for both compressors as 70%. The hydrogen supply pressure is 120 kN/m² and the temperature 25°C. The hydrogen chloride is cooled to 50°C after leaving the burner. Assume that the compressor intercooler cools the gas to 50°C, for both duties.

Which process would you select and why?

3.9. Estimate the work required to compress ethylene from 32 to 250 MPa in a two-stage reciprocating compressor where the gas is initially at 30°C and leaves the intercooler at 30°C.

3.10. Determine the pinch temperature and the minimum utility requirements for the process set out below. Take the minimum approach temperature as 15°C. Devise a heat-exchanger network to achieve maximum energy recovery.

Stream Number	Type	Heat capacity kW/°C	Source Temp. °C	Target Temp.°C
1	hot	13.5	180	80
2	hot	27.0	135	45
3	cold	53.5	60	100
4	cold	23.5	35	120

3.11. Determine the pinch temperature and the minimum utility requirements for the process set out below. Take the minimum approach temperature as 15°C. Devise a heat-exchanger network to achieve maximum energy recovery.

Stream Number	Type	Heat Capacity kW/°C	Source Temp. °C	Target Temp. °C
1	hot	10.0	200	80
2	hot	20.0	155	50
3	hot	40.0	90	35
4	cold	30.0	60	100
5	cold	8.0	35	90

3.12. To produce a high-purity product two distillation columns are operated in series. The overhead stream from the first column is the feed to the second column. The overhead from the second column is the purified product. Both columns are conventional distillation columns fitted with reboilers and total condensers. The bottom products are passed to other processing units, which do not form part of this problem. The feed to the first column passes through a preheater. The condensate from the second column is passed through a product cooler. The duty for each stream is summarized below:

No.	Stream	Type	Source temp. °C	Target Temp. °C	Duty, kW
1	Feed preheater	cold	20	50	900
2	First condenser	hot	70	60	1350
3	Second condenser	hot	65	55	1100
4	First reboiler	cold	85	87	1400
5	Second reboiler	cold	75	77	900
6	Product cooler	hot	55	25	30

Find the minimum utility requirements for this process, for a minimum approach temperature of 10°C.

Note: the stream heat capacity is given by dividing the exchanger duty by the temperature change.

3.13. At what value of the minimum approach temperature does the problem in Example 3.16 become a threshold problem? Design a heat-exchanger network for the resulting threshold problem. What insights does this give into the design proposed in Example 3.16?

4 FLOW-SHEETING

Chapter Contents

Key Learning Objectives

- How to prepare and present a process flow diagram
- How to use commercial process simulation software to build a process heat and material balance model
- How to use user-specified models and components when the simulator does not have what you need
- How to converge flow-sheets that include recycles

4.1. INTRODUCTION

This chapter covers the preparation and presentation of the process flow-sheet, also known as the Process Flow Diagram (PFD). The flow-sheet is the key document in process design. It shows the arrangement of the equipment selected to carry out the process; the stream connections; stream flow rates and compositions; and the operating conditions. It is a diagrammatic model of the process.

The flow-sheet is used by specialist design groups as the basis for their designs. These include piping, instrumentation, and equipment design and plant layout. It is also used by operating personnel for the preparation of operating manuals and operator training. During plant start-up and subsequent operation, the flow-sheet forms a basis for comparison of operating performance with design.

The flow-sheet is drawn up from material balances made over the complete process and each individual unit operation. Energy balances are also made to determine the energy flows and the utility requirements.

Most flow-sheet calculations are carried out using commercial process simulation programs. The process simulation programs contain models for most unit operations as well as thermodynamic and physical property models. All the commercial programs feature some level of custom modelling capability that allows the designer to add models for non-standard operations.

Many companies developed proprietary flow-sheeting programs between 1960 and 1980. The cost of maintaining and updating proprietary software is high; consequently, very few of the proprietary flow-sheeting programs are still in use, and most companies now rely entirely on commercially available software. Each of the commercial process simulation programs has its own unique idiosyncrasies, but they share many common features. The discussion in this chapter addresses general problems of process simulation and flow-sheeting rather than software-specific issues. The latter are usually thoroughly documented in the user manuals and on-line help that come with the software. Examples have been provided in this chapter using both Aspen Plus® (Aspen Technology Inc.) and UniSim Design™ (Honeywell Inc.). UniSim Design is based on the HYSYS™ software that was originally developed by Hyprotech Ltd. and is now owned and licensed by Honeywell. An older version of this software is also available from Aspen Technology Inc. as Aspen HYSYS®.

Because flow-sheeting is usually carried out using computer programs, it is necessary for the design engineer to have a good understanding of how to set up and solve computer models. The flow-sheet model that is solved on the computer to generate a mass and energy balance is often not an exact representation of the process flow diagram. The designer may need to use combinations of simulation library models and user models to capture the performance of process equipment. Spreadsheet or hand calculations are also often helpful in setting up process simulation models and providing good initial estimates, so as to accelerate convergence.

The next step in process design after the flow-sheet is the preparation of Piping and Instrumentation diagrams (abbreviated to P & I diagrams), often also called the Engineering Flow-sheet or Mechanical Flow-sheet. The P & I diagrams, as the name implies, show the engineering details of the process, and are based on the process flow-sheet. The preparation and presentation of P & I diagrams are discussed in

Chapter 5. The abbreviation PFD (for Process Flow Diagram) is often used for process flow-sheets and PID for Piping and Instrumentation Diagrams.

4.2. FLOW-SHEET PRESENTATION

As the process flow-sheet is the definitive document on the process, the presentation must be clear, comprehensive, accurate and complete. The various types of flow-sheet are discussed below.

4.2.1. Block Diagrams

A block diagram is the simplest form of presentation. Each block can represent a single piece of equipment or a complete stage in the process. Block diagrams were used to illustrate the examples in Chapters 2 and 3. They are useful for showing simple processes. With complex processes, their use is limited to showing the overall process, broken down into its principal stages; as in Example 2.13 (vinyl chloride). In that example, each block represented the equipment for a complete reaction stage: the reactor, separators and distillation columns.

Block diagrams are useful for representing a process in a simplified form in reports, textbooks and presentations, but have only a limited use as engineering documents.

The stream flow rates and compositions can be shown on the diagram adjacent to the stream lines, when only a small amount of information is to be shown, or tabulated separately.

Block diagrams are often drawn using simple graphics programs such as Visio™ or Microsoft PowerPoint™.

4.2.2. Pictorial Representation

On the detailed flow-sheets used for design and operation, the equipment is normally drawn in a stylized pictorial form. For tender documents or company brochures, actual scale drawings of the equipment are sometimes used, but it is more usual to use a simplified representation. There are several international standards for PFD symbols, but most companies use their own standard symbols, as the cost of converting all of their existing drawings would be excessive. ISO 10628 is the international standard for PFD drawing symbols, but few British or American companies apply this standard. The symbols given in British Standard, BS 1553 (1977) 'Graphical Symbols for General Engineering' Part 1, 'Piping Systems and Plant' are more typical of those in common use and a selection of symbols from BS 1553 is given in Appendix A. The symbols in BS 1553 are used in the UK and Commonwealth countries. A library of common symbols is available in Microsoft Visio, and these are similar to the symbols in Appendix A. Most European countries have adopted ISO 10628 as their standard.

4.2.3. Presentation of Stream Flow Rates

The data on the flow rate of each individual component, on the total stream flow rate, and the percentage composition, can be shown on the flow-sheet in various ways. The simplest method, suitable for simple processes with few pieces of equipment, is to tabulate the data in blocks alongside the process stream lines, as shown in Figure 4.1. Only a limited amount of information can be shown this way, and it is difficult to make neat alterations or to add additional data.

A better method for the presentation of data on flow-sheets is shown in Figure 4.2. In this method, each stream line is numbered and the data are tabulated at the bottom of the sheet. Alterations and additions can be easily made. This is the method generally used by professional design offices. A typical commercial flow-sheet is shown in Figure 4.3. Guide rules for the layout of this type of flow-sheet presentation are given in Section 4.2.5.

4.2.4. Information to be Included

The amount of information shown on a flow-sheet will depend on the custom and practice of the particular design office. The list given below has therefore been divided into essential items and optional items. The essential items must always be shown; the optional items add to the usefulness of the flow-sheet but are not always included.

Essential Information

1. Stream composition, either:
   i the flow rate of each individual component, kg/h, which is preferred, or
   ii the stream composition as a weight fraction.

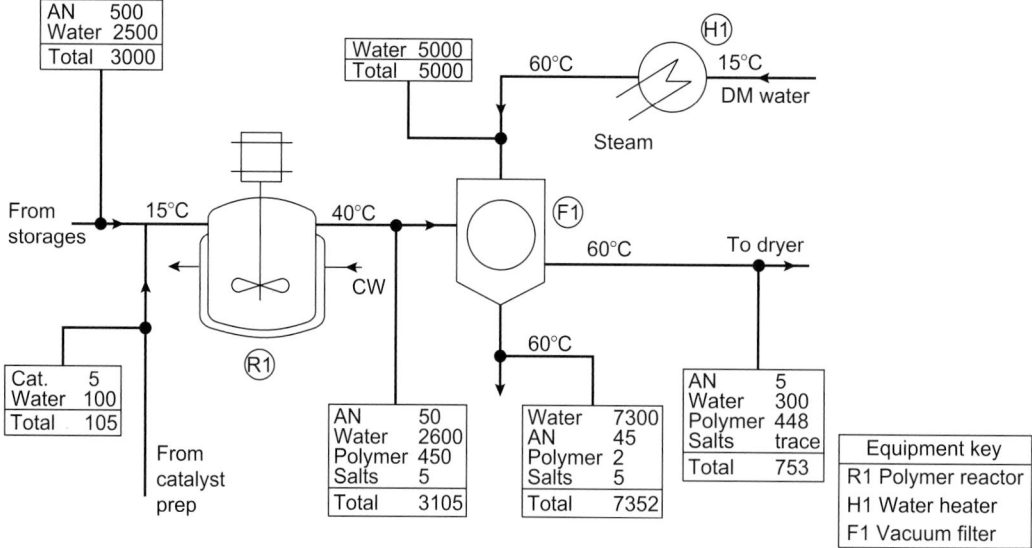

Figure 4.1.   Flow-sheet: polymer production.

Flows kg/h Pressures nominal

Line no.	1	1A	2	2A	3	4	5	6	7	8	9	10	11	12	13
Stream	Ammonia feed	Ammonia vapour	Filtered air	Oxidiser air	Oxidiser feed	Oxidiser outlet	W.H.B. outlet	Condenser gas	Condenser acid	Secondary air	Absorber feed	Tail(2) gas	Water feed	Absorber acid	Product acid
Component															
NH_3	731.0	731.0	—	—	731.0	Nil	—	—	—	—	—	—	—	—	—
O_2	—	—	3036.9	2628.2	2628.2	935.7[1]	(935.7)[1]	275.2	Trace	408.7	683.9	371.5	—	Trace	Trace
N_2	—	—	9990.8	8644.7	8644.7	8668.8	8668.8	8668.8	Trace	1346.1	10,014.7	10,014.7	—	Trace	Trace
NO_2	—	—	—	—	—	1238.4[1]	Trace (1238.4)[1]	202.5	—	—	202.5	21.9	(Trace)[1]	Trace	Trace
HNO_3	—	—	—	—	—	—	Nil	Nil	850.6	—	—	—	—	1704.0	2554.6
H_2O	—	—	Trace	—	—	1161.0	1161.0	996.6	1010.1	—	996.6	26.3	1376.9	1136.0	2146.0
Total	731.0	731.0	13,027.7	11,272.9	12,003.9	12,003.9	12,003.9	10,143.1	1860.7	1754.8	11,897.7	10,434.4	1376.9	2840.0	4700.6
Press bar	8	8	1	8	8	8	8	8	8	8	8	8	8	1	1
Temp. °C	15	20	15	230	204	907	234	40	40	40	40	25	25	40	43

C & R Construction Inc

Nitric acid 60 percent
100,000 t/y
Client BOP Chemicals
SLIGO   Sheet no. 9316

Dwg by   Date
Checked   25/7/1980

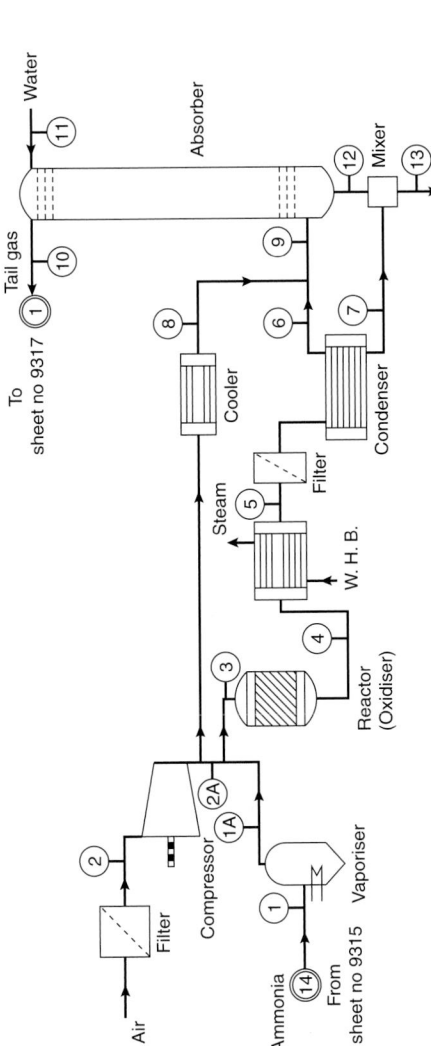

Figure 4.2. Flow-sheet: simplified nitric acid process.

C & R Construction Inc.	Title : Nitric Acid 60 per cent	Dwg by : GRM	DATE :
	Client : BOP Chemicals SILGO	Chk by : RKS	14/12/1992

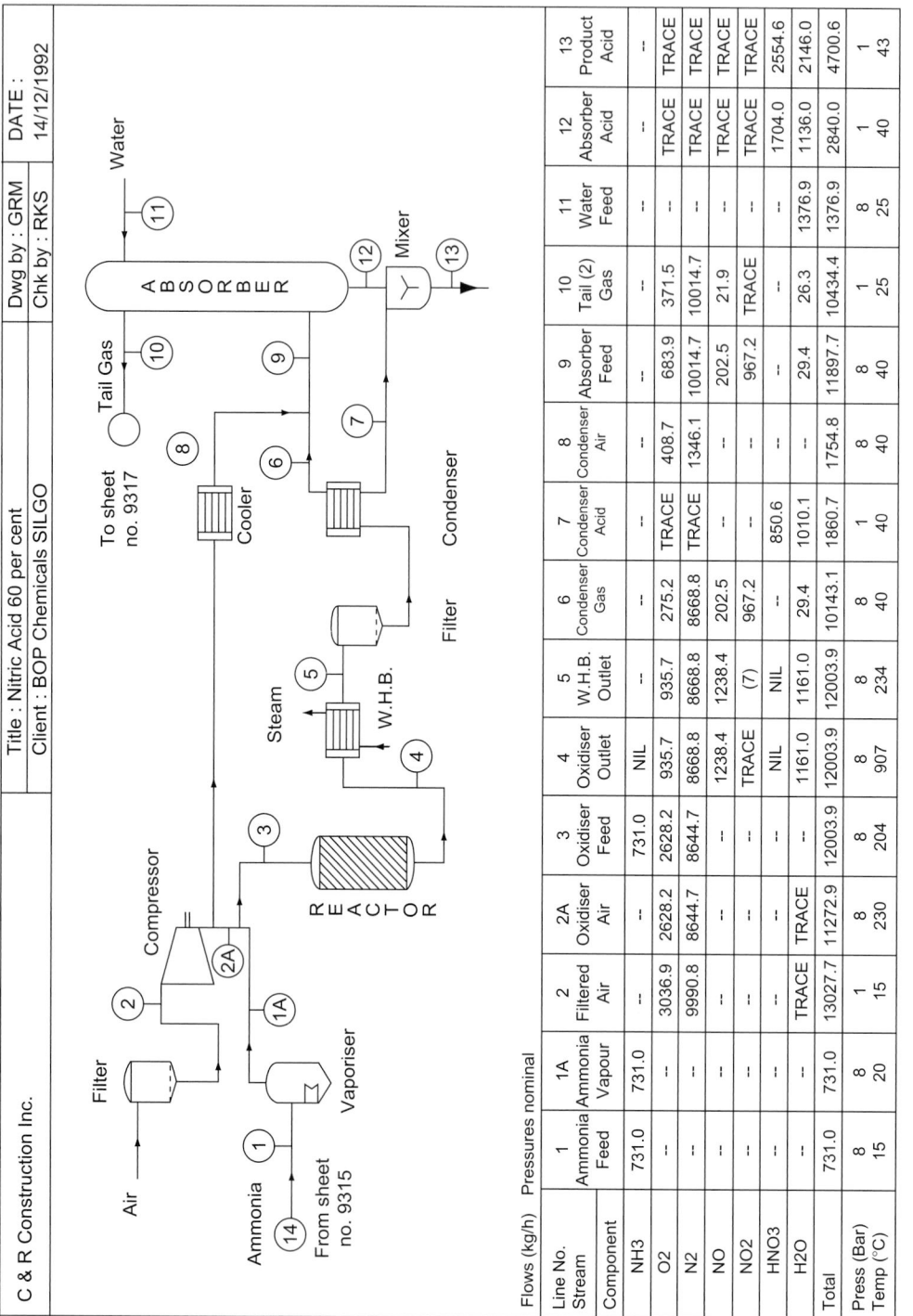

Flows (kg/h)   Pressures nominal

Line No. Stream Component	1 Ammonia Feed	1A Ammonia Vapour	2 Filtered Air	2A Oxidiser Air	3 Oxidiser Feed	4 Oxidiser Outlet	5 W.H.B. Outlet	6 Condenser Gas	7 Condenser Acid	8 Condenser Air	9 Absorber Feed	10 Tail (2) Gas	11 Water Feed	12 Absorber Acid	13 Product Acid
NH_3	731.0	731.0	--	--	731.0	NIL	--	--	--	--	--	--	--	--	--
O_2	--	--	3036.9	2628.2	2628.2	935.7	935.7	275.2	TRACE	408.7	683.9	371.5	--	TRACE	TRACE
N_2	--	--	9990.8	8644.7	8644.7	8668.8	8668.8	8668.8	TRACE	1346.1	10014.7	10014.7	--	TRACE	TRACE
NO	--	--	--	--	--	1238.4	1238.4	202.5	--	--	202.5	21.9	--	TRACE	TRACE
NO_2	--	--	--	--	--	TRACE	(7)	967.2	--	--	967.2	TRACE	--	TRACE	TRACE
HNO_3	--	--	--	--	--	NIL	NIL	--	850.6	--	--	--	--	1704.0	2554.6
H_2O	--	--	TRACE	TRACE	--	NIL	NIL	29.4	1010.1	1754.8	29.4	26.3	1376.9	1136.0	2146.0
Total	731.0	731.0	13027.7	11272.9	12003.9	12003.9	12003.9	10143.1	1860.7	1754.8	11897.7	10434.4	1376.9	2840.0	4700.6
Press (Bar)	8	8	1	8	8	8	8	8	1	8	8	1	8	1	1
Temp (°C)	15	20	15	230	204	907	234	40	40	40	40	25	25	40	43

Figure 4.2A.   Alternative presentation.

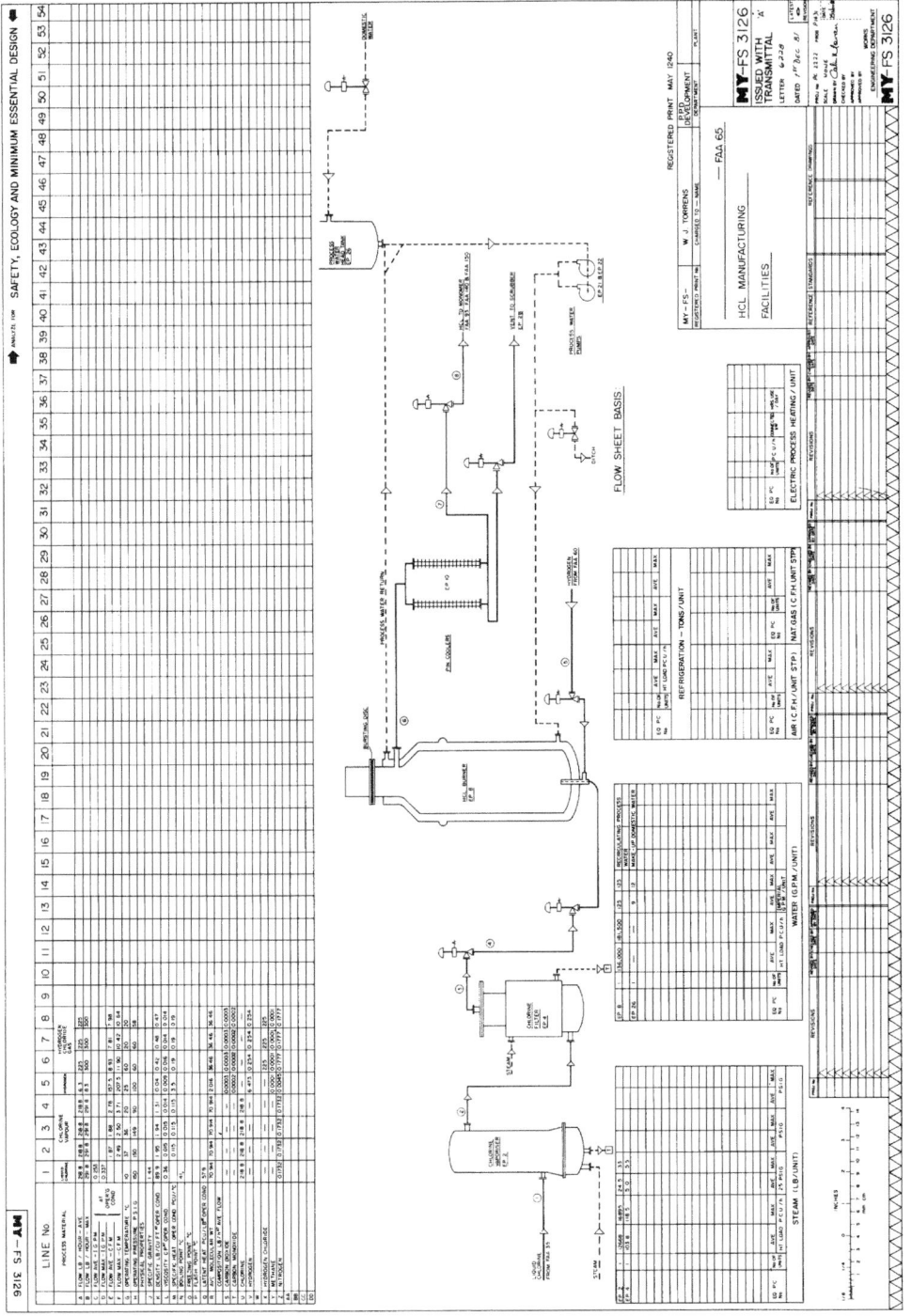

Figure 4.3. A typical flow-sheet.

2. Total stream flow rate, kg/h.
3. Stream temperature, degrees Celsius preferred.
4. Nominal operating pressure (the required operating pressure).
5. Stream enthalpy, kJ/h.

Optional Information

1. Molar percentage composition and/or molar flow rates.
2. Physical property data, mean values for the stream, such as:
    i density, kg/m^3
    ii viscosity, $mN\ s/m^2$.
3. Stream name, a brief, one- or two-word, description of the nature of the stream, for example 'ACETONE COLUMN BOTTOMS'.

4.2.5.   Layout

The sequence of the main equipment items shown symbolically on the flow-sheet follows that of the proposed plant layout. Some license must be exercised in the placing of ancillary items, such as heat exchangers and pumps, or the layout will be too congested. The aim should be to show the flow of material from stage to stage as it will occur, and to give a general impression of the layout of the actual process plant.

The equipment should be drawn approximately to scale. Again, some license is allowed for the sake of clarity, but the principal equipment items should be drawn roughly in the correct proportion. Ancillary items can be drawn out of proportion. For a complex process, with many process units, several sheets may be needed, and the continuation of the process streams from one sheet to another must be clearly shown. One method of indicating a line continuation is shown in Figure 4.2; those lines that are continued over to another drawing are indicated by a double concentric circle around the line number and the continuation sheet number is written below. An alternative method is to extend lines to the side of the page and then indicate the drawing sheet on which the line is continued.

The table of stream flows and other data can be placed above or below the equipment layout. Normal practice is to place it below. The components should be listed down the left-hand side of the table, as in Figure 4.2. For a long table, it is good practice to repeat the list at the right-hand side, so the components can be traced across from either side.

The stream line numbers should follow consecutively from left to right of the layout, as far as is practicable; so that when reading the flow-sheet it is easy to locate a particular line and the associated column containing the data.

All the process stream lines shown on the flow-sheet should be numbered and the data for the stream given. There is always a temptation to leave out the data on a process stream if it is clearly just formed by the addition of two other streams, as at a junction, or if the composition is unchanged when flowing through a process unit, such as a heat exchanger; this should be avoided. What may be clear to the process designer is not necessarily clear to the others who will use the flow-sheet. Complete, unambiguous information on all streams should be given, even if this involves some repetition. The purpose of the flow-sheet is to show the function of each process unit; even when the function has no discernible impact on the mass and energy balance.

4.2.6. Precision of Data

The total stream and individual component flows do not normally need to be shown to a high precision on the process flow-sheet; three or four significant figures are all that is usually justified by the accuracy of the flow-sheet calculations, and will typically be sufficient. The flows should, however, balance to within the precision shown. If a stream or component flow is so small that it is less than the precision used for the larger flows, it can be shown to a greater number of places, if its accuracy justifies this and the information is required. Imprecise small flows are best shown as 'TRACE'. If the composition of a trace component is specified as a process constraint, as, say, for an effluent stream or product quality specification, it can be shown in parts per million, ppm.

A trace quantity should not be shown as zero, or the space in the tabulation left blank, unless the process designer *is sure* that it has no significance. Trace quantities can be important. Only a trace of an impurity is needed to poison a catalyst, and trace quantities can determine the selection of the materials of construction; see Chapter 7. If the space in the data table is left blank opposite a particular component, the quantity may be assumed to be zero by the specialist design groups who take their information from the flow-sheet.

4.2.7. Basis of the Calculation

It is good practice to show on the flow-sheet the basis used for the flow-sheet calculations. This includes: the operating hours per year, the reaction and physical yields, and the datum temperature used for energy balances. It is also helpful to include a list of the principal assumptions used in the calculations. This alerts the user to any limitations that may have to be placed on the flow-sheet information.

If the amount of information that needs to be presented is excessive, then it can be summarized in a separate document that is referenced on the flow-sheet.

In some cases, mass and energy balances are prepared for multiple scenarios. These might include: winter and summer operating conditions, start of catalyst life and end of catalyst life, manufacture of different products or product grades, etc. Usually these different scenarios are shown as several tables on the same flow-sheet, but occasionally different flow-sheets are drawn for each case.

4.2.8. Batch Processes

Flow-sheets drawn up for batch processes normally show the quantities required to produce one batch. If a batch process forms part of an otherwise continuous process, it can be shown on the same flow-sheet, providing a clear break is made when tabulating the data between the continuous and batch sections; i.e., the change from kg/h to kg/batch.

A continuous process may include batch make-up of minor reagents, such as the catalyst for a polymerization process. Batch flows into a continuous process are usually labelled 'Normally no flow' and show the flow rates that will be obtained when the stream is flowing. It is these instantaneous flow rates that govern the equipment design, rather than the much lower time-averaged flow rates.

4.2.9.   Utilities

To avoid cluttering up the flow-sheet, it is not normal practice to show the utility (service) headers and lines on the process flow-sheet. The utility connections required on each piece of equipment should be shown and labelled, for example, 'CTW' for cooling tower water. The utility requirements for each piece of equipment should be tabulated on the flow-sheet.

4.2.10.   Equipment Identification

Each piece of equipment shown on the flow-sheet must be identified with a code number and name. The identification number (usually a letter and some digits) is normally that assigned to a particular piece of equipment as part of the general project control procedures, and is used to identify it in all the project documents.

If the flow-sheet is not part of the documentation for a project, then a simple, but consistent, identification code should be devised. The easiest code is to use an initial letter to identify the type of equipment, followed by digits to identify the particular piece. For example, H — heat exchangers, C — columns, R — reactors. Most companies have a standard convention that should be followed, but if there is no agreed standard then the key to the code should be shown on the flow-sheet.

4.2.11.   Computer-Aided Drafting

Most design offices use drafting software for the preparation of flow-sheets and other process drawings. With drafting software, standard symbols representing the process equipment, instruments and control systems are held in files, and these symbols are called up as required when drawing flow-sheets and piping and instrumentation diagrams (see Chapter 5). Final flow-sheet drawings are usually produced by professional drafters, who are experienced with the drafting software and conventions, rather than by the design engineer. The design engineer has to provide the required numbers, sketch the flow-sheet and review the final result.

To illustrate the use of a commercial computer-aided design program, Figure 4.2 has been redrawn using the program FLOSHEET and is shown as Figure 4.2a. FLOSHEET is part of a suite of programs called PROCEDE, described by Preece *et al.* (1991).

Although most process simulation programs feature a graphical user interface (GUI) that creates a drawing that resembles a PFD, printouts of these drawings are very seldom used as actual process flow diagrams. The unit operations shown in the process simulation usually do not exactly match the unit operations of the process. The simulation may include dummy items that do not physically exist and may omit some equipment that is needed in the plant but is not part of the simulation.

4.3.   PROCESS SIMULATION PROGRAMS

The most commonly used commercial process simulation programs are listed in Table 4.1. Most of these programs can be licensed by universities for educational purposes at nominal cost.

TABLE 4.1. Simulation Packages

Name	Type	Source	Internet Address http//www.—
Aspen Plus	steady-state	Aspen Technology Inc.Ten Canal Park, Cambridge, MA 02141-2201, USA	Aspentech.com
CHEMCAD	steady-state	Chemstations Inc.2901 Wilcrest, Suite 305,Houston, TX 77042 USA	Chemstations.net
DESIGN II	steady-state	WinSim Inc.P.O. Box 1885,Houston,TX 77251-1885, USA	Winsim.com
Aspen HYSYS	steady-state and dynamic	Aspen Technology Inc.Ten Canal Park,Cambridge, MA 02141-2201, USA	Aspentech.com
PRO/II and DYNSIM	steady-state and dynamic	SimSci-Esscor 5760 Fleet Street,Suite 100, Carlsbad,CA 92009, USA	Simsci.com
UniSim Design	steady-state and dynamic	Honeywell 300-250 York Street London, Ontario N6A 6K2, Canada	Honeywell.com

Note: Contact the web site to check the full features of the most recent versions of the programs.

Detailed discussion of the features of each of these programs is beyond the scope of this book. For a general review of the requirements, methodology and application of process simulation programs the reader is referred to the books by: Husain (1986), Wells and Rose (1986), Leesley (1982), Benedek (1980), and Westerberg *et al.* (1979). The features of the individual programs are described in their user manuals and on-line help. Two of these simulators have been used to generate the examples in this chapter: Aspen Plus® (v.11.1) and UniSim Design™ (R360.1).

Process simulation programs can be divided into two basic types:

Sequential-modular programs: in which the equations describing each process unit operation (module) are solved module-by-module in a stepwise manner. Iterative techniques are then used to solve the problems arising from the recycle of information.

Simultaneous (also known as *equation-oriented*) programs: in which the entire process is described by a set of equations, and the equations are solved simultaneously, not stepwise as in the sequential approach. Simultaneous programs can simulate the unsteady-state operation of processes and equipment, and can give faster convergence when multiple recycles are present.

In the past, most simulation programs available to designers were of the sequential-modular type. They were simpler to develop than the equation-oriented programs, and required only moderate computing power. The modules are processed sequentially, so essentially only the equations for a particular unit are in the computer memory at one time. Also, the process conditions, temperature, pressure, flow rate, etc., are fixed in time. With the sequential-modular approach, computational difficulties can arise due to the iterative methods used to solve recycle problems and obtain convergence. A major limitation of sequential-modular simulators is the inability to simulate the dynamic, time dependent, behaviour of a process.

Simultaneous, dynamic, simulators require appreciably more computing power than steady-state simulators to solve the thousands of differential equations needed to describe a process, or even a single item of equipment. With the development of fast, powerful computers, this is no longer a restriction. By their nature, simultaneous programs do not experience the problems of recycle convergence inherent in sequential simulators; however, as temperature, pressure and flow rate are not fixed and the input of one unit is not determined by the calculated output from the previous unit in the sequence, simultaneous programs demand more computer time. This has led to the development of hybrid programs in which the steady-state simulator is used to generate the initial conditions for the equation-oriented or dynamic simulation.

The principal advantage of simultaneous, dynamic, simulators is their ability to model the unsteady-state conditions that occur at start-up and during fault conditions. Dynamic simulators are being increasingly used for safety studies and in the design of control systems, as discussed in Section 4.9.

The structure of a typical simulation program is shown in Figure 4.4.

The program consists of:

1. A main executive program that controls and keeps track of the flow-sheet calculations and the flow of information to and from the subroutines.
2. A library of equipment performance subroutines (modules) that simulate the equipment and enable the output streams to be calculated from information on the inlet streams.
3. A data bank of physical properties. To a large extent, the utility of a sophisticated flow-sheeting program depends on the comprehensiveness of the physical

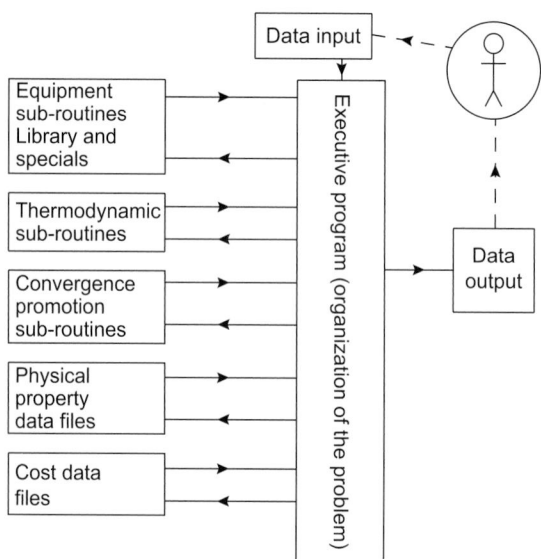

Figure 4.4.   A typical simulation program.

property data bank. The collection of the physical property data required for the design of a particular process and its transformation into a form suitable for a particular flow-sheeting program can be very time-consuming.

4. Subroutines for thermodynamics, such as the calculation of vapour-liquid equilibrium and stream enthalpies.

5. Sub-programs and data banks for equipment sizing and costing. Process simulation programs enable the designer to consider alternative processing schemes, and the cost routines allow quick economic comparisons to be made. Some programs include optimization routines. To make use of a costing routine, the program must be capable of producing at least approximate equipment designs.

In a sequential-modular program, the executive program sets up the flow-sheet sequence, identifies the recycle loops, and controls the unit operation calculations, while interacting with the unit operations library, physical property data bank and the other subroutines. The executive program also contains procedures for the optimum ordering of the calculations and routines to promote convergence.

In an equation-oriented simulator, the executive program sets up the flow-sheet and the set of equations that describe the unit operations, and then solves the equations using data from the unit operations library and the physical property data bank and calling on the file of thermodynamics subroutines.

All process simulators use graphical user interfaces to display the flow-sheet and facilitate the input of information to the package. The entry of data is usually intuitive to anyone familiar with the MS Windows™ operating systems.

In an industrial context, it is very important to ensure that the simulation model is properly benchmarked against operating plant data whenever such data are available. A considerable amount of time is usually spent validating plant models before they can be used for evaluation of new designs or plant expansions.

4.4. SPECIFICATION OF COMPONENTS AND PHYSICAL PROPERTY MODELS

The first step in building a process simulation is usually establishing the chemical basis for the model. This consists of choosing the components that will be included in the mass balance and deciding which models to use for the prediction of physical properties and phase equilibrium. The correlation of physical properties and prediction of phase equilibrium are described in detail in Chapter 8. This section thus focuses on the selection of suitable components.

4.4.1. Pure Components

Each of the commercial process simulation programs contains a large data bank of pure component compounds. Most of the pure components are organic compounds, but inorganic compounds and electrolytes are also included.

The fact that a pure component is listed in a simulator data bank does not guarantee that any of the properties given for that component are based on measured data. If the properties of a compound are critical to process performance, then the scientific literature should be consulted to confirm that the values used in the simulation are realistic.

The most important decision when building a pure component model is choosing the right number of components. The design engineer needs to consider carefully which components will have a significant impact on process design, operation and economics. If too few components are used, then the model will be inadequate for process design, as it will not correctly predict the performance of reactors and separation equipment. Conversely, if too many components are used, then the model can become difficult to converge, particularly if there are multiple recycles in the design.

Some guidelines to keep in mind when building a component list include:

1. Always include any component that has a specified limit in any of the products if that component is present in any of the feeds or could be formed in the process. This is critical to determining whether the separations are meeting product specifications.

2. Always include any component that has a specified limit in any of the feeds. These components can be a source of by-products or can act as catalyst or enzyme inhibitors. They must be tracked to ensure that they do not accumulate in the process or make it difficult to meet product specifications. In some cases, an additional separation may be needed to remove a feed contaminant.

3. Always include components that are expected to be formed in side reactions or consecutive reactions. It is important to understand where these components will accumulate or leave the process, even if their yield is not yet known.

4. Always include any compounds that are expected to be present and are known to have significant health, safety or environmental concerns, such as compounds with high toxicity or explosivity, known carcinogens, or listed hazardous air pollutants (see Chapters 9 and 14). These compounds must be tracked to make sure that they do not reach unsafe levels in any stream and to understand where they might be released to the environment.

5. Usually include any compound that might be present at a mass or mole fraction greater than 2% in any stream in the process.

6. Do not include isomers unless the process specifically requires distinction between isomers (for example, if the process is selective for one isomer, gives different products for different isomers, or is designed to separate isomers). Considering all of the possible isomers of organic compounds becomes combinatorially explosive at high carbon numbers. For fuels and bulk petrochemical processes that are carried out at relatively high temperatures it is often reasonable to assume an equilibrium distribution of isomers. For fine chemical and pharmaceutical processes it is usually important to track isomers separately, particularly enantiomers, as the desired product is often only one of the isomers.

In general, pure component models solve more efficiently with less than about 40 components. If the number of components becomes too large and there are many recycles, then it may be necessary to build two models. The first is a high-level model that contains only the main bulk components. This model is then used to initialize a second, more detailed model that has the full component list.

4.4.2. Pseudocomponents

Pseudocomponents (hypocomponents) are components created by the design engineer or automatically generated by the simulation program to match the boiling curves of petroleum mixtures.

Crude oil, fuels such as gasoline, kerosene and diesel, and most intermediate streams in an oil refinery consist of many different hydrocarbon compounds. The number of possible hydrocarbon isomers present depends on the carbon number, and both increase with boiling range. For diesel, crude oil and heavy fuel oils, the number of possible compounds can be from 10^4 to $>10^6$. At the time of writing, there is no analytical method that can identify all of these compounds, so it would be impossible to include them all in a model even if the resulting model could be solved. Instead, a large number of possible compounds with boiling points in a given range are 'lumped' together and represented by a single pseudocomponent with a boiling point in the middle of that range. A set of 10 to 30 pseudocomponents can then be fitted to any petroleum assay and used to model that oil.

Pseudocomponent models are very useful for oil fractionation and blending problems. They can also be used to characterize heavy products in some chemical processes such as ethane cracking. Pseudocomponents are treated as inert in most of the reactor models, but they can be converted or produced in yield shift reactors (see Section 4.5.1).

Some of the commercial simulation programs use a standard default set of pseudocomponents and fit the composition of each to match a boiling curve of the oil that is entered by the user. This can sometimes lead to errors when predicting ASTM D86 or D2887 curves for products from a feed that has been defined based on a true boiling point (TBP) curve, or when making many sub-cuts or cuts with tight distillation specifications. It is often better to work back from the product distillation curves and add extra pseudocomponents around the cut points to make sure that the recoveries and 5% and 95% points on the product distillation curves are predicted properly. All of the simulators have the option to add pseudocomponents to the default set or use a user-generated curve.

4.4.3. Solids and Salts

Most chemical and pharmaceutical processes involve some degree of solids handling. Examples of solids that must be modelled include:

- Components that are crystallized for separation, recovery or purification.
- Pharmaceutical products that are manufactured as powders or tablets.
- Insoluble salts formed by the reaction of acids and bases or other electrolytes.
- Hydrates, ice and solid carbon dioxide that can form in cryogenic processes.
- Cells, bacteria and immobilized enzymes in biological processes.
- Pellets or crystals of polymer formed in polymerization processes.
- Coal and ash particles in power generation.
- Catalyst pellets in processes in which the catalyst is fluidized or transported as a slurry.
- Mineral salts and ores that are used as process feeds.

- Fertilizer products.
- Fibres in paper processing.

Some solid-phase components can be characterized as pure components and can interact with other components in the model through phase and reaction equilibrium. Others, such as cells and catalysts, are unlikely to equilibrate with other components, although they can play a vital role in the process.

In Aspen Plus, solid components are identified as different types. Pure materials with measurable properties, such as molecular weight, vapour pressure and critical temperature and pressure, are known as conventional solids and are present in the MIXED sub-stream with other pure components. They can participate in any of the phase or reaction equilibria specified in any unit operation. If the solid phase participates only in reaction equilibrium but not in phase equilibrium (for example, when the solubility in the fluid phase is known to be very low), then it is called a conventional inert solid and is listed in a sub-stream CISOLID. If a solid is not involved in either phase or reaction equilibrium, then it is a non-conventional solid and is assigned to sub-stream NC. Non-conventional solids are defined by attributes rather than molecular properties and can be used for coal, cells, catalysts, bacteria, wood pulp and other multicomponent solid materials.

In UniSim Design, non-conventional solids can be defined as hypothetical components (see Section 4.4.4). The solid phases of pure components are predicted in the phase and reaction equilibrium calculations and do not need to be identified separately.

Many solids-handling operations have an effect on the particle size distribution (PSD) of the solid phase. The particle size distribution can also be an important product property. Aspen Plus allows the user to enter a particle size distribution as an attribute of a solid sub-stream. In UniSim Design, the particle size distribution is entered on the 'PSD Property' tab, which appears under 'worksheet' on the stream editor window for any stream that contains a pure or hypothetical solid component. Unit operations such as yield shift reactor, crusher, screen, cyclone, electrostatic precipitator and crystallizer can then be set up to modify the particle size distribution; typically by using a conversion function or a particle capture efficiency in each size range.

When inorganic solids and water are present, an electrolyte phase equilibrium model must be selected for the aqueous phase, to properly account for the dissolution of the solid and formation of ions in solution.

4.4.4.   User Components

The process simulators were originally developed for petrochemical and fuels applications; consequently, many molecules that are made in specialty chemical and pharmaceutical processes are not listed in the component data banks. All of the simulators allow the designer to overcome this drawback by adding new molecules to customize the data bank.

In UniSim Design, new molecules are added as hypothetical components. The minimum information needed to create a new hypothetical pure component is the normal boiling point, although the user is encouraged to provide as much information as is available. If the boiling point is unknown, then the molecular weight and density are used instead. The input information is used to tune the UNIFAC correlation to predict the physical and phase equilibrium properties of the molecule, as described in Chapter 8.

User-defined components are created in Aspen Plus using a 'user-defined component wizard'. The minimum required information is the molecular weight and normal boiling point. The program also allows the designer to enter molecular structure, specific gravity, enthalpy and Gibbs energy of formation, ideal gas heat capacity and Antoine vapour pressure coefficients, but for complex molecules usually only the molecular structure is known.

It is often necessary to add user components to complete a simulation model. The design engineer should always be cautious when interpreting simulation results for models that include user components. Phase equilibrium predictions for flashes, decanters, extraction, distillation and crystallization operations should be carefully checked against laboratory or plant data to ensure that the model is correctly predicting the component distribution between the phases. If the fit is poor, the binary interaction parameters in the phase equilibrium model can be tuned to improve the prediction.

4.5. SIMULATION OF UNIT OPERATIONS

A process simulation is built up from a set of unit operation models connected by mass and energy streams. The commercial simulators include many unit operation subroutines, sometimes referred to as library models. These operations can be selected from a palette or menu and then connected together using the simulator graphical user interface. Table 4.2 gives a list of the main unit operation models available in Aspen Plus and UniSim Design. Details of how to specify unit operations are given in the simulator manuals. This section provides general advice on unit operations modelling and modelling of non-standard unit operations.

4.5.1. Reactors

The modelling of real industrial reactors is usually the most difficult step in process simulation. It is usually easy to construct a model that gives a reasonable prediction of the yield of main product, but the simulator library models are not sophisticated enough to fully capture all the details of hydraulics, mixing, mass transfer, catalyst and enzyme inhibition, cell metabolism and other effects that often play a critical role in determining the reactor outlet composition, energy release or consumption, rate of catalyst deactivation and other important design parameters.

In the early stages of process design, the simulator library models are usually used with simplistic reaction models that give the design engineer a good enough idea of yields and enthalpy changes to allow design of the rest of the process. If the design seems economically attractive, then more detailed models can be built and substituted into the flow-sheet. These detailed models are usually built as user models, as described in Section 4.6.

TABLE 4.2.   Unit Operation Models in Aspen Plus™ and UniSim Design™

Unit Operation	Aspen Plus Models	UniSim Design Models
Stream mixing	Mixer	Mixer
Component splitter	Sep, Sep2	Component Splitter
Decanter	Decanter	3-Phase Separator
Flash	Flash2, Flash3	Separator, 3-Phase Separator
Piping components		
Piping	Pipe, Pipeline	Pipe Segment, Compressible Gas Pipe
Valves & fittings	Valve	Valve, Tee, Relief Valve
Hydrocyclone	HyCyc	Hydrocyclone
Reactors		
Conversion reactor	RStoic	Conversion Reactor
Equilibrium reactor	REquil	Equilibrium Reactor
Gibbs reactor	RGibbs	Gibbs Reactor
Yield reactor	RYield	Yield Shift Reactor
CSTR	RCSTR	Continuous Stirred Tank Reactor
Plug flow reactor	RPlug	Plug Flow Reactor
Columns		
Shortcut distillation	DSTWU, Distl, SCFrac	Shortcut column
Rigorous distillation	RadFrac, MultiFrac	Distillation, 3-Phase Distillation
Liquid-liquid extraction	Extract	Liquid-Liquid Extractor
Absorption and stripping	RadFrac	Absorber, Refluxed Absorber, Reboiled
Fractionation	PetroFrac	Absorber
Rate-based distillation	RATEFRAC™	3 Stripper Crude, 4 Stripper Crude, Vaccum
Batch distillation	BatchFrac	Resid
		Column, FCCU Main Fractionator
Heat-transfer equipment		
Heater or cooler	Heater	Heater, Cooler
Heat exchanger	HeatX, HxFlux, Hetran, HTRI-Xist	Heat Exchanger
Air cooler	Aerotran	Air Cooler
Fired heater	Heater	Fired Heater
Multi-stream exchanger	MheatX	LNG Exchanger
Rotating equipment		
Compressor	Compr, MCompr	Compressor
Turbine	Compr, MCompr	Expander
Pump, hydraulic turbine	Pump	Pump
Solids handling		
Size reduction	Crusher	Screen
Size selection	Screen	Crystallizer, Precipitation
Crystallizer	Crystallizer	Neutralizer
Neutralization		
Solids washing	SWash	
Filter	Fabfl, CFuge, Filter	Rotary Vacuum Filter
Cyclone	HyCyc, Cyclone	Hydrocyclone, Cyclone
Solids decanting	CCD	Simple Solid Separator
Solids transport		Conveyor
Secondary recovery	ESP, Fabfl, VScrub	Baghouse Filter
User models	User, User2, User3	User Unit Op

Most of the commercial simulation programs have variants on the following reactor models.

Conversion Reactor (Stoichiometric Reactor)

A conversion reactor requires a reaction stoichiometry and an extent of reaction, which is usually specified as an extent of conversion of a limiting reagent. No reaction kinetics information is needed, so it can be used when the kinetics are unknown (which is often the case in the early stages of design) or when the reaction is known to proceed to full conversion. Conversion reactors can handle multiple reactions, but care is needed in specifying the order in which they are solved if they use the same limiting reagent.

Equilibrium Reactor

An equilibrium reactor finds the equilibrium product distribution for a specified set of stoichiometric reactions. Phase equilibrium is also solved. The engineer can enter the outlet temperature and pressure and let the reactor model calculate the duty needed to reach that condition, or else enter a heat duty and let the model predict the outlet conditions from an energy balance.

An equilibrium reactor only solves the equations specified, so it is useful in situations where one or more reactions equilibrate rapidly while other reactions proceed much more slowly. An example is the steam reforming of methane to hydrogen. In this process, the water–gas shift reaction between water and carbon monoxide equilibrates rapidly at temperatures above $450°C$, while methane conversion requires catalysis even at temperatures above $800°C$. This process chemistry is explored in Example 4.2.

In some simulation programs, the equilibrium reactor model requires the designer to specify both liquid and vapour phase products, even though one of the streams may be calculated to have zero flow. If the real reactor has a single outlet, then the two product streams in the model should be mixed back together.

Gibbs Reactor

The Gibbs reactor solves the full reaction (and optionally phase) equilibrium of all species in the component list by minimization of the Gibbs free energy, subject to the constraints of the feed mass balance and either isothermal or adiabatic operation. A Gibbs reactor can be specified with restrictions such as a temperature approach to equilibrium or a fixed conversion of one species.

The Gibbs reactor is very useful when modelling a system that is known to come to equilibrium, in particular, high-temperature processes involving simple molecules. It is less useful when complex molecules are present, as these usually have high Gibbs energy of formation; consequently, very low concentrations of these species are predicted unless the number of components in the model is very restricted.

The designer must specify the components carefully when using a Gibbs reactor in the model, as the Gibbs reactor can only solve for specified components. If a component that is actually formed is not listed in the component set, then the Gibbs reactor results will be meaningless. Furthermore, if some of the species have high Gibbs free energy, their concentrations may not be properly predicted by the model. An example is aromatic hydrocarbon compounds such as benzene, toluene and xylenes, which have Gibbs free energy of formation greater than zero. If these species are in a model component set that also contains hydrogen and carbon, then a Gibbs reactor will predict that only carbon and hydrogen are formed. Although hydrogen and coke are indeed the final equilibrium products, the aromatic hydrocarbons are kinetically stable and there are many processes that convert aromatic hydrocarbon compounds without significant coke yields. In this situation, the designer must either omit carbon from the component list or else use an equilibrium reactor in the model.

Continuous Stirred Tank Reactor (CSTR)

The CSTR is a model of the conventional well-mixed reactor. It can be used when a model of the reaction kinetics is available and the reactor is believed to be well mixed; i.e., the conditions everywhere in the reactor are the same as the outlet conditions. By specifying forward and reverse reactions, the CSTR model can model equilibrium and rate-based reactions simultaneously. The main drawback of using the CSTR model is that a detailed understanding of kinetics is necessary if by-products are to be predicted properly.

Plug-Flow Reactor (PFR)

A plug-flow reactor models the conventional plug-flow behaviour, assuming radial mixing, but no axial dispersion. The reaction kinetics must be specified and the model has the same limitations as the CSTR model.

Most of the simulators allow heat input or removal from a plug-flow reactor. Heat transfer can be with a constant wall temperature (as encountered in a fired tube, steam-jacketed pipe or immersed coil) or with counter-current flow of a utility stream (as in a heat-exchanger tube or jacketed pipe with cooling water).

Yield-Shift Reactor

The yield-shift reactor overcomes some of the drawbacks of the other reactor models by allowing the designer to specify a yield pattern. Yield-shift reactors can be used when there is no model of the kinetics, but some laboratory or pilot plant data are available, from which a yield correlation can be established.

Yield-shift reactors are particularly useful when modelling streams that contain pseudocomponents, solids with a particle size distribution or processes that form small amounts of many by-products. These can all be described easily in yield correlations, but can be difficult to model with the other reactor types.

The main difficulty in using the yield-shift reactor is in establishing the yield correlation. If a single point, for example from a patent, is all that is available, then entering the yield distribution is straightforward. If, on the other hand, the purpose is to optimize the reactor conditions, then a substantial set of data must be collected to build a response surface model that accurately predicts yields over a wide enough range of conditions. If different catalysts can be used, the underlying reaction mechanism may be different for each, and each will require its own yield model. The development of yield models can be an expensive process and is often not undertaken until corporate management has been satisfied that the process is likely to be economically attractive.

Modelling Real Reactors

Industrial reactors are usually more complex than the simple simulator library models. Real reactors usually involve multiple phases and have strong mass transfer, heat transfer and mixing effects. The residence time distributions of real reactors can be determined by tracer studies, and seldom exactly match the simple CSTR or PFR models.

Sometimes a combination of library models can be used to model the reaction system. For example, a conversion reactor can be used to establish the conversion of main feeds, followed by an equilibrium reactor that establishes an equilibrium distribution among specified products. Similarly, reactors with complex mixing patterns can be modelled as networks of CSTR and PFR models, as described in Section 1.9.10 and illustrated in Figure 1.19.

When using a combination of library models to simulate a reactor, it is a good idea to group these models in a sub-flow-sheet. The sub-flow-sheet can be given a suitable label such as 'reactor' that indicates that all the unit operations it contains are modelling a single piece of real equipment. This makes it less likely that someone else using the model will misinterpret it as containing additional distinct operations.

Detailed models of commercial reactors are usually written as user models. These are described in Section 4.6.

Example 4.1

When heavy oils are cracked in a catalytic or thermal cracking process, lighter hydrocarbon compounds are formed. Most cracking processes on heavy oil feeds form products with carbon numbers ranging from two to greater than 20. How does the equilibrium distribution of hydrocarbon compounds with five carbons (C_5 compounds) change as the temperature of the cracking process is increased at 200 kPa?

Solution

This problem was solved using UniSim Design.

The problem asks for an equilibrium distribution, so the model should contain either a Gibbs reactor or an equilibrium reactor.

A quick glance at the component list in UniSim Design shows that there are 22 hydrocarbon species with five carbons. To model the equilibrium among these species, we also need to include hydrogen to allow for the formation of alkenes, dienes and alkynes. Although it would be possible to enter 21 reactions and use an equilibrium reactor, it is clearly easier to use a Gibbs reactor for this analysis. Figure 4.5 shows the Gibbs reactor model.

To specify the feed we must enter the temperature, pressure, flow rate and composition. The temperature, pressure and flow rate are entered in the stream editor window (Figure 4.6). The feed composition can be entered as 100% of any of the C_5 paraffin species, for example normal pentane. The results from a Gibbs reactor would be the same if 100% isopentane was entered. It should be noted, however, that if a mixture of a pentane and a pentene was specified then the overall ratio of hydrogen to carbon would be different and different results would be obtained.

A spreadsheet was also added to the model, as illustrated in Figure 4.5, to make it easier to capture and download the results. The spreadsheet was set up to import

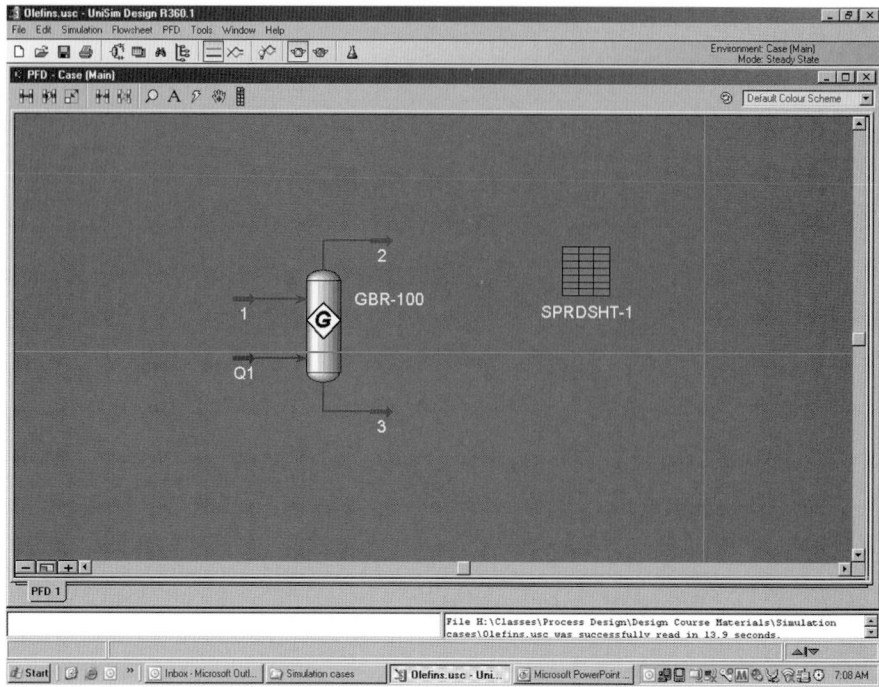

Figure 4.5.   Gibbs reactor model.

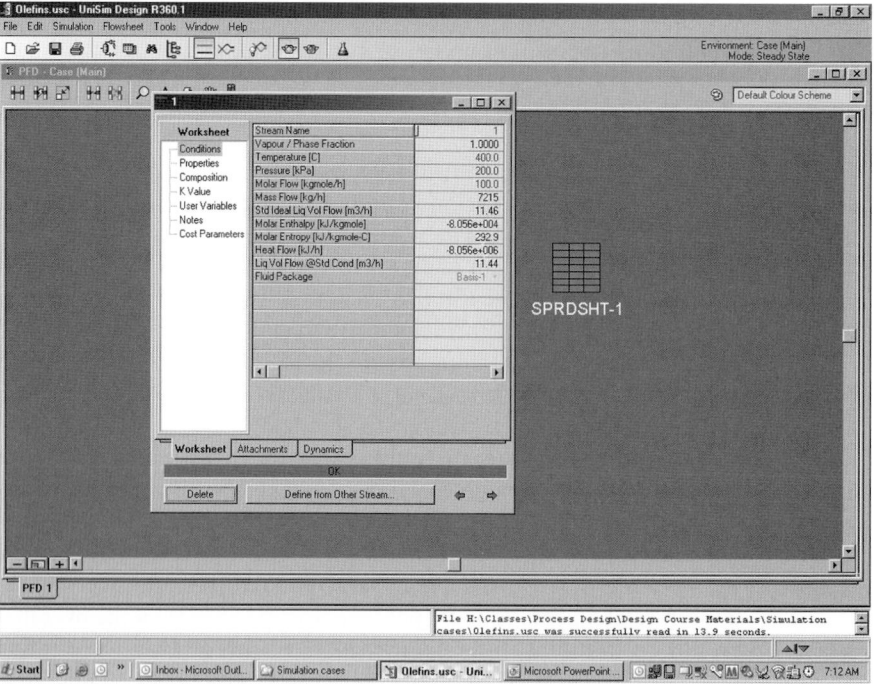

Figure 4.6.   Stream entry.

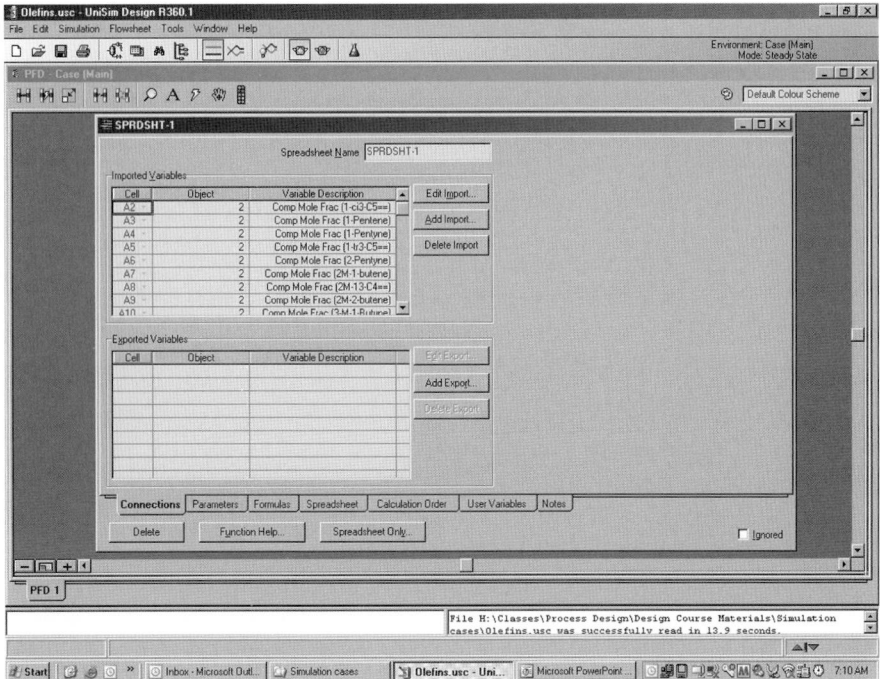

Figure 4.7.   Product composition spreadsheet.

component mole fractions from the simulation (Figure 4.7). The simulation was then run for a range of temperatures, and after each run a new column was entered in the spreadsheet (Figure 4.8).

When the results are examined, many of the individual species are present at relatively low concentrations. It thus makes sense to group some compounds together by molecular type, for example, adding all the dienes together and adding all the alkynes (acetylenes) together.

The spreadsheet results were corrected to give the distribution of C_5 compounds by dividing by one minus the mole fraction of hydrogen, and then plotted to give the graph in Figure 4.9.

It can be seen from the graph that the equilibrium products at temperatures below $500°C$ are mainly alkanes (also known as paraffins or saturated hydrocarbons), with the equilibrium giving roughly a 2:1 ratio of isopentane to normal pentane. As the temperature is increased from $500°C$ to $600°C$, there is increased formation of alkene compounds (also known as olefins). At $700°C$, we see increased formation of cyclopentene and of dienes, and above $800°C$ dienes are the favoured product.

Of course this is an incomplete picture, as the relative fraction of C_5 compounds would be expected to decrease as the temperature is raised and C_5 species are cracked to lighter compounds in the C_2 and C_3 range. The model also did not contain carbon (coke), and so could not predict the temperature at which coke would become the preferred product. A more rigorous equilibrium model of a cracking process might include all of the possible hydrocarbon compounds up to C_7 or higher.

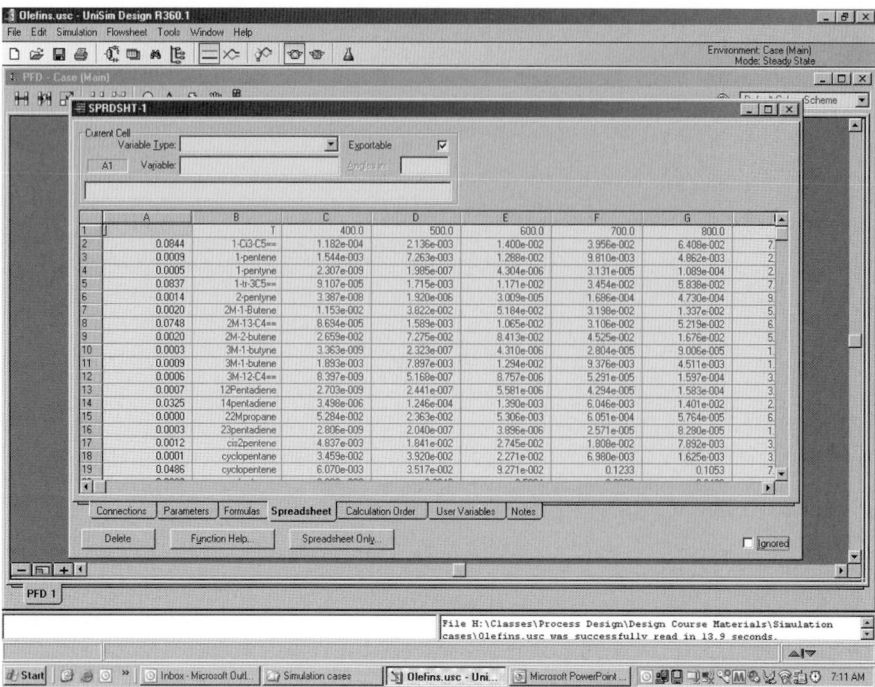

Figure 4.8. Spreadsheet results.

A real reactor might give a very different distribution of C_5 compounds from that calculated using the Gibbs reactor model. Dienes formed at high temperatures might recombine with hydrogen during cooling, giving a mixture that looked more like the equilibrium product at a lower temperature. There might also be formation of C_5 compounds by condensation reactions of C_2 and C_3 species during cooling, or loss of dienes and cyclopentene due to coke formation.

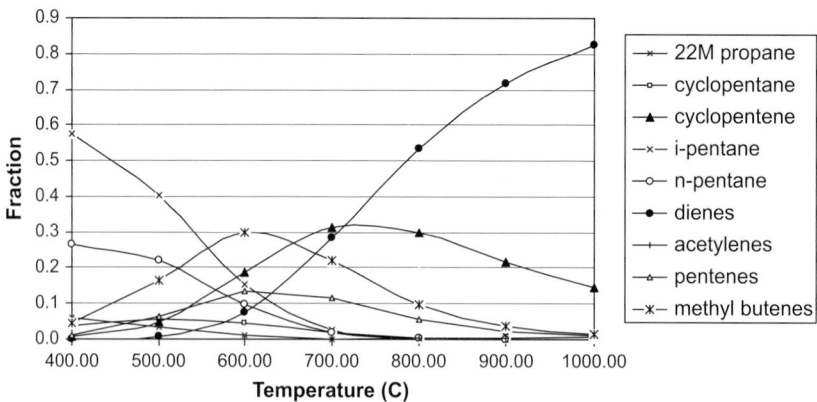

Figure 4.9. Product distribution.

Example 4.2

Hydrogen can be made by steam reforming of methane, which is a highly endo-thermic process:

$$CH_4 + H_2O \leftrightarrow CO + 3H_2 \qquad \Delta H^\circ_{rxn} = 2.1 \times 10^5 \text{ kJ/kgmol}$$

Steam reforming is usually carried out in fired tubular reactors, with catalyst packed inside the tubes and fuel fired on the outside of the tubes to provide the heat of reaction. The product gas mixture contains carbon dioxide and water vapour as well as carbon monoxide and hydrogen and is conventionally known as synthesis gas or syngas.

Hydrogen can also be made by partial oxidation of methane, which is an exothermic process, but yields less product per mole of methane feed:

$$CH_4 + \tfrac{1}{2}O_2 \rightarrow CO + 2H_2 \qquad \Delta H^\circ_{rxn} = -7.1 \times 10^4 \text{ kJ/kgmol}$$

When steam, oxygen and methane are combined, heat from the partial oxidation reaction can be used to provide the heat for steam reforming. The combined process is known as autothermal reforming. Autothermal reforming has the attraction of requiring less capital investment than steam reforming (because it does not need a fired-heater reactor), but giving higher yields than partial oxidation.

The yield of hydrogen can be further increased by carrying out the water–gas shift reaction:

$$CO + H_2O \rightarrow CO_2 + H_2 \qquad \Delta H^\circ_{rxn} = -4.2 \times 10^4 \text{ kJ/kgmol}$$

The water–gas shift reaction equilibrates rapidly at temperatures above about 450°C. At high temperatures this reaction favours the formation of carbon monoxide, whilst at low temperatures more hydrogen is formed. When hydrogen is the desired product, the shift reaction is promoted at lower temperatures by using an excess of steam and providing a medium- or low-temperature shift catalyst.

In an autothermal reforming process, 1000 kmol/h of methane at 20°C is compressed to 10 bar, mixed with 2500 kmol/h of saturated steam and reacted with pure oxygen to give 98% conversion of the methane. The resulting products are cooled and passed over a medium-temperature shift catalyst that gives an outlet composition corresponding to equilibrium at 350°C.

(i) How much heat is required to vaporize the steam?
(ii) How much oxygen is needed?
(iii) What is the temperature at the exit of the autothermal reforming reactor?
(iv) What is the final molar flow rate of each component of the synthesis gas?

Solution

This problem was solved using Aspen Plus. The model must simulate the high temperature reforming reaction and also the re-equilibration of the water–gas shift reaction as the product gas is cooled. A Gibbs reactor can be used for the high-temperature reaction, but an equilibrium reactor must be specified for the shift reactor, as only the water–gas shift reaction will re-equilibrate at 350°C. Because the methane compressor supplies some heat to the feed, it should be included in the model. Since the question asks how much heat is needed to vaporize the steam, a steam boiler should also be included. The oxygen supply system can also be included, giving the model shown in Figure 4.10.

The heat duty to the reforming reactor is specified as zero. The oxygen flow rate can then be adjusted until the desired methane conversion is achieved. For 98% conversion, the flow rate of methane in the autothermal reactor product (stream 502) is 2% of the flow rate in the reactor feed (stream 501); i.e., 20 kmol/h. For the purpose of this example, the oxygen flow rate was adjusted manually, although a controller could have been used, as described in Section 4.8. The results are shown in Figure 4.11.

When the simulation model was run, the following values were calculated:

(i) The steam heater requires 36 MW of heat input.
(ii) 674 kmol/h of oxygen are needed.
(iii) The temperature at the exit of the reforming reactor is 893°C.
(iv) The molar flow rates at the outlet of the shift reactor (stream 504) are:

H_2	2504
H_2O	1956
CO	68
CO_2	912
CH_4	20

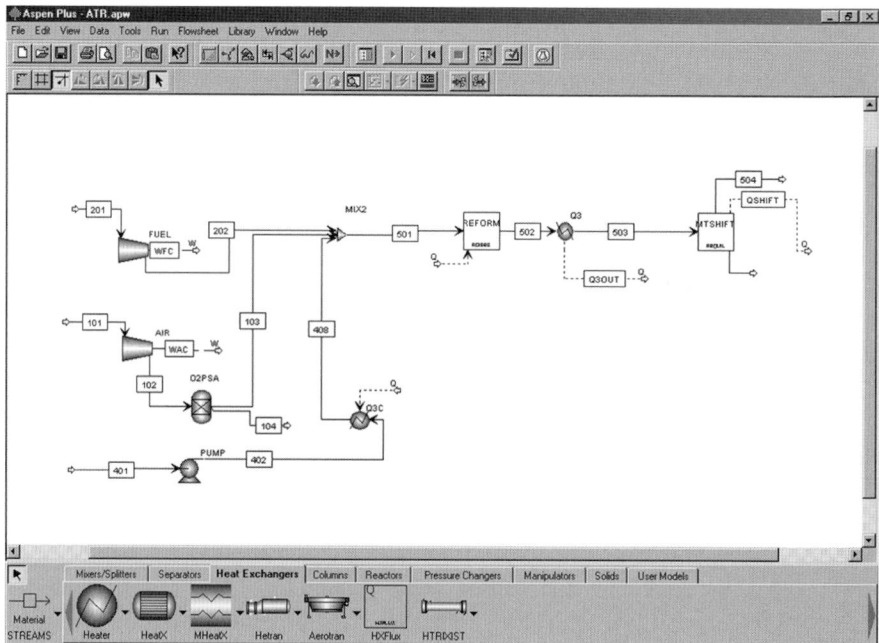

Figure 4.10.   Autothermal reforming model.

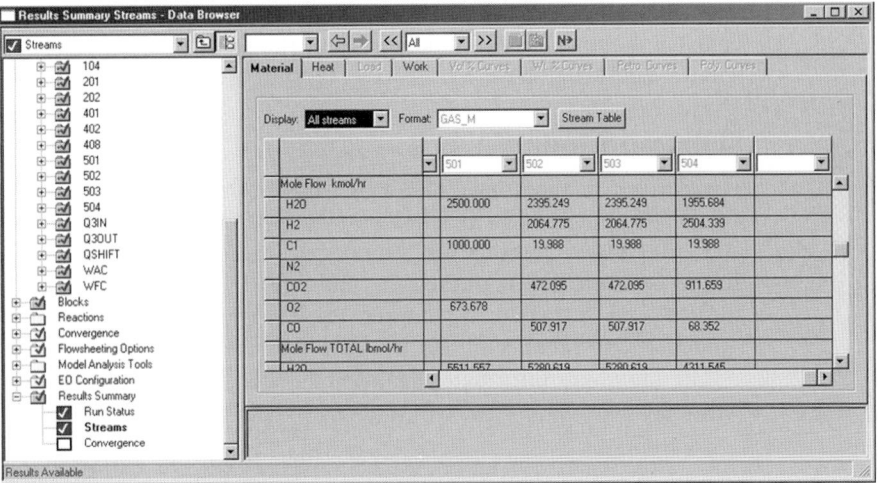

Figure 4.11. Autothermal reactor model results.

It should be immediately apparent from the model output that the process as simulated is far from optimal. The oxygen consumption is larger than the 500 kmol/h that would have been needed for partial oxidation. The excess oxygen is needed because the additional steam that is being fed must also be heated to the reactor outlet temperature, which requires more of the feed methane to be burned. The corollary of this result is that the hydrogen yield, at roughly 2.5 moles per mole methane, is not much better than could have been obtained with partial oxidation followed by shift, despite the large excess of steam used.

The designer has several options that could be examined to improve this process:

1. Increase heat recovery from the product gas to the feed streams to preheat the reactor feed and reduce the amount of oxygen that is needed.
2. Reduce the amount of steam fed with the methane.
3. Bypass a part of the steam from the reformer feed to the shift reactor feed, so as to obtain the benefit of driving the equilibrium in the shift reactor without the cost of providing extra heat to the reformer.
4. Reduce the conversion of methane so that a lower reactor conversion and lower outlet temperature are required.

In practice, all of these options are implemented to some extent to arrive at the optimal autothermal reforming conditions. This optimization is explored further in problem 4.13.

4.5.2. Distillation

The commercial process simulators contain a range of distillation models with different degrees of sophistication. The design engineer must choose a model that is suitable for the purpose, depending on the problem type, the extent of design information available and the level of detail required in the solution. In some cases, it may make sense to build different versions of the flow-sheet, using different levels of detail in the distillation models, so that the simpler model can be used to initialize a more detailed model.

Shortcut Models

The simplest distillation models to set up are the shortcut models. These models use the Fenske-Underwood-Gilliland or Winn-Underwood-Gilliland method to determine the minimum reflux and number of stages or to determine the required reflux given a number of trays or the required number of trays for a given reflux ratio. These methods are described in Chapter 11. The shortcut models can also estimate the condenser and reboiler duties and determine the optimum feed tray.

The minimum information needed to specify a shortcut distillation model is as follows:

- The component recoveries of the light and heavy key components.
- The condenser and reboiler pressures.
- Whether the column has a total or partial condenser.

In some cases, the designer can specify the purities of the light and heavy key components in the bottoms and distillate respectively. Care is needed when using purity as a specification, as it is easy to specify purities or combinations of purity and recovery that are infeasible; see Section 11.6.

The easiest way to use a shortcut distillation model is to start by estimating the minimum reflux and number of stages. The optimum reflux ratio is usually between 1.05 and 1.25 times the minimum reflux ratio, R_{min}, so $1.15 \times R_{min}$ is often used as an initial estimate. Once the reflux ratio is specified, the number of stages and optimum feed stage can be determined. The shortcut model results can then be used to set up and initialize a rigorous distillation simulation.

Shortcut models can also be used to initialize fractionation columns (complex distillation columns with multiple products), as described below.

Shortcut distillation models are robust and are solved quickly. They do not give an accurate prediction of the distribution of non-key components, and they do not perform well when there is significant liquid–phase non-ideality, but they are an efficient way of generating a good initial design for a rigorous distillation model. In processes that have a large number of recycle streams, it is often worthwhile to build one model with shortcut columns and a second model with rigorous columns. The simple model will converge more easily and can be used to provide good initial estimates of column conditions and recycle streams for the detailed model.

The main drawback of shortcut models is that they assume constant relative volatility, usually calculated at the feed condition. If there is significant liquid- or vapour-phase non-ideality, then constant relative volatility is a very poor assumption and shortcut models should not be used.

Rigorous Models

Rigorous models carry out full stage-by-stage mass and energy balances. They give better predictions of the distribution of components than shortcut models, particularly when the liquid phase behaves non-ideally, as the flash calculation is made on each stage. Rigorous models allow many more column configurations, including use of side streams, intermediate condensers and reboilers, multiple feeds and side strippers and rectifiers. Rigorous models can be much harder to converge, particularly if poor initial estimates are used or if the column is improperly specified.

The two main types of rigorous distillation models are equilibrium-stage models and rate-based models. Equilibrium-stage models assume either full vapour-liquid equilibrium on each stage or else an approach to equilibrium based on a stage efficiency entered by the designer. When using an equilibrium-stage model for column sizing, the stage efficiencies must be entered. Stage efficiency is typically less than 0.8, and is discussed in more detail in Section 11.10. Rate-based models do not assume phase equilibrium, except at the vapour–liquid interface, and instead solve the inter-phase mass-transfer and heat-transfer equations. Rate-based models are more realistic than the idealized equilibrium-stage models, but because it can be difficult to predict the interfacial area and mass-transfer coefficients, rate-based models are less widely used in practice.

Rigorous distillation models can be used to model absorber columns, stripper columns, refluxed absorbers, three-phase systems such as extractive distillation columns, many possible complex column configurations, and columns that include reactions such as reactive distillation and reactive absorption columns. The formation of a second liquid phase (usually a water phase) in the column can be predicted if the designer has selected a liquid-phase activity model that allows for the prediction of two liquid phases.

One of the most useful features of the rigorous distillation models in the commercial simulation programs is that most include a tool for plotting column profiles. The design engineer can generate plots showing the molar composition of each species in either phase versus tray number. These plots can be helpful in troubleshooting column designs.

For example, Figures 4.12 to 4.17 show column profiles for the distillation problem introduced in Example 1.1, which is described in more detail in Examples 4.3 and 4.4. The column was simulated in UniSim Design.

- In Figure 4.12, the feed stage was moved up to tray 10, which is too high. The column profiles show a broad flat region between trays 20 and 45, indicating that nothing much is going on over this part of the column. There are too many trays in the stripping section and the feed tray should be moved lower. Sections with very small change in composition can also be indicative of pinched regions where an azeotropic mixture is being formed.
- In Figure 4.13, the feed tray has been moved down to tray 63, which is too low. The column profiles for benzene and toluene, the light components, are flat between trays 30 and 60 in the rectifying section, indicating that the feed tray should be moved higher.
- In Figure 4.14, the column specification was changed from toluene recovery to reflux ratio and a low value of reflux ratio (2.2) was entered. This is less than the minimum reflux required for the specified separation; consequently, the desired recovery of toluene cannot be achieved. The recovery of toluene is reduced to 72%.

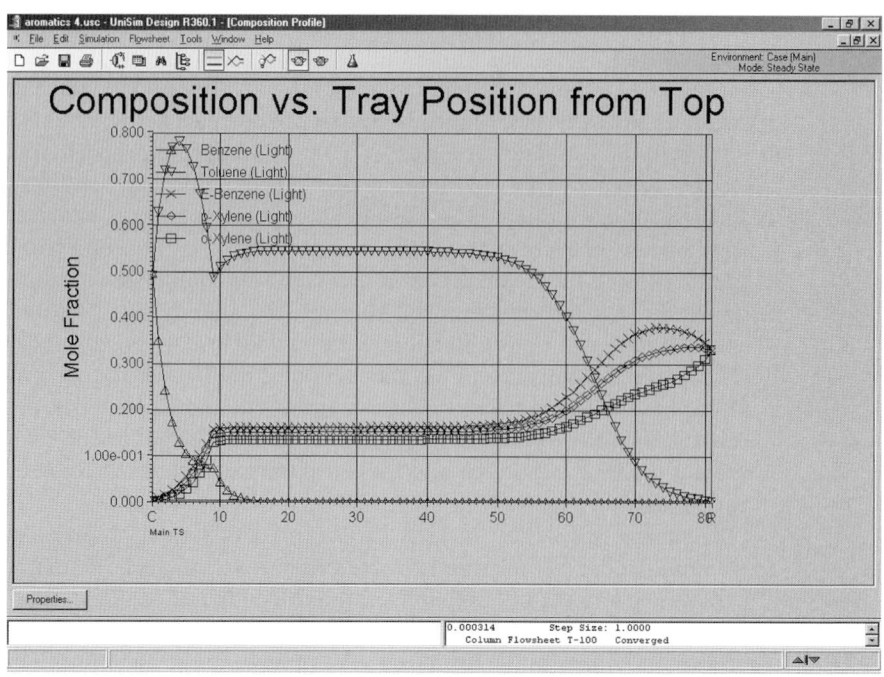

Figure 4.12.   Feed tray too high.

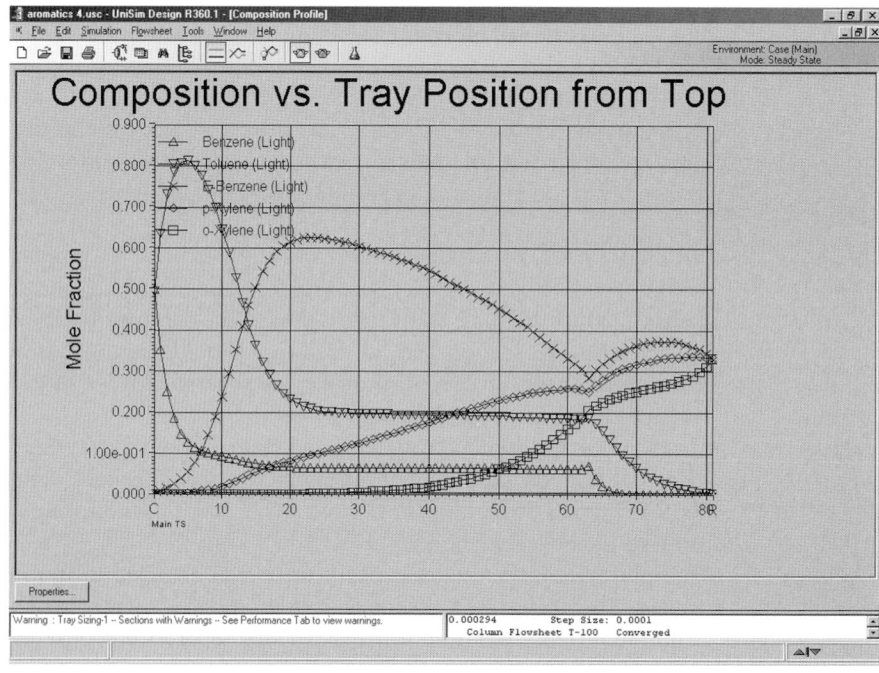

Figure 4.13.   Feed tray too low.

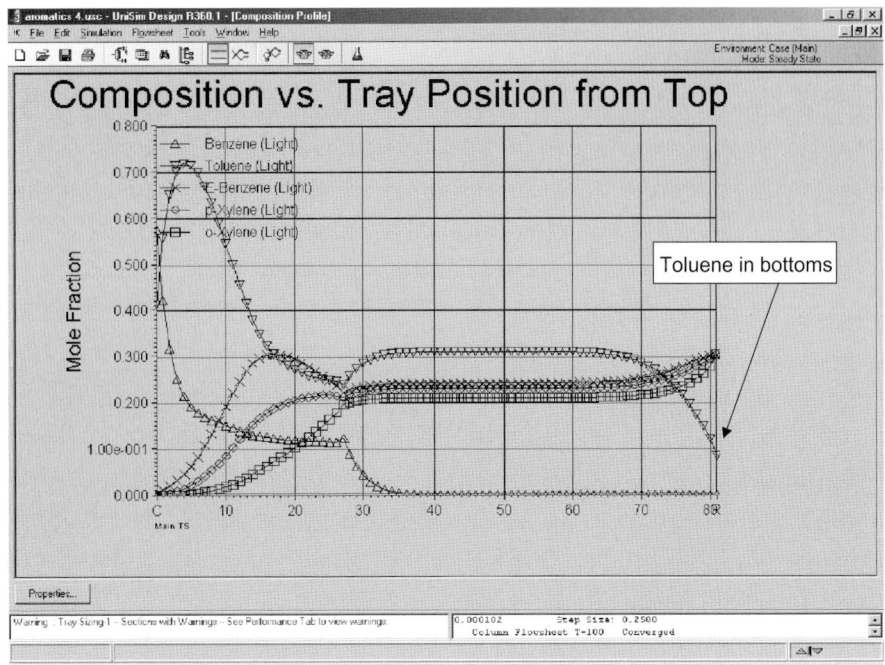

Figure 4.14. Reflux ratio too low: toluene recovery 72%.

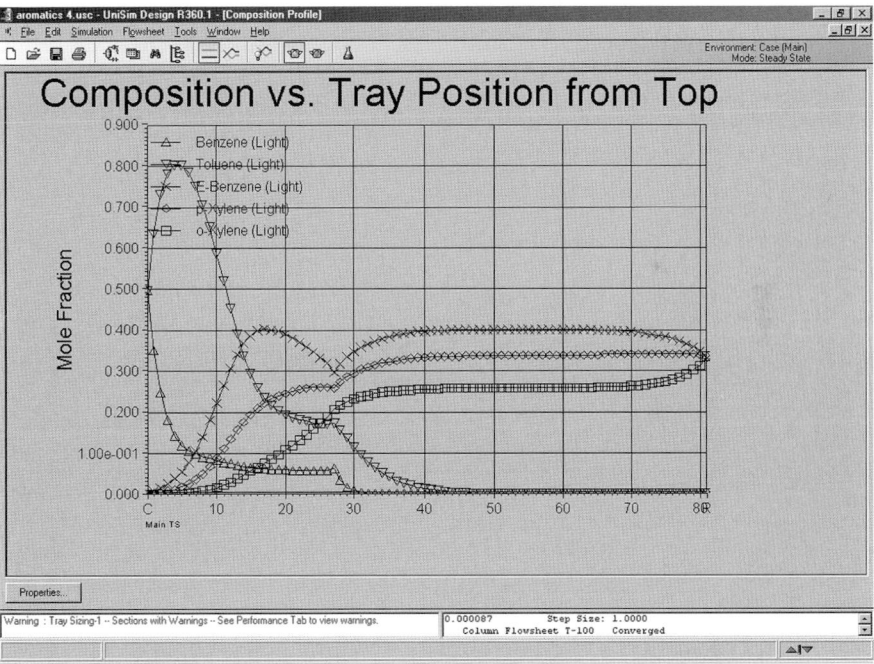

Figure 4.15. Reflux ratio too high: toluene recovery 100%.

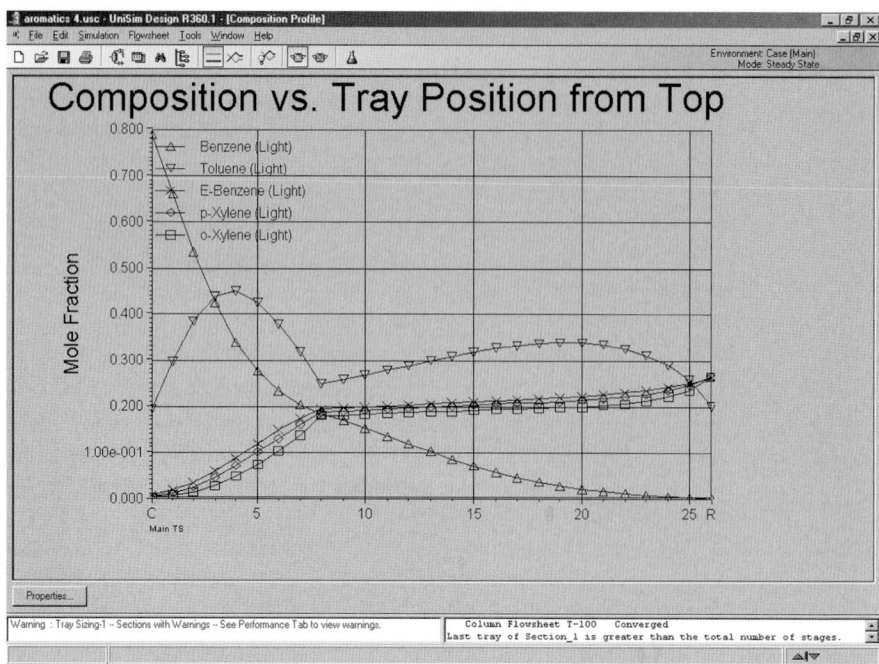

Figure 4.16.   Too few trays: toluene recovery 24.5%.

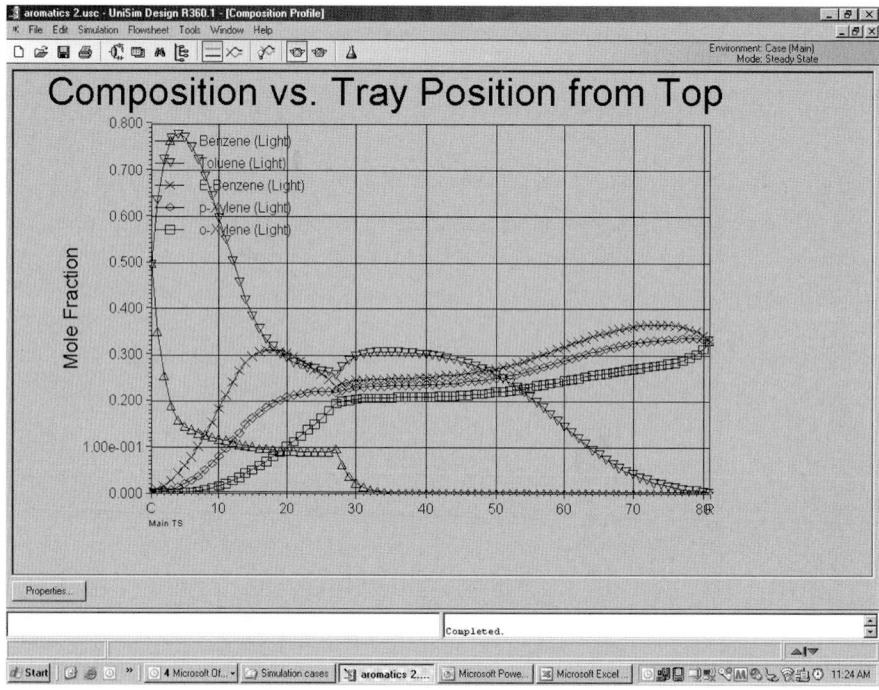

Figure 4.17.   Optimized column profiles.

- In Figure 4.15, the reflux ratio was increased to 4.0. The recovery of toluene is now 100%, which is greater than the 99% required. This represents a sub-optimal use of energy and capital.
- Figure 4.16 shows the column profiles when the number of trays was reduced to 25, with the feed on tray 8. The column profile for toluene shows that there are insufficient stages (and/or reflux). Although the profile is changing smoothly, the recovery in the distillate is only 24.5%.
- The column profiles with the optimum conditions determined in Example 1.1 are shown in Figure 4.17. The poor features shown in the other profiles are absent.

Complex Columns for Fractionation

Several of the commercial simulation programs offer pre-configured complex column rigorous models for petroleum fractionation. These models include charge heaters, several side strippers, and one or two pump-around loops. These fractionation column models can be used to model refinery distillation operations such as crude oil distillation, vacuum distillation of atmospheric residue oil, fluidized catalytic cracking (FCC) process main columns and hydrocracker or coker main columns. Aspen Plus also has a shortcut fractionation model, SCFrac, which can be used to configure fractionation columns in the same way that shortcut distillation models are used to initialize multicomponent rigorous distillation models.

A typical crude oil distillation column is illustrated in Figure 4.18, which shows a simulation using an Aspen Plus PetroFrac model. The crude oil is preheated in

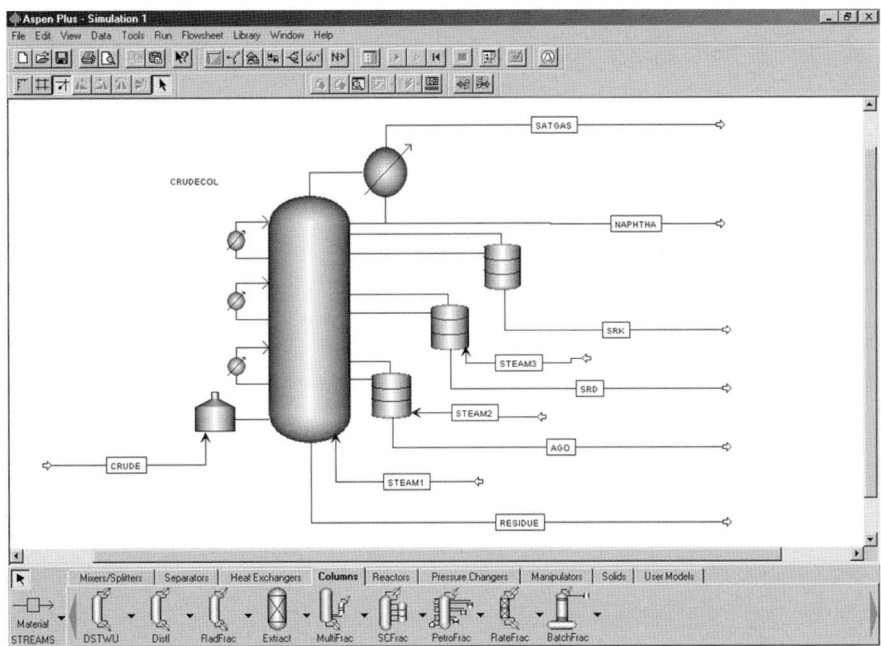

Figure 4.18. Crude oil fractionation.

a heat-exchange network and charge heater and is then fed to the flash zone at the bottom of the column. Stripping steam is also added at the bottom of the column to provide additional vapour flow. Products with different boiling ranges are withdrawn from the column. The intermediate products are withdrawn from the bottom of side-stripper columns, so as to minimize loss of lighter products in the side stream. Although the exact distillation ranges can vary depending on the local fuels specifications and the sophistication of the refinery, the typical products taken in a crude oil distillation unit are (from the bottom up):

1. Atmospheric residue oil (Residue), containing compounds that boil above about 340°C (650°F). This is normally sent to a vacuum distillation unit to recover more light products, but parts of it may be blended into high sulphur fuels such as heating oil or bunker fuel (marine fuel).

2. Atmospheric gas oil (AGO), containing compounds that boil in the range 275°C to 340°C (530°F to 650°F). This material is too high-boiling for use as a transportation fuel and is usually sent to a hydrocracker or FCC unit for conversion to lighter products.

3. Heavy distillate (straight-run distillate or SRD), containing compounds that boil in the range 205°C to 275°C (400°F to 530°F). This material is hydrotreated to remove sulphur compounds and can then be blended into heating oils and diesel fuels for trucks, railroad engines and off-road applications such as tractors and mining equipment.

4. Light distillate (straight-run kerosene or SRK), containing compounds that boil in the range 175°C to 230°C (350°F to 450°F). Light distillate is hydrotreated to remove sulphur and can then be blended into jet fuel or sold as kerosene (sometimes called paraffin) for lamp and cooking fuel.

5. Naphtha, boiling in the range 25°C to 205°C (80°F to 400°F). Naphtha is usually sent to an additional column for separation into a light naphtha boiling below 80°C (180°F) and a heavy naphtha. Heavy naphtha has the right boiling range for gasoline, but usually has very low octane number. It is typically upgraded by catalytic reforming using noble metal catalysts, to increase the concentration of aromatic hydrocarbons in the naphtha and raise the octane number. Catalytic reforming is also the first step in the production of aromatic hydrocarbons for petrochemicals manufacture. Light naphtha also boils in a suitable range for blending into gasoline, and often has an acceptable octane number. It is usually treated to oxidize odorous mercaptan sulphur compounds. Light naphtha is also widely used as a petrochemical feedstock for steam cracking to produce olefin compounds such as ethylene and propylene.

6. The overhead product of the crude unit contains hydrogen, methane, carbon dioxide, hydrogen sulphide and hydrocarbons up to butanes and some pentanes. It is usually sent to a set of distillation columns known as a 'saturate gas plant' for recovery of propane and butane for sale. The lighter gases are then used as refinery fuel.

The design of refinery fractionation columns can be complex. The pump-around streams function as intermediate condensers and remove surplus heat from the column. This heat is usually recovered by heat exchange with the cold crude oil feed.

Oil refineries are often designed to handle many different crude oils with different boiling assays. The refinery may make different product slates at different times of the year, or in response to market conditions. The crude oil distillation and associated heat-exchange network must be flexible enough to handle all of these variations, while still achieving tight specifications on the boiling point curves of every product.

Column Sizing

The rigorous column models allow the design engineer to carry out tray-sizing and hydraulics calculations for the basic types of distillation trays and for some types of random and structured packing. Different commercial simulators use different tray-sizing correlations, but they all follow a method similar to that described in Chapter 11.

The tray-sizing tools are not always enabled when running the distillation models. In some of the simulation programs, the design engineer must enable a tray-sizing program and/or enter default values for tray type and tray spacing before the sizing algorithm will work properly. If the column diameter does not change when the reflux rate is significantly changed (or if all the columns in the simulation appear to have the same diameter), then the designer should check to make sure that the tray-sizing part of the program is properly configured.

The tray-sizing options in the simulators are restricted to standard internals such as sieve trays, valve trays, bubble-cap trays, random packings and structured packings. They do not include high-capacity trays, high-efficiency trays or the latest packing designs. When designing a column that has many stages or a large diameter, it is always worth contacting the column internals vendors for estimates, as use of high-capacity, high-efficiency internals can lead to substantial savings. Advanced internals are also usually used when revamping an existing column to a higher throughput or tighter product specifications.

The design engineer should always allow for tray inefficiency when using column-sizing tools in conjunction with an equilibrium-stage model. Failure to do so would under-predict number of stages and hence have an impact on the column pressure drop and hydraulics. Estimation of stage efficiency is discussed in Chapter 11. For initial design purposes, a stage efficiency of 0.5 to 0.7 is usually used. For detailed design, stage efficiencies depend on the type of tray used and are often provided by the column internals vendor.

The design engineer must remember to allow a suitable design factor or design margin when sizing columns. Design factors are discussed in Section 1.7. It may be necessary to create two versions of the flow-sheet. One version will have the design basis flow rates for producing the mass and energy balances, while the second will have flow rates that are 10% larger for purposes of sizing equipment.

The simulation of distillation processes is discussed in more detail by Luyben (2006).

Example 4.3

This example provides more detail on the solution of the problem that was introduced as Example 1.1. The original problem statement was to optimize the design of a distillation column to separate 225 metric tons per hour of an equimolar mixture of benzene, toluene, ethylbenzene, paraxylene and orthoxylene with minimum total annualized cost. The feed is a saturated liquid at 330 kPa. The recovery of toluene in the distillate should be greater than 99%, and the recovery of ethylbenzene in the bottoms should be greater than 99%.

In this example, a column simulation should be set up using a shortcut model. The shortcut model results will be used to initialize a rigorous model in the example that follows. Determine:

  (i) The minimum reflux ratio.
 (ii) The minimum number of trays.
(iii) The actual number of trays when the reflux is 1.15 R_{min}.
(iv) The optimum feed tray.

Solution

This problem was solved using UniSim Design. The problem was set up as a shortcut column as shown in Figure 4.19.

UniSim Design requires the designer to specify the mole fraction of the light key component in the bottoms and the heavy key component in the distillate. We have an equimolar feed, so if we take a basis of 100 mol/h of feed, then the molar flow rate of each component is 20 mol/h. A 99% recovery of each key component corresponds to allowing 0.2 mol/h of that component into the other stream. The mole fractions are then:

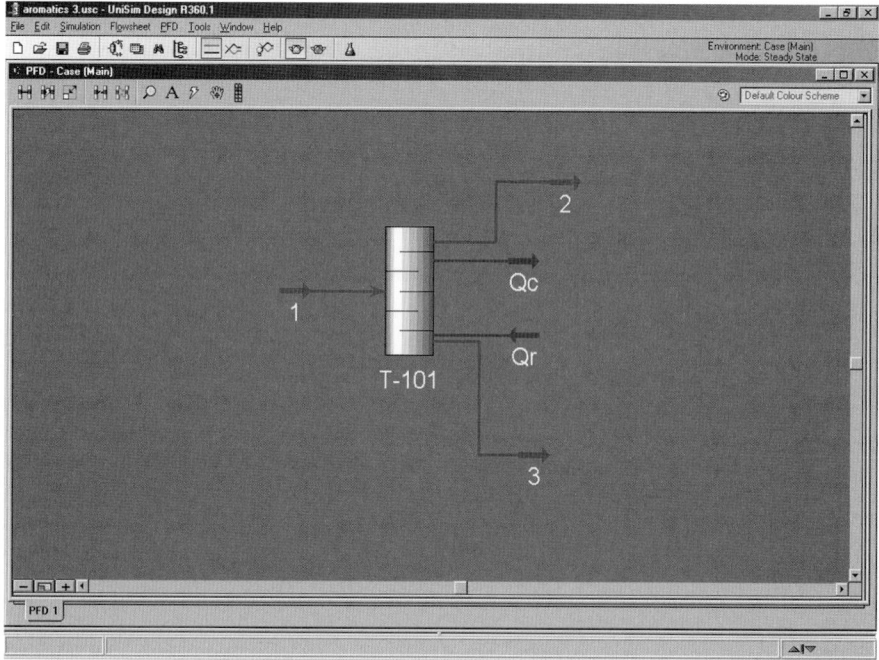

Figure 4.19.   Shortcut distillation.

$$\text{Ethylbenzene in distillate} = 0.2/40 = 0.005$$
$$\text{Toluene in bottoms} = 0.2/60 = 0.00333$$

When these are entered into the shortcut column as specifications, the minimum reflux is calculated to be $R_{min} = 2.130$. The actual reflux ratio can then be specified as $2.13 \times 1.15 = 2.45$, as shown in Figure 4.20.

The shortcut column results are shown in Figure 4.21. The minimum number of stages is calculated as 16.4, which should be rounded up to 17. The actual number of trays required is 39, with feed at stage 18.

Example 4.4

Continuing the problem defined in Example 4.3, use a rigorous simulation to carry out tray sizing and estimate the required column diameter.

Solution

Since we are now sizing the column, the first step is to increase the flow rate to allow for a design factor. The process design basis is 225 metric tons per hour of feed. The equipment design should include at least a 10% safety factor, so the equipment design basis was set at 250 metric tons per hour of feed (rounding up from 247.5 for convenience).

Figure 4.22 shows the rigorous column simulation. UniSim Design allows the designer to enter any two specifications for the column, so instead of entering the reflux ratio as a specification, we can enter the required recoveries and provide the value of reflux ratio found in the shortcut model as an initial estimate, as shown in Figure 4.23.

The column converges quickly with the good estimate provided from the shortcut model. The column profiles can be checked by selecting the 'performance' tab in the column environment and then selecting 'plots' from the menu on the left and

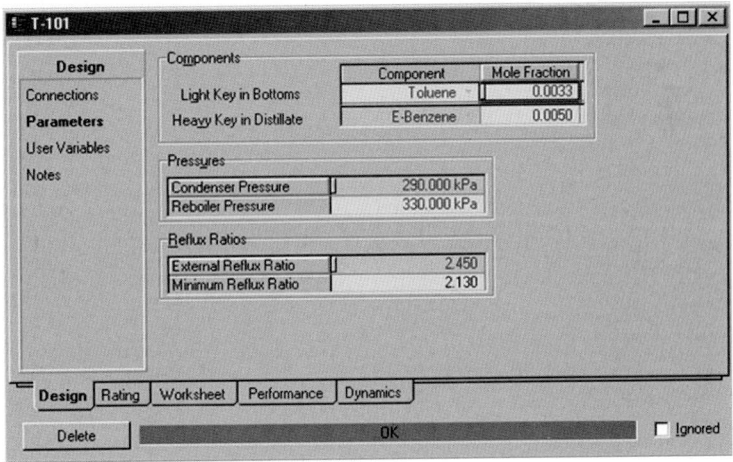

Figure 4.20. Shortcut column specifications.

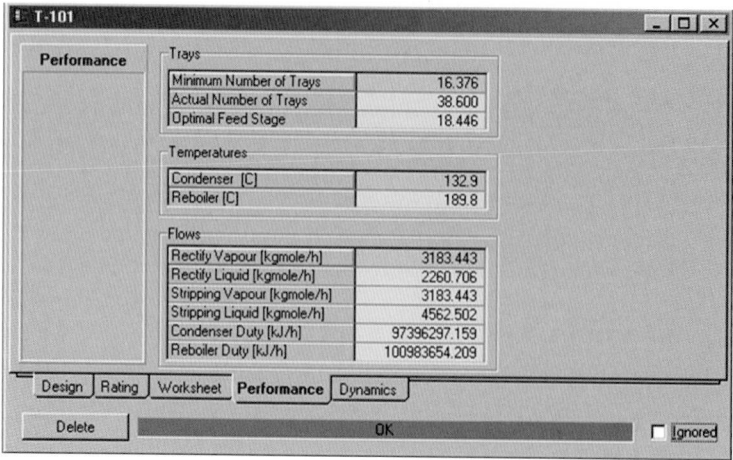

Figure 4.21.   Shortcut column results.

'composition' from the list of possible plots, as shown in Figure 4.24. This generates composition profiles like those presented in Figures 4.12 to 4.17.

To size the trays in UniSim Design, the tray-sizing utility must be activated (from the tools menu via tools/utilities/tray sizing). When sieve trays are selected with the default spacing of 609.6 mm (2 ft) and the other default parameters shown in

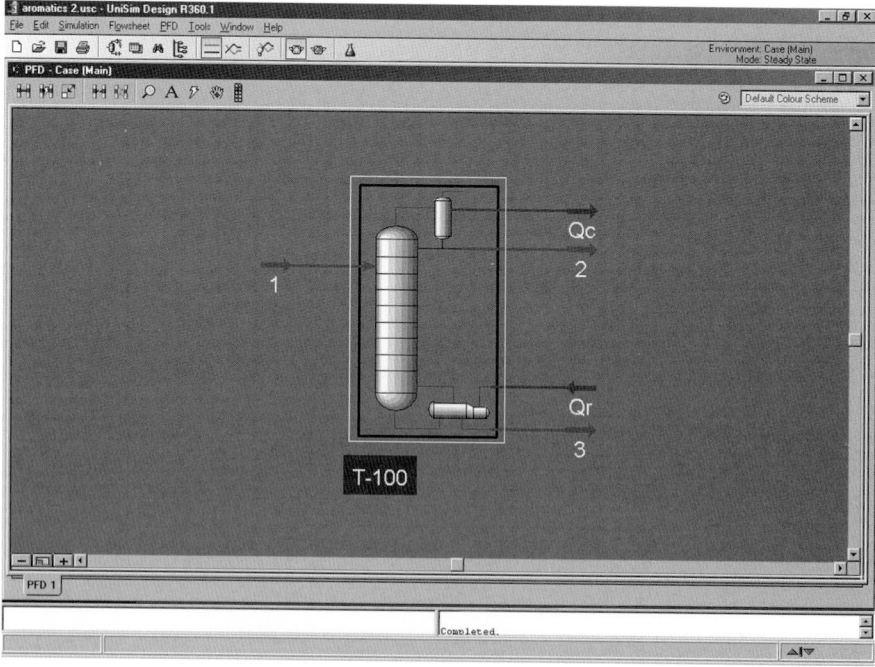

Figure 4.22.   Rigorous distillation.

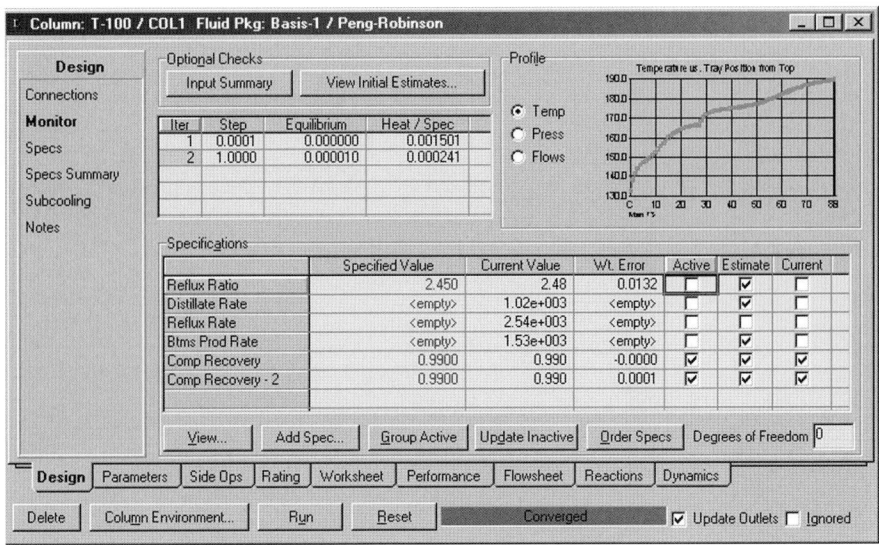

Figure 4.23.   Rigorous column specifications.

Figure 4.25, then the results in Figure 4.26 are obtained. The column diameter is found to be 4.42 m.

The data on column size, number of trays, reboiler and condenser duty can then be extracted from the simulation and put into a cost model or spreadsheet to carry out optimization of the total annual cost of production. The results of the optimization are described in Example 1.1.

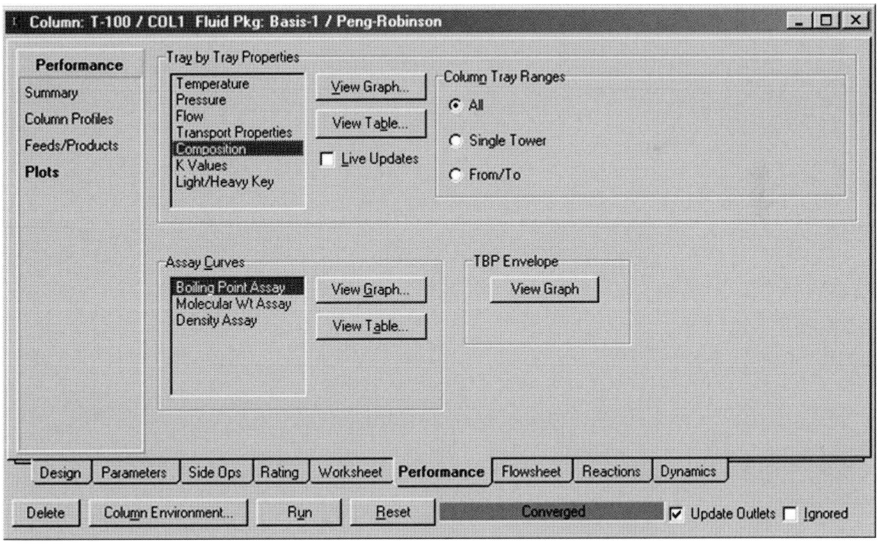

Figure 4.24.   Generating column profiles.

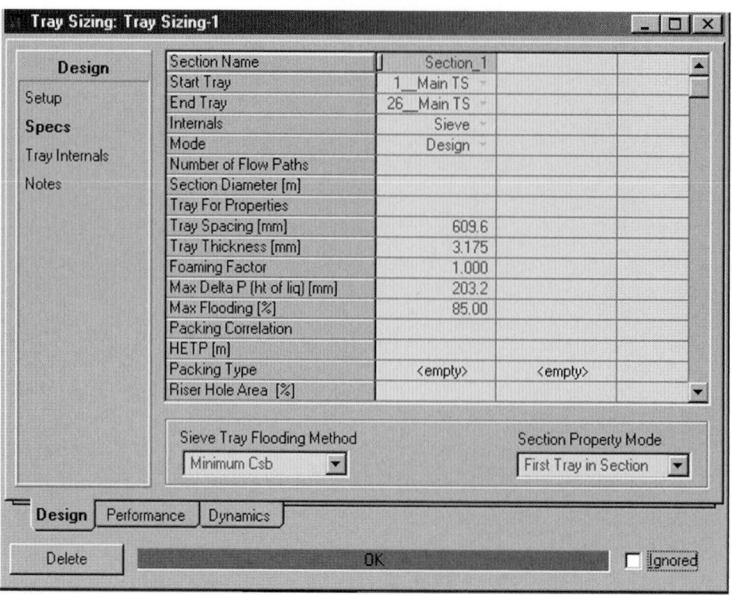

Figure 4.25.   Default tray-sizing options.

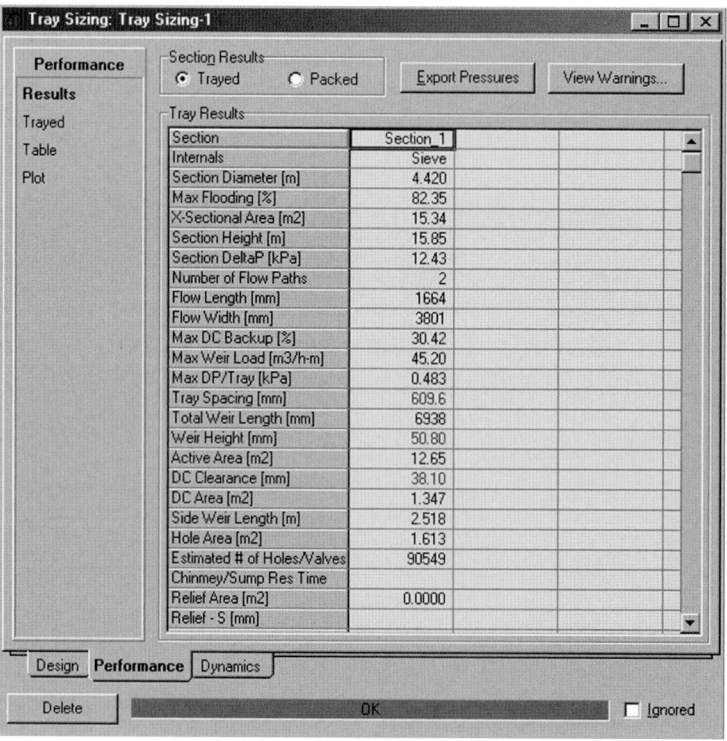

Figure 4.26.   Tray-sizing results.

4.5.3. **Other Separations**

Other multi-stage vapour–liquid separations such as absorption and stripping can be modelled using variations of the rigorous distillation models, as can multi-stage liquid–liquid extraction.

Single-stage liquid–liquid or vapour–liquid separation can be modelled as a flash vessel, but some caution is needed. The simulation programs assume perfect separation in a flash unless the designer specifies otherwise. If there is entrainment of droplets or bubbles, then the outlet compositions of a real flash vessel will be different from those predicted by the simulation. If the flash is critical to process performance, then the designer should make an allowance for entrainment. Most of the simulation programs allow the designer to specify a fraction of each phase that is entrained with the other phases. This is illustrated in Figure 4.27, which shows the data entry sheet for entrained flows for UniSim Design. In UniSim Design, the entrained fractions are entered on the 'Rating' tab of the flash model window. Users can also use built-in correlation models with their specified information such as vessel dimensions and nozzle locations. More sophisticated real separator modelling can be found in the three-phase separator model in UniSim Design. The fraction that is entrained depends on the design of the vessel, as described in Chapter 10.

Most of the simulators contain several models for fluid–solid separation. These models can be used to manipulate the particle size distribution when solids are present.

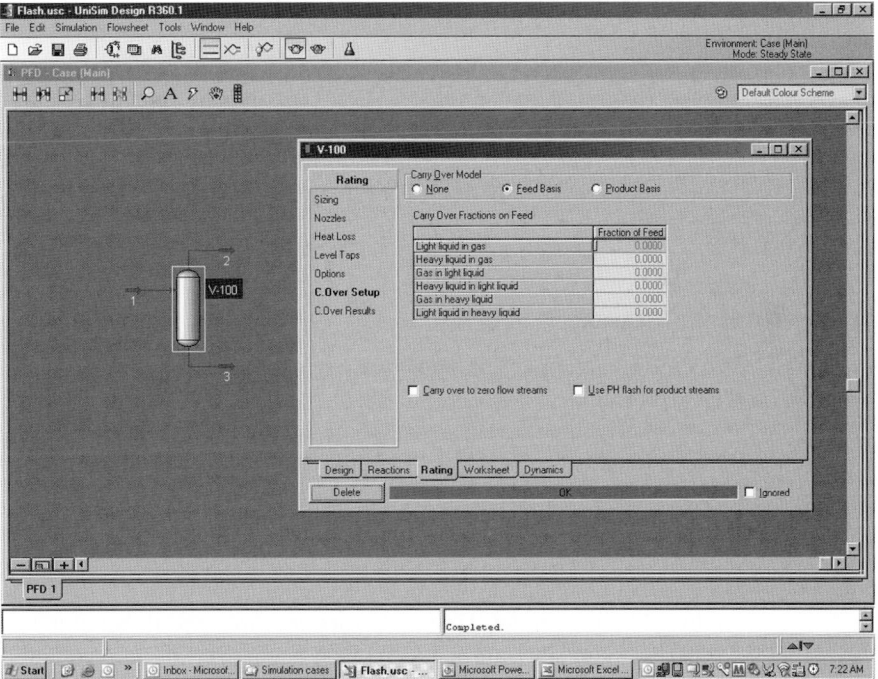

Figure 4.27.   Flash model with entrainment.

None of the commercial process simulators contains a good library model for adsorptive separations or membrane separations at the time of writing. These separation methods are important for gas–gas separations, chromatographic separations and size-exclusion or permeation-based separations. All of these processes must be modelled using component splitters, as described below.

Component Splitter Models

A component splitter is a sub-routine in the simulation that allows a set of components from a stream to be transferred into another stream with a specified recovery. Component splitters are convenient for modelling any separation process that cannot be described using one of the library models. Examples of real operations that are usually modelled as component splitters include:

- Pressure-swing adsorption
- Temperature-swing adsorption
- Chromatography
- Simulated-moving-bed adsorption
- Membrane separation
- Ion exchange
- Guard beds (irreversible adsorption).

When a component splitter is used in a model, it is a good practice to give the splitter a label that identifies the real equipment that is being modelled.

Component splitters are sometimes used in place of distillation columns when building simple models to provide initial estimates for processes with multiple recycles. There is little advantage to this approach compared with using shortcut distillation models, as the component splitter will not calculate the distribution of non-key components unless a recovery is entered for each. Estimating and entering the recoveries for every component is difficult and tedious and poor estimates of recoveries can lead to poor estimates of recycle flows, so the use of component splitters in this context effectively adds another layer of iteration to the model.

4.5.4. Heat Exchange

All of the commercial simulators include models for heaters, coolers, heat exchangers, fired heaters and air coolers. The models are easy to configure and the only inputs that are usually required on the process side are the estimated pressure drop and either the outlet temperature or the duty. A good initial estimate of pressure drop is 0.3 to 0.7 bar (5 to 10 psi).

The heater, cooler and heat-exchanger models allow the design engineer to enter estimates of film transfer coefficients, and hence estimate the exchanger area. As with distillation columns, the designer must remember to add a design factor to the sizes predicted by the model. Design factors are discussed in Section 1.7.

Problems often arise when using heat-exchanger models to simulate processes that have a high degree of process-to-process heat exchange. Whenever a process-to-process heat exchanger is included in a simulation, it sets up an additional recycle

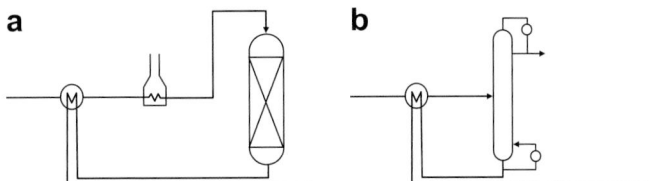

Figure 4.28.   Common feed heating schemes. (a) Feed–effluent exchange. (b) Feed–bottoms exchange.

of information; consequently, an additional loop must be converged. A common situation is where the effluent from a reactor or the bottoms from a distillation column is used to preheat the reactor or column feed, as illustrated in Figure 4.28. If these process flow schemes are simulated using heat exchangers, a recycle of energy is set up between the product and the feed. This recycle must be converged every time the flow-sheet is calculated (i.e., at every iteration of any other recycle loop in the process). If more than a few of these exchangers are present, then the overall flow-sheet convergence can become difficult.

Instead, it is usually a good practice to model the process using only heaters and coolers, and then set up sub-problems to model the heat exchangers. This facilitates data extraction for pinch analysis, makes it easier for the designer to recognize when exchangers might be internally pinched or have low F_t factors (see Chapter 12), and improves convergence.

Another problem that is often encountered when simulating heat exchangers and heat exchange networks is temperature cross. A temperature cross occurs when the cold stream outlet temperature is hotter than the hot stream outlet temperature (Section 12.6). When temperature cross occurs, many types of shell and tube heat exchanger give a very poor approximation of counter-current flow, and consequently have low F_t factors and require large surface areas. In some of the commercial simulation programs, the heat exchanger models will indicate if the F_t factor is low. If this is the case, then the designer should split the exchanger into several shells in series so that temperature cross is avoided. Some of the simulation programs allow the designer to plot profiles of temperature versus heat flow in the exchanger. These plots can be useful in identifying temperature crosses and internal pinches.

Example 4.5

A mixture of 100 kmol/h of 80 mol% benzene and 20 mol% ethylene at 40°C and 100 kPa is fed to a feed–effluent exchanger, where it is heated to 300°C and fed to a reactor. The reaction of ethylene and benzene to form ethylbenzene proceeds to 100% conversion of ethylene, and the reactor products are withdrawn, cooled by heat exchange with the feed, and sent to further processing. Estimate the outlet temperature of the product after heat exchange and the total surface area required if the average heat transfer coefficient is $200\,\mathrm{Wm^{-2}K^{-1}}$.

Solution

This problem was solved using UniSim Design. The reaction goes to full conversion, so a conversion reactor can be used. The simulation model is shown in Figure 4.29.

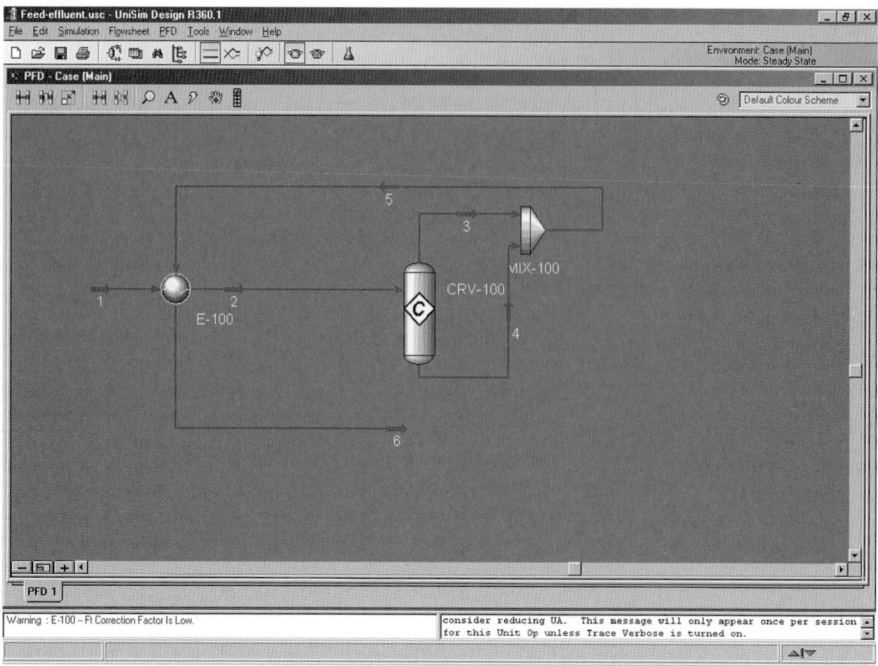

Figure 4.29.   Feed–effluent heat-exchange model for Example 4.5.

When the temperature at the outlet of the exchanger on the feed side is specified, the duty of the exchanger is defined and there is no recycle of information. The model thus solves very quickly, but it is necessary to check the results to see that the exchanger design makes sense.

The outlet temperature of the product (stream 6) is found to be 96.9°C, so there is enough heat in the product mixture to give an approach temperature of nearly 60°C, which seems perfectly adequate. If we open the exchanger worksheet though, there is a warning that the F_t factor is too low. Figure 4.30 shows the exchanger worksheet, and the F_t factor is only 0.2, which is not acceptable. When we examine the temperature–heat duty plot shown in Figure 4.31 (generated from the Performance tab of the exchanger worksheet), it is clear that there is a substantial temperature cross. This temperature cross causes the exchanger to have such a low F_t factor and gives a UA value of 78.3×10^3 WK^{-1}, where U is the overall heat transfer coefficient in Wm^{-2}K^{-1} and A is the area in m^2.

If $UA = 78.3 \times 10^3$ WK^{-1} and $U = 200$ Wm^{-2}K^{-1}, then the exchanger area is $A = 392$ m^2. This would be a feasible size of exchanger, but is large for the duty and is not acceptable because of the low F_t factor. We should add more shells in series.

By examining the temperature–heat duty plot in Figure 4.31, we can see that if we break the exchanger into two shells, with the first shell heating the feed up to the dew point (the kink in the lower curve), then the first shell will not have a temperature cross. This design corresponds to an outlet temperature of about 70°C for the first exchanger. The second exchanger would still have a tempera-ture cross though. If we break this second exchanger into two more exchangers,

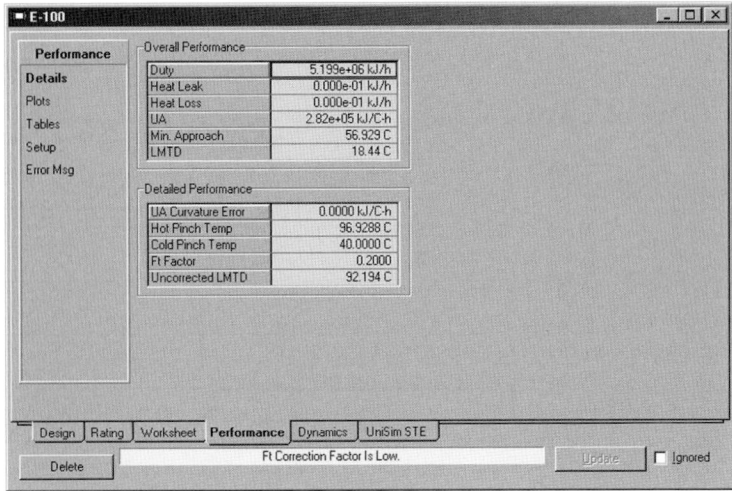

Figure 4.30. Exchanger worksheet for a single-shell design.

then the temperature cross is eliminated. We thus need at least three heat exchangers in series to avoid the temperature cross. This result could have been obtained by 'stepping off' between the temperature–duty plots, as illustrated in Figure 4.32.

Figure 4.33 shows a modified flow-sheet with two additional heat exchangers added in series. The outlet temperature of the second exchanger was specified as 200°C, to divide the duty of the second and third exchangers roughly equally. The

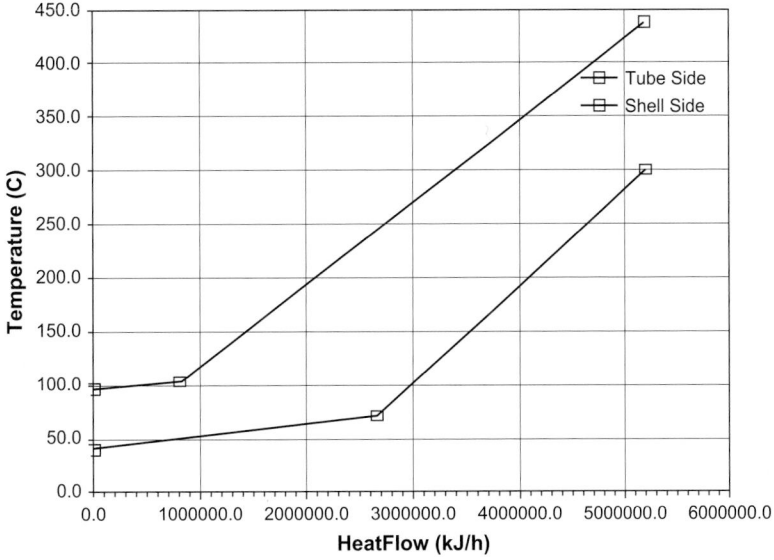

Figure 4.31. Temperature–heat flow plot for a single-shell design.

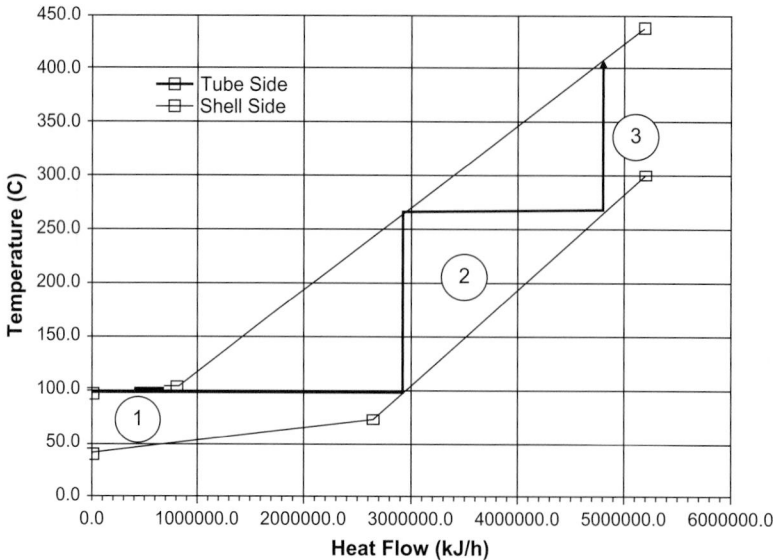

Figure 4.32.   Stepping between heat profiles to avoid temperature cross.

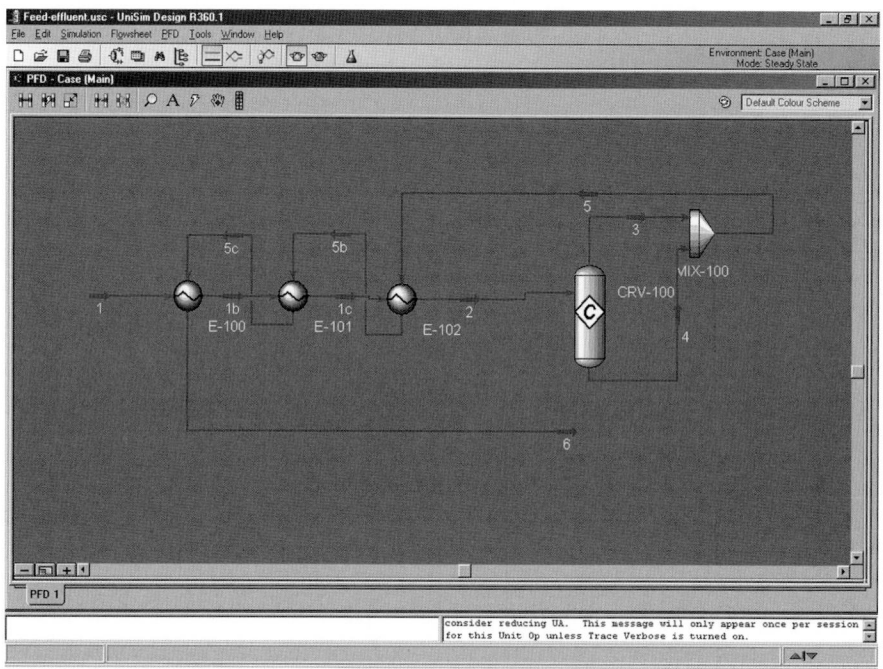

Figure 4.33.   Feed–effluent heat exchange with three shells in series.

TABLE 4.3. Heat-Exchanger Results

Design Case	Original (Single Shell)	Modified (Multiple Shell)		
Exchanger	E100	E100	E101	E102
Duty (MW)	1.44	0.53	0.57	0.35
UA (W/K)	78,300	6310	4780	2540
F	0.2	0.93	0.82	0.93
ΔT_{min}	56.9	56.9	134.3	139.7
ΔT_{lmtd}	18.4	83.6	118.7	138.4
A (m^2)	392	32	24	13
Total area (m^2)	392	68		

results are given in Table 4.3. Temperature–heat flow plots for the three exchangers are given in Figure 4.34.

The modified design achieved a reduction in surface area from 392 m^2 to 68 m^2 at the price of having three shells instead of the original one. More importantly, the modified design is more practical than the original design and is less likely to suffer from internal pinch points. The modified design is not yet optimized. Optimization of this problem is explored in problem 4.11.

4.5.5. Hydraulics

Most of the commercial simulation programs contain models for valves, pipe segments, tees and elbows. These models can be used to make an initial estimate of system pressure drop for the purposes of sizing pumps and compressors.

If a process hydraulic model is built, then care must be taken to specify pressure drop properly in the unit operation models. Rules of thumb are adequate for initial estimates, but in a hydraulic model these should be replaced with rigorous pressure drop calculations.

A hydraulic model will not be accurate unless some consideration has been given to plant layout and piping layout. Ideally, the hydraulic model should be built after the piping isometric drawings have been produced, when the designer has a good idea of pipe lengths and bends. The designer should also refer to the piping and instrumentation diagram for isolation valves, flow meters and other obstructions that cause increased pressure drop. These subjects are discussed in Chapter 5 and Chapter 14.

Care is needed when modelling compressible gas flows, flows of vapour–liquid mixtures, slurry flows and flows of non-Newtonian liquids. Some simulators use different pipe models for compressible flow. The prediction of pressure drop in multiphase flow is inexact at best and can be subject to very large errors if the extent of vaporization is unknown. In most of these cases, the simulation model should be replaced by a computational fluid dynamics (CFD) model of the important parts of the plant.

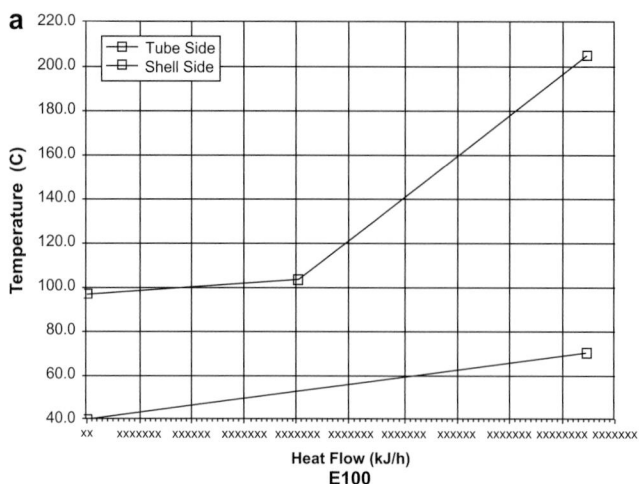

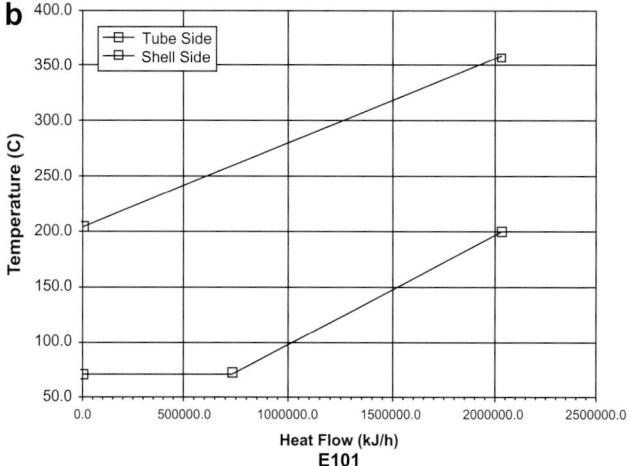

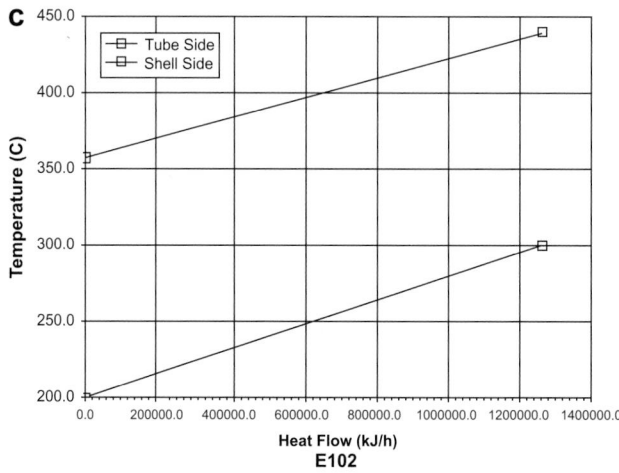

Figure 4.34.   Temperature–heat flow profiles for the three exchangers in series (a) E100, (b) E101, (c) E102.

4.5.6. Solids Handling

The commercial simulation programs were originally developed mainly for petro-chemical applications and none of them has a complete set of solids-handling operations. Although models for filters, crystallizers, decanters and cyclones are present in most of the simulators, the designer may have to add user models for operations such as:

- hoppers
- belt conveyors
- elevators
- pipe conveyors
- screw conveyors
- kneaders
- extruders
- slurry pumps
- fluidized bed heaters
- fluidized bed reactors

- washers
- flocculators
- spray driers
- prill towers
- rotary driers
- rotary kilns
- belt driers
- centrifuges
- falling film evaporators
- moving bed reactors

- crushers and pulverizers
- jet mills
- ball mills
- agglomerators
- granulators
- tableting presses
- paper machines
- classifiers
- electrostatic precipitators
- homogenizers

Because solids are handled in many commodity chemical processes as well as pharmaceuticals, polymers and biological processes, the simulation software vendors are under pressure from their customers to enhance the capability of the programs for modelling solids operations. This continues to be an area of evolution of the commercial software.

4.6. USER MODELS

When the design engineer needs to specify a unit operation that is not represented by a library model and cannot be approximated by a simple model such as a component splitter or a combination of library models, then it is necessary to construct a user model. All of the commercial simulators allow the user to build add-in models of varying sophistication.

4.6.1. Spreadsheet Models

Models that require no internal iteration are easily coded as spreadsheets. Most of the simulators offer some degree of spreadsheet capability, ranging from simple calculation blocks to full Microsoft Excel™ functionality.

In UniSim Design, spreadsheets can be created by selecting the spreadsheet option on the unit operations palette. The spreadsheet is easy to configure and allows data to be imported from streams and unit operations. The functionality of the UniSim Design spreadsheet is rather basic at the time of writing, but is usually adequate for simple input–output models. Values calculated by the spreadsheet can be exported back to the simulation model. The spreadsheet can thus be set up to act as a unit operation. The use of a spreadsheet as a unit operation is illustrated in Example 4.6. Aspen Plus has a similar simple spreadsheet capability using Microsoft Excel, which can be specified as a calculator block (via Data/Flowsheet Options/Calculator). The Excel calculator block in Aspen Plus requires a little more time to configure than the UniSim Design spreadsheet, but at the time of writing it can perform all of the functions available in MS Excel 97.

For more sophisticated spreadsheet models, Aspen Plus allows the user to link a spreadsheet to a simulation via a user model known as a USER2 block. The designer can create a new spreadsheet or customize an existing spreadsheet to interact with an Aspen Plus simulation. The USER2 block is much easier to manipulate when handling large amounts of input and output data, such as streams with many components or unit operations that involve multiple streams. The procedure for setting up a USER2 MS Excel model is more complex than using a calculator block, but avoids having to identify every number required from the flow-sheet individually. Instructions on how to build USER2 spreadsheet models are given in the Aspen Plus manuals and on-line help (Aspen Technology, 2001).

4.6.2.   User Sub-Routines

Models that require internal convergence are best written as sub-routines rather than spreadsheets, as more efficient solution algorithms can be used. Most user sub-routines are written in FORTRAN or Visual Basic, though some of the simulators allow other programming languages to be used.

It is generally a good practice to compile and test a user model in a simplified flow-sheet or as a stand-alone program before adding it to a complex flow-sheet with recycles. It is also a good practice to check the model carefully over a wide range of input values, or else constrain the inputs to ranges where the model is valid.

Detailed instructions on how to write user models to interface with commercial simulation programs can be found in the simulator manuals. The manuals also contain specific requirements for how the models should be compiled and registered as extensions or shared libraries (.dll files in Microsoft Windows). In Aspen Plus, user models can be added as USER or USER2 blocks, following the instructions in the Aspen Plus manuals. In UniSim Design, it is very easy to add user models using the User Unit Operation, which can be found on the object palette or under the Flowsheet/Add Operation menu. The UniSim Design User Unit Operation can be linked to any program without requiring an extension file to be registered. The User Unit Operation is not documented in the UniSim Design manual, but instructions on setting it up and adding code are given in the on-line help.

Example 4.6

A gas turbine engine is fuelled with 3000 kg/h of methane at 15°C and 1000 kPa, and supplied with ambient air at 15°C. The air and fuel are compressed to 2900 kPa and fed to a combustor. The air flow rate is designed to give a temperature of 1400°C at the outlet of the combustor. The hot gas leaving the combustor is expanded in the turbine. Shaft work produced by the turbine is used to power the two compressors and run a dynamo for generating electricity.

If the efficiency of the compressors is 98% and that of the turbine is 88% and 1% of the shaft work is lost due to friction and losses in the dynamo, estimate the rate of power production and the overall cycle efficiency.

Solution

This problem was solved using UniSim Design.

A gas turbine engine should run with a large excess of air to provide full combustion of the fuel, so the combustor can be modelled as a conversion reactor. There is no model for a dynamo in UniSim Design, so the dynamo and shaft losses can be modelled using a spreadsheet operation, as shown in Figure 4.35.

Figure 4.35 also illustrates the use of an 'Adjust' controller to set the air flow rate so as to give the desired reactor outlet temperature. The specifications for the Adjust are shown in Figures 4.36 and 4.37. The Adjust was specified with a minimum air flow rate of 60,000 kg/h to ensure that the solver did not converge to a solution in which the air flow did not give full conversion of methane. The stoichiometric requirement is $3000 \times 2 \times (32/16) /0.21 = 57,000$ kg/h of air.

The spreadsheet model of the dynamo is relatively simple, as illustrated in Figure 4.38. The model takes the turbine shaft work and compressor duties as inputs. The friction losses are estimated as 1% of the turbine shaft work. The friction losses and compressor duties are then subtracted from the shaft work to give the net power from the dynamo, which is calculated to be 17.7 MW.

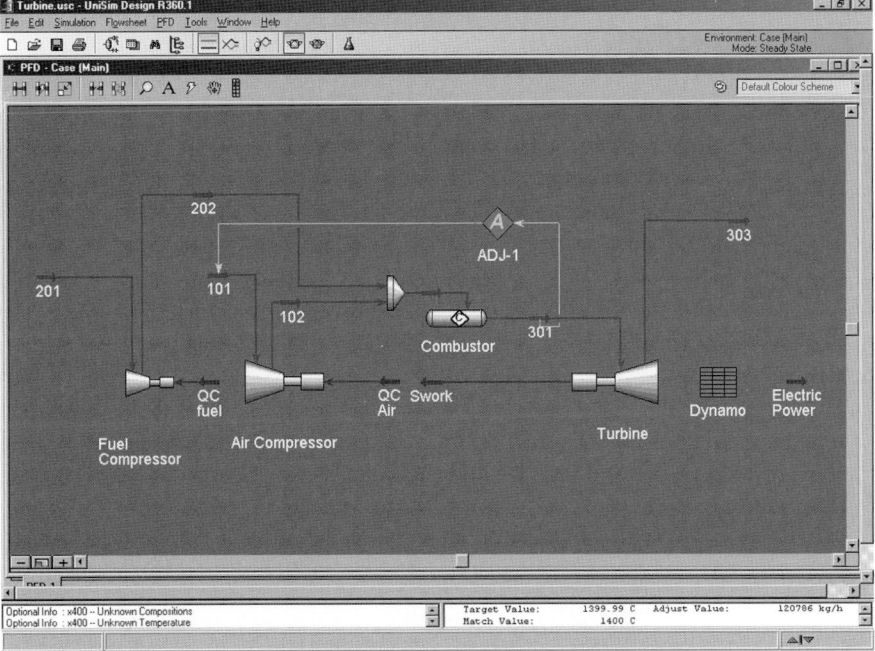

Figure 4.35.   Gas turbine model.

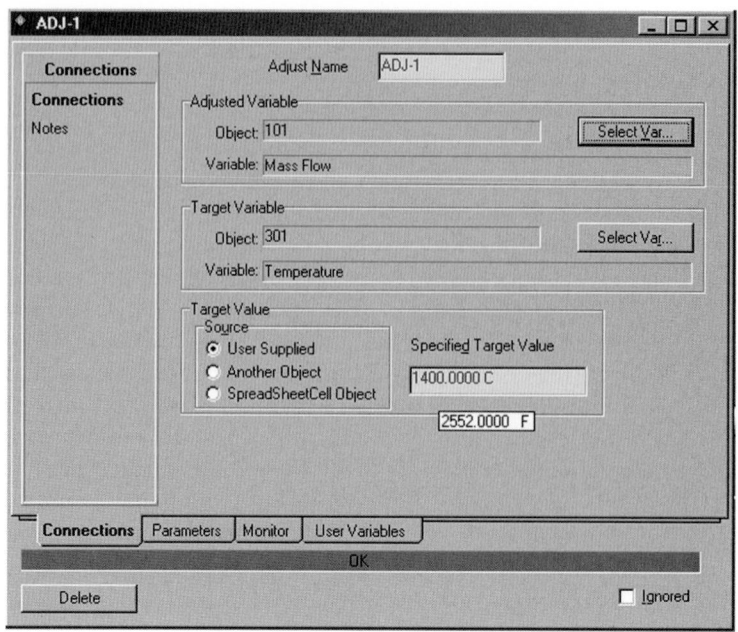

Figure 4.36.   Adjust specifications.

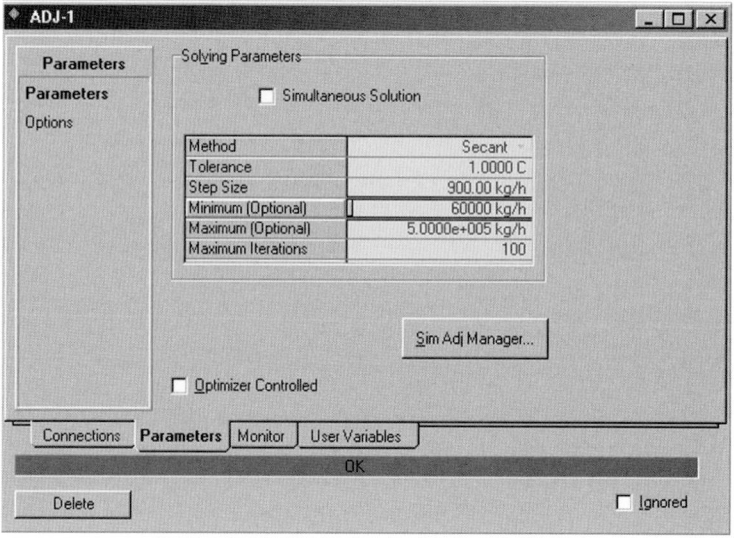

Figure 4.37.   Adjust solving parameters.

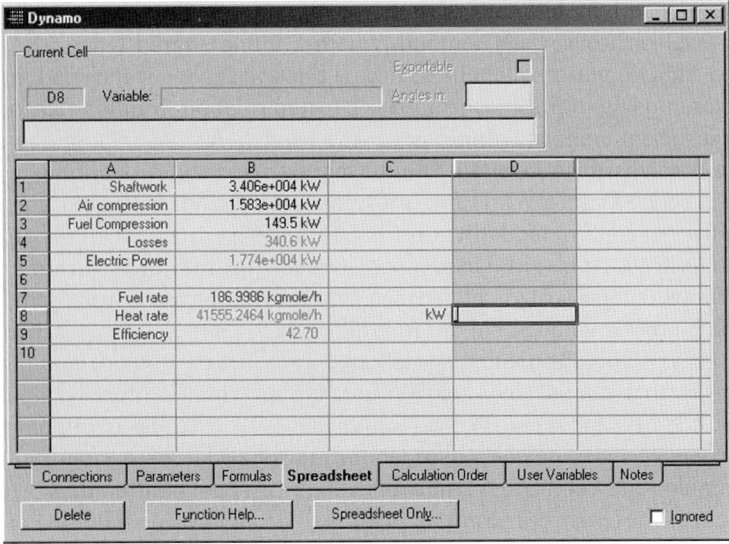

Figure 4.38. Spreadsheet model of dynamo.

The cycle efficiency is the net power produced divided by the heating rate of the fuel. The heating rate is the molar flow of fuel multiplied by the standard molar heat of combustion:

$$\text{Heating rate (kW)} = \text{molar flow (mol/h)} \times \Delta H_c^\circ \text{(kJ/mol)}/3600 \qquad (4.1)$$

The cycle efficiency is calculated to be 42.7%.

4.7. FLOW-SHEETS WITH RECYCLE

Recycles of solvents, catalysts, unconverted feed materials and by-products are found in many processes. Most processes contain at least one material recycle, and some may have six or more. Furthermore, when energy is recovered by process-to-process heat transfer then energy recycles are created, as discussed in Section 4.5.4.

4.7.1. Tearing the Flow-Sheet

For a sequential-modular simulation program to be able to solve a flow-sheet with a recycle, the design engineer needs to provide an initial estimate of a stream somewhere in the recycle loop. This is known as a 'tear' stream, as the loop is 'torn' at that point. The program can then solve and update the tear stream values with a new estimate. The procedure is repeated until the difference between values at each iteration becomes less than a specified tolerance, at which point the flow-sheet is said to be converged to a solution.

The procedure for tearing and solving a simulation can be illustrated by a simple example. Figure 4.39 shows a process in which two feeds, A and B, are combined and

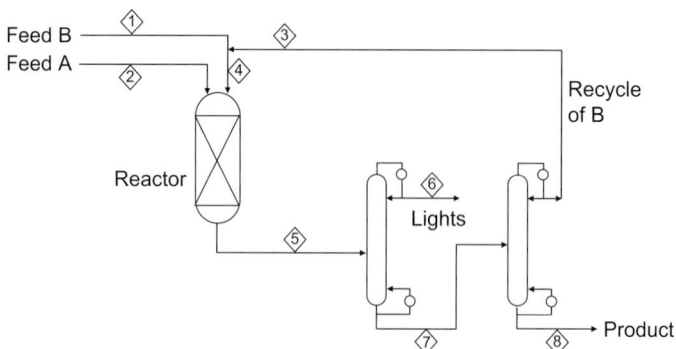

Figure 4.39.   Sample process with recycle.

fed to a fixed bed reactor. The reactor product is sent to a stripping column to remove light ends and is then sent to a column that separates heavy product from unreacted feed B. The unreacted feed B is recycled to the reactor.

To solve the reactor model, we need to specify the reactor feeds, streams 2 and 4. Stream 4 is made by adding fresh feed stream 1 to recycle stream 3, so a logical first approach might be to make an estimate of the recycle stream, in which case stream 3 is the tear stream. Figure 4.40 shows the flow-sheet torn at stream 3. The designer provides an initial estimate of stream 3a. The flow-sheet then solves and calculates stream 3b. The design engineer specifies a recycle operation connecting streams 3a and 3b, and the simulator then updates stream 3a with the values from stream 3b (or with other values if an accelerated convergence method is used, as discussed below). The calculation is then repeated until the convergence criteria are met.

The choice of tear stream can have a significant impact on the rate of convergence. For example, if the process of Figure 4.39 were modelled with a yield shift reactor, then tearing the flowsheet at stream 5 would probably give faster convergence. Some of the simulation programs automatically identify the best tear stream.

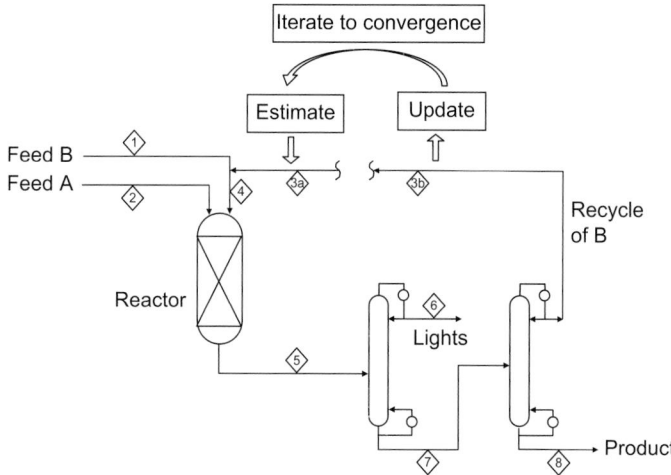

Figure 4.40.   Tearing the recycle loop.

4.7.2. Convergence Methods

The methods used to converge recycle loops in the commercial process simulation programs are similar to the methods described in Section 1.9. Most of the commercial simulation programs include the methods described below.

Successive Substitution (Direct Substitution)

In this method, an initial estimate, x_k, is used to calculate a new value of the parameter, $f(x_k)$. The estimate is then updated using the calculated value:

$$x_{k+1} = f(x_k)$$
$$x_{k+2} = f(x_{k+1}), \text{ etc.}$$
(4.2)

This method is simple to code, but is computationally inefficient and convergence is not guaranteed.

Bounded Wegstein

The bounded Wegstein method is the default method in most of the simulation programs. It is a linear extrapolation of successive substitution.

The Wegstein method initially starts out with a direct substitution step:

$$x_1 = f(x_0)$$
(4.3)

An acceleration parameter, q, can then be calculated:

$$q = \frac{s}{s-1}$$
(4.4)

where

$$s = \frac{f(x_k) - f(x_{k-1})}{x_k - x_{k-1}}$$
(4.5)

and the next iteration is then:

$$x_{k+1} = q\, x_k + (1-q)f(x_k)$$
(4.6)

If $q = 0$, the method is the same as successive substitution. If $0 < q < 1$, then convergence is damped, and the closer q is to 1.0, the slower convergence becomes. If q is less than 0, then the convergence is accelerated. The bounded Wegstein method sets bounds on q, usually keeping it in the range $-5 < q < 0$, so as to guarantee acceleration without overshooting the solution too widely.

The bounded Wegstein method is usually fast and robust. If convergence is slow, then the designer should consider reducing the bounds on q. If convergence oscillates, then consider damping the convergence by setting bounds such that $0 < q < 1$.

Newton and Quasi-Newton Methods

The Newton method uses an estimate of the gradient at each step to calculate the next iteration, as described in Section 1.9.6. Quasi-Newton methods such as Broyden's method use linearized secants rather than gradients. This approach reduces the number of calculations per iteration, although the number of iterations may be increased.

Newton and quasi-Newton methods are used for more difficult convergence problems, for example, when there are many recycle streams, or many recycles that include operations that must be converged at each iteration, such as distillation columns. The Newton and quasi-Newton methods are also often used when there are many recycles and control blocks (see Section 4.8.1). The Newton method should not normally be used unless the other methods have failed, as it is more computationally intensive and can be slower to converge for simple problems.

4.7.3. Manual Calculations

The convergence of recycle calculations is almost always better if a good initial estimate of the tear stream is provided.

If the tear stream is chosen carefully, then it may be easy for the design engineer to generate a good initial estimate. This can be illustrated by returning to the problem of Figure 4.39. We can tear the recycle loop at the reactor effluent, as shown in Figure 4.41. We can then state the following about the reactor effluent:

1. The reactor effluent must contain the net production rate of product (which is known), plus any product that is in the recycle. Recycling product to the reactor is not a good idea, as it is likely to lead to by-product formation. A reasonable estimate of product recovery in the separation section is probably 99% or greater, so a good initial estimate of the amount of product in stream 5b is the net production rate divided by the separation recovery, or roughly 101% of the net production rate.

2. Since feed B is recycled and feed A is not, it looks like we are using an excess of B to drive full conversion of A. A good initial estimate of the flow rate of component A in stream 5b is therefore zero. If we have conversion data in terms of A, then we could produce a better estimate.

3. Feed B is supplied to the reactor in excess. The amount of B consumed in the reactor must be equal to the amount required by stoichiometry to

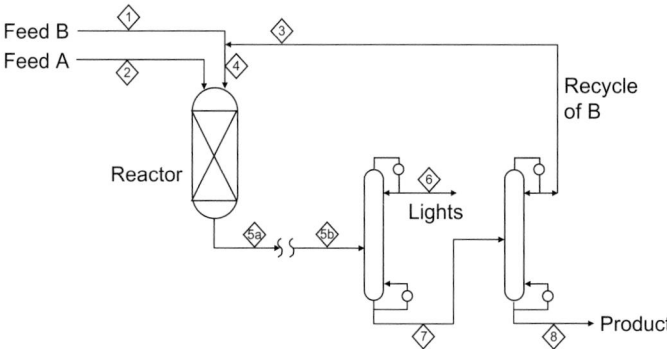

Figure 4.41.   Tearing at the reactor outlet.

produce the product. The amount of B remaining in the reactor effluent is given by:

$$\frac{\text{moles B remaining}}{\text{per mole product}} = \frac{\text{moles B fed}}{\text{stoichiometric moles B per mole product}} - 1$$

$$= \frac{1}{\text{conversion of B}} - 1 \qquad (4.7)$$

So, knowing the flow rate of product, we can get a good initial estimate of the flow rate of B if we know either the conversion of B or the ratio in excess of the stoichiometric feed rate of B that we want to supply.

We can thus make good estimates of the three major components that are present in stream 5b. If light or heavy by-products are formed in the reactor but not recycled, then a single successive substitution step will provide good estimates for these components, as well as a better estimate of the conversion of B and the amount of A that is required in excess of stoichiometric requirements.

Manual calculations are also very useful when solving flow-sheets that use recycle and purge. Purge streams are often withdrawn from recycles to prevent the accumulation of species that are difficult to separate, as described in Section 2.15. A typical recycle and purge flow scheme is illustrated in Figure 4.42. A liquid feed and a gas are mixed, heated, reacted, cooled and separated to give a liquid product. Unreacted gas from the separator is recycled to the feed. A make-up stream is added to the gas recycle to make up for consumption of gas in the process. If the make-up gas contains any inert gases, then over time these would accumulate in the recycle and eventually the reaction would be slowed down when the partial pressure of reactant gas fell. To prevent this situation from occurring, we withdraw a purge stream to maintain the inerts at an acceptable level. We can provide a good initial estimate of the recycle stream by noting:

1. The flow rate of inerts in the purge is equal to the flow rate of inerts in the make-up gas.
2. The required partial pressure of reactant gas at the reactor outlet sets the concentration of reactant gas and inerts in the recycle and the unconverted gas flow rate if the reactor pressure is specified.

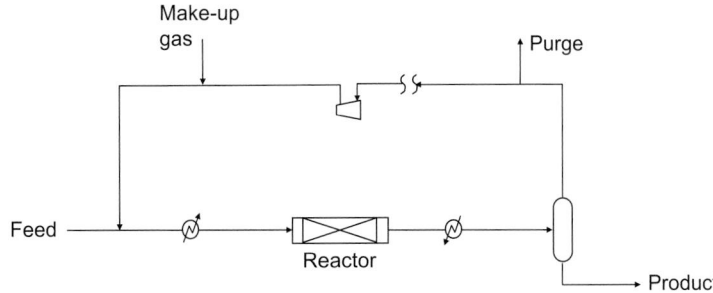

Figure 4.42. Process with gas recycle and purge.

We can then write a mass balance on inerts:

$$My_M = Py_R \tag{4.8}$$

and on reactant gas:

$$M(1 - y_M) = G + P(1 - y_R) \tag{4.9}$$

Hence

$$M(1 - y_M) = G + M\frac{y_M}{y_R}(1 - y_R)$$

where

M = make-up molar flow rate
P = purge molar flow rate
y_M = mole fraction of inerts in make-up
y_R = mole fraction of inerts in recycle and purge
G = molar rate of consumption of gas in reactor.

Hence we can solve for M and P if G is known.

The temperature of the recycle gas at the outlet of the compressor is not easily estimated, so the logical place to tear the recycle is between the purge and the compressor, as indicated in Figure 4.42.

4.7.4. Convergence Problems

If a flow-sheet is not converged, or if the process simulation software runs and gives a statement 'converged with errors', then the results *cannot be used* for design. The designer must take steps to improve the simulation so that a converged solution can be found.

The first steps that an experienced designer would usually take would be to:

1. Make sure that the specifications are feasible.
2. Try increasing the number of iterations.
3. Try a different convergence algorithm.
4. Try to find a better initial estimate.
5. Try a different tear stream.

If one or more unit operation has been given infeasible specifications, then the flow-sheet will never converge. This problem also occurs with multicomponent distillation columns, particularly when purity specifications or flow rate specifications are used, or when non-adjacent key components are chosen. A quick manual mass balance around the column can usually determine whether the specifications are feasible. Remember that all the components in the feed must exit the column somewhere. The use of recovery specifications is usually more robust, but care is still needed to make sure that the reflux ratio and number of trays are greater than the minimum required. A similar problem is encountered in recycle loops if a component accumulates because of the separation specifications that have been set. Adding a purge stream usually solves this problem.

For large problems with multiple recycles, it may be necessary to increase the number of iterations to allow the flow sheet time to converge. This strategy can be effective, but is obviously inefficient if underlying problems in the model are causing the poor convergence.

In some cases, it may be worthwhile to develop a simplified simulation model to arrive at a first estimate of tear stream composition, flow rate and conditions (temperature and pressure). Models can be simplified by using faster and more robust unit operation models, for example, substituting shortcut column models for rigorous distillation models. Models can also be simplified by reducing the number of components in the model. Reducing the number of components often leads to a good estimate of the bulk flows and stream enthalpies, which can be useful if there are interactions between the mass and energy balances. Another simplification strategy that is often used is to model heat exchangers using a dummy stream on one side (usually the side that is downstream in the process). The recycle of energy from downstream to upstream is then not converged until after the rest of the flow sheet has been converged. Alternatively, heaters and coolers can be used in a simplified model, or even in the rigorous model, as long as the stream data is then extracted and used to design the real exchangers.

Another approach that is widely used is to 'creep up on' the converged solution. This entails building up the model starting from a simplified version and successively adding detail while re-converging at each step. As more complexity is added, the values from the previous run are used to initialize the next version. This is a slow, but effective, method. The design engineer must remember to save the intermediate versions every so often, in case later problems are encountered. A similar strategy is often used when running sensitivity analyses or case studies that require perturbations of a converged model. The designer changes the relevant parameters in small steps to reach the new conditions, while re-converging at each step. The results of each step then provide a good initial estimate for the next step and convergence problems are avoided.

When there are multiple recycles present, it is sometimes more effective to solve the model in a simultaneous (equation-oriented) mode rather than in a sequential-modular mode. If the simulation problem allows simultaneous solution of the equation set, this can be attempted. If the process is known to contain many recycles, then the designer should anticipate convergence problems and should select a process simulation program that can be run in a simultaneous mode.

Example 4.7

Light naphtha is a mixture produced by distillation of crude oil. Light naphtha primarily contains alkane compounds (paraffins) and it can be blended into gasoline. The octane value of methyl-substituted alkanes (iso-paraffins) is higher than that of straight-chain compounds (normal paraffins), so it is often advantageous to isomerize the light naphtha to increase the proportion of branched compounds.

A simple naphtha isomerization process has a feed of 10,000 barrels per day (bpd) of a 50 wt% mixture of n-hexane and methyl pentane. The feed is heated and sent to a reactor where it is brought to equilibrium at 1300 kPa and 250°C. The reactor products are cooled to the dew point and fed to a distillation column operated at

300 kPa. The bottoms product of the distillation is rich in n-hexane and is recycled to the reactor feed. An overall conversion of n-hexane of 95% is achieved.

Simulate the process to determine the recycle flow rate and composition.

Solution

This problem was solved using UniSim Design. The first step is to convert the volumetric flow rate into a mass flow rate in metric units. We can set up a stream that has a 50:50 mixture by weight of n-hexane and methyl pentane. This stream has a density of 641 kg/m^3 at 40°C, so the required flow rate is:

$$10{,}000 \text{ bpd} = 10{,}000 \times 641 \text{ (kg/m}^3) \times 0.1596 \text{ (m}^3/\text{bbl})/24 = 42.627 \text{ t/h}$$

In a real isomerization process, a part of the feed will be lost due to cracking reactions; however, in our simplified model the only reactions that occur are isomerization reactions. Because we only consider isomerization reactions, all of the product and feed components have the same molecular weight (C_6H_{14}, $M_w = 86$). The feed flow rate of n-hexane is thus $42.627 \times 0.5 = 21.31$ t/h. So for 95% conversion of n-hexane, the amount of n-hexane in the product is $0.05 \times 21.31 = 1.0655$ t/h, or $1065.5/86 = 12.39$ kmol/h. The mole fraction of n-hexane in the product is 5% of 50%, or 2.5 mol%.

To get an initial estimate of the distillation column conditions, the process was first simulated using a shortcut column model, as shown in Figure 4.43. If we assume that no cyclic compounds are formed in the process, then the component list includes all

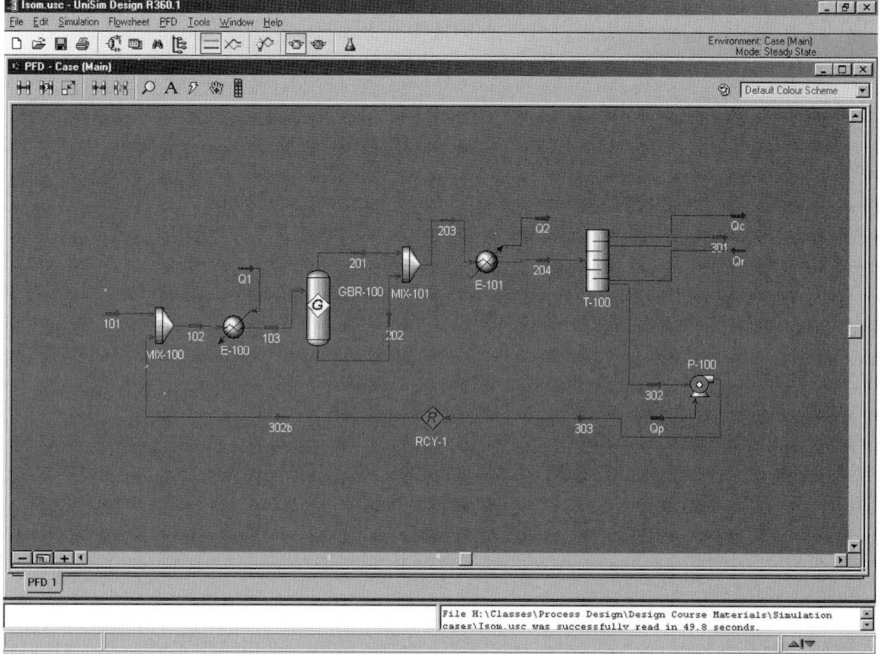

Figure 4.43.   Isomerization process model using shortcut distillation.

of the available C_6 paraffin compounds; i.e., n-hexane, 2-methyl pentane, 3-methyl pentane, 2,3-methyl butane and 2,2-methyl butane. The reactor achieves complete equilibrium between these species and so can be modelled using a Gibbs reactor.

The shortcut column model requires a second specification, given in terms of the heavy key component. We can define either of the methyl pentane species as the heavy key. In the simplified model that we have built, the level of methyl pentane in the recycle is not important to the process performance. Increasing the recycle of methyl pentane species increases the process yield of dimethyl butane species, which would lead to an improvement in the product octane number. In reality, the presence of side reactions that caused cracking to less valuable light hydrocarbons would establish a trade-off that would set the optimum level of methyl pentane recycle. For now, we will assume that the mole fraction of 2-methyl pentane in the bottoms is 0.2.

With these conditions, and with the recycle not closed, the shortcut column model predicts a minimum reflux of 3.75. The reflux ratio is then set at $1.15 \times R_{min} = 4.31$, as shown in Figure 4.44. The shortcut model then calculates that we need 41 theoretical trays, with optimal feed tray 26, as shown in Figure 4.45. The column bottoms flow rate is 18,900 kg/h, which can be used as an initial estimate for the recycle flow. The recycle loop can now be closed and run. The converged solution still has $R_{min} = 3.75$, so the reflux ratio does not need to be adjusted. The converged recycle flow rate is 18.85 t/h or 218.7 kmol/h, as shown in Figure 4.46. The shortcut column design of the converged flowsheet still has 41 trays with the feed on tray 26.

The results from the shortcut model can now be used to provide a good initial estimate for a rigorous model. The shortcut column is replaced with a rigorous column, as shown in Figure 4.47. The rigorous column model can be set up with the number of stages and feed stage predicted by the shortcut model, Figure 4.48. If we specify the reflux ratio and bottoms product rate as column specifications, as in Figure 4.49, then the flow-sheet converges quickly.

The results from the rigorous model with the inputs specified as above show a flow rate of 1084.5 kg/h of n-hexane in the distillate product. This exceeds the requirements calculated from the problem statement (1065.5 kg/h). The simplest way to get

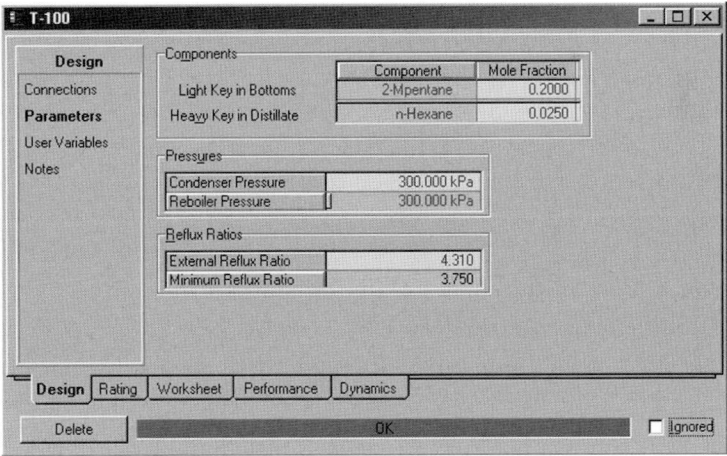

Figure 4.44. Shortcut column specifications.

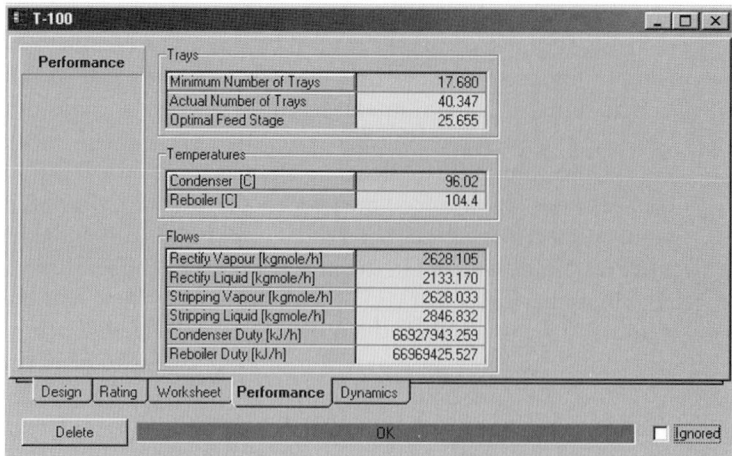

Figure 4.45.   Shortcut column results.

back to the required specification is to use it directly as a specification for the column. From the 'Design' tab on the column window we can select 'Monitor' and then 'Add spec' to add a specification on the distillate flow rate of n-hexane, as shown in Figure 4.50. This specification can then be made active and the bottoms flow rate specification can be relaxed. When the simulation is re-converged, the bottoms flow rate increases to 19,350 kg/h and the n-hexane in the distillate meets the specification flow rate of 1065.5 kg/h.

The column profiles for the rigorous distillation model are shown in Figure 4.51. The profiles do not show any obvious poor design of the column, although the design is not yet optimized.

The simulation was converged to achieve the target conversion of n-hexane with a recycle of 19.35 t/h. The recycle composition is 50.0 mol% n-hexane, 21.1 mol% 2-methyl pentane, 25.1 mol% 3-methyl pentane, 3.6 mol% 2,3-methyl butane and 0.2 mol% 2,2-methyl butane. This is a converged solution, but it is only one of many

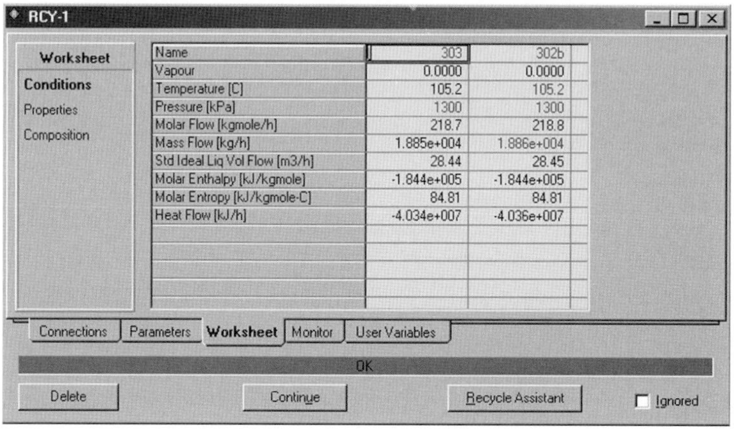

Name	303	302b
Vapour	0.0000	0.0000
Temperature [C]	105.2	105.2
Pressure [kPa]	1300	1300
Molar Flow [kgmole/h]	218.7	218.8
Mass Flow [kg/h]	1.885e+004	1.886e+004
Std Ideal Liq Vol Flow [m3/h]	28.44	28.45
Molar Enthalpy [kJ/kgmole]	-1.844e+005	-1.844e+005
Molar Entropy [kJ/kgmole-C]	84.81	84.81
Heat Flow [kJ/h]	-4.034e+007	-4.036e+007

Figure 4.46.   Converged recycle results for the shortcut column model.

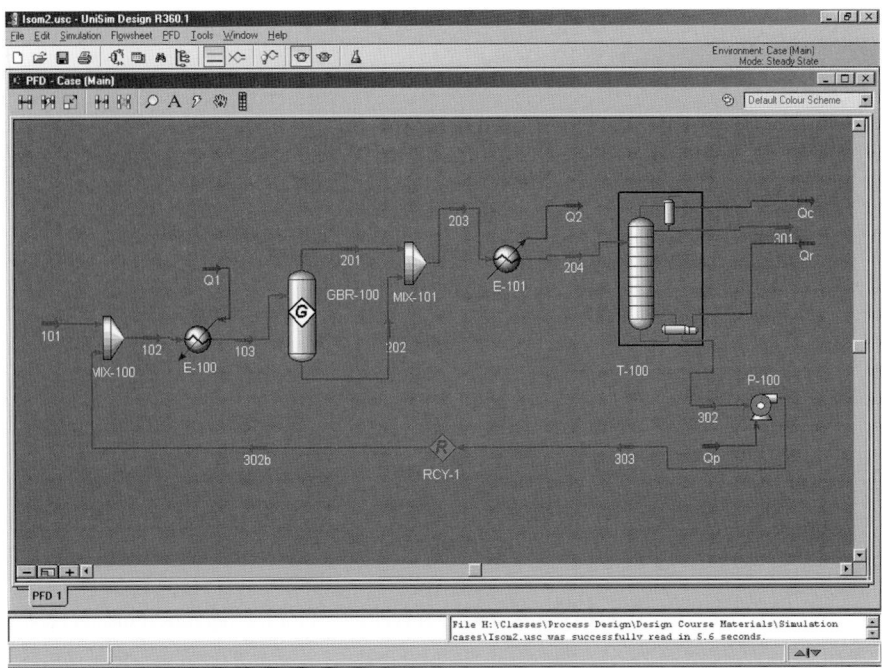

Figure 4.47. Isomerization process model using rigorous distillation.

possible converged solutions. No attempt has yet been made to optimize the design. The optimization of this process is examined in problem 4.14. For more realistic information on isomerization process conditions, the reader should consult Meyers (2003).

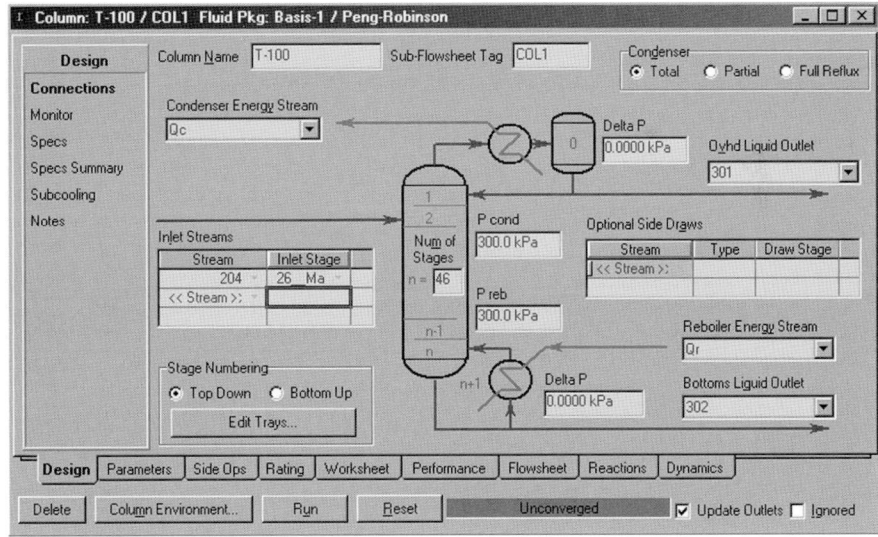

Figure 4.48. Design parameters for the rigorous distillation column.

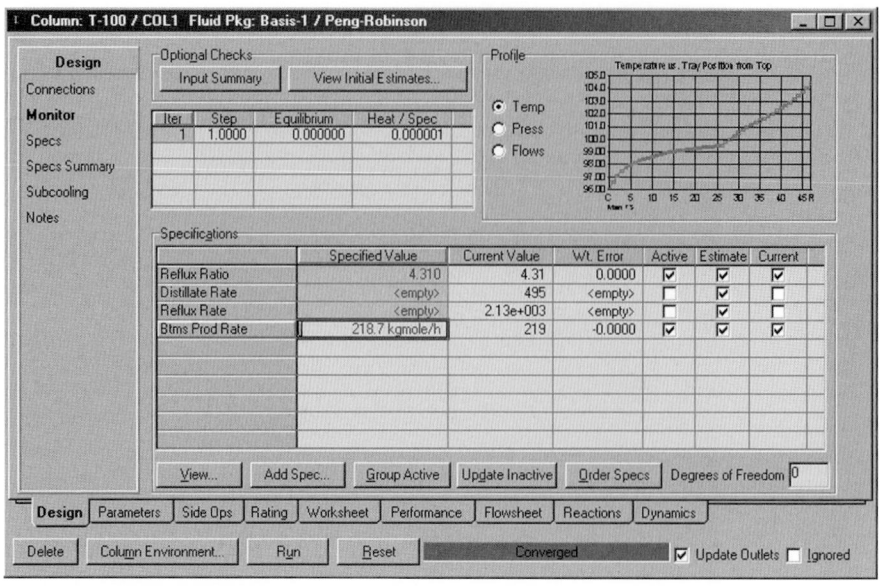

Figure 4.49.   Specifications for the rigorous distillation column.

4.8. FLOW-SHEET OPTIMIZATION

After achieving a converged simulation of the process, the designer will usually want to carry out some degree of optimization. The commercial simulation programs have a limited optimization capability that can be used with suitable caution.

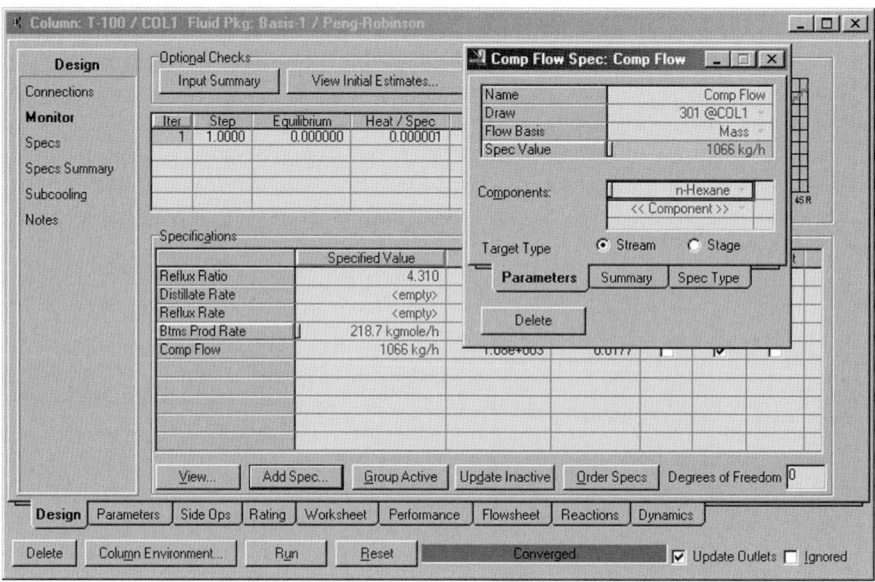

Figure 4.50.   Adding a specification on n-hexane mass flow.

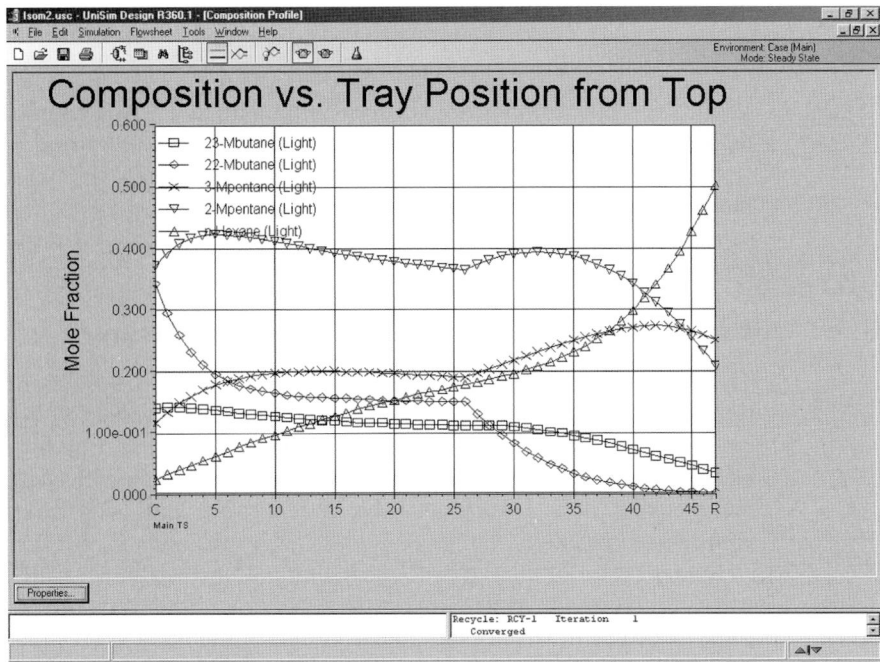

Figure 4.51. Column profiles for the rigorous distillation model.

4.8.1. Use of Controllers

The simplest form of optimization is to impose additional constraints on the simulation so that it meets requirements specified by the designer. For example, if the designer made estimates of the feed rates, then the production rate of product that is predicted by the model may be less (or more) than the desired rate. The designer could correct this by calculating the appropriate ratio, multiplying all the feed streams by this ratio, and then re-converging the model, but this approach would soon become tedious.

Instead, the simulation programs allow the designer to impose constraints on the model. In the example above, this would be a constraint that the product flow rate is equal to a target value. Constraints are imposed using controller functions, known as a 'Design Spec' in Aspen Plus or a 'Set' or 'Adjust' in UniSim Design. Controllers are specified either as:

Set variable x to value z

or

Adjust variable x to value y by manipulating variable z

where z is an unknown variable or set of variables that will be calculated by the simulation and x is the variable that the designer wants to specify.

Controllers can be used to capture all kinds of design constraints and specifications. They are particularly useful for setting feed ratios and controlling purge rates and recycle ratios to achieve target compositions. Some care is needed to ensure that they are used sparingly, otherwise too many recycles of information can be introduced and convergence becomes difficult.

Controllers behave much like recycles, and it is usually a good idea to generate a converged simulation to act as a good initial estimate before adding controllers. This does not apply to simple controller functions such as feed ratio controllers.

In a dynamic simulation, controllers are used to model the real control valves of the process. When converting a steady-state simulation to a dynamic simulation, some care is needed to ensure that the controller functions correspond to physically achievable control structures.

4.8.2.   Optimization Using Process Simulation Software

The commercial process simulation programs all have the ability to solve optimization problems that can be posed as non-linear programming (NLP) problems. At the time of writing, only Aspen Plus allows the designer to carry out discreet optimization using integer variables. It is, therefore, not always possible to optimize integer parameters such as number of trays or feed tray, while simultaneously optimizing continuous variables.

Optimization of a large process simulation model is intrinsically difficult, particularly if there are multiple recycles. As noted in Section 1.9.9, the solution algorithms for NLP problems require multiple solutions of the model, which must be converged at each solution.

An additional complication of flow-sheet optimization is the formulation of the objective function. The objective function for industrial design is always a measure of economic performance. The design parameters calculated by the simulation program can be used to give relatively good estimates of equipment cost, but this typically requires exporting the parameters into a specialized cost-estimating program, such as Aspen ICARUS, as described in Section 1.8. Furthermore, the equipment must usually be oversized by a suitable design factor compared to the design flow rates, as discussed in Section 1.7. The simplest way to address this problem is to generate two or three simulation runs with variations of the key design parameters. These designs can then be costed to develop approximate cost curves, which can then be used in the optimization tool of the simulation program.

The Aspen Plus manual provides several useful recommendations for specifying optimization problems (Aspen Technology, 2001):

1. Start by converging a simulation of the flow-sheet. This helps the designer detect errors, ensures that specifications are feasible and provides good estimates for tear streams.
2. Carry out a sensitivity analysis to determine which variables have the most impact on the objective function. These are the variables that should be used as decision variables. It is also important to determine reasonable ranges for these variables and set upper and lower bound constraints. If the ranges set are too narrow, then the optimum may not be found. If they are too wide, then convergence may be difficult.
3. While carrying out the sensitivity analysis, see if the optimum is broad or sharp. If there are only small changes in the objective function, then further optimization may not be justified.

Another approach that is often used is to carry out optimization using simplified models to fix the process structure and determine the approximate values of key decision variables. A final NLP optimization can then be carried out using a rigorous model.

4.9. DYNAMIC SIMULATION

Most continuous processes are only simulated in steady-state mode. Some of the simulation programs allow a steady-state simulation to be converted to run in a dynamic mode. Dynamic simulation is useful for:

1. Simulating batch and semi-continuous processes to determine rate-controlling steps and investigate batch-to-batch recycles and heat recovery.
2. Simulating process start-up and shut-down.
3. Simulating cyclic processes.
4. Simulating process disturbances to evaluate control system performance and tune controllers.
5. Simulating emergency conditions to evaluate alarm system and relief system responses and ensure that they are adequate.
6. Developing operator training programs.

For a good dynamic simulation, the designer must specify the actual control system from the piping and instrumentation diagram (see Chapter 5) and also all of the vessel designs so that hold-ups can be calculated. Mass transfer rates and reaction rates must also be known or assumed.

Dynamic simulation is more computationally intensive than steady-state simulation. Dynamic simulation is usually applied to parts of a process (or even single-unit operations) rather than an entire process. Different simulation strategies are needed to give a robust dynamic model. Good introductions to dynamic simulation are given in the books by Luyben (2006), Ingham *et al.* (2007), Seborg *et al.* (2003) and Asprey and Machietto (2003), and the paper by Pantelides (1988).

4.10 REFERENCES

Aspen Technology (2001) Aspen Plus® 11.1 User Guide (Aspen Technology Inc.).

Asprey, S. P. and Machietto, S. (2003) *Dynamic Model Development: Methods, Theory and Applications* (Elsevier).

Benedek, P. (ed.) (1980) *Steady-state Flow-sheeting of Chemical Plants* (Elsevier).

Husain, A. (1986) *Chemical Process Simulation* (Wiley).

Ingham, J., Dunn, I. J., Heinzle, E., Prenosil, J. E., and Snape, J. B. (2007) *Chemical Engineering Dynamics*, 3rd. edn (Wiley-VCH).

Leesley, M. E. (ed.) (1982) *Computer Aided Process Plant Design* (Gulf).

Luyben, W. L. (2006) *Distillation Design and Control Using Aspen™ Simulation* (Wiley).

Meyers, R. A. (2003) *Handbook of Petroleum Refining Processes*, 3rd edn (McGraw-Hill),

Pantelides, C. C. (1988) *Comp. and Chem. Eng.*, **12**, 745. SpeedUp—recent advances in process engineering.

Preece, P. E., Kift, M. H. and Grills, D. M. (1991) *Computer-Orientated Process Design*, Proceedings of COPE, Barcelona, Spain, Oct. 14–16, 209, A graphical user interface for computer aided process design.

Seborg, D. E., Edgar, T. F., and Mellichamp, D. A. (2003) *Process Dynamics and Control* (Prentice Hall).

Wells, G. L. and Rose, L. M. (1986) *The Art of Chemical Process Design* (Elsevier).

Westerberg, A. W., Hutchinson, H. P., Motard, R. L., and Winter, P. (1979) *Process Flow-sheeting* (Cambridge U.P.).

British Standards

BS 1553-1 (1977) Specification for graphical symbols for general engineering. Part 1: Piping systems and plant (British Standards Institute).

American and International Standards

ASTM D 86 (2007) Standard test method for distillation of petroleum products at atmospheric pressure (ASTM International).

ASTM D 2887 (2006) Standard test method for boiling range distribution of petroleum fractions by gas chromatography (ASTM International).

ISO 10628 (1997) Flow diagrams for process plants – General rules, 1st edn (International Organization for Standardization).

4.11. NOMENCLATURE

		Dimensions in **MLT**
A	Heat-exchanger area	L^2
F_t	Shell and tube exchanger factor (for non-counter-current flow)	—
G	Molar rate of consumption of gas in reactor	MT^{-1}
M	Make-up gas molar flow rate	MT^{-1}
P	Purge gas molar flow rate	MT^{-1}
q	Wegstein method acceleration parameter	—
s	Wegstein method estimate of gradient	—
U	Overall heat transfer coefficient	$MT^{-3}\theta^{-1}$
x_k	Estimate of parameter x at k^{th} iteration	—
x	Controlled parameter	—
y	Target value	—
y_M	Mole fraction of inerts in make-up	—
y_R	Mole fraction of inerts in recycle and purge	—
z	Target value or unknown variable calculated by simulation program	—

4.12. **PROBLEMS**

4.1. Monochlorobenzene is produced by the reaction of benzene with chlorine. A mixture of monochlorobenzene and dichlorobenzene is produced, with a small amount of trichlorobenzene. Hydrogen chloride is produced as a by-product. Benzene is fed to the reactor in excess to promote the production of monochlorobenzene.

The reactor products are fed to a condenser where the chlorobenzenes and unreacted benzene are condensed. The condensate is separated from the non-condensable gases in a separator. The non-condensable gases, hydrogen chloride and unreacted chlorine, pass to an absorption column where the hydrogen chloride is absorbed in water. The chlorine leaving the absorber is recycled to the reactor.

The liquid phase from the separator, containing chlorobenzenes and unreacted benzene, is fed to a distillation column, where the chlorobenzenes are separated from the unreacted benzene. The benzene is recycled to the reactor.

Using the data given below, calculate the stream flows and draw up a preliminary flow-sheet for the production of 1.0 tonne (metric ton) of monochlorobenzene per day.

Data:

Reactor

Reactions:

$$C_6H_6 + Cl_2 \rightarrow C_6H_5 + HCl$$
$$C_6H_6 + 2Cl_2 \rightarrow C_6H_4\,Cl_2 + 2HCl$$

Mol ratio $Cl_2 : C_6H_6$ at inlet to reactor = 0.9
Overall conversion of benzene = 55.3%
Yield of monochlorobenzene = 73.6%
Yield of dichlorobenzene = 27.3%
Production of other chlorinated compounds can be neglected.

Separator

Assume 0.5% of the liquid stream is entrained with the vapor.

Absorber

Assume 99.99% absorption of hydrogen chloride, and that 98% of the chlorine is recycled, the remainder being dissolved in the water. The water supply to the absorber is set to produce a 30% w/w strength hydrochloric acid.

Distillation column

Take the recovery of benzene to be 95%, and 99.99% recovery of the chlorobenzenes.

Note: This problem can be solved without using process simulation software. Start the mass balance at the reactor inlet (after the recycle streams have been added) and assume 100 kmol/h of benzene at this point.

4.2. Methyl tertiary butyl ether (MTBE) is used as an anti-knock additive in gasoline.

It is manufactured by the reaction of isobutene with methanol. The reaction is highly selective and practically any C_4 stream containing isobutene can be used as a feedstock:

$$CH_2 = C(CH_3)_2 + CH_3OH \rightarrow (CH_3)_3 - C - O - CH_3$$

A 10% excess of methanol is used to suppress side reactions.

In a typical process, the conversion of isobutene in the reactor stage is 97%.

The product is separated from the unreacted methanol and any C_4 compounds by distillation.

The essentially pure, liquid, MTBE leaves the base of the distillation column and is sent to storage. The methanol and C_4 compounds leave the top of the column as vapour and pass to a column where the methanol is separated by absorption in water. The C_4 compounds leave the top of the absorption column, saturated with water, and are used as a fuel gas. The methanol is separated from the water solvent by distillation and recycled to the reactor stage. The water, which leaves the base of the column, is recycled to the absorption column. A purge is taken from the water recycle stream to prevent the build-up of impurities.

1. Draw up a block flow diagram for this process.
2. Estimate the feeds for each stage.
3. Draw a flow-sheet for the process.

Treat the C_4 compounds, other than isobutene, as one component.

Data:

1. Feedstock composition, mol%: n-butane = 2, butene-1 = 31, butene-2 = 18, isobutene = 49.
2. Required production rate of MTBE, 7000 kg/h.
3. Reactor conversion of isobutene, 97%.
4. Recovery of MTBE from the distillation column, 99.5%.
5. Recovery of methanol in the absorption column, 99%.
6. Concentration of methanol in the solution leaving the absorption column, 15%.
7. Purge from the water recycle stream, to waste treatment, 10% of the flow leaving the methanol recovery column.
8. The gases leave the top of the absorption column saturated with water at 30°C.
9. Both columns operate at essentially atmospheric pressure.

4.3. Ethanol can be produced by fermentation of sugars and is used as a gasoline blending component. Because the sugars can be derived from biomass, ethanol is potentially a renewable fuel. In the fermentation of cane sugar to ethanol, sucrose ($C_{12}H_{22}O_{11}$) is converted by yeast (*Saccharomyces cerevisae*) to yield ethanol and CO_2. Some sucrose is also consumed in maintaining the cell culture in the fermentation reactor. The fermentation reaction can be carried out in a continuous reactor as long as the ethanol concentration does not exceed about 8 wt%, at which point the productivity of the yeast declines significantly. The sucrose is fed as a 12.5 wt% solution in water, which must be sterilized before it can be fed to the reactor. The sterilization is usually accomplished by heating with steam. Carbon dioxide is vented from the fermentation reactor. The liquid product of the fermentation reactor is sent to a hydrocyclone to concentrate the yeast for recycle to the reactor. The remaining liquid is sent to a distillation column known as a 'beer column', which concentrates the alcohol to about 40 mol% ethanol and 60 mol% water in the distillate. The recovery of ethanol in the beer column is 99.9%. The bottoms stream from the beer column contains the remaining components of the fermentation broth and can be processed for use as animal feed.

1. Draw a flow-sheet for this process.
2. Estimate the stream flow rates and compositions for a production rate of 200,000 US gal/d of dry (100%) ethanol.
3. Estimate the ethanol lost in the CO_2 vent gas.
4. Estimate the reboiler duty of the beer column.

Data:

1. Yield per kg sucrose: ethanol 443.3 g, CO_2 484 g, non-sugar solids 5.3 g, yeast 21 g, fermentation by-products 43.7 g, higher alcohols (fuel oil) 2.6 g.
2. Conversion of sucrose, 98.5%.
3. Yeast concentration in fermentation reactor at steady state, 3 wt%.
4. Fermenter temperature, 38°C.

4.4. In an ethanol plant, the mixture of water and ethanol from the beer column distillate contains about 40% ethanol (molar basis) in water, together with the fusel oils described in the previous problem. This mixture is distilled to give an azeotropic mixture of ethanol and water (89% ethanol) overhead, with 99.9% recovery of ethanol. The fusel oil can cause blending problems if it is allowed to accumulate in the distillate. Fusel oil is a mixture of higher alcohols and ethers that can be approximated as a mixture of n-butanol and diethyl ether. This mixture is usually removed as a side stream from the column. When the side stream is contacted with additional water a two-phase mixture can be formed and the oil phase can be decanted to leave an ethanol–water phase that is returned to the column.

1. Draw a flow-sheet for this process.
2. Estimate the stream flow rates and compositions for a production rate of 200,000 US gal/d of dry (100%) ethanol.
3. Optimize the distillation column using the cost correlations given in Section 6.3 and assuming that reboiler heat costs $5/MMBtu. Minimize the total annualized cost of the column.

4.5. Water and ethanol form a low boiling point azeotrope; hence, water cannot be completely separated from ethanol by conventional distillation. To produce absolute (100%) ethanol, it is necessary to add an entraining agent to break the azeotrope. Benzene is an effective entrainer and is used where the product is not required for food products. Three columns are used in the benzene process.

Column 1. This column separates the ethanol from the water. The bottom product is essentially pure ethanol. The water in the feed is carried overhead as the ternary azeotrope of ethanol, benzene and water (roughly 24% ethanol, 54% benzene, 22% water). The overhead vapour is condensed and the condensate separated in a decanter into, a benzene-rich phase (22% ethanol, 74% benzene, 4% water) and a water-rich phase (35% ethanol, 4% benzene, 61% water). The benzene-rich phase is recycled to the column as reflux. A benzene make-up stream is added to the reflux to make up any loss of benzene from the process. The water-rich phase is fed to the second column.

Column 2. This column recovers the benzene as the ternary azeotrope and recycles it as vapour to join the overhead vapour from the first column. The bottom product from the column is essentially free of benzene (29% ethanol, 51% water). This stream is fed to the third column.

Column 3. In this column, the water is separated and sent to waste treatment. The overhead product consists of the azeotropic mixture of ethanol and water (89% ethanol, 11% water). The overheads are condensed and recycled to join the feed to the first column. The bottom product is essentially free of ethanol.

1. Draw a flow-sheet for this process.
2. Estimate the stream flow rates and compositions for a production rate of 200,000 US gal/d of dry (100%) ethanol.

Take the benzene losses to total 0.1 kmol/h. All the compositions given are molar percentages.

4.6. A plant is required to produce 10,000 metric tons per year of anhydrous hydrogen chloride from chlorine and hydrogen. The hydrogen source is impure: 90 mol% hydrogen, balance nitrogen.

The chlorine is essentially pure chlorine, supplied in rail tankers.

The hydrogen and chlorine are reacted in a burner at 1.5 bar pressure:

$$H_2 + Cl_2 \rightarrow 2HCl$$

Hydrogen is supplied to the burner in 3% excess over the stoichiometric amount. The conversion of chlorine is essentially 100%. The gases leaving the burner are cooled in a heat exchanger.

The cooled gases pass to an absorption column where the hydrogen chloride gas is absorbed in dilute hydrochloric acid. The absorption column is designed to recover 99.5% of the hydrogen chloride in the feed.

The unreacted hydrogen and inerts pass from the absorber to a vent scrubber where any hydrogen chloride present is neutralized by contact with a dilute, aqueous solution of sodium hydroxide. The solution is recirculated around the scrubber. The concentration of sodium hydroxide is maintained at 5% by taking a purge from the recycle loop and introducing a make-up stream of 25% concentration. The maximum concentration of hydrogen chloride discharged in the gases vented from the scrubber to atmosphere must not exceed 200 ppm (parts per million) by volume.

The strong acid from the absorption column (32% HCl) is fed to a stripping column where the hydrogen chloride gas is recovered from the solution by distillation. The diluted acid from the base of this column (22% HCl), is recycled to the absorption column.

The gases from the top of the stripping column pass through a partial condenser, where the bulk of the water vapour present is condensed and returned to the column as reflux. The gases leaving the column will be saturated with water vapour at 40°C.

The hydrogen chloride gas leaving the condenser is dried by contact with concentrated sulphuric acid in a packed column. The acid is recirculated over the packing. The concentration of sulphuric acid is maintained at 70% by taking a purge from the recycle loop and introducing a make-up stream of strong acid (98% H_2SO_4).

The anhydrous hydrogen chloride product is compressed to 5 bar and supplied as a feed to another process.

Using the information provided, calculate the flow rates and compositions of the main process streams, and draw a flow-sheet for this process. All compositions are wt%, except where indicated.

4.7. Ammonia is synthesized from hydrogen and nitrogen. The synthesis gas is usually produced from hydrocarbons. The most common raw materials are oil or natural gas, though coal and even peat can be used.

When produced from natural gas, the synthesis gas will be impure, containing up to 5% inerts, mainly methane and argon. The reaction equilibrium and rate are favoured by high pressure. The conversion is low, about 15%, and so, after removal of the ammonia produced, the gas is recycled to the converter inlet. A typical process consists of: a converter (reactor) operating at 350 bar, a refrigerated system to condense out the ammonia product from the recycle loop, and compressors to compress the feed and recycle gas. A purge is taken from the recycle loop to keep the inert concentration in the recycle gas at an acceptable level.

Using the data given below, draw a flow diagram of the process and calculate the process stream flow rates and compositions for the production of 600 t/d ammonia.

Data:

Composition of synthesis gas, mol fraction:

N_2	H_2	CH_4	A
24.5	73.5	1.7	0.3

Temperature and operating pressure of liquid ammonia–gas separator: -28°C and 340 bar.

Inert gas concentration in recycle gas, not greater than 15 mol%.

4.8. Methyl ethyl ketone (MEK) is manufactured by the dehydrogenation of 2-butanol.

A simplified description of the process listing the various units used is given below:

1. A reactor in which the butanol is dehydrated to produce MEK and hydrogen, according to the reaction:

$$CH_3CH_2CH_3CHOH \rightarrow CH_3CH_2CH_3CO + H_2$$

The conversion of alcohol is 88% and the selectivity to MEK can be taken as 100%.

2. A cooler-condenser, in which the reactor off-gases are cooled and most of the MEK and unreacted alcohol are condensed. Two exchangers are used but they can be modelled as one unit. Of the MEK entering the unit, 84% is condensed, together with 92% of the alcohol. The hydrogen is non-condensable. The condensate is fed forward to the final purification column.

3. An absorption column, in which the uncondensed MEK and alcohol are absorbed in water. Around 98% of the MEK and alcohol can be considered to be absorbed in this unit, giving a 10 wt% solution of MEK. The water feed to the absorber is recycled from the next unit, the extractor. The vent stream from the absorber, containing mainly hydrogen, is sent to a flare stack.

4. An extraction column, in which the MEK and alcohol in the solution from the absorber are extracted into trichloroethylane (TCE). The raffinate, water containing around 0.5 wt% MEK, is recycled to the absorption column.

   The extract, which contains around 20 wt% MEK, and a small amount of butanol and water, is fed to a distillation column.

5. A distillation column, which separates the MEK and alcohol from the solvent TCE. The recovery of MEK is 99.99%.

   The solvent containing a trace of MEK and water is recycled to the extraction column.

6. A second distillation column, which produces a 99.9% pure MEK product from the crude product from the first column. The residue from this column, which contains the bulk of the unreacted 2-butanol, is recycled to the reactor.

For a production rate of 1250 kg/h MEK:

1. Draw a flow-sheet for the process.
2. Estimate the stream flow rates and compositions.
3. Estimate the reboiler and condenser duties of the two distillation columns.
4. Estimate the number of theoretical trays required in each column.

4.9. In the problem of Example 4.1, the feed was specified as pentane (C_5H_{12}) with a hydrogen to carbon ratio of 2.4:1. If the feed to the process were a heavy oil, the hydrogen to carbon ratio would be more like 2:1. How would the distribution of C_5 carbon compounds change if the feed had a 2:1 carbon ratio?

4.10. Example 4.1 examined the equilibrium distribution of hydrocarbon compounds within a single carbon number (C_5). In reality, cracking reactions to ethylene, propylene and other light alkenes and alkynes will have a significant effect on the yield of a cracking process.

1. What is the effect of including C_2 and C_3 compounds on the equilibrium distribution?
2. What is the effect of including coke (carbon) as well as the C_2 and C_3 compounds?
3. What do these results tell you about cracking processes?

4.11. Optimize the heat exchanger design of Example 4.5 to minimize the total surface area required.

4.12. A stream containing 4 metric tons/h of a 20 wt% mixture of benzene in toluene is heated from 20°C to the bubble point at 4 atm pressure. The mixture is separated in a distillation column to give 99.9% recovery of benzene overhead and toluene in the bottoms.
  1. If the toluene product must be cooled to 20°C, how much of the feed heat can be supplied by heat exchange with the bottoms?
  2. How many heat-exchange shells are needed?
  3. What is the minimum total heat-exchange area?
  4. What is the distillation column diameter?
  5. How many sieve trays are needed if the tray efficiency is 70%?

4.13. The autothermal reforming of methane to hydrogen was described in Example 4.2. The solution in the example was not optimized, and suggestions were given for how to improve the results. Optimize the process to minimize the cost of production of hydrogen, assuming:
  1. Cost of methane = 16 ¢/lb
  2. Cost of oxygen = 2 ¢/lb
  3. Cost of water = 25 ¢/1000 lb
  4. Annualized cost of heat exchangers = \$30,000 + 3A, where A is the area in ft^2
  5. Cost of electric power = 6 ¢/kWh
  6. Reactor and catalyst costs are the same in all cases.

  Hint: first determine the optimal heat recovery and steam and oxygen to methane ratios for a given methane conversion. Repeat for different methane conversions to find the overall optimum.

4.14. The light naphtha isomerization process is more complex than the description given in Example 4.7.
  1. Hydrogen is flowed through the plant to reduce catalyst deactivation. The hydrogen flow rate is typically about 2 moles per mole of hydrocarbon on a pure hydrogen basis. The hydrogen make-up gas is typically about 90 mol% hydrogen, with the balance methane.
  2. Light hydrocarbon compounds are formed by cracking reactions. These compounds accumulate in the hydrogen recycle and are controlled by taking a purge stream. A stabilizer column is also required, upstream of the distillation column, to remove light hydrocarbons and hydrogen before the distillation.
  3. Each of the C$_6$ isomers has a different blending octane value. The total octane value of the product can be found by summing the products of the mole fraction of each component and the component blending value. The blending values are: n-hexane 60; 2-methyl pentane 78.5; 3-methyl pentane 79.5; 2,2-dinethyl butane 86.3; 2,3-dimethyl butane 93.
  Optimize the design of Example 4.7, subject to the following:

  1. The selectivity loss due to cracking reactions can be approximated as 1% conversion of C$_6$ compounds to propane per reactor pass.

2. The wholesale value of gasoline can be assumed to be $2.0 + 0.05$ (octane number $- 87$) $/US gal.
3. The cost of hydrogen is $6/1000 scf, and the fuel value of the hydrogen and propane purge stream is $5/MMBtu.
4. The reactor plus catalyst total installed cost can be taken as $0.5 MM per 1000 bpd of liquids processed.
5. Other costs can be estimated using the cost correlations given in Section 6.3.

Additional flow-sheeting problems are given in the form of design projects in Appendices E and F.

5 PIPING AND INSTRUMENTATION

Chapter Contents

Key Learning Objectives

- How to read a piping and instrument diagram drawn using ISA-5.1 symbols
- How valves and controllers work
- How to calculate line pressure drop and size and select pumps
- How to design control schemes for common unit operations and whole processes

5.1. INTRODUCTION

The process flow-sheet shows the arrangement of the major pieces of equipment and their interconnection. It is a description of the nature of the process.

The Piping and Instrument diagram (P and I diagram or PID) shows the engineering details of the equipment, instruments, piping, valves and fittings; and their arrangement. It is often called the Engineering Flow-sheet or Engineering Line Diagram. This chapter covers the preparation of the preliminary P and I diagrams at the process design stage of the project.

The design of piping systems, and the specification of the process instrumentation and control systems, is usually done by specialist design groups, and a detailed discussion of piping design and control systems is beyond the scope of this book. Only general guide rules are given. The piping handbook edited by Nayyar *et al.* (2000) is particularly recommended for the guidance on the detailed design of piping systems and process instrumentation and control. The references cited in the text and listed at the end of the chapter should also be consulted.

5.2. THE P AND I DIAGRAM

The P and I diagram shows the arrangement of the process equipment, piping, pumps, instruments, valves and other fittings. It should include:

1. All process equipment identified by an equipment number. The equipment should be drawn roughly in proportion, and the location of nozzles shown.
2. All pipes, identified by a line number. The pipe size and material of construction should be shown. The material may be included as part of the line identification number.
3. All valves, control and block valves, with an identification number. The type and size should be shown. The type may be shown by the symbol used for the valve or included in the code used for the valve number.
4. Ancillary fittings that are part of the piping system, such as inline sight-glasses, strainers and steam traps; with an identification number.
5. Pumps, identified by a suitable code number.
6. All control loops and instruments, with an identification number.

For simple processes, the utility (service) lines can be shown on the P and I diagram. For complex processes, separate diagrams should be used to show the service lines, so the information can be shown clearly, without cluttering up the diagram. The service connections to each unit should, however, be shown on the P and I diagram.

The P and I diagram will resemble the process flow-sheet, but the process information is not shown. The same equipment identification numbers should be used on both diagrams.

5.2.1. Symbols and Layout

The symbols used to show the equipment, valves, instruments and control loops will depend on the practice of the particular design office. The equipment symbols are usually more detailed than those used for the process flow-sheet. A typical example of a P and I diagram is shown in Figure 5.34 at the end of this chapter.

At the time of writing, there is no ISO or European Standard for P and I symbols. The most widely used international standard symbols for instruments, controllers and valves are those given by the Instrumentation Systems and Automation Society design code ISA-5.1-1984 (R1992). Some companies use their own symbols though, and different standards are followed in some countries, such as BS 1646 in the UK and DIN 19227 and DIN 2429 in Germany.

When laying out the diagram, it is only necessary to show the relative elevation of the process connections to the equipment where these affect the process operation; for example, the net positive suction head (NPSH) of pumps, barometric legs, syphons and the operation of thermosyphon reboilers. Full details of pipe layout are usually shown in a different drawing, known as a piping isometric drawing. See Figure 5.20 for an example.

Computer-aided drafting programs are available for the preparation of P and I diagrams; see the reference to the PROCEDE package in Chapter 4.

5.2.2. Basic Symbols

The symbols illustrated below are those given in ISA-5.1-1984 (R1992).

Control Valves

| General | Three-way | Globe | Diaphragm |

Figure 5.1. Control valves.

Different types of valves are discussed in Section 5.3.

Actuators

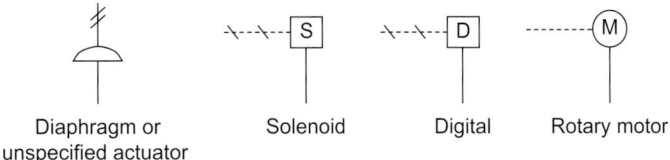

Diaphragm or unspecified actuator   Solenoid   Digital   Rotary motor

Figure 5.2. Actuators.

Most modern control valves (final control elements) are actuated by electric motors, but older valves are actuated by pneumatic signals using instrument air. Pneumatic actuators are preferred in situations where electronic controllers might cause a process hazard or where electric power is not available or reliable. Pneumatic controllers are also found in many older plants where replacement with electronic controllers has not yet occurred. Motor actuators are used for larger valves, while digital and solenoid actuators are used for valves that are switched from open to closed, as often occurs in batch processing. Many newer controllers use a combination of these approaches. For example, a digital signal can be sent to a solenoid that opens or shuts an instrument air line that then actuates a pneumatically driven control valve.

Instrument Lines

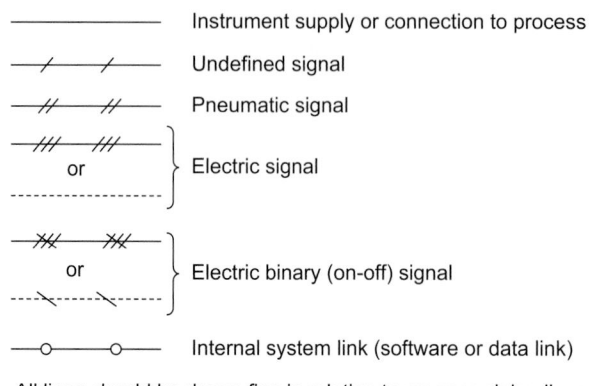

All lines should be drawn fine in relation to process piping lines

Figure 5.3.   Instrument lines.

The instrument connecting lines are drawn in a manner to distinguish them from the main process lines. Process lines are drawn as solid lines and are usually drawn thicker.

Failure Mode

The direction of the arrow shows the position of the valve on failure of the power supply.

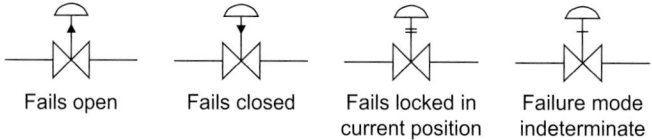

Figure 5.4.   Valve failure modes.

General Instrument and Controller Symbols

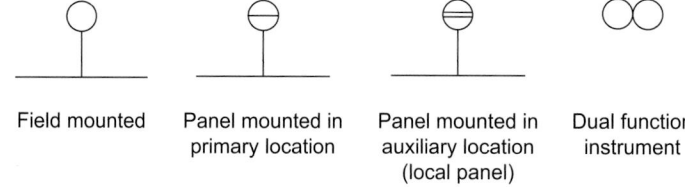

Figure 5.5.   General instrument and controller symbols.

Locally mounted means that the controller and display are located out on the plant near to the sensing instrument location. *Main panel* means that they are located on a panel in the control room. Except on small plants, most controllers would be mounted in the control room.

Distributed Control: Shared Display Symbols

Field mounted shared display device with limited access to adjustments

Shared display device with operator access to adjustments

*AH
AL Shared display device with software alarms (is measured variable)

Programmable logic controller accessible to operator

Field mounted programmable logic controller

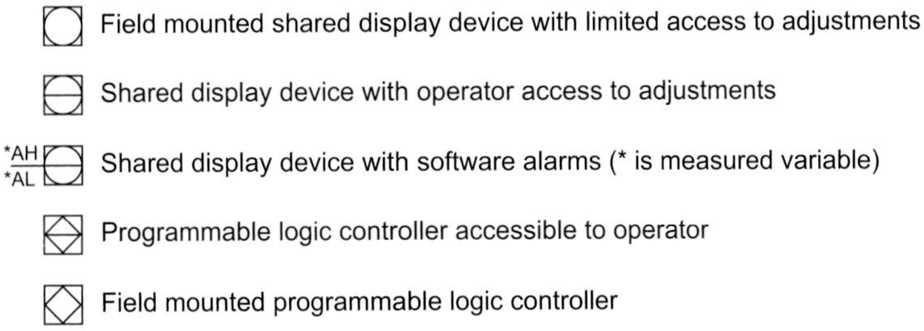

Figure 5.6. Shared display symbols for distributed control and logic control.

A distributed control system is a system that is functionally integrated, but consists of subsystems that may be physically separate and remotely located from one another. A shared display is an operator interface device such as a computer screen or video screen that is used to display process control information from a number of sources at the command of the operator. Most plants built since 1990 (and many older plants) use shared displays instead of instrument panels.

Programmable logic controllers are used to control discrete operations, such as steps in a batch or semi-continuous process, and to program interlock controls that guard against unsafe or uneconomic conditions. For example, a logic controller could be used to ensure that an operator cannot open an air vent line to a vessel unless the feed valves are closed and nitrogen purge is open.

Other Common Symbols

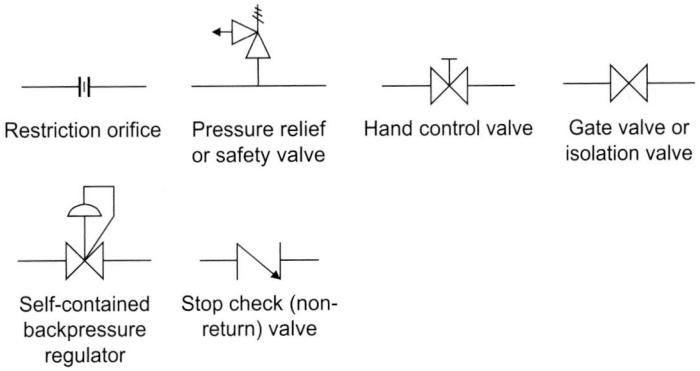

Restriction orifice   Pressure relief or safety valve   Hand control valve   Gate valve or isolation valve

Self-contained backpressure regulator   Stop check (non-return) valve

Figure 5.7. Other common symbols.

TABLE 5.1. Letter Code for Instrument Symbols (Based on ISA-5.1-1984 (R1992))

Initiating or Measured Variable	First Letter	Indicating Only	Controllers Recording	Controllers Indicating	Controllers Blind	Transmitters	Final Control Element
Analysis (composition)	A	AI	ARC	AIC	AC	AT	AV
Flow rate	F	FI	FRC	FIC	FC	FT	FV
Flow ratio	FF	FFI	FFRC	FFIC	FFC	FFT	FFV
Power	J	JI	JRC	JIC		JT	JV
Level	L	LI	LRC	LIC	LC	LT	LV
Pressure, vacuum	P	PI	PRC	PIC	PC	PT	PV
Pressure differential	PD	PDI	PDRC	PDIC	PDC	PDT	PDV
Quantity	Q	QI	QRC	QIC		QT	QZ
Radiation	R	RI	RRC	RIC	RC	RT	RZ
Temperature	T	TI	TRC	TIC	TC	TT	TV
Temperature differential	TD	TDI	TDRC	TDIC	TDC	TDT	TDV
Weight	W	WI	WRC	WIC	WC	WT	WZ

Notes:
(1) The letters C, D, G, M, N and O are not defined and can be used for any user-specified property.
(2) The letter S as second or subsequent letter indicates a switch.
(3) The letter Y as second or subsequent letter indicates a relay or a compute function.
(4) The letter Z is used for the final control element when this is not a valve.
Consult the standard for the full set of letter codes.

Type of Instrument

This is indicated on the circle representing the instrument-controller by a letter code (see Table 5.1).

The first letter indicates the property measured; for example, F = flow. Subsequent letters indicate the function; for example,

$$I = \text{indicating}$$
$$RC = \text{recorder controller}$$

The letters AH or AL indicate high or low alarms.

The P and I diagram shows all the components that make up a control loop. For example, Figure 5.8 shows a field-located pressure transmitter connected to a shared display pressure indicator-controller with operator access to adjustments and high and low alarms. The pressure controller sends an electric signal to a fail-closed diaphragm-actuated pressure control valve.

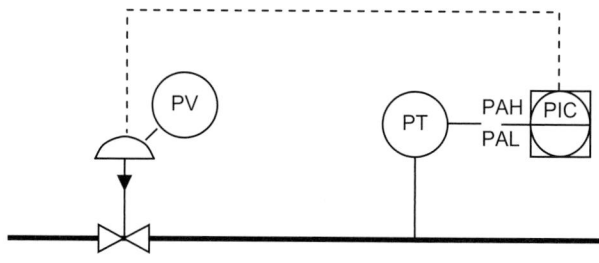

Figure 5.8. A typical control loop.

5.3. VALVE SELECTION

The valves used for a chemical process plant can be divided into two broad classes, depending on their primary function:

1. Shut-off valves (block valves or isolation valves), whose purpose is to close off the flow.
2. Control valves, both manual and automatic, used to regulate flow.

The main types of valves used are:

Gate	Figure 5.9a
Plug	Figure 5.9b
Ball	Figure 5.9c
Globe	Figure 5.9d
Diaphragm	Figure 5.9e
Butterfly	Figure 5.9f

A valve selected for shut-off purposes should give a positive seal in the closed position and minimum resistance to flow when open. Gate, plug and ball valves are most frequently used for this purpose. Gate valves are available in the widest range of sizes and can be operated manually or by a motor. They have a straight-through flow channel and low pressure drop when fully open. Several turns of the valve handle are usually required to close the valve, so they are best used when operated infrequently. Gate valves should not be operated partially open, as the valve seals can become deformed, causing the valve not to seal properly. Plug valves and ball valves have the advantage that they only require a quarter turn to open or close. These valves are often actuated by solenoids and are used where quick on–off switching is needed. The selection of valves is discussed by Merrick (1986) (1990), Smith and Vivian (1995) and Smith and Zappe (2003).

If flow control is required, the valve should be capable of giving smooth control over the full range of flow, from fully open to closed. Globe valves are normally used, though diaphragm valves are also common. Butterfly valves are often used for the control of gas and vapour flows. Automatic control valves are usually globe valves with special trim designs (see Peacock and Richardson, 1994; Chapter 7).

The careful selection and design of control valves is important; good flow control must be achieved, whilst keeping the pressure drop as low as possible. The valve must also be sized to avoid the flashing of hot liquids and the super-critical flow of gases and vapours. Control valve sizing is discussed by Chaflin (1974).

Non-return valves are used to prevent back-flow of fluid in a process line. They do not normally give an absolute shut-off of the reverse flow. A typical design is shown in Figure 5.9g. Since swing-type check valves depend on gravity to close the valve, care must be taken to orient the valve properly when locating and installing it.

Most of the British Standards for valves have now been brought into conformance with European Standards. The more widely used British Standards are listed in the references at the end of this chapter. American Standards are set by the ASME B16 Standards Committee and can be ordered from the American Society of Mechanical Engineers. General standards are described in ASME B16.34-2004 (ASME, 2004),

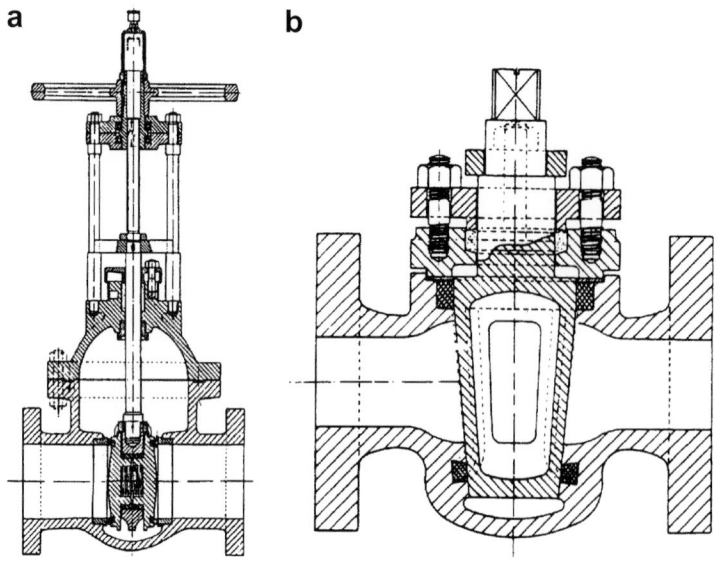

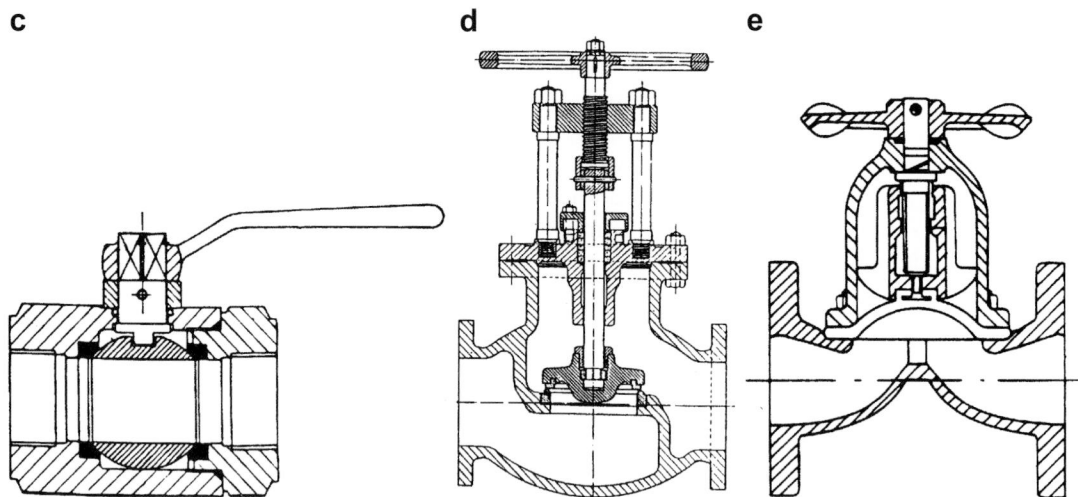

Figure 5.9.   (a) Gate valve (slide valve). (b) Plug valve. (c) Ball valve. (d) Globe valve. (e) Diaphragm valve.

f   **g**

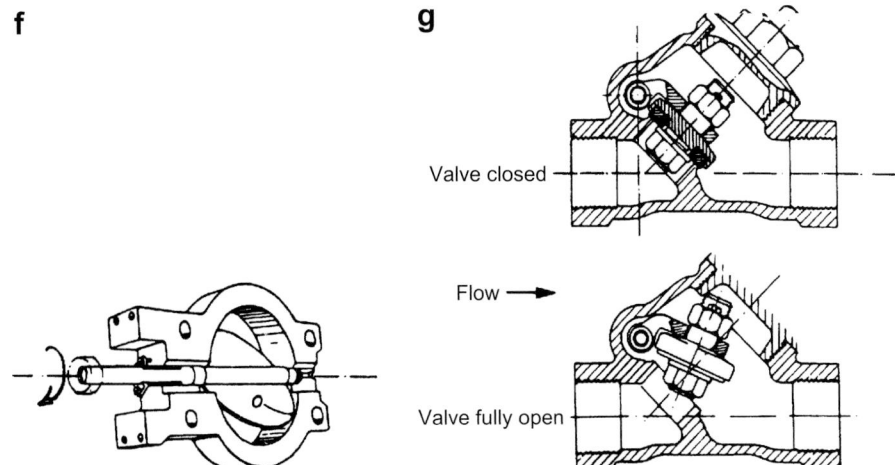

Valve closed

Flow ⟶

Valve fully open

Figure 5.9. Cont'd.   (f) Butterfly valve. (g) Non-return valve, check valve, hinged disc type.

while valve dimensions are given in ASME B16.10-2000 (ASME, 2000). Valve design is covered by Pearson (1978).

5.4. PUMPS AND COMPRESSORS

5.4.1. Pump Selection

The pumping of liquids is covered by Coulson *et al.* (1999), Chapter 8. Reference should be made to that chapter for a discussion of the principles of pump design and illustrations of the more commonly used pumps.

Pumps can be classified into two general types:

1. Dynamic pumps, such as centrifugal pumps.
2. Positive displacement pumps, such as reciprocating and diaphragm pumps.

The single-stage, horizontal, overhung, centrifugal pump is by far the most commonly used type in the chemical process industry. Other types are used where a high head or other special process considerations are specified. For example, when small flow rates of additives must be added to a process positive displacement metering pumps are often used.

Pump selection is made on the flow rate and head required, together with other process considerations, such as corrosion or the presence of solids in the fluid.

The chart shown in Figure 5.10 can be used to determine the type of pump required for a particular head and flow rate. This figure is based on one published by Doolin (1977).

Centrifugal pumps are characterized by their specific speed (see Coulson *et al.*, 1999; Chapter 8). In the dimensionless form, specific speed is given by:

$$N_s = \frac{NQ^{1/2}}{(gh)^{3/4}} \tag{5.1}$$

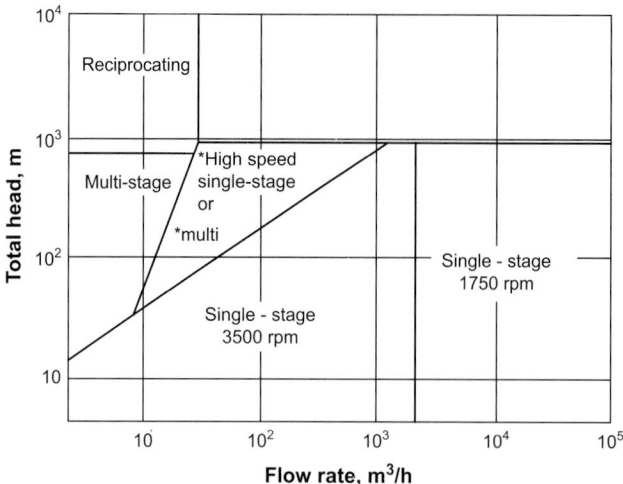

Figure 5.10.   Centrifugal pump selection guide. *Single-stage >1750 rpm, multi-stage 1750 rpm.

where

> N = revolutions per second
> Q = flow, m^3/s
> h = head, m
> g = gravitational acceleration, m/s^2.

Pump manufacturers do not generally use the dimensionless specific speed, but define the impeller specific speed by the equation:

$$N'_S = \frac{N'Q^{1/2}}{h^{3/4}} \tag{5.2}$$

where

> N' = revolutions per minute (rpm),
> Q = flow, US gal/min
> h = head, ft.

Values of the non-dimensional specific speed, as defined by equation 5.1, can be converted to the form defined by equation 5.2 by multiplying by 1.72×10^4.

The impeller specific speed for centrifugal pumps (equation 5.2) usually lies between 400 and 20,000, depending on the type of impeller. Generally, pump impellers are classified as centrifugal or radial for specific speeds between 400 and 4000, mixed flow between 4000 and 9000, and axial above 9000 (Heald, 1996). Doolin (1977) states that below a specific speed of 1000 the efficiency of single-stage centrifugal pumps is low and multi-stage pumps should be considered.

For a detailed discussion of the factors governing the selection of the best centrifugal pump for a given duty the reader should refer to the article by De Santis (1976), Neerkin (1974), Jacobs (1965) or Walas (1990).

Positive displacement, reciprocating, pumps are normally used where a high head is required at a low flow rate. Holland and Chapman (1966) review the various types of positive displacement pumps available and discuss their applications.

A general guide to the selection, installation and operation of pumps for the process industries is given by Davidson and von Bertele (1999) or Jandiel (2000).

The selection of the pump cannot be separated from the design of the complete piping system. The total head required will be the sum of the dynamic head due to friction losses in the piping, fittings, valves and process equipment, and any static head due to differences in elevation.

The pressure drop required across a control valve will be a function of the valve design. Sufficient pressure drop must be allowed for when sizing the pump to ensure that the control valve operates satisfactorily over the full range of flow required. If possible, the control valve and pump should be sized together, as a unit, to ensure that the optimum size is selected for both. As a rough guide, if the characteristics are not specified, the control valve pressure drop should be taken as at least 30% of the total dynamic pressure drop through the system, with a minimum value of 50 kPa (7 psi). A good rule of thumb in the early stages of process design is to allow 70 kPa (10 psi) pressure drop for each control valve. The valve should be sized for a maximum flow rate 30% above the normal stream flow rate. Some of the pressure drop across the valve will be recovered downstream, the amount depending on the type of valve used.

Methods for the calculation of pressure drop through pipes and fittings are given in Section 5.4.2 and Coulson *et al.* (1999), Chapter 3. It is important that a proper analysis is made of the system and the use of a calculation form (work sheet) to standardize pump-head calculations is recommended. A standard calculation form ensures that a systematic method of calculation is used, and provides a check list to ensure that all the usual factors have been considered. It is also a permanent record of the calculation. A template for a standard pump and line calculation is given in Appendix G and can be downloaded in MS Excel format from http://books.elsevier. com/companions. Example 5.8 has been set out using this calculation form. The calculation should include a check on the net positive suction head (NPSH) available; see Section 5.4.3.

Kern (1975) discusses the practical design of pump suction piping, in a series of articles on the practical aspects of piping system design published in the journal *Chemical Engineering* from December 1973 to November 1975. A detailed presentation of pipe-sizing techniques is also given by Simpson (1968), who covers liquid, gas and two-phase systems. Line sizing and pump selection are also covered in a comprehensive article by Ludwig (1960).

5.4.2. Pressure Drop in Pipelines

The pressure drop in a pipe due to friction is a function of the fluid flow rate, fluid density and viscosity, pipe diameter, pipe surface roughness, and the length of the pipe. It can be calculated using the following equation:

$$\Delta P_f = 8f(L/d_i)\frac{\rho u^2}{2} \tag{5.3}$$

where

ΔP_f = pressure drop, N/m^2
f = friction factor
L = pipe length, m
d_i = pipe inside diameter, m
ρ = fluid density, kg/m^3
u = fluid velocity, m/s.

The friction factor is dependent on the Reynolds number and pipe roughness. The friction factor for use in equation 5.3 can be found from Figure 5.11.

The Reynolds number is given by:

$$Re = (\rho \times u \times d_i)/\mu \tag{5.4}$$

Values for the absolute surface roughness of commonly used pipes are given in Table 5.2. The parameter to use with Figure 5.11 is the relative roughness, e, given by:

$$e = \text{absolute roughness/pipe inside diameter}$$

Note: the friction factor used in equation 5.3 is related to the shear stress at the pipe wall, R, by the equation $f = (R/\rho\, u^2)$. Other workers use different relationships. Their charts for friction factor will give values that are multiples of those given by Figure 5.11. So, it is important to make sure that the pressure drop equation used matches the friction factor chart. One of the most commonly used is that of Fanning, which defines the coefficient of friction as $C_f = (2R/\rho\, u^2)$; i.e., $C_f = 2f$, in which case equation 5.3 becomes:

$$\Delta P_f = 4C_f(L/d_i)\frac{\rho u^2}{2} \tag{5.3a}$$

Non-Newtonian Fluids

In equation 5.3, and when calculating the Reynolds number for use with Figure 5.11, the fluid viscosity and density are taken to be constant. This will be true for Newtonian liquids but not for non-Newtonian liquids, where the apparent viscosity is a function of the shear stress.

More complex methods are needed to determine the pressure drop of non-Newtonian fluids in pipelines. Suitable methods are given in Richardson *et al.* (2002), Chapter 4, and in Chabbra and Richardson (1999); see also Darby (2001).

Gases

When a gas flows through a pipe, the gas density is a function of the pressure and so is determined by the pressure drop. Equation 5.3 and Figure 5.11 can be used to estimate the pressure drop, but it may be necessary to divide the pipeline into short sections and sum the results.

Two-Phase Mixtures

For vapour–liquid mixtures the pressure drop in horizontal pipes can be found using the correlation of Lockhart and Martinelli (1949), which relates the

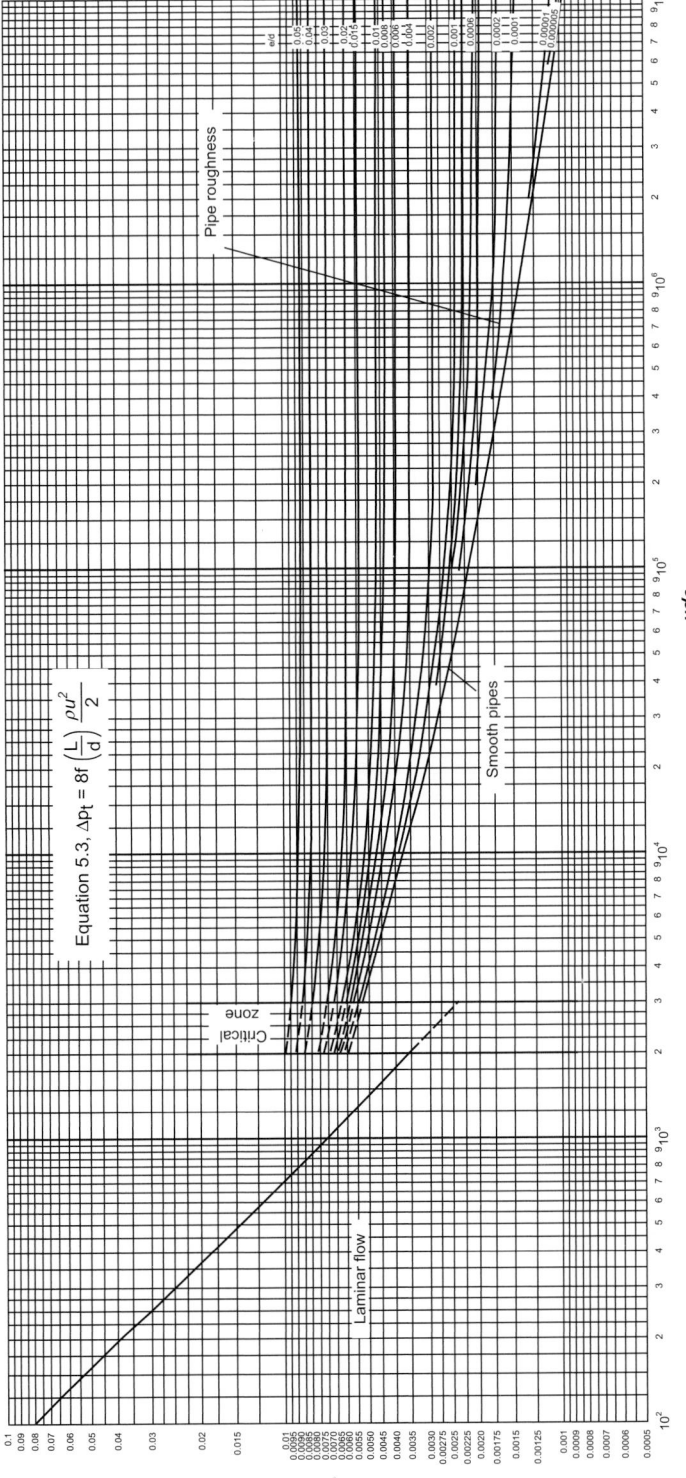

Figure 5.11. Pipe friction versus Reynolds number and relative roughness.

TABLE 5.2. Pipe Roughness

Material	Absolute Roughness (mm)
Drawn tubing	0.0015
Commercial steel pipe	0.046
Cast-iron pipe	0.26
Concrete pipe	0.3 to 3.0

two-phase pressure drop to the pressure drop that would be calculated if each phase was flowing separately in the pipe. Details of the correlation and methods for two-phase flow in vertical pipes are given in Green and Perry (2007).

Liquid-solid mixtures that do not settle out rapidly are usually treated as non-Newtonian fluids. This will usually be the case if the solid particle size is less than about 200 microns (0.2mm). Larger particle sizes form settling slurries and require a critical velocity to maintain the solids in suspension. Correlations for critical velocity and pressure drop are given in Green and Perry (2007).

Gas–solid mixtures are commonly encountered in pneumatic conveying. This is discussed in Coulson *et al.* (1999), Chapter 5, and by Mills (2004) and Mills, Jones and Agarwal (2004).

Miscellaneous Pressure Losses

Any obstruction to flow will generate turbulence and cause a pressure drop. So, pipe fittings, such as bends, elbows, reducing or enlargement sections and tee junctions, will increase the pressure drop in a pipeline.

There will also be a pressure drop due to the valves used to isolate equipment and control the fluid flow. The pressure drop due to these miscellaneous losses can be estimated using either of two methods:

1. As the number of velocity heads, K, lost at each fitting or valve. A velocity head is $u^2/2g$, metres of the fluid, equivalent to $(\rho u^2/2)$, N/m^2. The total number of velocity heads lost due to all the fittings and valves is added to the pressure drop due to pipe friction.
2. As a length of pipe that would cause the same pressure loss as the fitting or valve. As this will be a function of the pipe diameter, it is expressed as the number of equivalent pipe diameters. The length of pipe to add to the actual pipe length is found by multiplying the total number of equivalent pipe diameters by the diameter of the pipe being used.

The number of velocity heads lost, or equivalent pipe diameter, is a characteristic of the particular fitting or type of valve used. Values can be found in handbooks and manufacturers' literature. The values for a selected number of fittings and valves are given in Table 5.3.

The two methods used to estimate the miscellaneous losses are illustrated in Example 5.1.

TABLE 5.3. Pressure Loss in Pipe Fittings and Valves (for Turbulent Flow)

Fitting or Valve	K, Number of Velocity Heads	Number of Equivalent Pipe Diameters
45° standard elbow	0.35	15
45° long radius elbow	0.2	10
90° standard radius elbow	0.6–0.8	30–40
90° standard long elbow	0.45	23
90° square elbow	1.5	75
Tee-entry from leg	1.2	60
Tee-entry into leg	1.8	90
Union and coupling	0.04	2
Sharp reduction (tank outlet)	0.5	25
Sudden expansion (tank inlet)	1.0	50
Gate valve:		
fully open	0.15	7.5
1/4 open	16	800
1/2 open	4	200
3/4 open	1	40
Globe valve, bevel seat:		
fully open	6	300
1/2 open	8.5	450
Globe valve, plug disk:		
fully open	9	450
1/2 open	36	1800
1/4 open	112	5600
Plug valve: open	0.4	18

Pipe fittings are discussed in Section 5.5.4, see also Green and Perry (2007). Valve types and applications are discussed in Section 5.3.

Example 5.1

A pipeline connecting two tanks contains four standard elbows, a globe valve that is fully open and a gate valve that is half open. The line is commercial steel pipe, 25 mm internal diameter, length 120 m.

The properties of the fluid are: viscosity $0.99 \, \text{mNm}^{-2}\,\text{s}$, density $998 \, \text{kg/m}^3$. Calculate the total pressure drop due to friction when the flow rate is 3500 kg/h.

Solution

$$\text{Cross-sectional area of pipe} = \frac{\pi}{4}(25 \times 10^{-3})^2 = 0.491 \times 10^{-3} \, \text{m}^2$$

$$\text{Fluid velocity}, u = \frac{3500}{3600} \times \frac{1}{0.491 \times 10^{-3}} \times \frac{1}{998} = 1.98 \, \text{m/s}$$

$$\text{Reynolds number}, Re = (998 \times 1.98 \times 25 \times 10^{-3})/0.99 \times 10^{-3}$$

$$= 49,900 = 5 \times 10^4 \tag{5.4}$$

Absolute roughness commercial steel pipe, Table 5.2 = 0.046 mm
Relative roughness $= 0.046/(25 \times 10^{-3}) = 0.0018$, round to 0.002
From friction factor chart, Figure 5.11, $f = 0.0032$

Miscellaneous Losses

Fitting/valve	Number of velocity heads, K	Equivalent pipe diameters
Entry	0.5	25
Elbows	(0.8×4)	(40×4)
Globe valve, open	6.0	300
Gate valve, 1/2 open	4.0	200
Exit	1.0	50
Total	14.7	735

Method 1, Velocity Heads

$$\text{A velocity head} = u^2/2g = 1.98^2/(2 \times 9.8) = 0.20 \text{ m of liquid.}$$
$$\text{Head loss} = 0.20 \times 14.7 = 2.94 \text{ m}$$
$$\text{as pressure} = 2.94 \times 998 \times 9.8 = 28{,}754 \text{ N/m}^2$$
$$\text{Friction loss in pipe, } \Delta P_f = 8 \times 0.0032 \frac{(120)}{(25 \times 10^{-3})} 988 \times \frac{1.98^2}{2}$$
$$= 240{,}388 \text{ N/m}^2$$
$$\text{Total pressure} = 28{,}754 + 240{,}388 = 269{,}142 \text{ N/m}^2$$
$$= \underline{\underline{270 \text{ kN/m}^2}}$$

(5.3)

Method 2, Equivalent Pipe Diameters

Extra length of pipe to allow for miscellaneous losses

$$= 735 \times 25 \times 10^{-3} = 18.4 \text{ m.}$$

So, total length for ΔP calculation $= 120 + 18.4 = 138.4 \text{ m.}$

$$\Delta P_f = 8 \times 0.0032 \frac{(138.4)}{(25 \times 10^{-3})} 998 \times \frac{1.98^2}{2} = 277{,}247 \text{ N/m}^2$$
$$= \underline{\underline{277 \text{ kN/m}^2}}$$

(5.3)

Note: The two methods will not give exactly the same result. The method using velocity heads is the more fundamentally correct approach, but the use of equivalent diameters is easier to apply and sufficiently accurate for use in preliminary design calculations.

5.4.3. Power Requirements for Pumping Liquids

To transport a liquid from one vessel to another through a pipeline (see Figure 5.12), energy has to be supplied to:

1. overcome the friction losses in the pipes;
2. overcome the miscellaneous losses in the pipe fittings (e.g. bends), valves, instruments etc.;

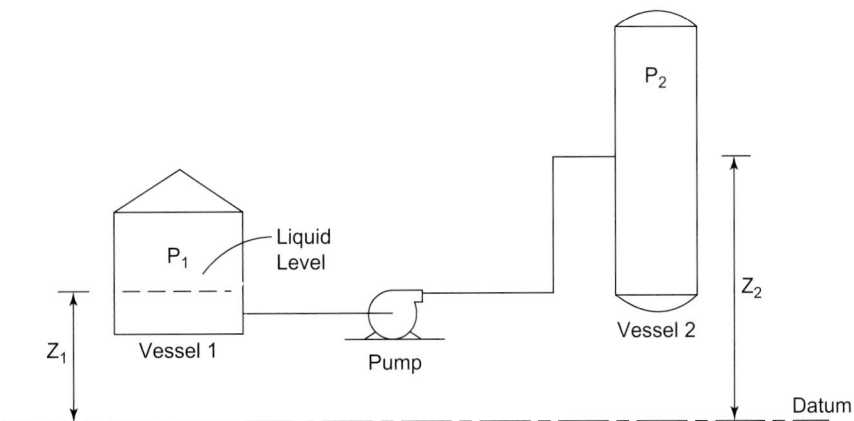

Figure 5.12.   Piping system.

3. overcome the losses in process equipment (e.g. heat exchangers, packed beds);
4. overcome any difference in elevation from end to end of the pipe;
5. overcome any difference in pressure between the vessels at each end of the pipeline.

The total energy required can be calculated from the energy equation:

$$g\Delta z + \Delta P/\rho - \Delta P_f/\rho - W = 0$$

where

W = work done by the fluid, J/kg
Δz = difference in elevations $(z_1 - z_2)$, m
ΔP = difference in system pressures $(P_1 - P_2)$, N/m^2
ΔP_f = pressure drop due to friction, including miscellaneous losses, and equipment losses, (see Section 5.4.2), N/m^2
ρ = liquid density, kg/m^3
g = acceleration due to gravity, m/s^2.

If W is negative a pump is required; if it is positive a turbine could be installed to extract energy from the system.

$$\text{The head required from the pump} = \Delta P_f/\rho g - \Delta P/\rho g - \Delta z \qquad (5.5a)$$

The power is given by:

$$(W \times m)/\eta, \text{ for a pump} \qquad (5.6a)$$

$$(W \times m) \times \eta, \text{ for a turbine} \qquad (5.6b)$$

where

m = mass flow rate, kg/s
η = efficiency = power out/power in.

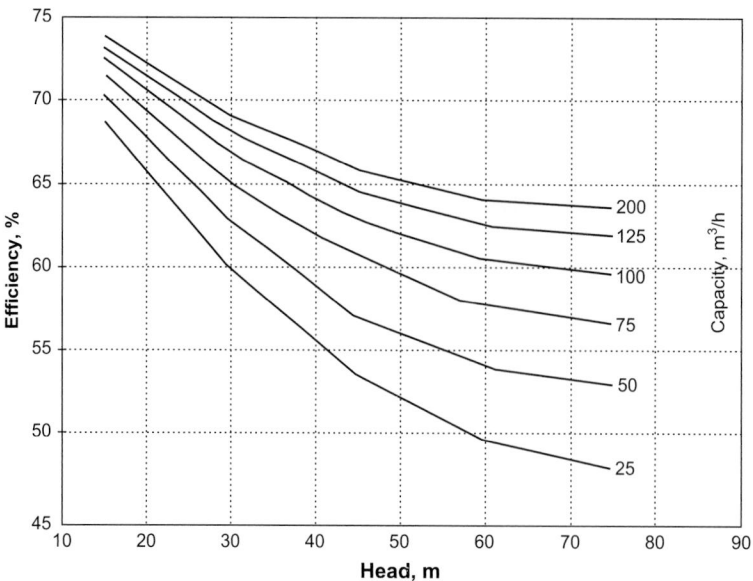

Figure 5.13.   Centrifugal pump efficiency.

The efficiency will depend on the type of pump used and the operating conditions. For preliminary design calculations, the efficiency of centrifugal pumps can be estimated using Figure. 5.13.

Example 5.2

A tanker carrying toluene is unloaded, using the ship's pumps, to an on-shore storage tank. The pipeline is 225 mm internal diameter and 900 m long. Miscellaneous losses due to fittings, valves, etc., amount to 600 equivalent pipe diameters. The maximum liquid level in the storage tank is 30 m above the lowest level in the ship's tanks. The ship's tanks are nitrogen blanketed and maintained at a pressure of 1.05 bar. The storage tank has a floating roof, which exerts a pressure of 1.1 bar on the liquid.

The ship must unload 1000 metric tons within 5 h to avoid demurrage charges. Estimate the power required by the pump. Take the pump efficiency as 70%.

Physical properties of toluene: density 874 kg/m^3, viscosity 0.62 mNm^{-2} s.

Solution

$$\text{Cross-sectional area of pipe} = \frac{\pi}{4}(225 \times 10^{-3})^2 = 0.0398\,\text{m}^2$$

$$\text{Minimum fluid velocity} = \frac{1000 \times 10^3}{5 \times 3600} \times \frac{1}{0.0398} \times \frac{1}{874} = 1.6\,\text{m/s}$$

$$\text{Reynolds number} = (874 \times 1.6 \times 225 \times 10^{-3})/0.62 \times 10^{-3}$$
$$= 507,484 = 5.1 \times 10^5 \tag{5.4}$$

Absolute roughness commercial steel pipe, Table 5.2 = 0.046 mm
Relative roughness = 0.046/225 = 0.0002

Friction factor from Figure 5.11, $f = 0.0019$
Total length of pipeline, including miscellaneous losses,

$$= 900 + 600 \times 225 \times 10^{-3} = 1035 \text{ m}$$

Friction loss in pipeline, $\Delta P_f = 8 \times 0.0019 \times \left(\frac{1035}{225 \times 10^{-3}}\right) \times 874 \times \frac{1.62^2}{2}$

$$= 78,221 \text{ N/m}^2 \tag{5.3}$$

Maximum difference in elevation, $(z_1 - z_2) = (0 - 30) = \underline{-30 \text{ m}}$

Pressure difference, $(P_1 - P_2) = (1.05 - 1.1)10^5 = \underline{-5 \times 10^3 \text{ N/m}^2}$

Energy balance

$$9.8(-30) + (-5 \times 103)/874 - (78,221)/874 - W = 0 \tag{5.5}$$

$$W = \underline{-389.2} \text{ J/kg}$$

$$\text{Power} = (389.2 \times 55.56)/0.7 = 30,981 \text{ W, say } \underline{31 \text{ kW}} \tag{5.6a}$$

Note that this is the maximum power required by the pump, at the end of the unloading when the ship's tank is nearly empty and the storage tank is nearly full. Initially, the difference in elevation is lower and the power required is reduced. For design purposes, the maximum power case would be the governing case and would be used to size the pump and motor.

5.4.4. Characteristic Curves for Centrifugal Pumps

The performance of a centrifugal pump is characterized by plotting the head developed against the flow rate. The pump efficiency can be shown on the same curve. A typical plot is shown in Figure 5.14. The head developed by the pump falls as the flow rate is increased. The efficiency rises to a maximum and then falls.

For a given type and design of pump, the performance will depend on the impeller diameter, the pump speed, and the number of stages. Pump manufacturers publish families of operating curves for the range of pumps they sell. These can be used to select the best pump for a given duty. A typical set of curves is shown in Figure 5.15.

5.4.5. System Curve (Operating Line)

There are two components to the pressure head that has to be supplied by the pump in a piping system:

1. The static pressure, to overcome the differences in head (height) and pressure.
2. The dynamic loss due to friction in the pipe, the miscellaneous losses, and the pressure loss through equipment.

The static pressure difference will be independent of the fluid flow rate. The dynamic loss will increase as the flow rate is increased. It will be roughly proportional

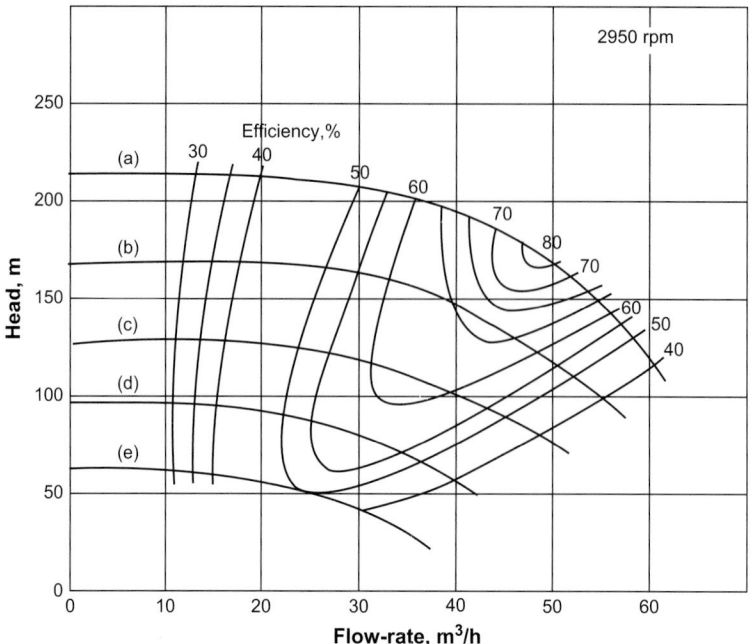

Figure 5.14.   Pump characteristic for a range of impeller sizes: (a) 250 mm, (b) 225 mm, (c) 200 mm, (d) 175 mm, (e) 150 mm.

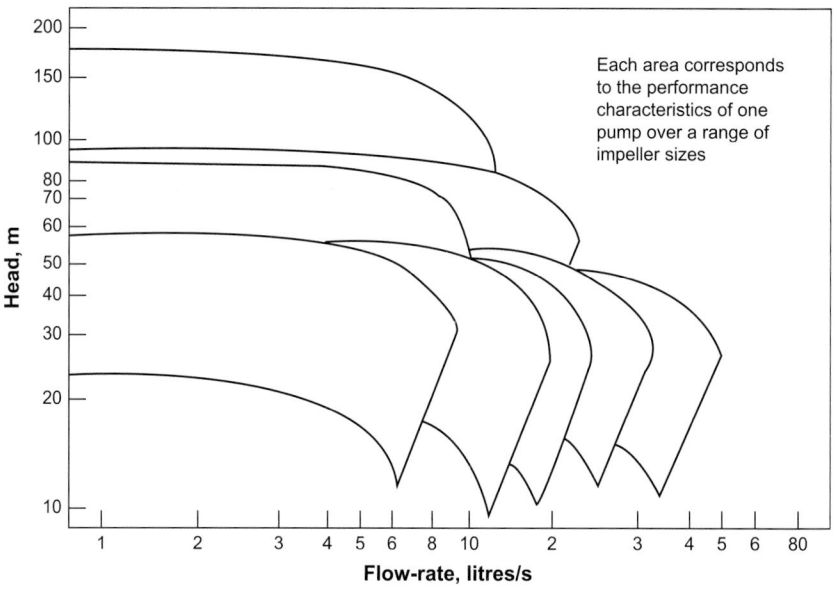

Figure 5.15.   Family of pump curves.

to the flow rate squared; see equation 5.3. The system curve, or operating line, is a plot of the total pressure head versus the liquid flow rate. The operating point of a centrifugal pump can be found by plotting the system curve on the pump's characteristic curve; see Example 5.3. The operating point is the point where the system curve and pump curve intersect.

When selecting a centrifugal pump for a given duty, it is important to match the pump characteristic with the system curve. The operating point should be as close as is practical to the point of maximum pump efficiency, allowing for the range of flow rate over which the pump may be required to operate. This requires a good understanding of the pressure drop across the control valve if a valve is used in the line downstream of the pump. The control valve pressure drop will be proportional to the velocity squared, and can range from as low as 6 to over 100 velocity heads over the range of operation of the valve, depending on the type of valve chosen (see Table 5.3). The system curve should be plotted for the case when the valve is fully open and the case when the valve is quarter-open (or at the minimum fraction open recommended by the valve manufacturer) to determine the range over which flow can be controlled with a given combination of valve and pump. Details of valve pressure drop can be obtained from manufacturers. The equations for design of a valve are given in Coulson *et al.* (1999) and in Green and Perry (2007).

Most centrifugal pumps are controlled by throttling the flow with a valve on the pump discharge; see Section 5.8.3. This varies the dynamic pressure loss, and so the position of the operating point on the pump characteristic curve.

Throttling the flow causes an energy loss. This energy loss is acceptable in most applications; however, when the flow rates are large, the use of variable speed control on the pump drive should be considered to conserve energy.

A more detailed discussion of the operating characteristics of centrifugal and other types of pump is given by Walas (1990) and Karassik *et al.* (2001).

Example 5.3

A process liquid is pumped from a storage tank to a distillation column, using a centrifugal pump. The pipeline is 80-mm internal diameter commercial steel pipe, 100 m long. Miscellaneous losses are equivalent to 600 pipe diameters. The storage tank operates at atmospheric pressure and the column at 1.7 bara. The lowest liquid level in the tank will be 1.5 m above the pump inlet, and the feed point to the column is 3 m above the pump inlet.

Plot the system curve on the pump characteristic given in Figure 5.16 and determine the operating point and pump efficiency.

Properties of the fluid: density $900 \, \text{kg/m}^3$, viscosity $1.36 \, \text{mNm}^{-2} \, \text{s}$.

Solution

Static Head

Difference in elevation, $\Delta z = 3.0 - 1.5 = 1.5 \, \text{m}$

Difference in pressure, $\Delta P = (1.7 - 1.013)10^5 = 0.7 \times 10^5 \, \text{N/m}^2$

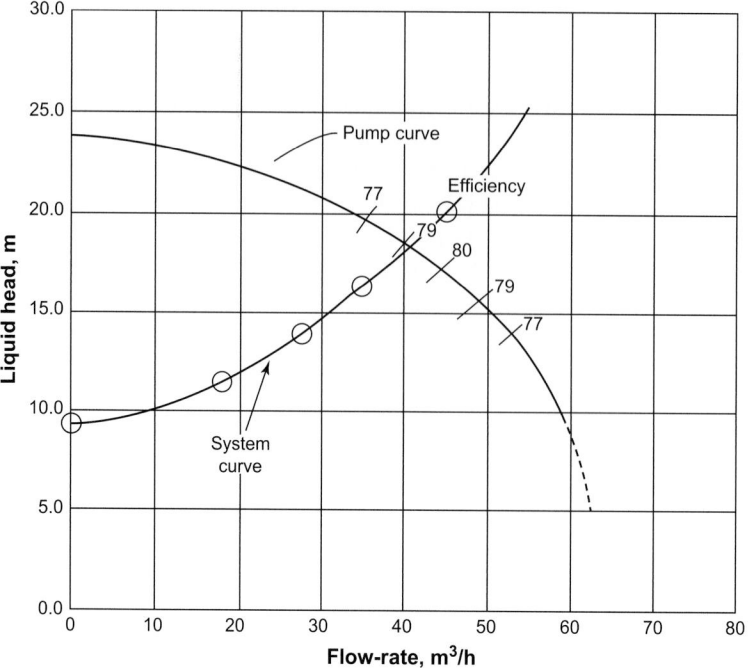

Figure 5.16.   Example 5.3.

$$\text{as head of liquid} = (0.7 = 10^5)/(900 \times 9.8) = 7.9 \text{ m}$$

$$\text{Total static head} = 1.5 + 7.9 = \underline{9.4 \text{ m}}$$

Dynamic Head

As an initial value, take the fluid velocity as 1 m/s, a reasonable value.

$$\text{Cross-sectional area of pipe} = \frac{\pi}{4}(80 \times 10^{-3})^2 = 5.03 \times 10^{-3} \text{ m}^2$$

$$\text{Volumetric flow rate} = 1 \times 5.03 \times 10^{-3} \times 3600 = 18.1 \text{ m}^3/\text{h}$$

$$\text{Reynolds number} = \frac{900 \times 1 \times 80 \times 10^{-3}}{1.36 \times 10^{-3}} = 5.3 \times 10^4 \quad (5.4)$$

$$\text{Relative roughness} = 0.046/80 = 0.0006$$

Friction factor from Figure 5.11, $f = 0.0027$

$$\text{Length including miscellaneous losses} = 100 + (600 \times 80 \times 10^{-3}) = 148 \text{ m}$$

$$\text{Pressure drop, } \Delta P_f = 8 \times 0.0027 \frac{(148)}{(80 \times 10^{-3})} \times 900 \times \frac{1^2}{2} = \underline{\underline{17,982 \text{ N/m}^2}}$$

$$= 17,982/(900 \times 9.8) = \underline{2.03} \text{ m liquid} \quad (5.3)$$

Total head = 9.4 + 2.03 = 11.4 m

To find the system curve the calculations were repeated for the velocities shown in the table below:

Velocity m/s	Flow rate m³/h	Static head m	Dynamic head m	Total head m
1	18.1	9.4	2.0	11.4
1.5	27.2	9.4	4.3	14.0
2.0	36.2	9.4	6.8	16.2
2.5	45.3	9.4	10.7	20.1
3.0	54.3	9.4	15.2	24.6

Plotting these values on the pump characteristic gives the operating point as 18.5 m at 41 m³/h and the pump efficiency as 79%.

5.4.6. Net Positive Suction Head (NPSH)

The pressure at the inlet to a pump must be high enough to prevent cavitation occurring in the pump. Cavitation occurs when bubbles of vapour, or gas, form in the pump casing. Vapour bubbles will form if the pressure falls below the vapour pressure of the liquid.

The net positive suction head available ($NPSH_{avail}$) is the pressure at the pump suction, above the vapour pressure of the liquid, expressed as head of liquid.

The net positive head required ($NPSH_{reqd}$) is a function of the design parameters of the pump, and will be specified by the pump manufacturer. As a general guide, the NPSH should be above 3 m for pump capacities up to 100 m³/h, and 6 m above this capacity. Special impeller designs can be used to overcome problems of low suction head; see Doolin (1977).

The net positive head available is given by the following equation:

$$NPSH_{avail} = P/\rho.g + H - P_f/\rho.g - P_v/\rho.g \qquad (5.7)$$

where

$NPSH_{avail}$ = net positive suction head available at the pump suction, m
P = the pressure above the liquid in the feed vessel, N/m²
H = the height of liquid above the pump suction, m
P_f = the pressure loss in the suction piping, N/m²
P_v = the vapour pressure of the liquid at the pump suction, N/m²
ρ = the density of the liquid at the pump suction temperature, kg/m³
g = the acceleration due to gravity, m/s².

The inlet piping arrangement must be designed to ensure that $NPSH_{avail}$ exceeds $NPSH_{reqd}$ under all operating conditions.

The calculation of $NPSH_{avail}$ is illustrated in Example 5.4.

Example 5.4

Liquid chlorine is unloaded from rail tankers into a storage vessel. To provide the necessary NPSH, the transfer pump is placed in a pit below ground level. Given the following information, calculate the NPSH available at the inlet to the pump, at a maximum flow rate of 16,000 kg/h.

The total length of the pipeline from the rail tanker outlet to the pump inlet is 50 m. The vertical distance from the tank outlet to the pump inlet is 10 m. Commercial steel piping, 50-mm internal diameter, is used.

Miscellaneous friction losses due to the tanker outlet constriction and the pipe fittings in the inlet piping are equivalent to 1000 equivalent pipe diameters. The vapour pressure of chlorine at the maximum temperature reached at the pump is 685 kN/m^2 and its density and viscosity 1286 kg/m^3 and 0.364 mNm^{-2} s. The pressure in the tanker is 7 bara.

Solution

Friction Losses

$$\text{Miscellaneous losses} = 1000 \times 50 \times 10^{-3} = 50 \text{ m of pipe}$$

$$\text{Total length of inlet piping} = 50 + 50 = 100 \text{ m}$$

$$\text{Relative roughness, } e/d = 0.046/50 = 0.001$$

$$\text{Pipe cross-sectional area} = \frac{\pi}{4}(50 \times 10^{-3})^2 = 1.96 \times 10^{-3} \text{ m}^2$$

$$\text{Velocity, } u = \frac{16,000}{3600} \times \frac{1}{1.96 \times 10^{-3}} \times \frac{1}{1286} = 1.76 \text{ m/s}$$

$$\text{Reynolds number} = \frac{1286 \times 1.76 \times 50 \times 10^{-3}}{0.364 \times 10^{-3}} = 3.1 \times 10^5 \tag{5.4}$$

$$\text{Friction factor from Figure 5.11, } f = 0.00225$$

$$\Delta P_f = 8 \times 0.00225 \frac{(100)}{(50 \times 10^{-3})} \times 1286 \times \frac{1.76^2}{2} = 71,703 \text{ N/m}^2 \tag{5.3}$$

$$\text{NPSH} = \frac{7 \times 10^5}{1286 \times 9.8} + 10 - \frac{71.703}{1286 \times 9.8} - \frac{685 \times 10^3}{1286 \times 9.8} \tag{5.7}$$

$$= 55.5 + 10 - 5.7 - 54.4 = \underline{5.4 \text{ m}}$$

5.4.7. Pump and Other Shaft Seals

A seal must be made where a rotating shaft passes through the casing of a pump or the wall of a vessel. The seal must serve several functions:

1. To keep the liquid contained.
2. To prevent ingress of incompatible fluids, such as air.
3. To prevent escape of flammable or toxic materials.

Packed Glands

The simplest, and oldest, form of seal is the packed gland, or stuffing box (Figure 5.17). Its applications range from sealing the stems of the water faucets in every home, to proving the seal on industrial pumps, agitator and valve shafts.

The shaft runs through a housing (gland) and the space between the shaft and the wall of the housing is filled with rings of packing. A gland follower is used to apply pressure to the packing to ensure that the seal is tight. Proprietary packing materials are used. A summary of the factors to be considered in the selection of packing materials for packed glands is given by Hoyle (1975). To make a completely tight seal, the pressure on the packing must be two to three times the system pressure. This can lead to excessive wear on rotating shafts and lower pressures are used, allowing some leakage which lubricates the packing. So, packed glands should only be specified for fluids that are not toxic, corrosive or flammable.

To provide positive lubrication, a lantern ring is often incorporated in the packing and lubricant forced through the ring into the packing (Figure 5.18). With a pump seal, a flush is often take from the pump discharge and returned to the seal, through the lantern ring, to lubricate and cool the packing. If any leakage to the environment must be avoided, a separate flush liquid can be used. A liquid must be selected that is compatible with the process fluid and the environment; water is often used.

Mechanical Seals

In the process industries the conditions at the pump seal are often harsh and more complex seals are needed. Mechanical face seals are used (Figure 5.19). They are generally referred to simply as mechanical seals, and are used only on rotating shafts.

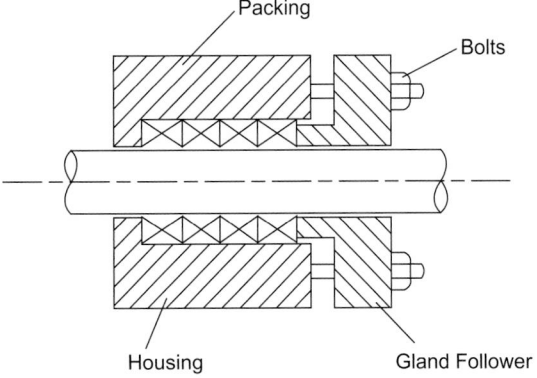

Figure 5.17. Packed gland.

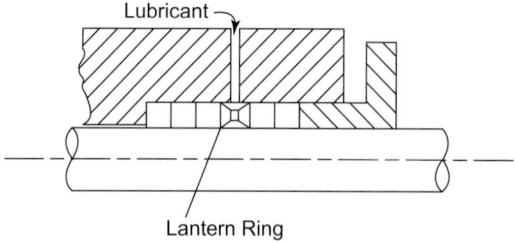

Figure 5.18.   Packed gland with lantern ring.

The seal is formed between two flat faces, set perpendicular to the shaft. One face rotates with the shaft, the other is stationary. The seal is made, and the faces lubricated, by a very thin film of liquid, about 0.0001 μm thick. A particular advantage of this type of seal is that it can provide a very effective seal without causing any wear on the shaft. The wear is transferred to the special seal faces. Some leakage will occur but it is small, normally only a few drops per hour.

Unlike a packed gland, a mechanical seal, when correctly installed and maintained, can be considered leak tight.

A great variety of mechanical seal designs are available, and seals can be found to suit virtually all applications. Only the basic mechanical seal is described below. Full details, and specifications, of the range of seals available and their applications can be obtained from manufacturers' catalogues.

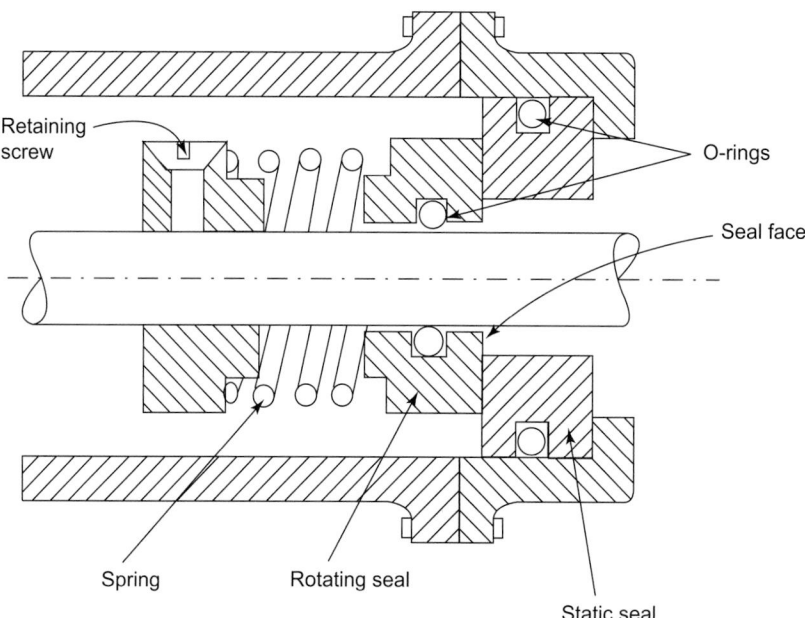

Figure 5.19.   Basic mechanical seal.

The Basic Mechanical Seal

The components of a mechanical seal (Figure 5.19) are:

1. A stationary sealing ring (mating ring).
2. A seal for the stationary ring, O-rings or gaskets.
3. A rotating seal ring (primary ring), mounted so that it can slide along the shaft to take up wear in the seal faces.
4. A secondary seal for the rotating ring mount, usually O-rings or chevron seals.
5. A spring to maintain contact pressure between the seal faces, to push the faces together.
6. A thrust support for the spring, either a collar keyed to the shaft or a step in the shaft.

The assembled seal is fitted into a gland housing (stuffing box) and held in place by a retaining ring (gland plate).

Mechanical seals are classified as inside or outside, depending on whether the primary (rotating ring) is located inside the housing, running in the fluid, or outside. Outside seals are easier to maintain, but inside seals are more commonly used in the process industries, as it is easier to lubricate and flush this type.

Double Seals

Where it is necessary to prevent any leakage of fluid to the atmosphere, a double mechanical seal is used. The space between the two seals is flushed with a harmless fluid, compatible with the process fluid, and provides a buffer between the two seals.

Seal-Less Pumps (Canned Pumps)

Pumps that have no seal on the shaft between the pump and the drive motor are available. They are used for severe duties, where it is essential that there is no leakage into the process fluid, or the environment.

The drive motor and pump are enclosed in a single casing and the stator windings and armature are protected by metal cans; they are usually referred to as canned pumps. The motor runs in the process fluid. The use of canned pumps to control environmental pollution is discussed by Webster (1979).

5.4.8. Gas Compressors

The equipment used to compress a gas through a process piping system is different from that used for liquids. At low pressure drops a simple fan may be adequate. At higher pressure drops multi-stage compressors are commonly used. The different types of compressor and guidelines for their selection are discussed in Chapter 10, Section 10.12. Equations for the power consumed in gas compression are given in Chapter 3, Section 3.13.

5.5. MECHANICAL DESIGN OF PIPING SYSTEMS

5.5.1. Piping System Design Codes

The most widely used international design codes for pressure piping are those set by the ASME B31 Committee. Different standards are required for different services, as

TABLE 5.4.   ASME Pipe Codes

Code No.	Scope	Latest Revision
B31.1	Power piping	2004
B31.2	Fuel gas piping	1968
B31.3	Process piping	2004
B31.4	Pipeline transportation systems for liquid hydrocarbons and other liquids	1997
B31.5	Refrigeration piping and heat transfer components	2001
B31.8	Gas transmission and distribution piping systems	2003
B31.9	Building services piping	2004
B31.11	Slurry transportation piping systems	2002

shown in Table 5.4. Most chemical plant and oil refinery piping is designed in accordance with ASME B31.3, which will be used as reference in the following sections. The ASME B31.3 code applies to piping for raw, intermediate and finished chemicals; petroleum products; gas, steam, air and water; fluidized solids; refrigerants and cryogenic fluids. It does not apply to:

1. Piping systems designed for pressures less than 15 psi gage that handle fluids that are non-flammable, non-toxic and not damaging to human tissues, and are at temperatures between $-29°C$ $(-20°F)$ and $186°C$ $(366°F)$.
2. Power boiler piping that conforms to ASME B31.1 and power boilers that conform to ASME Boiler and Pressure Vessel Code Section I.
3. Fired heater tubes, tube headers and manifolds that are internal to the heater enclosure.
4. Internal piping and external pipe connections of pressure vessels, heat exchangers, pumps, compressors and other fluid handling or process equipment.

Note though that different design standards apply for refrigeration plants, fuel gas piping, power plant and slurry-handling systems.

The British Standards for piping were largely developed from the ASME B31 standards. For example, BS 3799 for pipe fittings is based on ASME B16.11. The older British Standards are now being superseded by European Standards; some of those that are now active are listed in the references at the end of this chapter.

5.5.2.   Wall Thickness: Pipe Schedule

The pipe wall thickness is selected to resist the internal pressure, with allowances for corrosion, erosion, and other mechanical allowances for pipe threads, etc. Process pipes can normally be considered as thin cylinders; only high-pressure pipes, such as high-pressure steam lines, are likely to be classified as thick cylinders and must be given special consideration (see Chapter 13).

The ASME B31.3 code gives the following formula for pipe thickness:

$$t_m = t_p + c$$
$$t_p = \frac{Pd}{2(SE + P\gamma)} \tag{5.8}$$

where

t_m = minimum required thickness
t_p = pressure design thickness
c = sum of mechanical allowances (thread depth) plus corrosion and erosion allowances
P = internal design gauge pressure, lb/in^2 (or N/mm^2)
d = pipe outside diameter
S = basic allowable stress for pipe material, lb/in^2 (or N/mm^2)
E = casting quality factor
γ = temperature coefficient.

Allowable stresses and values of the coefficients for different materials are given in Appendix A of the design code. Standard dimensions for stainless steel pipe are given in ASME B36.19 and for wrought steel and wrought iron pipe in ASME B36.10M. Standard pipe dimensions are also summarized by Green and Perry (2007).

Pipes are often specified by a schedule number (based on the thin cylinder formula), defined by:

$$\text{Schedule number} = \frac{P_s \times 1000}{\sigma_s} \qquad (5.9)$$

where

P_s = safe working pressure, lb/in^2 (or N/mm^2)
σ_s = safe working stress, lb/in^2 (or N/mm^2).

Schedule 40 pipe is commonly used for general purpose applications at low pressure.

Example 5.5

Estimate the safe working pressure for a 4-in (100-mm) diameter, schedule 40 pipe, SA53 carbon steel, butt welded, working temperature 100°C. The maximum allowable stress for butt-welded steel pipe up to 120°C is 11,700 lb/in^2 (79.6 N/mm^2).

Solution

$$P_S = \frac{(\text{schedule no.}) \times \sigma_S}{1000} = \frac{40 \times 11700}{1000} = \underline{\underline{468\,lb/in^2}} = \underline{\underline{3180\,kN/m^2}}$$

5.5.3. Pipe Supports

Over long runs, between buildings and equipment, pipes are usually carried on pipe racks. These carry the main process and service pipes, and are laid out to allow easy access to the equipment.

Various designs of pipe hangers and supports are used to support individual pipes. Details of typical supports can be found in the books by Green and Perry (2007) and Nayyar *et al.* (2000). Pipe supports frequently incorporate provision for thermal expansion.

5.5.4. Pipe Fittings

Pipe runs are normally made up from lengths of pipe, incorporating standard fittings for joints, bends and tees. Joints are usually welded but small sizes may be screwed. Flanged joints are used where this is a more convenient method of assembly, or if the joint will have to be frequently broken for maintenance. Flanged joints are normally used for the final connection to the process equipment, valves and ancillary equipment.

Details of the standard pipe fittings, welded, screwed and flanged, can be found in manufacturers' catalogues and in the appropriate national standards. The European Standards for pipe fittings are BS EN 10241 and BS EN 10253. American Standards for pipe fittings are set by the ASME B16 committee. The standards for metal pipes and fittings are discussed by Masek (1968). Flanges and flange standards are discussed in Chapter 13, Section 13.10.

5.5.5. Pipe Stressing

Piping systems must be designed so as not to impose unacceptable stresses on the equipment to which they are connected.

Loads arise from:

1. Thermal expansion of the pipes and equipment.
2. The weight of the pipes, their contents, insulation and any ancillary equipment.
3. The reaction to the fluid pressure drop.
4. Loads imposed by the operation of ancillary equipment, such as relief valves.
5. Vibration.

Thermal expansion is a major factor to be considered in the design of piping systems. The reaction load due to pressure drop will normally be negligible. The dead-weight loads can be carried by properly designed supports.

Flexibility is incorporated into piping systems to absorb the thermal expansion. A piping system will have a certain amount of flexibility due to the bends and loops required by the layout. If necessary, expansion loops, bellows and other special expansion devices can be used to take up expansion.

A discussion of the methods used for the calculation of piping flexibility and stress analysis are beyond the scope of this book. Manual calculation techniques, and the application of computers in piping stress analysis, are discussed in the handbook edited by Nayyar *et al.* (2000).

5.5.6. Layout and Design

An extensive discussion of the techniques used for piping system design and specification is beyond the scope of this book. The subject is covered thoroughly in the books by Sherwood (1991), Kentish (1982a) (1982b) and Lamit (1981).

5.6. PIPE SIZE SELECTION

If the motive power to drive the fluid through the pipe is available free, for instance when pressure is let down from one vessel to another or if there is sufficient head for gravity flow, the smallest pipe diameter that gives the required flow rate would normally be used.

If the fluid has to be pumped through the pipe, the size should be selected to give the least total annualized cost.

Typical pipe velocities and allowable pressure drops that can be used to estimate pipe sizes are given below:

	Velocity, m/s	ΔP, kPa/m
Liquids, pumped (not viscous)	1–3	0.5
Liquids, gravity flow	—	0.05
Gases and vapours	15–30	0.02% of line pressure
High-pressure steam, >8 bar	30–60	—

Rase (1953) gives expressions for design velocities in terms of the pipe diameter. His expressions, converted to SI units, are:

Pump discharge	$0.06d_i + 0.4$ m/s
Pump suction	$0.02d_i + 0.1$ m/s
Steam or vapour	$0.2d_i$ m/s

where d_i is the internal diameter in mm.

Simpson (1968) gives values for the optimum velocity in terms of the fluid density. His values, converted to SI units and rounded, are:

Fluid density, kg/m^3	Velocity, m/s
1600	2.4
800	3.0
160	4.9
16	9.4
0.16	18.0
0.016	34.0

The maximum velocity should be kept below that at which erosion is likely to occur. For gases and vapours the velocity cannot exceed the critical velocity (sonic velocity) (see Coulson *et al.* (1999), Chapter 4) and would normally be limited to 30% of the critical velocity.

Economic Pipe Diameter

The capital cost of a pipe run increases with diameter, whereas the pumping costs decrease with increasing diameter. The most economic pipe diameter will be the one that gives the lowest total annualized cost. Several authors have published formulae and nomographs for the estimation of the economic pipe diameter: Genereaux (1937), Peters and Timmerhaus (1968) (1991), Nolte (1978) and Capps (1995).

A rule of thumb for the economic pipe diameter that is widely used in oil refining is:

$$\text{Economic diameter in inches} = (\text{flow rate in gpm})^{0.5}$$

In metric units this converts to:

$$d_i, \text{optimum} = 0.33(G/\rho)^{0.5}$$

where

$G = $ flow rate, kg/s
$\rho = $ density, kg/m^3
$d_i = $ pipe id, m.

The formulae developed in this section are presented as an illustration of a simple optimization problem in design, and to provide an estimate of economic pipe diameter in SI units. The method used is essentially that first published by Genereaux (1937).

The cost equations can be developed by considering a 1-m length of pipe.

The purchase cost will be roughly proportional to the diameter raised to some power.

$$\text{Purchase cost} = Bd^n \$/m$$

The value of the constant B and the index n depend on the pipe material and schedule.

The installed cost can be calculated by using the factorial method of costing discussed in Chapter 6.

$$\text{Installed cost} = Bd^n(1 + F)$$

where the factor F includes the cost of valves, fittings and erection, for a typical run of the pipe.

The capital cost can be included in the operating cost as an annual capital charge. There will also be an annual charge for maintenance, based on the capital cost.

$$C_C = Bd^n(1 + F)(a + b) \tag{5.10}$$

where

$C_C = $ annualized capital cost of the piping, \$/m.yr
$a = $ capital annualization factor, yr^{-1}
$b = $ maintenance costs as fraction of installed capital, yr^{-1}.

The power required for pumping is given by:

$$\text{Power} = \text{Volumetric flow rate} \times \text{pressure drop}$$

Only the friction pressure drop need be considered, as any static head is not a function of the pipe diameter.

To calculate the pressure drop the pipe friction factor needs to be known. This is a function of Reynolds number, which is in turn a function of the pipe diameter. Several expressions have been proposed for relating friction factor to Reynolds number. For simplicity the relationship proposed by Genereaux (1937) for turbulent flow in clean commercial steel pipes will be used.

$$C_f = 0.04\, Re^{-0.16}$$

where C_f is the Fanning friction factor $= 2(R/\rho u^2)$.

Substituting this into the Fanning pressure drop equation gives:

$$\Delta P = 0.125 G^{1.84} \mu^{0.16} \rho^{-1} d_i^{-4.84} \tag{5.11}$$

where

ΔP = pressure drop, N/m^2 (Pa)
μ = viscosity, Nm^{-2} s.

The annual pumping costs will be given by:

$$C_w = \frac{A\ p}{1000\ \eta}\, \Delta P\, \frac{G}{\rho}$$

where

A = plant attainment, hours/year
p = cost of power, \$/kWh
η = pump efficiency.

Substituting from equation 5.11:

$$C_w = \frac{A\ p}{\eta}\, 1.25 \times 10^{-4}\, G^{2.84}\, \mu^{0.16}\, \rho^{-2}\, d_i^{-4.84} \tag{5.12}$$

The total annual operating cost $C_t = C_C + C_w$.

Adding equations 5.10 and 5.12, differentiating, and equating to zero to find the pipe diameter to give the minimum cost gives:

$$d_i,\ \text{optimum} = \left(\frac{6.05 \times 10^{-4}\ A\ p\ G^{2.84}\ \mu^{0.16}\ \rho^{-2}}{\eta\ n\ B\ (1+F)(a+b)} \right)^{1/(4.84+n)} \tag{5.13}$$

Equation 5.13 is a general equation and can be used to estimate the economic pipe diameter for any particular situation. It can be set up on a spreadsheet and the effect of the various factors investigated.

The equation can be simplified by substituting typical values for the constants.

A The normal attainment for a chemical process plant will be 90–95%, so take the operating hours per year as 8000.
η Pump and compressor efficiencies will be 50–70%, so take 0.6.
p A typical wholesale cost of electric power for a large user is 0.06 \$/kWh (mid-2008).

F This is the most difficult factor to estimate. Other authors have used values ranging from 1.5 (Peters and Timmerhaus, 1968) to 6.75 (Nolte, 1978). It is best taken as a function of the pipe diameter, as has been done to derive the simplified equations given below.

B, n Can be estimated from the current cost of piping.

    a Will depend on the current cost of capital, and could range from 0.1 to 0.25, but is typically around 0.16. See Chapter 6 for a detailed discussion.

    b A typical figure for process plant will be 5%; see Chapter 6.

F, *B*, and *n* are best established from a recent correlation of piping costs and should include the costs of fittings, paint or insulation and installation. For initial estimates the following correlations (basis January 2006) can be used.

$$\text{A106 Carbon steel 1 to 8 inches, } \$/\text{ft} = 17.4 \ (d_i \text{ in inches})^{0.74}$$

$$\text{10 to 24 inches, } \$/\text{ft} = 1.03 \ (d_i \text{ in inches})^{1.73}$$

$$\text{304 Stainless steel 1 to 8 inches, } \$/\text{ft} = 24.5 \ (d_i \text{ in inches})^{0.9}$$

$$\text{10 to 24 inches, } \$/\text{ft} = 2.74 \ (d_i \text{ in inches})^{1.7}$$

In metric units (d_i in metres), these become:

$$\text{A106 Carbon steel 25 to 200 mm, } \$/\text{m} = 880 \ d_i^{0.74}$$

$$\text{250 to 600 mm, } \$/\text{m} = 1900 \ d_i^{1.73}$$

$$\text{304 Stainless steel 25 to 200 mm, } \$/\text{m} = 2200 \ d_i^{0.94}$$

$$\text{250 to 600 mm, } \$/\text{m} = 4700 \ d_i^{1.7}$$

For small-diameter carbon steel pipes, substitution in equation 5.12 gives:

$$d_i, \text{ optimum} = 0.830 G^{0.51} \mu^{0.03} \rho^{-0.36}$$

Because the exponent of the viscosity term is small, its value will change very little over a wide range of viscosity

at
$$\mu = 10^{-5} \text{ Nm}^{-2}\text{s} \ (0.01 \text{ cp}), \mu^{0.03} = 0.71$$

$$\mu = 10^{-2} \text{ Nm}^{-2}\text{s} \ (10 \text{ cp}), \mu^{0.03} = 0.88$$

Taking a mean value of 0.8 gives the following equations for the optimum diameter, for turbulent flow:

A106 Carbon steel pipe:

$$\text{25 to 200 mm, } d_i, \text{optimum} = 0.664 \ G^{0.51} \ \rho^{-0.36}$$

$$\text{250 to 600 mm, } d_i, \text{optimum} = 0.534 \ G^{0.43} \ \rho^{-0.30} \tag{5.14}$$

304 Stainless steel pipe:

$$\text{25 to 200 mm, } d_i, \text{optimum} = 0.550 \ G^{0.49} \ \rho^{-0.35}$$

$$\text{250 to 600 mm, } d_i, \text{optimum} = 0.465 \ G^{0.43} \ \rho^{-0.31} \tag{5.15}$$

Note that the optimum diameter for stainless steel is smaller than for carbon steel, as would be expected given the higher materials cost of the pipe. Note also that equations 5.14 and 5.15 predict optimum pipe diameters that are roughly double those given by the rule of thumb at the start of this section. This most likely reflects a change in the relative values of capital and energy since the period when the rule of thumb was deduced.

Equations 5.14 and 5.15 can be used to make an approximate estimate of the economic pipe diameter for normal pipe runs. For a more accurate estimate, or if the fluid or pipe run is unusual, the method used to develop equation 5.13 can be used, taking into account the special features of the particular pipe run.

For very long pipe systems, such as transportation pipelines, the capital costs of the required pumps should also be included.

For gases, the capital cost of compression is much more significant and should always be included in the analysis.

Equations for the optimum pipe diameter with laminar flow can be developed by using a suitable equation for pressure drop in the equation for pumping costs.

The approximate equations should not be used for steam, as the quality of steam depends on its pressure, and hence the pressure drop.

Nolte (1978) gives detailed methods for the selection of economic pipe diameters, taking into account all the factors involved. He gives equations for liquids, gases, steam and two-phase systems. He includes in his method an allowance for the pressure drop due to fittings and valves, which was neglected in the development of equation 5.12, and by most other authors.

The use of equations 5.14 and 5.15 are illustrated in Examples 5.6 and 5.7, and the results compared with those obtained by other authors. The older correlations give lower values for the economic pipe diameters, probably due to changes in the relative values of capital and energy. Note, however, that the refining rule of thumb given at the start of this section gives optimum pipe diameters roughly twice those predicted from equations 5.14 and 5.15, perhaps reflecting more conservative pipe sizing in that industry.

Example 5.6

Estimate the optimum pipe diameter for a water flow rate of 10 kg/s, at 20°C. Carbon steel pipe will be used. Density of water, 1000 kg/m^3.

Solution

$$d_i, \text{ optimum} = 0.664 \times (10)^{0.51} 1000^{-0.36} \tag{5.14}$$

$$= \underline{177 \text{ mm}}$$

This resulting size is 6.97 inches, which is not a standard pipe size. We can choose either 6-in or 8-in pipe, so try 6-in pipe, sch 40, inside diameter 6.065 in (154 mm). Viscosity of water at 20°C = 1.1 × 10^{-3} Ns/m^2,

$$Re = \frac{4\,G}{\pi\,\mu\,d} = \frac{4 \times 10}{\pi \times 1.1 \times 10^{-3} \times 154 \times 10^{-3}} = 7.51 \times 10^4$$

> 4000, so flow is turbulent.

Comparison of methods:

	Economic Diameter
Equation 5.14	180 mm
Peters and Timmerhaus (1991)	4 in (100 mm)
Nolte (1978)	80 mm

Example 5.7

Estimate the optimum pipe diameter for a flow of HCl of 7000 kg/h at 5 bar, 15°C, stainless steel pipe. Molar volume 22.4 m^3/kmol, at 1 bar, 0°C.

Solution

Molecular weight HCl = 36.5.

$$\text{Density at operating conditions} = \frac{36.5}{22.4} \times \frac{5}{1} \times \frac{273}{288} = \underline{\underline{7.72 \ \text{kg/m}^3}}$$

$$\text{Optimum diameter} = 0.465 \left(\frac{7000}{3600}\right)^{0.43} \times 7.72^{-0.31} = \underline{\underline{328.4 \ \text{mm}}} \quad (5.15)$$

which is 12.9 inches, so we can use 14-in pipe, sch 40, with inside diameter 13.124 inches (333 mm).

Viscosity of HCl 0.013 mNs/m^2

$$Re = \frac{4}{\pi} \times \frac{7000}{3600} \times \frac{1}{0.013 \times 10^{-3} \times 333 \times 10^{-3}} = \underline{\underline{5.71 \times 10^5}}, \text{ turbulent}$$

Comparison of methods:

	Economic Diameter
Equation 5.15	14 in (333 mm)
Peters and Timmerhaus (1991)	9-in (220-mm) carbon steel
Nolte (1978)	7-in (180-mm) carbon steel

Example 5.8

Calculate the line size and specify the pump required for the line shown in Figure 5.20; material ortho-dichlorobenzene (ODCB), flow rate 10,000 kg/h, temperature 20°C, pipe material carbon steel.

Solution

$$\text{ODCB density at } 20°\text{C} = 1306 \ \text{kg/m}^3$$

$$\text{Viscosity: } 0.9 \ \text{mNs/m}^2 (0.9 \ \text{cp})$$

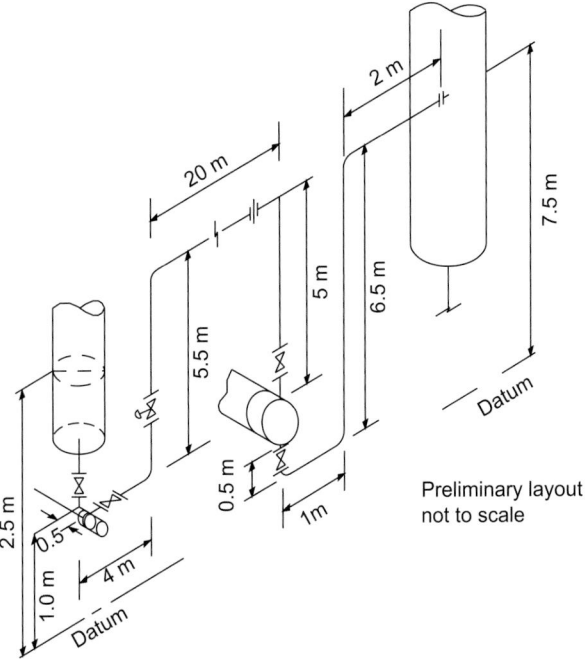

Figure 5.20. Piping isometric drawing (Example 5.8).

Estimation of Pipe Diameter Required

typical velocity for liquid 1 m/s

$$\text{mass flow} = \frac{10^4}{3600} = 2.78 \text{ kg/s}$$

$$\text{volumetric flow} = \frac{2.78}{1306} = 2.13 \times 10^{-3} \text{ m}^3/\text{s}$$

$$\text{area of pipe} = \frac{\text{volumetric flow}}{\text{velocity}} = \frac{2.13 \times 10^{-3}}{1} = 2.13 \times 10^{-3} \text{ m}^2$$

$$\text{diameter of pipe} = \sqrt{\left(2.13 \times 10^{-3} \times \frac{4}{\pi}\right)} = 0.052 \text{ m} = 52 \text{ mm}$$

Or, use economic pipe diameter formula:

$$d, \text{ optimum} = 0.664 \times 2.78^{0.51} \times 1306^{-0.36} \tag{5.14}$$

$$= 78.7 \text{ mm}$$

Take diameter as 77.9 mm (3-in sch 40 pipe)

$$\text{cross-sectional area} = \frac{\pi}{4} (77.9 \times 10^{-3})^2 = 4.77 \times 10^{-3} \text{ m}^2$$

Pressure Drop Calculation

$$\text{fluid velocity} = \frac{2.13 \times 10^{-3}}{4.77 \times 10^{-3}} = 0.45 \text{ m/s}$$

Friction loss per unit length, Δf_1:

$$Re = \frac{1306 \times 0.45 \times 77.9 \times 10^{-3}}{0.9 \times 10^{-3}} = 5.09 \times 10^4 \qquad (5.5)$$

Absolute roughness commercial steel pipe, Table 5.2 = 0.46 mm
Relative roughness, $e/d = 0.046/80 = 0.0005$
Friction factor from Figure 5.11, $f = 0.0025$

$$\Delta f_1 = 8 \times 0.0025 \times \frac{1}{77.9 \times 10^{-3}} \times 1306 \times \frac{0.45^2}{2} = 33.95 \text{ N/m}^2 \quad (5.3)$$

Design for a maximum flow rate of 20% above the average flow.

$$\text{Friction loss} = 0.0339 \times 1.2^2 = 0.0489 \text{ kPa/m}$$

Miscellaneous Losses

Take as equivalent pipe diameters. All bends will be taken as 90° standard radius elbow.

Line to pump suction:

$$\text{length} = 1.5 \text{ m}$$
$$\text{bend, } 1 \times 30 \times 80 \times 10^{-3} = 2.4 \text{ m}$$
$$\text{valve, } 1 \times 18 \times 80 \times 10^{-3} = 1.4 \text{ m}$$
$$\text{total} = 5.3 \text{ m}$$

$$\text{entry loss} = \frac{\rho u^2}{2} \quad \text{(see Section 5.4.2)}$$

$$\text{at maximum design velocity} = \frac{1306 \, (0.45 \times 1.2)^2}{2 \times 10^3} = 0.19 \text{ kPa}$$

Control valve pressure drop, allow normal	140 kPa
($\times 1.2^2$) maximum	200 kPa
Heat exchanger, allow normal	70 kPa
($\times 1.2^2$) maximum	100 kPa
Orifice, allow normal	15 kPa
($\times 1.2^2$) maximum	22 kPa

TABLE 5.5. Line Calculation Form (Example 5.8)

Company Name Address		Project Name									
		Project Number					Sheet	1	of		1
		REV	DATE	BY	APVD	REV	DATE		BY		APVD
Pump and Line Calculation Sheet		1	8.7.06	GPT							
Form XXXXX-YY-ZZ											

Owner's Name					
Plant Location					
Case Description	Chapter 5 Example 5.4				
Equipment label	P101		Equipment name	Stripper bottoms pump	
Plant section					
Process service					
Fluid	ODCB		Density		1306 kg/m^3
Operating temperature	Normal	20 ºC	Viscosity		0.9 N.s/m^2
	Min	15 ºC	Normal flow rate		2.78 kg/s
	Max	30 ºC	Design flow rate		3.34 kg/s

LINE PRESSURE DROP

	SUCTION						DISCHARGE				
	Line size		77.9	mm			Line size		77.9	mm	
Note		Normal	Max.	Units		Note	Flow	Normal	Max.	Units	
u_1	Velocity	0.4	0.5	m/s		u_2	Velocity	0.4	0.5	m/s	
Δf_1	Friction loss	0.03	0.05	kPa/m		Δf_2	Friction loss	0.03	0.05	kPa/m	
L_1	Line length	5.30	5.30	m		L_2	Line length	63.4	63.4	m	
$\Delta f_1 L_1$	Line loss	0.18	0.26	kPa		$\Delta f_2 L_2$	Line loss	2.15	3.09	kPa	
$\rho u_1^2/2$	Entrance loss	0.130	0.188	kPa			Orifice / Flow meter	15	22	kPa	
(40 kPa)	Strainer			kPa			Control valve	140	200	kPa	
	(1) Sub-total	0.310	0.446	kPa			Equipment				
						S&THX	H 205	70	100	kPa	
z_1	Static head	1.5	1.5	m						kPa	
$\rho g z_1$		19.2	19.2	kPa							
	Upstream equipment pressure	100	100	kPa		Total	(6) Dynamic loss	227	325	kPa	
	(2) Sub-total	119.2	119.2	kPa							
						z_2	Static head	6.5	6.5	m	
(2) − (1)	(3) Suction pressure	118.9	111.8	kPa		$\rho g z_2$		83.3	83.3	kPa	
	(4) Vapor pressure	0.1	0.1	kPa			Equip. press (max)	200	200	kPa	
(3) − (4)	(5) NPSH available	118.8	111.7	kPa			Contingency	0	0	kPa	
(5)/ρg	NPSH available	9.3	8.7	m			(7) Sub-total	283.3	283.3	kPa	
	NPSH available	12.1	11.4	m water		(7) + (6)	Discharge pressure	510.4	608.4	kPa	
						(3)	Suction pressure	118.9	111.8	kPa	
							(8) Differential pressure	391.5	496.6	kPa	
						(8)/ρg	Pump head	30.6	38.8	m	
							Control valve				
						Valve/(6)	% Dyn. loss		62%		

PUMP DATA

Pump manufacturer			Driver type	Electric 3-phase
Catalog No.			Power supply	440 V
Pump flow rate	normal	7.7 m^3/h	Seal type	Mechanical, external flush
	max.	9.2 m^3/h	Hydraulic power	0.833 kW
Differential pressure		391.5 kPa	Rated power	kW
		30.6 m	Efficiency	%
		39.9 m water	Suction specific speed	
NPSH required		m		
Pump type			Casing design pressure	610 kPa
No. of stages	single		Casing design temperature	30 ºC
Impeller type	closed		Casing type	
Mounting	horizontal		Casing material	

SKETCH

```
C 201          C 203
1 bar          2 bar

        LT   LIC              H 205

2.5 m                    7.5 m

 M        1.0 m    Z₁ = 2.5-1 = 1.5 m
                   Z₂ = 7.5-1 = 6.5 m
```

NOTES

1. Process data completed, remaining information to be filled in after equipment selection
2.
3.
4.
5.

Line from pump discharge:

$$\text{length} = 4 + 5.5 + 20 + 5 + 0.5 + 1 + 6.5 + 2 = 44.5 \text{ m}$$
$$\text{bends, } 6 \times 30 \times 80 \times 10^{-3} = 14.4 = 14.4 \text{ m}$$
$$\text{valves, } 3 \times 18 \times 80 \times 10^{-3} = 4.4 = \underline{4.4 \text{ m}}$$
$$\text{total} = 63.4 \text{ m}$$

The line pressure-drop calculation is set out on the calculation sheet shown in Table 5.5. A blank version of this calculation sheet can be found in Appendix G and is also available in MS Excel format at http://elsevierdirect.com/companions.

Pump selection:

flow rate $= 2.13 \times 10^{-3} \times 3600 = 7.7 \text{ m}^3/\text{h}$

differential head, maximum, $\underline{38 \text{ m}}$

select single-stage centrifugal (Figure 5.10)

5.7. CONTROL AND INSTRUMENTATION

5.7.1. Instruments

Instruments are provided to monitor the key process variables during plant operation. They may be incorporated in automatic control loops, or used for the manual monitoring of the process operation. In most modern plants, the instruments will be connected to a computer control and data logging system. Instruments monitoring critical process variables will be fitted with automatic alarms to alert the operators to critical and hazardous situations.

Details of process instruments and control equipment can be found in various handbooks; see Green and Perry (2007) and Liptak (2003). Reviews of process instruments and control equipment are published periodically in the journals *Chemical Engineering* and *Hydrocarbon Processing*. These reviews give details of instruments and control hardware available commercially.

It is desirable that the process variable that is to be monitored should be measured directly; however, this is often impractical and some dependent variable that is easier to measure is monitored in its place. For example, in the control of distillation columns the continuous, on-line, analysis of the overhead product is desirable but difficult and expensive to achieve reliably, so temperature is often monitored as an indication of composition. The temperature instrument may form part of a control loop controlling, say, reflux flow; with the composition of the overheads checked frequently by sampling and laboratory analysis.

5.7.2. Instrumentation and Control Objectives

The primary objectives of the designer when specifying instrumentation and control schemes are:

1. Safe plant operation:
   (a) To keep the process variables within known safe operating limits.

(b) To detect dangerous situations as they develop and to provide alarms and automatic shut-down systems.

(c) To provide interlocks and alarms to prevent dangerous operating procedures.

2. Production rate:
   To achieve the design product output.

3. Product quality:
   To maintain the product composition within the specified quality standards.

4. Cost:
   To operate at the lowest production cost, commensurate with the other objectives.

5. Stability:
   To maintain steady, automatic plant operation with minimal operator intervention.

These are not separate objectives and must be considered together. The order in which they are listed is not meant to imply the precedence of any objective over another, other than that of putting safety first. Product quality, production rate and the cost of production will be dependent on sales requirements. For example, it may be a better strategy to produce a better quality product at a higher cost.

In a typical chemical processing plant these objectives are achieved by a combination of automatic control, manual monitoring and laboratory and on-line analysis.

5.7.3. Automatic Control Schemes

The detailed design and specification of the automatic control schemes for a large project is usually done by specialists. The basic theory underlying the design and specification of automatic control systems is covered in several texts: Coughanowr (1991), Shinskey (1984) (1996), Luyben *et al.* (1999), Henson *et al.* (1996), Seborg *et al.* (2004) and Green and Perry (2007). The books by Murrill (1988), Shinskey (1996) and Kalani (2002) cover many of the more practical aspects of process control system design, and are recommended.

In this chapter only the first step in the specification of the control systems for a process will be considered: the preparation of a preliminary scheme of instrumentation and control, developed from the process flow-sheet. This can be drawn up by the process designer based on experience with a similar plant and critical assessment of the process requirements. Many of the control loops will be conventional and a detailed analysis of the system behaviour will not be needed, or justified. Judgment, based on experience, must be used to decide which systems are critical and need detailed analysis and design.

Some examples of typical (conventional) control systems used for the control of specific process variables and unit operations are given in the next section, and can be used as a guide in preparing preliminary instrumentation and control schemes.

Guide Rules

The following procedure can be used when drawing up preliminary P and I diagrams:

1. Identify and draw in those control loops that are obviously needed for steady plant operation, such as:
   (a) level controls
   (b) flow controls
   (c) pressure controls
   (d) temperature controls.
2. Identify the key process variables that need to be controlled to achieve the specified product quality. Include control loops using direct measurement of the controlled variable, where possible; if not practicable, select a suitable dependent variable.
3. Identify and include those additional control loops required for safe operation, not already covered in steps 1 and 2.
4. Decide and show those ancillary instruments needed for the monitoring of the plant operation by the operators, and for trouble-shooting and plant development. It is well worth including additional connections for instruments that may be needed for future trouble-shooting and development, even if the instruments are not installed permanently. These would include extra thermowells, pressure tappings, orifice flanges and sample points.
5. Decide on the location of sample points.
6. Decide on the type of control instrument that will be used, including whether it will be a local instrument or tied into the plant computer control system. Also decide on the type of actuator that can be used, the signal system and whether the instrument will record data. This step should be done in conjunction with steps 1 to 4.
7. Decide on the alarms and interlocks needed; this should be done in conjunction with step 3 (see Chapter 9).

In step 1 it is important to remember the following basic rules of process control:

- There can only be a single control valve on any given stream between unit operations.
- A level controller is needed anywhere where a vapour–liquid or liquid–liquid interface is maintained.
- Pressure control is more responsive when the pressure controller actuates a control valve on a vapour stream.
- Two operations cannot be controlled at different pressures unless there is a valve or other restriction (or a compressor or pump) between them.
- Temperature control is usually achieved by controlling the flow of a utility stream (such as steam or cooling water) or a bypass around an exchanger.
- The overall plant material balance is usually set by flow controllers or flow ratio controllers on the process feeds. There cannot be an additional flow controller on an intermediate stream unless there is provision for accumulation (surge), such as an intermediate storage tank.

Some simple examples of control schemes for common unit operations are given in the next section.

5.8. TYPICAL CONTROL SYSTEMS

5.8.1. Level Control

In any equipment where an interface exists between two phases (e.g. a liquid and a vapour), some means of maintaining the interface at the required level must be provided. This may be incorporated in the design of the equipment, as is usually done for decanters, or by automatic control of the flow from the equipment. Figure 5.21 shows a typical arrangement for level control at the base of a column. The control valve should be placed on the discharge line from the pump.

5.8.2. Pressure Control

Pressure control will be necessary for most systems handling vapour or gas. The method of control depends on the nature of the process. Typical schemes are shown in Figure 5.22a–d. The scheme shown in Figure 5.22a would not be used where the vented gas was toxic or valuable. In these circumstances the vent should be taken to a vent recovery system, such as a scrubber. The controls shown in Figure 5.22b, c and d are commonly used for controlling the pressure of distillation columns.

In processes that have a high-pressure reaction section and low-pressure separation section, the high-pressure section is usually pressure controlled by expanding the product from the high-pressure section across a control valve. If the process fluid does not change phase then a more economical scheme is to expand the product through a turbine or turbo-expander and recover shaft work from the expansion.

5.8.3. Flow Control

Flow control is usually associated with inventory control in a storage tank or other equipment or with feeds to the process. There must be a reservoir to take up the changes in flow rate.

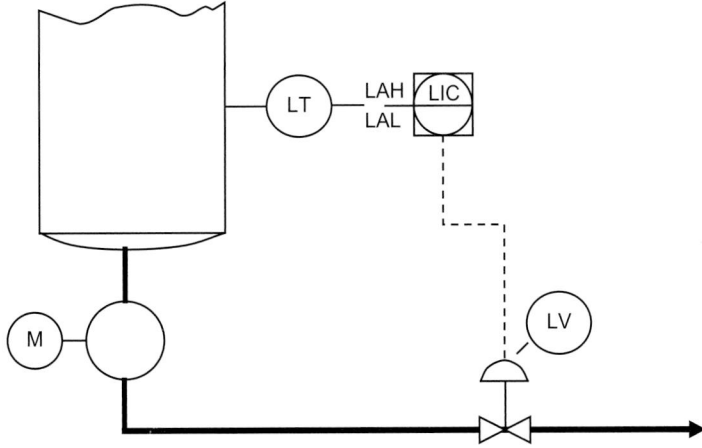

Figure 5.21.   Level control.

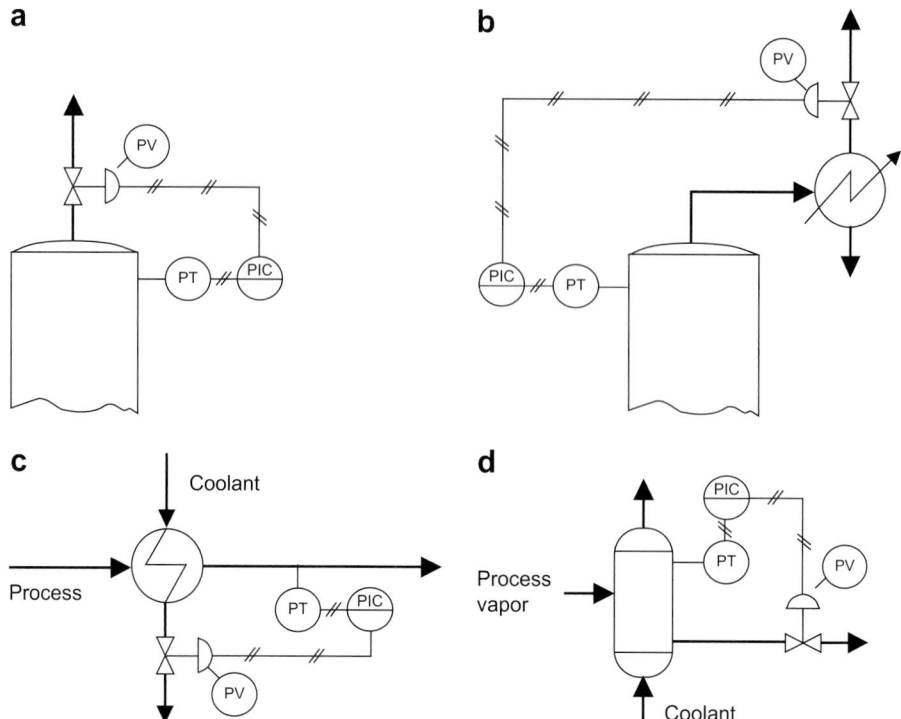

Figure 5.22.   (a) Pressure control by direct venting. (b) Venting of non-condensibles after a condenser. (c) Condenser pressure control by controlling coolant flow. (d) Pressure control of a condenser by varying the heat-transfer area, area dependent on liquid level.

To provide flow control on a compressor or pump running at a fixed speed and supplying a near constant volume output, a bypass control would be used, as shown in Figure 5.23a. The use of variable speed motors as shown in Figure 5.23c is more energy efficient than the traditional arrangement shown in Figure 5.23b, and is becoming increasingly common.

The overall process material balance is usually set by flow controllers on the feed streams. These will often control feeds in ratio to a flow of valuable feed, a solid stream flow (which is difficult to change quickly) or a measured flow of process mixture. Flow rates of small streams are often controlled using special metering pumps that deliver a constant mass flow rate.

5.8.4.   Heat Exchangers

Figure 5.24a shows the simplest arrangement, the temperature being controlled by varying the flow of the cooling or heating medium.

If the exchange is between two process streams whose flows are fixed, bypass control will have to be used, as shown in Figure 5.24b.

For air coolers, the coolant temperature may vary widely on a seasonal (or even hourly) basis. A bypass on the process side can be used as shown in Figure 5.24c, or else a variable speed motor can be used as shown in Figure 5.24d.

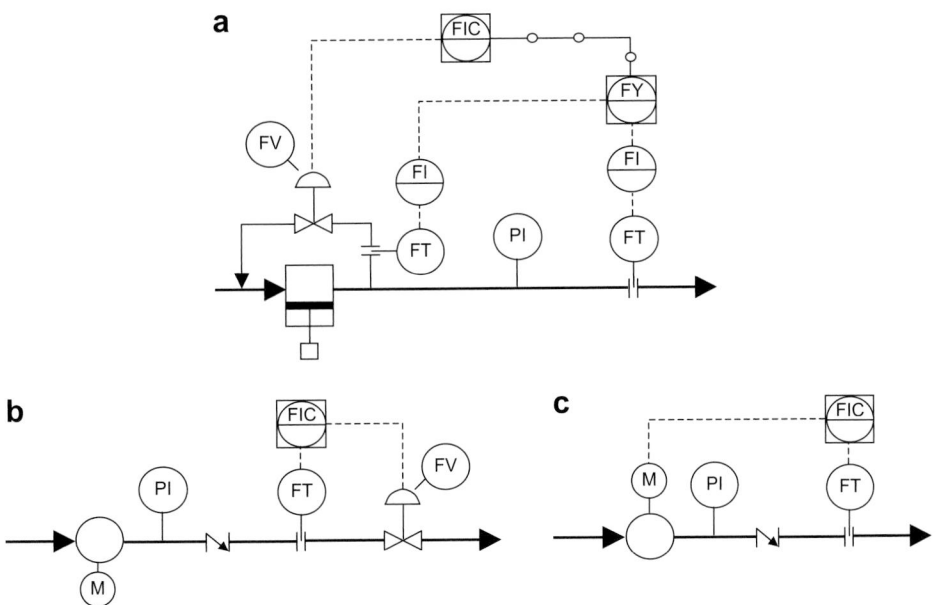

Figure 5.23. (a) Spill-back flow control for a reciprocating pump. (b) Flow control for a centrifugal pump. (c) Centrifugal pump with variable speed drive.

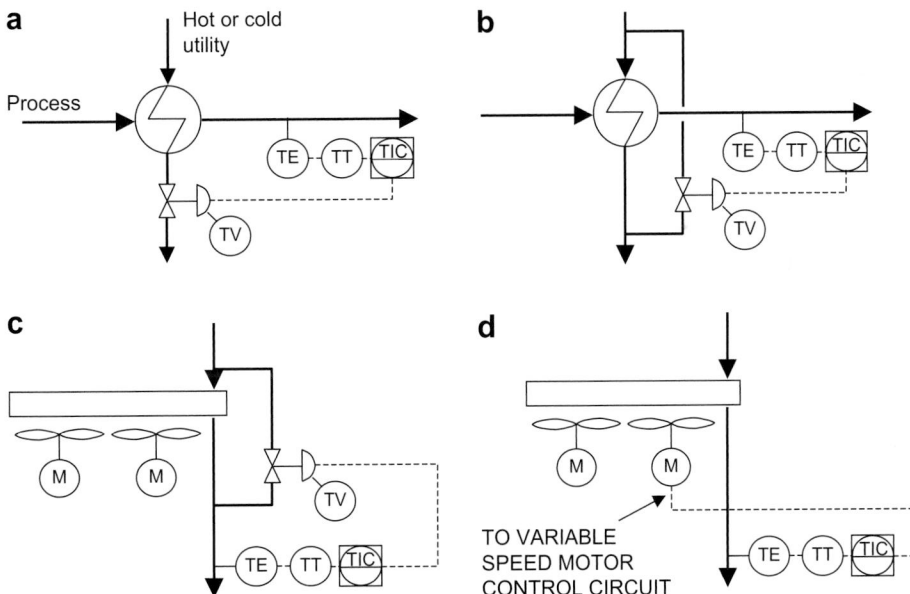

Figure 5.24. (a) Temperature control of one fluid stream. (b) Bypass control. (c) Air cooler with bypass control. (d) Air cooler with variable speed drive.

Condenser Control

Temperature control is unlikely to be effective for condensers, unless the liquid stream is sub-cooled. Pressure control is often used, as shown in Figure 5.22d, or control can be based on the outlet coolant temperature.

Reboiler and Vaporizer Control

As with condensers, temperature control is not effective, as the saturated vapour temperature is constant at constant pressure. Level control is often used for vaporizers; the controller controlling the steam supply to the heating surface, with the liquid feed to the vaporizer on flow control, as shown in Figure 5.25. An increase in the feed results in an automatic increase in steam to the vaporizer, to vaporize the increased flow and maintain the level constant.

Reboiler control systems are selected as part of the general control system for the distillation column and are discussed in Section 5.8.7.

5.8.5. Cascade Control

With this arrangement, the output of one controller is used to adjust the set point of another. Cascade control can give smoother control in situations where direct control of the variable would lead to unstable operation. The 'slave' controller can be used to compensate for any short-term variations in, say, a utility stream flow, which would upset the controlled variable; the primary ('master') controller controlling long-term variations. Typical examples are shown in Figures 5.30 and 5.31 below.

5.8.6. Ratio Control

Ratio control can be used where it is desired to maintain two flows at a constant ratio; for example, reactor feeds or distillation column reflux. A typical scheme for ratio control is shown in Figure 5.26.

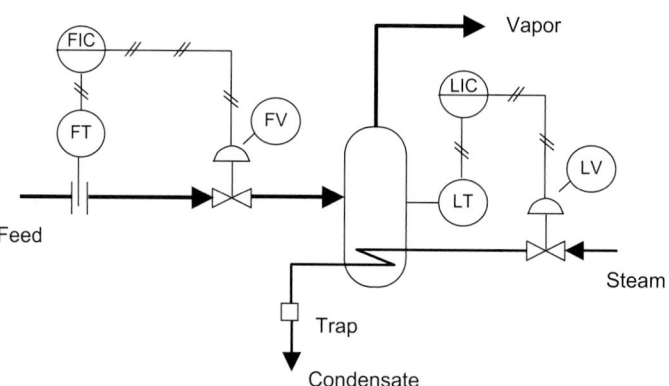

Figure 5.25.   Vaporizer control.

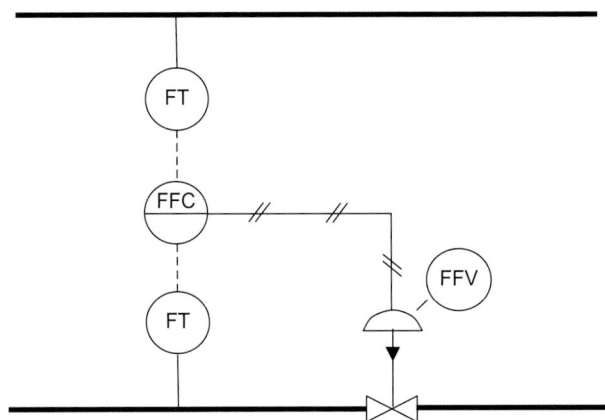

Figure 5.26. Ratio control.

5.8.7. Distillation Column Control

The primary objective of distillation column control is to maintain the specified composition of the top and bottom products, and any side streams; correcting for the effects of disturbances in:

1. Feed flow rate, composition and temperature.
2. Steam or other hot utility supply.
3. Cooling water or air cooler conditions.
4. Ambient conditions, which can cause cooling of the column shell and changes in internal reflux (see Chapter 11).

The feed flow rate is often set by the level controller on a preceding column. It can be independently controlled if the column is fed from a storage or surge tank. Feed temperature is not normally controlled, unless a feed preheater is used.

In the usual case where the feed rate is set by upstream operations and the column produces a liquid distillate product, there are five control valves, and hence five degrees of freedom; see Figure 5.27. One degree of freedom is used to set the column pressure, usually by control of the condenser using one of the schemes shown in Figure 5.22. Column pressure is normally controlled at a constant value, which then sets the vapour inventory in the column. The use of variable pressure control to conserve energy has been discussed by Shinskey (1976). Two degrees of freedom are needed to control the liquid inventories by controlling the vapour–liquid level in the column sump and the reflux drum (or condenser if no reflux drum is used).

The remaining two degrees of freedom can be used to achieve the desired separation, either in terms of product purity or recovery, by adjusting two flow rates. One of these flows is controlled by a flow or flow ratio controller to achieve the desired split between distillate and bottoms, while the other is usually controlled by a column temperature to achieve a desired composition in one of the products. The flow controller cannot be on the distillate or bottoms stream if the designer intends to control composition, as it would then be impossible to maintain product composition if there were changes in feed composition. The temperature controller can, however,

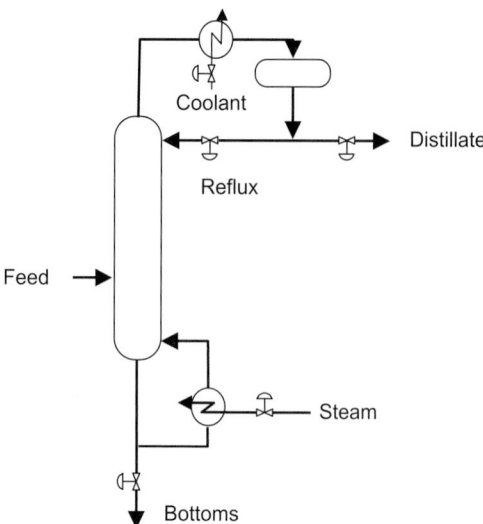

Figure 5.27.   Control valves and degrees of freedom for a simple distillation column.

control either the distillate or bottoms flow rate. The usual practice is to control a top temperature by varying the reflux ratio or distillate flow rate if the overhead product purity is more important (Figure 5.28), or control a bottom temperature by varying the boil-up rate or bottoms flow if bottoms purity is more important (Figure 5.29).

Control schemes of this type are commonly referred to as *material balance* control schemes, as they achieve the desired product purity by manipulating the column material balance. These schemes are very robust for processes where the feed flow rate to the column is relatively constant but the composition varies and close control must be maintained on one product composition.

Temperature is usually used as an indication of composition. The temperature sensor should be located at a position in the column where the rate of change of temperature with change in composition of the key component is a maximum; see Parkins (1959). Near the top and bottom of the column the change is usually small. When designing the column, it is a good idea to allow for thermowells on several trays, so that the best control point can be found when the column is actually operating. If reliable on-line analysers are available they can be incorporated in the control loop, but more complex control equipment will be needed. With multicomponent systems, temperature is not a unique function of composition.

Flow ratio controllers are sometimes used in distillation control, controlling the reflux or boil-up in ratio to the feed, distillate or bottoms rate. The same effect can be accomplished using cascade control, with the feed rate adjusting the set point of the flow controller on reflux or boil-up.

Shinskey (1984) has shown that there are 120 ways of connecting the five main pairs of measured and controlled variables, in single loops. A variety of control schemes has been devised for distillation column control. Some typical schemes are shown in Figures 5.28 to 5.30; ancillary control loops and instruments are not shown.

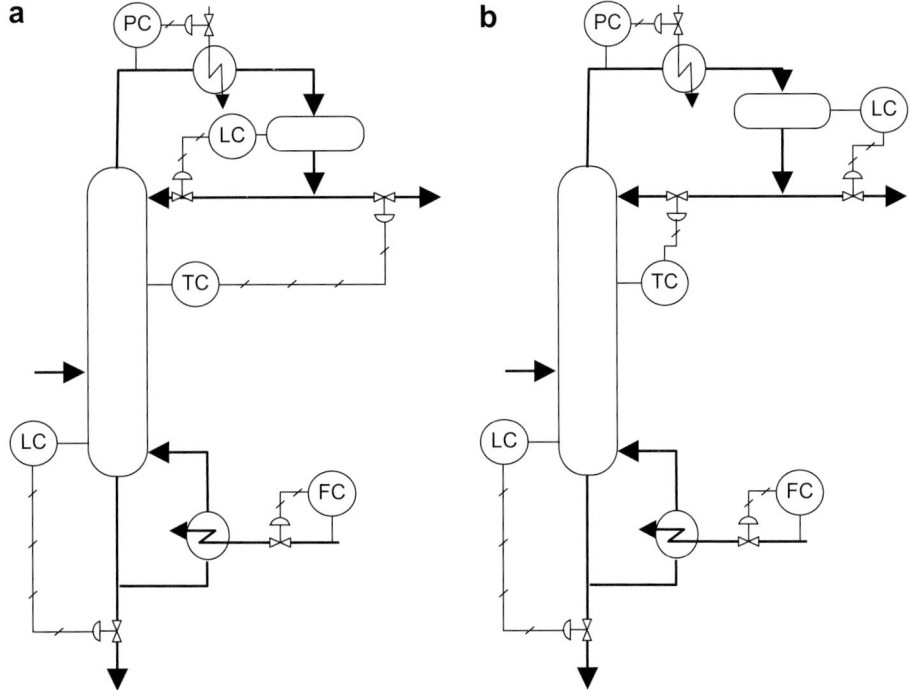

Figure 5.28. Material balance control schemes for controlling overhead product composition. Flow control on reboiler can be in ratio to feed if feed rate varies. (a) Direct control of distillate by composition. (b) Indirect control of distillate, composition controls reflux.

The choice of control scheme may be influenced by many other factors. For example, the control scheme of Figure 5.29b controls boil-up by composition and gives the fastest control response to variations in composition of any of the schemes. Kister (1990) discusses the advantages and drawbacks of the material balance control schemes shown in Figures 5.28 and 5.29.

An older control scheme that is often encountered is similar to Figure 5.28b, but has the steam to the reboiler controlled by a temperature in the stripping section of the column. This scheme is known as temperature-pattern control or dual composition control, and in principle allows both top and bottom compositions to be controlled. The drawback of this scheme is that there is a tendency for the controllers to fight each other, leading to unstable operation.

Distillation column control is discussed in detail by Parkins (1959), Bertrand and Jones (1961), Shinskey (1984) and Buckley *et al.* (1985).

Additional temperature indicating or recording points should be included up the column for monitoring column performance and for trouble-shooting.

5.8.8. Reactor Control

The schemes used for reactor control depend on the process and the type of reactor. If a reliable on-line analyser is available, and the reactor dynamics are suitable, the

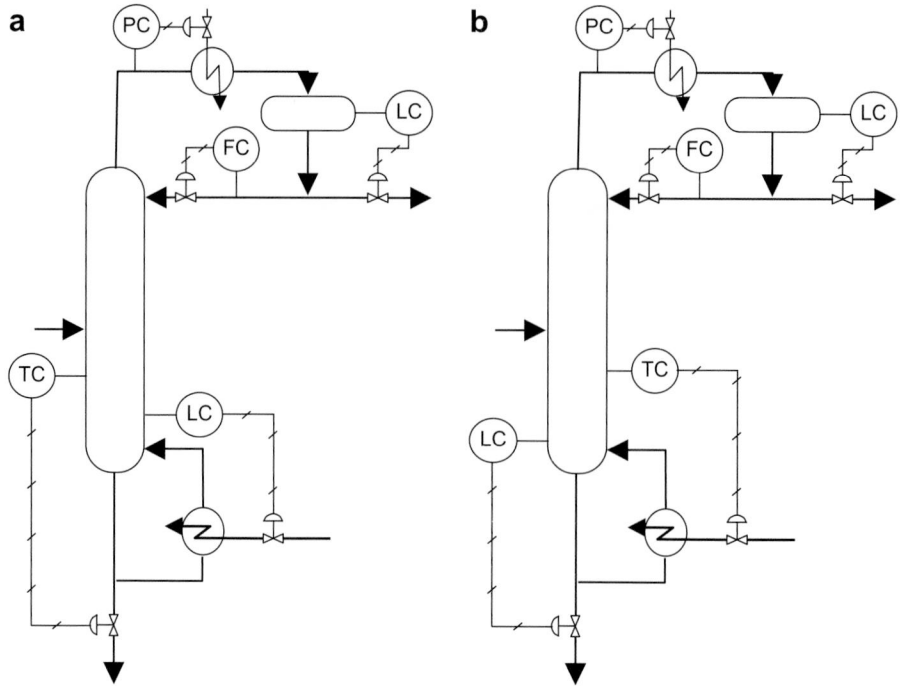

Figure 5.29. Material balance control schemes for controlling bottoms product composition. Flow control on reflux can be in ratio to feed if feed rate varies. (a) Direct control of bottoms by composition. (b) Indirect control of bottoms, composition controls boil-up.

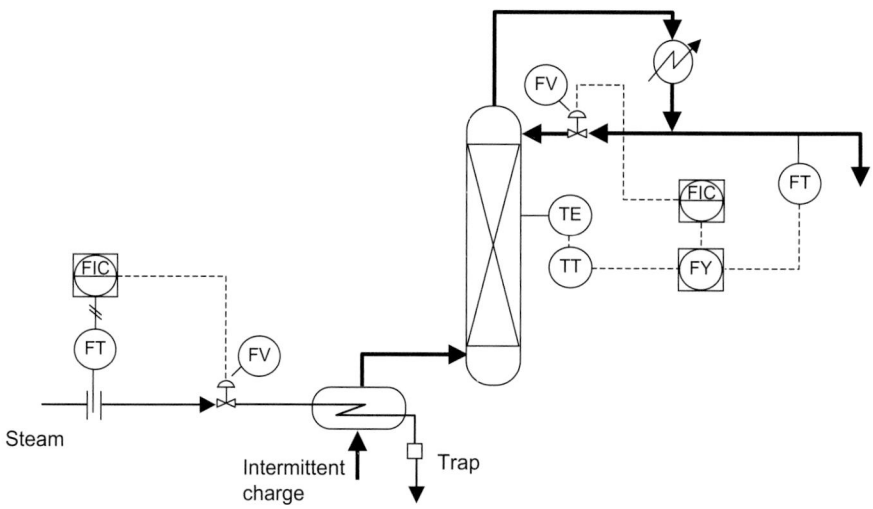

Figure 5.30. Batch distillation, reflux flow controlled based on temperature to infer composition.

product composition can be monitored continuously and the reactor conditions and feed flows controlled automatically to maintain the desired product composition and yield. More often, the operator is the final link in the control loop, adjusting the controller set points to maintain the product within specification, based on periodic laboratory analyses.

For small stirred-tank reactors, temperature will normally be controlled by regulating the flow of the heating or cooling medium. For larger reactors, temperature is often controlled by recycling a part of the product stream or adding inert material to the feed to act as a heat sink. Pressure is usually held constant. For liquid-phase reactors, pressure is often controlled by maintaining a vapour space above the liquid reagents. This space can be pressurized with nitrogen or other suitable gases. Material balance control will be necessary to maintain the correct flow of reactants to the reactor and the flow of products and unreacted materials from the reactor. A typical control scheme for a simple liquid-phase reactor is shown in Figure 5.31.

5.9. ALARMS, SAFETY TRIPS AND INTERLOCKS

Alarms are used to alert operators of serious, and potentially hazardous, deviations in process conditions. Key instruments are fitted with switches and relays to operate audible and visual alarms on the control panels and annunciator panels. Where delay, or lack of response, by the operator is likely to lead to the rapid development of a hazardous situation, the instrument would be fitted with a trip system to take action automatically to avert the hazard; such as shutting down pumps, closing valves and operating emergency systems.

The basic components of an automatic trip system are:

1. A sensor to monitor the control variable and provide an output signal when a preset value is exceeded (the instrument).

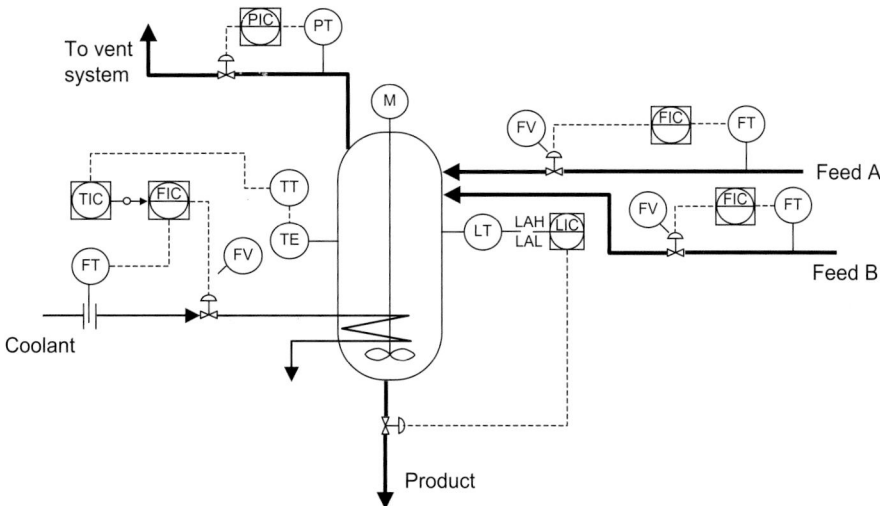

Figure 5.31. A typical stirred tank reactor control scheme, with temperature cascade control of coolant flow and flow control of reagents.

2. A link to transfer the signal to the actuator, usually consisting of a system of pneumatic or electric relays.
3. An actuator to carry out the required action: close or open a valve, switch off a motor.

A description of some of the equipment (hardware) used is given by Rasmussen (1975).

A safety trip can be incorporated in a control loop, as shown in Figure 5.32a. In this system the level control instrument has a built-in software alarm that alerts the operator if the level is too low and a programmed trip set for a level somewhat lower than the alarm level. However, the safe operation of such a system will be dependent on the reliability of the control equipment, and for potentially hazardous situations it is better practice to specify a separate trip system, such as that shown in Figure 5.32b, in which the trip is activated by a separate low-level switch. Provision must be made for the periodic checking of the trip system to ensure that the system operates when needed.

More information on the design of safety instrumented systems is given in Chapter 9, Section 9.8.

Interlocks

Where it is necessary to follow a fixed sequence of operations — for example, during a plant start-up and shut-down, or in batch operations — interlocks are included to prevent operators departing from the required sequence. They may be incorporated in the control system design, as pneumatic or electric relays, or may be mechanical interlocks. Various proprietary special lock and key systems are also available. In most plants, programmable logic controllers are used and the interlocks are coded into the control algorithms. Care should be taken to test all of the interlocks in the plant

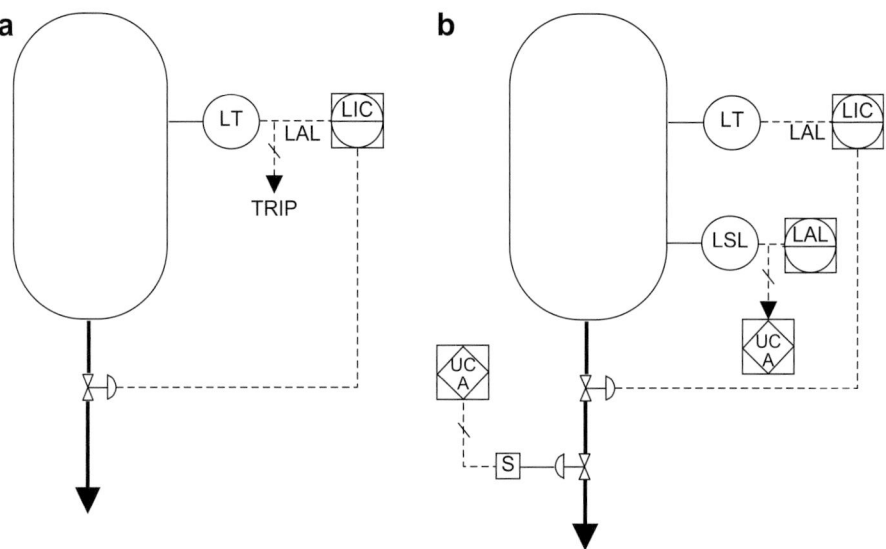

Figure 5.32.   (a) Trip as part of control system. (b) Separate shut-down trip.

automation during commissioning or whenever changes are made to the plant control and automation.

5.10. COMPUTERS IN PROCESS CONTROL

Almost all process control systems installed on new plants use programmable electronic devices based on microprocessors. These range from simple digitally actuated single-loop controllers that produce a single-output signal (Single Input – Single Output or SISO devices) up to complex distributed control systems that carry out control, real-time optimization and data logging and archiving for multiple process plants across a site or even an enterprise (Multiple Input – Multiple Output or MIMO devices).

A simple example of a multiple input device is a gas mass-flow controller (Figure 5.33), in which the gas mass flow is computed based on inputs from temperature, pressure and flow instruments.

The control schemes described in Section 5.8 mainly make use of SISO controllers, since the schemes were developed for single-unit operations. At the unit operation level, the primary focus of process control is usually on safe and stable operation and it is difficult to take advantage of the capability of advanced microprocessor-based control systems. When several unit operations are put together to form a process then the scope for use of MIMO devices increases, particularly when the devices are able to communicate with each other rapidly. The digital control system can then make use of more complex algorithms and models that enable feed-forward control (model-based or multivariable predictive control) and allow data collected from upstream in the process to guide the selection of operating conditions and controller set points for downstream operations. This allows for better response to process dynamics and more rapid operation of batch, cyclic and other unsteady-state processes. Model-based predictive control is also often used as a means of controlling product quality. This is because devices for measuring product quality typically require analytical procedures that take several minutes to hours to run, making effective feedback control difficult to accomplish.

The use of instruments that log and archive data facilitates remote monitoring of process performance and can improve plant trouble-shooting and optimization, as well as providing high-level data for enterprise-wide supply-chain management.

The electronic equipment and systems technology available for process control continues to evolve rapidly. Because of the pace of innovation, industry-wide

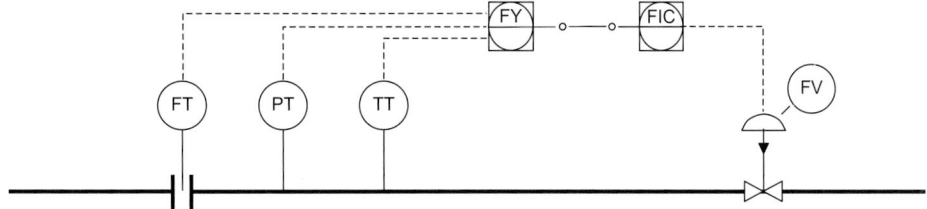

Figure 5.33. Gas mass-flow controller.

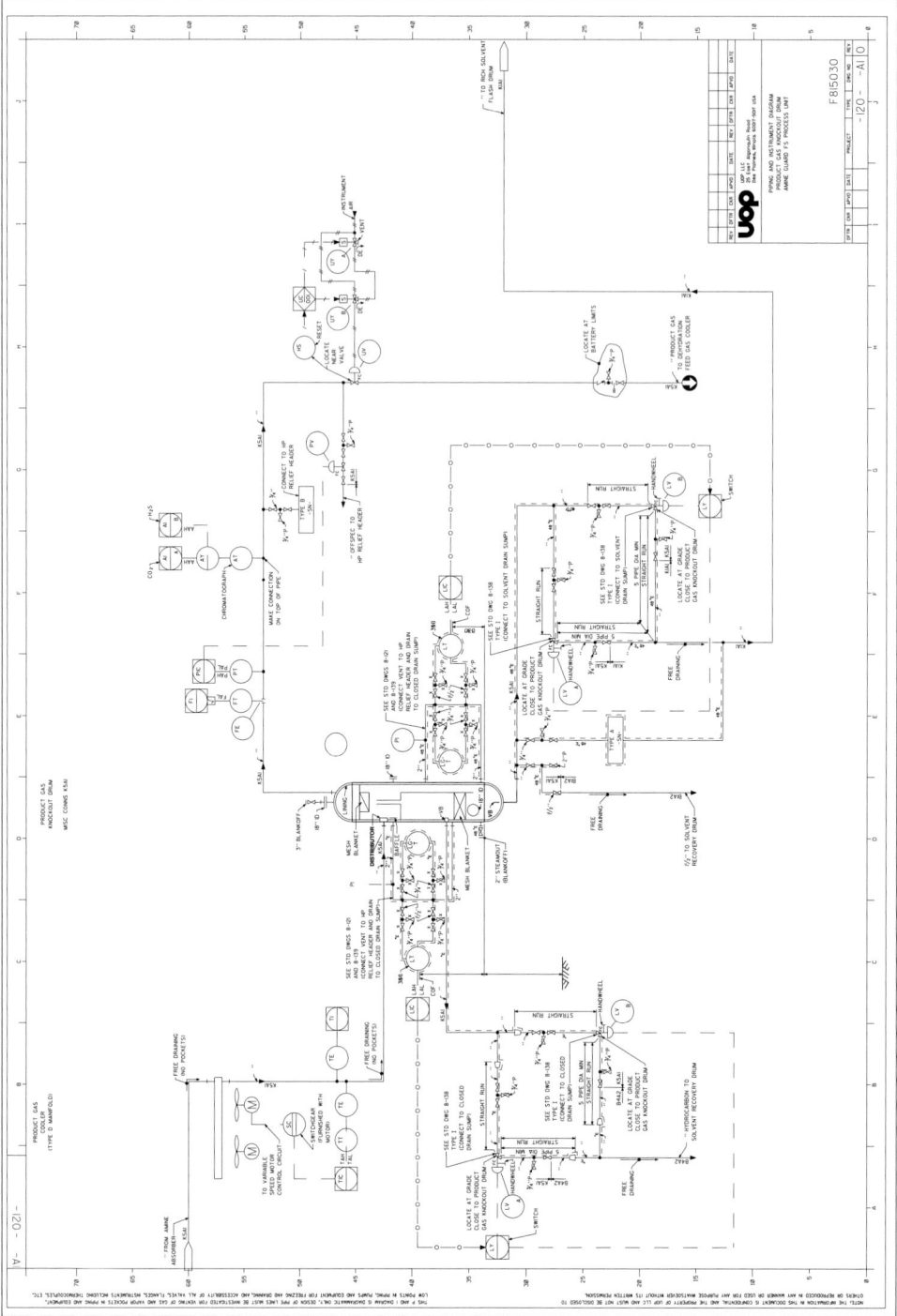

Figure 5.34. Piping and instrumentation diagram.

standards have not been able to keep up, and consequently different manufacturers' systems usually use proprietary technology and are often not fully compatible with each other. The implementation of the ISA SP50 and Hart Foundation Fieldbus standards has substantially improved digital communications between control devices, leading to improved control, faster set-up, better reliability through higher redundancy and even greater distribution of functions between devices. The ISA is also developing the SP100 standard for wireless transmission. Wireless systems are beginning to be used in inventory control and maintenance management, but are not widely used yet in plant control, mainly because of problems with interference, signal blocking and signal loss. The development of more robust error checking and transmission protocols, and the adoption of mesh networking approaches will probably lead to greater use of wireless instrumentation in the future.

A detailed treatment of digital technology for process control is beyond the scope of this book. Kalani (1988), Edgar *et al.* (1997) and Liptak (2003) all provide excellent reviews of the subject. Mitchell and Law (2003) give a good overview of digital bus technologies.

5.11. REFERENCES

Bertrand, L. and Jones, J. B. (1961) *Chem. Eng., NY* **68** (Feb. 20th) 139. Controlling distillation columns.

Buckley, P. S., Luyben, W. L., and Shunta, J. P. (1985) *Design of Distillation Column Control Systems* (Arnold).

Capps, R. W. (1995) *Chem. Eng., NY,* **102** (July) 102. Select the optimum pipe diameter.

Chabbra, R. P. and Richardson, J. F. (1999) *Non-Newtonian Flow in the Process Industries* (Butterworth-Heinemann).

Chaflin, S. (1974) *Chem. Eng., NY* **81** (Oct. 14th) 105. Specifying control valves.

Coughanowr, D. R. (1991) *Process Systems Analysis and Control*, 2nd edn. (MacGraw-Hill).

Coulson, J. M., Richardson, J. F., Backhurst, J., and Harker, J. H. (1999) *Chemical Engineering: Volume 1*, 6th edn. (Butterworth-Heinemann).

Darby, R. (2001) *Chem. Eng., NY* **108** (March) 66. Take the mystery out of non-Newton fluids.

Davidson, J. and von Bertele, O. (1999) *Process Pump Selection—A Systems Approach* (I. Mech E.).

De Santis, G. J. (1976) *Chem. Eng., NY* **83** (Nov. 22nd) 163. How to select a centrifugal pump.

Doolin, J. H. (1977) *Chem. Eng., NY* (Jan. 17th) **137**. Select pumps to cut energy cost.

Eckert, J. S. (1964) *Chem. Eng., NY* **71** (Mar. 30th) 79. Controlling packed-column stills.

Eder, H. (2003) *Chem. Eng., NY* **110**(6) (Jun 1) 63. For good process control, understand the process.

Edgar, T. F., Smith, C. L., Shinskey, F. G. *et al.* (1997) Process control. In *Perry's Chemical Engineers Handbook*, 7th edn (McGraw-Hill).

Genereaux, R. P. (1937) *Ind. Eng., Chem.* **29**, 385. Fluid-flow design methods.

Green, D. W. and Perry, R. H. (eds) (2007) *Perry's Chemical Engineers' Handbook*, 8th edn (McGraw-Hill).

Heald, C. C. (1996) 1-20. *Cameron Hydraulic Data* (Ingersoll-Dresser Pumps).

Henson, M., Seporg, D. E. and Hempstead, H. (1996) *Nonlinear Process Control* (Prentice Hall).

Holland, F. A. and Chapman, F. S. (1966) *Chem. Eng., NY* **73** (Feb. 14th) 129. Positive displacement pumps.

Hoyle, R. (1978) *Chem. Eng., NY*, **85** (Oct 8th) 103. How to select and use mechanical packings.

Jacobs, J. K. (1965) *Hydrocarbon Proc.* **44** (June) 122. How to select and specify process pumps.

Jandiel, D. G. (2000) *Chem. Eng., Prog.* **96** (July) 15. Select the right compressor.

Kalani, G. (1988) *Microprocessor Based Distributed Control Systems* (Prentice Hall).

Kalani, G. (2002) *Industrial Process Control: Advances and Applications* (Butterworth Heinemann).

Karassik, I. J., *et al.* (2001) *Pump Handbook*, 3rd edn. (McGraw-Hill).

Kentish, D. N. W. (1982a) *Industrial Pipework* (McGraw-Hill).

Kentish, D. N. W. (1982b) *Pipework Design Data* (McGraw-Hill).

Kern, R. (1975) *Chem. Eng., NY* **82** (April 28th) 119. How to design piping for pump suction conditions.

Kister, H. Z. (1990) *Distillation Operation* (McGraw-Hill).

Lamit, L. G. (1981) *Piping Systems: Drafting and Design* (Prentice Hall).

Liptak, B. G. (2003) *Instrument Engineers' Handbook, Vol. 1: Process Measurement and Analysis*, 4th edn. (CRC Press).

Lockhart, R. W. and Martinelli, R. C. (1949) *Chem. Eng., Prog.* **45**(1), 39. Proposed correlation of data for isothermal two-component flow in pipes.

Ludwg, E. E. (1960) *Chem. Eng., NY* **67** (June 13th) 162. Flow of fluids.

Luyben, W. L., Tyreus, b.d., and Luyben, m.l. (1999) *Plantwide Process Control* (McGraw-Hill).

Masek, J. A. (1968) *Chem. Eng., NY* **75** (June 17th) 215. Metallic piping.

Merrick, R. C. (1986) *Chem. Eng., NY* **93** (Sept. 1st) 52. Guide to the selection of manual valves.

Merrick, R. C. (1990) *Valve Selection and Specification Guide* (Spon.).

Mills, D. (2004) *Pneumatic Conveying Design Guide*, 2nd edn. (Butterworth-Heinemann).

Mills, D., Jones, M. G., and Agarwal, V. K. (2004) *Handbook of Pneumatic Conveying* Marcel Dekker.

Mitchell, J. A. and Law, G. (2003) *Chem. Eng., NY* **110**(2) (Feb 1). Get up to speed on digital buses.

Murrill, P. W. (1988) *Application Concepts of Process Control* (Instrument Society of America).

Nayyar, M. L., et al. (2000) *Piping Handbook*, 7th edn. ((McGraw-Hill)).

Neerkin, R. F. (1974) *Chem. Eng., NY* **81** (Feb. 18) 104. Pump selection for chemical engineers.

Nolte, C. B. (1978) *Optimum Pipe Size Selection* (Trans. Tech. Publications).

Parkins, R. (1959) *Chem. Eng., Prog.* **55** (July) 60. Continuous distillation plant controls.

Peacock, D. G. and Richardson, J. F. (1994) *Chemical Engineering: Volume 3*, 3rd edn. (Butterworth-Heinemann).

Pearson, G. H. (1978) *Valve Design* (Mechanical Engineering Publications).

Peters, M. S. and Timmerhaus, K. D. (1968) *Plant Design and Economics for Chemical Engineers*, 2nd edn. (McGraw-Hill).

Peters, M. S. and Timmerhaus, K. D. (1991) *Plant Design and Economics*, 4th edn. (McGraw-Hill).

Rase, H. F. (1953) *Petroleum Refiner* **32** (Aug.) 14. Take another look at economic pipe sizing.

Rasmussen, E. J. (1975) *Chem. Eng., NY* **82** (May 12th) 74. Alarm and shut down devices protect process equipment.

Richardson, J. F., Harker, J. H., and Backhurst, J. (2002) *Chemical Engineering: Volume 2*, 5th edn. (Butterworth-Heinemann).

Sherwood, D. R. (1991) *The Piping Guide*, 2nd edn. (Spon.).

Shinskey, F. G. (1976) *Chem. Eng. Prog.* **72** (May) 73. Energy-conserving control systems for distillation units.

Shinskey, F. G. (1984) *Distillation Control*, 2nd edn. (McGraw-Hill).

Shinskey, F. G. (1996) *Process Control Systems*, 4th edn. (McGraw-Hill).

Simpson, L. L. (1968) *Chem. Eng., NY* **75** (June 17th) 1923. Sizing piping for process plants.

Smith, E. and Vivian, B. E. (1995) *Valve Selection* (Mechanical Engineering Publications).

Smith, P. and Zappe, R. W. (2003) *Valve Selection Handbook*, 5th edn. (Gulf Publishing).

Walas, S. M. (1990) *Chemical Process Equipment* (Butterworth-Heinemann).

Webster, G.R. (1979) *Chem. Engr. London* No. 341 (Feb.) 91. The canned pump in the petrochemical environment.

British and European Standards

BS 1646:1984 (1984) Symbolic representation for process measurement control functions and instrumentation.

BS 5353:1989 (1989) Steel plug valves.

BS EN 558:1996 (1996) Industrial valves: face-to-face and centre-to-face dimensions of metal valves for use in flanged pipe systems.

BS EN 593:2004 (2004) Industrial valves: metallic butterfly valves.

BS EN 736-1:1995 (1995) Valves: terminology. Type of valve.

BS EN 736-2:1995 (1995) Valves: terminology. Definition of components of valves.

BS EN 1349:2002 (2002) Industrial process control valves.

BS EN 1983:2006 (2006) Industrial valves: steel ball valves.

BS EN 1984:2000 (2000) Industrial valves: steel gate valves (replaces BS 5157).

BS EN 10241:2000 (2000) Steel threaded pipe fittings.

BS EN 10253:1999 (1999) Butt-welding pipe fittings.

BS EN 12570:2000 (2000) Industrial valves: method for sizing the operating element.

BS EN 12982:2000 (2000) Industrial valves: end-to-end and centre-to-end dimensions for butt welding end valves.

BS EN 13397:2002 (2002) Industrial valves: diaphragm valves made of metallic materials.

BS EN 13709:2003 (2003) Industrial valves: steel globe and globe stop and check valves.

BS EN 14341:2006 (2006) Industrial valves: steel check valves.

BS EN ISO 10434 (2004) Bolted steel bonnet gate valves for the petroleum, petrochemical and allied industries, 2nd edn.

BS EN ISO 17292 (2004) Metal ball valves for the petroleum, petrochemical and allied industries, 2nd edn (replaces BS 5351).

BS ISO 7121:2006 (2006) Steel ball valves for general-purpose industrial applications.

DIN 2429-2 (1988) Symbolic representation of pipework components for use on engineering drawings; functional representation.

DIN 19227-1 (1993) Control technology; graphical symbols and identifying letters for process control engineering; symbolic representation for functions.

DIN 19227-2 (1991) Control technology; graphical symbols and identifying letters for process control engineering; representation of details.

American and International Standards

ASME B16.34-2004 Valves, flanged, threaded and welding end.
ASME B16.10-2000 Face-to-face and end-to-end dimensions of valves.
ASME B31.1-2004 Power piping.
ASME B31.2-1968 Fuel gas piping.
ASME B31.3-2004 Process piping.
ASME B31.4-1997 Pipeline transportation systems for liquid hydrocarbons and other liquids.
ASME B31.5-2001 Refrigeration piping and heat transfer components.
ASME B31.8-2003 Gas transmission and distribution piping systems.
ASME B31.9-2004 Building services piping.
ASME B31.11-2002 Slurry transportation piping systems.
ASME B36.19M-2004 Stainless steel pipe.
ISA-5.1-1984 (R1992) Instrumentation symbols and identification.
ISO 5209 (1977) General purpose industrial valves – marking, 1st edn.

5.12. NOMENCLATURE

		Dimensions in MLT$
A	Plant attainment (hours operated per year)	—
B	Purchased cost factor, pipes	L^{-1}
a	Capital charges factor, piping	T^{-1}
b	Maintenance cost factor, piping	T^{-1}
c	Sum of mechanical, corrosion and erosion allowances	L
C_f	Fanning friction factor	—
C_C	Annual capital charge, piping	$L^{-1}T^{-1}$
C_t	Total annual cost, piping	$L^{-1}T^{-1}$
C_w	Annual pumping cost, piping	$L^{-1}T^{-1}$
d	Pipe diameter	L
d_i	Pipe inside diameter	L
E	Pipe casting quality factor	—
e	Relative roughness	—
F	Installed cost factor, piping	—
f	Friction factor	—
G	Mass flow rate	MT^{-1}
g	Gravitational acceleration	LT^{-2}
H	Height of liquid above the pump suction	L
h	Pump head	L
K	Number of velocity heads	—
L	Pipe length	L
m	Mass flow rate	MT^{-1}
N	Pump speed, revolutions per unit time	T^{-1}
N_s	Pump specific speed	—
n	Index relating pipe cost to diameter	—
P	Pressure	$ML^{-1}T^{-2}$
P_f	Pressure loss in suction piping	$ML^{-1}T^{-2}$

P_s	Safe working pressure	$ML^{-1}T^{-2}$
P_v	Vapour pressure of liquid	$ML^{-1}T^{-2}$
ΔP	Difference in system pressures $(P_1 - P_2)$	$ML^{-1}T^{-2}$
ΔP_f	Pressure drop	$ML^{-1}T^{-2}$
p	Cost of power, pumping	$\$M^{-1}L^{-2}T^2$
Q	Volumetric flow rate	L^3T^{-1}
R	Shear stress on surface, pipes	$ML^{-1}T^{-2}$
S	Basic allowable stress for pipe material	$ML^{-1}T^{-2}$
t	Pipe wall thickness	L
t_m	Minimum required thickness	L
t_p	Pressure design thickness	L
u	Fluid velocity	LT^{-1}
W	Work done	L^2T^{-2}
z	Height above datum	L
Δz	Difference in elevation $(z_1 - z_2)$	L
γ	Temperature coefficient	—
η	Pump efficiency	—
ρ	Fluid density	ML^{-3}
μ	Viscosity of fluid	$ML^{-1}T^{-1}$
σ_d	Design stress	$ML^{-1}T^{-2}$
σ_s	Safe working stress	$ML^{-1}T^{-2}$
Re	Reynolds number	—
$NPSH_{avail}$	Net positive suction head available at the pump suction	L
$NPSH_{reqd}$	Net positive suction head required at the pump suction	L

5.13. PROBLEMS

5.1. Select suitable valve types for the following applications:
1. Isolating a heat exchanger.
2. Manual control of the water flow into a tank used for making up batches of sodium hydroxide solution.
3. The valves to isolate a pump and provide emergency manual control on a bypass loop.
4. Isolation valves in the line from a vacuum column to the steam ejectors producing the vacuum.
5. Valves in a line where cleanliness and hygiene are an essential requirement.
State the criterion used in the selection for each application.

5.2. Crude dichlorobenzene is pumped from a storage tank to a distillation column.

The tank is blanketed with nitrogen and the pressure above the liquid surface is held constant at 0.1-bar gauge pressure. The minimum depth of liquid in the tank is 1 m.

The distillation column operates at a pressure of 500 mmHg (500 mm of mercury, absolute). The feed point to the column is 12 m above the base of the tank. The tank and column are connected by a 50-mm internal diameter commercial steel pipe, 200 m long. The pipe run from the tank to the column contains the following valves and fittings: 20 standard radius 90° elbows, two gate valves to isolate the pump (operated fully open), an orifice plate and a flow-control valve.

If the maximum flow rate required is 20,000 kg/h, calculate the pump motor rating (power) needed. Take the pump efficiency as 70% and allow for a pressure drop of 0.5 bar across the control valve and a loss of 10 velocity heads across the orifice.

Density of dichlorobenzene 1300 kg/m^3, viscosity 1.4 cp.

5.3. A liquid is contained in a reactor vessel at 115-bar absolute pressure. It is transferred to a storage vessel through a 50-mm internal diameter commercial steel pipe. The storage vessel is nitrogen blanketed and pressure above the liquid surface is kept constant at 1500-N/m^2 gauge. The total run of pipe between the two vessels is 200 m. The miscellaneous losses due to entry and exit losses, fittings, valves, etc., amount to 800 equivalent pipe diameters. The liquid level in the storage vessel is at an elevation 20 m *below* the level in the reactor.

A turbine is fitted in the pipeline to recover the excess energy that is available, over that required to transfer the liquid from one vessel to the other. Estimate the power that can be taken from the turbine, when the liquid transfer rate is 5000 kg/h. Take the efficiency of the turbine as 70%.

The properties of the fluid are: density 895 kg/m^3, viscosity 0.76 mNm^{-2} s.

5.4. A process fluid is pumped from the bottom of one distillation column to another, using a centrifugal pump. The line is standard commercial steel pipe, 75 mm internal diameter. From the column to the pump inlet the line is 25 m long and contains six standard elbows and a fully open gate valve. From the pump outlet to the second column the line is 250 m long and contains 10 standard elbows, four gate valves (operated fully open) and a flow-control valve. The fluid level in the first column is 4 m above the pump inlet. The feed point of the second column is 6 m above the pump inlet. The operating pressure in the first column is 1.05 bara and that of the second column 0.3 barg.

Determine the operating point on the pump characteristic curve when the flow is such that the pressure drop across the control valve is 35 kN/m^2. The physical properties of the fluid are: density 875 kg/m^3, viscosity 1.46 mNm^{-2} s.

Also, determine the NPSH, at this flow rate, if the vapour pressure of the fluid at the pump suction is 25 kN/m^2.

Pump characteristic							
Flow rate, m^3/h	0.0	18.2	27.3	36.3	45.4	54.5	63.6
Head, m of liquid	32.0	31.4	30.8	29.0	26.5	23.2	18.3

5.5. Revisiting Problem 5.4, suppose the flow was controlled using a plug-disk globe valve and the initial design in the example assumed that the valve is fully open. What range of flow rates can be achieved if the valve can be throttled down to quarter-open? When the valve is quarter-open, what fraction of the pump work is lost across the valve?

5.6. Estimate the shaft work required to pump 65 gal/min of sugar solution in water (specific gravity = 1.05) if the pump inlet pressure is 25 psig and the outlet pressure required is 155 psig.

5.7. A shell and tube cooler in an aromatics complex cools 26,200 lb/h of naphtha (specific gravity 0.78, viscosity 0.007 cP). The cooler has 347 tubes, 16 ft long, $\frac{3}{4}$ inch diameter. If the naphtha is on the tube side, estimate the tube side pressure drop.

5.8. In a detergent making process, 1400 gal/h of water flows through a 2-in pipe system as follows:

Exit from pump, 2 ft vertical, open gate valve, 14 ft vertical, 90° bend, 12 ft horizontal, ¼ open globe valve, 20 ft horizontal, 90° bend, 6 ft horizontal, 90° bend, 12 ft vertical, 90° bend, 14 ft horizontal, 90° bend, 4 ft vertical, 90° bend, 28 ft horizontal, open gate valve, 3 ft horizontal, entry to tank containing 30 ft of liquid.

a) If the pump and tank are both at grade level, estimate the head that the pump must deliver.

b) If the pump inlet pressure is 25 psig, what is the outlet pressure?

c) Estimate the pump shaft work.

d) If the pump is powered by an electric motor with 85% efficiency, what is the annual electricity consumption?

5.9. A polymer is produced by the emulsion polymerization of acrylonitrile and methyl methacrylate in a stirred vessel. The monomers and an aqueous solution of catalyst are fed to the polymerization reactor continuously. The product is withdrawn from the base of the vessel as a slurry.

Devise a control system for this reactor, and draw up a preliminary piping and instrument diagram. The follow points need to be considered:

1. Close control of the reactor temperature is required.

2. The reactor runs 90% full.

3. The water and monomers are fed to the reactor separately.

4. The emulsion is a 30% mixture of monomers in water.

5. The flow of catalyst will be small compared with the water and monomer flows.

6. Accurate control of the catalyst flow is essential.

5.10. Devise a control system for the distillation column described in Chapter 11, Example 11.2. The flow to the column comes from a storage tank. The product, acetone, is sent to storage and the waste to an effluent pond. It is essential that the specifications on product and waste quality are met.

6 COSTING AND PROJECT EVALUATION

Chapter Contents

Key Learning Objectives

- How to estimate process capital and operating costs
- How to find and forecast prices for use in economic analysis
- How corporations finance projects
- Different criteria that companies use to compare the financial attractiveness of alternative projects, and other factors that are also taken into account in project selection
- How to allow for error in cost estimates

6.1. INTRODUCTION

Most chemical engineering design projects are carried out to provide information from which estimates of capital and operating costs can be made. Chemical plants are built to make a profit, and an estimate of the investment required and the cost of production are needed before the profitability of a project can be assessed. Cost estimation is a specialized subject and a profession in its own right, but the design engineer must be able to make rough cost estimates to decide between project alternatives and optimize the design.

This chapter introduces the components of capital and operating costs and the techniques used for estimating. Simple costing methods and some cost data are given, which can be used to make preliminary estimates of capital and operating costs in the early stages of design. Sources of cost data and methods for updating cost estimates are described. The main methods used for economic evaluation of projects are introduced, together with an overview of factors that influence project selection.

Most cost-estimating and economic analysis calculations are easily carried out using spreadsheets. Templates are introduced in the examples throughout the chapter. Blank templates are given in Appendix G and in the on-line material at http://elsevierdirect.com/companions. The more sophisticated software that is used in industry for preliminary estimating is discussed in Section 6.3.

For a more detailed treatment of the subject the reader should refer to the numerous specialized texts that have been published on cost estimation. The following books are particularly recommended: Happle and Jordan (1975), Guthrie (1974), Page (1996), Garrett (1989), Humphreys (1991) and Humphreys (2005).

Several companies regularly publish economic analyses of chemical processes. Nexant publishes the Process Evaluation and Research Planning (PERP) reports (www.nexant.com/products). Roughly 10 new reports are issued each year and almost 200 processes have been analyzed. The PERP reports provide estimates of capital and operating costs, usually for two or three process alternatives, as well as an overview of the market. Access Intelligence publishes the SRI Chemical Economics Handbook (CEH) series, which contains 288 reports on a range of commodity and specialty chemicals. The CEH reports provide an overview of production technologies and analyses of several regional markets, but do not provide the level of production cost detail given in the PERP reports. Various consulting firms also carry out paid economic studies of 'state of the art' technology. Although there are minor variations in methodology, most of these studies estimate production costs using similar assumptions. The conventions used will be introduced in the following sections and should be followed when making preliminary economic analyses and when accurate cost information is not available.

6.2. COSTS, REVENUES AND PROFITS

This section introduces the components of project costs and revenues.

6.2.1. Fixed Capital Investment

The fixed capital investment is the total cost of designing, constructing and installing a plant and the associated modifications needed to prepare the plant site. The fixed capital investment is made up of:

1. The inside battery limits (ISBL) investment – the cost of the plant itself.
2. The modifications and improvements that must be made to the site infrastructure, known as off-site or OSBL investment.
3. Engineering and construction costs.
4. Contingency charges.

ISBL Plant Costs

The ISBL plant cost includes the cost of procuring and installing all the process equipment that makes up the new plant.

The direct field costs include:

1. All the major process equipment, such as vessels, reactors, columns, furnaces, heat exchangers, coolers, pumps, compressors, motors, fans, turbines, filters, centrifuges, driers, conveyors, etc., including field fabrication and testing if necessary.
2. Bulk items, such as piping, valves, wiring, instruments, structures, insulation, paint, lube oils, solvents, catalysts, etc.
3. Civil works such as roads, foundations, piling, buildings, sewers, ditches, bunds, etc.
4. Installation labour and supervision.

In addition to the direct field costs there will be indirect field costs including:

1. Construction costs such as construction equipment rental, temporary construction (rigging, trailers, etc.), temporary water and power, construction workshops, etc.
2. Field expenses and services such as field canteens, specialists' costs, overtime pay and adverse weather costs.
3. Construction insurance.
4. Labour benefits and burdens (social security, workers compensation, etc.).
5. Miscellaneous overhead items such as agent's fees, legal costs, import duties, special freight costs, local taxes, patent fees or royalties, corporate overheads, etc.

In the early stages of a project it is important to define the ISBL scope carefully, as other project costs are often estimated from ISBL cost. The overall project economics can be badly miscalculated if the ISBL scope is poorly defined. Methods for estimating ISBL costs are given in Section 6.3.

Off-Site Costs

Off-site cost or OSBL investment includes the costs of the additions that must be made to the site infrastructure to accommodate adding a new plant or increasing the capacity of an existing plant. Off-site investments may include:

- Electric main substations, transformers, switchgear and power lines.
- Power generation plants, turbine engines, standby generators.
- Boilers, steam mains, condensate lines, boiler feed water treatment plant, supply pumps.
- Cooling towers, circulation pumps, cooling water mains, cooling water treatment.
- Water pipes, water demineralization, waste water treatment plant, site drainage and sewers.
- Air separation plants to provide site nitrogen for inert gas, nitrogen lines.
- Driers and blowers for instrument air, instrument air lines.
- Pipe bridges, feed and product pipelines.
- Tanker farms, loading facilities, conveyors, docks, warehouses, railroads, lift trucks.
- Laboratories, analytical equipment, offices, canteens, changing rooms, central control rooms.
- Workshops and maintenance facilities.
- Emergency services, fire fighting equipment, fire hydrants, medical facilities, etc.
- Site security, fencing, gatehouses, landscaping.

Off-site investments often involve interactions with utility companies such as electricity or water suppliers. They may be subject to equal or greater scrutiny than ISBL investments, because of their impact on the local community through water consumption and discharge, traffic, etc.

Off-site costs are typically estimated as a proportion of ISBL costs in the early stages of design. Off-site costs are usually in the range from 10% to 100% of ISBL costs, depending on the project scope and its impact on site infrastructure. For typical petrochemical projects, off-site costs are usually between 20% and 50% of ISBL cost, and 40% is usually used as an initial estimate if no details of the site are known. Off-site costs will generally be lower for an established site with well developed infrastructure. This is particularly true of sites that have undergone contraction, where some plants have closed, leaving underutilized infrastructure ('brown-field' sites). On the other hand, if the site infrastructure is in need of repair or upgrading to meet new regulations, or if the plant is built on a completely new site (a 'green-field' site) then off-site costs will be higher.

Once a site has been chosen for the project then the modifications to the site infrastructure that are needed can be designed in detail in the same manner as the ISBL investments. Infrastructure upgrades are usually the first part of a project to be implemented as they usually need to be commissioned before the plant can begin operation.

Engineering Costs

The engineering costs, sometimes referred to as home office costs or contractor charges, include the costs of detailed design and other engineering services required to carry out the project:

1. Detailed design engineering of process equipment, piping systems, control systems and off-sites, plant layout, drafting, cost engineering, scale models and civil engineering.

2. Procurement of main plant items and bulks.
3. Construction supervision and services.
4. Administrative charges, including engineering supervision, project management, expediting, inspection, travel and living expenses and home office overheads.
5. Bonding.
6. Contractor's profit.

Very few operating companies retain a large enough engineering staff to carry out all of these activities internally, except for very small projects. In most cases, one or more of the major engineering contracting firms will be brought in.

Engineering costs are best estimated individually based on project scope, as they are not directly proportional to project size. A rule of thumb for engineering costs is 30% of ISBL plus OSBL cost for smaller projects and 10% of ISBL plus OSBL cost for larger projects. The actual charges paid for real industrial projects vary considerably from customer to customer and are strongly influenced by long-term client–contractor relationships and overall market demand for engineering services. Customers usually have to pay premiums or surcharges if they want to complete a project on an accelerated timeline or if they make a lot of changes once a project is underway.

Contingency Charges

Contingency charges are extra costs added into the project budget to allow for variation from the cost estimate. All cost estimates are uncertain (see Section 6.3.1) and the final installed cost of many items is not known until installation has been successfully completed. Apart from errors in the cost estimate, contingency costs also help cover:

- changes in project scope,
- changes in prices (e.g., prices of steel, copper, catalyst, etc.),
- currency fluctuations,
- labour disputes,
- subcontractor problems, and
- other unexpected problems.

A minimum contingency charge of 10% of ISBL plus OSBL cost should be used on all projects. If the technology is uncertain then higher contingency charges (up to 50%) are used. Contingency charges are discussed in more detail in Section 6.8.4.

6.2.2. Working Capital

Working capital is the additional money needed, above what it cost to build the plant, to start the plant up and run it until it starts earning income. Working capital typically includes:

1. Value of raw material inventory – usually estimated as 2 weeks' delivered cost of raw materials.
2. Value of product and by-product inventory – estimated as 2 weeks' cost of production.

3. Cash on hand – estimated as 1 week's cost of production.
4. Accounts receivable – products shipped but not yet paid for – estimated as 1 month's cost of production.
5. Credit for accounts payable – feedstocks, solvents, catalysts, packaging, etc. received but not yet paid for – estimated as 1 month's delivered cost.
6. Spare parts inventory – estimated as 1% to 2% of ISBL plus OSBL investment cost.

It can be seen that the sum of items 1 through 5 is roughly 7 weeks' cost of production minus 2 weeks' feedstock costs (item 5 is a credit).

Working capital can vary from as low as 5% of the fixed capital for a simple, single-product process, with little or no finished product storage, to as high as 30% for a process producing a diverse range of product grades for a sophisticated market, such as synthetic fibres. A typical figure for petrochemical plants is 15% of the fixed capital (ISBL plus OSBL cost).

Working capital is better estimated from the cost of production rather than capital investment. It is recovered at the end of the plant life.

Other methods for estimating the working capital requirement are given by Bechtel (1960), Lyda (1972) and Scott (1978).

6.2.3.   Variable Costs of Production

Variable costs of production are costs that are proportional to the plant output or operation rate. These include the costs of:

1. Raw materials consumed by the process.
2. Utilities – fuel burned in process heaters, steam, cooling water, electricity, raw water, instrument air, nitrogen and other services brought in from elsewhere on the site.
3. Consumables – solvents, acids, bases, inert materials, corrosion inhibitors, additives, catalysts and adsorbents that require continuous or frequent replacement.
4. Effluent disposal.
5. Packaging and shipping – drums, bags, tankers, freight charges, etc.

Variable costs can usually be reduced by more efficient design or operation of the plant. Methods for estimating variable costs are discussed in Section 6.4.

6.2.4.   Fixed Costs of Production

Fixed production costs are costs that are incurred regardless of the plant operation rate or output. If the plant cuts back its production these costs are not reduced. Fixed costs include:

1. Operating labour – see Section 6.4.7.
2. Supervision – usually taken as 25% of operating labour.
3. Direct salary overhead – costs of fringe benefits, payroll taxes, health insurance, etc., usually 40% to 60% of operating labour plus supervision.
4. Maintenance, which includes both materials and labour, and is typically estimated as 3% to 5% of ISBL investment, depending on the expected plant

reliability. Plants with more moving equipment or more solids handling usually require higher maintenance.

5. Property taxes and insurance – typically 1% to 2% of ISBL fixed capital.

6. Rent of land (and/or buildings) – typically estimated as 1% to 2% of ISBL plus OSBL investment. Most projects assume land is rented rather than purchased, but in some cases the land is bought and the cost is added to the fixed capital investment and recovered at the end of the plant life.

7. General plant overhead – charges to cover corporate overhead functions such as human resources, research and development (R&D), information technology, finance, legal, etc. Corporate overhead varies widely depending on the industry sector. Oil refining companies that carry out minimal R&D have much lower overhead than pharmaceuticals manufacturers. Plant overhead is typically taken as 65% of total labour (including supervision and direct overhead) plus maintenance.

8. Allocated environmental charges; for example, to cover Superfund payments in the United States, or costs associated with REACH in the European Union (see Chapter 9, Section 9.1.1) – typically 1% of ISBL plus OSBL cost.

9. Running license fees and royalty payments – i.e., those not capitalized at the start of the project.

10. Capital charges – these include interest payments due on any debt or loans used to finance the project, but *do not* include expected returns on invested equity capital – see Section 6.6.

11. Sales and marketing costs – in some cases these are considered part of general plant overhead. They can vary from almost zero for some commodities to millions of dollars a year for branded items such as foods, toiletries, drugs and cosmetics.

Fixed costs should never be neglected, even in the earliest stages of design, as they can have a significant impact on project economics. Very few chemical plants in developed economies carry less than US$1 million (US$1 MM) of fixed costs.

Fixed costs are also a strong disincentive for building small plants. As plant size is increased, labour, supervision and overhead costs usually do not increase, and hence the fixed cost per kilogram of product decreases. This, together with economies of scale in capital investment (see Section 6.3), gives larger plants more flexibility to reduce prices and hence force smaller plants out of business during downturns in the business cycle.

Fixed costs are not easily influenced by better design or operation of the plant, other than improvements that allow the plant to be operated safely with a smaller workforce. Fixed costs are more amenable to control at the corporate level than the plant level.

6.2.5. Revenues, Margins and Profits

Revenues

The revenues for a project are the incomes earned from sales of main products and by-products.

The production rate of main product is usually specified in the design basis and is determined based on predictions of overall market growth.

Determining which by-products to recover, purify and sell is usually more difficult than determining the main product. Some by-products are produced by the main reaction stoichiometry and are unavoidable unless new chemistry can be found. These stoichiometric by-products must usually be sold for whatever price they can get, otherwise waste-disposal costs will be excessive. Some examples of stoichiometric by-products are given in Table 6.1. Other by-products are produced from feed impurities or by non-selective reactions. The decision to recover, purify and sell; recycle or otherwise attenuate; or dispose of them as wastes is an important design optimization problem and is discussed in Section 6.4.8.

Margins

The sum of product and by-product revenues minus raw material costs is known as the gross margin (or sometimes product margin or just margin).

$$\text{Gross margin} = \text{Revenues} - \text{Raw materials costs} \qquad (6.1)$$

Gross margin is a useful concept, as raw materials costs are almost always the largest contributor to production costs (typically 80% to 90% of total cost of production). Raw materials and product prices of commodities are often subject to high variability and can be difficult to forecast, but margins suffer less variability if producers are able to pass feedstock price increases on to their customers. Margins are therefore often used in price forecasting, as described in Section 6.4.2.

Margins vary widely between different sectors of the chemical industry. For commodities such as bulk petrochemicals and fuels, margins are typically very low (less than 10% of revenues) and may even occasionally be negative. Commodity businesses are usually cyclical because of investment cycles and experience higher margins when supply is short, as described in Section 6.4. When a product is tightly regulated (making market entry difficult) or subject to patent protection, then margins can be much higher. For example, margins on food additives, pharmaceutical products and biomedical implants are typically more than 40% of revenues and often higher than 80% of revenues.

Table 6.1.   Some Stoichiometric By-Products

Feeds	Main Product	By-Product
cumene + air	phenol	acetone
propylene + ethylbenzene + air	propylene oxide	styrene
ethylene + chlorine	vinyl chloride monomer	HCl
allyl chloride + HOCl + NaOH	epichlorohydrin	NaCl
methane + steam	hydrogen	carbon dioxide
glucose	ethanol (by fermentation)	carbon dioxide
acetone cyanohydrin + methanol + H_2SO_4	methyl methacrylate	ammonium sulphate
sodium chloride + electricity	chlorine	sodium hydroxide

Profits

The cash cost of production (CCOP) is the sum of the fixed and variable production costs:

$$CCOP = VCOP + FCOP \qquad (6.2)$$

where

VCOP = sum of all the variable costs of production minus by-product revenues
FCOP = sum of all the fixed costs of production.

The cash cost of production is the cost of making product, not including any return on the equity capital invested. By convention, by-product revenues are usually taken as a credit and included in the VCOP. This makes it easier to determine cost per kilogram of producing the main product.

The gross profit is:

$$\text{Gross profit} = \text{Main product revenues} - CCOP \qquad (6.3)$$

Gross profit should not be confused with gross margin, as gross profit includes all the other variable costs in addition to raw materials, and also includes fixed costs.

The profit made by the plant is usually subject to taxation. Different tax codes apply in different countries and locations, and the taxable income may not be the full gross profit. Taxes are discussed in more detail in Section 6.5. The net profit (or cash flow after tax) is the amount left after taxes are paid:

$$\text{Net profit} = \text{gross profit} - \text{taxes} \qquad (6.4)$$

The net profit from the project is the money that is available as a return on the initial investments. Methods for evaluating the economic performance of investments are introduced in Sections 6.6 and 6.7.

It is sometimes useful to calculate a total cost of production (TCOP), assuming that a plant generates a specified return on investment. In this case an annual capital charge (ACC) is added to the cash cost of production:

$$TCOP = CCOP + ACC \qquad (6.5)$$

Methods for calculating the annual capital charge are discussed in Section 6.7.6.

6.2.6. Cash Flows at the End of the Project

If a plant ceases operation or is 'mothballed' (shut down on a semi-permanent basis) then the working capital is recovered, but must be reinvested if the plant is restarted. When a plant is shut down permanently then it can be sold in its entirety or else broken up and sold as scrap. There are several companies that specialize in buying and reselling second-hand plant, and advertisements for used plants and equipment can usually be found in the classified sections of the trade journals. The scrap value can be estimated based on the equipment weight and is usually less than 10% of the ISBL investment. OSBL investments are not recovered unless the entire site is shut down. If land was purchased for the plant, which is increasingly uncommon, then the land can be sold as an additional end of life credit. These cash flows at the end of the project are often not included in profitability analysis, as their timing is uncertain and they are often far enough in the future that they have negligible impact on any of the measures of profitability.

6.3. ESTIMATING CAPITAL COSTS

6.3.1. Accuracy and Purpose of Capital Cost Estimates

The accuracy of an estimate depends on the amount of design detail available, the accuracy of the cost data available, and the time spent on preparing the estimate. In the early stages of a project only an approximate estimate will be required, and justified, by the amount of information available.

The Association for the Advancement of Cost Estimating International (AACE International) is the professional association representing the cost-engineering profession in the USA. AACE International classifies capital cost estimates into five types according to their accuracy and purpose:

1. Order of magnitude estimates ('ballpark estimate', 'guesstimate', 'Class 5 estimate'), accuracy typically ±30–50%, usually based on the costs of similar processes and requiring essentially no design information. These are used in initial feasibility studies and for screening purposes.
2. Preliminary ('approximate', 'study', 'feasibility', 'Class 4') estimates, accuracy typically ±30%, which are used to make coarse choices between design alternatives. They are based on limited cost data and design detail.
3. Definitive ('authorization', 'budgeting', 'control', 'Class 3') estimates, accuracy typically ±10–15%. These are used for the authorization of funds to proceed with the design to the point where an accurate and more detailed estimate can be made. Authorization may also include funds to cover cancellation charges on any long-delivery equipment ordered at this stage of the design to avoid delay in the project. In a contracting organization this type of estimate could be used with a large contingency factor to obtain a price for tendering. Normally, however, an accuracy of about ±5% would be needed and a more detailed estimate would be made, if time permitted. With experience, and where a company has cost data available from similar projects, estimates of acceptable accuracy can be made at the flow-sheet stage of the project. A rough P and I diagram and the approximate sizes of the major items of equipment would also be needed.
4. Detailed estimates ('quotation', 'tender', 'firm estimate', 'contractor's estimate', 'Class 2 estimate'), accuracy ±5–10%, which are used for project cost control and estimates for fixed price contracts. These are based on the completed (or near complete) process design, firm quotes for equipment, and a detailed breakdown and estimation of the construction cost. By this stage the contractor can usually present a list of all the items that must be purchased and can make a firm commitment to the client.
5. Check estimates ('tender', 'as-bid', 'Class 1 estimate'), accuracy ±5–10%. This is based on a completed design and concluded negotiations on procurement of specialized items and long lead-time items.

The cost of preparing an estimate increases from about 0.1% of the total project cost for ±30% accuracy, to about 3% for a detailed estimate with an accuracy of ±5%.

As a project proceeds from initial concept through detailed design to start-up, costs begin to be accumulated, particularly once procurement and construction get underway (Figure 6.1a). At the same time, the ability of the design engineer to influence project cost decreases, and is minimal by the time construction begins (Figure 6.1b). There is therefore a strong incentive to try to estimate project costs at as early a stage as possible, even if the design information is incomplete, so that the project can be optimized, evaluated and abandoned if it is not attractive.

6.3.2. Rapid Cost Estimates

Historic Cost Data

The quickest way to make an order-of-magnitude estimate of plant cost is to scale it from the known cost of an earlier plant that used the same technology or from published data. This requires no design information other than the production rate.

The capital cost of a plant is related to capacity by the equation

$$C_2 = C_1 \left(\frac{S_2}{S_1}\right)^n \tag{6.6}$$

where

C_2 = ISBL capital cost of the plant with capacity S_2
C_1 = ISBL capital cost of the plant with capacity S_1.

The exponent n is typically 0.8 to 0.9 for processes that use a lot of mechanical work or gas compression (e.g., methanol, paper pulping, solids handling plants). For typical petrochemical processes n is usually about 0.7. For small-scale, highly

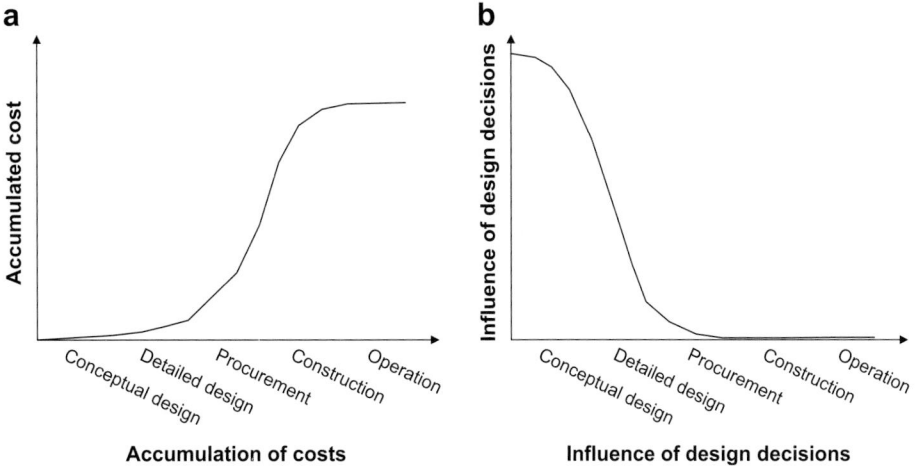

Figure 6.1. Influence of design decisions on project cost.

instrumented processes n is in the range 0.4 to 0.5. Averaged across the whole chemical industry, n is about 0.6, and hence equation 6.6 is commonly referred to as the 'six-tenths rule'. This value can be used to get a rough estimate of the capital cost if there are not sufficient data available to calculate the index for the particular process. Estrup (1972) gives a critical review of the six-tenths rule. Equation 6.6 is only an approximation, and if sufficient data are available the relationship is best represented on a log–log plot. Garrett (1989) has published capital cost–plant capacity curves for over 250 processes.

The journal *Hydrocarbon Processing* publishes supplements on refining, petrochemical and gas processing processes every other year. These supplements are available in print or CD format to subscribers and give approximate capital cost data for various licensed processes, which can be fitted using a rearranged form of equation 6.6:

$$C_2 = \frac{C_1}{S_1^n} \times S_2^n = a \; S_2^n \tag{6.7}$$

Values of the parameters a and n for some fuels and commodity chemical processes are given in Table 6.2. The costs in the *Hydrocarbon Processing* supplements are supplied by the technology vendors and are suitable for ballpark estimates only.

Step-Count Method

If cost data for a similar process is not available then an order-of-magnitude estimate can sometimes be made by adding contributions for different plant sections or functional units.

Experienced design engineers can often figure out costs of plant sections from historic total plant costs. For example, in many petrochemical processes roughly 20% of ISBL capital cost is in the reactor section and 80% is in the distillation and product purification sections.

An alternative approach is Bridgewater's method, which correlates plant cost against number of processing steps (Bridgewater and Mumford, 1979). For plants primarily processing liquids and solids:

$$Q \geq 60,000: \quad C = 3200 \; N \left(\frac{Q}{s}\right)^{0.675} \tag{6.8}$$

$$Q < 60,000: \quad C = 280,000 \; N \left(\frac{Q}{s}\right)^{0.3} \tag{6.9}$$

where

C = ISBL capital cost in US$, US Gulf Coast, 2000 basis
Q = plant capacity in metric tons per year
s = reactor conversion (= mass of desired product per mass fed to the reactor)
N = number of functional units.

(*Note*: the correlations have been updated from the original reference.)

Table 6.2. Process Cost Correlations

Process	Licensor	Capacity Units	S_{lower}	S_{upper}	a	n
ABS Resin (15% Rubber) by emulsion polymerization	Generic	MMlb/y	50	300	12.146	0.6
Acetic Acid by Cativa process	BP	MMlb/y	500	2000	3.474	0.6
Acetic Acid by Low Water Methanol Carbonylation	Celanese	MMlb/y	500	2000	2.772	0.6
Acrolein by propylene oxidation with Bi/Mo catalyst	Generic	MMlb/y	30	150	6.809	0.6
Adipic acid from phenol	Generic	MMlb/y	300	1000	3.533	0.6
Alkylation (sulphuric acid effluent refrigeration process)	Stratco/ DuPont	bpd	4,000	20,000		
Alkylation (HF process)	UOP	bpd	5000	12,000	0.153	0.6
Allyl chloride by propylene chlorination	Generic	MMlb/y	80	250	7.581	0.6
Alpha olefins (full range process)	Chevron Phillips	MMlb/y	400	1200	5.240	0.6
Alpha olefins (full range process)	Shell	MMlb/y	400	1000	8.146	0.6
Benzene by Sulpholane extraction	UOP/Shell	MMgal/y	50	200	7.793	0.6
Benzene by toluene hydrodealkylation	Generic	MMgal/y	50	200	7.002	0.6
Benzene reduction by Bensat	UOP	bpd	8000	15,000	0.0275	0.6
Biodiesel (FAME) from vegetable oil	Generic	MMlb/y	100	500	2.747	0.6
bis-HET by Eastman Glycolysis	Eastman	MMlb/y	50	200	0.500	0.6
BTX Aromatics by Cyclar process	BP/UOP	tpy	200,000	800,000	0.044	0.6
BTX Aromatics by CCR Platforming	UOP	tpy	200,000	800,000	0.015	0.6
Butadiene by extractive distillation	UOP/BASF	MMlb/y	100	500	5.514	0.6
Butadiene by Oxo-D plus extractive distillation	Texas Petrochem.	MMlb/y	100	500	11.314	0.6
Butene-1 by Alphabutol ethylene dimerization	Axens	tpy	5000	30,000	0.0251	0.6
Butene-1 by BP Process	BP	tpy	20,000	80,000	0.169	0.6
Caprolactam from nitration-grade toluene	SNIA BPD S.p.A.	tpy	40,000	120,000	0.321	0.6
Carbon monoxide by steam methane reforming	Generic	MMscf/y	2000	6000	0.363	0.6
Catalytic Condensation for Gasoline Production	UOP	bpd	10,000	30,000	0.222	0.6
Catalytic reforming by CCR Platforming	UOP	bpd	15,000	60,000	0.179	0.6
Coking by Flexicoking including Fluid Coking	ExxonMobil	bpd	15,000	40,000	0.343	0.6
Coking by Selective Yield Delayed Coking	Foster Wheeler/UOP	bpd	15,000	60,000	0.109	0.68
Copolymer polypropylene by INNOVENE	BP	MMlb/y	300	900	3.430	0.6
Copolymer polypropylene by Unipol	Dow	MMlb/y	300	900	3.641	0.6
Copolymer polypropylene by SPHERIPOL Bulk	Basell	MMlb/y	300	900	3.649	0.6
Copolymer polypropylene by BORSTAR	Borealis	MMlb/y	300	900	4.015	0.6
Crude distillation by D2000	TOTAL/Technip	bpd	150,000	300,000	0.151	0.6
Cumene by Q-Max	UOP	tpy	150,000	450,000	0.0120	0.6
Cyclic Olefin Copolymer by Mitsui Process	Mitsui	MMlb/y	60	120	12.243	0.6
Cyclohexane by liq-phase hydrogenation of benzene	Axens	tpy	100,000	300,000	0.0061	0.6
Dewaxing by ISODEWAXING	Chevron Lummus	bpd	6000	15,000	0.256	0.6
2,6-Dimethylnaphthalene by MeOH alkylation	Exxon Mobil/Kobe	MMlb/y	50	100	7.712	0.6
Dimethyl terephthalate by methanolysis	Generic	MMlb/y	30	80	5.173	0.6
Dimethyl terephthalate by Huels Oxidation	Huels	MMlb/y	300	800	7.511	0.6
Ethanol by ethylene hydration	Generic	Mgal/y	30	90	9.643	0.6
Ethanol (fuel grade) by Corn Dry Milling	Generic	tpy	100,000	300,000	0.0865	0.6
Ethylbenzene by EBOne	ABB Lummus/UOP	tpy	300,000	700,000	0.0085	0.6

Table 6.2.   Process Cost Correlations—Cont'd

Process	Licensor	Capacity Units	S_{lower}	S_{upper}	a	n
Ethylene by ethane cracking	Generic	MMlb/y	500	2000	9.574	0.6
Ethylene by UOP Hydro MTO	UOP/Norsk Hydro	MMlb/y	500	2000	8.632	0.6
Ethylene: light naphtha cracker (max ethylene)	Generic	MMlb/y	1000	2000	16.411	0.6
Ethylene by ethane/propane cracker	Generic	MMlb/y	1000	2000	7.878	0.6
Ethylene by gas oil cracker	Generic	MMlb/y	1000	2000	17.117	0.6
Ethylene glycol via ethylene oxide hydrolysis	Shell	MMlb/y	500	1000	5.792	0.6
Expandable polystyrene by suspension process	Generic	MMlb/y	50	100	3.466	0.6
Fischer Tropsch Process	ExxonMobil	tpy	200,000	700,000	0.476	0.6
Fluid catalytic cracking	KBR	bpd	20,000	60,000	0.210	0.6
Fluid catalytic cracking with power recovery	UOP	bpd	20,000	60,000	0.302	0.6
Gas to liquids by Syntroleum Process	Syntroleum	bpd	30,000	100,000	2.279	0.6
Gas sweetening by Amine Guard FS to pipeline spec	UOP	MMscf/d	300	800	0.386	0.6
Gasification by GE Gasification Process Maya crude	GE Energy	bpd	7000	15,000	0.681	0.6
Gasoline desulphurization, ultra-deep by Prime-G+	Axens	bpd	7000	15,000	0.0420	0.58
Glucose (40% Solution) by basic wet corn milling	Generic	MMlb/y	300	800	3.317	0.6
HDPE Pellets by BP Gas Phase Process	BP Amoco	MMlb/y	300	700	3.624	0.6
HDPE Pellets by Phillips Slurry Process	Phillips	MMlb/y	300	700	3.370	0.6
HDPE Pellets by Zeigler Slurry Process	Zeigler	MMlb/y	300	700	4.488	0.6
High impact polystyrene by bulk polymerization	Dow	MMlb/y	70	160	2.970	0.6
Hydrocracking by ISOCRACKING	Chevron Lummus	bpd	20,000	45,000	0.221	0.6
Hydrocracking by Unicracking, distillate	UOP	bpd	20,000	45,000	0.136	0.66
Hydrocracking	Axens	bpd	20,000	45,000	0.198	0.6
Hydrogen by steam methane reforming	Foster Wheeler	MMscf/d	10	50	1.759	0.79
Hydrotreating by Unionfining	UOP	bpd	10,000	40,000	0.0532	0.68
Isomerization by Once-through Penex	UOP	bpd	8000	15,000	0.0454	0.6
Isomerization by Penex-Molex	UOP	bpd	8000	15,000	0.120	0.6
Isophthalic acid by m-Xylene oxidation	Generic	MMlb/y	160	300	9.914	0.6
Isoprene via isobutylene carbonylation	IFP	MMlb/y	60	200	10.024	0.6
Isoprene by propylene dimerization and pyrolysis	Generic	MMlb/y	60	200	6.519	0.6
Linear alkylbenzene by PACOL/DeFine/PEP/Detal	UOP	MMlb/y	100	250	4.896	0.6
Linear alpha olefins	Chevron	MMlb/y	300	700	5.198	0.6
Linear alpha olefins by Linear-1	UOP	tpy	200,000	300,000	0.122	0.6
Maleic anhydride by fluid bed process	Generic	MMlb/y	70	150	7.957	0.6
Methacrylic acid by isobutylene oxidation	Generic	MMlb/y	70	150	7.691	0.6
Methanol via steam reforming & synthesis	Davy Process Tech.	tpd	3000	7000	2.775	0.6
m-Xylene by MX Sorbex	UOP	MMlb/y	150	300	4.326	0.6
Naphthalene by 3-stage fractional crystallizer	Generic	MMlb/y	20	50	2.375	0.6
N-Butanol from crude C4s	BASF	MMlb/y	150	300	8.236	0.6
Norbornene by Diels-Alder reaction	Generic	MMlb/y	40	90	7.482	0.6
Pentaerythritol by condensation	Generic	MMlb/y	40	90	6.220	0.6
PET resin chip with comonomer by NG3	DuPont	MMlb/y	150	300	4.755	0.6
Phenol from cumene (zeolite catalyst)	UOP/ABB Lummus	MMlb/y	200	600	6.192	0.6
Phthalic anhydride by catalytic oxidation	Generic	MMlb/y	100	200	7.203	0.6
Polycarbonate by interfacial polymerization	Generic	MMlb/y	70	150	20.680	0.6
Polyethylene terephthalate (melt phase)	Generic	MMlb/y	70	200	5.389	0.6
Polystyrene by bulk polymerization, plug flow	Generic	MMlb/y	70	200	2.551	0.6
Propylene by Oleflex	UOP	tpy	150,000	350,000	0.0943	0.6
Propylene by metathesis	Generic	MMlb/y	500	1000	1.899	0.6
Purified terphthalic acid	EniChem/ Technimont	MMlb/y	350	700	10.599	0.6

Table 6.2. Process Cost Correlations—Cont'd

Process	Licensor	Capacity Units	S_{lower}	S_{upper}	a	n
p-Xylene by Isomar and Parex	UOP	tpy	300,000	700,000	0.0230	0.6
p-Xylene by Tatoray Process	UOP	bpd	12,000	20,000	0.0690	0.6
Refined Glycerine by distillation/adsorption	Generic	MMlb/y	30	60	2.878	0.6
Sebaccic Acid by cyclododecanone route	Sumitomo	MMlb/y	8	16	13.445	0.6
Sorbitol (70%) by continuous hydrogenation	Generic	MMlb/y	50	120	4.444	0.6
Styrene by SMART	ABB Lummus/UOP	tpy	300,000	700,000	0.0355	0.6
Vinyl acetate by Cativa Integrated Process	BP	MMlb/y	300	800	7.597	0.6
Vinyl acetate by Celanese VAntage Process	Celanese	MMlb/y	300	800	6.647	0.6
Visbreaking by coil-type visbreaker	Foster Wheeler/UOP	bpd	6000	15,000	0.278	0.48

Notes:

1. Values of a are in January 2006 million US\$ on a US Gulf Coast (USGC) basis (Nelson Farrer index = 1961.6, CE index = 478.6).
2. S_{lower} and S_{upper} indicate the bounds of the region over which the correlation can be applied.
3. S is based on product rate for chemicals, feed rate for fuels.
4. If the index n is 0.6 then the correlation is an extrapolation around a single cost point.
5. Flow units are: MMlb/y = million pounds per year; tpy = short tons per year; bpd = barrels per day.
6. Correlations are based on data taken from *Hydrocarbon Processing* (2003, 2004a and 2004b), except where the licensor is stated as 'Generic', in which cases the correlations are based on data from Nexant PERP reports (see www.Nexant.com/products for a full list of reports available).

A functional unit includes all the equipment and ancillaries needed for a significant process step or function, such as a reaction, separation or other major unit operation. Pumping and heat exchange are not normally considered as functional units unless they have substantial cost, for example, compressors, refrigeration systems or process furnaces.

Manufactured Products

Step-count methods such as Bridgewater's method were developed for chemical plants and do not extend well to other types of manufacturing. For large-scale production of manufactured items (>500,000 pieces per year) a rule of thumb is:

$$\text{TCOP} = 2 \times \text{materials cost} \tag{6.10}$$

This equation can be used to make a very approximate estimate of plant cost if fixed costs and utilities can be estimated.

Example 6.1

The process for making cyclohexane by saturation of benzene consists of a feed–effluent heat exchanger, a saturation reactor and a product stabilizer column. Estimate the cost of a plant that produces 200,000 metric tons per year (200 kt/y) of cyclohexane using the correlation in Table 6.2 and Bridgewater's method.

Solution

From Table 6.2, the cost correlation for the Axens process for benzene saturation gives:

$$C = 0.0061(S)^{0.6}$$
$$= 0.0061(2 \times 10^5)^{0.6}$$
$$= \$9.2 \text{ MM expressed on a Jan. 2006 USGC basis}$$

Using Bridgewater's method, we have two functional units (the reactor and product stabilizer – the heat exchanger doesn't count) and assuming that the reactor conversion is 1.0, we can substitute into equation 6.8:

$$C = 3200 \times 2 \times (Q)^{0.675}$$
$$= 3200 \times 2 \times (2 \times 10^5)^{0.675}$$
$$= \$24 \text{ MM expressed on a 2000 USGC basis.}$$

Note that we have obtained two very different answers. Bridgewater's correlation is known to be only an approximation; however, Table 6.2 is based on data from technology vendors that may be somewhat understated. With the level of information available it is probably safe to say that the cost is in the range $10 MM to $20 MM. Note also that the costs are not on the same time basis. Methods for correcting costs on different time bases will be discussed in Section 6.3.5, below.

6.3.3.   The Factorial Method of Cost Estimation

Capital cost estimates for chemical process plants are often based on an estimate of the purchase cost of the major equipment items required for the process, the other costs being estimated as factors of the equipment cost. The accuracy of this type of estimate will depend on what stage the design has reached at the time the estimate is made, and on the reliability of the data available on equipment costs. In the later stages of the project design, when detailed equipment specifications are available and firm quotes have been obtained from vendors, a relatively accurate estimate of the capital cost of the project can be made by this method.

Lang Factors

Lang (1948) proposed that the ISBL fixed capital cost of a plant is given as a function of the total purchased equipment cost by the equation:

$$C = F \left(\sum C_e \right) \tag{6.11}$$

where

C = total plant ISBL capital cost (including engineering costs)
$\Sigma\, C_e$ = total delivered cost of all the major equipment items: reactors, tanks, columns, heat exchangers, furnaces, etc.
F = an installation factor, later widely known as a Lang factor.

Lang originally proposed the following values of F, based on 1940s economics:

F = 3.1 for solids processing plant

$F = 4.74$ for fluids processing plant

$F = 3.63$ for mixed fluids–solids processing plant.

Hand (1958) suggested that better results are obtained by using different factors for different types of equipment. Examples of the factors proposed by Hand are given in Table 6.3. Hand also observed that this approach should only be used in the earliest stages of process design and in the absence of detailed design information.

Both Lang (1948) and Hand (1958) included home office costs but not off-site costs or contingency in their installation factors, so beware of double counting Engineering Procurement and Construction (EPC) costs when using this approach. The relative costs of materials and labour have changed substantially from when these factors were developed, and the accuracy of the correlation probably never warranted three significant figures for F. Most practitioners using this method therefore use a Lang factor of 3, 4 or 5, depending on the plant scale (larger plant = smaller factor) and type.

Detailed Factorial Estimates

Equation 6.11 can be used to make a preliminary estimate once the flow-sheet has been drawn up and the main plant equipment has been sized. When more detailed design information is available then the installation factor can be estimated somewhat more rigorously, by considering the cost factors that are compounded into the Lang factor individually.

The direct-cost items that are incurred in the construction of a plant, in addition to the cost of equipment are:

1. Equipment erection, including foundations and minor structural work.
2. Piping, including insulation and painting.
3. Electrical, power and lighting.
4. Instruments and automatic process control (APC) systems.
5. Process buildings and structures.
6. Ancillary buildings, offices, laboratory buildings, workshops.
7. Storage for raw materials and finished product.
8. Utilities (Services), provision of plant for steam, water, air, fire fighting services (if not costed separately as off-sites).
9. Site preparation.

Table 6.3. Installation Factors Proposed by Hand (1958)

Equipment Type	Installation Factor
Compressors	2.5
Distillation columns	4
Fired heaters	2
Heat exchangers	3.5
Instruments	4
Miscellaneous equipment	2.5
Pressure vessels	4
Pumps	4

The contribution of each of these items to the total capital cost is calculated by multiplying the total purchased equipment cost by an appropriate factor. As with the basic Lang factor, these factors are best derived from historical cost data for similar processes. Typical values for the factors are given in several references; see Happle and Jordan (1975) and Garrett (1989). Guthrie (1974) splits the costs into the material and labour portions and gives separate factors for each.

The accuracy and reliability of an estimate can be improved by dividing the process into sub-units and using factors that depend on the function of the sub-units; see Guthrie (1969). In Guthrie's detailed method of cost estimation the installation, piping and instrumentation costs for each piece of equipment are costed separately. Detailed costing is only justified if the cost data available are reliable and the design has been taken to the point where all the cost items can be identified and included. Gerrard (2000) gives factors for individual pieces of equipment as a function of equipment cost and complexity of installation.

Typical factors for the components of the capital cost are given in Table 6.4. These can be used to make an approximate estimate of capital cost using equipment cost data published in the literature.

The installation factors given in Tables 6.3 and 6.4 are for plants built from carbon steel. When more exotic materials are used then a materials factor f_m should also be introduced:

$$f_m = \frac{\text{purchased cost of item in exotic material}}{\text{purchased cost of item in carbon steel}} \qquad (6.12)$$

Note that f_m is not equal to the ratio of the metal prices, as the equipment purchased cost also includes labour costs, overheads, fabricator's profit and other

Table 6.4.   Typical factors for estimation of project fixed capital cost

Item	Process type Fluids	Fluids – Solids	Solids
Major equipment, total purchase cost	C_e	C_e	C_e
f_{er} Equipment erection	0.3	0.5	0.6
f_p Piping	0.8	0.6	0.2
f_i Instrumentation and control	0.3	0.3	0.2
f_{el} Electrical	0.2	0.2	0.15
f_c Civil	0.3	0.3	0.2
f_s Structures and buildings	0.2	0.2	0.1
f_l Lagging and paint	0.1	0.1	0.05
ISBL cost, $C = \Sigma C_e \times$	3.3	3.2	2.5
Offsites (OS)	0.3	0.4	0.4
Design and Engineering (D&E)	0.3	0.25	0.2
Contingency (X)	0.1	0.1	0.1
Total fixed capital cost $C_{FC} = C(1 + OS)(1 + D\&E + X)$			
$= C \times$	1.82	1.89	1.82
$= \Sigma C_e \times$	6.00	6.05	4.55

costs that do not scale directly with metal price. Equation 6.11 can then be expanded for each piece of equipment to give:

$$C = \sum_{i=1}^{i=M} C_{e,i,CS} \; [(1+f_p)f_m + (f_{er} + f_{el} + f_i + f_c + f_s + f_l)] \tag{6.13}$$

or

$$C = \sum_{i=1}^{i=M} C_{e,i,A} \; [(1+f_p) + (f_{er} + f_{el} + f_i + f_c + f_s + f_l)/f_m] \tag{6.14}$$

where

$C_{e,i,CS}$ = purchased equipment cost of equipment i in carbon steel
$C_{e,i,A}$ = purchased equipment cost of equipment i in alloy
M = total number of pieces of equipment
f_p = installation factor for piping
f_{er} = installation factor for equipment erection
f_{el} = installation factor for electrical work
f_i = installation factor for instrumentation and process control
f_c = installation factor for civil engineering work
f_s = installation factor for structures and buildings
f_l = installation factor for lagging, insulation or paint.

Failure to properly correct installation factors for materials of construction is one of the most common sources of error with the factorial method. Typical values of the materials factor for common engineering alloys are given in Table 6.5.

Summary of the Factorial Method

Many variations on the factorial method are used. The method outlined below can be used with the data given in this chapter to make a quick, approximate estimate of the fixed capital investment needed for a project.

1. Prepare material and energy balances; draw up preliminary flow-sheets; size major equipment items and select materials of construction.
2. Estimate the purchased cost of the major equipment items. See next section.

Table 6.5. Materials Cost Factors, f_m, Relative to Plain Carbon Steel

Material	f_m
Carbon steel	1.0
Aluminium and bronze	1.07
Cast steel	1.1
304 stainless steel	1.3
316 stainless steel	1.3
321 stainless steel	1.5
Hastelloy C	1.55
Monel	1.65
Nickel and Inconel	1.7

3. Calculate the ISBL installed capital cost, using the factors given in Table 6.4 and correcting for materials of construction using equation 6.13 or 6.14 with the materials factors given in Table 6.5.
4. Calculate the OSBL, engineering and contingency costs using the factors given in Table 6.4.
5. The sum of ISBL, OSBL, engineering and contingency costs is the fixed capital investment.
6. Estimate the working capital as a percentage of the fixed capital investment; 10 to 20% is typical (or better, calculate it from the cost of production if this has been estimated – see Section 6.4).
7. Add the fixed and working capital to get the total investment required.

6.3.4.   Estimating Purchased Equipment Costs

The factorial method of cost estimation is based on purchased equipment costs and therefore requires good estimates for equipment costs. Costs of single pieces of equipment are also often needed for minor revamp and de-bottlenecking projects.

The best source of purchased equipment costs is recent data on actual prices paid for similar equipment. Engineers working for EPC companies (often referred to as Contractors) have access to large amounts of high-quality data, as these companies carry out many projects globally every year. Engineers working in operating companies may have access to data from recent projects, but unless they work for a large company that carries out many capital projects they are unlikely to be able to develop and maintain current cost correlations for more than a few basic equipment types. Most large companies recognize the difficulty of making reliable cost estimates and employ a few experienced cost engineering specialists who collect data and work closely with the EPC companies on project budgets.

Actual prices paid for equipment and bulk items may differ substantially from catalogue or list prices, depending on the purchasing power of the contractor or client and the urgency of the project. Discounts and surcharges are highly confidential business information and will be closely guarded even within EPC companies.

Those design engineers who are outside the EPC sector and do not have the support of a cost estimating department must rely on cost data from the open literature or use cost estimating software. The most widely used software for estimating chemical plant costs is the ICARUS™ suite of tools licensed by Aspen Technology Inc. ICARUS™ does not use the factorial method, but instead estimates equipment costs, bulk costs and installation costs from the costs of materials and labour, following the practice used by cost engineers for detailed estimating. The models in ICARUS™ are developed by a team of cost engineers based on data collected from EPC companies and equipment manufacturers. The models are updated annually. The ICARUS Process Estimator software is included in the standard Aspen/Hsysys academic package and is available in most universities. The ICARUS™ software can give reasonably good estimates when used properly and is described in more detail in Section 6.3.8.

There is an abundance of equipment cost data and cost correlations in the open literature, but much of it is of very poor quality. The relationship between size and cost given in equations 6.6 and 6.7 can also be used for equipment if a suitable size parameter is used. If the size range spans several orders of magnitude, then log–log plots usually give a better representation of the relationship than simple equations.

Some of the most reliable information on equipment costs can be found in the professional cost engineering literature. Correlations based on recent data are occasionally published in *Cost Engineering*, which is the journal of the Association for the Advancement of Cost Engineering International (AACE International). AACE International also has an excellent web site, www.aacei.org, which has cost models that can be used by members. There is also an extensive listing of other web resources for cost estimating at www.aacei.org/resources. The UK Association of Cost Engineers (ACostE) publishes the journal *The Cost Engineer*, and also prints a guide to capital cost estimating (Gerrard, 2000), which gives cost curves for the main types of process equipment based on recent data. The prices are given in British pounds sterling on a UK basis, and are useful for making estimates of prices in Northwest Europe. The International Cost Engineering Council web site (www.icoste.org) provides links to 46 international cost-engineering societies, several of which maintain databases of local costs.

Many cost correlations can be found in chemical engineering textbooks; for example, Douglas (1988), Garrett (1989), Turton *et al.* (2003), Peters *et al.* (2003) and Ulrich and Vasudevan (2004). The references for such correlations should always be checked carefully. When they are properly referenced they are often found to be based on data published by Guthrie (1969, 1974) and updated using either cost indices (as described in Section 6.3.6) or a few recent data points. Guthrie's correlations were reasonably good when published, but there have been substantial changes in the relative contributions of material and fabrication costs of most process equipment since then. Academic authors usually do not have access to sufficient high-quality cost data to be able to make reliable correlations, and most of the academic correlations predict lower costs than would be obtained using Aspen ICARUS™ or other detailed estimating methods. These correlations are adequate for the purposes of university design projects but are not useful for assessing the costs of real projects. It is to be hoped that the authors of these publications will benchmark the correlations against Aspen ICARUS™ in future editions, which will improve the accuracy of the correlations and make them more useful to those who do not have access to costing software.

Detailed estimates are usually made by costing the materials and labour required for each item in the plant, making a full analysis of the work breakdown structure (WBS) to arrive at an accurate estimate of the labour. This method must be followed whenever cost or price data is not available, for example, when making an estimate of the cost of specialized equipment that cannot be found in the literature. For example, a reactor design is usually unique for a particular process but the design can be broken down into standard components (vessels, heat-exchange surfaces, spargers, agitators, etc.) the cost of which can be found in the literature and used to build up an estimate of the reactor cost. This method is described by Dysert (2007) and Woodward and Chen (2007) in sections of the AACE International training manual (Amos, 2007). Breakdowns of the materials and labour components for many types of process equipment are given by Page (1996). Pikulik and Diaz (1977) give a method of costing major equipment items from cost data on the basic components: shells, heads, nozzles and internal fittings. Purohit (1983) gives a detailed procedure for estimating the cost of heat exchangers.

A large amount of vendor information is now available on-line and can easily be found using any of the major search engines or by starting from directories such as www.purchasing.com. On-line costs are usually manufacturer's catalogue prices for small-order quantities. Large order sizes (as filled by contractors) are often steeply discounted. Items requiring special fabrication, for example large vessels or compressors, may experience discounts or surcharges depending on the state of the manufacturer's order books and the purchasing power of the customer.

For those design engineers who lack access to reliable cost data or estimating software, the correlations given in Table 6.6 can be used for preliminary estimates. The correlations in Table 6.6 are of the form:

$$C_e = a + bS^n \tag{6.15}$$

where

C_e = purchased equipment cost on a US Gulf Coast basis, Jan. 2007 (CE index (CEPCI) = 509.7, NF refinery inflation index = 2059.1)
a, b = cost constants in Table 6.6
S = size parameter, units given in Table 6.6
N = exponent for that type of equipment.

The correlations in Table 6.6 are only valid between the lower and upper values of S indicated. The prices are all for carbon steel equipment except where noted in the table. Costs calculated from Table 6.6 can be updated and converted to international locations by following the methods set out in Sections 6.3.5 and 6.3.6.

Example 6.2

A plant modification has been proposed that will allow recovery of a by-product. The modification consists of adding the following equipment:

Distillation column, height 30 m, diameter 3 m, 50 sieve trays, operating pressure 10 bar
U-tube heat exchanger, area 60 m^2
Kettle reboiler, area 110 m^2
Horizontal pressure vessel, volume 3 m^3, operating pressure 10 bar
Storage tank, volume 50 m^3
Two centrifugal pumps, flow rate 3.6 m^3/h, driver power 500 W
Three centrifugal pumps, flow rate 2.5 m^3/h, driver power 1 kW (two installed plus one spare).

Estimate the installed ISBL capital cost of the modification if the plant is to be built from type 304 stainless steel. Estimate the cost using both Hand's method and the factors given in Table 6.4.

Solution

The first step is to convert the units to those required for the correlations and determine any missing design information. The distillation column can be costed as a combination of a vertical pressure vessel and internals. For both pressure vessels we need to know the wall thickness. The details of how to calculate vessel wall thickness

Table 6.6. Purchased Equipment Cost for Common Plant Equipment

Equipment	Units for size, S	S_{lower}	S_{upper}	a	b	n	Note
Agitators & mixers							
Propeller	driver power, kW	5.0	75	15,000	990	1.05	
Spiral ribbon mixer	driver power, kW	5.0	35	27,000	110	2.0	
Static mixer	L/s	1.0	50	500	1030	0.4	
Boilers							
Packaged, 15 to 40 bar	kg/h steam	5000	200,000	106,000	8.7	1.0	
Field erected, 10 to 70 bar	kg/h steam	20,000	800,000	110,000	45	0.9	
Centrifuges							
High-speed disk	diameter, m	0.26	0.49	50,000	423,000	0.7	
Atmospheric suspended basket	power, kW	2.0	20	57,000	660	1.5	
Compressors							
Blower	m^3/h	200	5000	3800	49	0.8	
Centrifugal	driver power, kW	75	30,000	490,000	16,800	0.6	
Reciprocating	driver power, kW	93	16,800	220,000	2300	0.75	
Conveyors							
Belt, 0.5 m wide	length, m	10	500	36,000	640	1.0	
Belt, 1.0 m wide	length, m	10	500	40,000	1160	1.0	
Bucket elevator, 0.5 m bucket	height, m	10	30	15,000	2300	1.0	
Crushers							
Reversible hammer mill	t/h	30	400	60,000	640	1.0	
Pulverisers	kg/h	200	4000	14,000	590	0.5	
Crystallizers							
Scraped surface crystallizer	length, m	7	280	8400	11,300	0.8	
Distillation columns							
See pressure vessels, packing and trays							
Dryers							
Direct contact rotary	area, m^2	11	180	13,000	9100	0.9	1
Atmospheric tray batch	area, m^2	3.0	20	8700	6800	0.5	2
Spray dryer	evap. rate, kg/h	400	4,000	350,000	1900	0.7	
Evaporators							
Vertical tube	area, m^2	11	640	280	30,500	0.55	
Agitated falling film	area, m^2	0.5	12	75,000	56,000	0.75	
Exchangers							
U-tube shell and tube	area, m^2	10	1000	24,000	46	1.2	
Double pipe	area, m^2	1.0	80	1600	2100	1.0	
Thermosyphon reboiler	area, m^2	10	500	26,000	104	1.1	
U-tube kettle reboiler	area, m^2	10	500	25,000	340	0.9	
Plate and frame	area, m^2	1.0	500	1350	180	0.95	3
Filters							
Plate and frame	capacity, m^3	0.4	1.4	110,000	77,000	0.5	
Vacuum drum	area, m^2	10	180	−63,000	80,000	0.3	
Furnaces							
Cylindrical	duty, MW	0.2	60	68,500	93,000	0.8	
Box	duty, MW	30	120	37,000	95,000	0.8	

Table 6.6.   Purchased Equipment Cost for Common Plant Equipment—Cont'd

Equipment	Units for size, S	S_{lower}	S_{upper}	a	b	n	Note
Packings							
304 ss Raschig rings	m^3			0	7300	1.0	
Ceramic intalox saddles	m^3			0	1800	1.0	
304 ss Pall rings	m^3			0	7700	1.0	
PVC structured packing	m^3			0	500	1.0	
304 ss structured packing	m^3			0	6900	1.0	4
Pressure vessels							
Vertical, cs	shell mass, kg	160	250,000	10,000	29	0.85	5
Horizontal, cs	shell mass, kg	160	50,000	8800	27	0.85	
Vertical, 304 ss	shell mass, kg	120	250,000	15,000	68	0.85	5
Horizontal, 304 ss	shell mass, kg	120	50,000	11,000	63	0.85	
Pumps and drivers							
Single-stage centrifugal	flow, L/s	0.2	126	6900	206	0.9	
Explosion proof motor	power, kW	1.0	2500	-950	1770	0.6	
Condensing steam turbine	power, kW	100	20,000	-12,000	1630	0.75	
Reactors							
Jacketed, agitated	volume, m^3	0.5	100	53,000	28,000	0.8	3
Jacketed, agitated, glass lined	volume, m^3	0.5	25	11,000	76,000	0.4	
Tanks							
Floating roof	capacity, m^3	100	10,000	97,000	2800	0.65	
Cone roof	capacity, m^3	10	4000	5000	1400	0.7	
Trays							
Sieve trays	diameter, m	0.5	5.0	110	380	1.8	6
Valve trays	diameter, m	0.5	5.0	180	340	1.9	6
Bubble cap trays	diameter, m	0.5	5.0	290	550	1.9	6
Utilities							
Cooling tower & pumps	flow, L/s	100	10,000	150,000	1300	0.9	7
Packaged mechanical refrigerator	evaporator duty, kW	50	1500	21,000	3100	0.9	
Water ion exchange plant	flow, m^3/h	1	50	12,000	5400	0.75	

Notes

1. Direct heated.
2. Gas fired.
3. Type 304 stainless steel.
4. With surface area 350 m^2/m^3.
5. Not including heads, ports, brackets, internals, etc. (see Chapter 13 for how to calculate wall thickness).
6. Cost per tray, based on a stack of 30 trays.
7. Field assembly.
8. All costs are U.S. Gulf Coast basis, Jan. 2007 (CE index (CEPCI) = 509.7, NF refinery inflation index = 2059.1).

in accordance with the ASME Boiler and Pressure Vessel Code are given in Section 13.5, and the equation to use is equation 13.41.

The design pressure of the vessels should be 10% above the operating pressure (see Chapter 13), so the design pressure is 11 bar or roughly 1.1×10^6 N/m^2. The maximum allowable stress for type 304 stainless steel at 500°F (260°C) is 12.9 ksi or roughly 89 N/mm^2 (Table 13.2). Assuming the welds will be fully radiographed the weld efficiency is 1.0. Substituting in equation 13.41 for the column wall thickness, t_w, then gives:

$$t_w = \frac{1.1 \times 10^6 \times 3}{(2 \times 89 \times 10^6 \times 1.0) - (1.2 \times 1.1 \times 10^6)} \qquad (13.41)$$
$$= 0.0187 \text{ m}$$

say 20 mm.

We can now calculate the shell mass, using the density of 304 stainless steel (= 8000 kg/m^3, from Table 7.2).

$$\text{Shell mass} = \pi D_c L_c t_w \rho$$

where

D_c = vessel diameter, m
L_c = vessel length, m
t_w = wall thickness, m
ρ = metal density, kg/m^3.

So the shell mass for the distillation column is:

$$\text{Shell mass} = \pi \times 3.0 \times 30 \times 0.02 \times 8000 = 46{,}685 \text{ kg}$$

For the horizontal pressure vessel we need to convert the volume into a length and diameter. Assuming that the vessel is a cylinder with $L_c = 2D_c$ then we can follow the same method as for the column and find $t_w = 8$ mm and shell mass = 636 kg.

Using the correlations in Table 6.6, we obtain the following purchase costs for the stainless steel pressure vessels:

Distillation column shell, cost = $15{,}000 + 68 \, (46{,}685)^{0.85} = \$650{,}000$
Horizontal pressure vessel, cost = $11{,}000 + 63 \, (636)^{0.85} = \$26{,}000$

For the remaining equipment we obtain the following purchase costs from the correlations in Table 6.6 based on carbon steel construction:

Distillation column trays
        cost per tray = $110 + 380 \, (3.0)^{1.8} = \$2855$
        cost for 50 trays = \$143,000
U-tube heat exchanger, cost = $24{,}000 + 46 \, (60)^{1.2} = \$30{,}300$
        Kettle reboiler, cost = $25{,}000 + 340 \, (110)^{0.9} = \$48{,}400$
        Storage tank (conical head), cost = $5000 + 1400 \, (50)^{0.7} = \$27{,}000$
Centrifugal pump, 3.6 m^3/h = 1 L/s, so:
        cost each = $6900 + 206 \, (1.0)^{0.9} = \$7100$
        cost for two pumps = \$14,200

driver (electric motor), cost each $= -950 + 1770 \, (0.5)^{0.6} = \$220$
cost for two drivers $= \$440$
Centrifugal pump, 2.5 m^3/h $= 0.694$ L/s, so:
cost each $= 6900 + 206 \, (0.694)^{0.9} = \$7050$
cost for three $= \$21,100$
driver (electric motor), cost each $= -950 + 1770 \, (1.0)^{0.6} = \$820$
cost for three drivers $= \$2460$

Note that the pumps and drivers are at the lower end of the range of validity of the cost correlations, but their costs are small compared to the other costs and the error introduced is therefore negligible given the overall accuracy of $\pm 30\%$.

Following Hand's method, the installed cost of the distillation column is then:

$$C = 4 \times 650,000 = \$2,600,000$$

The cost of the trays can be converted to type 304 stainless steel by multiplying by the appropriate materials factor from Table 6.5, giving:

$C = 1.3 \times 143,000 = \$185,900$

This then gives a total cost for the column plus internals of $2,600,000 + 185,900 = \$2,790,000$.

The installed cost of the horizontal pressure vessel is $4 \times 26,000 = \$104,000$.

The installed cost for the exchangers and storage tank in carbon steel construction is:

$C = 3.5 \, (30,300 + 48,400) + 2.5 \, (27,000) = \$343,000$

so the cost in type 304 stainless steel is $1.3 \times 343,000 = \$446,000$.

For the pumps, we need to add the cost of the pump and driver before determining the installed cost. Only the cost of the pump needs to be converted to stainless steel. For the first set of pumps:

$$C = 4 \times (440 + (1.3 \times 14,200)) = \$75,600$$

For the second set of pumps only two are installed (the other is a warehouse spare), so the total installed cost is:

$$C = (1.3 \times 7050) + 820 + (4 \times 2 \times (820 + (1.3 \times 7,050))) = \$90,000$$

The total installed ISBL cost of the plant is then:

$$C = 2,790,000 + 104,000 + 446,000 + 75,600 + 90,000 = \$3,506,000$$

or \$3.5 MM $\pm 30\%$ within the accuracy of the method.

If instead we use the factors given in Table 6.4, then using equation 6.13, the installed cost for the exchangers, tank and pumps is equal to:

$$C = (30,300 + 48,400 + 27,000 + 14,200 + 14,100)[(1 + 0.8) \times 1.3 + (0.3$$
$$+ \, 0.3 + 0.2 + 0.3 + 0.2 + 0.1)]$$

$$C = (134,000)[3.74] = \$501,200$$

The installed cost for the pressure vessels and pump drivers (which do not require a materials conversion factor) is:

$$C = (650,000 + 26,000 + 440 + 1640)[1 + 0.8 + 0.3 + 0.3 + 0.2 + 0.3 + 0.2 + 0.1]$$

$$C = (678,080)[3.2] = \$2,170,000$$

In addition to this, we require the cost of the trays in stainless steel and the cost of the spare pump and driver:

$$C = 820 + 1.3(185,900 + 7050) = \$251,700$$

The total installed ISBL cost of the plant is then:
$C = 501,200 + 2,170,00 + 251,700 = \$2,920,000$
or $2.9 MM $\pm$ 30% within the accuracy of the method.

Note that although the answers obtained by the two methods are different, each is well within the range of accuracy of the other. Both estimates should be stated as being on a US Gulf Coast basis, January 2007, as this is the basis for the correlations in Table 6.6.

6.3.5. Cost Escalation

All cost-estimating methods use historical data, and are themselves forecasts of future costs. The prices of the materials of construction and the costs of labour are subject to inflation. Some method has to be used to update old cost data for use in estimating at the design stage, and to forecast the future construction cost of the plant.

The method usually used to update historical cost data makes use of published cost indices. These relate present costs to past costs, and are based on data for labour, material and energy costs published in government statistical digests.

$$\text{Cost in year A} = \text{Cost in year B} \times \frac{\text{Cost index in year A}}{\text{Cost index in year B}} \qquad (6.16)$$

To get the best estimate, each job should be broken down into its components and separate indices should be used for labour and materials. It is often more convenient to use the composite indices published for various industries in the trade journals. These are weighted average indices combining the various components of costs in proportions considered typical for the particular industry.

In the UK, the two main cost indices are the ACE index published by the Association of Cost Engineers (ACostE) in *The Cost Engineer* and the PREDICT Plant Cost Index published in the journal *Process Engineering*. *Process Engineering* also publishes monthly cost indices for several countries, including the USA, UK, Japan, Australia and many of the EU countries.

A composite index for the US process plant industry is published monthly in the journal *Chemical Engineering*; this is the Chemical Engineering Plant Cost Index (CEPCI), usually referred to as the CE index. *Chemical Engineering* also publishes the Marshall and Swift index (M&S equipment cost index).

For oil refinery and petrochemicals projects, the *Oil and Gas Journal* publishes the Nelson-Farrer Refinery Construction Index (NF index). This index is updated monthly and indices for forty types of equipment are updated quarterly. The Nelson-Farrer index is on a US Gulf Coast basis rather than US average, and is more reliable than the CE index for the types of equipment used in hydrocarbon processing.

The journal *Engineering News Record* publishes a monthly construction cost index. This is based on civil engineering projects and is sometimes used for updating off-sites costs. This index has been published since 1904 and is the oldest of all the indices.

All cost indices should be used with caution and judgment. They do not necessarily relate the true make-up of costs for any particular piece of equipment or plant, nor the effect of supply and demand on prices. The longer the period over which the

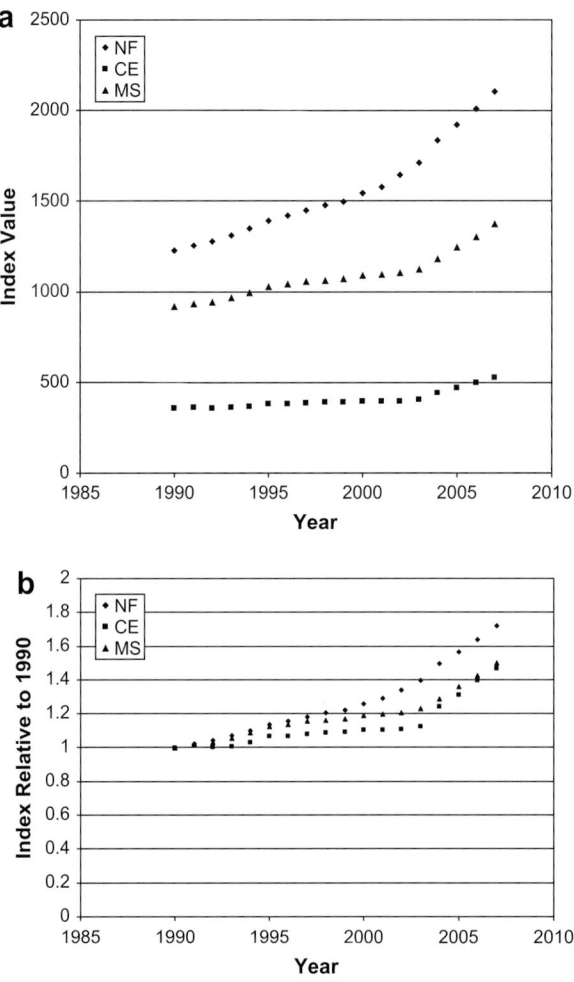

Figure 6.2.   a. Variation of major cost indices. b. Variation of major cost indices relative to 1990 = 1.0.

correlation is made the more unreliable the estimate. Between 1970 and 1990 prices rose dramatically. Prices then grew at a more or less steady 2% to 3% per year until 2003, when high demand for fuels projects and high energy prices caused another period of steeper price inflation. The major cost indices for the USA are plotted in Figure 6.2a. Figure 6.2b shows the same data plotted relative to the 1990 value of each index. Figure 6.2b clearly shows the NF index starting to accelerate ahead of the M&S and CE indices as fuels sector activity led price inflation from 2000 onwards.

To estimate the future cost of a plant, some prediction has to be made of the future annual rate of inflation. This can be based on the extrapolation of one of the published indices, tempered by the engineer's own assessment of what the future may hold. Inflation is difficult to forecast, and allowance for inflation is often included in the contingency charges added to the project cost.

The time basis of a cost estimate is sometimes identified by writing the year as a subscript to the currency; for example $\$_{2004}$ or $€_{2008}$.

Example 6.3

The purchased cost of a shell and tube heat exchanger, carbon shell, 316 stainless steel tubes, heat-transfer area 500 m^2, was \$64,000 in January 2003; estimate the cost in January 2010. Use the M&S Equipment Cost Index.

Solution

From Figure 6.2a (or by looking up the index in *Chemical Engineering*):

index in 2003 = 1123.6
index in 2007 = 1373.3.

By extrapolation from the period 2003 to 2005, the M&S index for 2010 will be about 1560.
So, estimated cost in January 2010 = \$64,000 × 1560/1124 = \$89,000.

6.3.6. Location Factors

Most plant and equipment cost data is given on a US Gulf Coast (USGC) or Northwest Europe basis, as these are historically the main centres of the chemical industry, for which the most data are available. The cost of building a plant in any other location will depend on:

- Local fabrication and construction infrastructure.
- Local labour availability and cost.
- Costs of shipping or transporting equipment to site.
- Import duties or other local tariffs.
- Currency exchange rates, which affect the relative cost of locally purchased items such as bulk materials, when converted to a conventional pricing basis such as Euros or US dollars.

These differences are often captured in cost estimating by using a location factor:

$$\text{Cost of plant in location A} = \text{cost of plant on USGC} \times LF_A \qquad (6.17)$$

where LF_A = location factor for location A relative to USGC basis.

Location factors for international locations are a strong function of currency exchange rates and hence fluctuate with time. Cran (1976a,b), Bridgewater (1979), Soloman (1990) and Gerrard (2000) give location factors for international locations from which this variation can be seen. It can be argued that as a result of globalization all international installation factors are trending closer to 1.0 (Gerrard, 2000). Location factors within a country are somewhat easier to predict and Bridgewater (1979) suggested a simple rule of thumb: add 10% for every 1000 miles from the nearest major industrial centre.

Table 6.7 gives example location factors relative to a USGC installation. These are based on data from Aspen Richardson's *International Construction Cost Factor Location Manual (2003)*. More recent versions of this manual can be found by searching for Richardson Engineering Services at www.aspentech.com. The values in Table 6.7 give costs on a local basis in US dollars. The location factors in Table 6.7 are based on 2003 data and can be updated by dividing by the ratio US dollar/local currency in 2003 and multiplying by the ratio U.S. dollar/local currency in the year of interest. If a cost estimate for a future year is being made then currency variation will have to be forecasted. Currency exchange rates are published in the financial press and on foreign exchange web sites. Several web sites have excellent historic currency converter programs; see, for example, www.xe.com/ict/, www.x-rates.com/cgi-bin/hlookup.cgi and www.oanda.com/convert/fxhistory.

Table 6.7. Location Factors

Country	Region	Location Factor
UK		1.02
France		1.13
Germany		1.11
Italy		1.14
Netherlands		1.19
Russia		1.53
India		1.02
Middle East		1.07
China	imported	1.12
	indigenous	0.61
Japan		1.26
SE Asia		1.12
Australia		1.21
US	Gulf coast	1.00
	East coast	1.04
	West Coast	1.07
	Midwest	1.02
Canada	Ontario	1.00
	Fort McMurray	1.60
Mexico		1.03
Brazil		1.14

Example 6.4

The cost of constructing a 30,000 metric tons per year (30 kMTA) acrolein plant was estimated as $80 million ($80 MM) on a 2006 US Gulf Coast basis. What would be the cost in Euros and in US dollars on a 2006 Germany basis?

Solution

From Table 6.7, the 2003 location factor for Germany was 1.11.

The exchange rate in 2003 averaged about €1 = US$1.15 and in 2006 it averaged about €1 = US$1.35.

The 2006 location factor for Germany is thus $1.11 \times 1.35/1.15 = 1.30$.

The cost of building the acrolein plant in Germany in 2006 is US$80 MM × 1.30 = US$104 MM.

The cost of building the plant in Euros (€$_{2006}$) is US$104/1.35 = €77 million.

6.3.7. Off-Site Costs

Improvements to the site infrastructure are almost always needed when a new plant is added to a site or a major expansion is carried out. The cost of such improvements is known as the off-site or OSBL investment, as described in Section 6.2.1.

In the early stages of designing a new process, the off-site requirements are usually not precisely known and an allowance for off-site costs is made by assuming that they will be a ratio of the ISBL investment. A typical number is 20% to 50% of ISBL investment, depending on the process and site conditions. As the design details are established and the requirements for utilities such as steam, electricity and cooling water are determined, the site requirements can also be determined. Potential modifications to the infrastructure can then be designed to accommodate the new plant.

Many of the off-site items are designed as 'packaged' plants or systems that are purchased from specialized suppliers. In some cases, the supplier may even offer an *over-the fence* contract, in which the supplier builds, owns and operates the off-site plant and contracts to supply the site with the desired utility stream or service. Over-the-fence contracts are widely used for industrial gases such as nitrogen, oxygen and hydrogen, and most plants also import electricity from the local utility company. Over-the-fence contracts for steam, cooling water and effluent treatment are less common, but are sometimes used in smaller plants or where several companies share a site.

The question of whether to build a self-contained infrastructure for a plant or contract for off-site services is an example of a *make or buy* problem. The over-the-fence price will usually be higher than the cost of producing the utility or service internally, since the supplier needs to make a profit and recover their capital investment. On the other hand, contracting for the service reduces the project capital investment and fixed costs, since the supplier must take on the costs of labour, maintenance and overheads. The make or buy decision is usually made by comparing annualized costs, as described in Section 6.7.6. Correlations for costs of utility plants and other off-sites are given in the sources listed in Section 6.3.4.

6.3.8. Computer Tools for Cost Estimation

It is difficult for engineers outside the EPC sector to collect recent cost data from a large set of real projects and maintain accurate and up-to-date cost correlations. Instead, the most common method for making preliminary estimates in industry is to use commercial cost estimating software.

A wide variety of cost-estimating programs are available. These include CostLink/ CM (Building Systems Design, Inc.), Cost Track™ (OnTrack Engineering Ltd.), ICARUS™ (Aspen Technology Inc.), PRISM Project Estimator (ARES Corp.), Success Estimator (U.S. Cost), Visual Estimator (CPR Internation Inc.), WinEst® (Win Estimator®), and others that can be found by searching on the web or looking at the listings provided by AACE International at www.aacei.org. The discussion in this section will focus on Aspen Technology's ICARUS Process Evaluator™ (IPE) software, as this is probably the most widely used program and is the one with which the author is most familiar. This software is made available as part of the standard Aspen academic license and so is available in any University that licenses Aspen Technology products. It is also available in most chemical companies.

The ICARUS™ cost estimating tools are simple to use and give quick, defensible estimates without requiring a lot of design data. Design information can be uploaded from any of the major flow-sheet simulation programs, or else entered manually in the ICARUS programs. The program allows the design to be updated as more information on design details becomes available, so that a more accurate estimate can be developed. Costs can be estimated for a whole plant or for one piece of equipment at a time. Over 250 types of equipment are included and these can be designed in a broad range of materials, including US, UK, German and Japanese standard alloys.

The ICARUS software uses a combination of mathematical models and expert systems to develop cost estimates. Costs are based on the materials and labour required (following the practice used for detailed estimates) rather than installation factors. If design parameters are not specified by the user then they are calculated or set to default values by the program. The user should always review the design details carefully to make sure that the default values make sense for the application. If any values are not acceptable they can be manually adjusted and a more realistic estimate can be generated.

A detailed description of how to run the ICARUS software is beyond the scope of this book, and is unnecessary, as the program is extensively documented (AspenTech 2002a, 2002b). Some of the common issues that arise in using the software are discussed below. These or similar problems are also faced when using other cost estimating programs.

Mapping Simulation Data

Instructions on loading data from a process simulation are given in the Aspen ICARUS Process Evaluator™ User's Guide (AspenTech, 2002a). When a simulator report file is loaded, IPE generates a block-flow diagram with each unit operation of the simulation shown as a block. These blocks must then be 'mapped' to Icarus project components (pieces of equipment or bulk items).

Unless the user specifies otherwise, each simulator block is mapped to a default ICARUS project component. The mapping defaults need to be understood properly, as large errors can be introduced if unit operations are mapped incorrectly. The default mapping specifications are given in section 3 of the user's guide (AspenTech, 2002a). Some mappings that commonly cause problems include:

1. Reactors: plug-flow reactor models (PLUG in HYSYS and ProII, RPLUG in AspenPlus) are mapped to a packed tower, which is fine for fixed bed catalytic reactors, but not for other types of plug-flow reactor. All other reactor models (Gibbs, stoichiometric, equilibrium and yield) are mapped to agitated tank reactors. Reactors that are not suitable for these mappings can be mapped to other ICARUS project components or set up as user models (see below).

2. Heaters, coolers and heat exchangers: the default mapping for all heat transfer equipment is the floating head heat exchanger. ICARUS contains several different heat-exchanger types, including a generic TEMA heat exchanger that can be customized to the other types, as well as fired heater and air cooler components. It is often worthwhile to change the default mapping to the TEMA exchanger to allow the exchangers to be customized in ICARUS.

3. Distillation columns: the simulator column models include not just the column itself, but also the reboiler, condenser, overhead receiver drum and reflux pump (but not bottoms pump). ICARUS has 10 possible configurations to which a column can be mapped. Alternatively, the column can be mapped to a packed or trayed tower and the ancillary items can be created as separate ICARUS project components.

4. Dummy items: process simulations often contain models of items that are not actual plant equipment (see Chapter 4). For example, heat exchangers are sometimes modelled as a series of heaters and coolers linked by a calculator block as a means of checking for internal pinch points or allowing for heat losses to ambient. When the simulation is mapped into ICARUS, dummy items should be excluded from the mapping process. In the above example, only the heaters should be mapped, so as to avoid double counting the heat transfer area.

The default mapping can be edited by right-clicking on 'Project Component Map Specifications' in the Project Basis/Process Design folder. A simulator model can be excluded from the mapping by selecting the item and then selecting 'delete all mappings'. New mappings can be specified by selecting a simulator item and adding a new mapping.

To map loaded simulator data, click the map button on the toolbar (which maps all items) or right-click on an area or plant item in the process view window (which allows items to be mapped individually). If individual items are selected then the user is given an option to use simulator data to override the default mapping in the Component Map Specs file. This is useful for heat exchangers and other equipment where the simulator allows the equipment type to be specified.

Design Factors

All good designs include an appropriate degree of over-design to allow for uncertainties in the design data and method (see Chapter 1). For some equipment the design factor or margin is specified by design codes and standards, for example, in the design of pressure vessels, as described in Chapter 13. In other cases, the design engineer must specify the degree of over-design or margin based on experience, judgement or company policy.

The equipment sizes calculated by a process simulator will be at the design flow rate unless a higher throughput was specified by the user, and hence include no design margin. The IPE software adds an 'equipment design allowance' to the equipment cost to allow for the design factor that will be introduced when the equipment is designed in detail. The equipment design allowance is based on the process description as follows:

New and unproven process	15%
New process	10%
Redesigned process	7%
Licensed process	5%
Proven process	3%

The process description is entered by right-clicking on 'General Specs' in the Project Basis/Basis for Capital Costs folder.

The equipment design allowance is only applied to system-developed costs. If different design margins are needed for different equipment types, then the default should be set to 'proven process' and the equipment can be over-sized appropriately. Design margins can also be added to components using the IPE custom model tool. Care should be taken to avoid adding more design margin than is necessary.

Pressure Vessels

When costing pressure vessels such as reactors and distillation columns, care must be taken to ensure that the wall thickness is adequate. The default method in IPE calculates the wall thickness required based on the ASME Boiler and Pressure Vessel Code Section VIII Division 1 method for the case where the wall thickness is governed by containment of internal pressure (see Chapter 13 for details of this method). The program can also allow for wind and seismic loads and design for under-pressure (vacuum). If other loads govern the design then the IPE software can significantly underestimate the vessel cost. This is particularly important for vessels that operate at pressures below 5 bara, where the required wall thickness is likely to be influenced by dead weight loads and bending moments from the vessel supports, and for tall vessels such as distillation columns and large packed-bed reactors. Similarly, if the vessel is designed under a different section of the Boiler and Pressure Vessel Code, which is usually the case for vessels operated at high pressures, then IPE can overestimate the vessel cost. It is important to always remember to enter the design pressure and temperature of the vessel, not the operating pressure and temperature.

The best approach to costing pressure vessels using the IPE software is to enter all of the dimensions after completing the mechanical design of the vessel using the methods given in Chapter 13, or using suitable pressure vessel design software.

Non-Standard Components

Although IPE contains over 250 equipment types, many processes require equipment that is not on the list of available project components. Also, in some cases the user will want to specify a certain make or model of equipment that may only be available in discrete sizes (for example, gas turbine engines or large pumps and compressors). In these situations, the non-standard equipment can be included by setting up an Equipment Model Library (EML). Many companies maintain standard EMLs listing equipment that they often specify.

A new EML can be created by selecting the 'Libraries' tab in the palette and opening the folder Cost Libraries/Equipment Model Library. Right-clicking on either of the sub-folders then allows the user to create a new EML in the appropriate set of units. Once an EML has been created, equipment items can be added to it. When a new item is added, a dialog box opens in which the user has to specify the sizing or costing method (linear, log-log, semi-log or discrete) and primary sizing parameters. Two costs and sizes must also be entered to establish the cost correlation.

Equipment model libraries are useful for completing an IPE model of a process that contains non-standard items. Care must be taken to update the EML costs so that they remain current.

Example 6.5

Estimate the cost of a waste heat boiler designed to produce 4000 lb/h of steam. The exchanger area has been estimated as 1300 ft^2.

Solution

Starting from the IPE project explorer window (on the far left of the screen), right-click on the Main Area and select Add Project Component (Figure 6.3a).

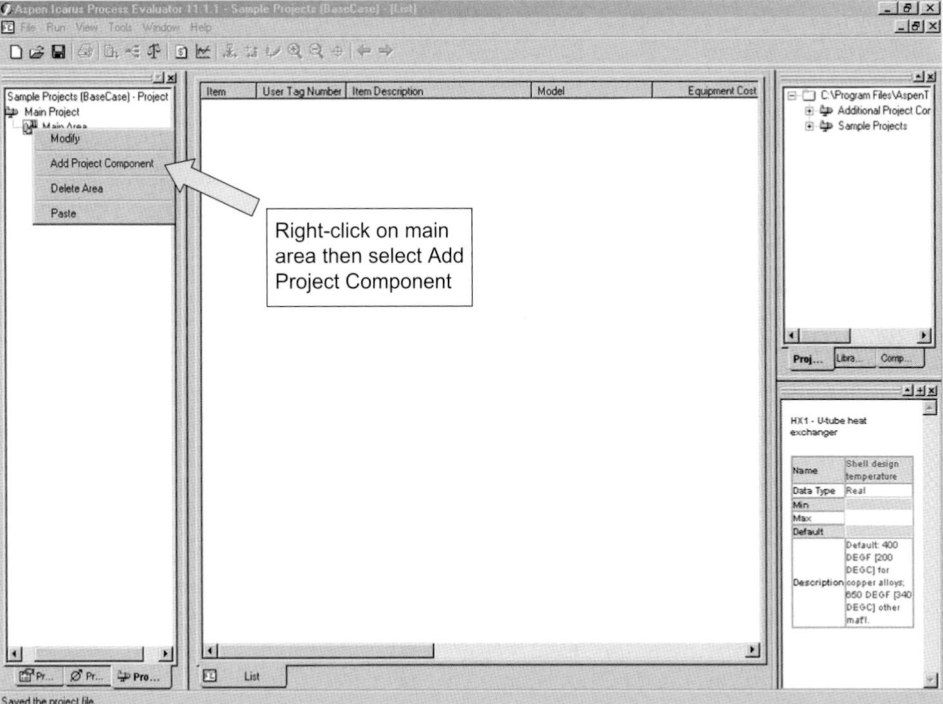

Figure 6.3a. Aspen ICARUS example.

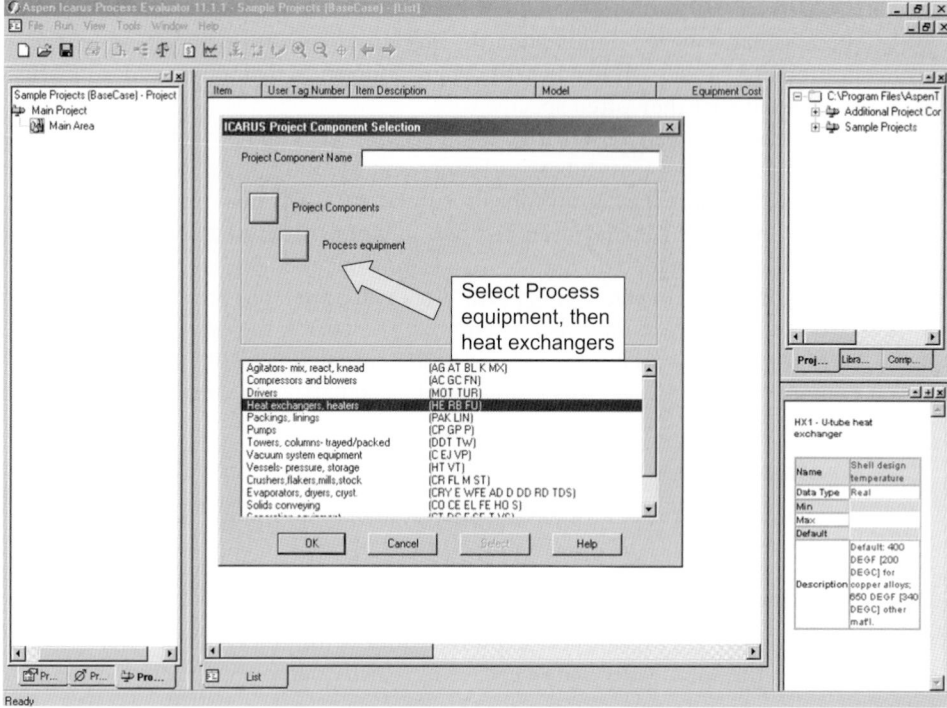

Figure 6.3b. Aspen ICARUS example.

Select Process Equipment, then Heat Exchangers (Figure 6.3b). Select Waste Heat Boiler and enter a name (Figure 6.3c).

Enter the size parameters and then click the Evaluate button (Figure 6.3d). This runs the evaluator program and gives the results screen shown in Figure 6.3e. The purchased equipment cost is $145,900 on a Jan 2006 USGC basis. The installed cost is $196,225. Note that the installed cost is calculated directly by estimating bulk materials and labour rather than using an installation factor.

6.3.9. Validity of Cost Estimates

It should always be remembered that cost estimates are only estimates and are subject to error. An estimate should always indicate the margin of error. The error in a cost estimate is primarily determined by the degree of design detail that is available, and even a skilled estimator cannot estimate an accurate cost for a sketchy design.

When more design information has been developed a professional cost engineer will be able to develop a more accurate estimate. The process design engineer should compare this estimate with the preliminary estimate to gain a better understanding of where the preliminary estimate could have been improved (either through capturing missing plant items or using better costing methods). This will help the design engineer to produce better preliminary estimates in future.

Additional resources for cost estimating are available from the various cost estimating associations: the UK Association of Cost Engineers (www.acoste.org.uk); the

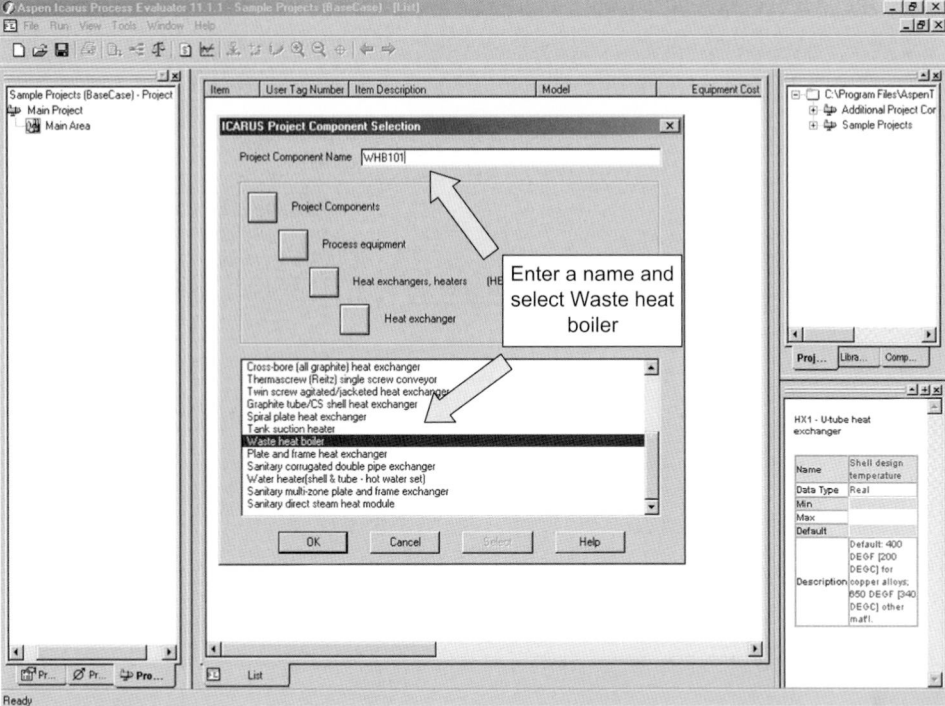

Figure 6.3c. Aspen ICARUS example.

Association for the Advancement of Cost Engineering International (www.aacei.org); the Project Management Institute (www.pmi.org); and the International Cost Engineering Council (www.icoste.org). The ICEC web site has links to cost engineering societies in 46 countries.

6.4. ESTIMATING PRODUCTION COSTS AND REVENUES

The revenues and variable costs of production are obtained by multiplying the product, feed or utility flow rates from the flow-sheet by the appropriate prices. The difficult step is usually finding good price data.

6.4.1. Sources of Price Data

This section describes the most widely used sources of price data. Some pricing terminology is given in Table 6.8.

Internal Company Forecasts

In many large companies the marketing or planning department develops official forecasts of prices for use in internal studies. These forecasts sometimes include multiple price scenarios, and projects must be evaluated under every scenario. Company forecasts are occasionally made available to the public. See for example

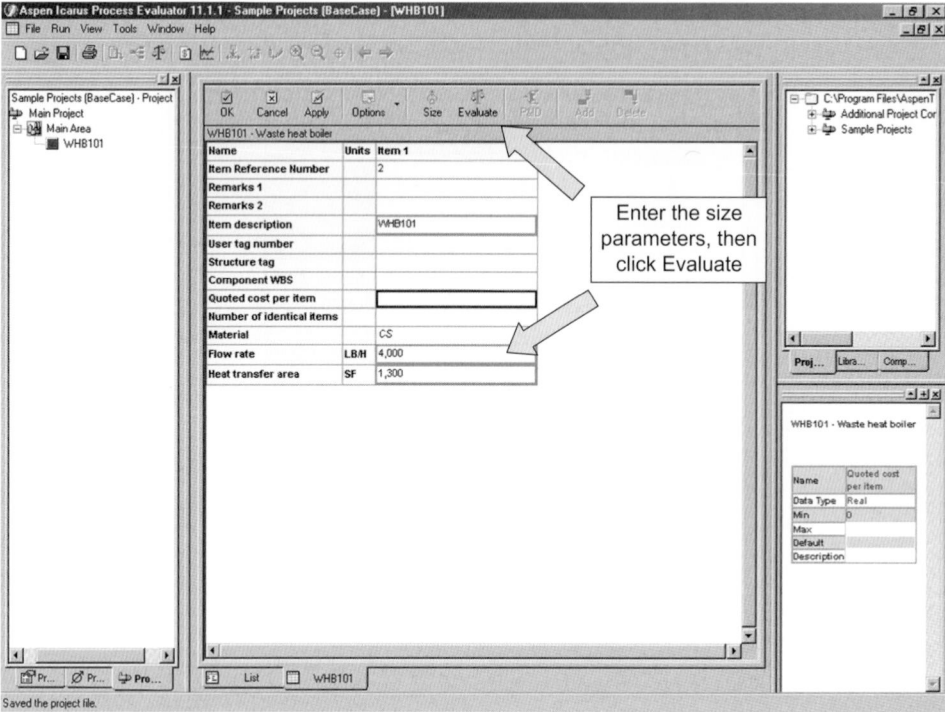

Figure 6.3d. Aspen ICARUS example.

Shell (2002) or Shell (2005), which can be downloaded from www.Shell.com. When an officially-approved price set exists, the design engineer should use it. The main concern is then ensuring that prices for feeds, products or consumables that are not part of the standard forecast are put on a consistent basis.

Trade Journals

Several journals publish chemicals and fuel prices on a weekly basis:

ICIS Chemical Business Americas, formerly known as *Chemical Marketing Reporter* (ICIS Publications), used to list prices for 757 chemicals with multiple locations and product grades for some. This list was reduced to only 85 compounds in 2006, with most of the remaining set being natural extracts. Data for 80 chemicals, 44 fuels and 11 base oils is now provided on-line through the subscription service www.icispricing. com. At the time of writing this service was very expensive compared to some of the alternatives listed below. ICIS also publishes *ICIS Chemical Business Europe* (formerly *European Chemical News*) and *ICIS Chemical Business Asia*, which provide regional price data for a smaller set of compounds.

The *Oil and Gas Journal* (Pennwell) publishes prices for several crude oils and a range of petroleum products on US, NW Europe and SE Asia bases, as well as natural gas prices for the USA.

Chemical Week (Access Intelligence) gives spot and contract prices for 22 commodity chemicals in US and NW Europe markets.

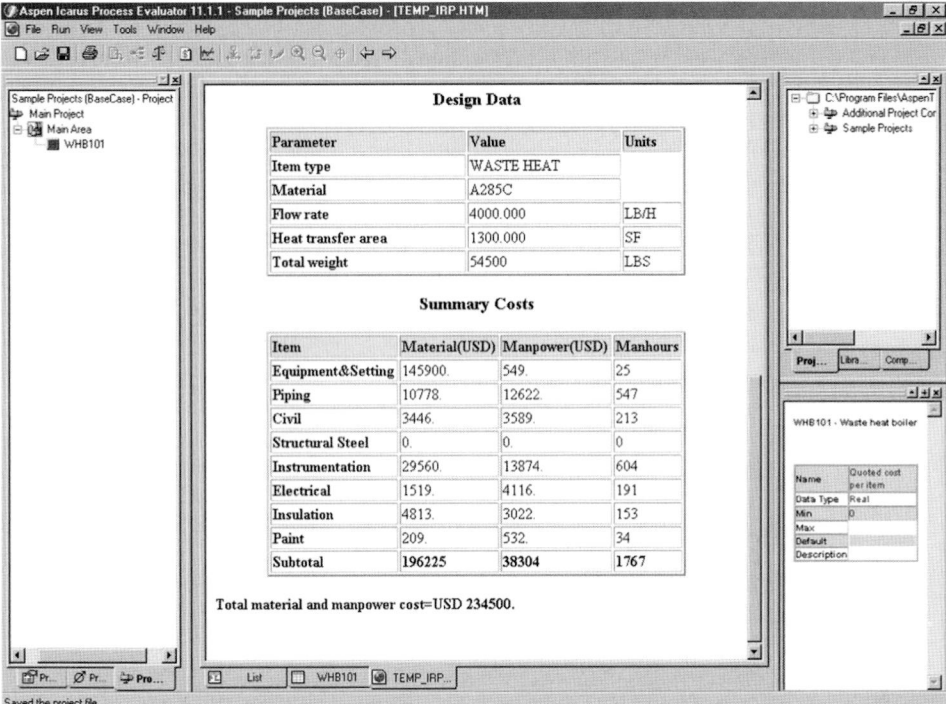

Figure 6.3e. Aspen ICARUS example.

Consultants

There are many companies that can be hired as consultants to provide economic and marketing information, or that allow access to such information on a subscription basis. The information provided generally includes market surveys and technical and economic analyses of competing technologies as well as price data and forecasts. There is not room here to list all of these companies, but some of the most widely used are:

Table 6.8. Pricing Terminology

Abbreviation	Meaning
c.i.f.	Cost, insurance and freight
dlvd.	Delivered
f.o.b.	Free on board
frt. alld.	Freight allowed
dms.	Drums
bgs.	Bags
refy.	Refinery gate
syn.	Synthetic
t.t.	Tank truck
t.c.	Tank car (rail)
t.l.	Truck load
imp.	Imported

- *Purvin and Gertz*: Provides quarterly forecasts of oil, gas and fuels prices that are widely used in the oil industry. They have a 10-year archive of historic data and forecast prices of most fuel products as well as crude oils on US, NW Europe, Middle East and Asia bases.
- *Cambridge Energy Research Associates*: Publishes forecasts of crude oil prices based on macroeconomics and industry trends (drilling rates, etc.).
- *Chemical Market Associates Inc.* (CMAI): Maintains a large archive of historic data and future price forecasts for 70 commodity chemicals, including multiple grades, US, NW Europe, Middle East, NE and SE Asia. Spot and contract prices are given for some compounds and in some cases margins are also estimated by formula.
- *SRI*: The *Chemical Economics Handbook* series of reports published by SRI provides overviews of the markets for 288 compounds. These reports are not updated as frequently as the others, but are useful for less commoditized compounds.

On-Line Brokers and Suppliers

Much price data is available on-line from suppliers' web sites that can be found through directory sites such as www.purchasing.com and www.business.com/directory/chemicals.

Some caution is needed when using price data from the web. The prices quoted are generally for spot sale of small quantity orders, and are thus much higher than the market rates for large order sizes under long-term contract. The prices listed on-line are also often for higher quality material such as analytical, laboratory or USP pharmaceutical grades, which have much higher prices than bulk grades.

Reference Books

Prices for some of the more common commodity chemicals are sometimes given in process economics textbooks. These prices are usually single data points rather than forecasts. They are only suitable for undergraduate design projects.

6.4.2. Forecasting Prices

In most cases, it will take between 1 and 3 years for a project to go through the phases of design, procurement and construction before the plant can begin operation. The plant will then operate for the project life of 10 to 20 years. The design engineer thus needs to carry out the economic analysis using prices forecasted over the next 20 or so years rather than the current price when the design is carried out.

For some compounds the only variation in price over time is minor adjustments to allow for inflation. This is the case for some specialty compounds that have relatively high prices and are not subject to competitive pressure (which tends to drive prices down). Prices can also be stable if they are controlled by governments, but this is increasingly rare. In most cases, however, prices are determined largely by feedstock prices, which are ultimately determined by fluctuations in the prices of commodity fuels and chemicals. The prices of these commodities are set by markets in response to variations in supply and demand, and vary widely over time.

Most price forecasts are based on an analysis of historic price data. Several methods are used, as illustrated in Figure 6.4. The simplest method is to use the current price, Figure 6.4a, but this is unsatisfactory for most commodities. Linear regression of past prices is a good method for capturing long-term trends (>10 years), but can give very different results depending on the start date chosen, as shown in Figure 6.4b. This method can be very misleading if the data set is too small.

Many commodity prices exhibit cyclic behaviour due to the investment cycle, so in some cases non-linear models can be used (Figure 6.4c). Unfortunately, both the amplitude and the frequency of the price peaks usually vary somewhat erratically, making it difficult to fit the cyclic price behaviour with simple wave models or even advanced Fourier transform methods.

A fourth approach, illustrated in Figure 6.4d is to recognize that feed and product prices are usually closely linked, since increases in feed costs are passed on to customers whenever possible via increases in product price. Although feed and product prices may both be variable, the gross margin is therefore subject to much less variation and can be forecasted more reliably. Forecasting of margins is the method used widely in the fuels and petrochemicals industry as it is much easier to predict the variation in margins than the underlying variation in the prices of crude oil and natural gas. The drawbacks of this method are that it does not work so well when there are multiple routes to the same product, and it involves making assumptions about yields that may not hold true throughout the forecast period. In cases where the gross margin is high, it can be more difficult for the manufacturer to pass on the full

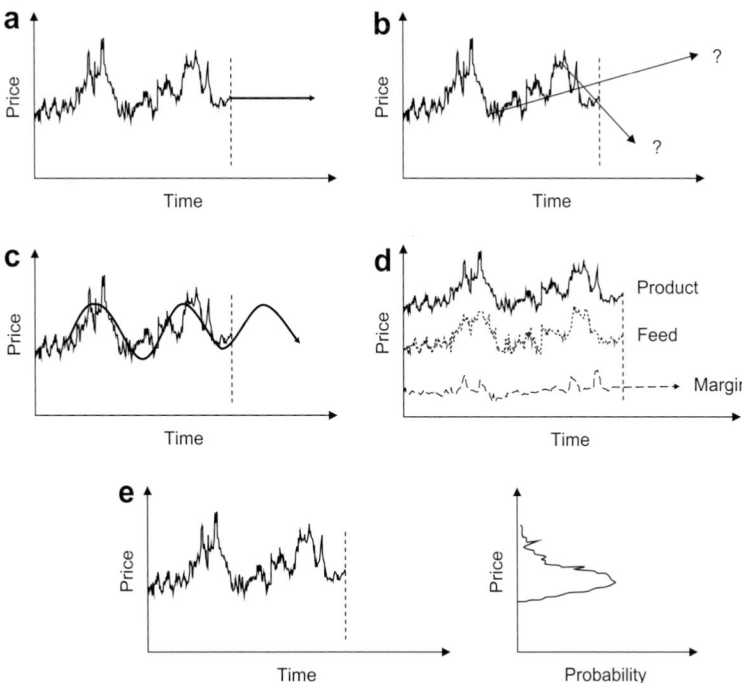

Figure 6.4a–e. Forecasting commodity prices.

impact of feedstock price increases in the form of increased product prices. In such cases, when feed prices rise rapidly there is a drop in margins while producers wait for the market to absorb the impact of higher prices.

Another method is to model the statistical distribution of the price (or margin) as illustrated in Figure 6.4e. At its simplest, this method involves taking the average price, adjusted for inflation, over a recent period. This method can miss long-term trends in the data and few prices follow any of the more commonly used distributions. It is useful, however, in combination with sensitivity analysis methods such as Monte Carlo Simulation (see Section 6.8).

Figure 6.5 shows North American prices from CMAI data for polyethylene terephthalate resin (PET), which is made from terephthalic acid (TPA), which in turn is made from paraxylene (PX). Several things are apparent from Figure 6.5:

1. The spot prices of PX and TPA show more volatility than the contract prices, as would be expected.
2. All the prices follow the same broad trends, with a major peak in 1995 and long recovery leading to a second peak in 2006.
3. The sharp peak in PX spot price in 1995 was not passed on to the other prices.

Figure 6.6 shows the simple margins TPA-PX and PET-PX over the same time period, all based on contract prices. The degree of variation in margins is clearly less than the variation in the base prices. There also appears to be a long-term decline in TPA margins relative to PX.

A similar examination of feed and product prices along the value chain of a given chemical can usually provide valuable insights into the best method of forecasting. No method is perfect, and anyone capable of accurately predicting commodity prices would

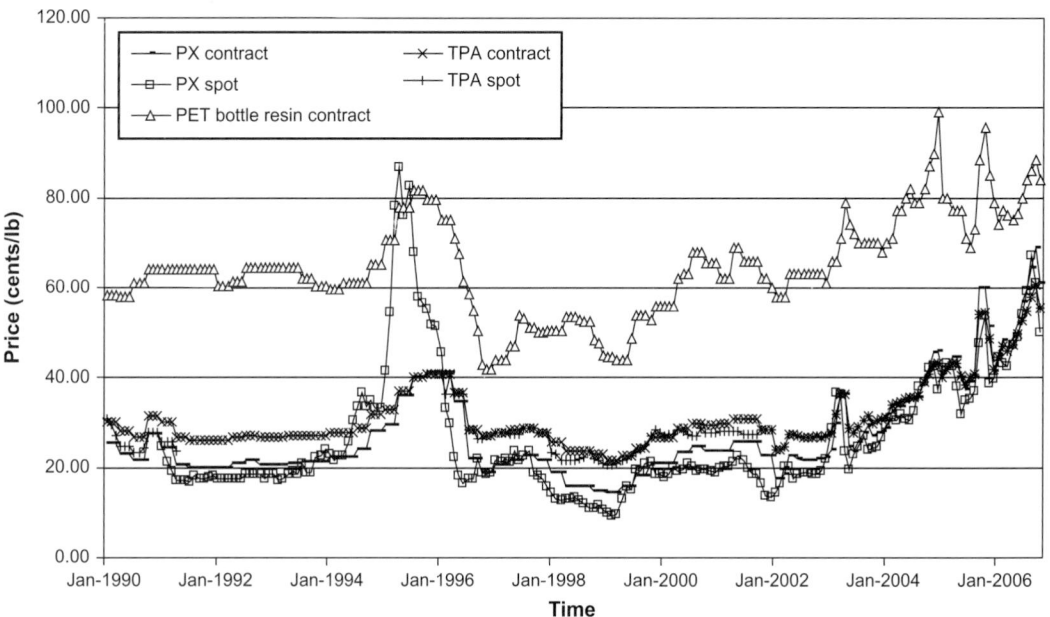

Figure 6.5. North American prices for the PET value chain.

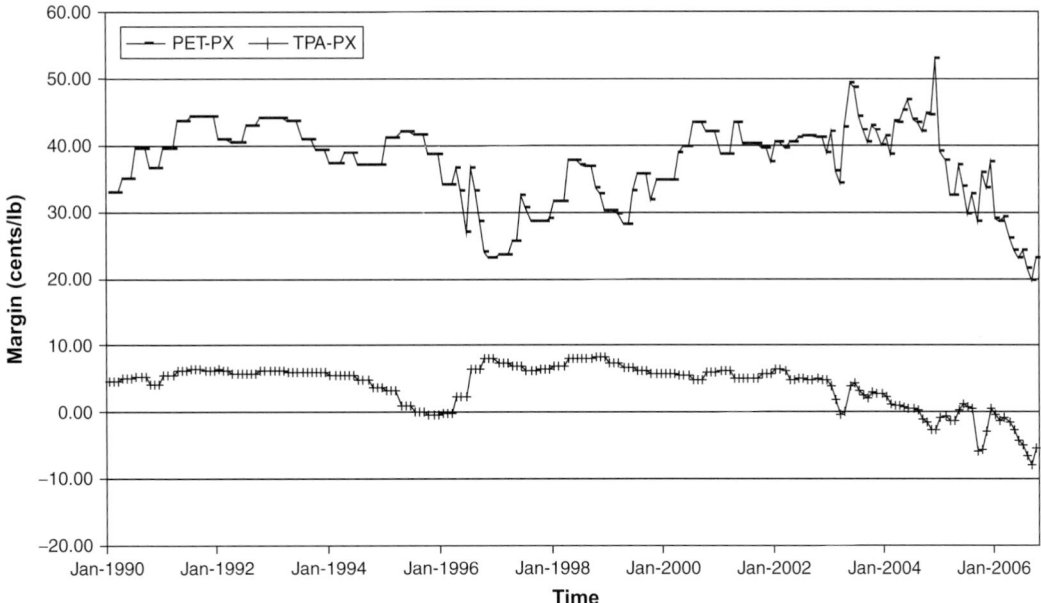

Figure 6.6. Simple margins for the PET value chain.

be well advised to pursue a more lucrative career than chemical engineering. For process design purposes it is usually sufficient to show that the prices used for optimization and economic analysis are realistic and consistent with consensus views of the market.

6.4.3. Transfer Pricing

If the raw material for plant B is the product of plant A on the same site and owned by the same company, then the price that plant B pays to plant A is known as a 'transfer price'. Whenever realistic, transfer prices should be set by open market prices. This reflects the reality that plant A could sell its product on the open market or plant B could similarly buy its feed. Some cases when transfer prices do not match market prices include:

When plant A produces material that is suitable for internal consumption but does not meet specifications for traded product. In this case, the transfer price to plant B should be discounted to allow for the added costs incurred in plant B from handling the less pure feed.

When plant A is underutilized or cannot sell its product and has recovered all of its initial capital investment, then the transfer price to plant B can be set at the cash cost of production of plant A (see Section 6.2.5). This encourages maximum use of the existing asset of plant A and thereby reduces the fixed costs per kilogram of product from plant A and makes it more competitive.

When the pricing of product from the upstream plant is set to drive capacity utilization or conservation, for example, by using a sliding price scale based on the amount of material used.

When transfer pricing is used it is important to keep in mind which processes actually bring in money from customers and which do not. If unrealistic transfer prices are used, uneconomic projects may seem attractive and poor investment decisions may be made.

6.4.4. Utility Costs

The utility consumption of a process cannot be estimated accurately without completing the material and energy balances and carrying out a pinch analysis, as described in Chapter 3. The pinch analysis gives targets for the minimum requirements of hot and cold utility. More detailed optimization then translates these targets into expected demands for fired heat, steam, electricity, cooling water and refrigeration. In addition to the utilities required for heating and cooling, the process may also need process water and air for applications such as washing, stripping and instrument air supply. A good overview of methods for design and optimization of utility systems is given by Smith (2005).

The electricity demand of the process is mainly determined by the work required for pumping, compression, air coolers and solids-handling operations, but also includes the power needed for instruments, lights and other small users. Some plants generate their own electricity using a gas-turbine co-generation plant with a heat recovery steam generator (waste heat boiler) to raise steam (Figure 6.7). The co-generation plant can be sized to meet or exceed the plant electricity requirement, depending on whether the export of electricity is an attractive use of capital.

Most plants are located on sites where the utilities are provided by the site infrastructure. The price charged for a utility is mainly determined by the operating cost of generating and transmitting the utility stream. Some companies also include a capital recovery charge in the utility cost, but if this is done then the off-site (OSBL) capital cost must be reduced to avoid double counting and biasing the project capital-energy trade-off, leading to poor use of capital.

Some smaller plants purchase utilities 'over the fence' from a supplier such as a larger site or a utility company, in which case the utility prices are set by contract and are typically pegged to the price of natural gas or fuel oil.

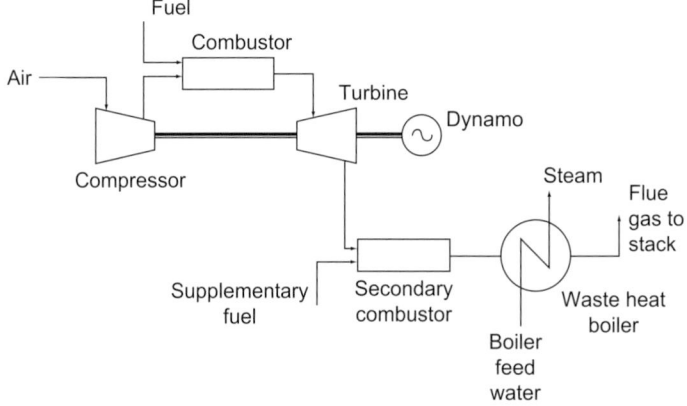

Figure 6.7. Gas-turbine based cogeneration plant.

Fired Heat

Fired heaters are used for process heating above the highest temperatures that can be reached using high pressure steam, typically about 250°C (480°F). Process streams may be heated directly in the furnace tubes, or indirectly using a hot oil circuit. The design of fired heaters is described in Chapter 12. The cost of fired heat can be calculated from the price of the fuel fired. Most fired process heaters use natural gas, as it is cleaner burning than fuel oil and therefore easier to fit NO_x control systems and obtain permits. Natural gas also requires less maintenance of burners and fuel lines and natural gas burners can often co-fire process waste streams such as hydrogen, light hydrocarbons, or air saturated with hydrocarbons or solvents. In recent years, North American and European prices for natural gas have had very high mid-winter peaks. This has caused some plants to revert to using heating oil as fuel.

Natural gas and heating oil are traded as commodities and prices can be found at on-line trading sites or business news sites (e.g., http://markets.ft.com or www.cnn.money.com). Historic prices for forecasting can be found in the *Oil and Gas Journal* or from the US Energy Information Agency (www.eia.doe.gov). The EIA also has an extensive database of historic international prices for natural gas, heating oil and electricity (www.eia.doe.gov/emeu/international).

The fuel consumed in a fired heater can be estimated from the fired heater duty divided by the furnace efficiency. The furnace efficiency will typically be about 0.85 if both the radiant and convective sections are used (see Chapter 12) and about 0.6 if the process heating is in the radiant section only.

Steam

Steam is the most widely used heat source on most chemical plants. Steam has a number of advantages as a hot utility:

- The heat of condensation of steam is high, giving a high heat output per pound of utility at constant temperature (compared to other utilities such as hot oil and flue gas that release sensible heat over a broad temperature range).
- The temperature at which heat is released can be precisely controlled by controlling the pressure of the steam. This enables tight temperature control, which is important in many processes.
- Condensing steam has very high heat-transfer coefficients, leading to cheaper heat exchangers.
- Steam is non-toxic, non-flammable, visible if it leaks externally and inert to many (but not all) process fluids.

Most sites have a pipe network supplying steam at three or more pressure levels for different process uses. A typical steam system is illustrated in Figure 6.8. Boiler feed water at high pressure is preheated and fed to boilers where high pressure steam is raised and superheated above the dew point to allow for heat losses in the piping. Boiler feed water preheat can be accomplished using process waste heat or convective section heating in the boiler plant. High-pressure (HP) steam is typically at about 40 bar, corresponding to a condensing temperature of 250°C, but every site is different. Some of the HP steam is used for process heating at high temperatures. The remainder of the HP steam is expanded either through steam turbines known as *back-pressure turbines* or through let-down valves to form medium-pressure (MP) steam. The pressure of the MP steam mains varies widely from site to site, but is typically about 20 bar, corresponding to a condensing temperature of 212°C. MP steam is used for intermediate temperature heating or expanded to form low-pressure (LP) steam, typically at about 3 bar, condensing at 134°C. Some of the LP steam may be used for process heating if there are low temperature heat requirements. LP (or MP or HP) steam can also be expanded in condensing turbines to generate shaft work for process drives or electricity production. A small amount of LP steam is used to strip dissolved non-condensable gases such as air from the condensate and make-up water. LP steam is also often used as 'live steam' in the process; for example, as stripping vapour or for cleaning, purging or sterilizing equipment.

When steam is condensed without coming into contact with process fluids, then the hot condensate can be collected and returned to the boiler feed water system. Condensate can also sometimes be used as a low-temperature heat source if the process requires low-temperature heat.

The price of HP steam can be estimated from the cost of boiler feed water treatment, the price of fuel and the boiler efficiency:

$$P_{HPS} = P_F \times \frac{dH_b}{\eta_B} + P_{BFW} \qquad (6.18)$$

where

P_{HPS} = price of HP steam (£/t)
P_F = price of fuel (£/GJ)
dH_b = heating rate (GJ/t steam)
η_B = boiler efficiency
P_{BFW} = price or cost of boiler feed water (£/t).

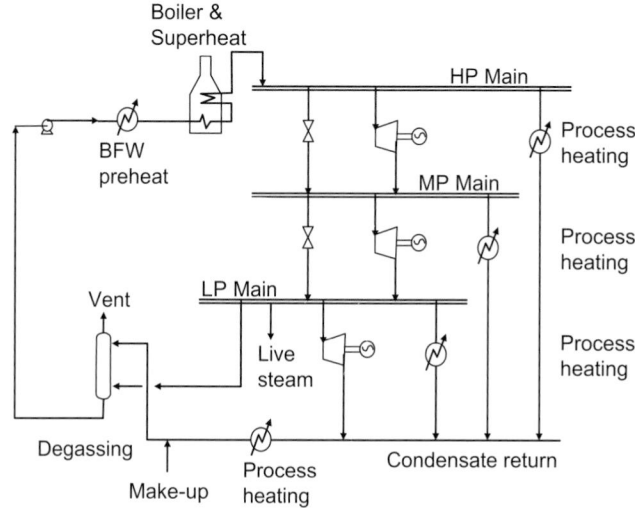

Figure 6.8. Steam system.

Package boilers typically have efficiencies similar to fired heaters, in the range 0.8 to 0.9.

The heating rate should include boiler feed water preheat, the latent heat of vaporization and the superheat specified.

The cost of boiler feed water includes allowances for water make-up, chemical treatment and degassing, and is typically about twice the cost of raw water (see below). If no information on the price of water is available, then 0.50 £/t (~0.50 $/1000 lb) can be used as an initial estimate. If the steam is condensed and the condensate is returned to the boiler feed water (which will normally be the case), then the price of steam should include a credit for the condensate. The condensate credit will often be close enough to the boiler feed water cost that the two terms cancel each other out and can be neglected.

The prices of MP and LP steam are usually discounted from the HP steam price, to allow for the shaft work credit that can be gained by expanding the steam through a turbine, and also to encourage process heat recovery by raising steam at intermediate levels and using low-grade heat when possible. Several methods of discounting are used. The most rational of these is to calculate the shaft work generated by expanding the steam between levels and price this as equivalent to electricity (which could be generated by attaching the turbine to a dynamo or else would be needed to run a motor to replace the turbine if it is used as a driver). The value of the shaft work then sets the discount between steam at different levels. This is illustrated in the following example.

Example 6.6

A site has steam levels at 40 bar, 20 bar and 6 bar. The price of fuel is £4/GJ and electricity costs £0.05/kWh. If the boiler efficiency is 0.8 and the steam turbine efficiency is 0.85, suggest prices for HP, MP and LP steam.

Solution

The first step is to look up the steam conditions, enthalpies and entropies in steam tables:

Steam level	HP	MP	LP
Pressure (bar)	40	20	6
Saturation temperature (°C)	250	212	159

The steam will be superheated above the saturation temperature to allow for heat losses in the pipe network. The following superheat temperatures were set to give an adequate margin above the saturation temperature for HP steam and also to give (roughly) the same specific entropy for each steam level. The actual superheat temperatures of MP and LP steam will be higher, due to the non-isentropic nature of the expansion.

	HP	MP	LP
Superheat temperature (°C)	400	300	160
Specific entropy, s_g, (kJ/kg.K)	6.769	6.768	6.761
Specific enthalpy, h_g, (kJ/kg)	3214	3025	2757

We can then calculate the difference in enthalpy between levels for isentropic expansion:

Isentropic delta enthalpy (kJ/kg)	189	268

Multiplying by the turbine efficiency gives the non-isentropic enthalpy of expansion:

Actual delta enthalpy (kJ/kg)	161	228

This can be converted to give the shaft work in kWh/t:

Shaft work (kWh/t)	44.7	63.3

Multiplying by the price of electricity converts this into a shaft work credit:

Shaft work credit (£/t)	2.24	3.17

The price of HP steam can be found from equation 6.18, assuming that the boiler feed water cost is cancelled out by a condensate credit and the heating rate is 2.2 GJ/t steam (allowing for some boiler feed water preheat). The other prices can then be estimated by subtracting the shaft work credits.

Steam price (£/t)	11.0	8.76	5.59

For quick estimates, this example can easily be coded into a spreadsheet and updated with the current prices of fuel and power. A sample steam costing spreadsheet is available in the on-line material at http://books.elsevier.com/companions.

Cooling

The cost of process cooling usually depends strongly on the cost of power (electricity).

- Air coolers use electric power to run the fans. The power requirement is determined as part of the cooler design, as described in Chapter 12.
- Cooling water systems use power for pumping the cooling water through the system and for running fans (if installed) in the cooling towers. They also have costs for water make-up and chemical treatment. The power used in a typical recirculating cooling water system is usually between 1 and 2 kWh/1000 US gal (3.8 m^3) of circulating water. The costs of water make-up and chemical treatment usually add about $0.02/1000 gal circulated (£2.60/1000 metric tons).

- Refrigeration systems use power to compress the refrigerant. The power can be estimated using the cooling duty and the refrigerator coefficient of performance (COP).

$$COP = \frac{\text{Refrigeration produced (Btu/hr or MW)}}{\text{Shaft work used (Btu/hr or MW)}} \qquad (6.19)$$

The COP is a strong function of the temperature range over which the refrigeration cycle operates. For an ideal refrigeration cycle (a reverse Carnot cycle), the COP is:

$$COP = \frac{T_1}{(T_2 - T_1)} \qquad (6.20)$$

where

T_1 = evaporator absolute temperature (K)
T_2 = condenser absolute temperature (K).

The COP of real refrigeration cycles is always less than the Carnot efficiency. It is usually about 0.6 times the Carnot efficiency for a simple refrigeration cycle, but can be as high as 0.9 times the Carnot efficiency if complex cycles are used. Good overviews of refrigeration cycle design are given by Dincer (2003), Stoecker (1998) and Trott and Welch (1999).

Electricity

Chemical plants consume large enough amounts of electricity that it is often economically attractive for them to install gas-turbine engines or steam turbines and generate their own electric power. This 'make or buy' scenario gives chemical producers strong leverage when negotiating electric power contracts and they are usually able to purchase electricity at or close to wholesale prices. Wholesale electricity prices vary internationally (see www.eia.doe.gov/emeu/international for details), but are typically in the range $0.05 to $0.12/kWh in most developed economies at the time of writing.

Water

Raw water is brought in to make up for losses in the steam and cooling water systems and is also treated to generate demineralized and deionized water for process use. The price of water varies strongly by location, depending on fresh water availability. Water prices are often set by local government bodies and often include a charge for waste water rejection. This charge is usually applied on the basis of the water consumed by the plant, regardless of whether that water is actually rejected as a liquid (as opposed to being lost as vapour or incorporated into a product by reaction).

A very rough estimate of water costs can be made by assuming $2 per 1000 gal (0.26 £/t). Demineralized water typically costs about double the price of raw water, but this obviously varies strongly with the mineral content of the water and the disposal cost of effluents from the demineralization system. Water demineralization is discussed in the sections on ion exchange and reverse osmosis in Chapter 10.

Air and Nitrogen

Air at 1 atmosphere pressure is freely available in most chemical plants. Compressed air can be priced based on the power needed for compression (see Chapter 3). Drying the air, for example for instrument air, typically adds about $0.005 per standard m^3 ($0.14/1000 scf). Nitrogen and oxygen are usually purchased from one of the industrial gas companies via pipeline or a small dedicated over-the-fence plant. The price varies depending on local power costs, but is typically in the range $20 to $70 per metric ton ($0.01 to $0.03 per lb) for large facilities.

Example 6.7

Estimate the annual cost of providing refrigeration to a condenser with duty 1.2 MW operating at -5°C. The refrigeration cycle rejects heat to cooling water that is available at 40°C, and has an efficiency of 80% of the Carnot cycle efficiency. The plant operates for 8000 h per year and electricity costs €0.06/kWh.

Solution

The refrigeration cycle needs to operate with an evaporator temperature below $-5°C$, say at $-10°C$ or 263 K. The condenser must operate above 40°C, say at 45°C (318 K).

For this temperature range the Carnot cycle efficiency is:

$$COP = \frac{T_1}{(T_2 - T_1)} = \frac{263}{318 - 263} = 4.78 \qquad (6.20)$$

If the cycle is 80% efficient then the actual coefficient of performance $= 4.78 \times 0.8 = 3.83$.

The shaft work needed to supply 1.2 MW of cooling is given by:

$$\text{Shaft work required} = \frac{\text{Cooling duty}}{COP} = \frac{1.2}{3.83} = 0.313 \text{ MW}$$

The annual cost is then $= 313$ kW $\times$ 8000 h/y $\times$ 0.06 €/kWh $= 150,000$€/y.

6.4.5. Consumables Costs

Consumables include materials such as acids, bases, sorbents, solvents and catalysts that are used in the process. Over time these become depleted or degraded and require replacement. In some cases a continuous purge and make-up is used (for example, for acids and bases), while in other cases an entire batch is periodically replaced (for example, for sorbents and catalysts).

The prices of acids, bases and solvents can be found from the same sources used for raw materials prices. Whenever possible, the cheapest base (NaOH) or acid (H_2SO_4) would be used in the process, but for neutralizing spent sulphuric acid, lime (CaO) or ammonia (NH_3) are often used, as these bases react with sulphuric acid to form insoluble sulphates that can be recovered and sold as by-products. Acids and bases are also consumed in ion-exchange units that are used to treat process feed water and boiler feed water; see Chapter 10. The cost of process acid or base must always include the costs of neutralizing the spent stream.

The price of adsorbents and catalysts varies very widely depending on the nature of the material. The cheapest catalysts and adsorbents cost less than $2/kg, while more expensive catalysts containing noble metals such as platinum and palladium have costs that are mainly determined by the amount of precious metal on the catalyst. In some cases, the value of the noble metal on a load of catalyst is so high that the chemical plant rents the catalyst rather than buying it and when the catalyst is spent it is returned to the manufacturer for precious metal recovery.

Although small in quantity, consumables can add a lot of cost and complexity to a plant. The plant must be designed with systems for handling, storing, metering and disposing of all the consumables used. In many chemical plants over half of the total pieces of equipment are associated with consumables handling.

6.4.6. Waste-Disposal Costs

Materials produced by the process that cannot be recycled or sold as by-products must be disposed of as waste. In some cases additional treatment is required to concentrate the waste stream before sending it to final disposal.

Hydrocarbon waste streams such as off-spec products, slop oils, spent solvents and off-gases (including hydrogen-rich gases) can often be incinerated or used as process fuel. This allows the fuel value of the stream to be recovered, and the waste stream can be assigned a value based on its heat of combustion:

$$P_{WFV} = P_F \times \Delta H_C^o \qquad (6.21)$$

where

P_{WFV} = waste value as fuel ($/lb or $/kg)
P_F = price of fuel ($/MMBtu or $/GJ)
$\Delta H^{\circ}{}_C$ = heat of combustion (MMBtu/lb or GJ/kg).

If additional systems such as flue gas scrubbers must be fitted to allow the waste to be combusted, then the waste stream value should be discounted to recover the extra cost.

Dilute aqueous streams are sent to waste-water treatment unless the contaminants are toxic to the bacteria in the waste-water plant. Acidic or basic wastes are neutralized prior to treatment. Neutralization is usually carried out using a base or acid that will form a solid salt that can be precipitated from the water, so that the total dissolved solids (TDS) load on the waste-water plant is not excessive. The cost of waste-water treatment is typically about $6 per 1000 gal ($1.5/t), but there may also be local charges for spent water discharge.

Inert solid wastes can be sent to landfill at a cost of about $50/t, or in some cases used to make roads. Wastes from neutralizing spent sulphuric acid are typically calcium sulphate (gypsum) which can be used as road fill or ammonium sulphate, which can be sold as fertilizer.

Concentrated liquid streams that cannot be incinerated locally (for example, compounds containing halogens) and non-inert solids must be disposed of as hazardous waste. This entails shipping the material to a hazardous waste company for incineration in a specialized plant or long-term storage in a suitable facility. The costs of hazardous waste disposal depend strongly on the plant location, proximity to waste-disposal plants and the nature of the hazardous waste, and must be evaluated on a case-by-case basis.

Additional information on waste-disposal considerations is given in Chapter 14.

6.4.7. Labour Costs

The wages paid to plant operators and supervisors are a fixed cost of production, as described in Section 6.2.4. Almost all plants are operated on a shift-work basis (even batch plants), with typically 4.8 operators per shift position. This gives a four-shift rotation with allowance for weekends, vacations and holidays and some use of overtime. Most plants require at least three shift positions: one operator in the control room, one outside, and one in the tank farm or other feed/product shipping and receiving area. Plants that use more mechanical equipment, particularly solids-handling plants, typically require more shift positions, as do plants that involve batch operations. More shift positions are also needed when handling highly toxic compounds. In some cases two or more smaller plants may be grouped together with a common control room and tank farm to reduce the number of operators needed. Very few plants run entirely unattended though, with the exception of gas-processing plants, which hold no inventories of feed or product and are usually automated to allow a single control room operator to watch over several plants. A chart for estimating the minimum number of shift positions is given in Figure 6.9, but it should be emphasized that this only gives a rough guide. The design engineer should always carefully think through the operations required per shift, particularly for processes that handle solids or involve batch operations or frequent sampling.

Operator salaries vary by region and experience level. For initial estimates, an average salary of $60,000 per shift position per year on a USGC basis, not including direct or corporate overhead can be used. Supervision and overhead costs are discussed in Section 6.2.4.

6.4.8. By-Product Revenues

A good deal of process design effort is often spent analysing by-product recovery. Potentially valuable by-products include:

1. Materials produced in stoichiometric quantities by the reactions that form the main product (see Table 6.1 for examples). If these are not recovered as by-products, then the waste-disposal costs will be excessive.
2. Components that are produced in high yield by side reactions. Some examples include propylene, butylenes and butadiene, all of which are by-products of ethylene from steam cracking of naphtha feed. Orthoxylene and metaxylene are by-products of paraxylene manufacture by catalytic reforming of naphtha.

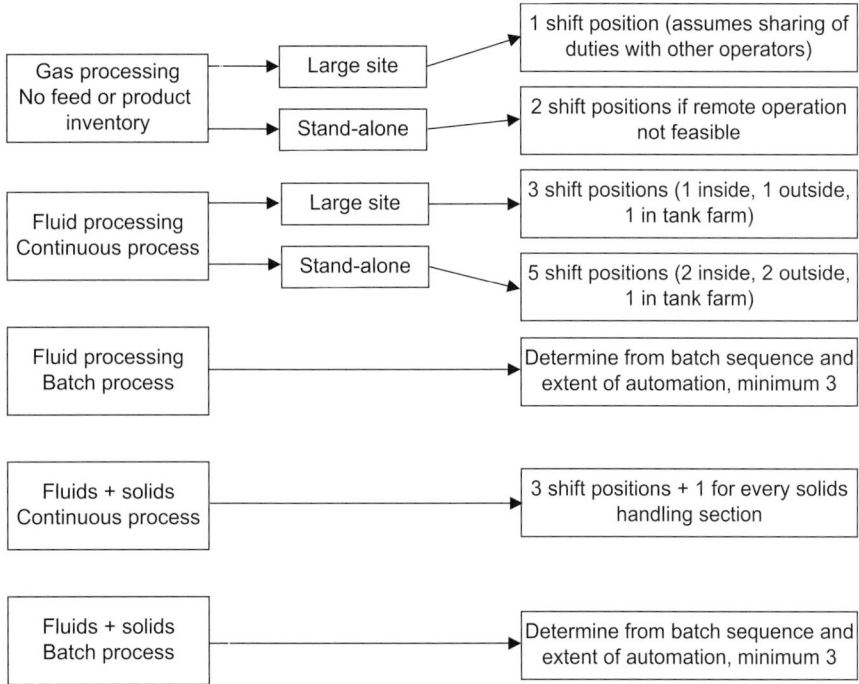

Figure 6.9. Algorithm for estimating the minimum number of shift positions.

3. Components formed in high yield from feed impurities. Most sulphur is produced as a by-product of fuels manufacture. Crude oil and natural gas contain sulphur compounds that are converted to H_2S during refining or gas treatment. The H_2S is then converted to elemental sulphur by the Claus process. Mannitol (a valuable hexose) is made from fructose that is present in the glucose feed to the sorbitol process.

4. Components produced in low yield that have high value. Dicyclopentadiene can be recovered from the products of steam naphtha cracking. Acetophenone is recovered as a by-product of phenol manufacture, although it can also be made by oxidation of ethylbenzene or fermentation of cinnamic acid.

5. Degraded consumables such as solvents that have re-use value.

Prices for by-products can be found in the same sources used for prices of main products. The difficult part is deciding whether it is worthwhile to recover a by-product. For the by-product to have value, it must meet the specifications for that material, which may entail additional processing costs. The design engineer must therefore assess whether the additional cost of recovering and purifying the by-product is justified by the by-product value and avoided waste-disposal cost, before deciding whether to value the material as a by-product or as a waste stream.

An algorithm for assessing the economic viability of recovering a by-product X is given in Figure 6.10. Note that it is important to consider not only the cost of purifying the by-product, but also whether it can be converted into something more valuable. This would include recycling the by-product within the process if that might be expected to lead to a higher yield of main product or formation of a more valuable by-product. Note also that when analysing whether to recover a by-product the value created by recovering the by-product includes not only the revenue from by-product sales, but also the avoided by-product disposal cost. If the by-product has fuel value then the fuel value should be subtracted from the revenue instead.

A rule of thumb that can be used for preliminary screening of by-products for large plants is that for by-product recovery to be economically viable the net benefit must be greater than US$200,000 per year. The net benefit is the by-product revenue plus the avoided waste-disposal cost. (This is based on the assumption that recovering by-product is going to add at least one separation to the process, which will cost at least $0.5 million of capital, or an annualized cost of about $170,000, as described in Section 6.7.6.).

6.4.9. Summarizing Production Costs and Revenues

It is useful to create a single-page summary of all of the production costs and revenues associated with a project, as this makes it easier to review the project economics and understand the relative contribution of different components to the overall cost of production. The summary sheet usually lists the quantity per year and per unit production of product, the price, the cost per year and the cost per unit production of product for each of the raw materials, by-products, consumables and utilities, as well as fixed costs and capital charges.

Most chemical companies have a preferred format for summarizing costs of production, and often use standard spreadsheets. Good examples are given in the PERP reports published by Nexant (www.nexant.com/products). A template for summarizing production costs is given in Appendix G and can be downloaded in MS Excel format from the on-line material at http://books.elsevier.com/companions. The use of this template is illustrated in Example 6.11.

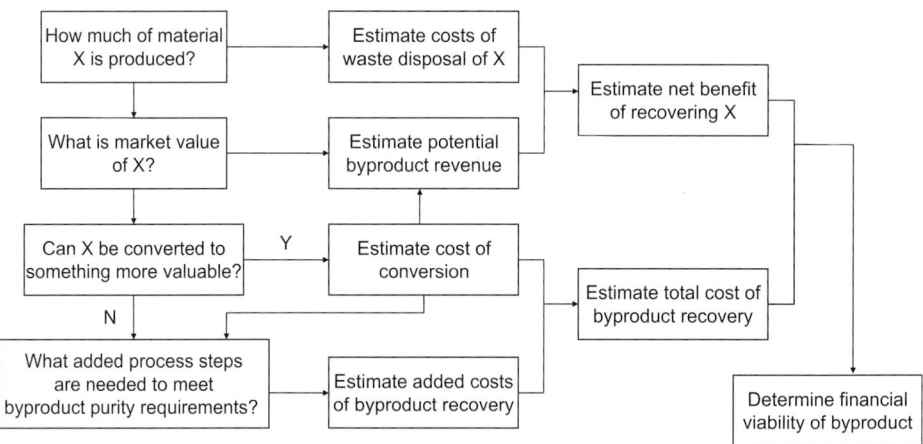

Figure 6.10. Algorithm for assessing the economic viability of by-product recovery.

6.5. **TAXES AND DEPRECIATION**

The profits generated by most chemical plants are subject to taxation. Taxes can have a significant impact on the cash flows from a project. The design engineer needs to have a basic understanding of taxation and tax allowances such as depreciation in order to make an economic evaluation of the project.

6.5.1. **Taxes**

Individuals and corporations must pay income tax in most countries. The details of tax law can be complicated and governments enact changes almost every year. Companies generally retain tax specialists, either as employees or as consultants, who have deep expertise in the intricacies of the field. Such specialized knowledge is not required for engineering design projects, which are usually compared on a relatively simple after-tax basis. The design engineer may occasionally need to consult a tax expert though, particularly when comparing projects in different countries with different tax laws and investment incentives.

Information on corporate taxes can usually be found from government web sites. In the UK the main rate of corporation tax is 28% for companies with profits greater than £1.5 million per year at the time of writing. Companies with profits below this threshold are taxed at the lower Small Companies Rate (SCR) of 21%. Information on current UK taxes can be found on the HM Revenue and Customs web site at www.hmrc.gov.uk.

Information on taxes in the United States is given on the Internal Revenue Service web site at www.irs.gov. At the time of writing, the top marginal rate of federal income tax on corporations in the United States is 35%, which applies to all incomes greater than $18,333,333 (IRS Publication 542). Since almost all companies engaged in building chemical plants substantially exceed this income threshold, it is common to assume that all profits will be taxed at the marginal rate. In many locations corporations must also pay state or local income taxes.

The amount of tax that must be paid in a given year is calculated by multiplying the taxable income by the tax rate. The taxable income is given by:

$$\text{Taxable income} = \text{gross profit} - \text{tax allowances} \qquad (6.22)$$

Various types of tax allowances are permitted in the tax laws of different countries, the most common of which is depreciation, discussed in Section 6.5.3. The after tax cash flow is then:

$$\begin{aligned} CF &= P - (P - D)t_r \\ &= P(1 - t_r) + Dt_r \end{aligned} \qquad (6.23)$$

where

CF = after tax cash flow
P = gross profit
D = sum of tax allowances
t_r = rate of taxation.

It can be seen from equation 6.23 that the effect of tax allowances is to reduce taxes paid and increase cash flow.

In some countries, taxes are paid in a given year based on the previous year's income. This is true for the USA, where corporate taxes are based on a calendar year of operations and are due by March 15 of the following year. This complicates the calculations somewhat, but is easily coded into a spreadsheet.

6.5.2. Investment Incentives

National and regional governments often provide incentives to encourage companies to make capital investments, since these investments create employment, generate taxation revenue and provide other benefits to politicians and the communities they represent.

The most common incentives used are tax allowances. Most countries allow some form of depreciation charge as a tax allowance, by which the fixed capital investment can be deducted from taxable income over a period of time, as described in Section 6.5.3. Other incentives that are often used include:

1. Tax waivers or vacations, in which no taxes are paid for a fixed period of time, typically 2 to 5 years after the project begins generating revenue.
2. Investment grants or credits, in which the government makes a cash contribution towards the initial investment.
3. Low-cost loans, in which the government either loans capital directly or else subsidizes the interest due on a commercial loan.
4. First year allowances (FYA), in which a high proportion of the investment can be depreciated in the first year. For example, at the time of writing the UK government allows a 100% FYA for energy-saving or water-efficient plant and machinery (www.hmrc.gov.uk/capital_allowances/).
5. Loan guarantees, in which the government agrees to underwrite loans for the project, making it easier to secure financing on advantageous terms.

An economic comparison between different process alternatives for the same site should usually be made using the same assumptions on investment incentives. This might not always be the case though, for example, if one project is eligible for a government grant because of using renewable energy and another project is not. It should also be noted that differences in incentives can have a significant impact on investment decisions when comparing investments at a company-wide level in a global context.

6.5.3. Depreciation Charges

Depreciation charges are the most common type of tax allowance used by governments as an incentive for investment. Depreciation charges are also sometimes referred to as writing down allowances (WDA). Depreciation is a non-cash charge reported as an expense, which reduces income for taxation purposes. There is no cash outlay for depreciation, and no money is transferred to any fund or account, so the depreciation charge is added back to the net income after taxes to give the total cash flow from operations.

$$\begin{aligned} CF &= I - (I \times t_r) + D \\ &= (P - D) - ((P - D) \times t_r) + D \\ &= P(1 - t_r) + Dt_r \end{aligned} \qquad (6.24)$$

where

I = taxable income

D = depreciation tax allowance.

It can be seen that equations 6.23 and 6.24 are equivalent.

Depreciation charges can be thought of as an allowance for the 'wear and tear, deterioration or obsolescence of the property' as a result of its use (IRS publ. 946).

The book value or 'written down' value of an asset is the original cost paid minus the accumulated depreciation charged. The book value has no connection to the resale value or current market value of the asset.

$$\text{Book value} = \text{initial cost} - \text{accumulated depreciation} \qquad (6.25)$$

Note that the law usually only allows depreciation of fixed capital investments, and not total capital, since working capital is not consumed and can be recovered at the end of the project. If land was purchased for the project, then the cost of the land must be deducted from the fixed capital cost as land is assumed to retain its value and cannot be depreciated.

Over a period of time the book value of the asset or fixed investment decreases until it is fully 'paid off' or 'written off', at which point depreciation can no longer be charged. The schedule of how depreciation charges are taken is set by the tax law. In a globalized economy it is necessary for design engineers to have familiarity with several different methods of depreciation, as each country has different rules. For example, the UK uses a declining balance method, while the US uses the Modified Accelerated Cost Recovery System (MACRS) described below (IRS publ. 946). When analysing international projects the appropriate national and regional tax laws must be checked to ensure that the correct depreciation rules are followed.

There are several other less widely used depreciation methods that are not discussed here. A good overview of these is given by Humphreys (1991).

Straight-Line Depreciation

Straight-line depreciation is the simplest method. The depreciable value, C_d, is depreciated over n years with annual depreciation charge D_i in year i, where:

$$D_i = \frac{C_d}{n} \quad \text{and} \quad D_j = D_i \forall j \qquad (6.26)$$

The depreciable value of the asset is the initial cost of the fixed capital investment, C, minus the salvage value (if any) at the end of the depreciable life. For chemical plants the salvage value is often taken as zero, as the plant usually continues to operate for many years beyond the end of the depreciable life.

The book value of the asset after m years of depreciation, B_m is:

$$B_m = C - \sum_{i=1}^{m} D_i$$

$$= C - \frac{m \, C_d}{n} \tag{6.27}$$

When the book value is equal to the salvage value (or zero) then the asset is fully depreciated and no further depreciation charge can be taken.

Straight-line depreciation must be used in the US for software (with a 36-month depreciable life), patents (with life equal to the patent term remaining) and other depreciable intangible property (IRS publ. 946).

Declining-Balance Depreciation

The declining-balance method is an accelerated depreciation schedule that allows higher charges in the early years of a project. This helps improve project economics by giving higher cash flows in the early years. In the declining-balance method, the annual depreciation charge is a fixed fraction, F_d, of the book value:

$$D_1 = CF_d \tag{6.28}$$

$$B_1 = C - D_1 = C(1 - F_d)$$

$$D_2 = B_1 F = C(1 - F_d)F_d$$

$$B_2 = B_1 - D_2 = C(1 - F_d)(1 - F_d) = C(1 - F_d)^2$$

Hence

$$D_m = C(1 - F_d)^{m-1} F_d \tag{6.29}$$

$$B_m = C(1 - F_d)^m \tag{6.30}$$

The fraction F_d must be less than $2/n$, where n is the depreciable life in years. When $F_d = 2/n$, this method is known as double declining-balance depreciation.

At the time of writing, the UK uses declining-balance depreciation with rates of 4% for buildings and structures, 25% for plant and machinery and 6% for long-life assets (see www.hmrc.gov.uk/capital_allowances for details).

Modified Accelerated Cost Recovery System (MACRS)

The MACRS depreciation method was established by the US Tax Reform Act of 1986 and is the depreciation method used for most tangible assets in the USA. The details of the MACRS depreciation method are given in IRS publication 946, which is available on-line at www.irs.gov/publications. The method is basically a combination of the double declining-balance method and the straight-line method. The double declining-balance method is used until the depreciation charge becomes less than it would be under the straight-line method, at which point the MACRS method switches to charge the same amount as the straight-line method.

Under MACRS depreciation, different recovery periods are assigned to different kinds of asset, based on a usable life ('class life') designated by the US Internal Revenue Service (IRS). For chemical plants the class life is 9.5 years and the recovery period is 5 years. The class life for other process industries ranges from 7.5 years for

Table 6.9. MACRS Depreciation Charges

Recovery year	Depreciation rate ($F_i = D_i / C_d$)	
	5-year recovery	**15-year recovery**
1	20.00	5.00
2	32.00	9.50
3	19.20	8.55
4	11.52	7.70
5	11.52	6.93
6	5.76	6.23
7		5.90
8		5.90
9		5.91
10		5.90
11		5.91
12		5.90
13		5.91
14		5.90
15		5.91
16		2.95

offshore oil platforms to 18 years for coal gasification; see Appendix B of IRS publication 946. It should also be noted that for roads, docks and other civil infrastructure a 15-year recovery period is used, so some off-site investments are depreciated on a different schedule from that used for the ISBL investment.

Another important convention within MACRS depreciation is that the method assumes that all property is acquired mid-year and hence assigns half of the full year depreciation in the first and last years of the recovery period. The result is the schedule of depreciation charges given in Table 6.9.

There are other details of MACRS depreciation that are not discussed here, and at the time of writing the tax law also allows assets to be depreciated by the straight-line method (over the class life, not the recovery period and still following the half-year convention). The US tax law is revised frequently and the most recent version of IRS publication 946 should be consulted for the current regulations.

Example 6.8

A chemical plant with a fixed capital investment of $100 million generates an annual gross profit of $50 million. Calculate the depreciation charge, taxes paid and after-tax cash flows for the first 10 years of plant operation using straight-line depreciation over 10 years and using MACRS depreciation with a 5-year recovery period. Assume the plant is built at time zero and begins operation at full rate in year 1. Assume the rate of corporate income tax is 35% and taxes must be paid based on the previous year's income.

Solution

The solution is easily coded into a spreadsheet. The results are shown in the tables below:

Year	Gross profit (MM$)	Depreciation charge (MM$)	Taxable income (MM$)	Taxes paid (MM$)	Cash Flow (MM$)
0	0	0	0	0	−100
1	50	10	40	0	50
2	50	10	40	14	36
3	50	10	40	14	36
4	50	10	40	14	36
5	50	10	40	14	36
6	50	10	40	14	36
7	50	10	40	14	36
8	50	10	40	14	36
9	50	10	40	14	36
10	50	10	40	14	36

Year	Gross profit (MM$)	Depreciation charge (MM$)	Taxable income (MM$)	Taxes paid (MM$)	Cash Flow (MM$)
0	0	0	0	0	−100
1	50	20	30	0	50
2	50	32	18	10.50	39.50
3	50	19.2	30.8	6.30	43.70
4	50	11.52	38.48	10.78	39.22
5	50	11.52	38.48	13.47	36.53
6	50	5.76	44.24	13.47	36.53
7	50	0	50	15.48	34.52
8	50	0	50	17.50	32.50
9	50	0	50	17.50	32.50
10	50	0	50	17.50	32.50

6.6.   PROJECT FINANCING

The construction and operation of chemical plants require large amounts of capital. Corporations engaged in the production of chemicals must raise the finances to support such investments. Like taxation, corporate financing is a specialized subject with many intricacies that require expert knowledge. The design engineer needs a superficial awareness of this subject to carry out economic analysis and optimization of the design.

6.6.1.   Basics of Corporate Accounting and Finance

The purpose of financial accounting is to report the economic performance and financial condition of a company to its owners (shareholders), lenders, regulatory agencies and other stakeholders. The primary means for financial reporting is the annual report to shareholders. The annual reports for companies in the chemical, life sciences and fuels industries generally contain:

1. A letter from the Chief Executive Officer (CEO) describing the past year's operations, significant acquisitions, divestitures and restructuring, and plans for the short and long term.
2. Financial information:
   a. Balance sheet
   b. Income statement
   c. Cash flow statement
   d. Notes to the financial statements
   e. Comments from the independent auditors.
3. Information on the directors and executive management of the company.
4. A report on the health, safety and environmental performance of the company (sometimes published separately).

The annual report of any publicly traded company will usually be available on-line and can easily be found by visiting the company's web site. The site will usually have a prominent link to 'information for investors' or something similar. No attempt has been made to create fictitious financial statements for the purposes of this book as an abundance of real examples is readily available on-line. The reader is encouraged to search the web for real examples.

Balance Sheet

The balance sheet is a snapshot of the financial condition of the company. It lists all the assets owned by the company and all the liabilities or amounts owed by the company. The difference between assets and liabilities is the stockholder's equity, i.e., notionally the amount of money the stockholders would have available to share out if they decided to liquidate the company.

$$\text{Stockholder's equity} = \text{assets} - \text{liabilities} \qquad (6.31)$$

Assets are typically listed in order of decreasing liquidity. Liquidity is a measure of how easily the asset could be turned into cash. Assets include:

- Cash and cash equivalents.
- Notes and accounts receivable, i.e., money owed to the company for goods shipped but not yet paid for.
- Inventories of raw materials, products, spare parts and other supplies.
- Prepaid taxes and expenses.
- Investments such as equity stakes in other companies or joint ventures.
- Property, plant and equipment. This is listed at book value, i.e., cost less accumulated depreciation. The actual market value of these assets may be considerably higher.
- Intangible assets such as patents, trademarks, goodwill, etc.

Liabilities are usually listed in the order in which they are due, starting with current liabilities. Liabilities include:

- Accounts payable, i.e., payment owed on goods already received by the company.
- Notes and loans due for repayment.

- Accrued liabilities and expenses such as legal settlements, amounts set aside for warranties, guarantees, etc.
- Deferred income taxes.
- Long-term debt.

The difference between assets and liabilities is the shareholder's equity. This consists of the capital paid in by the owners of common and preferred stocks, together with earnings retained and reinvested in the business. The capital paid in by the shareholders is often listed as the par value of the stock (typically 25¢ to $1 per share) plus the additional capital paid in when the stock was initially sold by the company. Note that this reflects only the capital raised by the company and has no relation to subsequent increases or decreases in the value of the stock that may have resulted from trading.

Income Statement

The income statement or consolidated statement of operations is a summary of the incomes, expenditures and taxes paid by the company over a fixed period of time. Results are usually presented for the past 3 calendar years.

The income statement lists the following items:

1. Sales and operating revenues (positive).
2. Income from equity holdings in other companies (positive).
3. Cost of goods sold (negative).
4. Selling, general and administrative expenses (negative).
5. Depreciation (negative on the income statement, but will be added back on the cash flow statement).
6. Interest paid on debt (negative).
7. Taxes other than income tax, such as excise duties (negative).
8. Income taxes (negative).

The sum of items 1 through 5 is sometimes listed as earnings before interest and taxes (EBIT). The sum of items 1 through 7 is listed as income before taxes or taxable income, and is usually positive. The net income is the sum of items 1 through 8, i.e., income before taxes minus taxes paid. Net income is also usually expressed as earnings per share of common stock.

The income statement gives a good insight into the overall profitability and margins of a business. It has to be read carefully though, as several items listed are non-cash charges such as depreciation, that do not affect the cash flow of the business. Corrections for these items are made in the cash flow statement.

Cash-Flow Statement

The cash-flow statement gives a summary of overall cash flows into and out of the business as a result of operating activities, investments and financing activities. It is also usually reported for the past 3 calendar years.

The cash flow from operating activities section starts with the net income. Adjustments are made for non-cash transactions (depreciation and deferred taxes are added back in), and changes in assets and liabilities.

The cash flow from investing activities section lists the cash spent on acquiring fixed assets such as property, plant and equipment, less any revenues from sale of fixed assets. It also lists acquisitions or divestitures of subsidiary businesses.

The cash flow from financing activities section summarizes changes in the company's long-term and short-term debt, proceeds from issues of common stock, repurchase of stocks, and dividends paid to stockholders.

The sum of cash flows from operations, investments and financing gives the net change in cash and cash equivalents. This is then added to the cash and cash equivalents from the beginning of the year to give the cash and cash equivalents at the end of the year, which appears on the balance sheet.

Summary

The business and accounting literature contains a wealth of information on how to read and analyse corporate financial statements. Most engineers work for or with corporations and have a direct personal interest in understanding financial performance; however, a detailed treatment of the subject is beyond the scope of this book. Excellent introductions to finance and accounting are given in the books by Spiro (1996) and Shim and Henteleff (1995).

6.6.2. Debt Financing and Repayment

Most debt capital is raised by issuing long-term bonds. A mortgage is a bond that is backed by pledging a specific real asset as security against the loan. An unsecured bond is called a debenture. The ratio of total debt divided by total assets is known as the debt ratio (DR) or leverage of the company.

All debt contracts require payment of interest on the loan and repayment of the principal (either at the end of the loan period or amortized over the period of the loan). Interest payments are a fixed cost, and if a company defaults on these payments then its ability to borrow money will be drastically reduced. Since interest is deducted from earnings, the greater the leverage of the company, the higher is the risk to future earnings, and hence to future cash flows and the financial solvency of the company. In the worst case, the company could be declared bankrupt and the assets of the company sold off to repay the debt. Finance managers therefore carefully adjust the amount of debt owed by the company so that the cost of servicing the debt (the interest payments) does not place an excessive burden on the company.

The rate of interest owed on debt depends on the bond markets, government central banks and the credit worthiness of the company. When new bonds are issued, they must be offered at a competitive interest rate, otherwise they will not sell. If the bond issuer has a high credit rating then they will be able to issue bonds at close to the interest rates set by the government. (US Treasury bonds are not rated, as it is assumed that they will be backed by the United States Federal Government. Some other countries' bonds are also not rated.) If the credit rating of the issuer is lower, then there is a higher chance that the debt may not be repaid, in which case it must be offered at a higher interest rate to offset this risk. Credit rating services such as Moody's and Standard and Poor's study the finances of corporations and publish credit ratings. These ratings are usually not advertised by issuers unless they are very high, but they are published in the financial papers. The difference in interest rate between low-rated and high-rated bonds issued at the same time is typically 2% to 3%.

Once they have been issued, bonds are traded on financial exchanges such as the London Stock Exchange, the New York Stock Exchange or the American Stock Exchange. Although the price of the bond in subsequent trading may vary from the offer price (or face value), the interest rate remains fixed. The financial newspapers report prices daily for the most actively traded corporate bonds. Bond prices can also be found at www.investinginbonds.com, together with much other useful information on bond markets. The interest rate is listed as the 'coupon' and the date on which the bond expires is the 'maturity'. Bonds are also assigned a unique nine-digit identification number by the American Bankers' Association Committee on Uniform Security Identification Procedures (CUSIP). For example, in 2006 Honeywell Inc. issued a 30-year bond CUSIP #438516AR7 with coupon 5.700 and maturity 03/15/2036.

6.6.3. Equity Financing

Equity capital consists of the capital contributed by stockholders, together with earnings retained for reinvestment in the business. Stockholders purchase stocks in the expectation of getting a return on their investment. This return can come from the dividends paid annually to stockholders (the part of earnings returned to the owners) or from growth of the company that is recognized by the stock market and leads to an increase in the price of the stock. Most stock is usually held by sophisticated institutional investors such as banks, mutual funds, insurance companies and pension funds. These investors employ expert analysts to assess the performance of companies relative to other companies in the same sector, and to the market as a whole. If the management of a company does not effectively deliver the financial return expected by investors, the stock price will suffer and the management will soon be replaced.

Simple measures of the effectiveness of management are the return on equity and earnings per share. Return on equity (ROE) is defined as:

$$\text{ROE} = \frac{\text{net annual profit}}{\text{stockholders' equity}} \times 100\% \tag{6.32}$$

The stockholders' expectation of return on their equity can be expressed as an interest rate and is known as the cost of equity capital. The cost of equity required to meet the expectations of the market is usually substantially higher than the interest rate owed on debt, because of the riskier nature of equity finance (since debt holders are paid first and hence have the primary right to any profit made by the business). For most corporations in the European Union and United States at the time of writing the cost of equity is in the range 25% to 30%.

6.6.4. Cost of Capital

Very few companies operate entirely on debt or equity financing alone and most use a balance of both. The overall cost of capital is simply the weighted average of the cost of debt and the cost of equity.

$$i_c = (DR \times i_d) + ((1 - DR) \times i_e) \tag{6.33}$$

where

$$i_c \quad = \text{cost of capital}$$
$$DR \quad = \text{debt ratio}$$
$$i_d \quad = \text{interest rate due on debt}$$
$$i_e \quad = \text{cost of equity.}$$

For example, if a company were financed 55% with debt at an average 8% interest and 45% with equity that carried an expectation of a 25% return, then the overall cost of capital would be:

$$i_c = (0.55 \times 0.08) + (0.45 \times 0.25)$$

$$= 0.1565$$

Since the equity is by definition (equation 6.31) the assets minus the liabilities (debt), the overall return on assets (ROA) can be expressed as:

$$\text{ROA} = \frac{\text{net annual profit}}{\text{total assets}} \times 100\% \tag{6.34}$$

It follows that:

$$\frac{\text{ROA}}{\text{ROE}} = \frac{\text{stockholders' equity}}{\text{total assets}} = 1 - DR \tag{6.35}$$

The overall cost of capital sets the interest rate that is used in economic evaluation of projects. The total portfolio of projects funded by a company must meet or exceed this interest rate if the company is to achieve its targeted return on equity and hence satisfy the expectations of its owners.

6.7. ECONOMIC EVALUATION OF PROJECTS

As the purpose of investing money in a chemical plant is to earn money, some means of comparing the economic performance of projects is needed. Before a company agrees to spend a large amount of capital on a proposed project, the management must be convinced that the project will provide a sound investment compared to other alternatives. This section introduces the principal methods used for making economic comparisons between projects.

6.7.1. Cash Flow and Cash-Flow Diagrams

During any project, cash initially flows out of the company to pay for the costs of engineering, equipment procurement and plant construction. Once the plant is constructed and begins operation, then the revenues from sale of product begin to flow into the company. The 'net cash flow' at any time is the difference between the earnings and expenditure. A cash-flow diagram, such as that shown in Figure 6.11, shows the forecast cumulative net cash flow over the life of a project. The cash flows are based on the best estimates of investment, operating costs, sales volume and sales

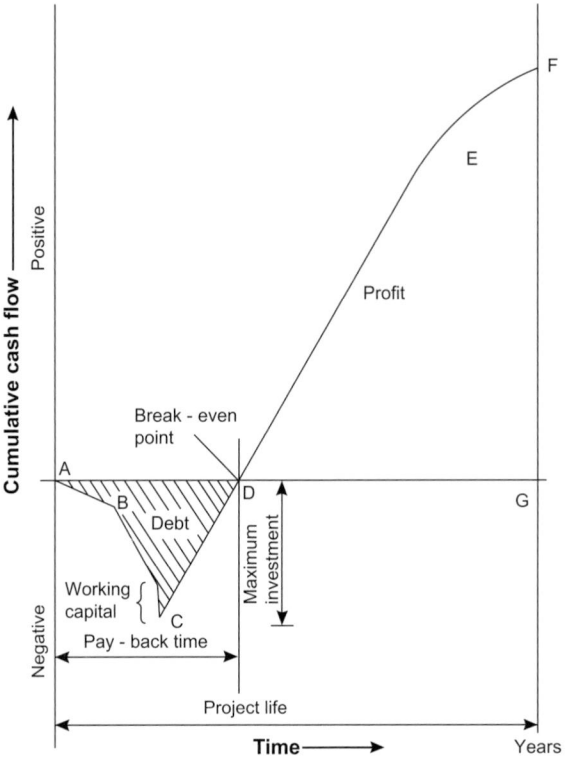

Figure 6.11. Project cash-flow diagram.

price that can be made for the project. A cash-flow diagram gives a clear picture of the resources required for a project and the timing of the earnings. The diagram can be divided into the following characteristic regions:

A–B The investment required to design the plant.

B–C The heavy flow of capital to build the plant and provide funds for start-up, including working capital.

C–D The cash-flow curve turns up at C, as the process comes on stream and income is generated from sales. The net cash flow is now positive but the cumulative amount remains negative until the investment is paid off, at point D.

Point D is known as the *break-even point* and the time to reach the break-even point is called the *pay-back time*. (In a different context, the term *break-even point* is also sometimes used for the percentage of plant capacity at which the income equals the cost of production).

D–E In this region the cumulative cash flow is positive. The project is earning a return on the investment.

E–F Toward the end of project life the rate of cash flow may tend to fall off, due to increased operating costs and falling sales volume and price due to obsolescence of the plant, and the slope of the curve changes.

The point F gives the final cumulative net cash flow at the end of the project life.

Net cash flow is a relatively simple and easily understood concept, and forms the basis for the calculation of other, more complex, measures of profitability. Taxes and the effect of depreciation are usually not considered in cash flow diagrams.

6.7.2. Simple Methods for Economic Analysis

Pay-Back Time

A simple method for estimating the pay-back time is to divide the total initial capital (fixed capital plus working capital) by the average annual cash flow:

$$\text{simple pay-back time} = \frac{\text{total investment}}{\text{average annual cash flow}} \tag{6.36}$$

This is not the same pay-back time indicated by the cash-flow diagram, as it assumes that all the investment is made in year zero and revenues begin immediately. For most chemical plant projects, this is not realistic as investments are typically spread over 1 to 3 years and revenues may not reach 100% of design basis until the second year of operation. The simple pay-back time also neglects taxes and depreciation.

Return on Investment

Another simple measure of economic performance is the return on investment (ROI). The ROI is defined in a similar manner to ROA and ROE:

$$\text{ROI} = \frac{\text{net annual profit}}{\text{total investment}} \times 100\% \tag{6.37}$$

If ROI is calculated as an average over the whole project then:

$$\text{ROI} = \frac{\text{cumulative net profit}}{\text{plant life} \times \text{initial investment}} \times 100\% \tag{6.38}$$

Calculation of the after-tax ROI is complicated if the depreciation term is less than the plant life and if an accelerated method of depreciation such as declining balance or MACRS is used. In such cases, it is just as easy to calculate one of the more meaningful economic criteria such as net present value or discounted cash-flow rate of return, described below. Because of this complication, a pre-tax ROI is often used instead:

$$\text{pre-tax ROI} = \frac{\text{pre-tax cash flow}}{\text{total investment}} \times 100\% \tag{6.39}$$

Note that pre-tax ROI is based on cash flow, not profit or taxable income, and therefore does not include a depreciation charge.

Return on investment is also sometimes calculated for incremental modifications to a large project, as described in Section 6.9.3.

6.7.3.   Time Value of Money

In Figure 6.11 the net cash flow is shown at its value in the year in which it occurred. So the numbers on the ordinate show the 'future worth' of the project. The cumulative value is the 'net future worth' (NFW).

The money earned in any year can be reinvested as soon as it is available and can start to earn a return. So money earned in the early years of the project is more valuable than that earned in later years. This 'time value of money' can be allowed for by using a variation of the familiar compound interest formula. The net cash flow in each year of the project is brought to its 'present value' at the start of the project by discounting it at some chosen compound interest rate.

The future worth of an amount of money, P, invested at interest rate, i, for n years is:

$$\text{Future worth in year } n = P(1+i)^n$$

Hence the present value of a future sum is:

$$\text{present value of future sum} = \frac{\text{future worth in year } n}{(1 + i)^n} \qquad (6.40)$$

The interest rate used in discounting future values is known as the discount rate and is chosen to reflect the earning power of money. In most companies the discount rate is set at the cost of capital (see Section 6.6.4).

Discounting of future cash flows should not be confused with allowing for price inflation. Inflation is a general increase in prices and costs, usually caused by imbalances between supply and demand. Inflation raises the costs of feed, products, utilities, labour and parts, but does not affect depreciation charges, which are based on original cost. Discounting, on the other hand, is a means of comparing the value of money that is available now (and can be reinvested) with money that will become available at some time in the future. All of the economic analysis methods can be modified to allow for inflation. See, for example, Humphreys (1991), Chapter 6. In practice, most companies assume that although prices may suffer inflation, margins and hence cash flows will be relatively insensitive to inflation. Inflation can therefore be neglected for the purposes of comparing the economic performance of projects.

6.7.4.   Net Present Value

The net present value (NPV) of a project is the sum of the present values of the future cash flows:

$$\text{NPV} = \sum_{n=1}^{n=t} \frac{CF_n}{(1 + i)^n} \qquad (6.41)$$

where

CF_n = cash flow in year n
t = project life in years
i = interest rate (= cost of capital, percent /100).

Table 6.10. Typical Start-up Schedule

Year	Costs	Revenues	Explanation
1st year	30% of fixed capital	0	Engineering + long lead-time items
2nd year	40–60% of fixed capital	0	Procurement and construction
3rd year	10–30% of fixed capital + working capital + FCOP + 30% VCOP	30% of design basis revenue	Remaining construction Initial production
4th year	FCOP + 50–90% VCOP	50 – 90% of design basis revenue	Shake-down of plant
5th year +	FCOP + VCOP	100% of design basis revenue	Full production at design rates

The net present value is always less than the total future worth of the project because of the discounting of future cash flows. Net present value is easily calculated using spreadsheets and most spreadsheet programs have a NPV function.

The net present value is a strong function of the interest rate used and the time period studied. When different time periods are analyzed the time period is sometimes denoted by a subscript. For example, NPV_{10} would denote the NPV over a 10-year period.

Net present value is a more useful economic measure than simple pay-back and ROI, since it allows for the time value of money and also for annual variation in expenses and revenues. Few large projects are completed in a single year and immediately begin production at full capacity. A more typical start-up schedule is given in Table 6.10. Net present value is also a more appropriate method to use when considering after-tax income using an accelerated depreciation method such as declining balance or MACRS.

6.7.5. Discounted Cash-Flow Rate of Return (DCFROR)

By calculating the NPV at various interest rates, it is possible to find an interest rate at which the cumulative net present value at the end of the project is zero. This particular rate is called the 'discounted cash-flow rate of return' (DCFROR) and is a measure of the maximum interest rate that the project could pay and still break even by the end of the project life.

$$\sum_{n=1}^{n=t} \frac{CF_n}{(1 + i')^n} = 0 \qquad (6.42)$$

where

CF_n = cash flow in year n
t = project life in years
i' = the discounted cash flow rate of return (percent /100).

The value of i' is found by trial-and-error calculations or by using the appropriate function in a spreadsheet. A more profitable project will be able to pay a higher DCFROR.

DCFROR provides a useful way of comparing the performance of capital for different projects, independent of the amount of capital used, the life of the plant, or the actual interest rates prevailing at any time. DCFROR is a more useful method than NPV when comparing projects of very different size. The NPV of large projects is usually greater than that of small projects, but then the investment is also much greater. DCFROR is independent of project size and the project with the highest DCFROR always provides the best 'bang for the buck'. When DCFROR is used as an investment criterion, companies usually expect projects to have a DCFROR greater than the cost of capital.

DCFROR can also be compared directly with interest rates. Because of this, it is sometimes known as the interest rate of return or internal rate of return (IRR).

Example 6.9

Estimate the NPV at a 12% interest rate and the DCFROR for the project described in Example 6.8, using the MACRS depreciation method.

Solution

Calculating the present values of the cash flows from the previous example requires adding two columns to the spreadsheet. We first calculate the discount factor $(1 + i)^{-n}$, and then multiply this by the cash flow in year n to give the present value of the cash flow. The present values can then be summed to give the net present value:

Year	Gross profit (MM$)	Depreciation charge (MM$)	Taxable income (MM$)	Taxes paid (MM$)	Cash Flow (MM$)	Discount factor	Present value of CF (MM$)
0	0	0	0	0	−100	1	−100
1	50	20	30	0	50	0.893	44.64
2	50	32	18	10.50	39.50	0.797	31.49
3	50	19.2	30.8	6.30	43.70	0.712	31.10
4	50	11.52	38.48	10.78	39.22	0.636	24.93
5	50	11.52	38.48	13.47	36.53	0.567	20.73
6	50	5.76	44.24	13.47	36.53	0.507	18.51
7	50	0	50	15.48	34.52	0.452	15.61
8	50	0	50	17.50	32.50	0.404	13.13
9	50	0	50	17.50	32.50	0.361	11.72
10	50	0	50	17.50	32.50	0.322	10.46
				Interest rate		12.00%	
				Total = Net present value =			122.32

Note that we could also have calculated NPV directly using the NPV function. In MS Excel, the NPV function starts at the end of year 1, so any cash flows in year 0 should not be included in the function range.

The DCFROR can then be found by adjusting the interest rate until the NPV is equal to zero. This is easily accomplished in the spreadsheet using the 'Goal Seek' tool, giving DCFROR = 40.9%.

6.7.6. Annualized Cost Methods

An alternative method of comparing the magnitude of a capital investment in current dollars with a revenue stream in the future is to convert the capital cost into a future annual capital charge.

If an amount P is invested at an interest rate i, then after n years of compound interest it matures to the sum $P(1+i)^n$.

If, instead, an amount A is invested each year, also at interest rate i, then it matures to a sum, S, where:

$$S = A + A(1+i) + A(1+i)^2 + \ldots + A(1+i)^{n-1} \tag{6.43}$$

so

$$S(1+i) = A(1+i) + A(1+i)^2 + \ldots + A(1+i)^n \tag{6.44}$$

Hence subtracting equation 6.43 from equation 6.44:

$$Si = A[(1+i)^n - 1] \tag{6.45}$$

If the annual payments A have matured to give the same final sum that would have been obtained by investing the principal P at the same interest rate then:

$$S = P(1+i)^n = \frac{A}{i}[(1+i)^n - 1]$$

Hence

$$A = P\frac{[i(1+i)^n]}{[(1+i)^n - 1]} \tag{6.46}$$

and we can define an annual capital charge ratio, $ACCR$, as:

$$ACCR = \frac{A}{P} = \frac{[i(1+i)^n]}{[(1+i)^n - 1]} \tag{6.47}$$

The annual capital charge ratio is the fraction of the principal that must be paid out each year to fully repay the principal and all accumulated interest over the life of the investment. This is the same formula used for calculating payments on home mortgages and other loans where the principal is amortized over the loan period.

If the cost of capital is used as the interest rate (see Section 6.6.4), then the annual capital charge ratio can be used to convert the initial capital expense into an annual capital charge, or annualized capital cost, as described in Section 6.2.5.

$$\text{Annual capital charge (ACC)} = ACCR \times \text{total capital cost} \tag{6.48}$$

The annual capital charge can be added to the operating costs to give a total annualized cost, TAC:

$$\text{TAC} = \text{operating costs} + ACCR \times \text{total capital cost} \tag{6.49}$$

The TAC can be compared with forecasted future revenues. The TAC is also sometimes referred to as total cost of production or TCOP.

Table 6.11 shows values of $ACCR$ for different values of i and n. For a typical cost of capital of about 15% and a plant life of 10 years the value of $ACCR$ is 0.199, or about one-fifth of the capital investment.

Table 6.11. Values of Annual Capital Charge
Ratio (*ACCR*) for Different Interest Rates

Interest Rate, *i*	ACCR: 10-year Life	ACCR: 20-year Life
0.1	0.163	0.117
0.12	0.177	0.134
0.15	0.199	0.16
0.2	0.239	0.205
0.25	0.280	0.253
0.3	0.323	0.302

There are a few important things that should be noted when using the annualized cost method:

1. The method assumes investment and cash flows begin immediately, and so does not capture information on the timing of early expenditures and revenues.
2. The method does not take into account taxes or depreciation, and assumes that all of the revenue from the project is available to provide a return on the initial investment.
3. Working capital is recovered at the end of the project and so strictly only the fixed capital should be annualized. Equations 6.46 and 6.47 can be modified for the case where an additional sum becomes available at the end of the investment term, but this modified version is seldom used in practice and working capital is often either neglected in the annualized cost method or else (wrongly) thrown in with fixed capital. A simple way around this problem is to assume that the working capital is entirely funded by debt, in which case the cost of carrying the working capital is reduced to an interest payment that appears as part of the fixed costs of production. At the end of the project life the working capital will be released and will be available to repay the principal on the debt.
4. As described in Section 6.2.4, several of the fixed costs of production are proportional to the fixed capital invested (*FC*). If we assume annual charges of 3% of *FC* for maintenance, 2% of *FC* for property tax and 65% plant overhead then the annual capital charge ratio is increased by $0.02 + (1.65 \times 0.03) = 0.07$.
5. If we also assume engineering costs are 10% of (ISBL + OSBL) capital investment and add 15% of (ISBL + OSBL) capital as contingency, then with a 10-year plant life and a 15% interest rate the annual capital charge ratio is:

$$ACCR = [0.199 \times (1.0 + 0.1 + 0.15) + 0.07] \times [\text{Installed ISBL}$$
$$+ \text{OSBL capital cost}]$$

$$= 0.32 \times [\text{Installed ISBL} + \text{OSBL capital cost}] \qquad (6.50)$$

Equation 6.50 is the basis for the widely used rule of thumb of annualizing capital cost by dividing by three. When using this rule of thumb, it is important to remember that some, but not all, of the fixed costs have been counted in the annual capital charge.

The annualized cost method involves more assumptions than calculating NPV or DCFROR, but it is widely used as a quick way of comparing investments with the resulting benefits. Annualized cost is also useful as a method for analysing small projects and modifications that lead to reduced operating costs (for example, heat recovery projects), since the annualized capital outlay can be directly traded off against the expected annual savings and there is usually no change in working capital, operating labour or other fixed costs of production. Small projects usually can be executed quickly, so the error introduced by neglecting the timing of investments and revenues is less important than it is when designing a new plant.

The annualized cost method is also used when comparing the costs of equipment with different expected operating life. Annualization of the costs allows equipment with different service life to be compared on the same annual basis. This is illustrated in the example that follows.

Example 6.10

A carbon steel heat exchanger that costs €140,000 is expected to have a service life of 5 years before it requires replacement. If type 304 stainless steel is used then the service life will be increased to 10 years. Which exchanger is the most economical if the cost of capital is 12%?

Solution

With a 12% interest rate and 5-year life, the annual capital charge ratio is:

$$ACCR = \frac{[i\ (1\ +\ i)^n]}{[(1\ +\ i)^n\ -\ 1]} = \frac{[0.12\ (1.12)^5]}{[(1.12)^5\ -\ 1]} = 0.277 \quad (6.47)$$

The annualized capital cost of the carbon steel exchanger is then

$$= €140,000 \times 0.277 = €38,780/y$$

From Table 6.5, we can estimate the cost of the type 304 stainless steel exchanger to be €140,000 × 1.3 = €182,000. From Table 6.11 (or equation 6.47), with a 10-year life and 12% interest rate the annual capital charge ratio is 0.177, so the annualized cost of the stainless steel exchanger is:

$$= €182,000 \times 0.177 = €32,210/y$$

In this case, it would be more economical to buy the stainless steel heat exchanger.

6.7.7. Summary

There is no single best criterion for economic evaluation of projects. Each company uses its own preferred methods and sets criteria for the minimum performance that will allow a project to be funded (see Section 6.9). The design engineer must be careful to ensure that the method and assumptions used are in accordance with company policy, and that projects are compared on a fair basis. Projects should always be compared using the same economic criterion, but do not have to be compared on *the exact same* basis, since in a global economy there may be significant regional advantages in feed and product pricing, capital costs, financing or investment incentives.

As well as economic performance, many other factors have to be considered when evaluating projects; such as those listed below:

1. Safety.
2. Environmental problems (waste disposal).
3. Political considerations (government policies).
4. Location of customers and suppliers (supply chain).
5. Availability of labour and supporting services.
6. Corporate growth strategies.
7. Company experience in the particular technology.

Project selection is discussed in more detail in Section 6.9.

Example 6.11

Adipic acid is used in the manufacture of nylon 6,6. It is made by hydrogenation of phenol to a mixture of cyclohexanol and cyclohexanone (known as KA oil – ketone and alcohol), followed by oxidation with nitric acid. The overall reaction can be written approximately as:

$$C_6H_5OH + 2H_2 \rightarrow C_6H_{10}O$$

$$C_6H_{10}O + H_2 \rightarrow C_6H_{11}OH$$

$$C_6H_{11}OH + 2HNO_3 \rightarrow HOOC(CH_2)_4COOH + N_2O + 2H_2O$$

The actual process requirements of phenol, hydrogen, nitric acid and utilities and consumables have been determined to be:

Material	Amount	Units
Phenol	0.71572	lb/lb product
Hydrogen	0.0351	lb/lb product
Nitric acid (100% basis)	0.71778	lb/lb product
By-product off gas	0.00417	lb/lb product
Various catalysts and chemicals	32.85	$/metric ton product
Electric power	0.0939	kWh/lb product
Cooling water	56.1	gal/lb product
HP steam	0.35	lb/lb product
MP steam	7.63	lb/lb product
Boiler feed water	0.04	gal/lb product

These yields were taken from Chem Systems PERP report 98/99-3 Adipic acid (Chem Systems, 1999). The nitric acid consumption is given on a 100% basis, but 60% nitric acid is used in the process.

Estimate the fixed capital cost, the working capital, the cash cost of production and total cost of production for a new 400,000 metric ton per year (400 kt/y) adipic acid plant located in Northeast Asia. The prices of adipic acid, phenol, hydrogen and nitric acid have been forecasted for Northeast Asia as $1400/t, $1000/t, $1100/t and $380/t respectively. Assume a 15% cost of capital and a 10-year project life.

Solution

It is convenient to summarize costs of production in a spreadsheet, as discussed in Section 6.4.9. The template from Appendix G has been used in this example and is given in Figure 6.12. In addition to entering the information from the problem statement into the spreadsheet (with any necessary conversion of units) a few additional calculations are needed, as described below.

Estimating Capital Cost

The capital cost of the process can be estimated based on historic data using the correlation given in Table 6.2. The correlation is based on the plant capacity in MMlb/y, so we need to convert the capacity: 400 kt/y is equal to 880 MMlb/y:

$$\text{ISBL capital cost} = 3.533 \, S^{0.6} = 3.533 \, (880)^{0.6} = \$206.5 \text{ MM}$$

The ISBL cost is on a US Gulf Coast basis, so we need to convert to a Northeast Asia basis. If we look up the location factor in Table 6.7, then it is not clear what factor we should use. The location factor for Japan is 1.26, while for China it varies from 0.6 to 1.1, depending on the amount of indigenous versus imported equipment used. Since the exact location of the plant has not yet been specified, we are not able to make a definitive assessment of what the location factor should be. As a first approximation we therefore assume it is 1.0 and note that this should be revisited as part of the sensitivity analysis.

The OSBL capital cost is estimated as 40% of ISBL cost. The engineering cost and contingency are estimated as 10% and 15% of the sum (ISBL + OSBL) cost respectively, giving a total fixed capital cost of $361.3 MM.

Closing Mass Balance

The first thing that is apparent when entering the yield data is that the mass balance for the process does not close properly with the information given. This suggests that we still need to account for some waste streams.

The first waste stream is apparent from the process stoichiometry. Nitric acid is recycled in the process until it is eventually converted to N_2O and vented to the atmosphere. The yield of N_2O can therefore be found by a mass balance on nitrogen:

$$\text{Nitrogen fed} = \text{nitrogen purged}$$

$$400,000 \times 0.71778 \times \frac{14}{63} = m_{N2O} \times \frac{2 \times 14}{44}$$

where m_{N2O} is the flow rate of N_2O, which can be calculated as 100,261 MT/y. As a first approximation, there is no cost for handling this stream, although we might

Company Name	Project Name	Adipic acid from phenol						
Address **COST OF PRODUCTION** **Adipic Acid from Phenol**	Project Number					Sheet	1	
	REV	DATE	BY	APVD	REV	DATE	BY	APVD
	1	1.1.07	GPT					
Form XXXXX-YY-ZZ								

Owner's Name			Capital Cost Basis Year	2006
Plant Location	Northeast Asia		Units	Metric
Case Description			On Stream	8,000 hr/yr	333.33 day/yr

YIELD ESTIMATE | **CAPITAL COSTS**

Yield information taken from ChemSystems PERP report 98/99-3, Adipic Acid, p. 89
Yields input for phenol, nitric acid, hydrogen, off-gas, utilities and consumables

Scale of production set to 400 t/y = 880 MMlb/yr

	$MM
ISBL Capital Cost	206.5
OSBL Capital Cost	82.6
Engineering Costs	28.9
Contingency	43.4
Total Fixed Capital Cost	361.3
Working Capital	59.5

REVENUES AND RAW MATERIAL COSTS

MASS BALANCE   MB closure   101%

Key Products	Units	Units/Unit product	Units/yr	Price $/unit	$MM/yr	$/unit main product
Adipic acid	t	1	400,000	1400	560.00	1400.00
Total Key Product Revenues (REV)	t	1	400,000		560.00	1400.00

By-products & Waste Streams						
Nitrous oxide (vented)	t		100,261	0	0.00	0.00
Off-gas	t	0.00417	1,670	700	1.17	2.92
Organic Waste (Fuel value)	t	0.03072	12,288	300	3.69	9.22
Aqueous Waste	t		273,440	-1.5	-0.41	-1.03
Total Byproducts and Wastes (BP)	t	0.0348939	387,659		4.44	11.11

Raw Materials						
Phenol	t	0.71572	286,288	1000	286.29	715.72
Nitric acid 60% (100% basis)	t	0.71778	287,112	380	109.10	272.76
water with nitric acid	t		191,408	0	0.00	0.00
Hydrogen, 99%	t	0.0351	14,040	1100	15.44	38.61
Total Raw Materials (RM)		1	778,848		410.83	1027.09

Gross Margin (GM = REV + BP - RM)   153.61   384.03

CONSUMABLES

	Units	Units/Unit product	Units/yr	Price $/unit	$MM/yr	$/unit product
Various catalyst and chemicals	kg	32.85	13,138,263	1.00	13.14	32.85
Other	kg	0	0	0.00	0.00	0.00
Total Consumables (CONS)					13.14	32.85

UTILITIES

	Units	Units/Unit product	Units/hr	Price $/unit	$MM/yr	$/unit product
Electric	kWh	206.0	10,300	0.05	4.120	10.30
HP Steam	t	0.4	18	14.30	2.002	5.01
MP Steam	t	7.6	382	12.00	36.624	91.56
LP Steam	t	0.0	0	8.90	0.000	0.00
Boiler Feed	t	0.3	17	1.10	0.145	0.36
Condensate	t	0.0	0	0.80	0.000	0.00
Cooling Water	t	463.0	23,150	0.024	4.445	11.11
Fuel Fired	GJ	0.0	0	6.00	0.000	0.00
Total Utilities (UTS)					47.336	118.340

Variable Cost of Production (VCOP = RM - BP + CONS + UTS)   466.86   1167.16

FIXED OPERATING COSTS

			$MM/yr	$/unit product
Labor				
	4.8 Operators per Shift Position			
Number of shift positions	9			
		30,000 $/yr each	1.30	3.24
Supervision		25% of Operating Labor	0.32	0.81
Direct Ovhd.		45% of Labor & Superv.	0.73	1.82
Maintenance		3% of ISBL Investment	10.84	27.10
Overhead Expense				
Plant Overhead		65% of Labor & Maint.	8.57	21.43
Tax & Insurance		2% of Fixed Investment	5.42	13.55
Interest on Debt Financing		0% of Fixed Capital	0.00	0.00
		6% of Working Capital	3.57	8.93
		Fixed Cost of Production (FCOP)	30.75	76.88

ANNUALIZED CAPITAL CHARGES

	$MM	Interest Rate	Life (yr)	ACCR	$MM/yr	$/unit product
Fixed Capital Investment	361.303	15%	10	0.199	71.99	179.98
Royalty Amortization	15.000	15%	10	0.199	2.99	7.47
Inventory Amortization						
Catalyst 1	0.000	15%	3	0.438	0.00	0.00
Catalyst 2	0.000	15%	3	0.438	0.00	0.00
Adsorbent 1	0.000	15%	3	0.438	0.00	0.00
Equipment 1	0.000	15%	5	0.298	0.00	0.00
Equipment 2	0.000	15%	5	0.298	0.00	0.00
			Total Annual Capital Charge		74.98	187.45

SUMMARY

	$MM/yr	$/unit product
Variable Cost of Production	466.86	1167.16
Fixed Cost of Production	30.75	76.88
Cash Cost of Production	497.61	1244.04
Gross Profit	62.39	155.96
Total Cost of Production	572.59	1431.48

Figure 6.12. Cost of production worksheet for Example 6.11.

revisit this at a more detailed design stage if we need to fit vent scrubbers or other equipment to handle this off-gas.

The second waste stream is also apparent from the overall stoichiometry. Phenol has a molecular weight of 100 and adipic acid has a molecular weight of 146, so the stoichiometric requirement of phenol is $= 100/146 = 0.68493$ lb/lb product. The actual process consumption has been estimated as 0.71572 lb/lb product, so the difference $(0.71572 - 0.68493 = 0.03079$ lb/lb) must be converted into organic by-products. It is possible that some of the organic by-product may be material that is lost with the hydrogen-rich fuel gas, but as a first approximation we can assume that we recover an organic liquid waste product from the process. It is also possible (in fact quite likely) that some of the material that we are calling organic by-product is actually losses of organics in the nitrous oxide vent stream. Since this stream probably must be scrubbed before discharge, it is fair to assume as a first approximation that any organic material in it would be collected as an organic waste. This assumption should be revisited at a later stage in the design process when better information on process yields is available. The organic waste stream is priced at a typical fuel value of $300/t, assuming that it can be burned as process fuel.

The third waste stream is an aqueous waste. This consists of the water that is brought in with the nitric acid, the water formed by the reaction stoichiometry and any other water consumed, for example in vent scrubbers or process water washes.

The water brought in with the nitric acid is easily found by mass balance, since it is equal to the mass flow rate of nitric acid (100% basis) $\times 40/60 = 400,000 \times 0.71778 \times 4/6 = 191,408$ t/y.

The water formed by reaction stoichiometry can be estimated as 1 mole per mole nitric acid consumed, i.e., 18 t per 63 t consumed, giving $400,000 \times 0.71778 \times 18/63 = 82,032$ t/y. Note that we could also have estimated this as 2 moles per mole product, but that would give an overestimate of the water production as the amount of nitric acid consumed is less than the apparent stoichiometric requirement. This is because the overall reaction given above is only an approximation and does not include the reaction of cyclohexanone.

The water consumed in process washes and scrubbers is harder to estimate, but since no process water consumption was listed under utilities, we can assume as a first approximation that all the process water needs are met by internal recycles. This gives a total waste-water flow of $191,408 + 82,032 = 273,440$ t/y. The waste-water stream is assigned a cost of $1.5/t (see Section 6.4.6)

When the values above for nitrous oxide, organic waste and aqueous waste are entered in the spreadsheet, the mass balance shows 101 t of product for every 100 t of feed. This is not perfectly closed, but is good enough at this stage in the analysis. The error is most likely in the organic or aqueous waste streams and will have little impact on the economic analysis. This should of course be revisited when better process yield data and a converged process simulation are available.

Estimating Utility Costs

The amounts of utilities consumed are easily estimated from the production rate and the information in the problem statement (with conversion to metric units).

The prices of steam at different levels can be taken from Example 6.6, since the costs of fuel and natural gas are the same.

The prices of boiler feed water, condensate and cooling water are estimated as described in Section 6.4.4.

The utility cost is about 10% of the variable cost of production. This is typical for many commodity chemical processes.

Estimating Fixed Costs

The adipic acid process is a relatively complex process and essentially contains two plants – phenol hydrogenation and KA oil oxidation. We should therefore assume at least four shift positions for each plant, say nine in total. For a Northeast Asia basis we expect that the salary cost per shift position will be lower than the typical $50,000 per year that we would assume for a US Gulf Coast plant. As a first approximation this is estimated as $30,000/y. The remaining salary and overhead costs are fixed following the assumptions given in Section 6.2.4.

Interest charges are not included for the fixed capital (since we will calculate an annualized charge based on overall cost of capital below). An interest charge is included for the working capital, as working capital is recovered at the end of the project and so should not be amortized, as discussed in Section 6.7.6.

The total fixed cost of production is calculated to be $31 MM/y, which is low, compared to the variable cost of production ($467 MM/y). It is not uncommon for fixed costs to make a relatively minor contribution to the total cost of production for a world-scale plant.

Estimating Working Capital

The working capital is estimated as 7 weeks' cash cost of production minus 2 weeks' feedstock costs plus 1% of the fixed capital investment, as described in Section 6.2.2. Because the cash cost of production includes the interest payable on the working capital, this sets up a circular reference in the spreadsheet. The spreadsheet options must be adjusted to ensure that the calculation iterates to convergence. The converged result is $59.5 MM. Note that the value calculated is about 10% greater than it would have been had we estimated the working capital as 15% of fixed capital investment.

Estimating Annualized Capital Costs

The fixed capital investment is to be annualized over 10 years at a 15% interest rate. For this interest rate and recovery period the annual capital charge ratio is 0.199, so the annual capital charge is = 0.199 × 361.3 = $71.99 MM/y, or $179.98/t of product. As a quick check, we can see that this is roughly 10% of the total cost of production, which is typical for commodity chemical processes.

In addition to the fixed capital investment, we should also make an allowance for a process royalty. The problem statement did not specify whether the plant was to be built using proprietary technology, but it is reasonable to assume that a royalty will need to be paid. If a $15 MM royalty is added then this annualizes to a cost of

$3 MM/y, or roughly 0.5% of revenues, which is a reasonable initial estimate. This should be revisited during more detailed design when discussions with technology vendors take place.

Estimating Cost of Production

The cash cost of production is the sum of the fixed and variable production costs (equation 6.2):

$$CCOP = VCOP + FCOP = 466.86 + 30.75 = \$497.61 \text{ MM/y}$$

The total cost of production is the sum of the cash cost of production and the annual capital charge (equation 6.5):

$$TCOP = CCOP + ACC = 497.61 + 74.98 = \$572.59 \text{ MM/y}$$

It is worth noting that the calculated total cost of production is greater than the projected annual revenue of $560 MM/y. This suggests that the project would not earn the expected 15% interest rate. This is explored further in the following example and in problems 6.14 and 6.15.

Example 6.12

The adipic acid plant in Example 6.11 is built with 30% of the fixed investment in year 1 and 70% in year 2, and the plant operates at 50% of capacity in year 3 before reaching full capacity in year 4. The plant can be depreciated by the straight-line method over 10 years and profits can be assumed to be taxed at 35% per year, payable the next year. Assume that losses cannot be offset against revenues from other operations for tax purposes (i.e., no tax credits in years when the plant makes a loss). Estimate the following:

1. The cash flow in each year of the project.
2. The simple pay-back period.
3. The net present value with a 15% cost of capital for 10 years and 15 years of production at full capacity.
4. The DCFROR for 15 years of production at full capacity.

Is this an attractive investment?

Solution

The solution requires calculating the cash flows in each year of the project. This is easily coded into a spreadsheet, as illustrated in Figure 6.13. A blank template of this spreadsheet is given in Appendix G and is available in MS Excel format in the on-line material at http://books.elsevier.com/companions.

Cash-Flow Table

In years 1 and 2 of the project there are capital expenses but no revenues or operating costs. The capital expenses are not operating losses and so they have no effect on taxes or depreciation. They are negative cash flows.

| Company Name | | | | | | Project Name | Adipic acid from phenol | | | | | |
| Address | | | | | | Project Number | | | | Sheet | | 1 |

						REV	DATE	BY	APVD	REV	DATE	BY	APVD
ECONOMIC ANALYSIS						1	1.1.07	GPT					
Adipic Acid from Phenol													

Form XXXXX-YY-ZZ

Owner's Name			Capital Cost Basis Year	2006	
Plant Location	Northeast Asia		Units	Metric	
Case Description			On Stream	8,000 hr/yr	333.33 day/yr

REVENUES AND PRODUCTION COSTS | **CAPITAL COSTS** | **CONSTRUCTION SCHEDULE**

	$MM/yr			$MM		Year	% FC	% WC	% FCOP	% VCOP
Main product revenue	560.0		ISBL Capital Cost	206.5		1	30%	0%	0%	0%
Byproduct revenue	4.4		OSBL Capital Cost	82.6		2	70%	0%	0%	0%
Raw materials cost	410.8		Engineering Costs	28.9		3	0%	100%	100%	50%
Utilities cost	47.3		Contingency	43.4		4	0%	0%	100%	100%
Consumables cost	13.1		Total Fixed Capital Cost	361.3		5	0%	0%	100%	100%
VCOP	466.8					6	0%	0%	100%	100%
Salary and overheads	16.4		Working Capital	59.5		7+	0%	0%	100%	100%
Maintenance	10.8									
Interest	3.6									
Royalties	3.0									
FCOP	33.8									

ECONOMIC ASSUMPTIONS

Cost of equity	25%	Debt ratio	0.5	Tax rate	35%	
Cost of debt	5%			Depreciation method	Straight-line	
Cost of capital	15.0%			Depreciation period	10	years

CASH FLOW ANALYSIS

All figures in $MM unless indicated

Project year	Cap Ex	Revenue	CCOP	Gr. Profit	Deprcn	Taxbl Inc	Tax Paid	Cash Flow	PV of CF	NPV
1	108.4	0.0	0.0	0.0	0.0	0.0	0.0	-108.4	-94.3	-94.3
2	252.9	0.0	0.0	0.0	0.0	0.0	0.0	-252.9	-191.2	-285.5
3	59.5	280.0	267.2	12.8	36.1	-23.3	0.0	-46.7	-30.7	-316.2
4	0.0	560.0	500.6	59.4	36.1	23.3	0.0	59.4	34.0	-282.2
5	0.0	560.0	500.6	59.4	36.1	23.3	8.1	51.3	25.5	-256.8
6	0.0	560.0	500.6	59.4	36.1	23.3	8.1	51.3	22.2	-234.6
7	0.0	560.0	500.6	59.4	36.1	23.3	8.1	51.3	19.3	-215.3
8	0.0	560.0	500.6	59.4	36.1	23.3	8.1	51.3	16.8	-198.6
9	0.0	560.0	500.6	59.4	36.1	23.3	8.1	51.3	14.6	-184.0
10	0.0	560.0	500.6	59.4	36.1	23.3	8.1	51.3	12.7	-171.3
11	0.0	560.0	500.6	59.4	36.1	23.3	8.1	51.3	11.0	-160.3
12	0.0	560.0	500.6	59.4	36.1	23.3	8.1	51.3	9.6	-150.7
13	0.0	560.0	500.6	59.4	0.0	59.4	8.1	51.3	8.3	-142.4
14	0.0	560.0	500.6	59.4	0.0	59.4	20.8	38.6	5.5	-136.9
15	0.0	560.0	500.6	59.4	0.0	59.4	20.8	38.6	4.7	-132.2
16	0.0	560.0	500.6	59.4	0.0	59.4	20.8	38.6	4.1	-128.1
17	0.0	560.0	500.6	59.4	0.0	59.4	20.8	38.6	3.6	-124.5
18	0.0	560.0	500.6	59.4	0.0	59.4	20.8	38.6	3.1	-121.4
19	0.0	560.0	500.6	59.4	0.0	59.4	20.8	38.6	2.7	-118.7
20	-59.5	560.0	500.6	59.4	0.0	59.4	20.8	98.1	6.0	-112.7

ECONOMIC ANALYSIS

Average cash flow	44.7 $MM/yr	NPV	10 years	-171.3 $MM	IRR	10 years	-2.0%
Simple pay-back period	9.4 yrs		15 years	-132.2 $MM		15 years	5.6%
Return on investment (10 yrs)	3.32%		20 years	-112.7 $MM		20 years	8.4%
Return on investment (15 yrs)	5.77%	NPV to yr	19	-118.7 $MM			

Figure 6.13. Economic analysis worksheet for Example 6.12.

In year 3 the plant operates at 50% capacity and generates 50% of the design basis revenue. All of the working capital must be invested. The plant incurs 100% of the fixed cost of production but only 50% of the variable cost. Because the plant makes a profit, depreciation can be charged. Using the straight-line method of depreciation with a 10-year recovery period, the annual depreciation charge is one-tenth of the total fixed capital investment = 361.3/10 = $36.1 MM. Since the gross profit in year 3 is only $12.8 MM, the effect of charging depreciation is that the taxable income is negative and so no taxes are owed in year 4 (taxes are paid based on the previous year's income).

In year 4 the plant operates at full capacity and generates 100% of the design basis revenues with 100% of the VCOP. From here onwards the plant makes a gross profit of $59.4 MM each year.

Depreciation is charged for 10 years, i.e., until year 12. The taxable income therefore increases in year 13 and the taxes paid increase in year 14, giving a reduction in cash flow from $51.3 MM to $38.6 MM.

In the final year of the project, the working capital is released and should be taken as a positive increment to the cash flow. This is shown as occurring in year 20 in Figure 6.13, but should be adjusted when the length of the project is varied, as described below.

The present value of the cash flow in year n can be found by multiplying by $(1 + i)^{-n}$, as described in equation 6.40. The net present value up to year n is the cumulative sum of all the present values of cash flow up to that year.

Simple Pay-Back Period

The simple pay-back is calculated from the fixed investment and the average annual cash flow (equation 6.36). The average annual cash flow should be based only on the years in which the plant generates revenue, i.e., years 3 to 20, and is found to be $44.7 MM/y. Note that it does not matter if this range includes the year in which working capital is invested, as long as it also includes the year in which working capital is recovered. The working capital thereby cancels out and is not included in the average cash flow.

The simple pay-back period is then found from:

$$\text{simple pay-back time} = \frac{\text{total investment}}{\text{average annual cash flow}} = \frac{361.3}{44.7} = 8.08 \text{ years}$$

$$(6.36)$$

Net Present Value

The net present value with a 15% cost of capital after 10 years of production is the NPV at the end of year 13. This can be looked up in the cash-flow table and is $-142.4 MM. If the plant is closed after 10 years of production and the working capital is released, then there would be an additional cash flow of $59.1 MM in year 13, increasing the NPV to $-132.7 MM.

The net present value after 15 years of production is the NPV at the end of year 18, which can also be found from the cash-flow table and is $-121.4 MM. If the plant is closed after 15 years of production and the working capital is released, then there would be an additional cash flow of $59.1 MM in year 18, increasing the NPV to $-116.6 MM.

In all cases the NPV for this project is negative, so it is not an attractive investment with a 15% cost of capital. We already knew this would be the case based on the cost of production analysis in Example 6.11, which had shown that the TCOP with capital recovered at a 15% interest rate was greater than the expected revenue.

Internal Rate of Return (DCFROR)

The DCFROR (IRR) of the project after 15 years of production at full capacity can be found by either adjusting the interest rate (manually or using the goal seek function) until the NPV at the end of year 18 is equal to zero, or by using the IRR function in the spreadsheet over the range year 1 to year 18. The working capital should be included as a recovered cost in year 18.

The answer obtained in either case is DCFROR = 7.85%. This is the maximum interest rate at which this project can be financed to break even in 15 years of production.

Summary

None of the economic measures indicates that this is an attractive project with the projected costs, revenues and capital expenses. It should perhaps be noted though, that this analysis was based on a class 5 estimate of the capital cost ($\pm$ 50%). If we had any technical improvement in mind that could reduce either the capital investment or the cost of production, then we might want to develop the design further to assess if the economic analysis was sufficiently improved.

Example 6.13

A plant is producing 10,000 metric tons per year (10 kMTA) of a product. The overall yield is 70%, on a mass basis (kg of product per kg raw material). The raw material costs $500/metric ton, and the product sells for $900/metric ton. A process modification has been devised that will increase the yield to 75%. The additional investment required is $1,250,000, and the additional operating costs are negligible. Is the modification worth making?

Solution

There are two ways of looking at the earnings to be gained from the modification:

1. If the additional production given by the yield increase can be sold at the current price, the earnings on each additional ton of production will equal the sales price less the raw material cost.
2. If the additional production cannot be readily sold, the modification results in a reduction in raw material requirements, rather than increased sales, and the earnings (savings) are from the reduction in annual raw material costs.

The second way gives the lowest figures and is the safest basis for making the evaluation. At 10 kMTA production:

$$\text{Raw material requirements at 70\% yield} = \frac{10,000}{0.7} = 14,286$$

at 75% yield

$$= \frac{10,000}{0.75} = 13,333$$

Cost savings $= 953$ metric tons/y, which is worth $953 \times 500 = \$476,500/\text{y}$

$$\text{Pre} - \text{tax ROI} = \frac{476,500}{1,250,000} = 38\%$$

As the annual savings are constant, the simple pay-back period is the inverse of the pre-tax ROI:

$$\text{Simple pay} - \text{back period} = \frac{1,250,000}{476,500} = 2.62 \text{ years}$$

Based on the attractive ROI and pay-back period, this investment would seem to be worth pursuing further. Whether or not it was implemented would depend on the hurdle rate set for investments by the company.

6.8. SENSITIVITY ANALYSIS

6.8.1. Simple Sensitivity Analysis

The economic analysis of a project can only be based on the best estimates that can be made of the investment required and the cash flows. The actual cash flows achieved in any year will be affected by changes in raw materials costs and other operating costs, and will be very dependent on the sales volume and price. A sensitivity analysis is a way of examining the effects of uncertainties in the forecasts on the viability of a project. To carry out the analysis, the investment and cash flows are first calculated using what are considered the most probable values for the various factors; this establishes the base case for analysis. Various parameters in the cost model are then adjusted, assuming a range of error for each factor in turn. This will show how sensitive the cash flows and economic criteria are to errors in the forecast figures. A sensitivity analysis gives some idea of the degree of risk involved in making judgments on the forecast performance of the project.

The results of a sensitivity analysis are usually presented as plots of an economic criterion such as NPV or DCFROR versus the parameter studied. Several plots are sometimes shown on the same graph using a scale from $0.5 \times$ base value to $2 \times$ base value as the abscissa.

6.8.2. Parameters to Study

The purpose of sensitivity analysis is to identify those parameters that have a significant impact on project viability over the expected range of variation of the parameter. Typical parameters investigated and the range of variation that is usually assumed are given in Table 6.12.

Varying the production rate (while keeping investment and fixed costs constant) investigates the effects of unexpectedly high down time due to maintenance or operations problems, as well as unexpected difficulties in selling the full volume of product that could be produced. An increase in production rate beyond the design capacity might also be possible if the plant design margins allow some extra capacity or if the yields can be improved by use of a better catalyst, etc.

The choice of which feed and product prices to use in the sensitivity analysis depends on the method of price forecasting that has been used. Typically, total raw material cost is studied rather than treating each feed separately, but if raw material costs are found to be the dominant factor then they may be broken out into the costs of individual raw materials.

6.8.3. Statistical Methods for Risk Analysis

In a simple sensitivity analysis, each parameter is varied individually and the output is a qualitative understanding of which parameters have the most impact on project

Table 6.12.   Sensitivity Analysis Parameters

Parameter	Range of variation
Sales price	$\pm$ 20% of base (larger for cyclic commodities)
Production rate	$\pm$ 20% of base
Feed cost	$-$ 10% to $+$ 30% of base
Fuel cost	$-$ 50% to $+$ 100% of base
Fixed costs	$-$ 20% to $+$ 100% of base
ISBL capital investment	$-$ 20% to $+$ 50% of base
OSBL capital investment	$-$ 20% to $+$ 50% of base
Construction time	$-$ 6 months to $+$ 2 years
Interest rate	base to base $+$ 2 percentage points

viability. In a more formal risk analysis, statistical methods are used to examine the effect of variation in all of the parameters simultaneously and hence quantitatively determine the range of variability in the economic criteria. This allows the design engineer to estimate the degree of confidence with which the chosen economic criterion can be said to exceed a given threshold.

A simple method of statistical analysis was proposed by Piekarski (1984) and is described in Humphreys (2005). Each item in the estimate is expressed as a most likely value, ML, an upper value, H, and a lower value, L. The upper and lower values can be estimated using the ranges of variation given in Table 6.12. The mean and standard deviation are then estimated as:

$$\text{mean value,} \quad \bar{x} = \frac{(H + 2ML + L)}{4} \tag{6.51}$$

$$\text{standard deviation,} \quad S_x = \frac{(H - L)}{2.65} \tag{6.52}$$

Note that the mean is not necessarily equal to the most likely value if the distribution is skewed. This is often the case for cost functions.

The mean and standard deviation of other parameters can then be estimated by combination of the individual means and standard deviations using the mathematics of statistics given in Table 6.13.

This allows relatively easy estimation of the overall error in a completed cost estimate, and with a little more difficulty can be extended to economic criteria such as NPV, TAC or ROI.

Rather than build the simple method above into a spreadsheet, a more sophisticated approach is to take the economic model and subject it to analysis using Monte Carlo simulation. In Monte Carlo simulation, random numbers are generated and used to establish the value of each parameter within its allowed range. For example, each parameter could be set equal to $L + (R \times (H - L)/10)$, where R is a random number between 0 and 10. The overall probability distribution in the calculated parameter (economic criterion) can be estimated by performing a large number of such simulations. Several commercial programs for Monte Carlo simulation are

Table 6.13. Mathematics of Statistics
If: $y = f(\bar{x}, \bar{z})$, then the standard deviation of y, S_y is given as a function of S_x and S_z

Function y of $\bar{x}$, $\bar{z}$	Standard deviation S_y
$y = a\bar{x} + b\bar{z}$	$S_y = \sqrt{a^2 S_x^2 + b^2 S_z^2}$
$y = \bar{x}\,\bar{z}$	$S_y = \bar{x}\,\bar{z}\sqrt{\dfrac{S_x^2}{\bar{x}^2} + \dfrac{S_z^2}{\bar{z}^2}}$
$y = \dfrac{\bar{x}}{\bar{z}}$	$S_y = \dfrac{\bar{x}}{\bar{z}}\sqrt{\dfrac{S_x^2}{\bar{x}^2} + \dfrac{S_z^2}{\bar{z}^2}}$

Notes:

1. These formulae are strictly true only when the covariance of x and z is zero, i.e., there is no statistical interrelation between x and z, and when x and z have been estimated from a small set of data points.
2. For a more general description of the formulae, see Ku (1966).

available, for example REP/PC (Decision Sciences Corp.), @RISK (Palisade Corp.) and CRYSTAL BALL® (Decisioneering® Corp.).

Care must be taken in formulating Monte Carlo simulation problems. The Monte Carlo method implicitly assumes that all parameters vary randomly and independently. If two parameters are correlated (for example feedstock and product prices or feedstock and energy prices) then they should not be varied independently. The correct approach is to vary one of the parameters and then predict the other by correlation, imposing a random error on the predicted parameter to reflect the accuracy of the correlation.

The cost-estimating literature contains a lot of information on risk analysis. Good introductions to the use of statistics in risk analysis are given by Humphreys (2005) and Sweeting (1997).

6.8.4. Contingency Costs

The concept of a contingency charge to allow for variation in the capital cost estimate was introduced in Section 6.2.1, where it was suggested that a minimum contingency charge of 10% of ISBL plus OSBL fixed capital should be used.

If the confidence interval of the estimate is known, then the contingency charges can also be estimated based on the desired level of certainty that the project will not over-run the projected cost. For example, if the cost estimate is normally distributed then the estimator has the following confidence levels:

- 90% confidence that the cost is less than $\bar{x} + 1.3S_x$.
- 95% confidence that the cost is less than $\bar{x} + 1.65S_x$.
- 98% confidence that the cost is less than $\bar{x} + 2.05S_x$.
- 99% confidence that the cost is less than $\bar{x} + 2.33S_x$.

Although many of the components of a cost estimate are skewed distributions, when these are combined the resulting distribution is often approximately normal. The above guidelines can thus be used to determine the amount of contingency charge needed for a given level of confidence.

Note also that a 10% contingency charge gives 98% confidence of the cost coming in under estimate if the estimate has accuracy $\pm$ 6.5% (using the approximate method of calculating S_x given in equation 6.52). This illustrates that a 10% contingency charge should really be viewed as a minimum level and is only appropriate for detailed estimates (Class 1 and Class 2), when the technology is well understood.

Example 6.14

A preliminary (Class 4) estimate of the ISBL capital cost of building a 200,000 ton per year ethanol plant by corn dry milling has been stated as $130 MM -30% / $+50\%$. The plant is to be built on a green-field site and off-site costs are estimated to be between $40 MM and $60 MM. Estimate a value for the total project cost that will give 98% confidence that the project can be carried out within the amount estimated.

Solution

For the ISBL cost, $H = \$195$ MM, $L = \$91$ MM and $ML = \$130$ MM, so:

$$\overline{x}_{ISBL} = \frac{(H + 2ML + L)}{4} = \frac{(195 + 260 + 91)}{4} = \$136.5 \text{ MM} \quad (6.51)$$

$$S_{x,ISBL} = \frac{(H - L)}{2.65} = \frac{195 - 91}{2.65} = \$39.2 \text{ MM} \quad (6.52)$$

Similarly, for the OSBL costs, assuming the most likely value is in the middle of the range given:

$$\overline{x}_{OSBL} = \frac{(H + 2ML + L)}{4} = \frac{(40 + 100 + 60)}{4} = \$50 \text{ MM} \quad (6.51)$$

$$S_{x,OSBL} = \frac{(H - L)}{2.65} = \frac{60 - 40}{2.65} = \$7.55 \text{ MM} \quad (6.52)$$

Both mean values should be increased by 10% to allow for engineering costs, so combining the means gives:

$$\overline{x}_{Total} = 1.1\,\overline{x}_{ISBL} + 1.1\,\overline{x}_{OSBL} = \$205.2 \text{ MM}$$

$$S_{x,Total} = \sqrt{(1.1\,S_{x,ISBL})^2 + (1.1\,S_{x,OSBL})^2} = \$43.9 \text{ MM}$$

We have 98% confidence that the cost is less than $\overline{x} + 2.05 S_x = 205.2 + (2.05 \times 43.9) = \$295$ MM. If we budget the project for this amount (or tender a contract) then we are accepting a 1 in 50 risk that the project will exceed the given budget.

6.9. PROJECT PORTFOLIO SELECTION

A typical company involved in the chemical, pharmaceutical or fuels industries will evaluate many projects each year. Only a few of these projects are selected for implementation. This section discusses some of the criteria and methods used in making that selection.

6.9.1. Types of Projects

Investment projects are carried out for a variety of reasons.

Regulatory compliance projects are often required as a result of changes in environmental or other legislation. If the government changes the rules on plant safety, emissions or product specifications, then unless an exemption can be obtained the plant must be modified or closed down. Regulatory compliance projects often have poor financial performance unless the costs of going out of business are considered.

Cost-reduction projects are aimed at reducing the cost of production of an existing plant. The most common cost-reduction investments are for *preventive maintenance*, in which equipment is replaced, repaired or cleaned after a planned interval and before the equipment deteriorates to the point where it could impact process performance or safety. Most preventive maintenance projects are small and are handled through the plant maintenance budget, but some can be very large expensive projects requiring a major plant shutdown, for example, replacing the fired tubes in a main plant furnace. Another common type of cost-reduction project is a *Heat recovery* or *Heat-integration project*, in which the plant heat-exchange network or utility system is upgraded to reduce energy costs.

Whenever possible, companies also seek to fund *Growth projects* that can be expected to give high returns on the capital invested. Growth projects include expansions of existing units, often referred to as *Debottlenecking* or *Revamp projects* as well as construction of entirely new plants in *Grassroots projects*.

In all cases except grassroots projects, a large amount of information about the existing plant, site and products is usually needed before the project can be designed. Much effort is usually spent on reconciling simulation or other models to the plant performance so as to be useful for designing the plant modifications. Grassroots projects are typically used as undergraduate design projects because they are self-contained and do not require model reconciliation; however, in industrial practice they make up less than 10% of all projects.

6.9.2. Limits on the Project Portfolio

The most obvious limit on the portfolio of projects that can be funded is the availability of capital, which is in turn limited by the financing arrangements of the company (see Section 6.6).

Capital spending is often set in proportion to sales, operating profit or total assets. Table 6.14 shows recent information on capital spending for some of the largest chemical and pharmaceutical companies in the world. It can be seen from Table 6.14 that most of the companies' capital spending was between 4% and 8% of sales and

Table 6.14.   Capital Spending of Large Chemical and Pharmaceutical Companies

Company	Sales (MM$)	Net Profit (MM$)	Total Assets (MM$)	Capital Spending (MM$)	Capital/ Sales	Capital/ Net Profit	Capital/ Assets
BASF	79,457	5574	64,170	3516	0.044	0.631	0.055
Johnson & Johnson	61,095	10,576	52,191	2942	0.048	0.278	0.056
Dow	53,513	2887	44,448	2075	0.039	0.719	0.047
Pfizer	48,418	8213	73,388	1880	0.039	0.229	0.026
GlaxoSmithKline	45,477	11,153	62,068	3035	0.067	0.272	0.049
Bayer	44,403	6459	70,444	2593	0.058	0.401	0.037
Novartis	39,800	11,968	75,452	2549	0.064	0.213	0.034
Sanofi-Aventis	38,462	7216	98,601	2207	0.057	0.306	0.022
Roche	38,447	9532	65,158	3040	0.079	0.319	0.047
AstraZeneca	29,559	5627	47,957	1130	0.038	0.201	0.024
DuPont	29,378	2988	29,201	1585	0.054	0.530	0.054
Abbott Laboratories	25,914	3606	23,864	1656	0.064	0.459	0.069
Mitsubishi Chemical	24,880	1393	23,487	1444	0.058	1.037	0.061
Merck & Co.	24,197	3275	46,183	1011	0.042	0.309	0.022
Wyeth	22,400	4616	38,198	1391	0.062	0.301	0.036
Bristol-Myers Squibb	19,348	1968	19,844	843	0.044	0.428	0.042
Eli Lilly & Co	18,634	2953	24,333	1082	0.058	0.366	0.044
Linde	16,873	1305	34,216	1419	0.084	1.087	0.041
Air Liquide	16,317	1540	18,454	1863	0.114	1.210	0.101
Sumitomo Chemical	16,105	536	20,032	1210	0.075	2.257	0.060
Mitsui Chemical	15,172	211	12,477	719	0.047	3.408	0.058
Asahi Kasei	14,409	594	12,104	704	0.049	1.185	0.058
Akzo Nobel	14,009	795	26,384	492	0.035	0.619	0.019
Solvay	13,124	1071	15,329	1065	0.081	0.994	0.069
DSM	12,232	588	13,475	779	0.064	1.325	0.058
PPG Industries	11,206	815	10,541	2426	0.216	2.977	0.230
Air Products	10,038	1043	11,154	1055	0.105	1.012	0.095

Notes:

1. Source: Storck *et al.* (2008).
2. Numbers are based on 2007 financial data.

also between 4% and 6% of assets. On average, capital spending was 6.6% of sales and 5.6% of assets.

A second important constraint on the number of projects that can be carried out is the availability of critical resources. Companies with small engineering staffs will only be able to carry out a few projects at one time. Even if extensive use is made of EPC contractors, the owners will still need to provide some engineering support to each project. The availability of EPC contractors can also be an issue during times of peak industry construction. Projects that require extensive research and development work may be delayed because of constraints on the availability of researchers and pilot plant facilities.

Often the most important constraint is set by regulatory timelines. Regulatory compliance projects must be completed in time for the plant or product to comply with the new law. This may dictate a narrow window of typically less than 5 years, in which the project must be planned, designed and constructed, giving the company little choice on when the project must be begun.

Regulatory timelines are extremely important for pharmaceutical products. A new drug is protected by patent for 20 years from the date the patent is filed. Beyond that time, competitors are able to sell generic versions of the drug and the price usually falls significantly. Before a new drug can be marketed, both the product and the manufacturing process must be approved by regulatory agencies such as the U.S. Food and Drug Administration (FDA). Pharmaceutical manufacturers thus seek to maximize the revenue that they can obtain from a drug between FDA approval and patent expiration. This requires making advance preparations during the approvals process so that the rate of production can be ramped up quickly when final approval is obtained. The portfolio of investment projects for a pharmaceutical company will be strongly influenced by the expected outcomes of the regulatory approval process for new products.

6.9.3. Decision Criteria

Different types of projects are often judged using different economic criteria.

At the plant or site scale, management may have been given a small discretionary capital budget that can be used for preventive maintenance and cost reduction projects (if it is not swallowed up by regulatory compliance projects). These projects are often ranked using simple measures such as pay-back, ROI or total annualized cost. For a project to be considered for funding, it must meet a minimum (or maximum) criterion, known as a 'hurdle rate'. For example, a company may dictate that projects should not be funded unless the pay-back period is less than 2 years. Regulatory compliance projects are often evaluated based on minimum incremental total annual cost, since it is implicitly assumed that there will be no additional revenue. If there is additional revenue, for example, from sale of a by-product, then this can be off-set against the costs. If the cost of compliance is excessive, then the alternative costs of closing down or selling the site will also be evaluated.

Small projects or modifications to ongoing projects are often evaluated based on an 'incremental ROI' defined as:

$$\text{Incremental ROI} = \frac{\text{incremental profit}}{\text{incremental investment}} \times 100\% \qquad (6.53)$$

A separate hurdle rate is set for incremental ROI to ensure that modifications to a large project pay out in their own right and do not get funded just because of the attractiveness (or size) of the base project. This helps to prevent creep of project expenses.

Major growth and expansion projects that require significant investment are usually evaluated at the corporate level. Most companies look at the internal rate of return (IRR or DCFROR), the fixed and working capital, and the NPV with the interest rate set equal to the cost of capital. The selection of projects is constrained by the factors described in Section 6.9.2. The set of projects chosen may also be strongly influenced by strategic factors such as the desire to expand a particular business or product line, or a desire to expand the presence of the company in a region that is experiencing rapid economic growth, such as India or China.

Two means of simplifying the selection problem are usually used so that the company senior management is not faced with a list of thousands of potential projects. The first is to set internal hurdle rates based on simple measures such as IRR or pay-back so that unattractive projects are weeded out at an early stage of the evaluation process. The second method is to divide the available capital budget into categories (sometimes referred to as 'buckets') so as to balance the competing needs of different regions and businesses, growth areas versus established products, etc. The various strategic business units or regional subsidiaries (depending on how the company is organized) each submit their proposed capital budgets and a ranked list of projects. Corporate senior management then makes strategic adjustments between the different categories, and determines where to draw the line in each list such that the overall portfolio is balanced in accordance with the strategic objectives that they have set for the company. In a large corporation, this process may be repeated at two or more levels of management, with the list of selected projects being passed up to a higher level for further review and approval before the capital is authorized.

The problem of portfolio selection is easily expressed numerically as a constrained optimization: maximize economic criterion subject to constraint on available capital. This is a form of the 'knapsack problem', which can be formulated as a mixed-integer linear program (MILP), as long as the project sizes are fixed. (If not, then it becomes a mixed-integer non-linear program). In practice, numerical methods are very rarely used for portfolio selection, as many of the strategic factors considered are difficult to quantify and relate to the economic objective function.

6.10. REFERENCES

Amos, S. J. (2007) *Skills and Knowledge of Cost Engineering*, 5th edn. (AACE International). revised.

Aspen Richardosn (2003) *International Construction Cost Factor Location Manual* (Aspen Technology Inc.).

Aspentech (2002a) *Aspen ICARUS Process Evaluator User's Guide* (Aspen Technology Inc.)

Aspentech (2002b) *ICARUS Reference: ICARUS Evaluation Engine (IEE) 30.0* (Aspen Technology Inc.).

Bechtel, L. B. (1960) *Chem. Eng., NY* **67**, (Feb. 22nd) 127. Estimate working capital needs.

Bridegewater, A. V. and Mumford, C. J. (1979) *Waste Recycling and Pollution Control Handbook,* Ch. 20 (George Godwin).

Bridegewater, A. V. (1979) *Chem. Eng.,* **86**(24), 119. International construction cost location factors.

Chem Systems (1999) *Adipic Acid: PERP Report 98/99-3* (Chem Systems Inc.).

Cran, J. (1976a) *Engineering and Process Economics*, 1, *109–112.* EPE Plant cost indices international *(1970=100).*

Cran, J. (1976b) *Engineering and Process Economics*, 1, *321–323.* EPE Plant cost indices international

Dincer, I. (2003) *Refrigeration Systems and Applications* (Wiley).

Douglas, J. M. (1988) *Conceptual Design of Chemical Processes* (McGraw-Hill).

Dysert, L. R. (2007) *Chapter 9 Estimating* in *Skills and Knowledge of Cost Engineering*, 5th edn revised (AACE International).

Estrup, C. (1972) *Brit. Chem. Eng., Proc. Tech.* **17**, 213. The history of the six-tenths rule in capital cost estimation.

Garrett, D. E. (1989) *Chemical Engineering Economics* (Van Norstrand Reinhold).

Gerrard, A. M. (2000) *Guide to Capital Cost Estimation*, 4th edn. (Institution of Chemical Engineers).

Guthrie, K. M. (1969) *Chem. Eng.* **76**(6), 114. Capital cost estimating.

Guthrie, K. M. (1974) *Process Plant Estimating, Evaluation, and Control* (Craftsman Book Co.).

Hand, W. E. (1958) *Petrol. Refiner* **37**(9), 331. From flow sheet to cost estimate.

Happle, J. and Jordan, D. G. (1975) *Chemical Process Economics*, 2nd edn. (Marcel Dekker).

Humphreys, K. K. (1991) *Jelen's Cost and Optimization Engineering*, 3rd edn. (McGraw-Hill).

Humphreys, K. K. (2005) *Project and Cost Engineers' Handbook*, 4th edn. (AACE International).

Hydrocarbon Processing (2003) *Petrochemical Processes 2003* (Gulf Publishing Co).

Hydrocarbon processing. (2004a) *Gas Processes 2004*. Gulf Publishing Co.

Hydrocarbon processing. (2004b) *Refining Processes 2004*. Gulf Publishing Co.

Ku, H. (1966) *J. Res. Nat. Bur. Standards-C. Eng. And Instr.* **70**C(4), 263–273. Notes on the use of propagation of error formulas.

Lang, H. J. (1948) *Chem. Eng.*, **55**(6), 112. Simplified approach to preliminary cost estimates.

Lyda, T. B. (1972) *Chem. Eng.* **79**, 182. How much working capital will the new project need? (Sept. 18th).

Page, J. S. (1996) *Conceptual Cost Estimating Manual*, 2nd edn. (Gulf).

Peters, M. S., Timmerhaus, K. D., and West, R. E. (2003) *Plant Design and Economics*, 5th edn. (McGraw-Hill).

Piekarski, J. A. (1984) *AACE Transactions* **1984**, D5. Simplified risk analysis in project economics.

Pikulik, A. and Diaz, H. E. (1977) *Chem. Eng., NY* **84**, 106. Cost estimating for major process equipment. Oct. 10th.

Purohit, G. P. (1983) *Chem. Eng., NY* **90**, 56. Estimating the cost of heat exchangers (Aug. 22nd).

Scott, R. (1978) *Eng. and Proc. Econ.* **3**, 105. Working capital and its estimation for project evaluation.

SHELL *People and Connections – Global Scenarios to 2020. www.Shell.com.*

SHELL *Shell Global Scenarios to 2025. www.Shell.com.*

Shim, J. K. and Henteleff, N. (1995) *What Every Engineer Should Know About Accounting and Finance* (Marcel Dekker).

Smith, R. (2005) *Chemical Process Design and Integration* (Wiley).

Soloman, G. (1990) *The Cost Engineer* **28**(2). Location factors.

Spiro, H. T. (1996) *Finance for the Non-financial Manager*, 4th edn. (Wiley).

Stoecker, W. F. (1998) *Industrial Refrigeration Handbook* (McGraw-Hill).

Storck, W. J., Voith, M., McCoy, M., Reisch, M. S., Tullo, A. H., Short, P. L., and Tremblay, J.-F. (2008) *Chem. & Eng. News* **86**(27), 35. Facts and figures of the chemical industry.

Sweeting, J. (1997) *Project Cost Estimating – Principles and Practice* (Institution of Chemical Engineers).

Trott, A. R. and Welch, T. C. (1999) *Refrigeration and Air Conditioning* (Butterworth-Heinemann).

Turton, R., Bailie, R. C., Whiting, W. B., and Shaeiwitz, J. A. (2003) *Analysis, Synthesis and Design of Chemical Processes*, 2nd edn. (Prentice Hall).

Ulrich, G. D. and Vasudevan, P. T. (2004) *Chemical Engineering Process Design and Economics: A Practical Guide*, 2nd edn. (Process Publishing).

Woodward, C. P. and Chen, M. T. (2007) *Appendix C: Estimating Reference Material in Skills and Knowledge of Cost Engineering*, 5th edn. (AACE International). revised.

AMERICAN LAWS AND STANDARDS

IRS Publication (2006) *Corporations (United States Department of the Treasury Internal Revenue Service)* 542.

IRS Publication (2006) *How to depreciate property (United States Department of the Treasury Internal Revenue Service)* 946.

6.11.   NOMENCLATURE

		Dimensions in $MLT\theta$
A	Annual amount invested in equations 6.43 to 6.47	$
ACC	Annual capital charge	$
ACCR	Annual capital charge ratio	—
a	Constant in equation 6.7 or equation 6.15	$
B_m	Book value in after m years of depreciation	$
b	Constant in equation 6.15	$
C	Capital cost	$
C_d	Depreciable value	$
C_e	Purchased equipment cost	$
$C_{e,i,A}$	Purchased cost of equipment i in alloy	$
$C_{e,i,CS}$	Purchased cost of equipment i in carbon steel	$
C_1	Capital cost of plant with capacity S_1	$
C_2	Capital cost of plant with capacity S_2	$
CCOP	Cash cost of production	M^{-1} or T^{-1}
CF	Cash flow	$
CF_n	Cash flow in year n	$
COP	Coefficient of performance of a refrigeration cycle	—
D	Sum of tax allowances, depreciation	$
D_c	Diameter of distillation column	L
D_i	Depreciation charge in year i	$
DR	Debt ratio (leverage)	—
dH_b	Boiler heating rate	$L^{-2}T^2$
F	Installation (Lang) factor	—
FCOP	Fixed cost of production	M^{-1} or T^{-1}
F_d	Fraction of book value depreciated each year in declining balance method	—

f_c	Installation factor for civil engineering work	—
f_{el}	Installation factor for electrical work	—
f_{er}	Installation factor for equipment erection	—
f_i	Installation factor for instrumentation and control	—
f_l	Installation factor for lagging, insulation and paint	—
f_m	Materials factor	—
f_p	Installation factor for piping	—
f_s	Installation factor for structures and buildings	—
H	High value of range (equation 6.51)	$
I	Taxable income	$
i	Interest rate	—
i'	Discounted cash flow rate of return (internal rate of return)	—
i_c	Cost of capital	—
i_d	Interest rate due on debt	—
i_e	Cost of equity	—
L	Low value of range (equation 6.51)	$
L_c	Vessel length	L
LF_A	Location factor for location A relative to US Gulf Coast basis	—
M	Total number of pieces of equipment	—
ML	Most likely value of range (equation 6.51)	$
m	Number of years	T
N	Number of significant processing steps (functional units)	—
NPV	Net present value	S
n	Capital cost exponent in equations 6.6 and 6.15	—
n	Number of years	T
P	Gross profit, or principle invested in equations 6.43 to 6.47	$
P_{BFW}	Price of boiler feed water	$M^{-1}
P_F	Price of fuel	$M^{-1}L^{-2}T^2
P_{HPS}	Price of HP steam	$M^{-1}
P_{WFV}	Value of waste as fuel	$M^{-1}
Q	Plant capacity	MT^{-1}
ROA	Return on assets	—
ROE	Return on equity	—
ROI	Return on investment	—
S	Plant or equipment capacity	*
S	Matured sum in equations 6.43 to 6.47	$
S_x	Standard deviation	$
S_1	Capacity of plant 1	*
S_2	Capacity of plant 2	*
s	Reactor conversion	—
T_1	Evaporator absolute temperature	θ
T_2	Condenser absolute temperature	θ

TAC	Total annualized cost	$
TCOP	Total cost of production	$M^{-1} or $T^{-1}
t	Time, project life in years	T
t_r	Tax rate	—
t_w	Vessel wall thickness	L
VCOP	Variable cost of production	$M^{-1} or $T^{-1}
$\bar{x}$	Mean value	$
$\Delta H^{\circ}{}_C$	Heat of combustion	$L^{-2}T^2$
η_B	Boiler efficiency	—
ρ	Metal density	ML^{-3}

Asterisk (*) indicates that the dimensions are dependent on the type of equipment or process.

6.12. PROBLEMS

6.1. Estimate the capital cost of a plant that produces 80,000 metric tons per year of caprolactam.

6.2. The process used in the manufacture of aniline from nitrobenzene is described in Appendix F, design problem F.8. The process involves six significant stages:
Vaporization of the nitrobenzene
Hydrogenation of the nitrobenzene
Separation of the reactor products by condensation
Recovery of crude aniline by distillation
Purification of the crude nitrobenzene
Recovery of aniline from waste water streams.
Estimate the capital cost of a plant to produce 20,000 metric tons per year.

6.3. A reactor vessel cost $365,000 in June 1998; estimate the cost in January 2010.

6.4. The cost of a distillation column was €225,000 in early 1998; estimate the cost in January 2011.

6.5. Using the data on equipment costs given in this chapter or commercial cost estimating software, estimate the cost of the following equipment:
1. A shell and tube heat exchanger, heat-transfer area 50 m^2, floating head type, carbon steel shell, stainless steel tubes, operating pressure 25 bar.
2. A kettle reboiler: heat-transfer area 25 m^2, carbon steel shell and tubes, operating pressure 10 bar.
3. A horizontal, cylindrical, storage tank, 3 m in diameter, 12 m long, used for liquid chlorine at 10 bar, material carbon steel.
4. A plate column: diameter 2 m, height 25 m, stainless clad vessel, 20 stainless steel sieve plates, operating pressure 5 bar.

6.6. Compare the cost of the following types of heat exchangers, each with a heat-transfer area of 10 m^2. Take the construction material as carbon steel.

1. Shell and tube, fixed head.
2. Double-pipe.

6.7. Estimate the cost of the following items of equipment:
1. A packaged boiler to produce 20,000 kg/h of steam at 40 bar.
2. A centrifugal compressor, driver power 75 kW.
3. A plate and frame filter press, filtration area 10 m^2.
4. A floating roof storage tank, capacity 50,000 m^3.
5. A cone roof storage tank, capacity 35,000 m^3.

6.8. A storage tank is purged continuously with a stream of nitrogen. The purge stream leaving the tank is saturated with the product stored in the tank. A major part of the product lost in the purge could be recovered by installing a scrubbing tower to absorb the product in a solvent. The solution from the tower could be fed to a stage in the production process, and the product and solvent recovered without significant additional cost. A preliminary design of the purge recovery system has been made. It would consist of:
1. A small tower 0.5 m in diameter, 4.0 m high, packed with 25-mm ceramic saddles, packed height 3.0 m.
2. A small storage tank for the solution, 5 m^3 capacity.
3. The necessary pipe work, pump, and instrumentation.
All the equipment can be constructed from carbon steel.

Using the following data, evaluate whether it would be economical to install the recovery system:
1. Cost of product $5 per lb.
2. Cost of solvent $0.5 per lb.
3. Additional solvent make-up 10 kg/day.
4. Current loss of product 0.7 kg/h.
5. Anticipated recovery of product 80%.
6. Additional utility costs, negligible.
Other operating costs will be insignificant.

6.9. Make a rough estimate of the cost of steam per metric ton, produced from a packaged boiler. At 15 bar, 10,000 kg per hour of steam are required. Natural gas will be used as the fuel, calorific value 39 MJ/m^3 (roughly 1 MMBtu/1000 scf). Take the boiler efficiency as 80%. No condensate will be returned to the boiler.

6.10. The production of methyl ethyl ketone (MEK) is described in Appendix F, problem F.3. A preliminary design has been made for a plant to produce 10,000 metric tons per year. The major equipment items required are listed below. The plant operating rate will be 8000 h per year.
Estimate the capital required for this project, and the cash cost of production.
The plant will be built on an existing site with adequate infrastructure to provide the ancillary requirements of the new plant (no off-site investment is needed).

Major Equipment Items

1. Butanol vaporizer: shell and tube heat exchanger, kettle type, heat-transfer area 15 m², design pressure 5 bar, material carbon steel.
2. Reactor feed heaters (two): shell and tube, fixed head, heat-transfer area 25 m², design pressure 5 bar, material stainless steel.
3. Reactors, (three): shell and tube construction, fixed tube sheets, heat-transfer area 50 m², design pressure 5 bar, material stainless steel.
4. Condenser: shell and tube heat exchanger, fixed tube sheets, heat-transfer area 25 m², design pressure 2 bar, material stainless steel.
5. Absorption column: packed column, diameter 0.5 m, height 6.0 m, packing height 4.5 m, packing 25-mm ceramic saddles, design pressure 2 bar, material carbon steel.
6. Extraction column: packed column, diameter 0.5 m, height 4 m, packed height 3 m, packing 25-mm stainless steel pall rings, design pressure 2 bar, material carbon steel.
7. Solvent recovery column: plate column, diameter 0.6 m, height 6 m, 10 stainless steel sieve plates, design pressure 2 bar, column material carbon steel.
8. Recovery column reboiler: thermosyphon, shell and tube, fixed tube sheets, heat transfer area 4 m², design pressure 2 bar, material carbon steel.
9. Recovery column condenser: double-pipe, heat-transfer area 1.5 m², design pressure 2 bar, material carbon steel.
10. Solvent cooler: double-pipe exchanger, heat-transfer area 2 m², material stainless steel.
11. Product purification column: plate column, diameter 1 m², height 20 m, 15 sieve plates, design pressure 2 bar, material stainless steel.
12. Product column reboiler: kettle type, heat-transfer area 4 m², design pressure 2 bar, material stainless steel.
13. Product column condenser: shell and tube, floating head, heat-transfer area 15 m², design pressure 2 bar, material stainless steel.
14. Feed compressor: centrifugal, rating 750 kW.
15. Butanol storage tank: cone roof, capacity 400 m³, material carbon steel.
16. Solvent storage tank: horizontal, diameter 1.5 m, length 5 m, material carbon steel.
17. Product storage tank: cone roof, capacity 400 m³, material carbon steel.

Raw Materials

1. 2-butanol, 1.045 kg per kg of MEK, price $800 per metric ton.
2. Solvent (trichloroethane) make-up 7000 kg per year, price $1.0/kg.

Utilities

1. Fuel oil, 3000 metric tons per year, heating value 45 GJ/metric ton.
2. Cooling water, 120 metric tons per hour.
3. Steam, low pressure, 1.2 metric tons per hour.
4. Electrical power, 1 MW.

The fuel oil is burnt to provide flue gases for heating the reactor feed and the reactor. Some of the fuel requirements could be provided by using the by-product hydrogen. Also, the exhaust flue gases could be used to generate steam. The economics of these possibilities need not be considered.

6.11. A plant is proposing to install a combined heat and power system to supply electrical power and process steam. Power is currently taken from a utility company and steam is generated using on-site boilers.

The capital cost of the CHP plant is estimated to be $23 million. Combined heat and power is expected to give net savings of $10 million per year. The plant is expected to operate for 10 years after the completion of construction.

Calculate the cumulative net present value of the project, at a discount rate of 12%, using MACRS depreciation with a 5-year recovery term. Also, calculate the discounted cash flow rate of return.

Construction will take 2 years, and the capital will be paid in two equal increments, at the end of the first and second years. The savings (income) can be taken as paid at the end of each year. Production will start on the completion of construction.

6.12. A process heat-recovery study identifies five potential modifications, none of which are mutually exclusive, with the costs and energy savings given below.

Project	Capital Cost (MM$)	Fuel savings (MMBtu/h)
A	1.5	15
B	0.6	9
C	1.8	16
D	2.2	17
E	0.3	8

If fuel costs $6/MMBtu and the plant operates for 350 days/year, which projects have a simple pay-back period less than 1 year?

What is the maximum 10-year NPV that can be achieved with a 15% interest rate and a 35% tax rate? Assume all the projects can be built immediately, and use MACRS depreciation with a 5-year recovery term. What combination of projects is selected to meet the maximum NPV?

6.13. An electronics company wants to fit a solvent-recovery system on the vent gas from its circuit board manufacturing line. The solvent-recovery system consists of a chiller, a knockout drum and an adsorbent bed. The adsorbent is periodically regenerated by circulating hot air over the bed and to the chiller and knockout. After consultation with equipment vendors, the following purchased prices are estimated for the major plant equipment:

Item	Cost ($)
Chiller	4000
Knockout drum	1000
Packaged refrigeration plant	3000
Adsorbent vessel (x2)	1500 each
Air blower	4000
Air heater	3000

Estimate the ISBL cost of the plant and the total project cost. If the annual operating costs are $38,000 and the annual savings in recovered solvent are $61,500, what is the IRR of this project?

6.14. Carry out a sensitivity analysis of the adipic acid project described in Examples 6.11 and 6.12.

6.15. The adipic acid plant described in Examples 6.11 and 6.12 is to be built in China, with a location factor of 0.85. Up to 45% of the total investment can be secured as a low-cost loan at an interest rate of 1%.
1. What is the cost of capital if the cost of equity is 40%?
2. What is the NPV for 15 years of production?
3. What is the IRR if the debt must be amortized over 15 years as a fixed cost of production?

7 MATERIALS OF CONSTRUCTION

Chapter Contents

Key Learning Objectives

- Mechanical and chemical properties that must be considered when selecting materials of construction for a chemical plant
- Relative costs of common materials of construction
- Properties of alloys commonly used in engineering
- When to use polymers or ceramic materials

7.1. INTRODUCTION

This chapter covers the selection of materials of construction for process equipment and piping.

Many factors have to be considered when selecting engineering materials, but for chemical process plants the overriding considerations are usually high temperature strength and the ability to resist corrosion. The process designer will be responsible for recommending materials that will be suitable for the process conditions. The process engineer must also consider the requirements of the mechanical design engineer; the material selected must have sufficient strength and be easily worked. The most economical material that satisfies both process and mechanical requirements should be selected; this will be the material that gives the lowest cost over the working life of the plant, allowing for maintenance and replacement. Other factors, such as product contamination and process safety, must also be considered. The mechanical properties that are important in the selection of materials are discussed briefly in this chapter. Several books have been published on the properties of materials, and the metal-working processes used in equipment fabrication; a selection suitable for further study is given in the list of references at the end of this chapter. The mechanical design of process equipment is discussed in Chapter 13.

A detailed discussion of the theoretical aspects of corrosion is not given in this chapter, as this subject is covered comprehensively in several books: Revie (2005), Fontana (1986), Dillon (1986) and Schweitzer (1989).

An extensive set of corrosion data for different materials is given by Craig and Anderson (1995).

7.2. MATERIAL PROPERTIES

The most important characteristics to be considered when selecting a material of construction are:

1. Mechanical properties
   a Strength: tensile strength
   b Stiffness: elastic modulus (Young's modulus)
   c Toughness: fracture resistance
   d Hardness: wear resistance
   e Fatigue resistance
   f Creep resistance.
2. The effect of high temperature, low temperature and thermal cycling on the mechanical properties.
3. Corrosion resistance.
4. Any special properties required; such as, thermal conductivity, electrical resistance, magnetic properties.
5. Ease of fabrication: forming, welding, casting (see Table 7.1).
6. Availability in standard sizes: plates, sections, tubes
7. Cost.

TABLE 7.1. A Guide to the Fabrication Properties of Common Metals and Alloys

	Machining	Cold Working	Hot Working	Casting	Welding	Annealing Temp. (°C)
Mild steel	S	S	S	D	S	750
Low-alloy steel	S	D	S	D	S	750
Cast iron	S	U	U	S	D/U	—
Stainless steel (18Cr, 8Ni)	S	S	S	D	S	1050
Nickel	S	S	S	S	S	1150
Monel	S	S	S	S	S	1100
Copper (deoxidized)	D	S	S	S	D	800
Brass	S	D	S	S	S	700
Aluminium	S	S	S	D	S	550
Dural	S	S	S	—	S	350
Lead	—	S	—	—	S	—
Titanium	S	S	U	U	D	—

S = satisfactory, D = difficult, special techniques needed, U = unsatisfactory.

7.3. MECHANICAL PROPERTIES

Typical values of the mechanical properties of the more common materials used in the construction of chemical process equipment are given in Table 7.2.

7.3.1. Tensile Strength

The tensile strength (tensile stress) is a measure of the basic strength of a material. It is the maximum stress that the material will withstand, measured by a standard tensile

TABLE 7.2. Mechanical Properties of Common Metals and Alloys (Typical Values at Room Temperature)

	Tensile Strength (N/mm²)	0.1% Proof Stress (N/mm²)	Modulus of Elasticity (kN/mm²)	Hardness, Brinell	Specific Gravity
Mild steel	430	220	210	100–200	7.9
Low-alloy steel	420–660	230–460	210	130–200	7.9
Cast iron	140–170	—	140	150–250	7.2
Stainless steel (18Cr, 8Ni)	>540	200	210	160	8.0
Nickel (>99% Ni)	500	130	210	80–150	8.9
Monel	650	170	170	120–250	8.8
Copper (deoxidized)	200	60	110	30–100	8.9
Brass (Admiralty)	400–600	130	115	100–200	8.6
Aluminium (>99%)	80–150	—	70	30	2.7
Dural	400	150	70	100	2.7
Lead	30	—	15	5	11.3
Titanium	500	350	110	150	4.5

Note: Tensile stress and proof stress are not the same as the maximum allowable stress permitted by design code. See Tables 7.5 and 7.7 for maximum allowable stress values.

test. The older name for this property, which is more descriptive of the property, was Ultimate Tensile Strength (UTS).

Proof stress is the stress to cause a specified permanent extension, usually 0.1%.

The maximum allowable stresses specified by the ASME boiler and pressure vessel (BPV) code and the European EN 13445 pressure vessel code are calculated from these and other material properties at the design temperature, and allowing for suitable safety factors. The basis for establishing maximum allowable stress values is discussed in Chapter 13 and is described in detail in the ASME BPV Code Section II Part D, Mandatory Appendix 1 or in BS EN 13445-3.

7.3.2. Stiffness

Stiffness is the ability to resist bending and buckling. It is a function of the elastic modulus of the material and the shape of the cross-section of the member (the second moment of area).

7.3.3. Toughness

Toughness is associated with tensile strength, and is a measure of the material's resistance to crack propagation. The crystal structure of ductile materials, such as steel, aluminium and copper, is such that they stop the propagation of a crack by local yielding at the crack tip. In other materials, such as the cast irons and glass, the structure is such that local yielding does not occur and the materials are brittle. Brittle materials are weak in tension but strong in compression. Under compression any incipient cracks present are closed up. Various techniques have been developed to allow the use of brittle materials in situations where tensile stress would normally occur. For example, the use of pre-stressed concrete, and glass-fibre-reinforced plastics in pressure vessel construction.

A detailed discussion of the factors that determine the fracture toughness of materials can be found in the books by Institute of Metallurgists (1960) and Boyd (1970). Gordon (1976) gives an elementary, but very readable, account of the strength of materials in terms of their macroscopic and microscopic structure.

7.3.4. Hardness

The surface hardness, as measured in a standard test such as the Brinell hardness test, is an indication of a material's ability to resist wear. This will be an important property if the equipment is being designed to handle abrasive solids, or liquids containing suspended solids that are likely to cause erosion.

7.3.5. Fatigue

Fatigue failure is likely to occur in equipment subject to cyclic loading; for example, rotating equipment, such as pumps and compressors, and equipment subjected to temperature or pressure cycling. A comprehensive treatment of this subject is given by Harris (1976).

7.3.6. Creep

Creep is the gradual extension of a material under a steady tensile stress, over a prolonged period of time. It is usually only important at high temperatures; for instance, with steam and gas turbine blades. For a few materials, notably lead, the rate of creep is significant at moderate temperatures. Lead will creep under its own weight at room temperature and lead linings must be supported at frequent intervals.

The creep strength of a material is usually reported as the stress to cause rupture in 100,000 h, at the test temperature.

7.3.7. Effect of Temperature on the Mechanical Properties

The tensile strength and elastic modulus of metals decrease with increasing temperature. For example, the tensile strength of mild steel (low carbon steel, $C < 0.25\%$) is 450 N/mm^2 at 25°C falling to 210 at 500°C, and the value of Young's modulus 200,000 N/mm^2 at 25°C falling to 150,000 at 500°C. The ASME BPV Code Section II Part D specifies maximum temperatures for each material. For example, SA-285 plain carbon steel plate cannot be used to construct a pressure vessel that meets the specifications of ASME BPV Code Section VIII Div. 1 with a design temperature greater than 900°F (482°C). Any pressure vessel that is designed for use above this temperature must be made from killed steel or alloy. The maximum allowable stress used in design is always based on the design temperature. Materials must be chosen that have sufficient strength at the design temperature to give an economic and mechanically feasible wall thickness. The stainless steels are superior in this respect to plain carbon steels.

Creep resistance will be important if the material is subjected to high stresses at elevated temperatures. Special alloys, such as Inconel 600 (UNS N06600) or Incoloy 800 (UNS N08800) (both trademarks of International Nickel Co.) are used for high temperature equipment such as furnace tubes in environments that do not contain sulphur. The selection of materials for high-temperature applications is discussed by Day (1979) and Lai (1990).

At low temperatures, less than 10°C, metals that are normally ductile can fail in a brittle manner. Serious disasters have occurred through the failure of welded carbon steel vessels at low temperatures. The phenomenon of brittle failure is associated with the crystalline structure of metals. Metals with a body-centred-cubic (bcc) lattice are more liable to brittle failure than those with a face-centred-cubic (fcc) or hexagonal lattice. For low-temperature equipment, such as cryogenic plant and liquefied-gas storage, austenitic stainless steel (fcc) or aluminium alloys (hex) should be specified; see Wigley (1978).

V-notch impact tests, such as the Charpy test, are used to test the susceptibility of materials to brittle failure: see Wells (1968) and ASME BPV Code Sec. VIII Div. 1 Part UG-84.

The brittle fracture of welded structures is a complex phenomenon and is dependent on plate thickness and the residual stresses present after fabrication, as well as the operating temperature. A comprehensive discussion of brittle fracture in steel structures is given by Boyd (1970).

7.4. CORROSION RESISTANCE

The conditions that cause corrosion can arise in a variety of ways. For this brief discussion on the selection of materials it is convenient to classify corrosion into the following categories:

1. General wastage of material: uniform corrosion
2. Galvanic corrosion: dissimilar metals in contact
3. Pitting: localized attack
4. Intergranular corrosion
5. Stress corrosion
6. Erosion–corrosion
7. Corrosion fatigue
8. High-temperature oxidation and sulphidation
9. Hydrogen embrittlement.

Metallic corrosion is essentially an electrochemical process. Four components are necessary to set up an electrochemical cell:

1. Anode: the corroding electrode
2. Cathode: the passive, non-corroding electrode
3. The conducting medium: the electrolyte, the corroding fluid
4. Completion of the electrical circuit: through the material.

Cathodic areas can arise in many ways:

i. Dissimilar metals
ii. Corrosion products
iii. Inclusions in the metal, such as slag
iv. Less well aerated areas
v. Areas of differential concentration
vi. Differentially strained areas.

7.4.1. Uniform Corrosion

This term describes the more or less uniform wastage of material by corrosion, with no pitting or other forms of local attack. If the corrosion of a material can be considered to be uniform, the life of the material in service can be predicted from experimentally determined corrosion rates.

Corrosion rates are usually expressed as a penetration rate in inches per year (ipy), or mills per year (mpy) (where a mill $= 10^{-3}$ in). They are also expressed as a weight loss in milligrams per square decimetre per day (mdd). In corrosion testing, the corrosion rate is measured by the reduction in weight of a specimen of known area over a fixed period of time.

$$\text{ipy} = \frac{12w}{tA\rho} \tag{7.1}$$

where

> w = mass loss in time t, lb
> t = time, years
> A = surface area, ft^2
> ρ = density of material, lb/ft^3

as most of the published data on corrosion rates are in imperial units.

In SI units, 1 ipy = 25 mm per year.

When judging corrosion rates expressed in mdd it must be remembered that the penetration rate depends on the density of the material. For ferrous metals 100 mdd = 0.02 ipy.

What can be considered as an acceptable rate of attack will depend on the cost of the material; the duty, particularly as regards to safety; and the economic life of the plant. For the more commonly used inexpensive materials, such as the carbon and low-alloy steels, a guide to what is considered acceptable is given in Table 7.3. For the more expensive alloys, such as the high-alloy steels, the brasses and aluminium, the figures given in Table 7.3 should be divided by 2.

If the predicted corrosion rate indicates only short exposures then the design engineer should allow for frequent inspection of the plant and periodic replacement of the affected equipment. This affects process economics in two ways, as it reduces the on-stream factor (number of days of production per year) and increases the maintenance costs. Usually the economic impact of frequent shut down and replacement is so negative that use of a more expensive alloy with better corrosion resistance can be justified.

Allowances for expected corrosion over the plant life or time between replacements must be added to the minimum vessel wall thicknesses calculated to comply with the pressure vessel codes. These corrosion allowances can be economically or mechanically prohibitive if the corrosion rate is high. Guidance on corrosion allowances is given in the ASME BPV Code Sec. VIII Div. 1 Non-mandatory Appendix E. The corrosion allowance should at least equal the expected corrosion loss during the desired life of the vessel.

The corrosion rate will depend on the temperature and concentration of the corrosive fluid. An increase in temperature usually results in an increased rate of corrosion; though not always. The rate will depend on other factors that are affected by temperature, such as oxygen solubility.

TABLE 7.3. Acceptable Corrosion Rates

	Corrosion Rate	
	ipy	mm/y
Completely satisfactory	<0.01	0.25
Use with caution	<0.03	0.75
Use only for short exposures	<0.06	1.5
Completely unsatisfactory	>0.06	1.5

The effect of concentration can also be complex. For example, the corrosion of mild steel in sulphuric acid, where the rate is unacceptably high in dilute acid and at concentrations above 70%, but is acceptable at intermediate concentrations.

7.4.2. Galvanic Corrosion

If dissimilar metals are placed in contact in an electrolyte, the corrosion rate of the anodic metal will be increased, as the metal lower in the electrochemical series will readily act as a cathode. The galvanic series in sea water for some of the more commonly used metals is shown in Table 7.4. Some metals under certain conditions form a natural protective film; for example, stainless steel in oxidizing environments. This state is denoted by 'passive' in the series shown in Table 7.4. Active indicates the absence of the protective film; for example, where the surface of the metal is subject to wear due to moving parts or abrasion by the fluid. Minor shifts in position in the series can be expected in other electrolytes, but the series for sea water is a good indication of the combinations of metals to be avoided. If metals that are widely separated in the galvanic series have to be used together, they should be electrically insulated from each other, breaking the conducting circuit. Alternatively, if sacrificial loss of the anodic material can be accepted, the thickness of this material can be increased to allow for the increased rate of corrosion. The corrosion rate will depend on the relative areas of the anodic and cathodic metals. A high cathode to anode area should be avoided. Sacrificial anodes are used to protect underground steel pipes.

7.4.3. Pitting

Pitting is the term given to highly localized corrosion that forms pits in the metal surface. If a material is liable to pitting, penetration can occur prematurely and corrosion rate data are not a reliable guide to the equipment life.

TABLE 7.4.   Galvanic Series in Sea Water

Noble end (protected end)	18/8 stainless steel (passive)
	Monel
	Inconel (passive)
	Nickel (passive)
	Copper
	Aluminium bronze (Cu 92%, Al 8%)
	Admiralty brass (Cu 71%, Zn 28%, Sn 1%)
	Nickel (active)
	Inconel (active)
	Lead
	18/8 stainless steel (active)
	Cast iron
	Mild steel
	Aluminium
	Galvanized steel
	Zinc
	Magnesium

Pitting can be caused by a variety of circumstances; any situation that causes a localized increase in corrosion rate may result in the formation of a pit. In an aerated medium the oxygen concentration will be lower at the bottom of a pit, and the bottom will be anodic to the surrounding metal, causing increased corrosion and deepening of the pit. A good surface finish will reduce this type of attack. Pitting can also occur if the composition of the metal is not uniform; for example, the presence of slag inclusions in welds. The impingement of bubbles can also cause pitting. This occurs during cavitation in pumps, which is an example of erosion–corrosion.

7.4.4. Intergranular Corrosion

Intergranular corrosion is the preferential corrosion of material at the grain (crystal) boundaries. Though the loss of material will be small, intergranular corrosion can cause the catastrophic failure of equipment. Intergranular corrosion is a common form of attack on alloys but occurs rarely with pure metals. The attack is usually caused by a differential couple being set up between impurities existing at the grain boundary. Impurities will tend to accumulate at the grain boundaries after heat treatment. The classic example of intergranular corrosion in chemical plant is the weld decay of unstabilized stainless steel. This is caused by the precipitation of chromium carbides at the grain boundaries in a zone adjacent to the weld, where the temperature has been $500-800°C$ during welding. Weld decay can be avoided by annealing after welding, if practical (post-weld heat treatment); or by using low carbon grades ($<0.3\%$ C); or grades stabilized by the addition of titanium or niobium.

7.4.5. Effect of Stress

Corrosion rate and the form of attack can be changed if the material is under stress. Generally, the rate of attack will not change significantly within normal design stress values. However, for some combinations of metal, corrosive media and temperature, the phenomenon called stress corrosion cracking can occur. This is the general name given to a form of attack in which cracks are produced that grow rapidly, and can cause premature, brittle failure of the metal. The conditions necessary for stress corrosion cracking to occur are:

1. Simultaneous stress and corrosion
2. A specific corrosive substance; in particular the presence of Cl^-, OH^-, NO_3^-, or NH_4^+ ions.

Mild stress can cause cracking; the residual stresses from fabrication and welding are sufficient.

For a general discussion of the mechanism of stress corrosion cracking see Fontana (1986).

Some classic examples of stress corrosion cracking are:

The seasonal cracking of brass cartridge cases
Caustic embrittlement of steel boilers
The stress corrosion cracking of stainless steels in the presence of chloride ions.

Stress corrosion cracking can be avoided by selecting materials that are not susceptible in the specific corrosion environment; or, less certainly, by stress relieving by post-weld heat treatment.

Comprehensive tables of materials susceptible to stress corrosion cracking in specific chemicals are given by Moore (1979). Moore's tables are taken from the corrosion data survey published by NACE (1974). See also ASME BPV Code Sec. II Part D Appendix A-330.

The term corrosion fatigue is used to describe the premature failure of materials in corrosive environments caused by cyclic stresses. Even mildly corrosive conditions can markedly reduce the fatigue life of a component. Unlike stress corrosion cracking, corrosion fatigue can occur in any corrosive environment and does not depend on a specific combination of corrosive substance and metal. Materials with a high resistance to corrosion must be specified for critical components subjected to cyclic stresses.

7.4.6. Erosion–Corrosion

The term erosion–corrosion is used to describe the increased rate of attack caused by a combination of erosion and corrosion. If a fluid stream contains suspended particles or where there is high velocity or turbulence, erosion will tend to remove the products of corrosion and any protective film, and the rate of attack will be markedly increased. If erosion is likely to occur, more resistant materials must be specified, or the material surface protected in some way. For example, plastic inserts can be used to prevent erosion–corrosion at the inlet to heat-exchanger tubes.

7.4.7. High-Temperature Oxidation and Sulphidation

Corrosion is normally associated with aqueous solutions but oxidation can occur in dry conditions. Carbon and low-alloy steels will oxidize rapidly at high temperatures and their use is limited to temperatures below 480°C (900°F).

Chromium is the most effective alloying element to give resistance to oxidation, forming a tenacious oxide film. Chromium alloys should be specified for equipment subject to temperatures above 480°C in oxidizing atmospheres. For example, type 304L stainless steel (18% Cr) can be used up to 650°C (1200°F). For temperatures above 700°C additional stabilization is needed. Type 347 stainless steel is stabilized with niobium and can be used up to 850°C. High-nickel alloys can also be used as long as sulphur is not present, and high-chromium-content Ni alloys are used at the highest temperatures. For example, Inconel 600 (15.5% Cr) can be used up to 650°C (1200°F) and Incoloy 800 (21% Cr) can be used up to 850°C (1500°F).

Sulphur is a very common corrosive contaminant in gas processing, oil refining and energy conversion. In reducing environments sulphur is present as H_2S, which causes sulphidation of metals. The metal chosen must often withstand a sulphiding environment on one side and an oxidizing environment on the other side, all at high temperature (for example in a furnace tube). Sulphur can attack the chromium oxide scale that protects the alloy, causing breakaway corrosion, particularly for high-nickel alloys. Lai (1990) gives high-temperature corrosion data for various sulphiding and mixed-gas environments and recommends the use of high-chromium high-silicon alloys such as HR-160 in this service.

7.4.8. Hydrogen Embrittlement

Hydrogen embrittlement is the name given to the loss of ductility caused by the absorption (and reaction) of hydrogen in a metal. It is of particular importance when specifying steels for use in hydrogen reforming plant. Alloy steels have a greater resistance to hydrogen embrittlement than the plain carbon steels. A chart showing the suitability of various alloy steels for use in hydrogen atmospheres, as a function of hydrogen partial pressure and temperature, is given in the NACE (1974) corrosion data survey. Below 500°C plain carbon steel can be used.

7.5. SELECTION FOR CORROSION RESISTANCE

In order to select the correct material of construction, the process environment to which the material will be exposed must be clearly defined. In addition to the main corrosive chemicals present, the following factors must be considered:

1. Temperature: affects corrosion rate and mechanical properties
2. Pressure
3. pH
4. Presence of trace impurities: stress corrosion
5. The amount of aeration: differential oxidation cells
6. Stream velocity and agitation: erosion–corrosion
7. Heat-transfer rates: differential temperatures.

The conditions that may arise during abnormal operation, such as at start-up and shutdown, must be considered, in addition to normal, steady-state, operation.

Corrosion Charts

The resistance of some commonly used materials to a range of chemicals is shown in Appendix B. More comprehensive corrosion data, covering most of the materials used in the construction of process plant, in a wide range of corrosive media, are given by, Rabald (1968), NACE (1974), Hamner (1974), Green and Perry (2007), Lai (1990) and Schweitzer (1976, 1989, 1998).

The 12-volume *Dechema Corrosion Handbook* is an extensive guide to the interaction of corrosive media with materials (Dechema, 1987). The ASM *Handbook of Corrosion Data* also has extensive data (Craig and Anderson, 1995).

These corrosion guides can be used for the preliminary screening of materials that are likely to be suitable, but the fact that published data indicate that a material is satisfactory cannot be taken as a guarantee that it will be suitable for the process environment being considered. Slight changes in the process conditions, or the presence of unsuspected trace impurities, can markedly change the rate of attack or the nature of the corrosion. The guides will, however, show clearly those materials that are manifestly unsuitable. Judgment, based on experience with the materials in similar process environments, must be used when assessing published corrosion data.

Pilot plant tests, and laboratory corrosion tests under simulated plant conditions, will help in the selection of suitable materials if actual plant experience is not available. Preliminary tests can be carried out by inserting coupons of different materials into an apparatus that is known to resist corrosion before testing plant components. This reduces the likelihood of component failure and possible release of chemicals during testing. Care is needed in the interpretation of laboratory tests.

The advice of the technical service department of the company supplying the materials should also be sought.

7.6.   MATERIAL COSTS

An indication of the cost of some commonly used metals is given in Table 7.5. The actual cost of metals and alloys will fluctuate quite widely, depending on movements in the world metal exchanges.

Current metals prices can be found at:

www.steelonthenet.com	free site with monthly carbon steel prices
www.steelbb.com	steel business briefing: subscription site with weekly carbon steel and stainless steel prices
www.steelweek.com	subscription site with weekly international prices
www.lme.com	London Metals Exchange: free site for major commodity metals
http://metalprices.com/freesite/ metals	great alloy calculator: 3-month-old prices are free, current prices subscription only

TABLE 7.5.   Relative Cost of Metals (February 2008)

Metal	Type or Grade	Price ($/lb)	Max. Allowable Stress (ksi = 1000 psi)	Relative Cost Rating
Carbon steel	A-285	0.46	12.9	1
Austenitic stainless steel	304	1.44	20	2.0
	316	2.47	20	3.5
Aluminium alloy	A03560	1.19	8.6	1.3
Copper	C10400	3.92	6.7	18.8
Nickel	99% Ni	12.87	10	41.1
Incoloy	N08800	4.66	20	6.7
Monel	N04400	9.16	18.7	15.6
Titanium	R50250	3.63	10	5.9

Notes: 1. The maximum allowable stress values are at 40°C (100°F) and are taken from ASME BPV Code Sec. II Part D. The code should be consulted for values at other temperatures.
2. The European EN 13445 pressure vessel code uses different maximum allowable stresses; see BS EN 13445-2 and BS EN 13445-3.
3. Several other grades exist for most of the materials listed.

The quantity of a material used will depend on the material density and strength (maximum allowable stress) and these must be taken into account when comparing material costs. Moore (1970) compares costs by calculating a cost rating factor defined by the equation:

$$\text{Cost rating} = \frac{C \times \rho}{\sigma_d} \tag{7.2}$$

where

C = cost per unit mass, \$/kg
ρ = density, kg/m^3
σ_d = maximum allowable stress, N/mm^2.

Cost ratings, relative to the rating for mild steel (low carbon), are shown in Table 7.5 for February 2008 prices. Materials with a relatively high maximum allowable stress, such as stainless and low-alloy steels, can be used more efficiently than carbon steel. Note that the simplified formula given in equation 7.2 does not take into account different corrosion allowances for the different materials.

The relative cost of equipment made from different materials will depend on the cost of fabrication, as well as the basic cost of the material. Unless a particular material requires special fabrication techniques, the relative cost of the finished equipment will be lower than the relative bare material cost. For example; the purchased cost of a stainless steel storage tank will be two to three times the cost of the same tank in carbon steel, whereas the relative cost of the metals is from five to eight.

If the corrosion rate is uniform, then the optimum material can be selected by calculating the annual costs for the possible candidate materials. The annual cost will depend on the predicted life, calculated from the corrosion rate, and the purchased cost of the equipment. In a given situation, it may prove more economic to install a cheaper material with a high corrosion rate and replace it frequently, rather than select a more resistant but more expensive material. This strategy would only be considered for relatively simple equipment with low fabrication costs, and where premature failure would not cause a serious hazard. For example, carbon steel could be specified for an aqueous effluent line in place of stainless steel, accepting the probable need for replacement. The pipe wall thickness would be monitored *in situ* frequently to determine when replacement was needed.

The more expensive, corrosion-resistant, alloys are frequently used as a cladding on carbon steel. If a thick plate is needed for structural strength, as for pressure vessels, the use of clad materials can substantially reduce the cost. The design requirements for pressure vessels with cladding or applied internal linings are given in ASME BPV Code Sec. VIII Div. 1 Part UCL.

7.7. CONTAMINATION

With some processes, the prevention of the contamination of a process stream, or a product, by certain metals, or the products of corrosion, overrides any other considerations when selecting suitable materials. For instance, in textile processes, stainless steel or aluminium is often used in preference to carbon steel, which would be quite suitable except that any slight rusting will mark the textiles (iron staining).

With processes that use catalysts, care must be taken to select materials that will not cause contamination and poisoning of the catalyst.

Some other examples that illustrate the need to consider the effect of contamination by trace quantities of other materials are:

1. For equipment handling acetylene, the pure metals, or alloys containing copper, silver, mercury or gold, must be avoided to prevent the formation of explosive acetylides.
2. The presence of trace quantities of mercury in a process stream can cause the catastrophic failure of brass heat-exchanger tubes, from the formation of a mercury–copper amalgam. Incidents have occurred where the contamination has come from unsuspected sources, such as the failure of mercury-in-steel thermometers.
3. In the Flixborough disaster (see Chapter 9), there was evidence that the stress corrosion cracking of a stainless steel pipe had been caused by zinc contamination from galvanized wire supporting lagging.

7.7.1. Surface Finish

In industries such as the food, pharmaceutical, biochemical, and textile industries, the surface finish of the material is as important as the choice of material, to avoid contamination.

Stainless steel is widely used, and the surfaces, inside and out, are given a high finish by abrasive blasting and mechanical polishing. This is done for the purposes of hygiene; to prevent material adhering to the surface; and to aid cleaning and sterilization. The surface finishes required in food processing are discussed by Timperley (1984) and Jowitt (1980).

A good surface finish is important in textile fibre processing to prevent the fibres snagging.

7.8. COMMONLY USED MATERIALS OF CONSTRUCTION

The general mechanical properties, corrosion resistance and typical areas of use of some of the materials commonly used in the construction of chemical plant are given in this section. The values given are for a typical, representative, grade of the material or alloy. The alloys used in chemical plant construction are known by a variety of trade names, and code numbers are designated in the various national standards. With the exception of the stainless steels, no attempt has been made in this book to classify the alloys discussed by using one or other of the national standards; the commonly used, generic, names for the alloys have been used. For the full details of the properties and compositions of the different grades available in a particular class of alloy, and the designated code numbers, reference should be made to the appropriate national code, to the various handbooks, or to manufacturers' literature. See, for example, ASME BPV Code Sec. II Part D for a full listing of materials' properties and ASME BPV Code Sec. VIII Div.1 for material-specific fabrication guidelines.

The US trade names and codes are given by Green and Perry (2007). A comprehensive review of the engineering materials used for chemical and process plant can be found in the book by Evans (1974).

7.8.1. Iron and Steel

Low carbon steel (mild steel) is the most commonly used engineering material. It is cheap, is available in a wide range of standard forms and sizes, and can be easily worked and welded. It has good tensile strength and ductility.

The carbon steels and iron are not resistant to corrosion, except in certain specific environments, such as concentrated sulphuric acid and the caustic alkalis. They are suitable for use with most organic solvents, except chlorinated solvents, but traces of corrosion products may cause discoloration.

Mild steel is susceptible to stress corrosion cracking in certain environments.

The corrosion resistance of the low-alloy steels (less than 5% of alloying elements), where the alloying elements are added to improve the mechanical strength and not for corrosion resistance, is not significantly different from that of the plain carbon steels.

A comprehensive reference covering the properties and application of steels, including the stainless steels, is the book by Llewellyn (1992). The use of carbon steel in the construction of chemical plant is discussed by Clark (1970).

The high silicon irons (14% to 15% Si) have a high resistance to mineral acids, except hydrofluoric acid. They are particularly suitable for use with sulphuric acid at all concentrations and temperatures. They are, however, very brittle.

7.8.2. Stainless Steel

The stainless steels are the most frequently used corrosion-resistant materials in the chemical industry.

To impart corrosion resistance the chromium content must be above 12%, and the higher the chromium content, the more resistant is the alloy to corrosion in oxidizing conditions. Nickel is added to improve the corrosion resistance in non-oxidizing environments.

Types

A wide range of stainless steels are available, with compositions tailored to give the properties required for specific applications. They can be divided into three broad classes according to their microstructure:

1. Ferritic: 13–20% Cr, <0.1% C, with no nickel
2. Austenitic: 18–20% Cr, >7% Ni
3. Martensitic: 12–14% Cr, 0.2–0.4% C, up to 2% Ni.

The uniform structure of Austenite (face-centred cubic, with the carbides in solution) is the structure desired for corrosion resistance, and it is these grades that are widely used in the chemical industry. The composition of the main grades of austenitic steel is shown in Table 7.6. Their properties are discussed below.

TABLE 7.6.   Commonly Used Grades of Austenitic Stainless Steel

Specification no. AISI No.	Composition %							
	C max.	Si max.	Mn max.	Cr Range	Ni Range	Mo Range	Ti	Nb
304	0.08	—	2.00	17.5 20.0	8.0 11.0	—	—	—
304L	0.03	1.00	2.00	17.5 20.0	8.0 12.0	—	—	—
321	0.12	1.00	2.00	17.0 20.0	9.0 12.0	—	$4 \times C$	—
347	0.08	1.00	2.00	17.0 20.0	9.0 13.0	—	—	$10 \times C$
316	0.08	1.00	2.00	16.0 18.0	10.0 14.0	2.0 3.0	—	—
316L	0.03	1.0	2.0	16.0 18.0	10.0 14.0	2.0 3.0	—	—
309	0.20	—	—	22.0 24.0	12.0 15.0	—	—	—
310	0.25	—	—	24.0 26.0	19.0 22.0	—	—	—

S and P = 0.045% all grades.
AISI = American Iron and Steel Institute.

Type 304 (the so-called 18/8 stainless steels): the most generally used stainless steel. It contains the minimum Cr and Ni that give a stable austenitic structure. The carbon content is low enough for heat treatment not to be normally needed with thin sections to prevent weld decay (see Section 7.4.4).

Type 304L: low-carbon version of type 304 (<0.03% C) used for thicker welded sections, where carbide precipitation would occur with type 304.

Type 321: a stabilized version of 304, stabilized with titanium to prevent carbide precipitation during welding. It has a slightly higher strength than 304L, and is more suitable for high-temperature use.

Type 347: stabilized with niobium.

Type 316: in this alloy, molybdenum is added to improve the corrosion resistance in reducing conditions, such as in dilute sulphuric acid, and, in particular, to solutions containing chlorides.

Type 316L: a low carbon version of type 316, which should be specified if welding or heat treatment is liable to cause carbide precipitation in type 316.

Types 309/310: alloys with a high chromium content, to give greater resistance to oxidation at high temperatures. Alloys with greater than 25% Cr are susceptible to embrittlement due to sigma phase formation at temperatures above 500°C. Sigma phase is an intermetallic compound, FeCr. The formation of the sigma phase in austenitic stainless steels is discussed by Hills and Harries (1960).

Mechanical Properties

The austenitic stainless steels have greater strength than the plain carbon steels, particularly at elevated temperatures (see Table 7.7).

TABLE 7.7. Comparative Strength of Carbon Steel and Stainless Steel

Temperature (°F)		100	300	500	700	900
Maximum allowable	Carbon steel (A285 plate)	12.9	12.9	12.9	11.5	5.9
stress (1000 psi)	Stainless steel (304L plate)	16.7	16.7	14.7	13.5	11.9

Maximum allowable stress values from ASME BPV Code Sec. II Part D.

As was mentioned in Section 7.3.7, the austenitic stainless steels, unlike the plain carbon steels, do not become brittle at low temperatures. It should be noted that the thermal conductivity of stainless steel is significantly lower than that of mild steel.

Typical values at 100°C are: type 304 (18/8)   16 W/m°C
                                 mild steel   60 W/m°C

Austenitic stainless steels are non-magnetic in the annealed condition.

General Corrosion Resistance

The higher the alloying content, the better the corrosion resistance over a wide range of conditions, strongly oxidizing to reducing, but the higher the cost. A ranking in order of increasing corrosion resistance, taking type 304 as 1, is given below:

304	304L	321	316	316L	310
1.0	1.1	1.1	1.25	1.3	1.6

Intergranular corrosion (weld decay) and stress corrosion cracking are problems associated with the use of stainless steels, and must be considered when selecting types suitable for use in a particular environment. Stress corrosion cracking in stainless steels can be caused by a few ppm of chloride ions (see Section 7.4.5).

In general, stainless steels are used for corrosion resistance when oxidizing conditions exist. Special types, or other high-nickel alloys, should be specified if reducing conditions are likely to occur. The properties, corrosion resistance and uses of the various grades of stainless steel are discussed fully by Peckner and Bernstein (1977). A comprehensive discussion of the corrosion resistance of stainless steels is given in Sedriks (1979).

Stress corrosion cracking in stainless steels is discussed by Turner (1989).

High-Alloy-Content Stainless Steels

Super austenitic, high-nickel stainless steels, containing between 29% and 30% nickel and 20% chromium, have a good resistance to acids and acid chlorides. They are more expensive than the lower alloy content, 300 series, of austenitic stainless steels.

Duplex and super-duplex stainless steels contain high percentages of chromium. They are called duplex because their structure is a mixture of the austenitic and ferritic phases. They have a better corrosion resistance than the austenitic stainless steels and are less susceptible to stress corrosion cracking. The chromium content of duplex stainless steels is around 20%, and around 25% in the super-duplex grades. The super-duplex steels were developed for use in aggressive off-shore environments.

The duplex range of stainless steels can be readily cast, wrought and machined. Problems can occur in welding, due to the need to keep the correct balance of ferrite and austenite in the weld area, but this can be overcome using the correct welding materials and procedures.

The cost of the duplex grades is comparable with the 316 steels. Super-duplex costs around 50% more than duplex.

The selection and properties of duplex stainless steels are discussed by Bendall and Guha (1990) and Warde (1991).

7.8.3. Nickel

Nickel has good mechanical properties and is easily worked. The pure metal (>99%) is not generally used for chemical plant, its alloys being preferred for most applications. The main use is for equipment handling caustic alkalis at temperatures above that at which carbon steel could be used; above 70°C. Nickel is not subject to corrosion cracking like stainless steel.

7.8.4. Monel

Monel, the classic nickel–copper alloy with the metals in the ratio 2:1, is probably, after the stainless steels, the most commonly used alloy for chemical plants. It is easily worked and has good mechanical properties up to 500°C. It is more expensive than stainless steel but is not susceptible to stress corrosion cracking in chloride solutions. Monel has good resistance to dilute mineral acids and can be used in reducing conditions where the stainless steels would be unsuitable. It may be used for equipment handling alkalis, organic acids and salts, and sea water.

7.8.5. Inconel and Incoloy

Inconel (typically 76% Ni, 7% Fe, 15% Cr) is used primarily for acid resistance at high temperatures. It maintains its strength at elevated temperature and is resistant to furnace gases, if sulphur free. It is not suitable for use in sulphiding environments. Nickel alloys with higher chromium content such as Incoloy 800 (21% Cr) and RA-33 (25% Cr) have better oxidation resistance at higher temperatures.

7.8.6. The Hastelloys

The trade name Hastelloy covers a range of nickel, chromium, molybdenum, iron alloys that were developed for corrosion resistance to strong mineral acids, particularly HCl. The corrosion resistance and use of the two main grades, Hastelloy B (65% Ni, 28% Mo, 6% Fe) and Hastelloy C (54% Ni, 17% Mo, 15% Cr, 5% Fe), are discussed in papers by Weisert (1952a,b).

7.8.7. Copper and Copper Alloys

Pure copper is not widely used for chemical equipment. It has been used traditionally in the food industry, particularly in brewing. Copper is a relatively soft, very easily worked metal, and is used extensively for small-bore pipes and tubes.

The main alloys of copper are the brasses, alloyed with zinc, and the bronzes, alloyed with tin. Other, so-called bronzes are the aluminium bronzes and the silicon bronzes.

Copper is attacked by mineral acids, except cold, dilute, un-aerated sulphuric acid. It is resistant to caustic alkalis, except ammonia, and to many organic acids and salts. The brasses and bronzes have a similar corrosion resistance to the pure metal. Their main use in the chemical industry is for valves and other small fittings, and for heat-exchanger tubes and tube sheets. If brass is used, a grade must be selected that is resistant to dezincification.

The cupro-nickel alloys (70% Cu) have a good resistance to erosion–corrosion and are used for heat-exchanger tubes, particularly where sea water is used as a coolant.

7.8.8. Aluminium and Its Alloys

Pure aluminium lacks mechanical strength but has higher resistance to corrosion than its alloys. The main structural alloys used are the Duralumin (Dural) range of aluminium–copper alloys (typical composition 4% Cu, with 0.5% Mg) which have a tensile strength equivalent to that of mild steel. The pure metal can be used as a cladding on Dural plates, to combine the corrosion resistance of the pure metal with the strength of the alloy. The corrosion resistance of aluminium is due to the formation of a thin oxide film (as with the stainless steels). It is therefore most suitable for use in strong oxidizing conditions. It is attacked by mineral acids, and by alkalis, but is suitable for concentrated nitric acid, greater than 80%. It is widely used in the textile and food industries, where the use of mild steel would cause contamination. It is also used for the storage and distribution of demineralized water.

7.8.9. Lead

Lead was one of the traditional materials of construction for chemical plants but has now, due to its price, been largely replaced by other materials, particularly plastics. It is a soft, ductile material, and is mainly used in the form of sheets (as linings) or pipe. It has a good resistance to acids, particularly sulphuric.

7.8.10. Titanium

Titanium is now used quite widely in the chemical industry, mainly for its resistance to chloride solutions, including sea water and wet chlorine. It is rapidly attacked by dry chlorine, but the presence of as low a concentration of moisture as 0.01% will prevent attack. Like the stainless steels, titanium depends for its resistance on the formation of an oxide film. Titanium is also used in other halide services; for example, in liquid-phase oxidation processes, such as the manufacture of terephthalic acid, that use bromide as catalyst or promoter.

Alloying with palladium (0.15%) significantly improves the corrosion resistance, particularly to HCl. Titanium is being increasingly used for heat exchangers, both shell and tube, and plate exchangers; replacing cupro-nickel for use with sea water.

The use of titanium for corrosion resistance is discussed by Deily (1997).

7.8.11. Tantalum

The corrosion resistance of tantalum is similar to that of glass, and it has been called a metallic glass. It is expensive, about five times the cost of stainless steel, and is used for special applications, where glass or a glass lining would not be suitable. Tantalum plugs are used to repair glass-lined equipment.

The use of tantalum as a material of construction in the chemical industry is discussed by Fensom and Clark (1984) and Rowe (1994, 1999).

7.8.12. Zirconium

Zirconium and zirconium alloys are used in the nuclear industry, because of their low neutron absorption cross-section and resistance to hot water at high pressures.

In the chemical industry, zirconium is finding use where resistance to hot and boiling acids is required: nitric, sulphuric and particularly hydrochloric. Its resistance is equivalent to that of tantalum but zirconium is less expensive, similar in price to high-nickel steel. Rowe (1999) gives a brief review of the properties and use of zirconium for chemical plants.

7.8.13. Silver

Silver linings are used for vessels and equipment handling hydrofluoric acid. It is also used for special applications in the food and pharmaceutical industries where it is vital to avoid contamination of the product.

7.8.14. Gold

Gold is rarely used as a material of construction because of its high cost. It is highly resistant to attack by dilute nitric acid and hot concentrated sulphuric acid, but is dissolved by aqua regia (a mixture of concentrated nitric and sulphuric acids). It is attacked by chlorine and bromine, and forms an amalgam with mercury.

It has been used as thin plating on condenser tubes and other surfaces.

7.8.15. Platinum

Platinum has a high resistance to oxidation at high temperature. One of its main uses has been in the form of an alloy with copper, in the manufacture of the spinnerets used in synthetic textile spinning processes.

7.9. PLASTICS AS MATERIALS OF CONSTRUCTION FOR CHEMICAL PLANTS

Plastics are being increasingly used as corrosion-resistant materials for chemical plant construction. They are also widely used in food processing and biochemical plants. They can be divided into two broad classes:

1. Thermoplastic materials, which soften with increasing temperature; for example, polyvinyl chloride (PVC) and polyethylene.
2. Thermosetting materials, which have a rigid, cross-linked structure; for example, the polyester and epoxy resins.

Details of the chemical composition and properties of the wide range of plastics used as engineering materials can be found in the books by Butt and Wright (1980), Evans (1974) and Harper (2001).

The biggest use of plastics is for piping; sheets are also used for lining vessels and for fabricated ducting and fan casings. Mouldings are used for small items, such as pump impellers, valve parts and pipe fittings.

The mechanical strength and operating temperature of plastics are low compared with metals. The mechanical strength, and other properties, can be modified by the addition of fillers and plasticizers. When reinforced with glass or carbon fibres, thermosetting plastics can have a strength equivalent to mild steel, and are used for pressure vessels and pressure piping. Guidelines for the design of fibre-reinforced plastic pressure vessels are given in the ASME BPV Code Sec. X Part RD and in European standards BS EN 13121 and BS EN 13923. Unlike metals, plastics are flammable. Plastics can be considered to complement metals as corrosion-resistant materials of construction. They generally have good resistance to dilute acids and inorganic salts, but suffer degradation in organic solvents that would not attack metals. Unlike metals, plastics can absorb solvents, causing swelling and softening. The properties and typical areas of use of the main plastics used for chemical plant are reviewed briefly in the following sections. A comprehensive discussion of the use of plastics as corrosion-resistant materials is given in a book by Fontana (1986). Information on selection of plastics for different applications is also given by Harper (2001). The mechanical properties and relative cost of plastics are given in Table 7.8.

TABLE 7.8. Mechanical Properties and Relative Cost of Polymers

Material	Tensile Strength (N/mm^2)	Elastic Modulus (kN/mm^2)	Density (kg/m^3)	Relative Cost
PVC	55	3.5	1400	1.5
Polyethylene (low density)	12	0.2	900	1.0
Polypropylene	35	1.5	900	1.5
PTFE	21	1.0	2100	30.0
GRP polyester	100	7.0	1500	3.0
GRP epoxy	250	14.0	1800	5.0

Approximate cost relative to polyethylene, volumetric basis.

7.9.1. Poly-Vinyl Chloride (PVC)

PVC is probably the most commonly used thermoplastic material in the chemical industry. Of the available grades, rigid (unplasticized) PVC is the most widely used. It is resistant to most inorganic acids, except strong sulphuric and nitric, and inorganic salt solutions. It is unsuitable, due to swelling, for use with most organic solvents. The maximum operating temperature for PVC is low, 60°C. The use of PVC as a material of construction in chemical engineering is discussed in a series of articles by Mottram and Lever (1957).

7.9.2. Polyolefins

Low-density polyethylene is a relatively cheap, tough, flexible plastic. It has a low softening point and is not suitable for use above about 60°C. The higher density polymer (950 kg/m^3) is stiffer, and can be used at higher temperatures. Polypropylene is a stronger material than the polyethylenes and can be used at temperatures up to 120°C.

The chemical resistance of the polyolefins is similar to that of PVC.

7.9.3. Polytetrafluoroethylene (PTFE)

PTFE, known under the trade names Teflon and Fluon, is resistant to all chemicals, except molten alkalis and fluorine, and can be used at temperatures up to 250°C. It is a relatively weak material, but its mechanical strength can be improved by the addition of fillers (glass and carbon fibres). It is expensive and difficult to fabricate. PTFE is used extensively for gaskets, gland packings (for example on valve stems) and demister pads. As a coating, it is used to confer non-stick properties to surfaces, such as filter plates. It can also be used as a liner for vessels.

7.9.4. Polyvinylidene Fluoride (PVDF)

PVDF has properties similar to PTFE but is easier to fabricate. It has good resistance to inorganic acids and alkalis, and organic solvents. It is limited to a maximum operating temperature of 140°C.

7.9.5. Glass-Fibre-Reinforced Plastics (GRP)

The polyester resins, reinforced with glass fibre, are the most common thermosetting plastics used for chemical plant. Complex shapes can be easily formed using the techniques developed for working with reinforced plastics. Glass-reinforced plastics are relatively strong and have a good resistance to a wide range of chemicals. The mechanical strength depends on the resin used, the form of the reinforcement (chopped mat or cloth), and the ratio of resin to glass.

By using special techniques, in which the reinforcing glass fibres are wound on in the form of a continuous filament, high strength can be obtained, and this method is used to produce pressure vessels.

The polyester resins are resistant to dilute mineral acids, inorganic salts and many solvents. They are less resistant to alkalis.

Glass-fibre-reinforced epoxy resins are also used for chemical plant but are more expensive than the polyester resins. In general they are resistant to the same range of chemicals as the polyesters, but are more resistant to alkalis.

The chemical resistance of GRP is dependent on the amount of glass reinforcement used. High ratios of glass to resin give higher mechanical strength but generally lower resistance to some chemicals. The design of chemical plant equipment in GRP is the subject of a book by Malleson (1969); see also Shaddock (1971), Baines (1984), ASME BPV Code Sec. X, and BS EN 13121 and 13923.

7.9.6. Rubber

Rubber, particularly in the form of linings for tanks and pipes, has been extensively used in the chemical industry for many years. Natural rubber is most commonly used, because of its good resistance to acids (except concentrated nitric) and alkalis. It is unsuitable for use with most organic solvents.

Synthetic rubbers are also used for particular applications. Hypalon (trademark, E. I. du Pont de Nemours) has a good resistance to strongly oxidizing chemicals and can be used with nitric acid. It is unsuitable for use with chlorinated solvents. Viton (trademark, E. I. du Pont de Nemours) has a better resistance to solvents, including chlorinated solvents, than other rubbers. Both Hypalon and Viton are expensive compared with other synthetic, and natural, rubbers.

The use of natural rubber lining is discussed by Saxman (1965) and the chemical resistance of synthetic rubbers by Evans (1963).

Butt and Wright (1984) give an authoritative account of the application and uses of rubber and plastic linings and coatings.

7.10. CERAMIC MATERIALS (SILICATE MATERIALS)

Ceramics are compounds of non-metallic elements and include the following materials used for chemical plants:

Glass, the borosilicate glasses (hard glass)
Stoneware
Acid-resistant bricks and tiles
Refractory materials
Cements and concrete.

Ceramic materials have a cross-linked structure and are therefore brittle.

7.10.1. Glass

Borosilicate glass (known by several trade names, including Pyrex) is used for chemical plants as it is stronger than the soda glass used for general purposes; it is more resistant to thermal shock and chemical attack. Glass equipment is often used in the small-scale manufacture of specialty chemicals. Glass can be used up to moderately high temperatures (700°C) but is not suitable for pressures above atmospheric unless used as a lining.

Glass equipment is available from several specialist manufacturers. Pipes and fittings are produced in a range of sizes, up to 0.5 m. Special equipment, such as heat exchangers, is available and, together with the larger sizes of pipe, can be used to construct distillation and absorption columns. Teflon gaskets are normally used for jointing glass equipment and pipes.

Where failure of the glass could cause injury, pipes and equipment should be protected by external shielding or wrapping with plastic tape. Glass apparatus should allow adequate venting to the atmosphere to handle anticipated relief scenarios without accumulating high pressure.

Glass linings, also known as glass enamel, have been used on steel and iron vessels for many years. Borosilicate glass is used, and the thickness of the lining varies from about 1 mm down to a few microns. The techniques used for glass lining, and the precautions to be taken in the design and fabrication of vessels to ensure a satisfactory lining, are discussed by Landels and Stout (1970) and in the ASME BPV Code Sec. VIII Div. 1, Mandatory Appendix 27. Borosilicate glass is resistant to acids, salts and organic chemicals. It is attacked by the caustic alkalis and fluorine.

7.10.2. Stoneware

Chemical stoneware is similar to the domestic variety but of higher quality: stronger and with a better glaze. It is available in a variety of shapes for pipe runs and columns. As for glass, it is resistant to most chemicals, except alkalis and fluorine. The composition and properties of chemical stoneware are discussed by Holdridge (1961). Stoneware and porcelain shapes are used for packing absorption and distillation columns (see Chapter 11).

7.10.3. Acid-Resistant Bricks and Tiles

High-quality bricks and tiles are used for lining vessels and ditches, and to cover floors. The linings are usually backed with a corrosion-resistant membrane of rubber or plastic, placed behind the tiles, and special acid-resistant cements are used for the joints. Brick and tile linings are covered in a book by Falcke and Lorentz (1985).

7.10.4. Refractory Materials (Refractories)

Refractory bricks and cements are needed for equipment operating at high temperatures, such as fired heaters, high-temperature reactors and boilers.

The refractory bricks in common use are composed of mixtures of silica (SiO_2) and alumina (Al_2O_3). The quality of the bricks is largely determined by the relative amounts of these materials and the firing temperature. Mixtures of silica and alumina form a eutectic (94.5% SiO_2, 1545°C) and for a high refractoriness under load (the ability to resist distortion at high temperature) the composition must be well removed from the eutectic composition. The highest quality refractory bricks, for use in load-bearing structures at high temperatures, contain high proportions of silica or alumina. 'Silica bricks', containing greater than 98% SiO_2, are used for general furnace construction. High alumina bricks, 60% Al_2O_3, are used for special furnaces where resistance to attack by alkalis is important, such as lime and cement kilns. Fire bricks, typical composition 50% SiO_2, 40% Al_2O_3, balance CaO and Fe_2O_3, are used for general furnace construction. Silica can exist in a variety of allotropic forms, and bricks containing a high proportion of silica undergo reversible expansion when heated up to working temperature. The higher the silica content, the greater the expansion and this must be allowed for in furnace design and operation.

Ordinary fire bricks, fire bricks with a high porosity, and special bricks composed of diatomaceous earths are used for insulating walls.

Full details of the refractory materials used for process and metallurgical furnaces can be found in the books by Norton (1968) and Lyle (1947). Additional information on refractories can be found in the books by Schacht (1995, 2004) and Routschka (1997).

7.11. CARBON

Impervious carbon, impregnated with chemically resistant resins, is used for specialized equipment, particularly heat exchangers. It has a high thermal conductivity and a good resistance to most chemicals, except oxidizing acids of concentrations greater than 30%. Carbon tubes can be used in conventional shell and tube exchanger arrangements, or proprietary designs can be used, in which the fluid channels are formed in blocks of carbon; see Hilland (1960) and Denyer (1991).

7.12. PROTECTIVE COATINGS

A wide range of paints and other organic coatings are used for the protection of mild steel structures. Paints are used mainly for protection from atmospheric corrosion. Special chemically resistant paints have been developed for use on chemical process equipment. Chlorinated rubber paints and epoxy-based paints are used. In the application of paints and other coatings, good surface preparation is essential to ensure good adhesion of the paint film or coating.

Brief reviews of the paints used to protect chemical plant are given by Ruff (1984) and Hullcoop (1984).

7.13. DESIGN FOR CORROSION RESISTANCE

The life of equipment subjected to corrosive environments can be increased by proper attention to design details. Equipment should be designed to drain freely and completely. The internal surfaces should be smooth and free from crevices where corrosion products and other solids can accumulate. Butt joints should be used in preference to lap joints. The use of dissimilar metals in contact should be avoided, or care taken to ensure that they are effectively insulated to avoid galvanic corrosion. Fluid velocities and turbulence should be high enough to avoid the deposition of solids, but not so high as to cause erosion–corrosion. The design and operating procedures should make allowance for changes in the environment to which the materials are exposed. For example, heating and cooling rates should be slow enough to prevent thermal shocks and care should be taken during maintenance not to damage corrosion-resistant films that have developed during operation.

7.14. REFERENCES

Baines, D. (1984) *Chem. Engr., London* No. 161 (July) 24. Glass reinforced plastics in the process industries.

Bendall, K. and Guha, P. (1990) *Process Industry Journal* (Mar.) 31. Balancing the cost of corrosion resistance.

Boyd, G. M. (1970) *Brittle Fracture of Steel Structures* (Butterworths).

Butt, L. T. and Wright, D. C. (1980) *Use of Polymers in Chemical Plant Construction* (Applied Science).

Clark, E. E. (1970) Chem. Engr., London No. 242 (Oct.) 312. Carbon steels for the construction of chemical and allied plant.

Craig, B. D. and Anderson, D. B. (1995) *Handbook of Corrosion Data* (ASM International).

Day, M. F. (1979) *Materials for High Temperature Use, Engineering Design Guide* No. 28 (Oxford University Press).

Dechema (1987) *Corrosion Handbook* (VCH).

Deily, J. E. (1997) *Chem. Eng. Prog.* **93** (June) 50. Use titanium to stand up to corrosives.

Denyer, M. (1991) *Processing* (July) 23. Graphite as a material for heat exchangers.

Dillon, C. P. (1986) *Corrosion Control in the Chemical Industry* (McGraw-Hill).

Evans, L. S. (1963) *Rubber and Plastics Age* **44**, 1349. The chemical resistance of rubber and plastics.

Evans, L. S. (1974) *Selecting Engineering Materials for Chemical and Process Plant* (Business Books); see also 2nd edn (Hutchinson, 1980).

Evans, L. S. (1980) *Chemical and Process Plant: a Guide to the Selection of Engineering Materials*, 2nd edn (Hutchinson).

Falcke, F. K. and Lorentz, G. (eds) (1985) *Handbook of Acid Proof Construction* (VCH).

Fensom, D. H. and Clark, B. (1984) *Chem. Engr., London* No. 162 (Aug.) 46. Tantalum: Its uses in the chemical industry.

Fontana, M. G. (1986) *Corrosion Engineering*, 3rd edn (McGraw-Hill).

Gordon, J. E. (1976) *The New Science of Strong Materials*, 2nd edn. (Penguin Books).

Green, D. W. and Perry, R. H. (eds) (2007) *Perry's Chemical Engineers' Handbook*, 8th edn (McGraw-Hill).

Hamner, N. E. (1974) *Corrosion Data Survey*, 5th edn (National Association of Corrosion Engineers).

Harper, C. A. (2001) *Handbook of Materials for Product Design* (McGraw-Hill).

Harris, W. J. (1976) *The Significance of Fatigue* (Oxford University Press).

Hilland, A. (1960) *Chem. and Proc. Eng* **41**, 416. Graphite for heat exchangers.

Hills, R. F. and Harries, D. P. (1960) *Chem. and Proc. Eng* **41**, 391. Sigma phase in austenitic stainless steel.

Holdridge, D. A. (1961) *Chem. and Proc. Eng.* **42**, 405. Ceramics.

Hullcoop, R. (1984) Processing (April) 13. The great cover up.

Institute of Metallurgists (1960) *Toughness and Brittleness of Metals* (Iliffe).

Jowitt, R. (ed.) (1980) *Hygienic design and operation of food plant* (Ellis Horwood).

Lai, G. Y. (1990) *High Temperature Corrosion of Engineering Alloys* (ASM International).

Landels, H. H. and Stout, E. (1970) *Brit. Chem. Eng* **15**, 1289. Glassed steel equipment: a guide to current technology.

Llewellyn, D. T. (1992) *Steels: Metallurgy and Applications* (Butterworth-Heinemann).

Lyle, O. (1947) *Efficient Use of Steam* (HMSO).

Malleson, J. H. (1969) *Chemical Plant Design with Reinforced Plastics* (McGraw-Hill).

Moore, D. C. (1970) *Chem. Engr. London* No. 242 (Oct.) 326. Copper.

Moore, R. E. (1979) *Chem. Eng., NY* **86** (July 30th) 91. Selecting materials to resist corrosive conditions.

Mottram, S. and Lever, D. A. (1957) *The Ind. Chem.* **33**, 62, 123, 177 (in three parts). Unplasticized P. V. C. as a constructional material in chemical engineering.

NACE (1974) *Standard TM-01-69 Laboratory Corrosion Testing of Metals for the Process Industries* (National Association of Corrosion Engineers).

Norton, F. H. (1968) *Refractories*, 4th edn (McGraw-Hill).

Peckner, D. and Bernstein, I. M. (1977) *Handbook of Stainless Steels* (McGraw-Hill).

Rabald, E. (1968) *Corrosion Guide*, 2nd edn (Elsevier).

Revie, R. W. (2005) *Uhlig's Corrosion Handbook*, 2nd edn (Wiley).

Routschka, G. (1997) *Pocket Manual of Refractory Materials*, 1st edn (Vulkan-Verlag)

Rowe, D. (1994) *Process Industry Journal* (March) 37. Tempted by tantalum.

Rowe, D. (1999) *Chem. Engr., London* No. 683 (June 24) 19. Tantalising Materials.

Ruff, C. (1984) *Chem. Engr., London* No. 409 (Dec.) 27. Paint for Plants.

Saxman, T. E. (1965) *Materials Protection* **4** (Oct.) 43. Natural rubber tank linings.

Schacht, C. A. (1995) *Refractory Linings: Thermochemical Design and Applications* (Marcel Dekker).

Schacht, C. A. (2004) *Refractories Handbook* (Marcel Dekker).

Schweitzer, P. A. (1976) *Corrosion Resistance Tables* (Marcel Dekker).

Schweitzer, P. A. (1989) (ed.) *Corrosion and Corrosion Protection Handbook*, 2nd edn (Marcel Dekker).

Schweitzer, P. A. (1998) *Encyclopedia of Corrosion Protection* (Marcel Dekker).

Sedriks, A. J. (1979) *Corrosion Resistance of Stainless Steel* (Wiley).

Shaddock, A. K. (1971) *Chem. Eng., NY* **78** (Aug. 9th) 116. Designing for reinforced plastics.

Timperley, D. A. (1984) *Inst. Chem. Eng. Sym. Ser.* No. 84, 31. Surface finish and spray cleaning of stainless steel.

Turner, M. (1989) *Chem. Engr., London* No. 460 (May) 52. What every chemical engineer should know about stress corrosion cracking.

Warde, E. (1991) *Chem. Engr., London* No. 502 (Aug. 15th) 35. Which super-duplex?

Weisert, E. D. (1952a) *Chem. Eng., NY* **59** (June) 267. Hastelloy alloy C.

Weisert, E. D. (1952b) *Chem. Eng., NY* **59** (July) 314. Hastelloy alloy B.

Wells, A. A. (1968) *British Welding Journal* **15**, 221. Fracture control of thick steels for pressure vessels.

Wigley, D. A. (1978) *Materials for Low Temperatures, Engineering Design Guide* No. 28 (Oxford University Press).

British and European Standards

BS EN 13121 (2003) GRP tanks and vessels for use above ground.

BS EN 13445-1 (2002) Unfired pressure vessels.

BS EN 13923 (2006) Filament-wound FRP pressure vessels. Materials, design, manufacturing and testing.

American and International Standards

ASME Boiler and Pressure Vessel Code Section II Part D. Materials Properties. 2005.
ASME Boiler and Pressure Vessel Code Section VIII Division 1. Rules for the Construction of Pressure Vessels. 2006.
ASME Boiler and Pressure Vessel Code Section X. Fibre-reinforced Plastic Vessels. 2005.

Bibliography

Further reading on materials, materials selection and equipment fabrication.
Callister, W. D. (1991) *Materials Science and Engineering, an Introduction* (Wiley).
Champion, F. A. (1967) *Corrosion Testing Procedures*, 3rd edn (Chapman Hall).
Crane, F. A. A. and Charles, J. A. (1989) *Selection and Use of Engineering Materials*, 2nd edn (Butterworths).
Ewalds, H. L. (1984) *Fracture Mechanics* (Arnold).
Flinn, R. A. and Trojan, P. K. (1990) *Engineering Materials and Their Applications*, 4th edn (Houghton Mifflin).
Ray, M. S. (1987) *The Technology and Application of Engineering Materials* (Prentice Hall).
Rolfe, S. T. (1987) *Fracture Mechanics and Fatigue Control in Structures*, 2nd edn (Prentice Hall).

7.15. NOMENCLATURE

		Dimensions in MLT$
A	Area	L^2
C	Cost of material	$/M$
t	Time	T
w	Mass loss	M
ρ	Density	ML^{-3}
σ_d	Maximum allowable stress	$ML^{-1}T^{-2}$

7.16. PROBLEMS

7.1. A pipeline constructed of carbon steel failed after 3 years of operation. On examination it was found that the wall thickness had been reduced by corrosion to about half the original value. The pipeline was constructed of nominal 100 mm (4 in) schedule 40, pipe, inside diameter 102.3 mm (4.026 in), outside diameter 114.3 mm (4.5 in). Estimate the rate of corrosion in ipy and mm per year.

7.2. The pipeline described in question 7.1 was used to carry wastewater to a hold-up tank. The effluent is not hazardous. A decision has to be made on what material to use to replace the pipe. Three suggestions have been made:
1. Replace with the same schedule carbon steel pipe and accept renewal at 3-year intervals.
2. Replace with a thicker pipe, schedule 80, outside diameter 114.3 mm (4.5 in), inside diameter 97.2 mm (3.826 in).

3. Use stainless steel pipe, which will not corrode.

The estimated cost of the pipes, per unit length is: schedule 40 carbon steel $5, schedule 80 carbon steel $8.3, stainless steel (304) schedule 40 $24.8.

Installation and fittings for all the materials adds $16.5 per unit length.

The downtime required to replace the pipe does not result in a loss of production.

If the expected future life of the plant is 7 years, recommend which pipe to use.

7.3. Choose a suitable material of construction for the following duties:
1. 98% w/w sulphuric acid at 70°C
2. 5% w/w sulphuric acid at 30°C
3. 30% w/w hydrochloric acid at 50°C
4. 5% aqueous sodium hydroxide solution at 30°C
5. Concentrated aqueous sodium hydroxide solution at 50°C
6. 5% w/w nitric acid at 30°C
7. Boiling concentrated nitric acid
8. 10% w/w sodium chloride solution
9. A 5% w/w solution of cuprous chloride in hydrochloric acid
10. 10% w/w hydrofluoric acid

In each case, select the material for a 50-mm pipe operating at approximately 2-bar pressure.

7.4. Suggest suitable materials of construction for the following applications:
1. A 10,000-m^3 storage tank for toluene.
2. A 5.0-m^3 tank for storing a 30% w/w aqueous solution of sodium chloride.
3. A 2-m diameter, 20-m-high distillation column, distilling acrylonitrile.
4. A 100-m^3 storage tank for strong nitric acid.
5. A 500-m^3 aqueous waste hold-up tank. The wastewater pH can vary from 1 to 12. The wastewater will also contain traces of organic material.
6. A packed absorption column 0.5 m diameter, 3 m high, absorbing gaseous hydrochloric acid into water. The column will operate at essentially atmospheric pressure.

7.5. Aniline is manufactured by the hydrogenation of nitrobenzene in a fluidized bed reactor. The reactor operates at 250°C and 20 bar. The reactor vessel is approximately 3 m diameter and 9 m high. Suggest suitable materials of construction for this reactor.

7.6. Methyl ethyl ketone is manufactured by the dehydrogenation of 2-butanol using a shell and tube type of reactor. Flue gases are used for heating and pass though the tubes. The flue gases will contain traces of sulphur dioxide. The reaction products include hydrogen.

The reaction takes place in the shell at a pressure of 3 bar and temperature of 500°C. Select suitable materials for the tubes and shell.

7.7. In the manufacture of aniline by the hydrogenation of nitrobenzene, the off-gases from the reactor are cooled and the products and unreacted nitro-benzene condensed in a shell and tube exchanger. A typical composition of the condensate is, in kmol/h: aniline 950, cyclo-hexylamine 10, water 1920, nitrobenzene 40. The gases enter the condenser at 230°C and leave at 50°C. The cooling water enters the tubes at 20°C and leaves at 50°C. Suggest suitable materials of construction for the shell and the tubes.

7.8. A slurry of acrylic polymer particles in water is held in storage tanks prior to filtering and drying. Plain carbon steel would be a suitable material for the tanks, but it is essential that the polymer does not become contaminated with iron in storage. Suggest some alternative materials of construction for the tanks.

7.9. Coal gasification is carried out at 850°C and 40 atmospheres pressure, by reaction of coal with steam and oxygen. The empirical formula of the coal is roughly $CH_{0.8}S_{0.013}$. Recommend materials of construction for:
1. The coal addition system
2. The oxygen injection system
3. The gasification reactor
4. The product gas transfer line.

8 DESIGN INFORMATION AND DATA

Chapter Contents

Key Learning Objectives

- How to obtain the chemical and physical properties needed for design calculations
- How commercial process simulators predict properties for compounds when little or no data are available
- How to select a suitable phase equilibrium model

8.1.   INTRODUCTION

Information on manufacturing processes, equipment parameters, materials of construction, costs and the physical properties of process materials are needed at all stages of design, from the initial screening of possible processes to the plant start-up and production.

Sources of data on costs were discussed in Chapter 6 and materials of construction in Chapter 7. This chapter covers sources of information on manufacturing processes and physical properties, and the estimation of physical property data. Information on the types of equipment (unit operations) used in chemical process plants is given in Volume 2, and in the chapters concerned with equipment selection and design in this volume, Chapters 10, 11 and 12.

When a project is largely a repeat of a previous project, the data and information required for the design will be available in the company's process files, if proper detailed records are kept. For a new project or process, the design data must be obtained from the literature, or by experiment (research laboratory and pilot plant), or be purchased from other companies. The information on manufacturing processes available in the general literature can be of use in the initial stages of process design for screening potential processes, but is usually mainly descriptive and too superficial to be of much use for detailed design and evaluation.

The literature on the physical properties of elements and compounds is extensive, and reliable values for common materials can usually be found. The principal sources of physical property data are listed in the references at the end of this chapter.

Where values cannot be found, the data required must be measured experimentally or estimated. Methods of estimating (predicting) the more important physical properties required for design are given in this chapter. A physical property data bank is given in Appendix C and is available in MS Excel format in the on-line material at http://books.elsevier.com/companions.

Readers who are unfamiliar with the sources of information, and the techniques used for searching the literature, should consult one of the many guides to the technical literature that have been published, such as those by Lord (2000) and Maizell (1998).

8.2.   SOURCES OF INFORMATION ON MANUFACTURING PROCESSES

In this section, the sources of information available in the open literature on commercial processes for the production of chemicals and related products are reviewed.

The chemical process industries are competitive, and the information that is published on commercial processes is restricted. The articles on particular processes published in the technical literature and in textbooks invariably give only a superficial account of the chemistry and unit operations used. They lack the detailed information on reaction kinetics, process conditions, equipment parameters and physical properties that is needed for process design. The information that can be found in the general literature is, however, useful in the early stages of a project, when searching for possible process routes. It is often sufficient for a flow-sheet of the process to be drawn up and a rough estimate of the capital and production costs made.

The most comprehensive collection of information on manufacturing processes is probably the *Encyclopaedia of Chemical Technology* edited by Kirk and Othmer (2001) (2003), which covers the whole range of chemical and associated products. Another encyclopaedia covering manufacturing processes is that edited by McKetta (2001). Several books have also been published that give brief summaries of the production processes used for the commercial chemicals and chemical products. The best known of these is probably Shreve's book on the chemical process industries, now updated by Austin and Basta (1998). Comyns (1993) lists named chemical manufacturing processes, with references.

The extensive German reference work on industrial processes, *Ullman's Encyclopaedia of Industrial Technology*, is now available in an English translation, Ullman (2002).

Specialized texts have been published on some of the more important bulk industrial chemicals, such as that by Miller (1969) on ethylene and its derivatives; these are too numerous to list but should be available in the larger reference libraries and can be found by reference to the library catalogue. Myers (2003) gives a good introduction to the processes used in oil refining. Kohl and Nielsen (1997) provide an excellent overview of the processes used for gas treating and sulphur recovery.

Books quickly become outdated, and many of the processes described are obsolete, or at best obsolescent. More up-to-date descriptions of the processes in current use can be found in the technical journals. The journal *Hydrocarbon Processing* publishes an annual review of petrochemical processes, which was entitled *Petrochemical Developments* and is now called *Petrochemicals Notebook*; this gives flow-diagrams and brief process descriptions of new process developments.

Patents

Patents can be a useful source of information, but some care is needed in extracting information from them. To obtain a patent, an inventor is legally obliged to disclose the best mode of practice of the invention; failure to do so could render the patent invalid if it were contested. Most patents therefore include one or more examples illustrating how the invention is practiced and differentiating it from the prior art. The examples given in a patent often give an indication of the process conditions used, although they are frequently examples of laboratory preparations rather than of the full-scale manufacturing processes. Many process patents also include examples based on computer simulations, in which case the data should be viewed with suspicion. When using data from patents, it is important to carefully read the section that describes the experimental procedure to be sure that the experiments were run under appropriate conditions.

A patent gives its owner the right to sue anyone who practices the technology described in the patent claims without a license from the patent owner. Patent attorneys generally try to write patents to claim broad ranges of process conditions, so as to maximize the range of validity and make it hard for competitors to avoid the patent by making a slight change in temperature, pressure or other process parameters. Very often, a patent will say something along the lines of: 'the reaction is carried out at a temperature in the range 50 to 500°C, more preferably in the range 100 to 300°C and most preferably in the range 200 to 250°C'. It is usually possible to use engineering judgment to determine the optimal conditions from such ranges. The best conditions will usually be at or near the upper or lower end of the narrowest defined range. The examples in the patent will often indicate the best operating point.

Patents can be downloaded for free from the web site of the US patent office, www. uspto.gov. The US PTO web site also has limited search capability. Most large companies subscribe to more sophisticated patent search services such as Delphion (www.delphion. com), PatBase (www.patbase.com) and GetthePatent (www.getthepatent.com).

Several guides have been written to help engineers understand the use of patents for the protection of inventions and as sources of information, such as those by Auger (1992) and Gordon and Cookfair (2000).

The Internet

It is worthwhile searching on-line for information on processes, equipment, products and physical properties. Many manufacturers and government departments maintain web sites. In particular, up-to-date information can be obtained on the health and environmental effects of products.

Many university libraries or engineering departments provide information guides for students and these are available on the internet. A search using the key words such as 'chemical engineering information' will usually find them. Some examples are:

- The University of Manchester, UK, Heriot-Watt University, Edinburgh, UK and the Joint Information Systems Committee: www.intute.ac.uk/sciences/
- University of Florida Web Virtual Library: www.che.ufl.edu/www-che/
- Karlsburg University, Germany: International Directory of Chemical Engineering URLs: www.ciw.uni-karlsruhe.de/links.php

Many of the important sources of engineering information are subscription services. The American Chemical Society's Chemical Abstracts Service is the best source for chemical properties and reaction kinetics data. Chemical abstracts can be searched on-line through the SciFinder subscription service (www.cas.org). This is available in most university libraries.

Another important source of information is Knovel. This provides on-line access to most standard reference books. It is a subscription service but can be accessed through many libraries, including those of the professional engineering institutions and some universities. At the time of writing, Knovel is available for free to members of the AIChE and IChemE. In addition to having many reference books in .pdf format, Knovel has interactive graphs and look-up tables for books such as *Perry's Chemical Engineers Handbook* and the *International Critical Tables*.

8.3. GENERAL SOURCES OF PHYSICAL PROPERTIES

In this section, those references that contain comprehensive compilations of physical property data are reviewed. Sources of data on specific physical properties are given in the remaining sections of the chapter.

International Critical Tables (1933) is still probably the most comprehensive compilation of physical properties, and is available in most reference libraries. Though it was first published in 1933, physical properties do not change, except in as much as experimental techniques improve, and ICT is still a useful source of engineering data. ICT is now available as an e-book and can be referenced on the internet through Knovel (2003).

Tables and graphs of physical properties are given in many handbooks and textbooks on chemical engineering and related subjects. Many of the data given are duplicated from book to book, but the various handbooks do provide quick, easy access to data on the more commonly used substances.

An extensive compilation of thermophysical data has been published by Plenum Press (Touloukian, 1970–1977). This multiple-volume work covers conductivity, specific heat, thermal expansion, viscosity and radiative properties (emittance, reflectance, absorptance and transmittance).

Elsevier have published a series of volumes on physical property and thermodynamic data. Those of use in design are included in the Bibliography at the end of this chapter.

The Engineering Sciences Data Unit (ESDU, www.ihsesdu.com) was set up to provide validated data for engineering design, developed under the guidance and approval of engineers from industry, the universities and research laboratories. ESDU data include equipment design data and software and extensive high-quality physical property data – mostly for pure fluids that are in use in the oil and process industries.

Caution should be exercised when taking data from the literature, as typographical errors often occur. If a value looks doubtful it should be cross-checked in an independent reference, or by estimation.

The values of some properties are dependent on the method of measurement; for example, surface tension and flash point, and the method used should be checked, by reference to the original paper if necessary, if an accurate value is required.

The results of research work on physical properties are reported in the general engineering and scientific literature. The *Journal of Chemical Engineering Data* specializes in publishing physical property data for use in chemical engineering design. A quick search of the literature for data can be made by using the abstracting journals; such as *Chemical Abstracts* (American Chemical Society) and *Engineering Index* (Engineering Index Inc., New York). *Engineering Index* is now called Engineering Information (EI) and is a web-based reference source owned by Elsevier information (www.ei.org).

Computerized physical property data banks have been set up by various organizations to provide a service to the design engineer. They can be incorporated into computer-aided design programs and are increasingly being used to provide reliable, authenticated, design data. Examples of such programs are the PPDS and the DIPPR™ databases.

PPDS (Physical Property Data Service) was originally developed in the UK by the Institution of Chemical Engineers and the National Physical Laboratory. It is now available as a Microsoft™ Windows version from NEL, a division of the TUV Suddeutschland Group (www.tuvnel.com/content/ppds.aspx). PPDS is made available to universities at a discount.

The DIPPR™ databases were developed in the USA by the Design Institute for Physical Properties of the American Institute of Chemical Engineers. The DIPPR™ projects are aimed at providing evaluated process design data for the design of chemical processes and equipment (www.aiche.org/TechnicalSocieties/DIPPR/index. aspx). The DIPPR Project 801 has been made available to university departments; see Rowley *et al.* (2004).

8.4. ACCURACY REQUIRED OF ENGINEERING DATA

The accuracy needed depends on the use to which the data will be put. Before spending time and money searching for the most accurate value, or arranging for special measurements to be made, the designer must decide what accuracy is required; this will depend on several factors:

1. The level of design: less accuracy is obviously needed for rough screening calculations, made to sort out possible alternative designs, than in the final stages of design, when money will be committed to purchase equipment.
2. The reliability of the design methods: if there is some uncertainty in the techniques to be used, it is clearly a waste of time to search out highly accurate physical property data that will add little or nothing to the reliability of the final design.
3. The sensitivity to the particular property: how much will a small error in the property affect the design calculation? For example, it was shown in Chapter 5 that the estimation of the optimum pipe diameter is insensitive to viscosity. The sensitivity of a design method to errors in physical properties, and other data, can be checked by repeating the calculation using slightly altered values.

It is often sufficient to estimate a value for a property (sometimes even to make an intelligent guess) if the value has little effect on the final outcome of the design calculation. For example, in calculating the heat load for a reboiler or vaporizer an accurate value of the liquid specific heat is seldom needed, as the latent heat load is usually many times the sensible heat load and a small error in the sensible heat calculation will have little effect on the design. The designer must, however, exercise caution when deciding to use less reliable data, and must be sure that they are sufficiently accurate for the design purpose. For example, it would be correct to use an approximate value for density when calculating the pressure drop in a pipe system where a small error could be tolerated, considering the other probable uncertainties in the design; but it would be quite unacceptable in the design of a decanter, where the operation depends on small differences in density.

Consider the accuracy of the equilibrium data required to calculate the number of equilibrium stages needed for the separation of a mixture of acetone and water by distillation (see Chapter 11, Example 11.2). Several investigators have published vapour–liquid equilibrium (VLE) data for this system: Othmer *et al.* (1952), York and Holmes (1942), Kojima *et al.* (1968), Reinders and De Minjer (1947).

If the purity of the acetone product required is less than 95%, inaccuracies in the VLE plot have little effect on the estimate of the number of stages required, as the relative volatility is very high. If a high purity is wanted, say >99%, then reliable data are needed in this region as the equilibrium line approaches the operating line (a pinch point occurs). Of the references cited, none gives values in the region above 95%, and only two give values above 90%; more experimental values are needed to design with confidence. There is a possibility that the system forms an azeotrope in this region. An azeotrope does form at higher pressure (Othmer *et al.*, 1952).

8.5. PREDICTION OF PHYSICAL PROPERTIES

Whenever possible, experimentally determined values of physical properties should be used. If reliable values cannot be found in the literature and if time or facilities are not available for their determination, then to proceed with the design the designer must resort to estimation. Techniques are available for the prediction of most physical properties with sufficient accuracy for use in process and equipment design. A detailed review of all the different methods available is beyond the scope of this book; selected methods are given for the more commonly needed properties. The criterion used for selecting a particular method for presentation in this chapter was to choose the most easily used, simplest, method that had sufficient accuracy for general use. If highly accurate values are required, then specialized texts on physical property estimation should be consulted, such as those by Reid *et al.* (1987), Poling *et al.* (2000), Bretsznajder (1971), Sterbacek *et al.* (1979), and AIChE (1983, 1985).

A quick check on the probable accuracy of a particular method can be made by using it to estimate the property for an analogous compound, for which experimental values are available.

The techniques used for prediction are also useful for the correlation, and extrapolation and interpolation, of experimental values.

Group contribution techniques are based on the concept that a particular physical property of a compound can be considered to be made up of contributions from the constituent atoms, groups and bonds; the contributions being determined from experimental data. They provide the designer with simple, convenient, methods for physical property estimation; requiring only a knowledge of the structural formula of the compound.

Also useful, and convenient to use, are prediction methods based on the use of reduced properties (corresponding states), providing that values for the critical properties are available or can be estimated with sufficient accuracy; see Sterbacek *et al.* (1979).

In most cases, the methods described in the following sections, or their equivalents, are available in commercial process simulation programs such as AspenPlus™, ChemCAD™, Pro II™ and UniSim™. The easiest way to estimate mixture properties is usually to set up a stream with the desired temperature, pressure and composition in one of the simulators. The design engineer should always check the results from the simulation against any available data. If no experimental data are available then it is usually a good idea to make an independent estimate of any parameters that have a strong influence on the design, to be satisfied that the results from the simulator are credible. If the independent estimate does not agree with the simulation result then it may be worthwhile to conduct some experiments to collect real data.

8.6. DENSITY

8.6.1. Liquids

Values for the density of pure liquids can usually be found in the handbooks. It should be noted that the density of most organic liquids, other than those containing a halogen or other 'heavy atom', usually lies between 700 and 1000 kg/m³. Liquid densities are given in Appendix C.

An approximate estimate of the density at the normal boiling point can be obtained from the molar volume (see Table 8.6)

$$\rho_b = \frac{M}{V_m} \tag{8.1}$$

where

ρ_b = density, kg/m³
M = molecular mass
V_m = molar volume, m³/kmol.

For mixtures, it is usually sufficient to take the specific volume of the components as additive; even for non-ideal solutions, as is illustrated by Example 8.1.

The densities of many aqueous solutions are given by Green and Perry (2007).

Example 8.1

Calculate the density of a mixture of methanol and water at 20°C, composition 40% w/w methanol.

$$\text{Density of water at 20°C} \qquad 998.2 \text{ kg/m}^3$$
$$\text{Density of methanol at 20°C} \quad 791.2 \text{ kg/m}^3$$

Solution

Basis: 1000 kg

$$\text{Volume of water} = \frac{0.6 \times 1000}{998.2} = 0.601 \text{ m}^3$$

$$\text{Volume of methanol} = \frac{0.4 \times 1000}{791.2} = \underline{0.506 \text{ m}^3}$$

$$\text{Total} \quad 1.107 \text{ m}^3$$

$$\text{Density of mixture} = \frac{1000}{1.107} = \underline{\underline{903.3 \text{ kg/m}^3}}$$

$$\text{Experimental value} = 934.5 \text{ kg/m}^3$$

$$\text{Error} = \frac{934.5 - 903.3}{903.3} = 3\%, \text{ which would be}$$

acceptable for most engineering purposes

If data on the variation of density with temperature cannot be found, they can be approximated for non-polar liquids from Smith's equation for thermal expansion (Smith *et al.*, 1954).

$$\beta = \frac{0.04314}{(T_c - T)^{0.641}} \tag{8.2}$$

where

β = coefficient of thermal expansion, K^{-1}
T_c = critical temperature, K
T = temperature, K.

8.6.2. Gas and Vapour Density (Specific Volume)

For general engineering purposes it is often sufficient to consider that real gases and vapours behave ideally, and to use the gas law:

$$PV = nRT \tag{8.3}$$

where

P = absolute pressure, N/m^2 (Pa)
V = volume, m^3
n = mols of gas
T = absolute temperature, K
R = universal gas constant, $8.314 \, J \, K^{-1} \, mol^{-1}$ (or $kJ \, K^{-1} \, kmol^{-1}$).

$$\text{Specific volume} = \frac{RT}{P} \tag{8.4}$$

These equations will be sufficiently accurate up to moderate pressures, in circumstances where the value is not critical. If greater accuracy is needed, the simplest method is to modify equation 8.3 by including the compressibility factor z:

$$PV = znRT \tag{8.5}$$

The compressibility factor can be estimated from a generalized compressibility plot, which gives z as a function of reduced pressure and temperature (Chapter 3, Figure 3.8).

For mixtures, the pseudocritical properties of the mixture should be used to obtain the compressibility factor.

$$P_{c,m} = P_{c,a} y_a + P_{c,b} y_b + \cdots \tag{8.6}$$

$$T_{c,m} = T_{c,a} y_a + T_{c,b} y_b + \cdots \tag{8.7}$$

where

P_c = critical pressure
T_c = critical temperature
y = mol fraction

suffixes
$$m = \text{mixture}$$
$$a, b, \text{etc.} = \text{components.}$$

8.7. VISCOSITY

Viscosity values are needed for any design calculations involving the transport of fluids or heat. Values for pure substances can usually be found in the literature; see Yaws (1993–1994). Liquid viscosities are given in Appendix C. Methods for the estimation of viscosity are given below.

8.7.1. Liquids

A rough estimate of the viscosity of a pure liquid at its boiling point can be obtained from the modified Arrhenius equation:

$$\mu_b = 0.01 \rho_b^{0.5} \tag{8.8}$$

where

μ_b = viscosity, mNs/m^2
ρ_b = density at boiling point, kg/m^3.

A more accurate value can be obtained if reliable values of density are available, or can be estimated with sufficient accuracy from Souders' equation (Souders, 1938):

$$\log(\log\ 10\mu) = \frac{I}{M} \rho \times 10^{-3} - 2.9 \tag{8.9}$$

where

μ = viscosity, mNs/m^2
M = molecular mass
I = Souders' index, estimated from the group contributions given in Table 8.1
ρ = density at the required temperature, kg/m^3.

Example 8.2

Estimate the viscosity of toluene at 20°C.

Solution

Toluene

TABLE 8.1. Contributions for Calculating the Viscosity Constant I in Souders' Equation

Atom	H	O	C	N	Cl	Br	I
Contribution	+2.7	+29.7	+50.2	+37.0	+60	+79	+110

Contributions of groups and bonds

Double bond	−15.5	$H-\overset{\|\|}{\underset{O}{C}}-R$	+10
Five-member ring	−24		
Six-member ring	−21		
Side groups on a six-member ring:		$-CH=CH-CH_2-X^+$	+4
Molecular weight < 17	−19	$\overset{R}{\underset{R}{>}}CH-X$	+6
Molecular weight > 16	−17		
Ortho or *para* position	+3		
Meta position	−1		
$\overset{R}{\underset{R}{>}}CH-CH\overset{R}{\underset{R}{<}}$	+8	OH	+57.1
		COO	+90
		COOH	+104.4
		NO_2	+80
$R-\overset{R}{\underset{R}{C}}-R$	+10		
$-CH_2-$	+55.6		

[†] X is a negative group.

Contributions from Table 8.1:

7 carbon atoms	7 × 50.2	= 351.4
8 hydrogen atoms	8 × 2.7	= 21.6
3 double bonds	3(−15.5)	= −46.5
1 six-membered ring		−21.1
1 side group		−9.0
	Total, I	= 296.4

Density at 20°C = 866 kg/m^3
Molecular weight 92

$$\log(\log 10\,\mu) = \frac{296.4 \times 866 \times 10^{-3}}{92} - 2.9 = -0.11$$

$$\log 10\mu = 0.776$$

$$\mu = 0.597, \text{ rounded} = 0.6 \text{ mNs/m}^2$$

Experimental value, 0.6 cp = 0.6 mNs/m^2.

Author's note: the fit obtained in this example is rather fortuitous; the usual accuracy of the method for organic liquids is around ±10%.

Variation with Temperature

If the viscosity is known at a particular temperature, the value at another temperature can be estimated with reasonable accuracy (within ±20%) by using the generalized plot of Lewis and Squires (1934); see Figure 8.1. The scale of the temperature ordinate is obtained by plotting the known value, as illustrated in Example 8.3.

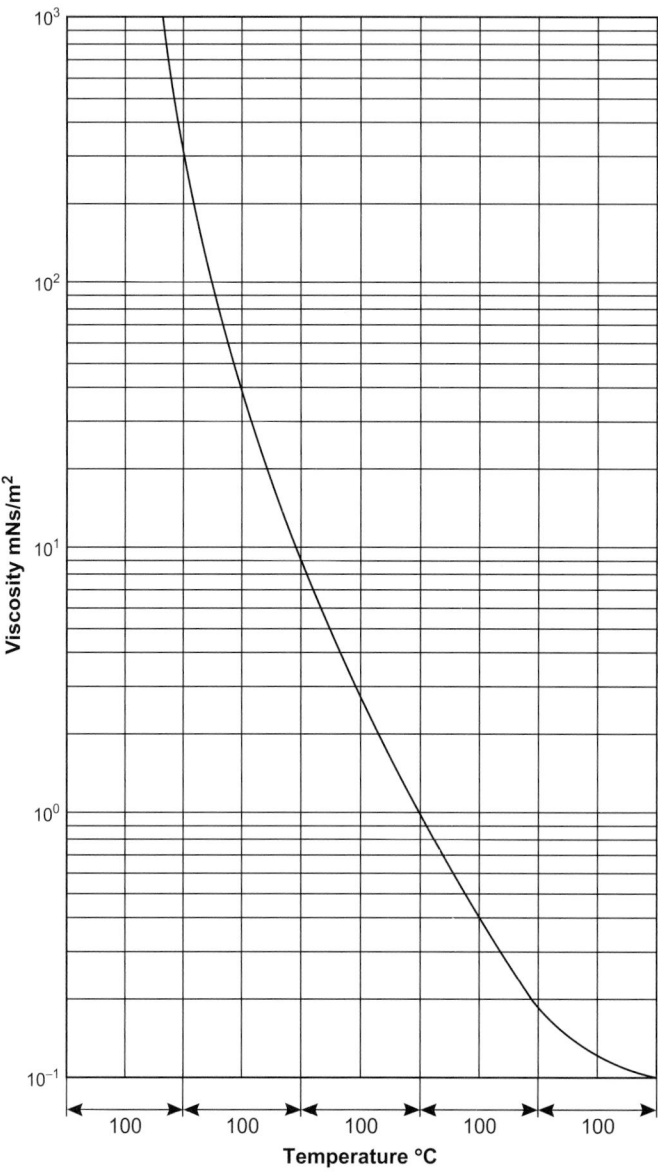

Figure 8.1.   Generalized viscosity vs temperature curve for liquids.

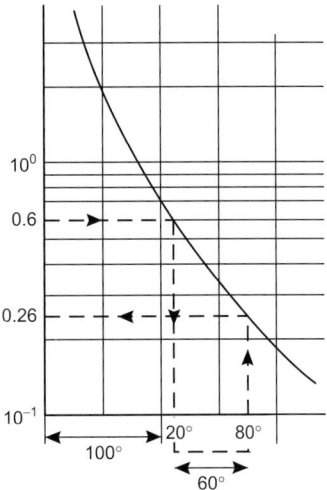

Figure 8.1a.

Example 8.3

Estimate the viscosity of toluene at 80°C, using the value at 20°C given in Example 8.2.

Solution

Temperature increment $80 - 20 = 60$°C.
From Figure 8.1a, viscosity at 80°C = 0.26 mNs/m².

Effect of Pressure

The viscosity of a liquid is dependent on pressure as well as temperature, but the effect is not significant except at very high pressures. A rise in pressure of 300 bar is roughly equivalent to a decrease in temperature of 1°C.

Mixtures

It is difficult to predict the viscosity of mixtures of liquids. Viscosities are rarely additive, and the shape of the viscosity–concentration curve can be complex. The viscosity of the mixture may be lower or, occasionally, higher than that of the pure components. A rough check on the magnitude of the likely error in a design calculation, arising from uncertainty in the viscosity of a mixture, can be made by using the smallest and largest values of the pure components in the calculation, and noting the result.

As an approximation, the variation can be assumed to be linear, if the range of viscosity is not very wide, and a weighted average viscosity calculated. For organic liquid mixtures a modified form of Souders' equation can be used, using a mol fraction weighted average value for the viscosity constant for the mixture I_m and the average molecular weight.

For a binary mixture equation 8.9 becomes:

$$\log(\log 10 \, \mu_m) = \rho_m \left[\frac{x_1 I_1 + x_2 I_2}{x_1 M_1 + x_2 M_2} \right] \times 10^{-3} - 2.9 \qquad (8.10)$$

where

μ_m = viscosity of mixture
ρ_m = density of mixture
x_1, x_2 = mol fraction of components
M_1, M_2 = molecular masses of components.

Bretsznajder (1971) gives a detailed review of the methods that have been developed for estimating the viscosity of mixtures, including methods for aqueous solutions and dispersions.

For heat-transfer calculations, Kern (1950) gives a rough rule of thumb for organic liquid mixtures:

$$\frac{1}{\mu_m} = \frac{w_1}{\mu_1} + \frac{w_2}{\mu_2} \qquad (8.11)$$

where

w_1, w_2 = mass fractions of the components 1 and 2
μ_1, μ_2 = viscosities of components 1 and 2.

8.7.2. Gases

Reliable methods for the prediction of gas viscosities, and the effect of temperature and pressure, are given by Bretsznajder (1971) and Reid et al. (1987).

Where an estimate of the viscosity is needed to calculate Prandtl numbers the methods developed for the direct estimation of Prandtl numbers should be used.

For gases at low pressure Bromley (1952) has suggested the following values:

	Prandtl number
Monatomic gases (e.g. Ar, He)	$0.67 \pm 5\%$
Non-polar, linear molecules (e.g. O_2, Cl_2)	$0.73 \pm 15\%$
Non-polar, non-linear molecules (e.g. CH_4, C_6H_6)	$0.79 \pm 15\%$
Strongly polar molecules (e.g. CH_3OH, SO_2, HCl)	$0.86 \pm 8\%$

The Prandtl number for gases varies only slightly with temperature.

8.8. THERMAL CONDUCTIVITY

The experimental methods used for the determination of thermal conductivity are described by Tsederberg (1965), who also lists values for many substances. The four-volume handbook by Yaws (1995–1997) is a useful source of thermal conductivity data for hydrocarbons and inorganic compounds.

8.8.1. Solids

The thermal conductivity of a solid is determined by its form and structure, as well as composition. Values for the commonly used engineering materials are given in various handbooks, and in Table 12.6.

8.8.2. Liquids

The data available in the literature up to 1973 have been reviewed by Jamieson *et al.* (1975). The Weber equation (Weber, 1880) can be used to make a rough estimate of the thermal conductivity of organic liquids, for use in heat-transfer calculations.

$$k = 3.56 \times 10^{-5} C_p \left(\frac{\rho^4}{M}\right)^{1/3} \tag{8.12}$$

where

k = thermal conductivity. W/m°C
M = molecular mass
C_p = specific heat capacity, kJ/kg°C
ρ = density, kg/m^3.

Bretsznajder (1971) gives a group contribution method for estimating the thermal conductivity of liquids.

Example 8.4

Estimate the thermal conductivity of benzene at 30°C.

Solution

Density at 30°C = 875 kg/m^3
Molecular mass = 78
Specific heat capacity = 1.75 kJ/kg°C

$$k = 3.56 \times 10^{-5} \times 1.75 \left(\frac{875^4}{78}\right)^{1/3} = \underline{\underline{0.12 \text{ W/m°C}}} \tag{8.12}$$

Experimental value, 0.16 W/m°C, error 25%.

8.8.3. Gases

Approximate values for the thermal conductivity of pure gases, up to moderate pressures, can be estimated from values of the gas viscosity, using Eucken's equation (Eucken, 1912):

$$k = \mu \left(C_p + \frac{10.4}{M}\right) \tag{8.13}$$

where

μ = viscosity, mNs/m^2
C_p = specific heat capacity, kJ/kg°C
M = molecular mass.

Example 8.5

Estimate the thermal conductivity of ethane at 1 bar and 450°C.

Solution

$$\text{Viscosity} = 0.0134 \, \text{mNs/m}^2$$
$$\text{Specific heat capacity} = 2.47 \, \text{kJ/kg°C}$$

$$k = 0.0134 \left(2.47 + \frac{10.4}{30} \right) = \underline{\underline{0.038 \ \text{W/m°C}}} \tag{8.13}$$

Experimental value, 0.043 W/m°C, error 12%.

8.8.4.   Mixtures

In general, the thermal conductivities of liquid mixtures, and gas mixtures, are not simple functions of composition and the thermal conductivity of the components. Bretsznajder (1971) discusses the methods that are available for estimating the thermal conductivities of mixtures from knowledge of the thermal conductivity of the components.

If the components are all non-polar, a simple weighted average is usually sufficiently accurate for design purposes.

$$k_m = k_1 w_1 + k_2 w_2 + \cdots \tag{8.14}$$

where

k_m = thermal conductivity of mixture
k_1, k_2 = thermal conductivity of components
w_1, w_2 = component mass fractions.

8.9.   SPECIFIC HEAT CAPACITY

The specific heats of the most common organic and inorganic materials can usually be found in the handbooks.

8.9.1.   Solids and Liquids

Approximate values can be calculated for solids, and liquids, by using a modified form of Kopp's law, which is given by Werner (1941). The heat capacity of

TABLE 8.2. Heat Capacities of the Elements, J/mol°C

Element	Solids	Liquids
C	7.5	11.7
H	9.6	18.0
B	11.3	19.7
Si	15.9	24.3
O	16.7	25.1
F	20.9	29.3
P and S	22.6	31.0
all others	26.0	33.5

a compound is taken as the sum of the heat capacities of the individual elements of which it is composed. The values attributed to each element, for liquids and solids, at room temperature, are given in Table 8.2; the method is illustrated in Example 8.6.

Example 8.6

Estimate the specific heat capacity of urea, CH_4N_2O.

Solution

Element	mol. mass	Heat capacity
C	12	$7.5 = 7.5$
H	4	$4 \times 9.6 = 38.4$
N	28	$2 \times 26.0 = 52.0$
O	16	$16.7 = 16.7$
	60	114.6 J/mol°C

$$\text{Specific heat capacity} = \frac{114.6}{60} = \underline{\underline{1.91 \text{ J/g°C}}} \text{ (kJ/kg°C)}$$

Experimental value, 1.34 kJ/kg°C, error 43%.

Kopp's rule does not take into account the arrangement of the atoms in the molecule, and, at best, gives only very approximate, 'ball-park' values.

For organic liquids, the group contribution method proposed by Chueh and Swanson (1973a, b) gives reasonably accurate predictions. The contributions to be assigned to each molecular group are given in Table 8.3 and the method is illustrated in Examples 8.7 and 8.8.

Liquid specific heats do not vary much with temperature, at temperatures well below the critical temperature (reduced temperature <0.7).

TABLE 8.3. Group Contributions for Liquid Heat Capacities at 20°C, kJ/kmol°C (Chueh and Swanson, 1973a, b)

Group	Value	Group	Value
Alkane		$\underset{\|}{\overset{O}{\overset{\|}{C}}}$ —C—O—	60.71
—CH$_3$	36.84	—CH$_2$OH	73.27
—CH$_2$—	30.40	$\|$ —CHOH	76.20
$\|$ —CH—	20.93	$\|$	
$\|$ —C— $\|$	7.37	$\|$ —COH $\|$	111.37
		—OH	44.80
Olefin		—ONO$_2$	119.32
=CH$_2$	21.77	**Halogen**	
$\|$ =C—H	21.35	—Cl (first or second on a carbon)	36.01
		—Cl (third or fourth on a carbon)	25.12
$\|$ =C—	15.91	—Br	37.68
Alkyne		—F	16.75
—C≡H	24.70	—I	36.01
—C≡	24.70	**Nitrogen**	
In a ring		H $\|$ H—N—	58.62
$\|$ —CH=	18.42	H $\|$ —N—	43.96
$\|$ —C= or —C— $\|$	12.14	$\|$ —N—	31.40
—C=	22.19	—N=(in a ring)	18.84
—CH$_2$—	25.96	—C≡N	58.70
Oxygen		**Sulphur**	
—O—	35.17	—SH	44.80
$\underset{/}{\overset{\backslash}{C}}$=O	53.00	—S—	33.49
—C—O $\|$ H	53.00	**Hydrogen**	
$\overset{O}{\overset{\|}{C}}$ —C—OH	79.97	H— (for formic acid, formates, hydrogen cyanide, etc.)	14.65

Add 18.84 for any carbon group which fulfills the following criterion: a carbon group which is joined by a single bond to a carbon group connected by a double or triple bond with a third carbon group. In some cases a carbon group fulfills the above criterion in more ways than one; 18.84 should be added each time the group fulfills the criterion.

Exceptions to the above 18.84 rule:
  1. No such extra 18.84 additions for –CH3 groups.
  2. For a –CH2– group fulfilling the 18.84 addition criterion add 10.47 instead of 18.84. However, when the –CH2– group fulfils the addition criterion in more ways than one, the addition should be 10.47 the first time and 18.84 for each subsequent addition.
  3. No such extra addition for any carbon group in a ring.

The specific heats of liquid mixtures can be estimated, with sufficient accuracy for most technical calculations, by taking heat capacities as the mass (or mole) weighted sum of the pure component heat capacities.

For dilute aqueous solutions it is usually sufficient to take the specific heat of the solution as that of water.

Example 8.7

Using Chueh and Swanson's method, estimate the specific heat capacity of ethyl bromide at 20°C.

Solution

Ethyl bromide CH_3CH_2Br

Group	Contribution	No. of	
—CH_3	36.84	1	= 36.84
—CH_2—	30.40	1	= 30.40
—Br	37.68	1	= 37.68
		Total	104.92 kJ/kmol°C

$$\text{Specific heat capacity} = \frac{104.92}{109} = 0.96 \text{ kJ/kg}°\text{C}$$

Experimental value, 0.90 kJ/kg°C.

Example 8.8

Estimate the specific heat capacity of chlorobutadiene at 20°C, using Chueh and Swanson's method.

Solution

Structural formula $CH_2{=}C{-}CH{=}CH_2$, mol. wt. 88.5
                              |
                             Cl

Group	Contribution	No. of	Addition rule	Total
$={CH_2}$	21.77	2	—	= 43.54
$={C}-$	15.91	1	18.84	= 34.75
|				
|				
$={CH}$	21.35	1	18.84	= 40.19
$-{Cl}$	36.01	1	—	= 36.01
			Total	154.49 kJ/kmol°C

$$\text{Specific heat capacity} = \frac{154.49}{88.5} = 1.75 \text{ kJ/kg}°\text{C}$$

8.9.2. Gases

The dependence of gas specific heats on temperature was discussed in Chapter 3, Section 3.5. For a gas in the ideal state the specific heat capacity at constant pressure is given by:

$$C_p^o = a + bT + cT^2 + dT^3 \qquad \text{(equation 3.19)}$$

Values for the constants in this equation for the more common gases can be found in the handbooks, and in Appendix C.

Several group contribution methods have been developed for the estimation of the constants, such as that by Rihani and Doraiswamy (1965) for organic compounds. Their values for each molecular group are given in Table 8.4, and the method illustrated in Example 8.9. The values should not be used for acetylenic compounds.

The correction of the ideal gas heat capacity to account for real conditions of temperature and pressure was discussed in Chapter 3, Section 3.7.

Example 8.9

Estimate the specific heat capacity of isopropyl alcohol at 500 K.

Solution

Structural formula

$$
\begin{array}{c}
CH_3 \\
| \\
CH_3-CH-OH
\end{array}
$$

Group	No. of	a	$b \times 10^2$	$c \times 10^4$	$d \times 10^6$
$-CH_3$	2	5.0970	17.9480	−0.7134	0.0095
$-\overset{|}{\underset{|}{CH}}$	1	−14.7516	14.3020	−1.1791	0.03356
$-OH$	1	27.2691	−0.5640	0.1733	−0.0068
Total		17.6145	31.6860	−1.7190	0.0363

$$C_p^o = 17.6145 + 31.6860 \times 10^{-2}T - 1.7192 \times 10^{-4}T^2 + 0.0363 \times 10^{-6}T^3.$$

At 500 K, substitution gives:

$$C_p = \underline{\underline{137.6 \text{ kJ/kmol}^\circ C}}$$

Experimental value, 31.78 cal/mol°C = 132.8 kJ/kmol°C, error 4%.

TABLE 8.4. Group Contributions to Ideal Gas Heat Capacities, kJ/kmol°C (Rihani and Doraiswamy, 1965)

Group	a	$b \times 10^2$	$c \times 10^4$	$d \times 10^6$
Aliphatic hydrocarbon groups				
—CH₃	2.5485	8.9740	−0.3567	0.004752
—CH₂	1.6518	8.9447	−0.5012	0.0187
=CH₂	2.2048	7.6857	−0.3994	0.008264
—C—H	−14.7516	14.3020	−1.1791	0.03356
—C—	−24.4131	18.6493	−1.7619	0.05288
H\C=CH₂	1.1610	14.4786	−0.8031	0.01792
\C=CH₂	−1.7472	16.2694	−1.1652	0.03083
H\C=C\H	−13.0676	15.9356	−0.9877	0.02305
H\C=C\H	3.9261	12.5208	−0.7323	0.01641
\C=C\H	−6.161	14.1696	−0.9927	0.02594
\C=C\	1.9829	14.7304	−1.3188	0.03854
H\C=C=CH₂	9.3784	17.9597	−1.07433	0.02474
\C=C=CH₂	11.0146	17.4414	−1.1912	0.03047
H\C=C=C\H	−13.0833	20.8878	−1.8018	0.05447
Aromatic hydrocarbon groups				
HC<	−6.1010	8.0165	−0.5162	0.01250
—C<	−5.8125	6.3468	−0.4476	0.01113
↔C<	0.5104	5.0953	−0.3580	0.00888
Contributions due to ring formation				
Three-membered ring	−14.7878	−0.1256	0.3129	−0.02309
Four-membered ring	−36.2368	4.5134	0.1779	−0.00105
Five-membered ring:				
Pentane	−51.4348	7.7913	−0.4342	0.00898
Pentene	−28.8106	3.2732	−0.1445	0.00247
Six-membered ring:				
Hexane	−56.0709	8.9564	−0.1796	−0.00781
Hexene	−33.5941	9.3110	−0.80118	0.02291

(continued)

TABLE 8.4.   Group Contributions to Ideal Gas Heat Capacities, kJ/kmol°C (Rihani and Doraiswamy, 1965)—cont'd

Group	a	$b \times 10^2$	$c \times 10^4$	$d \times 10^6$
	Oxygen-containing groups			
—OH	27.2691	−0.5640	0.1733	−0.00680
—O—	11.9161	−0.04187	0.1901	−0.01142
—CH=O	14.7308	3.9511	0.2571	−0.02922
>C=O	4.1935	8.6931	−0.6850	0.01882
—C(=O)—O—H	5.8846	14.4997	−1.0706	0.02883
—C(=O)—O—	11.4509	4.5012	0.2793	−0.03864
O<	−15.6352	5.7472	−0.5296	0.01586
	Nitrogen-containing groups			
—C≡N	18.8841	2.2864	0.1126	−0.01587
—N≡C	21.2941	1.4620	0.1084	−0.01020
—NH₂	17.4937	3.0890	0.2843	−0.03061
>NH—	−5.2461	9.1825	−0.6716	0.01774
>N—	−14.5186	12.3230	−1.1191	0.03277
N<	10.2401	1.4386	0.07159	−0.01138
—NO₂	4.5638	11.0536	−0.7834	0.01989
	Sulphur-containing groups			
—SH	10.7170	5.5881	−0.4978	0.01599
—S—	17.6917	0.4719	−0.0109	−0.00030
S<	17.0922	−0.1260	0.3061	−0.02546
—SO₃H	28.9802	10.3561	0.7436	−0.09397
	Halogen-containing groups			
—F	6.0215	1.4453	−0.0444	−0.00014
—Cl	12.8373	0.8885	−0.0536	0.00116
—Br	11.5577	1.9808	−0.1905	0.0060
—I	13.6703	2.0520	−0.2257	0.00746

8.10. ENTHALPY OF VAPORIZATION (LATENT HEAT)

The latent heats of vaporization of the more commonly used materials can be found in the handbooks and in Appendix C.

A very rough estimate can be obtained from Trouton's rule (Trouton, 1884), one of the oldest prediction methods.

$$\frac{L_v}{T_b} = \text{constant} \tag{8.15}$$

where

L_v = latent heat of vaporization, kJ/kmol
T_b = normal boiling point, K.

For organic liquids the constant can be taken as 100.

More accurate estimates, suitable for most engineering purposes, can be made from knowledge of the vapour pressure–temperature relationship for the substance. Several correlations have been proposed; see Reid *et al.* (1987).

The equation presented here, due to Haggenmacher (1946), is derived from the Antoine vapour pressure equation (see Section 8.11).

$$L_v = \frac{8.32\, BT^2 \Delta z}{(T + C)^2} \tag{8.16}$$

where
L_v = latent heat at the required temperature, kJ/kmol
T = temperature, K
B, C = coefficients in the Antoine equation (equation 8.20)
$\Delta z = z_{gas} - z_{liquid}$ (where z is the compressibility constant), calculated from the equation:

$$\Delta z = \left[1 - \frac{P_r}{T_r^3}\right]^{0.5} \tag{8.17}$$

P_r = reduced pressure
T_r = reduced temperature.

If an experimental value of the latent heat at the boiling point is known, the Watson equation (Watson, 1943), can be used to estimate the latent heat at other temperatures.

$$L_v = L_{v,b} \left[\frac{T_c - T}{T_c - T_b}\right]^{0.38} \tag{8.18}$$

where
L_v = latent heat at temperature T, kJ/kmol
$L_{v,b}$ = latent heat at the normal boiling point, kJ/kmol
T_b = boiling point, K
T_c = critical temperature, K
T = temperature, K.

Over a limited range of temperature, up to 100°C, the variation of latent heat with temperature can usually be taken as linear.

8.10.1. Mixtures

For design purposes it is usually sufficiently accurate to take the latent heats of the components of a mixture as additive:

$$L_v \text{ mixture} = L_{v1}x_1 + L_{v2}x_2 + \ldots \tag{8.19}$$

where

L_{v1}, L_{v2} = latent heats of the components kJ/kmol
x_1, x_2 = mol fractions of components.

Example 8.10

Estimate the latent heat of vaporization of acetic anhydride, $C_4H_6O_3$, at its boiling point, 139.6°C (412.7 K), and at 200°C (473 K).

Solution

For acetic anhydride, $T_c = 569.1$ K, $P_c = 46$ bar,

Antoine constants $A = 16.3982$
$B = 3287.56$
$C = -75.11$

Experimental value at the boiling point, 41,242 kJ/kmol.
From Trouton's rule:

$$L_{v,b} = 100 \times 412.7 = \underline{\underline{41,270 \ \text{kJ/kmol}}}$$

Note: the close approximation to the experimental value is fortuitous; the rule normally gives only a very approximate estimate.

From Haggenmacher's equation:

$$\text{at the b.p. } P_r = \frac{1}{46} = 0.02124$$

$$T_r = \frac{412.7}{569.1} = 0.7252$$

$$\Delta z = \left[1 - \frac{0.02124}{0.7252^3}\right]^{0.5} = 0.972$$

$$L_{v,b} = \frac{8.32 \times 3287.6 \times (412.7)^2 \times 0.972}{(412.7 - 75.11)^2} = \underline{\underline{39,733 \ \text{kJ/mol}}}$$

At 200°C, the vapour pressure must first be estimated from the Antoine equation:

$$\ln P = A - \frac{B}{T + C}$$

$$\ln P = 16.3982 - \frac{3287.56}{473 - 75.11} = 8.14$$

$$P = 3421.35 \text{ mmHg} = 4.5 \text{ bar}$$

$$P_c = \frac{4.5}{46} = 0.098$$

$$T_c = \frac{473}{569.1} = 0.831$$

$$\Delta z = \left[1 - \frac{0.098}{0.831^3}\right]^{0.5} = 0.911$$

$$L_v = \frac{8.32 \times 3287.6 \times (473)^2 \times 0.911}{(473 - 75.11)^2} = \underline{\underline{35,211 \text{ kJ/kmol}}}$$

Using Watson's equation and the experimental value at the boiling point:

$$L_v = 41,242 \left[\frac{569.1 - 473}{569.1 - 412.7}\right]^{0.38} = \underline{\underline{34,260 \text{ kJ/kmol}}}$$

8.11. VAPOUR PRESSURE

If the normal boiling point (vapour pressure = 1 atm) and the critical temperature and pressure are known, then a straight line drawn through these two points on a plot of log-pressure versus reciprocal absolute temperature can be used to make a rough estimation of the vapour pressure at intermediate temperatures.

Several equations have been developed to express vapour pressure as a function of temperature. One of the most commonly used is the three-term Antoine equation (Antoine, 1888):

$$\ln P = A - \frac{B}{T + C} \tag{8.20}$$

where

$$P = \text{vapour pressure, mmHg}$$
$$A, B, C = \text{the Antoine coefficients}$$
$$T = \text{temperature, K.}$$

Vapour pressure data, in the form of the constants in the Antoine equation, are given in several references; the compilations by Ohe (1976), Dreisbach (1952), Hala et al. (1968) and Hirata et al. (1975) give values for several thousand compounds. Antoine vapour pressure coefficients for the elements are given by Nesmeyanov (1963). Care must be taken when using Antoine coefficients taken from the literature in equation 8.20, as the equation is often written in different and ambiguous forms; the logarithm of the pressure may be to the base 10, instead of the natural logarithm, and the temperature may be degrees Celsius, not absolute temperature. Also, occasionally, the minus sign shown in equation 8.20 is included in the constant B and the equation written with a plus sign. The pressure may also be in units other than mmHg. Always check the actual form of the equation used in the particular reference. Antoine constants for use in equation 8.20 are given in Appendix C. A spreadsheet for calculating vapour pressure is available in MS Excel format in the on-line material at http://elsevierdirect.com/companions. Vapour pressure data for hydrocarbons can be found in the four-volume handbook by Yaws (1994–1995).

8.12. DIFFUSION COEFFICIENTS (DIFFUSIVITIES)

Diffusion coefficients are needed in the design of mass-transfer processes; such as gas absorption, distillation and liquid–liquid extraction, as well as in catalytic reactions where mass transfer can limit the rate of reaction.

Experimental values for the more common systems can be often found in the literature, but for most design work the values must be estimated.

8.12.1. Gases

The equation developed by Fuller *et al.* (1966) is easy to apply and gives reliable estimates:

$$D_v = \frac{1.013 \times 10^{-7} T^{1.75} \left(\frac{1}{M_a} + \frac{1}{M_b} \right)^{1/2}}{P \left[\left(\sum_a v_i \right)^{1/3} + \left(\sum_b v_i \right)^{1/3} \right]^2} \tag{8.21}$$

where

$$D_v = \text{diffusivity, m}^2/\text{s}$$
$$T = \text{temperature, K}$$
$$M_a, M_b = \text{molecular masses of components } a \text{ and } b$$
$$P = \text{total pressure, bar}$$
$$\sum_a v_i, \sum_b v_i = \text{the summation of the special diffusion volume coefficients for}$$
components a and b, given in Table 8.5.

The method is illustrated in Example 8.11.

Example 8.11

Estimate the diffusivity of methanol in air at atmospheric pressure and 25°C.

Solution

Diffusion volumes from Table 8.5; methanol:

Element	v_i	No. of	
C		$16.50 \times 1 =$	16.50
H		$1.98 \times 4 =$	7.92
O		$5.48 \times 1 =$	5.48
		$\sum_a v_i$	29.90

TABLE 8.5. Special atomic diffusion volumes (Fuller *et al.*, 1966)

Atomic and Structural Diffusion Volume Increments			
C	16.5	Cl	19.5*
H	1.98	S	17.0*
O	5.48	Aromatic or hetrocyclic rings	−20.0
N	5.69*		

Diffusion Volumes of Simple Molecules			
H_2	7.07	CO	18.9
D_2	6.70	CO_2	26.9
He	2.88	N_2O	35.9
N_2	17.9	NH_3	14.9
O_2	16.6	H_2O	12.7
Air	20.1	CCL_2F_2	114.8*
Ne	5.59	SF_6	69.7*
Ar	16.1	Cl_2	37.7*
Kr	22.8	Br_2	67.2*
Xe	37.9*	SO_2	41.1*

* Value based on only a few data points.

Diffusion volume for air = 20.1
1 standard atmosphere = 1.013 bar
Molecular mass CH_3OH = 32, air = 29

$$D_v = \frac{1.013 \times 10^{-7} \times 298^{1.75}(1/32 + 1/29)^{1/2}}{1.013[(29.90)^{1/3} + (20.1)^{1/3}]^2}$$

$$= \underline{\underline{16.2 \times 10^{-6} \text{m}^2/\text{s}}} \tag{8.21}$$

Experimental value, 15.9×10^{-6} m^2/s.

8.12.2. Liquids

The equation developed by Wilke and Chang (1955), given below, can be used to predict liquid diffusivity.

$$D_L = \frac{1.173 \times 10^{-13}(\phi M)^{0.5}T}{\mu V_m^{0.6}} \tag{8.22}$$

where

D_L = liquid diffusivity, m^2/s
ϕ = an association factor for the solvent
 = 2.6 for water (some workers recommend 2.26)
 = 1.9 for methanol
 = 1.5 for ethanol
 = 1.0 for unassociated solvents

M = molecular mass of solvent

μ = viscosity of solvent, mN s/m^2

T = temperature, K

V_m = molar volume of the solute at its boiling point, m^3/kmol. This can be estimated from the group contributions given in Table 8.6.

The method is illustrated in Example 8.12.

The Wilke-Chang correlation is shown graphically in Figure 8.2. This figure can be used to determine the association constant for a solvent from experimental values for D_L in the solvent.

The Wilke-Chang equation gives satisfactory predictions for the diffusivity of organic compounds in water but not for water in organic solvents.

Example 8.12

Estimate the diffusivity of phenol in ethanol at 20°C (293 K).

Solution

Viscosity of ethanol at 20°C, 1.2 mNs/m^2

Molecular mass, 46

Molar volume of phenol   from Table 8.6:

TABLE 8.6.   Structural Contributions to Molar Volumes, m^3/kmol (Gambill, 1958)

Molecular Volumes							
Air	0.0299	CO_2	0.0340	H_2S	0.0329	NO	0.0236
Br_2	0.0532	COS	0.0515	I_2	0.0715	N_2O	0.0364
Cl_2	0.0484	H_2	0.0143	N_2	0.0312	O_2	0.0256
CO	0.0307	H_2O	0.0189	NH_3	0.0258	SO_2	0.0448

Atomic Volumes							
As	0.0305	F	0.0087	P	0.0270	Sn	0.0423
Bi	0.0480	Ge	0.0345	Pb	0.0480	Ti	0.0357
Br	0.0270	H	0.0037	S	0.0256	V	0.0320
C	0.0148	Hg	0.0190	Sb	0.0342	Zn	0.0204
Cr	0.0274	I	0.037	Si	0.0320		
Cl, terminal, as in RCl			0.0216	in higher esters, ethers		0.0110	
medial, as in R—CHCl—R			0.0246	in acids		0.0120	
Nitrogen, double-bonded			0.0156	in union with S, P, N		0.0083	
triply bonded, as in nitriles			0.0162	three-membered ring		−0.0060	
in primary amines, RNH_2			0.0105	four-membered ring		−0.0085	
in secondary amines, R_2NH			0.012	five-membered ring		−0.0115	
in tertiary amines, R_3N			0.0108	six-membered ring as in benzene, cyclohexane, pyridine		−0.0150	
Oxygen, except as noted below			0.0074				
in methyl esters			0.0091	Naphthalene ring		−0.0300	
in methyl ethers			0.0099	Anthracene ring		−0.0475	

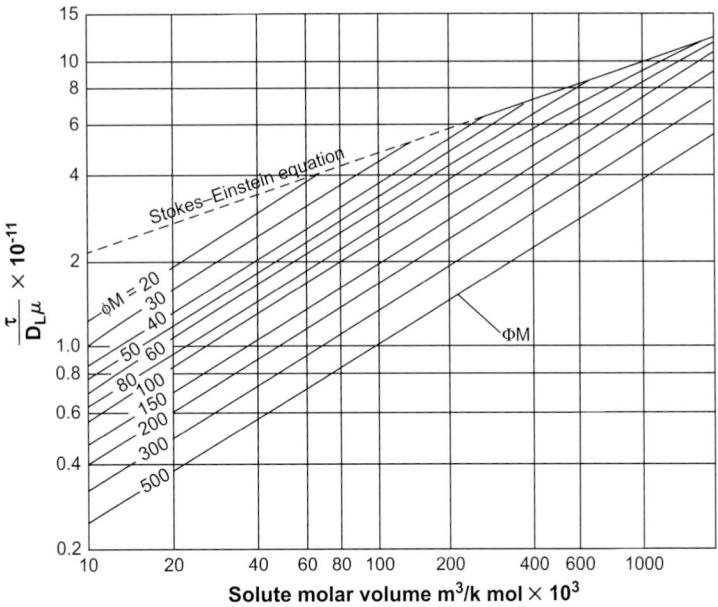

Figure 8.2. The Wilke-Chang correlation.

Atom	Vol.	No. of		
C	0.0148×6		$=$	0.0888
H	0.0037×6		$=$	0.0222
O	0.0074×1		$=$	0.0074
Ring	-0.015×1		$=$	-0.015
		Total	$=$	$0.1034 \, \text{m}^3/\text{k mol}$

$$D_L = \frac{1.173 \times 10^{-13}(1.5 \times 46)^{0.5} 293}{1.2 \times 0.1034^{0.6}} = 9.28 \times 10^{-10} \, \text{m}^2/\text{s} \qquad (8.22)$$

Experimental value, $8 \times 10^{-10} \, \text{m}^2/\text{s}$, error 16%.

8.13. SURFACE TENSION

It is usually difficult to find experimental values for surface tension for any but the more commonly used liquids. A useful compilation of experimental values is that by Jasper (1972), which covers over 2000 pure liquids. Othmer *et al.* (1968) give a nomograph covering about 100 compounds.

If reliable values of the liquid and vapour density are available, the surface tension can be estimated from the Sugden parachor; which can be estimated by a group contribution method (Sugden, 1924).

$$\sigma = \left[\frac{P_{ch}(\rho_L - \rho_v)}{M}\right]^4 \times 10^{-12} \tag{8.23}$$

where

σ = surface tension, mJ/m^2 (dyne/cm)
P_{ch} = Sugden's parachor
ρ_L = liquid density, kg/m^3
ρ_v = density of the saturated vapour, kg/m^3
M = molecular mass
with σ, ρ_L, ρ_v evaluated at the system temperature.

The vapour density can be neglected when it is small compared with the liquid density.

The parachor can be calculated using the group contributions given in Table 8.7. The method is illustrated in Example 8.13.

8.13.1.   Mixtures

The surface tension of a mixture is rarely a simple function of composition. However, for hydrocarbons a rough value can be calculated by assuming a linear relationship.

$$\sigma_m = \sigma_1 x_1 + \sigma_2 x_2 \ldots \tag{8.24}$$

where

TABLE 8.7.   Contribution to Sugdens's Parachor for Organic Compounds (Sugden, 1924)

Atom, Group or Bond	Contribution	Atom, Group or Bond	Contribution
C	4.8	Si	25.0
H	17.1	Al	38.6
H in (OH)	11.3	Sn	57.9
O	20.0	As	50.1
O$_2$ in esters, acids	60.0	Double bond: terminal	
N	12.5	2,3-position	23.2
S	48.2	3,4-position	
P	37.7	Triple bond	46.6
F	25.7	Rings	
Cl	54.3	3-membered	16.7
Br	68.0	4-membered	11.6
I	91.0	5-membered	8.5
Se	62.5	6-membered	6.1

σ_m = surface tension of mixture
σ_1, σ_2 = surface tension of components
x_1, x_2 = component mol fractions.

Example 8.13

Estimate the surface tension of pure methanol at 20°C, density 791.7 kg/m³, molecular weight 32.04.

Solution

Calculation of parachor, CH_3OH, Table 8.7.

Group	Contribution	No.	
C	4.8	× 1	= 4.8
H—O	11.3	× 1	= 11.3
H—C	17.1	× 3	= 51.3
O	20.0	× 1	= 20.0
		Total	= 87.4

$$\sigma = \left[\frac{87.4 \times 791.7}{32.04}\right]^4 \times 10^{-12} = \underline{\underline{21.8 \text{ mJ/m}^2}} \tag{8.23}$$

Experimental value, 22.5 mJ/m².

8.14. CRITICAL CONSTANTS

Values of the critical temperature and pressure are needed for prediction methods that correlate physical properties with the reduced conditions. It is also important to know the critical conditions when applying equation of state methods, as some of the equation of state models are unreliable close to the critical point. Experimental values for many substances can be found in various handbooks, and in Appendix C. Critical reviews of the literature on critical constants, and summaries of selected values, have been published by Kudchadker *et al.* (1968) for organic compounds and by Mathews (1972) for inorganic compounds. An earlier review was published by Kobe and Lynn (1953).

If reliable experimental values cannot be found, techniques are available for estimating the critical constants with sufficient accuracy for most design purposes. For organic compounds, Lydersen's method is normally used (Lydersen, 1955):

$$T_c = \frac{T_b}{[0.567 + \Sigma\Delta T - (\Sigma\Delta T)^2]} \tag{8.25}$$

$$P_c = \frac{M}{(0.34 + \Sigma\Delta P)^2} \tag{8.26}$$

where
$$V_c = 0.04 + \Sigma \Delta V \qquad (8.27)$$

T_c = critical temperature, K
P_c = critical pressure, atm (1.0133 bar)
V_c = molar volume at the critical conditions, m³/kmol
T_b = normal boiling point, K
M = relative molecular mass
ΔT = critical temperature increments, Table 8.8
ΔP = critical pressure increments, Table 8.8
ΔV = molar volume increments, Table 8.8.

Fedons (1982) gives a simple method for the estimation of critical temperature that does not require knowledge of the boiling point of the compound.

Example 8.14

Estimate the critical constants for diphenylmethane using Lydersen's method; normal boiling point 537.5 K, molecular mass 168.2, structural formula:

Solution

		Total contribution		
Group	No. of	ΔT	ΔP	ΔV
H—C—(ring)	10	0.11	1.54	0.37
—C—(ring)	2	0.022	0.308	0.072
—CH²—	1	0.02	0.227	0.055
		Σ 0.152	2.075	0.497

$$T_c = \frac{537.5}{(0.567 + 0.152 - 0.152^2)} = \underline{\underline{772\ k}}$$

experimental value 767 K,

$$P_c = \frac{168.2}{(0.34 + 2.075)^2} = \underline{\underline{28.8\ atm}}$$

experimental value 28.2 atm,

$$V_c = 0.04 + 0.497 = \underline{\underline{0.537\ m^3/kmol}}$$

TABLE 8.8. Critical Constant Increments (Lydersen, 1955)

	ΔT	ΔP	ΔV		ΔT	ΔP	ΔV
Non-ring increments							
—CH₃	0.020	0.227	0.055	=C—	0.0	0.198	0.036
—CH₂	0.020	0.227	0.055	=C=	0.0	0.198	0.036
—CH	0.012	0.210	0.051	≡CH	0.005	0.153	0.036*
				≡C—	0.005	0.153	0.036*
—C—	0.00	0.210	0.041	H	0	0	0
=CH₂	0.018	0.198	0.045				
=CH	0.018	0.198	0.045				
Ring increments							
—CH₂—	0.013	0.184	0.0445	=CH	0.011	0.154	0.037
—CH	0.012	0.192	0.046	=C—	0.011	0.154	0.036
—C—	−0.007*	0.154*	0.031*	=C=	0.011	0.154	0.036
Halogen increments							
—F	0.018	0.224	0.018	—Br	0.010	0.50*	0.070*
—Cl	0.017	0.320	0.049	—I	0.012	0.83*	0.095*
Oxygen increments							
—OH (alcohols)	0.082	0.06	0.018*	—CO (ring)	0.033*	0.2*	0.050*
—OH (phenols)	0.031	−0.02*	0.030*	HC=O (aldehyde)	0.048	0.33	0.073
—O— (non-ring)	0.021	0.16	0.020	—COOH (acid)	0.085	0.4*	0.080
—O— (ring)	0.014*	0.12*	0.080*	—COO— (ester)	0.047	0.47	0.080
—C=O (non-ring)	0.040	0.29	0.060	=O (except for combinations above)	0.02*	0.12*	0.011*
Nitrogen increments							
—NH₂	0.031	0.095	0.028	—N— (ring)	0.007*	0.013*	0.032*
—NH (non-ring)	0.031	0.135	0.037*	—CN	0.060*	0.36*	0.080*
—NH (ring)	0.024*	0.09*	0.027*	—NO₂	0.055*	0.42*	0.078*
—N— (non-ring)	0.014	0.17	0.042*				

(continued)

TABLE 8.8.   Critical Constant Increments (Lydersen, 1955)—Cont'd

	ΔT	ΔP	ΔV		ΔT	ΔP	ΔV
Sulphur increments							
—SH	0.015	0.27	0.055	—S—(ring)	0.008*	0.24*	0.045*
—S—(non-ring)	0.015	0.27	0.055	S	0.003*	0.24*	0.047*
Miscellaneous							
$\mid$ —Si— $\mid$	0.03	0.54*		$\mid$ —B— $\mid$		0.03*	

Dashes represent bonds with atoms other than hydrogen.
Values marked with an asterisk are based on too few experimental points to be reliable.

8.15.   ENTHALPY OF REACTION AND ENTHALPY OF FORMATION

Enthalpies of reaction (heats of reaction) for the reactions used in the production of commercial chemicals can usually be found in the literature. Stephenson (1966) gives values for most of the production processes he describes in his book.

Heats of reaction can be calculated from the heats of formation of the reactants and products, as described in Chapter 3, Section 3.10. Values of the standard heats of formation for the more common chemicals are given in various handbooks; see also Appendix C. Care must be taken to correct the heat of reaction to the temperature and pressure of the process. A useful source of data on heats of formation, and combustion, is the critical review of the literature by Domalski (1972).

Benson has developed a detailed group contribution method for the estimation of heats of formation; see Benson (1976) and Benson *et al.* (1969). He estimates the accuracy of the method to be from ± 2.0 kJ/mol, for simple compounds, to about ± 12 kJ/mol, for highly substituted compounds. Benson's method and other group contribution methods for the estimation of heats of formation are described by Reid *et al.* (1987).

8.16.   PHASE EQUILIBRIUM DATA

Phase equilibrium data are needed for the design of all separation processes that depend on differences in concentration between phases.

8.16.1.   Experimental Data

Experimental data have been published for several thousand binary and many multicomponent systems. Virtually all the published experimental data have been collected together in the DETHERM database managed by DECHEMA (www. dechema.de/en/detherm.html). The DETHERM database allows free web access to search for whether a component or mixture is included in the data set. Older versions of this database were published in print form (DECHEMA, 1977). The books by

Chu *et al.* (1956), Hala *et al.* (1968, 1973), Hirata *et al.* (1975) and Ohe (1989, 1990) are also useful sources.

8.16.2. Phase Equilibrium

The criterion for thermodynamic equilibrium between two phases of a multicomponent mixture is that for every component, i:

$$f_i^v = f_i^L \tag{8.28}$$

where f_i^v is the vapour-phase fugacity and f_i^L the liquid-phase fugacity of component i:

$$f_i^v = P\phi_i y_i \tag{8.29}$$

and

$$f_i^L = f_i^{OL} \gamma_i x_i \tag{8.30}$$

where

P = total system pressure
ϕ_i = vapour fugacity coefficient
y_i = concentration of component i in the vapour phase
f_i^{OL} = standard state fugacity of the pure liquid
γ_i = liquid-phase activity coefficient
x_i = concentration of component i in the liquid phase.

Substitution from equations 8.29 and 8.30 into equation 8.28 and rearranging gives:

$$K_i = \frac{y_i}{x_i} = \frac{\gamma_i f_i^{OL}}{P\phi_i} \tag{8.31}$$

where

K_i is the distribution coefficient (the K value)
ϕ_i can be calculated from an appropriate equation of state (see Section 8.16.3)
f_i^{OL} can be computed from the following expression:

$$f_i^{OL} = P_i^o \phi_i^s \left\{ \exp\left\{ \frac{(P - P_i^o)}{RT} v_i^L \right\} \right\} \tag{8.32}$$

where

P_i^o = the pure component vapour pressure (which can be calculated from the Antoine equation, see Section 8.11), N/m^2
ϕ_i^s = the fugacity coefficient of the pure component i at saturation
v_i^L = the liquid molar volume, m^3/mol.

The exponential term in equation 8.32 is known as the *Poynting correction*, and corrects for the effects of pressure on the liquid-phase fugacity.

ϕ_i^s is calculated using the same equation of state used to calculate ϕ_i.

For systems in which the vapour phase imperfections are not significant, equation 8.32 reduces to the familiar Raoult's law equation:

$$K_i = \frac{\gamma_i P_i^o}{P} \tag{8.33}$$

Relative Volatility

The relative volatility of two components can be expressed as the ratio of their K values:

$$\alpha_{ij} = \frac{K_i}{K_j} \tag{8.34}$$

For ideal mixtures (obeying Raoult's law):

$$K_i = \frac{P_i^o}{P} \tag{8.35}$$

and

$$\alpha_{ij} = \frac{K_i^o}{K_j^o} = \frac{P_i^o}{P_j^o} \tag{8.36}$$

where K_i^o and K_j^o are the ideal K values for components i and j.

8.16.3.   Equations of State

An equation of state is an algebraic expression that relates temperature, pressure and molar volume, for a real fluid.

Many equations of state have been developed, of varying complexity. No one equation is sufficiently accurate to represent all real gases, under all conditions. The equations of state most frequently used in the design of multicomponent separation processes are given below. The actual equation is only given for one of the correlations, the Redlich-Kwong equation, as an illustration. Equations of state are normally solved using computer-aided design packages; see Chapter 11. For details of the other equations the reader should consult the reference cited, or the books by Reid *et al.* (1987), Prausnitz *et al.* (1998) and Walas (1985). To select the best equation to use for a particular process design refer to Table 8.10 below and Figure 8.4.

Redlich-Kwong Equation (R-K)

This equation is an extension of the more familiar Van der Waal's equation. The Redlich-Kwong equation is:

$$P = \frac{PT}{V - b} \times \frac{a}{T^{1/2}V(V + b)} \tag{8.37}$$

where

$$a = 0.427 \ R^2 \ T_c^{2.5}/P_c$$
$$b = 0.08664 \ RT_c/P_c$$
$$P = \text{pressure}$$
$$V = \text{volume.}$$

The R-K equation is not suitable for use near the critical pressure ($P_r > 0.8$), or for liquids; Redlich and Kwong (1949).

Redlich-Kwong-Soave Equation (R-K-S)

Soave (1972) modified the R-K equation to extend its usefulness to the critical region, and for use with liquids.

Benedict-Webb-Rubin (B-W-R) Equation

This equation has eight empirical constants and gives accurate predictions for vapour- and liquid-phase hydrocarbons. It can also be used for mixtures of light hydrocarbons with carbon dioxide and water; Benedict *et al.* (1951).

Lee-Kesler-Plocker (L-K-P) Equation

Lee and Kesler (1975) extended the B-W-R equation to a wider variety of substances, using the principle of corresponding states. The method was modified further by Plocker *et al.* (1978).

Chao-Seader Equation (C-S)

The Chao-Seader equation gives accurate predictions for light hydrocarbons and hydrogen, but is limited to temperatures below 530 K; Chao and Seader (1961).

Grayson-Streed Equation (G-S)

Grayson and Streed (1963) extended the C-S equation for use with hydrogen-rich mixtures, and for high-pressure and high-temperature systems. It can be used up to 200 bar and 4700 K.

Peng-Robinson Equation (P-R)

The Peng-Robinson equation is related to the R-K-S equation of state and was developed to overcome the instability in the R-K-S equation near the critical point; Peng and Robinson (1976).

Brown K_{10} Equation (B-K_{10})

Brown, see Cajander *et al.* (1960), developed a method which relates the equilibrium constant K to four parameters: component, pressure, temperature and the convergence pressure. The convergence pressure is the pressure at which all K values tend to 1. The B-K_{10} equation is limited to low pressure and its use is generally restricted to vacuum systems.

8.16.4. Correlations for Liquid Phase Activity Coefficients

The liquid-phase activity coefficient, γ_i, is a function of pressure, temperature and liquid composition. At conditions remote from the critical conditions it is virtually independent of pressure and, in the range of temperature normally encountered in distillation, can be taken as independent of temperature.

Several equations have been developed to represent the dependence of activity coefficients on liquid composition. Only those of most use in the design of separation processes will be given. For a detailed discussion of the equations for activity coefficients and their relative merits the reader is referred to the books by Reid *et al.* (1987), Prausnitz *et al.* (1998), Walas (1985) and Null (1970).

Wilson Equation

The equation developed by Wilson (1964) is convenient to use in process design:

$$\ln \gamma_k = 1.0 - \ln \left[\sum_{j=1}^{n} \left(x_j A_{kj} \right) \right] - \sum_{i=1}^{n} \left[\frac{x_i A_{ik}}{\sum_{j=1}^{n} \left(x_j A_{ij} \right)} \right] \qquad (8.38)$$

where

γ_k = activity coefficient for component k
A_{ij}, A_{ji} = Wilson coefficients (A values) for the binary pair i, j
n = number of components.

The Wilson equation is superior to the familiar Van-Laar and Margules equations for systems that are severely non-ideal, but like other three-suffix equations it cannot be used to represent systems that form two liquid phases in the concentration range of interest.

A significant advantage of the Wilson equation is that it can be used to calculate the equilibrium compositions for multicomponent systems using only the Wilson coefficients obtained for the binary pairs that comprise the multicomponent mixture. The Wilson coefficients for several hundred binary systems are given in the DETHERM database (www.dechema.de/en/detherm.html), and by Hirata (1975). Hirata gives methods for calculating the Wilson coefficients from VLE experimental data.

Non-Random Two-Liquid Equation (NRTL) Equation

The NRTL equation developed by Renon and Prausnitz overcomes the disadvantage of the Wilson equation in that it is applicable to immiscible systems. It can be used to predict phase compositions for vapour–liquid and liquid–liquid systems.

Universal Quasi-Chemical (UNIQUAC) Equation

The UNIQUAC equation developed by Abrams and Prausnitz is usually preferred to the NRTL equation in the computer-aided design of separation processes. It is suitable for miscible and immiscible systems, and so can be used for vapour-liquid and liquid-liquid systems. As with the Wilson and NRTL equations, the equilibrium compositions for a multicomponent mixture can be predicted from experimental data for the binary pairs that comprise the mixture. Also, in the absence of experimental data for the binary pairs, the coefficients for use in the UNIQUAC equation can be predicted by a group contribution method: UNIFAC, described below.

The UNIQUAC equation is not given here as its algebraic complexity precludes its use in manual calculations. It would normally be used as a sub-routine in a design or process simulation program. For details of the equation consult the text by Reid *et al.* (1987), Prausnitz *et al.* (1998) or Walas (1985).

The best source of data for the UNIQUAC constants for binary pairs is the DETHERM database: www.dechema.de/en/detherm/html.

8.16.5. Prediction of Vapour–Liquid Equilibrium

The designer will often be confronted with the problem of how to proceed with the design of a separation process without adequate experimentally determined equilibrium data. Some techniques are available for the prediction of VLE data and for the extrapolation of experimental values. Caution must be used in the application of these techniques in design and the predictions should be supported with experimentally determined values whenever practicable. The same confidence cannot be placed on the prediction of equilibrium data as that for many of the prediction techniques for other physical properties given in this chapter. Some of the techniques most useful in design are given in the following paragraphs.

Estimation of Activity Coefficients From Azeotropic Data

If a binary system forms an azeotrope, the activity coefficients can be calculated from knowledge of the composition of the azeotrope and the azeotropic temperature. At the azeotropic point the compositions of the liquid and vapour are the same, so from equation 8.31:

$$\gamma_i = \frac{P}{P_i^o}$$

where P_i^o is determined at the azeotropic temperature.

The values of the activity coefficients determined at the azeotropic composition can be used to calculate the coefficients in the Wilson equation (or any other of the three-suffix equations) and the equation can then be used to estimate the activity coefficients at other compositions.

Horsley (1973) and Gmehling (1994) give extensive collections of data on azeotropes.

Activity Coefficients at Infinite Dilution

The constants in any of the activity coefficient equations can be readily calculated from experimental values of the activity coefficients at infinite dilution. For the Wilson equation:

$$\ln \gamma_1^\infty = -\ln A_{12} - A_{21} + 1 \tag{8.39a}$$

$$\ln \gamma_2^\infty = -\ln A_{21} - A_{12} + 1 \tag{8.39b}$$

where

γ_1^∞, γ_2^∞ = the activity coefficients at infinite dilution for components 1 and 2 respectively

A_{12} = the Wilson A-value for component 1 in component 2

A_{21} = the Wilson A-value for component 2 in component 1.

Relatively simple experimental techniques, using ebulliometry and chromatography, are available for the determination of the activity coefficients at infinite dilution. The methods used are described by Null (1970) and Conder and Young (1979).

Pieratti *et al.* (1955) have developed correlations for the prediction of the activity coefficients at infinite dilution for systems containing water, hydrocarbons and some other organic compounds. Their method, and the data needed for predictions, is described by Treybal (1963) and Reid *et al.* (1987).

Calculation of Activity Coefficients from Mutual Solubility Data

For systems that are only partially miscible in the liquid state, the activity coefficient in the homogeneous region can be calculated from experimental values of the mutual solubility limits. The methods used are described by Reid *et al.* (1987), Treybal (1963), Brian (1965) and Null (1970). Treybal (1963) has shown that the Van-Laar equation should be used for predicting activity coefficients from mutual solubility limits.

Group Contribution Methods

Group contribution methods have been developed for the prediction of liquid-phase activity coefficients. The objective has been to enable the prediction of phase equilibrium data for the tens of thousands of possible mixtures of interest to the process designer to be made from the contributions of the relatively few functional groups that made up the compounds. The UNIFAC method (Fredenslund *et al.*, 1977a) is probably the most useful for process design. Its use is described in detail in a book by Fredenslund *et al.* (1977b). A method was also developed to predict the parameters required for the NRTL equation: the ASOG method (Kojima and Tochigi, 1979). More extensive work has been done to develop the UNIFAC method, to include a wider range of functional groups; see Gmehling *et al.* (1982) and Magnussen *et al.* (1981).

The UNIFAC method is the preferred group contribution method for use in design, and it is included in all the commercial simulation and design programs.

Care must be exercised in applying the UNIFAC method. The specific limitations of the method are:

1. Pressure not greater than a few bar (say, limit to 5 bar)
2. Temperature below 150°C
3. No non-condensable components or electrolytes
4. Components must not contain more than 10 functional groups.

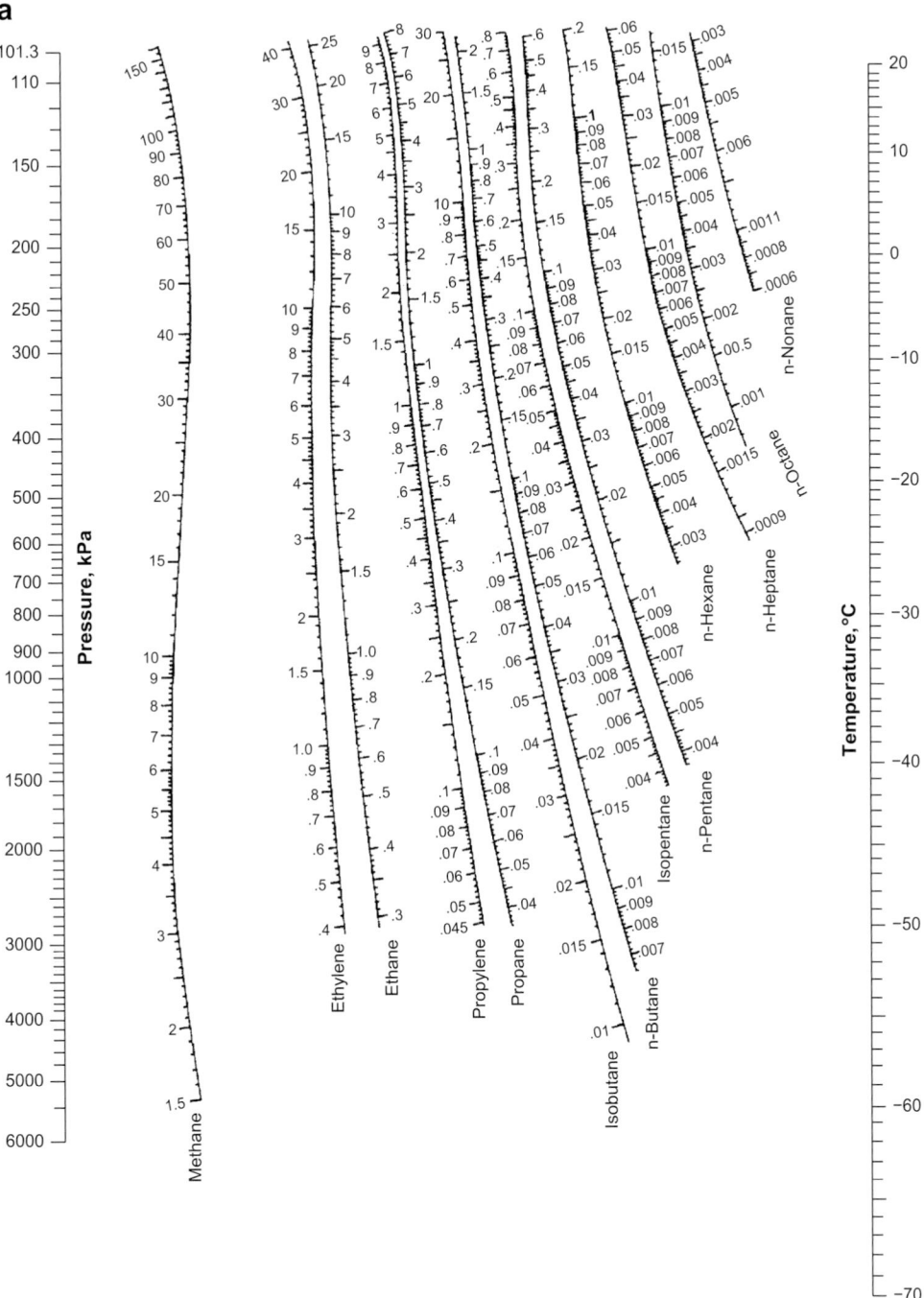

Figure 8.3. (a) De Priester chart: *K*-values for hydrocarbons, low temperature.

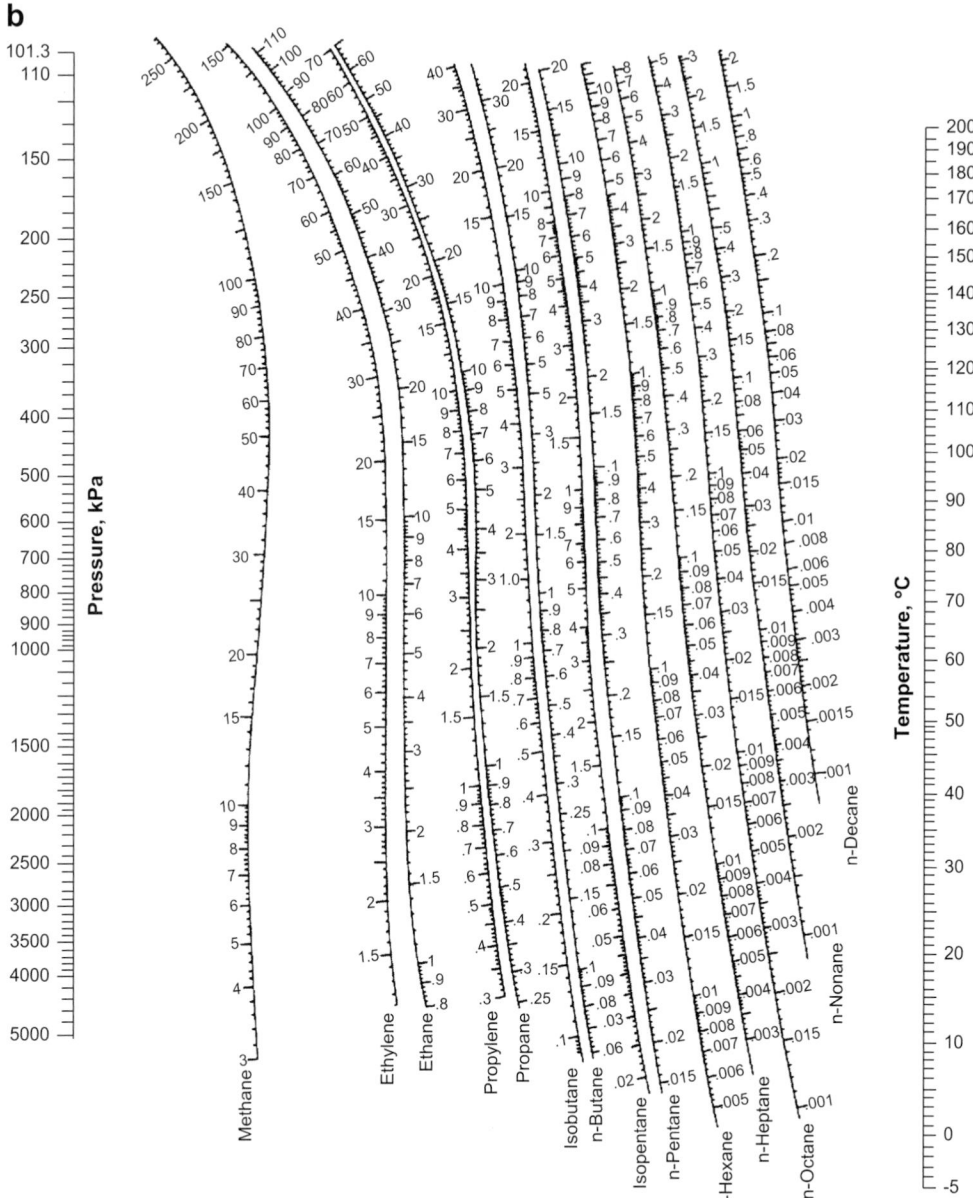

Figure 8.3 — cont'd. (b) De Priester chart: *K*-values for hydrocarbons, high temperature.

8.16.6. *K*-Values for Hydrocarbons

A useful source of *K*-values for light hydrocarbons is the well known 'De Priester charts' (Dabyburjor, 1978), which are reproduced as Figure 8.3a and b. These charts give the *K*-values over a wide range of temperature and pressure.

8.16.7. Sour-Water Systems

The term 'sour water' is used for water containing carbon dioxide, hydrogen sulphide and ammonia encountered in refinery operations.

Special correlations have been developed to handle the VLE of such systems, and these are incorporated in most design and simulation programs.

Newman (1991) gives the equilibrium data required for the design of sour-water systems, as charts.

8.16.8. Vapour–Liquid Equilibrium at High Pressures

At pressures above a few atmospheres, the deviations from ideal behaviour in the gas phase will be significant and must be taken into account in process design. The effect of pressure on the liquid-phase activity coefficient must also be considered. A discussion of the methods used to correlate and estimate VLE data at high pressures is beyond the scope of this book. The reader should refer to the texts by Null (1970), Prausnitz *et al.* (1998) and Prausnitz and Chueh (1968).

Prausnitz and Chueh also discuss phase equilibrium in systems containing components above their critical temperature (super-critical components).

8.16.9. Liquid–Liquid Equilibrium

Experimental data, or predictions, that give the distribution of components between the two solvent phases are needed for the design of liquid–liquid extraction processes, and mutual solubility limits are needed for the design of decanters, and other liquid–liquid separators.

Green and Perry (2007) give a useful summary of solubility data. Liquid–liquid equilibrium (LLE) compositions can be predicted from VLE data, but the predictions are seldom accurate enough for use in the design of liquid–liquid extraction processes.

Null (1970) gives a computer program for the calculation of ternary diagrams from VLE data, using the Van-Laar equation.

The DETHERM data collection includes LLE data for several hundred mixtures: www.dechema.de/en/detherm.html.

The UNIQUAC equation can be used to estimate activity coefficients and liquid compositions for multicomponent liquid–liquid systems. The UNIFAC method can be used to estimate UNIQUAC parameters when experimental data are not available; see Section 8.16.5.

It must be emphasized that extreme caution should be exercised when using predicted values for liquid–liquid activity coefficients in design calculations.

8.16.10. Choice of Phase Equilibrium Model for Design Calculations

The choice of the best method for deducing VLE and LLE for a given system will depend on three factors:

1. The composition of the mixture (the class of system)
2. The operating pressure (low, medium or high)
3. The experimental data available.

TABLE 8.9.   Classification of Mixtures

Class	Principal Interactions	Examples
I. Simple molecules	Dispersion forces	H_2, N_2, CH_4
II. Complex non-polar molecules	Dispersion forces	CCl_4, iC_5H_{10}
III. Polarisable	Induction dipole	CO_2, C_6H_6
IV. Polar molecules	Dipole moment	dimethyl formamide, chloroethane
V. Hydrogen bonding	Hydrogen bonds	alcohols, water

Classes of Mixtures

For the purpose of deciding which phase equilibrium method to use, it is convenient to classify components into the classes shown in Table 8.9.

Using the classification given in Table 8.9, Table 8.10 can be used to select the appropriate VLE or LLE method.

Flow-Chart for Selection of Phase Equilibrium Method

The flow-chart shown in Figure 8.4 has been adapted from a similar chart published by Wilcon and White (1986). The abbreviations used in the chart for the equations of state correspond to those given in Section 8.16.3.

8.16.11.   Gas Solubility

At low pressures, most gases are only sparingly soluble in liquids, and at dilute concentrations the systems obey Henry's law. Markham and Kobe (1941) and Battino and Clever (1966) give comprehensive reviews of the literature on gas solubility.

TABLE 8.10.   Selection of Phase Equilibrium Method

Class of Mixture	Pressure Low <3 bar		Pressure Moderate <15 bar		Pressure High >15 bar	
	f^L	f^V	f^L	f^V	f^L	f^V
I, II, III (none supercritical)	ES	I	ES	ES	ES	ES and K
I, II, III (supercritical)	ES	I	ES	ES	ES	ES and K
I, II, III, IV, V (vapour–liquid)	ACT	I	ACT	ES	ES	ES and K
I, II, III, IV, V (liquid–liquid)	ACT	I	ACT	ES	ES	ES
Hydrocarbons and water	ES	ES and K	ES	ES and K	ES	ES and K

I = ideal, vapour fugacity D partial pressure.
ES = appropriate equation of state.
K = equilibrium constant (K factor) derived from experimental data.
ACT = correlation for liquid-phase activity coefficient; such as, Wilson, NRTL, UNIQUAC, UNIFAC (see Section 8.16.4). Use UNIQUAC and UNIFAC v-l-e parameters for vapour–liquid systems and l-l-e parameters for liquid–liquid systems.

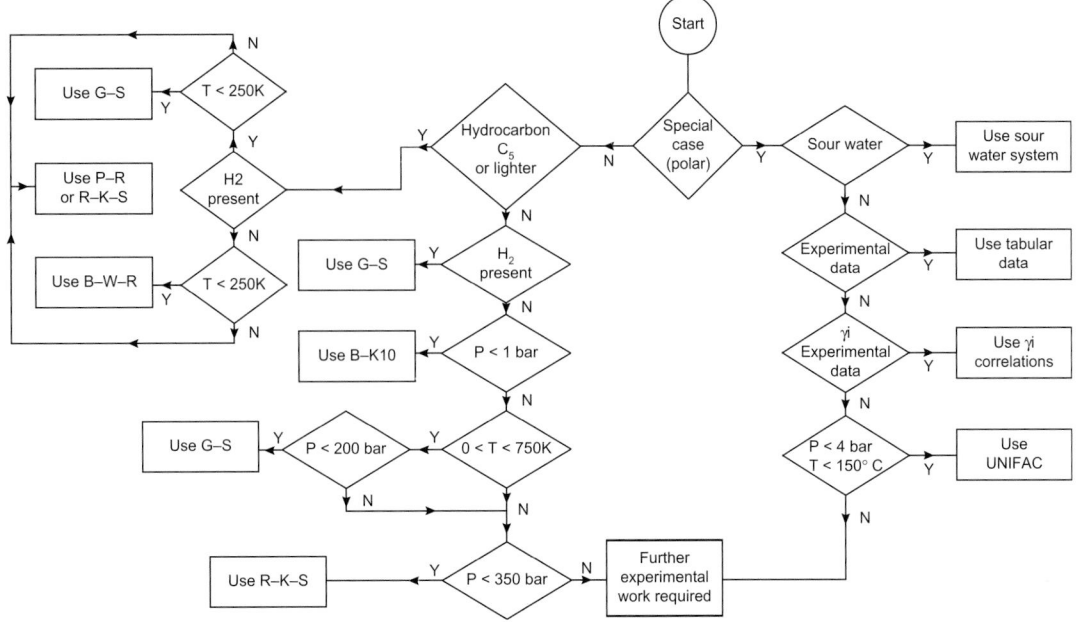

Figure 8.4. Flow-chart for the selection of phase equilibrium method.

8.16.12. Use of Equations of State to Estimate Specific Enthalpy and Density

Computer-aided packages for the design and simulation of separation processes will contain sub-routines for the estimation of excess enthalpy and liquid and vapour density from the appropriate equation of state.

Specific Enthalpy

For the vapour phase, the deviation of the specific enthalpy from the ideal state can be illustrated using the Redlich-Kwong equation, written in the form:

$$z^3 + z^2 + z(B^2 + B - A) = 0$$

where z = the compressibility factor

$$A = \frac{a \times P}{R^2 \times T^{2.5}}$$

$$B = \frac{b \times P}{R \times T}$$

The fugacity coefficient is given by:

$$\ln \phi = z - 1 - \ln(z - b) - \left(\frac{A}{B}\right) \ln \left(1 - \frac{B}{z}\right)$$

and the excess enthalpy $(H - H^\circ) = RT + \int_0^v \left[T\left(\frac{dP}{dT}\right)_v - P\right] dv$

where H is enthalpy at the system temperature and pressure and $H°$ enthalpy at the ideal state.

Unless liquid phase activity coefficients have been used, it is best to use the same equation of state for excess enthalpy that was selected for the VLE. If liquid-phase activity coefficients have been specified, then a correlation appropriate for the activity coefficient method should be used.

Density

For vapours, use the equation of state selected for predicting the VLE. For liquids, use the same equation if it is suitable for estimating liquid density.

8.17. REFERENCES

AIChE. (1983) *Design Institute for Physical Property Data, Manual for Predicting Chemical Process Design Data* (AIChE).

AIChE (1985) *Design Institute for Physical Property Data, Data Compilation, Part II* (AIChE).

Antoine, C. (1888) *Compte rend.* **107**, 681 and 836. Tensions des vapeurs: Nouvelle relation entre les tensions et les températures.

Auger, C. P. (ed.) (1992) *Information Sources in Patents* 1992 (Bower-Saur).

Austin, G. T. and Basta, N. (1998) *Shreve's Chemical Process Industries Handbook* (McGraw-Hill).

Battino, R. and Clever, H. L. (1966) *Chem. Rev.* **66**, 395. The solubility of gases in liquids.

Benedict, M., Webb, G. B., and Rubin, L. C. *Chem. Eng. Prog.* **47**, 419, 449, 517, 609 (in 4 parts). An experimental equation for thermodynamic properties of light hydrocarbons.

Benson, S. W., (1976) *Thermochemical Kinetics*, 2nd edn. (Wiley).

Benson, S. W., Cruickshank, F. R., Golden, D. M., Haugen, G. R., O'Neal, H. E., Rogers, A. S., Shaw, R., and Walsh, R. (1969) *Chem. Rev.* **69**, 279. Activity rules for the estimation of thermochemical properties.

Bretsznajder, S. (1971) *Prediction of Transport and other Physical Properties of Fluids* (Pergamon Press).

Brian, P. L. T. (1965) *Ind. Eng. Chem. Fundamentals* **4**, 100. Predicting activity coefficients from liquid phase solubility limits.

Bromley, L. A. (1952) *Thermal Conductivity of Gases at Moderate Pressure, University of California Radiation Laboratory Report UCRL—1852* (University of California, Berkeley).

Cajander, B. C., Hiplin, H. G., and Lenoir, J. M. (1960) *J. Chem. Eng. Data* 5, 251. Prediction of equilibrium ratios from nomograms of improved accuracy..

Chao, K. C. and Seader, J. D. (1961) *AIChEJ1* 7, 598. A generalized correlation for vapour-liquid equilibria in hydrocarbon mixtures.

Chueh, C. F. and Swanson, A. C. (1973a) *Can. J. Chem. Eng.* **51**, 576. Estimation of liquid heat capacity.

Chueh, C. F. and Swanson, A. C. (1973b) *Chem. Eng. Prog.* **69** (July) 83. Estimating liquid heat capacity.

Chu, J. C., Wang, S. L., Levy, S. L., and Paul, R. (1956) *Vapour-liquid Equilibrium Data* (J. W. Edwards Inc., Ann Arbor, MI).

Comyns, A. E. (1993) *Dictionary of Named Chemical Processes* (Oxford University Press).

Conder, J. R. and Young, C. L. (1979) *Physicochemical Measurement by Gas Chromatography* (Wiley).

Dabyburjor, D. B. (1978) *Chem. Eng. Prog.* **74** (April), 85. SI units for distribution coefficients.

DECHEMA (1977ff) DECHEMA *Chemistry Data Series* (DECHEMA).

Domalski, E. S. (1972) *J. Phys. Chem. Ref.* Data 1, 221. Selected values of heats of combustion and heats of formation of organic compounds containing the elements C, H, N, O, P, and S.

Dreisbach, R. R. (1952) *Pressure-volume-temperature Relationships of Organic Compounds*, 3rd edn. (Handbook Publishers).

Eucken, A. (1912) *Phys. Z.* **13**, 1101. The dependence on temperature of the thermal conductivity of certain gases.

Fedons, R. F. (1982) *Chem. Eng. Commns.* **16**, 149. A relationship between chemical structure and critical temperature.

Fredenslund, A., Gmehling, J., Michelsen, M. L., Rasmussen, P. and Prausnitz, J. M. (1977a) *Ind. Eng. Chem. Proc. Des. and Dev.* 16, 450. Computerized design of multicomponent distillation columns using the UNIFAC group contribution method for calculation of activity coefficients.

Fredenslund, A., Gmehling, J. and Rasmussen, P. (1977b) *Vapour-liquid Equilibria using UNIFAC: a Group Contribution Method* (Elsevier).

Fuller, E. N., Schettler, P. D., and Giddings, J. C. (1966) *Ind. Eng. Chem.* **58** (May), 19. A new method for the prediction of gas-phase diffusion coefficients.

Gambill, W. R. (1958) *Chem Eng NY.* **65**(6), 125. Predict diffusion coefficient, D.

Gmehling, J., Rasmussen, P., and Frednenslund, A. (1982) *Ind. Eng. Chem. Proc. Des. and Dev.* **21**, 118. Vapour liquid equilibria by UNIFAC group contribution, revision and extension.

Gmehling, J. (1994) *Azeotropic Data* (Wiley-VCH).

Gordon, T. T. and Cookfair, A. S. (2000) *Patent Fundamentals for Scientists and Engineers* (CRC Press).

Grayson, H. G. and Streed, C. W. (1963) *Proc. 6th World Petroleum Congress, Frankfurt, Germany*, paper 20, Sec. 7, 233. Vapour-liquid equilibrium for high temperature, high pressure hydrogen-hydrocarbon systems.

Green, D. W. and Perry, R. H. (eds) (2007) *Perry's Chemical Engineers' Handbook*, 8th edn (McGraw-Hill)

Haggenmacher, J. E. (1946) *J. Am. Chem. Soc.* **68**, 1633. Heat of vaporization as a function of temperature.

Hala, E., Wichterle, I., Polak, J., and Boublik, T. (1968) *Vapour-liquid Equilibrium Data at Normal Pressure* (Pergamon Press).

Hala, E.,Wichterle, I. and Linek, J. (1973) *Vapour-liquid Equilibrium Data Bibliography* (Elsevier). Supplements: 1, 1976; 2, 1979; 3, 1982, 4, 1985.

Hirata, M., Ohe, S., and Nagahama, K. (1975) *Computer Aided Data Book of Vapour-liquid Equilibria* (Elsevier).

Horsley, L. H. (1973) *Azeotropic Data III* (American Chemical Society).

Jamieson, D. T., Irving, J. B., and Tudhope, J. S. (1975) *Liquid Thermal Conductivity: A Data Survey to 1973* (HMSO).

Jasper, J. J. (1972) *J. Phys. Chem. Ref. Data* **1**, 841. The surface tension of pure liquids.

Kern, D. Q. (1950) *Process Heat Transfer* (McGraw-Hill).

Kirk, R. E. (2003) *Encyclopaedia of Chemical Technology: Concise Edition* (Wiley).

Encyclopaedia of Chemical Technology (Ed. by R. E. Kirk, D. F. Othmer), 4th edn. (Wiley).

Knovel (2003) *International Tables of Numerical Data, Physics, Chemistry and Technology*, 1st electronic edn. (Knovel).

Kobe, K. A. and Lynn, R. E. (1953) *Chem. Rev.* **52**, 177. The critical properties of elements and compounds.

Kojima, K. and Tochigi, K. (1979) *Prediction of Vapour-Liquid Equilibria by the ASOG Method* (Elsevier).

Kojima, K., Tochigi, K., Seki, H., and Watase, K. (1968) *Kagaku Kogaku* **32**, 149. Determination of vapourliquid equilibrium from boiling point curve.

Kohl, A. L. and Nielsen, R. B. (1997) *Gas Purification*, 5th edn. (Gulf Publishing).

Kudchadker, A. P., Alani, G. H., and Zwolinsk, B. J. (1968) *Chem. Rev.* **68**, 659. The critical constants of organic substances.

Lee, B. I. and Kesler, M. G. (1975) *AI Chem EJL* **21**, 510. A generalized thermodynamic correlation based on three-parameter corresponding states.

Lewis, W. K. and Squires, L. (1934) *Oil and Gas J.* (Nov. 15th) 92. The mechanism of oil viscosity as related to the structure of liquids.

Lord, C. R. (2000) *Guide to Information Sources in Engineering* (Libraries Unlimited).

Lydersen, A. L. (1955) *Estimation of Critical Properties of Organic Compounds, University of Wisconsin Coll. Eng. Exp. Stn.* Report 3 (University of Wisconsin).

Magnussen, T., Rasmussen, P., and Frednenslund, A. (1981) *Ind. Eng. Chem. Proc. Des. and Dev.* **20**, 331. UNIFAC parameter table for prediction of liquid-liquid equilibria.

Maizell, R. E. (1998) *How to find Chemical Information: A Guide for Practising Chemists, Educators and Students*, 3rd edn. (Wiley Interscience).

Markham, A. E. and Kobe, K. A. (1941) *Chem. Rev.* **28**, 519. The solubility of gases in liquids.

Mathews, J. F. (1972) *Chem. Rev.* **72**, 71. The critical constants of inorganic substances.

McKetta, J. J. (2001) *Encyclopaedia of Chemical Processes and Design* (Marcel Dekker).

Meyers, R. A. (2003) *Handbook of Petroleum Refining Processes*, 3rd edn. (McGraw-Hill).

Miller, S. A. (1969) *Ethylene and its Industrial Derivatives* (Benn).

Nesmeyanov, A. N. (1963) *Vapour Pressure of Elements* (Infosearch Ltd., London).

Newman, S. A. (1991) Hyd. Proc. **70** (Sept.) 145 (Oct.) 101 (Nov.) 139 (in 3 parts). Sour water design by charts.

Null, H. R. (1970) *Phase Equilibrium in Process Design* (Wiley).

Ohe, S. (1976) *Computer Aided Data Book of Vapour Pressure* (Data Book Publishing Co., Japan).

Ohe, S. (1989) *Vapour-Liquid Equilibrium* (Elsevier).

Ohe, S. (1990) *Vapour-Liquid Equilibrium at High Pressure* (Elsevier).

Othmer, D. F., Chudgar, M. M., and Levy, S. L. (1952) *Ind. Eng. Chem.* **44**, 1872. Binary and ternary systems of acetone, methyl ethyl ketone and water.

Othmer, D. F., Josefowitz, S., and Schmutzler, A. F. (1968) *Ind. Eng. Chem.* **40**, 886. Correlating surface tensions of liquids.

Peng, D. Y. and Robinson, D. B. (1976) *Ind. Eng. Chem. Fund* **15**, 59. A new two constant equation of state.

Pieratti, G. J., Deal, C. H., and Derr, E. L. (1955) *Ind. Eng. Chem.* **51**, 95. Activity coefficients and molecular structure.

Plocker, U., Knapp, H. and Prausnitz, J. (1978) *Ind. Eng. Chem. Proc. Des.* and Dev. **17**, 243. Calculation of high-pressure vapour-liquid equilibria from a corresponding-states correlation with emphasis on asymmetric mixtures.

Poling, B. E., Prausnitz, J. M., and O'Connell, J. P. (2000) *The Properties of Gases and Liquids*, 5th edn. (McGraw-Hill).

Prausnitz, J. M. and Chueh, P. L. (1968) *Computer Calculations for High-pressure Vapour-liquid-equilibria* (Prentice-Hall).

Prausnitz, J. M., Lichtenthaler, R. N., and Azevedo, E. G. (1998) *Molecular Thermo-dynamics of Fluid-phase Equilibria*, 3rd edn. (Prentice-Hall).

Redlich, O. and Kwong, J. N. S. (1949) *Chem. Rev.* **44**, 233. The thermodynamics of solutions, V. An equation of state. Fugacities of gaseous solutions.

Reid, R. C., Prausnitz, J. M., and Poling, B. E. (1987) *Properties of Liquids and Gases*, 4th edn. (McGraw-Hill).

Reinders, W. and De Minjer, C. H. (1947) *Trav. Chim. Pays-Bas* **66**, 573. Vapour-liquid equilibria in ternary systems VI. The system water-acetone-chloroform.

Rihani, D. N. and Doraiswamy, L. K. (1965) *Ind. Eng. Chem. Fundamentals* **4**, 17. Estimation of heat capacity of organic compounds from group contributions.

Rowley, R. L., Wilding, W. V., Oscarson, J. L., Yang, W., and Zundel, N. A. (2004) *DIPPR™ Data Compilation of Pure Chemical Properties* (Design Institute for Physical Properties, AIChE).

Smith, W. T., Greenbaum, S., and Rutledge, G. P. (1954) *J. Phys. Chem.* **58**, 443. Correlation of critical temperature with thermal expansion coefficients of organic liquids.

Soave, G. (1972) *Chem. Eng. Sci.* **27**, 1197. Equilibrium constants from modified Redlich-Kwong equation of state.

Souders, M. (1938) *J. Am. Chem. Soc.* **60**, 154. Viscosity and chemical constitution.

Stephenson, R. M. (1966) *Introduction to Chemical Process Industries* (Reinhold).

Sterbacek, Z., Biskup, B., and Tausk, P. (1979) *Calculation of Properties using Corre-sponding-state Methods* (Elsevier).

Sugden, S. (1924) *J. Chem. Soc.* **125**, 1177. A relation between surface tension, density, and chemical composition.

Touloukian, Y. S. (ed.) (1970-77) *Thermophysical Properties of Matter, TPRC Data Services* (Plenum Press).

Treybal, R. E. (1963) *Liquid Extraction*, 2nd edn. (McGraw-Hill).

Trouton, F. T. (1884) *Phil. Mag* **18**, 54. On molecular latent heat.

Tsederberg, N. V. (1965) *Thermal Conductivity of Gases and Liquids* (Arnold).

Ullman (2002) *Ullman's Encyclopaedia of Industrial Chemistry*, 5th edn. (VCH).

Walas, S. M. (1985) *Phase Equilibrium in Chemical Engineering* (Butterworths).

Watson, K. M. (1943) *Ind. Eng. Chem.* **35**, 398. Thermodynamics of the liquid state: generalized prediction of properties.

Weber, H. F. (1880) *Ann Phy. Chem.* **10**, 103. Untersuchungen über die wärmeleitung in flüssigkeiten.

Werner, R. R. (1941) *Thermochemical Calculations* (McGraw-Hill).

Wilke, C. R. and Chang, P. (1955) *A.I.Ch.E.Jl.* **1**, 264. Correlation of diffusion coefficients in dilute solutions.

Wilcon, R. F. and White, S. L. (1986) *Chem. Eng., NY* **93**, (Oct. 27th) 142. Selecting the proper model to stimulate vapour-liquid equilibrium.

Wilson, G. M. (1964) *J. Am. Chem. Soc.* **86**, 127. A new expression for excess energy of mixing.

Yaws, C. L. (1993–1994) *Handbook of Viscosity, 4 vols* (Gulf Publishing).

Yaws, C. L. (1994–1995) *Handbook of Vapour Pressure, 4 vols* (Gulf Publishing).

Yaws, C. L. (1995–1997) *Handbook of Thermal Conductivity, 4 vols* (Gulf Publishing).

York, R. and Holmes, R. C. (1942) *Ind. Eng. Chem.* **34**, 345. Vapour-liquid equilibria of the system acetoneacetic acid-water.

Bibliography: General Sources of Physical Properties

Boubik, T., Fried, V., and Hala, E. (1984) *The Vapour Pressures of Pure Substances*, 2nd edn. (Elsevier).

Boul, M., Nyvlt, J., and Sohnel, O. (1981) *Solubility of Inorganic Two-Component Systems* (Elsevier).

Christensen, J. J., Hanks, R. W., and Izatt, R. M. (1982) *Handbook of Heats of Mixing* (Wiley).

Dreisbach, R. R. (1955-61) *Physical Properties of Chemical Compounds, Vols. I, II, III* (American Chemical Society).

Dreisbach, R. R. (1952) *Pressure-Volume-Temperature Relationships of Organic Compounds*, 3rd edn. (Handbook Publishers).

Fenske, M., Braun, W. G., and Thompson, W. H. (1966) *Technical Data Book-Petroleum Refining* (American Petroleum Institute).

Flick, E. W. (1991) *Industrial Solvent Handbook*, 4th edn. (Noyes).

Gallant, R. W. (1968) (1970) *Physical Properties of Hydrocarbons*, Vols. 1 and 2 (Gulf).

Green, D. W. (2007) *Perry's Chemical Engineers' Handbook*, 8th edn. (McGraw-Hill).

Lange, N. A. (1961) *Handbook of Chemistry*, 10th edn. (McGraw-Hill).

Maxwell, J. B. (1950) *Data Book on Hydrocarbons* (Van Nostrand).

National Bureau of Standards (1951) *Selected Values of Thermodynamic Properties, Circular C500* (US Government Printing Office).

Renon, H. (1986) *Fluid Properties and Phase Equilibria for Chemical Engineers* (Elsevier).

Ross, T. K. and Freshwater, D. C. (1962) *Chemical Engineers Data Book* (Leonard Hill).

Rossini, F. D. (1953) *Selected Values of Physical and Thermodynamic Properties of Hydrocarbons and Related Compounds* (American Chemical Society).

Seidell, A. (1952) *Solubilities of Inorganic and Organic Compounds*, 3rd edn. (Van Nostrand).

Sohnel, O. and Novotny, P. (1985) *Densities of Aqueous Solutions in Organic Substances* (Elsevier).

Spiers, H. M. (1961) *Technical Data on Fuel*, 6th edn. (British National Committee, Conference on World Power).

Stephen, T. and Stephen, H. (1963) *Solubilities of Inorganic and Organic Compounds, 2 vols* (Macmillan).

Stephenson, R. M. (1966) *Introduction to Chemical Process Industries* (Reinhold).

Tamir, A., Tamir, E., and Stephan, K. (1983) *Heats on Phase Change of Pure Components and Mixtures* (Elsevier).

Timmermanns, J. (1950) *Physico-chemical Constants of Pure Organic Compounds* (Elsevier).

Timmermanns, J. (1959) *Physico-chemical Constants of Binary Systems, 4 vols* (Interscience).

Viswanath, D. S. and Natarajan, G. (1989) *Data Book on Viscosity* (Hemisphere).

Weast, R. C. (1972) *Handbook of Chemistry and Physics*, 53rd edn. (the Chemical Rubber Co.).

International Critical Tables of Numerical Data, Physics, Chemistry, and Technology, 8 vols (Ed. by E. W. Washburn) (McGraw-Hill).

Wisniak, J. and Tamir, A. (1980) *Liquid-liquid Equilibria and Extraction: A Literature Source Book*, Parts A and B.

Wisniak, J. and Herskowitz, M. (1984) *Solubility of Gases and Solids, 2 vols* (Elsevier).

Yaws, C. L. (1977) *Physical Properties* (McGraw-Hill).

Yaws, C. L. (1999) *Chemical Properties Handbook* (McGraw-Hill).

Yaw's Handbook of Thermodynamic and Physical Properties of Chemical Compounds (2003) Knovel.

8.18. NOMENCLATURE

		Dimensions in $\mathbf{MLT}\theta$
A	Coefficient in the Antoine equation	—
$A_{1,2}$	Coefficients in the Wilson equation for the binary pair 1, 2	—
a	Coefficient in the Redlich-Kwong equation of state	—
B	Coefficient in the Antoine equation	θ
B_i	Second viral coefficient for component i	$M^{-1}L^3$
b	Coefficient in the Redlich-Kwong equation of state	—
C	Coefficient in the Antoine equation	θ
C_p	Specific heat capacity at constant pressure	$L^2T^{-2}\theta^{-1}$
D_L	Liquid diffusivity	L^2T^{-1}
D_v	Gas diffusivity	L^2T^{-1}
f_i	Fugacity coefficient for component i	—
f_i^{OL}	Standard state fugacity coefficient of pure liquid	—
H	Specific enthalpy	L^2T^{-2}
H^0	Excess specific enthalpy	L^2T^{-2}
I	Souders' index (equation 8.9)	$M^{-1}L^3$
K	Equilibrium constant (ratio)	—
K^0	Equilibrium constant for an ideal mixture	—
k	Thermal conductivity	$MLT^{-3}\theta^{-1}$
k_m	Thermal conductivity of a mixture	$MLT^{-3}\theta^{-1}$
L_v	Latent heat of vaporization	L^2T^{-2}
$L_{v,b}$	Latent heat at normal boiling point	L^2T^{-2}
M	Molecular mass (weight)	M
n	Number of components	—
P	Pressure	$ML^{-1}T^{-2}$ or L
P_c	Critical pressure	$ML^{-1}T^{-2}$
P_{ch}	Sugden's parachor (equation 8.23)	—
P_i^0	Vapour pressure of component i	$ML^{-1}T^{-2}$ or L
P_k	Vapour pressure of component k	$ML^{-1}T^{-2}$ or L
P_r	Reduced pressure	—
ΔP_c	Critical constant increment in Lydersen equation (equation 8.26)	$M^{-1/2}L^{1/2}T$
R	Universal gas constant	$L^2T^{-2}\theta^{-1}$
T	Temperature, absolute scale	θ
T_b	Normal boiling point, absolute scale	θ
T_c	Critical temperature	θ
T_r	Reduced temperature	θ
ΔT_c	Critical constant increment in Lydersen equation (Equation 8.25)	—
t	Temperature, relative scale	θ
V_c	Critical volume	$M^{-1}L^3$
V_m	Molar volume at normal boiling point	$M^{-1}L^3$

ΔV_c	Critical constant increment in Lydersen equation (Equation 8.27)	$\mathbf{M^{-1}L^3}$
v_i	Special diffusion volume coefficient for component i (Table 8.5)	$\mathbf{L^3}$
v_i^0	Liquid molar volume	$\mathbf{M^{-1}L^3}$
w	Mass fraction (weight fraction)	—
x	Mol fraction, liquid phase	—
y	Mol fraction, vapour phase	—
z	Compressibility factor	—
α	Relative volatility	—
β	Coefficient of thermal expansion	$\boldsymbol{\theta^{-1}}$
γ	Liquid activity coefficient	—
γ^∞	Activity coefficient at infinite dilution	—
μ	Dynamic viscosity	$\mathbf{ML^{-1}T^{-1}}$
μ_b	Viscosity at boiling point	$\mathbf{ML^{-1}T^{-1}}$
μ_m	Viscosity of a mixture	$\mathbf{ML^{-1}T^{-1}}$
ρ	Density	$\mathbf{ML^{-3}}$
ρ_L	Liquid density	$\mathbf{ML^{-3}}$
ρ_v	Vapour (gas) density	$\mathbf{ML^{-3}}$
ρ_b	Density at normal boiling point	$\mathbf{ML^{-3}}$
σ	Surface tension	$\mathbf{MT^{-2}}$
σ_m	Surface tension of a mixture	$\mathbf{MT^{-2}}$
ϕ	Fugacity coefficient	—
ϕ^s	Fugacity coefficient of pure component	—
ϕ^L	Fugacity coefficient of pure liquid	—
ϕ^V	Fugacity coefficient of pure vapour	—

Suffixes

$\left.\begin{array}{l} a, b \\ i, j, k \\ 1, 2 \end{array}\right\}$	Components	
L	Liquid	—
V	Vapour	—

8.19.   PROBLEMS

8.1. Estimate the liquid density at their boiling points for the following:
1. 2-butanol
2. Methyl chloride
3. Methyl ethyl ketone
4. Aniline
5. Nitrobenzene.

8.2. Estimate the density of the following gases at the conditions given:
1. Hydrogen at 20 bara and 230°C
2. Ammonia at 1 bara and 50°C and at 100 bara and 300°C
3. Nitrobenzene at 20 bara and 230°C
4. Water at 100 bara and 500°C; check your answer using steam tables
5. Benzene at 2 barg and 250°C
6. Synthesis gas ($N_2 + 3H_2$) at 5 barg and 25°C.

8.3. Make a rough estimate of the viscosity of 2-butanol and aniline at their boiling points, using the modified Arrhenius equation. Compare your values with those given using the equation for viscosity in Appendix C.

8.4. Make a rough estimate of the thermal conductivity of n-butane both as a liquid at 20°C and as a gas at 5 bara and 200°C. Take the viscosity of the gaseous n-butane as $0.012 \, mN \, m^{-2}s$.

8.5. Estimate the specific heat capacity of liquid 1,4 pentadiene and aniline at 20°C.

8.6. For the compounds listed below, estimate the constants in the equation for ideal gas heat capacity, equation 3.19, using the method given in Section 8.9.2:
1. 3-methyl thiophene
2. Nitrobenzene
3. 2-methyl-2-butanethiol
4. Methyl-t-butyl ether.

8.7. Estimate the heat of vaporization of methyl-t-butyl ether, at 100°C.

8.8. Estimate the gaseous phase diffusion coefficient for the following systems, at 1 atmosphere and the temperatures given:
1. Carbon dioxide in air at 20°C
2. Ethane in hydrogen at 0°C
3. Oxygen in hydrogen at 0°C
4. Water vapour in air at 450°C
5. Phosgene in air at 0°C.

8.9. Estimate the liquid phase diffusion coefficient for the following systems at 25°C:
1. Toluene in n-heptane
2. Nitrobenzene in carbon tetrachloride
3. Chloroform in benzene
4. Hydrogen chloride in water
5. Sulphur dioxide in water.

8.10. Estimate the surface tension of pure acetone and ethanol at 20°C, and benzene at 16°C, all at 1 atmosphere pressure.

8.11. Using Lydersen's method, estimate the critical constants for isobutanol. Compare your values with those given in Appendix C.

8.12. The composition of the feed to a debutanizer is given below. The column will operate at 14 bar and below 750 K. The process is to be modelled using a commercial simulation program. Suggest a suitable phase equilibrium method to use in the simulation.

Feed composition:

		kg/h
propane	C_3	910
isobutane	$i\text{-}C_4$	180
n-butane	$n\text{-}C_4$	270
isopentane	$i\text{-}C_5$	70
normal pentane	$n\text{-}C_5$	90
normal hexane	$n\text{-}C_6$	20

8.13. In the manufacture of methyl ethyl ketone from butanol, the product is separated from unreacted butanol by distillation. The feed to the column consists of a mixture of methyl ethyl ketone, 2-butanol and trichloroethane. What would be a suitable phase equilibrium correlation to use in modelling this process?

9 SAFETY AND LOSS PREVENTION

Chapter Contents

Key Learning Objectives

- The importance of safety in the design and operation of chemical plants
- Safety legislation with which companies must comply
- Standards and codes of practice that help ensure safer designs
- Process and materials hazards that must be considered in design
- Methods such as HAZOP, FMEA and quantitative risk analysis that are used to analyze and quantify process hazards
- How relief valves are designed and used to prevent failure of vessels due to over-pressure

9.1.   INTRODUCTION

The safe design and operation of facilities is of paramount importance to every company that is involved in the manufacture of fuels, chemicals and pharmaceutical products.

Any organization has a legal and moral obligation to safeguard the health and welfare of its employees and the general public. Safety is also good business; the good management practices needed to ensure safe operation also ensure efficient operation.

The term 'loss prevention' is an insurance term, the loss being the financial loss caused by an accident. This loss will not only be the cost of replacing the damaged plant, paying fines and settling third party claims, but also the loss of earnings from lost production and lost sales opportunity. In the event of a major incident, such costs can be large enough to overwhelm a company.

All manufacturing processes are to some extent hazardous, but in chemical processes there are additional, special, hazards associated with the chemicals used and the process conditions. The designer must be aware of these hazards, and ensure, through the application of sound engineering practice, that the risks are reduced to tolerable levels.

In this chapter the discussion of safety in process design will of necessity be limited. A more complete treatment of the subject can be found in the books by Wells (1996, 1997), Mannan (2004), Fawcett and Wood (1982), Green (1982), Crowl and Louvar (2002), Cameron and Raman (2005) and Carson and Mumford (1988, 2002); and in the general literature, particularly the publications by the American Institute of Chemical Engineers and the Institution of Chemical Engineers. The proceedings of the symposia on safety and loss prevention organized by these bodies, and the European Federation of Chemical Engineering, also contain many articles of interest on general safety philosophy, techniques and organization, and the hazards associated with specific processes and equipment. A good general overview of safety issues in process design is given in the AIChE Center for Chemical Process Safety book Guidelines for Engineering Design for Process Safety (CCPS, 1993). The Institution of Chemical Engineers has published a book on safety of particular interest to students of Chemical Engineering (Marshall and Ruhemann, 2000).

While an effort has been made to provide a summary of the major legal requirements and relevant codes and standards, the area of process safety undergoes continuous improvement and many of the standards are re-written annually, so the information presented here may not be current by the time of publication. Updated information can be obtained from the standards organizations and regulatory agencies. The design engineer should always consult the most recent version of the laws, regulations or standards and should always make a thorough check to ensure that the design complies with all local regulations and current best practices.

9.1.1.   Safety Legislation

Because of the particular hazards associated with processing large quantities of chemicals and fuels, most governments have enacted legislation to ensure that best safety practices are followed. In the UK, the main safety laws are:

1. The Health and Safety at Work etc. Act (1974) (HSAW or HSW): Sets legal requirements for manufacturers, employers and employees to ensure the health

and safety of employees and the general public. Established the Health and Safety Executive (HSE) to enforce the act, set health and safety regulations, codes and standards, appoint inspectors and direct investigations and inquiries. Inspectors are given the power to serve prohibition notices directing an employer to cease activities that the inspector believes involve a risk of serious personal injury. Failure to comply with a prohibition notice, obstruction of inspectors or contravening health and safety regulations can be punished by fines, imprisonment, or both, depending on the seriousness of the offence. Additional details and the full text of the Act can be obtained from the HSE website: www.hse.gov.uk.

2. The Offshore Safety Act (1992): Passed following the Piper Alpha disaster of 1988. Extends the Health and Safety at Work etc. Act to cover offshore installations and pipelines, including safe decommissioning, dismantling and disposal of offshore facilities in parts of the sea in or adjacent to Great Britain.

In addition to national laws, countries that belong to the European Union (EU) are also subject to various European community directives regulating plant and product safety. The major European safety legislation includes:

1. Council Directive 88/379/EEC of 7 June 1988 on the approximation of the laws, regulations and administrative provisions of the member states relating to the classification, packaging and labelling of dangerous preparations (88/379/EEC): Sets common European standards for identifying hazardous materials. Article 10 requires potentially hazardous materials to have a materials safety data sheet, details of which are specified in Commission Directive 91/155/EEC of 5 March 1991.

2. Council Directive 89/391/EEC of 12 June 1989 on the introduction of measures to encourage improvements in the safety and health of workers at work (89/391/EEC): Employers have a duty to ensure the health and safety of workers in every aspect related to the work. Employers must avoid risks, carry out risk assessments, take steps to reduce risks, consult with and train workers, and provide suitable protective equipment and systems. Workers are responsible for making correct use of machinery, safety devices and protective equipment, and for informing employers of any situation that they believe represents a danger to health and safety. Under Article 16, the European Council is empowered to issue additional Directives in areas of specific hazards. Some of the areas relevant to chemical plant design that have been addressed under Article 16 include:

   - Directive 80/1107/EC: Chemical, physical and biological agents
   - Directive 1992/92/EC: Explosive atmospheres
   - Directive 98/24/EC: Chemical agents. This is implemented as the Control of Substances Hazardous to Health (COSHH) Regulations of 1992 in the UK; see www.hse.gov.uk/coshh
   - Directive 2000/54/EC: Biological agents
   - Directive 2003/10/EC: Physical agents (noise)
   - Directive 2004/37/EC: Carcinogens.

3. Council Directive 96/82/EC of 9 December 1996 on the control of major accident hazards involving dangerous substances (96/82/EC, known as the Seveso II Directive): Operators of facilities that contain or produce greater than prescribed quantities of hazardous chemicals must register with national authorities, establish and implement a major accident prevention policy, demonstrate that adequate safety and reliability have been incorporated into the design, construction, operation and maintenance of any facility linked to a major accident hazard, and draw up emergency plans that include internal and external responses to a major accident. Information on safety measures and required response must be supplied to anyone likely to be affected by a major accident, including the general public. Safety information must be updated when a plant modification, expansion or other change of use occurs. Facilities shall be inspected for compliance at least annually. In the UK, the Seveso II Directive is implemented as the Control of Major Accident Hazards Regulations 1999 (COMAH), administered by the HSE.

4. Regulation (EC) No 1907/2006 of the European Parliament and of the Council of 18 December 2006 concerning the Registration, Evaluation, Authorization and Restriction of Chemicals (REACH): Establishes a European Chemicals Agency (ECHA) and replaces several European Directives with a single set of laws to protect human health and the environment from the use of chemicals. Requires manufacturers and importers of chemicals to register them with ECHA and report information on properties and hazards. All registered substances will be evaluated by ECHA or national authorities for potential hazardous properties. Restrictions can be placed on the manufacture, import and use of substances of 'very high concern', including requiring manufacturers to obtain authorization from ECHA, or in extreme cases, a total ban of the substance. Substances of very high concern include carcinogens, mutagens, persistent pollutants, bio-accumulative compounds, neurotoxins and endocrine disruptors. One of the goals of REACH is to control use of such substances and encourage industry to develop safer materials to use instead. Because compliance with REACH is necessary to gain access to European markets, it is likely that ECHA will effectively set global standards for chemical product safety (Boxerman et al., 2008).

Other EU directives relevant to health and safety at work are described in the following sections. The European Parliament and Council frequently issue directives updating safety and environmental legislation, which are then implemented and enforced by the member states. More detailed information on all of the acts passed by the EU can be found on the European Union Occupational Safety and Health Agency website at http://osha.europa.eu. Full text versions of EU directives can be downloaded from http://eur-lex.europa.eu.

Information on health and safety legislation in the US and Canada is given in the North American version of this book (Towler and Sinnott, 2008).

In addition to safety legislation, releases of material to the environment as a result of loss of containment during an incident are also prohibited by various environmental laws. These are discussed in Chapter 14.

Various states, municipalities and other bodies may also enact legislation that regulates the safe operation of chemical plants (for example, local fire codes). Local regulations may place stricter requirements on the design and operation of facilities, but do not absolve the owner or designer from obligations under national or federal laws.

The most recent version of local, national and federal laws and standards must always be consulted during design. The design engineer should be aware of the relevant legislation, but is not usually expected to interpret the requirements of the law and will usually rely upon corporate lawyers and professional safety experts to set company policies, codes and standards that ensure legal compliance. If the design engineer has any concerns that corporate policies are not meeting regulatory requirements then these concerns should immediately be raised with management and, if there is no satisfactory response, with the regulatory agencies.

In this book only the particular hazards associated with chemical and allied processes will be considered. The more general hazards present in all manufacturing processes, such as the dangers from rotating machinery, falls, falling objects, use of machine tools and electrocution, will not be considered. General industrial safety and hygiene are covered in several books: King and Hirst (1998), Ashafi (2003) and Ridley (2003).

9.1.2. Layers of Plant Safety

Safety and loss prevention in process design can be considered under the following broad headings:

1. Identification and assessment of the hazards.
2. Control of the hazards: for example, by containment of flammable and toxic materials.
3. Control of the process: prevention of hazardous deviations in process variables (pressure, temperature, flow) by provision of automatic control systems, interlocks, alarms and trips; together with good operating practices and management.
4. Limitation of the loss, i.e., the damage and injury caused if an incident occurs: pressure relief, plant layout, provision of fire-fighting equipment.

Another way of expressing this is in terms of layers of plant safety, illustrated in Figure 9.1. Each of the layers in Figure 9.1 can be activated if the lower levels have all failed.

The most basic level of plant safety is safe process and equipment design. If the process is inherently safe (see Section 9.1.3) then incidents are much less likely to occur. The process equipment is the primary means for containing the chemicals that are being processed, as well as keeping out air, and containing high temperatures and pressures. Vessel design codes and standards incorporate safety margins for equipment to reduce the risk of it failing in operation (see Chapter 13). Most countries require chemical plants to be built and operated in accordance with national or industry standards.

Figure 9.1.   Layers of plant safety.

The basic process control system (BPCS) should be designed to maintain the plant under safe conditions of temperature, pressure, flow rates, levels and compositions. In most continuous plants the process control system will attempt to maintain the process within reasonable bounds of a steady-state condition. In batch or cyclic processes the variation of process parameters ('ramping') will be controlled to occur at a safe rate to prevent over-shooting.

If a process variable falls outside of the safe operating range, this should trigger an automatic alarm in the plant control room. The purpose of the alarm is to warn the process operators of the triggering condition so that the operators can intervene accordingly. Care should be taken when designing the plant control system not to include too many alarms and to clarify the necessary operator responses, since too many alarms can overwhelm the operators and increase the likelihood of human error (see Section 9.3.7). Alarms should be set so that they are not frequently triggered by normal process variability (in which case they will tend to be ignored) and also to allow time for the operator to respond before the next safety layer is activated. See Chapter 5 for more discussion of process control and instrumentation.

In the event that process operators are unable to bring the process back into control when there is a significant deviation of a variable that indicates a hazardous condition, an automatic shutdown of the process (also known as a 'trip') should be activated. Trip systems are sometimes activated by the plant control system and sometimes self-actuated, as described in Chapter 5. Emergency shutdown will usually involve shutting off feeds and sources of heat, depressurizing the process and purging the plant with inert material. When designing the emergency shutdown procedures and systems, care must be taken to ensure that unsafe conditions are not created or worsened. For example, in some high temperature or exothermic processes it may be safer to continue feed of one of the reagents while shutting off the others, so as to remove heat from the reactor. Closing all the valves in a plant is almost never the safest means of shutting it down. A good guide to the design of control, alarm and shutdown systems is given in the American Petroleum Institute Recommended Practice API RP 14c (2001). Although intended for offshore production platforms, this standard covers many unit operations found in chemical plants and refineries. Methods for quantifying the required reliability of shutdown systems are described in Section 9.8.

If the plant safety shutdown is not rapid enough and an over-pressure situation develops, then the pressure-relief system is activated. Pressure vessel design codes such as the ASME Boiler and Pressure Vessel Code require relief devices to be fitted on all pressure vessels. If the relief system has been properly designed and maintained, then in the event of an over-pressure incident the plant contents will be vented via relief valves or bursting disks into the relief system, where liquids are recovered for treatment and vapours are sent to flare stacks or discharged to atmosphere if it is safe to do so. The pressure-relief system should allow the plant to be relieved of any source of over-pressure before damage to process equipment (leaks, bursting or explosion) can occur. Pressure-relief systems are discussed in Section 9.9.

If a loss of containment does occur in a chemical plant, then an emergency response is required. A small-scale loss of containment might be a leak or a spill. Leaks of liquids are usually visible and obvious, while leaks of vapour can be much harder to detect and require special instrumentation for monitoring. If the material that escapes from the process is flammable, then the first manifestation of loss of containment might be a small or localized fire (often smouldering insulation is an early warning). Plant personnel should be trained to respond to such emergencies. Many large sites also have a dedicated emergency response staff to fight fires and clean up chemical spills. The emergency response in the process unit does not always cause the unit to be shut down, depending on the scale of the incident, but the root cause of every incident must always be determined and any deficiency in the plant must be corrected safely before normal operations are resumed.

In the event that an incident develops into a more serious accident, the resources required will be beyond those available on the plant or site. Local community emergency response providers will be brought into the site, and injured members of the workforce and local population will need treatment at local hospitals. The local community must be able to plan for such events and local emergency responders must be trained to cope with the hazards associated with the plant. In countries of the EU, the Seveso II Directive (Directive 96/82/EC) ensures that the local community has access to the necessary information.

9.1.3. Intrinsic and Extrinsic Safety

Processes can be divided into those that are intrinsically safe, and those for which the safety has to be engineered in. An intrinsically safe process is one in which safe operation is inherent in the nature of the process; a process that causes no danger, or negligible danger, under all foreseeable circumstances (all possible deviations from the design operating conditions). The term 'inherently safe' is often preferred to 'intrinsically safe', to avoid confusion with the narrower use of the term intrinsically safe as applied to electrical equipment (see Section 9.3.5). In the context of risk management, an inherently safe design is a design that has a very low likelihood of causing injury even in the absence of protective systems; see Section 9.8.

Clearly, the designer should always select a process that is inherently safe whenever it is practical and economic to do so; however, most chemical manufacturing processes are, to a greater or lesser extent, inherently unsafe, and dangerous situations can develop if the process conditions deviate from the design values. The safe operation of such processes depends on the design and provision of engineered safety devices and on good operating practices, to prevent a dangerous situation developing and to minimize the consequences of any incident that arises from the failure of these safeguards.

The term 'engineered safety' covers the provision in the design of control systems, alarms, trips, pressure-relief devices, automatic shutdown systems, duplication of key equipment services; and fire-fighting equipment, sprinkler systems and blast walls, to contain any fire or explosion.

The design of an inherently safe process plant is discussed by Kletz in a booklet published by the Institution of Chemical Engineers (Kletz, 1984), and in Kletz and Cheaper (1998), CCPS (1993) and Chapter 32 of Mannan (2004). Kletz makes the telling point that material that is not there cannot leak out; so cannot catch fire, explode or poison anyone. This is a plea to keep the inventory of dangerous material to the absolute minimum required for the operation of the process. The AIChE Center for Chemical Process Safety has published a checklist for inherently safer chemical reaction process design and operation, which can be downloaded from http://www.aiche.org/ccps/safetyalert. Additional information on inherently safe design is given in the CCPS book by Bollinger *et al.* (2008).

9.2. MATERIALS HAZARDS

In this section the special hazards of chemicals are reviewed (toxicity, flammability and reactivity). Hazards arising from process operation will be discussed in Section 9.3.

9.2.1. Toxicity

Most of the materials used in the manufacture of chemicals are poisonous, to some extent, and almost every chemical is toxic if someone is exposed to enough of it. The potential hazard will depend on the inherent toxicity of the material and the frequency and duration of any exposure.

It is usual to distinguish between the short-term effects (acute) and the long-term effects (chronic). Acute effects normally have symptoms that develop rapidly after exposure; for example, burns to the skin after direct contact, respiratory failure, renal failure, cardiac arrest, paralysis, etc. Acute effects are usually associated with a short exposure to a high concentration of toxin (although what constitutes a 'high concentration' depends on the toxicity). The chronic symptoms of poisoning develop over a long period of time (e.g. cancer) and often persist or recur frequently. Chronic effects may occur as the result of long-term exposure to low levels of a toxin, but may also occur as a delayed response to a short-term exposure to high levels of a toxin.

Highly toxic materials that cause immediate injury, such as phosgene and chlorine, are usually classified as safety hazards, whereas materials whose effects are only apparent after long exposure at low concentrations, for instance carcinogenic materials such as vinyl chloride, are usually classified as industrial health and hygiene hazards. The permissible limits and the precautions to be taken to ensure the limits are met will be very different for these two classes of toxic materials. Industrial hygiene is as much a matter of good operating practice and control as of good design.

The inherent toxicity of a material is measured by tests on animals. It is usually expressed as the lethal dose at which 50% of the test animals are killed, the LD_{50} (lethal dose fifty) value. The dose is expressed as the quantity in milligrams of the toxic substance per kilogram of body weight of the test animal.

Some values of LD_{50} for oral ingestion by rats are given in Table 9.1. Estimates of the LD_{50} for humans are based on tests on animals. The LD_{50} measures the acute effects; it gives only a crude indication of the possible chronic effects. The LD_{50} for humans should always be taken as the lowest measured value for other mammalian species. In some cases, LD_{50} data are given for different routes of ingestion. For example, ethanol has LD_{50} values 3450 (oral, mouse), 7060 (oral, rat) and 1440 (intravenous, rat).

There is no generally accepted definition of what can be considered toxic and non-toxic.

A system of classification is given in the *Classification, Packaging and Labelling of Dangerous Substances Regulations, 1984* (UK), which is based on EU guidelines; for example:

LD_{50}, absorbed orally in rats, mg/kg	
≤ 25	very toxic
25 to 200	toxic
200 to 2000	harmful

These definitions apply only to the short-term (acute) effects. In fixing permissible limits on concentration for the long-term exposure of workers to toxic materials, the exposure time must be considered together with the inherent toxicity of the material.

TABLE 9.1. Toxicity Data

Compound	PEL (ppm)	LD_{50} (mg/kg)
Carbon monoxide	50	1807
Carbon disulphide	20	3188
Chlorine	1	239
Chlorine dioxide	0.1	292
Chloroform	50	1188
Cyclohexane	300	
Dioxane	100	4200
Ethylbenzene	100	3500
Formic acid	5	1100
Furfural	5	260
Hydrogen chloride	5	4701
Hydrogen cyanide	10	3.7
Isopropyl alcohol	400	5045
Toluene	100	5000
Xylene	100	4300

Source: OSHA.

The 'threshold limit value' (TLV) is a commonly used guide for controlling the long-term exposure of workers to contaminated air. The TLV is defined as the concentration to which it is believed the average worker could be exposed, day by day, for 8 hours a day, 5 days a week, without suffering harm. It is expressed in ppm for vapours and gases, and in mg/m^3 (or grains/ft^3) for dusts and liquid mists. A comprehensive source of data on the toxicity of industrial materials is Sax's handbook (Lewis, 2004); which also gives guidance on the interpretation and use of the data. Recommended TLVs are given by the American Conference of Government Industrial Hygienists (www.acgih.org/home.htm).

Most governments require lower levels of chemicals exposure than the TLVs. In the UK the HSE sets workplace exposure limits under COSHH. These are published as EH40/2005, which is available from www.hse.gov.uk. Similar levels are allowed in other European countries under Directives 98/24/EC and 2000/39/EC. In the USA, Permissible Exposure Limits (PEL) for known toxins are set by the Occupational Safety and Health Administration (OSHA). Values can be found on the OSHA web site at www.osha.gov/SLTC/healthguidelines.

Fuller details of the methods used for toxicity testing, the interpretation of the results and their use in setting standards for industrial hygiene are given in the more specialized texts on the subject; see Carson and Mumford (1988) and Mannan (2004).

9.2.2. Flammability

The term 'flammable' is now more commonly used in the technical literature than 'inflammable' to describe materials that will burn, and will be used in this book. The hazard caused by a flammable material depends on a number of factors:

1. The flash-point of the material
2. The autoignition temperature of the material
3. The flammability limits of the material
4. The energy released in combustion.

Flash-Point

The flash-point is a measure of the ease of ignition of a liquid. It is the lowest temperature at which the material will ignite from an open flame. The flash-point is a function of the vapour pressure and the flammability limits of the material. It is measured in standard apparatus, following standard procedures (ASTM D92 and ASTM D93). Both open- and closed-cup apparatus is used. Closed-cup flash-points are lower than open cup, and the type of apparatus used should be stated clearly when reporting measurements. Flash-points are given in Sax's handbook (Lewis, 2004). The flash-points of many volatile materials are below normal ambient temperature; for example, ether $-45°C$, gasoline $-43°C$ (open cup).

Autoignition Temperature

The autoignition temperature of a substance is the temperature at which it will ignite spontaneously in air, without any external source of ignition. It is an indication of the maximum temperature to which a material can be heated in air; for example, in drying operations.

Flammability Limits

The flammability limits of a material are the lowest and highest concentrations in air, at normal pressure and temperature, at which a flame will propagate through the mixture. They show the range of concentration over which the material will burn in air, if ignited. At very low concentrations in air a flame will not propagate as there is insufficient fuel. Similarly, at very high concentrations a flame will not propagate due to insufficient oxidant. Flammability limits are characteristic of the particular material, and differ widely for different materials. For example, hydrogen has a lower limit of 4.1 and an upper limit of 74.2% by volume, whereas for gasoline the range is only from 1.3% to 7.0%. The flammability limits for a number of materials are given in Table 9.2. The limits for a wider range of materials are given in Sax's handbook (Lewis, 2004).

TABLE 9.2. Flammability Ranges

Material	Lower Limit	Upper Limit
Hydrogen	4.1	74.2
Ammonia	15.0	28.0
Hydrocyanic acid	5.6	40.0
Hydrogen sulphide	4.3	45.0
Carbon disulphide	1.3	44.0
Carbon monoxide	12.5	74.2
Methane	5.3	14.0
Ethane	3.0	12.5
Propane	2.3	9.5
Butane	1.9	8.5
Isobutane	1.8	8.4
Ethylene	3.1	32.0
Propylene	2.4	10.3
n-Butene	1.6	9.3
Isobutene	1.8	9.7
Butadiene	2.0	11.5
Benzene	1.4	7.1
Toluene	1.4	6.7
Cyclohexane	1.3	8.0
Methanol	7.3	36.0
Ethanol	4.3	19.0
Isopropanol	2.2	12.0
Formaldehyde	7.0	73.0
Acetaldehyde	4.1	57.0
Aetone	3.0	12.8
Methylethyl ketone	1.8	10.0
Dimethylamine (DEA)	2.8	184
Trimethylamine (TEA)	2.0	11.6
Gasoline	1.3	7.0
Kerosene (jet fuel)	0.7	5.6
Gas oil (diesel)	6.0	13.5

Volume percentage in air at ambient conditions.

A flammable mixture may exist in the space above the liquid surface in a storage tank. The vapour space above highly flammable liquids is usually purged with inert gas (nitrogen) or floating-head tanks are used. In a floating-head tank a 'piston' floats on top of the liquid, eliminating the vapour space.

9.2.3.   Materials Incompatibility

Some materials are naturally unstable and can spontaneously decompose, polymerize or undergo other reactions. These reactions can be initiated or accelerated by promoters such as light, heat, sources of free radicals or ions, or catalysts such as metal surfaces. These reactions can sometimes be retarded by adding inhibitors or diluents. Reactions of this kind are usually exothermic and if allowed to proceed will lead to a 'runaway' reaction with serious consequences.

Some materials are by their nature highly reactive and will react with many other compounds at low temperatures. Examples include strong oxidizing agents such as peroxides and chlorates, strong reducing agents, strong alkalis, strong acids and the metallic forms of alkali metals. In addition to being reactive with many other chemicals, these materials can also attack the materials from which the plant is constructed.

Other groups of compounds are known to react together rapidly and exothermically. These include mixtures such as acids and bases, acids and metals, fuels and oxidants, free radical initiators and epoxides, peroxides, or unsaturated molecules.

Another important class of incompatible materials are those that become more hazardous when contacted with water. For example, carbonyl sulphide (COS) and calcium sulphide (CaS) both release toxic H_2S on contact with water. Dry powders of sodium or potassium cyanide release toxic HCN in the presence of moisture. Care must be taken to prevent such materials from coming into contact with water during processing and storage. The 1985 Bhopal disaster was started by a runaway reaction involving a water-sensitive chemical.

Materials that are used to construct the process equipment and instrumentation must also be checked for compatibility with the process chemicals. This includes not only the metals or alloys from which the major vessels are built, but also welding, brazing or soldering materials, components of pumps, valves and instruments, gaskets, seals, linings and lubricants.

Information on incompatible materials can be found in most Materials Safety Data Sheets. The United States National Fire Protection Association (NFPA) also publishes standards NFPA 491 (1997) Guide to Hazardous Chemical Reactions and NFPA 49 (1994) Hazardous Chemicals Data, both of which provide data on incompatible materials.

Materials incompatibility is one of the most frequent causes of process incidents. Degradation of seals and gaskets that have become softened by solvent effects can lead to minor leaks or major loss of containment, and hence to fires, explosions or more serious accidents. If seal or gasket leakage is identified in a process then the plant engineer should consult with the manufacturer to confirm the material is suitable for the service. If necessary, all the seals or gaskets of that material should be replaced with something more resilient to the process conditions.

9.2.4. Ionizing Radiation

The radiation emitted by radioactive materials is harmful to living matter. Small quantities of radioactive isotopes are used in the process industry for various purposes; for example, in level and density-measuring instruments, and for the non-destructive testing of equipment.

The use of radioactive isotopes in industry is covered by government legislation. In the European Union this is Council Directive 96/29/Euratom, which sets safety standards for the protection of workers and the general public from ionizing radiation. Low levels of radiation may also be present in natural minerals. Care should be taken if these radioactive materials are concentrated or accumulated in the process, or dispersed into the environment.

A discussion of the particular hazards that arise in the chemical processing of nuclear fuels is outside the scope of this book.

9.2.5. Materials Safety Data Sheets

A Materials Safety Data Sheet (MSDS) is a document summarizing the hazards and health and safety information for a chemical. In EU countries, Commission Directive 91/155/EEC requires that chemical manufacturers must make an MSDS available to employees and customers for every chemical manufactured or sold. Other countries have similar requirements; for example, the OSHA Hazard Communication Standard (29 CFR 1910.1200) in the USA and the Canadian Hazardous Products Act (R.S., 1985, c. H-3).

The MSDS contains the information needed to begin analysing materials and process hazards; to understand the hazards to which the workforce is exposed; and to respond to a release of the material or other major incident where emergency response personnel may be exposed to the material.

The MSDS usually contains the following sections:

1. Chemical product and company information: chemical name and grade; catalogue numbers and synonyms; manufacturer's contact information, including 24-h contact numbers.
2. Composition and information of ingredients: chemical names, CAS numbers and concentration of major components of the product.
3. Hazards identification: summary of the major hazards and health effects.
4. First-aid measures: procedures for contact with eyes and skin or by ingestion or inhalation.
5. Fire-fighting measures: information on fire fighting, extinguishing media, flammability data, National Fire Protection Association ratings.
6. Accidental release measures: procedures for dealing with leaks or spills.
7. Handling and storage: procedures for transfer, storage and general use of the material.
8. Exposure controls and personal protection: required engineering controls such as eyewashes, safety showers, ventilation, etc.; OSHA PEL data; required personal protective equipment.
9. Physical and chemical properties.

10. Stability and reactivity: conditions that cause instability, known incompatible materials, hazardous decomposition products.
11. Toxicological information: acute effects, LD_{50} data, chronic effects, carcinogenicity, teratogenicity, mutagenicity.
12. Ecological information: ecotoxicity data for insects and fish, other known environmental impacts.
13. Disposal considerations: requirements for disposal.
14. Transport information: shipping information required by the US Department of Transport as well as other international bodies.
15. Regulatory information: US federal and state, European, Canadian and international regulations listing the material.
16. Additional information: date of creation and revisions, legal disclaimers.

Most MSDS forms are created by chemical manufacturers. They can be found in libraries, on manufacturer's web sites and by contacting manufacturers or suppliers directly. Web sites are available that catalogue MSDS forms from multiple sources (e.g. www.msdssearch.com). The Canadian Centre for Occupational Health and Safety also maintains an extensive collection at www.ccinfoweb.ccohs.ca. For legal reasons (limitation of liability), most MSDS forms contain a disclaimer stating that the user should also make their own evaluation of compatibility and fitness for use. An example MSDS form is given in Appendix I.

9.2.6.   Design for Materials Hazards

Under Council Directive 89/391/EEC, employers are required to carry out an assessment to evaluate the risk to health from any chemicals handled, and establish what precautions are needed to protect employees. A written record of the assessment should be kept, and details made available to employees. The design engineer should consider the preventative aspects of the use of hazardous substances. Points to consider are:

1. Substitution: of the processing route with one using less hazardous material, or of toxic process materials with non-toxic or less toxic materials.
2. Containment: sound design of equipment and piping, to avoid leaks. For example, specifying welded joints in preference to gasketed flanged joints that are liable to leak or suffer materials incompatibility problems.
3. Prevention of releases: by process and equipment design, operating procedures and design of disposal systems.
4. Ventilation: use open structures, or provide adequate ventilation systems.
5. Disposal: provision of effective vent stacks to disperse material vented from pressure-relief devices, or use of vent scrubbers. Collection and treatment of sewer and run-off waters and liquids collected from relief systems.
6. Emergency equipment and procedures: automated shutdown systems, escape routes, rescue equipment, respirators, antidotes (if appropriate), safety showers, eyebaths, emergency services.

In addition, good plant operating practice would include:

1. Written instruction in the use of the hazardous substances and the risks involved.
2. Adequate training of personnel.
3. Provision of protective clothing and equipment.
4. Good housekeeping and personal hygiene.
5. Monitoring of the environment to check exposure levels. Consider the installation of permanent instruments fitted with alarms.
6. Regular medical check-ups on employees, to check for the chronic effects of toxic materials.
7. Training of local emergency response personnel.

The process design engineer should always collect the MSDS of every component used in the process, including solvents, acids, bases, adsorbents, etc., at as early a stage in the design as possible. The information in the MSDS can be used to improve the inherent safety of the process; for example, by eliminating incompatible mixtures or substituting less hazardous chemicals as feeds, intermediates or solvents. The MSDS information can also be used to ensure that the design meets regulatory requirements on vapour recovery and other emissions.

9.3. PROCESS HAZARDS

In addition to the hazards caused by chemical or materials properties, hazards can arise from the conditions under which processes are carried out and the equipment that is used.

9.3.1. Pressure

Over-pressure, a pressure exceeding the system design pressure, is one of the most serious hazards in chemical plant operation. If the pressure exceeds the maximum allowable working pressure of the vessel by more than the safety margin allowed in the vessel design code, then failure of the vessel can occur, usually at a joint or flange. Failure of a vessel, or the associated piping, can precipitate a sequence of events that culminate in a disaster. Over-pressure occurs when mass, moles or energy accumulate in a contained volume or space with a restricted outflow. Specific causes of over-pressure are discussed in Section 9.9.1.

Pressure vessels are required to be fitted with some form of pressure-relief device, set at the maximum allowable working pressure, so that potential over-pressure is relieved in a controlled manner (ASME Boiler and Pressure Vessel Code, Section VIII Division 1, Part UG-125). See Section 9.9 for a more detailed discussion of pressure relief.

Process equipment must also be protected from under-pressure (vacuum), as this places compressive stresses on vessel walls that can lead to failure by bucking. Design for under-pressure is discussed in Section 9.9.6.

The factors to be considered in the design of relief systems are set out in a comprehensive paper by Parkinson (1979) and by Moore (1984). More extensive references, including design codes and standards and relief system design software, are given in Section 9.9.

9.3.2. Temperature Deviations

Excessively high temperature, over and above that for which the equipment was designed, can cause structural failure and initiate a disaster. High temperatures can arise from loss of control of reactors and heaters; and, externally, from open fires. In the design of processes where high temperatures are a hazard, protection against high temperatures is provided by:

1. Provision of high-temperature alarms and interlocks to shut down reactor feeds or heating systems if the temperature exceeds critical limits.
2. Use of additional temperature detectors to provide highly localized and redundant temperature monitoring. These include detectors on the wall of the vessel ('skin thermocouples') as well as in thermowells exposed to the process fluids. In some cases, temperature-sensitive paint that changes colour above a certain threshold is used on the outside of the vessel.
3. Provision of emergency cooling systems for reactors where heat continues to be generated after shutdown; for instance, in some polymerization systems.
4. Provision of quench systems for emergency shutdown, designed to flood the equipment with a cold, inert material.
5. Structural design of equipment to withstand the worst possible temperature excursion.
6. The selection of inherently safe heating systems for hazardous materials.

Steam and other vapour heating systems are inherently safer than fired heat and electric heat, as the temperature cannot exceed the saturation temperature at the supply pressure if the vapour is de-superheated and the supply line has a relief system that prevents it from becoming over-pressurized. Other heating systems rely on control of the heating rate to limit the maximum process temperature. Electrical heating systems can be particularly hazardous, since the heating rate is proportional to the resistance of the heating element, which increases with temperature.

Very low temperatures can also be hazardous. Low temperatures can be caused by ambient conditions, operation of cryogenic processes, expansion of gases and vapours, flashing of liquids (auto-refrigeration) and endothermic reactions. Low temperatures can cause embrittlement and stress cracking in metals. At very low temperatures some metals undergo micro-structural transformations that cause substantial changes in density (for example brass). When water freezes in a confined volume, the increase in specific volume can cause pipes or vessels to crack. Minimum design metal temperatures are specified in pressure vessel design (see Chapter 13).

9.3.3. Noise

Excessive noise is a hazard to health and safety. Long exposure to high noise levels can cause permanent damage to hearing. At lower levels, noise is a distraction and causes fatigue. Regulations on noise are set by EU Directive 2003/10/EC.

The unit of sound measurement is the decibel, defined by the expression:

$$\text{Sound level} = 20 \log_{10} \left[\frac{\text{RMS sound pressure (Pa)}}{2 \times 10^{-5}} \right], \text{dB} \qquad (9.1)$$

The subjective effect of sound depends on frequency as well as intensity.

Industrial sound meters include a filter network to give the meter a response that corresponds roughly to that of the human ear. This is termed the 'A' weighting network and the readings are reported as dB(A).

Permanent damage to hearing can be caused at sound levels above about 85 dB(A), and it is normal practice to provide ear protection in areas where the level is above 80 dB(A). In the UK, the 2005 Noise at Work Regulations require mandatory use of hearing protection and audiometric testing for workers who are exposed to noise levels greater than 85 dB(A).

Excessive plant noise can lead to complaints from neighbouring factories and local residents. Due attention should be given to noise levels when specifying, and when laying out, equipment that is likely to be excessively noisy, such as compressors, fans, burners and steam relief valves. This equipment should not be placed near the control room.

Several books are available on the general subject of industrial noise control (see Bias and Hansen, 2003) and on noise control in the process industries (Cheremisnoff, 1996; ASME, 1993).

9.3.4. Loss of Containment

The primary means for protecting employees and the public from exposure to toxic chemicals is the plant itself. Loss of containment can occur due to:

1. Pressure-relief events.
2. Operator errors such as leaving a sample point open or dripping.
3. Poor maintenance procedures, including failure to isolate, drain and purge properly before maintenance, leading to release when the equipment is opened, and failure to reconnect items properly and close drain valves when maintenance is complete.
4. Leaks from degraded equipment, including damaged seals, gaskets and packings, and corroded or eroded vessels and pipes.
5. Emissions from solid-handling operations (dust).
6. Internal equipment leaks (particularly in heat-exchanger tubesheets) that allow utility services such as cooling water to become contaminated with process chemicals.
7. Spills from drum or tanker loading and emptying.

Frequent loss of containment incidents are usually an indication that a plant has been poorly maintained, and are a leading indicator of major incidents.

If the potential impact of a loss of containment is high, then the design engineer should provide means of containment or mitigation. These means might include:

1. Secondary containment (bunding) to prevent run-off, but note that this can create a worse hazard if the chemicals are flammable and easily ignited.
2. Contained drainage and sewer systems to collect run-off and rainwater for waste treatment.
3. Use of concrete foundations to protect groundwater.
4. Containment of the plant inside a building with ventilation and vent scrubbers (used for hazardous dusts and very toxic compounds).

9.3.5. Fires and Ignition Sources

A fire occurs whenever sufficient amounts of fuel and oxidant are mixed and contacted with an ignition source. If a fuel is above its autoignition temperature then ignition can occur spontaneously in air. Though precautions are normally taken to eliminate sources of ignition on chemical plants, it is best to work on the principle that a leak of flammable material will ultimately find an ignition source. Guidelines for the control of ignition sources are given in section 7.9 of NFPA 30 (2003).

Electrical Equipment

The sparking of electrical equipment, such as motors, is a major potential source of ignition, and flame-proof equipment is normally specified. Electrically operated instruments, controllers and computer systems are also potential sources of ignition of flammable mixtures.

In the European Union, the use of electrical equipment in hazardous areas is regulated under the Seveso II Directive. In the UK, this is implemented as the Dangerous Substances and Explosive Atmospheres Regulations of 2002 (DSEAR), which forms part of the COMAH regulations. DSEAR requires employers to carry out hazardous area studies and classify areas as zones, depending on the level of hazard. Part 10 of BS EN 60079 (2003) defines hazardous areas as those where explosive gas–air mixtures are present, or may be expected to be present, in quantities such as to require special precautions for the construction and use of electrical apparatus. Non-hazardous areas are those where explosive gas–air mixtures are not expected to be present.

Three classifications are defined for hazardous areas:

Zone 0: explosive gas–air mixtures are present continuously or present for long periods. Specify: intrinsically safe equipment.
Zone 1: explosive gas–air mixtures likely to occur in normal operation. Specify: intrinsically safe equipment, or flame-proof enclosures: enclosures with pressurizing and purging.
Zone 3: explosive gas–air mixtures not likely to occur during normal operation, but could occur for short periods. Specify: intrinsically safe equipment, or total enclosure, or non-sparking apparatus.

The other parts of BS EN 60079 (2007) describe different methods of protecting equipment from explosive atmospheres, such as use of flameproof enclosures (part 1), pressurized enclosures (part 2) and intrinsically safe equipment (part 11). The standards should be consulted for the full specification before selecting equipment for use in the designated zones. Rules are also given for maintenance and inspection of explosion-proof equipment. Equipment for use in the presence of combustible dusts is described in BS EN 50281 (1999) and BS EN 61241 (2005). See also the references listed under explosions in Section 9.3.6.

In the USA, the use of electrical equipment in hazardous areas is covered by the National Electrical Code (NFPA 70, 2006), National Fire Protection Association standards NFPA 496 (2003) and NFPA 497 (2004), and OSHA standard 29 CFR 1910.307. The American Petroleum Institute Recommended Practices API RP 500 (2002) and API RP 505 (1997) should also be consulted.

The design and specification of intrinsically safe control equipment and systems is discussed by MacMillan (1998) and Cooper and Jones (1993).

Static Electricity

The movement of any non-conducting material, powder, liquid or gas, can generate static electricity, producing sparks. Precautions must be taken to ensure that all piping is properly earthed (grounded) and that electrical continuity is maintained around flanges. Escaping steam, or other vapours and gases, can generate a static charge. Gases escaping from a ruptured vessel can self-ignite from a static spark. For a review of the dangers of static electricity in the process industries, see the article by Napier and Russell (1974) and the books by Pratt (1999) and Britton (1999). Protection against static electricity, lightning and stray currents is discussed in API RP 2003 (1998). Protection against lightning is described in BS EN 62305 (2006). NFPA 77 (2000) is the national standard on static electricity protection in the USA.

Process Flames

Open flames from process furnaces, incinerators and flare stacks are obvious sources of ignition and must be sited well away from a plant containing flammable materials.

Miscellaneous Sources

It is the usual practice on plants handling flammable materials to control the entry on to the site of obvious sources of ignition; such as matches, cigarette lighters and battery-operated equipment. The use of portable electrical equipment and welding, cutting and spark-producing tools, and the movement of gasoline engine vehicles would also be subject to strict control. Exhaust gases from diesel engines are also a potential source of ignition.

Flame Traps

Flame arresters are fitted in the vent lines of equipment that contains flammable material to prevent the propagation of flame through the vents. Various types of proprietary flame arresters are used. In general, they work on the principle of providing a heat sink, usually expanded metal grids or plates, to dissipate the heat of the flame. Flame arrestors and their applications are discussed by Rogowski (1980), Howard (1992), Mendoza *et al.* (1998), API RP 2210 (2000) and ISO 16852 (2008).

Traps should also be installed in plant ditches to prevent the spread of flame. These are normally liquid U-legs, which block the spread of a flame along ditches.

Fire Protection

Recommendations on the fire precautions to be taken in the design of chemical plants are given in the standards NFPA 30 (2003), API RP 2001 (2005) and API PUBL 2218 (1999). Fire prevention and protection for machinery is discussed in BS EN 13478 (2001).

To protect against structural failure, water-deluge systems are usually installed to keep vessels and structural steelwork cool in a fire. Water mist fire protection systems are described in NFPA 750 (2006) and API PUBL 2030 (1998).

The lower section of structural steel columns are also often lagged with concrete or other suitable materials.

Plants that handle flammable liquids are usually designed to have slightly sloping ground or use drainage ditches or trenches to control run-off so that pools do not form. Drainage ditches and slopes should always direct flow away from sources of ignition.

9.3.6.   Explosions

An explosion is the sudden, catastrophic, release of energy causing a pressure wave (blast wave). An explosion can occur without fire, such as the failure through over-pressure of a steam boiler or an air receiver.

When discussing the explosion of a flammable mixture, it is necessary to distinguish between detonation and deflagration. If a mixture detonates, the reaction zone propagates at supersonic velocity (above approximately 300 m/s) and the principal heating mechanism in the mixture is shock compression. The pressure wave in a detonation can be up to 20 bar. In a deflagration, the combustion process is the same as in the normal burning of a gas mixture; the combustion zone propagates at subsonic velocity, and the pressure build-up is slow and usually less than 10 bar. Whether detonation or deflagration occurs in a gas–air mixture depends on a number of factors, including the concentration of the mixture and the source of ignition. Unless confined or ignited by a high-intensity source (a detonator), most materials will not detonate; however, the pressure wave (blast wave) caused by a deflagration can still cause considerable damage. In a confined space, such as a pipe, a deflagration can propagate into a detonation.

Certain materials, for example, acetylene and many peroxides, can decompose explosively in the absence of oxygen; such materials are particularly hazardous.

A good general introduction to explosions is given by Crowl (2003).

Confined Vapour Cloud Explosion (CVCE)

A relatively small amount of flammable material, a few kilograms, can lead to an explosion when released into the confined space of a building.

Unconfined Vapour Cloud Explosions (UCVCE)

This type of explosion results from the release of a considerable quantity of flammable gas, or vapour, into the atmosphere, and its subsequent ignition. Such an explosion can cause extensive damage, such as occurred at Flixborough (HMSO, 1975) and BP Texas City (CSHIB, 2005). Unconfined vapour explosions are discussed by Munday (1976) and Gugan (1979).

Boiling Liquid Expanding Vapour Explosions (BLEVE)

Boiling liquid expanding vapour explosions occur when there is a sudden release of vapour, containing liquid droplets, due to the failure of a vessel. A serious incident involving the failure of a LPG (liquefied petroleum gas) storage sphere occurred at Feyzin, France, in 1966, when the tank was heated by an external fire fuelled by a leak from the tank; see Mannan (2004) and Marshall (1987).

Dust Explosions

Finely divided combustible solids, if intimately mixed with air, can explode. Several disastrous explosions have occurred in grain silos.

Dust explosions usually occur in two stages: a primary explosion that disturbs deposited dust, followed by the second, severe, explosion of the dust thrown into the atmosphere. Any finely divided combustible solid is a potential explosion hazard. Particular care must be taken in the design of dryers, conveyors, cyclones, and storage hoppers for polymers and other combustible products or intermediates. The extensive literature on the hazard and control of dust explosions should be consulted before designing powder handling systems: Field (1982), Cross and Farrer (1982), Barton (2001), Eckhoff (2003), NFPA 61 (2007), NFPA 654 (2006), NFPA 664 (2006), BS EN 1127 (2007).

Explosivity Properties

Information on explosive materials is given in the standards NFPA 495 (2005), NFPA 491 (1997) and BS EN 1839 (2003). *Sax's Handbook of Hazardous Materials* (Lewis, 2004) is also a good general reference.

The expansion factor is defined as the molar density of the reagents divided by the molar density of the products in an explosive mixture. The expansion factor is a measure of the increase in volume resulting from combustion. The maximum value of the expansion factor is for adiabatic combustion.

The flame speed is the rate of propagation of a flame front through a flammable mixture, with respect to a fixed observer. Materials such as hydrogen and acetylene that have high flame speeds are more prone to detonation.

Values of these properties, autoignition temperature and adiabatic flame temperature are given for hydrogen and some hydrocarbons in Table 9.3, which is based on data from Dugdale (1985).

Design Implications

The usual approach in design is to prevent explosions from occurring; for example, by not allowing flammable mixtures to form in the process. If internal explosion is a possibility then it must be considered as a pressure-relief scenario and the pressure-relief devices must be sized to prevent detonation. This will usually require the use of large bursting disks; see Section 9.9. Flame arrestors should also be specified on process piping to prevent a deflagration event from propagating into a detonation. Particular care should be taken when designing plants that contain both pressurized fuels and pressurized oxidants. General guidelines on explosion protection are given in BS EN 1127 (2007), BS EN 14460 (2006), BS EN 60079 (2007), NFPA 69 (2007) and NFPA 68 (2006); see also the references listed under fires in Section 9.3.5. Explosion suppression and venting systems are discussed in BS EN 14373 (2005), BS EN 14797 (2006) and BS EN 14994 (2007). Specific guidelines for various sectors of the process industries are given in other standards, such as offshore production installations (BS EN ISO 13702, 1999), agricultural and food-processing plants (NFPA 61, 2007), wood-processing and woodworking facilities (NFPA 664, 2006), and plants handling sulphur (NFPA 655, 2006). In the European Union, regulations

TABLE 9.3. Explosivity Properties

Fuel	Formula	Maximum Flame Speed (m/s)	Adiabatic Flame Temperature (K)	Expansion Factor	Autoignition Temperature (°C)
Hydrogen	H_2	22.1	2318	6.9	400
Methane	CH_4	2.8	2148	7.5	601
Ethane	C_2H_6	3.4	2168	7.7	515
Propane	C_3H_8	3.3	2198	7.9	450
n-Butane	C_4H_{10}	3.3	2168	7.9	405
Pentane	C_5H_{12}	3.4	2232	8.1	260
Hexane	C_6H_{14}	3.4	2221	8.1	225
Acetylene	C_2H_2	14.8	2598	8.7	305
Ethylene	C_2H_4	6.5	2248	7.8	490
Propylene	C_3H_6	3.7	2208	7.8	460
Benzene	C_6H_6	5	2287	8.1	560
Cyclohexane	C_6H_{12}	4.2	2232	8.1	245

Adapted from Dugdale (1985).

on equipment for use in potentially explosive atmospheres are set under Directives 94/9/EC and 1999/92/EC.

9.3.7. Human Error

The intervention of well-trained process operators is a vital layer in process safety, as it is usually the last opportunity to restore the process to a safe condition before an emergency shutdown or incident occurs (Figure 9.1).

Even with capable, experienced and well-trained staff, there is always the possibility of human error. The likelihood of operator error is substantially increased if operating procedures are not clearly documented and followed or if there are lapses in training and supervision. Kletz (1999a) has suggested the following failure probabilities:

Action Required	Probability of Failure
A valve to be closed directly below an alarm	0.001
Simple action in a quiet environment	0.01
Simple action in a distracting environment	0.1
Complex and rapid action required	1.0

The Chemical Safety and Hazard Investigation Board preliminary report on their investigation of the explosion at the BP Texas City refinery on 23 March 2005, in which 15 people were killed and over 170 were injured, describes multiple failures in supervision, operating procedures and training that contributed to the accident (CSHIB, 2005). One of these was holding a plant safety training meeting in the control room while the operators were trying to start up the plant.

9.4. ANALYSIS OF PRODUCT AND PROCESS SAFETY

The analysis of the health, safety and environmental (HS&E) impact of technology is so important that it is carried out at every stage of a project, using the project technical information as it becomes available. As more design detail is developed, more quantitative methods can be used for analysing safety and environmental impact.

Table 9.4 shows typical steps in the evolution of a new product or process from initial concept to manufacture. In the early stages of process development the detail of the process has not been established, but qualitative assessments of major hazards can be made by collecting information from the MSDS forms for the chemicals involved. Once a conceptual flow-scheme has been developed, semi-quantitative methods such as Failure-Mode Effect Analysis (FMEA, see Section 9.5) and systematic procedures for identifying hazards such as HAZAN can be applied. An initial pollution prevention analysis can be made if the major process effluents are known. Some companies also calculate safety indices at this stage, to give a semi-quantitative comparison of the safety of the new process compared to existing processes (Section 9.6). When the process Piping and Instrumentation (P and I) diagram has been established and a full mass and energy balance has been completed then a full Hazard and Operability Study (HAZOP, see Section 9.7) can be carried out and the operating and emergency procedures can be updated. Safety check-lists (Section 9.4.1) are often completed at this stage and then updated and amended at subsequent stages. During detailed design and procurement, vendor information on instrument reliability becomes available. This information can be used to make a more quantitative analysis of likely failure

TABLE 9.4.   Health, Safety and Environmental Impact Analysis During the Evolution of a Project

Stage	Information Available	HS&E Analysis Methods
Research concept	Chemistry	MSDS review
	MSDS information	Major hazard review
Conceptual design	Process flow-diagram	Process FMEA/HAZAN
	Equipment list	Pollution prevention analysis
	Vessel designs	Preliminary operating procedures
	Reactor models	
Preliminary design	P and I diagrams	HAZOP
	Process control scheme	Emergency procedures
	Metallurgy	Safety indices
	Detailed mass and energy balance	Safety checklists
	Hydraulics	Effluent summary
	Off-sites	
Detailed design engineering	Mechanical designs	Quantitative risk analysis
	Instrument specs	Fault tree analysis
	Vendor details	
	Plot plans	
Procurement, construction	Piping isometrics	As-built HAZOP
	As-built specs	Operator training
Operation	Commissioning log	Ongoing training
	Operations log	Change management procedures
	Maintenance log	Revised operating procedures

rates, and hence determine whether duplicate or back-up systems are needed (Section 9.8). When the plant begins operation, any changes or modifications made during commissioning or in operation must also go through a detailed hazard analysis.

In the European Union, Directive 89/391/EEC requires employers to evaluate the risks to worker health and safety, in consultation with the workforce, and to take steps to improve worker health and safety (Articles 6.3, 9.1 and 11.1). A similar requirement exists in the USA under the OSHA standard 29 CFR 1910.119 *Process Safety Management of Highly Hazardous Chemicals,* available at www.osha.gov. The regulations do not specify the method of hazard analysis that must be used and most employers use several or all of the methods listed in the following sections, increasing the complexity of the analysis as more information is developed during the course of the project.

When the product that will be produced by the plant is a food, vitamin, cosmetic, medical implant, or human or veterinary drug then additional safety analysis must be carried out to comply with government regulations on food or drug safety. In the UK, food safety regulations are set by the Food Standards Agency (www.food.gov.uk), which is part of the Department of the Environment, Food and Rural Affairs (www.defra.gov.uk). Medicines and healthcare products are regulated by the Medicines and Healthcare Products Regulatory Agency (MHRA, www.mhra.gov.uk). These agencies require that plants that come under their jurisdiction must follow Good Manufacturing Practice (GMP) regulations, and such plants are subject to inspection and certification by the regulatory agencies.

In the European Union, food safety standards are set by the European Food Safety Authority (EFSA, www.efsa.europa.eu). Regulation (EC) No. 178/2002 of the European Parliament and Council established EFSA with authority to ensure free movement of food within the EU, while protecting consumers and ensuring that food in the EU is safe for consumption. Regulations from EFSA are implemented by the national authorities. In the United States, foods and healthcare products are regulated by the Food and Drug Administration (FDA). Details of FDA regulations can be found in the FDA Compliance Policy Guides, which are available at www.fda.gov.

Overviews of the different methods used for safety analysis are given in Crowl and Louvar (2002), Mannan (2004), CCPS (2008) and ISO 17776 (2000).

9.4.1. Safety Check Lists

Check lists are useful aids to memory. A check list that has been drawn up by experienced engineers can be a useful guide for the less experienced; however, too great a reliance should never be put on the use of check lists, to the exclusion of other considerations and techniques. No check list can be completely comprehensive, covering all the factors to be considered for any particular process or operation.

A short safety check list, covering the main items that should be considered in process design, is given below. More detailed check lists are given by Carson and Mumford (1988) and Wells (1980). Balemans (1974) gives a comprehensive list of guidelines for the safe design of chemical plant, drawn up in the form of a check list. A loss prevention check list is included in the Dow Fire and Explosion Index Hazard Classification Guide (Dow, 1994).

Design Safety Check List

Materials
  (a) flash-point
  (b) flammability range
  (c) autoignition temperature
  (d) composition
  (e) stability (shock sensitive?)
  (f) toxicity, TLV
  (g) corrosion
  (h) physical properties (unusual?)
  (i) heat of combustion/reaction

Process
  1. Reactors
      (a) exothermic – heat of reaction
      (b) temperature control – emergency systems
      (c) side reactions – dangerous?
      (d) effect of contamination
      (e) effect of unusual concentrations (including catalyst)
      (f) corrosion
  2. Pressure systems
      (a) need?
      (b) design to current codes
      (c) materials of construction – adequate?
      (d) pressure relief – adequate?
      (e) safe venting systems
      (f) flame arrestors

Control Systems
  (a) fail safe
  (b) back-up power supplies
  (c) high/low alarms and trips on critical variables
      (i) temperature
      (ii) pressure
      (iii) flow
      (iv) level
      (v) composition
  (d) back-up/duplicate systems on critical variables
  (e) remote operation of valves
  (f) block valves on critical lines
  (g) excess-flow valves
  (h) interlock systems to prevent mis-operation
  (i) automatic shutdown systems

Storages
  (a) limit quantity
  (b) inert purging/blanketing

(c) floating roof tanks
(d) dykeing
(e) loading/unloading facilities – safety
(f) earthing
(g) ignition sources – vehicles

General
(a) inert purging systems needed
(b) compliance with electrical codes
(c) adequate lighting
(d) lightning protection
(e) sewers and drains adequate, flame traps
(f) dust-explosion hazards
(g) build-up of dangerous impurities – purges
(h) plant layout
   (i) separation of units
   (ii) access
   (iii) siting of control rooms and offices
   (iv) services
(i) safety showers, eyebaths

Fire protection
(a) emergency water supplies
(b) fire mains and hydrants
(c) foam systems
(d) sprinklers and deluge systems
(e) insulation and protection of structures
(f) access to buildings
(g) fire-fighting equipment

The check list is intended to promote thought; to raise questions such as: is it needed, what are the alternatives, has provision been made for, checked for, has it been provided?

9.5. FAILURE-MODE EFFECT ANALYSIS

Failure-mode effect analysis (FMEA) is a method originally developed in manufacturing, which is used to determine the relative importance of different component failures within an overall system or product. It can be applied to analysis of chemical plant safety, as well as to design of products and even business plans and commercial projects. The method is semi-quantitative. It assigns numerical rankings to different failure modes based on the (qualitative) perceptions of the participants. Different groups or individuals will not necessarily reach the same conclusions, so the method is best used in the early stages of design as a means of brainstorming for safety issues. More rigorous methods such as HAZAN and HAZOP should be applied when more design details are available.

9.5.1. FMEA Procedure

An FMEA should ideally be carried out as a group brainstorming exercise. The group should include a diverse set of experts. When an FMEA is used for process safety analysis, these should include:

1. An expert on process chemistry
2. An expert on process equipment
3. An expert on process control
4. An expert on process operations
5. An expert on safety analysis
6. The process design engineer.

The analysis then proceeds as follows:

1. The group begins by reviewing the process and defining a set of process steps or key inputs.
2. For each step of input they then brainstorm for failure modes, i.e., ways in which the step or input might not perform its desired function.
3. For each failure mode, the group brainstorms for possible consequences. There may be multiple consequences for a given failure mode.
4. For each failure mode (and consequences) the group lists possible causes. Once again, there may be several causes that can trigger the same failure mode.
5. For each cause, the team lists the systems *that are currently in place* to prevent the cause from happening or allow the cause to be detected in time for operators to respond before the failure mode occurs. At this step, it is very important that the team considers the design as it currently exists. They must not assume that something will be added later to take care of any identified problem.
6. Once the brainstorming phase is completed (usually after several sessions), the team reviews the list of consequences and assigns each of them a 'severity' number, SEV. The severity is a measure of the impact of the consequence. Different scales can be used for severity, as discussed in Section 9.5.2.
7. The team then assigns a 'likelihood of occurrence' number, OCC, to each of the causes. The occurrence number is a measure of either the probability or frequency of the cause occurring.
8. For each of the current control methods or systems the team then assigns a 'detection' number, DET, that rates the probability that the existing systems will prevent the cause or failure mode from happening or detect the cause and allow an operator response before the failure mode occurs.
9. The three numbers, SEV, OCC and DET, are multiplied together to give an overall risk probability number, RPN.
10. Based on the RPN values, actions are assigned to each item in the FMEA. Low RPN items may require no action, while high RPN issues may require major changes to the process design and instrumentation.

An FMEA should always be associated with a particular revision of a design. Whenever a new revision is released the FMEA should be updated.

9.5.2. FMEA Rating Scales

The numbers assigned to the FMEA SEV, OCC and DET parameters are only a qualitative indication of the probability or impact. Because of this (and to reduce the time spent arguing over whether an item deserves a 4 or a 5 rating) most experienced practitioners use a 1, 4, 7, 10 scale to increase the granularity of the responses.

It is important that the team should agree on the meaning of each rating in the context of each of the FMEA parameters before starting to assign ratings. A suggested rating scale is given in Table 9.5, but other scales may be more appropriate in other cases.

It should be noted that the DET scale is inverse to the OCC scale. A high value of the DET number corresponds to a low probability of detection, while a high value of the OCC number corresponds to a high probability of occurrence.

When assigning ratings to the different FMEA parameters it is good practice to attempt to reach a consensus within the team. If no consensus can be developed (usually between a pair of values) then the best practice is to choose the higher value of the pair.

9.5.3. Interpretation of FMEA Scores

Once the RPN values have been calculated, the list should be ranked by RPN number and should be checked for consistency. This is particularly necessary when the FMEA has been completed during several sessions. FMEA is essentially a qualitative method, and the rankings based on RPN are at best only an indication of the team's assessment of the relative risk of the different failure modes. The team should not be overly concerned with the relative ranking of two issues as long as both are ranked appropriately high or low in the overall list.

Every item in the list should be reviewed to determine what follow-up action is required. If a 1, 4, 7, 10 scale is used then specific actions leading to changes in design or operating procedures are usually required for every item with RPN score greater than 100. This ensures that any item that scores 7 or 10 in one rating leads to an action unless it scores 1 in one of the other ratings.

Because FMEA is a qualitative method it is difficult to draw comparisons between FMEA studies of different processes. If a team studying process A identifies 70 items

TABLE 9.5.   Suggested Rating Scale for FMEA

Rating	SEV	OCC	DET
1	Effect is insignificant	Failure is very unlikely	Current safeguards will always prevent failure mode
4	Minor disruption, possible loss of production	Occasional failure possible	High probability that current safeguard will detect or prevent
7	Major disruption, possible damage to local equipment	Infrequent failure is likely	Low probability that current safeguard will detect or prevent
10	Severe disruption, major damage to plant, possible injury to personnel	Failure is very likely or frequent	No current method of detection

in an FMEA, while a team studying process B identifies 200 items, then either process B has more associated risks or else the team assigned to process B made a more thorough analysis. A short list of FMEA items (less than 50) is usually indicative of an incomplete analysis rather than a safe process.

9.5.4. Tools for FMEA

FMEA is easily carried out using spreadsheets. A Microsoft Excel template is available in the on-line material at http://elsevierdirect.com/companions and is given in Appendix G. Additional information on FMEA is given by Birolini (2004), Dodson and Nolan (1999) and Stamatis (1995).

9.6. SAFETY INDICES

Some companies make use of safety indices as a tool for assessing the relative risk of a new process or plant. The most widely used safety index is the Dow Fire and Explosion Index, developed by the Dow Chemical Company and published by the American Institute of Chemical Engineers (Dow, 1994; see www.aiche.org). A numerical 'fire and explosion index' (F & EI) is calculated, based on the nature of the process and the properties of the process materials. The larger the value of the F & EI, the more hazardous the process, see Table 9.6.

To assess the potential hazard of a new plant, the index can be calculated after the P and I and equipment layout diagrams have been prepared. In earlier versions of the guide the index was then used to determine what preventative and protection measures were needed; see Dow (1973). In the current version the preventative and protection measures that have been incorporated in the plant design to reduce the hazard are taken into account when assessing the potential loss, in the form of loss control credit factors.

It is worthwhile estimating the F & EI index at an early stage in the process design, as it will indicate whether alternative, less hazardous, process routes should be considered.

Only a brief outline of the method used to calculate the Dow F & EI will be given in this section. The full guide should be studied before applying the technique to a particular process. Judgment, based on experience with similar processes, is needed

TABLE 9.6. Assessment of Hazard

F & EI Range	Degree of Hazard
1–60	Light
61–96	Moderate
97–127	Intermediate
128–158	Heavy
>159	Severe

Adapted from the Dow F & EI guide (1994).

to decide the magnitude of the various factors used in the calculation of the index, and the loss control credit factors.

9.6.1. Calculation of the Dow F & EI

The procedure for calculating the index and the potential loss is set out in Figure 9.2.

The first step is to identify the units that would have the greatest impact on the magnitude of any fire or explosion. The index is calculated for each of these units.

The basis of the F & EI is a 'material factor' (MF). The MF is then multiplied by a 'unit hazard factor', F_3, to determine the F & EI for the process unit. The unit

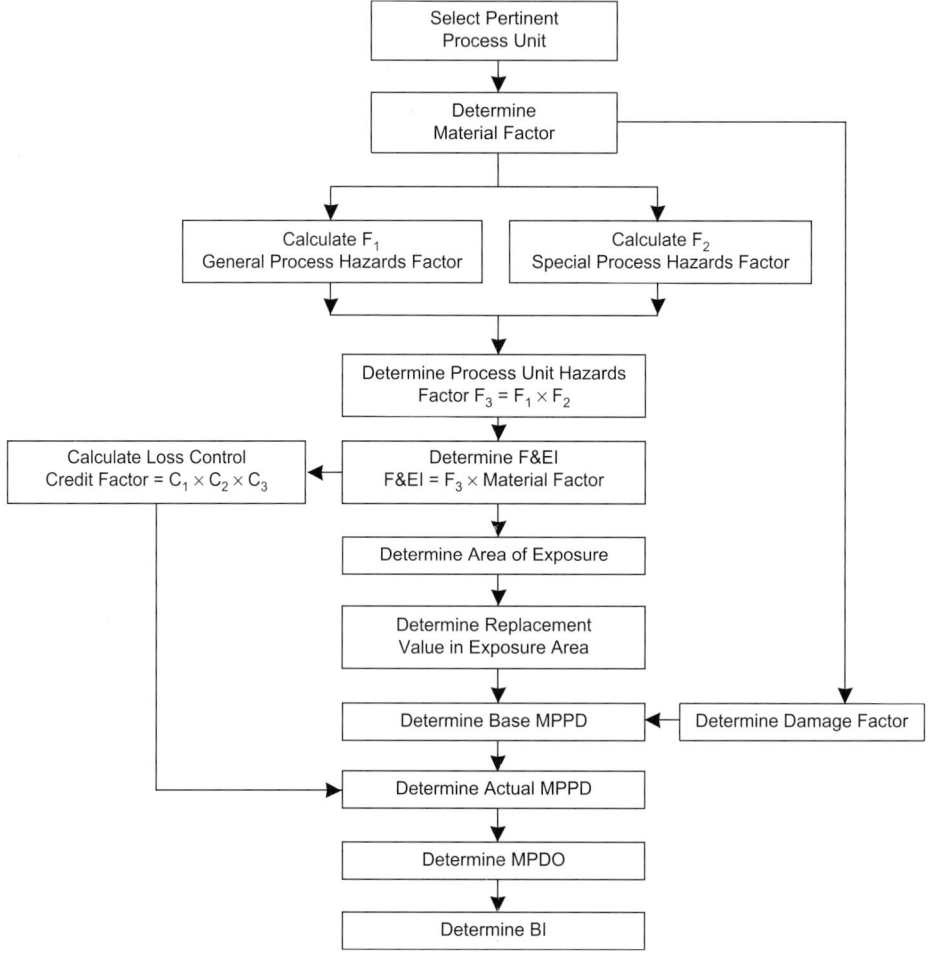

Figure 9.2. Procedure for calculating the F & EI index and other risk analysis information. From Dow (1994) reproduced by permission of the American Institute of Chemical Engineers. © 1994 AIChE. All rights reserved.

hazard factor is the product of two factors which take account of the hazards inherent in the operation of the particular process unit: the general and special process hazards (see Figure 9.3).

FIRE & EXPLOSION INDEX

AREA/COUNTRY	DIVISION	LOCATION	DATE
SITE	MANUFACTURING UNIT	PROCESS UNIT	
PREPARED BY:	APPROVED BY:(Superintendent)	BUILDING	
REVIEWED BY:(Management)	REVIEWED BY:(Technology Center)	REVIEWED BY:(Safety & Loss Prevention)	

MATERIALS IN PROCESS UNIT

STATE OF OPERATION   BASIC MATERIAL(S) FOR MATERIAL FACTOR

— DESIGN — START UP — NORMAL OPERATION —— SHUTDOWN

MATERIAL FACTOR (See Table 1 or Appendices A or B) Note requirements when unit temperature over 140 °F (60 °C)

		Penalty Factor Range	Penalty Factor Used(1)
1. General Process Hazards			
Base Factor		1.00	1.00
A. Exothermic Chemical Reactions		0.30 to 1.25	
B. Endothermic Processes		0.20 to 0.40	
C. Material Handling and Transfer		0.25 to 1.05	
D. Enclosed or Indoor Process Units	—	0.25 to 0.90	
E. Access		0.20 to 0.35	
F. Drainage and Spill Control gal or cu.m.		0.25 to 0.50	
General Process Hazards Factor (F₁)			
2. Special Process Hazards			
Base Factor		1.00	1.00
A. Toxic Material(s)		0.20 to 0.80	
B. Sub-Atmospheric Pressure (< 500 mm Hg)		0.50	
C. Operation In or Near Flammable Range ___ Inerted ___ Not Inerted			
1. Tank Farms Storage Flammable Liquids		0.50	
2. Process Upset or Purge Failure		0.30	
3. Always in Flammable Range		0.80	
D. Dust Explosion (See Table 3)		0.25 to 2.00	
E. Pressure (See Figure 2) Operating Pressure ____ psig or kPa gauge Relief Setting ____ psig or kPa gauge			
F. Low Temperature		0.20 to 0.30	
G. Quantity of Flammable/Unstable Material: Quantity ____ lb or kg H_C = ____ BTU/lb or kcal/kg			
1. Liquids or Gases in Process (See Figure 3)			
2. Liquids or Gases in Storage (See Figure 4)			
3. Combustible Solids in Storage, Dust in Process (See Figure 5)			
H. Corrosion and Erosion		0.10 to 0.75	
I. Leakage – Joints and Packing		0.10 to 1.50	
J. Use of Fired Equipment (See Figure 6)			
K. Hot Oil Heat Exchange System (See Table 5)		0.15 to 1.15	
L. Rotating Equipment		0.50	
Special Process Hazards Factor (F₂)			
Process Unit Hazards Factor (F₁ x F₂) = F₃			
Fire and Explosion Index (F₃ x MF = F&EI)			

(1) For no penalty use 0.00.

Figure 9.3. Dow F & EI calculation form. From Dow (1994) reproduced by permission of the American Institute of Chemical Engineers. © 1994 AIChE. All rights reserved. The figure numbers refer to the Dow guide. Gallons are US gallons.
Note: $1 \, m^3 = 264.2 \, US \, gal$; $1 \, kN/m^2 = 0.145 \, psi$; $1 \, kg = 2.2 \, lb$; $1 \, kJ/kg = 0.43 \, BTU/lb$.

Material Factor

The MF is a measure of the intrinsic rate of energy release from the burning, explosion, or other chemical reaction of the material. Values for the MF for over 300 of the most commonly used substances are given in the guide. The guide also includes a procedure for calculating the MF for substances not listed: from knowledge of the flash-points (for dusts, dust explosion tests) and a reactivity value, N_r. The reactivity value is a qualitative description of the reactivity of the substance, and ranges from 0 for stable substances to 4 for substances that are capable of unconfined detonation.

Some typical material factors are given in Table 9.7.

In calculating the F & EI for a unit, the value for the material with the highest MF, which is present in significant quantities, is used.

General Process Hazards

The general process hazards are factors that play a primary role in determining the magnitude of the loss following an incident.

Six factors are listed on the calculation form (Figure 9.3).

A. *Exothermic chemical reactions*: the penalty varies from 0.3 for a mild exotherm, such as hydrogenation, to 1.25 for a particularly sensitive exotherm, such as nitration.

B. *Endothermic processes*: a penalty of 0.2 is applied to reactors only. It is increased to 0.4 if the reactor is heated by the combustion of a fuel.

C. *Materials handling and transfer*: this penalty takes account of the hazard involved in the handling, transfer and warehousing of the material.

D. *Enclosed or indoor process units*: accounts for the additional hazard where ventilation is restricted.

E. *Access of emergency equipment*: areas not having adequate access are penalized. Minimum requirement is access from two sides.

TABLE 9.7.   Some Typical Material Factors

	MF	Flash-Point (°C)	Heat of Combustion (MJ/kg)
Acetaldehyde	24	−39	24.4
Acetone	16	−20	28.6
Acetylene	40	gas	48.2
Ammonia	4	gas	18.6
Benzene	16	−11	40.2
Butane	21	gas	45.8
Chlorine	1	−	0.0
Cyclohexane	16	−20	43.5
Ethyl alcohol	16	13	26.8
Hydrogen	21	gas	120.0
Nitroglycerine	40	−	18.2
Sulphur	4	−	9.3
Toluene	16	40	31.3
Vinyl chloride	21	gas	18.6

F. *Drainage and spill control*: penalizes design conditions that would cause large spills of flammable material adjacent to process equipment, such as inadequate design of drainage.

Special Process Hazards

The special process hazards are factors that are known from experience to contribute to the probability of an incident involving loss.

Twelve factors are listed on the calculation form (Figure 9.3).

A. *Toxic materials*: the presence of toxic substances after an incident will make the task of the emergency personnel more difficult. The factor applied ranges from 0 for non-toxic materials to 0.8 for substances that can cause death after short exposure.

B. *Sub-atmospheric pressure*: allows for the hazard of air leakage into equipment. It is only applied for pressure less than 500 mmHg (0.66 bara).

C. *Operation in or near flammable range*: covers the possibility of air mixing with material in equipment or storage tanks, under conditions where the mixture will be within the explosive range.

D. *Dust explosion*: covers the possibility of a dust explosion. The degree of risk is largely determined by the particle size. The penalty factor varies from 0.25 for particles above 175 μm to 2.0 for particles below 75 μm.

E. *Relief pressure*: this penalty accounts for the effect of pressure on the rate of leakage, should a leak occur. Equipment design and operation becomes more critical as the operating pressure is increased. The factor to apply depends on the relief device setting and the physical nature of the process material. It is determined from Figure 2 in the Dow Guide.

F. *Low temperature*: this factor allows for the possibility of brittle fracture occurring in carbon steel, or other metals, at low temperatures (see Chapter 7 of this book).

G. *Quantity of flammable material*: the potential loss will be greater, the greater the quantity of hazardous material in the process or in storage. The factor to apply depends on the physical state and hazardous nature of the process material, and the quantity of material. It varies from 0.1 to 3.0, and is determined from Figures 3, 4 and 5 in the Dow Guide.

H. *Corrosion and erosion*: despite good design and materials selection, some corrosion problems may arise, both internally and externally. The factor to be applied depends on the anticipated corrosion rate. The severest factor is applied if stress corrosion cracking is likely to occur (see Chapter 7 of this book).

I. *Leakage—joints and packing*: this factor accounts for the possibility of leakage from gaskets, pump and other shaft seals, and packed glands. The factor varies from 0.1 where there is the possibility of minor leaks to 1.5 for processes that have sight glasses, bellows or other expansion joints.

J. *Use of fired heaters*: the presence of boilers or furnaces, heated by the combustion of fuels, increases the probability of ignition should a leak of flammable material occur from a process unit. The risk involved will depend on the siting of the fired equipment and the flash-point of the process material. The factor to apply is determined with reference to Figure 6 in the Dow Guide.

K. *Hot oil heat-exchange system*: most special heat-exchange fluids are flammable and are often used above their flash-points; so their use in a unit increases the risk of fire or explosion. The factor to apply depends on the quantity and whether the fluid is above or below its flash-point; see Table 5 in the Guide.

L. *Rotating equipment*: this factor accounts for the hazard arising from the use of large pieces of rotating equipment: compressors, centrifuges and some mixers.

9.6.2. Potential Loss

The procedure for estimating the potential loss that would follow an incident is set out in Table 9.8: the Process Unit Risk Analysis summary.

The first step is to calculate the 'damage factor' for the unit. The damage factor depends on the value of the MF and the process unit hazards factor (F_3 in Figure 9.3). It is determined using Figure 8 in the Dow Guide.

An estimate is then made of the area (radius) of exposure. This represents the area containing equipment that could be damaged following a fire or explosion in the unit being considered. It is evaluated from Figure 7 in the Guide and is a linear function of the F & EI.

An estimate of the replacement value of the equipment within the exposed area is then made, and combined with the damage factor to estimate the 'base maximum probable property damage' (Base MPPD).

The 'maximum probable property damage' (MPPD) is then calculated by multiplying the Base MPPD by a 'loss control credit factor'. The loss control credit factors, see Table 9.9, allow for the reduction in the potential loss given by the preventative and protective measures incorporated in the design. The Dow Guide should be consulted for details of how to calculate the credit factors.

The MPPD is used to predict the maximum number of days which the plant will be down for repair, the 'maximum probable days outage' (MPDO). The MPDO is used to estimate the financial loss due to the lost production: the 'business interruption'

TABLE 9.8. Process Unit Risk Analysis Summary

1. Fire & Explosion Index (F& EI)	
2. Radius of Exposure	(Figure 7)* ft or m
3. Area of Exposure	ft² or m²
4. Value of Area of Exposure	$MM
5. Damage Factor	(Figure 8)*
6. Base Maximum Probable Property Damage (Base MPPD) [4 × 5]	$MM
7. Loss Control Credit Factor	(see above)
8. Actual Maximum Probable Property Damage (Actual MPPD) [6 × 7]	$MM
9. Maximum Probable Days Outage (MPDO)	(Figure 9)* days
10. Business Interruption (BI)	$MM

* Refer to Fire & Explosion Index Hazard Classification Guide for details.
From Dow (1994) reproduced by permission of the American Institute of Chemical Engineers. © 1994 AIChE. All rights reserved.

TABLE 9.9. Loss Control Credit Factors

Feature	Credit Factor Range	Credit Factor Used (2)
1. Process Control Credit Factor (C₁)		
a. Emergency Power	0.98	
b. Cooling	0.97 to 0.99	
c. Explosion Control	0.84 to 0.98	
d. Emergency Shutdown	0.96 to 0.99	
e. Computer Control	0.93 to 0.99	
f. Inert Gas	0.94 to 0.96	
g. Operating Instructions/Procedures	0.91 to 0.99	
h. Reactive Chemical Review	0.91 to 0.98	
i. Other Process Hazard Analysis	0.91 to 0.98	
C_1 value (3)		
2. Material Isolation Credit Factor (C₂)		
a. Remote Control Valves	0.96 to 0.98	
b. Dump/Blowdown	0.96 to 0.98	
c. Drainage	0.91 to 0.97	
d. Interlock	0.98	
C_2 value (3)		
3. Fire Protection Credit Factor (C₃)		
a. Leak Detection	0.94 to 0.98	
b. Structural Steel	0.95 to 0.98	
c. Fire Water Supply	0.94 to 0.97	
d. Special Systems	0.91	
e. Sprinkler Systems	0.74 to 0.97	
f. Water Curtains	0.97 to 0.98	
g. Foam	0.92 to 0.97	
h. Hand Extinguishers/Monitors	0.93 to 0.98	
i. Cable Protection	0.94 to 0.98	
C_3 value (3)		
Loss Control Credit Factor = $C_1 \times C_2 \times C_3$(3) = (enter on line 7, Table 9.6)		

(2) For no credit factor enter 1.00. (3) Product of all factors used.
From Dow (1994) reproduced by permission of the American Institute of Chemical Engineers. © 1994 AIChE. All rights reserved.

(BI). The financial loss due to lost business opportunity can often exceed the loss from property damage.

9.6.3. Basic Preventative and Protective Measures

The basic safety and fire protective measures that should be included in all chemical process designs are listed below. This list is based on that given in the Dow Guide, with some minor amendments.

    1. Adequate, and secure, water supplies for fire fighting.
    2. Correct structural design of vessels, piping, steel work.

3. Pressure-relief devices.
4. Corrosion-resistant materials, and/or adequate corrosion allowances.
5. Segregation of reactive materials.
6. Grounding of electrical equipment.
7. Safe location of auxiliary electrical equipment, transformers, switch gear.
8. Provision of back-up utility supplies and services.
9. Compliance with national codes and standards.
10. Fail-safe instrumentation.
11. Provision for access of emergency vehicles and the evacuation of personnel.
12. Adequate drainage for spills and fire-fighting water.
13. Insulation of hot surfaces.
14. No glass equipment used for flammable or hazardous materials, unless no suitable alternative is available.
15. Adequate separation of hazardous equipment.
16. Protection of pipe racks and cable trays from fire.
17. Provision of block valves on lines to main processing areas.
18. Protection of fired equipment (heaters, furnaces) against accidental explosion and fire.
19. Safe design and location of control rooms.

Note: the design and location of control rooms, particularly as regards protection against an unconfined vapour explosion, is covered in a publication of the Chemical Industries Association (CIA, 1979a).

9.6.4. Mond Fire, Explosion and Toxicity Index

The Mond index was developed from the Dow F &EI by personnel at the ICI Mond division. The third edition of the Dow index (Dow, 1973), was extended to cover a wider range of process and storage installations, the processing of chemicals with explosive properties, and the evaluation of a toxicity hazards index. Also included was a procedure to allow for the off-setting effects of good design, and of control and safety instrumentation. Their revised Mond fire, explosion and toxicity index was discussed in a series of papers by Lewis (1979a, 1979b), which included a technical manual setting out the calculation procedure. An extended version of the manual was issued in 1985, and an amended version published in 1993 (ICI, 1993).

Procedure

The basic procedures for calculating the Mond indices are similar to those used for the Dow index.

The process is first divided into a number of units that are assessed individually.

The dominant material for each unit is then selected and its MF determined. The MF in the Mond index is a function of the energy content per unit weight (the heat of combustion).

The MF is then modified to allow for the effect of general and special process and material hazards, the physical quantity of the material in the process step, the plant layout, and the toxicity of process materials.

Separate fire and explosion indices are calculated. An aerial explosion index can also be estimated, to assess the potential hazard of aerial explosions. An equivalent Dow index can also be determined.

The individual fire and explosion indexes are combined to give an overall index for the process unit. The overall index is the most important in assessing the potential hazard.

The magnitude of the potential hazard is determined by reference to rating tables, similar to that shown for the Dow index in Table 9.6.

After the initial calculation of the indices (the initial indices), the process is reviewed to see what measures can be taken to reduce the rating (the potential hazard).

The appropriate off-setting factors to allow for the preventative features included in the design are then applied, and final hazard indices calculated.

Preventative Measures

Preventative measures fall into two categories:

1. Those that reduce the number of incidents, such as sound mechanical design of equipment and piping, operating and maintenance procedures, and operator training.
2. Those that reduce the scale of a potential incident, such as measures for fire protection and fixed fire-fighting equipment.

Many measures will not fit neatly into individual categories but will apply to both.

Implementation

The Mond technique of hazard evaluation is fully explained in the ICI technical manual (ICI, 1993), to which reference should be made to implement the method. The calculations are made using a standard form, similar to that used for the Dow index.

9.6.5. Summary

The Dow and Mond indexes are useful techniques that can be used in the early stages of a project design to evaluate the hazards and risks of the proposed process.

Calculation of the indexes for the various sections of the process will highlight any particularly hazardous sections and indicate where a detailed study is needed to reduce the hazards.

Example 9.1

Evaluate the Dow F & EI for the nitric acid plant illustrated in Chapter 4, Figure 4.2.

Solution

The calculation is set out on the special form shown in Figure 9.3a. Notes on the decisions taken and the factors used are given below.

Unit: consider the total plant, no separate areas, but exclude the main storages.

a FIRE & EXPLOSION INDEX

AREA/COUNTRY -	DIVISION -	LOCATION SLIGO	DATE 20 JAN 1997
SITE -	MANUFACTURING UNIT NITRIC ACID	PROCESS UNIT COMPLETE PLANT	
PREPARED BY: RKS	APPROVED BY:(Superintendent) ANOTHER		BUILDING -
REVIEWED BY:(Management) -	REVIEWED BY:(Technology Center) -		REVIEWED BY:(Safety & Loss Prevention) -

MATERIALS IN PROCESS UNIT   AMMONIA, AIR, OXIDES OF NITROGEN, WATER

STATE OF OPERATION — DESIGN — START UP ✓ NORMAL OPERATION — SHUTDOWN	BASIC MATERIAL(S) FOR MATERIAL FACTOR AMMONIA

MATERIAL FACTOR (See Table 1 or Appendices A or B) Note requirements when unit temperature over 140 °F (60 °C)	4

1. General Process Hazards	Penalty Factor Range	Penalty Factor Used(1)
Base Factor	1.00	1.00
A. Exothermic Chemical Reactions	0.30 to 1.25	0.50
B. Endothermic Processes	0.20 to 0.40	
C. Material Handling and Transfer	0.25 to 1.05	
D. Enclosed or Indoor Process Units –	0.25 to 0.90	
E. Access	0.20 to 0.35	
F. Drainage and Spill Control gal or cu.m.	0.25 to 0.50	
General Process Hazards Factor (F₁)		1.50
2. Special Process Hazards		
Base Factor	1.00	1.00
A. Toxic Material(s)	0.20 to 0.80	0.60
B. Sub-Atmospheric Pressure (< 500 mm Hg)	0.50	
C. Operation In or Near Flammable Range ___ Inerted ___ Not Inerted		
1. Tank Farms Storage Flammable Liquids	0.50	
2. Process Upset or Purge Failure	0.30	0.80
3. Always in Flammable Range	0.80	
D. Dust Explosion (See Table 3)	0.25 to 2.00	
E. Pressure (See Figure 2) Operating Pressure ___103___ psig or kPa gauge Relief Setting ___125___ psig or kPa gauge		0.35
F. Low Temperature	0.20 to 0.30	
G. Quantity of Flammable/Unstable Material: Quantity ____ lb or kg H_C = ____ BTU/lb or kcal/kg		
1. Liquids or Gases in Process (See Figure 3)		
2. Liquids or Gases in Storage (See Figure 4)		
3. Combustible Solids in Storage, Dust in Process (See Figure 5)		
H. Corrosion and Erosion	0.10 to 0.75	0.10
I. Leakage – Joints and Packing	0.10 to 1.50	0.10
J. Use of Fired Equipment (See Figure 6)		
K. Hot Oil Heat Exchange System (See Table 5)	0.15 to 1.15	
L. Rotating Equipment	0.50	**0.50**
Special Process Hazards Factor (F₂)		3.45
Process Unit Hazards Factor (F₁ x F₂) = F₃		5.20
Fire and Explosion Index (F₃ x MF = F&EI)		21

(1) For no penalty use 0.00.

Figure 9.3a. F & EI calculation form, Example 9.1. From Dow (1994) reproduced by permission of the American Institute of Chemical Engineers. © 1994 AIChE. All rights reserved.

Material factor: for ammonia, from Dow Guide, and Table 9.6.

$$MF = 4.0$$

Note: Hydrogen is present, and has a larger material factor (21) but the concentration is too small for it to be considered the dominant material.

General Process Hazards

A. Oxidizing reaction, factor = 0.5.
B. Not applicable.
C. Not applicable.
D. Not applicable.
E. Adequate access would be provided, factor = 0.0.
F. Adequate drainage would be provided, factor = 0.0.

Special Process Hazards

A. Ammonia is highly toxic, likely to cause serious injury, factor = 0.6.
B. Not applicable.
C. Operation always is within the flammable limits, factor = 0.8.
D. Not applicable.
E. Operation pressure 8 atm = $8 \times 14.7 - 14.7 = 103$ psig. Set relief valve at 20% above the operating pressure (see Chapter 13 of this book) = 125 psig. From Figure 2 in the guide, factor = 0.35.

   Note: psig = pounds force per square inch, gauge.

F. Not applicable.
G. The largest quantity of ammonia in the process will be the liquid in the vaporizer, say around 500 kg.
   Heat of combustion, Table 9.3 = 18.6 MJ/kg
   Potential energy release = $500 \times 18.6 = 9300$ MJ
   $= 9300 \times 10^6/(1.05506 \times 10^3) = 8.81 \times 10^6$ Btu
   which is too small to register on Figure 3 in the Guide, factor = 0.0.
H. Corrosion-resistant materials of construction would be specified, but external corrosion is possible due to nitric oxide fumes; allow minimum factor = 0.1.
I. Welded joints would be used on ammonia service and mechanical seals on pumps. Use minimum factor as full equipment details are not known at the flow-sheet stage, factor = 0.1.
J. Not applicable.
K. Not applicable.
L. Large turbines and compressors used, factor = 0.5.

The index works out at 21: classified as 'Light'. Ammonia would not normally be considered a dangerously flammable material; the danger of an internal explosion in the reactor is the main process hazard. The toxicity of ammonia and the corrosiveness of nitric acid would also need to be considered in a full hazard evaluation. The Process Unit Risk Analysis would be completed when the site for the plant had been determined.

9.7. HAZARD AND OPERABILITY STUDIES

A hazard and operability study is a systematic procedure for critical examination of the operability of a process. When applied to a process design or an operating plant, it indicates potential hazards that may arise from deviations from the intended design conditions. The technique was developed by the Petrochemicals Division of Imperial Chemical Industries (see Lawley, 1974), and is now in general use in the chemical and process industries.

The term 'operability study' should more properly be used for this type of study, though it is usually referred to as a hazard and operability study, or HAZOP study. This can cause confusion with the term 'hazard analysis', or 'process hazard analysis' (PHA), which is a similar but somewhat less rigorous method. Numerous books have been written illustrating the use of HAZOP. Those by Hyatt (2003), CCPS (2000), Taylor (2000) and Kletz (1999a) give comprehensive descriptions of the technique, with examples.

A brief outline of the technique is given in this section to illustrate its use in process design. It can be used to make a preliminary examination of the design at the flow-sheet stage; and for a detailed study at a later stage, when a full process description, final flow-sheets, P and I diagrams, and equipment details are available. An 'as-built' HAZOP is often carried out after construction and immediately before commissioning a new plant.

9.7.1. Basic Principles

A formal operability study is the systematic study of the design, vessel by vessel and line by line, using 'guide words' to help generate thought about the way deviations from the intended operating conditions can cause hazardous situations.

The seven guide words recommended are given in Table 9.10. In addition to these words, the following words are also used in a special way, and have the precise meanings given below:

> *Intention*: the intention defines how the particular part of the process was intended to operate; the intention of the designer.
> *Deviations*: these are departures from the designer's intention that are detected by the systematic application of the guide words.

TABLE 9.10. A List of Guide Words

Guide Words	Meanings	Comments
No or Not	The complete negation of these intentions	No part of the intentions is achieved but nothing else happens
More/Less	Quantitative increases or decreases	These refer to quantities and properties such as flow rates and temperatures, as well as activities like 'Heat' and 'React'
As well as	A qualitative increase	All the design and operating intentions are achieved together with some additional activity
Part of	A qualitative decrease	Only some of the intentions are achieved; some are not
Reverse	The logical opposite of the intention	This is mostly applicable to activities, for example reverse flow or chemical reaction. It can also be applied to substances, e.g. 'Poison' instead of 'Antidote' or 'D' instead of 'L' optical isomers
Other than	Complete substitution	No part of the original intention is achieved. Something quite different happens

Causes: reasons why, and how, the deviations could occur. Only if a deviation can be shown to have a realistic cause is it treated as meaningful.

Consequences: the results that follow from the occurrence of a meaningful deviation.

Hazards: consequences that can cause damage (loss) or injury.

The use of the guide words can be illustrated by considering a simple example. Figure 9.4 shows a chlorine vaporizer, which supplies chlorine at 2 bar to a chlorination reactor. The vaporizer is heated by condensing steam.

Consider the steam supply line and associated control instrumentation. The designer's intention is that steam shall be supplied at a pressure and flow rate to match the required chlorine demand.

Apply the guide word No:

Possible deviation – no steam flow.

Possible causes – blockage, valve failure (mechanical or power), failure of steam supply (fracture of main, boiler shutdown).

Clearly this is a meaningful deviation, with several plausible causes.

Consequences – the main consequence is loss of chlorine flow to the chlorination reactor.

The effect of this on the reactor operation would have to be considered. This would be brought out in the operability study on the reactor; it would be a possible cause of no chlorine flow. Since the flow controller does not know that steam flow has been lost, chlorine will continue to be pumped into the vessel until the high-level alarm sounds and the high-level shutdown closes the control valve. A secondary consequence is that the vessel is now filled with liquid chlorine that must be drained to a safe level before operation can be resumed. The operating procedures must include instructions on how to deal with this scenario.

Apply the guide word More:

Possible deviation – more steam flow.

Possible cause – valve stuck open.

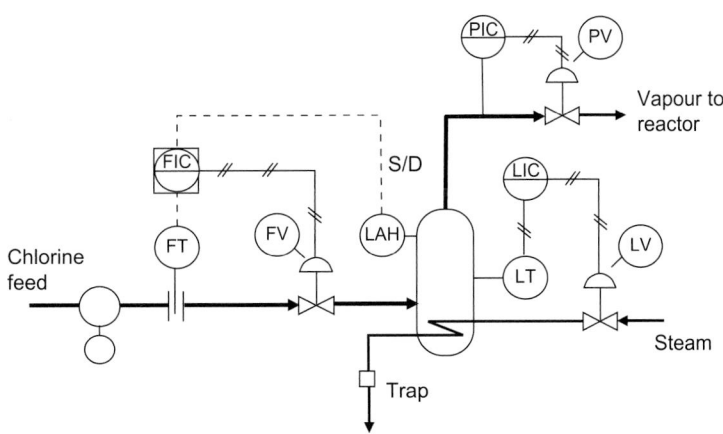

Figure 9.4. Chlorine vaporizer instrumentation.

Consequences – low level in vaporizer (this should activate the low-level alarm), higher rate of flow to the reactor.

Note: to some extent the level will be self-regulating, because as the level falls the heating surface is uncovered.

Hazard – depends on the possible effect of high flow on the reactor.

Possible deviation – more steam pressure (increase in mains pressure).

Possible causes – failure of pressure-regulating valves.

Consequences – increase in vaporization rate. Need to consider the consequences of the heating coil reaching the maximum possible steam system pressure.

Hazard – rupture of lines (unlikely), effect of sudden increase in chlorine flow on reactor.

A more detailed illustration of the HAZOP method is given in Example 9.2.

9.7.2. Explanation of Guide Words

It is important to understand the intended meaning of the guide words in Table 9.10. The meaning of the words No/Not, More and Less are easily understood; the No/Not, More and Less could, for example, refer to flow, pressure, temperature, level and viscosity. All circumstances leading to No flow should be considered, including reverse flow.

The other words need some further explanation:

As well as: something in addition to the design intention; such as impurities, side reactions, ingress of air, extra phases present.

Part of: something missing, only part of the intention realized; such as the change in composition of a stream, a missing component.

Reverse: the reverse of, or opposite to, the design intention. This could mean reverse flow if the intention was to transfer material. For a reaction, it could mean the reverse reaction. In heat transfer, it could mean the transfer of heat in the opposite direction to what was intended.

Other than: an important and far-reaching guide word, but consequently more vague in its application. It covers all conceivable situations other than that intended; such as, start-up, shutdown, maintenance, catalyst regeneration and charging, failure of plant services.

When referring to time, the guide words Sooner than and Later than can also be used.

9.7.3. Procedure

An operability study would normally be carried out by a team of experienced people, who have complementary skills and knowledge, led by a team leader who is experienced in the technique. The team would include a similar set of experts to an FMEA team, as described in Section 9.5.

The team examines the process vessel by vessel, and line by line, using the guide words to detect any hazards.

The information required for the study will depend on the extent of the investigation.

A preliminary study can be made from a description of the process and the process flow-diagrams. For a detailed, final, study of the design, the flow-sheets, piping and instrument diagrams, equipment specifications and layout drawings would be needed. For a batch process, information on the sequence of operation will also be required, such as that given in operating instructions, logic diagrams and flow-charts.

A typical sequence of events is shown in Figure 9.5. After each line has been studied it is marked on the flow-sheet as checked.

A written record is not normally made of each step in the study, only those deviations that lead to a potential hazard are recorded. If possible, the action needed to

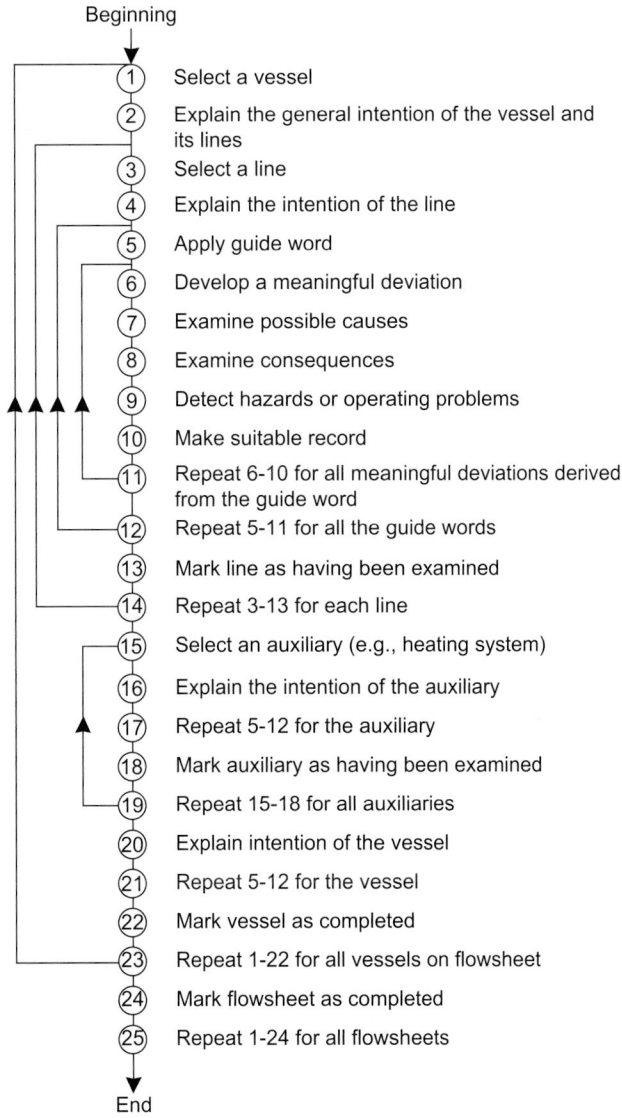

Figure 9.5. Detailed sequence of an operability study.

remove the hazard is decided by the team and recorded. If more information, or time, is needed to decide the best action, the matter is referred to the design group for action, or taken up at another meeting of the study team.

When using the operability study technique to vet a process design, the action to be taken to deal with a potential hazard will often be modifications to the control systems and instrumentation: the inclusion of additional alarms, trips or interlocks. If major hazards are identified, major design changes may be necessary, i.e., alternative processes, materials or equipment.

Example 9.2

This example illustrates how the techniques used in an operability study can be used to decide the instrumentation required for safe operation. Figure 9.6a shows the basic instrumentation and control systems required for the steady-state operation of the reactor section of the nitric acid process introduced in Figure 4.2. Figure 9.6b shows the additional instrumentation and safety trips added after making the operability study set out below. The instrument symbols used are explained in Chapter 5.

The most significant hazard of this process is the probability of an explosion if the concentration of ammonia in the reactor is inadvertently allowed to reach the explosive range, >14%. Note that this is a simplified flow diagram and a HAZOP based on the full P and I diagram would go into considerably more detail.

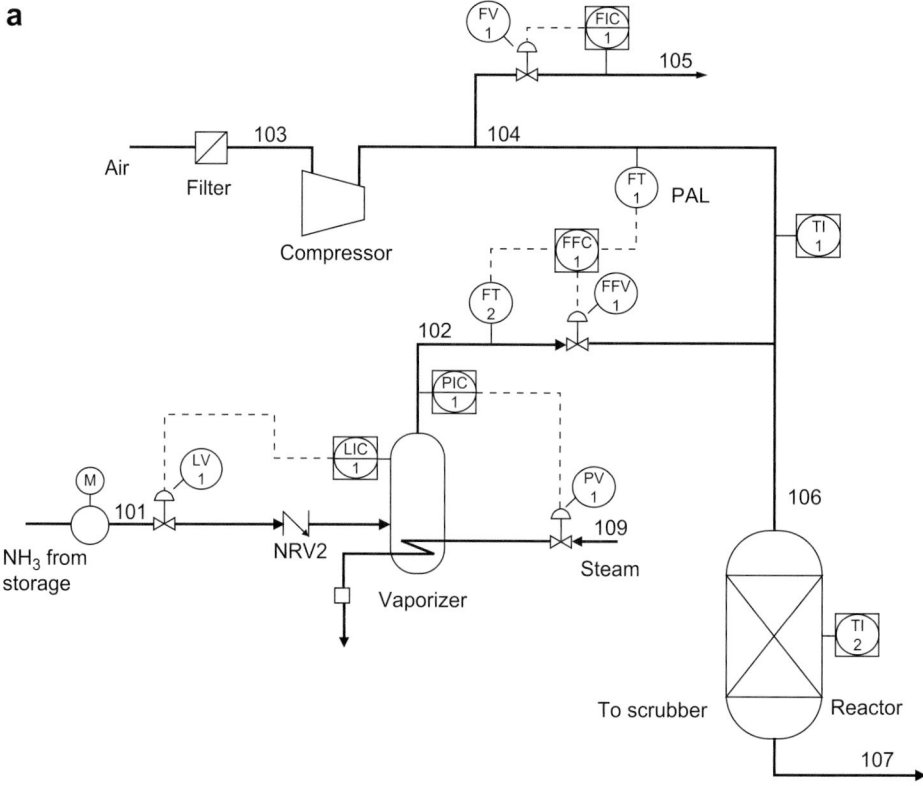

Figure 9.6a. Nitric acid plant reactor section before HAZOP.

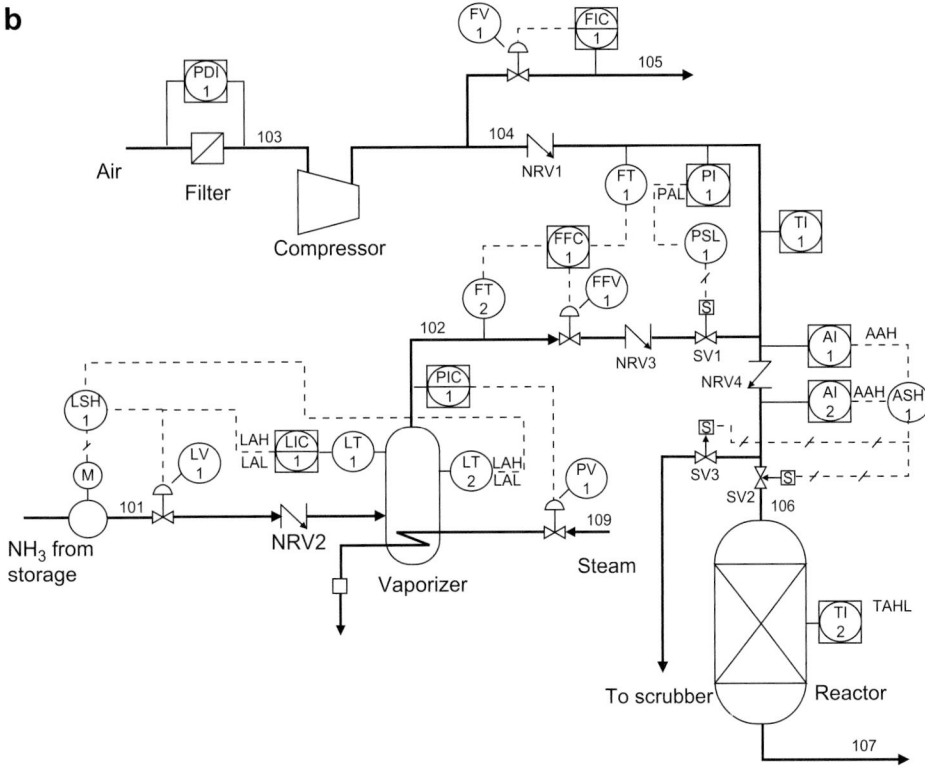

Figure 9.6b. Nitric acid plant reactor section after HAZOP.

Operability Study

The sequence of steps shown in Figure 9.4 is followed. Only deviations leading to action, and those having consequences of interest, are recorded.

*Vessel – **Air Filter***
Intention – to remove particles that would foul the reactor catalyst

Guide Word	Deviation	Cause	Consequences and Action
Line No. 103			
Intention – transfers clear air at atmospheric pressure and ambient temperature to compressor			
LESS OF	Flow	Partially blocked filter	Possible dangerous increase in NH_3 concentration: measure and log pressure differential
AS WELL AS	Composition	Filter damaged, incorrectly installed	Impurities, possible poisoning of catalyst: proper maintenance

Vessel – **Compressor**

Intention – to supply air at 8 bar, 12,000 kg/h, 250°C, to the mixing tee

Guide Word	Deviation	Cause	Consequences and Action
Line No. 104			
Intention – transfers air to reactor (mixing tee)			
No/None	Flow	Compressor failure	Possible dangerous NH_3 conc.: pressure indicator with low-pressure alarm (PI1) interlocked to shut down NH_3 flow
More	Flow	Failure of compressor controls	High rate of reaction, high reactor temperature: high- temperature alarms added to TI2
Reverse	Flow	Fall in line press (compressor fails); high pressure at reactor	NH_3 in compressor – explosion hazard: fit non-return valve (NRV1); hot wet acid gas-corrosion; fit second valve (NRV4)
Line No. 105			
Intention – transfer secondary air to absorber			
No	Flow	Compressor failure FV1 failure	Incomplete oxidation, air pollution from absorber vent: operating procedures
Less	Flow	FV1 plugging, FIC1 failure	As no flow

Vessel – **Ammonia vaporizer**

Intention – evaporate liquid ammonia at 8 bar, 25°C, 731 kg/h

Guide Word	Deviation	Cause	Consequences and Action
Line No. 101			
Intention – transfer liquid NH_3 from storage			
No	Flow	Pump failure LV1 fails	Level falls in vaporizer: fit low-level alarm on LIC1
Less	Flow	Partial failure pump/valve	LIC1 alarms

Guide Word	Deviation	Cause	Consequences and Action
MORE	Flow	LV1 sticking, LIC1 fails	Vaporizer floods, liquid to reactor: fit high-level alarm on LIC1 with automatic pump shutdown. Add independent level transmitter and alarm LT2
AS WELL AS	Water brine	Leakage into storages from refrigeration	Concentration of NH_4OH in vaporizer: routine analysis, maintenance
REVERSE	Flow	Pump fails, vaporizer pressure higher than delivery	Flow of vapour into storages: LIC1 alarms; fit non-return valve (NRV2)

Line No. 102
Intention – transfers vapour to mixing tee

NO	Flow	Failure of steam flow, FFV1 fails closed	LIC1 alarms, reaction ceases: considered low flow alarm, rejected – needs resetting at each rate
LESS	Flow	Partial failure or blockage FFV1	As no flow
	LEVEL	LIC1 fails	LT2 back-up system alarms
MORE	Flow	FT2/ratio control mis-operation	Danger of high ammonia concentration: fit alarm, fit analysers (duplicate) with high alarm 12% NH_3 (AI1, AI2)
	LEVEL	LIC1 fails	LT2 back-up system alarms
REVERSE	Flow	Steam failure	Hot, acid gases from reactor – corrosion: fit non-return valve (NRV3)

Line 109 (auxiliary)

NO	Flow	PV1 fails, trap frozen	High level in vaporizer: LIC1 actuated

Vessel – **Reactor**

Intention – oxidizes NH_3 with air, 8 bar, 900°C

Guide Word	Deviation	Cause	Consequences and Action
Line No. 106			
Intention – transfers mixture to reactor, 250°C			
No	Flow	NRV4 stuck closed	Fall in reaction rate: fit low temp. alarm on TI2
Less	Flow	NRV4 partially closed	As No
	NH_3 conc.	Failure of ratio control	Temperatures fall: TI2 alarms (consider low conc. alarm on AI1, AI2)
More	NH3 conc.	Failure of ratio control, air flow restricted	High reactor temp.: TI2 alarms 14% explosive mixture enters reactor – potential for disaster: include automatic shutdown bypass actuated by AI1, AI2, SV2 closes, SV3 opens
	Flow	Control systems failure	High reactor temp.: TI2 alarms
Line No. 107			
Intention – transfers reactor products to waste-heat boiler			
As well as	Composition	Refractory particles from reactor	Possible plugging of boiler tubes: install filter up-stream of boiler

9.8. QUANTITATIVE HAZARD ANALYSIS

Methods such as FMEA, HAZOP and use of safety indices will identify potential hazards, but give only qualitative guidance on the likelihood of an incident occurring and the loss suffered; these are left to the intuition of the team members. In a quantitative hazard analysis, the engineer attempts to determine the probability of an event occurring and the potential cost in terms of injuries, financial loss, etc. The international standard IEC 61508 (1998) (adopted in the UK as BS EN 61508:2000) defines the risk of a hazard as the probable rate of occurrence (typically expressed as events per year) multiplied by the degree of severity of the harm caused. If there are no protective systems in place then the inherent risk, R_{np}, is:

$$R_{np} = F_{np} \times C \qquad (9.1)$$

where

F_{np} is the inherent frequency of the hazard with no protective system (number of events per year)

C is the impact of the hazard (impact per loss event).

The impact can be stated in terms of injuries, serious injuries, emissions, financial loss or other measures. The analysis is sometimes repeated with different measures of impact, as the organization may have a different tolerance for risk of injuries than for financial risk.

In most designs, protective systems are added to reduce the risk of a hazard to a tolerable level. The tolerable risk is defined as:

$$R_t = F_t \times C \tag{9.2}$$

where

F_t is the tolerable frequency of the hazard (number of events per year)

R_t is the tolerable risk, also sometimes called the acceptable risk.

The risk reduction factor of the protective system, ΔR, is defined as the ratio of inherent frequency to tolerable frequency:

$$\Delta R = \frac{F_{np}}{F_t} \tag{9.3}$$

It can be seen that the risk reduction factor is the inverse of the average probability that the protective system will fail when it is called upon to operate, PFD_{av}:

$$PFD_{av} = \frac{F_t}{F_{np}} = \frac{1}{\Delta R} \tag{9.4}$$

The quantitative hazard analysis can thus be used to set targets for the reliability of the protective system. The reliability of the protective system can then be increased until the desired risk reduction factor is attained. Methods that are used to improve the reliability of the protective system are discussed in Section 9.8.2.

The design of safety-instrumented systems for maintaining safe operation of processes is discussed in the international standard IEC 61511, which is adopted as BS EN 61511 in the UK and ANSI/ISA-84.00.01-2004 in the USA. For a safety system that must operate on demand (i.e., in response to an initiating event), IEC 61511 states that each safety-instrumented function (SIF) must have a safety integrity level (SIL) specified to give the required risk reduction or average probability of failure on demand shown in Table 9.11. A good introduction to the application of the functional

TABLE 9.11. Safety Integrity Levels in Demand Mode of Operation (IEC 61511)

Safety Integrity Level (SIL)	Target average Probability of Failure on Demand (PFD_{av})	Target Risk Reduction Factor (ΔR)
4	$\geq 10^{-5}$ to $< 10^{-4}$	>10,000 to $\leq$ 100,000
3	$\geq 10^{-4}$ to $< 10^{-3}$	>1000 to $\leq$ 10,000
2	$\geq 10^{-3}$ to $< 10^{-2}$	>100 to $\leq$ 1000
1	$\geq 10^{-2}$ to $< 10^{-1}$	>10 to $\leq$ 100

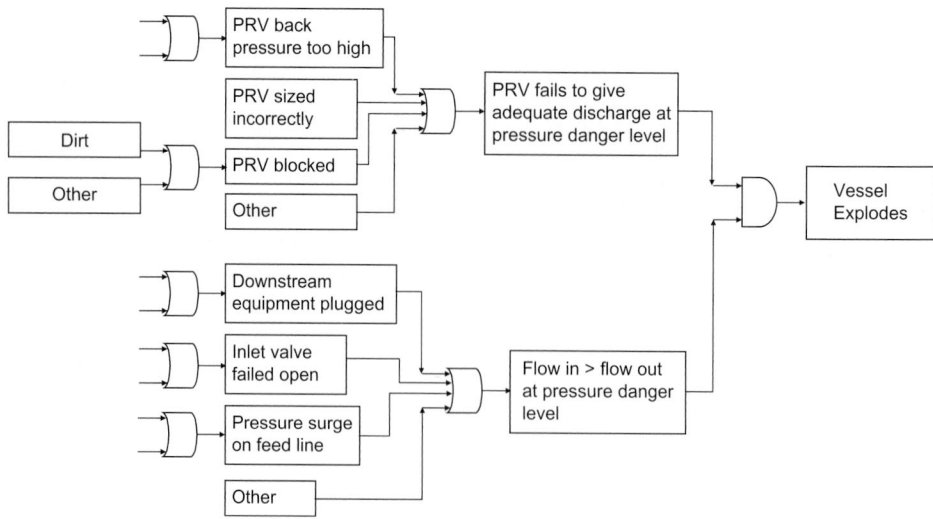

Figure 9.7.   Fault tree for failure of a pressure vessel.

safety standards IEC 61508 and 61511 and the development of safety-instrumented systems is given by King (2007). The book by Cameron and Raman (2005) is an extensive guide to risk management systems.

9.8.1.   Fault Trees

Incidents usually occur through the coincident failure of two or more items: failure of equipment, control systems and instruments, and mis-operation. The sequence of events leading to a hazardous incident can be shown as a fault tree (logic tree), such as that shown in Figure 9.7. This figure shows the set of events that could lead to a pressure vessel rupture. The AND symbol is used where all the inputs are necessary before the system fails, and the OR symbol where failure of any input, by itself, would cause failure of the system. A fault tree is analogous to the type of logic diagram used to represent computer operations, and the symbols are analogous to logic AND and OR gates (Figure 9.8). It can be seen from Figure 9.7 that failure of the vessel will only occur if there is a cause of over-pressure AND a failure of the pressure-relief valve (PRV) to respond adequately. These in turn have several possible causes, which may also have possible causes. Each chain of causality should be pursued to the root cause, and the diagram in Figure 9.7 is incomplete.

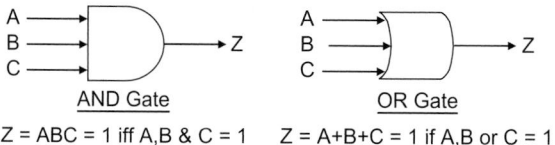

Figure 9.8.   Logic symbols for AND and OR.

The fault trees for even a simple process unit will be complex, with many branches. Fault trees are used to make a quantitative assessment of the likelihood of failure of a system, using data on the reliability of the individual components of the system.

Once the fault tree for a sub-section of the process has been developed, it can be used to improve the reliability of the design by introducing additional, redundant, instrumentation. Since a hazardous condition usually requires the failure of one or more devices, introducing additional parallel devices reduces the likelihood of a failure as long as the devices do not have a common mode of failure. The quantitative analysis of the likelihood of an event can be used to determine the level of system redundancy that is required to reduce the likelihood to an acceptably low value.

Event trees are a similar way of representing the same information, but there is not sufficient space to present both methods here; see Mannan (2004) or CCPS (2008).

9.8.2. Equipment Reliability

When a fault tree has been constructed, it can be used to estimate the probability of the system failing if the probabilities of the events in the fault tree can be estimated. In most cases, this requires a good understanding of the reliability of instruments, alarms and safety devices, since these devices would be expected to maintain the process in a safe condition.

If the failure rate, λ, is the number of occasions per year that a protective system develops a fault (year^{-1}) and the time interval between tests of the device is τ years, then intuitively, the device on average fails half way between tests. The probability that the device is inactive and will fail on demand (also known as the fractional dead time) is then approximately:

$$\phi = \frac{\lambda \tau}{2} \tag{9.5}$$

If the demand rate, δ, is the number of occasions per year that the protective device is actuated then the hazard rate, F, is:

$$F = \delta \phi = \frac{\delta \lambda \tau}{2} \tag{9.6}$$

The intuitive result in equation 9.6 is true if and only if $\delta \lambda$, $\lambda \tau$ and $\delta \tau$ are all $\ll 1$. For a more rigorous analysis of reliability see Chapters 7 and 13 of Mannan (2004). It can be seen that for a simple system with a single device, the demand rate is the inherent frequency of the hazard, $\delta = F_{np}$, and the probability of the device being inactive is the average probability of failure on demand, $\phi = PFD_{av}$. Equations 9.6 and 9.4 are therefore equivalent.

The hazard rate can be reduced by using more reliable equipment (lower value of λ), more frequent testing (lower τ) or making improvements that lead to steadier operation (lower δ). Alternatively, two protective systems in parallel can be installed, in which case the hazard rate becomes:

$$F = \frac{4}{3} \delta \phi_A \phi_B \tag{9.7}$$

where

ϕ_A is the fractional dead time for system A

ϕ_B is the fractional dead time for system B, and subject to the same conditions listed above.

Example 9.3

Laboratory test data for a trip system shows a failure rate of 0.2 per year. If the demand rate is once every 2 years and the test interval is 6 months, what is the hazard rate? Should a parallel system be installed?

$$\text{Fractional dead time, } \phi = \frac{\lambda \delta}{2} = \frac{0.2 \times 0.5}{2} = 0.05 \tag{9.5}$$

$$\text{Hazard rate for single system, } F = \delta \phi = 0.5 \times 0.05 = \underline{0.025}$$

i.e., once in every 40 years.

Many plants operate for more than 20 years, so this is probably too high a failure rate to be acceptable. If two systems are used in parallel then:

$$F = \frac{4}{3} \delta \phi_A \phi_B = \frac{4}{3} \times 0.5 \times 0.05 \times 0.05 = \underline{1.67 \times 10^{-3}}$$

or once in 600 years.

Two systems in parallel should be used, or alternatively, the test frequency could be increased to, say, once every 2 months, giving a more acceptable failure rate of once in every 120 years. Whether the test frequency could be increased will depend on the extent to which testing the device disrupts plant operations. On a large plant with many safety trips and interlocks it may not be possible to test every system on a frequent basis.

The data on probabilities given in this example are for illustration only, and do not represent actual data for these components. Some quantitative data on the reliability of instruments and control systems is given by Mannan (2004). Examples of the application of quantitative hazard analysis techniques in chemical plant design are given by Wells (1996) and Prugh (1980).

The Centre for Chemical Process Safety (CCPS) of the American Institute of Chemical Engineers has published a comprehensive and authoritative guide to quantitative risk analysis (CCPS, 1999). The CCPS has also collected extensive data on device reliability; see CCPS (1989).

Several other texts are available on the application of risk analysis techniques in the chemical process industries; see CCPS (2000), Frank and Whittle (2001), Cameron and Raman (2005), Crowl and Louvar (2002), Arendt and Lorenzo (2000), Kales (1997), Dodson and Nolan (1999), Green (1983) and Kletz (1999b).

9.8.3. Tolerable Risk and Safety Priorities

If the consequences of an incident can be predicted quantitatively (property loss and the possible number of fatalities), then a quantitative assessment can be made of the risk using equation 9.1.

If the loss can be measured in money, the cash value of the risk can be compared with the cost of safety equipment or design changes to reduce the risk. In this way, decisions on safety can be made in the same way as other design decisions: to give the best return of the money invested.

Hazards invariably endanger life as well as property, and any attempt to make cost comparisons will be difficult and controversial. It can be argued that no risk to life should be tolerated; however, resources are always limited and some way of establishing safety priorities is needed.

One approach is to compare the risks, calculated from a hazard analysis, with risks that are generally considered acceptable, such as the average risks in the particular industry, and the kind of risks that people accept voluntarily. One measure of the risk to life is the Fatal Accident Frequency Rate (FAFR), defined as the number of deaths per 10^8 working hours. This is equivalent to the number of deaths in a group of 1000 people over their working lives. The FAFR can be calculated from statistical data for various industries and activities; some of the published values are shown in Tables 9.12 and 9.13. Table 9.12 shows the relative position of the chemical industry compared with other industries; Table 9.13 gives values for some of the risks that people accept voluntarily.

In the chemical process industries, it is generally accepted that risks with an FAFR greater than 0.4 (one-tenth of the average for industry) should be eliminated as a matter of priority, the elimination of lesser risks depending on the resources available; see Kletz (1977a). This criterion is for risks to employees; for risks to the general public (undertaken involuntarily) a lower criterion must be used. In the UK, the HSE has developed the 'as low as reasonably practicable' (ALARP) principle, under which owners can operate a plant in a region defined as tolerable risk, as long as they can demonstrate that they have achieved the lowest risk possible taking into

TABLE 9.12. FAFR for Some Industries for the Period 1978–90

Industry	FAFR
Chemical industry	1.2
UK manufacturing	1.2
Deep sea fishing	4.2

TABLE 9.13. FAFR for Some Non-Industrial Activities

Activity	FAFR
Staying at home	3
Travelling by rail	5
Travelling by bus	3
Travelling by car	57
Travelling by air	240
Travelling by motor cycle	660
Rock climbing	4000

account cost versus risk reduction. The tolerable risk region is defined as a fatality frequency of 10^{-3} to 10^{-6} per person per year for workers and 10^{-4} to 10^{-6} per person per year for the general public; see Schmidt (2007). The level of risk to which the public outside the factory gate should be exposed by plant operations will always be a matter of debate and controversy. Kletz (1977b) suggested that a hazard can be considered acceptable if the average risk is less than one in 10 million, per person, per year. This is an order of magnitude lower than the HSE guideline and is equivalent to a FAFR of 0.001; about the same as the chance of being struck by lightning. Additional guidelines on setting tolerable risk are given by Schmidt (2007).

For further reading on the subject of tolerable risk and risk management, see Cox and Tait (1998) and Lowrance (1976).

9.8.4.   Computer Software for Quantitative Risk Analysis

The assessment of the risks and consequences involved in the planning and operation of a major plant site is a daunting task.

In industrial practice, the safety instrumented systems that are used are more complex than the simple systems described above. If two instruments are used in parallel, with either instrument able to activate a shutdown (a 'one out of two' system, denoted 1oo2) then the probability of failure on demand is reduced, as illustrated in Example 9.3, but the likelihood of a spurious shutdown due to instrument failure is doubled. The instrument engineer can overcome this problem by using three instruments with a voting policy that requires two out of three of the instruments to activate before a shutdown is caused (2oo3 voting). The use of programmable logic controllers, distributed control systems and device communication methods such as Fieldbus means that the reliability of many electrical, electronic and software components must also be considered in the analysis. Credit can also be taken for the performance of the basic process control system (BPCS) as a layer of safety protection (IEC 61511). The calculations soon become too complex to be carried out without using a computer.

The methodology of the classical method of quantitative risk analysis is shown in Figure 9.9. First, the likely frequency of failure of equipment, pipe-lines and storage vessels must be predicted, using the techniques described above. The probable magnitude of any discharges must then be estimated, and the consequences of failure evaluated: fire, explosion or toxic fume release. Other factors, such as site geography, weather conditions, site layout and safety management practices, must be taken into consideration. The dispersion of gas clouds can be predicted using suitable models. This methodology enables the severity of the risks to be assessed. Limits have to be agreed on the acceptable risks, such as the permitted concentrations of toxic gases. Decisions can then be made on the siting of plant equipment (see Chapter 14), on the suitability of a site location and on emergency planning procedures.

The comprehensive and detailed assessment of the risks required for a 'safety-case' can only be satisfactorily carried out for major installations with the aid of computer software. Programs for quantitative risk analysis have been developed by consulting firms specializing in safety and environmental protection. Typical of the software available is the SAFETI (Suite for Assessment of Flammability Explosion and Toxic Impact) suite of programs developed by DNV Technica Ltd (www.dnv.com). These

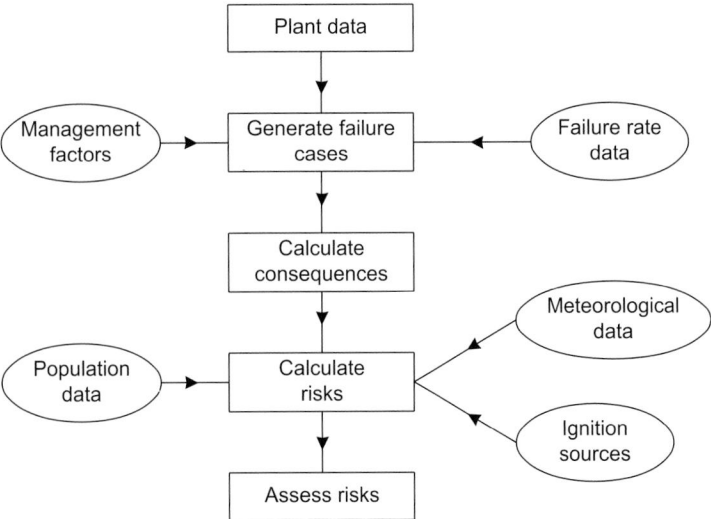

Figure 9.9. Quantitative risk assessment procedure.

programs were initially developed for the authorities in the Netherlands, as a response to the Seveso Directives of the EU. The programs have subsequently been developed further and extended, and are widely used in the preparation of safety cases; see Pitblado *et al.* (1990). Other examples include PHAWorks and FaultrEASE from Chempute Inc. (www.chempute.com); FTA-Pro and PHA-Pro from Dyadem (www.dyadem.com); RENO and BlockSim7 from Reliasoft (www.risk-analysis-software.org); and LOGAN from Reliass (www.reliability-safety-software.com). These and other programs for fault tree analysis, HAZOP, process hazard analysis and quantitative risk analysis are easily found by searching on-line, and most chemical companies license one of these programs. Some of the programs offer free trial versions and reduced cost student subscriptions.

Computer programs can be used to investigate a range of possible scenarios for a site, but as with all computer software used in design they should not be used without caution and judgment. They would normally be used with the assistance and guidance of the consulting firm supplying the software. With intelligent use, guided by experience, such programs can indicate the magnitude of the likely risks at a site, and allow sound decisions to be made when licensing a process operation or granting planning permission for a new installation.

9.9. PRESSURE RELIEF

Pressure-relief devices are an essential requirement for the safe use of pressure vessels. Pressure-relief devices provide a mechanical means of ensuring that the pressure inside a vessel cannot rise to an unsafe level. All pressure vessels within the scope of Section VIII of the ASME Boiler and Pressure Vessel Code must be fitted with a pressure-relief device. The purpose of the pressure-relief device is to prevent catastrophic failure of the vessel by providing a safe means of relieving over-pressure if the pressure inside the vessel exceeds the maximum allowable working pressure.

Three different types of relief device are commonly used:

Directly actuated valves: weight or spring-loaded valves that open at a pre-determined pressure, and that normally close after the pressure has been relieved. The system pressure provides the motive power to operate the valve.

Indirectly actuated valves: pneumatically or electrically operated valves that are activated by pressure-sensing instruments.

Bursting discs: thin discs of material that are designed and manufactured to fail at a predetermined pressure.

Relief valves are normally used to regulate minor excursions of pressure; and bursting discs as safety devices to relieve major over-pressure. Bursting discs are often used in conjunction with relief valves to protect the valve from corrosive process fluids during normal operation. The design and selection of relief valves is discussed by Morley (1989a,b), and is also covered by the pressure vessel standards; see below. Bursting discs are discussed by Mathews (1984), Asquith and Lavery (1990) and Murphy (1993). The discs are manufactured in a wide range of common engineering steels and alloys as well as a variety of materials for use in corrosive conditions, such as impervious carbon, gold and silver, and suitable discs can be found for use with all process fluids. Bursting discs and relief valves are proprietary items and the vendors should be consulted when selecting suitable types and sizes.

Selection and sizing of the relief device are the responsibility of the end user of the pressure vessel. Rules for the selection and sizing of pressure-relief devices are given in the ASME BPV Code Sec. VIII D.1 Parts UG-125 to UG-137 and D.2 Part AR or in the European standard BS EN ISO 4126 (2004).

Under the rules given in ASME BPV Code Sec. VIII D.1, the primary pressure-relief device must have a set pressure not greater than the maximum allowable working pressure of the vessel. The primary relief device must be sized to prevent the pressure from rising 10% or 3 psi (20 kPa), whichever is greater, above the maximum allowable working pressure. If secondary relief devices are used then their set pressure must be not greater than 5% above the maximum allowable working pressure. When multiple relief devices are used then their combined discharge must be adequate to prevent the vessel pressure from rising more than 16% or 4 psi (30 kPa), whichever is greater, above the maximum allowable working pressure. In a relief scenario where the pressure vessel is exposed to an external fire, the relief device or devices must prevent the vessel pressure from increasing to more than 21% above the maximum allowable working pressure.

Pressure-relief devices must be constructed, located and installed such that they can be easily inspected and maintained. They are normally located at the top of a vessel in a clean, free-draining location. They must be located on or close to the vessel that they are protecting.

9.9.1. Pressure-Relief Scenarios

Over-pressure will occur whenever mass, moles or energy accumulate in a contained volume or space with a restricted outflow. The rate at which material or energy accumulates determines the pressure rise. If the process control system is not able to respond quickly enough then the pressure-relief device must be activated before the vessel ruptures, explodes or suffers some other catastrophic loss of containment.

The first step in designing a pressure-relief system is to evaluate the possible causes of over-pressure to determine the rate of pressure accumulation associated with each, and hence estimate the relief load (the flow rate that must be discharged through the relief device). The API Recommended Practice (RP) 521 suggests the following causes:

Blocked outlet	Chemical reaction	Electric power loss
Utility failure	External fire	Accumulation of non-condensable species
Cooling or reflux failure	Abnormal heat input	Failure of automatic controls
Inadvertent valve opening	Operator error	Loss of heat in series fractionation
Loss of fans	Check valve failure	Volatile material entering system
Steam or water hammer	Internal explosion	Heat-exchanger tube failure
Adsorbent flow failure		Overheating a liquid full system

This list is not exhaustive, and the design engineers should always brainstorm for additional scenarios and review the results of FMEA, HAZOP, HAZAN or other process safety analyses.

In evaluating relief scenarios, the design engineer should consider sequential events that result from the same root cause event, particularly when these can increase the relief load. For example, the loss of electric power in a plant that carries out a liquid-phase exothermic reaction could have the following impacts:

1. Failure of all or part of the automatic control system.
2. Loss of cooling due to failure of cooling water pumps or air coolers.
3. Loss of mixing in reactor due to failure of stirrer, leading to localized runaway reaction.

Since these have a common cause they should be considered as simultaneous events for that cause. If two events do not share a common cause then the probability that they will occur simultaneously is remote and is not usually considered (API RP 521, 3.2). Root cause events such as power loss, utility loss and external fire will often cause multiple other events and hence large relief loads.

The rate at which pressure accumulates is also affected by the response of the process control system. API RP 521 recommends that instrumentation should be assumed to respond as designed if it increases the relieving requirement, but no credit should be taken for instrumentation response if it reduces the relieving requirement. For example in Figure 9.10a, if the outlet control valve becomes blocked and the pressure in the vessel rises, the flow from the pump will initially decrease because of the higher back-pressure. The flow controller will compensate for this by opening the flow control valve to try to maintain a constant flow rate, and will consequently increase the relieving load. The design engineer should assume that the instrumentation responds as designed and the flow rate remains constant. In Figure 9.10b, if the liquid outlet control valve becomes blocked the pressure controller will continue opening the pressure control valve until it is fully open. This provides an alternative outflow and reduces the relieving load, but according to API RP 521 this response should not be considered.

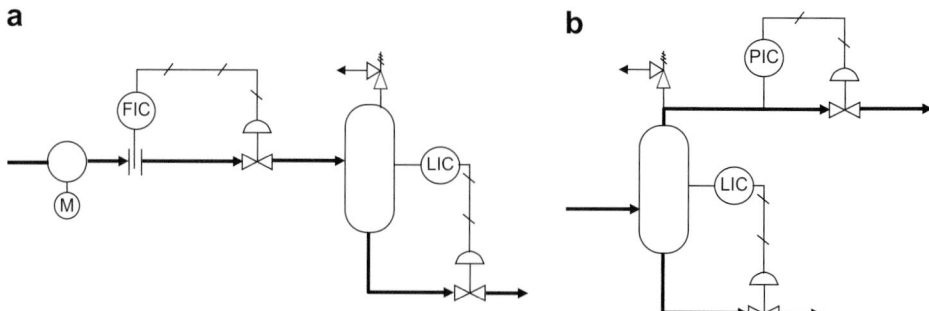

Figure 9.10. Instrumentation response to pressure-relief scenarios. (a) Instrumentation response increases relieving load. (b) Instrumentation response would reduce relieving load, but API RP 521 recommends taking no credit for instrumentation response.

Heat exchangers and other vessels with internal compartments must also be protected from over-pressure in the case of an internal failure. This is of particular importance for shell and tube type exchangers, as the common design practice is to put the higher pressure fluid on the tube side. This saves costs in constructing the shell and also obviates sizing the tubes to withstand a high compressive load due to external pressure. If the tube side is at higher pressure then in the event of a tube or tube-sheet failure the shell will be exposed to the higher tube-side pressure.

Both API RP 521 and ASME BPV Code Section VIII allow multiple vessels connected together to be considered as a single unit for relief scenarios, provided that there are no valves between the vessels and that the design considers the full relieving load of the system (ASME BPV Code Sec. VIII D.1 UG-133).

9.9.2. Pressure-Relief Loads

The rate at which pressure accumulates can be estimated by making non-steady state mass, mole and energy balances around the vessel or system:

$$\text{in} + \text{formed by reaction} = \text{out} + \text{accumulation} \tag{9.8}$$

Because liquids have very low compressibility, pressure vessels are seldom operated entirely filled with liquid, since small accumulations of material would cause large surges in pressure. Instead, it is common practice to operate with a 'bubble' of vapour (often nitrogen) at the top of the vessel. The mass balance equation can then be rearranged into an equation for the rate of change of pressure of this gas with time.

For example, consider a vessel of total volume V m^3 that is normally operated 80% full of liquid on level control (as in Figure 9.10a) and is fed with a flow rate v m^3/s of liquid. If the volume of liquid in the vessel is V_L, then if the outlet becomes blocked and the liquid is assumed to be incompressible, the change in the volume of the liquid is:

$$\frac{dV_L}{dt} = v \tag{9.9}$$

where t = time, s.

The volume occupied by vapour, $V_G = V - V_L$, so:

$$\frac{dV_G}{dt} = -\frac{dV_L}{dt} = -v \tag{9.10}$$

If there is no vapour flow in or out of the vessel then assuming the vapour behaves as an ideal gas:

$$V_G = nRT/P \tag{9.11}$$

where

n = number of moles of gas in the vessel, mol
R = ideal gas constant, J/mol.K
T = temperature, K
P = pressure, N/m^2.

If the temperature is constant (which is valid for a blocked outlet relief scenario) then until the relief valve opens:

$$\frac{dP}{dt} = nRT\frac{d}{dt}\left(\frac{1}{V_G}\right) = -\frac{nRT}{V_G^2}\frac{dV_G}{dt} = \frac{P^2 v}{nRT} \tag{9.12}$$

Equation 9.12 can be used to estimate the rate of pressure accumulation.

When the relief valve opens it allows vapour to discharge at a flow rate w kg/s. The number of moles of vapour in the vessel is then given by:

$$\frac{dn}{dt} = -\frac{1000\,w}{M_w} \tag{9.13}$$

where M_w is the average molecular weight of the vapour, g/mol.

The equation for the rate of change of pressure becomes:

$$\frac{dP}{dt} = RT\frac{d}{dt}\left(\frac{n}{V_G}\right) = \frac{RT}{V_G^2}\left(V_G\frac{dn}{dt} - n\frac{dV_G}{dt}\right)$$

$$= \frac{P^2}{nRT}\left(v - \frac{1000\,RTw}{M_w\,P}\right) \tag{9.14}$$

If the relief valve is sized correctly then the maximum pressure that can accumulate is 110% of the maximum allowable working pressure, P_m (ASME BPV Code Sec. VIII D.1 UG-125). At this point there is no further accumulation of pressure and $dp/dt = 0$, hence:

$$\frac{1000\,RTw}{M_w \times 1.1P_m} = v \tag{9.15}$$

and the required relief load is:

$$w = \frac{1.1P_m\,M_w v}{1000\,RT} \tag{9.16}$$

Equation 9.16 applies as long as only vapour is vented from the vessel. Once the vapour has been displaced by liquid then the relief load must be the liquid flow rate. If a two-phase mixture is vented then the calculation becomes more complex.

In most cases the governing relief scenario includes both material and heat input into the system and typically also includes vaporization of material, reaction and two-phase flow. Such systems are much more difficult to describe using simple differential algebraic models and the current industrial practice is to use dynamic simulation models for these cases. Dynamic models can be built in any of the commercial process simulators that have this capability. The AIChE Design Institute for Emergency Relief Systems (DIERS) also licenses software called SuperChems™ (formerly SAFIRE) that is written specifically for pressure-relief system design and incorporates the DIERS recommended methods and research findings for multiphase, reacting and highly non-ideal systems.

For some relief scenarios, correlations have been established for the relieving load. For the external fire case API RP 521 (Section 3.15.2) gives:

$$Q = 21000 \, F_e \, A_w^{\,0.82} = w_f \, \Delta H_{vap} \qquad (9.17)$$

where

Q = heat input due to fire, BTU/h
F_e = environmental factor
A_w = internal wetted surface area, ft^2
w_f = fire case relieving load, lb/h
ΔH_{vap} = heat of vaporization, BTU/lb.

The environmental factor F_e allows for insulation on the vessel. It is equal to 1.0 for a bare vessel or if the insulation can be stripped off by a liquid jet. The correlation in equation 9.17 assumes good general design practice and site layout, including use of sewers and trenches or the natural slope of the land to control run-off so that pools do not form. Other formulae for the rate of heat input and relief load are given by ROSPA (1971) and NFPA 30 (2003). Local safety regulations and fire codes should be consulted to determine the appropriate method to use in any particular design.

Design codes and standards such as API RP 521 and the DIERS Project Manual (Fisher *et al.*, 1993) should be consulted for other correlations and recommended methods for calculating relief loads. The DIERS Project Manual also discusses calculation of relief loads for under-pressure scenarios (Section 9.9.5).

9.9.3. Design of Pressure-Relief Valves

Spring-Loaded Relief Valves

The most commonly used relief device is the conventional spring-loaded relief valve shown in Figure 9.11. This design of valve is available in the widest range of sizes and materials (API Standard 526, BS EN ISO 4126-1:2004).

In a conventional relief valve the pressure force acts on a disk that is held against a seating surface by a spring. The compression of the spring can be adjusted using an adjusting screw so that the spring force is equal to the pressure force at the valve set pressure.

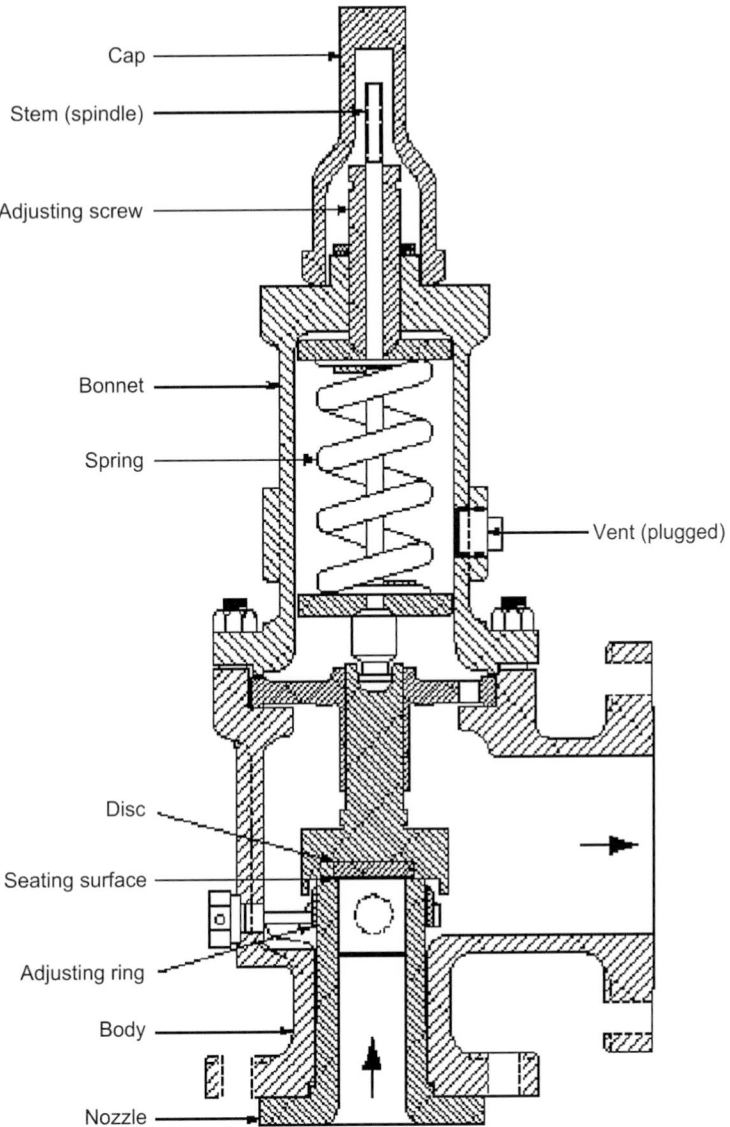

Cap

Stem (spindle)

Adjusting screw

Bonnet

Spring

Vent (plugged)

Disc

Seating surface

Adjusting ring

Body

Nozzle

Figure 9.11. Conventional spring-loaded relief valve. Reproduced with permission from API Recommended Practice 520.

The pressure flow response of a conventional relief valve is illustrated schematically in Figure 9.12. When the pressure in the vessel reaches 92% to 95% of the set pressure a spring-loaded relief valve in a gas or vapour service begins to 'simmer' and leak gas. Leakage can be reduced by lapping the disk and seating surface to a high degree of polish, using elastomeric seals (at low temperatures only) or using a high pressure differential between the operating pressure and set pressure. When the set pressure is reached, the valve 'pops' and the disk lifts from the seat. The disk and seat are shaped such that the force on the disk continues to increase until the valve is fully

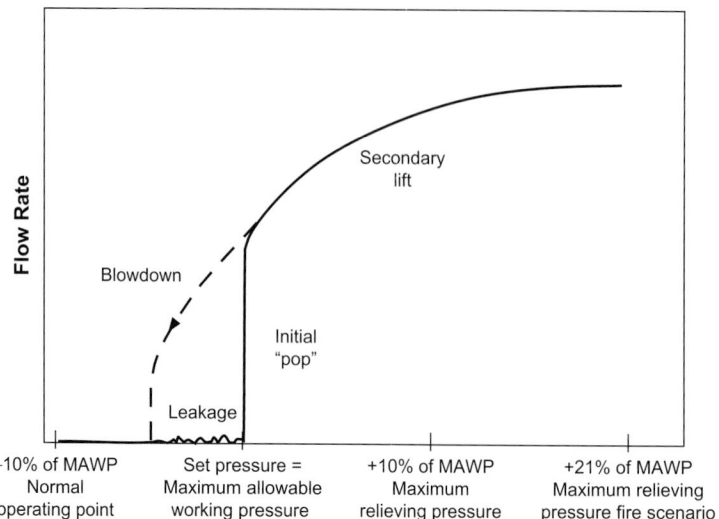

Figure 9.12.   Pressure–flow response of a conventional spring-loaded relief valve.

open, at which point the flow rate is limited only by the bore area of the seating surface and not by the gap between the seating surface and the disk. At this point the design flow rate is achieved and there should be no further pressure accumulation.

When the pressure falls sufficiently, the spring force can overcome the forces due to the flowing fluid and the valve re-seats. Re-seating usually occurs at a lower pressure than the set pressure, giving a different curve for blowdown.

The capacity and lift pressure of a conventional spring-loaded relief valve are affected by the back-pressure in the downstream relief system. The back-pressure exerts forces that are additive to the spring force. Where back-pressure is known to fluctuate or accumulate, balanced pressure-relief valves that incorporate a bellows or other means of compensating for back-pressure should be used (see API RP 520 for details). This is particularly important when multiple devices are relieved into the same vent or flare system, as common-cause relief scenarios such as power loss can trigger multiple relief events and send a lot of material into the vent or flare system, increasing the back-pressure acting on the relief valves.

Pilot-Operated Relief Valves

Pilot-operated relief valves are designed to overcome some of the major drawbacks of conventional spring-loaded relief valves. In a pilot-operated relief valve the spring and disk are replaced by a piston, as shown in Figure 9.13. A narrow-bore pipe known as a pilot supply line connects from the top of the piston to the relief valve inlet via a secondary (pilot) valve of the spring-loaded type. In normal operation both sides of the valve see the same pressure, but because the top surface area of the piston is greater than the area of the seat, the downward force is greater and the valve remains closed. When the pressure exceeds the set pressure the pilot valve opens and pressure above the piston is lost. This causes the piston to lift and the valve opens. The pilot valve vent can be exhausted to atmosphere or to the main valve outlet, depending on the containment requirements for the process fluid.

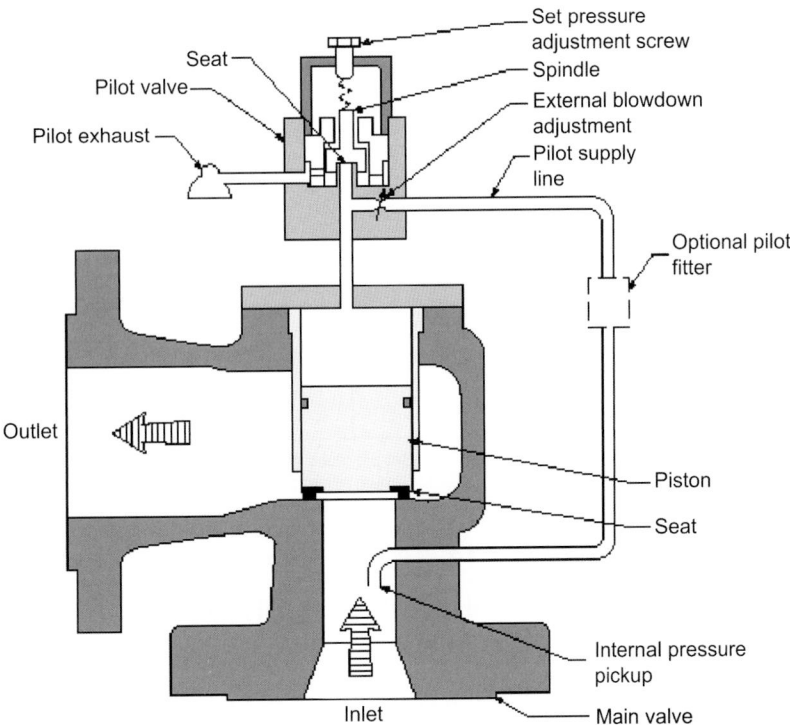

Figure 9.13. Pop-action pilot-operated relief valve. Reproduced with permission from API Recommended Practice 520.

The pressure-flow response of a pilot-operated relief valve is illustrated schematically in Figure 9.14. Leakage is eliminated and there is no blowdown.

Pilot-operated relief valves are used in applications that require a low differential between operating pressure and set pressure (for example, revamps where the vessel is now operated closer to the maximum allowable working pressure or vessels operating below 230 kPa or 20 psig), high-pressure services (above 69 bara or 1000 psig) and cases where low leakage is required. They are not available in the same range of metallurgies as spring-loaded relief valves. Pilot-operated relief valves are also restricted to lower temperature applications, as they typically use elastomeric materials to make a seal between the piston and its housing. More details of pilot-operated relief valves are given in BS EN ISO 4126-4:2004.

Sizing Relief Valves

Guidelines for sizing relief valves are given in API RP 520 and BS EN ISO 4126. Different design equations are recommended for vapour, liquid, steam or two-phase flows. Sizing methods are also discussed in the DIERS Project Manual (Fisher *et al.*, 1993) and the book by CCPS (1998).

When the fluid flowing through the valve is a compressible gas or a vapour, then the design must consider whether critical flow is achieved in the nozzle of the valve. The critical flow rate is the maximum flow rate that can be achieved and corresponds

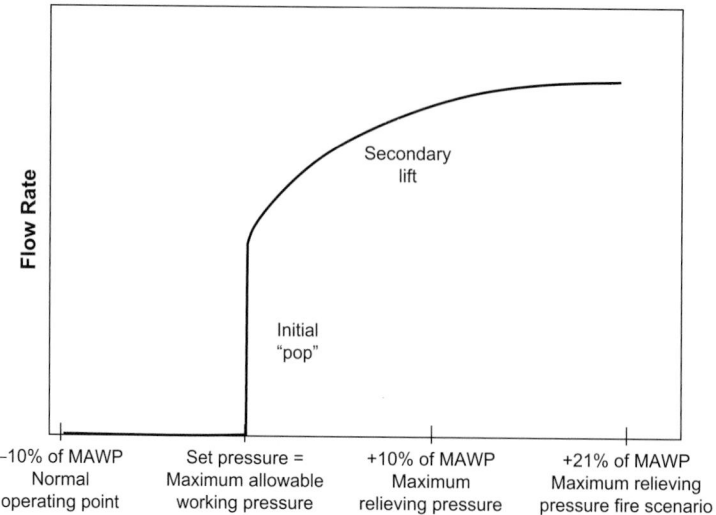

Figure 9.14.   Pressure–flow response of a pilot-operated relief valve.

to a sonic velocity at the nozzle. If critical flow occurs, then the pressure at the nozzle exit cannot fall below the critical flow pressure P_{cf}, even if a lower pressure exists downstream. The critical flow pressure can be estimated from the upstream pressure for an ideal gas using the equation:

$$\frac{P_{cf}}{P_1} = \left[\frac{2}{\gamma + 1}\right]^{\gamma / (\gamma - 1)} \tag{9.18}$$

where

γ = ratio of specific heats = C_p/C_v
P_1 = absolute pressure upstream
P_{cf} = critical flow pressure.

Any consistent set of units may be used for pressure as long as the absolute pressure is used, not the gauge pressure. The ratio P_{cf}/P_1 is called the critical pressure ratio. Typical values of this ratio are given in Table 9.14. If the downstream pressure is less than the critical flow pressure, critical flow will occur in the nozzle. It can be seen from the table that this will be the case whenever the upstream pressure is more than two times the downstream pressure. Since most relief systems are operated close to atmospheric pressure, critical flow is the usual case.

For critical flow, API RP 520 (Section 3.6.2) gives the following equation for valve area, A_d:

$$A_d = \frac{13,160 \, w}{C \, K_d \, P_1 \, K_b \, K_c} \sqrt{\frac{T \, Z}{M_w}} \tag{9.19}$$

where

A_d = discharge area, mm^2
w = required flow rate, kg/h

TABLE 9.14. Critical Flow Pressure Ratios (Adapted from API RP 520)

Gas	Specific Heat Ratio $\gamma = C_p/C_v$ at 60°F, 1 atm	Critical Flow Pressure Ratio at 60°F, 1 atm
Hydrogen	1.41	0.52
Air	1.40	0.53
Nitrogen	1.40	0.53
Steam	1.33	0.54
Ammonia	1.3	0.54
Carbon dioxide	1.29	0.55
Methane	1.31	0.54
Ethane	1.19	0.57
Ethylene	1.24	0.57
Propane	1.13	0.58
Propylene	1.15	0.58
n-Butane	1.19	0.59
n-Hexane	1.06	0.59
Benzene	1.12	0.58
n-Decane	1.03	0.60

Notes: 1. Taken from API RP 520, Table 7.

2. Some values of critical flow pressure ratio have been determined experimentally and do not necessarily agree with predictions from equation 9.18.

$$C = \text{coefficient} = 520 \sqrt{\gamma \left(\tfrac{2}{\gamma+1}\right)^{(\gamma+1)/(\gamma-1)}}$$

K_d = coefficient of discharge
P_1 = absolute pressure upstream, kPa
K_b = back-pressure correction factor
K_c = combination correction factor
T = relieving temperature, K
Z = compressibility at the inlet condition
M_w = molecular weight, g/mol.

For preliminary estimates, the coefficient K_d can be taken as 0.975 for a relief valve and 0.62 for a bursting disk. The back-pressure correction factor, K_b, can initially be assumed to be 1.0 for critical flow. The combination correction factor, K_c, is used when a rupture disk is used upstream of the relief valve (see next section), in which case it is 0.9. If no rupture disk is used then K_c, is 1.0. For vessels designed in accordance with ASME BPV Code Sec. VIII,, $P_1 = 1.1$ times the maximum allowable working pressure.

The relief valve selected should be one with equal or greater area than calculated using equation 9.19. Relief valve sizes are given in API Standard 526 or BS EN ISO 4126. Sizing equations for sub-critical flow of vapours, liquids, steam and two-phase mixtures are given in API RP 520.

9.9.4. Design of Non-Reclosing Pressure-Relief Devices

Two types of non-reclosing pressure-relief devices are used, rupture disks and breaking-pin devices.

A rupture disk device consists of a rupture disk and a clamp that holds the disk in position. The disk is made from a thin sheet of metal and is designed to burst if a set pressure is exceeded. Some rupture disks are scored so that they can burst without forming fragments that might damage downstream equipment.

Rupture disks are often used upstream of relief valves to protect the relief valve from corrosion or to reduce losses due to relief valve leakage. Large rupture disks are also used in situations that require very fast response time or high relieving load (for example, reactor runaway and external fire cases). They are also used in situations where pressure is intentionally reduced below the operating pressure for safety reasons. The use of bursting disc devices is described in BS EN ISO 4126-2:2004 and BS EN ISO 4126-6:2004.

If a rupture disk is used as the primary pressure-relief device then when it bursts the operators have no option but to shut down the plant so that the disk can be replaced before the vessel is re-pressured. Rupture disks are therefore most commonly used at the inlets of relief valves or as secondary relief devices. Rupture disks can be sized using equation 9.19 for compressible gases in sonic flow, with a value of $K_d = 0.62$. The combination of safety valves and rupture discs is discussed in BS EN ISO 4126-3:2004.

Breaking-pin devices have a similar construction to spring-loaded relief valves, except the valve disk is held against the seat by a pin that is designed to buckle or break when the set pressure is reached, as illustrated in Figure 9.15. Once the valve has opened the pin must be replaced before the valve can be re-set.

Both rupture disks and breaking-pin devices are sensitive to temperature. The manufacturer should always be consulted for applications that are not at ambient conditions. Since non-reclosing pressure-relief devices can only be used once, the set pressure is determined by testing a sample of the devices out of each manufactured batch. Pressure-relief valve test methods are specified in ASME PTC 25-2001.

9.9.5.   Design of Pressure-Relief Discharge Systems

When designing relief venting systems, it is important to ensure that flammable or toxic gases are vented to a safe location. This will normally mean venting at a sufficient height to ensure that the gases are dispersed without creating a hazard. For highly toxic materials it may be necessary to provide a scrubber to absorb and 'kill' the material; for instance, the provision of caustic scrubbers for chlorine and hydrochloric acid gases. If flammable materials have to be vented at frequent intervals, as for example in some refinery operations, flare stacks are used.

The rate at which material can be vented will be determined by the design of the complete venting system: the relief device and the associated piping. The maximum venting rate will be limited by the critical (sonic) velocity, whatever the pressure drop. The vent system must be designed such that sonic flow can only occur at the relief valve and not elsewhere in the system, otherwise the design relief load will not be attained. The design of venting systems to give adequate protection against over-pressure is a complex and difficult subject, particularly if two-phase flow is likely to occur. When two-phase flow can occur then the relief system must provide for disengagement of liquid from the vapour before the vapour is vented or sent to flare.

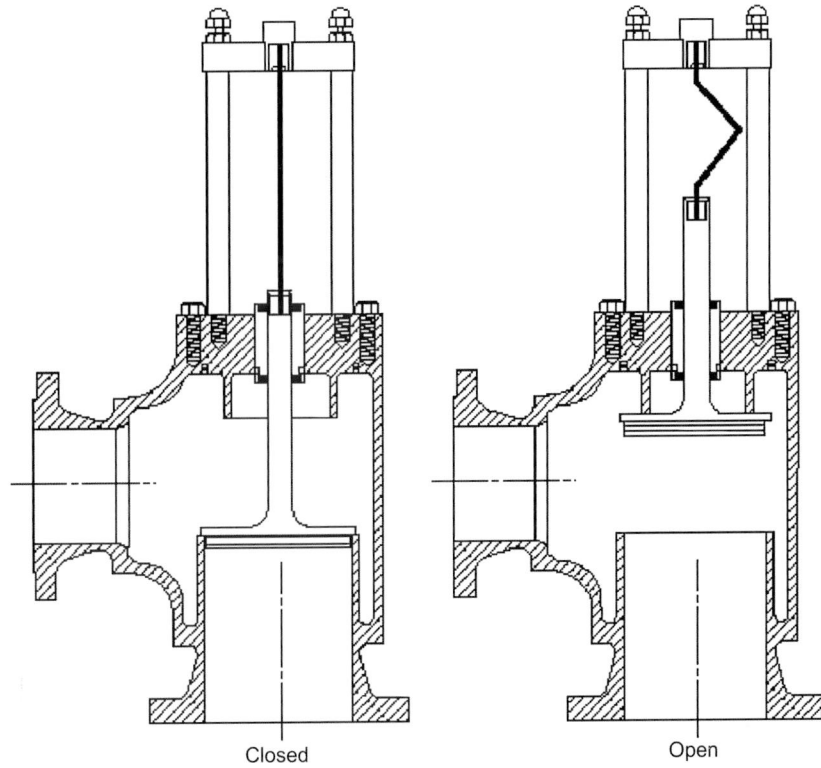

Closed   Open

Figure 9.15. Buckling-pin relief valve. Reproduced with permission from API Recommended Practice 520.

Guidelines for relief valve installation and relief systems design are given in API RP 520 Part II, API RP 521 Sections 4 and 5, and the DIERS Project Manual (Fisher *et al.*, 1993). API RP 521 also gives design methods for blowdown drums and flare systems. A typical relief system is shown in Figure 9.16. For a comprehensive discussion of the problem of vent system design, and the design methods available, see the papers by Duxbury (1976, 1979) and the guidelines by CCPS (1998).

9.9.6. Protection from Under-Pressure (Vacuum)

Unless designed to withstand external pressure (see Section 13.7) a vessel must be protected against the hazard of under-pressure, as well as over-pressure. Under-pressure will normally mean vacuum on the inside with atmospheric pressure on the outside. It requires only a slight drop in pressure below atmospheric pressure to collapse a storage tank. Though the pressure differential may be small, the force on the tank roof will be considerable. For example, if the pressure in a 10-m-diameter tank falls to 10 millibar below the external pressure, the total load on the tank roof will be around 80,000 N (equivalent to 8 tons in weight). It is not an uncommon occurrence for a storage tank to be sucked in (collapsed) by the suction pulled by the discharge pump, due to the tank vents having become blocked. Where practical,

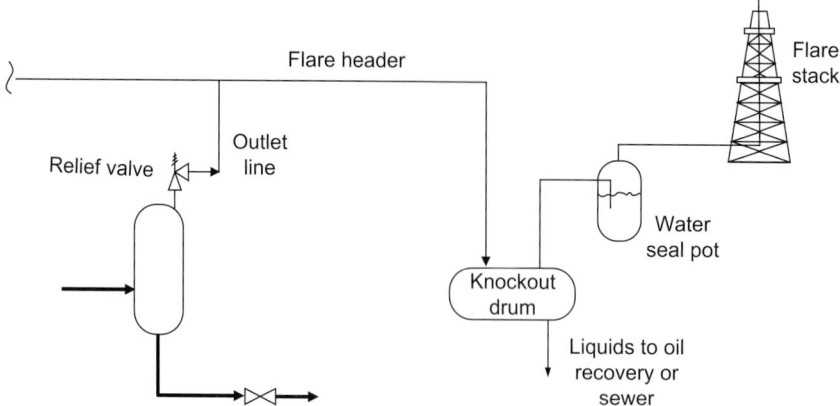

Figure 9.16.   Typical relief system design.

vacuum breakers (valves that open to atmosphere when the internal pressure drops below atmospheric) should be fitted.

Example 9.4

A gasoline surge drum has capacity 4 m³ (1060 gal) and is normally operated 50% full at 40°C (100°F) under 20 bar absolute pressure (280 psig) of hydrogen in the head space and using a level controlled outflow as shown in Figure 9.10a. Gasoline of specific gravity 0.7 is pumped into the surge drum at a normal flow rate of 130 m³/h.

Assuming the aspect ratio of the vessel (ratio length/diameter) is 3.0 and the heat of vaporization of gasoline is 180 BTU/lb, evaluate the relief loads for the blocked outflow and external fire cases, and hence determine the relief valve size. (In practice, gasoline contains many components that boil over a wide range of temperatures at the design pressure and a more complex calculation is needed than is given here.)

Blocked Outlet Case

$$w = \frac{1.1 P_m \, M_w v}{1000 \, RT} = \frac{1.1 \times \left(\dfrac{130}{3600}\right) \times \left(\dfrac{20 \times 10^5}{0.9}\right) \times 2}{1000 \times 8.314 \times 313} = \underline{\underline{67.8 \text{ g/s}}} \qquad (9.16)$$

External Fire Case

If the vessel has a hemispherical head then:

$$\text{volume} = \pi \left(\frac{D^2 L}{4} + \frac{D^3}{6} \right) = \frac{11 \pi D^3}{12}$$

$$\text{so} \quad D = 1.12 \text{ m}$$

$$\text{wetted area} = \pi \, (DL + D^2)/2 = 2\pi D^2$$

$$= 7.82 \text{ m}^2 = 84.2 \text{ ft}^2$$

Assume $F_e = 1$

$$w_f = \frac{21000 \, F_e \, A_w}{\Delta H_{vap}} = \frac{21000 \times 1 \times 84.2^{0.82}}{180}$$

$$= 4423 \text{ lb/hr}$$

$$= \underline{\underline{0.56 \text{ kg/s}}}$$

(9.17)

So the external fire case has the higher relieving load and governs the design.

If the vent line discharges to a flare system at atmospheric pressure then:

$$\frac{P_{outlet}}{P_1} = \frac{1}{20} << 0.52$$

so flow in the nozzle is critical.

For hydrogen:

$$C = 520 \sqrt{1.41 \left(\frac{2}{2.41}\right)^{(2.41/0.41))}}$$

$$= 356.9$$

Assume the valve will lift when the temperature reaches 60°C (333 K), $Z = 1.02$ for hydrogen:

$$A_d = \frac{13,160 \, w}{C \, K_d \, P_1 \, K_b \, K_c} \sqrt{\frac{T \, Z}{M_w}}$$

$$= \frac{13160 \times 0.56 \times 3600}{356.9 \times 0.975 \times 2000 \times 1.0 \times 1.0} \sqrt{\frac{333 \times 1.02}{2}}$$

(9.19)

$$= 496.8 \text{ mm}^2 \text{ or } 0.77 \text{ in}^2$$

From API Std. 526 we would select a 'H' orifice relief valve with effective orifice area 0.785 in^2. A size 2H3 carbon steel relief valve will allow a set pressure up to 740 psig in the expected range of operation temperature. In practice, however, we would have to consider two-phase flow due to entrainment of boiling liquid with the vapour and might select a larger orifice size after more detailed design.

9.10. REFERENCES

Arendt, J. S. and Lorenzo, D. K. (2000) *Evaluating Process Safety in the Chemical Industry: A User's Guide to Quantitative Risk Analysis.* (Wiley – AIChE).

Asquith, W. and Lavery, K. (1990) *Proc. Ind. Jl.* (Sept.) 15. Bursting discs – the vital element in relief.

ASME (1993) *Noise Control in the Process Industries* (ASME).

Ashafi, C. R. (2003) *Industrial Safety and Health Management*, 5th edn (Prentice Hall).

Balemans, A. W. M. (1974) Check-lists: guide lines for safe design of process plants. In: *Loss Prevention and Safety Promotion in the Process Industries*, C. H. Bushmann (ed.) (Elsevier).

Barton, J. (2001) *Dust Explosion, Prevention and Protection—A Practical Guide* (Institution of Chemical Engineers, London).

Bias, D. and Hansen, C. (2003) *Engineering Noise: Theory and Practice* (Spon Press).

Birolini, A. (2004) *Reliability Engineering Theory and Practice*, 4th ed. (Springer Verlag).

Bollinger, R. E., Clark, D. G., Dowell, R. M., Ewbank, R. M., Hendershot, D. C., Lutz, W. K., Meszaros, S. I., Park, D. E., Wixom, E. D. and Crowl, D. A. (ed.) (2008) *Inherently Safer Chemical Processes: A Life Cycle Approach (Center for Chemical Process Safety)*, 2nd edn (Wiley-AIChE).

Boxerman, S., Bell, C. and Nordlander, K. (2008) *Chem. Eng. NY,* 115(3), 38. Are you ready for REACH?.

Britton, L. G. (1999) *Avoiding Static Ignition Hazards in Chemical Processes* (AIChE).

Brown, D. (2004) *Chem. Engr* London No. 758 (August) 42. It's a risky business.

Cameron, I. T. and Raman, R. (2005) Process Systems Risk Management, Vol. 6 (Process Systems Engineering) (Academic Press).

Carson, P. A. and Mumford, C. J. (1988) *Safe Handling of Chemicals in Industry*, 2 vols (Longmans).

Carson, P. A. and Mumford, C. J. (2002) *Hazardous Chemicals Handbook*, 2nd edn (Newnes).

CCPS (2008) *Guidelines for Hazard Evaluation Procedures (Center for Chemical Process Safety)*, 3rd edn (Wiley – AIChE).

CCPS (2000) *Guidelines for Hazard Evaluation Procedures – with worked examples (Center for Chemical Process Safety)*, 2nd edn (Wiley – AIChE).

CCPS (1999) *Guidelines for Chemical Process Quantitative Risk Analysis (Center for Chemical Process Safety)*, 2nd edn (Wiley – AIChE).

CCPS (1998) *Guidelines for Pressure Relief and Effluent Handling Systems (Center for Chemical Process Safety)* (Wiley – AIChE).

CCPS (1993) *Guidelines for Engineering Design for Process Safety (Center for Chemical Process Safety)* (Wiley – AIChE).

CCPS (1989) *Guidelines for Process Equipment Reliability Data, with Data Tables (Center for Chemical Process Safety)* (Wiley – AIChE).

Cheremisnoff, N. P. (1996) *Noise Control in Industry: A Practical Guide* (Noyes).

CIA (1979) *Process Plant Hazards and Control Building Design* (Chemical Industries Association).

Cooper, W. F. and Jones, D. A. (1993) *Electrical Safety Engineering*, 3rd edn (Butterworth-Heinemann).

Cox, S. and Tait, R. (1998) *Safety, Reliability and Risk Management – An Integrated Approach* (Elsevier).

Cross, J. and Farrer, D. (1982) *Dust Explosions* (Plenum Press).

Crowl, D. A. (2003) *Understanding Explosions* (Wiley – AIChE).

Crowl, D. A. and Louvar, J. F. (2002) *Chemical Process Safety: Fundamentals with Applications*, 2nd edn (Prentice-Hall).

CSHIB (2005) Chemical Safety and Hazard Investigation Board preliminary findings on the BP Americas Texas City explosion. Oct. 27, available from www.chemsafety.gov.

Dodson, B. and Nolan, D. (1999) *Reliability Engineering Handbook* (Marcel Dekker).

Dow (1973) *Fire and Explosion Index Hazard Classification Guide*, 3rd edn (Dow Chemical Company).

Dow (1994) *Dow's Fire and Explosion Index Hazard Classification Guide* (American Institute of Chemical Engineers, New York).

Dugdale, D. (1985) *An Introduction to Fire Dynamics* (Wiley).

Duxbury, H. A. (1976) Loss Prevention No. 10 (AIChE) 147. Gas vent sizing methods.

Duxbury, H. A. (1979) Chem. Engr. London No. 350 (Nov.) 783. Relief line sizing for gases.

Eckhoff, R. K. (2003) *Dust Explosions* (Butterworth-Heinemann).

Fawcett, H. H. and Wood, W. S. (1982) *Safety and Accident Prevention in Chemical Operations* (Wiley).

Field, P. (1982) *Dust Explosions* (Elsevier).

Fisher, H. G., Forrest, H. S., Grossel, S. S., Huff, J. E., Muller, A. R., Noronha, J. A. Shaw, D. A., and Tilley, B. J. (1993) *Emergency Relief System Design Using DIERS Technology – The Design Institute for Emergency Relief Systems (DIERS) Project Manual* (Wiley – AIChE).

Frank, W. I. and Whittle, D. K. (2001) *Revalidating Process Hazard Analysis* (Wiley – AIChE).

Green, A. E. (ed.) (1982) *High Risk Technology* (Wiley).

Green, A. E. (ed.) (1983) *Safety System Reliablity* (Wiley).

Gugan, K. (1979) *Unconfined Vapour Cloud Explosions* (Gulf Publishing).

HMSO (1975) *The Flixborough Disaster, Report of the Court of Enquiry* (Stationery Office).

Howard, W. B. (1992) *Chem. Eng. Prog.* **88** (April) 69. Use precautions in selection, installation and operation of flame arresters.

Hyatt, N. (2003) *Guidelines for Process Hazard Analysis (PHA, Hazop), Hazard Identification and Risk Analysis* (CRC Press).

ICI (1993) *Mond Index: How to Identify, Assess and Minimise Potential Hazards on Chemical Plant Units for New and Existing Processes*, 2nd edn (ICI, Northwich).

Kales, P. (1997) *Reliability for Technology, Engineering and Management* (Prentice Hall).

King, R. and Hirst, R. (1998) *King's Safety in the Process Industries*, 2nd edn (Elsevier).

King, A. (2007) *The Chemical Engineer*, **790**, 46. Functional safety standards.

Kletz, T. A. (1977a) New Scientist (May 12th) 320. What risks should we run.

Kletz, T. A. (1977b) Hyd. Proc. 56 (May) 207. Evaluate risk in plant design.

Kletz, T. A. (1984) *Cheaper, Safer Plants or Wealth and Safety at Work* (Institution of Chemical Engineers, London).

Kletz, T. A. (1999a) *HAZOP and HAZAN*, 4th edn (Taylor and Francis).

Kletz, T. A. (1999b) *HAZOP and HAZAN: Identifying Process Industry Hazards* (Institution of Chemical Engineers, London).

Kletz, T. A. and Cheaper, T. A. (1998) *A Handbook for Inherently Safer Design*, 2nd edn (Taylor and Francis).

Lawley, H. G. (1974) Loss Prevention No. 8 (AIChE) 105. Operability Studies and hazard analysis.

Lewis, D. J. (1979a) AIChE Loss Prevention Symposium, Houston, April. The Mond fire, explosion and toxicity index: a development of the Dow index.

Lewis, D. J. (1979b) Loss Prevention No. 13 (AIChE) 20. The Mond fire, explosion and toxicity index applied to plant layout and spacing.

Lewis, R. J. (2004) *Sax's Dangerous Properties of Hazardous Materials*, 11th edn (Wiley).

Lowrance, W. W. (1976) *Of Acceptable Risk* (W Kaufmann, USA).

Macmillan, A. (1998) *Electrical Installations in Hazardous Areas* (Butterworth-Heinemann).

Mannan, S., ed. (2004) *Lees' Loss Prevention in the Process Industries*, 3rd edn, 2 vols (Butterworth-Heinemann).

Marshall, V. C. (1987) *Major Chemical Hazards* (Ellis Horwood).

Marshall, V. C. and Ruhemann, S. (2000) *Fundamentals of Process Safety* (Institution of Chemical Engineers, London).

Mathews, T. (1984) Chem. Engr., London No. 406 (Aug.-Sept.) 21. Bursting discs for over-pressure protection

Mendoza, V. A., Smolensky, V. G., and Straitz, J. F. (1998) *Hydrocarbon Proc.* 77(10), 63. Do your flame arrestors provide adequate protection?

Moore, A. (1984) Chem. Engr., London No. 407 (Oct.) 13. Pressure relieving systems.

Morley, P. G. (1989a) Chem. Engr., London No. 463 (Aug.) 21. Sizing pressure safety valves for gas duty.

Morley, P. G. (1989b) Chem. Engr., London No. 465 (Oct.) 47. Sizing pressure safety valves for flashing liquid duty.

Munday, G. (1976) *Chem. Engr.* London No. 308 (April) 278. Unconfined vapour explosions.

Murphy, G. (1993) *Processing* (Nov.) 6. Quiet life ends in burst of activity.

Napier, D. H. and Russell, D. A. (1974) *Proc. First Int. Sym.* on Loss Prevention (Elsevier). Hazard assessment and critical parameters relating to static electrification in the process industries.

Parkinson, J. S. (1979) *Inst. Chem. Eng. Sym.* Design 79, K1. Assessment of plant pressure relief systems.

Pitblado, R. M., Shaw, S. J., and Sevens, G. (1990) *Inst. Chem. Eng. Sym. Ser. No.* **120**, 51. The SAFETI risk assessment package and case study application.

Pratt, T. H. (1999) *Electrostatic Ignitions of Fires and Explosions* (AIChE).

Prugh, R. N. (1980) *Chem. Eng. Prog.* **76** (July) 59. Applications of fault tree analysis.

Ridley, J. (ed.) (2003) *Safety at Work* (Elsevier).

Rogowski, Z. W. (1980) *Inst. Chem. Eng. Sym. Ser. No.* 58, 53. Flame arresters in industry.

ROSPA (1971) *Liquid Flammable Gases: Storage and Handling* (Royal Society for the Prevention of Accidents, London).

Schmidt, M. (2007) *Chem. Eng. NY.* **114**(9), 69. Tolerable risk.

Stamatis, D. H. (1995) *Failure Mode and Effect Analysis: FMEA from Theory to Execution* (ASQC Quality Press).

Taylor, B. T. et al. (2000) *HAZOP: A Guide to Best Practice* (Institution of Chemical Engineers, London).

Towler, G. and Sinnott, R. (2008) *Chemical Engineering Design: Principles, Practice and Economics of Plant and Process Design* (Elsevier).

Wells, G. L. (1980) *Safety in Process Plant Design* (Institution of Chemical Engineers, London).

Wells, G. L. (1996) *Hazard Identification and Risk Assessment* (Institution of Chemical Engineers, London).

Wells, G. L. (1997) *Major Hazards and their Management* (Institution of Chemical Engineers, London).

Bibliography

Further reading on process safety:

Croner's Dangerous Goods Safety Advisor (Croner).

Fingas, M. (2002) (ed.) *Handbook of Hazardous Materials Spills and Technology* (McGraw-Hill).

Ghaival, S. (2004) (ed.) *Tolley's Health and Safety at Work Handbook* (Tolley publishing).

Johnson, R. W., Rudy, S. W., and Unwin, S. D. (2003) *Essential Practices for Managing Chemical Reactivity Hazards* (CCPS, American Institute of Chemical Engineers).

Martel, B. (2000) *Chemical Risk Analysis (English translation)* (Penton Press).

Redmilla, F., Chudleigh, M., and Catmur, J. (1999) *Systems Safety: HAZOP and Software HAZOP* (Wiley).

RSC (1991ff) *Dictionary of Substances and Their Effects*, 5 vols (Royal Society of Chemistry).

Smith, D. J. (2001) *Reliability, Maintainability and Risk – Practical Methods for Engineers*, 6th edn (Elsevier).

British and European Standards

BS EN 1127:2007 (2007) Explosive atmospheres. Explosion prevention and protection.

BS EN 1839:2003 (2003) Determination of explosion limits of gases and vapours.

BS EN ISO 4126-1:2004 (2004) Safety devices for protection against excessive pressure. Part 1: Safety valves.

BS EN ISO 4126-2:2004 (2004) Safety devices for protection against excessive pressure. Part 2: Bursting disc safety devices.

BS EN ISO 4126-3:2004 (2004) Safety devices for protection against excessive pressure. Part 3: Safety valves and bursting disc safety devices in combination.

BS EN ISO 4126-4:2004 (2004) Safety devices for protection against excessive pressure. Part 4: Pilot operated safety valves.

BS EN ISO 4126-5:2004 (2004) Safety devices for protection against excessive pressure. Part 5: Controlled safety pressure relief systems (CSPRS).

BS EN ISO 4126-6:2004 (2004) Safety devices for protection against excessive pressure. Part 6: Application, selection and installation of bursting disc safety devices.

BS EN ISO 4126-7:2004 (2004) Safety devices for protection against excessive pressure. Part 7: Common data.

BS EN 13478:2001 (2001) Safety of machinery – fire prevention and protection.

BS EN ISO 13702 (1999) Petroleum and natural gas industries – control and mitigation of fires and explosions on offshore production installations – requirements and guidelines.

BS EN 14373:2005 (2005) Explosion suppression systems.

BS EN 14460:2006 (2006) Explosion resistant equipment.

BS EN 14797:2006 (2006) Explosion venting devices.

BS EN 14994:2007 (2007) Gas explosion venting protective systems.

BS EN 50281 (1999) Electrical apparatus for use in the presence of combustible dust.

BS EN 60079-0:2006 (2006) Electrical apparatus for explosive gas atmospheres. General requirements.

BS EN 60079-1:2007 (2007) Explosive atmospheres. Equipment protection by flameproof enclosures.

BS EN 60079-2:2007 (2007) Explosive atmospheres. Equipment protection by pressurized enclosure.

BS EN 60079-10:2003 (2003) Electrical apparatus for explosive gas atmospheres. Classification of hazardous areas.

BS EN 60079-11:2007 (2007) Explosive atmospheres. Equipment protection by intrinsic safety.

BS EN 61241 (2005) Electrical apparatus with protection by enclosure for use in the presence of combustible dusts.

BS EN 61508 (2002) (7 parts) Functional safety of electrical/electronic/programmable electronic safety-related systems.

BS EN 61511 (2004) (3 parts) Functional safety: Safety instrumented systems for the process industry sector.

BS EN 62305:2006 (2006) Protection against lightning.

American and International Standards

ANSI/ISA – 84.00.01-2004 (2004) (3 parts) (IEC 61511 Mod) Functional safety: Safety instrumented systems for the process industry sector (American National Standards Institute/Instrumentation, Systems and Automation Society).

API Publication 2030 (1998) Application of water spray systems for fire protection in the petroleum industry, 2nd edn (American Petroleum Institute).

API Publication 2218 (1999) Fireproofing practices in petroleum and petrochemical processing plants, 2nd edn (American Petroleum Institute).

API Recommended Practice 14c (2001) Recommended practice for analysis, design, installation and testing of basic surface safety systems for offshore production platforms (American Petroleum Institute).

API Recommended Practice 500 (1997, R-2002) Recommended practice for classification of locations for electrical installations at petroleum facilities classified as Class I Division 1 and Division 2 (American Petroleum Institute).

API Recommended Practice 505 (1997) Recommended practice for classification of locations for electrical installations at petroleum facilities classified as Class I Zone 0, Zone 1 and Zone 2 (American Petroleum Institute).

API Recommended Practice 520 (2000) Sizing, selection, and installation of pressure-relieving devices in refineries, 7th edn (American Petroleum Institute).

API Recommended Practice 521 (1997) Guide for pressure-relieving and depressuring systems, 4th edn (American Petroleum Institute).

API Recommended Practice 2001 (2005) Fire protection in refineries, 8th edn (American Petroleum Institute).

API Recommended Practice 2003 (1998) Protection against ignitions arising out of static, lightning and stray currents, 6th edn (American Petroleum Institute).

API Recommended Practice 2210 (2000) Flame arrestors for vents of tanks storing petroleum products, 3rd edn (American Petroleum Institute).

API Standard 526 (2002) Flanged steel pressure relief valves, 5th edn (American Petroleum Institute).

API Standard 527 (1991) Seal tightness of pressure relief valves, 3rd edn (American Petroleum Institute).

ASME Boiler and Pressure Vessel Code Section VIII (2004) Rules for the construction of pressure vessels (ASME International).

ASME PTC 25-2001 Pressure relief devices – performance test codes (ASME International).

ASTM D92 (2005) Standard test method for fire and flash-points by Cleveland open cup tester (ASTM International).

ASTM D93 (2002) Standard test methods for flash-point by Pensky-Martens closed cup tester (ASTM International).

IEC 61508 (2000) see BS EN 61508.

IEC 61511 (2003) see BS EN 61511.

ISO 16852 (2008) Flame arrestors – performance requirements, test methods and limits for use, 1st edn (International Organization for Standardization).

ISO 17776 (2000) Petroleum and natural gas industries – offshore production installations: guidelines on tools and techniques for hazard identification and risk assessment.

NFPA 30 (2003) Flammable and combustible liquids code (National Fire Protection Association).

NFPA 49 (1994) Hazardous chemicals data (National Fire Protection Association).

NFPA 30 (2003) Flammable and combustible liquids code (National Fire Protection Association).

NFPA 61 (2007) Standard for the prevention of fires and dust explosions in agricultural and food processing facilities – 2008 edition (National Fire Protection Association).

NFPA 68 (2006) Standard on explosion protection by deflagration venting – 2007 edition (National Fire Protection Association).

NFPA 69 (2007) Standard on explosion prevention systems – 2008 edition (National Fire Protection Association).

NFPA 70 (2006) National electrical code (National Fire Protection Association).

NFPA 77 (2000) Recommended practice on static electricity (National Fire Protection Association).

NFPA 491 (1997) Guide to hazardous chemical reactions (National Fire Protection Association).

NFPA 495 (2006) Explosive materials code (National Fire Protection Association).

NFPA 496 (2003) Standard for purged and pressurized enclosures for electrical equipment (National Fire Protection Association).

NFPA 497 (2004) Recommended practice for the classification of flammable liquids, gases or vapours and of hazardous (classified) locations for electrical installations in chemical process areas (National Fire Protection Association).

NFPA 654 (2006) Standard for the prevention of fire and dust explosions from the manufacturing, processing and handling of combustible particulate solids (National Fire Protection Association).

NFPA 655 (2006) Standard for the prevention of sulphur fires and explosions (National Fire Protection Association).

NFPA 664 (2006) Standard for the prevention of fires and dust explosions in wood processing and woodworking facilities – 2007 edn (National Fire Protection Association).

NFPA 750 (2006) Standard on water mist fire protection systems (National Fire Protection Association).

OSHA Standard 29 CFR 1910.119 (2008) Process safety management of highly hazardous chemicals.

OSHA Standard 29 CFR 1910.307 (2008) Subpart S. Electrical.

OSHA Standard 29 CFR 1910.1200 (2008) Hazard communication.

9.11. NOMENCLATURE

		Dimensions in **MLT**
A_d	Discharge area	L^2
A_w	Internal wetted surface area	L^2
C	Impact of a hazard (impact per loss event) in equation 9.1	
C	Constant in equation 9.19	—
C_p	Specific heat capacity at constant pressure	$L^2T^{-2}\theta^{-1}$
C_v	Specific heat capacity at constant volume	$L^2T^{-2}\theta^{-1}$
D	Diameter	L
F	Hazard rate (frequency of a hazard in events per year)	T^{-1}
F_e	Environmental factor for external fire	—
F_{np}	Inherent frequency of a hazard with no protective system (number of events per year)	T^{-1}
F_t	Tolerable frequency of a hazard (number of events per year)	T^{-1}
F_3	Unit hazard factor in Dow Fire & Explosion Index	—
K_b	Back-pressure correction factor	—
K_c	Combination correction factor for a relief valve	—

K_d	Coefficient of discharge for a relief valve	—
L	Length of vessel	L
M_w	Molecular weight	—
N_r	Reactivity in the Dow Fire & Explosion Index	—
n	Number of moles of gas	M
P	Pressure	$ML^{-1}T^{-2}$
P_{cf}	Critical flow pressure	$ML^{-1}T^{-2}$
P_m	Maximum allowable working pressure	$ML^{-1}T^{-2}$
P_1	Upstream pressure	$ML^{-1}T^{-2}$
PFD_{av}	Average probability of failure on demand of a protective system	—
Q	Heat input due to fire	ML^2T^{-3}
R	Ideal gas constant	$L^2T^{-2}\theta^{-1}$
R_{np}	Inherent risk of a hazard with no protective system in place	
R_t	Tolerable risk, also sometimes called the acceptable risk	
T	Temperature	θ
t	Time	T
V	Vessel volume	L^3
V_G	Volume occupied by vapour	L^3
V_L	Volume occupied by liquid	L^3
v	Liquid volumetric flow rate	L^3T^{-1}
w	Relieving mass flow rate	MT^{-1}
w_f	Fire case relieving mass flow rate	MT^{-1}
Z	Vapour compressibility	—
γ	Ratio of specific heat capacities	—
ΔH_{vap}	Heat of vaporization	L^2T^{-2}
ΔR	Risk reduction factor, defined in equation 9.3	
δ	Demand rate (number of occasions per year that a safety system is actuated)	T^{-1}
ϕ	Fractional dead time (probability that a safety system is inactive)	—
λ	Failure rate (number of occasions per year that a safety system develops a fault)	T^{-1}
τ	Test interval	T

9.12. PROBLEMS

9.1. In the storage of flammable liquids, if the composition of the vapour–air mixture above the liquid surface falls within the flammability limits, a floating roof tank would be used or the tank blanketed with inert gas. Check if the vapour composition for the liquids listed below will fall within their flammability range, at atmospheric pressure and 25°C.
1. Toluene
2. Acrylonitrile
3. Nitrobenzene
4. Acetone.

9.2. Complete a failure mode effect analysis for the nitric acid plant reactor section described in Example 9.2 (This is best carried out as a group activity with a group size of three to six.)

9.3. Estimate the Dow Fire and Explosion Index, and determine the hazard rating, for the processes listed below.

Use the process descriptions given in Appendix F and develop the designs, as needed, to estimate the index.
1. Ethylhexanol from propylene and synthesis gas, F.1
2. Chlorobenzenes from benzene and chlorine, F.2
3. Methyl ethyl ketone from 2-butanol, F.3
4. Acrylonitrile from propylene and ammonia, F.4
5. Aniline from nitrobenzene and hydrogen. F.8.

9.4. Devise a preliminary control scheme for the sections of the nitric acid plant described in Chapter 4, flow-sheet Figure 4.2, which are listed below. Make a practice HAZOP study of each section and revise your preliminary control scheme.
1. Waste heat boiler (WHB)
2. Condenser
3. Absorption column.

9.5. A distillation column separates benzene from toluene using a control scheme similar to that shown in Figure 5.28b. Make a practice HAZOP study of the plant section and add any instrumentation that is needed to develop the full P and I diagram.

9.6. Develop a fault tree for events that could lead to release of benzene to the atmosphere for the distillation section designed in problem 9.5.

9.7. List the materials incompatibility problems you should be aware of in plants that handle:

1. hydrogen peroxide
2. ethylene oxide
3. chlorine
4. ammonium nitrate
5. styrene.

9.8. List possible relief scenarios for the vessel designed in problem 13.3.

9.9. Estimate the relieving load for the reactor designed in problem 13.7 for a blocked outlet scenario, and size the relief valve for this case.

9.10. A toluene surge drum has capacity 500 gal and is normally operated 60% full at 100°F under 300 psig of hydrogen in the head space using a level controlled outflow. The normal flow rate into the vessel is 30,000 lb/h. Determine the vessel dimensions if the vessel is vertically mounted. Evaluate the relief loads for the blocked outflow and external fire cases and hence determine the relief valve size.

10 EQUIPMENT SELECTION, SPECIFICATION AND DESIGN

Chapter Contents

Key Learning Objectives

- How to separate mixtures of gases
- How to size and design equipment for sizing, handling, transporting, mixing, separating and recovering solids
- How to design equipment for liquid–liquid and liquid–vapour contacting
- How to design mixers and reactors

10.1.   INTRODUCTION

The first chapters of this book covered process design: the synthesis of the complete process as an assembly of units, each carrying out a specific process operation. In this and the following chapters, the selection, specification and design of the equipment required to carry out the function of these process units (unit operations) is considered in more detail. The equipment used in the chemical process industries can be divided into two classes: proprietary and non-proprietary. Proprietary equipment, such as pumps, compressors, filters, centrifuges and dryers, is designed and manufactured by specialist firms. Non-proprietary equipment is designed as special, one-off, items for particular processes; for example, reactors, distillation columns and heat exchangers.

Unless employed by one of the specialist equipment manufacturers, the chemical engineer is not normally involved in the detailed design of proprietary equipment. The chemical engineer's job will be to select and specify the equipment needed for a particular duty; consulting with the vendors to ensure that the equipment supplied is suitable. Chemical engineers may be involved with the vendor's designers in modifying standard equipment for particular applications; for example, a standard tunnel dryer designed to handle particulate solids may be adapted to dry synthetic fibres. As was pointed out in Chapter 1, the use of standard equipment, whenever possible, will reduce costs.

Reactors, columns and other vessels are usually designed as special items for a given project. In particular, reactor designs are usually unique, except where more or less standard equipment is used; such as an agitated, jacketed, vessel. Distillation columns, vessels and tubular heat exchangers, though non-proprietary items, will be designed to conform to recognized standards and codes; this reduces the amount of design work involved.

The chemical engineer's part in the design of 'non-proprietary' equipment is usually limited to selecting and 'sizing' the equipment. For example, in the design of a distillation column the design engineer will typically determine the number of plates; type and design of plate; diameter of the column; and the position of the inlet, outlet and instrument nozzles. This information would then be transmitted, in the form of sketches and specification sheets, to the specialist mechanical design group, or the fabricator's design team, for detailed design.

In this chapter the emphasis is on equipment selection, rather than equipment design; as most of the equipment described is proprietary equipment. Design methods are given for some miscellaneous non-proprietary items. The main techniques used to separate phases, and the components within phases, are listed in Table 10.1 and discussed in Sections 10.2 to 10.9. Size reduction and enlargement, mixing and transport and storage of materials are discussed in Sections 10.10 to 10.12. A brief discussion of reactor design is included in Section 10.13. The design of two important classes of equipment, columns and heat exchangers, is covered separately in Chapters 11 and 12.

A great variety of equipment is used in the process industries, and it is only possible to give very brief descriptions of the main types in this book. Further details are given in Richardson *et al.* (2002) and McCabe *et al.* (2001). Descriptions and illustrations

TABLE 10.1. Separation Processes

		MINOR COMPONENT				
		Solid		**Liquid**		**Gas/Vapour**	

MAJOR COMPONENT		**Solid**		**Liquid**		**Gas/Vapour**	
Solid	Sorting	10.3	Pressing	10.4.5	Crushing	10.10
	Screening	10.3.1	Drying	10.4.6	Heating	—
	Hydrocyclones	10.3.2				
	Classifiers	10.3.3				
	Jigs	10.3.4				
	Tables	10.3.5				
	Centrifuges	10.3.6				
	Dense media	10.3.7				
	Flotation	10.3.8				
	Magnetic	10.3.9				
	Electrostatic	10.3.10				
Liquid	Thickeners	10.4.1	Decanters	10.6.1	Stripping	10.2.4
	Clarifiers	10.4.1	Coalescers	10.6.3		11.14
	Hydrocyclones	10.4.4	Solvent extraction	10.7.1		
	Filtration	10.4.2	Leaching	10.7.1		
	Centrifuges	10.4.3	Chromatography	10.7.2		
	Crystallizers	10.5.2	Distillation	Chapter 11		
	Evaporators	10.5.1				
	Precipitation	10.5.3				
	Membranes	10.5.4				
	Reverse osmosis	10.5.4				
	Ion exchange	10.5.5				
	Adsorption	10.5.6				
Gas/Vapour	Gravity Settlers	10.8.1	Separating Vessels	10.9	Adsorption	10.2.1
	Impingement Settlers	10.8.2	Demisting pads	10.9	Absorption	10.2.4
	Cyclones	10.8.3	Cyclones	10.8.3		11.14
	Filters	10.8.4	Wet scrubbers	10.8.5	Cryogenic	10.2.3
	Wet scrubbers	10.8.5	Electrostatic precipitators	10.8.6	distillation	Chapter 11
	Electrostatic precipitators	10.8.6			Membranes	10.2.2

Notes: Numbers refer to the sections in this chapter. The terms major and minor component only apply where different phases are to be separated, i.e., not to those on the diagonal.

of most of the equipment used can be found in various handbooks: Green and Perry (2007), Schweitzer (1997) and Walas (1990). Equipment manufacturers' advertisements in the technical press and on-line should also be studied. It is worthwhile building up a personal file of vendors' catalogues to supplement those that may be held in a firm's library. In the UK, a commercial organization, Technical Indexes Ltd, publishes the Process Engineering Index, which contains information from over 3000 manufacturers and suppliers of process equipment. Manufacturers' web sites are usually easily located using on-line search engines and often provide details of equipment construction, standard sizes, available metallurgies, specification sheets and performance information.

The scientific principles and theory that underlie the design and operation of processing equipment are covered in numerous textbooks such as Richardson *et al.* (2002), McCabe *et al.* (2001) and others cited in the relevant sections below.

10.2. GAS–GAS SEPARATIONS

Separation of gaseous or vapour components from the gas phase is important in many industrial processes:

1. Preparation of high-purity industrial gases such as oxygen, nitrogen, argon, neon, hydrogen, etc.
2. Purification of natural gas to meet specifications for pipelines and liquefaction plants.
3. Treatment of process gases to prevent accumulation of contaminants in recycles or to protect catalysts.
4. Removal of pollutants from vent gases to meet legislative requirements.
5. Drying and purification of air for use in sterile dryers, clean rooms, as instrument air, etc.

Bulk removal of condensable vapour components from a gas can usually be accomplished by cooling and condensation; see Section 12.10. However, even refrigerated condensers seldom give adequate recovery of the vapour and additional separation is usually needed to meet product gas specifications.

10.2.1. Adsorption

Adsorption is probably the most widely used method in gas separation. Adsorption can be expensive for bulk separation of large quantities of gas, but it has the advantage of allowing very high gas purities or high recoveries of contaminants to be achieved.

The basic principle of adsorption is to exploit differences in the strength of interaction between different species in the gas phase and the surface of a solid known as a sorbent or adsorbent. If one species is more strongly adsorbed, then it accumulates on the sorbent as the gas passes through a bed of sorbent, and hence is removed from the gas phase. If the gas flows down through the sorbent bed, as shown in Figure 10.1a, then it continues to contact fresh sorbent and all of the adsorbed component is removed, giving a purified gas. As the sorbent accumulates the adsorbed component, it eventually becomes saturated and can adsorb no more. A concentration profile is thus established in the sorbent bed (Figure 10.1b). The concentration profile moves down through the bed until it reaches the bottom, at which point 'breakthrough' of the adsorbed component occurs and the concentration in the exit gas begins to rise. Breakthrough occurs at the time labelled t_B in Figure 10.1. The thermodynamics and kinetics of adsorption that lead to the formation of the concentration profiles shown in Figure 10.1b are discussed in detail in the books by Ruthven (1984), Yang (1997), Richardson *et al.* (2002) and Ruthven *et al.* (1993). Adsorption is also covered in the books by Suziki (1990) and Crittenden and Thomas (1998).

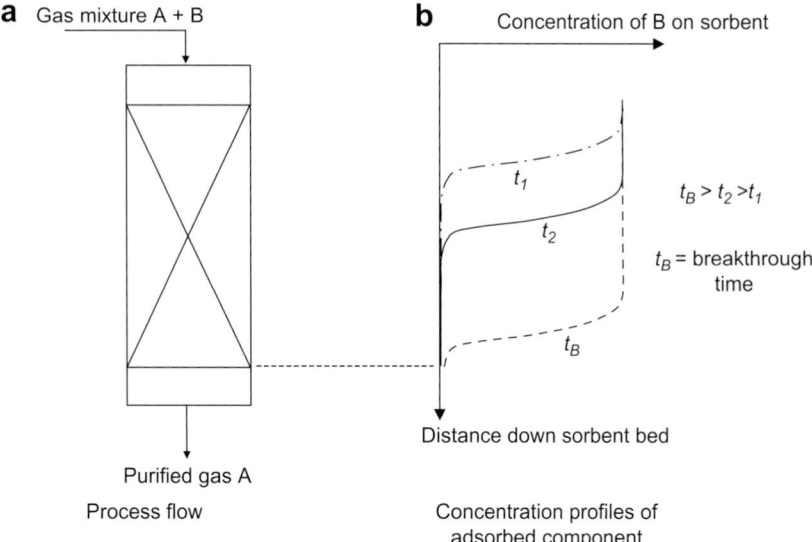

a Gas mixture A + B

Purified gas A

Process flow

b Concentration of B on sorbent

t_1

t_2

t_B

$t_B > t_2 > t_1$

t_B = breakthrough time

Distance down sorbent bed

Concentration profiles of adsorbed component

Figure 10.1.   Adsorption.

Irreversible Adsorption

When adsorption is used to remove a low level of a contaminant from a gas stream, a sorbent can be selected that reacts irreversibly with the contaminant. In this case, the process is termed irreversible adsorption and the adsorption vessel is sometimes referred to as a guard bed. An example of irreversible adsorption is the use of zinc oxide to remove trace amounts of hydrogen sulphide from natural gas and petrochemical processes.

When a bed of irreversible sorbent becomes saturated it must be replaced. A common arrangement is to use two beds in parallel, with one in service and the other isolated to allow for sorbent change-out (Figure 10.2). This scheme is simple, but can be wasteful of sorbent, as the bed must usually be replaced before breakthrough occurs.

An alternative is to use a 'lead-lag' arrangement (Figure 10.3). In this flow scheme, the gas normally flows through both beds in series, as illustrated in the figure. When

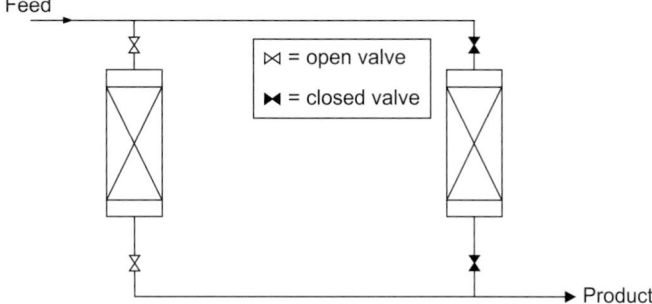

Feed

⋈ = open valve

▶◀ = closed valve

Product

Figure 10.2.   Parallel flow through guard beds.

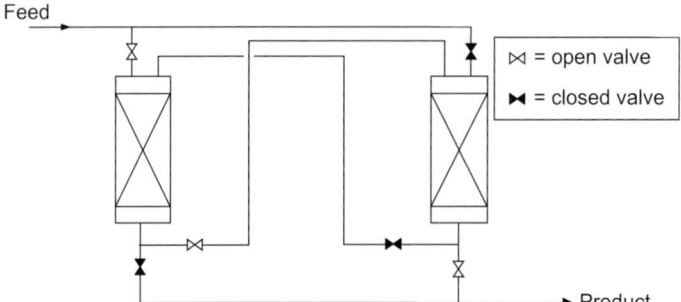

Figure 10.3.   Lead-lag guard beds.

the first (lead) bed becomes saturated and breakthrough occurs then the second (lag) bed removes the contaminant, so there is no breakthrough to the product gas. The first bed is then removed from service for sorbent change-out while the feed gas flows through the second bed alone. The first bed is then placed back into service and becomes the lag bed, while the second bed now operates as lead bed. This arrangement allows the sorbent to be completely used up and reduces spent sorbent waste disposal costs.

The sorbents for irreversible adsorption are usually cheap, but the process can only be justified when the amount of contaminant removed is small, because the inconvenience and labour costs of changing out sorbent and the costs of waste disposal must also be taken into account.

Reversible Adsorption

In most cases, irreversible adsorption is not practical, as the adsorbed component either has a high flow rate or is desired as a product. When the bed reaches or nears breakthrough is must be taken out of process service so that the sorbent can be regenerated and the adsorbed components recovered. Multiple beds are usually used, with an arrangement of isolation valves that allows the beds to be sequenced so that a fresh bed of sorbent is brought into service whenever a bed is switched to regeneration.

Regeneration of the sorbent is usually carried out by increasing temperature or reducing pressure to give a lower equilibrium concentration of the adsorbed component on the solid surface. The equilibrium between the partial pressure of a component in the vapour phase and the adsorbent loading, expressed as grams of adsorbed material per gram of adsorbent, can be plotted at constant temperature as an isotherm; see Figure 10.4. Various expressions can be used to model isotherms, but for design purposes they can usually be fitted with an equation of the form:

$$m = k\,p^n \tag{10.1}$$

where
  m = adsorbent loading, g/g adsorbent
  k = equilibrium constant
  p = partial pressure of adsorbed component
  n = exponent.

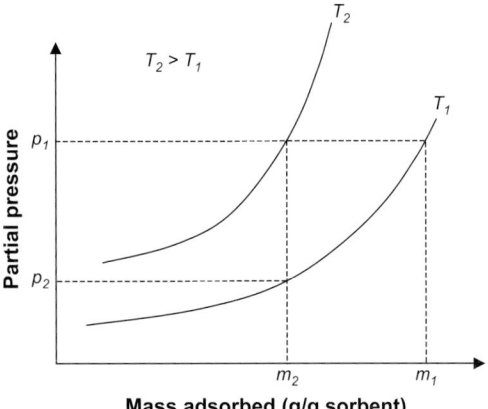

Figure 10.4. Adsorption isotherms.

Figure 10.4 shows isotherms at two temperatures, T_2 and T_1, where T_2 is greater than T_1. It can be seen from the figure that a change in adsorbent loading from m_1 to m_2 could be accomplished either by decreasing the partial pressure from p_1 to p_2 at constant temperature T_1, or by increasing temperature from T_1 to T_2 at constant pressure p_1. Cycling the bed over a range of temperature or pressure thus allows material to be adsorbed under process conditions and then recovered from the bed under regeneration conditions.

Pressure Swing Adsorption

When the bed is regenerated by reducing pressure, the process is known as pressure swing adsorption, or PSA. Pressure swing adsorption processes are nominally isothermal, although heat of adsorption and desorption effects must be allowed for in the design. Because the process is not limited by heat-transfer rates, very rapid cycles as short as a few minutes can be used. Rapid cycling gives efficient use of adsorbent and leads to smaller vessel sizes and lower capital cost.

If the process feed is under high pressure, then carrying out the regeneration at or near atmospheric pressure may be adequate. Alternatively, a vacuum pump can be used to generate a lower pressure for regeneration, in which case the process is known as vacuum swing adsorption (VSA or VPSA).

A small flow of sweep gas is usually passed through the bed during regeneration to promote desorption and clear away the desorbed components. The sweep gas is typically a slip stream from the purified product, although other gases are occasionally used. When the product gas is used as sweep gas, very high purity product gas can be obtained, but the recovery of the product gas is typically less than 95%.

Typical industrial flow-sheets for PSA use 4 to 12 adsorbent vessels. With multiple vessels, the designer can sequence the switching valves to compensate for heating and cooling effects due to heat of adsorption and desorption, and also to economize on losses of product gas during depressurization, regeneration and repressurization. Examples of typical commercial configurations are given by Ruthven *et al.* (1993), Cassidy and Doshi (1984) and Kumar *et al.* (1994).

Pressure swing adsorption is used in many processes; for example, air separation, hydrogen manufacture by steam reforming and dehydration of ethanol–water azeotropic mixtures (see Section 11.6.5).

Temperature Swing Adsorption

Regeneration of the adsorbent by raising temperature is known as temperature swing adsorption (TSA). In TSA processes, heat is required to raise the temperature of the adsorbent, the vessel and piping, and to provide the heat of desorption. This heat is usually provided by flowing a hot stripping gas over the adsorbent. The stripping gas is typically steam, dry air, nitrogen or a slip stream of purified product. The stripping gas must be chosen carefully to ensure that flammable atmospheres are not formed and the adsorbent is not deactivated. The stripping gas is usually heated in a fired heater, but electric or steam heaters are used in smaller plants.

After desorption is completed, the bed is usually flushed with cold purified product to bring the temperature back down to the adsorption temperature. Temperature swing processes typically operate on longer time cycles than pressure swing adsorption, as the cycle time is usually governed by the maximum attainable rates of heating and cooling.

Temperature swing adsorption is widely used for drying air and for removing trace amounts of organic compounds from vent gases using activated carbon as adsorbent. The activated carbon can then be regenerated with steam.

Adsorbent Selection

Adsorbents for a reversible process such as PSA or TSA are chosen to give a wide range of adsorbent loading over an acceptable range of either temperature or pressure. Many adsorbents are proprietary materials that have been designed to have high specific surface area and hence give high loadings; however, isotherm data for common adsorbents such as activated carbon, alumina, silica gel and some of the more common zeolites are available in the open literature and in books such as Breck (1974), Ruthven (1984) and Yang, R.T. (2003).

Care must be taken to ensure that the adsorbent is not irreversibly poisoned by any component of the gas phase. When multiple contaminants must be removed from a gas, then several layers of different adsorbents with selectivity for different contaminants can be loaded in the same vessel so that each contaminant is removed in turn.

Prices for some of the more common adsorbents are listed in Aspen ICARUS and other costing programs. Current prices for commodity adsorbents such as activated carbon, silica gel and alumina can be found online or by contacting manufacturers.

Adsorption Equipment Design

Adsorption plants are often purchased from an industrial gas company as packaged modular plants that include the adsorbent, vessels and valve skids. For TSA plants, the heater may also be included in the packaged plant scope. The design and optimization of PSA and TSA processes require detailed understanding of the adsorbent properties and heat of adsorption effects and are best left to specialist suppliers. The simplified method that follows is suitable for undergraduate design projects and for generating preliminary estimates.

When a plant is designed from scratch, the first step is to estimate the amount of adsorbent required. The amount of adsorbent can be determined from the flow rate of the adsorbed species and the change in the bed loading during the cycle. A mass balance on the bed gives:

$$(F_1 y_1 - F_2 y_2) M_w t_a = 1000 (m_1 - m_2) M_a f_L \qquad (10.2)$$

where

F_1 = feed molar flow rate (mol/s)
F_2 = product molar flow rate (mol/s)
y_1 = feed mole fraction of adsorbed component
y_2 = product mole fraction of adsorbed component
M_w = molecular weight of adsorbed component (g/mol)
t_a = time the bed is in the adsorption stage of the cycle (s)
m_1 = maximum adsorbent loading (g/g adsorbent)
m_2 = minimum adsorbent loading (g/g adsorbent)
M_a = mass of adsorbent per bed (kg)
f_L = fraction of bed that is fully loaded at end of adsorption phase of cycle.

The fraction of the adsorbent bed, f_L, that reaches loading m_1 at the end of the adsorption step of the cycle depends on the process arrangement and the number of beds used. For a simple two-bed system it will usually be less than 0.7 unless a very sharp front is formed or a lead-lag arrangement can be used. With four or more beds, the cycle can usually be designed to give fractional loading close to 1.0 unless multiple components are being removed.

The time that the bed spends in the adsorption phase of the cycle is less than the total cycle time, which also includes time for regeneration, depressurizing, heating, cooling, etc. The rates of these processes are governed by intrinsic rates of desorption, mass transfer and heat transfer. The development of rate-based models of these processes is discussed by Ruthven (1984) and Richardson *et al.* (2002). For a preliminary analysis, a time in the range 10 to 60 min can be assumed for PSA and 60 to 200 min for TSA. The total cycle time is equal to the time spent in adsorption multiplied by the number of beds in the sequence.

The volume of each adsorbent bed can be estimated from the mass of adsorbent and the adsorbent bulk density. Fixed beds of adsorbent are usually used, to give a sharp adsorption concentration profile. The adsorption vessel can then be sized as a cylinder that contains the adsorbent volume. The head space is usually left empty and about 20% of the volume between the tangent lines of the vessel is packed with inert material to ensure that a uniform flow profile is established at the entry and exit of the bed and to prevent 'fingering' of contaminant through the bed (Figure 10.5). The aspect ratio of the bed is usually at least 3:1 to ensure that high fractional loading is obtained and the adsorbent is used efficiently.

The total plant cost will include at least two adsorbent vessels, the set of switching valves and any other equipment, such as blowers, vacuum pumps or heaters, that is required for regeneration.

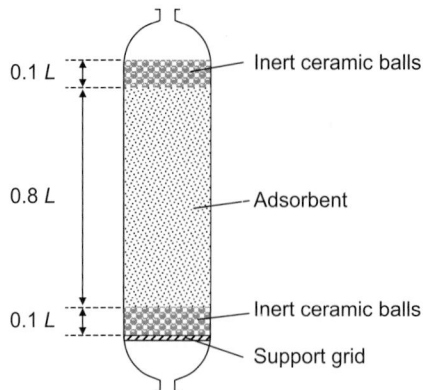

Figure 10.5.   Adsorption vessel internals.

Example 10.1

Yang *et al.* (1997) give isotherm data for adsorption of methane on zeolite 5A in the presence of hydrogen:

Pressure (atm)	0	1.5	2.5	4.5	6.5	9	12.5	16.5	19.5
Adsorbed amount (mmol/g)	0.0	0.4	0.7	1.0	1.25	1.45	1.6	1.7	1.75

It has been determined that a four-bed PSA plant using this adsorbent can be used to purify a stream containing 40% CH_4 in H_2 to give 90% recovery of hydrogen at purity 99.99 mol%. The adsorption time is 300 s and the beds are 85% loaded at the end of the adsorption step. If the feed gas is available at 25 atm and the off-gas can be discharged at 2 atm, then estimate the amount of adsorbent per bed and the dimensions of each adsorbent vessel for a feed rate of 1000 kmol/h.

Data (Yang *et al.*, 1997): Adsorbent bulk density = 795 kg/m^3.

Solution

Feed flow rate = 1000 kmol/h

$$\text{hydrogen in feed} = 1000 \times 0.60 = 600 \text{ kmol/h}$$

$$\text{methane in feed} = 1000 \times 0.40 = 400 \text{ kmol/h}, \; y_{CH4,feed} = 0.4$$

90% recovery at 99.99% purity:

$$\text{hydrogen in product gas} = 600 \times 0.9 = 540 \text{ kmol/h}$$

$$\text{methane in product gas} = \frac{(1 - 0.9999)}{0.9999} \times 540 = 5.4 \times 10^{-2} \text{ kmol/h}$$

$$\text{hydrogen in off-gas} = 600 - 540 = 60 \text{ kmol/h}$$

$$\text{methane in off-gas} \approx 400 \text{ kmol/h}$$

So, methane mole fraction in off-gas, $y_{CH4, \text{off-gas}} = \dfrac{400}{460} = 0.870$

Partial pressures of methane:

$$\text{in feed at 25 atm} = 25 \times 0.4 = 10 \text{ atm}$$

$$\text{in off-gas at 2 atm} = 2 \times 0.87 = 1.74 \text{ atm}$$

By fitting an isotherm through the data, plotting or interpolating, we find:

$$\text{at 10 atm, adsorbent loading} = 1.5 \text{ mmol/g}$$

$$\text{at 1.74 atm, adsorbent loading} = 0.5 \text{ mmol/g}$$

Substituting into equation 10.2 (noting that we don't need the molecular weight, as the adsorbent loadings were given in mmol/g = mol/kg):

$$\{(1000 \times 0.4) - (540 \times 0.0001)\} \times \frac{1000}{3600} \times 300$$

$$= (1.5 - 0.5) \times M_a \times 0.85 \quad (\text{mol/s})$$

Hence

$$\text{mass of adsorbent } M_a = \frac{400 \times 300}{3.6 \times 0.85 \times 1.0} = \underline{\underline{39,200 \text{ kg.}}}$$

$$\text{volume of adsorbent bed} = \frac{39,200}{795} = 49 \text{ m}^3.$$

Allow 20% of total volume for inert packing for good flow distribution, so tangent-to-tangent volume = 49/0.8 = 61.7 m³.

Assume 4:1 cylinder,

$$\text{volume} = \frac{\pi D_T^2 L_v}{4} = \pi D_T^3$$

where

D_T = vessel diameter
L_v = tangent-to-tangent length.

So, $D_T^3 = 61.7/\pi$, vessel diameter = 2.70 m.

Round up to nearest standard head size (see Chapter 13):

diameter = <u>2.74 m</u> (9 ft)
tangent length = <u>11.0 m</u> (36 ft).

We now have sufficient information to complete the pressure vessel design and costing (see Chapters 13 and 6). Note that four vessels will be needed, and the total quantity of adsorbent is 4 × 39,200 kg = 156.8 kg.

10.2.2. Membrane Separation

A membrane is a thin layer of material that allows species to pass through by permeation. The material that flows through the membrane is known as the permeate, while the material that does not is called the retentate. If some species pass though the membrane faster than others, then these species will accumulate in the permeate and the membrane can be used to separate these components from the gas mixture.

The flux through a membrane is defined as the flow rate through a unit area of membrane per unit time. The flux is proportional to the applied partial pressure gradient:

$$M_i = \frac{P_i}{\delta} (p_{i,f} - p_{i,p}) \qquad (10.3)$$

where

   M_i = molar flux of component i (mol/m^2.s)
   P_i = permeability of membrane for component i (mol/m.s.bar)
   δ = membrane thickness (m)
   $p_{i,f}$ = local partial pressure of component i on feed side (bar)
   $p_{i,p}$ = local partial pressure of component i on permeate side (bar).

The average flux across a long cylindrical membrane is given by:

$$M_{i,ave} = \frac{\int_0^{L_m} M_i \ dx}{L_m} \qquad (10.4)$$

where

   $M_{i,ave}$ = average molar flux of component i (mol/m^2.s)
   L_m = length of membrane (m)
   x = length (m).

If the feed or permeate partial pressure varies substantially along the membrane length, then an expression for the variation of partial pressure with length must be substituted into equation 10.4 before the average flux can be found.

The ratio of permeability of two components is called the membrane selectivity, or ideal separation factor:

$$S_{ij} = \frac{P_i}{P_j} \qquad (10.5)$$

where S_{ij} = selectivity of the membrane for component i over component j.

If the selectivity of the membrane for a component is very high then that component can be recovered at high purity in the permeate.

At the retentate end of the membrane, the partial pressure of the permeating component is reduced; consequently the driving force for permeation is lower and the flux is lower. If the partial pressure for the permeating species becomes too low, then the flux becomes very low and an uneconomically large membrane area is required. Because of this effect, membrane processes for gas separation usually cannot achieve high purity in the retentate or high recovery of permeate. Techniques for maintaining an adequate flux at the retentate end are discussed below.

Membrane Selection and Construction

Membranes are described as 'microporous' if they have an open pore structure and 'dense' if there are no holes or pores. Microporous membranes perform essentially

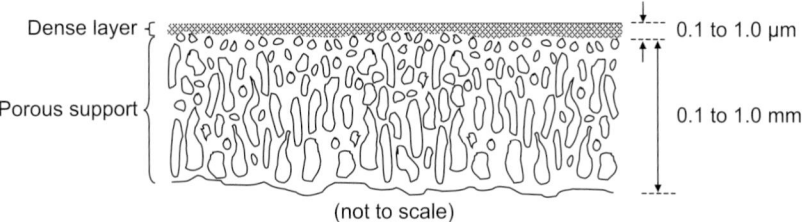

Figure 10.6. Asymmetric membrane structure.

like sieves and allow small molecules to move through the pores while large molecules are blocked. The separation of $^{235}UF_6$ from $^{238}UF_6$ in uranium enrichment is the only commercial gas separation process that uses a microporous membrane. Most gas separations use 'asymmetric' membranes, consisting of a very thin dense layer supported on a thick porous layer that gives strength to the membrane; see Figure 10.6.

Gas separation membranes are usually made from elastomeric or glassy polymers, although very thin layers of metal or ceramic materials are used in some special applications. The use of polymers restricts membrane processes to operating at low temperatures; typically below 100°C. For a dense membrane, the mechanism of permeation is for the gas to dissolve in the membrane on the feed side and then evaporate from the membrane on the permeate side. The membrane permeability and selectivity therefore depend on the solubility of different species in the membrane. With careful design and selection of membrane material, high selectivities can be obtained.

In addition to high selectivity, it is desirable to have high membrane permeability, as this will determine the area required, and hence the membrane cost. Table 10.2 gives permeability coefficients for some membrane materials. Other important parameters in membrane selection include mechanical stability, chemical resistance, thermal stability, ease of manufacture and cost.

Gas separation membranes are usually manufactured as hollow fibres or else cast as flat sheets that are then spiral wound into modular assemblies. Details of the manufacturing methods are given by Rautenbach and Albrecht (1989) and Scott and Hughes (1996). Hollow-fibre membranes are glued ('potted') into a resin to form

TABLE 10.2. Membrane Permeability Coefficients $(cm^3(STP).cm/cm^2.s.cmHg) \times 10^{10}$

Membrane	Temperature (°C)	CO$_2$	O$_2$	N$_2$
Natural rubber	25	99.6	17.7	6.12
Ethyl cellulose	25	113	15	3.0
Polystyrene	20	10.0	2.01	0.32
Polycarbonate	25	8.0	1.4	0.3
Poly(dimethylsiloxane)	25	3240	605	300
Poly(ethylene phthalate)	25	0.15	0.03	0.006
Poly(vinyl alcohol)	20	0.0005	0.00052	0.00045

Taken from Osada and Nakagawa (1992).

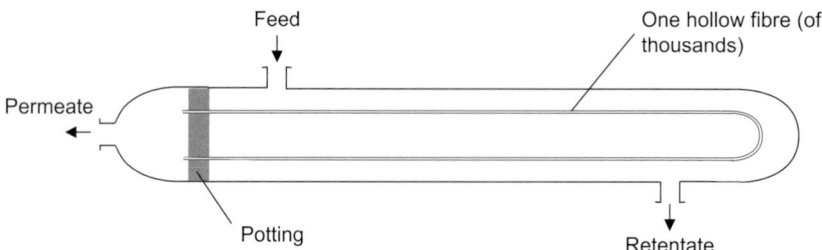

Figure 10.7. Hollow-fibre membrane module.

a closure similar to the tube-sheet of a shell and tube exchanger (Figure 10.7). In a spiral-wound membrane, two membrane sheets are placed back to back and sealed along three edges to form an envelope. The fourth edge is attached to a perforated tube, from which the permeate can be withdrawn. Several membrane envelopes can be attached to one tube, with net-like spacers between them. The entire assembly is then wrapped into a spiral and inserted into a tubular shell with a suitable head assembly through which the permeate can be withdrawn (Figure 10.8).

Both the hollow-fibre and spiral-wound designs use dead-ended membranes that do not allow the use of a sweep gas on the permeate side. The permeate stream can be withdrawn from the feed end or retentate end, giving counter-current or co-current flow (Figure 10.9). With spiral-wound membranes, the permeate stream can be withdrawn from both ends, giving an approximation of cross flow (Figure 10.9c). The choice of flow arrangement determines the boundary conditions for integrating equation 10.4; see Chapter 5 of Scott and Hughes (1996).

In both the hollow-fibre and spiral-wound designs, the cost of the membrane itself is typically 80% to 90% of the total module cost.

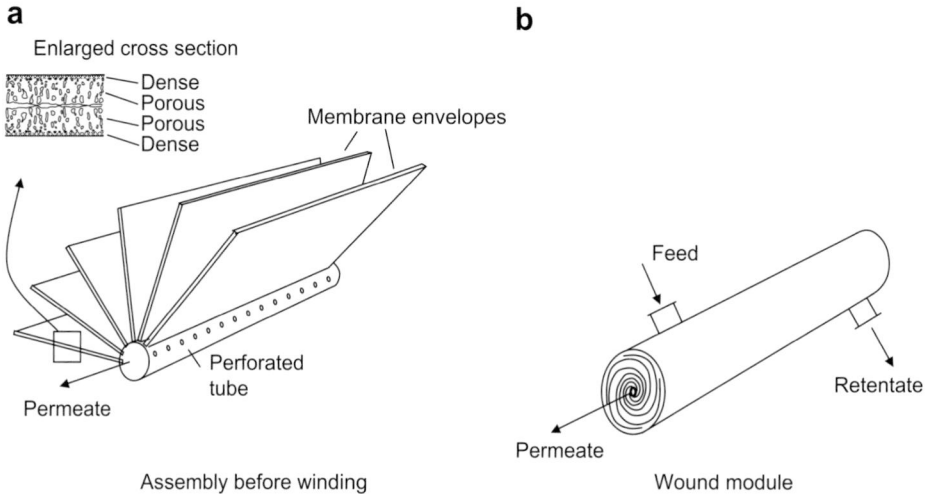

Figure 10.8. Spiral-wound membrane module.

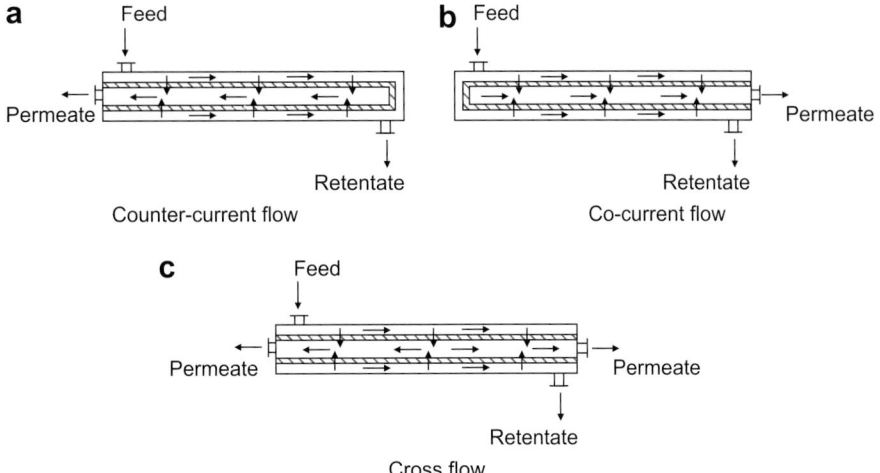

Figure 10.9.   Membrane flow patterns.

Membrane Process Design

A typical membrane process uses a large number of membrane modules to achieve the overall separation. The modules can be arranged in various networks, depending on the processing objective and membrane performance. Some of the more common arrangements are illustrated in Figure 10.10 and described below.

1. Tapered cascade (Figure 10.10a). This flow scheme is used when the final retentate flow is a small fraction of the feed flow rate. It ensures that the

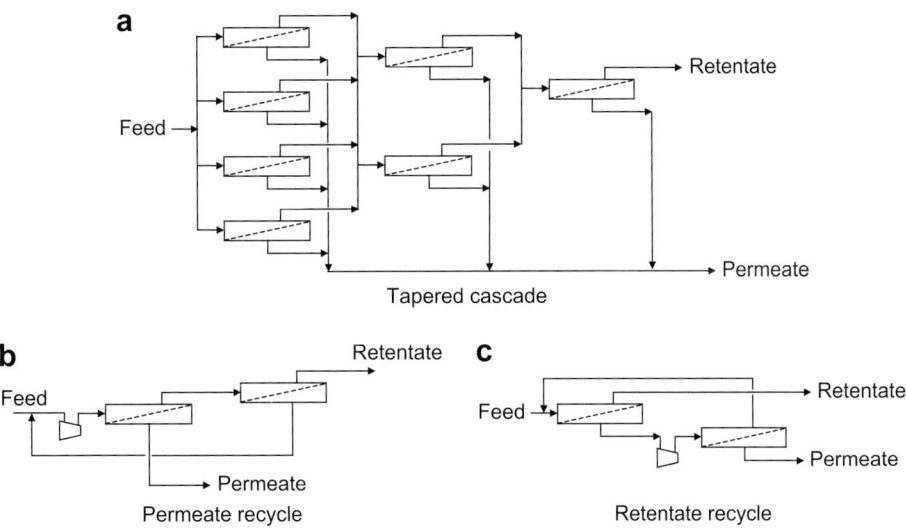

Figure 10.10.   Membrane flow schemes.

retentate side velocities are maintained in downstream modules as the flow rate decreases. Tapered cascades with retentate or permeate recycle can also be used.

2. Permeate recycle (Figure 10.10b). The use of a permeate recycle increases both the recovery and the purity of the permeate. The permeate stream from the downstream module is recycled to the feed of the upstream module. This increases the concentration and partial pressure of the permeating species in the first module, leading to better permeate purity and higher flux. The downstream module can be run to higher recovery, even if this causes the purity obtained in the downstream module to be lower.

3. Retentate recycle (Figure 10.10c). In this scheme, the permeate stream from a first module is sent to a second module. Retentate from the second module is recycled to the first module. This flow scheme is used to increase permeate purity when the membrane average selectivity is low. In some cases, several modules can be placed in series, with retentate recycle from the downstream module to the preceding module.

In addition to the membrane modules, membrane separation units often incorporate chillers, condensers, heaters and guard beds. If the feed contains condensable components, these are usually knocked out by chilling and condensing upstream of the membrane so that they do not condense on the membrane and 'blind' the surface. The gas is then reheated to give a comfortable margin of temperature above the dew point. When a network of membrane modules is used, intermediate chillers, condensers and reheaters may be needed if the condensable components build up to the dew-point concentration in either the permeate or the retentate of an intermediate module.

Sometimes, condensable or reactive species can poison or permanently damage the membrane. In such cases, these components are removed upstream of the membrane unit, usually by using an adsorption process.

The design of membrane separation plants is usually left to specialist vendors. Details of the membrane design and construction are usually proprietary, and optimal process design requires careful selection of flow pattern, flow scheme and ancillary equipment. Accurate prediction of the overall purity and recovery requires a good knowledge of the permeability of all species present, so that equation 10.4 can be integrated for every species simultaneously. Design methods for membrane units are described in Scott and Hughes (1996), Noble and Stern (1995), Mulder (1996), Hoffman (2003) and Baker (2004). Short-cut design methods are given by Hogsett and Mazur (1983) and Fleming and Dupuis (1993), but both papers caution that final design should be left to experts. A simplified membrane design example is given in Example 10.4 in Section 10.5.4, for the purification of water by reverse osmosis. Gas separations are more complex in that there are usually multiple species that permeate through the membrane.

The proprietary nature of membrane technology for gas separations means that there is little reliable cost information in the open literature. The *Handbook of Gas Processing Processes* published biennially as a supplement to *Hydrocarbon Processing* usually provides some information from vendors that can be used to make preliminary estimates.

10.2.3. **Cryogenic Distillation**

In cryogenic distillation, the feed gas is compressed to high pressure, cooled, and then chilled until the gas is partially liquefied and distillation can be carried out. The refrigeration load for chilling the gas may be met by expansion of the products or by using an external refrigeration cycle. The distillation column is designed to operate at very low temperatures and usually has many stages, but other than that is much like any conventional distillation; see Chapter 11.

The efficiency of cryogenic separations depends critically on heat recovery (cold recovery) during chilling of the feed. Very close temperature approaches of the order of 1°C to 5°C are used, to minimize the amount of low temperature refrigeration that is needed. Complex multi-level refrigeration cycles are also used, to improve the coefficient of performance and reduce the compression work.

If the pressure of a cryogenic process is increased, the process can be carried out closer to ambient temperature, giving a smaller temperature range for the refrigeration cycle and hence a more efficient refrigeration (see Section 6.4.4). It also becomes easier to generate the required refrigeration by expansion of the products; however, the feed compression becomes more costly and the distillation becomes more difficult, as relative volatility generally decreases with increased pressure. One of the main trade-offs in design of cryogenic processes is therefore between feed compression and refrigeration plant compression.

Cryogenic distillation is usually the lowest cost process when handling large quantities of gases heavier than hydrogen. It is widely used industrially for air separation, natural gas liquids recovery, ethylene recovery and propylene recovery; see Flynn (2004).

10.2.4. **Absorption and Stripping**

In an absorption process, a component is removed from a gas by contacting the gas with a solvent that dissolves the component selectively. The solvent is then regenerated in a stripping process and recycled to the absorber. Absorption and stripping are usually carried out in vapour–liquid contacting columns; see Section 11.14.

10.3. **SOLID–SOLID SEPARATIONS**

Processes and equipment are required to separate valuable solids from unwanted material, and for size grading (classifying) solid raw materials and products.

The equipment used for solid–solid separation processes was developed primarily for the minerals processing and metallurgical industries for the benefication (upgrading) of ores. The techniques used depend on differences in physical, rather than chemical, properties, though chemical additives may be used to enhance separation. The principal techniques used are shown in Figure 10.11, which can be used to select the type of processes likely to be suitable for a particular material and size range.

Sorting material by appearance, by hand, is now rarely used due to the high cost of labour.

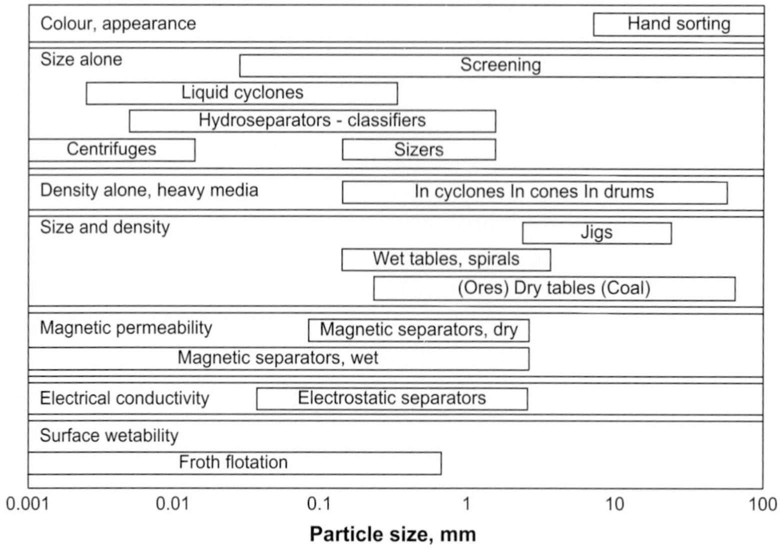

Figure 10.11.   A particle size selection guide to solid–solid separation techniques and equipment (after Roberts *et al.* 1971).

10.3.1. Screening (sieving)

Screens separate particles on the basis of size. Their main application is in grading raw materials and products into size ranges, but they are also used for the removal of trash (over- and under-sized contaminants) and for dewatering. Industrial screening equipment is used over a wide range of particle sizes, from fine powders to large rocks. For small particles woven cloth or wire screens are used, and for larger sizes perforated metal plates or grids.

Screen sizes are defined in two ways: by a mesh size number for small sizes and by the actual size of opening in the screen for the larger sizes. There are several different standards in use for mesh size, and it is important to quote the particular standard used when specifying particle size ranges by mesh size. In the European Union the appropriate standards are BS EN 933 (1996), BS 7792 (ISO 10630) (1995) and BS ISO 14315 (1997). In the USA the appropriate ASTM Standards should be used (ASTM E11).

The simplest industrial screening equipment are stationary screens, over which the material to be screened flows. Typical of this type are 'Grizzly' screens, which consist of rows of equally spaced parallel bars, and which are used to 'scalp' off over-sized rocks in the feed to crushers.

Dynamic screening equipment can be categorized according to the type of motion used to shake-up and transport the material on the screen. The principal types used in the chemical process industries are described briefly below.

Vibrating screens: horizontal and inclined screening surfaces vibrated at high frequencies (1000 to 7000 Hz). These are high-capacity units, with good separating efficiency, and are used for a wide range of particle sizes.

Oscillating screens: operated at lower frequencies than vibrating screens (100–400 Hz) with a longer, more linear, stroke.

Reciprocating screens: operated with a shaking motion, a long stroke at low frequency (20–200 Hz). Used for conveying with size separation.

Sifting screens: operated with a circular motion in the plane of the screening surface. The actual motion may be circular, gyratory, or circularly vibrated. Used for the wet and dry screening of fine powders.

Revolving screens: inclined, cylindrical screens, rotated at low speeds (10–20 rpm). Used for the wet screening of relatively coarse material, but have now been largely replaced by vibrating screens.

Figure 10.12, which is based on a similar chart given by Matthews (1971), can be used to select the type of screening equipment likely to be suitable for a particular size range. Equipment selection will normally be based on laboratory and pilot scale screening tests, conducted by the equipment vendors or with their co-operation. The main factors to be considered, and the information that would be required by the firms supplying proprietary screening equipment, are listed below:

1. Rate, throughput required.
2. Size range (test screen analysis).
3. Characteristics of the material: free-flowing or sticky, bulk density, abrasiveness.
4. Hazards: flammability, toxicity, dust explosion.
5. Wet or dry screening to be used.

10.3.2. Liquid–Solid Cyclones

Cyclones can be used for the classification of solids, as well as for liquid–solid, and liquid–liquid separations. The design and application of liquid cyclones (hydrocyclones) is discussed in Section 10.4.4. A typical unit is shown in Figure 10.13.

Liquid cyclones can be used for the classification of solid particles over a size range from 5 to 100 μm. Commercial units are available in a wide range of materials of construction and sizes, from as small as 10 mm to up to 30 m diameter. The separating efficiency of liquid cyclones depends on the particle size and density, and the density and viscosity of the liquid medium.

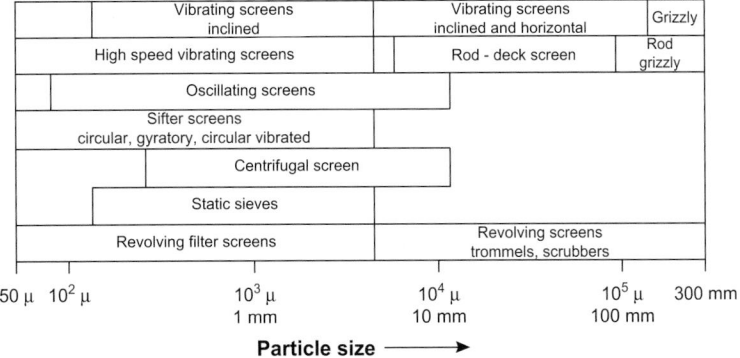

Figure 10.12. Screen selection by particle size range.

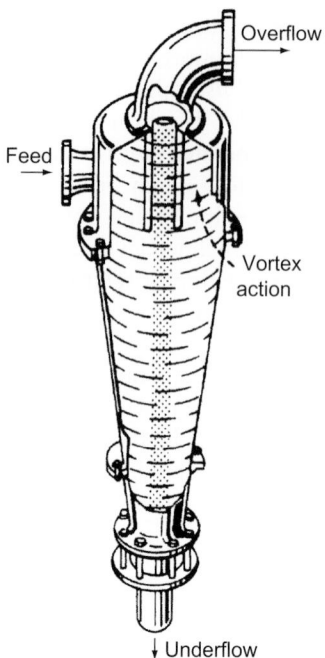

Figure 10.13.   Liquid–solid cyclone (hydrocyclone).

10.3.3. Hydroseparators and Sizers (Classifiers)

Classifiers that depend on the difference in the settling rates of different size particles in water are frequently used for separating fine particles, in the 50 to 300 μm range. Various designs are used. The principal ones used in the chemical process industries are described below.

Thickeners: are primarily used for liquid–solid separation (see Section 10.4). When used for classification, the feed rate is such that the overflow rate is greater than the settling rate of the slurry, and the finer particles remain in the overflow stream.

Rake classifiers: are inclined, shallow, rectangular troughs, fitted with mechanical rakes at the bottom to rake the deposited solids to the top of the incline (Figure 10.14). Several rake classifiers can be used in series to separate the feed into different size ranges.

Bowl classifiers: are shallow bowls with concave bottoms, fitted with rakes. Their operation is similar to that of thickeners.

10.3.4. Hydraulic Jigs

Jigs separate solids by difference in density and size. The material is immersed in water, supported on a screen (Figure 10.15). Pulses of water are forced through the bed of material, either by moving the screen or by pulsating the water level. The flow of water fluidizes the bed and causes the solids to stratify with the lighter material at the top and the heavier at the bottom.

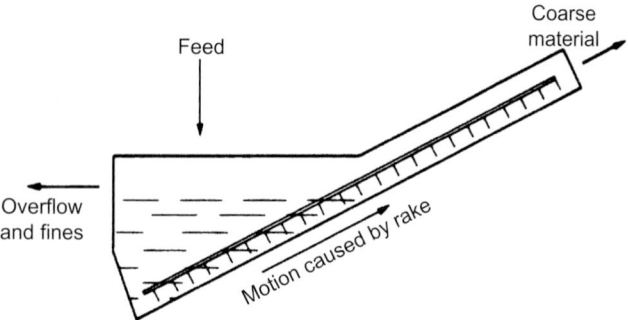

Figure 10.14.   Rake classifier.

10.3.5. Tables

Tables are used wet and dry. The separating action of a wet table resembles that of the traditional miner's pan. Riffled tables (Figure 10.16) are basically rectangular decks, inclined at a shallow angle to the horizontal (2° to 5°), with shallow slats (riffles) fitted to the surface. The table is mechanically shaken, with a slow stroke in the forward direction and a faster backward stroke. The particles are separated into different size ranges under the combined action of the vibration, water flow and the resistance to flow over the riffles.

10.3.6. Classifying Centrifuges

Centrifuges are used for the classification of particles in size ranges below 10 μm. Two types are used: solid bowl centrifuges, usually with a cylindrical, conical bowl, rotated about a horizontal axis; and 'nozzle' bowl machines, fitted with discs.

These types are described in Section 10.4.3.

10.3.7. Dense-Medium Separators (Sink and Float Processes)

Solids of different densities can be separated by immersing them in a fluid of intermediate density. The heavier solids sink to the bottom and the lighter float to the

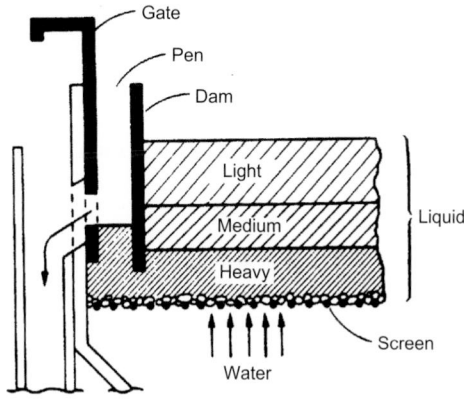

Figure 10.15.   A hydraulic jig.

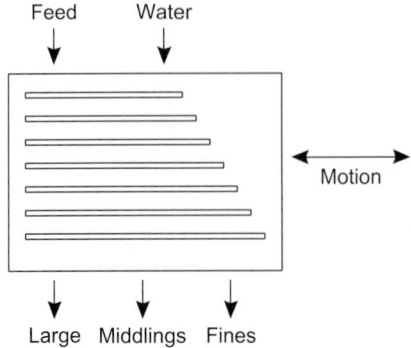

Figure 10.16. Wilfley riffled table.

surface. Water suspensions of fine particles are often used as the dense liquid (heavy-medium). The technique is used extensively for the benefication (concentration) of mineral ores.

10.3.8. Flotation Separators (Froth-Flotation)

Froth-flotation processes are used extensively for the separation of finely divided solids. Separation depends on differences in the surface properties of the materials. The particles are suspended in an aerated liquid (usually water), and air bubbles adhere preferentially to the particles of one component and bring them to the surface. Frothing agents are used so that the separated material is held on the surface as froth and can be removed.

Froth-flotation is an extensively used separation technique, having a wide range of applications in the minerals processing industries and other industries. It can be used for particles in the size range from 50 to 400 μm.

10.3.9. Magnetic Separators

Magnetic separators can be used for materials that are affected by magnetic fields; the principle is illustrated in Figure 10.17. Rotating-drum magnetic separators are used for a wide range of materials in the minerals processing industries. They can be designed to

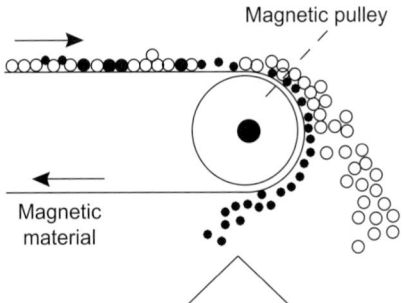

Figure 10.17. Magnetic separator.

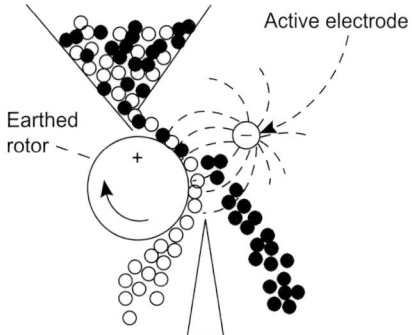

Figure 10.18.   Electrostatic separator.

handle relatively high throughputs, up to 3000 kg/h per meter length of drum. Simple magnetic separators are often used for the removal of iron from the feed to a crusher.

The various types of magnetic separators used and their applications are described by Bronkala (1988).

10.3.10.   Electrostatic Separators

Electrostatic separation depends on differences in the electrical properties (conductivity) of the materials to be treated. In a typical process the particles pass through a high-voltage electric field as it is fed on to a revolving drum, which is at earth potential (Figure 10.18). Those particles that acquire a charge adhere to the drum surface and are carried further around the drum before being discharged.

10.4.   LIQUID–SOLID (SOLID–LIQUID) SEPARATORS

The need to separate solid and liquid phases is probably the most common phase separation requirement in the process industries, and many techniques are used (Figure 10.19). The most suitable technique to use will depend on the solids concentration and feed rate, as well as the size and nature of the solid particles. The

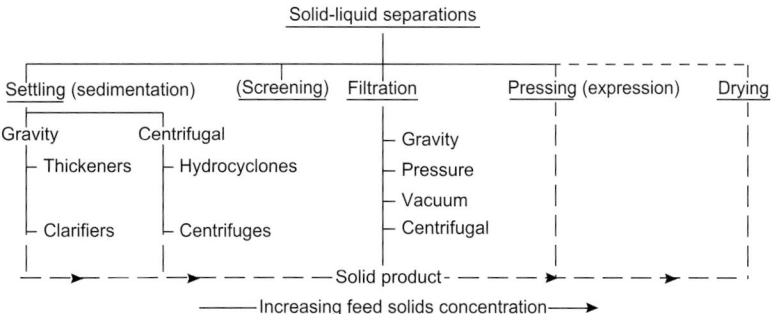

Figure 10.19.   Solid–liquid separation techniques.

range of application of various techniques and equipment, as a function of slurry concentration and particle size, is shown in Figure 10.20.

The choice of equipment also depends on whether the prime objective is to obtain a clear liquid or a solid product, and on the degree of dryness of the solid required.

The design, construction and application of thickeners, centrifuges and filters is a specialized subject, and firms that have expertise in these fields should be consulted when selecting and specifying equipment for new applications. Several specialist texts on the subject are available: Svarovsky (2001), Ward *et al.* (2000) and Wakeman and Tarleton (1998). The theory of sedimentation processes is covered in Richardson *et al.* (2002), Chapter 5 and filtration in Chapter 7.

10.4.1. Thickeners and Clarifiers

Thickening and clarification are sedimentation processes, and the types of equipment used for the two techniques are similar. The primary purpose of thickening is to increase the concentration of a relatively large quantity of suspended solids; whereas that of clarifying, as the name implies, is to remove a small quantity of fine solids to

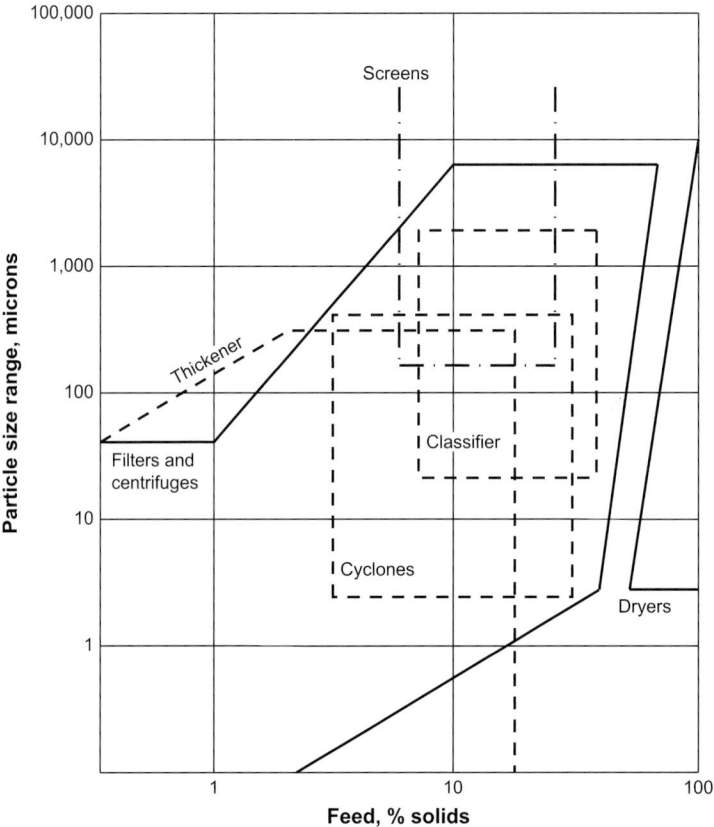

Figure 10.20. Solid–liquid separation techniques (after Dahlstrom and Cornell, 1971).

produce a clear liquid effluent. Thickening and clarification are relatively cheap processes when used for the treatment of large volumes of liquid.

A thickener, or clarifier, consists essentially of a large circular tank with a rotating rake at the base. Rectangular tanks are also used, but the circular design is preferred. They can be classified according to the way the rake is supported and driven. The three basic designs are shown in Figure 10.21. Various designs of rake are used, depending on the nature of the solids.

The design and construction of thickeners and clarifiers is described by Dahlstrom and Cornell (1971).

Flocculating agents are often added to promote the separating performance of thickeners.

10.4.2. Filtration

In filtration processes, the solids are separated from the liquid by passing (filtering) the slurry through some form of porous filter medium. Filtration is a widely used separation process in the chemical and other process industries. Many types of equipment and filter media are used, designed to meet the needs of particular applications. Descriptions of the filtration equipment used in the process industries and their fields of application can be found in various handbooks: Green and Perry (2007), Dickenson (1997), Schweitzer (1997); and in specialist texts on the subject: Cheremisnoff (1998), Orr (1977), Sutherland (2008) and Wakeman and Tarleton

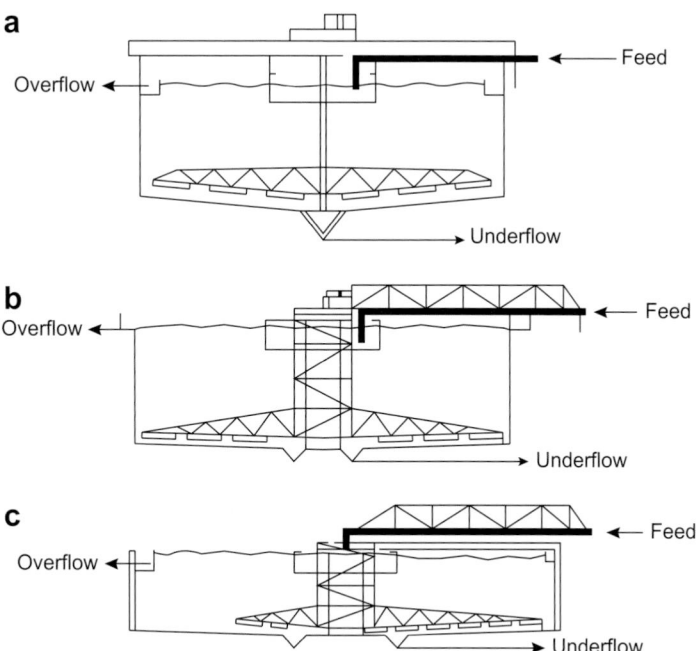

Figure 10.21. Types of thickener and clarifier. (a) Bridge supported (up to <40 m dia.). (b) Centre column supported (<30 m dia.). (c) Traction driven (<60 m dia.).

(2005). Biopharmaceutical applications are discussed in Meltzer and Jornitz (1997). A short discussion of filtration theory and descriptions of the principal types of equipment is given in Richardson *et al.* (2002), Chapter 7.

The most commonly used filter medium is woven cloth, but many other media are also used. The main types are listed in Table 10.3. A comprehensive discussion of the factors to be considered when selecting filter media is given by Purchas (1971) and Mais (1971); see also Purchas and Sutherland (2001). Filter aids are often used to increase the rate of filtration of difficult slurries. They are either applied as a precoat to the filter cloth or added to the slurry and deposited with the solids, assisting in the formation of a porous cake.

Industrial filters use vacuum, pressure, or centrifugal force to drive the liquid (filtrate) through the deposited cake of solids. Filtration is essentially a discontinuous process. With batch filters, such as plate and frame presses, the equipment has to be shut down to discharge the cake; and even with those filters designed for continuous operation, such as rotating-drum filters and cross-flow filters, periodic stoppages are necessary to change the filter cloths. Batch filters can be coupled to continuous plant by using several units in parallel, or by providing buffer storage capacity for the feed and product.

The principal factors to be considered when selecting filtration equipment are:

1. The nature of the slurry and the cake formed.
2. The solids concentration in the feed.
3. The throughput required.
4. The nature and physical properties of the liquid: viscosity, flammability, toxicity, corrosiveness.
5. Whether cake washing is required.
6. The cake dryness required.
7. Whether contamination of the solid by a filter aid is acceptable.
8. Whether the valuable product is the solid or the liquid, or both.

TABLE 10.3.   Filter Media

Type	Examples	Min. Size Particle Trapped (μm)
1. Solid fabrications	Scalloped washers	
	Wire-wound tubes	5
2. Rigid porous media	Ceramics, stoneware	1
	Sintered metal	3
3. Metal	Perforated sheets	100
	Woven wire	5
4. Porous plastics	Pads, sheets	3
	Membranes	0.005
5. Woven fabrics	Natural and synthetic fibre cloths	10
6. Non-woven sheets	Felts, lap	10
	Paper, cellulose	5
7. Cartridges	Yarn-wound spools, graded fibres	2
8. Loose solids	Fibres, asbestos, cellulose	sub-micron

The overriding factor will be the filtration characteristics of the slurry; whether it is fast filtering (low specific cake resistance) or slow filtering (high specific cake resistance). The filtration characteristics can be determined by laboratory or pilot plant tests. A guide to filter selection by the slurry characteristics is given in Table 10.4, which is based on a similar selection chart given by Porter *et al.* (1971).

The principal types of industrial scale filter used are described briefly below.

Nutsche (Gravity and Vacuum Operation)

This is the simplest type of batch filter. It consists of a tank with a perforated base, which supports the filter medium.

Plate and Frame Press (Pressure Operation) (Figure 10.22)

These are the oldest and most commonly used batch filters. The equipment is versatile, made in a variety of materials, and capable of handling viscous liquids and cakes with a high specific resistance.

Leaf Filters (Pressure and Vacuum Operation)

Various types of leaf filter are used, with the leaves arranged in horizontal or vertical rows. The leaves consist of metal frames over which filter cloths are draped. The cake is removed either mechanically or by sluicing it off with jets of water. Leaf filters are used for similar applications as plate and frame presses, but generally have lower operating costs.

TABLE 10.4. Guide to Filter Selection

Slurry Characteristics	Fast Filtering	Medium Filtering	Slow Filtering	Dilute	Very Dilute
Cake formation rate	Cm/s	mm/s	0.02–0.12 mm/s	0.02 mm/s	No cake
Normal concentration	>20%	10–20%	1–10%	<5%	<0.1%
Settling rate	Very fast	Fast	Slow	Slow	—
Leaf test rate, kg/h m^2	>2500	250–2500	25–250	<25	—
Filtrate rate, m^3/h m^2	>10	5–10	0.02–0.05	0.02–5	0.02–5

Filter application

Continuous vacuum filters

Multi compartment drum

Single compartment drum

Top feed drum

Scroll discharge drum

Tilting pan

Belt

Disc

Batch vacuum leaf

Batch nutsche

Batch pressure filters

Plate and frame

Vertical leaf

Horizontal plate

Cartridge edge

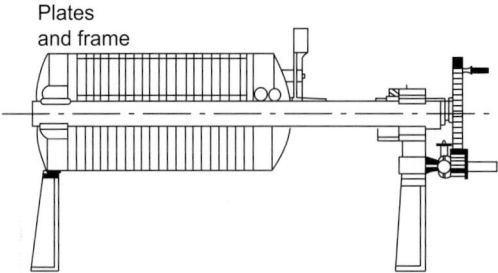

Figure 10.22. Plate and frame filter press.

Rotary Drum Filters (Usually Vacuum Operation) (Figure 10.23)

A drum filter consists essentially of a large hollow drum round which the filter medium is fitted. The drum is partially submerged in a trough of slurry, and the filtrate is sucked through the filter medium by vacuum inside the drum. Wash water can be sprayed on to the drum surface and multi-compartment drums are used so that the wash water can be kept separate from the filtrate. A variety of methods can be used to remove the cake from the drum: knives, strings, air jets and wires. Rotating drum filters are essentially continuous in operation. They can handle large throughputs, and are widely used for free-filtering slurries.

Disc Filters (Pressure and Vacuum Operation)

Disc filters are similar in principle to rotary filters, but consist of several thin discs mounted on a shaft, in place of the drum. This gives a larger effective filtering area on a given floor area, and vacuum disc filters are used in preference to drum filters where

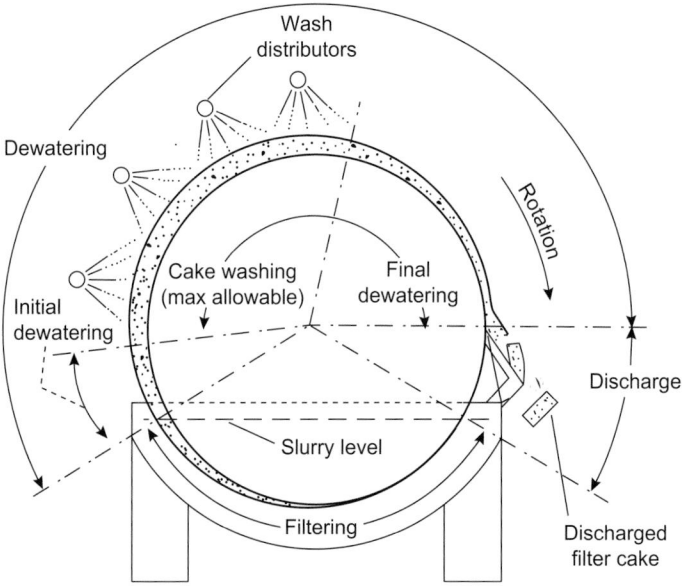

Figure 10.23. Drum filter.

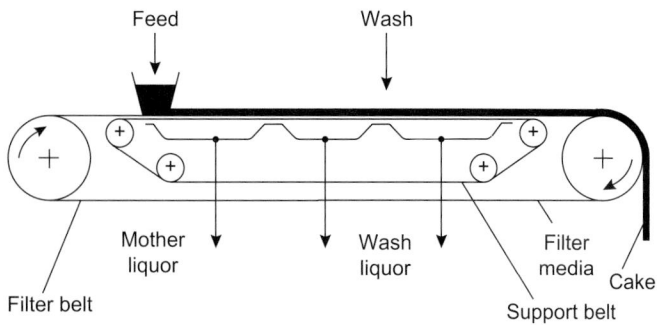

Figure 10.24. Belt filter.

space is restricted. At sizes above approximately 25 m^2 filtration area, disc filters are cheaper; but their applications are more restricted, as they are not as suitable for the application of wash water, or precoating.

Belt Filters (Vacuum Operation) (Figure 10.24)

A belt filter consists of an endless reinforced rubber belt, with drainage holes along its centre, which supports the filter medium. The belt passes over a stationary suction box, into which the filtrate is sucked. Slurry and wash water are sprayed on to the top of the belt.

Horizontal Pan Filters (Vacuum Operation) (Figure 10.25)

This type is similar in operation to a vacuum Nutsche filter. It consists of shallow pans with perforated bases, which support the filter medium. By arranging a series of pans

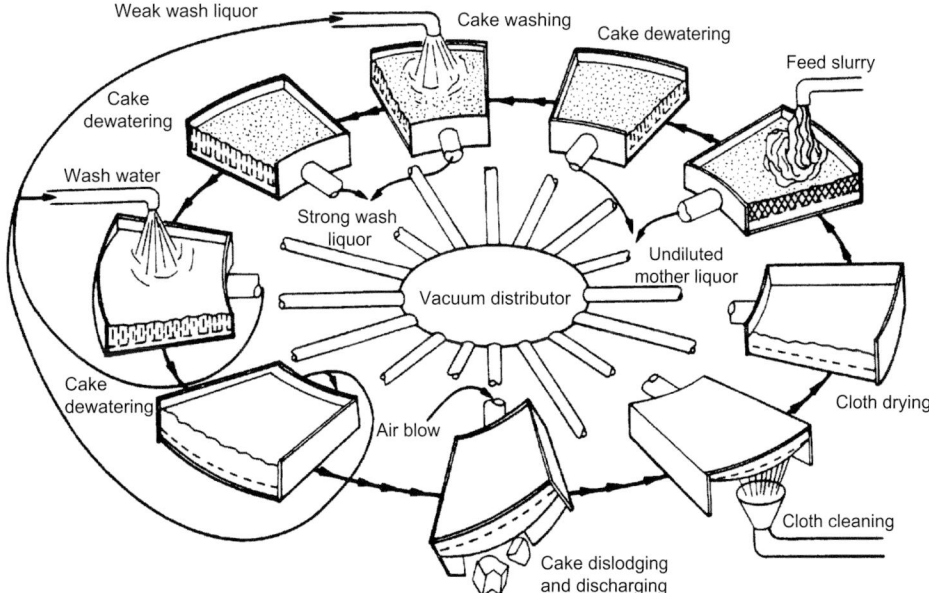

Figure 10.25. Pan filters.

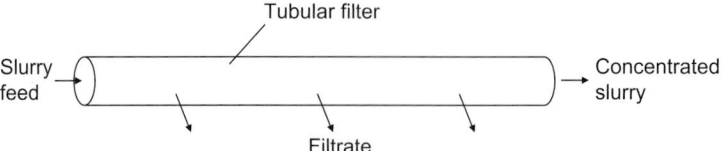

Figure 10.26.   Cross-flow filtration.

around the circumference of a rotating wheel, the operation of filtering, washing, drying and discharging can be made automatic.

Centrifugal Filters

Centrifugal filters use centrifugal force to drive the filtrate through the filter cake. The equipment used is described in the next section.

Cross-Flow Filters

Cross-flow filters are used to reject liquid from a slurry, similar to thickeners. The filter is arranged as tubular modules, and is often a porous membrane; see Section 10.5.4. The slurry is usually fed through the tubes and clear liquid is withdrawn through the tube wall as filtrate, as illustrated in Figure 10.26. The flow on the inside of the tube prevents the accumulation of solids and the concentrated slurry can then be sent to further processing.

Cross-flow filtration for solvent rejection is often carried out in a 'feed and bleed' flow scheme, shown in Figure 10.27. A portion of the retentate from the filtration unit is recycled to the feed. This flow scheme allows higher velocities to be used on the retentate side, and hence higher concentrations can be reached without fouling the filter surface.

Cross-flow filtration processes such as microfiltration and ultrafiltration are widely used in food processing; for example, in concentrating orange juice, and also in the recovery of products from fermentation processes.

10.4.3. Centrifuges

Centrifuges are classified according to the mechanism used for solids separation:

(a) Sedimentation centrifuges: in which the separation is dependent on a difference in density between the solid and liquid phases (solid heavier).

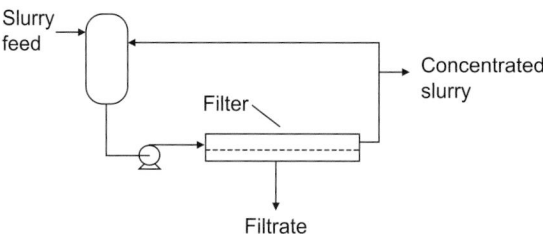

Figure 10.27.   Feed and bleed filtration.

TABLE 10.5.   Selection of Sedimentation or Filter Centrifuge

Factor	Sedimentation	Filtration
Solids size, fine		x
Solids size, >150 μm	x	
Compressible cakes	x	
Open cakes		x
Dry cake required		x
High filtrate clarity	x	
Crystal breakage problems		x
Pressure operation		
High-temperature operation	will depend on the type of centrifuge used	

(b) Filtration centrifuges: which separate the phases by filtration. The walls of the centrifuge basket are porous, and the liquid filters through the deposited cake of solids and is removed.

The choice of centrifuge for a particular application will depend on the nature of the feed and the product requirements.

The main factors to be considered are summarized in Table 10.5. As a general rule, sedimentation centrifuges are used when it is required to produce a clarified liquid, and filtration centrifuges to produce a pure, dry, solid.

A variety of centrifugal filter and sedimenter designs can be used. The main types are listed in Table 10.6. They can be classified by a number of design and operating features, such as:

TABLE 10.6.   Centrifuge Types (After Sutherland, 1970)

Sedimentation	Filtration-fixed bed
Laboratory	Vertical basket
Bottle	Manual discharge
Ultra	Bag discharge
	Knife discharge
Tubular bowl	Horizontal basket
	Inclined basket
Disc	
Batch bowl	
Nozzle discharge	
Valve discharge	Filtration-moving bed
Opening bowl	
Imperforate basket	Conical bowl
Manual discharge	Wide angle
Skimmer discharge	Vibrating
	Torsional
	Tumbling
Scroll discharge	Scroll discharge
Horizontal	
Cantilevered	Cylindrical bowl
Vertical	Scroll discharge
Screen bowl	Pusher

1. Mode of operation: batch or continuous
2. Orientation of the bowl/basket: horizontal or vertical
3. Position of the suspension and drive: overhung or underhung
4. Type of bowl: solid, perforated basket, disc bowl
5. Method of solids cake removal
6. Method of liquid removal.

Descriptions of the various types of centrifuges and their fields of application can be found in various handbooks, in a book by Leung (1998) and articles by Ambler (1971) and Linley (1984).

The fields of application of each type, classified by the size range of the solid particles separated, are given in Figure 10.28. A similar selection chart is given by Schroeder (1998).

Sedimentation Centrifuges

There are four main types of sedimentation centrifuge:

1.   Tubular Bowl (Figure 10.29)
High-speed, vertical axis, tubular bowl centrifuges are used for the separation of immiscible liquids, such as water and oil, and for the separation of fine solids. The bowl is driven at speeds of around 15,000 rpm (250 Hz) and the centrifugal force generated exceeds 130,000 N.

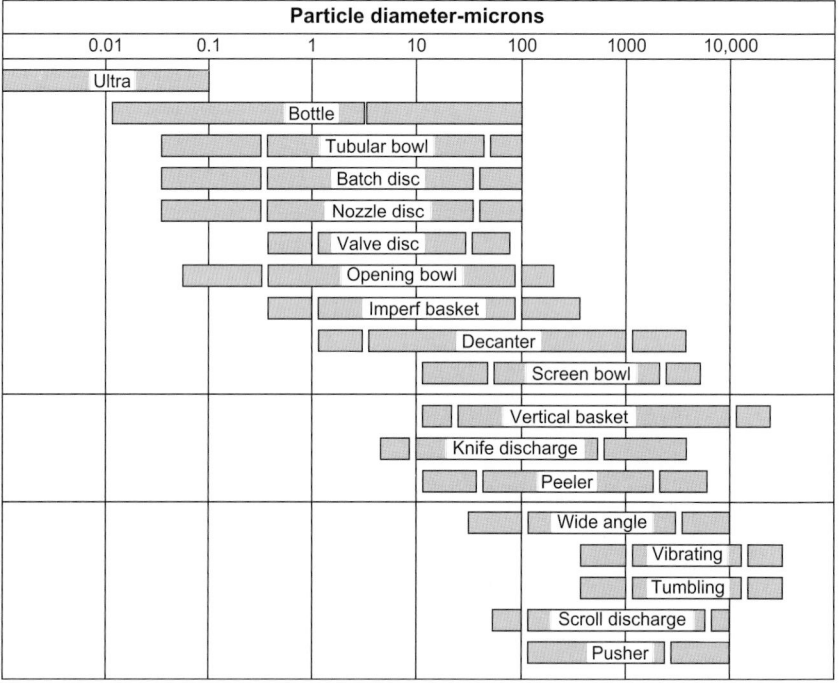

Figure 10.28.   Classification of centrifuges by particle size (after Sutherland, 1970).

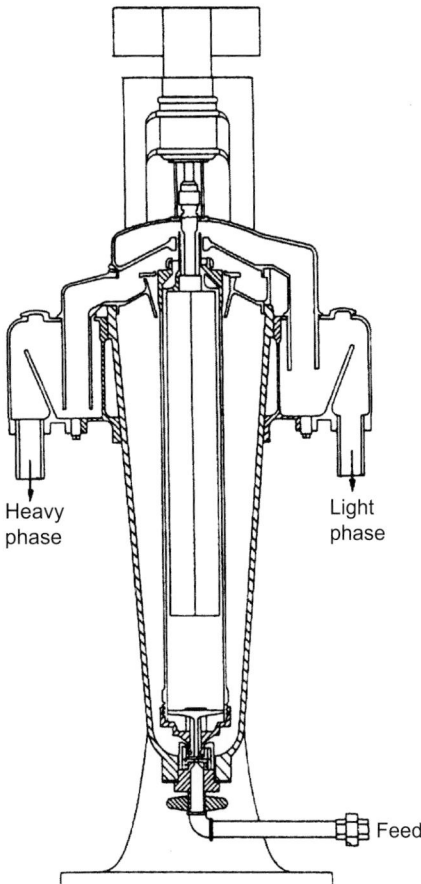

Figure 10.29. Tubular bowl centrifuge.

2. Disc Bowl (Figure 10.30)

The conical discs in a disc bowl centrifuge split the liquid flow into a number of very thin layers, which greatly increases the separating efficiency. Disc bowl centrifuges are used for separating liquids and fine solids, and for solids classification.

3. Scroll Discharge

In this type of machine, the solids deposited on the wall of the bowl are removed by a scroll (a helical screw conveyer) which revolves at a slightly different speed from the bowl. Scroll discharge centrifuges can be designed so that solids can be washed and relatively dry solids can be discharged.

4. Solid Bowl Batch Centrifuge

The simplest type; similar to the tubular bowl machine, but with a smaller bowl length to diameter ratio (less than 0.75).

The tubular bowl type is rarely used for solids concentrations above 1% by volume. For concentrations between 1% and 15%, any of the other three types can be used.

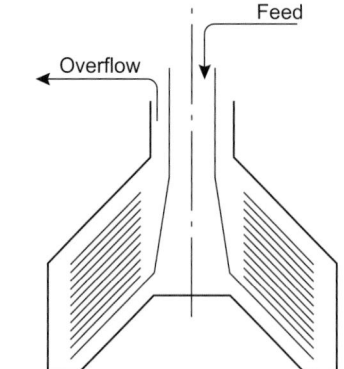

Figure 10.30.   Disc bowl centrifuge.

Above 15%, either the scroll discharge type or the batch type may be used, depending on whether continuous or intermittent operation is required.

Sigma Theory for Sedimentation Centrifuges

The basic equations describing sedimentation in a centrifugal field have been developed in Richardson *et al.* (2002), Chapter 9. In that discussion the term 'sigma' ($\sum$) is introduced, which can be used to define the performance of a centrifuge independently of the physical properties of the solid–fluid system. The sigma value of a centrifuge, normally expressed in cm^2, is equal to the cross-sectional area of a gravity settling tank having the same clarifying capacity.

This approach to describing centrifuge performance has become known as the 'sigma theory'. It provides a means for comparing the performance of sedimentation centrifuges and for scaling up from laboratory and pilot scale tests; see Ambler (1952) and Trowbridge (1962).

In the general case, it can be shown that:

$$Q = 2u_g\Sigma \tag{10.6}$$

and (where Stokes' law applies)

$$u_g = \frac{\Delta\rho d_s^2 g}{18\mu} \tag{10.7}$$

Note: The factor of 2 is included in equation 10.6 as d_s is the 'cut-off' size: 50% of particles of this size will be removed in passage through the centrifuge.

where

Q = volumetric flow of liquid through the centrifuge, m^3/s
u_g = terminal velocity of the solid particle settling under gravity through the liquid, m/s
Σ = sigma value of the centrifuge, m^2
$\Delta\rho$ = density difference between solid and liquid, kg/m^3

TABLE 10.7. Selection of Sedimentation Centrifuges

Type	Approximate Efficiency (%)	Normal Operating Range Q, m³/h at Q/Σ m/s
Tubular bowl	90	0.4 at 5×10^{-8} to 4 at 3.5×10^{-7}
Disc	45	0.1 at 7×10^{-8} to 110 at 4.5×10^{-7}
Solid bowl (scroll discharge)	60	0.7 at 1.5×10^{-6} to 15 at 1.5×10^{-5}
Solid bowl (basket)	75	0.4 at 5×10^{-6} to 4 at 1.5×10^{-4}

d_s = the diameter of the solid particle, *the cut-off* size, m ($\mu m \times 10^{-6}$)
μ = viscosity of the liquid, $Nm^{-2}s$
g = gravitational acceleration, 9.81 m/s².

Morris (1966) gives a method for the selection of the appropriate type of sedimentation centrifuge for a particular application based on the ratio of the liquid overflow to sigma value (Q/Σ). His values for the operating range of each type, and their approximate efficiency rating, are given in Table 10.7. The efficiency term is used to account for the different amounts by which the various designs differ from the theoretical sigma values given by equation 10.1. Sigma values depend solely on the geometrical configuration and speed of the centrifuge. Details of the calculation for various types are given by Ambler (1952). To use Table 10.7, it is necessary to know the feed rate of slurry (and hence the liquid overflow Q), the density of the liquid and solid, the liquid viscosity, and the diameter of the particle for, say, a 98% size removal. The use of Table 10.7 is illustrated in Example 10.2.

A selection guide for sedimentation centrifuges by Lavanchy *et al.* (1964), which includes other types of solid–liquid separators, is shown in Figure 10.31, adapted to SI units.

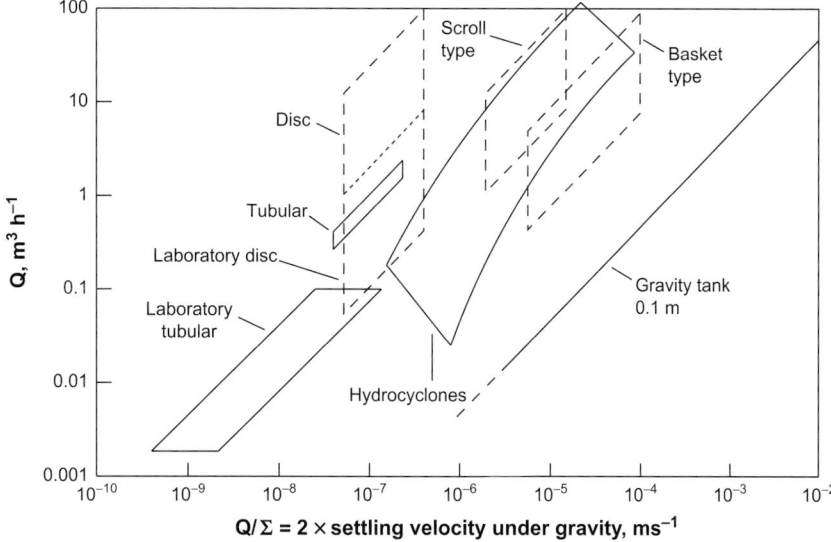

Figure 10.31. Performance of sedimentation equipment (after Lavanchy *et al.*, 1964).

Example 10.2

A precipitate is to be continuously separated from a slurry. The solids concentration is 5% and the slurry feed rate 5.5 m³/h. The relevant physical properties at the system operating temperature are:

liquid density 1050 kg/m³, viscosity 4 cp (mNm⁻²s)
solid density 2300 kg/m³, cut-off particle size 10 μm = 10×10^{-6} m.

Solution

$$\text{Overflow rate, } Q = 0.95 \times 5.5 = 5.23 \text{ m}^3/\text{h}$$
$$= \frac{5.13}{3600} = 1.45 \times 10^{-3} \text{ m}^3/\text{s}$$
$$\Delta\rho = 2300 - 1050 = 1250 \text{ kg/m}^3$$

From equations 10.6 and 10.7

$$\frac{Q}{\Sigma} = 2 \times \frac{1250(10 \times 10^{-6})^2}{18 \times 4 \times 10^{-3}} \times 9.81 = 3.4 \times 10^{-5}$$

From Table 10.7 for a Q of 5.23 m³/h at a Q/Σ of 3.4×10^{-5} a solid bowl basket type should be used.

To obtain an idea of the size of the machine needed, the sigma value can be calculated using the efficiency value from Table 10.7.

From equation 10.6:

$$\Sigma = \frac{Q}{eff. \times 2u_g} = \frac{1.45 \times 10^{-3}}{0.75 \times 3.4 \times 10^{-5}}$$
$$= \underline{56.9 \text{ m}^2}$$

The sigma value is the equivalent area of a gravity settler that would perform the same separation as the centrifuge.

Filtration Centrifuges (Centrifugal Filters)

It is convenient to classify centrifugal filters into two broad classes, depending on how the solids are removed: fixed bed or moving bed.

In the fixed-bed type, the cake of solids remains on the walls of the bowl until removed manually, or automatically by means of a knife mechanism. It is essentially cyclic in operation. In the moving-bed type, the mass of solids is moved along the bowl by the action of a scroll (similar to the solid-bowl sedimentation type), a ram (pusher type), a vibration mechanism, or by the bowl angle. Washing and drying zones can be incorporated into the moving bed type.

Bradley (1965) has grouped the various types into the family tree shown in Figure 10.32.

Schematic diagrams of the various types are shown in Figure 10.33. The simplest machines are the basket types (Figure 10.33a,b,c), and these form the basic design from which the other types have been developed (Figure 10.33d–o).

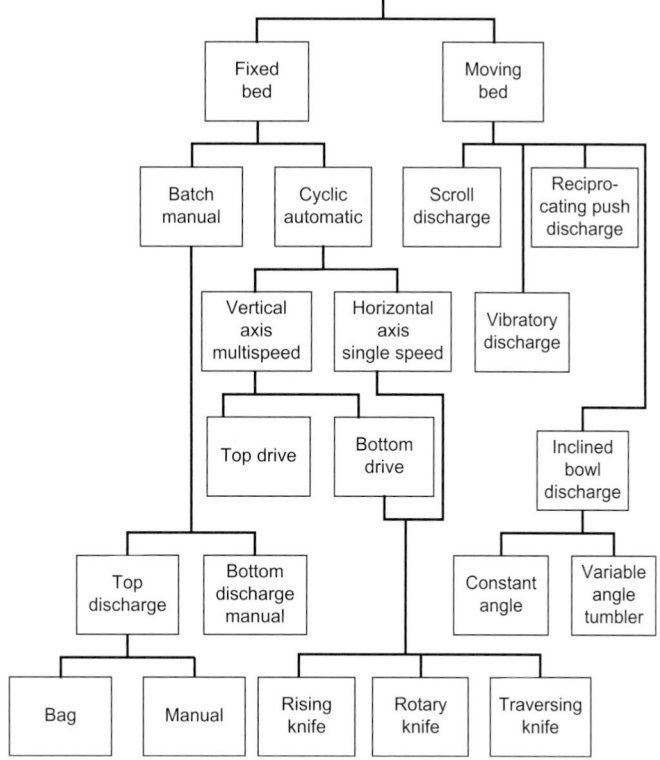

Figure 10.32.   Filtration centrifuge family tree (after Bradley, 1965).

The various arrangements of knife mechanisms used for automatic removal of the cake are shown in Figures 10.33d–h. The bottom discharge-type machines (Figure 10.33d,e) can be designed for variable speed, automatic discharge, and are suitable for use with fragile, or plate or needle-shaped crystals, where it is desirable to avoid breakage or compaction of the bed. They can be loaded and discharged at low speeds, which reduces breakage and compaction of the cake. The single-speed machines (Figures 10.33f,g,h) are used where cakes are thin, and short cycle times are required. They can be designed for high-temperature and high-pressure operation. When continuous operation is required, the scroll, pusher or other self-discharge types are used (Figures 10.33i–o). The scroll discharge centrifuge is a low-cost, flexible machine, capable of a wide range of applications, but is not suitable for handling fragile materials. It is normally used for coarse particles, where some contamination of the filtrate with fines can be tolerated.

The capacity of filtration centrifuges is very dependent on the solids concentration in the feed. For example, at 10% feed slurry concentration 9 kg of liquid will be centrifuged for every 1 kg of solids separated, whereas with a 50% solids concentration the quantity will be less than 1 kg. For dilute slurries it is well worth considering using some form of pre-concentration, such as gravity sedimentation, hydrocyclones or cross-flow filtration.

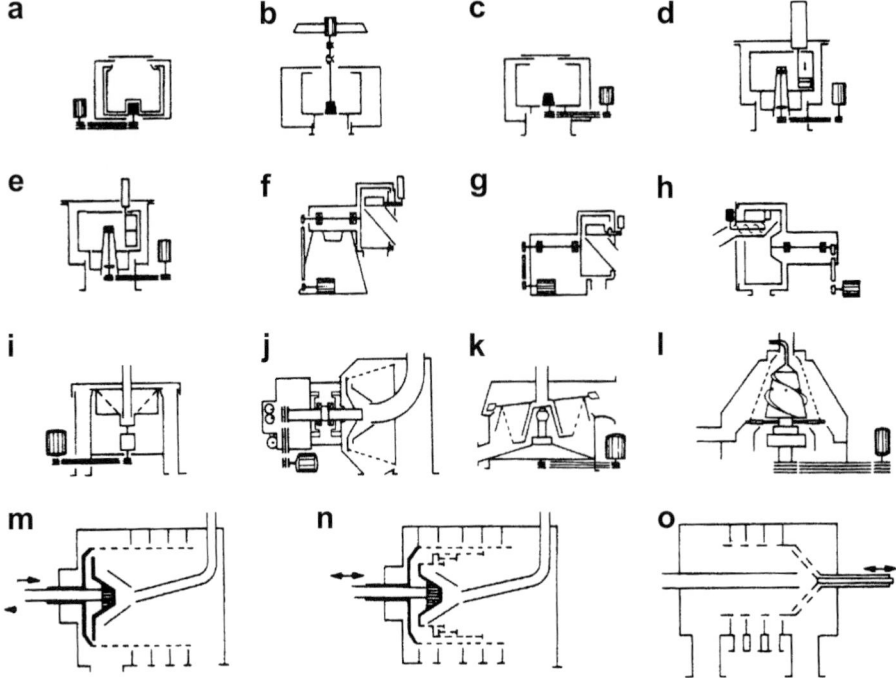

Figure 10.33. Schematic diagrams of filtration centrifuge types (Bradley, 1965). (a) Bottom drive batch basket with bag. (b) Top drive bottom discharge batch basket. (c) Bottom drive bottom discharge batch basket. (d) Bottom drive automatic basket, rising knife. (e) Bottom drive automatic basket, rotary knife. (f) Single reversing knife, rising knife. (g) Single-speed automatic rotary knife. (h) Single-speed automatic traversing knife. (i) Inclined wall self-discharge. (j) Inclined vibrating wall self-discharge. (k) Inclined 'tumbling' wall self-discharge. (l) Inclined wall scroll discharge. (m) Traditional single-stage pusher. (n) Traditional multi-stage pusher. (o) Conical pusher with de-watering cone.

10.4.4. Hydrocyclones (Liquid-Cyclones)

Hydrocyclones are used for solid–liquid separations, as well as for solids classification and liquid–liquid separation. A hydrocyclone is a centrifugal device with a stationary wall, the centrifugal force being generated by the liquid motion. The operating principle is basically the same as that of the gas cyclone described in Section 10.8.3, and in Richardson *et al.* (2002), Chapter 1. Hydrocyclones are simple, robust, separating devices that can be used over the particle size range from 4 to 500 μm. They are often used in groups, as illustrated in Figure 10.34. The design and application of hydrocyclones is discussed fully in books by Abulnaga (2002) and Svarovsky and Thew (1992). Design methods and charts are also given by Zanker (1977), Day *et al.* (1997) and Moir (1985).

The nomographs by Zanker can be used to make a preliminary estimate of the size of cyclone needed. The specialist manufacturers of hydrocyclone equipment should be consulted to determine the best arrangements and design for a particular application.

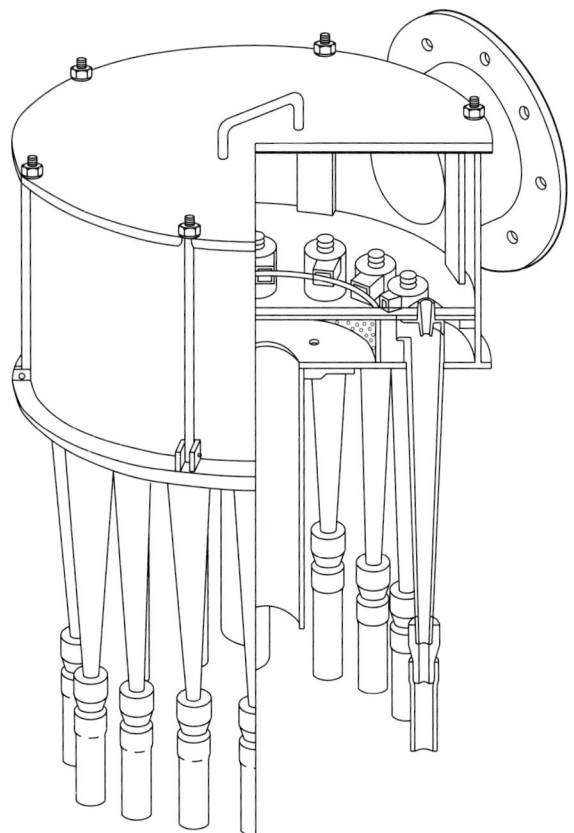

Figure 10.34. A 'Clog' assembly of 16 × 2 in (50 mm) diameter hydrocyclone. (Courtesy of Richard Mozley Ltd.)

Zanker's method is outlined below and illustrated in Example 10.3. Figure 10.35 is based on an empirical equation by Bradley (1960):

$$d_{50} = 4.5 \left[\frac{D_c^3 \mu}{L^{1.2}(\rho_s - \rho_L)} \right] \tag{10.8}$$

where

d_{50} = the particle diameter for which the cyclone is 50% efficient, μm
D_c = diameter of the cyclone chamber, cm
μ = liquid viscosity, centipoise (mN s/m^2)
L = feed flow rate, l/min
ρ_L = density of the liquid, g/cm^3
ρ_s = density of the solid, g/cm^3.

The equation gives the chamber diameter required to separate the so-called d_{50} particle diameter, as a function of the slurry flow rate and the liquid and solid physical

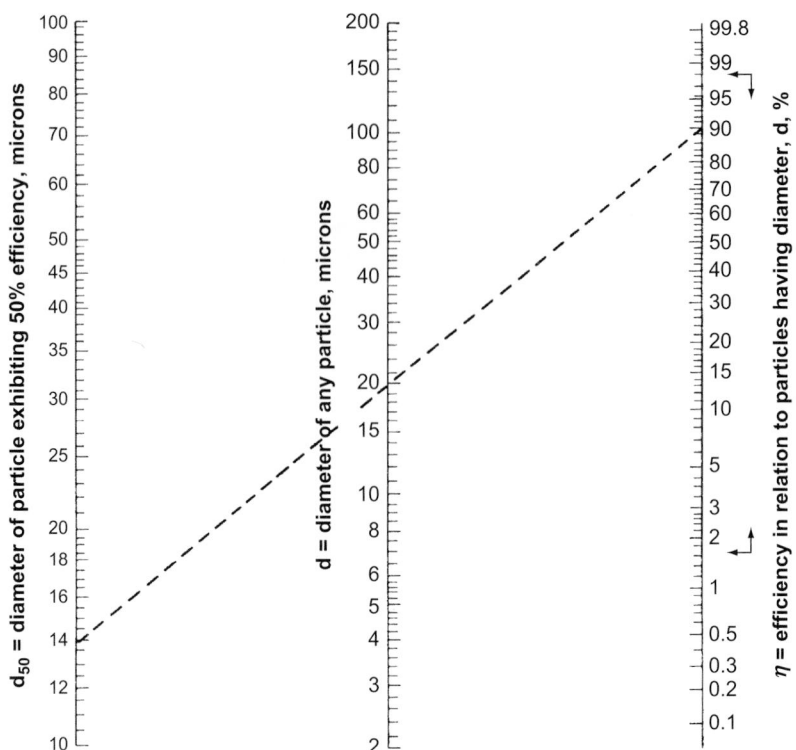

Figure 10.35. Determination of d_{50} from the desired particle separation (equation 10.8; Zanker, 1977) (Example 10.3).

properties. The d_{50} particle diameter is the diameter of the particles, 50% of which will appear in the overflow and 50% in the underflow. The separating efficiency for other particles is related to the d_{50} diameter by Figure 10.36, which is based on a formula by Bennett (1936).

$$\eta = 100 \left[1 - e^{-(d/d_{50}-0.115)^3}\right] \tag{10.9}$$

where

η = the efficiency of the cyclone in separating any particle of diameter d, %
d = the selected particle diameter, μm.

The method applies to hydrocyclones with the proportions shown in Figure 10.37.

Example 10.3

Estimate the size of hydrocyclone needed to separate 90% of particles with a diameter greater than 20 μm from 10 m³/h of a dilute slurry.

Physical properties: solid density 2000 kg/m³, liquid density 1000 kg/m³, viscosity 1 mN s/m².

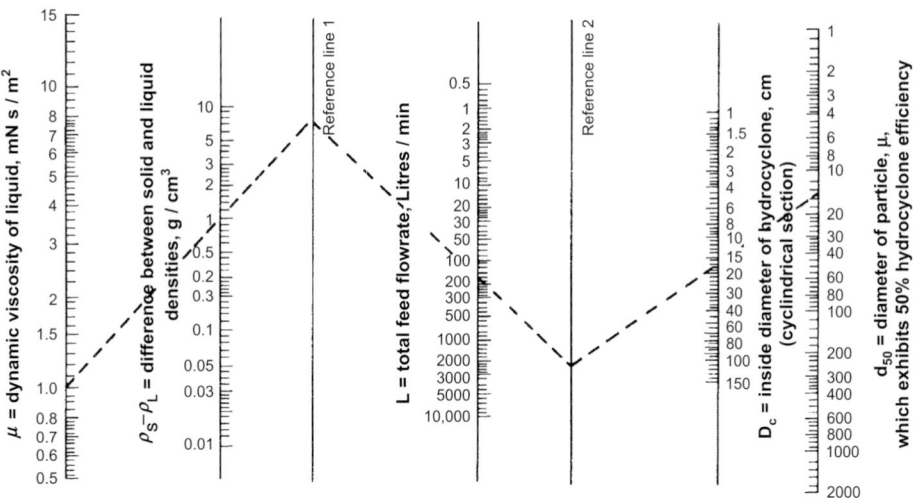

Figure 10.36. Chamber dia. D_c from flow rate, physical properties, and d_{50} particle size (equation 10.9; Zanker, 1977) (Example 10.3).

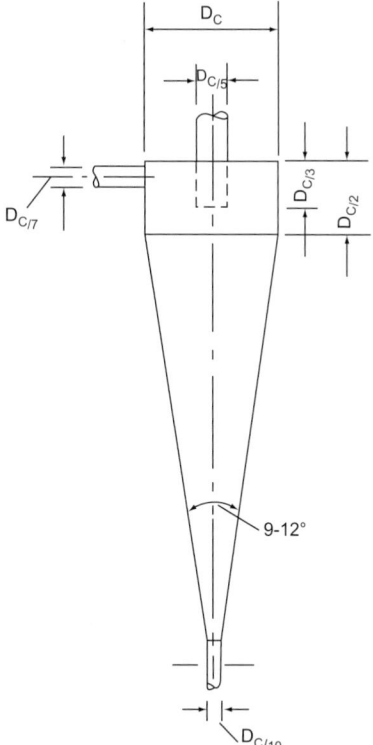

Figure 10.37. Hydrocyclone-typical proportions.

Solution

$$\text{Flow-rate} = \frac{10 \times 10^3}{60} = 1.66.7 \text{l/min}$$
$$(\rho_s - \rho_L) = 2.0 - 1.0 = 1.0 \text{ g/cm}^3$$

From Figure 10.35, for 90% removal of particles above 20 μm

$$d_{50} = 40 \text{ μm}$$

From Figure 10.36, for $\mu = 1$ mN s/m^2, $(\rho_s - \rho_L) = 1.0$ g/cm^3, $L = 167$/min

$$D_c = \underline{16 \text{ cm}}$$

10.4.5. Pressing (Expression)

Pressing, in which the liquid is squeezed (expressed) from a mass of solids by compression, is used for certain specialized applications. Pressing consumes a great deal of energy and should not be used unless no other separating technique is suitable; however, in some applications dewatering by pressing can be competitive with drying.

Presses are of two basic types: hydraulic batch presses and screw presses. Hydraulic presses are used for extracting fruit juices, and screw presses for dewatering materials such as paper pulp, rubbish and manure. The equipment used is described in the handbooks; see Green and Perry (2007).

10.4.6. Solids Drying

Drying is the removal of water, or other volatile liquids, by evaporation. Most solid materials require drying at some stage in their production. The choice of suitable drying equipment cannot be separated from the selection of the upstream equipment feeding the drying stage.

The overriding consideration in the selection of drying equipment is the nature and concentration of the feed. Drying is an energy-intensive process, and the removal of liquid by thermal drying will be more costly than by mechanical separation techniques.

Drying equipment can be classified according to the following design and operating features:

1. Batch or continuous
2. Physical state of the feed: liquid, slurry, wet solid
3. Method of conveyance of the solid: belt, rotary, fluidized
4. Heating system: conduction, convection, radiation.

Hot air is usually used as the heating and mass-transfer medium in industrial dryers unless there are concerns about solvent flammability, in which case nitrogen, depleted air or recirculating flue gas is used. The air may be directly heated by the products of combustion of the fuel used (oil, gas or coal) or indirectly heated, usually by banks of steam-heated finned tubes. The heated air is usually propelled through the dryer by electrically driven fans.

Table 10.8, adapted from a similar selection guide by Parker (1963a), shows the basic features of the various types of solids dryer used in the process industries, and Table 10.9, by Williams-Gardner (1965), shows typical applications.

TABLE 10.8. Dryer Selection

Mode of operation	Generic type	Feed condition 1	2	3	Specific dryer types	Jacketed	Suitable for heat-sensitive materials	Suitable for vacuum service	Retention or cycle time	Heat transfer method	Capacity	Typical evaporation capacity
Batch	Stationary				1. Shelf 2. Cabinet 3. Compartment	Yes	Yes	Yes	6.48 h	Radiant and conduction	Limited	0.15–1.0
					Truck	No	Yes	No	6.48 h	Convection	Limited	0.15–1.0
					1. Kettle 2. Pan	Yes	No	Yes	3.12 h	Conduction	Limited	1.5–15
					Rotary shell	Yes	Yes	Yes	4.48 h	Conduction	Limited	0.5–12
					Rotary internal	Yes	Yes	Yes	4.48 h	Conduction	Limited	0.5–12
					Double cone	Yes	Yes	Yes	3.12 h	Conduction	Limited	0.5–12
Continuous	Drum				1. Single drum 2. Double drum 3. Twin drum	No	Yes	Yes	Very short	Conduction	Medium	5–50
	Rotary				Rotary direct heat	No	No	No	Long	Convection	High	3–110
					Rotary, indirect heat	No	No	No	Long	Conduction	Medium	15–200
					Rotary, steam tube	No	Depends on material	No	Long	Conduction	High	15–200
					Rotary, direct-indirect heat	No	No	No	Long	Conduction Convection	High	50–150
					Louver	No	Depends on material	No	Long	Convection	High	5–240
	Conveyor				Tunnel belt, screen	No	Yes	No	Long	Convection	Medium	1.5–35
					Rotary shelf	Yes	Depends on material	No	Medium	Conduction Convection	Medium	0.5–10
					Trough	Yes	Depends on material	Yes	Varies	Conduction	Medium	0.5–15
					Vibrating	Yes	Depends on material	No	Medium	Convection Conduction	Medium	0.5–100
					Turbo	No	Depends on material	No	Medium	Convection	Medium	1–10
	Suspended particle				Spray	No	Yes	No	Short	Convection	High	1.5–50
					Flash	No	Yes	No	Short	Convection	High	–
					Fluid bed	No	Yes	No	Short	Convection	Medium	–

⟵⟶ = applicable to feed conditions noted
Key to feed conditions:

1. Solutions, colloidal suspensions and emulsions, pumpable solids suspensions, pastes and sludges.
2. Free-flowing powders, granular, crystalline or fibrous solids that can withstand mechanical handling.
3. Solids incapable of withstanding mechanical handling.

TABLE 10.9.   Dryer Applications

Dryer Type	System	Feed Form	Typical Products
Batch ovens	Forced convection	Paste, granules, extrude cake	Pigment dyestuffs, pharmaceuticals, fibres
	Vacuum	Extrude cake	Pharmaceuticals
pan (agitated)	Atmospheric and vacuum	Crystals, granules, powders	Fine chemicals, food products
rotary	Vacuum	Crystals, granules solvent recovery	Pharmaceuticals
fluid bed	Forced convection	Granular, crystals	Fine chemicals, pharmaceuticals, plastics
infra-red	Radiant	Components sheets	Metal products, plastics
Continuous rotary	Convection Direct/indirect Direct Indirect Conduction	Crystals, coarse powders, extrudes, preformed cake lumps, granular paste and fillers, cakes back-mixedwith dry product	Chemical ores, food products, clays, pigments, chemicals Carbon black
film drum	Conduction	Liquids, suspensions	Foodstuffs, pigment
trough	Conduction		Ceramics, adhesives
spray	Convection	Liquids, suspensions	Foodstuffs, pharmaceuticals, ceramics, fine chemicals, detergents, organic extracts
band	Convection	Preformed solids	Foodstuffs, pigments, chemicals, rubber, clays, ores, textiles
fluid bed	Convection	Preformed solids granules, crystals	Ores, coal, clays, chemicals
pneumatic	Convection	Preformed pastes, granules, crystals, coarse products	Chemicals, starch, flour, resins, wood-products, food
infra-red	Radiant	Components sheets	Metal products, moulded fibre articles, painted surfaces

Batch dryers are normally used for small-scale production and where the drying cycle is likely to be long. Continuous dryers require less labour and floor space, and produce a more uniform quality product.

When the feed is solids, it is important to present the material to the dryer in a form that will produce a bed of solids with an open, porous, structure.

For pastes and slurries, some form of pre-treatment equipment will normally be needed, such as extrusion or granulation.

The main factors to be considered when selecting a dryer are:

1. Feed condition: solid, liquid, paste, powder, crystals
2. Feed concentration, the initial liquid content
3. Product specification: dryness required, physical form
4. Throughput required
5. Heat sensitivity of the product
6. Nature of the vapour: toxicity, flammability
7. Nature of the solid: flammability (dust explosion hazard), toxicity.

The drying characteristics of the material can be investigated by laboratory and pilot plant tests, which are best carried out in consultation with the equipment vendors.

The theory of drying processes is discussed in Richardson *et al.* (2002), Chapter 16. Full descriptions of the various types of dryer and their applications are given in that chapter and in Green and Perry (2007), Majumdar (2006) and Walas (1990). Only brief descriptions of the principal types will be given in this section.

The basic types used in the chemical process industries are: tray, band, rotary, fluidized, pneumatic, drum and spray dryers.

Tray Dryers (Figure 10.38)

Batch tray dryers are used for drying small quantities of solids, and are used for a wide range of materials.

The material to be dried is placed in solid-bottomed trays over which hot air is blown; or perforated-bottom trays through which the air passes.

Batch dryers have high labour requirements, but close control can be maintained over the drying conditions and the product inventory, and they are suitable for drying valuable products.

Conveyor Dryers (Continuous Circulation Band Dryers) (Figure 10.39)

In this type, the solids are fed on to an endless, perforated, conveyor belt, through which hot air is forced. The belt is housed in a long rectangular cabinet, which is divided up into zones, so that the flow pattern and temperature of the drying air can be controlled. The relative movement through the dryer of the solids and drying air can be parallel or, more usually, counter-current.

This type of dryer is clearly only suitable for materials that form a bed with an open structure. High drying rates can be achieved, with good product quality control. Thermal efficiencies are high and, with steam heating, steam usage can be as low as 1.5 kg per kg of water evaporated. The disadvantages of this type of dryer are high initial cost and, due to the mechanical belt, high maintenance costs.

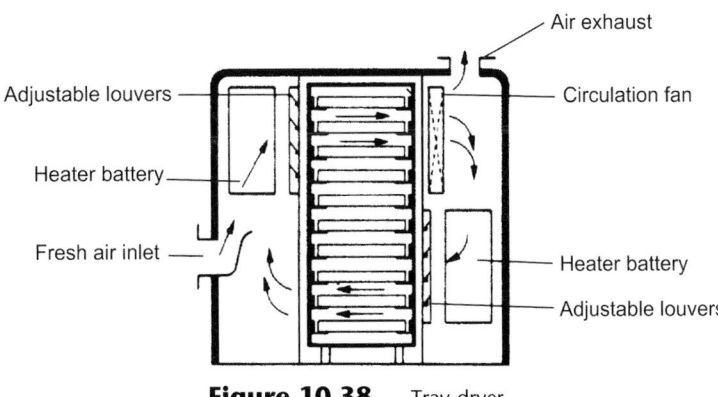

Figure 10.38. Tray dryer.

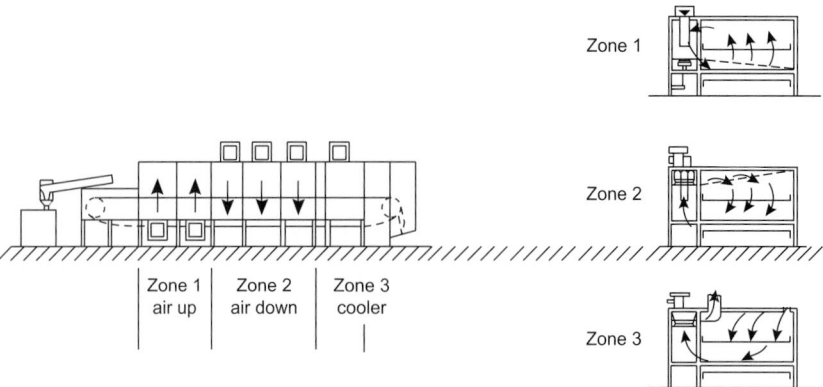

Figure 10.39.   Conveyor dryer.

Rotary Dryers (Figure 10.40)

In rotary dryers, the solids are conveyed along the inside of a rotating, inclined, cylinder and are heated and dried by direct contact with hot air or gases flowing through the cylinder. In some, the cylinders are indirectly heated.

Rotating dryers are suitable for drying free-flowing granular materials. They are suitable for continuous operation at high throughputs, have a high thermal efficiency and relatively low capital cost and labour costs. Some disadvantages of this type are: a non-uniform residence time, dust generation and high noise levels.

Fluidized-Bed Dryers (Figure 10.41)

In this type of dryer, the drying gas is passed through the bed of solids at a velocity sufficient to keep the bed in a fluidized state; which promotes high heat transfer and drying rates.

Fluidized-bed dryers are suitable for granular and crystalline materials within the particle size range 1 to 3 mm. They are designed for continuous and batch operation.

The main advantages of fluidized dryers are: rapid and uniform heat transfer, short drying times with good control of the drying conditions, and low floor area requirements. The power requirements are high compared with other types.

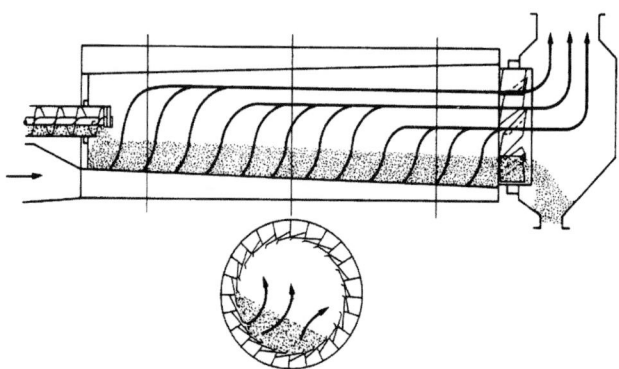

Figure 10.40.   Rotary dryer.

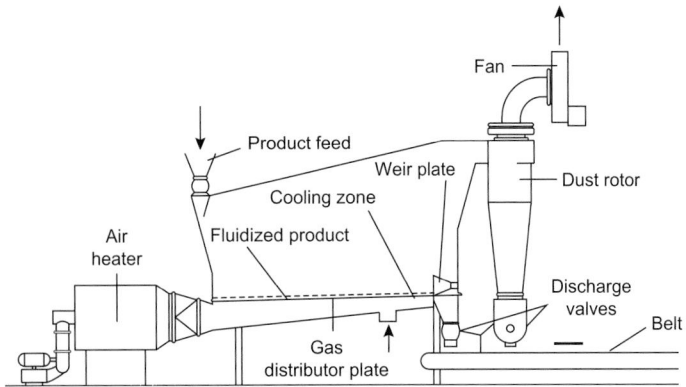

Figure 10.41. Fluidized bed dryer.

Pneumatic Dryers (Figure 10.42)

Pneumatic dryers, also called flash dryers, are similar in their operating principle to spray dryers. The product to be dried is dispersed into an upward-flowing stream of hot gas by a suitable feeder. The equipment acts as a pneumatic conveyor and dryer.

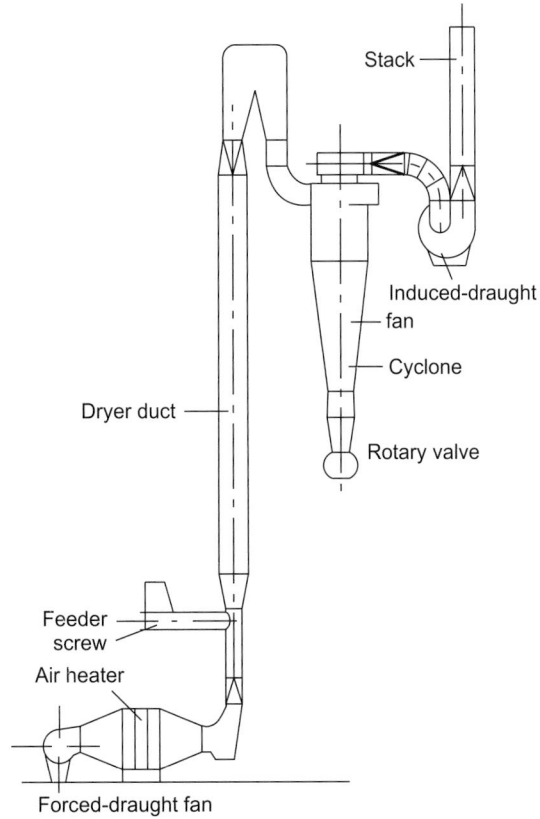

Figure 10.42. Pneumatic dryer.

Contact times are short, and this limits the size of particle that can be dried. Pneumatic dryers are suitable for materials that are too fine to be dried in a fluidized bed dryer but which are heat sensitive and must be dried rapidly. The thermal efficiency of this type is generally low.

Spray Dryers (Figure 10.43)

Spray dryers are normally used for liquid and dilute slurry feeds, but can be designed to handle any material that can be pumped. The material to be dried is atomized in a nozzle, or by a disc-type atomizer, positioned at the top of a vertical cylindrical vessel. Hot air flows up the vessel (in some designs downward) and conveys and dries the droplets. The liquid vaporizes rapidly from the droplet surface and open, porous particles are formed. The dried particles are removed in a cyclone separator or bag filter.

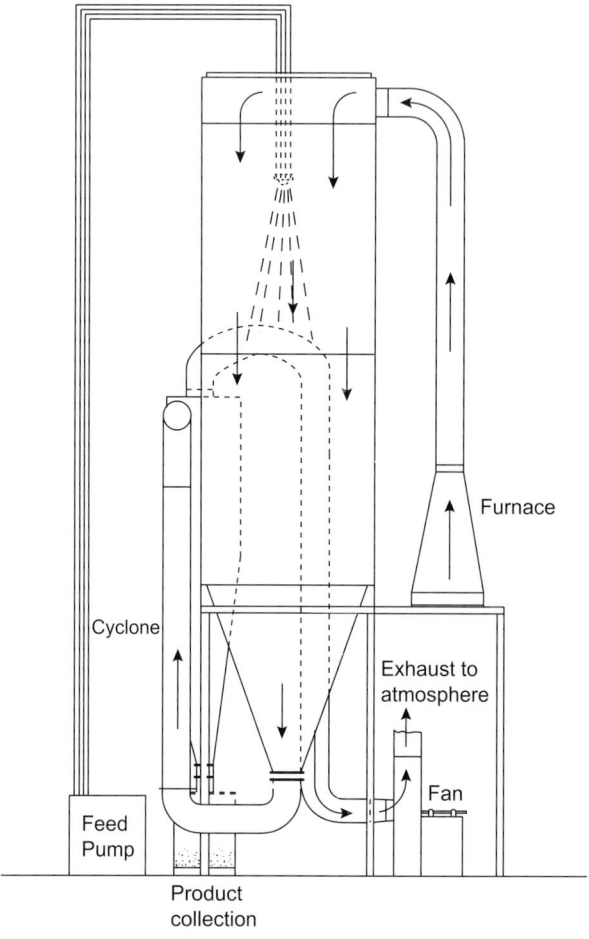

Figure 10.43.   Spray dryer.

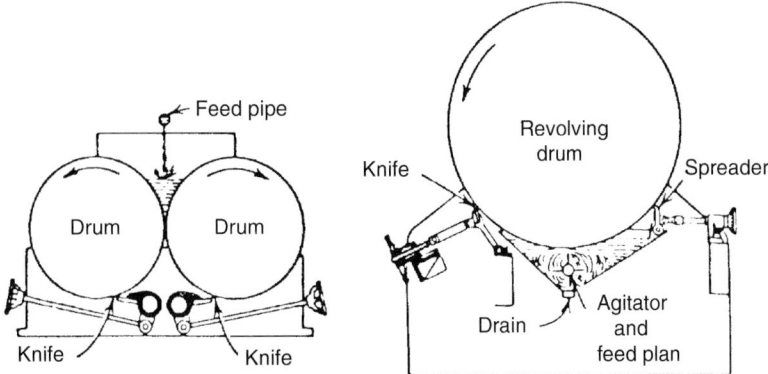

Figure 10.44. Rotary drum dryers.

The main advantages of spray drying are the short contact time, making it suitable for drying heat-sensitive materials, and good control of the product particle size, bulk density and form. Because the solids concentration in the feed is low, the heating requirements will be high. Spray drying is discussed in a book by Masters (1991).

Rotary Drum Dryers (Figure 10.44)

Drum dryers are used for liquid and dilute slurry feeds. They are an alternative choice to spray dryers when the material to be dried will form a film on a heated surface, and is not heat sensitive.

A drum dryer consists essentially of a revolving, internally heated, drum, on which a film of the solids is deposited and dried. The film is formed either by immersing part of the drum in a trough of the liquid or by spraying, or splashing, the feed on to the drum surface; double drums are also used in which the feed is fed into the 'nip' formed between the drums.

The drums are usually heated with steam, and steam economies of 1.3 kg steam per kg of water evaporated are typically achieved.

10.5. SEPARATION OF DISSOLVED SOLIDS

On an industrial scale, evaporation and crystallization are the main processes used for the recovery of dissolved solids from solutions. Membrane filtration processes, such as reverse osmosis, and micro- and ultra-filtration, are used to 'filter out' dissolved solids in certain applications. Ion exchange is used to substitute one ion for another in electrolyte solutions; for example, replacing a metal cation with H^+ to form an acid that can then be recovered by distillation.

10.5.1. Evaporators

Evaporation is the removal of a solvent by vaporization, from solids that are not volatile. It is normally used to produce a concentrated liquid, often prior to crystallization, but a dry solid product can be obtained with some specialized designs. The general subject of evaporation is covered in Richardson *et al.* (2002), Chapter 14. The selection of the appropriate type of evaporator is discussed by Cole (1984). Evaporation is the subject of a book by Billet (1989).

Many evaporator designs have been developed for specialized applications in particular industries. The designs can be grouped into the following basic types.

Direct-Heated Evaporators

This type includes solar pans and submerged combustion units. Submerged combustion evaporators can be used for applications where contamination of the solution by the products of combustion is acceptable.

Long-Tube Evaporators (Figure 10.45)

In this type, the liquid flows as a thin film on the walls of a long, vertical, heated, tube. Both falling film and rising film types are used. They are high-capacity units; suitable for low-viscosity solutions.

Forced-Circulation Evaporators (Figure 10.46)

In forced circulation evaporators the liquid is pumped through the tubes. They are suitable for use with materials that tend to foul the heat transfer surfaces, and where crystallization can occur in the evaporator.

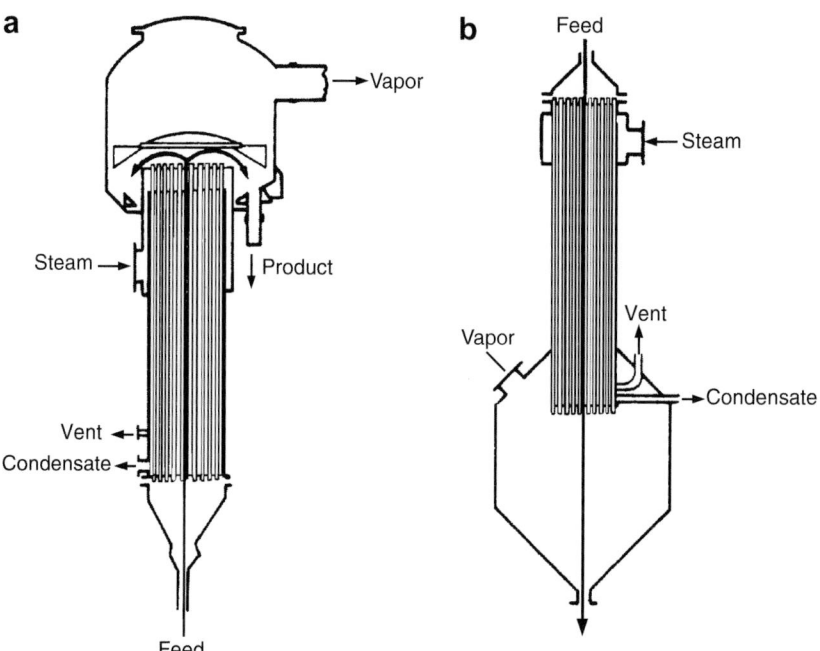

Figure 10.45.   Long-tube evaporators. (a) Rising film. (b) Falling film.

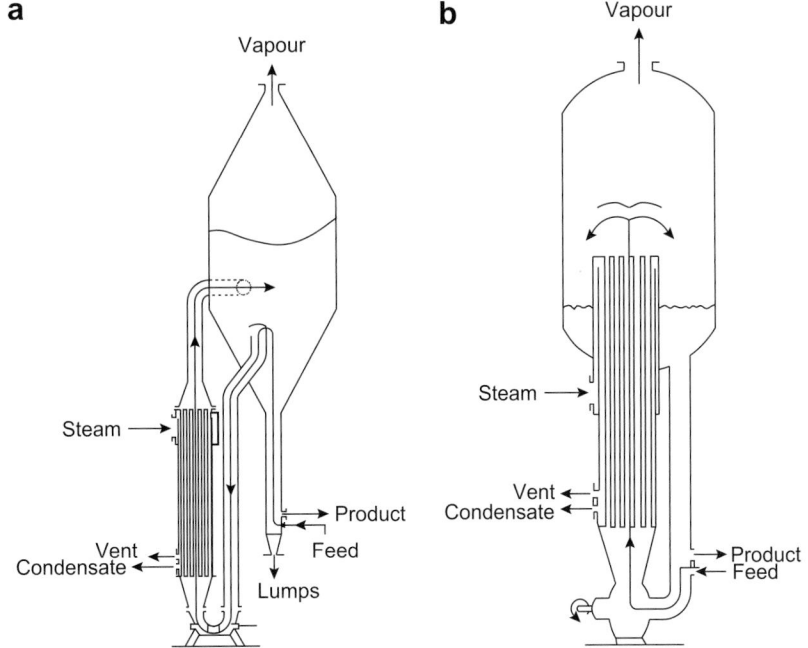

Figure 10.46. Forced-circulation evaporators. (a) Submerged tube. (b) Boiling tube.

Agitated Thin-Film Evaporators (Figure 10.47)

In this design, a thin layer of solution is spread on the heating surface by mechanical means. Wiped-film evaporators are used for very viscous materials and for producing solid products. The design and applications of this type of evaporator are discussed by Mutzenburg (1965), Parker (1965) and Fischer (1965).

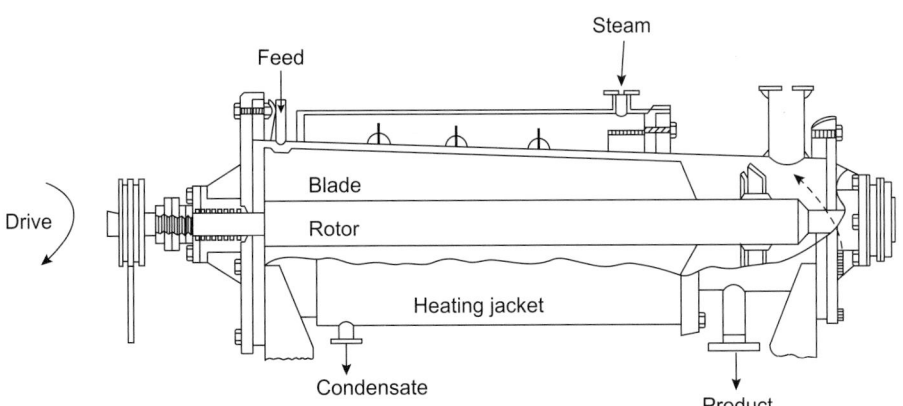

Figure 10.47. Horizontal wiped-film evaporator.

Short-Tube Evaporators

Short-tube evaporators, also called callandria evaporators, are used in the sugar industry; see Richardson *et al.* (2002).

Evaporator Selection

The selection of the most suitable evaporator type for a particular application will depend on the following factors:

1. The throughput required
2. The viscosity of the feed and the increase in viscosity during evaporation
3. The nature of the product required; solid, slurry, or concentrated solution
4. The heat sensitivity of the product
5. Whether the materials are fouling or non-fouling
6. Whether the solution is likely to foam
7. Whether direct heating can be used.

A selection guide based on these factors is given in Figure 10.48; see also Parker (1963b).

Auxiliary Equipment

Condensers and vacuum pumps will be needed for evaporators operated under vacuum. For aqueous solutions, steam ejectors and jet condensers are normally used. Jet condensers are direct-contact condensers, where the vapour is condensed by contact with jets of cooling water. Indirect, surface condensers are used where it is necessary to keep the condensed vapour and cooling water effluent separate.

| Evaporator type | Feed conditions | | | | | | | Suitable for heat-sensitive materials |
| | Viscosity, mN s/m^2 | | | | | | | |
	Very viscous > 1000	Medium viscosity < 1000 max	Low viscosity < 100	Foaming	Scaling or fouling	Crystals produced	Solids in suspension	
Recirculating Calandria (short vertical tube)		←					→	No
Forced circulation		←				→		Yes
Falling film			← →					No
Natural circulation			← →					No
Single pass wiped film	←						→	Yes
Tubular (long tube) Falling film			← →					Yes
Rising film			← →					Yes

Figure 10.48.　Evaporator selection guide.

10.5.2. Crystallization

Crystallization is used for the production, purification and recovery of solids. Crystalline products have an attractive appearance, are free flowing, and easily handled and packaged. The process is used in a wide range of industries: from the small-scale production of specialized chemicals, such as pharmaceutical products, to the high tonnage production of products such as sugar, common salt and fertilizers.

Crystallization theory is covered in Richardson *et al.* (2002), Chapter 15 and in other texts: Mullin (2001) and Jones (2002). Descriptions of the various crystallizers used commercially can be found in these texts and in handbooks: Mersmann (2001), Myerson (1993), Green and Perry (2007) and Schweitzer (1997). Procedures for the scale-up and design of crystallizers are given by Mersmann (2001) and Mersham (1988, 1984).

Crystallization equipment can be classified by the method used to obtain super-saturation of the liquor, and also by the method used to suspend the growing crystals. Super-saturation is obtained by cooling or evaporation. There are four basic types of crystallizer; these are described briefly below.

Tank Crystallizers

Tank crystallizers are the simplest type of industrial crystallizing equipment. Crystallization is induced by cooling the mother liquor in tanks, which may be agitated and equipped with cooling coils or jackets. Tank crystallizers are operated batch-wise, and are generally used for small-scale production.

Scraped-Surface Crystallizers

Scraped-surface crystallizers are similar in principle to the tank type, but the cooling surfaces are continually scraped or agitated to prevent fouling by deposited crystals and to promote heat transfer. They are suitable for processing high-viscosity liquors. Scraped-surface crystallizers can be operated batch-wise, with recirculation of the mother liquor, or continuously. A disadvantage of this type is that they tend to produce very small crystals.

Circulating Magma Crystallizers (Figure 10.49)

In this type, both the liquor and growing crystals are circulated through the zone in which super-saturation occurs. Circulating magma crystallizers are probably the most important type of large-scale crystallizers used in the chemical process industry. Designs are available in which super-saturation is achieved by direct cooling, evaporation or evaporative cooling under vacuum.

Circulating Liquor Crystallizers (Figure 10.50)

In a circulating liquor crystallizer only the liquor is circulated through the heating or cooling equipment; the crystals are retained in suspension in the crystallizing zone by the up-flow of liquor. Circulating liquor crystallizers produce crystals of regular size. The basic design consists of three components: a vessel in which the crystals are suspended and grow and are removed; a means of producing super-saturation, by cooling or evaporation; and a means of circulating the liquor.

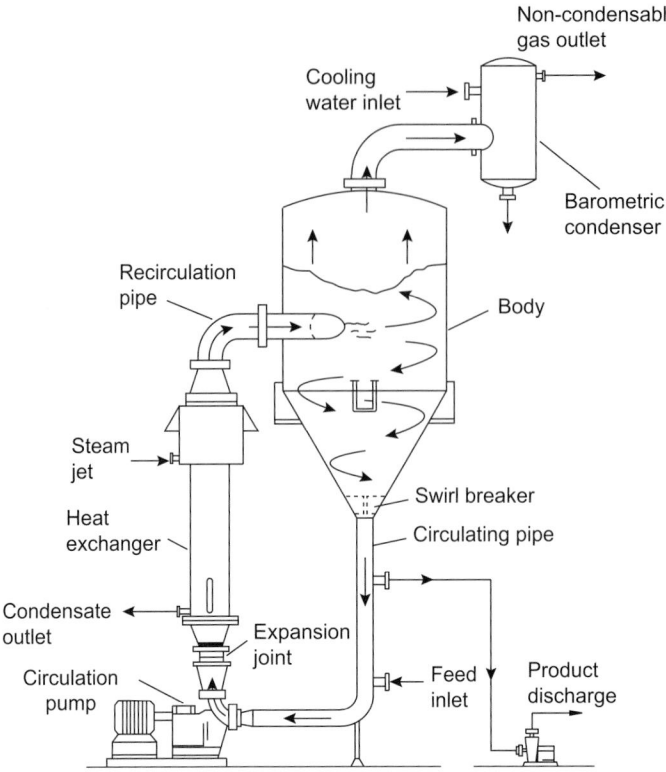

Figure 10.49. Circulating magma crystallizer (evaporative type).

The Oslo crystallizer (Figure 10.50) is the archetypical design for this type of crystallizing equipment.

Circulating liquor crystallizers and circulating magma crystallizers are used for the large-scale production of a wide range of crystal products.

Typical applications of the main types of crystallizer are summarized in Table 10.10; see also Larson (1978).

Crystallizer Design

Crystallizers are normally sized in consultation with a specialist equipment vendor. The important design parameters are:

1. Process throughput
2. Feed concentration
3. Target solids yield (recovery)
4. Target particle size distribution
5. Product purity (particularly in fractional crystallization)
6. Heat addition or removal requirements, including latent heat as well as sensible heat.

The yield of a crystallizer is limited by both solid–liquid equilibrium and the desire to maintain a low enough solids fraction to allow slurry flow of the product. If the

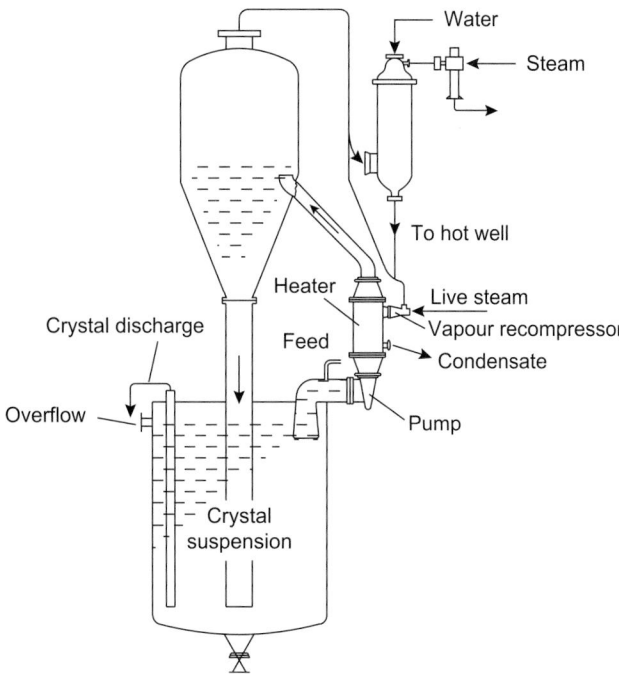

Figure 10.50.   Oslo evaporative crystallizer.

solvent is removed by evaporation, then the mother liquor can be recycled to the crystallizer after the crystals have been removed by filtration, as long as impurities do not accumulate in the recycle.

10.5.3. Precipitation

Precipitation can be considered as a branch of crystallization, although the solid phase that is formed need not be crystalline.

TABLE 10.10.   Selection of Crystallizers

Crystallizer Type	Applications	Typical Uses
Tank	Batch operation, small-scale production	Fatty acids, vegetable oils, sugars
Scraped surface	Organic compounds, where fouling is a problem, viscous materials	Chlorobenzenes, organic acids, paraffin waxes, naphthalene, urea
Circulating magma	Production of large-sized crystals. High throughputs	Ammonium and other inorganic salts, sodium and potassium chlorides
Circulating liquor	Production of uniform crystals (smaller size than circulating magma). High throughputs.	Gypsum, inorganic salts, sodium and potassium nitrates, silver nitrates

The solubility of organic solutes can be influenced by temperature, composition, pH, solvent polarity and ionic strength. If something is added to the solvent to change one or more of these properties then the solute can sometimes be precipitated out of solution. If the solvent volume is not significantly changed, the recovery of solute is equal to the change in solute solubility divided by the initial solubility:

$$\text{Solute recovery} = \frac{\text{solute precipitated}}{\text{total solute fed}} \approx \frac{\text{initial solubility - final solubility}}{\text{initial solubility}}$$

(10.10)

Precipitation is widely used in the recovery of large organic molecules; such as, specialty chemicals, pharmaceutical and food compounds, proteins, and other biological products. Some of the common techniques that are used include:

1. 'Salting out', in which a salt such as calcium citrate, calcium chloride or ammonium sulphate is added to an aqueous solution to raise the ionic strength and cause precipitation.
2. Changing solution polarity by adding methanol, ethanol, acetone, acetonitrile or other suitable solvent.
3. Changing the pH by adding acid (sometimes known as acidulation) or base.
4. Heat treatment ('cooking') to thermally degrade an unwanted solute that is then precipitated.
5. Adsorption precipitation by adding diatomaceous earth, casein, gelatine, activated carbon, clay or other large particles that can adsorb the organic species and then settle as precipitate.

Precipitation operations usually do not require evaporation of solvent or cooling of a saturated solution, and so can be carried out in simpler equipment than crystallization. The process usually consists of a mixing tank or in-line mixer, followed by a solid–liquid separation device such as a hydrocyclone or centrifuge, as described in Section 10.4.

Precipitation is discussed in detail by Sohnel and Garside (1992).

10.5.4. Membrane Separations

Membranes are widely used for concentration of solutions of dissolved solids, as well as suspensions of particulates. A tubular membrane that is permeable to the solvent but not the solute can be used to remove solvent from a solution in the same manner as cross-flow filtration (Section 10.4.2). These processes are classified as microfiltration, ultrafiltration or nanofiltration depending on the size of particulate or molecule that is retained; see Table 10.11. The use of a membrane for solvent removal is usually preferred over solvent evaporation if the solute is sensitive to high temperatures. This is often the case for biologically active large molecules such as proteins and enzymes and for flavour compounds found in foods and beverages.

Membranes can also be found that are selective for the solute over the solvent. Such membranes allow the solute to be transferred to a different solvent without intimate mixing of the solvents, which can be beneficial for solvent or product degradation or when the two solvents are mutually miscible. In the pharmaceuticals industry, the use of solute-selective membranes is known as diafiltration, because of the close analogy

TABLE 10.11. Membrane Filtration Process

Process	Approximate Size Range (m)	Applications
Microfiltration	10^{-8} to 10^{-4}	pollen, bacteria, blood cells
Ultrafiltration	10^{-9} to 10^{-8}	proteins and virus
Nanofiltration	5×10^{-9} to 15×10^{-9}	water softening
Reverse osmosis	10^{-10} to 10^{-9}	desalination
Dialysis	10^{-9} to molecules	blood purification
Electrodialysis	10^{-9} to molecules	separation of electrolytes
Pervaporation	10^{-9} to molecules	dehydration of ethanol
Gas permeation	10^{-9} to molecules	hydrogen recovery, dehydration

to dialysis in organisms. A recent variant on diafiltration is to use a charged membrane to increase the selectivity for solute; see Mehta *et al.* (2008) for a good example.

When membranes are used for filtration or solute concentration, the limit on solvent recovery is set by the need to avoid fouling or scaling of the membrane, and maintain a pumpable slurry. Mass transfer at the membrane wall causes the fluid closest to the membrane to be depleted in solvent; a phenomenon known as concentration polarization. Because the fluid at the membrane wall is depleted in solvent and enriched in solute, precipitation or crystallization can occur if the solute solubility is exceeded. Either of these processes can lead to membrane fouling and loss of throughput. Solvent recovery membranes are therefore usually operated at an outlet concentration well below saturation.

The design of membranes for solute transfer is similar to the design of gas separation membranes, as described in Section 10.2.2, with the exception that a second solvent is usually introduced on the permeate side to remove the solute. Hollow-fibre or tubular membranes are most commonly used.

The design of membrane separations for liquids is discussed in Scott and Hughes (1996), Cheryan (1986), McGregor (1986), Rautenbach and Albrecht (1989), Noble and Stern (1995), Mulder (1996), Porter (1997), Hoffman (2003) and Baker (2004). Applications of membranes to biological systems are described in the book by Wang (2001). The special case of recovery of purified water from salt solutions by reverse osmosis is discussed below.

Reverse Osmosis

Reverse osmosis (RO) is by far the most widely used membrane process. In a reverse osmosis plant, water passes through a membrane, while dissolved minerals and other solids are rejected in the retentate. Reverse osmosis is used to generate deionized process feed water, purify boiler feed water, recover water from waste streams and desalinate sea water or brackish water for drinking and irrigation.

In a reverse osmosis process, the feed water is pressurized to provide an adequate pressure gradient to overcome the difference in osmotic potential between the briny retentate and the purer permeate. Under the applied pressure gradient, water flows through the membrane against the concentration gradient. The membranes are usually designed as spiral-wound modules and are operated in cross-flow; see Section 10.2.2.

Reverse osmosis plants are usually purchased as modular plants designed by one of the major water-treatment companies. Cost correlations have been developed and can be found in Aspen ICARUS and other cost-estimating programs.

The recovery of water from an RO plant depends on the feed water quality, the product specifications and the need to prevent membrane fouling. Highly pure water is usually not obtained in a single stage. The membrane typically rejects 96% to 98% of the salts per stage, and several stages may be used to achieve the desired purity, with recycle of retentate as described in Section 10.2.2.

As with any solvent-rejection membrane, the designer must ensure that the retentate will not reach concentrations of solute that will cause the membrane to be fouled. Allowance must also be made for concentration polarization near the membrane. At high recovery of water, the osmotic pressure of the solution on the retentate side increases dramatically. The limit on recovery is often set by an economic trade-off between the cost of consuming additional feed and the cost of pumping to a higher pressure. Table 10.12 gives the osmotic pressure of NaCl and sea salt solutions at different concentrations at 40°C. Some guidelines on membrane fluxes and retentate flows are given by Kucera (2008). A typical product recovery when generating purified water from town water is 50% to 75%, but the recovery depends strongly on the factors described above and is often less than 30% in

TABLE 10.12. Osmotic Pressure of Sea Salt and Sodium Chloride Solutions at 40°C (Stoughton and Lietzke, 1965)

Molality of NaCl (mol/kg)	Osmotic Pressure (atm)
0.01	0.49
0.10	4.76
0.50	23.60
1.0	48.08
1.50	73.93
2.00	101.3
3.00	161.6
4.00	230.5
5.00	309.4

Weight % Sea Salts (wt%)	Osmotic Pressure (atm)
1.00	7.41
2.00	14.88
3.45*	26.17
5.00	38.96
7.50	61.40
10.00	86.46
15.00	146.6
20.00	225.1
25.00	331

*3.45 wt% solids is taken as the value for standard sea water.

desalination plants. When specifying the use of an RO system for process feed water, the design engineer must allow for the cost of the extra feed water that is needed. This is illustrated in Example 10.4, below.

The performance of RO plants can often be substantially improved by pre-treatment of the water feed. Common pre-treatment processes include filtration, softening of the water by cation exchange, activated carbon adsorption of chlorine and organics, and addition of chemicals to prevent biological fouling and suppress precipitation.

Because of its widespread use, there is an abundant literature on reverse osmosis. It is covered in all of the general books on membranes listed above, and also in the specialist books by Amjad (1993), Byrne (1995), Wilf *et al.* (2007) and AWWA (2007).

Example 10.4

A reverse osmosis plant is to be designed to produce 50 kg/s of boiler feed water with less than 20 ppmw of NaCl starting from sea water that contains 3.5 wt% NaCl. The 40-m^2 membrane modules operate at 60 atm and achieve a flux of 0.4 m^3/m^2.day when the permeate pressure is 2 atm. If each membrane module has a rejection of 96% of the salt fed to it, determine the overall membrane sequence and the water feed rate required.

Solution

Target concentration = 20 ppmw = 0.002 wt%.

If the permeate pressure is 2 atm and the retentate is at 60 atm then the pressure drop across the membrane is $60 - 2 = 58$ atm.

From Table 10.12, an osmotic pressure of 58 atm corresponds to a concentration of 7.44 wt%. We need to make an allowance for concentration polarization at the membrane, so assume that the retentate bulk concentration is 70% of this concentration, i.e., 5.21 wt%.

Mass balance on salt across first stage, basis 100 kg/s of feed, 96% rejection of salts:

Feed	=	Permeate	+	Retentate
3.5	=	3.5×0.04	+	3.5×0.96
	=	0.14	+	3.36

So if the retentate water flow rate is x, then:

$$\frac{3.36}{x + 3.36} = 0.0521, \text{ hence } x = \frac{3.36}{0.0521} - 3.36 = 61.1 \text{kg/s}$$

So, by difference, the water flow in the permeate is $96.5 - 61.1 = 35.4$ kg/s, and the salt concentration in the permeate is $0.14/(0.14 + 61.1) = 0.39$ wt%, which does not meet the target.

If we assume a simple cascade with no retentate recycle, then the second stage can be calculated in the same way:

$$\text{Salt retained in permeate} = 0.14 \times 0.04 = 0.0056 \text{ kg/s}$$

$$\text{Salt rejected in retentate} = 0.14 \times 0.96 = 0.1344 \text{ kg/s}$$

$$\text{Retentate water flow} = \frac{0.1344}{0.0521} - 0.1344 = 2.44 \text{ kg/s}$$

$$\text{Permeate water flow} = 35.4 - 2.44 = 32.94 \text{ kg/s}$$

Permeate salt concentration $= 0.0056/(0.0056 + 32.94) = 0.017$ wt%, which still does not meet the target.

This calculation can be repeated for subsequent stages and is easily coded into a spreadsheet. The third-stage results are:

$$\text{Salt retained in premeate} = 0.0056 \times 0.04 = 0.000224 \text{ kg/s}$$

$$\text{Salt rejected in retentate} = 0.0056 \times 0.96 = 0.005376 \text{ kg/s}$$

$$\text{Retentate water flow} = \frac{0.005376}{0.0521} - 0.005376 = 0.098 \text{ kg/s}$$

$$\text{Permeate water flow} = 32.94 - 0.098 = 32.85 \text{ kg/s}$$

Permeate salt concentration $= 0.000224/(0.000224 + 32.85) = 0.00068$ wt%, which now exceeds the target specification, so a part of the second-stage product can bypass the third-stage; however, for design purposes we will assume that all the stages are fully used, as this provides some additional safety factor for fouling, concentration polarization, etc.

The overall recovery of water is 32.85 kg/s out of 96.5 kg/s fed, i.e., 34%, so the feed rate to produce 50 kg/s of RO water is $50/0.34 = 147$ kg/s of sea water.

Note that the retentate flow in the third stage has become very small. This is not a practical value and a higher flow rate would be necessary. A higher flow rate can be accomplished by reducing the outlet retentate concentration and recycling the retentate to an earlier point in the network; see Figure 10.10c. The network with recycles is more difficult to solve using hand calculations, but can easily be modelled using process simulation software. If the process simulation program does not have a membrane unit operation then the RO membrane can be modelled as a fixed-split separator; see Section 4.5.3.

By examination of the permeate flows calculated above, we see that the permeate flow is essentially unchanged and roughly 32.9 kg/s from stage 2 onwards ($\approx$ 50 kg/s on the design flow rate basis). The permeate from stage 1 is 35.4 kg/s, which when corrected to the 50 kg/s production basis would be $35.4 \times 147/100 = 52$ kg/s. So if we allow roughly 10% additional flow (relative to product rate) in each stage for recycles then we can assume 55 kg/s flow in each stage

$$55 \text{ kg/s} = \frac{55 \times 3600 \times 24}{1000} = 4752 \text{ m}^3/\text{day}$$

So if each 40-m^2 membrane module permeates 0.4 m^3/m^2.day, then the total number of modules per stage:

$$= \frac{4752}{40 \times 0.4} = 297$$

Hence the total number of modules $= 297 \times 3 = 891$.

With some allowance for spare modules, the sequence is then three stages, each containing 300 modules, with roughly 10% recycle from stages 2 and 3. These results would need to be confirmed by detailed simulation and discussion with a vendor.

10.5.5. Ion Exchange

Ion exchange is used for water softening, demineralization and separation and recovery of salts, including salts of organic acids and bases. In an ion-exchange process, the solution flows through a bed of resin beads. The resin is a polymer that has been functionalized by the addition of either acidic or basic groups. For example, sulphonated polystyrene contains $-SO_3$ groups that attach to cations from the solution, and hence can be used as a cation-exchange resin. The particular choice of acidic or basic groups allows the designer to modify the strength of interaction and hence the selectivity of the resin.

When a solution is passed over a cation-exchange resin, the cations in the solution equilibrate with the cations that were attached to the resin, and hence effectively become adsorbed onto the resin. When the resin nears breakthrough, it can be regenerated by washing with a solution of counter-ion; typically H^+, Na^+ or Ca^{2+} for cation exchange or Cl^-, HO^-, or NO_3^- for anion exchange.

Full deionization can be accomplished by carrying out cation exchange using H^+ as the counter-ion, followed by anion exchange using HO^- as the counter-ion.

The most common ion-exchange process is water softening, in which Ca^{2+} and Mg^{2+} that occur naturally in hard water are exchanged with Na^+ using a cation-exchange resin that is regenerated with NaCl. Water softening is used for boiler feed water and to prepare water for reverse osmosis units. Small units are also used for domestic water softening in regions that have high water hardness.

The capacity of an ion-exchange resin depends on the extent of functionalization of the polymer, and is normally expressed in mmol/g or mmol/mL of resin. The units millimoles (mmol) are sometimes written as milliequivalents (meq). For cation-exchange resins the loading is per g or mL of dry hydrogen-form resin, and for anion-exchange resins it is usually per g or mL of dry chlorine-form resin. Capacities for some of the more commonly used resins are given in Green and Perry (2007). A rough estimate of the total bed volume required can be made by assuming that the bed operates to 70% of breakthrough; a more detailed analysis would usually be made by a specialist designer.

As with adsorption (see Section 10.2.1), a continuous ion-exchange system requires at least two beds of resin, so that one can be in regeneration while the other is in process service.

When specifying an ion-exchange system, the design engineer must allow for the regenerant and effluent treatment systems. The regenerant is a salt of the counter-ion, usually in aqueous solution. The regenerant is required in an amount that is somewhat greater than the stoichiometric equivalent of the exchanged ion, to provide an adequate difference in chemical potential to drive the exchanged ion off the resin and ensure that regeneration is completed. A good initial estimate is 150% to 200% of the stoichiometric equivalent. The regenerant is often fed at high concentration, to minimize the amount of effluent formed during regeneration. The spent regenerant may require neutralization or other additional treatment before it can be sent to a waste-water plant.

Many pharmaceutical products and intermediates are organic salts that can be recovered by ion exchange. If the ion loaded on the resin is the desired product, then the regenerant should be chosen to give the product in a suitable form for further processing.

An introduction to the theory of ion exchange is given in Chapter 18 of Richardson *et al.* (2002). A more detailed discussion of the technology is given in the book by Helfferich (1995). Wachinski and Etzel (1997) discuss the application of ion exchange to waste recovery.

10.5.6. Adsorption

Fixed beds of adsorbent are occasionally used for removing small amounts of dissolved solids or liquids from a liquid stream. Common adsorbents are silica, alumina, activated carbon, zeolites and clays.

When removing a dissolved solid, the process is usually treated as irreversible adsorption; see Section 10.2.1. Reversible adsorption of dissolved components with regeneration of the sorbent using a different solvent is a form of chromatography, and is discussed in Section 10.7.2.

10.6. LIQUID–LIQUID SEPARATION

Separation of two liquid phases, immiscible or partially miscible liquids, is a common requirement in the process industries. For example, in the unit operation of liquid–liquid extraction, the liquid-contacting step must be followed by a separation stage (Chapter 11, Section 11.16). It is also frequently necessary to separate small quantities of entrained water from process streams. The simplest form of equipment used to separate liquid phases is the gravity settling tank, the decanter. Various types of proprietary equipment are also used to promote coalescence and improve separation in difficult systems, or where emulsions are likely to form. Centrifugal separators are also used.

10.6.1. Decanters (Settlers)

Decanters are used to separate liquids where there is a sufficient difference in density between the liquids for the droplets to settle readily. Decanters are essentially tanks that give sufficient residence time for the droplets of the dispersed phase to rise (or settle) to the interface between the phases and coalesce. In an operating decanter there will be three distinct zones or bands: clear heavy liquid, separating dispersed liquid (the dispersion zone) and clear light liquid.

Decanters are normally designed for continuous operation, but the same design principles apply to batch-operated units. Many vessel shapes are used for decanters, but for most applications a cylindrical vessel will be suitable, and will be the cheapest shape. Typical designs are shown in Figures 10.51 and 10.52. The position of the interface can be controlled, with or without the use of instruments, by use of a siphon take-off for the heavy liquid, Figure 10.51.

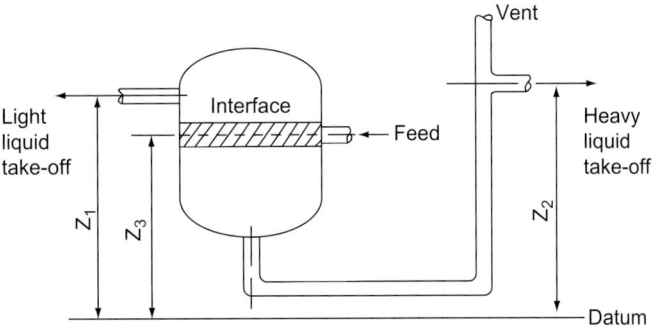

Figure 10.51. Vertical decanter.

The height of the take-off can be determined by making a pressure balance. Neglecting friction loss in the pipes, the pressure exerted by the combined height of the heavy and light liquid in the vessel must be balanced by the height of the heavy liquid in the take-off leg, Figure 10.51.

$$(z_1 - z_3)\rho_1 g + z_3 \rho_2 g = z_2 \rho_2 g$$

hence

$$z_2 = \frac{(z_1 - z_3)\rho_1}{\rho_2} + z_3 \qquad (10.11)$$

where

ρ_1 = density of the light liquid, kg/m^3
ρ_2 = density of the heavy liquid, kg/m^3
z_1 = height from datum to light liquid overflow, m
z_2 = height from datum to heavy liquid overflow, m
z_3 = height from datum to the interface, m.

The height of the liquid interface should be measured accurately when the liquid densities are close, when one component is present only in small quantities, or when the throughput is very small. A typical scheme for the automatic control of the interface, using a level instrument that can detect the position of the interface, is

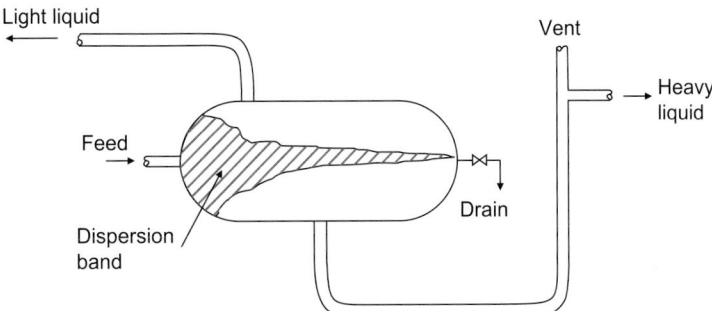

Figure 10.52. Horizontal decanter.

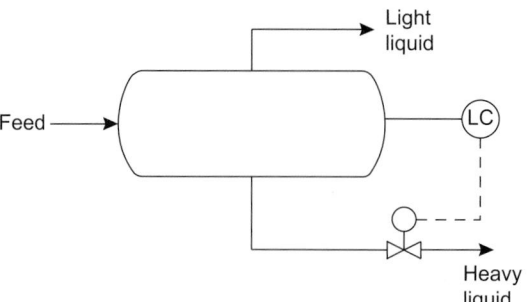

Figure 10.53.   Automatic control, level controller detecting interface.

shown in Figure 10.53. Where one phase is present only in small amounts it is often recycled to the decanter feed to give more stable operation.

Decanter Design

A rough estimate of the decanter volume required can be made by taking a hold-up time of 5 to 10 min, which is usually sufficient where emulsions are not likely to form. Methods for the design of decanters are given by Hooper (1997) and Signales (1975). The general approach taken is outlined below and illustrated by Example 10.5.

The decanter vessel is sized on the basis that the velocity of the continuous phase must be less than the settling velocity of the droplets of the dispersed phase. Plug flow is assumed, and the velocity of the continuous phase is calculated using the area of the interface:

$$u_c = \frac{L_c}{A_i} < u_d \qquad (10.12)$$

where

u_d = settling velocity of the dispersed phase droplets, m/s
u_c = velocity of the continuous phase, m/s
L_c = continuous phase volumetric flow rate, m^3/s
A_i = area of the interface, m^2.

Stokes' law (see Richardson *et al.*, 2002, Chapter 3) is used to determine the settling velocity of the droplets:

$$u_d = \frac{d_d^2 g(\rho_d - \rho_c)}{18\mu_c} \qquad (10.13)$$

where

d_d = droplet diameter, m
u_d = settling (terminal) velocity of the dispersed phase droplets with diameter d, m/s
ρ_c = density of the continuous phase, kg/m^3
ρ_d = density of the dispersed phase, kg/m^3
μ_c = viscosity of the continuous phase, N s/m^2
g = gravitational acceleration, 9.81 m/s^2.

Equation 10.13 is used to calculate the settling velocity with an assumed droplet size of 150 μm, which is well below the droplet sizes normally found in decanter feeds. If the calculated settling velocity is greater than 4×10^{-3} m/s, then a figure of 4×10^{-3} m/s is used.

For a horizontal, cylindrical, decanter vessel, the interfacial area will depend on the position of the interface.

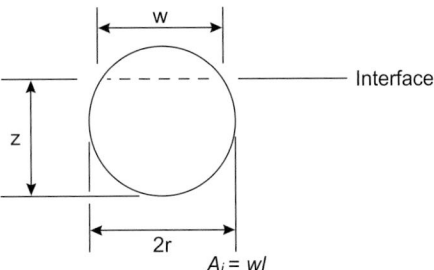

and

$$w = 2(2rs - z^2)^{1/2}$$

where

w = width of the interface, m
z = height of the interface from the base of the vessel, m
l = length of the cylinder, m
r = radius of the cylinder, m.

For a vertical, cylindrical decanter:

$$A_i = \pi r^2$$

The position of the interface should be such that the band of droplets that collect at the interface waiting to coalesce and cross the interface does not extend to the bottom (or top) of the vessel. Ryon *et al.* (1959) and Mizrahi and Barnea (1973) have shown that the depth of the dispersion band is a function of the liquid flow rate and the interfacial area. A value of 10% of the decanter height is usually taken for design purposes. If the performance of the decanter is likely to be critical to process performance, the design can be investigated using scale models. The model should be scaled to operate at the same Reynolds number as the proposed design, so that the effect of turbulence can be investigated; see Hooper (1975).

Example 10.5

Design a decanter to separate a light oil from water.
   The oil is the dispersed phase.
   Oil, flow rate 1000 kg/h, density 900 kg/m^3, viscosity 3 mN s/m^2.
   Water, flow rate 5000 kg/h, density 1000 kg/m^3, viscosity 1 mN s/m^2.

Solution

Take $d_d = 150 \, \mu m$

$$u_d = \frac{(150 \times 10^{-6})^2 9.81(900 - 1000)}{18 \times 1 \times 10^{-3}} \tag{10.13}$$

$$= -0.0012 \text{ m/s}, \; -1.2 \text{ mm/s (rising)}$$

As the flow rate is small, use a vertical, cylindrical vessel.

$$L_c = \frac{5000}{1000} \times \frac{1}{3600} = 1.39 \times 10^{-3} \text{ m}^3/\text{s}$$

$$u_c \leq u_d, \text{ and } u_c = \frac{L_c}{A_i}$$

Hence

$$A_i = \frac{1.39 \times 10^{-3}}{0.0012} = 1.16 \text{ m}^2$$

$$r = \sqrt{\frac{1.16}{\pi}} = 0.61 \text{ m}$$

$$\text{diameter} = \underline{1.2 \text{ m}}$$

Take the height as twice the diameter, a reasonable value for a decanter:

$$\text{height} = \underline{2.4 \text{ m}}$$

Take the dispersion band as 10% of the height = $\underline{0.24 \text{ m}}$
  Check the residence time of the droplets in the dispersion band

$$\frac{0.24}{u_d} = \frac{0.24}{0.0012} = 200 \text{ s } (\sim 3 \text{ min})$$

This is satisfactory; a time of 2 to 5 min is normally recommended for control purposes. Check the size of the water (continuous, heavy phase) droplets that could be entrained with the oil (light phase).

$$\text{Velocity of oil phase} = \frac{1000}{900} \times \frac{1}{3600} \times \frac{1}{1.16}$$

$$= 2.7 \times 10^{-4} \text{ m/s } (0.27 \text{ mm/s})$$

From equation 10.13

$$d_d = \left[\frac{u_d 18 \mu_c}{g(\rho_d - \rho_c)}\right]^{1/2}$$

so the entrained droplet size will be:

$$= \left[\frac{2.7 \times 10^{-4} \times 18 \times 3 \times 10^{-3}}{9.81(1000 - 900)} \right]^{1/2}$$

$$= 1.2 \times 10^{-4} \text{ m} = 120 \, \mu\text{m}$$

which is satisfactory; below $150 \, \mu\text{m}$.

Piping Arrangement

To minimize entrainment by the jet of liquid entering the vessel, the inlet velocity for a decanter should be kept below 1 m/s.

$$\text{Flow-rate} = \left[\frac{1000}{900} + \frac{5000}{1000} \right] \frac{1}{3600} = 1.7 \times 10^{-3} \text{ m}^3/\text{s}$$

$$\text{Area of pipe} = \frac{1.7 \times 10^{-3}}{1} = 1.7 \times 10^{-3} \text{ m}^2$$

$$\text{Pipe diameter} = \sqrt{\frac{1.7 \times 10^{-3} \times 4}{\pi}} = 0.047 \text{ m, say } \underline{50 \text{ mm}}$$

Take the position of the interface as half-way up the vessel and the light liquid off-take as at 90% of the vessel height, then

$$z_1 = 0.9 \times 2.4 = 2.16 \text{ m}$$
$$z_3 = 0.5 \times 2.4 = 1.2 \text{ m}$$
$$z_2 = \frac{(2.16 - 1.2)}{1000} \times 900 + 1.2 = \underline{\underline{2.06 \text{ m}}}$$
$$\text{say } \underline{2.0 \text{ m}}$$

Proposed Design

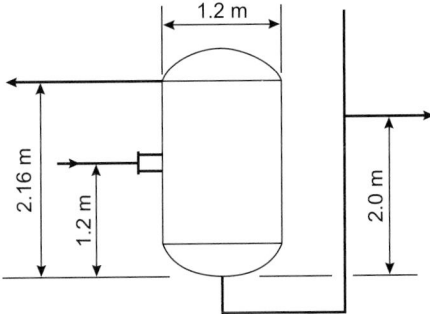

Drain valves should be fitted at the interface so that any tendency for an emulsion to form can be checked, and the emulsion accumulating at the interface drained off periodically as necessary.

10.6.2. Plate Separators

Stacks of horizontal, parallel, plates are used in some proprietary decanter designs to increase the interfacial area per unit volume and to reduce turbulence. They effectively convert the decanter volume into several smaller separators connected in parallel.

10.6.3. Coalescers

Proprietary equipment, in which the dispersion is forced through some form of coalescing medium, is often used for the coalescence and separation of finely dispersed droplets. A medium is chosen that is preferentially wetted by the dispersed phase; knitted wire or plastic mesh, beds of fibrous material, or special membranes are used. The coalescing medium works by holding up the dispersed droplets long enough for them to form globlets of sufficient size to settle. A typical unit is shown in Figure 10.54; see Redmon (1963). Coalescing filters are suitable for separating small quantities of dispersed liquids from large throughputs.

Electrical coalescers, in which a high voltage field is used to break down the stabilizing film surrounding the suspended droplets, are used for desalting crude oils and for similar applications; see Waterman (1965).

10.6.4. Centrifugal Separators

Sedimentation Centrifuges

For difficult separations, where simple gravity settling is not satisfactory, sedimentation centrifuges should be considered. Centrifuging will give a cleaner separation than that obtainable by gravity settling. Centrifuges can be used where the difference in gravity between the liquids is very small, as low as 100 kg/m^3, and they can handle high throughputs, up to around 100 m^3/h. Also, centrifuging will usually break any emulsion that may form. Bowl or disc centrifuges are normally used (see Section 10.4.3).

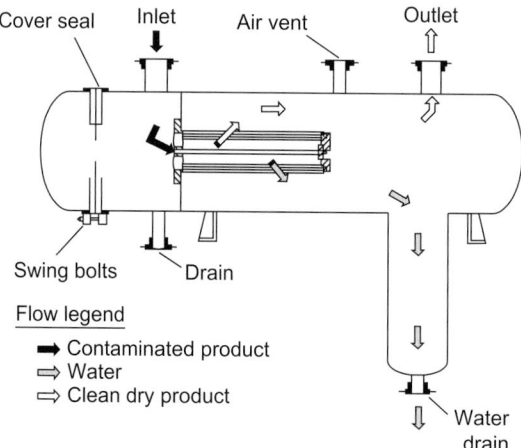

Figure 10.54.   Typical coalescer design.

Hydrocyclones

Hydrocyclones are used for some liquid–liquid separations, but are not as effective in this application as in separating solids from liquids.

10.7. SEPARATION OF DISSOLVED LIQUIDS

The most commonly used techniques for the separation and purification of miscible liquids are distillation and solvent extraction. In recent years, adsorption, ion exchange and chromatography have become practical alternatives to distillation or solvent extraction in many special applications.

Distillation is probably the most widely used separation technique in the chemical process industries, and is covered in Chapter 11 of this book. Solvent extraction and the associated technique, leaching (solid–liquid extraction), are described below and in Section 11.16. Ion exchange, membrane separations and precipitation are discussed under separation of dissolved solids in Section 10.5.

Adsorption can be used to selectively remove a dissolved liquid or solid component from solution. When the adsorbent is regenerated by elution with a solvent, the process is known as preparative chromatography (for small production volumes) or production chromatography (for large volumes), which is discussed in Section 10.7.2. Adsorption with regeneration by cycling temperature or pressure is less commonly used for liquids and is discussed in Section 10.2.1.

10.7.1. Solvent Extraction and Leaching

Solvent Extraction (Liquid–Liquid Extraction)

Solvent extraction, also called liquid–liquid extraction, can be used to separate a substance from a solution by extraction into another solvent. It can be used either to recover a valuable substance from the original solution, or to purify the original solvent by removing an unwanted component. Examples of solvent extraction are the extraction of uranium and plutonium salts from solution in nitric acid in the nuclear industry using kerosene as solvent, and the extraction of benzene from reformed naphtha using sulfolane as solvent.

The process depends on the substance being extracted, the solute, having a greater solubility in the solvent used for the extraction than in the original feed solvent. The two solvents must be essentially immiscible.

The solvents are mixed in a contactor, to effect the transfer of solute, and then the phases are separated. The depleted feed solvent leaving the extractor is called the raffinate, and the solute-rich extraction solvent, the extract. The solute is normally recovered from the extraction solvent by distillation, and the extraction solvent recycled.

The simplest form of extractor is a mixer-settler, which consists of an agitated tank and a decanter. For multi-stage extraction processes, liquid–liquid contacting columns are used. The design of extraction columns is discussed in Chapter 11, Section 11.16. See also Richardson *et al.* (2002), Chapter 13, Treybal (1963), Walas (1990), and Green and Perry (2007).

Leaching

Liquids can be extracted from solids by leaching. As the name implies, the soluble component contained in a solid is leached out by contacting the solid with a suitable solvent. A principal application of leaching is in the extraction of valuable oils from nuts and seeds, such as palm oil and rape seed oil.

The equipment used to contact the solids with the solvent is usually a special design to suit the type of solid being processed, and is to an extent unique to the particular industry. General details of leaching equipment are given in Richardson *et al.* (2002), Chapter 10, and in Green and Perry (2007).

The leaching is normally done using a number of stages. In this respect, the process is similar to liquid–liquid extraction, and the methods used to determine the number of stages required are similar.

For a detailed discussion of the procedures used to determine the number of stages required for a particular process, see Richardson *et al.* (2002), Chapter 10, or Prabhudesai (1997).

10.7.2.　Chromatography

The term chromatography is broadly applied to separation processes in which a fluid is separated into components by passing it over a bed of adsorbent in a continuous flow of carrier fluid. Gas chromatography (GC) is widely used as an analytical method, but is only rarely used for product recovery because of the high-volume flow rates and pressure-drop requirements. Liquid-phase chromatography, on the other hand, is extensively used for product recovery and purification, particularly for fine chemicals and biological products.

Most chromatographic separations are carried out in batch or semi-batch mode, but continuous chromatography can be carried out using flow schemes, such as simulated moving-bed (SMB) chromatography described below.

The general principles of chromatography are discussed by Ruthven (1984), Ganetsos and Barker (1992), Richardson *et al.* (2002), Chapter 19, and Hagel *et al.* (2007).

Chromatographic separations rely on different components in the feed having different adsorption equilibria with the solid phase. The solid-phase material can be an inorganic or organic adsorbent, resin or gel, and is sometimes referred to as the stationary phase. The liquid phase is called the mobile phase, and consists of the feed liquid as well as the carrier liquid, which is also sometimes called the eluent or desorbent. The process performance is strongly influenced by the choice of stationary phase and mobile phase.

Batch Chromatography

Batch chromatography processes operate in a very similar mode to laboratory-scale chromatography. They are consequently favoured by chemists for preparative chromatography and smaller scale production chromatography, as the scale-up from lab methods is more straightforward than for continuous chromatography.

In batch chromatography, a pulse of feed is injected into a continuously flowing stream of mobile phase that passes over a long column packed with a suitable stationary phase, as shown in Figure 10.55a. The more strongly adsorbed species pass through the

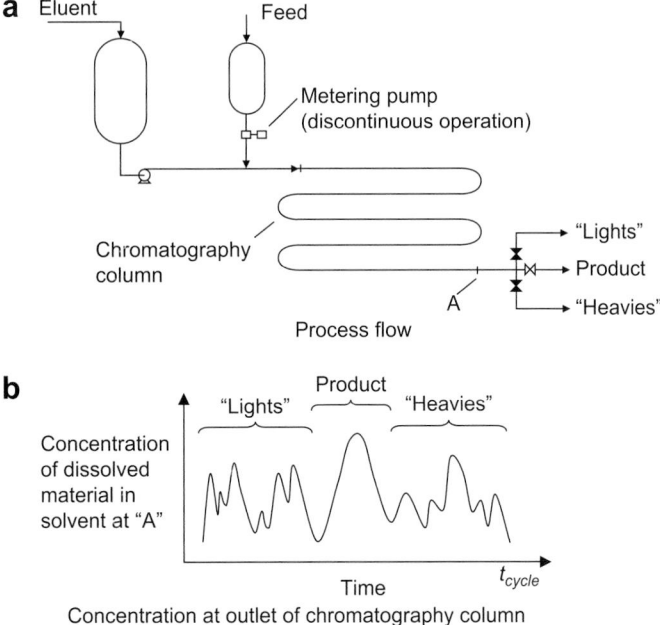

Figure 10.55. Batch chromatography.

column more slowly than less strongly adsorbed species. If a long enough column is used, then the different species can be resolved into fractions (or 'peaks') that are enriched in different components of the feed; Figure 10.55b. If the column effluent composition is monitored, the first fractions recovered will contain material that adsorbs less strongly than the desired product ('lights'). This material can be sent for eluent recovery or discarded as waste. The next fractions are rich in product and are sent to product recovery. Finally, material that is more strongly adsorbed than the desired product ('heavies') is eluted and this material is also sent to solvent recovery or waste disposal.

If all the material can be desorbed from the column within a reasonable time and carrier flow rate, then the column is clean enough to accept a new injection of feed and the process is repeated. Occasionally, the cost in time and desorbent to recover the heavier material is greater than the cost of emptying the column and reloading with fresh stationary phase, in which case the cycle is ended after collection of the product and the spent stationary phase is sent to waste disposal.

The eluent can be recovered from the product and waste streams and recycled to the feed; however, in many fine chemical and pharmaceutical processes the spent solvent is discarded or sold as a by-product, to eliminate concerns about components accumulating in the solvent recycle.

Batch-chromatography columns can be designed as dynamic processes or by using an equilibrium stage analogy; see Chapter 8 of Ruthven (1984).

When a batch column is scaled up to a larger diameter, there will be greater dispersion of the components, leading to broader peaks, and hence the column length must be increased. Batch chromatography is therefore inefficient in the use of stationary phase, particularly if the stationary phase is discarded frequently. When very long columns are used, the pressure drop can be very high, hence the process is sometimes known as high-pressure liquid chromatography or high-performance liquid chromatography (HPLC).

Gel Permeation Chromatography

Gel permeation chromatography is a variation of batch chromatography. The stationary phase is selected to have a pore structure that excludes the desired product. The larger molecules that cannot enter the pores are eluted first, while smaller molecules are eluted later. The separation order is thus reversed from typical batch chromatography. With gel permeation chromatography it is much easier to completely regenerate the sorbent and the cycle time can be reduced. Other aspects of the design are similar to conventional batch chromatography.

Affinity Chromatography

Affinity chromatography is probably the most widely used preparative and production chromatography method in industry, particularly for recovery of biochemicals and large molecules.

In affinity chromatography, the stationary phase is selected or designed to have a highly specific interaction with the desired component. Commonly used interactions include enzyme–inhibitor, antibody–antigen, and lectin–cell-wall. For example, monoclonal antibodies (mAb) can be synthesized that have very highly specific binding to a desired protein. If the mAb is chemically bound to beads of agarose, polyacrylamide or other suitable material, then the beads can be used to pack a column with specific selectivity for the target protein.

Affinity chromatography is in many respects more like an adsorption–desorption process (Section 10.2.1) than a chromatography process. The feed liquid can be passed through the bed without requiring additional eluent. The high selectivity of the stationary phase allows the bed to be kept on stream until the sorbent is fully loaded or nearly fully loaded. The sorbent is then regenerated by the eluent flow. The regeneration step usually involves a change in solvent properties to disrupt the affinity between the sorbent and adsorbed species; for example, a change in solvent polarity, pH, ionic strength or occasionally even temperature. When multiple species are adsorbed, the eluent properties may be changed over time so that different species are eluted in sequence. This is known as applying a solvent gradient.

The cost of an affinity chromatography packing is usually much greater than the cost of conventional chromatography media, even though the adsorbent is used more efficiently and column sizes are generally smaller. The column is therefore more likely to be fully regenerated so that the stationary phase can be reused. Even so, the sorbent performance usually deteriorates over multiple cycles and stationary phase replacement can be a significant consumable cost in biochemical and pharmaceutical processes. For example, Follman and Fahrner (2004) state that protein A affinity chromatography accounts for 35% of the downstream purification costs for monoclonal antibody production.

A good introduction to affinity chromatography is given in the book by Mohr and Pommerening (1986). More recent work is described in Ganetsos and Barker (1992) and Hagel *et al.* (2007).

Continuous Chromatography

A truly continuous chromatography process would employ counter-current flow of the solid phase and the desorbent, as shown in Figure 10.56a. If the liquid feed is introduced at a height h_F above the base and desorbent liquid is introduced at the top, then a component that is more strongly adsorbed on the sorbent will tend to move up the column with the solid phase, giving the composition profile labelled A in Figure 10.56b. Conversely, the component or components that are less strongly adsorbed will tend to move down the column with the desorbent, giving a concentration profile like that labelled B in Figure 10.56b. At height h_E, above the feed point, the liquid phase is substantially free of component B and a product stream containing only desorbent and A can be withdrawn. This stream is known as the extract. Similarly, at height h_R, below the feed point, all of the component A has been adsorbed by the solid and a liquid stream containing only desorbent and B can be recovered. This stream is known as the raffinate.

Above h_E, in the region labelled zone I, the desorbent flow washes A off the solid phase, regenerating a clean sorbent that can be recycled to the bottom of the column. Likewise, below h_R, in the region labelled zone IV, the adsorbent adsorbs remaining B from the liquid, generating a clean desorbent that can be recycled to the top of the column. In practice, this bottom zone is not always used, as it is often cheaper to separate B from desorbent by other methods such as distillation or crystallization.

A true counter-current continuous chromatography process is analogous to a series of absorbers and strippers:

Zone I: desorption (stripping) of A
Zone II: desorption of B

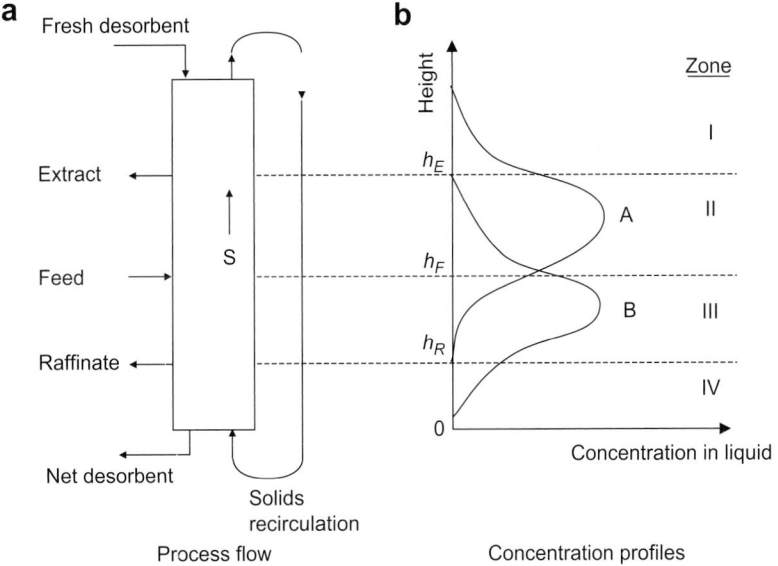

Figure 10.56. Continuous counter-current chromatography.

Zone III: adsorption (absorption) of A, desorption of B
Zone IV: adsorption of B

If the equilibrium constants are known, then the process can be approximately modelled in terms of theoretical stages using the Kremser equations or McCabe Thiele analysis; see Ruthven (1984).

A true moving-bed counter-current chromatography process ('Hypersorption') was commercially demonstrated by Dow Chemical and Union Oil Co. in 1947; see Kehde *et al.* (1948). Unfortunately, most good sorbents are not strong enough to withstand circulation at high solids flow rates and suffer unacceptably high attrition, so this process is no longer practised.

Instead of circulating the sorbent, the movement of solids can be simulated by using a large number of beds and periodically switching the bed location at which the net flows (feed, extract, raffinate and desorbent) are fed or removed. In the UOP Sorbex™ process, illustrated in Figure 10.57, this is accomplished using a rotary valve; however, the same effect can be obtained using a large number of solenoid switching valves. When the rotary valve is moved to the next position, the net flow that was moving through the pipe between beds n and n+1 is switched to move through the pipe between beds n+1 and n+2. Bed n thus effectively becomes bed n-1 relative to the net flows and movement of the bed is simulated.

Although the solid phase undergoes periodic discrete movement rather than continuous movement relative to the liquid and a steady state is never really established, the performance is nonetheless close to that of true counter-current chromatography; see Ruthven (1984) and Menet and Thibaut (1999).

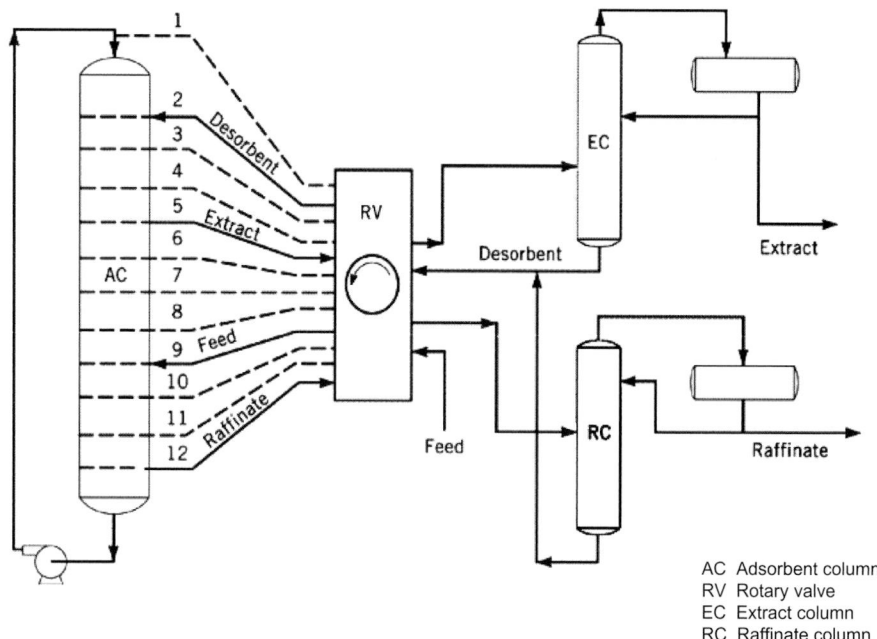

Figure 10.57.   UOP Sorbex™ Process (reproduced with permission of UOP LLC).

Simulated moving-bed (SMB) chromatography can be used to obtain products with high purity and high recovery if a suitable combination of sorbent and desorbent can be found. The desorbent can be a solvent or a liquid that has similar adsorption properties to the target molecule and competes for sorbent sites. The sorbent is used much more efficiently than in batch chromatography and the production rate per kilogram of sorbent is usually much higher.

Although SMB chromatography has been very successful in a number of applications, its use is not as widespread as affinity chromatography. The development and scale-up of SMB processes can be tricky, as the ratio of liquid to solid flow rates in each zone has to be chosen to give the desired adsorption or desorption for that zone; see Mazzotti *et al.* (1997) and Jupke *et al.* (2002). If the effective stripping factor or absorption factor is too close to 1.0, then a large number of effective stages are needed or separation will be poor. It is also much more difficult to apply a solvent gradient or vary solvent properties in a SMB process. Because of its higher process development costs, SMB chromatography tends to only be applied to products that are produced in high volume or that require a large amount of expensive sorbent. The largest applications are in recovery of paraxylene from mixed xylenes and production of high fructose corn syrup; see Ruthven (1984) and Meyers (2003).

10.8. GAS–SOLIDS SEPARATIONS (GAS CLEANING)

The primary need for gas–solid separation processes is for gas cleaning: the removal of dispersed finely divided solids (dust) and liquid mists from gas streams. Process gas streams must often be cleaned up to prevent contamination of catalysts or products, and to avoid damage to equipment, such as compressors. Also, effluent gas streams must be cleaned to comply with air-pollution regulations and for reasons of hygiene, to remove toxic and other hazardous materials; see IChemE (1992).

There is also often a need for clean, filtered, air for processes that use air as a raw material, and where clean working atmospheres are needed; for instance, in the pharmaceutical and electronics industries.

The particles to be removed may range in size from large molecules, measuring a few hundredths of a micrometre, to the coarse dusts arising from the attrition of catalysts or the fly ash from the combustion of pulverized fuels.

A variety of equipment has been developed for gas cleaning. The principal types used in the process industries are listed in Table 10.13, which is adapted from a selection guide given by Sargent (1971). Table 10.13 shows the general field of application of each type in terms of the particle size separated, the expected separation efficiency and the throughput. It can be used to make a preliminary selection of the type of equipment likely to be suitable for a particular application. Descriptions of the equipment shown in Table 10.13 can be found in various handbooks: Green and Perry (2007), Schweitzer (1997); and in specialist texts: Strauss (1975). Gas cleaning is also covered in Richardson *et al.* (2002), Chapter 1.

Gas-cleaning equipment can be classified according to the mechanism employed to separate the particles: gravity settling, impingement, centrifugal force, filtering, washing and electrostatic precipitation.

TABLE 10.13.   Gas-Cleaning Equipment

Type of Equipment	Min. Particle size (μm)	Min. Loading (mg/m³)	Approx. Efficiency (%)	Typical Gas Velocity (m/s)	Maximum Capacity (m³/s)	Gas Pressure Drop (mmH₂O)	Liquid Rate (m³/10³ m³ gas)	Space Required (relative)
Dry collectors								
Settling chamber	50	12,000	50	1.5–3	none	5	—	Large
Baffle chamber	50	12,000	50	5–10	none	3–12	—	Medium
Louver	20	2500	80	10–20	15	10–50	—	Small
Cyclone	10	2500	85	10–20	25	10–70	—	Medium
Multiple cyclone	5	2500	95	10–20	100	50–150	—	Small
Impingement	10	2500	90	15–30	none	25–50	—	Small
Wet scrubbers								
Gravity spray	10	2500	70	0.5–1	50	25	0.05–0.3	Medium
Centrifugal	5	2500	90	10–20	50	50–150	0.1–1.0	Medium
Impingement	5	2500	95	15–30	50	50–200	0.1–0.7	Medium
Packed	5	250	90	0.5–1	25	25–250	0.7–2.0	Medium
Jet	0.5–5 (range)	250	90	10–100	50	none	7–14	Small
Venturi	0.5	250	99	50–200	50	250–750	0.4–1.4	Small
Others								
Fabric filters	0.2	250	99	0.01–0.1	100	50–150	—	Large
Electrostatic precipitators	2	250	99	5–30	1000	5–25	—	Large

10.8.1.   Gravity Settlers (Settling Chambers)

Settling chambers are the simplest form of industrial gas-cleaning equipment, but have only a limited use; they are suitable for coarse dusts, particles larger than 50 μm. They are essentially long, horizontal, rectangular chambers through which the gas flows. The solids settle under gravity and are removed from the bottom of the chamber. Horizontal plates or vertical baffles are used in some designs to improve the separation. Settling chambers offer little resistance to the gas flow, and can be designed for operation at high temperature and high pressure, and for use in corrosive atmospheres.

The length of chamber required to settle a given particle size can be estimated from the settling velocity (calculated using Stokes' law) and the gas velocity. A design procedure is given by Jacob and Dhodapkar (1997).

10.8.2.   Impingement Separators

Impingement separators employ baffles to achieve the separation. The gas stream flows easily round the baffles, whereas the solid particles, due to their higher momentum, tend to continue in their line of flight, strike the baffles and are collected. Many different baffle designs are used in commercial equipment; a typical example is shown in Figure 10.58. Impingement separators cause a higher pressure drop than settling chambers, but are capable of separating smaller particle sizes, 10–20 μm.

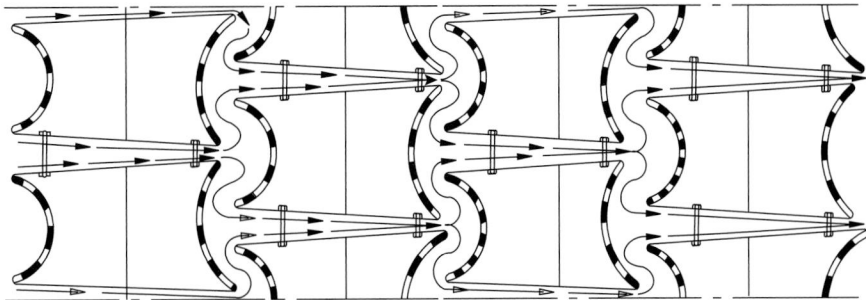

Figure 10.58. Impingement separator (section showing gas flow).

10.8.3. Centrifugal Separators (Cyclones)

Cyclones are the principal type of gas–solids separator employing centrifugal force, and are widely used. They are basically simple constructions, can be made from a wide range of materials, and can be designed for high temperature and pressure operation.

Cyclones are suitable for separating particles above about 5 μm diameter; smaller particles, down to about 0.5 μm, can be separated where agglomeration occurs.

The most commonly used design is the reverse-flow cyclone (Figure 10.59); other configurations are used for special purposes. In a reverse-flow cyclone the gas enters

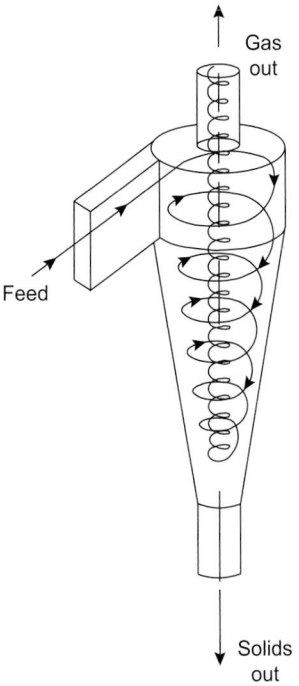

Figure 10.59. Reverse-flow cyclone.

the top chamber tangentially and spirals down to the apex of the conical section; it then moves upward in a second, smaller diameter, spiral, and exits at the top through a central vertical pipe. The solids move radially to the walls, slide down the walls and are collected at the bottom. Design procedures for cyclones are given by Constantinescu (1984), Strauss (1975), Koch and Licht (1977), and Stairmand (1951). The theoretical concepts and experimental work on which the design methods are based are discussed in Richardson *et al.* (2002), Chapter 1. Stairmand's method is outlined below and illustrated in Example 10.6.

Cyclone Design

Stairmand developed two standard designs for gas–solid cyclones: a high-efficiency cyclone (Figure 10.60a) and a high-throughput design (Figure 10.60b). The performance curves for these designs, obtained experimentally under standard test conditions, are shown in Figure 10.61a and b. These curves can be transformed to other cyclone sizes and operating conditions by use of the following scaling equation, for a given separating efficiency:

$$d_2 = d_1 \left[\left(\frac{D_c}{D_c} \right)^3 \times \frac{Q_1}{Q_2} \times \frac{\Delta\rho_1}{\Delta\rho_2} \times \frac{\mu_2}{\mu_1} \right]^{1/2} \tag{10.14}$$

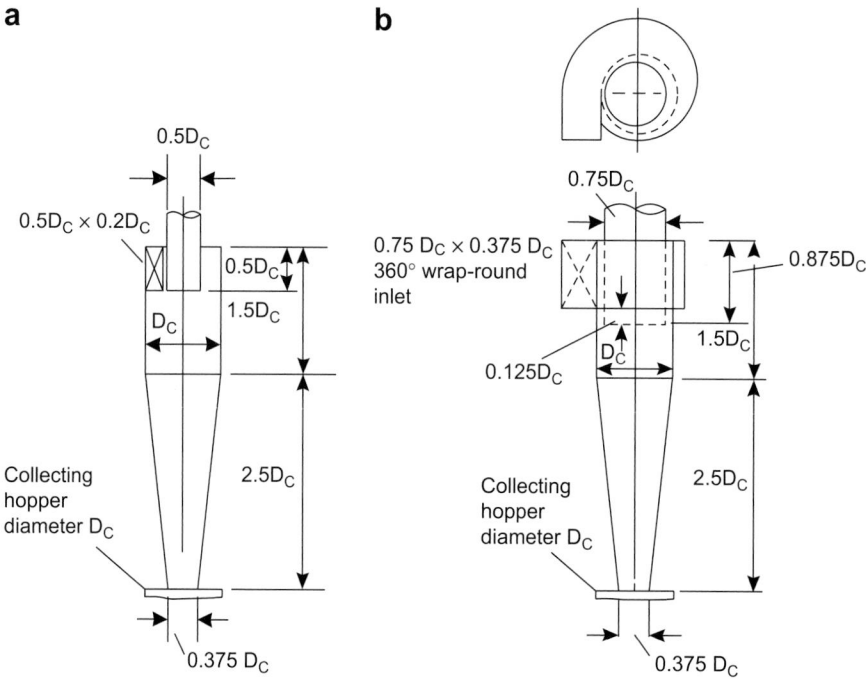

Figure 10.60. Standard cyclone dimension. (a) High-efficiency cyclone. (b) High-throughput cyclone.

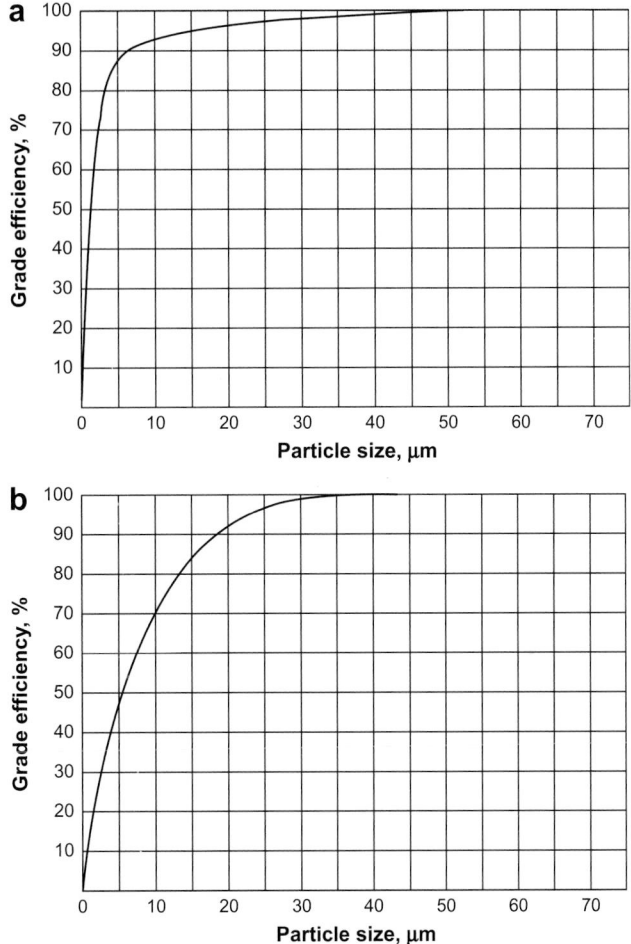

Figure 10.61. Performance curves, standard conditions. (a) High-efficiency cyclone performance curves, standard conditions. (b) High-throughput cyclone.

where

d_1 = mean diameter of particle separated at the standard conditions, at the chosen separating efficiency (Figure 10.61a or b)

d_2 = mean diameter of the particle separated in the proposed design, at the same separating efficiency

D_{c_1} = diameter of the standard cyclone = 8 in (203 mm)

D_{c_2} = diameter of proposed cyclone, mm

Q_1 = standard flow rate:

for high-efficiency design = 223 m³/h

for high-throughput design = 669 m³/h

Q_2 = proposed flow rate, m³/h

$\Delta\rho_1$ = solid–fluid density difference in standard conditions = 2000 kg/m³

$\Delta\rho_2$ = density difference, proposed design

μ_1 = test fluid viscosity (air at 1 atm, 20°C) = 0.018 mN s/m^2

μ_2 = viscosity, proposed fluid.

A performance curve for the proposed design can be drawn up from Figure 10.61a or b by multiplying the grade diameter at, say, each 10% increment of efficiency, by the scaling factor given by equation 10.14; as shown in Figure 10.62.

An alternative method of using the scaling factor, that does not require redrawing the performance curve, is used in Example 10.6. The cyclone should be designed to give an inlet velocity of between 9 and 27 m/s (30 to 90 ft/s); the optimum inlet velocity has been found to be 15 m/s (50 ft/s).

Pressure Drop

The pressure drop in a cyclone will be due to the entry and exit losses, and friction and kinetic energy losses in the cyclone. The empirical equation given by Stairmand (1949) can be used to estimate the pressure drop:

$$\Delta P = \frac{\rho_f}{203}\left\{ u_1^2\left[1 + 2\phi^2\left(\frac{2r_t}{r_e} - 1\right)\right] + 2u_2^2 \right\} \tag{10.15}$$

where

ΔP = cyclone pressure drop, millibars
ρ_f = gas density, kg/m^3
u_1 = inlet duct velocity, m/s
u_2 = exit duct velocity, m/s
r_t = radius of circle to which the centre line of the inlet is tangential, m
r_e = radius of exit pipe, m
ϕ = factor from Figure 10.63
ψ = parameter in Figure 10.63, given by:
$\psi = f_c\frac{A_s}{A_1}$
f_c = friction factor, taken as 0.005 for gases

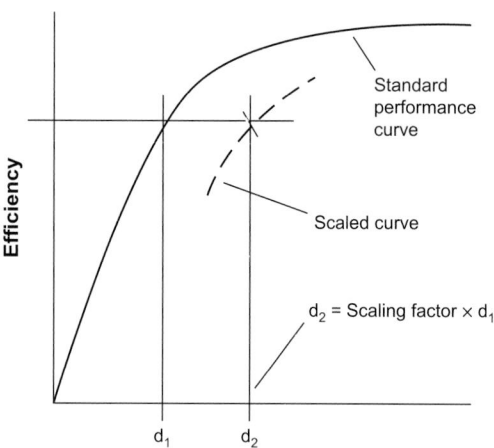

Figure 10.62.   Scaled performance curve

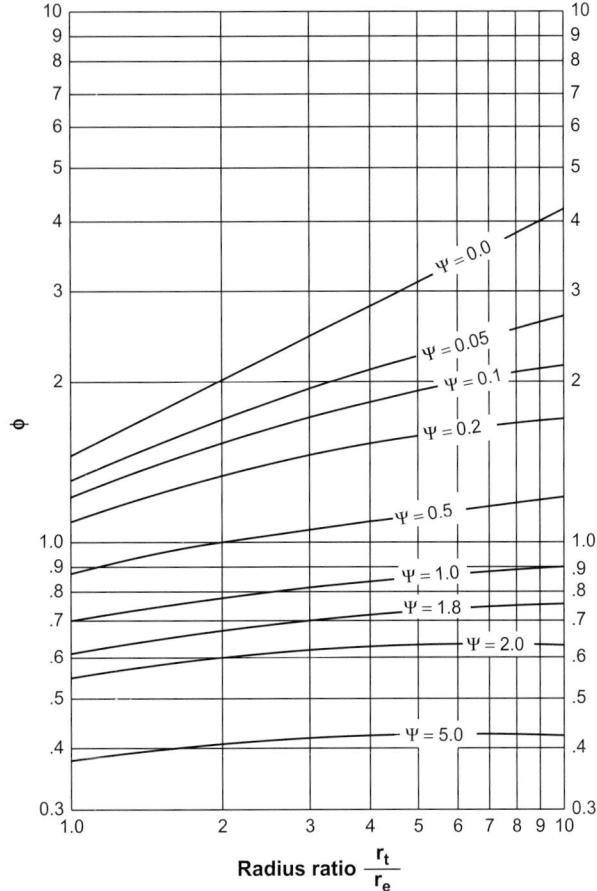

Figure 10.63.   Cyclone pressure-drop factor.

A_s = surface area of cyclone exposed to the spinning fluid, m^2 (for design purposes this can be taken as equal to the surface area of a cylinder with the same diameter as the cyclone and length equal to the total height of the cyclone; barrel plus cone)

A_1 = area of inlet duct, m^2.

Stairmand's equation is for the gas flowing alone, containing no solids. The presence of solids will normally increase the pressure drop over that calculated using equation 10.15, depending on the solids loading. Alternative design methods for cyclones, which include procedures for estimating the true pressure drop, are given by Green and Perry (2007) and Yang (1999); see also Zenz (2001).

General Design Procedure

1. Select either the high-efficiency or high-throughput design, depending on the performance required.
2. Obtain an estimate of the particle size distribution of the solids in the stream to be treated.

3. Estimate the number of cyclones needed in parallel.
4. Calculate the cyclone diameter for an inlet velocity of 15 m/s (50 ft/s). Scale the other cyclone dimensions from Figure 10.60a or b.
5. Calculate the scale-up factor for the transposition of Figures 10.61a or b.
6. Calculate the cyclone performance and overall efficiency (recovery of solids). If unsatisfactory, try a smaller diameter.
7. Calculate the cyclone pressure drop and, if required, select a suitable blower.
8. Cost the system and optimize to make the best use of the pressure drop available, or, if a blower is required, to give the lowest operating cost.

Example 10.6

Design a cyclone to recover solids from a process gas stream. The anticipated particle size distribution in the inlet gas is given below. The density of the particles is 2500 kg/m^3, and the gas is essentially nitrogen at 150°C. The stream volumetric flow rate is 4000 m^3/h, and the operation is at atmospheric pressure. An 80% recovery of the solids is required.

Particle size (μm)	50	40	30	20	10	5	2
Percentage by weight less than	90	75	65	55	30	10	4

Solution

As 30% of the particles are below 10 μm the high-efficiency design will be required to give the specified recovery.

$$\text{Flow-rate} = \frac{4000}{3600} = 1.11 \text{ m}^3/\text{s}$$

$$\text{Area of inlet duct, at 15 m/s} = \frac{1.11}{15} = 0.07 \text{ m}^2$$

From Figure 10.60a

$$\text{duct area} = 0.5D_c \times 0.2D_c$$

so

$$D_c = 0.84 \text{ m}$$

This is clearly too large compared with the standard design diameter of 0.203 m. Try four cyclones in parallel, $D_c = 0.42$ m:

$$\text{Flow-rate per cyclone} = 1000 \text{ m}^3/\text{h}$$

$$\text{Density of gas at 150°C} = \frac{28}{22.4} \times \frac{273}{423} = 0.81 \text{ kg/m}^2,$$

negligible compared with the solids density.

$$\text{Viscosity of } N_2 \text{ at } 150°C = 0.023 \text{ cp(mN s/m}^2)$$

From equation 10.14

$$\text{scaling factor} = \left[\left(\frac{0.42}{0.203}\right)^3 \times \frac{223}{1000} \times \frac{2000}{2500} \times \frac{0.023}{0.018}\right]^{1/2} = \underline{\underline{1.42}}$$

The performance calculations, using this scaling factor and Figure 10.61a, are set out in Table 10.14. The collection efficiencies shown in column 4 of the table were read from Figure 10.61a at the scaled particle size, column 3. The overall collection efficiency satisfies the specified solids recovery. The proposed design with dimensions in the proportions given in Figure 10.60a is shown in Figure 10.64.

Pressure-Drop Calculation

$$\text{Area of inlet duct, } A_1, = 210 \times 80 = 16,800 \text{ mm}^2$$

$$\text{Cyclone surface area, } A_s = \pi \times 420 \times (630 + 1050)$$
$$= 2.218 \times 10^6 \text{ mm}^2$$

f_c taken as 0.005

$$\psi = \frac{f_c, A_s}{A_1} = \frac{0.005 \times 2.218 \times 10^6}{16,800} = 0.66$$

$$\frac{r_t}{r_e} = \frac{(420 - (80/2))}{210} = 1.81$$

From Figure 10.63, $\phi = 0.9$.

TABLE 10.14. Calculated Performance of Cyclone Design (Example 10.6)

Particle Size (μm)	% in Range	Mean Particle Size ÷ Scaling Factor	Efficiency at Scaled Size % (Figure 10.61a)	Collected (2)×(4) 100	Grading at Exit (2)−(5)	% at Exit
>50	10	35	98	9.8	0.2	1.8
50–40	15	32	97	14.6	0.4	3.5
40–30	10	25	96	9.6	0.4	3.5
30–20	10	18	95	9.5	0.5	4.4
20–10	25	11	93	23.3	1.7	15.1
10–5	20	5	86	17.2	2.8	24.8
5–2	6	3	72	4.3	1.7	15.1
2–0	4	1	10	0.4	3.6	31.8
	100		Overall collection efficiency	88.7	11.3	100.0

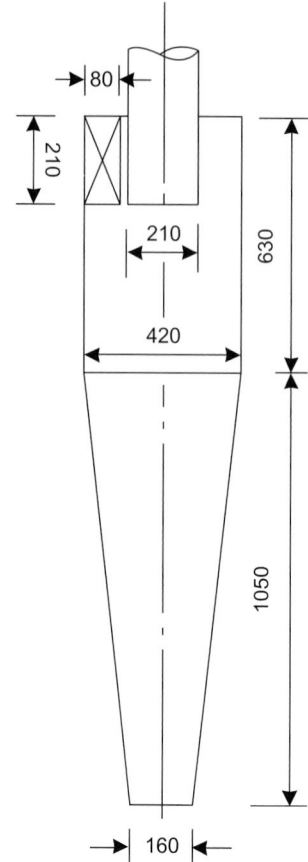

Figure 10.64.   Proposed cyclone design, all dimensions in mm (Example 10.6).

$$u_1 = \frac{1000}{3600} \times \frac{10^6}{16,800} = 16.5 \text{ m/s}$$

$$\text{Area of exit pipe} = \frac{\pi \times 210^2}{4} = 34,636 \text{ mm}^2$$

$$u_2 = \frac{1000}{3600} \times \frac{10^6}{34,636} = 8.0 \text{ m/s}$$

From equation 10.15

$$\Delta P = \frac{0.81}{203}[16.5^2[1 + 2 \times 0.9^2(2 \times 1.81 - 1)] + 2 \times 8.0^2]$$

$$= \underline{16.4 \text{ millibar}}\,(67 \text{ mm H}_2\text{O})$$

This pressure drop looks reasonable.

10.8.4. Filters

The filters used for gas cleaning separate the solid particles by a combination of impingement and filtration; the pore sizes in the filter media used are too large simply to filter out the particles. The separating action relies on the precoating of the filter medium by the first particles separated, which are separated by impingement on the filter medium fibres. Woven or felted cloths of cotton and various synthetic fibres are commonly used as the filter media. Glass-fibre mats and paper filter elements are also used.

A typical example of this type of separator is the bag filter, which consists of a number of bags supported on a frame and housed in a large rectangular chamber (Figure 10.65). The deposited solids are removed by mechanically vibrating the bag, or by periodically reversing the gas flow. Bag filters can be used to separate small particles, down to around 1 μm, with a high separating efficiency. Commercial units are available to suit most applications and should be selected in consultation with the vendors.

The design and specification of bag filters (baghouses) is covered by Kraus (1979).

Air Filters

Dust-free air is required for many process applications. The requirements of air filtration differ from those of process gas filtration mainly in that the quantity of dust

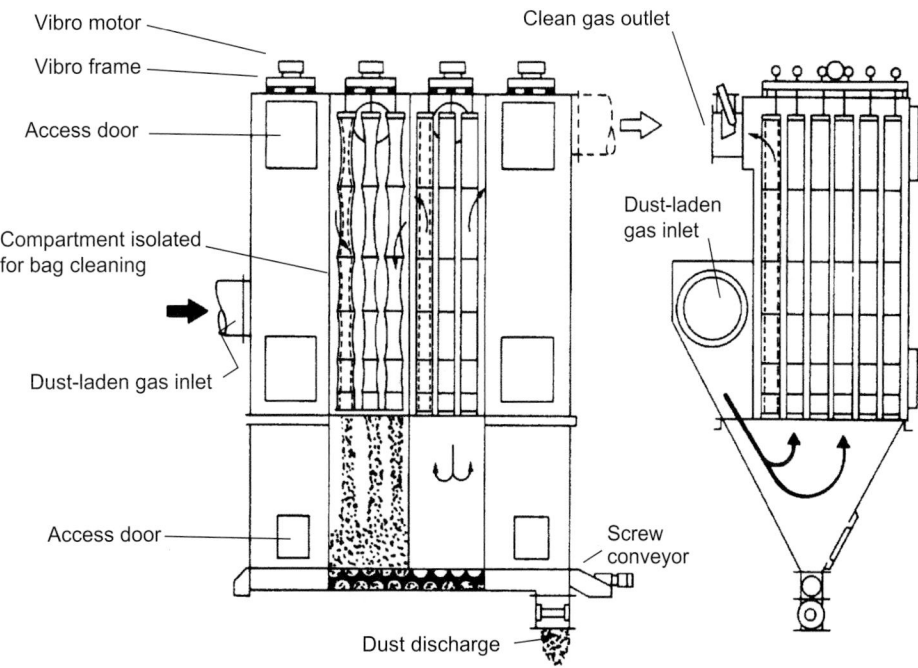

Figure 10.65. Multi-compartment vibro bag filter.

to be removed will be lower, typically less than $10 \, mg/m^3$ (~ 5 grains per $1000 \, ft^3$), and also in that there is no requirement to recover the material collected.

Three basic types of air filter are used: viscous, dry and continuous. Viscous and dry units are similar in construction, but the filter medium of the viscous type is coated with a viscous material, such as a mineral oil, to retain the dust. The filters are made up from standard, preformed, sections, supported on a frame in a filter housing. The sections are removed periodically for cleaning or replacement. Various designs of continuous filtration equipment are also available, employing either viscous or dry filter elements, but in which the filter is cleaned continuously. A comprehensive description of air-filtration equipment is given by Strauss (1975).

10.8.5. Wet Scrubbers (Washing)

In wet scrubbing, the dust is removed by counter-current washing with a liquid, usually water, and the solids are removed as a slurry. The principal mechanism involved is the impact (impingement) of the dust particles and the water droplets. Particle sizes down to $0.5 \, \mu m$ can be removed in suitably designed scrubbers. In addition to removing solids, wet scrubbers can be used to simultaneously cool the gas and neutralize any corrosive constituents.

Spray towers, plate and packed columns are used, as well as a variety of proprietary designs. Spray towers have a low pressure drop but are not suitable for removing very fine particles, below $10 \, \mu m$. The collecting efficiency can be improved by the use of plates or packing but at the expense of a higher pressure drop.

Venturi and orifice scrubbers are simple forms of wet scrubbers. The turbulence created by the venturi or orifice is used to atomize water sprays and promote contact between the liquid droplets and dust particles. The agglomerated particles of dust and liquid are then collected in a centrifugal separator, usually a cyclone.

10.8.6. Electrostatic Precipitators

Electrostatic precipitators are capable of collecting very fine particles, $<2 \, \mu m$, at high efficiencies; however, their capital and operating costs are high. Electrostatic precipitation should only be considered in place of alternative processes, such as filtration, where the gases are hot or corrosive.

Electrostatic precipitators are used extensively in the metallurgical, cement and electrical power industries. Their main application is probably in the removal of the fine fly ash formed in the combustion of pulverized coal in power station boilers. The basic principle of operation is simple. The gas is ionized in passing between a high-voltage electrode and an earthed (grounded) electrode; the dust particles become charged and are attracted to the earthed electrode. The precipitated dust is removed from the electrodes mechanically, usually by vibration, or by washing. Wires are normally used for the high-voltage electrode, and plates or tubes for the earthed electrode. A typical design is shown in Figure 10.66. A full description of the construction, design and application of electrostatic precipitators is given by Schneider *et al.* (1975) and Parker (2002).

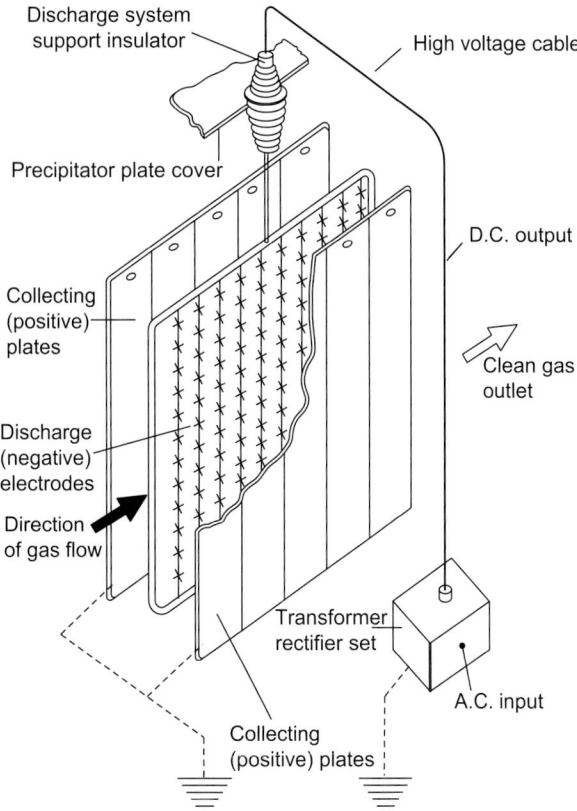

Discharge system
support insulator

High voltage cable

Precipitator plate cover

D.C. output

Collecting
(positive)
plates

Clean gas
outlet

Discharge
(negative)
electrodes

Direction
of gas flow

Transformer
rectifier set

A.C. input

Collecting
(positive) plates

Figure 10.66. Electrostatic precipitator.

10.9. GAS–LIQUID SEPARATORS

The separation of liquid droplets and mists from gas or vapour streams is analogous to the separation of solid particles and, with the possible exception of filtration, the same techniques and equipment can be used.

Where the carryover of some fine droplets can be tolerated, it is often sufficient to rely on gravity settling in a vertical or horizontal separating vessel (known as a knockout pot).

Knitted mesh demisting pads are frequently used to improve the performance of separating vessels where the droplets are likely to be small, down to 1 μm, and where high separating efficiencies are required. Proprietary demister pads are available in a wide range of materials (metals and plastics), thicknesses and pad densities. For liquid separators, stainless-steel pads around 100 mm thick and with a nominal density of 150 kg/m^3 would generally be used. Use of a demister pad allows a smaller vessel to be used. Separating efficiencies above 99% can be obtained with low pressure drop. The design and specification of demister pads for gas–liquid separators is discussed by Pryce Bailey and Davies (1973).

The design methods for horizontal separators given below are based on a procedure given by Gerunda (1981).

Cyclone separators are also frequently used for gas–liquid separation. They can be designed using the same methods for gas–solids cyclones. The inlet velocity should be kept below 30 m/s to avoid pick-up of liquid from the cyclone surfaces.

10.9.1. Settling Velocity

Equation 10.16 can be used to estimate the settling velocity of the liquid droplets, for the design of separating vessels.

$$u_t = 0.07(\rho_L - \rho_v)/\rho_v^{\frac{1}{2}} \tag{10.16}$$

where

u_t = settling velocity, m/s
ρ_L = liquid density, kg/m^3
ρ_v = vapour density, kg/m^3.

If a demister pad is not used, the value of u_t obtained from equation 10.16 should be multiplied by a factor of 0.15 to provide a margin of safety and to allow for flow surges.

10.9.2. Vertical Separators

The layout and typical proportions of a vertical liquid–gas separator are shown in Figure 10.67.

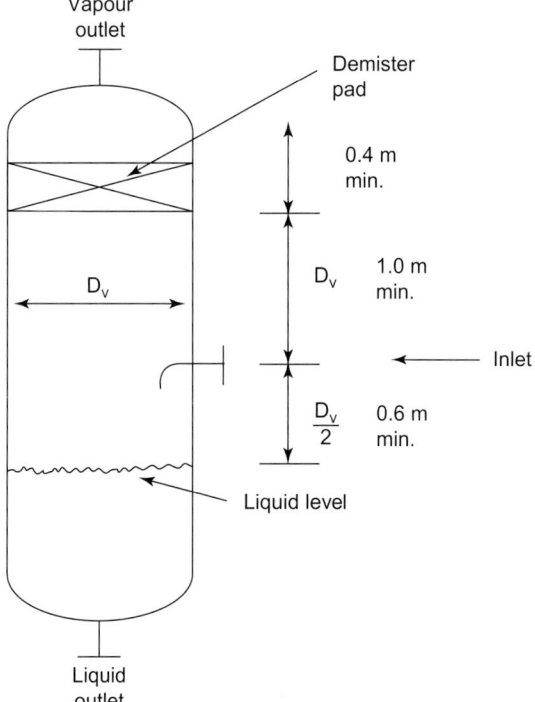

Figure 10.67.   Vertical liquid–vapour separator.

The diameter of the vessel must be large enough to slow the gas down to below the velocity at which the droplets will settle out. So the minimum allowable diameter will be given by:

$$D_v = \sqrt{\left(\frac{4\ V_v}{\pi\ u_s}\right)} \tag{10.17}$$

where

D_v = minimum vessel diameter, m
V_v = gas, or vapour volumetric flow rate, m³/s
$u_s = u_t$, if a demister pad is used, and 0.15 u_t for a separator without a demister pad; u_t from equation (10.16), m/s.

The diameter is usually rounded up to the nearest standard vessel size so that standard vessel closures can be used; see Section 13.5.

The height of the vessel outlet above the gas inlet should be sufficient to allow for disengagement of the liquid drops. A height equal to the diameter of the vessel or 1 m, whichever is the greatest, should be used; see Figure 10.67.

The liquid level will depend on the hold-up time necessary for smooth operation and control; typically 10 min would be allowed.

Example 10.7

Make a preliminary design for a separator to separate a mixture of steam and water; flow rates: steam 2000 kg/h, water 1000 kg/h; operating pressure 4 bar.

Solution

From steam tables, at 4 bar: saturation temperature 143.6°C, liquid density 926.4 kg/m³, vapour density 2.16 kg/m3.

$$u_t = 0.07[(926.4 - 2.16)/2.16]^{\frac{1}{2}} = 1.45 \text{ m/s} \tag{10.16}$$

As the separation of condensate from steam is unlikely to be critical, a demister pad will not be specified.
So

$$u_t = 0.15 \times 1.45 = 0.218 \text{ m/s}$$

$$\text{Vapour volumetric flow rate} = \frac{2000}{3600 \times 2.16} = 0.257 \text{ m}^3/\text{s}$$

$$D_v = \sqrt{[(4 \times 0.257)/(\pi \times 0.218)]}$$
$$= 1.23 \text{ m, round to nearest standard vessel size, 1.25 m (4 ft).} \tag{10.17}$$

$$\text{Liquid volumetric flow rate} = \frac{1000}{3600 \times 926.14} = 3.0 \times 10^{-4} \text{ m}^3/\text{s}$$

Allow a minimum of 10 min hold-up.

$$\text{Volume held in vessel} = 3.0 \times 10^{-4} \times (10 \times 60) = 0.18 \text{ m}^3$$

$$\text{Liquid depth required, } h_v = \frac{\text{volume held-up}}{\text{vessel cross-sectional area}}$$

$$= \frac{0.18}{(\pi \times 1.25^2/4)} = 0.15 \text{ m}$$

Increase to 0.3 m to allow space for positioning the level controller.

10.9.3. Horizontal Separators

The layout of a typical horizontal separator is shown in Figure 10.68.

A horizontal separator would be selected when a long liquid hold-up time is required.

In the design of a horizontal separator, the vessel diameter cannot be determined independently of its length, unlike for a vertical separator. The diameter and length, and the liquid level, must be chosen to give sufficient vapour residence time for the liquid droplets to settle out, and for the required liquid hold-up time to be met.

The most economical length to diameter ratio will depend on the operating pressure (see Chapter 13). As a general guide the following values can be used:

Operating pressure, bar	Length/diameter, L_v/D_v
0–20	3
20–35	4
>35	5

The relationship between the area for vapour flow, A_v, and the height above the liquid level, h_v, can been found from tables giving the dimensions of the segments of circles (see Green and Perry, 2007), or from Figures 11.40 and 11.41 in Chapter 11.

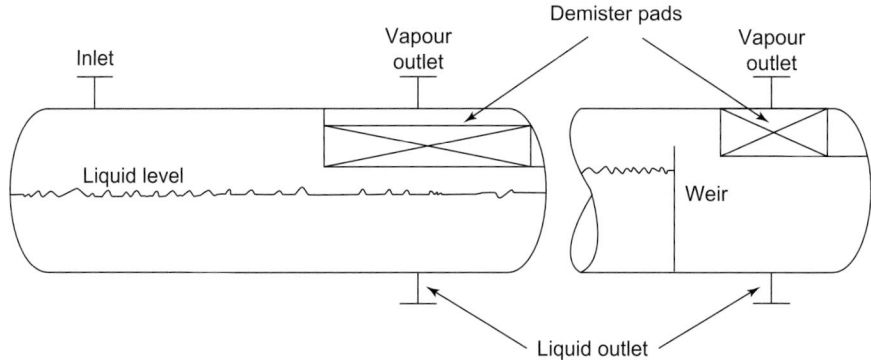

Figure 10.68.   Horizontal liquid–vapour separator.

For preliminary designs, set the liquid height at half the vessel diameter

$$h_v = D_v/2 \quad and \quad f_v = 0.5$$

where f_v is the fraction of the total cross-sectional area occupied by the vapour.

The design procedure for horizontal separators is illustrated in the following example.

Example 10.8

Design a horizontal separator to separate 10,000 kg/h of liquid, density 962.0 kg/m³, from 12,500 kg/h of vapor, density 23.6 kg/m3. The vessel operating pressure will be 21 bar.

Solution

$$u_t = 0.07[(962.0 - 23.6)/23.6]^{1/2} = 0.44 \text{ m/s}$$

Try a separator without a demister pad.

$$u_s = 0.15 \times 0.44 = 0.066 \text{ m/s}$$

$$\text{Vapour volumetric flow rate} = \frac{12,500}{3600 \times 23.6} = 0.147 \text{ m}^3/\text{s}$$

Take $h_v = 0.5D_v$ and $L_v/D_v = 4$

$$\text{Cross} - \text{sectional area for vapour flow} = \frac{\pi D_v^2}{4} \times 0.5 = 0.393 \, D_v^2$$

$$\text{Vapour velocity, } u_v = \frac{0.147}{0.393 D_v^2} = 0.374 \, D_v^{-2}$$

Vapour residence time required for the droplets to settle to liquid surface

$$h_v/u_s = 0.5D_v/0.066 = 7.58D_v$$

Actual residence time = vessel length/vapour velocity

$$= \frac{L_v}{u_v} = \frac{4 \, D_v}{0.374 \, D_v^{-2}} = 10.70 \, D_v^3$$

For satisfactory separation, required residence time = actual.
So,

$$7.58D_v = 10.70D_v^3$$

$D_v = 0.84$ m, say 0.92 m (3 ft, standard pipe size). Liquid hold-up time

$$\text{liquid volumetric flow rate} = \frac{10,000}{3600 \times 962.0} = 0.00289 \text{ m}^3/\text{s}$$

$$\text{liquid cross-sectional area} = \frac{\pi \times 0.92^2}{4} \times 0.5 = 0.332 \text{ m}^2$$

Length, $L_v = 4 \times 0.92 = 3.7$ m

Hold-up volume $= 0.332 \times 3.7 = 1.23$ m^3

Hold-up time = liquid volume/liquid flow rate $= 1.23/0.00289 = 426$ s $= 7$ min.

This is unsatisfactory; 10 min minimum required.

Need to increase the liquid volume. This is best done by increasing the vessel diameter. If the liquid height is kept at half the vessel diameter, the diameter must be increased by a factor of roughly $(10/7)^{0.5} = 1.2$.

$$\text{New } D_v = 0.92 \times 1.2 = 1.1 \text{ m}.$$

Check liquid residence time

$$\text{new liquid volume} = \frac{\pi \times 1.1^2}{4} \times 0.5 \times (4 \times 1.1) = 2.09 \text{ m}^3$$

$$\text{new residence time} = 2.09/0.00289 = 723 \text{ s} = 12 \text{ minutes, satisfactory}$$

Increasing the vessel diameter will have also changed the vapour velocity and the height above the liquid surface. The liquid separation will still be satisfactory as the velocity, and hence the residence time, is inversely proportional to the diameter squared, whereas the distance the droplets have to fall is directly proportional to the diameter.

In practice, the distance travelled by the vapour will be less than the vessel length, L_v, as the vapour inlet and outlet nozzles will be set in from the ends. This could be allowed for in the design but will make little difference.

10.10.　SIZE REDUCTION AND SIZE ENLARGEMENT

10.10.1.　Crushing and Grinding (Comminution) Equipment

Crushing is the first step in the process of size reduction, reducing large lumps to manageably sized pieces. For some processes crushing is sufficient, but for chemical processes it is usually followed by grinding to produce a fine-sized powder. Though many articles have been published on comminution and Marshall (1974) mentions over 4000, the subject remains essentially empirical. The designer must rely on experience, and the advice of the equipment manufacturers, when selecting and sizing crushing and grinding equipment, and to estimate the power requirements. Several models have been proposed for the calculation of the energy consumed in size reduction, some of which are discussed in Richardson *et al.* (2002), Chapter 2. For a fuller treatment of the subject, the reader should refer to the books by Lowrison (1974), Prasher (1987) and Fuerstenau and Han (2003).

The main factors to be considered when selecting equipment for crushing and grinding are:

1. The size of the feed
2. The size reduction ratio
3. The required particle size distribution of the product
4. The throughput

5. The properties of the material: hardness, abrasiveness, stickiness, density, toxicity, flammability
6. Whether wet grinding is permissible.

The selection guides given by Lowrison (1974) and Marshall (1974), which are reproduced in Tables 10.15 and 10.16, can be used to make a preliminary selection based on particle size and material hardness. Descriptions of most of the equipment listed in these tables are given in Richardson *et al.* (2002), Chapter 2, or can be found in the literature: Green and Perry (2007), Hiorns (1970), Lowrison (1974). The most commonly used equipment for coarse size reduction are jaw crushers and rotary crushers; and for grinding, ball mills or their variants: pebble, roll and tube mills.

10.10.2. Size Enlargement (Agglomeration) Equipment

Size enlargement processes are used to form larger agglomerates out of small particles. Size enlargement is a common step in processes that produce solid products; for example, in the manufacture of foods, tablets, fertilizers, catalysts, adsorbents, soap powders, solid fuels and ceramics.

The agglomerates may be held together by natural cohesive forces between the particles or by partially melting the solids and sintering the particles together; however, in most cases a 'binder' is added to the particles to stick them together. The choice of binder depends on the process temperature and product requirements. After the right-size particles have been formed, the binder is sometimes driven off by drying or calcining the particles. Common binders for pharmaceutical products include

TABLE 10.15. Selection of Comminution Equipment (After Lowrison, 1974)

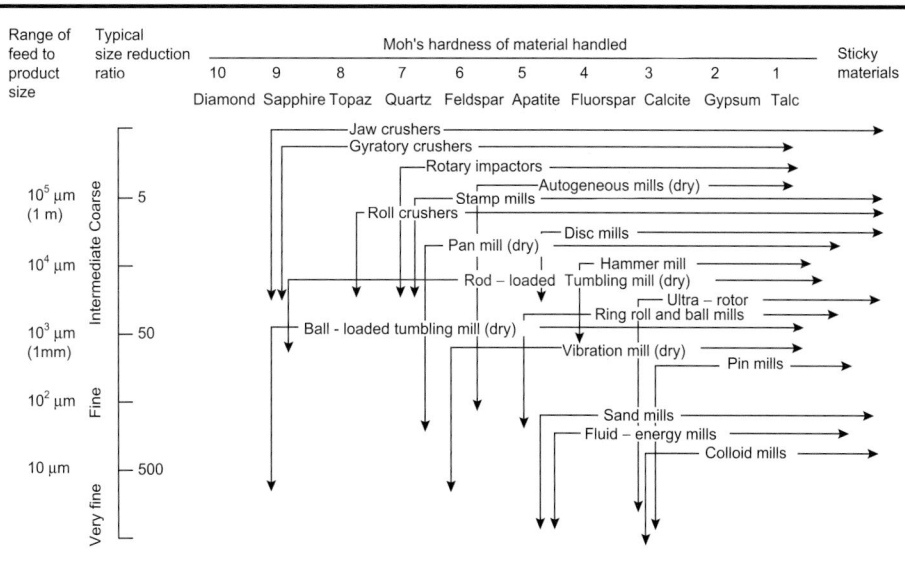

TABLE 10.16. Selection of Comminution Equipment for Various Materials (After Marshall, 1974)

Material Class No	Material Classification	Typical Materials in Class	Suitable Equipment for Product Size Classes			Remarks
			Down to 5 Mesh	Between 5 and 300 Mesh	Less than 300 Mesh	
1	Hard and tough	Mica Scrap and powdered metals	Jaw crushers Gyratory crushers Cone crushers Autogeneous mills	Ball, pebble, rod and cone mills Tube mills Vibration mills	Ball, pebble and cone mills Tube mills Vibration and vibro-energy mills Fluid-energy mills	Moh's hardness 5–10, but includes other tough materials of lower hardness
2	Hard, abrasive and brittle	Coke, quartz, granite	Jaw crushers Gyratory and cone crushers Roll crushers	Ball, pebble, rod and cone mills Vibration mills Roller mills	Ball, pebble and cone mills Tube mills Vibration and vibro-energy mills Fluid-energy mills	Moh's hardness 5–10 High wear rate/ contamination in high-speed machinery Use machines with abrasion resistant linings
3	Intermediate hard, and friable	Barytes, fluorspar, limestone	Jaw crushers Gyratory crushers Roll crushers Edge runner mills Impact breakers Autogeneous mills Cone crushers	Ball, pebble, rod and cone mills Tube mills Ring roll mills Ring ball mills Roller mills Peg and disc mills Cage mills Impact breakers Vibration mills	Ball, pebble and cone mills Tube mills Perl mills Vibration and vibro-energy mills Fluid-energy mills	Moh's hardness 3–5

		Coarse crushing	Intermediate	Fine grinding	Remarks
4	Fibrous, low abrasion and possibly tough	Wood, asbestos	Cone crushers Roll crushers Edge runner mills Autogeneous mills Impact breakers	Ball, pebble, rod and cone mills Tube mills Roller mills Peg and disc mills Cage mills Impact breakers Vibration mills Rotary cutters and dicers	Ball, pebble and cone mills Tube mills Sand mills Perl mills Vibration and vibro-energy mills Colloid mills	Wide range of hardness Low-temperature, liquid nitrogen, useful to embrittle soft but tough materials
5	Soft and friable	Sulphur, gypsum rock salt	Cone crushers Roll crushers Edge runner mills Impact breakers Autogeneous mills	Ball, pebble and cone mills Tube mills Ring roll mills Ring ball mills Roller mills Peg and disc mills Cage mills Impact breakers Vibration mills	Ball, pebble and cone mills Tube mills Sand mills Perl mills Vibration and vibro-energy mills Colloid mills Fluid-energy mills Peg and disc mills	Moh's hardness 1–3
6	Sticky	Clays, certain organic pigments	Roll crushers Impact breakers Edge runner mills	Ball, pebble, rod and cone mills* Tube mills* Peg and disc mills Cage mills Ring roll mills	Ball, pebble and cone mills* Tube mills* Sand mills Perl mills Vibration and vibro-energy mills Colloid mills	Wide range of Moh's hardness although mainly less than 3 Tends to clog *Wet grinding employed except for certain exceptional cases

Note: Moh's scale of hardness is given in Table 10.15.

*All ball, pebble, rod and cone mills, edge runner mills, tube mills, vibration mills and some ring ball mills may be used wet or dry except where stated. The perl mills, sand mills and colloid mills may be used for wet milling only.

dextrose, starch, glucose, gelatine and gums, while for catalysts or fertilizers resins or water are used.

When a uniform size product is required, a pressure compaction method is used. Rotary tableting presses or roll presses allow uniform shape and size solids to be produced at high throughputs, up to of the order 10,000 tablets per minute. Roll presses give less uniform product than tableting presses, but allow larger briquettes to be formed. Alternatively, the solid and binder can be formed into a paste and extruded through a die. The resulting extrudate may either break up naturally, giving product of uniform cross-section but varying length, or else may be periodically cut with a knife edge or wire.

If a broader particle size distribution is acceptable, then particles can be agglomerated cheaply using a rotary agglomerator, also sometimes known as a granulator, pelletizer, tumbling drum or balling drum. An agglomerator consists of an inclined drum into which the solids and binder are fed. Material leaving the drum is sent to a screen and undersize particles are recycled to the feed to act as seed particles; see Figure 10.69. Occasionally, oversize material is also separated, ground up and returned to the feed. Inclined open pans can also be used as rotary agglomerators; these are cheaper, but produce more dust and require more recycle.

Once the desired size and shape particles have been formed, they will often be coated. Various types of coatings are applied; for example, pharmaceutical tablets may be coated with sugar or gels to make them more palatable or easier to swallow, while many solids are coated with waxes to suppress dust formation during handling. Coatings are usually applied in a tumbler or spray coating drum that is essentially the same a rotary agglomerator. The particles must be kept moving during coating to ensure a uniform coating and prevent unwanted agglomeration. If the coating is applied in a solvent then a rotary dryer is sometimes used as the coating drum.

The equipment used for size enlargement is usually of proprietary design and must be specified in consultation with the vendor. The equipment vendors are easy to locate on the web and usually have technically proficient sales engineers who can provide advice on device selection, dust handling, etc. In many cases, the vendor will be prepared to run demonstration trials to prove that their device forms an acceptable product. Some of the common designs are described in Green and Perry (2007), and Richardson *et al.* (2002), Chapter 2.

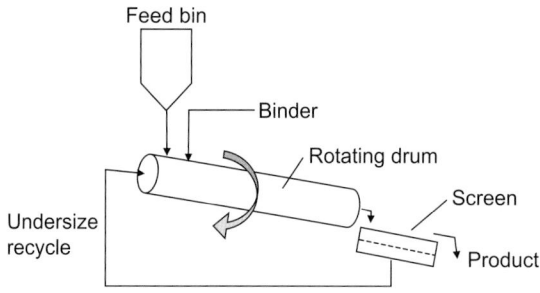

Figure 10.69.   Rotary agglomerator.

10.11. MIXING EQUIPMENT

The preparation of mixtures of solids, liquids and gases is an essential part of most production processes in the chemical and allied industries, covering all processing stages from the preparation of reagents through to the final blending of products. The equipment used depends on the nature of the materials and the degree of mixing required. Mixing is often associated with other operations, such as reaction and heat transfer. Liquid and solids mixing operations are frequently carried out as batch processes.

In this section, mixing processes will be considered under three separate headings: gases, liquids and solids.

10.11.1. Gas Mixing

Specialized equipment is seldom needed for mixing gases, which because of their low viscosities mix easily. The mixing given by turbulent flow in a length of pipe is usually sufficient for most purposes. Turbulence promoters, such as orifices or baffles, can be used to increase the rate of mixing. The piping arrangements used for inline mixing are discussed in the section on liquid mixing.

10.11.2. Liquid Mixing

The following factors must be taken into account when choosing equipment for mixing liquids:

1. Batch or continuous operation
2. Nature of the process: miscible liquids, preparation of solutions or dispersion of immiscible liquids
3. Degree of mixing required
4. Physical properties of the liquids, particularly the viscosity
5. Whether the mixing is associated with other operations: reaction, heat transfer.

In-line mixers can be used for the continuous mixing of low-viscosity fluids. For other mixing operations, stirred vessels or proprietary mixing equipment will be required.

In-line Mixing

Static devices that promote turbulent mixing in pipelines provide an inexpensive way of continuously mixing fluids. Some typical designs are shown in Figure 10.70a–c. A simple mixing tee, Figure 10.70a, followed by a length of pipe equal to 10 to 20 pipe diameters, is suitable for mixing low-viscosity fluids (≤ 50 mNs/m^2) providing the flow is turbulent, and the densities and flow rates of the fluids are similar.

With injection mixers (Figure 10.70b,c), in which the one fluid is introduced into the flowing stream of the other through a concentric pipe or an annular array of jets, mixing will take place by entrainment and turbulent diffusion. Such devices should be used where one flow is much lower than the other, and will give a satisfactory blend in about 80 pipe diameters. The inclusion of baffles or other flow restrictions will reduce the mixing length required.

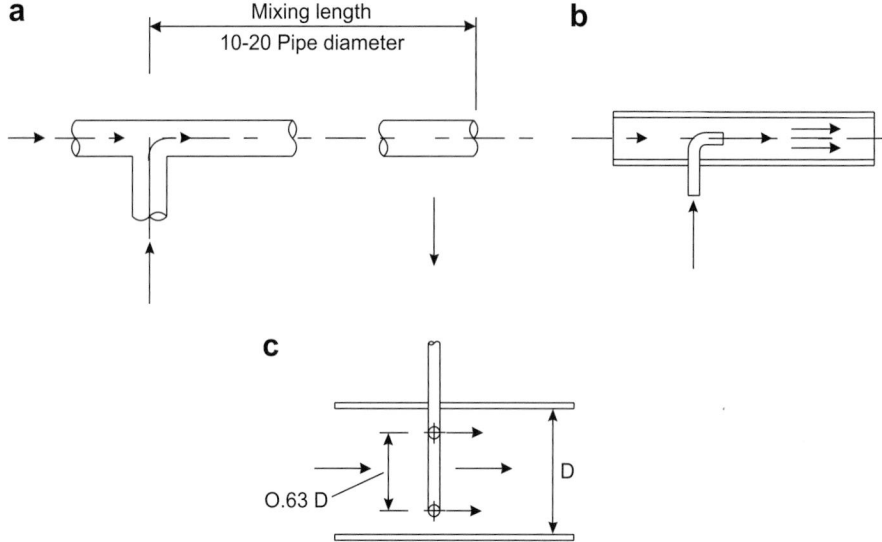

Figure 10.70. Inline mixers. (a) Tee. (b) Injection. (c) Annular.

The static in-line mixer shown in Figure 10.71 is effective in both laminar and turbulent flow, and can be used to mix viscous mixtures. The division and rotation of the fluid at each element causes rapid radial mixing; see Rosenzweig (1977) and Baker (1991). There is a great variety of different proprietary designs for static mixers and they are easily found by searching on the internet. The dispersion and mixing of liquids in pipes is discussed by Zughi et al. (2003) and Lee and Brodkey (1964).

Centrifugal pumps are effective in-line mixers for blending and dispersing liquids. Various proprietary motor-driven in-line mixers are also used for special applications; see Green and Perry (2007).

Stirred Tanks

Mixing vessels fitted with some form of agitator are the most commonly used type of equipment for blending viscous liquids and preparing solutions of dissolved solids.

Liquid mixing in stirred tanks is covered in Coulson et al. (1999), Chapter 7, and in several textbooks: Uhl and Gray (1967), Harnby et al. (1997) and Tatterson (1991), (1993).

A typical arrangement of the agitator and baffles in a stirred tank, and the flow pattern generated, is shown in Figure 10.72. Mixing occurs through the bulk flow of the liquid and, on a microscopic scale, by the motion of the turbulent eddies created

Figure 10.71. Static mixer (Kenics Corporation).

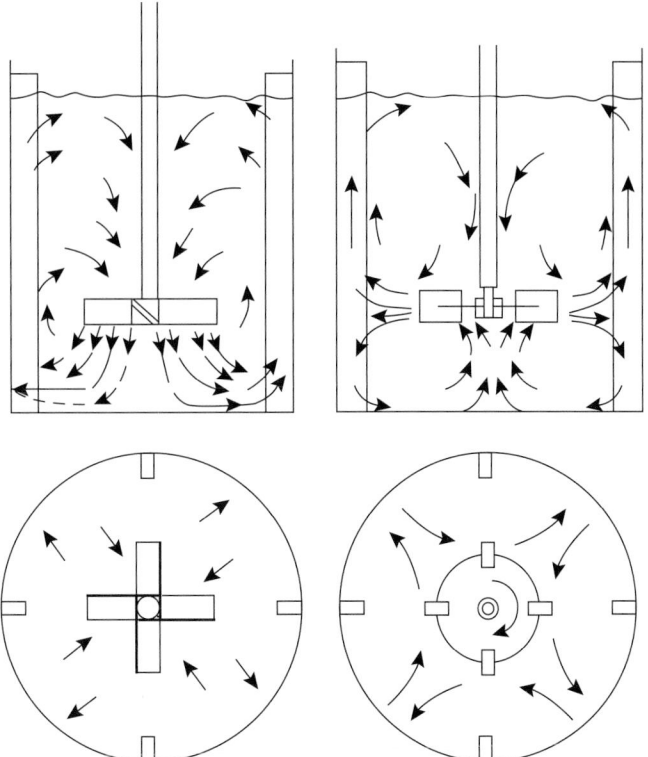

Figure 10.72. Agitator arrangements and flow patterns.

by the agitator. Bulk flow is the predominant mixing mechanism required for the blending of miscible liquids and for solids suspension. Turbulent mixing is important in operations involving mass and heat transfer, which can be considered as shear-controlled processes.

The most suitable agitator for a particular application will depend on the type of mixing required, the capacity of the vessel and the fluid properties, mainly the viscosity.

The three basic types of impeller that are used at high Reynolds numbers (low viscosity) are shown in Figure 10.73a–c. They can be classified according to the predominant direction of flow leaving the impeller. The flat-bladed (Rushton) turbines are essentially radial-flow devices, suitable for processes controlled by turbulent mixing (shear-controlled processes). The propeller and pitched-bladed turbines are essentially axial-flow devices, suitable for bulk fluid mixing.

Paddle, anchor and helical ribbon agitators (Figure 10.74a–c), and other special shapes, are used for more viscous fluids.

The selection chart given in Figure 10.75, which has been adapted from a similar chart given by Penney (1970), can be used to make a preliminary selection of the agitator type, based on the liquid viscosity and tank volume.

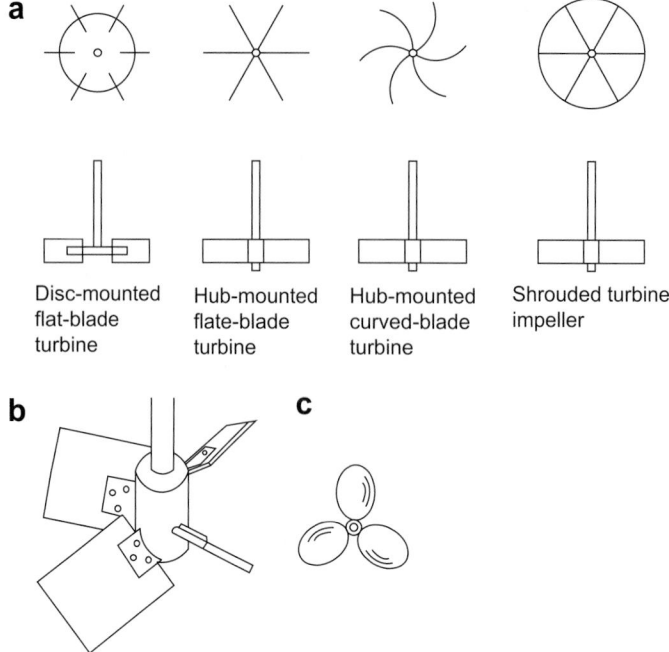

Figure 10.73.   Basic impeller types. (a) Turbine impeller. (b) Pitched bladed turbine. (c) Marine propeller.

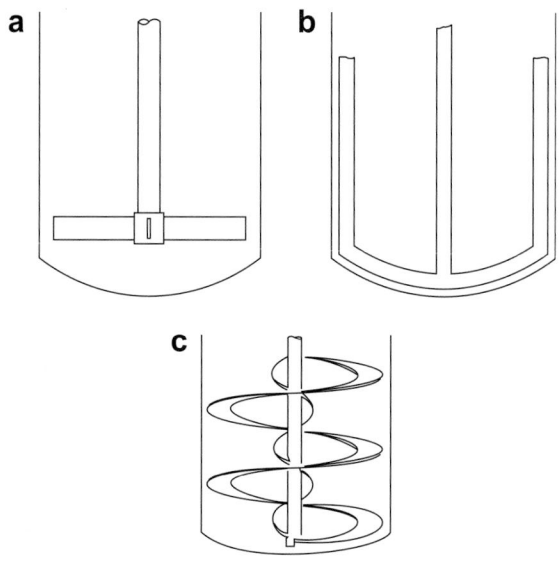

Figure 10.74.   Low-speed agitators. (a) Paddle. (b) Anchor. (c) Helical ribbon.

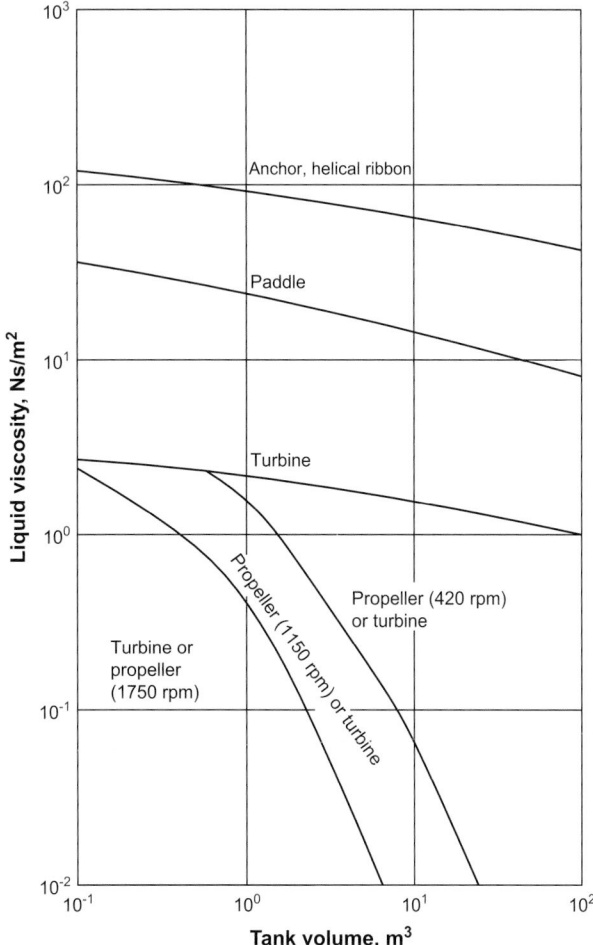

Figure 10.75. Agitator selection guide.

For turbine agitators, impeller to tank diameter ratios of up to about 0.6 are used, with the depth of liquid equal to the tank diameter. Baffles are normally used, to improve the mixing and reduce problems from vortex formation. Anchor agitators are used with close clearance between the blades and vessel wall, anchor to tank diameter ratios of 0.95 or higher. The selection of agitators for dispersing gases in liquids is discussed by Hicks (1976).

Agitator Power Consumption

The shaft power required to drive an agitator can be estimated using the following generalized dimensionless equation, the derivation of which is given in Coulson *et al.* (1999), Chapter 7.

$$N_p = K Re^b Fr^c \tag{10.18}$$

where

N_p = power number = $\frac{P}{D^5 N^3 \rho}$

Re = Reynolds number = $\frac{D^2 N \rho}{\mu}$,

Fr = Froude number = $\frac{DN^2}{g}$,

P = shaft power, W

K = a constant, dependent on the agitator type, size, and agitator-tank geometry

ρ = fluid density, kg/m^3

μ = fluid viscosity, Ns/m^2

N = agitator speed, s^{-1} (revolutions per second) (rps)

D = agitator diameter, m

g = gravitational acceleration, 9.81 m/s^2.

Values for the constant K and the indices b and c for various types of agitator, tank-agitator geometries and dimensions, can be found in the literature; Rushton *et al.* (1950). A useful review of the published correlations for agitator power consumption and heat transfer in agitated vessels is given by Wilkinson and Edwards (1972); they include correlations for non-Newtonian fluids. Typical power curves for propeller and turbine agitators are given in Figures 10.76 and 10.77. In the laminar flow region, the index 'b' = 1, and at high Reynolds number the power number is independent of the Froude number; index 'c' = 0.

An estimate of the power requirements for various applications can be obtained from Table 10.17.

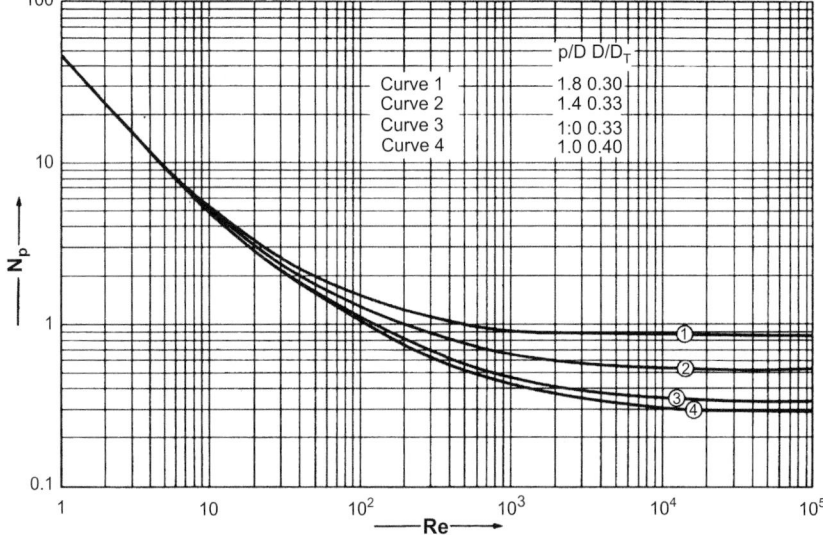

Figure 10.76.   Power correlation for single three-bladed propellers baffled (from Uhl and Gray,1967; with permission). p = blade pitch, D = impeller diameter, D_T = tank diameter.

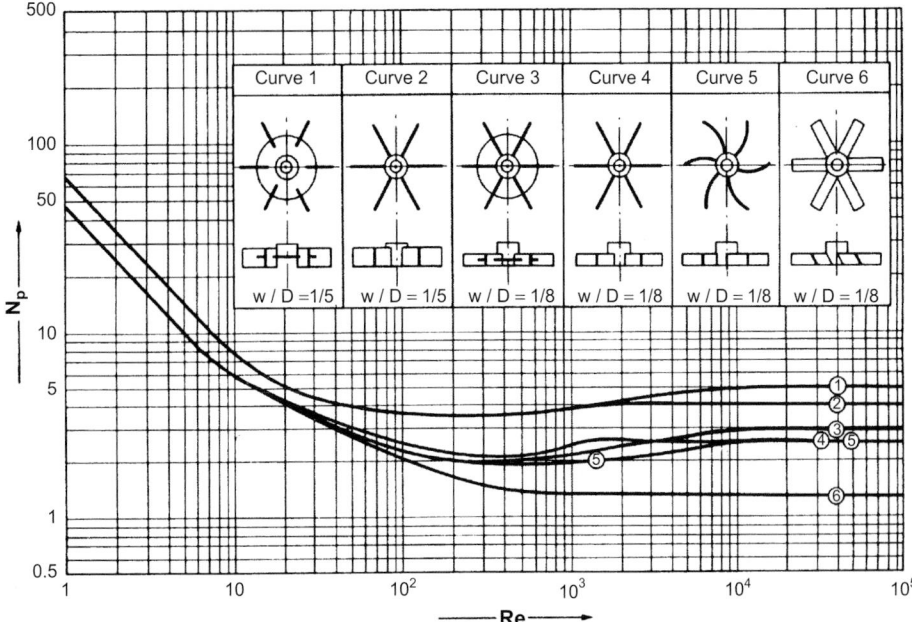

Figure 10.77. Power correlations for baffled turbine impellers, for tank with four baffles (from Uhl and Gray, 1967; with permission). w = impeller width, D = impeller diameter.

Side-entering Agitators

Side-entering agitators are used for blending low-viscosity liquids in large tanks, where it is impractical to use conventional agitators supported from the top of the tank; see Oldshue *et al.* (1956).

Where they are used with flammable liquids, particular care must be taken in the design and maintenance of the shaft seals, as any leakage may cause a fire.

For blending flammable liquids, the use of liquid jets should be considered as an 'intrinsically' safer option; see Fossett and Prosser (1949).

TABLE 10.17. Power Requirements: Baffled Agitated Tanks

Agitation	Applications	Power (kW/m³)
Mild	Blending, mixing	0.04–0.10
	Homogeneous reactions	0.01–0.03
Medium	Heat transfer	0.03–1.0
	Liquid–liquid mixing	1.0–1.5
Severe	Slurry suspension	1.5–2.0
	Gas absorption	1.5–2.0
	Emulsions	1.5–2.0
Violent	Fine slurry suspension	>2.0

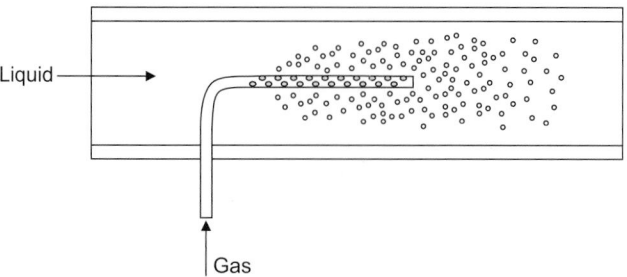

Figure 10.78. Gas sparger.

Gas–Liquid Mixing

Gases can be mixed into liquids using either in-line mixing, stirred vessels, or the vapour–liquid contacting devices described in Chapter 11.

When a small amount of gas is fed or the gas dissolves completely, in-line mixing can be used. The most common arrangement is an injection mixer (Figure 10.70b) followed by a static mixer. In some cases, a long injection tube with multiple holes drilled in it is used. This is known as a sparger; Figure 10.78.

If a gas is injected into a stirred tank, the location of the gas injection must be chosen based on the mixing pattern obtained with the impeller that has been selected. The gas injection device is usually an annular ring with multiple small openings, and the openings are oriented to promote the desired circulation of gas bubbles. Methods such as computational fluid dynamics (CFD) are used to analyse the gas bubble flow pattern and ensure that the gas hold-up and interfacial area are adequate.

A small flow of liquid can be dispersed into a gas stream using a spray nozzle; Figure 10.79. Many different proprietary spray nozzles are available and the nozzle is usually selected in consultation with a vendor.

When large flow rates of vapour and liquid are to be contacted to carry out mass transfer or direct heat transfer, plate or packed columns are usually used; these are discussed in detail in Chapter 11.

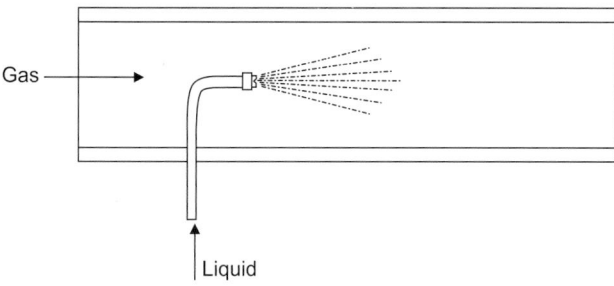

Figure 10.79. Liquid injection into gas.

TABLE 10.18. Solids and Paste Mixers

Type of Equipment	Mixing Action	Applications	Examples
Rotating: cone, double cone, drum	Tumbling action	Blending dry, free-flowing powders, granules, crystals	Pharmaceuticals, food, chemicals
Air blast fluidization	Air blast lifts and mixes particles	Dry powders and granules	Milk powder, detergents, chemicals
Horizontal trough mixer, with ribbon blades, paddles or beaters	Rotating element produces contra-flow movement of materials	Dry and moist powders	Chemicals, food, pigments, tablet granulation
Z-blade mixers	Shearing and kneading by the specially shaped blades	Mixing heavy pastes, creams and doughs	Bakery industry, rubber doughs, plastic dispersions
Pan mixers	Vertical, rotating paddles, often with planetary motion	Mixing, whipping and kneading of materials ranging from low-viscosity pastes to stiff doughs	Food, pharmaceuticals and chemicals, printing inks and ceramics
Cylinder mixers, single and double	Shearing and kneading action	Compounding of rubbers and plastics	Rubbers, plastics, and pigment dispersion

10.11.3. Solids and Pastes

A great variety of specialized equipment has been developed for mixing dry solids and pastes (wet solids). The principal types of equipment and their fields of application are given in Table 10.18. Descriptions of the equipment can be found in the literature: Green and Perry (2007) and Reid (1979). Cone blenders are used for free-flowing solids. Ribbon blenders can be used for dry solids and for blending liquids with solids. Z-blade mixers and pan mixers are used for kneading heavy pastes and doughs. Most solid and paste mixers are designed for batch operation.

A selection chart for solids mixing equipment is given by Jones (1985).

10.12. TRANSPORT AND STORAGE OF MATERIALS

In this section the principal means used for the transport and storage of process materials: gases, liquids and solids are discussed briefly. Further details and full descriptions of the equipment used can be found in various handbooks. Pumps and compressors are also discussed in Chapters 3 and 5.

10.12.1. Gases

The type of equipment best suited for the pumping of gases in pipelines depends on the flow rate, the differential pressure required, and the operating pressure.

In general, fans are used where the pressure drop is small, less than $35 \, cmH_2O$ (0.03 bar); axial flow compressors for high flow rates and moderate differential pressures; centrifugal compressors for high flow rates and, by staging, high

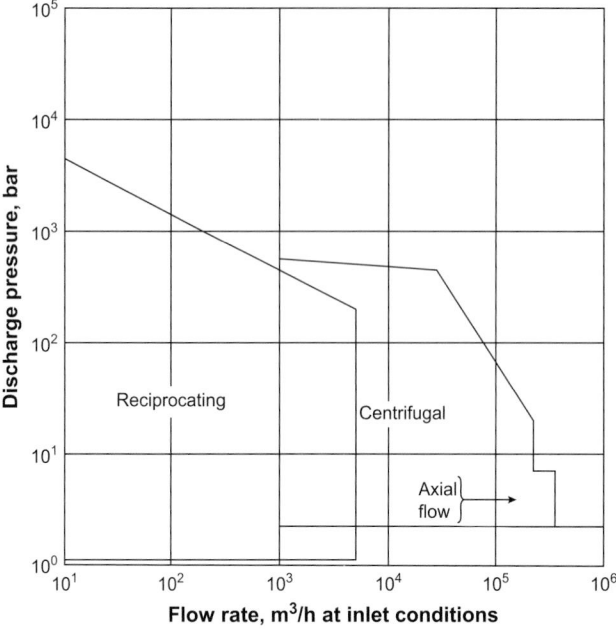

Figure 10.80.   Compressor operating ranges.

differential pressures. Reciprocating compressors can be used over a wide range of pressures and capacities, but are normally only specified in preference to centrifugal compressors where high pressures are required at relatively low flow rates.

Reciprocating, centrifugal and axial flow compressors are the principal types used in the chemical process industries, and the range of application of each type is shown in Figure 10.80 which has been adapted from a diagram by Dimoplon (1978). A more

TABLE 10.19.   Operating Range of Compressors and Blowers (After Begg, 1966)

Type of Compressor	Normal Maximum Speed (rpm)	Normal Maximum Capacity (m³/h)	Normal Maximum Pressure (differential) (bar)	
			Single Stage	**Multiple Stage**
Displacement				
1. Reciprocating	300	85,000	3.5	5000
2. Sliding vane	300	3400	3.5	8
3. Liquid ring	200	2550	0.7	1.7
4. Rootes	250	4250	0.35	1.7
5. Screw	10,000	12,750	3.5	17
Dynamic				
6. Centrifugal fan	1000	170,000		0.2
7. Turbo blower	3000	8500	0.35	1.7
8. Turbo compressor	10,000	136,000	3.5	100
9. Axial flow fan	1000	170,000	0.35	2.0
10. Axial flow blower	3000	170,000	3.5	10

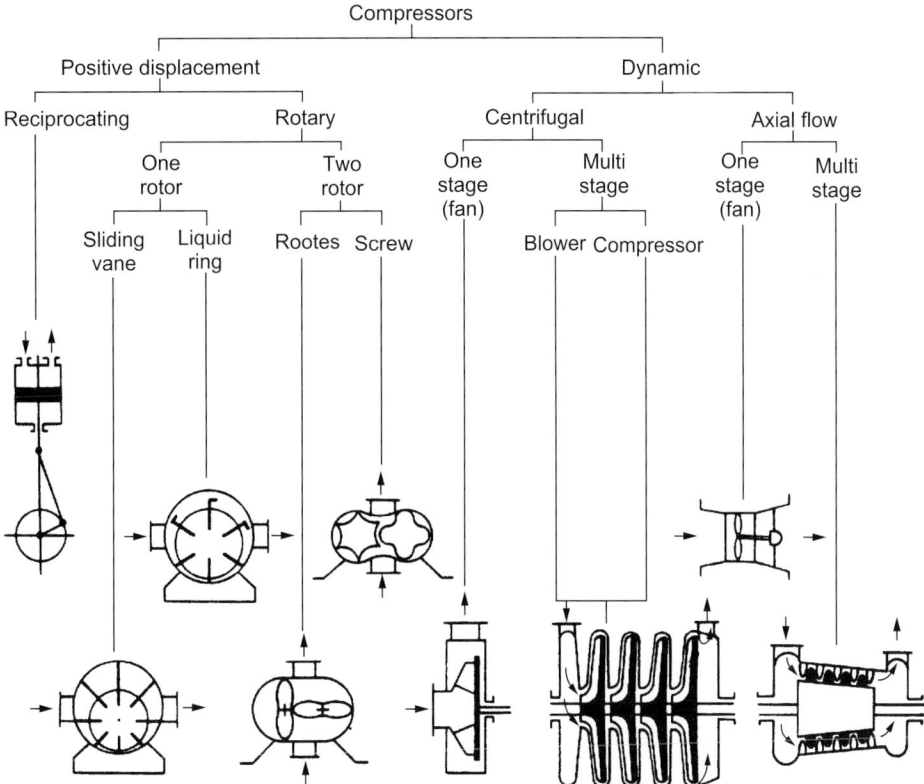

Figure 10.81.   Type of compressor (Begg, 1966).

comprehensive selection guide is given in Table 10.19. Diagrammatic sketches of the compressors listed are given in Figure 10.81.

Several textbooks are available on compressor design, selection and operation: Bloch *et al.* (1982), Brown (1990) and Aungier (1999, 2003).

Vacuum Production

The production of vacuum (sub-atmospheric pressure) is required for many chemical engineering processes; for example, vacuum distillation, drying and filtration. The type of vacuum pump needed will depend on the degree of vacuum required, the capacity of the system and the rate of air in-leakage.

Reciprocating and rotary positive displacement pumps are commonly used where moderately low vacuum is required, about 10 mmHg (0.013 bar), at moderate to high flow rates; such as in vacuum filtration.

Steam-jet ejectors are versatile and economic vacuum pumps and are frequently used, particularly in vacuum distillation. They can handle high vapour flow rates and, by using several ejectors in series, can produce low pressures, down to about 0.1 mmHg (0.13 mbar).

The operating principle of steam-jet ejectors is explained in Coulson *et al.*, Chapter 8. Their specification, sizing and operation are covered in a comprehensive series of papers by Power (1964). Diffusion pumps are used where very low pressures are required (hard vacuum) for processes such as molecular distillation.

For a general reference on the design and application of vacuum systems see Ryan and Roper (1986).

Storage

Gases are stored at low pressure in gas holders similar to those used for town gas. The liquid-sealed type is most commonly used. These consist of a number of telescopic sections (lifts) that rise and fall as gas is added to or withdrawn from the holder. The dry-sealed type is used where the gas must be kept dry. In this type the gas is contained by a piston moving in a large vertical cylindrical vessel. Water-seal holders are intrinsically safer for use with flammable gases than the dry seal type, as any leakage through the piston seal may form an explosive mixture in the closed space between the piston and the vessel roof. Details of the construction of gas holders can be found in text books on Gas Engineering: Meade (1921) and Smith (1945).

Gases are stored at high pressures where this is a process requirement and to reduce the storage volume. For some gases, the volume can be further reduced by liquefying the gas by pressure or refrigeration. Cylindrical and spherical vessels (Horton spheres) are used. The design of pressure vessels is discussed in Chapter 13.

10.12.2. Liquids

The selection of pumps for liquids is discussed in Chapter 5. Descriptions of most of the types of pumps used in the chemical process industries are given in Coulson *et al.*, Chapter 8, and several textbooks and handbooks on this subject: Garay (1997), Karassik (2001) and Parmley (2000).

The principal types used and their operating pressures and capacity ranges are summarized in Table 10.20 and Figure 10.82. Centrifugal pumps will normally be the first choice for pumping process fluids, the other types only being used for special applications, such as the use of reciprocating and gear pumps for metering.

Pump Shaft Power

The power required for pumping an incompressible fluid is given by:

$$\text{Power} = \frac{\Delta P Q_p}{\eta_p} \times 100 \tag{10.19}$$

TABLE 10.20. Normal Operating Range of Pumps

Type	Capacity Range (m³/h)	Typical Head (m of Water)
Centrifugal	$0.25-10^3$	10–50 300 (multistage)
Reciprocating	0.5–500	50–200
Diaphragm	0.05–50	5–60
Rotary gear and similar	0.05–500	60–200
Rotary sliding vane or similar	0.25–500	7–70

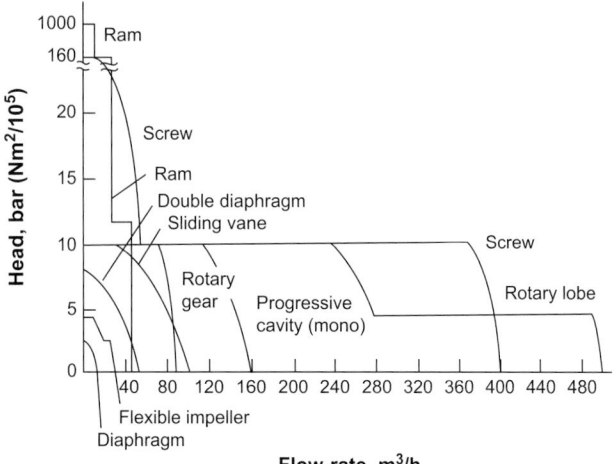

Figure 10.82. Selection of positive displacement pumps (adapted from Marshall, 1985). Descriptions of the types mentioned are given in Coulson *et al.* (1999), Chapter 8.

where

ΔP = pressure differential across the pump, N/m^2
Q_p = flow rate, m^3/s
η_p = pump efficiency, %.

See also, Chapter 5, Section 5.4.

The efficiency of centrifugal pumps depends on their size. The values given in Figure 10.83 can be used to estimate the power and energy requirements for preliminary design purposes. The efficiency of reciprocating pumps is usually around 90%.

Storage

Liquids are usually stored in bulk in vertical cylindrical steel tanks. Fixed and floating-roof tanks are used. In a floating-roof tank a movable piston floats on the surface of the liquid and is sealed to the tank walls. Floating-roof tanks are used to eliminate evaporation losses and, for flammable liquids, to obviate the use of inert gas blanketing to prevent an explosive mixture forming above the liquid, as would be the situation with a fixed-roof tank.

Horizontal cylindrical tanks and rectangular tanks are also used for storing liquids, usually for relatively small quantities.

The design of fixed roof, vertical tanks is discussed in Chapter 13, Section 13.16.

10.12.3. Solids

The movement and storage of solids is usually more expensive than the movement of liquids and gases, which can be easily pumped down a pipeline. The best equipment to use will depend on a number of factors:

1. The throughput
2. Length of travel

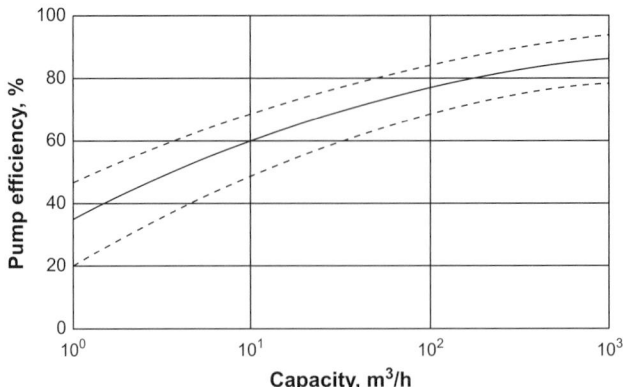

Figure 10.83.   Efficiencies of centrifugal pumps.

3. Change in elevation
4. Nature of the solids: size, bulk density, angle of repose, abrasiveness, corrosiveness, wet or dry.

Belt conveyors are the most commonly used type of equipment for the continuous transport of solids. They can carry a wide range of materials economically over long and short distances; either horizontally or at an appreciable angle, depending on the angle of repose of the solids. A belt conveyor consists of an endless belt of a flexible material, supported on rollers (idlers), and passing over larger rollers at each end, one of which is driven. The belt material is usually fabric-reinforced rubber or plastics; segmental metal belts are also used. Belts can be specified to withstand abrasive and corrosive materials; see BS EN ISO 21183, BS EN ISO 14890 and BS EN ISO 15236.

Screw conveyors, also called worm conveyors, are used for materials that are free flowing. The basic principle of the screw conveyor has been known since the time of Archimedes. The modern conveyor consists of a helical screw rotating in a U-shaped trough. They can be used horizontally or, with some loss of capacity, at an incline to lift materials. Screw conveyors are less efficient than belt conveyors, due to the friction between the solids and the flights of the screw and the trough, but are cheaper and easier to maintain. They are used to convey solids over short distances, and when some elevation (lift) is required. They can also be used for delivering a metered flow of solids.

Pneumatic conveying is used for movement of solids over relatively short distances. It is generally suitable only for free-flowing particles in the range 20 μm to 50 mm, as finer dusts tend to stick to the pipes, while larger particles are hard to entrain. In pneumatic conveying, the solids are transported in suspension in a gas or liquid. The solids may be either dilute phase, with void fraction typically greater than 95%, or dense phase, with void fraction as low as 50%. The velocity of carrier fluid must be large enough to keep the particles suspended; see Coulson *et al.* (1999),

Chapter 5. Pneumatic conveying can be used for both horizontal and vertical transport of solids, including making pipe turns. Sharp turns are generally avoided, as these cause solids attrition and pipe abrasion.

Pipe conveyors are a relatively new technology that is becoming widely used in minerals handling. A pipe conveyor is similar to a belt conveyor in that solids are dropped onto a flexible belt. After loading the solids, the belt passes through rollers that fold the sides over to cover the solid; see Figure 10.84. The rolled-up tube containing the solids then passes through idlers that keep the tube rolled until the destination, where the tube is unrolled and the solids are discharged. Pipe conveyors cost only slightly more than belt conveyors and have a number of advantages. They cause less dust formation, there is no need to cover the conveyor to keep material dry, and there is no need for the pipes to follow straight flights so a pipe conveyor can more easily traverse difficult terrain.

The most widely used equipment where a vertical lift is required is the bucket elevator. This consists of buckets fitted to an endless chain or belt, which passes over a driven roller or sprocket at the top end. Bucket elevators can handle a wide range of solids, from heavy lumps to fine powders, and are suitable for use with wet solids and slurries.

The mechanical conveying of solids is the subject of books by Colijn (1985), Fayed and Skocir (1996), Levy and Kalman (2001) and McGlinchey (2008). Screw conveyors are described in detail in the book by Forcade (1999). Pneumatic and hydraulic conveying are discussed in a book by Mills (2004); see also Mills *et al.* (2004).

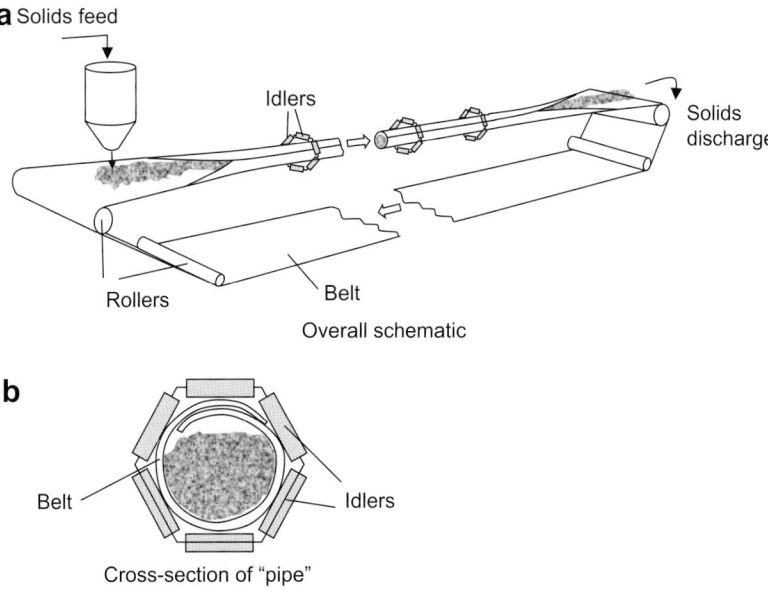

Figure 10.84. Pipe conveyor.

Storage

The simplest way to store solids is to pile them on the ground in the open air. This is satisfactory for the long-term storage of materials that do not deteriorate on exposure to the elements; for example, the seasonal stock piling of coal at collieries and power stations. For large stockpiles, permanent facilities are usually installed for distributing and reclaiming the material; travelling gantry cranes, grabs and drag scrapers feeding belt conveyors are used. For small, temporary, storage, mechanical shovels and trucks can be used. Where the cost of recovery from the stockpile is large compared with the value of the stock held, storage in silos or bunkers should be considered.

Overhead bunkers, also called bins or hoppers, are normally used for the short-term storage of materials that must be readily available for the process. They are arranged so that the material can be withdrawn at a steady rate from the base of the bunker on to a suitable conveyor. Bunkers must be carefully designed to ensure the free flow of material within the bunker, to avoid packing and bridging. Jenike (1967) and Jenike and Johnson (1970), have studied the flow of solids in containers and developed design methods. All aspects of the design of bins and hoppers, including feeding and discharge systems, are covered in books by Reisner (1971) and Brown and Nielsen (1998). See also the British Material Handling Board's code of practice on the design of silos and bunkers (BMHB, 1992).

The storage and transport of wet solids are covered by Heywood (1991).

10.13.   REACTORS

The reactor is the heart of a chemical process. It is the only place in the process where raw materials are converted into products, and reactor design is a vital step in the overall design of the process.

Numerous textbooks have been published on reactor design, and a selection is given in the bibliography at the end of this chapter. The volumes by Rase (1977, 1990) cover the practical aspects of reactor design and include case studies of industrial reactors. The design of electrochemical reactors is covered by Rousar *et al.* (1985) and Scott (1991).

The treatment of reactor design in this section will be restricted to a discussion of the selection of the appropriate reactor type for a particular process, and an outline of the steps to be followed in the design of a reactor.

The design of an industrial chemical reactor must satisfy the following requirements:

1. The chemical factors: the kinetics of the reaction. The design must provide sufficient residence time for the desired reaction to proceed to the required degree of conversion. When the reaction is promoted by a catalyst, enzyme or organism, then the reaction conditions must maintain the activity of the promoter, or allow activity to be periodically restored.
2. The mass-transfer factors: with heterogeneous reactions the reaction rate may be controlled by the rates of diffusion of the reacting species, rather than the chemical kinetics.
3. The heat-transfer factors: the removal, or addition, of the heat of reaction.

4. The safety factors: the confinement of hazardous reactants and products, and the control of the reaction and the process conditions.

The need to satisfy these interrelated, and often contradictory factors, makes reactor design a complex and difficult task. However, in many instances one of the factors will predominate and will determine the choice of reactor type and the design method.

10.13.1. Principal Types of Reactor

The following characteristics are normally used to classify reactor designs:

1. Mode of operation: batch or continuous
2. Phases present: homogeneous or heterogeneous
3. Reactor geometry: flow pattern and manner of contacting the phases
   (i) stirred tank reactor
   (ii) tubular reactor
   (iii) packed bed, fixed and moving
   (iv) fluidized bed.

Batch or Continuous Processing

In a batch process, some of the reagents are added at the start; the reaction proceeds, the compositions changing with time. Additional reagents may be added as the reaction proceeds, and changes in temperature may also be made. At the end of the recipe the reaction is stopped and the product withdrawn when the required conversion has been reached. Batch processes are suitable for small-scale production and for processes where a range of different products, or grades, is to be produced in the same equipment; for instance, pigments, dyestuffs and polymers.

In continuous processes the reactants are fed to the reactor and the products withdrawn continuously; the reactor operates under steady-state conditions. Continuous production will normally give lower production costs than batch production, but lacks the flexibility of batch production. Continuous reactors will usually be selected for large-scale production.

Processes that do not fit the definition of batch or continuous are often referred to as semi-continuous or semi-batch. In a semi-batch reactor some of the products may be withdrawn as the reaction proceeds. A semi-continuous process can be one which is interrupted periodically for some purpose; for instance, for the regeneration of catalyst.

Homogeneous and Heterogeneous Reactions

Homogeneous reactions are those in which the reactants, products, and any catalyst used form one continuous phase: gaseous or liquid.

Homogeneous gas-phase reactors are always operated continuously, whereas liquid-phase reactors may be batch or continuous. Tubular (pipe-line) reactors are normally used for homogeneous gas-phase reactions; for example, in the thermal cracking of petroleum crude oil fractions to ethylene, and the thermal decomposition of dichloroethane to vinyl chloride. Both tubular and stirred tank reactors are used for homogeneous liquid-phase reactions.

In a heterogeneous reaction two or more phases exist, and the overriding problem in the reactor design is to promote mass transfer between the phases. The possible combinations of phases are:

1. *Liquid–liquid*: immiscible liquid phases; reactions such as the nitration of toluene or benzene with mixed acids, emulsion polymerizations, and liquid acid catalysed alkylation. These are usually carried out in stirred tank reactors so that agitation can be used to generate a high liquid–liquid area for mass transfer.

2. *Liquid–solid*: with one, or more, liquid phases in contact with a solid. The solid may be a reactant or catalyst. Any of the reactor geometries can be used.

3. *Liquid–solid–gas*: where the solid is normally a catalyst; such as in the hydrogenation of amines using a slurry of platinum on activated carbon as a catalyst, or an organism; as in fermentation reactors, where the gas is air and carbon dioxide. Liquid–solid–gas reactions are usually carried out in slurry phase with the solids suspended in the liquid if the solids must be added to or removed from the reactor. If the solid phase is a catalyst or enzyme that deactivates slowly, a trickle bed can be used, in which the liquid and gas flow over a packed bed of solids, usually in co-current down flow.

4. *Gas–solid*: where the solid may take part in the reaction or act as a catalyst. The reduction of iron ores in blast furnaces and the combustion of solid fuels are examples where the solid is a reactant. Packed bed, moving bed or fluidized bed reactors are normally used.

5. *Gas–liquid*: where the liquid may take part in the reaction or act as a catalyst. Vapour–liquid contacting columns are preferred if the residence time requirements are short enough, because of the high area for mass transfer; see Chapter 11. When long residence time is needed for the liquid phase, stirred tanks or tubular reactors are used.

Reactor Geometry (type)

The reactors used for established processes are usually complex designs that have been developed (have evolved) over many years to suit the requirements of the process, and are unique designs; however, it is convenient to classify reactor designs into the following broad categories.

Stirred Tank Reactors

Stirred tank (agitated) reactors consist of a tank fitted with a mechanical agitator. Stirred tank reactors often also have a heating or cooling jacket or coils. They are operated as batch reactors or continuously. Several reactors may be used in series or parallel.

The stirred tank reactor can be considered the basic chemical reactor; modelling on a large scale the conventional laboratory flask. Tank sizes range from a few litres to several thousand litres. They are used for homogeneous and heterogeneous liquid–liquid and liquid–gas reactions, and for reactions that involve finely suspended solids which are held in suspension by the agitation. As the degree of agitation is under the designer's control, stirred tank reactors are particularly suitable for reactions where good mass transfer or heat transfer is required.

When operated as a continuous process, the composition in the reactor is constant and the same as the product stream, and, except for very rapid reactions, this will limit the conversion that can be obtained in one stage.

The power requirements for agitation will depend on the degree of agitation required and will range from about $0.2\,kW/m^3$ for moderate mixing to $2\,kW/m^3$ for intense mixing. Agitation of stirred tanks and power requirements are discussed in Section 10.11.2. Heat transfer to or from stirred tanks is discussed in Section 12.18.

Tubular Reactors

Tubular reactors are generally used for gaseous reactions, but are also suitable for some liquid-phase reactions.

If high heat-transfer rates are required, small-diameter tubes are used to increase the surface area to volume ratio. Several tubes may be arranged in parallel, connected to a manifold or fitted into a tube sheet in a similar arrangement to a shell and tube heat exchanger. Highly exothermic reactions are sometimes carried out inside the tubes of a heat exchanger, with boiling water on the shell side to provide a high rate of heat removal. Highly endothermic reactions at high temperature can be carried out inside tubes arranged in a furnace, as in ethylene cracking.

The pressure-drop and heat-transfer coefficients in empty tube reactors can be calculated using the methods for flow in pipes given in Chapters 5 and 12.

Packed-Bed Reactors

There are two basic types of packed-bed reactor: those in which the solid is a reactant and those in which the solid is a catalyst. Many examples of the first type can be found in the extractive metallurgical industries.

In the chemical process industries the designer will normally be concerned with the second type: catalytic reactors. Industrial packed-bed catalytic reactors range in size from small tubes, a few centimetres in diameter, to large-diameter packed beds. Packed-bed reactors are used for gas and gas–liquid reactions. When large-diameter gas–liquid packed bed reactors are used, care is needed to ensure a good distribution of liquid over the catalyst bed.

The practical aspects of designing packed-bed reactors are similar to designing packed columns for vapour–liquid contacting; see Section 11.14, although the liquid and gas usually flow co-currently down the bed, so there is less chance of flooding. When up-flow is used, the designer must ensure that the fluid velocity is not high enough to cause fluidization of the bed.

Heat-transfer rates in large diameter packed beds are poor. For relatively low heat addition or removal a packed bed reactor is usually operated adiabatically. Higher heating or cooling requirements can sometimes be accommodated by recirculating reactor effluent to the feed to act as heat carrier. In some cases, such as some selective oxidation processes, the catalyst is packed inside the tubes of a shell and tube heat exchanger. Where high heat-transfer rates are required, fluidized beds should be considered.

Moving-Bed Reactors

When a catalyst requires periodic regeneration the designer can use two or more packed bed reactors with one in operation and the other in regeneration mode (a cyclic or swing-bed system). Alternatively, it may be cheaper to use a moving-bed reactor, in which the catalyst is slowly circulated between a reaction vessel and a regeneration vessel. Moving-bed reactors are most commonly applied to gas–solid reactions, although they can also be used for liquid–solid reactions. Moving-bed reactors allow flow of solids without the high rates of attrition that are encountered in slurries and fluidized beds, and are therefore attractive when the catalyst is expensive or prone to attrition.

In a moving-bed reactor the solid flows downwards while fluid flows either radially or downwards through the catalyst bed. The catalyst is usually retained by screens that have slots or perforations through which the fluid can flow but the solid particles can not. A radial-flow moving-bed reactor is illustrated in Figure 10.85. With radial-flow reactors, the fluid velocity must be low enough to not cause 'pinning' of the catalyst against the screens.

Fluidized-Bed Reactors

The essential feature of a fluidized-bed reactor is that the solids are held in suspension by the upward flow of the reacting fluid; this promotes high mass-transfer and heat-transfer rates and good mixing. Heat-transfer coefficients on the order of $200 \text{ W/m}^2 °\text{C}$ to jackets and internal coils are typically obtained. The solids may be a catalyst, a reactant in fluidized combustion processes, or an inert powder added to promote heat transfer.

Though the principal advantage of a fluidized bed over a fixed bed is the higher heat transfer rate, fluidized beds are also useful where it is necessary to transport large quantities of solids as part of the reaction processes, such as where catalysts are transferred to another vessel for regeneration.

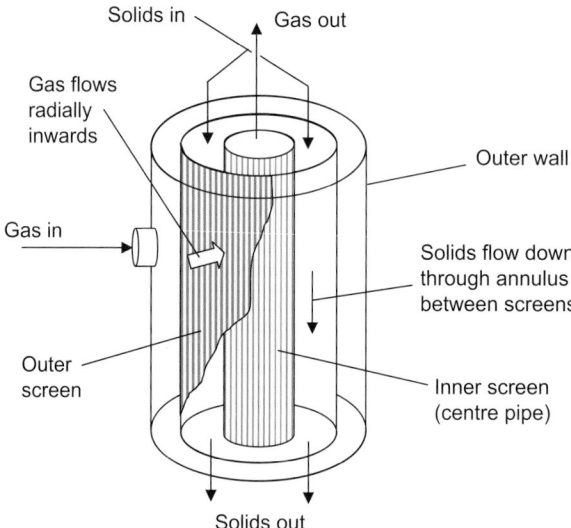

Figure 10.85.   Radial-flow moving-bed reactor.

Fluidization can only be used with relatively small sized particles, $< 300\ \mu m$ with gases. The solid material must be strong enough to withstand attrition in the fluidized bed and cheap enough to allow for make-up to replace attrition losses.

The gases that leave the fluidized bed are usually passed through one or more stages of cyclones to recover solids and return them to the reactor. After bulk solids recovery, methods such as fabric filters or electrostatic precipitation are used for control of fine particulates. Any equipment downstream of the particulate control equipment must still be designed for the presence of dust.

A great deal of research and development work has been done on fluidized-bed reactors in recent years, but the design and scale-up of large-diameter reactors is still an uncertain process and design methods are largely empirical.

The principles of fluidization processes are covered in Richardson *et al.* (2002), Chapter 6, and Yang, W.-C. (1999, 2003). The design of fluidized-bed reactors is discussed by Rase (1977).

10.13.2. Design Procedure

A general procedure for reactor design is outlined below.

1. Collect the kinetic and thermodynamic data on the desired reaction and the side reactions. It is unlikely that much useful information will be gleaned from a literature search, as little is published in the open literature on commercially attractive processes. The kinetic data required for reactor design are normally obtained from laboratory and pilot plant studies. Values will be needed for the rate of reaction over a range of operating conditions: pressure, temperature, flow rate and catalyst concentration (space velocity). The design of experimental reactors and scale-up are discussed by Rase (1977) and Bisio and Kabel (1985).

2. Collect the physical property data required for the design; either from the literature by estimation or, if necessary, by laboratory measurements.

3. Identify the predominant rate-controlling mechanism: kinetic, mass or heat transfer. Choose a suitable reactor type, based on experience with similar reactions, or from the laboratory and pilot plant work.

4. Make an initial selection of the reactor conditions to give the desired conversion and yield.

5. Size the reactor and estimate its performance. Exact analytical solutions of the design relationships are rarely possible; semi-empirical methods based on the analysis of idealized reactors will normally have to be used. Methods such as computational fluid dynamics (CFD) are becoming increasingly useful for modelling reactor performance, but unless the designer has a very good model of the reaction kinetics, skill and judgement will still be needed.

6. Select suitable materials of construction.

7. Make a preliminary mechanical design for the reactor: the vessel design, heat-transfer surfaces, internals and general arrangement.

8. Cost the proposed design, capital and operating, and repeat steps 4 to 8, as necessary, to optimize the design.

In choosing the reactor conditions, particularly the conversion, and optimizing the design, the interaction of the reactor design with the other process operations must not be overlooked. The degree of conversion of raw materials in the reactor will determine the size and cost of any equipment needed to separate and recycle unreacted materials. In these circumstances, the reactor and associated equipment must be optimized as a unit.

In most cases, the reactor design will require testing and validation at the pilot plant scale before a full-scale unit is built. Experience with similar reaction systems is usually the best way to ensure successful scale-up.

10.14. REFERENCES

Abulnaga, B. (2002) *Slurry Systems Handbook* (McGraw-Hill).

Ambler, C. M. (1952) *Chem. Eng. Prog.* **48** (March) 150. Evaluating the performance of centrifuges.

Ambler, C. M. (1971) *Chem. Eng., NY* **78** (Feb. 15th) 55. Centrifuge selection.

Amjad, Z. (ed) (1993) *Reverse Osmosis: Membrane Technology, Water Chemistry and Industrial Applications* (Van Nostrand Reinhold).

Aungier, R. H. (1999) *Centrifugal Compressors: A Strategy for Aerodynamic Design and Analysis* (American Society of Mechanical Engineers).

Aungier, R. H. (2003) *Axial-Flow Compressors: A Strategy for Aerodynamic Design and Analysis* (American Society of Mechanical Engineers).

AWWA (2007) *Reverse Osmosis and Nanofiltration, 2nd edn, AWWA Manual* (American Waterworks Association).

Baker, J. R. (1991) *Chem. Eng. Prog.* **87**(6), 32. Motionless mixtures stir up new uses.

Baker, R. W. (2004) *Membrane Technology and Applications*, 2nd edn (Wiley).

Begg, G. A. J. (1966) *Chem. & Process Eng.* **47**, 153. Gas compression in the chemical industry.

Bennett, J. G. (1936) *J. Inst. Fuel* **10**, 22. Broken coal.

Billet, R. (1989) *Evaporation Technology: Principles, Applications, Economics* (Wiley).

Bisio, A. and Kabel, R. L. (1985) *Scale Up of Chemical Processes: Conversion from Laboratory Scale Tests to Successful Commercial Size Design* (Wiley).

Bloch, H. P., Cameron, J. A., Danowsky, F. M., James, R., Swearingen, J. S. and Weightman, M. E. (1982) *Compressors and Expanders: Selection and Applications for the Process Industries* (Dekker).

BMHB (1992) *Draft Code of Practice for the Design of Hoppers, Bins, Bunkers and Silos*, 3rd edn (British Standards Institute).

Bradley, D. (1960) Institute of Minerals and Metals, International Congress, London, April, Paper 7, Group 2. Design and performance of cyclone thickeners.

Bradley, D. (1965) *Chem. & Process Eng.* 595. Medium-speed centrifuges.

Breck, D. W. (1974) *Zeolite Molecular Sieves* (Wiley).

Bronkala, W. J. (1988) *Chem. Eng., NY* **95** (March 14th) 133. Purification: doing it with magnets.

Brown, R. L. (1990) *Compressors: Sizing and Selection* (Gulf).

Brown, C. J. and Nielsen, J. (1998) *Silos: Fundamentals of Theory, Behaviour and Design* (Taylor & Francis).

Byrne, W. (1995) *Reverse Osmosis: A Practical Guide for Industrial Users* (Tall Oaks).

Cassidy, R. T. and Doshi, K. J. (1984) US 4,461,630 Product recovery in pressure swing adsorption process and system.

Cheryan, M. (1986) *Ultrafiltration Handbook* (Techonomonic).

Cheremisnoff, N. P. (1998) *Liquid Filtration* 2nd edn (Butterworth-Heinemann).

Cole, J. (1984) *Chem. Engr., London* No. 404 (June) 20. A guide to the selection of evaporation plant.

Constantinescu, S. (1984) *Chem. Eng., NY* **91** (Feb. 20th) 97. Sizing gas cyclones.

Colijn, H. (1985) *Mechanical Conveyors for Bulk Solids* (Elsevier).

Coulson, J. M., Richardson, J. F., Backhurst, J., and Harker, J. H. (1999) *Chemical Engineering: Volume 1*, 6th edn (Butterworth-Heinemann).

Crittenden, B. and Thomas, W. J. (1998) *Adsorption Design and Technology* (Butterworth-Heinemann).

Dahlstrom, D. A. and Cornell, C. F. (1971) *Chem. Eng., NY* **78** (Feb. 15th) 63. Thickening and clarification.

Day, R. W., Gricher, G. N., and Bier, T. H. (1997) Hydrocyclone separation. In *Handbook of Separation Processes for Chemical Engineers*, 3rd edn, Schweitzer, P. A. (ed.) (McGraw-Hill).

Dickenson, T. C. (1997) *Filters and Filtration Handbook* (Elsevier).

Dimoplon, W. (1978) *Hyd. Proc.* **57** (May) 221. What process engineers need to know about compressors.

Fayed, M. E. and Skocir, T. S. (1996) *Mechanical Conveyors: Selection and Operation* (CRC).

Fischer, R. (1965) *Chem. Eng., NY* **72** (Sept. 13th) 179. Agitated evaporators, Part 2, equipment and economics.

Fleming, G. K. and Dupuis, G. E. (1993) *Hyd. Proc.* **72**(4), 61. Hydrogen membrane recovery estimates.

Flynn, T. (2004) *Cryogenic Engineering*, 2nd edn (CRC).

Follman, D. K. and Fahrner, R. L. (2004) *J. Chromatogr. A.* **1024**, 79. Factorial screening of antibody purification processes using three chromatography steps without protein.

Forcade, M. P. (1999) *Screw Conveyor 101* (Goodman Conveyor Co).

Fossett, H. and Prosser, L. E. (1949) *Proc. Inst. Mech. Eng* **160**, 224. The application of free jets to the mixing of fluids in tanks.

Fuerstenau, M.C. and Han, K.N. (eds) (2003) *Principles of Mineral Processing* (Society for Mining, Metallurgy and Exploration).

Garay, P. N. (1997) *Pump Applications Desk Book* (Prentice Hall).

Ganetsos, G. and Barker, P. E. (eds) (1992) *Preparative and Production Scale Chromatography* (CRC).

Gerunda, A. (1981) *Chem. Eng., NY* **74** (May 4) 81. How to size liquid–vapour separators.

Green, D. W and Perry, R. H. (eds) (2007) *Perry's Chemical Engineers Handbook*, 8th edn (McGraw-Hill).

Hagel, L., Jagschies, G., and Sofer, G. K. (2007) *Handbook of Process Chromatography, 2nd edn: Development, Manufacturing, Validation and Economics* (Academic Press).

Harnby, N., Edwards, M. F., and Nienow, A. W. (1997) (eds) *Mixing in the Process Industries*, 2nd edn (Butterworths).

Helfferich, F. (1995) *Ion Exchange* (Dover Science).

Heywood, N. (1991) *The Storage and Conveying of Wet Granular Solids in the Process Industries* (Royal Society of Chemistry).

Hicks, R. W. (1976) *Chem. Eng., NY* **83** (July 19th) 141. How to select turbine agitators for dispersing gas into liquids.

Hiorns, F. J. (1970) *Brit. Chem. Eng.* **15**, 1565. Advances in comminution.

Hoffman, E. J. (2003) *Membrane Separations Technology: Single Stage, Multistage and Differential Permeation* (Gulf).

Hogsett, J. E. and Mazur, W. H. (1983) *Hyd. Proc.* **62**(8), 52. Estimate membrane system area.

Hooper, W. B. (1975) *Chem. Eng., NY* **82** (Aug. 4th) 103. Predicting flow patterns in plant equipment.

Hooper, W. B. (1997) Decantation. In *Handbook of Separation Processes for Chemical Engineers*, 3rd edn, Schweitzer, P. A. (ed.) (McGraw-Hill).

IChemE (1992) *Dust and Fume Control: a User Guide*, 2nd edn (Institution of Chemical Engineers, London).

Jacob, K. and Dhodapkar, S. (1997) Gas–Solid Separations. In: *Handbook of Separation Processes for Chemical Engineers*, 3rd edn, Schweitzer, P. A. (ed.) (McGraw-Hill).

Jenike, A. W. (1967) *Powder Technology* **1**, 237. Quantitive design of mass flow in bins.

Jenike, A.W. and Johnson, J.R. (1970) *Chem. Eng. Prog.* **66** (June) 31. Solids flow in bins and moving beds.

Jones, R. L. (1985) *Chem. Engr., London* **419**(9) 41. Mixing equipment for powders and pastes.

Jones, A. G. (2002) *Crystallization Process Systems* (Butterworth-Heinemann).

Jupke, A., Epping, A., and Schmidt-Traub, H. (2002) *J. Chromatog. A* **944**, 93. Optimal design of batch and simulated moving bed chromatographic separation processes.

Karassik, I. J. (ed.) (2001) *Pump Handbook*, 3rd edn (McGraw-Hill)

Kehde, H., Fairfield, R. G., Frank, J. C., and Zahnstecher, L. W. (1948) *Chem. Eng. Prog.* **44**(8), 575. Ethylene recovery. Commercial Hypersorption operation.

Koch, W. H. and Licht, W. (1977) *Chem. Eng., NY* **84** (Nov. 7th) 80. New design approach boosts cyclone efficiency.

Kraus, M. N. (1979) *Chem. Eng., NY* **86** (April 9th) 94. Separating and collecting industrial dusts. (April 23rd) 133. Baghouses: selecting, specifying and testing of industrial dust collectors.

Kucera, J. (2008) *Chem. Eng. Prog* **104**(5), 30. Understanding RO membrane performance.

Kumar, R., Naheiri, T. and Watson, C. F. (1994) US 5,328,503 Adsorption process with mixed repressurization and purge/equalization.

Larson, M.A. (1978) *Chem. Eng., NY* **85** (Feb. 13th) 90. Guidelines for selecting crystallizers.

Lavanchy, A. C., Keith, F. W. and Beams, J. W. (1964) Centrifugal separation. In *Kirk-Othmer Encyclopedia of Chemical Technology*, 2nd edn (Interscience).

Lee, J. and Brodkey, R. S. (1964) *AIChEJ* **10**, 187. Turbulent motion and mixing in a pipe.

Leung, W. W.-F. (1998) *Industrial Centrifugation Technology* (McGraw-Hill).

Levy, A. and Kalman, H. (2001) *Handbook of Conveying and Handling of Particulate Solids, Vol. 10 (Handbook of Powder Technology)* (Eslevier).

Linley, J. (1984) *Chem. Engr., London* No. 409 (Dec.) 28. *Centrifuges, Part 1: Guidelines on selection.*

Lowrison, G. C. (1974) *Crushing and Grinding* (Butterworths).

Mais, L. G. (1971) *Chem. Eng., NY* **78** (Feb. 15th) 49. Filter media.

Majumdar, A. S. (2006) *Handbook of Industrial Drying*, 3rd edn (CRC).

Marshall, P. (1985) *Chem. Engr., London*, No. 418 (Oct.) 52. *Positive displacement pumps: a brief survey.*

Marshall, V. C. (1974) *Comminution* (IChemE, London).

Masters, K. (1991) *Spray Drying Handbook*, 5th edn ((Longmans)).

Matthews, C. W. (1971) *Chem. Eng., NY* **78** (Feb. 15th) 99. Screening.

Mazzotti, M., Storti, G., and Morbidelli, M. (1997) *J. Chromatog. A* **769**, 3. Optimal operation of simulated moving bed units for nonlinear chromatographic separations.

McCabe, W. L., Smith, J. C., and Harriott, P. (2001) *Unit Operations of Chemical Engineering*, 6th edn (McGraw-Hill).

McGlinchey, D. (2008) *Bulk Solids Handling* (Wiley-Blackwell).

McGregor, W. C. (ed.) (1986) *Membrane Separation Processes in Biotechnology* (Dekker).

Meade, A. (1921) Modern Gasworks Practice, 2nd edn (Benn Bros.).

Mehta, A., Lovato Tse, M., Fogle, J., Len, A., Shrestha, R., Fontes, N., Lebreton, B., Wolk, B., and van Reis, R. (2008) *Chem. Eng. Prog* **104**(5), S14. Purifying therapeutic monoclonal antibodies.

Meltzer, T. H. and Jornitz, M. W. (eds.) (1997) *Filtration in the Biopharmaceutical Industry* (Informa Healthcare).

Menet, J.-M. and Thiebaut, D. (1999) *Countercurrent Chromatography* (CRC).

Mersham, A. (1984) *Int. Chem. Eng.*, **24** (3) 401. Design and scale-up of crystallizers.

Mersham, A. (1988) *Chem. Eng. & Proc.* **23**(4), 213. Design of crystallizers.

Mersmann, A, (ed.) (2001) *Crystallization Technology Handbook*, 2nd edn (CRC).

Meyers, R. A. (2003) *Handbook of Petroleum Refining Processes*, 3rd edn (McGraw-Hill).

Mills, D. (2004) *Pneumatic Conveying Design Guide*, 2nd edn (Butterworth-Heinemann).

Mills, D., Jones, M. G., and Agarwal, V. K. (2004) *Handbook of Pneumatic Conveying* (Marcel Dekker).

Mizrahi, J. and Barnea, E. (1973) *Process Engineering* (Jan.) 60. Compact settler gives efficient separation of liquid-liquid dispersions.

Mohr, P. and Pommerening, K. (1986) *Affinity Chromatography: Practical and Theoretical Aspects* (CRC).

Moir, D. N. (1985) *Chem. Engr., London* No. 410 (Jan.) 20. Selection and use of hydrocyclones.

Morris, B. G. (1966) *Brit. Chem. Eng.* **11**, 347, 846. Application and selection of centrifuges.

Mulder, M. (1996) *Basic principles of membrane technology* (Springer).

Mullin, J. W. (2001) *Crystallization*, 4th edn (Butterworth-Heinemann).

Mutzenburg, A. B. (1965) *Chem. Eng., NY* **72** (Sept. 13th) 175. Agitated evaporators, Part 1, thin-film technology.

Myerson, A. S. (1993) *Handbook of Industrial Crystallization* (Butterworth-Heinemann).

Noble, R. D. and Stern, S. A. (1995) *Membrane Separations Technology: Principles and Applications* (Elsevier).

Oldshue, J. Y., Hirshland, H. E. and Gretton, A. T. (1956) *Chem. Eng. Prog.* **52** (Nov.) 481. Side-entering mixers.

Orr, C. (ed.) (1977) *Filtration: Principles and Practice*, 2 vols (Marcel Dekker).

Osada, Y. and Nakagawa, T. (1992) *Membrane Science and Technology* (Marcel Dekker).

Parker, N. H. (1963a) *Chem. Eng., NY* **70** (June 24th) 115. Aids to dryer selection.

Parker, N. H. (1963b) *Chem. Eng., NY* **70** (July 22nd) 135. How to specify evaporators.

Parker, N. (1965) *Chem. Eng., NY* **72** (Sept. 13th) 179. Agitated evaporators, Part 2, equipment and economics.

Parker, K. (2002) *Electrostatic Precipitators* (Institution of Electrical Engineers).

Parmley, R. O. (2000) *Illustrated Source Book of Mechanical Components* (McGraw-Hill).

Porter, H. F., Flood, J. E. and Rennie, F. W. (1971) *Chem. Eng., NY* **78** (Feb. 15th) 39. Filter selection.

Porter, M. C. (1997) Membrane filtration. In *Handbook of Separation Processes for Chemical Engineers*, 3rd edn, Schweitzer, P. A. (ed.) (McGraw-Hill).

Power, R. B. (1964) *Hyd. Proc.* **43** (March) 138. Steam jet air ejectors.

Prabhudesai, R. K. (1997) Leaching. In *Handbook of Separation Processes for Chemical Engineers*, 3rd edn, Schweitzer, P. A. (ed.) (McGraw-Hill).

Prasher, C. L. (1987) *Crushing and Grinding Process Handbook* (Wiley).

Pryce Bayley, D. and Davies, G. A. (1973) *Chemical Processing* **19** (May) 33. Process applications of knitted mesh mist eliminators.

Purchas, D. B. (1971) *Chemical Processing* **17** (Jan.) 31, (Feb.) 55 (in two parts). Choosing the cheapest filter medium.

Purchas, D. B. and Sutherland, K. (2001) *Handbook of Filter Media*, 2nd edn (Elsevier).

Rase, H. F. (1977) *Chemical Reactor Design for Process Plants*, 2 vols (Wiley).

Rase, H. F. (1990) *Fixed-bed Reactor Design and Diagnostics* (Butterworths).

Rautenbach, R. and Albrecht, R. (1989) *Membrane Processes* (Wiley).

Redmon, O. C. (1963) *Chem. Eng. Prog.* **59** (Sept.) 87. Cartridge type coalescers.

Reid, R. W. (1979) *Mixing and kneading equipment*. In *Solids Separation and Mixing*, Bhatia, M. V. and Cheremisinoff, P. E. (eds) (Technomic).

Reisner, W. (1971) *Bins and Bunkers for Handling Bulk Materials* (Trans. Tech. Publications).

Richardson, J. F., Harker, J. H., and Backhurst, J. (2002) *Chemical Engineering: Volume 2*, 5th edn (Butterworth-Heinemann).

Roberts, E. J., Stavenger, P., Bowersox, J. P., Walton, A. K. and Mehta, M. (1971) *Chem. Eng., NY* **78** (Feb. 15th) 89. Solid/solid separation.

Rosennzweig, M. D. (1977) *Chem. Eng., NY* **84** (May 9th) 95. Motionless mixers move into new processing roles.

Rousar, I., Micha, K. and Kimla, A. (1985) *Electrochemical Engineering*, 2 vols. (Butterworths).

Rushton, J. H., Costich, E. W., and Everett, H. J. (1950) *Chem. Eng. Prog* **46**, 467. Power characteristics of mixing impellers.

Ruthven, D. M. (1984) *Principles of Adsorption and Adsorption Processes* (Wiley).

Ruthven, D. M., Farooq, S. and Knaebel, K. S. (1993) *Pressure Swing Adsorption* (VCH).

Ryan, D. L. and Roper, D. L. (1986) *Process Vacuum System Design and Operation* (McGraw-Hill).

Ryon, A. D., Daley, F. L. and Lowrie, R. S. (1959) *Chem. Eng. Prog.* **55** (Oct.) 70. Scale-up of mixer-settlers.

Sargent, G. D. (1971) *Chem. Eng., NY* **78** (Feb. 15) 11. Gas/solid separations.

Schneider, G. G., Horzella, T. I., Spiegel, P. J. and Cooper, P. J. (1975) *Chem. Eng., NY* **82** (May 26th) 94. Selecting and specifying electrostatic precipitators.

Schroeder, T. (1998) *Chem. Eng., NY* **105** (Sept.) 82. Selecting the right centrifuge.

Schweitzer, P. A. (eds) (1997) *Handbook of Separation Techniques for Chemical Engineers* 3rd edn 1997 (McGraw Hill)

Scott, K. (1991) *Electrochemical Reaction Engineering* (Academic Press).

Scott, K. S. and Hughes, R. (1996) *Industrial Membrane Separation Processes* (Kluwer).

Signales, B. (1975) *Chem. Eng., NY* **82** (June 23rd) 141. How to design settling drums.

Smith, N. (1945) *Gas Manufacture and Utilisation* (British Gas Council).

Sohnel, O. and Garside, J. (1992) *Precipitation* (Butterworth-Heinemann).

Stairmand, C. J. (1949) *Engineering* **168**, 409. Pressure drop in cyclone separators.

Stairmand, C. J. (1951) *Trans. Inst. Chem. Eng* **29**, 356. Design and performance of cyclone separators.

Stoughton, R. W. and Lietzke,, M. H. (1965) *J. Chem. Eng. Data* **10**(3), 254. Calculation of some thermodynamic properties of sea salt solutions at elevated temperatures from data on NaCl solutions.

Strauss, N. (1975) *Industrial Gas Cleaning* (Pergamon).

Sutherland, K. S. (1970) *Chemical Processing* **16** (May) How to specify a centrifuge 10.

Sutherland, K. (2008) *Filters and Filtration Handbook*, 5th edn (Elsevier).

Suziki, M. (1990) *Adsorption Engineering* (Elsevier).

Svarovsky, L. (ed.) (2001) *Solid-Liquid Separation*, 4th edn (Butterworth-Heinemann).

Svarovsky, L. and Thew, M. T. (1992) *Hydrocyclones: Analysis and Applications* (Kluwer).

Tatterson, G. B. (1991) *Fluid Mixing and Gas Dispersion in Agitated Tanks* (McGraw-Hill).

Tatterson, G. B. (1993) *Scale-up and Design of Industrial Mixing Processes* (McGraw-Hill).

Treybal, R. E. (1963) *Liquid Extraction*, 2nd edn ((McGraw Hill)).

Trowbridge, M. E. O'K. (1962) *Chem. Engr., London* No. 162 (Aug.) 73. Problems in scaling-up of centrifugal separation equipment.

Uhl, W. W. and Gray, J. B. (eds) (1967) *Mixing, Theory and Practice*, 2 vols (Academic Press).

Wachinski, A. M. and Etzel, J. E. (1997) *Environmental Ion Exchange: Principles and Design* (CRC).

Wakeman, R. and Tarleton, S. (1998) *Filtration Equipment Selection, Modelling and Process Simulation* (Elsevier).

Wakeman, R. and Tarleton, S. (2005) *Solid/Liquid Separation: Principles of Industrial Filtration* (Elsevier).

Walas, S. M. (1990) *Chemical Process Equipment: Selection and Design* (Butterworths).

Wang, W. K. (2001) *Membrane Separations in Biotechnology*, 2nd edn (CRC).

Ward, A. S., Rushton, A. and Holdrich, R. G. (2000) *Solid–Liquid Filtration and Separation Technology*, 2nd edn (Wiley – VCH).

Waterman, L. L. (1965) *Chem. Eng. Prog.* 61 (Oct.) 51. Electrical coalescers.

Wilf, M., Awerbuch, L., Bartels, C., Mickley, M., Pearce, G. and Voutchkov, N. (2007) *The Guidebook to Membrane Desalination Technology: Reverse Osmosis, Nanofiltration and Hybrid Systems Process, Design, Applications and Economics* (Balaban).

Wilkinson, W. L. and Edwards, M. F. (1972) *Chem. Engr., London* No. 264 (Aug.) 310; No. 265 (Sept.) 328 (in two parts). Heat transfer in agitated vessels.

Williams-Gardner, A. (1965) *Chem & Process Eng.* **46**, 609. Selection of industrial dryers.

Yang, J., Lee, C. H., and Chang, J. W. (1997) *Ind. Eng. Chem. Res.* **36**(7), 2789. Separation of hydrogen mixtures by a two-bed pressure swing adsorption process using zeolite 5A.

Yang, R. T. (1997) *Gas Separation by Adsorption Processes* (World Scientific Publishing).

Yang, R. T. (2003) *Adsorbents: Fundamentals and Applications* (Wiley).

Yang, W.-C. (ed.) (1999) *Fluidisation, Solids Handling and Processing—Industrial Applications* (Noyes).

Yang, W.-C. (ed.) (2003) *Handbook of Fluidization and Fluid-Particle Systems* (CRC).

Zanker, A. (1977) *Chem. Eng., NY* **84** (May 9th) 122. Hydrocyclones: dimensions and performance.

Zenz, F. A. (2001) *Chem. Eng., NY* **108** (Jan.) 60. Cyclone design tips.

Zughi, H. D., Khokar, Z. H. and Sharna, R. H. (2003) *Ind. Eng. Chem. Research* **42** (Oct. 15th), 2003. Mixing in pipelines with side and opposed tees.

Bibliography

Books on reactor design (not cited in text):

Aris, R. *Elementary Chemical Reactor Analysis* (Dover Publications, 2001).

Carberry, J. J. *Chemical and Catalytic Reactor Engineering* (McGraw Hill, 1976).

Chen, N. H. *Process Reactor Design* (Allyn and Bacon, 1983).

Doraiswamy, L. K. and Sharma, M. M. *Heterogeneous Reactions: analysis, examples, and reactor design* (Wiley, 1983):

Volume 1: *Gas-solid and solid-solid reactions*

Volume 2: *Fluid-fluid-solid reactions.*

Fogler, H. S. *Elements of Chemical Reactor Design* (Pearson Educational, 1998).

Froment, G. F. and Bischoff, K. B. *Chemical Reactor Analysis and Design*, 2nd edn (Wiley, 1990).

Levenspiel, O. *Chemical Reaction Engineering*, 3rd edn (Wiley, 1998).

Levenspiel, O. *The Chemical Reactor Omnibook* (Corvallis: OSU book centre, 1979).

Nauman, E. B. *Handbook of Chemical Reactor Design, Optimization and Scaleup* (McGraw-Hill, 2001).

Rose, L. M. *Chemical Reactor Design in Practice* (Elsevier, 1981).

Smith, J. M. *Chemical Engineering Kinetics* (McGraw-Hill, 1970).

Westerterp, K. R., Van Swaaij, W. P. M. and Beenackers, A. A. C. M. *Chemical Reactor Design and Operation*, 2nd edn (Wiley, 1988).

British and European Standards

BS 7792 (1995) (ISO 10630) Industrial plate screens.

BS EN 933-2 (1996) Tests for geometrical properties of aggregates. Part 2: Determination of particle size distribution – Test sieves, nominal size of apertures.

BS EN ISO 14890:2003 (2003) Conveyor belts. Specification for rubber or plastics covered conveyor belts of textile construction for general use.

BS EN ISO 15236-1:2005 (2005) Steel cord conveyor belts. Design, dimensions and mechanical requirements of conveyor belts for general use.

BS EN ISO 21183-1:2006 (2006) Light conveyor belts. Principal characteristics and applications.

BS ISO 14315 (1997) Specifications for industrial wire screens.

American and International Standards

ASTM E11 (2004) Standard specification for wire cloth and sieves for testing purposes (ASTM International).

ISO 10630 (1993) Industrial plate screens.

10.15. NOMENCLATURE

		Dimensions in **MLT**
A_i	Area of interface	L^2
A_s	Surface area of cyclone	L^2
A_v	Area for vapour flow	L^2
A_1	Area of cyclone inlet duct	L^2
b	Index in equation 10.18	—
c	Index in equation 10.18	—
D	Agitator diameter	L
D_c	Cyclone diameter	L
D_{c1}	Diameter of standard cyclone	L
D_{c2}	Diameter of proposed cyclone design	L
D_T	Tank or vessel diameter	L
D_v	Minimum vessel diameter for separator	L
d	Particle diameter	L
d_d	Droplet diameter	L
d_s	Diameter of solid particle removed in a centrifuge	L

d_1	Mean diameter of particles separated in cyclone under standard conditions	L
d_2	Mean diameter of particles separated in proposed cyclone design	L
d_{50}	Particle diameter for which cyclone is 50% efficient	L
F_1	Feed molar flow rate	MT^{-1}
F_2	Product molar flow rate	MT^{-1}
f_c	Friction factor for cyclones	—
f_L	Fraction of bed that is fully loaded at end of adsorption phase of cycle	—
f_v	Fraction of cross-sectional area occupied by vapour.	—
g	Gravitational acceleration	LT^{-2}
h_v	Height above liquid level	L
K	Constant in equation 10.18	—
k	Equilibrium constant for adsorption	$M^{-n}L^nT^{2n}$
L	Cyclone feed volumetric flow-rate	L^3T^{-1}
L_c	Continuous phase volumetric flow-rate	L^3T^{-1}
L_m	Length of membrane	L
L_v	Length of separator, vessel tangent-to-tangent length	L
l	Length of decanter vessel	L
M_a	Mass of adsorbent per bed	M
M_i	Molar flux of component i	$ML^{-2}T^1$
$M_{i,ave}$	Average molar flux of component i	$ML^{-2}T^1$
M_w	Molecular weight of adsorbed component	—
m	Adsorbent loading, g/g adsorbent	—
m_1	Maximum adsorbent loading (g/g adsorbent)	—
m_2	Minimum adsorbent loading (g/g adsorbent)	—
N	Agitator speed	T^{-1}
n	Exponent in equation 10.1	—
P	Pressure	$ML^{-1}T^{-2}$
P	Agitator shaft power in equation 10.18	ML^2T^{-3}
P_i	Permeability of membrane for component i (mol/m.s.bar)	T
ΔP	Pressure differential (pressure drop)	$ML^{-1}T^{-2}$
p	Partial pressure of adsorbed component	$ML^{-1}T^{-2}$
p	Agitator blade pitch (Figure 10.76)	L
$p_{i,f}$	Local partial pressure of component i on feed side	$ML^{-1}T^{-2}$
$p_{i,p}$	Local partial pressure of component i on permeate side	$ML^{-1}T^{-2}$
Q	Volumetric flow rate of liquid through a centrifuge	L^3T^{-1}
Q_p	Volumetric liquid flow through a pump	L^3T^{-1}
Q_1	Standard flow rate in cyclone	L^3T^{-1}
Q_2	Proposed flow rate in cyclone	L^3T^{-1}
r	Radius of decanter vessel	L
r_e	Radius of cyclone exit pipe	L

r_t	Radius of circle to which centre line of cyclone inlet duct is tangential	L
S_{ij}	Selectivity of a membrane for component i over component j	—
T	Temperature	θ
t_a	Time an adsorption bed is in the adsorption stage of the cycle	T
t_B	Breakthrough time	T
u_c	Velocity of continuous phase in a decanter	LT^{-1}
u_d	Settling (terminal) velocity of dispersed phase in a decanter	LT^{-1}
u_g	Terminal velocity of solid particles settling under gravity	LT^{-1}
u_s	velocity in a separator	LT^{-1}
u_t	Settling velocity	LT^{-1}
u_v	Maximum allowable vapour velocity in a separating vessel	LT^{-1}
u_1	Velocity in cyclone inlet duct	LT^{-1}
u_2	Velocity in cyclone exit duct	LT^{-1}
V_v	Gas, or vapour volumetric flow-rate	L^3T^{-1}
w	Width of interface in a decanter	L
x	Length	L
y_1	Feed mole fraction of adsorbed component	—
y_2	Product mole fraction of adsorbed component	—
z	Height of the interface from the base of the vessel	L
z_1	Height to light liquid overflow from a decanter	L
z_2	Height to heavy liquid overflow from a decanter	L
z_3	Height to the interface in a decanter	L
δ	Membrane thickness	L
η	Separating efficiency of a centrifuge	—
η	Efficiency of a cyclone in separating any particle of diameter d	—
η_p	Pump efficiency	—
μ	Liquid viscosity	$ML^{-1}T^{-1}$
μ_c	Viscosity of continuous phase	$ML^{-1}T^{-1}$
μ_1	Cyclone test fluid viscosity	$ML^{-1}T^{-1}$
μ_2	Viscosity of fluid in proposed cyclone design	$ML^{-1}T^{-1}$
ρ	Liquid density	ML^{-3}
ρ_c	Density of the continuous phase	ML^{-3}
ρ_d	Density of the dispersed phase	ML^{-3}
ρ_f	Gas density	ML^{-3}
ρ_L	Liquid density	ML^{-3}
ρ_s	Density of solid	ML^{-3}
ρ_v	Vapour density	ML^{-3}
ρ_1	Light liquid density in a decanter	ML^{-3}
ρ_2	Heavy liquid density in a decanter	ML^{-3}

$\Delta\rho$	Difference in density between solid and liquid	ML^{-3}
$\Delta\rho_1$	Density difference under standard conditions in standard cyclone	ML^{-3}
$\Delta\rho_2$	Density difference in proposed cyclone design	ML^{-3}
Σ	Sigma value for centrifuges, defined by equation 10.6	L^2
ϕ	Factor in Figure 10.63	—
ψ	Parameter in Figure 10.63	—

Dimensionless Numbers

Fr	Froude number
N_p	Power number, defined by equation 10.18
Re	Reynolds number

10.16. PROBLEMS

10.1. An electronics plant uses an adsorption system to recover solvent from a vent gas stream and prevent volatile organic compound (VOC) emissions. The vent gas flow rate is 20 m^3/s of dry air at 293 K, 1.5 atm. The initial solvent loading is 1.5 mol%, which must be reduced to 20 ppm to comply with emissions permits. Activated carbon is used as adsorbent, and has a capacity of 20 mol/kg of adsorbent and a heat of adsorption of 8 kcal/mol solvent. The sorbent can be regenerated by raising the temperature to 363 K. Design a suitable TSA system for this process. Estimate the amount of adsorbent needed, the vessel volumes and the minimum regenerant heat requirement for your design. The activated carbon adsorbent has average bulk density 120 kg/m^3 and heat capacity 0.7 J/g.°C. The solvent flammability limits in air are 2.5 to 12.0 vol% at 293 K and 1.2 to 16.0 vol% at 363 K.

10.2. It has been suggested that carbon dioxide can be recovered from flue gas using a membrane process. The flue gas has composition: nitrogen 73.9 mol%, oxygen 3.1 mol%, carbon dioxide 7.7 mol%, water vapour 15.3 mol%. Using the data in Table 10.2, determine the best membrane material for this process. What are the advantages and disadvantages of the proposed membrane process?

10.3. The product from a crystallizer is to be separated from the liquor using a centrifuge. The concentration of the crystals is 6.5% and the slurry feed rate to the centrifuge will be 5.0 m^3/h. The density of the liquor is 995 kg/m^3 and that of the crystals 1500 kg/m^3. The viscosity of the liquor is 0.7 mN m^{-2}s. The cut-off crystal size required is 5 μm.

Select a suitable type of centrifuge to use for this duty.

10.4. Dissolved solids in the tar from the bottom of a distillation column are precipitated by quenching the hot tar in oil. The solids are then separated from the oil and burnt. The density of the solids is 1100 kg/m^3. The density of the liquid phase after addition of the tar is 860 kg/m^3 and its viscosity, at the temperature of the mixture, 1.7 mN m^{-2}s. The solid content of the oil and tar mixture is 10% and the flow rate of the liquid phase leaving the separator will be 1000 kg/h. The cut-off particle size required is 0.1 mm.

List the types of separator that could be considered for separating the solids from the liquid. Bearing mind the nature of the process, what type of separator would you recommend for this duty?

10.5. The solids from a dilute slurry are to be separated using hydrocyclones. The density of the solids is 2900 kg/m^3, and liquid is water. A recovery of 95% of particles greater than 100 μm is required. The minimum operating temperature will be 10°C and the maximum 30°C. Design a hydrocyclone system to handle 1200 L/min of this slurry.

10.6. A fluidized bed is used in the production of aniline by the hydrogenation of nitrobenzene. Single-stage cyclones, followed by candle filters, are used to remove fines from the gases leaving the fluidized bed.

The reactor operates at a temperature 270°C and a pressure of 2.5 bara. The reactor diameter is 10 m. Hydrogen is used in large excess in the reaction, and for the purposes of this exercise the properties of the gas may be taken as those of hydrogen at the reactor conditions. The density of the catalyst particles is 1800 kg/m^3.

The estimated particle size distribution of the fines is:

Particle size, μm	50	40	30	20	10	5	2
Percentage by weight less than	100	70	40	20	10	5	2

A 70% recovery of the solids is required in the cyclones.

For a gas flow rate of 100,000 m^3/h, at the reactor conditions, determine how many cyclones operating in parallel are need and design a suitable cyclone. Estimate the size distribution of the particles entering the filters.

10.7. In a process for the production of acrylic fibres by the emulsion polymerization of acrylonitrile, the unreacted monomer is recovered from water by distillation. Acrylonitrile forms an azeotrope with water and the overhead product from the column contain around 5 mol% water. The overheads are condensed and the recovered acrylonitrile separated from the water in a decanter. The decanter operating temperature will be 20°C.

Size a suitable decanter for a feed rate of 3000 kg/h.

10.8. In the production of aniline by the hydrogenation of nitrobenzene, the reactor products are separated from unreacted hydrogen in a condenser. The condensate, which is mainly water and aniline, together with a small amount of unreacted nitrobenzene and cyclohexylamine, is fed to a decanter to separate the water and aniline. The separation will not be complete, as aniline is slightly soluble in water, and water in aniline. A typical material balance for the decanter is given below:

Basis 100 kg feed.

	Feed	Aqueous Stream	Organic Stream
water	23.8	21.4	2.4
aniline	72.2	1.1	71.1
nitrobenzene	3.2	trace	3.2
cyclohexylamine	0.8	0.8	trace
total	100	23.3	76.7

Design a decanter for this duty, for a feed rate of 3500 kg/h. Concentrate on the separation of the water and aniline. The densities of water–aniline solutions are given in Appendix F, problem F.8. The decanter will operate at a maximum temperature of 30°C.

10.9. Water droplets are to be separated from air in a simple separation drum. The flow-rate of the air is 1000 m^3/h, at s.t.p., and it contains 75 kg of water. The drum will operate at 1.1 bara pressure and 20°C.

Size a suitable liquid–vapour separator.

10.10. The vapour from a chlorine vaporizer will contain some liquid droplets. The vaporizer consists of a vertical, cylindrical, vessel with a submerged bundle for heating. A vapour rate of 2500 kg/h is required and the vaporizer will operate at 6 bara. Size the vessel to restrict the carryover of liquid droplets. The liquid hold-up time need not be considered, as the liquid level will be a function of the thermal design.

11 SEPARATION COLUMNS (DISTILLATION, ABSORPTION AND EXTRACTION)

Chapter Contents

Key Learning Objectives

- How to design distillation columns
- How to size distillation columns and select and design distillation column trays
- How to design distillation columns using packing instead of trays
- How to design absorption and stripping columns
- How to design liquid-liquid extraction columns

11.1. INTRODUCTION

This chapter covers the design of separating columns. Though the emphasis is on distillation processes, the basic construction features, and many of the design methods, also apply to other multistage processes, such as stripping, absorption and extraction. Only a brief review of the fundamental principles that underlie the design procedures will be given; a fuller discussion can be found in Richardson *et al.* (2002), and in other text books: King (1980), Hengstebeck (1976), Kister (1992), Doherty and Malone (2001) and Luyben (2006).

Distillation is probably the most widely used separation process in the chemical and allied industries; its applications ranging from the rectification of alcohol, which has been practiced since antiquity, to the fractionation of crude oil. A good understanding of methods used for correlating vapour–liquid equilibrium data is essential to the understanding of distillation and other equilibrium-staged processes; this subject was covered in Chapter 8.

In recent years, much of the work done to develop reliable design methods for distillation equipment has been carried out by a commercial organization, Fractionation Research Inc. (FRI), an organization set up with the resources to carry out experimental work on full-size columns. Since their work is proprietary, it is not published in the open literature and it has not been possible to refer to their methods in this book. Fractionation Research's design manuals will, however, be available to design engineers whose companies are subscribing members of the organization. FRI has also produced an excellent training video that shows the physical phenomena that occur when a plate column is operated in different hydraulic regimes. This video can be ordered from FRI at www.fri.org.

Distillation Column Design

The design of a distillation column can be divided into the following steps:

1. Specify the degree of separation required: set product specifications.
2. Select the operating conditions: batch or continuous; operating pressure.
3. Select the type of contacting device: plates or packing.
4. Determine the stage and reflux requirements: the number of equilibrium stages.
5. Size the column: diameter, number of real stages.
6. Design the column internals: plates, distributors, packing supports.
7. Mechanical design: vessel and internal fittings.

The principal step is to determine the stage and reflux requirements. This is a relatively simple procedure when the feed is a binary mixture, but a complex and difficult task when the feed contains more than two components (multicomponent systems).

Almost all distillation design is carried out using commercial process simulation software, as introduced in Chapter 4. The process simulation programs allow the designer to determine the stage and reflux requirements that are needed to attain the desired separation, then size the column and design the column internals. Once the column size is known, the shell can be designed as a pressure vessel (see Chapter 13) and the condenser and reboiler can be designed as heat exchangers (see Chapter 12). The whole design can then be costed and optimized. An example of distillation column optimization was given in Chapter 1.

11.2. CONTINUOUS DISTILLATION: PROCESS DESCRIPTION

The separation of liquid mixtures by distillation depends on differences in volatility between the components. The greater the relative volatilities, the easier the separation. The basic equipment required for continuous distillation is shown in Figure 11.1. Vapour flows up the column and liquid counter-currently down the column. The vapour and liquid are brought into contact on plates or packing. Part of the condensate from the condenser is returned to the top of the column to provide liquid flow above the feed point (reflux), and part of the liquid from the base of the column is vaporized in the reboiler and returned to provide the vapour flow.

In the section below the feed, the more volatile components are stripped from the liquid and this is known as the stripping section. Above the feed, the concentration of the more volatile components is increased and this is called the enrichment, or more commonly, the rectifying section. Figure 11.1a shows a column producing two product streams, referred to as distillate and bottoms, from a single feed. Columns are occasionally used with more than one feed, and with side streams withdrawn at points up the column, Figure 11.1b. This does not alter the basic operation, but complicates the analysis of the process, to some extent.

If the process requirement is to strip a volatile component from a relatively non-volatile solvent, the rectifying section may be omitted, and the column would then be called a stripping column.

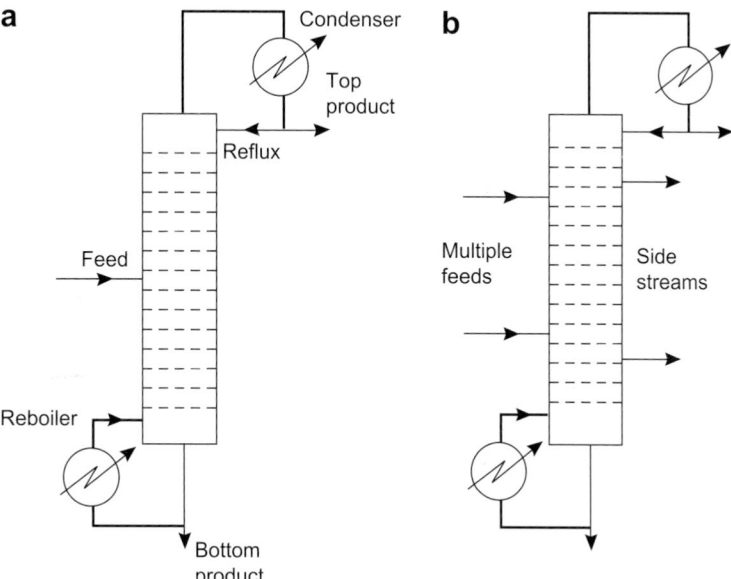

Figure 11.1. Distillation column. (a) Basic column. (b) Multiple feeds and side streams.

In some operations, where some or all of the top product is required as a vapour, only sufficient liquid is condensed to provide the reflux flow to the column, and the condenser is referred to as a partial condenser. When the liquid is totally condensed, the liquid returned to the column will have the same composition as the top product. In a partial condenser the reflux will be in equilibrium with the vapour leaving the condenser. Virtually pure top and bottom products can be obtained in a single column from a binary feed if no azeotrope is formed, but where the feed contains more than two components, only a single 'pure' product can be produced, either from the top or bottom of the column. Several columns will be needed to separate a multicomponent feed into its constituent parts.

11.2.1. Reflux Considerations

The reflux ratio, R, is normally defined as:

$$R = \frac{\text{flow returned as reflux}}{\text{flow of top product taken off}}$$

The number of stages required for a given separation will be dependent on the reflux ratio used.

In an operating column, the effective reflux ratio will be increased by vapour condensed within the column due to heat leakage through the walls. With a well lagged column the heat loss will be small and no allowance is normally made for this increased flow in design calculations. If a column is poorly insulated, changes in the internal reflux due to sudden changes in the external conditions, such as a sudden rain storm, can have a noticeable effect on the column operation and control.

Total Reflux

Total reflux is the condition when all the condensate is returned to the column as reflux: no product is taken off and there is no feed.

At total reflux the number of stages required for a given separation is the minimum at which it is theoretically possible to achieve the separation. Though not a practical operating condition, it is a useful guide to the likely number of stages that will be needed.

Columns are often started up with no product take-off and operated at total reflux until steady conditions are attained. The testing of columns is also conveniently carried out at total reflux.

Minimum Reflux

As the reflux ratio is reduced, a 'pinch point' will occur at which the separation can only be achieved with an infinite number of stages. This sets the minimum possible reflux ratio for the specified separation.

Optimum Reflux Ratio

Practical reflux ratios will lie somewhere between the minimum for the specified separation and total reflux. The designer must select a value at which the specified separation is achieved at minimum cost. Increasing the reflux reduces the number of stages required, and hence the capital cost, but increases the utility requirements (steam and cooling water) and the operating costs. The optimum reflux ratio will be that which gives the lowest total annualized cost or greatest net present value. No hard and fast rules can be given for the selection of the design reflux ratio, but for many systems the optimum will lie between 1.1 to 1.3 times the minimum reflux ratio. As a first approximation, 1.15 times minimum reflux is often used.

For new designs, where the ratio cannot be decided from past experience, the effect of reflux ratio on the number of stages can be investigated using a process simulation model.

At low reflux ratios the calculated number of stages will be very dependent on the accuracy of the vapour–liquid equilibrium data available. If the data or phase equilibrium model are suspect, a higher than normal ratio should be selected to give more confidence in the design.

11.2.2. Feed-Point Location

The precise location of the feed point will affect the number of stages required for a specified separation and the subsequent operation of the column. As a general rule, the feed should enter the column at the point that gives the best match between the feed composition (vapour and liquid if two phases) and the vapour and liquid streams in the column. In practice, it is wise to provide two or three feed-point nozzles located round the predicted feed point to allow for uncertainties in the design calculations and data, and possible changes in the feed composition after start-up.

11.2.3. Selection of Column Pressure

Except when distilling heat-sensitive materials, the main consideration when selecting the column operating pressure will be to ensure that the dew point temperature of the distillate is above that which can be easily obtained with the plant cooling water. The maximum, summer, temperature of cooling water is usually taken as $30°C$. If this means that high pressures will be needed, the provision of refrigerated cooling should be considered. Vacuum operation is used to reduce the column temperatures for the distillation of heat-sensitive materials and where very high temperatures would otherwise be needed to distil relatively non-volatile materials.

When calculating the stage and reflux requirements using short-cut methods it is usual to take the operating pressure as constant throughout the column. In vacuum columns, the column pressure drop will be a significant fraction of the total pressure and the change in pressure up the column should be allowed for when calculating the stage temperatures. When using rigorous simulation methods, a rough initial estimate of column pressure drop can be made by assuming a pressure drop per tray equal to twice the liquid static head on the tray; i.e., $2\rho_L g h_w$, where ρ_L is the liquid density (kg/m^3), g is the gravitational acceleration (m/s^2) and h_w is the weir height (m).

11.3. CONTINUOUS DISTILLATION: BASIC PRINCIPLES

11.3.1. Stage Equations

Material and energy balance equations can be written for any stage in a multistage process.

Figure 11.2 shows the material flows into and out of a typical stage n in a distillation column. The equations for this stage are set out below, for any component i.

Material Balance

$$V_{n+1}y_{n+1} + L_{n-1}x_{n-1} + F_n z_n = V_n y_n + L_n x_n + S_n x_n \qquad (11.1)$$

Energy Balance

$$V_{n+1}H_{n+1} + L_{n-1}h_{n-1} + Fh_f + q_n = V_n H_n + L_n h_n + S_n h_n \qquad (11.2)$$

where

$\quad V_n$ = vapour flow from the stage
$\quad V_{n+1}$ = vapour flow into the stage from the stage below
$\quad L_n$ = liquid flow from the stage
$\quad L_{n-1}$ = liquid flow into the stage from the stage above
$\quad F_n$ = any feed flow into the stage
$\quad S_n$ = any side stream from the stage
$\quad q_n$ = heat flow into, or removal from, the stage
$\quad n$ = any stage, numbered from the top of the column
$\quad z$ = mol fraction of component i in the feed stream (note, feed may be two-phase)
$\quad x$ = mol fraction of component i in the liquid streams
$\quad y$ = mol fraction component i in the vapour streams
$\quad H$ = specific enthalpy vapour phase
$\quad h$ = specific enthalpy liquid phase
$\quad h_f$ = specific enthalpy feed (vapour + liquid).

All flows are the total stream flows (mol/unit time) and the specific enthalpies are also for the total stream (J/mol).

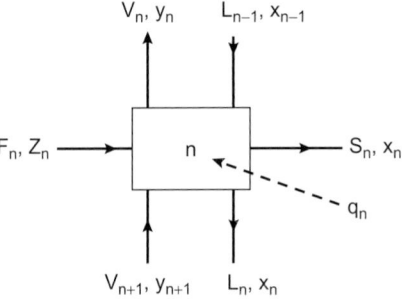

Figure 11.2.   Stage flows.

It is convenient to carry out the analysis in terms of 'equilibrium stages'. In an equilibrium stage (theoretical plate) the liquid and vapour streams leaving the stage are taken to be in equilibrium, and their compositions are determined by the vapour–liquid equilibrium relationship for the system (see Chapter 8). In terms of equilibrium constants:

$$y_i = K_i x_i \tag{11.3}$$

The performance of real stages is related to an equilibrium stage by the concept of plate or stage efficiencies for plate contactors, and 'height equivalent to a theoretical plate' for packed columns.

In addition to the equations arising from the material and energy balances over a stage, and the equilibrium relationships, there will be a fourth relationship, the summation equation for the liquid and vapour compositions:

$$\Sigma x_{i,n} = \Sigma y_{i,n} = 1.0 \tag{11.4}$$

These four equations are the so-called MESH equations for the stage: Material balance, Equilibrium, Summation and Heat (energy) balance, equations. MESH equations can be written for each stage, and for the reboiler and condenser. The solution of this set of equations forms the basis of the rigorous methods that have been developed for the analysis of staged separation processes and that are solved in the process simulation programs.

11.3.2. Dew Point and Bubble Point

To estimate the stage, and the condenser and reboiler temperatures, procedures are required for calculating dew and bubble points. By definition, a saturated liquid is at its bubble point (any rise in temperature will cause a bubble of vapour to form), and a saturated vapour is at its dew point (any drop in temperature will cause a drop of liquid to form).

Dew points and bubble points can be calculated from the vapour–liquid equilibrium for the system. In terms of equilibrium constants, they are defined by the equations

$$\text{bubble point: } \sum y_i = \sum K_i x_i = 1.0 \tag{11.5a}$$

$$\text{and dew point: } \sum x_i = \sum \frac{y_i}{K_i} = 1.0 \tag{11.5b}$$

For multicomponent mixtures, the temperature that satisfies these equations, at a given system pressure, must be found by iteration.

For binary systems the equations can be solved more readily because the component compositions are not independent; fixing one fixes the other.

$$y_a = 1 - y_b \tag{11.6a}$$

$$x_a = 1 - x_b \tag{11.6b}$$

11.3.3. Equilibrium Flash Calculations

In an equilibrium flash process, a feed stream is separated into liquid and vapour streams at equilibrium. The composition of the streams depends on the quantity of the feed vaporized (flashed). The equations used for equilibrium flash calculations are developed below and a typical calculation is shown in Example 11.1.

Flash calculations are often needed to determine the condition of the feed to a distillation column and, occasionally, to determine the flow of vapour from the reboiler, or condenser if a partial condenser is used.

Single-stage flash distillation processes are used to make a coarse separation of the light components in a feed; often as a preliminary step before a multicomponent distillation column.

Figure 11.3 shows a typical equilibrium flash process. The equations describing this process are:

Material balance, for any component, i

$$Fz_i = Vy_i + Lx_i \tag{11.7}$$

Energy balance, total stream enthalpies:

$$Fh_f = VH + Lh \tag{11.8}$$

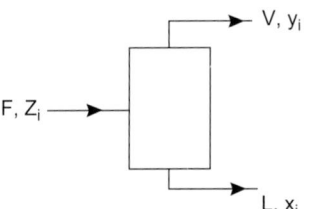

Figure 11.3.   Flash distillation.

If the vapour–liquid equilibrium relationship is expressed in terms of equilibrium constants, equation 11.7 can be written in a more useful form:

$$Fz_i = VK_i x_i + Lx_i$$
$$= Lx_i \left[\frac{V}{L} K_i + 1 \right]$$

from which

$$L = \sum_i \frac{Fz_i}{\left[\frac{VK_i}{L} + 1 \right]} \tag{11.9}$$

and, similarly,

$$V = \sum_i \frac{Fz_i}{\left[\frac{L}{VK_i} + 1 \right]} \tag{11.10}$$

The groups incorporating the liquid and vapour flow rates and the equilibrium constants have a general significance in separation process calculations.

The group L/VK_i is known as the absorption factor A_i, and is the ratio of the moles of any component in the liquid stream to the moles in the vapour stream.

The group VK_i/L is called the stripping factor S_i, and is the reciprocal of the absorption factor.

Efficient techniques for the solution of the trial-and-error calculations necessary in multicomponent flash calculations are given by several authors; see Hengstebeck (1976) and King (1980). Flash models are available in all the commercial process simulation programs and are very easy to configure. It is often a good idea to use flash models to check that the phase equilibrium model that has been selected makes an accurate prediction of any experimental data that are available. Flash models are also useful for checking for changes in volatility order or formation of second liquid phases within a distillation column.

Example 11.1

A feed to a column has the composition given in the table below, and is at a pressure of 14 bar and a temperature of 60°C. Calculate the flow and composition of the liquid and vapour phases. Take the equilibrium data from the Depriester charts given in Chapter 8.

		kmol/h	z_i
Feed	ethane (C$_2$)	20	0.25
	propane (C$_3$)	20	0.25
	isobutane (iC$_4$)	20	0.25
	n-pentane (nC$_5$)	20	0.25

Solution

For two phases to exist, the flash temperature must lie between the bubble point and dew point of the mixture.

From equations 11.5a and 11.5b:

$$\sum K_i z_i > 1.0$$

$$\sum \frac{z_i}{K_i} > 1.0$$

Check feed condition

	K_i	$K_i z_i$	z_i/K_i
C$_2$	3.8	0.95	0.07
C$_3$	1.3	0.33	0.19
iC$_4$	0.43	0.11	0.58
nC$_5$	0.16	0.04	1.56
		$\Sigma 1.43$	$\Sigma\, 2.40$

therefore the feed is a two-phase mixture.

Flash calculation

	K_i	Try $L/V = 1.5$		Try $L/V = 3.0$	
		$A_i = L/VK_i$	$V_i = Fz_i/(1 + A_i)$	A_i	V_i
C_2	3.8	0.395	14.34	0.789	11.17
C_3	1.3	1.154	9.29	2.308	6.04
iC_4	0.43	3.488	4.46	6.977	2.51
nC_5	0.16	9.375	1.93	18.750	1.01
		$V_{calc} = 30.02$		$V_{calc} = 20.73$	
		$L/V = \dfrac{80 - 30.02}{30.02} = 1.67$		$L/V = 2.80$	

Hengstebeck's method is used to find the third trial value for L/V. The calculated values are plotted against the assumed values and the intercept with a line at 45° (calculated = assumed) gives the new trial value, 2.4.

	Try $L/V = 2.4$			
	A_i	V_i	$y_i = V_i/V$	$x_i = (Fz_i - V_i)/L$
C_2	0.632	12.26	0.52	0.14
C_3	1.846	7.03	0.30	0.23
iC_4	5.581	3.04	0.13	0.30
nC_5	15.00	1.25	0.05	0.33
	$V_{calc} = 23.58$		1.00	1.00

$L = 80 - 23.58 = 56.42\,\text{kmol/h}$
L/V calculated $= 56.42/23.58 = 2.39$ close enough to the assumed value of 2.4

Adiabatic Flash

In many flash processes the feed stream is at a higher pressure than the flash pressure and the heat for vaporization is provided by the enthalpy of the feed. In this situation the flash temperature will not be known and must be found by iteration. A temperature must be found at which both the material and energy balances are satisfied. This is easily solved using process simulation software, by specifying the flash outlet pressure and specifying zero heat input. The program then calculates the temperature and stream flow rates that satisfy the MESH equations.

11.4. DESIGN VARIABLES IN DISTILLATION

It was shown in Chapter 1 that to carry out a design calculation the designer must specify values for a certain number of independent variables to define the problem completely, and that the ease of calculation will often depend on the judicious choice of these design variables.

The choice of design variables is particularly important in distillation, as the problem must be sufficiently well defined to find a feasible solution when simulated using a computer.

The total number of variables and equations required to describe a multicomponent distillation can be very large, since the MESH equations must be solved for every stage, including the reboiler and condenser. It becomes difficult for the designer to keep track of all the variables and equations, and mistakes are likely to be made, as the number of degrees of freedom will be the difference between two large numbers. Instead, a simpler procedure known as the 'description rule' given by Hanson *et al.* (1962) can be used. The description rule states that to determine a separation process completely the number of independent variables that must be set (by the designer) will equal the number that are set in the construction of the column or that can be controlled by external means in its operation. The method is best illustrated by considering the operation of the simplest type of column: with one feed, no side streams, a total condenser and a reboiler. The construction will fix the number of stages above and below the feed point (two variables). The feed rate, column pressure, and condenser and reboiler duties (cooling water and steam flows) will be controlled (four variables). There are therefore six variables in total.

To design the column this number of variables must be specified, but the same variables need not be selected. Typically, in a design situation the feed rate will be fixed by the upstream design. The column pressure will also usually be fixed early in the design. Distillation processes are usually operated at low pressure, where relative volatility is high, but the pressure is usually constrained to be high enough for the condenser to operate using cooling water rather than refrigeration. If the feed rate and pressure are specified then four degrees of freedom remain. Rigorous column models in process simulation programs require the designer to specify the number of stages above and below the feed, leaving the designer with two degrees of freedom. If two additional independent parameters are specified, then the problem is completely defined and has a single solution. For example, if the designer specifies a reflux ratio and a boil-up ratio or a reflux ratio and a distillate rate, then there will be a corresponding unique set of distillate and bottoms compositions for a given feed composition. If the designer chooses to specify the compositions of two key components in either the distillate or the bottoms then there will be a required reflux rate, boil-up rate, distillate flow rate, etc. Similarly, specifying the purity and recovery of a single component in one of the products will completely specify the problem.

When replacing variables identified by the application of the description rule it is important to ensure that those selected are truly independent, and that the values assigned to them lie within the range of possible, practical, values. This is particularly important when using purity or composition specifications in multicomponent distillation. It would clearly not be possible to obtain 99% purity of the light key component in the distillate if the feed contained 2% of components that boiled at lower temperatures than the light key component. The selection of key components and product specifications for multicomponent distillation are discussed in more detail in Section 11.6.

The number of independent variables that have to be specified to define a problem will depend on the type of separation process being considered. Some examples of the application of the description rule to more complex columns are given by Hanson *et al.* (1962).

11.5. DESIGN METHODS FOR BINARY SYSTEMS

The distillation of binary mixtures is a relatively simple problem. With a binary mixture, fixing the composition of one component fixes the composition of the other. The stage and reflux requirements can be determined using simple graphical methods developed in the 1920s, and iterative calculations are not required.

It must, however, be emphasized that the graphical methods for binary distillation are no longer used in any practical context. Very few industrial distillation problems involve true binary mixtures. There will usually be other components present even if the two main components constitute more than 99.9 mol% of the mixture. The design engineer will usually need to know how these other components distribute between the distillate and bottoms to ensure that product specifications can be met and to determine the contaminant loads on downstream operations. Furthermore, the distillation design problem is seldom solved in isolation from the overall process design and the widespread use of process simulation programs has made the graphical methods obsolete. The initialization of a rigorous simulation of a binary distillation uses the same methods used for a multicomponent distillation (as described in Section 4.5.2). The graphical methods give no insight into the solution procedures used for multicomponent distillation.

Despite the above considerations, many educators find the graphical methods for binary distillation to be useful as a means of explaining some of the phenomena that can occur in multistage separations, and these methods are part of the required chemical engineering curriculum in most countries. The graphical methods can be used to illustrate some problems that are common to binary and multicomponent distillation. Graphical methods are still useful in understanding and initializing other staged separation processes such as absorption, stripping and extraction.

The discussion of binary distillation methods in this chapter has been limited to a brief overview with emphasis on the insights that can be obtained from the graphs. For more details of the classical binary distillation methods, see Richardson *et al.* (2002) and earlier editions of this book.

11.5.1. Basic Equations

Sorel (1899) first derived and applied the basic stage equations to the analysis of binary systems. Figure 11.4a shows the flows and compositions in the top part of a column. Taking the system boundary to include the stage n and the condenser, gives the following equations:

Material Balance

Total flows

$$V_{n+1} = L_n + D \tag{11.11}$$

where D is the distillate flow rate, and for either component:

$$V_{n+1}y_{n+1} = L_n x_n + D x_d \tag{11.12}$$

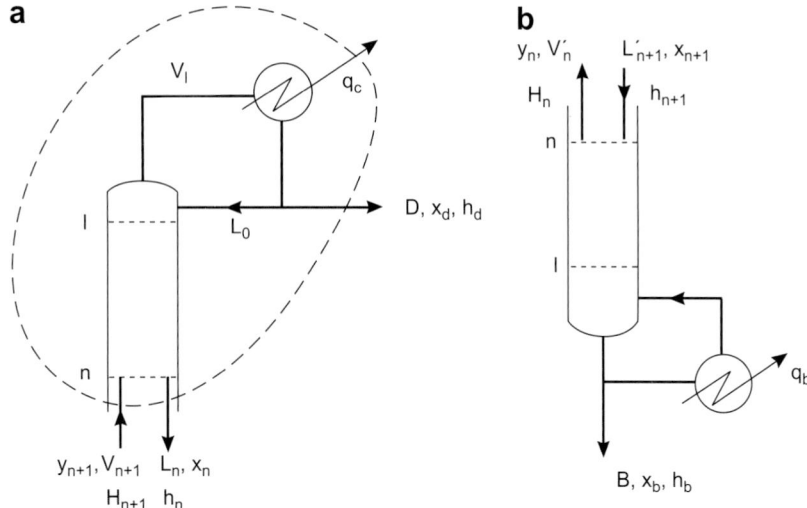

Figure 11.4. Column flows and compositions. (a) Above feed. (b) Below feed.

Energy Balance

Total stream enthalpies

$$V_{n+1}H_{n+1} = L_n h_n + D h_d + q_c \tag{11.13}$$

where q_c is the heat removed in the condenser.

Combining equations 11.11 and 11.12 gives

$$y_{n+1} = \frac{L_n}{L_n + D} x_n + \frac{D}{L_n + D} x_d \tag{11.14}$$

Combining equations 11.11 and 11.13 gives

$$V_{n+1}H_{n+1} = (L_n + D)H_{n+1} = L_n h_n + D h_d + q_c \tag{11.15}$$

Analogous equations can be written for the stripping section, Figure 11.4b.

$$x_{n+1} = \frac{V'_n}{V'_n + B} y_n + \frac{B}{V'_n + B} x_b \tag{11.16}$$

and

$$L'_{n+1} h_{n+1} = (V'_n + B)h_{n+1} = V'_n H_n + B h_b - q_b \tag{11.17}$$

where B is the bottoms flow rate.

At constant pressure, the stage temperatures will be functions of the vapour and liquid compositions only (dew and bubble points), and the specific enthalpies will therefore also be functions of composition

$$H = f(y) \qquad (11.18a)$$

$$h = f(x) \qquad (11.18b)$$

Lewis–Sorel Method (Equimolar Overflow)

For most distillation problems a simplifying assumption, first proposed by Lewis (1909), can be made that eliminates the need to solve the stage energy-balance equations. The molar liquid and vapour flow rates are taken as constant in the stripping and rectifying sections. This condition is referred to as equimolar overflow: the molar vapour and liquid flows from each stage are constant. This will only be true where the component molar latent heats of vaporization are the same and, together with the specific heats, are constant over the range of temperature in the column; there is no significant heat of mixing; and the heat losses are negligible. These conditions are substantially true for practical systems when the components form near-ideal liquid mixtures.

Even when the latent heats are substantially different, the error introduced by assuming equimolar overflow to calculate the number of stages is often small compared to the error in the stage efficiency, and is acceptable.

With equimolar overflow, equations 11.14 and 11.16 can be written without the subscripts to denote the stage number:

$$y_{n+1} = \frac{L}{L+D} x_n + \frac{D}{L+D} x_d \qquad (11.19)$$

$$x_{n+1} = \frac{V'}{V'+B} y_n + \frac{B}{V'+B} x_b \qquad (11.20)$$

where L = the constant liquid flow in the rectifying section = the reflux flow, L_0; and V' is the constant vapour flow in the stripping section.

Equations 11.19 and 11.20 can be written in an alternative form:

$$y_{n+1} = \frac{L}{V} x_n + \frac{D}{V} x_d \qquad (11.21)$$

$$y_n = \frac{L'}{V'} x_{n+1} - \frac{B}{V'} x_b \qquad (11.22)$$

where V is the constant vapour flow in the rectifying section = $(L + D)$; and L' is the constant liquid flow in the stripping section = $V' + B$.

These equations are linear, with slopes L/V and L'/V'. They are referred to as operating lines, and give the relationship between the liquid and vapour compositions between stages. For an equilibrium stage, the compositions of the liquid and vapour streams leaving the stage are given by the equilibrium relationship.

11.5.2. McCabe–Thiele Method

Equations 11.21 and 11.22 and the equilibrium relationship are conveniently solved by the graphical method developed by McCabe and Thiele (1925). A simple procedure for the construction of the diagram is given below and illustrated in Example 11.2.

Procedure

Refer to Figure 11.5; all compositions are those of the more volatile component.

1. Plot the vapour–liquid equilibrium curve from data available at the column operating pressure. In terms of relative volatility:

$$y = \frac{\alpha x}{(1 + (\alpha - 1)x)} \qquad (11.23)$$

where α is the geometric average relative volatility of the lighter (more volatile) component with respect to the heavier component (less volatile). It is usually more convenient, and less confusing, to use equal scales for the x and y axes.

2. Make a material balance over the column to determine the top and bottom compositions, x_d and x_b, from the data given.

3. The upper and lower operating lines intersect the diagonal at x_d and x_b respectively; mark these points on the diagram.

4. The point of intersection of the two operating lines is dependent on the phase condition of the feed. The line on which the intersection occurs is called the q line, found as follows:

(i) calculate the value of the ratio q given by

$$q = \frac{\text{heat to vaporize 1 mol of feed}}{\text{molar latent heat of feed}}$$

(ii) plot the q line, slope $= q/(q - 1)$, intersecting the diagonal at z_f (the feed composition).

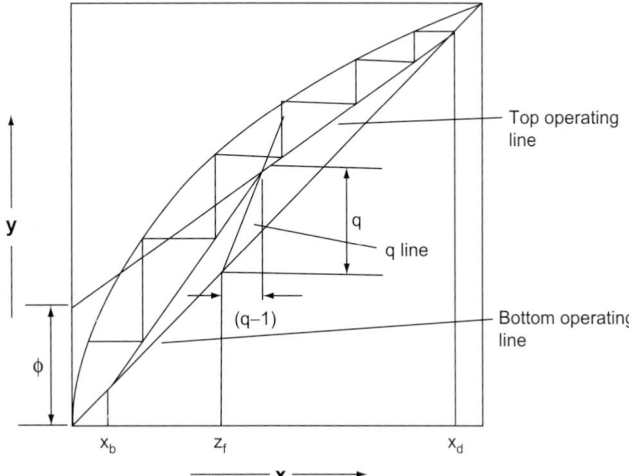

Figure 11.5. McCabe–Thiele diagram.

5. Select the reflux ratio and determine the point where the upper operating line extended cuts the y axis:

$$\phi = \frac{x_d}{1 + R} \tag{11.24}$$

6. Draw in the upper operating line (UOL), from x_d on the diagonal to ϕ.
7. Draw in the lower operating line (LOL), from x_b on the diagonal to the point of intersection of the top operating line and the q line.
8. Starting at x_d or x_b, step off the number of stages.

Note: The feed point should be located on the stage closest to the intersection of the operating lines.

The reboiler, and a partial condenser if used, act as equilibrium stages; however, when designing a column there is little point in reducing the estimated number of stages to account for this. Not counting the reboiler as a stage can be considered an additional design margin.

It can be seen from equation 11.24 and Figure 11.5 that as R increases, ϕ decreases, until the limit is reached where $\phi = 0$ and the upper and lower operating lines both lie along the diagonal, as in Figure 11.6. This is the total reflux condition, in which the minimum number of stages is needed for the separation.

Similarly, as R is reduced, the intersection between the upper and lower operating lines moves away from the diagonal until it reaches the equilibrium line, as illustrated in Figure 11.7. This is the minimum reflux condition. If the reflux ratio were to be reduced further, then there would be no feasible intersection of the operating lines.

It can also be seen that at minimum reflux the space between the operating and equilibrium lines becomes very small at the intersection point, which is known as a 'pinched' condition. An infinite number of trays are required at minimum reflux because of these pinch points, as can be seen in Figure 11.7. Pinch points also often occur when the relative volatility of the mixture is not constant, particularly when azeotropes or near azeotropes form, as illustrated in Figure 11.9. Pinch points also

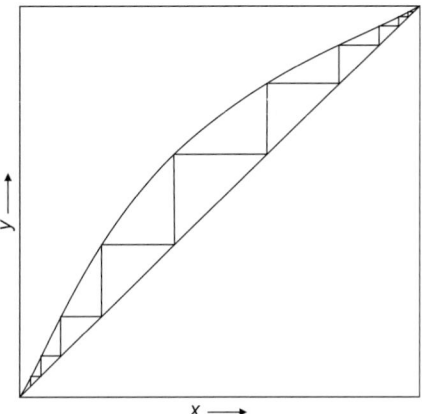

Figure 11.6.   Total reflux.

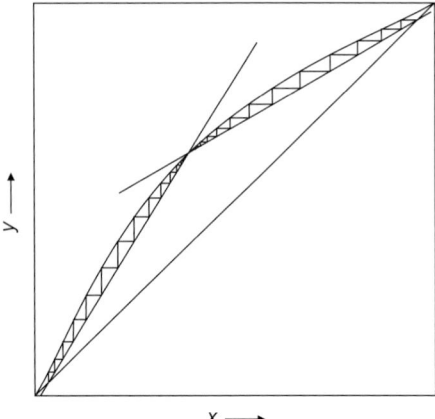

Figure 11.7.   Minimum reflux.

occur at the top or bottom of the column if very stringent purity specifications must be met. Pinch points are also found in multicomponent distillation and are easily visualised as regions where the composition profiles appear to be varying only very slightly from stage to stage. When a pinch point occurs, the solution is usually to increase the reflux or else change column pressure to obtain a more favourable equilibrium.

The efficiency of real contacting stages can be accounted for by reducing the height of the steps on the McCabe–Thiele diagram; see Figure 11.8. Stage efficiencies are discussed in Section 11.10.

The McCabe–Thiele method can be used for the design of columns with side streams and multiple feeds. The liquid and vapour flows in the sections between the feed and take-off points are calculated and operating lines drawn for each section.

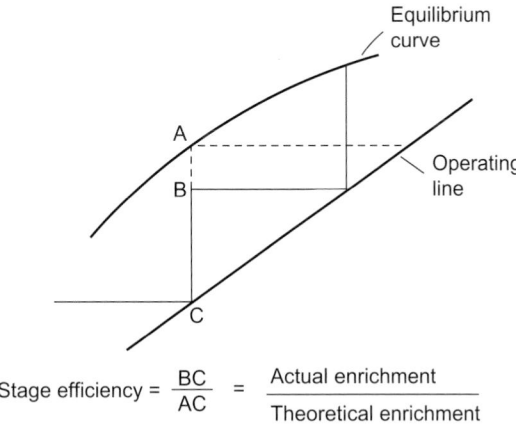

$$\text{Stage efficiency} = \frac{BC}{AC} = \frac{\text{Actual enrichment}}{\text{Theoretical enrichment}}$$

Figure 11.8.   Stage efficiency.

Example 11.2

Acetone is to be recovered from an aqueous waste stream by continuous distillation. The feed contains 10 mol% acetone. Acetone of at least 95 mol% purity is wanted, and the aqueous effluent must not contain more than 1 mol% acetone. The feed will be a saturated liquid. Estimate the number of ideal stages required.

Solution

There is no point in operating this column at other than atmospheric pressure. The equilibrium data available for the acetone–water system were discussed in Chapter 8, Section 8.4.

The data of Kojima *et al.* (1968) will be used.

Mol fraction x, liquid	0.00	0.05	0.10	0.15	0.20	0.25	0.30
Acetone y, vapour	0.00	0.6381	0.7301	0.7716	0.7916	0.8034	0.8124
Bubble point °C	100.0	74.80	68.53	65.26	63.59	62.60	61.87

x	0.35	0.40	0.45	0.50	0.55	0.60	0.65
y	0.8201	0.8269	0.8376	0.8387	0.8455	0.8532	0.8615
°C	61.26	60.75	60.35	59.95	59.54	59.12	58.71

x	0.70	0.75	0.80	0.85	0.90	0.95
y	0.8712	0.8817	0.8950	0.9118	0.9335	0.9627
°C	58.29	57.90	57.49	57.08	56.68	56.30

The equilibrium curve can be drawn with sufficient accuracy to determine the stages above the feed by plotting the concentrations at increments of 0.1.

Following the procedure given above, we can mark the product compositions.

Since the feed is a saturated liquid, $q = 1$, and the slope of the q line is $1/(1-1) = \infty$. The q line is thus plotted as a vertical line through the feed composition.

For this problem the condition of minimum reflux occurs where the top operating line just touches the equilibrium curve at the point where the q line cuts the curve.

From Figure 11.9

$$\phi \text{ for the operating line at minimum reflux } = 0.59$$

From equation 11.24

$$R_{min} = 0.95/0.59 - 1 = 0.62$$

Take

$$R = R_{min} \times 2 = 1.24$$

As the flows above the feed point will be small, a high reflux ratio is justified; the condenser duty will be small.

$$\phi = \frac{0.95}{1 + 1.24} = 0.42$$

We can then plot the upper operating line (UOL) and lower operating line (LOL).

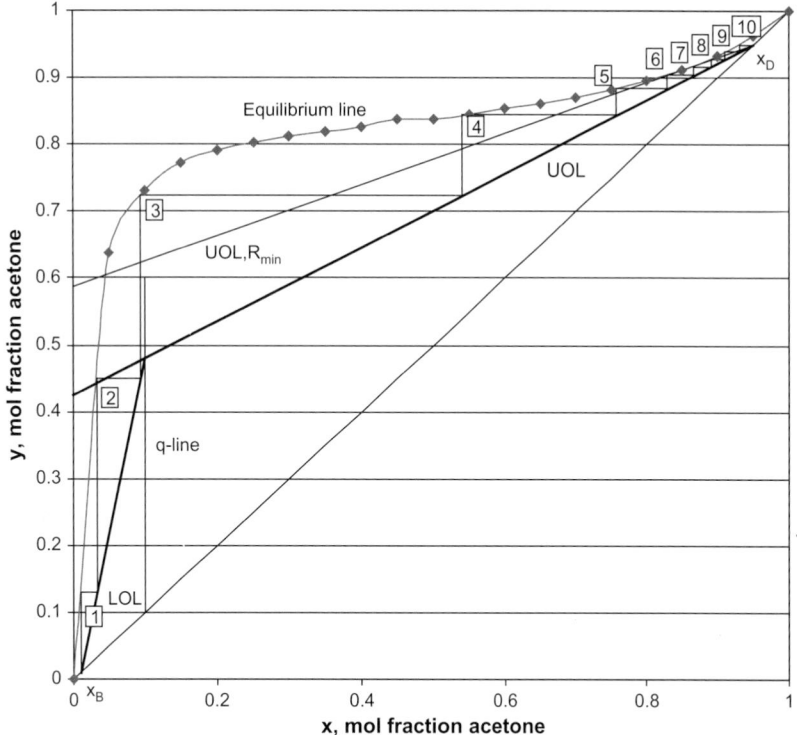

Figure 11.9.   McCabe–Thiele plot, Example 11.2.

Stepping off from the bottom, we require two stages in the stripping section and an additional eight stages in the rectifying section. The feed should be on the third stage from the bottom. Note that the number of stages below the feed is small, and since the reboiler counts as an equilibrium stage, if we feed onto the third plate from the bottom we would actually be feeding onto the stage marked 4 in Figure 11.9. This feed would be badly mismatched with the vapour and liquid compositions in the column, so it would be advisable to allow for possible feeds to the second and third plates from the bottom so that the feed point can be moved. Note also that the column is close to pinched at the top of the rectifying section. It would be prudent to add extra trays in this section.

11.6.   MULTICOMPONENT DISTILLATION: GENERAL CONSIDERATIONS

The problem of determining the stage and reflux requirements for multicomponent distillations is much more complex than for binary mixtures. With a multicomponent mixture, fixing one component composition does not uniquely determine the other component compositions and the stage temperature. Also when the feed contains more than two components it is not possible to specify the complete composition of the top and bottom products independently. The separation between the top and bottom products is usually specified by setting limits on two 'key' components, between which it is desired to make the separation.

The complexity of multicomponent distillation calculations can be appreciated by considering a typical problem. The normal procedure is to solve the MESH equations (Section 11.3.1) stage by stage, from the top and bottom of the column toward the feed point. For such a calculation to converge to a solution, the compositions obtained from both the bottom-up and top-down calculations must mesh at the feed point and match the feed composition. But the calculated compositions will depend on the compositions assumed for the top and bottom products at the start of the calculations. Though it is possible to match the key components, the other components will not match unless the designer was particularly fortunate in choosing the trial top and bottom compositions. For a completely rigorous solution, the compositions must be adjusted and the calculations repeated until satisfactory convergence at the feed point is obtained. Clearly, the greater the number of components considered, the more difficult the problem. As was shown in Section 11.3.2, iterative calculations will be needed to determine the stage temperatures. For other than ideal mixtures, the calculations will be further complicated by the fact that the component volatilities will be functions of the unknown stage compositions and temperatures. The relationship between volatility, composition and temperature may be highly non-linear, as discussed in Chapter 8. If more than a few stages are required, stage-by-stage calculations are complex and tedious and even with sophisticated process simulation programs convergence cannot be guaranteed.

Before the widespread availability of digital computers, various 'short-cut' methods were developed to simplify the task of designing multicomponent columns. A comprehensive summary of the methods used for hydrocarbon systems is given by Edmister (1947 to 1949) in a series of articles in the journal *The Petroleum Engineer*. Though computer programs will normally be available for the rigorous solution of the MESH equations, short-cut methods are still useful in preliminary design work, and as an aid in initializing problems for computer solution.

11.6.1. Key Components

Before starting the column design, the designer must select the two 'key' components between which it is desired to make the separation. The light key will be the component that it is desired to keep out of the bottom product, and the heavy key the component to be kept out of the top product. The keys are known as 'adjacent keys' if they are 'adjacent' in a listing of the components in order of volatility, and 'split keys' or 'non-adjacent keys' if some other component lies between them in the order. A separation between adjacent keys is known as a 'sharp split', while a separation with non-adjacent keys is known as a 'sloppy split'.

The choice of key components will normally be clear, but sometimes, particularly if close boiling isomers are present, judgment must be used in their selection. If any uncertainty exists, trial calculations should be made using different components as the keys to determine the pair that requires the largest number of stages for separation (the worst case). The Fenske equation can be used for these calculations; see Section 11.7.1.

The 'non-key' components that appear in both top and bottom products are known as 'distributed' components; and those that are not present, to any significant extent, in one or other product, are known as 'non-distributed' components.

11.6.2. Product Specifications

Specifications for the column will normally be set in terms of the purity or recovery of the key components.

A purity specification sets the mole (or mass) fraction of a component in one of the product streams. Purity specifications are easily understood and are easy to relate to the required product specifications. For example, if the standard specification for the desired grade of product is 99.5% pure then the designer could specify 99.5% purity of the product in the distillate of a finishing column. Similarly, if the purity of the heavy key component in the distillate must be less than 50 ppm then this can be used as the specification.

Although purity specifications are intuitively obvious, their use often leads to infeasible column specifications. The designer must check carefully to ensure that the amounts of lighter-than-light-key (or heavier-than-heavy-key) components are not large enough to render the purity specification infeasible. Consider, for example, a feed that contains 0.5 mol% A, 49.5 mol% B and 50 mol% C, where A is most volatile and C is least volatile. The highest purity of B that can be obtained in the distillate is 99%, and that would require complete recovery of B and complete rejection of C from the distillate, with a very high reflux ratio. If only 99% of the C is recovered in the bottoms product and only 99% of the B is recovered overhead, then the maximum feasible purity of B is $0.99(49.5)/[0.5 + 0.99(49.5) + 0.5] = 98\%$.

When a purity specification must be met in a column that forms part of a multi-component sequence of columns then it is important to set the specifications of the other columns so that the desired purity specification is feasible.

Instead of specifying purity, the designer can specify the recovery of one or more of the key components in either the distillate or bottoms. The recovery of a component in a product stream is defined as the fraction of the feed molar flow rate of the component that is recovered in that product stream. The relationship between purity and recovery is often not simple, particularly when many components are present or the key components are non-adjacent.

Recovery specifications are easily related to economic trade-offs, since the value of recovering an additional 0.1% or 0.01% of the desired product is easily assessed and can be traded off against the additional capital and operating costs of the column. Recovery of the desired product is usually set at 99% or greater. Recovery specifications for a distillation column are less likely to be infeasible than purity specifications; however, their use does not guarantee that the product will meet the specifications required for sale.

A combination of purity and recovery specifications can also be used. For example, in a finishing column in which the desired product is taken as distillate, the designer could specify the purity and recovery of the desired (light key) component in the distillate. No specifications on the heavy key or other components are needed, but the same feasibility checks must be made for the purity specification.

In mixtures that form azeotropes the volatility order changes with composition. This creates additional problems when setting product specifications. The design of azeotropic distillation sequences is discussed in Section 11.6.5.

Most process simulation programs allow the designer to select two specifications for a distillation column corresponding to the two remaining degrees of freedom once the feed rate, pressure, number of stages and feed stage have been selected. If the column is to be designed to achieve a given purity or recovery then it obviously makes sense to use this as a specification if the simulation program allows it, but the designer may need to provide an initial estimate of other parameters such as reflux ratio or distillate rate to ensure good convergence. Estimates of these parameters can be made using short-cut calculations or short-cut column models, as described in Section 11.7. The use of short-cut models to initialize a rigorous simulation model was discussed in Section 4.5.2 and illustrated in Example 4.3. Example 4.4 showed the effect of changing to a recovery specification in a rigorous solution of the same problem.

In some cases, the simulation program or model may not allow the use of purity or recovery specifications, in which case the designer must adjust other variables such as reflux rate, boil-up rate and distillate or bottoms flow rate until the specifications are met.

11.6.3. Number and Sequencing of Columns

In multicomponent distillation the production of a pure product usually requires at least two distillation columns. A common arrangement is to remove all components lighter than the desired product in a first column, then separate the desired product from heavier components in a second column. This arrangement is illustrated in Figure 11.10, and is known as a stripper and re-run column sequence. Since almost all processes produce some by-products that are lighter and some that are heavier than the desired product, this scheme is widely used.

If additional pure products are to be produced, then additional columns will be needed. The recovery of an additional pure component from the lights stream in Figure 11.10 would require one more column if the component was the least volatile component in the lights (i.e., the light key component of the first column in Figure 11.10), or two more columns if there were any additional components that boiled between the two desired products.

If a mixture contained N components and the designer wanted to separate it into pure components, then $N - 1$ columns would be needed, as each component could be

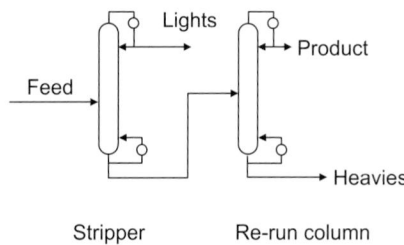

Figure 11.10. Stripper and re-run column.

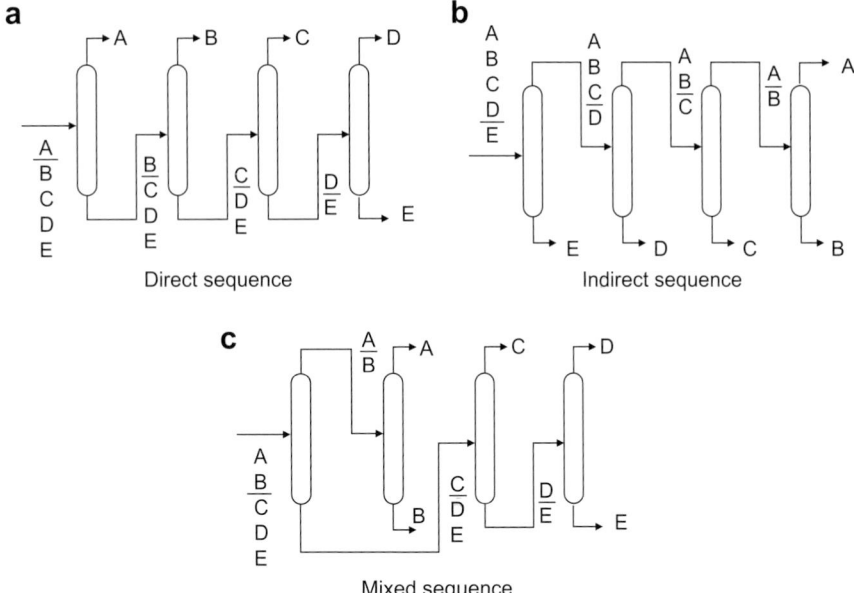

Figure 11.11. Column sequences for a five-component mixture.

removed in order of volatility until the final binary pair remained. If only M pure products are required then the number of columns needed is bounded between $M + 1$ and either $2M$ or $N - 1$, whichever is least.

The distillation sequence in which the components are separated in order of decreasing volatility is known as a direct sequence and is shown in Figure 11.11(a). There are many other possible distillation sequences. Figure 11.11(b) shows an indirect sequence in which the heaviest component is removed first and the distillate is fed to the second column. Components are then removed in the order from least volatile to most volatile. Figure 11.11(c) shows a mixed sequence in which the first separation is between components in the middle of the volatility order.

With five components, there are 14 possible column configurations. As the number of components increases, the number of possible column sequences increases combinatorially. With 10 components there are nearly 5000 possible schemes. The optimum scheme will be the one that has the best overall economic performance.

Various methods have been proposed for screening alternative designs to determine the optimum sequence; see Doherty and Malone (2001), Smith (2005) and Kumar (1982). These methods usually use short-cut column models and approximate costing relationships, so it is often worthwhile to complete detailed designs for a few of the best candidate schemes identified. There can also be strong interactions between the column sequence and the associated heat integration that can influence the final scheme that is selected.

Although distillation column sequencing is an interesting research problem, in practice there are very few processes that make more than two or three pure products. The optimum sequence can often be determined using heuristic rules such as:

1. Remove corrosive components first, to avoid using expensive metallurgy throughout the sequence.
2. Remove the heaviest component first if there are solids present in the feed. The presence of solids requires the use of special plates that are designed to resist plugging and have very low stage efficiency. It is best to get the solids out of the way as early as possible.
3. Split any components that cannot be condensed using cooling water from those that can early in the sequence. The lighter components can then be compressed to higher pressure, separated using absorption or adsorption, or separated in refrigerated columns. This rule avoids the use of refrigerated condensers, higher pressures or partial condensers elsewhere in the sequence.
4. Postpone the most difficult separation, such as between close-boiling compounds, until late in the sequence. A difficult separation will require many stages and high reflux and so the feed rate to that column should be made as low as possible so that the column handles less material.
5. Take the desired products as distillates whenever practical, to avoid any flushing of dirt or debris into the desired product. The same rule also applies to recycle streams.
6. Remove any components that are present in large excess early in the sequence, to make the downstream columns cheaper.

Tall Columns

Where a large number of stages are required, it may be necessary to split a column into two separate columns to reduce the height of the column, even though the required separation could, theoretically, have been obtained in a single column. This may also be done in vacuum distillations, to reduce the column pressure drop and limit the bottom temperatures.

11.6.4. Complex Columns

It is relatively easy to withdraw side streams from plate columns and to supply additional feeds to the column. If a liquid side stream is withdrawn from a tray above the feed, as shown in Figure 11.12(a), then it will be depleted in the heavier components of the feed (which preferentially stayed in the liquid phase and went down in the stripping section) and will also be depleted in the lighter components of the feed (which are preferentially in the vapour phase). Although the side stream will not be pure, it will be enriched in some of the components of mid-range volatility. In some cases, the purity of the side stream may be adequate, for example if the side stream is a process recycle.

The purity of a desired component in the side stream can be increased by sending the side stream to a small side stripper column that strips out any lighter components, as shown in Figure 11.12(b). The vapour from the side stripper is returned to the main column. A side stripper can be constructed as part of the main column by using a partitioned section of the main column, as shown in Figure 11.12(c). Side rectifiers are also used; see Figure 11.12(d).

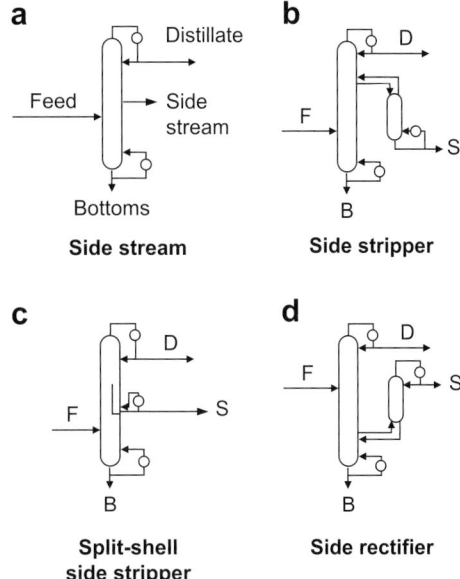

Figure 11.12. Side streams and side columns.

Side strippers and rectifiers allow up to three pure products to be made in one and a half columns, or in a single shell if a partition wall is used. Other complex column configurations are also possible, such as prefractionators and dividing-wall columns, illustrated in Figure 11.13. These complex columns generally have lower capital and operating costs than sequences of simple columns. More degrees of freedom are introduced into the design, so more care is needed in optimizing the columns. Smith (2005) gives an excellent introduction to the design of complex columns. Greene (2001), Schultz *et al.* (2002), Kaibel (2002) and Parkinson (2007) describe industrial applications of dividing-wall columns. Side strippers are widely used in petroleum refining; see Watkins (1979). Most process simulators allow the designer to add side strippers and rectifiers or select from a set of pre-built complex column models.

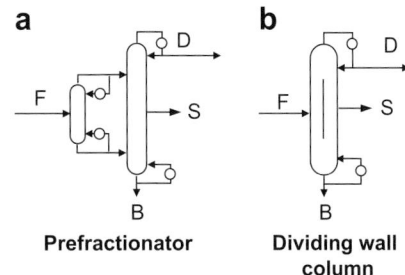

Figure 11.13. Complex column designs.

11.6.5. Distillation Column Sequencing for Azeotropic Mixtures

When a mixture forms an azeotrope then determining the best column sequence is not straightforward. Homogeneous azeotropes are mixtures of two or more components that have the same vapour and liquid phase composition at the boiling point. Heterogeneous azeotropes have two liquid phases that are in equilibrium with a vapour that has the same composition as the combined liquid composition at the boiling point. Different strategies are used for separation depending on the type of azeotrope.

The design of azeotropic distillation sequences has been the subject of much academic research and there is not sufficient space here to describe all of the techniques that have been developed. The reader should refer to Smith (2005) and Doherty and Malone (2001) for a more detailed treatment of the subject. The general strategy for separating an azeotropic mixture can be summarized as:

1. If the azeotrope is heterogeneous, then use a liquid–liquid split (decanter). The two liquid phases will usually have compositions on either side of the azeotrope. Each of these liquids can be distilled to give a pure product and the azeotrope, and the azeotropic mixtures can be recycled to the decanter. In some cases a third component, known as an entrainer, is added to cause the formation of a heterogeneous azeotropic mixture. The degree of separation in the liquid–liquid split can often be increased by lowering the temperature, which tends to increase the size of the two-phase region in the composition space.

    For example, Figure 11.14 shows the separation of an ethanol–water mixture using benzene as entrainer. The ethanol–water mixture is distilled to give a low-boiling azeotrope, which is then sent to a first column that is refluxed with the oil phase from a decanter. The first column produces ethanol as bottom product and a heterogeneous azeotrope as distillate. The distillate is sent to a decanter and separated into oil-rich and water-rich phases. The water phase is sent to a second column that produces water as bottoms product and heterogeneous azeotrope as distillate. The distillate from the second column is also sent to the decanter. This flow scheme was widely used for ethanol dehydration until cheaper and safer processes based on adsorbing the water using molecular sieves were introduced.

    This flow scheme is sometimes known as azeotropic distillation; however, the term heterogeneous azeotropic distillation is better, as all of the other methods also involve distilling azeotropes.

2. If the azeotrope is homogeneous, then the effect of varying pressure should be investigated. The composition of the azeotrope is always pressure dependent.

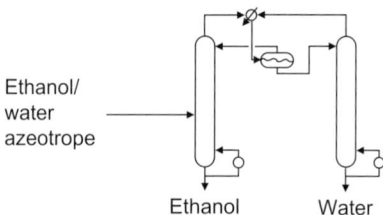

Figure 11.14. Dehydration of ethanol using benzene as entrainer.

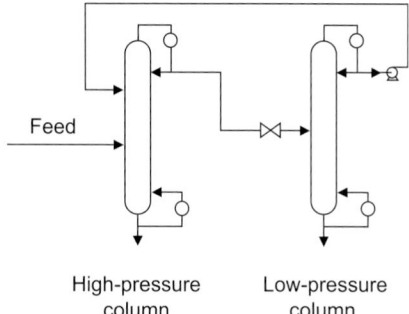

Figure 11.15. Pressure-swing distillation.

If there is a large change in composition over a reasonable range of pressure, two columns at different pressures can be used. Each column produces a pure product and a mixture corresponding to the azeotrope at the pressure of that column. The azeotropic mixture is then fed to the other column, as illustrated in Figure 11.15. The mixture from the low-pressure column must be pumped back up to the pressure of the high-pressure column. Note that the feed can be to either column and that it is also possible to produce the products as distillates if the azeotrope is maximum-boiling rather than minimum-boiling.

The pressure-swing distillation flow scheme is relatively simple and does not require any additional components to be added, but if the azeotrope composition is only weakly sensitive to pressure then the recycle from the low-pressure column to the high-pressure column will be large. The recycled material must be vaporized in the low-pressure column, so the low-pressure column can become very expensive.

3. If pressure-swing distillation is not economically attractive, then consider adding an entrainer. The preferred entrainers are usually those that form heterogeneous azeotropes, as discussed above, but homogeneous entrainers can also be used, in which case the process is known as extractive distillation.

The most commonly used type of entrainer is a higher boiling compound that does not form an azeotrope with either component of the azeotropic pair. If a high-boiling entrainer is used, it depresses the volatility of one component of the azeotrope, allowing the other component to be recovered in the distillate. The bottoms from the first column is then sent to a second column in which the other pure component is recovered as distillate and the entrainer is recovered as bottoms for recycle to the first column, as shown in Figure 11.16. Other schemes with low-boiling or mid-boiling entrainers are also possible, as described by Doherty and Malone (2001).

The entrainer should be selected from compounds that are already present within the process whenever possible. The use of compounds that are already present reduces consumables costs and waste formation and usually makes it easier to reach product specifications. The other compounds that are present in the process can be screened for suitability as entrainers by looking at boiling points and checking for the formation of additional azeotropes. If nothing

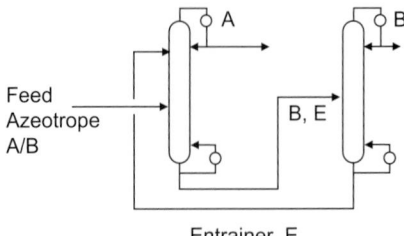

Figure 11.16.   Extractive distillation.

suitable is found then the more sophisticated methods for evaluating entrainers described by Doherty and Malone (2001) should be used.

4. If the azeotropic composition is close to the required purity specification, then consider removing the minor component by adsorption using a selective sorbent. If a regenerable sorbent can be found, then this process may be cheaper than a multi-column distillation. Pressure-swing adsorption using a molecular sieve sorbent as drying agent is now the most widely used method for breaking the ethanol–water azeotrope. Adsorption processes are discussed in Chapter 10.

5. If a suitable membrane material can be found that is permeable to one component of the mixture but impermeable to the other, then membrane separation can be used in combination with distillation. A typical flow sheet for the case where the light component is permeable is shown in Figure 11.17. A distillation column is used to separate the mixture of A and B into pure heavy component B and azeotrope. The azeotropic mixture is sent to the membrane unit, where pure light component A is recovered. It is usually not economical to operate a membrane unit at high recovery of permeate, so the retentate still contains a significant fraction of A, and is recycled to the distillation column. If the membrane is not impermeable to component B, then a permeate stream that is enriched in A can be sent to a second distillation column that then produces pure A and an azeotropic mixture for recycle to the membrane unit. The design of membrane processes is described in Chapter 10.

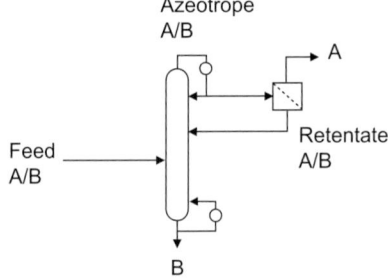

Figure 11.17.   Membrane distillation.

11.7. MULTICOMPONENT DISTILLATION: SHORT-CUT METHODS FOR STAGE AND REFLUX REQUIREMENTS

Some of the more useful short-cut procedures that can be used to estimate stage and reflux requirements without the aid of computers are given in this section. Most of the short-cut methods were developed for the design of separation columns for hydrocarbon systems in the petroleum and petrochemical systems industries, and caution must be exercised when applying them to other systems. They usually depend on the assumption of constant relative volatility, and should not be used for severely non-ideal systems. Short-cut methods for non-ideal and azeotropic systems are given by Featherstone (1971, 1973).

Although the short-cut methods were developed for hand calculations, they are easily coded into spreadsheets and are available as subroutines in all the commercial process simulation programs. The short-cut methods are useful when configuring rigorous distillation models, as described in Section 4.5.2.

The two most frequently used empirical methods for estimating the stage requirements for multicomponent distillations are the correlations published by Gilliland (1940) and by Erbar and Maddox (1961). These relate the number of ideal stages required for a given separation, at a given reflux ratio, to the number at total reflux (minimum possible) and the minimum reflux ratio (infinite number of stages). The Erbar–Maddox correlation is given in this section, as it is now generally considered to give more reliable predictions than Gilliland's correlation. The Erbar–Maddox correlation is shown in Figure 11.18, which gives the ratio of number of stages required to the number at total reflux, as a function of the reflux ratio, with the minimum reflux ratio as a parameter. To use Figure 11.18, estimates of the number of stages at total reflux and the minimum reflux ratio are needed.

11.7.1. Minimum Number of Stages (Fenske Equation)

The Fenske equation (Fenske, 1932) can be used to estimate the minimum stages required at total reflux. The derivation of this equation for a binary system is given in Richardson *et al.* (2002). The equation applies equally to multicomponent systems and can be written as:

$$\left[\frac{x_i}{x_r}\right]_d = \alpha_i^{N_{min}} \left[\frac{x_i}{x_r}\right]_b \qquad (11.25)$$

where

$[x_i/x_r]$ = the ratio of the concentration of any component i to the concentration of a reference component r, and the suffixes d and b denote the distillate (d) and the bottoms (b).

N_{min} = minimum number of stages at total reflux, including the reboiler.

α_i = average relative volatility of the component i with respect to the reference component.

Normally the separation required will be specified in terms of the key components, and equation 11.25 can be rearranged to give an estimate of the number of stages.

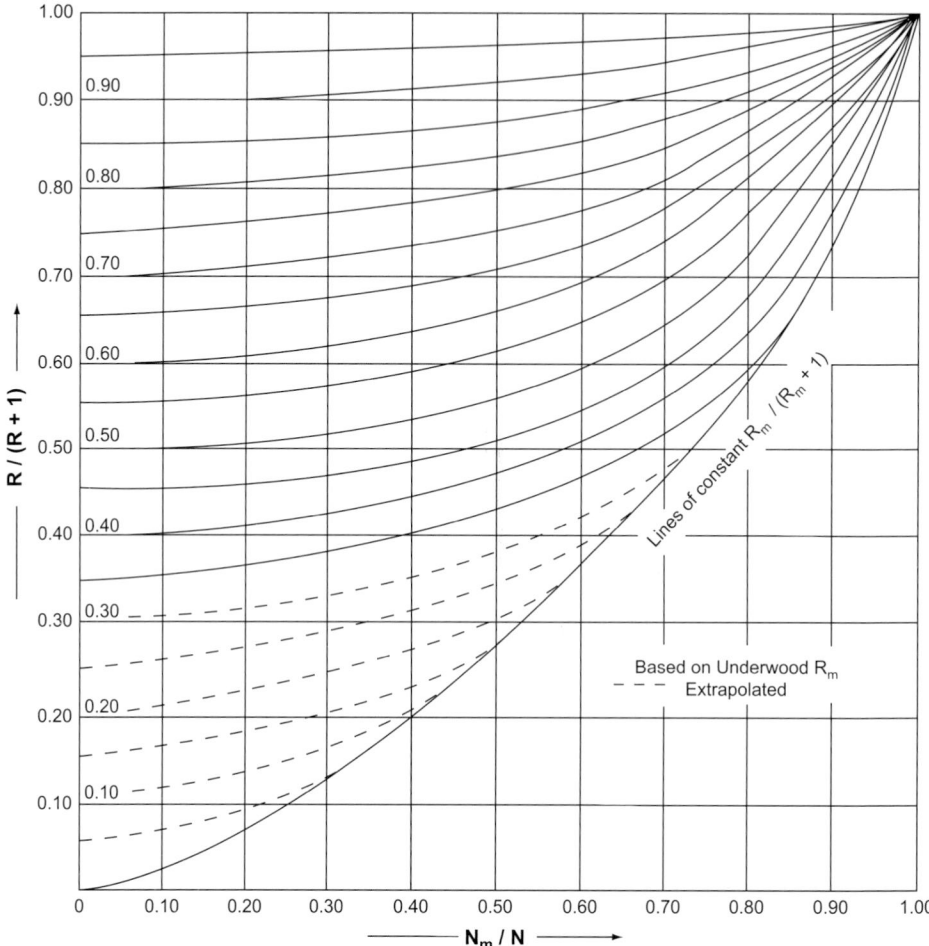

Figure 11.18.   Erbar–Maddox correlation (Erbar and Maddox, 1961).

$$N_{min} = \frac{\log\left[\dfrac{x_{LK}}{x_{HK}}\right]_d \left[\dfrac{x_{HK}}{x_{LK}}\right]_b}{\log \, \alpha_{LK}} \qquad (11.26)$$

where α_{LK} is the average relative volatility of the light key with respect to the heavy key, and x_{LK} and x_{HK} are the light and heavy key concentrations.

The relative volatility is taken as the geometric mean of the values at the column top and bottom temperatures. To calculate these temperatures initial estimates of the compositions must be made, so the calculation of the minimum number of stages by the Fenske equation is a trial-and-error procedure. The procedure is illustrated in Example 11.3. If there is a wide difference between the relative volatilities at the top and bottom of the column the use of the average value in the Fenske equation will underestimate the number of stages. In these circumstances, a better estimate can be made by calculating the number of stages in the rectifying and stripping sections separately; taking the feed concentration as the base concentration for the rectifying section and as the top concentration for the stripping section, and estimating the average relative volatilities separately for each section. This procedure will also give an estimate of the feed point location.

Winn (1958) has derived an equation for estimating the number of stages at total reflux, which is similar to the Fenske equation, but which can be used when the relative volatility cannot be taken as constant.

If the number of stages is known, equation 11.25 can be used to estimate the split of components between the top and bottom of the column at total reflux. It can be written in a more convenient form for calculating the split of components:

$$\frac{d_i}{b_i} = \alpha_i^{N_{min}} \left[\frac{d_r}{b_r} \right] \qquad (11.27)$$

where d_i and b_i are the flow rates of the component i in the distillate and bottoms; d_r and b_r are the flow rates of the reference component in the distillate and bottoms.

Note: from the column material balance:

$$d_i + b_i = f_i$$

where f_i is the flow rate of component i in the feed.

11.7.2. Minimum Reflux Ratio

Colburn (1941) and Underwood (1948) have derived equations for estimating the minimum reflux ratio for multicomponent distillations. As the Underwood equation is more widely used it is presented in this section. The equation can be stated in the form:

$$\sum \frac{\alpha_i x_{i,d}}{\alpha_i - \theta} = R_{min} + 1 \qquad (11.28)$$

where

$\quad \alpha_i$ = the relative volatility of component i with respect to some reference component, usually the heavy key

$\quad R_{min}$ = the minimum reflux ratio

$\quad x_{i,d}$ = concentration of component i in the distillate at minimum reflux,

and θ is the root of the equation

$$\sum \frac{\alpha_i x_{i,f}}{\alpha_i - \theta} = 1 - q \qquad (11.29)$$

where $x_{i,f}$ = the concentration of component i in the feed, and q depends on the condition of the feed and was defined in Section 11.5.2.

The value of θ must lie between the values of the relative volatility of the light and heavy keys, and is found by trial and error.

In the derivation of equations 11.28 and 11.29, the relative volatilities are taken as constant. The geometric average of values estimated at the top and bottom temperatures should be used. This requires an estimate of the top and bottom compositions. Though the compositions should strictly be those at minimum reflux, the values determined at total reflux, from the Fenske equation, can be used. A better estimate can be obtained by replacing the number of stages at total reflux in equation 11.27 by an estimate of the actual number; a value equal to $N_{min}/0.6$ is often used. The Erbar–Maddox method of estimating the stage and reflux requirements, using the Fenske and Underwood equations, is illustrated in Example 11.3.

11.7.3. Feed-Point Location

A limitation of the Erbar–Maddox, and similar empirical methods, is that they do not give the feed-point location. An estimate can be made by using the Fenske equation to calculate the number of stages in the rectifying and stripping sections separately, but this requires an estimate of the feed-point temperature. An alternative approach is to use the empirical equation given by Kirkbride (1944):

$$\log\left[\frac{N_r}{N_s}\right] = 0.206 \log\left[\left(\frac{B}{D}\right)\left(\frac{x_{f,\text{HK}}}{x_{f,\text{LK}}}\right)\left(\frac{x_{b,\text{LK}}}{x_{d,\text{HK}}}\right)^2\right] \tag{11.30}$$

where

N_r = number of stages above the feed, including any partial condenser
N_s = number of stages below the feed, including the reboiler
$x_{f,\text{HK}}$ = concentration of the heavy key in the feed
$x_{f,\text{LK}}$ = concentration of the light key in the feed
$x_{d,\text{HK}}$ = concentration of the heavy key in the top product
$x_{b,\text{LK}}$ = concentration of the light key if in the bottom product.

The use of this equation is illustrated in Example 11.4.

Example 11.3

Estimate the minimum number of ideal stages needed in the butane-pentane splitter defined by the compositions given in the table below. The column will operate at a pressure of 8.3 bar. Evaluate the effect of changes in reflux ratio on the number of stages required. This is an example of the application of the Erbar–Maddox method. The feed is at its boiling point.

	Feed (f)	Distillate (d)	Bottoms (b)
Propane, C_3	5	5	0
i-Butane, iC_4	15	15	0
n-Butane, nC_4	25	24	1
i-Pentane, iC_5	20	1	19
n-Pentane, nC_5	35	0	35
	100	45	55 kmol

Solution

The top and bottom temperatures (dew points and bubble points) were calculated by the method given in Section 11.3.2. Relative volatilities are given by equation 8.34:

$$\alpha_i = \frac{K_i}{K_{HK}}$$

Equilibrium constants were taken from the Depriester charts (Chapter 8).
Relative volatilities:

	Top	Bottom	Average
Temp. (°C)	65	120	
C_3	5.5	4.5	5.0
iC_4	2.7	2.5	2.6
(LK) nC_4	2.1	2.0	2.0
(HK) iC_5	1.0	1.0	1.0
nC_5	0.84	0.85	0.85

Minimum number of stages; Fenske equation, equation 11.26:

$$N_{min} = \frac{\log\left[\frac{24}{1}\right]\left[\frac{19}{1}\right]}{\log 2} = \underline{\underline{8.8}}$$

Minimum reflux ratio; Underwood equations 11.28 and 11.29.
This calculation is best tabulated.
As the feed is at its boiling point, $q = 1$

$$\sum \frac{\alpha_i x_{i,f}}{\alpha_i - \theta} = 0 \tag{11.29}$$

Try

$x_{i,f}$	α_i	$\alpha_i x_{i,f}$	$\theta = 1.5$	$\theta = 1.3$	$\theta = 1.35$
0.05	5	0.25	0.071	0.068	0.068
0.15	2.6	0.39	0.355	0.300	0.312
0.25	2.0	0.50	1.000	0.714	0.769
0.20	1	0.20	−0.400	−0.667	−0.571
0.35	0.85	0.30	−0.462	−0.667	−0.600
			$\sum = 0.564$	−0.252	0.022
					close enough
				$\theta = 1.35$	

Equation 11.28

$x_{i,d}$	α_i	$\alpha_i x_{i,d}$	$\alpha_i x_{i,d}/(\alpha_i - \theta)$
0.11	5	0.55	0.15
0.33	2.6	0.86	0.69
0.53	2.0	1.08	1.66
0.02	1	0.02	−0.06
0.01	0.85	0.01	−0.02
			$\sum = 2.42$

$$R_m + 1 = 2.42$$
$$R_m = \underline{\underline{1.42}}$$
$$\frac{R_m}{(R_m + 1)} = \frac{1.42}{2.42} = 0.59$$

Specimen calculation, for $R = 2.0$

$$\frac{R}{(R + 1)} = \frac{2}{3} = 0.66$$

from Figure 11.18

$$\frac{N_{min}}{N} = 0.56$$

$$N = \frac{8.8}{0.56} = \underline{\underline{15.7}}$$

for other reflux ratios

R	2	3	4	5	6
N	15.7	11.9	10.7	10.4	10.1

Note: The number of stages should be rounded up to the nearest integer. Above a reflux ratio of 4 there is little change in the number of stages required, but given the low number of theoretical stages needed the optimum reflux ratio is probably less than 2.0.

Example 11.4

Estimate the position of the feed point for the separation considered in Example 11.3, for a reflux ratio of 3.

Solution

Use the Kirkbride equation (11.30). Product distributions are taken from Example 11.3, though they could be confirmed using equation 11.27

$$x_{b,LK} = \frac{1}{55} = 0.018$$

$$x_{d,HK} = \frac{1}{45} = 0.022$$

$$\log\left(\frac{N_r}{N_s}\right) = 0.206 \ \log\left[\frac{55}{45}\left(\frac{0.2}{0.25}\right)\left(\frac{0.018}{0.022}\right)^2\right]$$

$$\log\left(\frac{N_r}{N_s}\right) = 0.206 \ \log\ (0.65)$$

$$\left(\frac{N_r}{N_s}\right) = \underline{\underline{0.91}}$$

for $R = 3$, $N = 12$.

Number of stages, excluding the reboiler $= \underline{\underline{11}}$

$$N_r + N_s = 11$$

$$N_s = 11 - N_r = 11 - 0.91N_s$$

$$N_s = \frac{11}{1.91} = 5.76, \ \text{say} \ \underline{\underline{6}}$$

11.8. MULTICOMPONENT SYSTEMS: RIGOROUS SOLUTION PROCEDURES (COMPUTER METHODS)

The rigorous column models in the commercial process simulation programs solve the full set of MESH equations (Section 11.3.1). A considerable amount of work has been done to develop efficient and reliable computer-aided design procedures for distillation and other staged processes. A detailed discussion of this work is beyond the scope of this book and the reader is referred to the specialist books that have been published on the subject, Smith (1963), Holland (1997) and Kister (1992), and to the numerous papers that have appeared in the chemical engineering literature. A good summary of the present state of the art is given by Haas (1992). In this section only a brief outline of the methods that have been developed will be given.

The basic steps in any rigorous solution procedure will be:

1. Specification of the problem; complete specification is essential for computer methods.
2. Selection of values for the iteration variables; for example, estimated stage temperatures, and liquid and vapour flows (the column temperature and flow profiles).
3. A calculation procedure for the solution of the stage equations.
4. A procedure for the selection of new values for the iteration variables for each set of trial calculations.
5. A procedure to test for convergence; to check if a satisfactory solution has been achieved.

Rating and Design Methods

All the methods described here require the specification of the number of stages below and above the feed point. They are therefore not directly applicable to design, where the designer wants to determine the number of stages required for a specified separation. They are strictly what are referred to as 'rating methods'; used to determine the performance of existing, or specified, columns. Given the number of stages, they can be used to determine product compositions. Iterative procedures are necessary to apply rating methods to the design of new columns. Short-cut models can be used to generate initial estimates of the number of stages and feed stage, as described above and in Section 4.5.2. If a good initial estimate is provided, then the rigorous model should converge faster and can be used to size and optimize the column.

11.8.1. Linear Algebra (Simultaneous) Methods

If the equilibrium relationships and flow rates are known (or assumed), the set of material balance equations for each component is linear in the component compositions. Amundson and Pontinen (1958) developed a method in which these equations are solved simultaneously and the results used to provide improved estimates of the temperature and flow profiles. The set of equations can be expressed in matrix form and solved using the standard inversion routines available in modern computer systems. Convergence can usually be achieved after a few iterations and can be improved by use of Newton's method.

This approach has been further developed by other workers, notably Wang and Henke (1966) and Naphtali and Sandholm (1971). The Naphtali and Sandholm method for solving rigorous columns is available in many commercial simulation programs.

11.8.2. Inside-Out Algorithms

The inside-out algorithms accelerate convergence by decomposing the solution of the MESH equations into two nested iteration loops. The method was initially proposed by Boston and Sullivan (1974) and has undergone many improvements; see Boston (1980).

The outer iteration loop determines local estimates of K values and stream enthalpies using models that depend on composition and temperature. The local model parameters are the iteration variables for the outside loop. The initial estimates for the outside loop come from the initial estimate of composition and temperature profile supplied by the user.

The inner iteration loop contains the MESH equations, expressed in terms of the local physical property parameters obtained from the outer loop. With simplified physical property models, the inner loop can be converged more quickly. Convergence methods such as bounded Wegstein and Broyden quasi-Newton are typically used, as described in Section 4.7.2.

When the inside loop is converged, the new estimates of composition and temperature are used to update the outer loop parameters. The convergence tolerance of the inside loop is usually tightened at each iteration of the outside loop. The outer loop converges when the changes in local model parameters are within a satisfactory tolerance from one iteration to the next.

All of the commercial process simulation programs offer inside-out algorithms, and some offer several variants that use different convergence methods. Inside-out algorithms are very effective if good initial estimates are provided. Because of their robust and rapid convergence, they are usually the default methods recommended by the simulation program.

Inside-out algorithms can be difficult to converge if no estimate of the temperature profile is provided, so the design engineer should always enter an estimated temperature profile. Short-cut methods can be used to obtain initial estimates of composition and temperature profiles. Another effective strategy is to initialize the model using specifications that are easily met, such as reflux rate and bottoms flow, and then use the resulting temperature and composition profiles as initial estimates for a simulation with the required purity or recovery specifications.

11.8.3. Relaxation Methods

With the exception of this method, all the methods described solve the stage equations for the steady-state design conditions. In an operating column other conditions will exist at start-up, and the column will approach the 'design' steady-state conditions after a period of time. The stage material balance equations can be written in a finite difference form, and procedures for the solution of these equations will model the unsteady-state behaviour of the column.

Rose *et al.* (1958) and Hanson and Sommerville (1963) have applied 'relaxation methods' to the solution of the unsteady-state equations to obtain the steady-state values. The application of this method to the design of multistage columns is described by Hanson and Sommerville (1963). They give a program listing and worked examples for a distillation column with side streams, and for a reboiled absorber.

Relaxation methods are not competitive with the 'steady-state' methods in the use of computer time, because of slow convergence. However, because they model the actual operation of the column, convergence should be achieved for all practical problems. Relaxation methods are used for dynamic simulation of distillation and for rate-based models such as Aspen Plus RateFrac™ and BatchFrac™. Dynamic models are very useful when attempting to understand the control and operation of distillation columns.

11.9. OTHER DISTILLATION PROCESSES

11.9.1. Batch Distillation

In batch distillation the mixture to be distilled is charged as a batch to the still and the distillation carried out until a satisfactory top or bottom product is achieved. The still usually consists of a vessel surmounted by a packed or plate column. The heater may be incorporated in the vessel or a separate reboiler used. Batch distillation should be considered under the following circumstances:

1. Where the quantity to be distilled is small
2. Where a range of products has to be produced
3. Where the feed is produced at irregular intervals
4. Where batch integrity is important
5. Where the feed composition varies over a wide range.

When the choice between batch and continuous is uncertain, an economic evaluation of both systems should be made.

Batch distillation is an unsteady-state process, the composition in the still (bottoms) varying as the batch is distilled.

Two modes of operation are used.

1. Fixed reflux, where the reflux rate is kept constant. The compositions will vary as the more volatile component is distilled off, and the distillation is stopped when the average composition of the distillate collected, or the bottoms left, meet the specification required.
2. Variable reflux, where the reflux rate is varied throughout the distillation to produce a fixed overhead composition. The reflux ratio will need to be progressively increased as the fraction of the more volatile component in the base of the still decreases.

The basic theory of batch distillation is given in Richardson *et al.* (2002) and in several other texts: Hart (1997), Green and Perry (2007) and Walas (1990). In the simple theoretical analysis of batch distillation columns the liquid hold-up in the column is usually ignored. This hold-up can have a significant effect on the separating efficiency and should be taken into account when designing batch distillation columns. The practical design of batch distillation columns is covered by Hengstebeck (1976), Ellerbe (1997) and Hart (1997).

11.9.2. Vacuum Distillation

Components that boil at high temperatures or suffer thermal degradation are sometimes distilled under vacuum to reduce the temperature required for distillation. Vacuum distillation is more expensive than steam distillation, but can be used for compounds that are miscible with water or for processes where the introduction of water might lead to problems such as the formation of azeotropes. The vacuum is usually generated using a vacuum pump or an ejector system on the column overhead product. Selection and design of vacuum pumps and ejectors is described in Section 10.12.1.

Vacuum columns have high capital and operating costs for the following reasons:

1. Low pressure decreases vapour density, so the column diameter is increased.
2. The vacuum production equipment has high capital and operating costs.
3. The column must be designed to withstand an external pressure. Thicker walls are required for vessels subject to external pressure, as described in Section 13.7.
4. Additional safety precautions and inspection are needed to ensure air cannot enter the equipment if the process fluids are flammable.

Because vacuum columns need low pressure drop per tray, low weir heights are used for plate columns, leading to low stage efficiency and a need for more trays. Packings are therefore often preferred for vacuum service.

11.9.3. Steam Distillation

In steam distillation, steam is introduced into the column to lower the partial pressure of the volatile components. Steam distillation is used for the distillation of heat sensitive products and for compounds with a high boiling point. It is an alternative to vacuum distillation. The products must be immiscible with water. Some steam will normally be allowed to condense to provide the heat required for the distillation. Live steam can be injected directly into the column base, or the steam can be generated by a heater in the still or in an external boiler.

The design procedure for columns employing steam distillation is essentially the same as that for conventional columns, making allowance for the presence of steam in the vapour.

Steam distillation is used extensively in the extraction of essential oils from plant materials.

11.9.4. Reactive Distillation

Reactive distillation is the name given to the process where the chemical reaction and product separation are carried out simultaneously in one unit. Carrying out the reaction, with separation and purification of the product by distillation, gives the following advantages:

1. Chemical equilibrium restrictions are overcome, as the product is removed as it is formed.
2. Energy savings can be obtained, as the heat of reaction can be used for the distillation.
3. Capital costs are reduced, as only one vessel is required.

The design of reactive distillation columns is complicated by the complex interactions between the reaction and separation processes. Detailed discussion of reactive distillation is given by Towler and Frey (2002) and Sundmacher and Kiene (2003).

Reactive distillation is used in the production of MTBE (methyl tertiary butyl ether) and methyl acetate.

11.9.5. Petroleum Distillation

Petroleum mixtures such as crude oil and the products of oil refining processes contain from 10^2 to greater than 10^5 components, typically including almost every possible hydrocarbon isomer in the boiling range of the mixture. It is usually neither necessary nor desirable to separate these mixtures into pure components, as the processing goal is to form mixtures with suitable properties, such as volatility and viscosity, for use as fuels. The mixture is distilled into 'fractions' or 'cuts' that have a suitable boiling range for blending into a fuel or sending for additional processing. The distillation of mineral oils is therefore known as fractionation, although this term is sometimes also applied to conventional multicomponent distillation.

The specifications for fractionation columns are usually not set in terms of key components. Instead, the designer specifies cut points for the product streams. The cut points are points on the product stream boiling curve, typically at 5% and 95% of the total material distilled. The sharpness of separation between two fractions is then measured by the overlap between the 95% cut temperature of the lighter fraction and the 5% cut temperature of the heavy fraction.

A good introduction to petroleum fractionation is given by Watkins (1979).

The simulation of petroleum fractionation columns is discussed in Sections 4.4.2 and 4.5.2. Most of the commercial process simulation programs incorporate pre-built complex column configurations for petroleum fractionation and also have standard sets of pseudocomponents that can be used to fit the feed and product boiling curves.

11.10. PLATE EFFICIENCY

The designer is concerned with real contacting stages; not the theoretical equilibrium stage assumed for convenience in the mathematical analysis of multistage processes. Equilibrium will rarely be attained in a real stage. The concept of stage efficiency is used to link the performance of practical contacting stages to the theoretical equilibrium stage.

Three principal definitions of efficiency are used:

1. Murphree plate efficiency (Murphree, 1925), defined in terms of the vapour compositions by:

$$E_{mV} = \frac{y_n - y_{n-1}}{y_e - y_{n-1}} \qquad (11.31)$$

where y_e is the composition of the vapour that would be in equilibrium with the liquid leaving the plate. The Murphree plate efficiency is the ratio of the actual separation achieved to that which would be achieved in an equilibrium stage (see Figure 11.8). In this definition of efficiency the liquid and the vapour stream are taken to be perfectly mixed; the compositions in equation 11.31 are the average composition values for the streams.

2. Point efficiency (Murphree point efficiency). If the vapour and liquid compositions are taken at a point on the plate, equation 11.31 gives the local or point efficiency, E_{mV}.

3. Overall column efficiency. This is sometimes confusingly referred to as the overall plate efficiency.

$$E_o = \frac{\text{number of ideal stages}}{\text{number of real stages}} \qquad (11.32)$$

An estimate of the overall column efficiency will be needed when the design method used gives an estimate of the number of ideal stages required for the separation.

In some methods, the Murphree plate efficiencies can be incorporated into the procedure for calculating the number of stages and the number of real stages determined directly.

For the idealized situation where the operating and equilibrium lines are straight, the overall column efficiency and the Murphree plate efficiency are related by an equation derived by Lewis (1936):

$$E_o = \frac{\log\left[1 + E_{mV}\left(\dfrac{mV}{L} - 1\right)\right]}{\log\left(\dfrac{mV}{L}\right)} \qquad (11.33)$$

where

m = slope of the equilibrium line
V = molar flow rate of the vapour
L = molar flow rate of the liquid.

Equation 11.33 is not of much practical use in distillation, as the slopes of the operating and equilibrium lines will vary throughout the column. It can be used by dividing the column into sections and calculating the slopes over each section. For most practical purposes, providing the plate efficiency does not vary too much, a simple average of the plate efficiency calculated at the column top, bottom and feed points will be sufficiently accurate.

11.10.1. Prediction of Plate Efficiency

Whenever possible, the plate efficiencies used in design should be based on measured values for similar systems, obtained on full-sized columns. There is no entirely satisfactory method for predicting plate efficiencies from the system physical properties and plate design parameters; however, the methods given in this section can be used to make a rough estimate where no reliable experimental values are available. They can also be used to extrapolate data obtained from small-scale experimental columns. If the system properties are at all unusual, experimental confirmation of the predicted values should always be obtained. The small, laboratory-scale, glass sieve plate column developed by Oldershaw (1941) has been shown to give reliable values for scale-up. The use of Oldershaw columns is described in papers by Swanson and Gester (1962), Veatch *et al.* (1960) and Fair *et al.* (1983).

Some typical values of plate efficiency for a range of systems are given in Table 11.1. More extensive compilations of experimental data are given by Vital *et al.* (1984) and Kister (1992).

Plate, and overall column, efficiencies will normally be between 30% and 80%, and as a rough guide a figure of 70% can be assumed for preliminary designs.

Efficiencies will be lower for vacuum distillations, as low weir heights are used to keep the pressure drop small (see Section 11.10.4).

Multicomponent Systems

The prediction methods given in the following sections, and those available in the open literature, are usually restricted to binary systems. It is clear that in a binary system the efficiency obtained for each component must be the same. This is not so for a multicomponent system; the heavier components will usually exhibit lower efficiencies than the lighter components.

The following guide rules, adapted from a paper by Toor and Burchard (1960), can be used to estimate the efficiencies for a multicomponent system from binary data:

1. If the components are similar, the multicomponent efficiencies will be similar to the binary efficiency.
2. If the predicted efficiencies for the binary pairs are high, the multicomponent efficiency will be high.
3. If the resistance to mass transfer is mainly in the liquid phase, the difference between the binary and multicomponent efficiencies will be small.

TABLE 11.1. Representative Efficiencies, Sieve Plates

System	Column dia. (m)	Pressure (kPa), abs	Efficiency (%)	
			E_{mV}	E_o
Water–methanol	1.0	—	80	
Water–ethanol	0.2	101	90	
Water–isopropanol	—	—		70
Water–acetone	0.15	90	80	
Water–acetic acid	0.46	101	75	
Water–ammonia	0.3	101	90	
Water–carbon dioxide	0.08	—	80	
Toluene–propanol	0.46	—	65	
Toluene–ethylene dichloride	0.05	101		75
Toluene–methylethylketone	0.15	—		85
Toluene–cyclohexane	2.4	—		70
Toluene–methylcyclohexane	—	27		90
Toluene–octane	0.15	101		40
Heptane–cyclohexane	1.2	165	95	85
	2.4	165		75
Propane–butane	—	—		100
Isobutane–n-butane	—	2070		110
Benzene–toluene	0.13	—	75	
Benzene–methanol	0.18	690	94	
Benzene–propanol	0.46	—	55	
Ethylbenzene–styrene	—	—	75	

E_{mV} = Murphree plate efficiency, E_o = overall column efficiency.

4. If the resistance is mainly in the vapour phase, as it normally will be, the difference between the binary and multicomponent efficiencies can be substantial.

The prediction of efficiencies for multicomponent systems is also discussed by Chan and Fair (1984b). For mixtures of dissimilar compounds, the efficiency can be very different from that predicted for each binary pair, and laboratory or pilot-plant studies should be made to confirm any predictions.

11.10.2. O'Connell's Correlation

A quick estimate of the overall column efficiency can be obtained from the correlation given by O'Connell (1946), which is shown in Figure 11.19. The overall column efficiency is correlated with the product of the relative volatility of the light key component (relative to the heavy key) and the molar average viscosity of the feed, estimated at the average column temperature. The correlation was based mainly on data obtained with hydrocarbon systems, but includes some values for chlorinated solvents and water–alcohol mixtures. It has been found to give reliable estimates of the overall column efficiency for hydrocarbon systems, and can be used to make an approximate estimate of the efficiency for other systems. The method takes no account of the plate design parameters, and includes only two physical property variables.

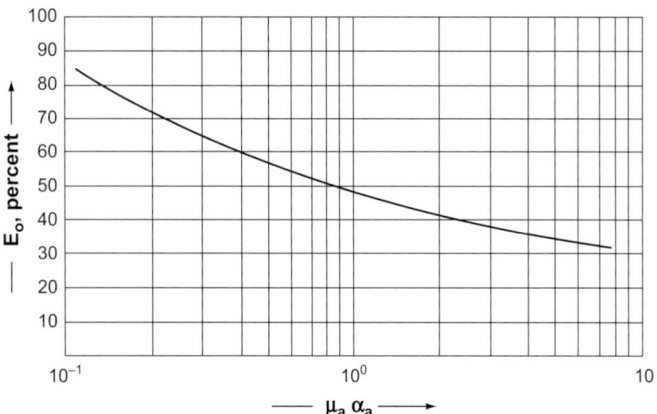

Figure 11.19. Distillation column efficiencies (bubble-caps) (after O'Connell, 1946).

Eduljee (1958) has expressed the O'Connell correlation in the form of an equation:

$$E_o = 51 - 32.5 \log (\mu_a \alpha_a) \qquad (11.34)$$

where

μ_a = the molar average liquid viscosity, mNs/m^2
α_a = average relative volatility of the light key.

Absorbers

O'Connell gave a similar correlation for the plate efficiency of absorbers (Figure 11.20). Appreciably lower plate efficiencies are obtained in absorption than in distillation.

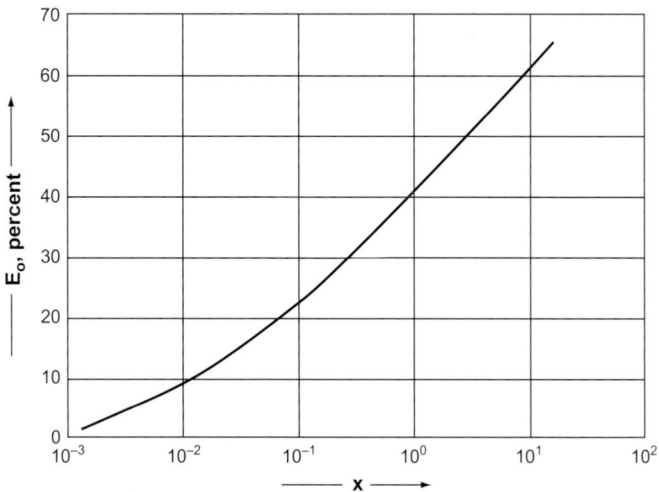

Figure 11.20. Absorber column efficiencies (bubble-caps) (after O'Connell, 1946).

In O'Connell's paper, the plate efficiency is correlated with a function involving Henry's constant, the total pressure, and the solvent viscosity at the operating temperature.

To convert the original data to SI units, it is convenient to express this function in the following form:

$$x = 0.062 \left[\frac{\rho_s P}{\mu_s H M_s} \right] = 0.062 \left[\frac{\rho_s}{\mu_s K M_s} \right] \tag{11.35}$$

where

H = the Henry's law constant, Nm^{-2}/mol fraction
P = total pressure, N/m^2
μ_s = solvent viscosity, mNs/m^2
M_s = molecular weight of the solvent
ρ_s = solvent density, kg/m^3
K = equilibrium constant for the solute.

Example 11.5

Using O'Connell's correlation, estimate the overall column efficiency and the number of real stages required for the separation given in Example 11.3, when the reflux ratio is 2.0.

Solution

From Example 11.3, feed composition, mol fractions:

propane 0.05, i-butane 0.15, n-butane 0.25, i-pentane 0.20, n-pentane 0.35.
Column: top temperature 65°C, bottom temperature 120°C.
Average relative volatility light key = 2.0. Take the viscosity at the average column temperature, 93°C,

viscosities:

propane = 0.03 mNs/m^2
butane = 0.12 mNs/m^2
pentane = 0.14 mNs/m^2.

For feed composition

$$\text{molar average viscosity} = 0.03 \times 0.05 + 0.12(0.15 + 0.25)$$
$$+ 0.14(0.20 + 0.35) = 0.13 \text{ mNs/m}^2$$
$$\alpha_a \mu_a = 2.0 \times 0.13 = 0.26$$

From Figure 11.19

$$E_o = \underline{\underline{70}}\%$$

From Example 11.3, when the reflux ratio is 2.0, number of ideal stages = 16, one ideal stage will be the reboiler, so number of actual stages (rounding up)

$$= \frac{(16 - 1)}{0.7} = \underline{\underline{22}}$$

11.10.3. Van Winkle's Correlation

Van Winkle *et al.* (1972) have published an empirical correlation for the plate efficiency that can be used to predict plate efficiencies for binary systems. Their correlation uses dimensionless groups that include those system variables and plate parameters that are known to affect plate efficiency. They give two equations, the simplest, and that which they consider the most accurate, is given below. The data used to derive the correlation covered both bubble-cap and sieve plates.

$$E_{mV} = 0.07 \mathrm{Dg}^{0.14} \mathrm{Sc}^{0.25} \mathrm{Re}^{0.08} \tag{11.36}$$

where

Dg = surface tension number = $(\sigma_L/\mu_L u_v)$
u_v = superficial vapour velocity
σ_L = liquid surface tension
μ_L = liquid viscosity
Sc = liquid Schmidt number = $(\mu_L/\rho_L D_{LK})$
ρ_L = liquid density
D_{LK} = liquid diffusivity, light key component
Re = Reynolds number = $(h_w u_v \rho_v/\mu_L(\mathrm{FA}))$
h_w = weir height
ρ_v = vapour density,

$$(\mathrm{FA}) = \text{fractional area} = \frac{(\text{area of holes or risers})}{(\text{total column cross-sectional area})}$$

The use of this method is illustrated in Example 11.8.

11.10.4. AIChE method

This method of predicting plate efficiency, published in 1958, was the result of a 5-year study of bubble-cap plate efficiency directed by the Research Committee of the American Institute of Chemical Engineers. The AIChE method is the most detailed method for predicting plate efficiencies that is available in the open literature. It takes into account all the major factors that are known to affect plate efficiency, including:

The mass transfer characteristics of the liquid and vapour phases
The design parameters of the plate
The vapour and liquid flow rates
The degree of mixing on the plate.

The method is well established, and in the absence of experimental values, or proprietary prediction methods, should be used when more than a rough estimate of efficiency is needed.

The approach taken is semi-empirical. Point efficiencies are estimated making use of the 'two-film theory', and the Murphree efficiency estimated allowing for the degree of mixing likely to be obtained on real plates.

The procedure and equations are given in this section without discussion of the theoretical basis of the method. The reader should refer to the AIChE manual, AIChE (1958), or to Smith (1963) who gives a comprehensive account of the method, and extends its use to sieve plates. Chan and Fair (1984a) published an alternative method for point efficiencies on sieve plates which they demonstrate gives closer predictions than the AIChE method. The Chan and Fair method follows the same overall methodology as the AIChE method but uses an improved correlation for vapour-phase mass transfer, given below.

AIChE method

The mass-transfer resistances in the vapour and liquid phases are expressed in terms of the number of transfer units, N_G and N_L. The point efficiency is related to the number of transfer units by the equation:

$$\frac{1}{\ln(1 - E_{mv})} = -\left[\frac{1}{N_G} + \frac{mV}{L} \times \frac{1}{N_L}\right] \tag{11.37}$$

where m is the slope of the equilibrium line, and V and L the vapour and liquid molar flow rates.

Equation 11.37 is plotted in Figure 11.21.

The number of gas-phase transfer units in the AIChE method is given by:

$$N_G = \frac{(0.776 + 4.57 \times 10^{-3}\, h_w - 0.24 F_v + 105 L_p)}{\left(\dfrac{\mu_v}{\rho_v\, D_v}\right)^{0.5}} \tag{11.38}$$

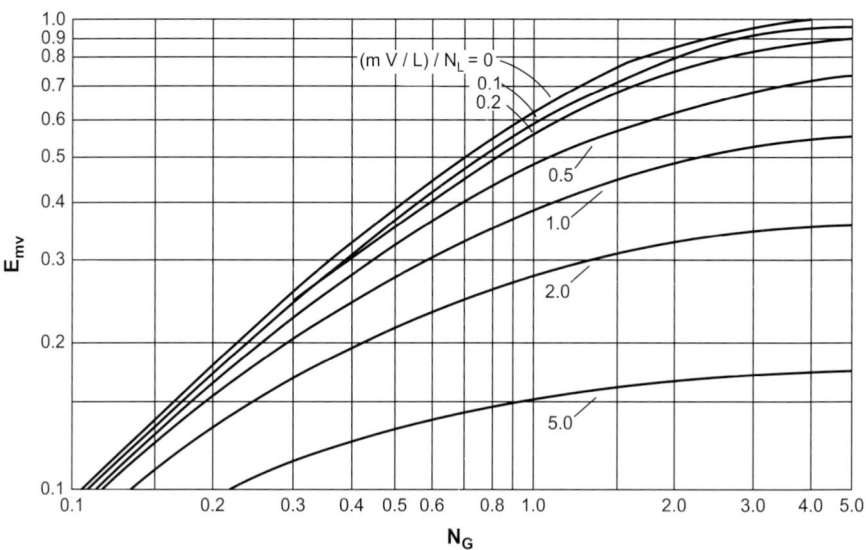

Figure 11.21.   Relationship between point efficiency and number of liquid and vapour transfer units (Equation 11.37).

where

h_w = weir height, mm

F_v = the column vapour 'F' factor = $u_a \, \rho_v^{0.5}$

u_a = vapour velocity based on the active tray area (bubbling area), see Section 11.13.2, m/s

L_p = the volumetric liquid flow rate across the plate, divided by the average width of the plate, m^3/sm. The average width can be calculated by dividing the active area by the length of the liquid path Z_L

μ_v = vapour viscosity, Ns/m^2

ρ_v = vapour density; kg/m^3

D_v = vapour diffusivity, m^2/s.

In the alternative method proposed by Chan and Fair (1984a), the number of gas-phase mass-transfer units is given by:

$$N_G = \frac{D_v^{0.5}(1030f - 867f^2)\, \overline{t_v}}{h_L^{0.5}} \tag{11.39}$$

where

h_L = liquid hold-up on tray, cm

$\overline{t_v}$ = average vapour residence time, s

$f = u_a/u_{af}$ = fractional approach to the vapour velocity based on active area at flooding, u_{af}.

The remainder of the Chan and Fair method is the same as the AIChE method. In both methods, the number of liquid phase transfer units is given by:

$$N_L = (4.13 \times 10^8 D_L)^{0.5}(0.21F_V + 0.15)t_L \tag{11.40}$$

where

D_L = liquid phase diffusivity, m^2/s

t_L = liquid contact time, s,

given by:

$$t_L = \frac{Z_c Z_L}{L_p} \tag{11.41}$$

where

Z_L = length of the liquid path, from inlet downcomer to outlet weir, m

Z_c = liquid hold-up on the plate, m^3 per m^2 active area,

given by:

for bubble-cap plates

$$Z_c = 0.042 + 0.19 \times 10^{-3}h_w - 0.014F_v + 2.5L_p \tag{11.42}$$

for sieve plates

$$Z_c = 0.006 + 0.73 \times 10^{-3}h_w - 0.24 \times 10 - 3F_v h_w + 1.22L_p \tag{11.43}$$

The Murphree efficiency E_{mV} is only equal to the point efficiency E_{mv} if the liquid on the plate is perfectly mixed. On a real plate this will not be so, and to estimate the plate efficiency from the point efficiency some means of estimating the degree of mixing is needed. The dimensionless Peclet number characterizes the degree of mixing in a system. For a plate the Peclet number is given by:

$$\text{Pe} = \frac{Z_L^2}{D_e \, t_L} \qquad (11.44)$$

where D_e is the 'eddy diffusivity', m^2/s.

A Peclet number of zero indicates perfect mixing and a value of ∞ indicates plug flow.

For bubble-cap and sieve plates the eddy diffusivity can be estimated from the equation:

$$D_e = (0.0038 + 0.017u_a + 3.86L_p + 0.18 \times 10^{-3}h_w)^2 \qquad (11.45)$$

The relation between the plate efficiency and point efficiency with the Peclet number as a parameter is shown in Figure 11.22a and b. The application of the AIChE method is illustrated in Example 11.7.

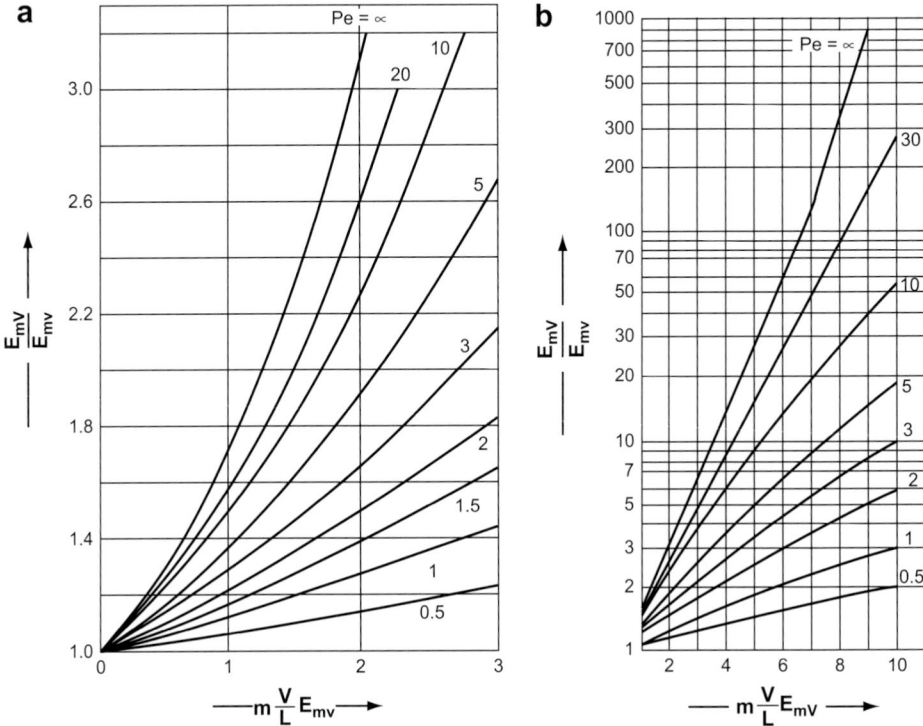

Figure 11.22. Relationship between plate and point efficiency.

Estimation of Physical Properties

To use the AIChE method or Van Winkle's correlation, estimates of the physical properties are required. It is unlikely that experimental values will be found in the literature for all systems that are of practical interest. The prediction methods given in Chapter 8, and in the references given in that chapter, can be used to estimate values.

The AIChE design manual recommends the Wilke and Chang (1955) equation for liquid diffusivities, and the Wilke and Lee (1955) modification to the Hirschfelder, Bird and Spotz equation for gas diffusivities.

Plate Design Parameters

The significance of the weir height in the AIChE equations should be noted. The weir height was the plate parameter found to have the most significant effect on plate efficiency. Increasing weir height will increase the plate efficiency, but at the expense of an increase in pressure drop and entrainment. Weir heights will normally be in the range 40 to 100 mm for columns operating at and above atmospheric pressure, but will be as low as 6 mm for vacuum columns. This, in part, accounts for the lower plate efficiencies obtained in vacuum columns.

The length of the liquid path Z_L is taken into account when assessing the plate mixing performance. The mixing correlation given in the AIChE method was not tested on large-diameter columns, and Smith (1963) states that the correlation should not be used for large-diameter plates; however, on a large plate the liquid path will normally be subdivided, and the value of Z_L will be similar to that in a small column. The assumption that the vapour space is well mixed across the tray may also not be valid for large column diameters.

The vapour 'F' factor F_v is a function of the active tray area. Increasing F_v decreases the number of gas-phase transfer units. The liquid flow term L_p is also a function of the active tray area, and the liquid path length. It will only have a significant effect on the number of transfer units if the path length is long. In practice, the range of values for F_v, the active area, and the path length will be limited by other plate design considerations.

Multicomponent Systems

The AIChE method was developed from measurements on binary systems. The AIChE manual should be consulted for advice on its application to multicomponent systems. See also the comments in Section 11.10.1.

11.10.5. Entrainment

The AIChE method, and that of Van Winkle, predicts the 'dry' Murphree plate efficiency. In operation some liquid droplets will be entrained and carried up the column by the vapour flow, and this will reduce the actual, operating, efficiency.

The dry-plate efficiency can be corrected for the effects of entrainment using the equation proposed by Colburn (1936):

$$E_a = \frac{E_{mV}}{1 + E_{mV}\left[\dfrac{\psi}{1 - \psi}\right]} \tag{11.46}$$

where

E_a = actual plate efficiency, allowing for entrainment

ψ = the fractional entrainment = $\dfrac{\text{entrained liquid}}{\text{gross liquid flow}}$.

A method for predicting the entrainment from sieve plates is given below in Section 11.13.5, Figure 11.36; a similar method for bubble-cap plates is given by Bolles (1963).

11.11. APPROXIMATE COLUMN SIZING

An approximate estimate of the overall column size can be made once the number of real stages required for the separation is known. This is often needed to make a rough estimate of the capital cost for project evaluation.

Plate Spacing

The overall height of the column will depend on the plate spacing. Plate spacings from 0.15 m (6 in) to 1 m (36 in) are normally used. The spacing chosen will depend on the column diameter and operating conditions. Close spacing is used with small-diameter columns and where head room is restricted, as it will be when a column is installed in a building. For columns above 1 m diameter, plate spacings of 0.3 to 0.6 m will normally be used, and 0.5 m (18 in) can be taken as an initial estimate. This would be revised, as necessary, when the detailed plate design is made.

A larger spacing will be needed between certain plates to accommodate feed and side-stream arrangements, and for manways.

Column Diameter

The principal factor that determines the column diameter is the vapour flow rate. The vapour velocity must be below that which would cause excessive liquid entrainment or a high pressure drop. The equation given below, which is based on the well known Souders and Brown equation (Lowenstein, 1961), can be used to estimate the maximum allowable superficial vapour velocity, and hence the column area and diameter,

$$\hat{u}_v = (-0.171 l_t^2 + 0.27 l_t - 0.047) \left[\frac{\rho_L - \rho_v}{\rho_v}\right]^{1/2} \qquad (11.47)$$

where

$\hat{u}_v$ = maximum allowable vapour velocity, based on the gross (total) column cross-sectional area, m/s

l_t = plate spacing, m (range 0.5–1.5).

The column diameter, D_c, can then be calculated:

$$D_c = \sqrt{\frac{4\hat{V}_w}{\pi \rho_v \hat{u}_v}} \qquad (11.48)$$

where $\hat{V}_w$ is the maximum vapour rate, kg/s.

This approximate estimate of the diameter would be revised when the detailed plate design is undertaken.

The column diameter estimated should then be rounded up to the nearest standard head size so that pre-formed heads can be used as vessel closures (see Section 13.5.2). The column sizing programs in most commercial process simulation programs use North American standard head sizes, which are available in 6-in (152.4-mm) increments.

11.12. PLATE CONTACTORS

Cross-flow plates are the most common type of plate contactor used in distillation and absorption columns. In a cross-flow plate the liquid flows across the plate and the vapour up through the plate. A typical layout is shown in Figure 11.23. The flowing liquid is transferred from plate to plate through vertical channels called 'down-comers'. A pool of liquid is retained on the plate by an outlet weir.

Other types of plate are used that have no downcomers (non-cross-flow plates), the liquid showering down the column through large openings in the plates (sometimes called shower plates). These, and, other proprietary non-cross-flow plates, are used for special purposes, particularly when a low pressure drop is required.

Four principal types of cross-flow tray are used, classified according to the method used to contact the vapour and liquid.

1. Sieve Plate (Perforated Plate) (Figure 11.24)
This is the simplest type of cross-flow plate. The vapour passes up through perforations in the plate, and the liquid is retained on the plate by the vapour flow. There is no positive vapour–liquid seal, and at low flow rates liquid will 'weep' through the holes, reducing the plate efficiency. The perforations are usually small holes, but larger holes and slots can be used.

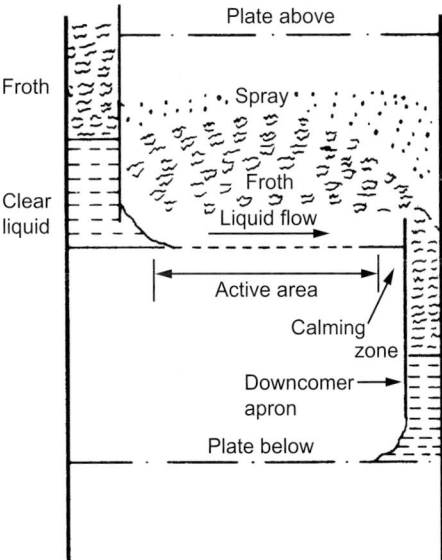

Figure 11.23. Typical cross-flow plate (sieve).

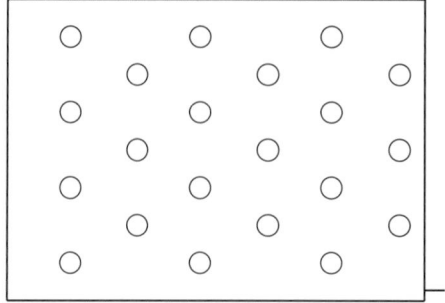

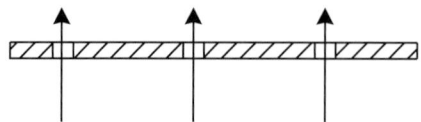

Figure 11.24.   Sieve plate.

2. Bubble-Cap Plates (Figure 11.25)
Bubble-cap plates are plates in which the vapour passes up through short pipes, called risers, covered by a cap with a serrated edge, or slots. The bubble-cap plate is the traditional, oldest, type of cross-flow plate, and many different designs have been developed. Standard cap designs would now be specified for most applications.

The most significant feature of the bubble-cap plate is that the use of risers ensures that a level of liquid is maintained on the tray at all vapour flow rates. Bubble-caps therefore have good turn-down performance at low flow rates. They are more expensive than sieve plates and more prone to corrosion, fouling and plugging, and so are usually only found on older columns.

3. Valve Plates (Floating-Cap Plates) (Figure 11.26)
Valve plates are proprietary designs. They are essentially sieve plates with large-diameter holes covered by movable flaps, which lift as the vapour flow increases.

As the area for vapour flow varies with the flow rate, valve plates can operate efficiently at lower flow rates than sieve plates, the valves closing at low vapour rates. The cost of valve plates is intermediate between sieve plates and bubble-cap plates.

Some very elaborate valve designs have been developed, but the simple type shown in Figure 11.26 is satisfactory for most applications.

4. Valve Plates (Fixed Valve Plates) (Figure 11.27)
A fixed valve plate is similar to a sieve plate, except the holes are only partially punched out, so that the hole remains partially covered, as shown in Figure 11.27. Fixed valve trays are almost as inexpensive as sieve trays and have improved turn-down performance. The relatively small cost difference between fixed valve trays and sieve trays can usually be justified by the improved turn-down performance, and fixed valve trays are the most common type specified in non-fouling applications.

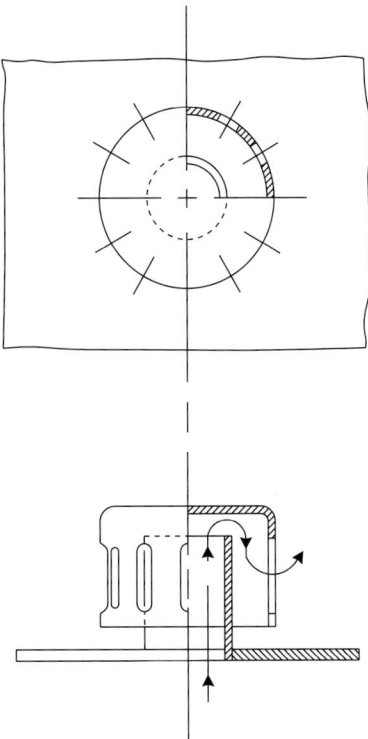

Figure 11.25. Bubble-cap.

Many different proprietary designs of fixed and floating valves have been developed. Performance details can be obtained from the tray vendors.

Liquid Flow Pattern

Cross-flow trays are also classified according to the number of liquid passes on the plate. The design shown in Figure 11.28a is a single-pass plate. For low liquid flow rates reverse flow plates are used (Figure 11.28b). In this type the plate is divided by a low central partition, and inlet and outlet downcomers are on the same side of the plate. Multiple-pass plates, in which the liquid stream is sub-divided by using several downcomers, are used for high liquid flow rates and large-diameter columns. A double-pass plate is shown in Figure 11.28c.

Selection of the liquid flow pattern is discussed in Section 11.13.4. An approximate criterion for selecting the liquid flow pattern is the liquid volumetric flow rate per unit weir length, which should ideally be in the range 5 to 8 L/s per m (2 to 3 gpm/in). Weir length is discussed in more detail in Section 11.13.8.

11.12.1. Selection of Plate Type

The principal factors to consider when comparing the performance of bubble-cap, sieve and valve plates are: cost, capacity, operating range, efficiency and pressure drop.

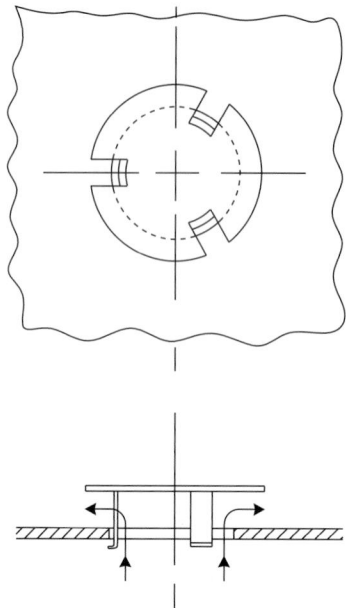

Figure 11.26.   Simple valve.

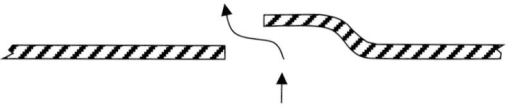

Figure 11.27.   Fixed valve.

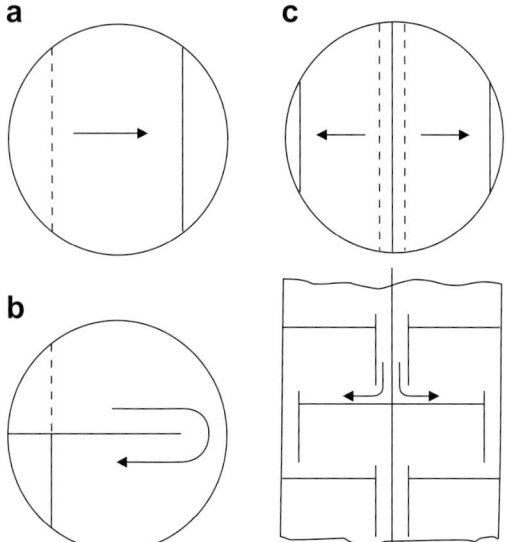

Figure 11.28. Liquid flow patterns on cross-flow trays. (*a*) Single pass. (*b*) Reverse flow. (*c*) Double pass.

Cost. Bubble-cap plates are appreciably more expensive than sieve or valve plates. The relative cost will depend on the material of construction used; for mild steel, the ratios bubble-cap:valve:fixed valve:sieve are approximately 3.0:1.2:1.1:1.0.

Capacity. There is little difference in the capacity rating of the three types (the diameter of the column required for a given flow rate); the ranking is sieve, valve, bubble-cap.

Operating range. This is the most significant factor. By operating range is meant the range of vapour and liquid rates over which the plate will operate satisfactorily (the stable operating range). Some flexibility will always be required in an operating plant to allow for changes in production rate, and to cover start-up and shutdown conditions. The ratio of the highest to the lowest flow rates is often referred to as the 'turn-down' ratio. Bubble-cap plates have a positive liquid seal and can therefore operate efficiently at very low vapour rates.

Sieve plates and fixed valve plates rely on the flow of vapour through the holes to hold the liquid on the plate, and cannot operate at very low vapour rates. With good design, sieve plates can give a satisfactory operating range; typically, from 50% to 120% of design capacity. Fixed valve plates have somewhat better turn-down performance. Valve plates are intended to give greater flexibility than sieve plates at a lower cost than bubble-caps.

Efficiency. The Murphree efficiency of the three types of plate will be virtually the same when operating over their design flow range, and no real distinction can be made between them; see Zuiderweg *et al.* (1960).

Pressure drop. The pressure drop over the plates can be an important design consideration, particularly for vacuum columns. The plate pressure drop will depend on the detailed design of the plate but, in general, sieve plates give the lowest pressure drop, followed by valves, with bubble-caps giving the highest.

Summary. Sieve plates are the cheapest and least prone to fouling and are satisfactory for most applications. Fixed valve plates are almost as cheap as sieve plates and have improved turn-down behaviour. The improved performance usually justifies the increased cost and this type is most commonly selected for non-fouling applications. Moving valve plates should be considered if the specified turn-down ratio cannot be met with sieve plates or fixed valve plates. Bubble-caps should only be used where very low vapour (gas) rates have to be handled and a positive liquid seal is essential at all flow rates.

11.12.2. Plate Construction

The mechanical design features of sieve plates are described in this section. The same general construction is also used for bubble-cap and valve plates. Details of the various types of bubble-cap used, and the preferred dimensions of standard cap designs, can be found in the books by Smith (1963) and Ludwig (1997). The manufacturers' design manuals should be consulted for details of valve plate design.

Two different types of plate construction are used. Large-diameter plates are normally constructed in sections, supported on beams. Small plates are installed in the column as a stack of pre-assembled plates.

Sectional Construction

A typical plate is shown in Figure 11.29. The plate sections are supported on a ring welded round the vessel wall, and on beams. The beams and ring are about 50 mm wide, with the beams set at around 0.6 m spacing. The beams are usually angle or channel sections, constructed from folded sheet. Special fasteners are used so the sections can be assembled from one side only. One section is designed to be removable to act as a manway. This reduces the number of manways needed on the vessel, which reduces the vessel cost.

Stacked Plates (Cartridge Plates)

The stacked type of construction is used where the column diameter is too small for a worker to enter to assemble the plates, say less than 1.2 m (4 ft). Each plate is fabricated complete with the downcomer, and joined to the plate above and below using screwed rods (spacers); see Figure 11.30. The plates are installed in the column shell as an assembly (stack) of 10, or so, plates. Tall columns have to be divided into flanged sections so that plate assemblies can be easily installed and removed. The weir and downcomer supports are usually formed by turning up the edge of the plate.

The plates are not fixed to the vessel wall, as they are with sectional plates, so there is no positive liquid seal at the edge of the plate and a small amount of leakage will occur. In some designs the plate edges are turned up round the circumference to make better contact at the wall. This can make it difficult to remove the plates for cleaning and maintenance, without damage.

Downcomers

The segmental or chord downcomer, shown in Figure 11.31a, is the simplest and cheapest form of construction and is satisfactory for most purposes. The downcomer

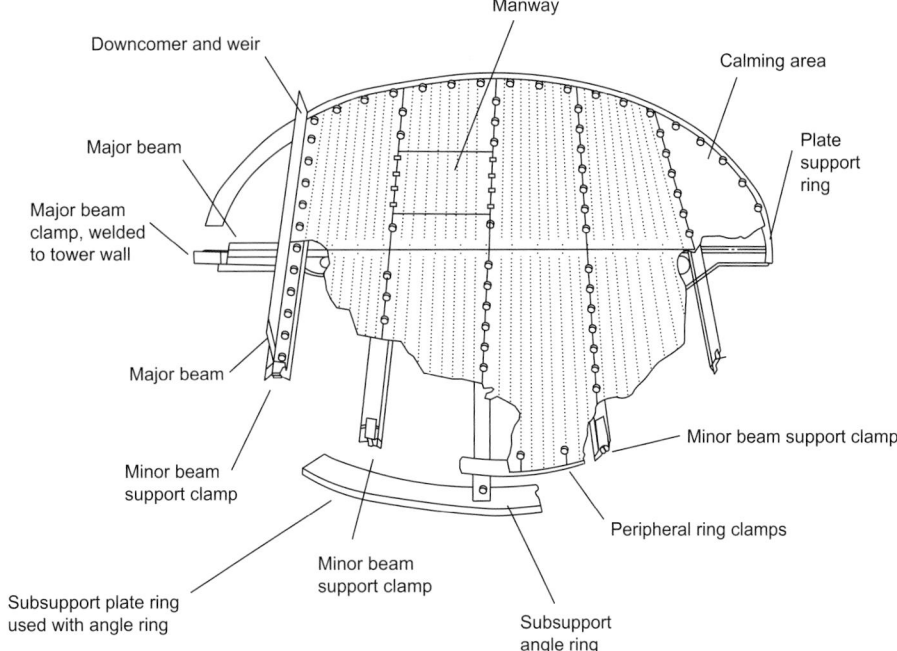

Figure 11.29. Typical sectional plate construction.

channel is formed by a flat plate, called an apron, which extends down from the outlet weir. The apron is usually vertical, but may be sloped (Figure 11.31b) to increase the plate area available for perforation. This design is commonly in high-capacity trays. If a more positive seal is required at the downcomer at the outlet, an inlet weir can be fitted (Figure 11.31c) or a recessed seal pan used (Figure 11.31d). Circular downcomers (pipes) are sometimes used for small liquid flow rates. Curved downcomers are often used in high-capacity trays for large columns. Truncated downcomers (Figure 11.31e) can be used to increase the plate area available for perforation and are also commonly used for high-capacity trays.

Side-Stream and Feed Points

Where a side stream is withdrawn from the column the plate design must be modified to provide a liquid seal at the take-off pipe. A typical design is shown in Figure 11.32a. Side-draw pipes and run-down lines must be sized for self-venting flow, and provision must be made for vapour to vent from the line in case vapour is entrained from the column or formed by flashing in the line. Sewell (1975) gives a correlation for the minimum pipe diameter that will allow self-venting flow.

When the feed stream is liquid, it will be normally introduced into the downcomer leading to the feed plate, and the plate spacing increased at this point; Figure 11.32b. This design should not be used if the feed is at the bubble point or is two-phase, as the feed may flash on entering the column, in which case downcomer flooding could occur.

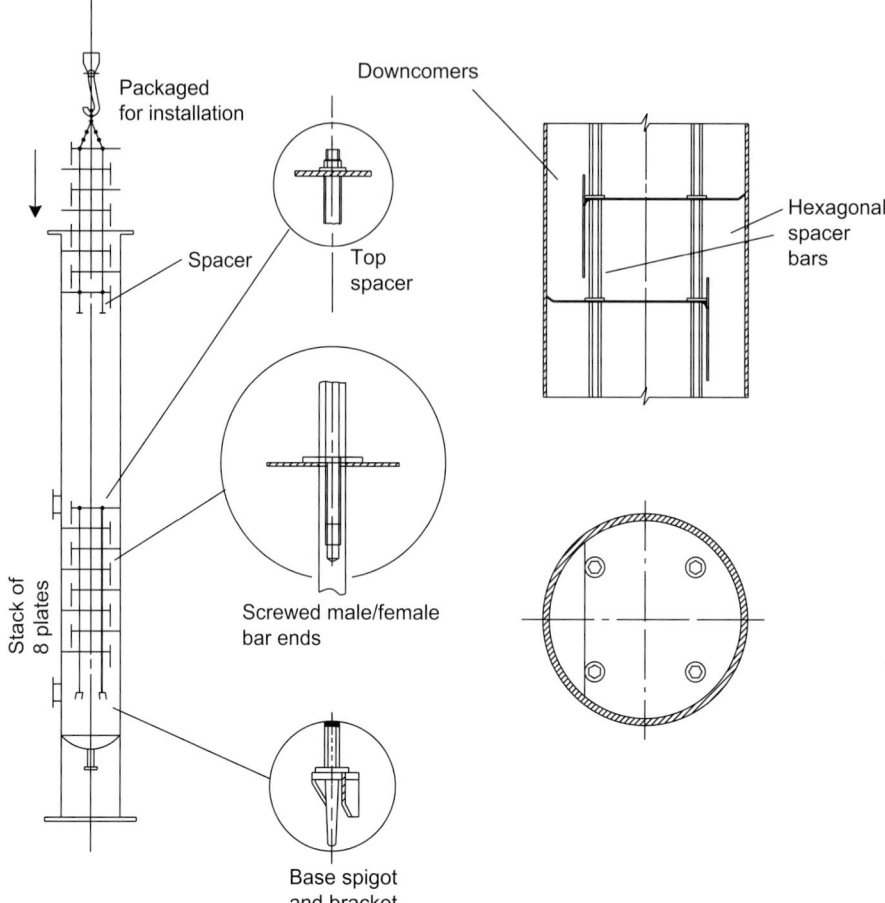

Figure 11.30.   Typical stacked-plate construction.

Structural Design

The plate structure must be designed to support the hydraulic loads on the plate during operation, and the loads imposed during construction and maintenance. Typical design values used for these loads are:

Hydraulic load: $600 \, \text{N/m}^2$ live load on the plate, plus $3000 \, \text{N/m}^2$ over the downcomer seal area.

Erection and maintenance: 1500 N concentrated load on any structural member.

It is important to set close tolerances on the weir height, downcomer clearance and plate flatness, to ensure an even flow of liquid across the plate. The tolerances specified will depend on the dimensions of the plate but will typically be about 3 mm.

The plate deflection under load is also important, and will normally be specified as not greater than 3 mm under the operating conditions for plates greater than 2.5 m, and proportionally less for smaller diameters.

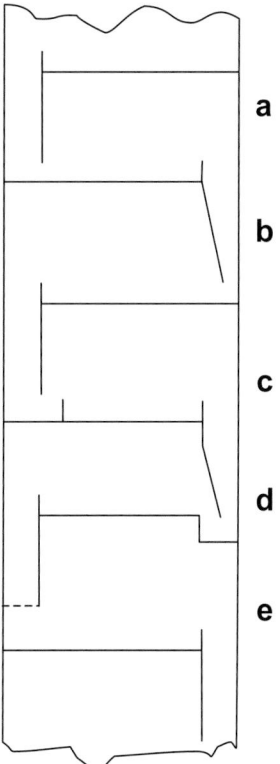

Figure 11.31.   Segment (chord) downcomer designs. (a) Vertical apron. (b) Inclined apron. (c) Inlet weir. (d) Recessed well. (e) Truncated downcomer.

The mechanical specification of bubble-cap, sieve and valve plates is covered in a series of articles by Glitsch (1960), McClain (1960), Thrift (1960a,b) and Patton and Pritchard (1960).

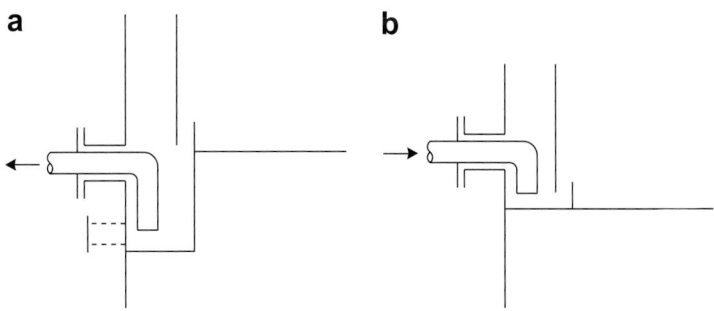

Figure 11.32.   Feed and take-off nozzles.

11.13.   PLATE HYDRAULIC DESIGN

The basic requirements of a plate contacting stage are that it should:

> Provide good vapour–liquid contact
> Provide sufficient liquid hold-up for good mass transfer (high efficiency)
> Have sufficient area and spacing to keep the entrainment and pressure drop within acceptable limits
> Have sufficient downcomer area for the liquid to flow freely from plate to plate.

Plate design, like most engineering design, is a combination of theory and practice. The design methods use semi-empirical correlations derived from fundamental research work combined with practical experience obtained from the operation of commercial columns. Proven layouts are used, and the plate dimensions are kept within the range of values known to give satisfactory performance.

A short procedure for the hydraulic design of sieve plates is given in this section. Design methods for bubble-cap plates are given by Bolles (1963) and Ludwig (1997). Valve plates are proprietary designs and will be designed in consultation with the vendors. Design manuals are available from some vendors.

A detailed discussion of the extensive literature on plate design and performance will not be given in this volume. Chase (1967) and Zuiderweg (1982) give critical reviews of the literature on sieve plates.

Several design methods have been published for sieve plates: Kister (1992), Barnicki and Davies (1989), Koch and Kuzniar (1966), Fair (1963), and Huang and Hodson (1958); see also the book by Lockett (1986).

Operating Range

Satisfactory operation will only be achieved over a limited range of vapour and liquid flow rates. A typical performance diagram for a sieve plate is shown in Figure 11.33.

The upper limit to vapour flow is set by the condition of flooding. At flooding there is a sharp drop in plate efficiency and increase in pressure drop. Flooding is caused by either the excessive carryover of liquid to the next plate by entrainment (entrainment or jet flooding), or by liquid backing up in the downcomers.

The lower limit of the vapour flow is set by the condition of weeping. Weeping occurs when the vapour flow is insufficient to maintain a level of liquid on the plate. 'Coning' occurs at low liquid rates, and is the term given to the condition where the vapour pushes the liquid back from the holes and jets upward, with poor liquid contact.

In the following sections, gas can be taken as synonymous with vapour when applying the method to the design of plates for absorption columns.

11.13.1.   Plate Design Procedure

A trial-and-error approach is necessary in plate design: starting with a rough plate layout, checking key performance factors and revising the design, as necessary, until a satisfactory design is achieved. A typical design procedure is set out below and

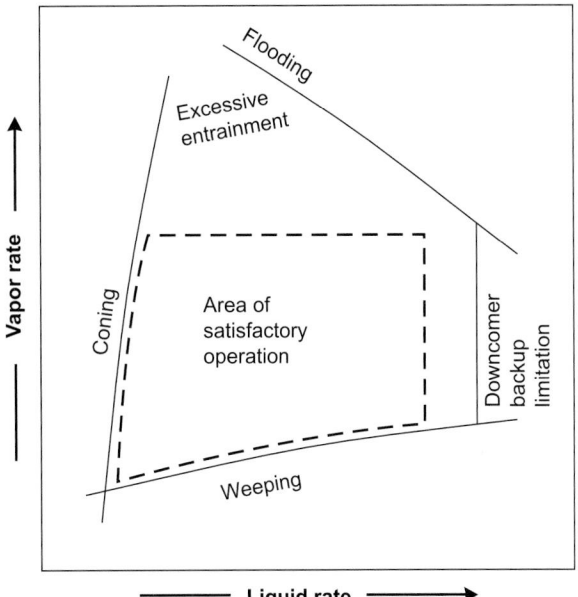

Figure 11.33. Sieve-plate performance diagram.

discussed in the following sections. The normal range of each design variable is given in the discussion, together with recommended values that can be used to start the design.

Most of the commercial process simulation programs offer tray design modules. These programs can be used for preliminary tray layout for costing purposes, but the default dimensions selected or calculated by these programs often do not give the best performance over the intended range of operation. An experienced designer will run several cases to confirm the tray performance is satisfactory over the whole range of operation. Hand calculations using the methods given in this section can also be used to guide the process simulation programs to a better design.

Procedure

1. Calculate the maximum and minimum vapour and liquid flow rates, for the turn-down ratio required.
2. Collect, or estimate, the system physical properties.
3. Select a trial plate spacing (Section 11.11).
4. Estimate the column diameter, based on flooding considerations (Section 11.13.3).
5. Decide the liquid flow arrangement (Section 11.13.4).
6. Make a trial plate layout: downcomer area, active area, hole area, hole size, weir height (Sections 11.13.8 to 11.13.10).
7. Check the weeping rate (Section 11.13.6), if unsatisfactory return to step 6.
8. Check the plate pressure drop (Section 11.13.14), if too high return to step 6.
9. Check downcomer back-up, if too high return to step 6 or 3 (Section 11.13.15).

10. Decide plate layout details: calming zones, unperforated areas. Check hole pitch, if unsatisfactory return to step 6 (Section 11.13.11).
11. Recalculate the percentage flooding based on chosen column diameter.
12. Check entrainment, if too high return to step 4 (Section 11.13.5).
13. Optimize design: repeat steps 3 to 12 to find smallest diameter and plate spacing acceptable (lowest cost).
14. Finalize design: draw up the plate specification and sketch the layout.

This procedure is illustrated in Example 11.6.

11.13.2. Plate Areas

The following area terms are used in the plate design procedure:

A_c = total column cross-sectional area
A_d = cross-sectional area of downcomer
A_n = net area available for vapour–liquid disengagement, normally equal to $A_c - A_d$ for a single-pass plate
A_a = active, or bubbling, area, equal to $A_c - 2A_d$ for single-pass plates
A_h = hole area, the total area of all the active holes
A_p = perforated area (including blanked areas)
A_{ap} = the clearance area under the downcomer apron.

11.13.3. Diameter

The flooding condition fixes the upper limit of vapour velocity. A high vapour velocity is needed for high plate efficiencies, and the velocity will normally be between 70 and 90% of that which would cause flooding. For design, a value of 80 to 85% of the flooding velocity should be used.

The flooding velocity can be estimated from the correlation given by Fair (1961):

$$u_f = K_1 \sqrt{\frac{\rho_L - \rho_v}{\rho_v}} \tag{11.49}$$

where

u_f = flooding vapour velocity, m/s, based on the net column cross-sectional area A_n (see Section 11.13.2)
K_1 = a constant obtained from Figure 11.34.

The liquid–vapour flow factor F_{LV} in Figure 11.34 is given by:

$$F_{LV} = \frac{L_w}{V_w} \sqrt{\frac{\rho_v}{\rho_L}} \tag{11.50}$$

where

L_w = liquid mass flow rate, kg/s
V_w = vapour mass flow rate, kg/s.

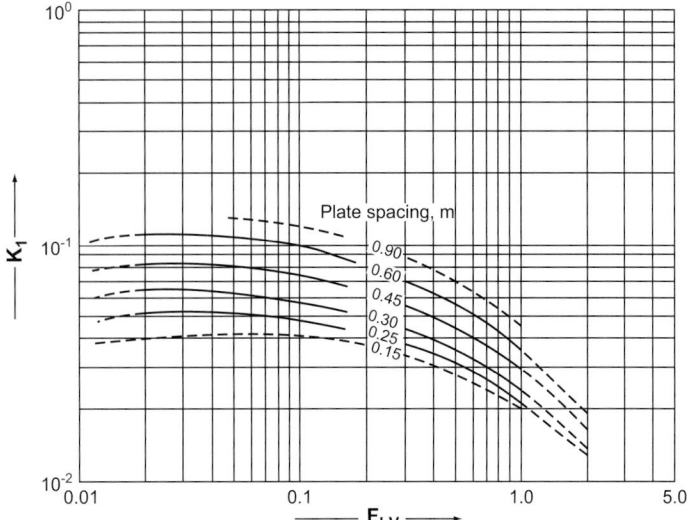

Figure 11.34. Flooding velocity, sieve plates.

The following restrictions apply to the use of Figure 11.34:

1. Hole size less than 6.5 mm. Entrainment may be greater with larger hole sizes
2. Weir height less than 15% of the plate spacing
3. Non-foaming systems
4. Hole: active area ratio greater than 0.10; for other ratios apply the following corrections:

hole: active area	multiply K_1 by
0.10	1.0
0.08	0.9
0.06	0.8

5. Liquid surface tension 0.02 N/m; for other surface tensions σ, multiply the value of K_1 by $[\sigma/0.02]^{0.2}$.

To calculate the column diameter an estimate of the net area A_n is required. As a first trial take the downcomer area as 12% of the total, and assume that the hole–active area is 10%.

Where the vapour and liquid flow rates, or physical properties, vary significantly throughout the column a plate design should be made for several points up the column. For distillation it will usually be sufficient to design for the conditions above and below the feed points. Changes in the vapour flow rate will normally be accommodated by adjusting the hole area, often by blanking off some rows of holes. Different column diameters would only be used where there is a considerable change in flow rate. Changes in liquid rate can be allowed for by adjusting the liquid downcomer areas.

11.13.4.   Liquid Flow Arrangement

The choice of plate type (reverse, single pass or multiple pass) will depend on the liquid flow rate and column diameter. An initial selection can be made using Figure 11.35, which has been adapted from a similar figure given by Huang and Hodson (1958).

11.13.5.   Entrainment

Entrainment can be estimated from the correlation given by Fair (1961) (Figure 11.36), which gives the fractional entrainment ψ (kg/kg gross liquid flow) as a function of the liquid–vapour factor F_{LV}, with the percentage approach to flooding as a parameter.

The percentage flooding is given by:

$$\text{percentage flooding} = \frac{u_n \ (\text{actual velocity based on net area})}{u_f(\text{from equation 11.49})} \qquad (11.51)$$

The effect of entrainment on plate efficiency can be estimated using equation 11.46.

As a rough guide the upper limit of ψ can be taken as 0.1; below this figure the effect on efficiency will be small. The optimum design value may be above this figure, see Fair (1963).

11.13.6.   Weep Point

The lower limit of the operating range occurs when liquid leakage through the plate holes becomes excessive. This is known as the weep point. The vapour velocity at the weep point is the minimum value for stable operation. The hole area must be chosen

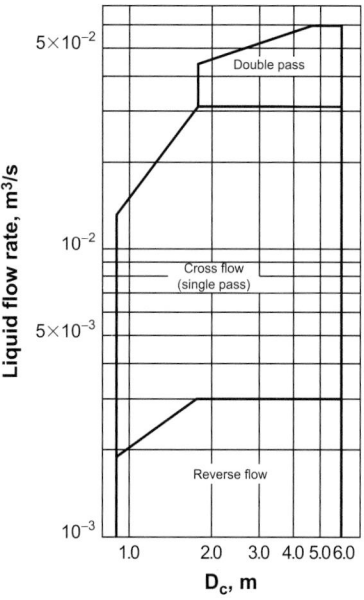

Figure 11.35.   Selection of liquid flow arrangement.

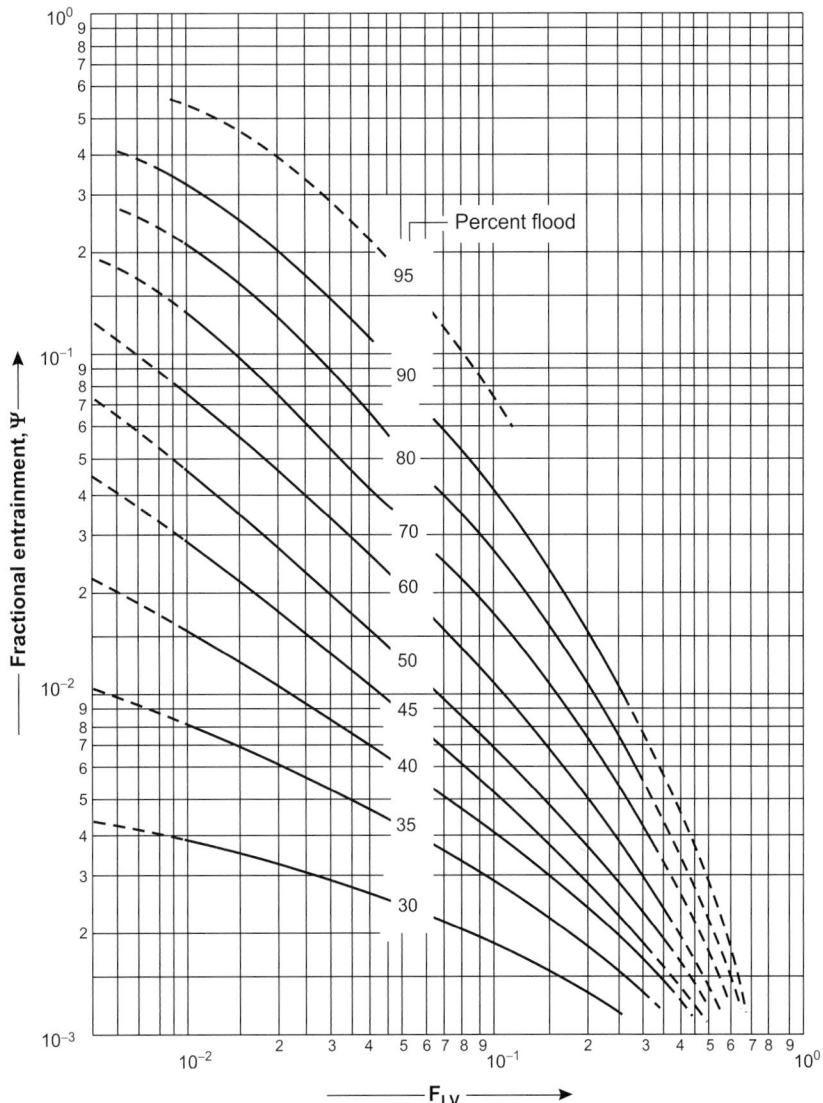

Figure 11.36. Entrainment correlation for sieve plates (Fair, 1961).

so that at the lowest operating rate the vapour flow velocity is still well above the weep point.

Several correlations have been proposed for predicting the vapour velocity at the weep point; see Chase (1967). That given by Eduljee (1959) is one of the simplest to use, and has been shown to be reliable.

The minimum design vapour velocity is given by:

$$u_h = \frac{[K_2 - 0.90(25.4 - d_h)]}{(\rho_v)^{1/2}} \qquad (11.52)$$

where

u_h = minimum vapour velocity through the holes (based on the hole area), m/s
d_h = hole diameter, mm
K_2 = a constant, dependent on the depth of clear liquid on the plate, obtained from Figure 11.37.

The clear liquid depth is equal to the height of the weir h_w plus the depth of the crest of liquid over the weir h_{ow}; this is discussed in the next section.

11.13.7. Weir Liquid Crest

The height of the liquid crest over the weir can be estimated using the Francis weir formula (see Coulson *et al.*, 1999). For a segmental downcomer this can be written as:

$$h_{ow} = 750 \left[\frac{L_w}{\rho_L l_w} \right]^{2/3} \tag{11.53}$$

where

l_w = weir length, m
h_{ow} = weir crest, mm liquid
L_w = liquid flow rate, kg/s.

With segmental downcomers the column wall constricts the liquid flow, and the weir crest will be higher than that predicted by the Francis formula for flow over an open weir. The constant in equation 11.53 has been increased to allow for this effect.

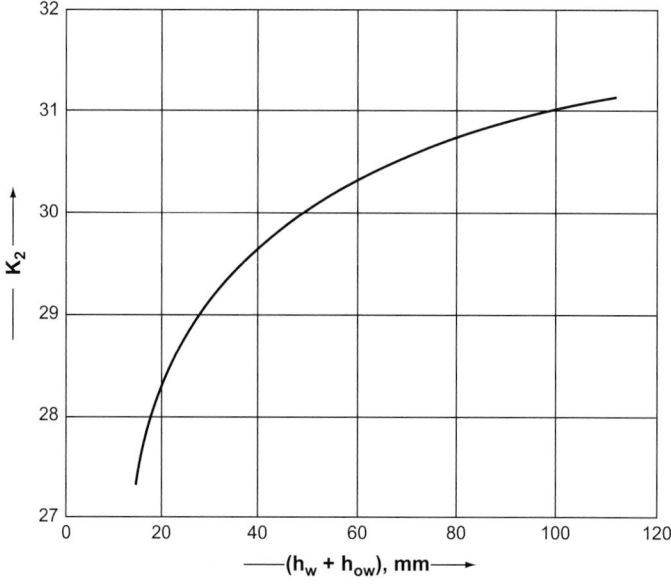

Figure 11.37.   Weep-point correlation (Eduljee, 1959).

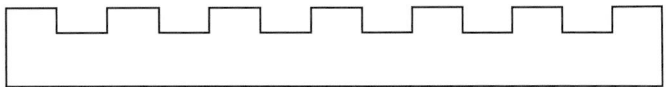

Figure 11.38. Picket-fence weir.

To ensure an even flow of liquid along the weir, the crest should be at least 10 mm at the lowest liquid rate. Serrated weirs known as picket-fence weirs are sometimes used for very low liquid rates, as illustrated in Figure 11.38.

11.13.8. Weir Dimensions

Weir Height

The height of the weir determines the volume of liquid on the plate and is an important factor in determining the plate efficiency (see Section 11.10.4). A high weir will increase the plate efficiency but at the expense of a higher plate pressure drop. For columns operating above atmospheric pressure, the weir heights will normally be between 40 mm and 90 mm (1.5 and 3.5 in); 40 to 50 mm is recommended. For vacuum operation lower weir heights are used to reduce the pressure drop; 6 to 12 mm (¼ to ½ in) is recommended.

Inlet Weirs

Inlet weirs, or recessed pans, are sometimes used to improve the distribution of liquid across the plate; but are seldom needed with segmental downcomers.

Weir Length

With segmental downcomers the length of the weir fixes the area of the downcomer. The chord length will normally be between 0.6 and 0.85 of the column diameter. A good initial value to use is 0.77, equivalent to a downcomer area of 12%. The liquid flow rate over the weir should ideally be in the range 5 to 8 L/s per m (2 to 3 gpm/in). If this is not feasible with a single-pass tray then reverse flow or multiple pass trays should be considered, as illustrated in Figure 11.28.

The relationship between weir length and downcomer area for segmental downcomers is given in Figure 11.39.

For double-pass plates the width of the central downcomer is normally 200 to 250 mm (8 to 10 in).

11.13.9. Perforated Area

The area available for perforation will be reduced by the obstruction caused by structural members (the support rings and beams), and by the use of calming zones.

Calming zones are unperforated strips of plate at the inlet and outlet sides of the plate. The width of each zone is usually made the same. Recommended values are: below 1.5 m diameter, 75 mm; above, 100 mm.

The width of the support ring for sectional plates will normally be 50 to 75 mm: the support ring should not extend into the downcomer area. A strip of unperforated plate will be left round the edge of cartridge-type trays to stiffen the plate.

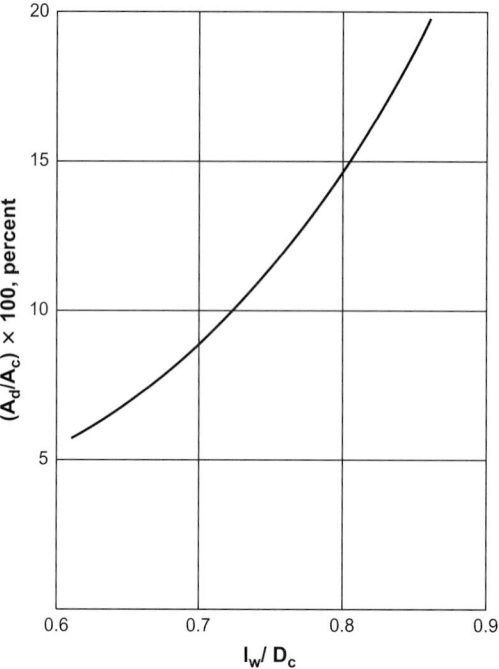

Figure 11.39.   Relation between downcomer area and weir length.

The unperforated area can be calculated from the plate geometry. The relationship between the weir chord length, chord height and the angle subtended by the chord is given in Figure 11.40.

11.13.10. Hole Size

The hole sizes used vary from 2.5 to 19 mm; 5 mm is the preferred size for non-fouling applications. Larger holes are recommended for fouling systems. The holes are drilled or punched. Punching is cheaper, but the minimum size of hole that can be punched will depend on the plate thickness. For carbon steel, hole sizes approximately equal to the plate thickness can be punched, but for stainless steel the minimum hole size that can be punched is about twice the plate thickness. Typical plate thicknesses used are: 5 mm (3/16 in) for carbon steel, and 3 mm (12 gauge) for stainless steel.

When punched plates are used, they should be installed with the direction of punching upward. Punching forms a slight nozzle, and reversing the plate will increase the pressure drop.

11.13.11. Hole Pitch

The hole pitch (distance between the hole centres) l_p should not be less than 2.0 hole diameters, and the normal range will be 2.5 to 4.0 diameters. Within this range, the pitch can be selected to give the number of active holes required for the total hole area specified.

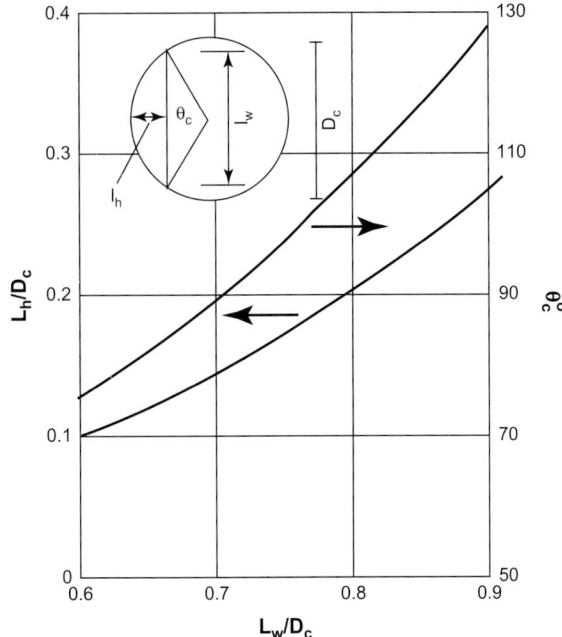

Figure 11.40. Relation between angle subtended by chord, chord height and chord length.

Square and equilateral triangular patterns are used; triangular is preferred. The total hole area as a fraction of the perforated area A_p is given by the following expression, for an equilateral triangular pitch:

$$\frac{A_h}{A_p} = 0.9 \left[\frac{d_h}{l_p}\right]^2 \qquad (11.54)$$

This equation is plotted in Figure 11.41.

11.13.12. Hydraulic Gradient

The hydraulic gradient is the difference in liquid level needed to drive the liquid flow across the plate. On sieve plates, unlike bubble-cap plates, the resistance to liquid flow will be small, and the hydraulic gradient is usually ignored in sieve-plate design. It can be significant in vacuum operation, as with the low weir heights used the hydraulic gradient can be a significant fraction of the total liquid depth. Methods for estimating the hydraulic gradient are given by Fair (1963).

11.13.13. Liquid Throw

The liquid throw is the horizontal distance travelled by the liquid stream flowing over the downcomer weir. It is only an important consideration in the design of multiple-pass plates. Bolles (1963) gives a method for estimating the liquid throw. If the liquid throw is excessive then anti-jump baffles can be used to ensure that liquid flows down and does not jump to the adjacent section.

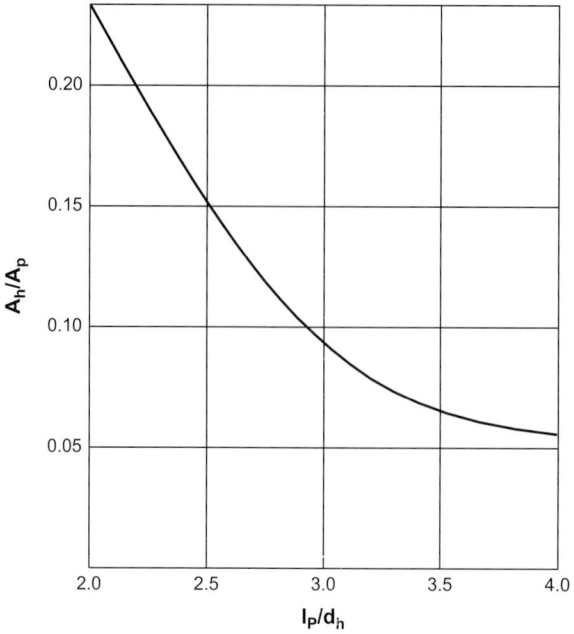

Figure 11.41.   Relation between hole area and pitch.

11.13.14. Plate Pressure Drop

The pressure drop over the plates is an important design consideration. There are two main sources of pressure loss: that due to vapour flow through the holes (an orifice loss), and that due to the static head of liquid on the plate.

A simple additive model is normally used to predict the total pressure drop. The total is taken as the sum of the pressure drop calculated for the flow of vapour through the dry plate (the dry-plate drop h_d), the head of clear liquid on the plate ($h_w + h_{ow}$), and a term to account for other, minor, sources of pressure loss, the so-called residual loss h_r. The residual loss is the difference between the observed experimental pressure drop and the simple sum of the dry-plate drop and the clear-liquid height. It accounts for the two effects: the energy to form the vapour bubbles and the fact that on an operating plate the liquid head will not be clear liquid but a head of 'aerated' liquid froth, and the froth density and height will be different from that of the clear liquid.

It is convenient to express the pressure drops in terms of millimetres of liquid. In pressure units:

$$\Delta P_t = 9.81 \times 10^{-3} h_t \rho_L \qquad (11.55)$$

where

ΔP_t = total plate pressure drop, Pa (N/m^2)
h_t = total plate pressure drop, mm liquid.

Dry-Plate Drop

The pressure drop through the dry plate can be estimated using expressions derived for flow through orifices.

$$h_d = 51 \left[\frac{u_h}{C_0}\right]^2 \frac{\rho_v}{\rho_L} \tag{11.56}$$

where the orifice coefficient C_0 is a function of the plate thickness, hole diameter, and the hole to perforated area ratio. C_0 can be obtained from Figure 11.42; which has been adapted from a similar figure by Liebson *et al.* (1957). u_h is the velocity through the holes, m/s.

Residual Head

Methods have been proposed for estimating the residual head as a function of liquid surface tension, froth density and froth height; however, as this correction term is

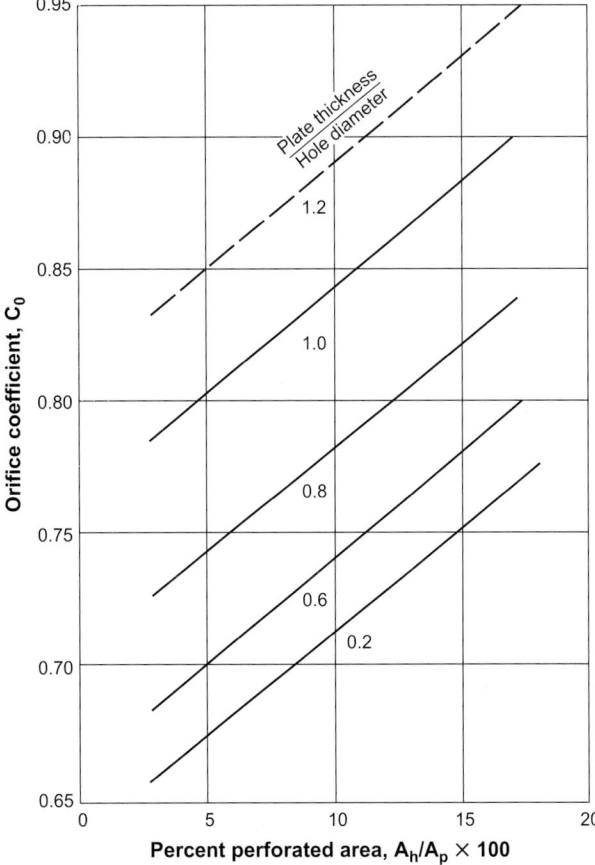

Figure 11.42. Discharge coefficient, sieve plates (Liebson *et al.*, 1957).

small, the use of an elaborate method for its estimation is not justified and the simple equation proposed by Hunt *et al.* (1955) can be used:

$$h_r = \frac{12.5 \times 10^3}{\rho_L} \tag{11.57}$$

Equation 11.57 is equivalent to taking the residual drop as a fixed value of 12.5 mm of water (½ in).

Total Drop

The total plate drop is given by:

$$h_t = h_d + (h_w + h_{ow}) + h_r \tag{11.58}$$

If the hydraulic gradient is significant, half its value is added to the clear liquid height.

11.13.15.   Downcomer Design (Back-Up)

The downcomer area and plate spacing must be such that the level of the liquid and froth in the downcomer is well below the top of the outlet weir on the plate above. If the level rises above the outlet weir the column will flood.

The back-up of liquid in the downcomer is caused by the pressure drop over the plate (the downcomer in effect forms one leg of a U-tube) and the resistance to flow in the downcomer itself; see Figure 11.43.

In terms of clear liquid, the downcomer back-up is given by:

$$h_b = (h_w + h_{ow}) + h_t + h_{dc} \tag{11.59}$$

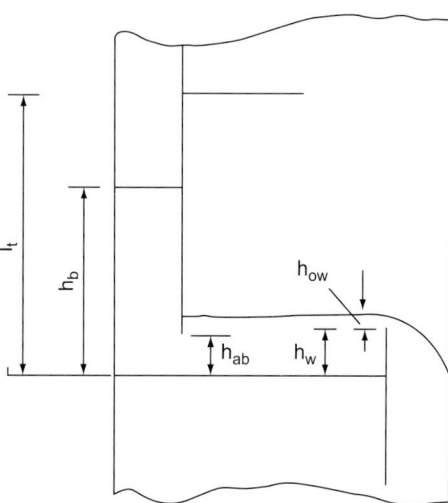

Figure 11.43.   Downcomer back-up.

where

h_b = downcomer back-up, measured from plate surface, mm
h_{dc} = head loss in the downcomer, mm.

The main resistance to flow will be caused by the constriction at the downcomer outlet, and the head loss in the downcomer can be estimated using the equation given by Cicalese *et al.* (1947)

$$h_{dc} = 166 \left[\frac{L_{wd}}{\rho_L A_m} \right]^2 \qquad (11.60)$$

where

L_{wd} = liquid flow rate in downcomer, kg/s
A_m = either the downcomer area A_d or the clearance area under the downcomer, A_{ap}; whichever is the smaller, m².

The clearance area under the downcomer is given by:

$$A_{ap} = h_{ap} l_w \qquad (11.61)$$

where h_{ap} is height of the bottom edge of the apron above the plate. This height is normally set at 5 to 10 mm (¼ to ½ in) below the outlet weir height:

$$h_{ap} = h_w - (5 \text{ to } 10 \text{ mm})$$

Froth Height

To predict the height of 'aerated' liquid on the plate, and the height of froth in the downcomer, some means of estimating the froth density is required. The density of the 'aerated' liquid will normally be between 0.4 and 0.7 times that of the clear liquid. A number of correlations have been proposed for estimating froth density as a function of the vapour flow rate and the liquid physical properties; see Chase (1967); however, none is particularly reliable, and for design purposes it is usually satisfactory to assume an average value of 0.5 of the liquid density.

This value is also taken as the mean density of the fluid in the downcomer, which means that for safe design the clear liquid back-up, calculated from equation 11.59, should not exceed half the plate spacing l_t, to avoid flooding.

Allowing for the weir height:

$$h_b \leq \tfrac{1}{2}(l_t + h_w) \qquad (11.62)$$

This criterion is, if anything, over-safe, and where close plate spacing is desired a better estimate of the froth density in the downcomer should be made. The method proposed by Thomas and Shah (1964) is recommended. Kister (1992) recommends that the froth height in the downcomer should not be greater than 80% of the tray spacing.

Downcomer Residence Time

Sufficient residence time must be allowed in the downcomer for the entrained vapour to disengage from the liquid stream, to prevent heavily 'aerated' liquid being carried under the downcomer. A time of at least 3 s is recommended.

The downcomer residence time is given by:

$$t_r = \frac{A_d h_{bc} \rho_L}{L_{wd}}$$ (11.63)

where

t_r = residence time, s
h_{bc} = clear liquid back-up, m.

Example 11.6

Design the plates for the column specified in Example 11.2. Take the minimum feed rate as 70% of the maximum (maximum feed 10,000 kg/h). Use sieve plates.

Solution

As the liquid and vapour flow rates and compositions will vary up the column, plate designs should be made above and below the feed point. Only the bottom plate will be designed in detail in this example.

From McCabe–Thiele diagram, Example 11.2:

Number of stages = 10
Top composition 95 mol%, bottom composition 1 mol%
Reflux ratio = 1.24.

Flow Rates

Mol. weight feed = $0.1 \times 58 + (1 - 0.1)18 = 22$
Feed = 10,000/22 = 454.5 kmol/h
Overall mass balance: $D + B = 454.5$

A mass balance on acetone gives:

$$0.95D + 0.01B = 0.1(454.5)$$

Hence

$$D = 43.5 \text{ kmol/h}, B = 411.0 \text{ kmol/h}$$

Vapour rate

$$V = D(1 + R) = 43.5(1 + 1.24) = 97.5 \text{ kmol/h}$$

The feed is saturated liquid, so:

liquid flow above feed, $L = RD = 1.24(43.52) = 54.0 \text{ kmol/h}$
liquid flow below feed, $L' = RD + F = 454.5 + 54 = 508.5 \text{ kmol/h}$

Physical Properties

Estimate base pressure, assume column efficiency of 60%, ignore reboiler.

$$\text{Number of real stages} = \frac{10}{0.6} = 17$$

Assume pressure drop per plate is 100 mm water.

$$\text{Column pressure drop} = 100 \times 10^{-3} \times 1000 \times 9.81 \times 17$$
$$= 16,677 \text{ Pa}$$

$$\text{Top pressure, 1 atm } (14.7 \text{ lb/in}^2) = 101.4 \times 10^3 \text{ Pa}$$

$$\text{Estimated bottom pressure} = 101.4 \times 10^3 + 16,677$$
$$= 118,077 \text{ Pa} = \underline{1.18 \text{ bar}}$$

From UniSim Design, base temperature 96.0°C.

$$\rho_v = 0.693 \text{ kg/m}^3, \rho_L = 944 \text{ kg/m}^3$$

Molecular weight = 18.4, surface tension = 58.9×10^{-3} N/m

Distillate, 95 mol% acetone, 56°C

$$\rho_v = 2.07 \text{ kg/m}^3, \rho_L = 748 \text{ kg/m}^3$$

Molecular weight = 56.1, surface tension = 22.7×10^{-3} N/m

Column Diameter

Neglecting differences in molecular weight between vapour and liquid:

$$F_{LV} \text{ bottom} = \frac{508.5}{97.5}\sqrt{\frac{0.693}{944}} = 0.141$$

$$F_{LV} \text{ top} = \frac{54}{97.5}\sqrt{\frac{2.07}{748}} = 0.0291$$

(11.50)

Take plate spacing as 0.5 m.
From Figure 11.34

$$\text{base } K_1 = 7.5 \times 10^{-2}$$

$$\text{top } K_1 = 9.0 \times 10^{-2}$$

Correction for surface tensions

$$\text{base } K_1 = \left(\frac{59}{20}\right)^{0.2} \times 7.5 \times 10^{-2} = 9.3 \times 10^{-2}$$

$$\text{top } K_1 = \left(\frac{23}{20}\right)^{0.2} \times 9.0 \times 10^{-2} = 9.3 \times 10^{-2}$$

(11.49)

$$\text{base } u_f = 9.3 \times 10^{-2}\sqrt{\frac{944 - 0.693}{0.693}} = 3.43 \text{ m/s}$$

$$\text{top } u_f = 9.3 \times 10^{-2}\sqrt{\frac{748 - 2.07}{2.07}} = 1.77 \text{ m/s}$$

Design for 85% flooding at maximum flow rate

$$\text{base } u_n = 3.43 \times 0.85 = 2.92 \text{ m/s}$$

$$\text{top } u_n = 1.77 \times 0.85 = 1.50 \text{ m/s}$$

Maximum volumetric flow rate

$$\text{base} = \frac{97.5 \times 18.4}{0.693 \times 3600} = 0.719 \text{ m}^3/\text{s}$$

$$\text{top} = \frac{97.5 \times 56.1}{2.07 \times 3600} = 0.734 \text{ m}^3/\text{s}$$

Net area required

$$\text{base} = \frac{0.719}{2.92} = 0.246 \text{ m}^2$$

$$\text{top} = \frac{0.734}{1.50} = 0.489 \text{ m}^2$$

As first trial take downcomer area as 12% of total.
    Column cross-sectioned area

$$\text{base} = \frac{0.246}{0.88} = 0.280 \text{ m}^2$$

$$\text{top} = \frac{0.489}{0.88} = 0.556 \text{ m}^2$$

Column diameter

$$\text{base} = \sqrt{\frac{0.28 \times 4}{\pi}} = 0.60 \text{ m}$$

$$\text{top} = \sqrt{\frac{0.556 \times 4}{\pi}} = 0.84 \text{ m}$$

Use same diameter above and below feed, reducing the perforated area for plates above the feed.
    This is too large to use standard pipe, so round up to nearest standard head size; inside diameter 914.4 mm (36 in).

Liquid Flow Pattern

$$\text{Maximum volumetric liquid rate} = \frac{508.5 \times 18.4}{3600 \times 944} = 2.75 \times 10^{-3} \text{ m}^3/\text{s}$$

The plate diameter is outside the range of Figure 11.35, but it is clear that a single-pass plate can be used.

Provisional Plate Design

Column diameter $D_c = 0.914$ m

Column area $A_c = 0.556$ m^2

Downcomer area $A_d = 0.12 \times 0.556 = 0.067$ m^2, at 12%

Net area $A_n = A_c - A_d = 0.556 - 0.067 = 0.489$ m^2

Active area $A_a = A_c - 2A_d = 0.556 - 0.134 = 0.422$ m^2

Hole area A_h take 10% A_a as first trial $= 0.042$ m^2

Weir length (from Figure 11.39) $= 0.76 \times 0.914 = 0.695$ m

Take weir height	50 mm
Hole diameter	5 mm
Plate thickness	5 mm

Check Weeping

Maximum liquid rate $= \dfrac{508.5 \times 18.4}{3600} = 2.60$ kg/s

Minimum liquid rate, at 70% turn-down $= 0.7 \times 2.6 = 1.82$ kg/s

$$\text{Maximum } h_{ow} = \left(\frac{2.6}{944 \times 0.695}\right)^{2/3} = 25.0 \text{ mm liquid} \qquad (11.53)$$

$$\text{Minimum } h_{ow} = \left(\frac{1.82}{944 \times 0.695}\right)^{2/3} = 19.7 \text{ mm liquid}$$

At minimum rate $h_w + h_{ow} = 50 + 19.7 = 69.7$ mm

From Figure 11.37

$$K_2 = 30.6$$

$$\hat{u}_h(\text{min}) = \frac{[30.6 - 0.90(25.4 - 5)]}{(0.693)^{1/2}} = 14.7 \text{ m/s} \qquad (11.52)$$

$$\text{Actual minimum vapour velocity} = \frac{\text{minimum vapour rate}}{A_h} = \frac{0.7 \times 0.719}{0.042}$$

$$= 12.0 \text{ m/s}$$

So the minimum operating rate will lead to weeping at the bottom of the column. Reduce hole area to 7% of active area $= 0.422 \times 0.07 = 0.0295 \text{ m}^2$.

$$\text{New actual minimum vapour velocity} = \frac{0.7 \times 0.719}{0.0295} = 17.1 \text{ m/s}$$

Which is now well above the weep point.

Plate Pressure Drop

Dry-plate drop
   Maximum vapour velocity through holes

$$\hat{u}_h(\text{max}) = \frac{0.719}{0.0295} = 24.4 \text{ m/s}$$

From Figure 11.42, for plate thickness/hole diameter $= 1$, and $A_h/A_p \simeq A_h/A_a = 0.07$, $C_0 = 0.82$

$$h_d = 51 \left[\frac{24.4}{0.82}\right]^2 \frac{0.693}{944} = 33.1 \text{ mm liquid} \tag{11.56}$$

residual head

$$h_r = \frac{12.5 \times 10^3}{944} = 13.2 \text{ mm liquid} \tag{11.55}$$

total plate pressure drop

$$h_t = 33 + (50 + 25) + 13 = 118 \text{ mm liquid}$$

Note: Here, 100 mm was assumed to calculate the base pressure. The calculation could be repeated with a revised estimate but the small change in physical properties will have little effect on the plate design; 118 mm per plate is considered acceptable.

Downcomer Liquid Back-Up

Downcomer pressure loss
   Take $h_{ap} = h_w - 10 = 40 \text{ mm}$
   Area under apron, $A_{ap} = 0.695 \times 40 \times 10^{-3} = 0.028 \text{ m}^2$
   As this is less than $A_d = 0.067 \text{ m}^2$ use A_{ap} in equation 11.60

$$h_{dc} = 166 \left[\frac{2.60}{944 \times 0.028}\right]^2 = 1.61 \text{ mm} \tag{11.60}$$

$$\text{say 2 mm.}$$

Back-up in downcomer

$$h_b = (50 + 25) + 118 + 2 = 195 \text{ mm} \tag{11.59}$$

195 mm $< \dfrac{1}{2}$ (plate spacing + weir height), so plate spacing is acceptable

Check residence time

$$t_r = \frac{0.067 \times 0.195 \times 944}{2.60} = 4.7 \text{ s} \tag{11.63}$$

$$> 3 \text{ s, satisfactory.}$$

Check Entrainment

$$u_v = \frac{0.719}{0.489} = 1.47 \text{ m/s}$$

$$\text{per cent flooding} = \frac{1.47}{3.43} = 42.8\%$$

$$F_{LV} = \underline{0.14}, \text{ so from Figure 11.36, } \Psi = 0.0038, \text{ well below 0.1}$$

As the per cent flooding is well below the design figure of 85, the column diameter could be reduced, but this would increase the pressure drop.

Trial Layout

Use cartridge-type construction. Allow 50-mm unperforated strip round plate edge; 50-mm-wide calming zones.

Perforated Area

From Figure 11.40, at $l_w/D_c = 0.695/0.914 = 0.76$, $\theta_c = 99°$

$$\text{angle subtended by the edge of the plate} = 180 - 99 = 81°$$

$$\begin{aligned} \text{mean length, unperforated edge strips} &= (0.914 - 50 \times 10^{-3})\pi \times 81/180 \\ &= 1.22 \text{ m} \end{aligned}$$

$$\text{area of unperforated edge strips} = 50 \times 10^{-3} \times 1.22 = 0.061 \text{ m}^2$$

$$\begin{aligned} \text{mean length of calming zone, approx.} &= \text{weir length} \\ &\quad + \text{ width of unperforated strip} \\ &= 0.695 + 50 \times 10^{-3} = 0.745 \text{ m} \end{aligned}$$

$$\text{area of calming zones} = 2(0.745 \times 50 \times 10^{-3}) = 0.0745 \text{ m}^2$$

$$\text{total area for perforations, } A_p = 0.422 - 0.061 - 0.075 = 0.286 \text{ m}^2$$

$$A_h/A_p = 0.0295/0.286 = 0.103$$

From Figure 11.41, $l_p/d_h = 2.9$; satisfactory, within 2.5 to 4.0.

Number of Holes

$$\text{Area of one hole} = 1.964 \times 10^{-5} \text{ m}^2$$

$$\text{Number of holes} = \frac{0.0295}{1.964 \times 10^{-5}} = 1502$$

Plate Specification

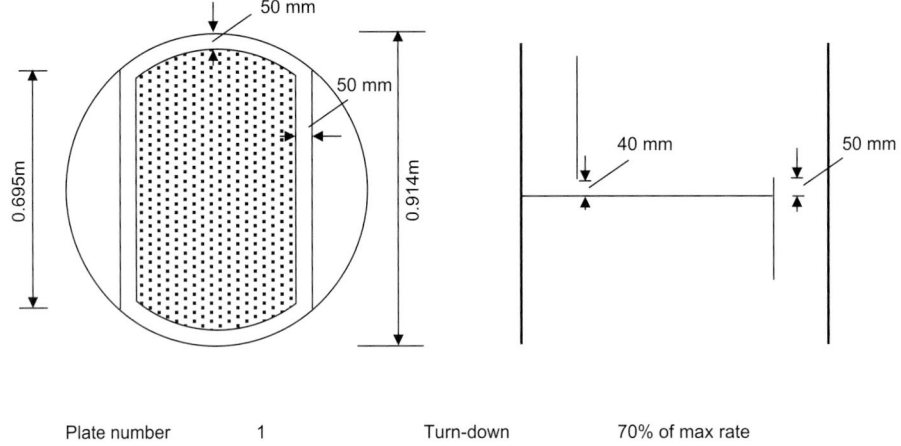

Plate number	1	Turn-down	70% of max rate
Plate inside dia.	0.914 m	Plate material	Mild steel
Hole size	5 mm	Downcomer material	Mild steel
Hole pitch	12.5 mm △	Plate spacing	0.5 m
Total holes	-	Plate thickness	5 mm
Active holes	1502	Plate pressure drop	120 mm liquid = 1.1 kPa
Blanking area	-		

Figure 11.44. Plate specification for Example 11.6.

Example 11.7

For the plate design in Example 11.6, estimate the plate efficiency for the plate on which the concentration of acetone is 5 mol%. Use the AIChE method.

Solution

Plate will be in the stripping section (see Figure 11.9).
  Plate dimensions:

active area $= 0.422 \text{ m}^2$
length between downcomers (Figure 11.40) (liquid path, Z_L) $= 0.914 \ (1 - 2 \times 0.175) = 0.594 \text{ m}$
weir height $= 50 \text{ mm}$.

Flow rates, check efficiency at minimum rates, at column base:

$$\text{Vapour} = 0.7 \times \frac{97.5}{3600} = 0.019 \text{ kmol/s}$$

$$\text{Liquid} = 0.7 \times \frac{508.5}{3600} = 0.099 \text{ kmol/s}$$

from the McCabe-Thiele diagram (Figure 11.9) at $x = 0.05$, assuming 60% plate efficiency, $y \approx 0.35$. The liquid composition, $x = 0.05$, will occur on around the third plate from the bottom (allowing for the reboiler and 60% efficiency per stage). The pressure on this plate will be approximately:

$$101.4 \times 10^3 + (14 \times 0.118 \times 9.81 \times 944) = 116.7 \text{ kPa}$$

$$\text{say, } 1.17 \text{ bar}$$

At this pressure the plate temperature will be about 92°C, and the liquid and vapour physical properties, from UniSim Process Design are:

Liquid

$$\text{molar weight} = 20, \rho_L = 932.7 \text{ kg/m}^3, \mu_L = 0.3544 \times 10^{-3} \text{ Nm}^{-2} \text{ s,}$$
$$\sigma = 60.2 \times 10^{-3} \text{ N/m}$$

Vapour

$$\text{molar weight} = 32, \rho_v = 1.233 \text{ kg/m}^3, \mu_v = 9.17 \times 10^{-6} \text{ Nm}^{-2} \text{ s}$$

$$D_L = 4.16 \times 10^{-9} \text{ m}^2/\text{s (estimated using Wilke-Chang equation, Chapter 8)}$$

$$D_v = 17.4 \times 10^{-6} \text{ m}^2/\text{s (estimated using Fuller equation, Chapter 8)}$$

$$\text{Vapour volumetric flow rate} = \frac{0.019 \times 32}{1.233} = 0.493 \text{ m}^3/\text{s}$$

$$\text{Liquid volumetric flow rate} = \frac{0.099 \times 20}{932.7} = 2.12 \times 10^{-3} \text{ m}^3/\text{s}$$

$$u_a = \frac{0.493}{0.422} = 1.17 \text{ m/s}$$

$$F_v = u_a\sqrt{\rho_v} = 2.365 \text{ kg}^{0.5}\text{m}^{-0.5}\text{s}^{-1}$$

Average width over active surface $= 0.422/0.594 = 0.71 \text{ m}$

$$L_p = \frac{2.12 \times 10^{-3}}{0.71} = 2.99 \times 10^{-3} \text{ m}^2/\text{s}$$

$$N_G = \frac{(0.776 + 4.57 \times 10^{-3} \times 50 - 0.24 \times 2.365 + 105 \times 2.99 \times 10^{-3})}{\left(\dfrac{9.17 \times 10^{-6}}{1.233 \times 17.4 \times 10^{-6}}\right)^{0.5}} = 1.15$$

$$(11.38)$$

$$Z_c = 0.006 + 0.73 \times 10^{-3} \times 50 - 0.24 \times 10^{-3} \times 2.365 \times 50 + 1.22 \times 2.99 \times 10^{-3}$$
$$= 17.8 \times 10^{-3} \tag{11.43}$$

$$t_L = \frac{17.8 \times 10^{-3} \times 0.594}{2.99 \times 10^{-3}} = 3.54 \text{ s} \tag{11.41}$$

$$N_L = (4.13 \times 10^8 \times 4.16 \times 10^{-9})^{0.5}(0.21 \times 2.365 + 0.15) \times 3.54 = 3.00 \tag{11.40}$$

$$D_e = (0.0038 + 0.017 \times 1.17 + 3.86 \times 2.99 \times 10^{-3} + 0.18 \times 10^{-3} \times 50)^2$$
$$= 1.96 \times 10^{-3} \tag{11.45}$$

$$\text{Pe} = \frac{(0.594)^2}{1.96 \times 10^{-3} \times 3.54} = 50.8 \tag{11.44}$$

From the McCabe–Thiele diagram, at $x = 0.05$, the slope of the equilibrium line ≈ 12.0, so

$$\frac{mV}{L} = \frac{12 \times 0.019}{0.099} = 2.30$$

$$\frac{\left(\dfrac{mV}{L}\right)}{N_L} = \frac{2.30}{3.00} = 0.767$$

From Figure 11.21

$$E_{mv} = 0.43$$

$$\frac{mV}{L} \times E_{mv} = 2.30 \times 0.43 = 0.989$$

From Figure 11.22

$$E_{mV}/E_{mv} = 1.62$$

$$E_{mV} = 0.43 \times 1.62 = 0.697$$

so plate efficiency $= \underline{\underline{70\%}}$.

Note: The slope of the equilibrium line is difficult to determine at $x = 0.05$, but any error will not greatly affect the value of E_{mV}.

Example 11.8

Calculate the plate efficiency for the plate design considered in Examples 11.6 and 11.7, using Van Winkle's correlation.

Solution

From Examples 11.6 and 11.7:

$$\rho_L = 932.7 \text{ kg/m}^3, \; \mu_L = 0.3544 \times 10^{-3} \text{ Nm}^{-2}\text{s}, \; D_{LK} = D_L$$
$$= 4.16 \times 10^{-9} \text{ m}^2/\text{s}, \; \sigma = 60.2 \times 10^{-3} \text{ N/m}$$

$$\rho_v = 1.233 \text{ kg/m}^3, \mu_v = 9.17 \times 10^{-6} \text{ Nm}^{-2} \text{ s},$$

$$h_w = 50 \text{ mm}$$

$$FA \text{ (fractional area)} = A_h/A_c = 0.0295/0.556 = 0.053$$

$$u_v = \text{superficial vapour velocity} = 0.493/0.556 = 0.887 \text{ m/s}$$

$$Dg = \left(\frac{0.0602}{0.3544 \times 10^{-3} \times 0.887} \right) = 191.6$$

$$Sc = \left(\frac{0.3544 \times 10^{-3}}{932.7 \times 4.16 \times 10^{-9}} \right) = 91.3$$

$$Re = \left(\frac{50 \times 10^{-3} \times 0.887 \times 932.7}{0.3544 \times 0.053} \right) = 2.2 \times 10^3$$

$$E_{mV} = 0.07(191.6)^{0.14}(91.3)^{0.25}(2.2 \times 10^3)^{0.08} \tag{11.36}$$

$$= \underline{0.836} \, (84\%)$$

This seems rather large compared to the value found using the AIChE method, so the value calculated in Example 11.7 is preferred.

11.14. PACKED COLUMNS

Packed columns are used for distillation, gas absorption, and liquid–liquid extraction; only distillation and absorption will be considered in this section. Stripping (desorption) is the reverse of absorption and the same design methods apply.

The gas–liquid contact in a packed bed column is continuous, not stage-wise, as in a plate column. The liquid flows down the column over the packing surface and the gas or vapour, counter-currently, up the column. In some gas-absorption columns co-current flow is used. The performance of a packed column is very dependent on the maintenance of good liquid and gas distribution throughout the packed bed, and this is an important consideration in packed-column design.

A schematic diagram, showing the main features of a packed absorption column, is given in Figure 11.45. A packed distillation column will be similar to the plate columns shown in Figure 11.1, with the plates replaced by packed sections.

The design of packed columns using random packings is covered in books by Kister (1992), Strigle (1994) and Billet (1995).

Choice of Plates or Packing

The choice between a plate or packed column for a particular application can only be made with complete assurance by costing each design; however, the choice can usually be made on the basis of experience by considering main advantages and disadvantages of each type, listed below:

1. Plate columns can be designed to handle a wider range of liquid and gas flow rates than packed columns.
2. Packed columns are not suitable for very low liquid rates.
3. The efficiency of a plate can be predicted with more certainty than the equivalent term for packing (HETP or HTU).
4. Plate columns can be designed with more assurance than packed columns. There is always some doubt that good liquid distribution can be maintained throughout a packed column under all operating conditions, particularly in large columns.

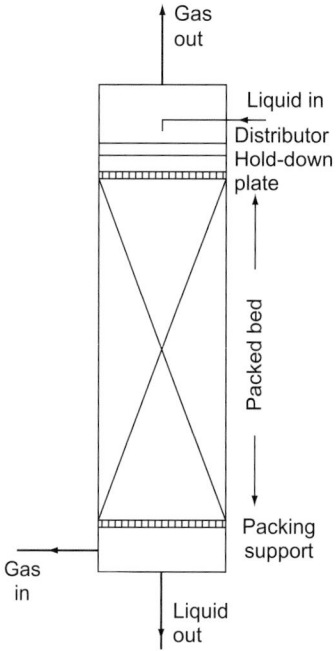

Figure 11.45. Packed absorption column.

5. It is easier to make provision for cooling in a plate column; coils can be installed on the plates.
6. It is easier to make provision for the withdrawal of side streams from plate columns.
7. If the liquid causes fouling, or contains solids, it is easier to make provision for cleaning in a plate column; manways can be installed on the plates. With small-diameter columns it may be cheaper to use packing and replace the packing when it becomes fouled.
8. For corrosive liquids a packed column will usually be cheaper than the equivalent plate column.
9. The liquid hold-up is appreciably lower in a packed column than a plate column. This can be important when the inventory of toxic or flammable liquids must be minimized for safety reasons.
10. Packed columns are more suitable for handling foaming systems.
11. The pressure drop per equilibrium stage (HETP) can be lower for packing than plates, and packing should be considered for vacuum columns.
12. Packing should always be considered for small-diameter columns, say less than 0.6 m, where plates would be difficult to install and expensive.

Packed-Column Design Procedures

The design of a packed column will involve the following steps:

1. Select the type and size of packing.
2. Determine the column height required for the specified separation.
3. Determine the column diameter (capacity), to handle the liquid and vapour flow rates.
4. Select and design the column internal features: packing support, liquid distributor, redistributors.

These steps are discussed in the following sections, and a packed-column design is illustrated in Example 11.9.

11.14.1. Types of Packing

The principal requirements of a packing are that it should:

Provide a large surface area: a high interfacial area between the gas and liquid
Have an open structure: low resistance to gas flow
Promote uniform liquid distribution on the packing surface
Promote uniform vapour or gas flow across the column cross-section.

Many diverse types and shapes of packing have been developed to satisfy these requirements.
They can be divided into two broad classes:

1. Packings with a regular geometry: such as stacked rings, grids and proprietary structured packings.
2. Random packings: rings, saddles and proprietary shapes, which are dumped into the column and take up a random arrangement.

Grids have an open structure and are used for high gas rates, where low pressure drop is essential; for example, in cooling towers. Random packings and structured packing elements are more commonly used in the process industries.

Random Packing

The principal types of random packings are shown in Figure 11.46. Design data for these packings are given in Table 11.2. The design methods and data given in this section can be used for the preliminary design of packed columns, but for detailed design it is advisable to consult the packing manufacturer's technical literature to obtain data for the particular packing that will be used. The packing manufacturers should be consulted for details of the many special types of packing that are available for special applications.

Raschig rings, Figure 11.46a, are one of the oldest specially manufactured types of random packing, and are still in general use. Pall rings, Figure 11.46b, are essentially Raschig rings in which openings have been made by folding strips of the surface into the ring. This increases the free area and improves the liquid distribution characteristics. Berl saddles, Figure 11.46c, were developed to give improved liquid distribution compared to Raschig rings. INTALOX® saddles, Figure 11.46d, can be considered to be an improved type of Berl saddle; their shape makes them easier to manufacture than Berl saddles. The HY-PAK® and SUPER INTALOX® packings shown in Figure 11.46e,f can be considered improved types of Pall ring and INTALOX® saddle respectively.

INTALOX® saddles, SUPER INTALOX® and HY-PAK® packings are proprietary designs, and registered trademarks of Koch-Glitsch, LP.

Ring and saddle packings are available in a variety of materials: ceramics, metals, plastics and carbon. Metal and plastics (polypropylene) rings are more efficient than ceramic rings, as it is possible to make the walls thinner.

Raschig rings are cheaper per unit volume than Pall rings or saddles but are less efficient, and the total cost of the column will usually be higher if Raschig rings are specified. For new columns, the choice will normally be between Pall rings and Berl or INTALOX® saddles.

The choice of material will depend on the nature of the fluids and the operating temperature. Ceramic packing will be the first choice for corrosive liquids, but ceramics are unsuitable for use with strong alkalis. Packings made of plastics are attacked by some organic solvents and can only be used up to moderate temperatures, so are unsuitable for distillation columns. Where the column operation is likely to be unstable, metal rings should be specified as ceramic packing is easily broken. The choice of packings for distillation and absorption is discussed in detail by Eckert (1963), Strigle (1994), Kister (1992) and Billet (1995).

Packing Size

In general, the largest size of packing that is suitable for the size of column should be used; up to 50 mm. Small sizes are appreciably more expensive than the larger sizes.

Above 50 mm the lower cost per cubic metre does not normally compensate for the lower mass transfer efficiency. Use of too large a size in a small column can cause poor liquid distribution.

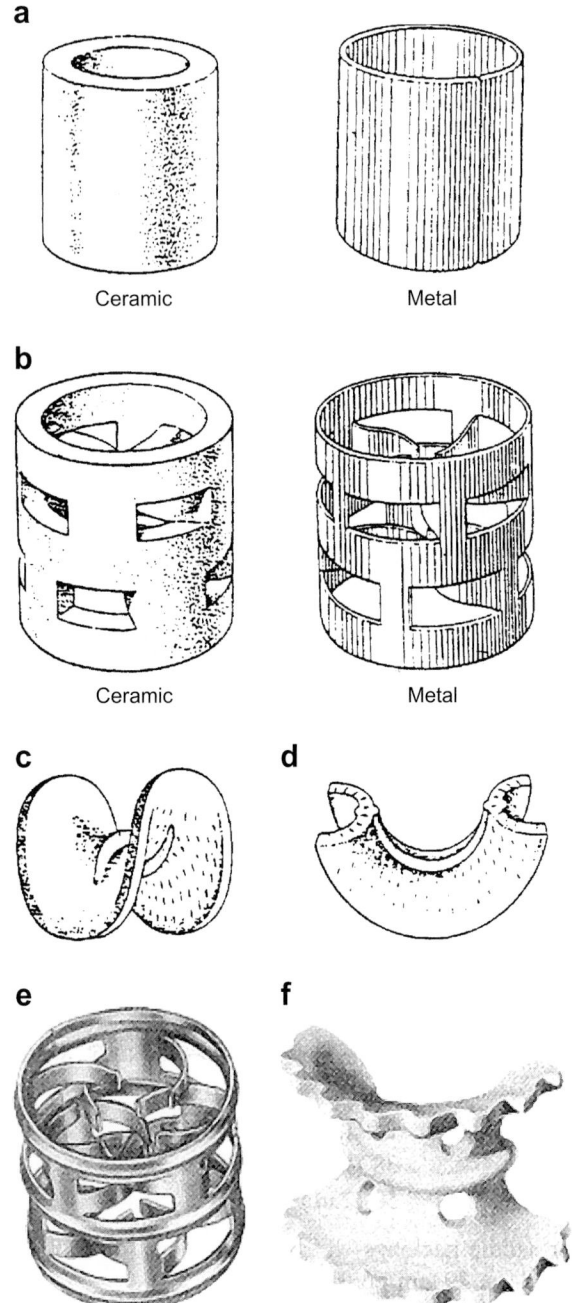

Figure 11.46. Types of packing (Koch-Glitsch, LP). (*a*) Raschig rings. (*b*) Pall rings. (*c*) Berl saddle ceramic. (*d*) INTALOX® saddle ceramic. (*e*) Metal HY-PAK®. (*f*) Ceramic, SUPER INTALOX®.

TABLE 11.2.   Design Data for Various Packings

| | Size | | Bulk density | Surface area a | Packing |
	in	mm	(kg/m³)	(m²/m³)	factor F_pm⁻¹
Raschig rings ceramic	0.50	13	881	368	2100
	1.0	25	673	190	525
	1.5	38	689	128	310
	2.0	51	651	95	210
	3.0	76	561	69	120
Metal (density for carbon steel)	0.5	13	1201	417	980
	1.0	25	625	207	375
	1.5	38	785	141	270
	2.0	51	593	102	190
	3.0	76	400	72	105
Pall rings metal	0.625	16	593	341	230
(density for carbon steel)	1.0	25	481	210	160
	1.25	32	385	128	92
	2.0	51	353	102	66
	3.5	76	273	66	52
Plastics (density for polypropylene)	0.625	16	112	341	320
	1.0	25	88	207	170
	1.5	38	76	128	130
	2.0	51	68	102	82
	3.5	89	64	85	52
INTALOX® saddles ceramic	0.5	13	737	480	660
	1.0	25	673	253	300
	1.5	38	625	194	170
	2.0	51	609	108	130
	3.0	76	577		72

Recommended size ranges are:

Column diameter	Use packing size
<0.3 m (1 ft)	<25 mm (1 in)
0.3 to 0.9 m (1 to 3 ft)	25 to 38 mm (1 to 1.5 in)
>0.9 m	50 to 75 mm (2 to 3 in)

Structured Packing

This refers to packing elements made up from wire mesh or perforated metal sheets. The material is folded and arranged with a regular geometry, to give a high surface area with a high void fraction. A typical example is shown in Figure 11.47.

Structured packings are produced by a number of manufacturers. The basic construction and performance of the various proprietary types available are similar. They are available in metal, plastics and stoneware. The advantage of structured packings over random packing is their low HETP (typically less than 0.5 m) and low

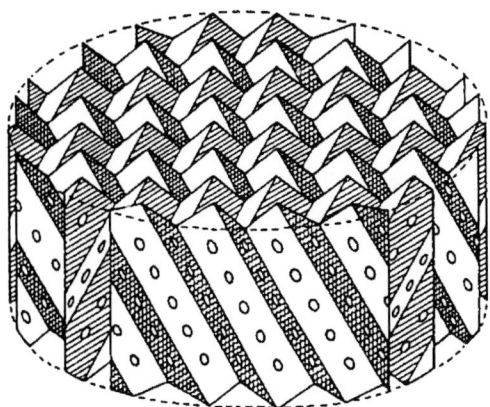

Figure 11.47. Make-up of structured packing. Reproduced from Butcher (1988) with permission.

pressure drop (around 100 Pa/m). They are being increasingly used in the following applications:

1. For difficult separations, requiring many stages: such as the separation of isomers.
2. High vacuum distillation.
3. For column revamps: to increase capacity and reduce reflux ratio requirements.

The applications have mainly been in distillation, but structured packings can also be used in absorption; in applications where high efficiency and low pressure drop are needed.

The cost of structured packings per cubic metre will be significantly higher than that of random packings, but this is offset by their higher efficiency.

The manufacturers' technical literature should be consulted for design data. A review of the types available is given by Butcher (1988). Generalized methods for predicting the capacity and pressure drop of structured packings are given by Fair and Bravo (1990) and Kister and Gill (1992). The use of structured packings in distillation is discussed in detail in the book by Kister (1992).

Structured packings have a high surface area of thin metal, and consequently can be susceptible to ignition of trapped hydrocarbons or pyrophoric corrosion products, as heat is not easily conducted away from local hot spots. The Fractionation Research Inc. (FRI) Design Practices Committee have provided guidelines on the design and maintenance of packed columns to reduce the likelihood of packing fires; see FRI Design Practices Committee (2007).

11.14.2. Packed-Bed Height

Distillation

For the design of packed distillation columns, it is simpler to treat the separation as a staged process and use the concept of the height of an equivalent equilibrium stage to convert the number of ideal stages required to a height of packing. The methods for estimating the number of ideal stages given in Sections 11.5 to 11.8 can then be applied to packed columns.

The height of an equivalent equilibrium stage, usually called the height equivalent to a theoretical plate (HETP), is the height of packing that will give the same separation as an equilibrium stage. It has been shown by Eckert (1975) that in distillation the HETP for a given type and size of packing is essentially constant, and independent of the system physical properties, providing good liquid distribution is maintained and the pressure drop is at least above 17 mm water per metre of packing height. The following values for Pall rings can be used to make an approximate estimate of the bed height required.

Size (mm)	HETP (m)
25 (1 in)	0.4–0.5
38 ($1\frac{1}{2}$ in)	0.6–0.75
50 (2 in)	0.75–1.0

The HETP for saddle packings will be similar to that for Pall rings providing the pressure drop is at least 29 mm/m.

The HETP for Raschig rings will be higher than for Pall rings or saddles, and the values given above will only apply at an appreciably higher pressure drop, greater than 42 mm/m.

The methods for estimating the heights of transfer units, HTU, given in Section 11.14.3 can be used for distillation. The relationship between transfer units and the height equivalent to a theoretical plate, HETP is given by:

$$\text{HETP} = \frac{\mathbf{H}_{OG}\ln\left(\dfrac{mG_m}{L_m}\right)}{\left(\dfrac{mG_m}{L_m} - 1\right)} \tag{11.64}$$

where $\mathbf{H}_{OG}$ is the height of an overall gas-phase transfer unit.

The slope of the equilibrium line m will normally vary throughout a distillation column, so it will be necessary to calculate the HETP for each plate or a series of plates.

Absorption

Though packed absorption and stripping columns can also be designed as staged processes, it is usually more convenient to use the integrated form of the differential equations set up by considering the rates of mass transfer at a point in the column. The derivation of these equations is given in Richardson *et al.* (2002).

Where the concentration of the solute is small, say less than 10%, the flow of gas and liquid will be essentially constant throughout the column, and the height of packing required, Z, is given by:

$$Z = \frac{G_m}{K_G a P} \int_{y_2}^{y_1} \frac{dy}{y - y_e} \qquad (11.65)$$

in terms of the overall gas-phase mass-transfer coefficient K_G and the gas composition.

Or,

$$Z = \frac{L_m}{K_L a C_t} \int_{x_2}^{x_1} \frac{dx}{x_e - x} \qquad (11.66)$$

in terms of the overall liquid-phase mass-transfer coefficient K_L and the liquid composition,

where

G_m = molar gas flow rate per unit cross-sectional area
L_m = molar liquid flow rate per unit cross-sectional area
a = interfacial surface area per unit volume
P = total pressure
C_t = total molar concentration
y_1 and y_2 = the mol fractions of the solute in the gas at the bottom and top of the column respectively
x_1 and x_2 = the mol fractions of the solute in the liquid at the bottom and top of the column respectively
x_e = the mole fraction in the liquid that would be in equilibrium with the gas concentration at any point
y_e = the mole fraction in the gas that would be in equilibrium with the liquid concentration at any point.

The relation between the equilibrium concentrations and actual concentrations is shown in Figure 11.48.

For design purposes it is convenient to write equations 11.65 and 11.66 in terms of 'transfer units' (HTU); where the value of the integral is the number of transfer units, and the group in front of the integral sign, which has units of length, is the height of a transfer unit.

$$Z = \mathbf{H}_{OG}\mathbf{N}_{OG} \qquad (11.67a)$$

or

$$Z = \mathbf{H}_{OL}\mathbf{N}_{OL} \qquad (11.67b)$$

where $\mathbf{H}_{OG}$ is the height of an overall gas-phase transfer unit

$$= \frac{G_m}{K_G a P} \qquad (11.68)$$

$\mathbf{N}_{OG}$ is the number of overall gas-phase transfer units

$$= \int_{y_2}^{y_1} \frac{dy}{y - y_e} \qquad (11.69)$$

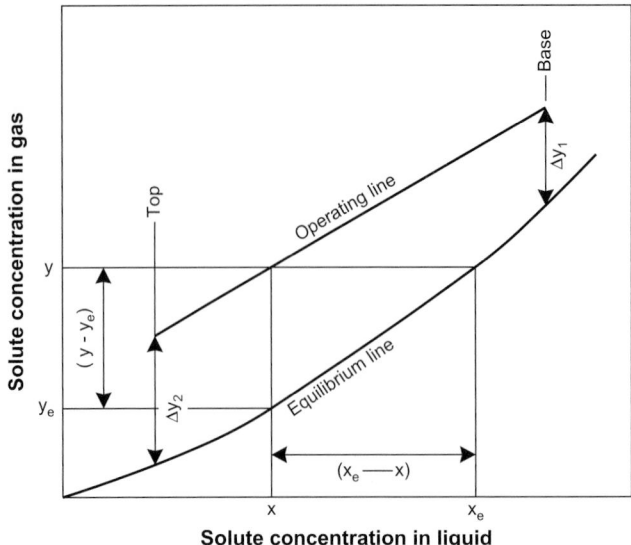

Figure 11.48.   Gas absorption concentration relationships.

H_{OL} is the height of an overall liquid-phase transfer unit

$$= \frac{L_m}{K_L a C_t} \tag{11.70}$$

N_{OL} is the number of overall liquid-phase transfer units

$$= \int_{x_2}^{x_1} \frac{dx}{x_e - x} \tag{11.71}$$

The number of overall gas-phase transfer units is often more conveniently expressed in terms of the partial pressure of the solute gas.

$$N_{OG} = \int_{p_1}^{p_2} \frac{dp}{p - p_e} \tag{11.72}$$

The relationship between the overall height of a transfer unit and the individual film transfer units H_L and H_G, which are based on the concentration driving force across the liquid and gas films, is given by

$$H_{OG} = H_G + m \frac{G_m}{L_m} H_L \tag{11.73}$$

$$H_{OL} = H_L + \frac{L_m}{m \, G_m} H_G \tag{11.74}$$

where m is the slope of the equilibrium line and G_m/L_m the slope of the operating line.

The number of transfer units is obtained by graphical or numerical integration of equations 11.69, 11.71 or 11.72.

Where the operating and equilibrium lines are straight, and they can usually be considered to be so for dilute systems, the number of transfer units is given by

$$\mathbf{N}_{OG} = \frac{y_1 - y_2}{\Delta y_{1m}} \qquad (11.75)$$

where Δy_{lm} is the log mean driving force, given by

$$y_{1m} = \frac{\Delta y_1 - \Delta y_2}{\ln\left(\dfrac{\Delta y_1}{\Delta y_2}\right)} \qquad (11.76)$$

where

$\Delta y_1 = y_1 - y_e$
$\Delta y_2 = y_2 - y_e.$

If the equilibrium curve and operating lines can be taken as straight and the solvent feed is essentially solute free, the number of transfer units is given by

$$\mathbf{N}_{OG} = \frac{1}{1 - \left(\dfrac{mG_m}{L_m}\right)} \ln\left[\left(1 - \frac{mG_m}{L_m}\right)\frac{y_1}{y_2} + \frac{mG_m}{L_m}\right] \qquad (11.77)$$

This equation is plotted in Figure 11.49, which can be used to make a quick estimate of the number of transfer units required for a given separation.

It can be seen from Figure 11.49 that the number of stages required for a given separation is very dependent on the flow rate L_m. If the solvent rate is not set by other process considerations, Figure 11.49 can be used to make quick estimates of the column height at different flow rates to find the most economic value. Colburn (1939) has suggested that the optimum value for the term mG_m/L_m will lie between 0.7 and 0.8.

Only physical absorption from dilute gases has been considered in this section. For a discussion of absorption from concentrated gases and absorption with chemical reaction, the reader should refer to Richardson *et al.* (2002), or to the book by Treybal (1980). The special case of absorption of acid gases is discussed extensively in the book by Kohl and Nielsen (1997). If the inlet gas concentration is not too high, the equations for dilute systems can be used by dividing the operating line up into two or three straight sections.

Stripping

In a stripping column, an absorbed solute is removed from a liquid solvent by counter-current contact with a vapour. Stripping and absorption are usually used together, with a stripping column regenerating the solvent for an absorber column. The analysis of stripping is similar to that for absorption; see Green and Perry (2007).

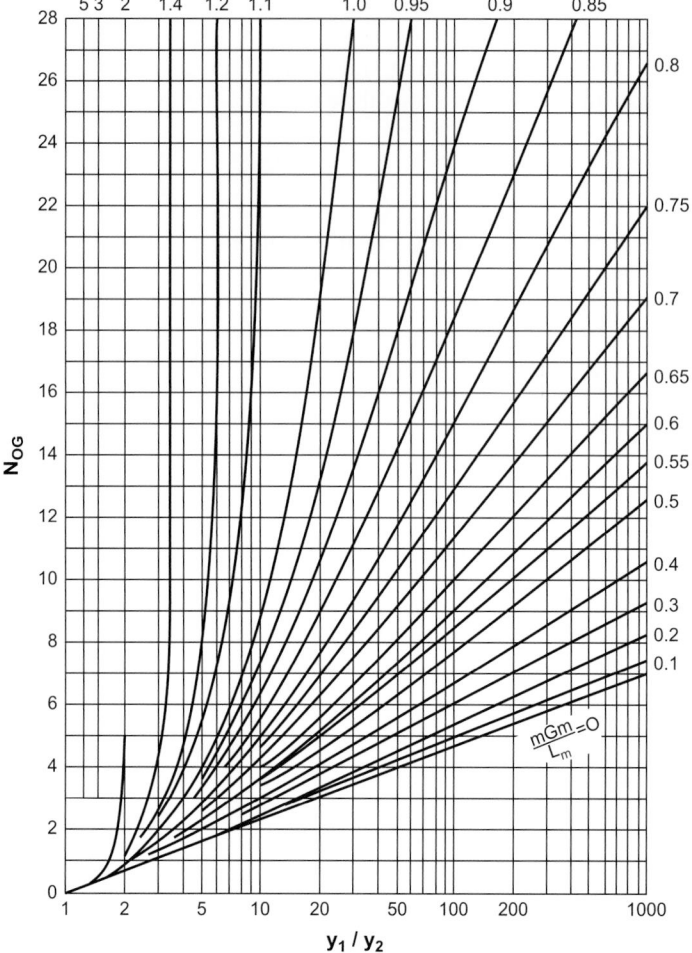

Figure 11.49.   Number of transfer units $\mathbf{N}_{OG}$ as a function of y_1/y_2 with mG_m/L_m as parameter.

Since the objective in stripping is to achieve a desired outlet liquid concentration, it is more customary to work in overall liquid-phase transfer units:

$$\mathbf{N}_{OL} = \int_{x_2}^{x_1} \frac{dx}{x_e - x} \qquad (11.71)$$

If the equilibrium and operating lines can be taken as straight, and the stripping vapour is essentially solute free, the number of transfer units is given by:

$$\mathbf{N}_{OL} = \frac{1}{\left(1 - \dfrac{L}{mG}\right)} \ln\left[\left(\frac{L}{mG}\right) + \left(1 - \frac{L}{mG}\right)\frac{x_2}{x_1}\right] \qquad (11.78)$$

The parameter (L/mG) is known as the stripping factor.

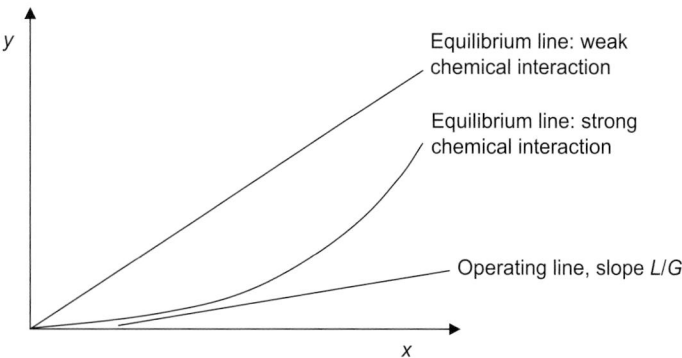

Figure 11.50. Equilibrium and operating lines for stripping.

When there are strong chemical interactions between the solute and solvent, the effect is usually to make the equilibrium line curve so that it is concave upwards, as shown in Figure 11.50. The operating line for stripping must lie below the equilibrium line, so if a low outlet concentration of solute is needed, as is often the case, then L/G must be less than the slope of the equilibrium line, m.

To achieve $L/G < m$, the designer can raise the stripping vapour rate, G, or increase temperature or reduce pressure to increase m.

As the equilibrium line becomes more concave it obviously becomes more difficult to achieve $L/G < m$ and obtain completely regenerated solvent. Absorber–stripper processes are therefore used for bulk separations, as it is too expensive to completely regenerate the solvent. If necessary, residual amounts of solute in the vapour leaving the absorber can be recovered using adsorption; see Chapter 10.

11.14.3. **Prediction of the Height of a Transfer Unit (HTU)**

There is no entirely satisfactory method for predicting the HTU. In practice, the value for a particular packing will depend not only on the physical properties and flow rates of the gas and liquid, but also on the uniformity of the liquid distribution throughout the column, which is dependent on the column height and diameter. This makes it difficult to extrapolate data obtained from small-size laboratory and pilot plant columns to industrial-size columns. Whenever possible, estimates should be based on actual values obtained from operating columns of similar size to that being designed.

Experimental values for several systems are given by Cornell *et al.* (1960), Eckert (1963), and Vital *et al.* (1984). A selection of values for a range of systems is given in Table 11.3. The composite mass transfer term $K_G a$ is normally used when reporting experimental mass-transfer coefficients for packing, as the effective interfacial area for mass transfer will be less than the actual surface area a of the packing.

Many correlations have been published for predicting the HTU and the mass-transfer coefficients; several are reviewed in Richardson *et al.* (2002). The two

TABLE 11.3. Typical Packing Efficiencies

System	Pressure (kPa)	Column dia, (m)	Packing		HTU (m)	HETP (m)
			type	size (mm)		
Absorption						
Hydrocarbons	6000	0.9	Pall	50		0.85
NH₃–Air–H₂O	101	—	Berl	50	0.50	
Air–water	101	—	Berl	50	0.50	
Acetone–water	101	0.6	Pall	50		0.75
Distillation						
Pentane–propane	101	0.46	Pall	25		0.46
IPA–water	101	0.46	Int.	25	0.75	0.50
Methanol–water	101	0.41	Pall	25	0.52	
	101	0.20	Int.	25		0.46
Acetone–water	101	0.46	Pall	25		0.37
	101	0.36	Int.	25		0.46
Formic acid–water	101	0.91	Pall	50		0.45
Acetone–water	101	0.38	Pall	38	0.55	0.45
	101	0.38	Int.	50	0.50	0.45
	101	1.07	Int.	38		1.22
MEK–toluene	101	0.38	Pall	25	0.29	0.35
	101	0.38	Int.	25	0.27	0.23
	101	0.38	Berl	25	0.31	0.31

MEK = methyl ethyl ketone, IPA = isopropyl alcohol.
Pall = Pall rings, Berl = Berl saddles, Int. = INTALOX® saddles.

methods given in this section have been found to be reliable for preliminary design work, and in the absence of practical values can be used for the final design with a suitable factor of safety.

The approach taken by the authors of the two methods is fundamentally different, and this provides a useful cross-check on the predicted values. Judgment must always be used when using predictive methods in design, and it is always worthwhile trying several methods and comparing the results.

Typical values for the HTU of random packings are:

25 nm (1 in)	0.3 to 0.6 m (1 to 2 ft)
38 mm ($1\frac{1}{2}$ in)	0.5 to 0.75 m ($1\frac{1}{2}$ to $2\frac{1}{2}$ ft)
50 mm (2 in)	0.6 to 1.0 m (2 to 3 ft)

Cornell's Method

Cornell *et al.* (1960) reviewed the previously published data and presented empirical equations for predicting the height of the gas- and liquid-film transfer units. Their correlation takes into account the physical properties of the system, the gas and liquid flow rates, and the column diameter and height. Equations and figures are given for a range of sizes of Raschig rings and Berl saddles. Only those for Berl saddles are given here, as it is unlikely that Raschig rings would be considered for a new column. Though the mass-transfer efficiency of Pall rings and INTALOX® saddles will be higher than that of the equivalent-size Berl saddle, the method can be used to make conservative estimates for these packings.

Bolles and Fair (1982) have extended the correlations given in the earlier paper to include metal Pall rings.

Cornell's equations are:

$$H_G = 0.011\psi_h(Sc)_v^{0.5}\left(\frac{D_c}{0.305}\right)^{1.11}\left(\frac{Z}{3.05}\right)^{0.33}\bigg/\left(L_w^*f_1f_2f_3\right)^{0.5} \tag{11.79}$$

$$H_L = 0.305\phi_h(Sc)_L^{0.5}K_3\left(\frac{Z}{3.05}\right)^{0.15} \tag{11.80}$$

where

H_G = height of a gas-phase transfer unit, m
H_L = height of a liquid-phase transfer unit, m
$(Sc)_v$ = gas Schmidt number = (μ_v/ρ_vD_v)
$(Sc)_L$ = liquid Schmidt number = (μ_L/ρ_LD_L)
D_c = column diameter, m
Z = column height, m
K_3 = percentage flooding correction factor, from Figure 11.51
ψ_h = H_G factor from Figure 11.52
ϕ_h = H_L factor from Figure 11.53
L_w^* = liquid mass flow rate per unit area column cross-sectional area, kg/m^{2}s
f_1 = liquid viscosity correction factor = $(\mu_L/\mu_w)^{0.16}$
f_2 = liquid density correction factor = $(\rho_w/\rho_L)^{1.25}$
f_3 = surface tension correction factor = $(\sigma_w/\sigma_L)^{0.8}$

where the suffix w refers to the physical properties of water at 20°C; all other physical properties are evaluated at the column conditions.

The terms $(D_c/0.305)$ and $(Z/3.05)$ are included in the equations to allow for the effects of column diameter and packed-bed height. The 'standard' values used by Cornell were 1 ft (0.305 m) for diameter, and 10 ft (3.05 m) for height. These correction terms will clearly give silly results if applied over too wide a range of values. For design purposes the diameter correction term should be taken as a fixed value of 2.3 for columns above 0.6 m (2 ft) diameter, and the height correction should only be included when the distance between liquid redistributors is greater than 3 m. To use Figures 11.51 and 11.52 an estimate of the column percentage flooding is needed. This can be obtained from Figure 11.54, where a flooding line has been included with the lines of constant pressure drop.

$$\text{Percentage flooding} = \left[\frac{K_4 \text{ at design pressure drop}}{K_4 \text{ at flooding}}\right]^{1/2} \tag{11.81}$$

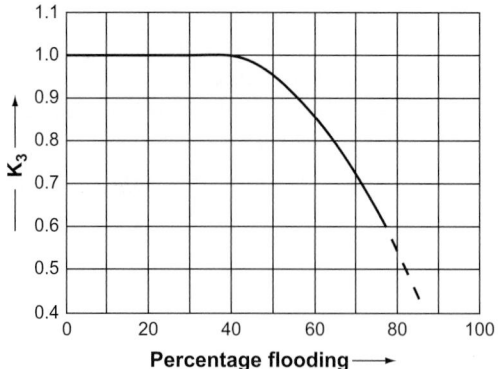

Figure 11.51.   Percentage flooding correction factor.

Onda's Method

Onda *et al.* (1968) published useful correlations for the film mass-transfer coefficients k_G and k_L and the effective wetted area of the packing a_w, which can be used to calculate H_G and H_L.

Their correlations were based on a large amount of data on gas absorption and distillation, with a variety of packings that included Pall rings and Berl saddles. Their method for estimating the effective area of packing can also be used with experimentally determined values of the mass-transfer coefficients, and values predicted using other correlations.

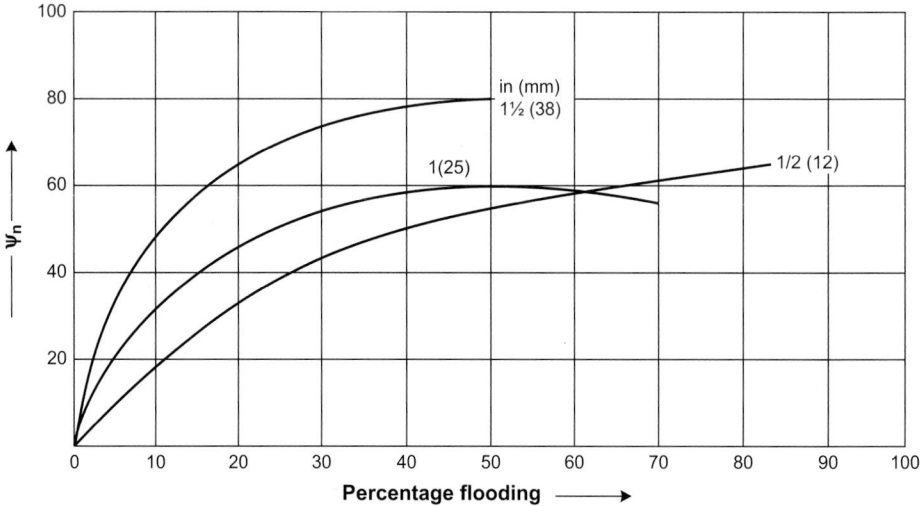

Figure 11.52.   Factor for H_G for Berl saddles.

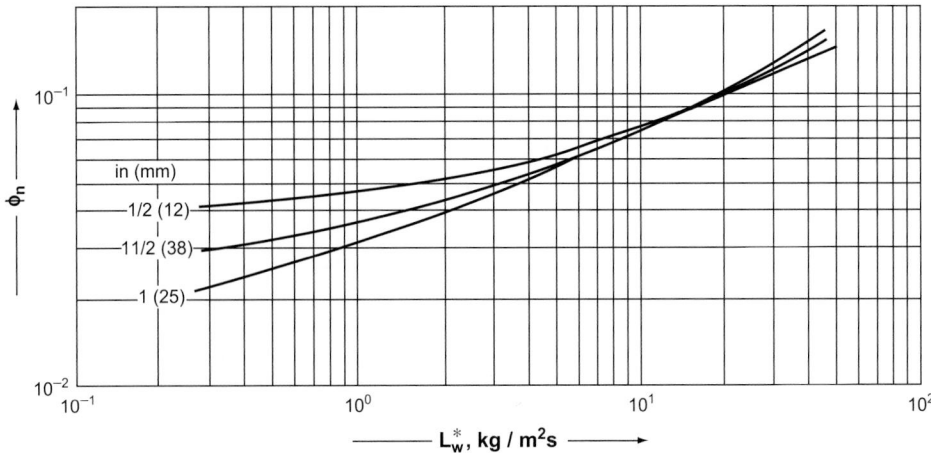

Figure 11.53. Factor for **H**$_L$ for Berl saddles.

The equation for the effective area is:

$$\frac{a_w}{a} = 1 - \exp\left[-1.45 \left(\frac{\sigma_c}{\sigma_L}\right)^{0.75} \left(\frac{L_w^*}{a\mu_L}\right)^{0.1} \left(\frac{L_w^{*2}a}{\rho_L^2 g}\right)^{-0.05} \left(\frac{L_w^{*2}}{\rho_L\sigma_L a}\right)^{0.2} \right]$$ (11.82)

and for the mass coefficients:

$$k_L \left(\frac{\rho_L}{\mu_L g}\right)^{1/3} = 0.0051 \left(\frac{L_w^*}{a_w\mu_L}\right)^{2/3} \left(\frac{\mu_L}{\rho_L D_L}\right)^{-1/2} (a d_p)^{0.4}$$ (11.83)

$$\frac{k_G}{a} \frac{RT}{D_v} = K_5 \left(\frac{V_w^*}{a\mu_v}\right)^{0.7} \left(\frac{\mu_v}{\rho_v D_v}\right)^{1/3} (a d_p)^{-2.0}$$ (11.84)

where

$K_5 = 5.23$ for packing sizes above 15 mm, and 2.00 for sizes below 15 mm
L_w^* = liquid mass flow rate per unit cross-sectional area, kg/m^{2}s
V_w^* = gas mass flow rate per unit column cross-sectional area, kg/m^{2}s
a_w = effective interfacial area of packing per unit volume, m^2/m^3
a = actual area of packing per unit volume (see Table 11.2), m^2/m^3
d_p = packing size, m
σ_c = critical surface tension for the particular packing material given below:

Material	σ_c (mN/m)
Ceramic	61
Metal (steel)	75
Plastic (polyethylene)	33
Carbon	56

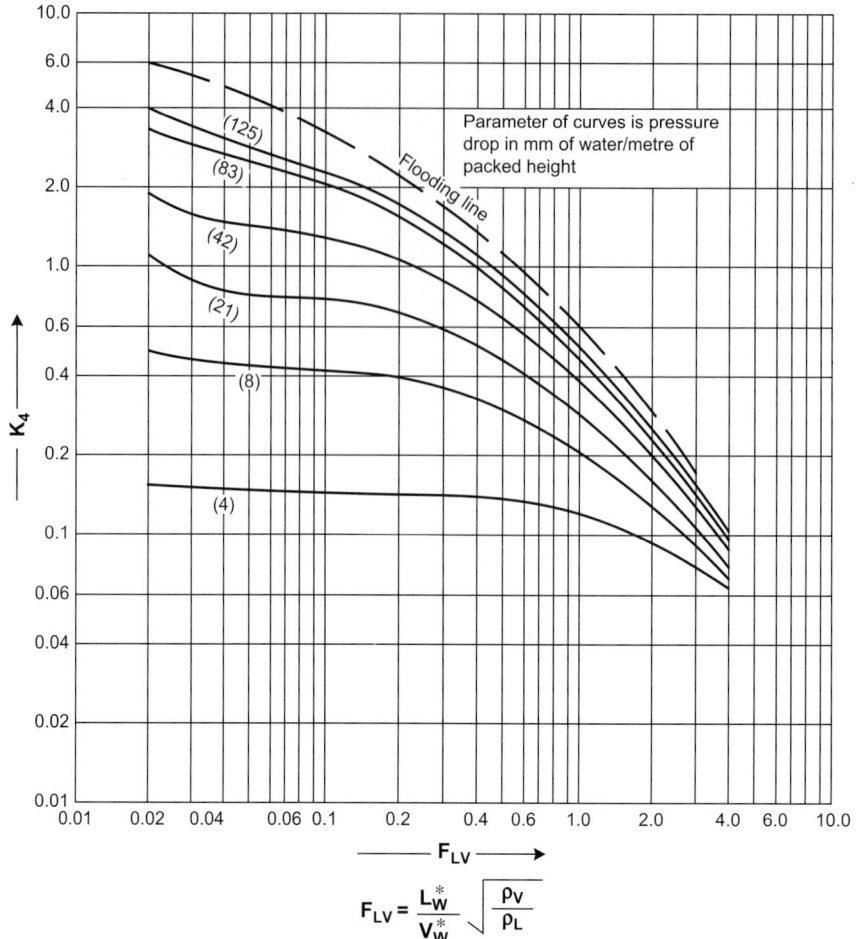

Figure 11.54. Generalized pressure-drop correlation, adapted from a figure by Koch-Glitsch, LP, with permission.

σ_L = liquid surface tension, N/m
k_G = gas-film mass-transfer coefficient, $kmol/m^2 s$ atm or $kmol/m^2 s$ bar
k_L = liquid-film mass-transfer coefficient, $kmol/m^2 s$ $(kmol/m^3)$ = m/s.

Note: all the groups in the equations are dimensionless.

The units for k_G will depend on the units used for the gas constant:

$$R = 0.08206 \text{ atm m}^3/\text{kmol K}$$

or

$$0.08314 \text{ bar m}^3/\text{kmol K}$$

The film transfer unit heights are given by:

$$H_G = \frac{G_m}{k_G a_w P} \qquad (11.85)$$

$$H_L = \frac{L_m}{k_L a_w C_t} \qquad (11.86)$$

where

P = column operating pressure, atm or bar
C_t = total concentration, kmol/m^3 = ρ_L/molecular weight solvent
G_m = molar gas flow rate per unit cross-sectional area, kmol/m^{2}s
L_m = molar liquid flow rate per unit cross-sectional area, kmol/m^{2}s.

11.14.4. Column diameter (capacity)

The capacity of a packed column is determined by its cross-sectional area. Normally, the column will be designed to operate at the highest economical pressure drop, to ensure good liquid and gas distribution. For random packings the pressure drop will not normally exceed 80 mm of water per metre of packing height. At this value the gas velocity will be about 80% of the flooding velocity. Recommended design values, millimetres water per metre packing, are:

Absorbers and strippers   15 to 50
Distillation, atmospheric and moderate pressure   40 to 80

Where the liquid is likely to foam, these values should be halved.

For vacuum distillations the maximum allowable pressure drop will be determined by the process requirements, but for satisfactory liquid distribution the pressure drop should not be less than 8 mm water per metre. If very low bottom pressures are required a special low-pressure-drop gauze packing should be considered, such as Hyperfil®, Multifil® or Dixon rings.

The column cross-sectional area and diameter for the selected pressure drop can be determined from the generalized pressure-drop correlation given in Figure 11.54. Figure 11.54 correlates the liquid and vapour flow rates, system physical properties and packing characteristics, with the gas mass flow rate per unit cross-sectional area; with lines of constant pressure drop as a parameter.

The term K_4 on Figure 11.54 is the function:

$$K_4 = \frac{13.1(V_w^*)^2 F_p \left(\dfrac{\mu_L}{\rho_L}\right)^{0.1}}{\rho_v(\rho_L - \rho_v)} \qquad (11.87)$$

where

V_w^* = gas mass flow rate per unit column cross-sectional area, kg/m^{2}s
F_p = packing factor, characteristic of the size and type of packing, see Table 11.2, m^{-1}

μ_L = liquid viscosity, Ns/m^2

ρ_L, ρ_v = liquid and vapour densities, kg/m^3.

The values of the flow factor F_{LV} given in Figure 11.54 cover the range that will generally give satisfactory column performance.

The ratio of liquid to gas flow will be fixed by the reflux ratio in distillation, and in gas absorption will be selected to give the required separation with the most economic use of solvent.

A new generalized correlation for pressure drop in packed columns, similar to Figure 11.54, has been published by Leva (1992, 1995). The new correlation gives a better prediction for systems where the density of the irrigating fluid is appreciably greater than that of water. It can also be used to predict the pressure drop over dry packing.

Example 11.9

Sulphur dioxide produced by the combustion of sulphur in air is absorbed in water. Pure SO_2 is then recovered from the solution by steam stripping. Make a preliminary design for the absorption column. The feed will be 5000 kg/h of gas containing 8% v/v SO_2. The gas will be cooled to 20°C. A 95% recovery of the sulphur dioxide is required.

Solution

As the solubility of SO_2 in water is high, operation at atmospheric pressure should be satisfactory. The feed-water temperature will be taken as 20°C, a reasonable design value.

Solubility Data

From Perry et al. (1973):

	per cent w/w solution	0.05	0.1	0.15	0.2	0.3	0.5	0.7	1.0	1.5
SO_2	Partial pressure gas mmHg	1.2	3.2	5.8	8.5	14.1	26	39	59	92

Partial pressure of SO_2 in the feed = $(8/100) \times 760 = 60.8$ mmHg.

These data are plotted in Figure 11.55.

Number of Stages

Partial pressure in the exit gas at 95% recovery = $60.8 \times 0.05 = 3.04$ mmHg.

Over this range of partial pressure the equilibrium line is essentially straight, so Figure 11.49 can be used to estimate the number of stages needed.

The use of Figure 11.49 will slightly overestimate the number of stages and a more accurate estimate would be made by graphical integration of equation 11.72, but this is not justified in view of the uncertainty in the prediction of the transfer unit height.

Molecular weights: $SO_2 = 64$, $H_2O = 18$, air $= 29$.

Slope of Equilibrium Line

From the data: partial pressure at 1.0% w/w $SO_2 = 59$ mmHg.

$$\text{Mole fraction in vapour} = \frac{59}{760} = 0.0776$$

$$\text{Mole fraction in liquid} = \frac{\frac{1}{64}}{\frac{1}{64} + \frac{99}{18}} = 0.0028$$

$$m = \frac{0.0776}{0.0028} = 27.4$$

To decide the most economic water flow rate, the stripper design should be considered together with the absorption design, but for the purpose of this example the absorption design will be considered alone. Using Figure 11.49, the number of stages required at different water rates will be determined and the 'optimum' rate chosen:

$$\frac{y_1}{y_2} = \frac{p_1}{p_2} = \frac{60.8}{3.04} = 20$$

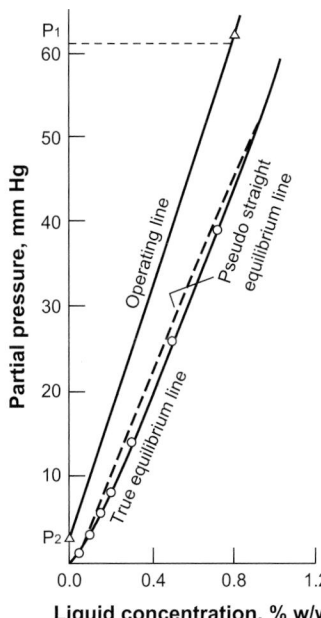

Figure 11.55. SO_2 absorber design (Example 11.9).

$m\frac{G_m}{L_m}$	0.5	0.6	0.7	0.8	0.9	1.0
N_{OG}	3.7	4.1	6.3	8	10.8	19.0

It can be seen that the 'optimum' will be between $mG_m/L_m = 0.6$ to 0.8, as would be expected. Below 0.6 there is only a small decrease in the number of stages required with increasing liquid rate; and above 0.8 the number of stages increases rapidly with decreasing liquid rate.

Check the liquid outlet composition at 0.6 and 0.8:

$$\text{Material balance } L_m x_1 = G_m(y_1 - y_2)$$

$$\text{so } x_1 = \frac{G_m}{L_m}(0.08 \times 0.95) = \frac{m}{27.4}\frac{G_m}{L_m}(0.076)$$

$$\text{at } \frac{mG_m}{L_m} = 0.6, x_1 = 1.66 \times 10^{-3} \text{ mol fraction}$$

$$\text{at } \frac{mG_m}{L_m} = 0.8, x_1 = 2.22 \times 10^{-3} \text{ mol fraction}$$

Use 0.8, as the higher concentration will favour the stripper design and operation, without significantly increasing the number of stages needed in the absorber.

$$N_{OG} = \underline{\underline{8}}$$

Column Diameter

The physical properties of the gas can be taken as those for air, as the concentration of SO_2 is low.

$$\text{Gas flowrate} = \frac{5000}{3600} = 1.39 \text{ kg/s}, = \frac{1.39}{29} = 0.048 \text{ kmol/s}$$

$$\text{Liquid flowrate} = \frac{27.4}{0.8} \times 0.048 = 1.64 \text{ kmol/s}$$

$$= 29.5 \text{ kg/s}$$

Select 38-mm ($1\frac{1}{2}$-in) ceramic INTALOX® saddles.
From Table 11.3, $F_p = 170 \text{ m}^{-1}$

$$\text{Gas density at } 20°C = \frac{29}{22.4} \times \frac{273}{293} = 1.21 \text{ kg/m}^3$$

$$\text{Liquid density} \simeq 1000 \text{ kg/m}^3$$

$$\text{Liquid viscosity} = 10^{-3} \text{ Ns/m}^2$$

$$\frac{L_W^*}{V_W^*}\sqrt{\frac{\rho_v}{\rho_L}} = \frac{29.5}{1.39}\sqrt{\frac{1.21}{10^3}} = 0.74$$

Design for a pressure drop of 20 mmH$_2$O per metre packing.

From Figure 11.54

$$K_4 = 0.35$$

At flooding $K_4 = 0.8$

$$\text{Percentage flooding} = \sqrt{\frac{0.35}{0.8}} \times 100 = 66\%, \text{ satisfactory.}$$

From equation 11.87

$$V_W^* = \left[\frac{K_4 \rho_V (\rho_L - \rho_v)}{13.1 F_p (\mu_L / \rho_L)^{0.1}}\right]^{1/2}$$

$$= \left[\frac{0.35 \times 1.21(1000 - 1.21)}{13.1 \times 170(10^{-3}/10^3)^{0.1}}\right]^{1/2} = 0.87 \text{ kg/m}^2\text{s}$$

$$\text{Column area required} = \frac{1.39}{0.87} = 1.6 \text{ m}^2$$

$$\text{Diameter} = \sqrt{\frac{4}{\pi} \times 1.6} = 1.43 \text{ m}$$

Round off to 1.50 m

$$\text{Column area} = \frac{\pi}{4} \times 1.5^2 = 1.77 \text{ m}^2$$

$$\text{Packing size to column diameter ratio} = \frac{1.5}{38 \times 10^{-3}} = 39$$

A larger packing size could be considered.
   Percentage flooding at selected diameter

$$= 66 \times \frac{1.6}{1.77} = 60\%,$$

Could consider reducing column diameter to the nearest standard pipe size.

Estimation of H_{OG}

Cornell's method

$$D_L = 1.7 \times 10^{-9} \text{m}^2/\text{s}$$

$$D_v = 1.45 \times 10^{-5} \text{m}^2/\text{s}$$

$$\mu_v = 0.018 \times 10^{-3} \text{Ns/m}^2$$

$$(Sc)_v = \frac{0.018 \times 10^{-3}}{1.21 \times 1.45 \times 10^{-5}} = 1.04$$

$$(Sc)_L = \frac{10^{-3}}{1000 \times 1.7 \times 10^{-9}} = 588$$

$$L_W^* = \frac{29.5}{1.77} = 16.7 \text{ kg/s m}^2$$

From Figure 11.51, at 60% flooding, $K_3 = 0.85$.

From Figure 11.52, at 60% flooding, $\psi_h = 80$.

From Figure 11.53, at $L_W^* = 16.7$, $\phi_h = 0.1$.

H_{OG} can be expected to be around 1 m, so as a first estimate Z can be taken as 8 m. The column diameter is greater than 0.6 m so the diameter correction term will be taken as 2.3.

$$H_L = 0.305 \times 0.1 (588)^{0.5} \times 0.85 \left(\frac{8}{3.05}\right)^{0.15} = 0.7 \text{ m} \qquad (11.80)$$

As the liquid temperature has been taken as 20°C, and the liquid is water

$$f_1 = f_2 = f_3 = 1$$

$$H_G = 0.011 \times 80 (1.04)^{0.5} (2.3) \left(\frac{8}{3.05}\right)^{0.33} \Big/ (16.7)^{0.5} = 0.7 \text{ m} \qquad (11.79)$$

$$H_{OG} = 0.7 + 0.8 \times 0.7 = 1.3 \text{ m}$$
$$(11.73)$$
$$Z = 8 \times 1.3 = \underline{10.4 \text{ m}}, \text{ close enough to the estimated value.}$$

Onda's method

$$R = 0.08314 \text{ bar m}^3/\text{kmol.K}$$

Surface tension of liquid, taken as water at 20°C $= 70 \times 10^{-3}$ N/m

$$g = 9.81 \text{ m/s}^2$$

$$d_p = 38 \times 10^{-3} \text{ m}$$

From Table 11.3, for 38-mm INTALOX® saddles

$$a = 194 \text{ m}^2/\text{m}^3$$

$$\sigma_c \text{ for ceramics} = 61 \times 10^{-3} \text{ N/m}$$

$$\frac{a_W}{a} = 1 - \exp\left[-1.45 \left(\frac{61 \times 10^{-3}}{70 \times 10^{-3}}\right)^{0.75} \left(\frac{17.6}{194 \times 10^{-3}}\right)^{0.1} \left(\frac{17.6^2 \times 194}{1000^2 \times 9.81}\right)^{-0.05}\right.$$
$$\left. \times \left(\frac{17.6^2}{1000 \times 70 \times 10^{-3} \times 194}\right)^{0.2}\right] = 0.71$$

$$a_W = 0.71 \times 194 = 138 \text{ m}^2/\text{m}^3 \qquad (11.82)$$

$$k_L \left(\frac{10^3}{10^{-3} \times 9.81} \right)^{1/3} = 0.0051 \left(\frac{17.6}{138 \times 10^{-3}} \right)^{2/3} \left(\frac{10^{-3}}{10^3 \times 1.7 \times 10^{-9}} \right)^{-1/2}$$

$$\times \, (194 \times 38 \times 10^{-3})^{0.4} \qquad (11.83)$$

$$k_L = 2.5 \times 10^{-4} \, \text{m/s}$$

$$V_W^* \text{ on actual column diameter} = \frac{1.39}{1.77} = 0.79 \, \text{kg/m}^2\text{s}$$

$$k_G \frac{0.08314 \times 293}{194 \times 1.45 \times 10^{-5}} = 5.23 \left(\frac{0.79}{194 \times 0.018 \times 10^{-3}} \right)^{0.7}$$

$$\times \, \left(\frac{0.018 \times 10^{-3}}{1.21 \times 1.45 \times 10^{-5}} \right)^{1/3} (194 \times 38 \times 10^{-3})^{-2.0}$$

$$k_G = 5.0 \times 10^{-4} \, \text{kmol/sm}^2 \, \text{bar} \qquad (11.84)$$

$$G_m = \frac{0.79}{29} = 0.027 \, \text{kmol/m}^2\text{s}$$

$$L_m = \frac{16.7}{18} = 0.93 \, \text{kmol/m}^2\text{s}$$

$$\mathbf{H}_G = \frac{0.027}{5.0 \times 10^{-4} \times 138 \times 1.013} = 0.39 \, \text{m}$$

$$C_T = \text{total concentration, as water,} \qquad (11.85)$$

$$= \frac{1000}{18} = 55.5 \, \text{kmol/m}^3$$

$$\mathbf{H}_L = \frac{0.93}{2.5 \times 10^{-4} \times 138 \times 55.6} = 0.49 \, \text{m} \qquad (11.86)$$

$$\mathbf{H}_{OG} = 0.39 + 0.8 \times 0.49 = \underline{0.78 \, \text{m}} \qquad (11.73)$$

Use higher value, estimated using Cornell's method, and round up packed-bed height to 11 m.

11.14.5. Column Internals

The internal fittings in a packed column are simpler than those in a plate column but must be carefully designed to ensure good performance. As a general rule, the standard fittings developed by the packing manufacturers should be specified. Some typical designs are shown in Figures 11.56 to 11.65, and their use is discussed in the following paragraphs.

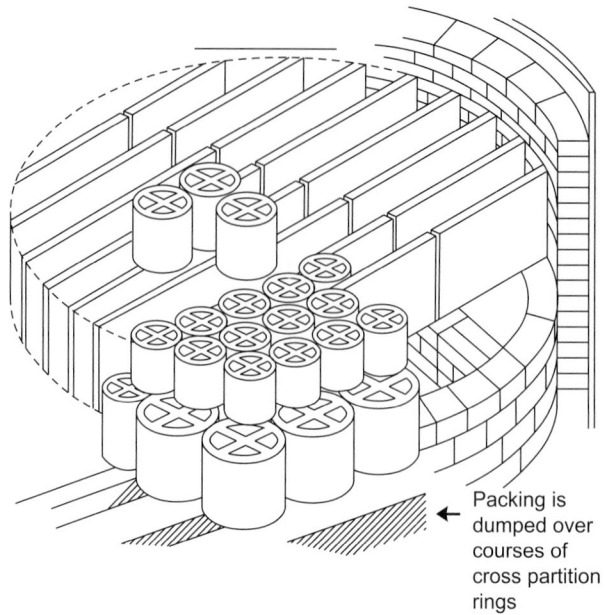

Packing is
dumped over
courses of
cross partition
rings

Figure 11.56. Stacked packing used to support random packing.

Packing support

The function of the support plate is to carry the weight of the wet packing, while allowing free passage of the gas and liquid. These requirements conflict; a poorly designed support will give a high pressure drop and can cause local flooding. Simple grid and perforated plate supports are used, but in these designs the liquid and gas have to vie for the same openings. Wide-spaced grids are used to increase the flow area; with layers of larger size packing stacked on the grid to support the small-size random packing (Figure 11.56).

The best design of packing support is one in which gas inlets are provided above the level where the liquid flows from the bed, such as the gas-injection type shown in Figures 11.57 and 11.58. These designs have a low pressure drop and no tendency to flooding. They are available in a wide range of sizes and materials: metals, ceramics and plastics.

Liquid Distributors

The satisfactory performance of a plate column is dependent on maintaining a uniform flow of liquid throughout the column, and good initial liquid distribution is essential. Various designs of distributors are used. For small-diameter columns a central open feed-pipe, or one fitted with a spray nozzle, may well be adequate; but for larger columns more elaborate designs are needed to ensure good distribution at all liquid flow rates. The two most commonly used designs are the orifice type, shown in Figure 11.59, and the weir type, shown in Figure 11.60. In the orifice type the liquid flows through holes in the plate and the gas through short stand pipes. The gas pipes should be sized to give sufficient area for gas flow without creating a significant pressure drop; the holes should be small enough to ensure that there is a level of liquid

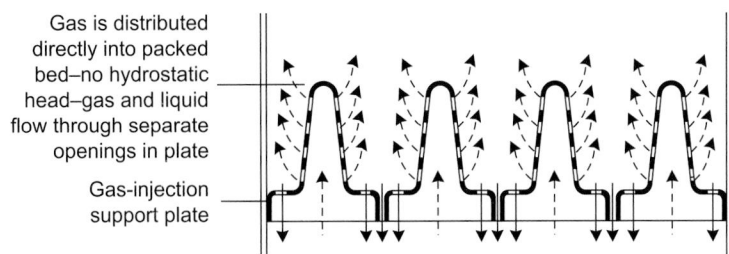

Gas is distributed directly into packed bed—no hydrostatic head—gas and liquid flow through separate openings in plate

Gas-injection support plate

Figure 11.57. The principle of the gas-injection packing support.

on the plate at the lowest liquid rate, but large enough to prevent the distributor overflowing at the highest rate. In the weir type the liquid flows over notched weirs in the gas stand-pipes. This type can be designed to cope with a wider range of liquid flow rates than the simpler orifice type.

For large-diameter columns, the trough-type distributor shown in Figure 11.61 can be used, and will give good liquid distribution with a large free area for gas flow.

All distributors that rely on the gravity flow of liquid must be installed in the column level, or maldistribution of liquid will occur.

A pipe manifold distributor, Figure 11.62, can be used when the liquid is fed to the column under pressure and the flow rate is reasonably constant. The distribution pipes and orifices should be sized to give an even flow from each element.

Liquid Redistributors

Redistributors are used to collect liquid that has migrated to the column walls and redistribute it evenly over the packing. They will also even out any maldistribution that has occurred within the packing.

A full redistributor combines the functions of a packing support and a liquid distributor; a typical design is shown in Figure 11.63.

The 'wall-wiper' type of redistributor, in which a ring collects liquid from the-column wall and redirects it into the centre packing, is occasionally used in

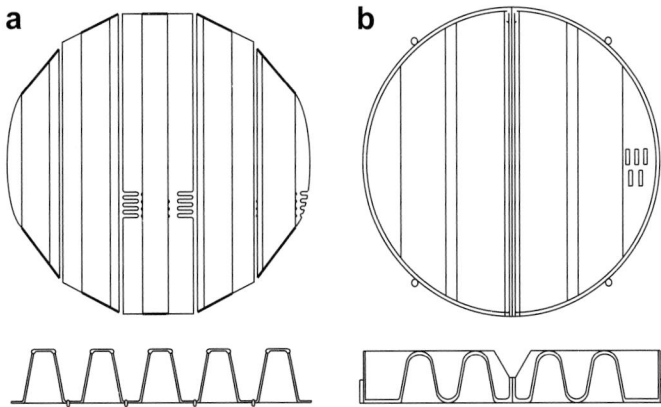

Figure 11.58. Typical designs of gas-injection supports (Koch-Glitsch, LP). (a) Small-diameter columns. (b) Large-diameter columns.

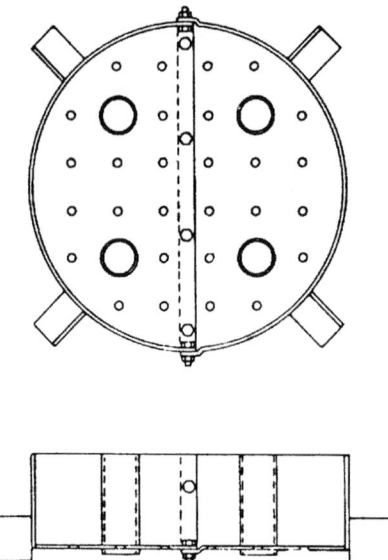

Figure 11.59.   Orifice-type distributor (Koch-Glitsch, LP).

small-diameter columns, less than 0.6 m. Care should be taken when specifying this type to select a design that does not unduly restrict the gas flow and cause local flooding. A good design is that shown in Figure 11.64.

The maximum bed height that should be used without liquid redistribution depends on the type of packing and the process. Distillation is less susceptible to maldistribution than absorption and stripping. As a general guide, the maximum bed height should not exceed three column diameters for Raschig rings, and 8 to 10 for

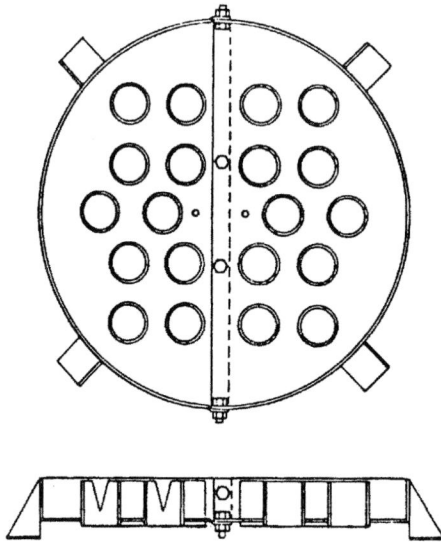

Figure 11.60.   Weir-type distributor (Koch-Glitsch, LP).

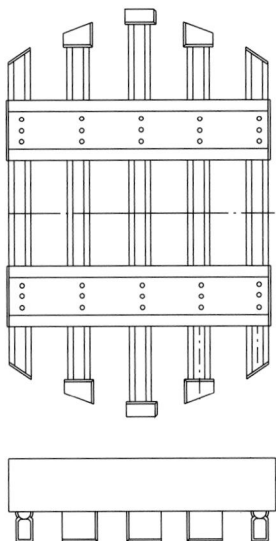

Figure 11.61.   Weir-trough distributors (Koch-Glitsch, LP).

Pall rings and saddles. In a large-diameter column the bed height will also be limited by the maximum weight of packing that can be supported by the packing support and column walls; this will be around 8 m.

Hold-Down Plates

At high gas rates, or if surging occurs through mis-operation, the top layers of packing can be fluidized. Under these conditions ceramic packing can break up and the pieces

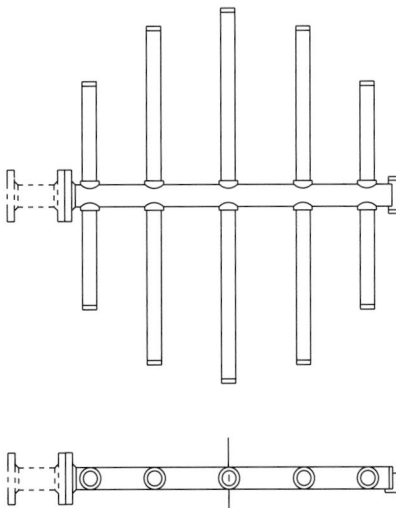

Figure 11.62.   Pipe distributor (Koch-Glitsch, LP).

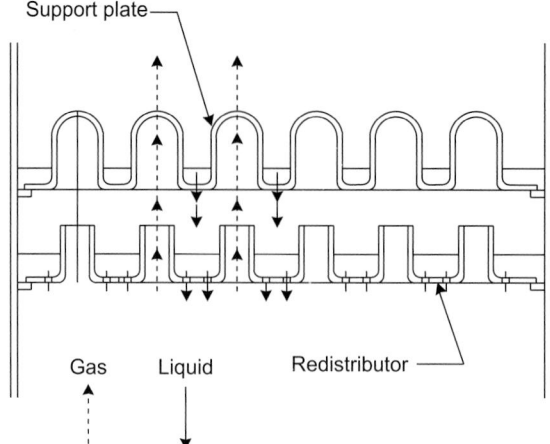

Figure 11.63. Full redistributor.

filter down the column and plug the packing; metal and plastic packing can be blown out of the column. Hold-down plates are used with ceramic packing to weigh down the top layers and prevent fluidization; a typical design is shown in Figure 11.65. Bed-limiters are sometimes used with plastics and metal packings to prevent expansion of the bed when operating at a high pressure drop. They are similar to hold-down plates

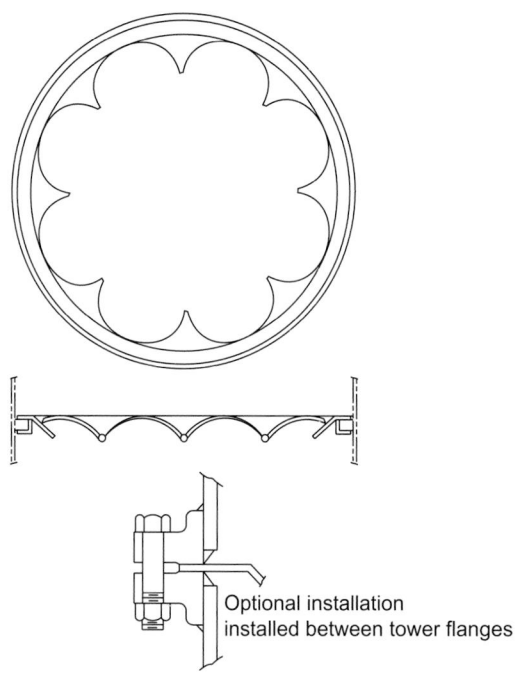

Figure 11.64. 'Wall wiper' redistributor (Koch-Glitsch, LP).

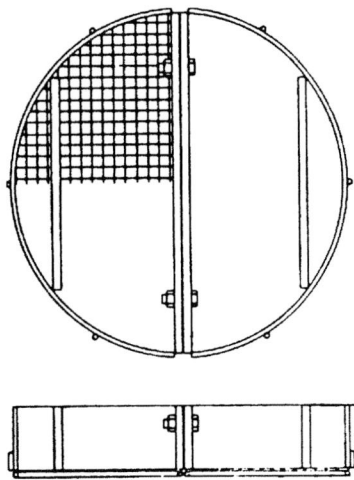

Figure 11.65. Hold-down plate design (Koch-Glitsch, LP).

but are of lighter construction and are fixed to the column walls. The openings in hold-down plates and bed-limiters should be small enough to retain the packing, but should not restrict the gas and liquid flow.

Installing Packing

Ceramic and metal packings are normally dumped into the column 'wet', to ensure a truly random distribution and prevent damage to the packing. The column is partially filled with water and the packing dumped into the water. A height of water must be kept above the packing at all times.

If the columns must be packed dry, for instance to avoid contamination of process fluids with water, the packing can be lowered into the column in buckets or other containers. Ceramic packings should not be dropped from a height of more than half a metre.

Liquid Hold-Up

An estimate of the amount of liquid held up in the packing under operating conditions is needed to calculate the total load carried by the packing support. The liquid hold-up will depend on the liquid rate and, to some extent, on the gas flow rate. The packing manufacturers' design literature should be consulted to obtain accurate estimates. As a rough guide, a value of about 25% of the packing weight can be taken for ceramic packings.

11.14.6. Wetting Rates

If very low liquid rates must be used, outside the range of F_{LV} given in Figure 11.54, the packing wetting rate should be checked to make sure it is above the minimum recommended by the packing manufacturer.

Wetting rate is defined as:

$$\text{wetting rate} = \frac{\text{volumetric liquid rate per unit cross-sectional area}}{\text{packing surface area per unit volume}}$$

A nomograph for the calculation of wetting rates is given in Richardson *et al.* (2002).

Wetting rates are frequently expressed in terms of mass or volume flow rate per unit column cross-sectional area. Kister (1992) gives values for minimum wetting rates of 0.5 to 2 gpm/ft^2 (0.35×10^{-3} to 1.4×10^{-3} m^3 s^{-1}/m^2) for random packing and 0.1 to 0.2 gpm/ft^2 (0.07×10^{-3} to 0.14×10^{-3} m^3 s^{-1}/m^2) for structured packing. Norman (1961) recommends that the liquid rate in absorbers should be kept above 2.7 kg/m^{2}s.

If the design liquid rate is too low, the diameter of the column should be reduced. For some processes liquid can be recycled to increase the flow over the packing.

A substantial factor of safety should be applied to the calculated bed height for processes where the wetting rate is likely to be low and less space should be allowed between liquid redistributors.

11.15. COLUMN AUXILIARIES

Intermediate storage tanks will normally be needed to smooth out fluctuations in column operation and process upsets. These tanks should be sized to give sufficient hold-up time for smooth operation and control. The hold-up time required will depend on the nature of the process and on how critical the operation is; some typical values for distillation processes are given below:

Operation	Time (min)
Feed to a train of columns	10 to 20
Between columns	5 to 10
Feed to a column from storage	2 to 5
Reflux drum	5 to 15

The time given is that for the level in the tank to fall from the normal operating level to the minimum operating level if the feed ceases.

Horizontal or vertical tanks are used, depending on the size and duty. Where only a small hold-up volume is required, this can be provided by extending the column base, or, for reflux accumulators, by extending the bottom header of the condenser.

The specification and sizing of surge tanks and accumulators is discussed in more detail by Mehra (1979) and Evans (1980).

11.16. SOLVENT EXTRACTION (LIQUID–LIQUID EXTRACTION)

Extraction should be considered as an alternative to distillation in the following situations:

1. Where the components in the feed have close boiling points. Extraction in a suitable solvent may be more economic if the relative volatility is below 1.2.

2. If the feed components form an azeotrope.
3. If the solute is heat sensitive, and can be extracted into a lower boiling solvent, to reduce the heat history during recovery.

Solvent Selection

The following factors must be considered when selecting a suitable solvent for a given extraction.

1. *Phase equilibrium*: the distribution coefficient or partition coefficient, K, is the ratio of the mole fractions in each phase: $x_\alpha = K x_\beta$. The distribution coefficient is analogous to the equilibrium constant in absorption. The relative separation or selectivity is a measure of the distribution of two solutes between the two solvents, and is the ratio of the distribution coefficients for the two solutes. Selectivity is analogous to relative volatility in distillation. The greater the difference in solubility of the preferred solute between the two solvents, the easier it will be to extract. This is particularly important when trying to preferentially extract one component from a mixture, as often occurs in the recovery of fermentation products and specialty chemicals.
2. *Partition ratio*: this is the weight fraction of the solute in the extract divided by the weight fraction in the raffinate. This determines the quantity of solvent and number of stages needed. The less solvent needed, the lower will be the solvent and solvent recovery costs.
3. *Density*: the greater the density difference between the feed and extraction solvents the easier it will be to separate the solvents.
4. *Miscibility*: ideally the two solvents should be immiscible. The greater the solubility of the extraction solvent in the feed solvent, the more difficult it will be to recover the solvent from the raffinate, and the higher the cost.
5. *Safety*: whenever it is possible and economical, a solvent should be chosen that is not toxic or dangerously flammable.
6. *Cost*: the purchase cost of the solvent is important but should not be considered in isolation from the total process costs. It may be worth considering a more expensive solvent if it is more effective and easier to recover.

11.16.1. Extraction Equipment

Extraction equipment can be divided into two broad groups:

1. Stage-wise extractors, in which the liquids are alternately contacted (mixed) and then separated, in a series of stages. The 'mixer-settler' contactor is an example of this type. Several mixer-settlers are often used in series to increase the effectiveness of the extraction.
2. Differential extractors, in which the phases are continuously in contact in the extractor and are only separated at the exits; for example, in packed column extractors.

Extraction columns can be further sub-divided according to the method used to promote contact between the phases: packed, plate, mechanically agitated or pulsed columns. Various types of proprietary centrifugal extractors are also used.

The following factors need to be taken into consideration when selecting an extractor for a particular application:

1. The number of stages required
2. The throughputs
3. The settling characteristics of the phases
4. The available floor area and head room.

Hanson (1968) has given a selection guide based on these factors, which can be used to select the type of equipment most likely to be suitable (Figure 11.66).

The basic principles of liquid–liquid extraction are covered in several specialist texts: Treybal (1980), Robbins (1997) and Humphrey and Keller (1997).

11.16.2. Extractor Design

Number of Stages

The primary task in the design of an extractor for a liquid–liquid extraction process is the determination of the number of stages needed to achieve the separation required.

The stages may be arranged in three ways:

1. Fresh solvent fed to each stage, the raffinate passing from stage to stage (cross-flow).
2. The extracting solvent fed co-currently with the raffinate, from stage to stage (co-current flow).
3. The exacting solvent fed counter-current to the raffinate (counter-current flow).

Counter-current flow is the most efficient method and the most commonly used. It will give the greatest concentration of the solute in the extract, and the least use of solvent.

Equilibrium Data

To determine the number of stages, it is best to plot the equilibrium data on a triangular diagram (Figure 11.67). Each corner of the triangle represents 100% of the feed solvent, solute or extraction solvent. Each side represents the composition of one of the binary pairs. The ternary compositions are shown in the interior of the triangle. Mixtures within the region bounded by the curve will separate into two phases. The tie-lines link the equilibrium compositions of the separate phases. The tie-lines reduce in length toward the top of the curve. The point where they disappear is called the plait point.

A fuller discussion of the various classes of diagram used to represent liquid–liquid equilibria is given in Richardson *et al.* (2002); see also Treybal (1980) and Humphrey *et al.* (1984).

The most comprehensive source of equilibrium data for liquid–liquid systems is the DECHEMA data series (Sorensen and Arlt, 1979). Equilibrium data for some systems are also given by Green and Perry (2007). Most commercial process simulations programs can be used to make initial estimates of liquid–liquid equilibria using a three-phase flash operation. Care is needed in selecting an appropriate phase equilibrium model. The UNIQUAC and UNIFAC equations can be used to estimate liquid–liquid equilibria; see Chapter 8.

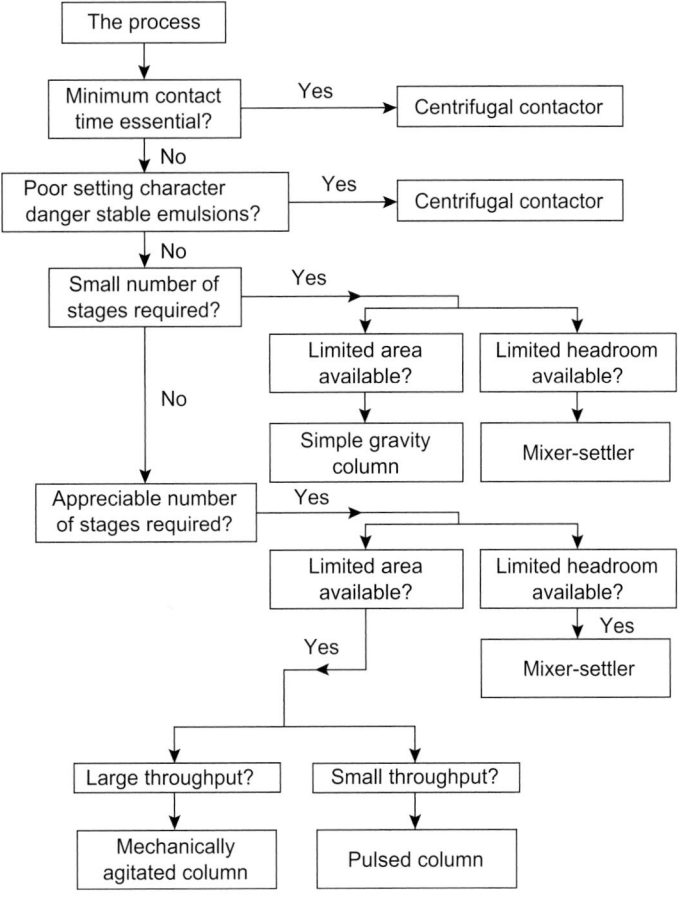

Figure 11.66.   Selection guide for liquid–liquid contactors (after Hanson, 1968).

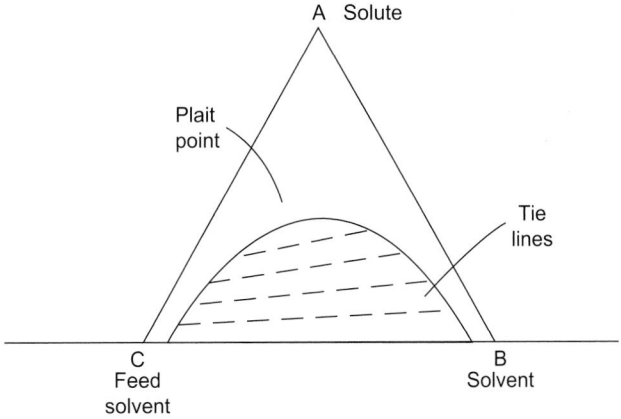

Figure 11.67.   Equilibrium diagram solute distributed between two solvents.

Number of Stages

The number of stages required for a given separation can be determined from the triangular diagram using a method analogous to the McCabe–Thiele method for distillation. The method set out below is for counter-current extraction.

Procedure

Refer to Figures 11.67 and 11.68.

Let the flow rates be

F = feed, of the solution to be extracted
E = extract
R = raffinate
S = the extracting solvent
   and the compositions:
r = raffinate
e = extract
s = solvent
f = feed.

Then a material balance over stage n gives:

$$F + E_{n+1} = R_n + E_1$$

It can be shown that the difference in flow rate between the raffinate leaving any stage, R_n, and the extract entering the stage, E_n, is constant. Also, that the difference between the amounts of each component entering and leaving a stage is constant. This means that if lines are drawn on the triangular diagram linking the composition of the raffinate from a stage and the extract entering from the next stage, they will pass through a common pole when extrapolated. The number of stages needed can be found by making use of this construction and the equilibrium compositions given by the tie-lines.

Construction

1. Draw the liquid–liquid equilibrium data on triangular graph paper. Show sufficient tie-lines to enable the equilibrium compositions to be determined at each stage.
2. Mark the feed-solvent and extraction-solvent compositions on the diagram. Join them with a line. The composition of a mixture of the feed and solvent will lie on this line.

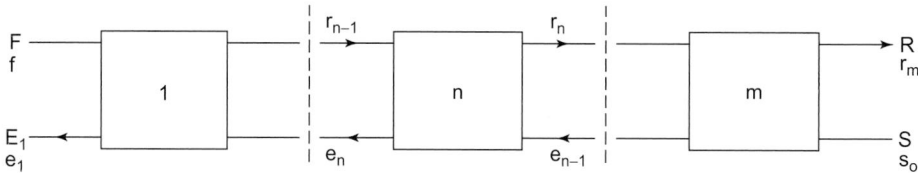

Figure 11.68.   Counter-current extraction.

3. Calculate the composition of the mixture given by mixing the feed with the extraction solvent. Mark this point, O, on the line drawn in step 2.
4. Mark the final raffinate composition, r_m, on the equilibrium curve.
5. Draw a line from r_m through the point 0. This will cut the curve at the final extract composition, e_1.

Note: if the extract composition is specified, rather than the raffinate, draw the line from e_1 through 0 to find r_m.

6. Draw a line from the solvent composition, S_0, through r_m and extend it beyond r_m.
7. Draw a line from e_1 through f and extend it to cross the line drawn in step 6, at the pole point, P.
8. Find the composition of the raffinate leaving the first stage, r_1, by judging the position of the tie-line from e_1. Draw a line from the pole point, P, through r_1 to cut the curve at e_2, the extract leaving stage 2.
9. Repeat this procedure until sufficient stages have been drawn to reach the desired raffinate final composition.

If an extended tie-line passes through the pole point P, an infinite number of stages will be needed. This condition sets the minimum flow of extraction solvent required. It is analogous to a pinch point in distillation.

The method is illustrated in Example 11.10.

Example 11.10

Acetone is to be extracted from a solution in water, using 1,1,2-trichloroethane. The feed concentration is 45.0% w/w acetone. Determine the number of stages required to reduce the concentration of acetone to below 10%, using 32 kg of extraction solvent per 100 kg feed.

The equilibrium data for this system are given by Treybal *et al. Ind. Eng. Chem.* **38**, 817 (1946).

Solution

Composition of feed + solvent, point $O = 0.45 \times 100/(100 + 32) = 0.34 = 34\%$.
Draw line from TCE (trichloroethane) $= 100\%$, point s_0, to feed composition, f, 45% acetone.
Mark point O on this line at 34% acetone.
Mark required final raffinate composition, r_m, on the equilibrium curve, at 10%.
Draw line from this point through point O to find final extract composition, e_1.
Draw line from this point though the feed composition, f, extend this line to cut a line extended from s_0 through r_m, at P.
Using the tie-lines plotted on the figure, judge the position that a tie-line would have from e_1 and mark it in, to find the point on the curve giving the composition of the raffinate leaving the first stage, r_1.

Draw a line through from the pole point P through r_1, to find the point on the curve giving the extract composition leaving the second stage, e_2.

Repeat these steps until the raffinate composition found is below 10%.

From the diagram, Figure 11.69, it can be seen that five stages are needed.

In this example, the fact that the raffinate composition from stage 5 passes through the specified raffinate composition of 10% is fortuitous. As the construction, particularly the judgment of the position of the tie-lines, is approximate, the number of stages will be increased to six. This should ensure that the specified raffinate composition of below 10% is met.

Immiscible Solvents

If the solvents are immiscible, the procedure for determining the number of stages required is simplified. The equilibrium curve can be drawn on regular, orthogonal, graph paper. An operating line, giving the relationship between the compositions of the raffinate and extracts entering and leaving each stage, can then be drawn, and the stages stepped off. The procedure is similar to the McCabe–Thiele construction for determining the number of stages in distillation; see Section 11.5.2. The slope of the operating line is the ratio of the final raffinate to fresh solvent flow rates. Alternatively, the methods given in Section 11.14.2 for design of absorption columns can be used, where G is the flow rate of the initial solvent phase, L is the flow rate of the extraction solvent, y is the mole fraction of solute in the initial solvent, and m is the partition coefficient.

For a full discussion of the methods that can be used to determine the stage requirements in liquid–liquid extraction refer to Treybal (1980), Green and Perry (2007) and Robbins (1997). Computer programs are available for the design of extraction processes and are included in the various commercial process simulation packages available; see Chapter 4. It is usually a good practice to make a few hand calculations to determine a good initialization for the process simulation model. As a minimum, the designer should estimate the partition coefficient and ensure that the absorption factor $mG/L \leq 1$ if the goal is to obtain a high recovery of solute.

11.16.3. Extraction Columns

The simplest form of extractor is a spray column. The column is empty; one liquid forms a continuous phase and the other liquid flows up or down the column in the form of droplets. Mass transfer takes places to or from the droplets to the continuous phase. The efficiency of a spray tower will be low, particularly with large-diameter columns, due to back mixing. The efficiency of the basic, empty, spray column can be improved by installing plates or packing.

Sieve plates are used, similar to those used for distillation and absorption. The plates can be designed so that either the heavy phase is dispersed in the light phase (sometimes called 'rain decks') or the light phase is dispersed in the heavy phase ('reverse rain decks').

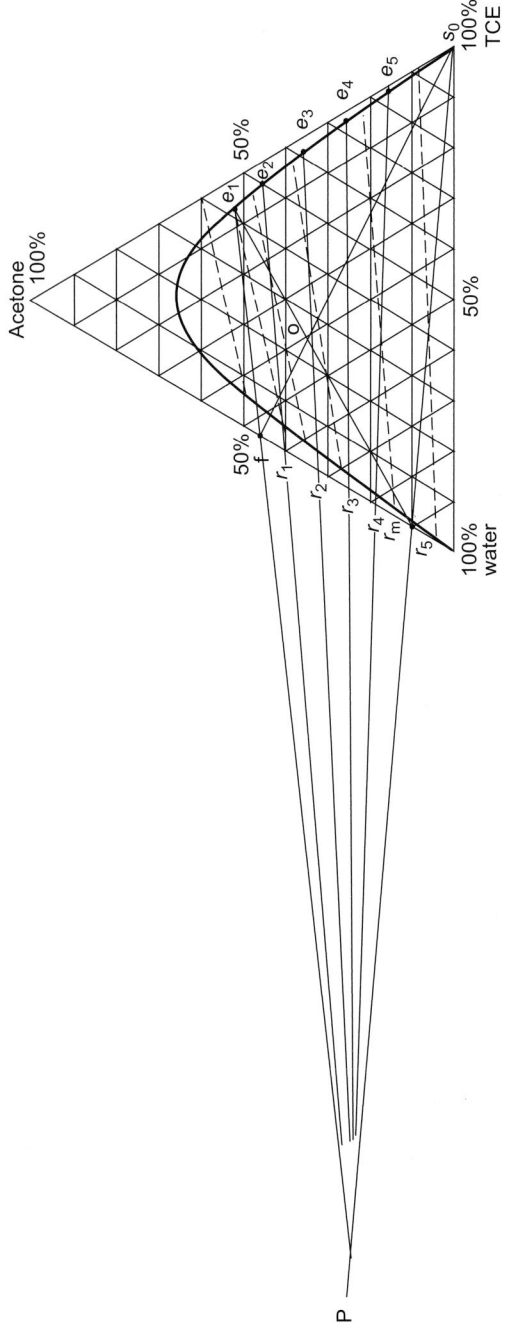

Figure 11.69.   Example 11.10.

Random packings are also used; they are the same as those used in packed distillation and absorption columns. The properties of random packings are given in Tables 11.2 and 11.3. Proprietary structured packings are also used.

Mass transfer in packed columns is a continuous, differential, process, so the transfer unit method should be used to determine the column height, as used in absorption; see Section 11.14.2. It is often convenient to treat packed extraction columns as staged processes and use the HETP for the packing employed. For random packings, the HETP in extraction will typically range from 0.5 to 1.5 m, depending on the type and size of packing used.

Flooding

No simple correlation is available to predict the flooding velocities in extraction columns, and hence the column diameter needed. The more specialized texts should be consulted to obtain guidance on the appropriate method to use for a particular problem; see Treybal (1980), Green and Perry (2007) and Humphrey and Keller (1997).

11.16.4.   Supercritical Fluid Extraction

A recent development in liquid–liquid extraction has been the use of supercritical fluids as the extraction solvent. Carbon dioxide at high pressure is the most commonly used fluid. It is used in processes for the decaffeination of coffee and tea. The solvent can be recovered from the extract solution as a gas, by reducing the pressure. Supercritical extraction processes are discussed by Humphrey and Keller (1997).

11.17.   REFERENCES

AIChE (1958) *Bubble-tray Design Manual* (American Institute of Chemical Engineers).

Amundson, N. R. and Pontinen, A. J. (1958) *Ind. Eng. Chem.* **50**, 730. Multicomponent distillation calculations on a large digital computer.

Barnicki, S. D. and Davies, J. F. (1989) *Chem. Eng., NY* **96** (Oct.) 140, (Nov.) 202. Designing sieve tray columns.

Billet, R. (1995) *Packed Towers* (VCH).

Bolles, W. L. (1963) Tray hydraulics: bubble-cap trays, in *Design of Equilibrium Stage Processes*, B. D. Smith (ed.) (McGraw-Hill).

Bolles, W. L. and Fair, J. R. (1982) *Chem. Eng., NY* **89** (July 12) 109. Improved mass transfer model enhances packed-column design.

Boston, J. F. and Sullivan, S. L. (1974) *Can. J. Chem. Eng.* **52**(1), 52. New class of solution methods for multicomponent, multistage separation processes.

Boston, J. F. (1980) *ACS Symp. Ser.* **124** (*Comput. Appl. Chem. Eng. Process Des. Simul.*) 135. Inside-out algorithms for multicomponent separation process calculations.

Butcher, C. (1988) *Chem. Engr., London* No. 451 (Aug.) 25. Structured packings.

Carey, J. S. and Lewis, W. K. (1932) *J. Ind. Eng. Chem.* **24**, 882. Studies in distillation. Liquid-vapour equilibria of ethyl alcohol-water mixtures.

Chan, H. and Fair, J. R. (1984a) *Ind. Eng. Chem. Proc. Des. Dev.* **23**, 814. Prediction of point efficiencies on sieve trays. 1. Binary systems.

Chan, H. and Fair, J. R. (1984b) *Ind. Eng. Chem. Proc. Des. Dev.* **23**, 820. Prediction of point efficiencies on sieve trays. 2. multicomponent systems.

Chase, J. D. (1967) *Chem. Eng., NY* **74** (July 31st) 105 (Aug. 28th) 139 (in two parts). Sieve-tray design.

Cicalese, J. J., Davis, J. A., Harrington, P. J., Houghland, G. S., Hutchinson, A. J. L. and Walsh, T. J. (1947) *Pet. Ref.* **26** (May) 495. Study of alkylation-plant isobutane tower performance.

Colburn, A. P. (1936) *Ind. Eng. Chem.* **28**, 520. Effect of entrainment on plate efficiency in distillation.

Colburn, A. P. (1939) *Trans. Am. Inst. Chem. Eng.* **35**, 211. The simplified calculation of diffusional processes.

Colburn, A. P. (1941) *Trans. Am. Inst. Chem. Eng.* **37**, 805. The calculation of minimum reflux ratio in the distillation of multicomponent mixtures.

Cornell, D., Knapp, W. G. and Fair, J. R. (1960) *Chem. Eng. Prog.* **56** (July) 68 (Aug.) 48 (in two parts). Mass transfer efficiency in packed columns.

Coulson, J. M., Richardson, J. F., Backhurst, J. and Harker, J. H. (1999) *Chemical Engineering: Volume 1*, 6th edn (Butterworth-Heinemann).

Doherty, M. F. and Malone, M. F. (2001) *Conceptual Design of Distillation Columns* (McGraw-Hill).

Eckert, J. S. (1963) *Chem. Eng. Prog.* **59** (May) 76. A new look at distillation — 4 tower packings — comparative performance.

Eckert, J. S. (1975) *Chem. Eng., NY* **82** (April 14th) 70. How tower packings behave.

Edmister, W. C. (1947) Hydrocarbon absorption and fractionation process design methods, a series of articles published in the *Petroleum Engineer* from May 1947 to March 1949 (19 parts). Reproduced in *A Sourcebook of Technical Literature on Distillation* (Gulf).

Eduljee, H. E. (1958) *Brit. Chem. Eng.* **53**, 14. Design of sieve-type distillation plates.

Eduljee, H. E. (1959) *Brit. Chem. Eng.* **54**, 320. Design of sieve-type distillation plates.

Ellerbe, R. W. (1997) Batch distillation, in *Handbook of Separation Processes for Chemical Engineers*, 3rd edn, P. A. Schweitzer (ed.) (McGraw-Hill).

Erbar, J. H. and Maddox, R. N. (1961) *Pet. Ref.* **40** (May) 183. Latest score: reflux vs. trays.

Evans, F. L. (1980) *Equipment Design Handbook for Refineries and Chemical Plants*, vol. 2, 2nd edn (Gulf).

Fair, J. R. (1961) *Petro/Chem. Eng.* **33** (Oct.) 45. How to predict sieve tray entrainment and flooding.

Fair, J. R. (1963) Tray hydraulics: perforated trays, in *Design of Equilibrium Stage Processes*, B. D. Smith (ed.) (McGraw-Hill).

Fair, J. R. and Bravo, J. L. (1990) *Chem. Eng. Prog.* **86**, (1) 19. Distillation columns containing structured packing.

Fair, J. R., Null, H. R. and Bolles, W. L. (1983) *Ind. Eng. Chem. Proc. Des. Dev.* **22**, 53. Scale-up of plate efficiency from laboratory Oldershaw data.

Featherstone, W. (1971) *Brit. Chem. Eng. & Proc. Tech.* **16** (12), 1121. Azeotropic systems, a rapid method of still design.

Featherstone, W. (1973) *Proc. Tech. Int.* **18** (April/May), 185. Non-ideal systems — A rapid method of estimating still requirements.

Fenske, M. R. (1932) *Ind. Eng. Chem.* **24**, 482. Fractionation of straight-run gasoline.

FRI Design Practices Committee (2007) *Chem. Eng. NY* **114**(7), 34. Causes and prevention of packing fires.

Gilliland, E. R. (1940) *Ind. Eng. Chem.* **32**, 1220. Multicomponent rectification, estimation of the number of theoretical plates as a function of the reflux ratio.

Glitsch, H. C. (1960) *Pet. Ref.* **39** (Aug) 91. Mechanical specification of trays.

Green, D. W. and Perry, R. H. (eds) (2007) *Perry's Chemical Engineers' Handbook*, 8th edn (McGraw-Hill).

Greene, R. (2001) *Chem. Eng. Prog.* **97**(6), 17. Update: Dividing-wall columns gain momentum.

Haas, J. R. (1992) Rigorous distillation calculations, in *Distillation Design*, H. Z. Kister (ed.) (McGraw-Hill).

Hanson, C. (1968) *Chem. Eng., NY* **75** (Aug. 26th) 76. Solvent extraction.

Hanson, D. N., Duffin, J. H. and Somerville, G. E. (1962) *Computation of Multistage Separation Processes* (Reinhold).

Hanson, D. N. and Somerville, G. F. (1963) *Advances in Chemical Engineering* **4**, 279. Computing multistage vapour-liquid processes.

Hart, D. R. (1997) Batch distillation, in *Distillation Design*, H. Z. Kister (ed.) (McGraw-Hill).

Hengstebeck, R. J. (1946) *Trans. Am. Inst. Chem. Eng.* **42**, 309. Simplified method for solving multicomponent distillation problems.

Hengstebeck, R. J. (1976) *Distillation: Principles and design procedures* (Kriger).

Holland, C. D. (1997) *Fundamentals of Multicomponent Distillation* (McGraw-Hill).

Huang, C-J. and Hodson, J. R. (1958) *Pet. Ref.* **37** (Feb.) 103. Perforated trays — designed this way.

Humphrey, J. L. and Keller, G. E. (1997) *Separation Process Technology.* (McGraw-Hill).

Humphrey, J. L., Rocha, J. A. and Fair, J. R. (1984) *Chem. Eng., NY* **91** (Sept. 17) 76. The essentials of extraction.

Hunt, C.d'A., Hanson, D. N. and Wilke, C. R. (1955) *AIChE Jl* **1**, 441. Capacity factors in the performance of perforated-plate columns.

Kaibel, G. (2002) *Computer-Aided Chem. Eng.* **10**, 9. Process synthesis and design in industrial practice.

King, C. J. (1980) *Separation Processes*, 2nd edn (McGraw-Hill).

Kirkbride, C. G. (1944) *Pet. Ref.* **23** (Sept.) 87(321). Process design procedure for multi-component fractionators.

Kister, H. Z. (1992) *Distillation Design* (McGraw-Hill).

Kister, H. Z. and Gill, D. R. (1992) *Chem. Engr., London* No. 524 (Aug.) s7. Flooding and pressure drop in structured packings.

Koch, R. and Kuzniar, J. (1966) *International Chem. Eng.* **6** (Oct.) 618. Hydraulic calculations of a weir sieve tray.

Kohl, A.L. and Nielsen, R. (1997) *Gas Purification* (Gulf).

Kojima, K., Tochigi, K., Seki, H. and Watase, K. (1968) *Kagaku Kogaku* **32**, 149. Determination of vapor-liquid equilibrium from boiling point curve.

Kumar, A. (1982) *Process Synthesis and Engineering Design* (McGraw-Hill).

Leva, M. (1992) *Chem. Eng. Prog.* **88**, 65. Reconsider Packed-Tower Pressure-Drop Correlations.

Leva, M. (1995) *Chem. Engr. London* No. 592 (July 27) 24. Revised GPDC applied.

Lewis, W. K. (1909) *Ind. Eng. Chem.* **1**, 522. The theory of fractional distillation.

Lewis, W. K. (1936) *Ind. Eng. Chem.* **28**, 399. Rectification of binary mixtures.

Liebson, I., Kelley, R. E. and Bullington, L. A. (1957) *Pet. Ref.* **36** (Feb.) 127. How to design perforated trays.

Lockett, M. J. (1986) *Distillation Tray Fundamentals* (Cambridge University Press).

Lowenstein, J. G. (1961) *Ind. Eng. Chem.* **53** (Oct.) 44A. Sizing distillation columns.

Ludwig, E. E. (1997) *Applied Process Design for Chemical and Petrochemical Plant*, Vol. 2, 3rd edn (Gulf).

Luyben, W.L. (2006) *Distillation Design and Control Using Aspen™ Simulation* (Wiley).

McCabe, W. L. and Thiele, E. W. (1925) *Ind. Eng. Chem.* **17**, 605. Graphical design of distillation columns.

McClain, R. W. (1960) *Pet. Ref.* **39** (Aug.) 92. How to specify bubble-cap trays.

Mehra, Y. R. (1979) *Chem. Eng., NY* **86** (July 2nd) 87. Liquid surge capacity in horizontal and vertical vessels.

Murphree, E. V. (1925) *Ind. Eng. Chem.* **17**, 747. Rectifying column calculations.

Naphtali, L. M. and Sandholm, D. P. (1971) *AIChE Jl* **17**, 148. Multicomponent separation calculations by linearisation.

Newman, M., Hayworth, C. B., Treybal, R. E. (1949) Ind Eng Chem., **41**, 2039. Dehydration of aqueous methyl ethyl ketone.

Norman, W. S. (1961) *Absorption, Distillation and Cooling Towers* (Longmans).

O'Connell, H. E. (1946) *Trans. Am. Inst. Chem. Eng.* **42**, 741. Plate efficiency of fractionating columns and absorbers.

Oldershaw, C. F. (1941) *Ind. Eng. Chem. (Anal. ed.)* **13**, 265. Perforated plate columns for analytical batch distillations.

Onda, K., Takeuchi, H. and Okumoto, Y. (1968) *J. Chem. Eng. Japan* **1**, 56. Mass transfer coefficients between gas and liquid phases in packed columns.

Othmer, D. F. (1943) *Ind Eng Chem.*, **35**, 614. Composition of vapors from boiling binary solutions.

Parkinson, G. (2007) *Chem. Eng. Prog.* **103**(5), 8. Dividing-wall columns find greater appeal.

Patton, B. A. and Pritchard, B. L. (1960) *Pet. Ref.* **39** (Aug.) 95. How to specify sieve trays.

Richardson, J. F., Harker, J. H. and Backhurst, J. (2002) *Chemical Engineering: Volume 2*, 5th edn (Butterworth-Heinemann).

Robbins, L. A. (1997) Liquid liquid extraction, in *Handbook of Separation Processes for Chemical Engineers*, 3rd edn, P. A. Schweitzer (ed.) (McGraw-Hill).

Rose, A., Sweeney, R. F. and Schrodt, V. N. (1958) *Ind. Eng. Chem.* **50**, 737. Continuous distillation calculations by relaxation method.

Schultz, M. A., Stewart, D. G., Harris, J. M., Rosenblum, S. P., Shakur, M. S. and O'Brien, D. E. (2002) *Chem. Eng. Prog.* **98**(5), 64. Reduce costs with dividing-wall columns.

Sewell, A. (1975) *Chem. Eng. (London)* **299/300**, 442. Practical aspects of distillation column design.

Smith, B. D. (1963) *Design of Equilibrium Stage Processes* (McGraw-Hill).

Smith, R. (2005) *Chemical Process Design*, 2nd edn. (McGraw-Hill).

Sorel, E. (1899) *Distillation et Rectification Industrielle* (G. Carr et C. Naud).

Sorensen, J. M. and Arlt, W. (1979) *Liquid–Liquid Equilibrium Data Collection*, Chemical Data Series Vols V/2, V/3 (DECHEMA).

Souders, M. and Brown, G. G. (1934) *Ind. Eng. Chem.* **26**, 98. Design of fractionating columns.

Strigle, R. F. (1994) *Random Packings and Packed Towers: design and applications* 2nd edn (Gulf).

Sundmacher, K. and Kiene, A. (eds) (2003) *Reactive Distillation: Status and Future Directions* (Wiley).

Swanson, R. W. and Gester, J. A. (1962) *J. Chem. Eng. Data* **7**, 132. Purification of isoprene by extractive distillation.

Thomas, W. J. and Shah, A. N. (1964) *Trans. Inst. Chem. Eng.* **42**, T71. Downcomer studies in a frothing system.

Thrift, C. (1960a) *Pet. Ref.* **39** (Aug.) 93. How to specify valve trays.

Thrift, C. (1960b) *Pet. Ref.* **39** (Aug.) 95. How to specify sieve trays.

Toor, H. L. and Burchard, J. K. (1960) *AIChE Jl* **6**, 202. Plate efficiencies in multicomponent systems.

Towler, G. P. and Frey, S. J. (2002) Reactive distillation, in *Reactive Separation Processes*, S. Kulprathipanja (ed.) (Taylor and Francis).

Treybal, R. E., Weber, L. D., and Daley, J. F. (1946) *Ind Eng Chem.* 38, 817. The System acetonewater-1,1,2-trichloroethane. Ternary liquid and binary vapor equilibria.

Treybal, R. E. (1980) *Mass Transfer Operations*, 3rd edn (McGraw-Hill).

Underwood, A. J. V. (1948) *Chem. Eng. Prog.* 44 (Aug.) 603. Fractional distillation of multicomponent mixtures.

Van Winkle, M., MacFarland, A. and Sigmund, P. M. (1972) *Hyd. Proc.* 51 (July) 111. Predict distillation efficiency.

Veatch, F., Callahan, J. L., Dol, J. D. and Milberger, E. C. (1960) *Chem. Eng. Prog.* 56 (Oct.) 65. New route to acrylonitrile.

Vital, T. J., Grossel, S. S. and Olsen, P. I. (1984) *Hyd. Proc.* 63 (Dec.) 75. Estimating separation efficiency.

Walas, S. M. (1990) *Chemical Process Equipment: Selection and Design* (Butterworth-Heinemann).

Wang, J. C. and Henke, G. E. (1966) *Hyd. Proc.* 48 (Aug) 155. Tridiagonal matrix for distillation.

Watkins, R.N. (1979) *Petroleum Refinery Distillation*, 2nd edn (Gulf).

Wilke, C. R. and Chang, P. (1955) *AIChE Jl* 1, 264. Correlation for diffusion coefficients in dilute solutions.

Wilke, C. R. and Lee, C. Y. (1955) *Ind. Eng. Chem.* 47, 1253. Estimation of diffusion coefficients for gases and vapours.

Winn, F. W. (1958) *Pet. Ref.* 37 (May) 216. New relative volatility method for distillation calculations.

Zuiderweg, F. J. (1982) *Chem. Eng. Sci.* 37, 1441. Sieve trays: A state-of-the-art review.

Zuiderweg, F. J., Verburg, H. and Gilissen, F. A. H. (1960) *First International Symposium on Distillation*, Inst. Chem. Eng. London, 201. Comparison of fractionating devices.

11.18. NOMENCLATURE

		Dimensions in **MLT** θ
A_a	Active area of plate	L^2
A_{ap}	Clearance area under apron	L^2
A_c	Total column cross-sectional area	L^2
A_d	Downcomer cross-sectional area	L^2
A_h	Total hole area	L^2
A_i	Absorption factor	—
A_m	Area term in equation 11.60	L^2
A_n	Net area available for vapour–liquid disengagement	L^2
A_p	Perforated area	L^2
a	Packing surface area per unit volume	L^{-1}
a_w	Effective interfacial area of packing per unit volume	L^{-1}
B	Moles of bottom product per unit time	MT^{-1}
b_i	Moles of component i in bottom product	M
C_o	Orifice coefficient in equation 11.56	—
C_T	Total molar concentration	ML^{-3}
D	Moles of distillate per unit time	MT^{-1}
D_c	Column diameter	L
D_e	Eddy diffusivity	L^2T^{-1}

D_L	Liquid diffusivity	L^2T^{-1}
D_{LK}	Diffusivity of light key component in liquid phase	L^2T^{-1}
D_v	Diffusivity of vapour	L^2T^{-1}
d_h	Hole diameter	L
d_i	Moles of component i in distillate per unit time	MT^{-1}
d_p	Size of packing	L
E	Extract flow rate	MT^{-1}
E_a	Actual plate efficiency, allowing for entrainment	—
E_{mV}	Murphree plate efficiency	—
E_{mv}	Murphree point efficiency	—
E_o	Overall column efficiency	—
e	Extract composition	—
FA	Fractional area, equation 11.69	—
F	Feed, of the solution to be extracted	MT^{-1}
F_n	Feed rate to stage n	MT^{-1}
F_p	Packing factor	L^{-1}
F_v	Column 'F' factor $= u_a\sqrt{\rho_v}$	$M^{1/2}L^{-1/2}T^{-1}$
F_{LV}	Column liquid-vapour factor in Figure 11.34	—
f	Feed composition (in extraction)	—
f	Fractional approach to flooding $= u_a / u_{af}$	—
f_i	Moles of component i in feed per unit time	MT^{-1}
f_1	Viscosity correction factor in equation 11.79	—
f_2	Liquid density correction factor in equation 11.79	—
f_3	Surface tension correction factor in equation 11.79	—
G_m	Molar flow rate of gas per unit area	$ML^{-2}T^{-1}$
g	Gravitational acceleration	LT^{-2}
H	Specific enthalpy of vapour phase	L^2T^{-2}
$\mathbf{H}_G$	Height of gas-film transfer unit	L
$\mathbf{H}_L$	Height of liquid-film transfer unit	L
$\mathbf{H}_{OG}$	Height of overall gas phase transfer unit	L
$\mathbf{H}_{OL}$	Height of overall liquid phase transfer unit	L
H	Henry's constant	$ML^{-1}T^{-2}$
h	Specific enthalpy of liquid phase	L^2T^{-2}
h_{ap}	Apron clearance	L
h_b	Height of liquid backed up in downcomer	L
h_{bc}	Downcomer back-up in terms of clear liquid head	L
h_d	Dry-plate pressure drop, head of liquid	L
h_{dc}	Head loss in downcomer	L
h_f	Specific enthalpy of feed stream	L^2T^{-2}
h_L	Liquid hold-up on tray	L
h_{ow}	Height of liquid crest over downcomer weir	L
h_r	Plate residual pressure drop, head of liquid	L
h_t	Total plate pressure drop, head of liquid	L
h_w	Weir height	L
K	Equilibrium constant	—
K_G	Overall gas-phase mass-transfer coefficient	$L^{-1}T$

K_i	Equilibrium constant for component i	—
K_L	Overall liquid-phase mass-transfer coefficient	$\mathbf{LT^{-1}}$
K_1	Constant in equation 11.49	$\mathbf{LT^{-1}}$
K_2	Constant in equation 11.52	—
K_3	Percentage flooding factor in equation 11.80	—
K_4	Parameter in Fig. 11.54, defined by equation 11.81	—
K_5	Constant in equation 11.84	—
k_G	Gas-film mass-transfer coefficient	$\mathbf{L^{-1}T}$
k_L	Liquid-film mass-transfer coefficient	$\mathbf{LT^{-1}}$
L	Liquid flow rate, mols per unit time	$\mathbf{MT^{-1}}$
L_m	Molar flow rate of liquid per unit area	$\mathbf{ML^{-2}T^{-1}}$
L_p	Volumetric flow rate across plate divided by average plate width	$\mathbf{L^2T^{-1}}$
L_w	Liquid mass flow rate	$\mathbf{MT^{-1}}$
L_w^*	Liquid mass flow rate per unit area	$\mathbf{MT^{-2}T^{-1}}$
L_{wd}	Liquid mass flow rate through downcomer	$\mathbf{MT^{-1}}$
l_h	Weir chord height	$\mathbf{L}$
l_p	Pitch of holes (distance between centers)	$\mathbf{L}$
l_t	Plate spacing in column	$\mathbf{L}$
l_w	Weir length	$\mathbf{L}$
M	Number of pure products required	—
M_s	Molecular weight of solvent	—
m	Slope of equilibrium line	—
N	Number of components	—
N	Number of stages	—
$\mathbf{N}_G$	Number of gas-film transfer units	—
$\mathbf{N}_L$	Number of liquid-film transfer units	—
N_{min}	Number of stages at total reflux	—
$\mathbf{N}_{OG}$	Number of overall gas-phase transfer units	—
$\mathbf{N}_{OL}$	Number of overall liquid-phase transfer units	—
N_r	Number of equilibrium stages above feed	—
N_s	Number of equilibrium stages below feed	—
n	Stage number	—
P	Total pressure	$\mathbf{ML^{-1}T^{-2}}$
ΔP_t	Total plate pressure drop	$\mathbf{ML^{-1}T^{-2}}$
p	Partial pressure	$\mathbf{ML^{-1}T^{-2}}$
q	Heat to vaporize one mol of feed divided by molar latent heat	—
q_b	Heat supplied to reboiler	$\mathbf{ML^2T^{-3}}$
q_c	Heat removed in condenser	$\mathbf{ML^2T^{-3}}$
q_n	Heat supplied to or removed from stage n	$\mathbf{ML^2T^{-3}}$
$\mathbf{R}$	Universal gas constant	$\mathbf{L^2T^{-2}\theta^{-1}}$
R	Reflux ratio	—
R	Raffinate flow rate (in extraction)	$\mathbf{MT^{-1}}$
R_{min}	Minimum reflux ratio	—
r	Raffinate composition	—
S	Extracting solvent flow rate	$\mathbf{MT^{-1}}$

S_i	Stripping factor	—
S_n	Side stream flow from stage n	MT^{-1}
s	Solvent composition	—
t_L	Liquid contact time	T
t_r	Residence time in downcomer	T
$\overline{t_v}$	Average vapour residence time	T
u_a	Vapour velocity based on active area	LT^{-1}
u_{af}	Vapour velocity at flooding point based on active area A_a	LT^{-1}
u_f	Vapour velocity at flooding point based on net area A_n	LT^{-1}
u_h	Vapour velocity through holes	LT^{-1}
u_n	Vapour velocity based on net cross-sectional area	LT^{-1}
u_v	Superficial vapour velocity (based on total cross-sectional area)	LT^{-1}
V	Vapour flow rate mols per unit time	MT^{-1}
V_w	Vapour mass flow rate	MT^{-1}
V_w^*	Vapour mass flow rate per unit area	$ML^{-2}T^{-1}$
x	Mol fraction of component in liquid phase	—
x_b	Mol fraction of component in bottom product	—
x_d	Mol fraction of component in distillate	—
x_e	Equilibrium mol fraction in liquid phase	—
x_i	Mol fraction of component i	—
x_r	Mol fraction of reference component (equation 11.25)	—
x_1	Mol fraction of solute in solution at column base	—
x_2	Mol fraction of solute in solution at column top	—
x_α	Mol fraction in 1st liquid phase	—
x_β	Mol fraction in 2nd liquid phase	—
y	Mol fraction of component in vapour phase	—
y_e	Equilibrium mol fraction in vapour phase	—
y_i	Mol fraction of component i	—
Δy	Concentration driving force in the gas phase	—
Δy_{lm}	Log mean concentration driving force	—
y_1	Concentration of solute in gas phase at column base	—
y_2	Concentration of solute in gas phase at column top	—
Z	Height of packing	L
Z_c	Liquid hold-up on plate	L
Z_L	Length of liquid path	L
z_i	Mol fraction of component i in feed stream	—
z_f	Mol fraction of component in feed stream	—
α	Relative volatility	—
α_i	Relative volatility of component i	—
α_a	Average relative volatility of light key	—
θ	Root of equation 11.29	—
μ	Dynamic viscosity	$ML^{-1}T^{-1}$
μ_a	Molar average liquid viscosity	$ML^{-1}T^{-1}$
μ_s	Viscosity of solvent	$ML^{-1}T^{-1}$

μ_w	Viscosity of water at 20°C	$\mathbf{ML^{-1}T^{-1}}$
ρ	Density	$\mathbf{ML^{-3}}$
ρ_s	Density of solvent	$\mathbf{ML^{-3}}$
ρ_w	Density of water at 20°C	$\mathbf{ML^{-3}}$
σ	Surface tension	$\mathbf{MT^{-2}}$
σ_c	Critical surface tension for packing material	$\mathbf{MT^{-2}}$
σ_w	Surface tension of water at 20°C	$\mathbf{MT^{-2}}$
Φ	Intercept of operating line on Y axis	—
Φ_n	Factor in equation 11.80	—
ψ	Fractional entrainment	—
ψ_h	Factor in equation 11.79	—
Dg	Surface tension number	
Pe	Peclet number	
Re	Reynolds number	
Sc	Schmidt number	

Subscripts

b	Bottoms
d	Distillate
e	At equilibrium with the other phase
f	Feed
HK	Heavy key
i	Component number
L	Liquid
LK	Light key
m	Last stage (in extraction)
min	Minimum
n	Stage number
r	Reference component
s	Solvent
v	Vapour
w	Water
α	1st liquid phase
β	2nd liquid phase
1	Base of packed column
2	Top of packed column

Superscript

$'$	Stripping section of column

11.19. PROBLEMS

All of the problems in this chapter can be solved using either hand calculations or process simulation programs. Where data are given, they can be used to confirm the validity of the phase equilibrium model used in a simulation.

11.1. At a pressure of 10 bar, determine the bubble and dew point of a mixture of hydrocarbons, composition, mol%: n-butane 21, n-pentane 48, n-hexane 31. The equilibrium K factors can be estimated using the De Priester charts in Chapter 8 or found using a process simulation program.

11.2. The feed to a distillation column has the following composition, mol%: propane 5.0, isobutane 15, n-butane 25, isopentane 20, n-pentane 35. The feed is preheated to a temperature of 90°C, at 8.3 bar pressure. Estimate the proportion of the feed that is vapour.

11.3. Propane is separated from propylene by distillation. The compounds have close boiling points and the relative volatility will be low. For a feed composition of 10% w/w propane, 90% w/w propylene, estimate the number of theoretical plates needed to produce propylene overhead with a minimum purity of 99.5 mol%. The column will operate with a reflux ratio of 20. The feed will be at its boiling point. Take the relative volatility as constant at 1.1.

11.4. The composition of the feed to a debutanizer is given below. Make a preliminary design for a column to recover 98% of the n-butane overhead and 95% of the isopentane from the column base. The column will operate at 14 bar and the feed will be at its boiling point. Use the short-cut methods and follow the procedure set out below. Use the De Priester charts to determine the relative volatility if solving by hand. The liquid viscosity can be estimated using the data given in Appendix C.
(a) Investigate the effect of reflux ratio on the number of theoretical stages
(b) Select the optimum reflux ratio
(c) Determine the number of stages at this reflux ratio
(d) Estimate the stage efficiency
(e) Determine the number of real stages
(f) Estimate the feed point
(g) Estimate the column diameter.

Feed composition:

		kg/h
propane	C_3	910
isobutane	$i\text{-}C_4$	180
n-butane	$n\text{-}C_4$	270
isopentane	$i\text{-}C_5$	70
normal pentane	$n\text{-}C_5$	90
normal hexane	$n\text{-}C_6$	20

11.5. In a process for the manufacture of acetone, acetone is separated from acetic acid by distillation. The feed to the column is 60 mol% acetone, the balance acetic acid.

The column is to recover 95% of the acetone in the feed with a purity of 99.5 mol% acetone. The column will operate at a pressure of 760 mmHg and the feed will be preheated to 70°C.

For this separation, determine:

(a) the number of minimum number of stages required
(b) the minimum reflux ratio
(c) the number of theoretical stages for a reflux ratio 1.5 times the minimum
(d) the number of actual stages if the plate efficiency can be taken as 60%.

Equilibrium data for the system acetone–acetic acid, at 760 mmHg, mol fractions acetone:

liquid phase	0.10	0.2	0.3	0.4	0.5	0.6	0.7	0.8	0.9
vapour phase	0.31	0.56	0.73	0.84	0.91	0.95	0.97	0.98	0.99
boiling point (°C)	103.8	93.1	85.8	79.7	74.6	70.2	66.1	62.6	59.2

Reference: Othmer (1943).

11.6. In the manufacture of absolute alcohol by fermentation, the product is separated and purified using several stages of distillation. In the first stage, a mixture of 5 mol% ethanol in water, with traces of acetaldehyde and fusel oil, is concentrated to 50 mol%. The concentration of alcohol in the wastewater is reduced to less than 0.1 mol%.

Design a sieve plate column to perform this separation, for a feed rate of 10,000 kg/h. Treat the feed as a binary mixture of ethanol and water.

Take the feed temperature as 20°C. The column will operate at 1 atmosphere.

Determine:

(a) the number of theoretical stages
(b) an estimate of the stage efficiency
(c) the number of actual stages needed.

Design a suitable sieve plate for conditions below the feed point.

Equilibrium data for the system ethanol–water, at 760 mmHg, mol fractions ethanol:

liquid phase	0.019	0.072	0.124	0.234	0.327	0.508	0.573	0.676	0.747	0.894
vapour phase	0.170	0.389	0.470	0.545	0.583	0.656	0.684	0.739	0.782	0.894
boiling point (°C)	95.5	89.0	85.3	82.7	81.5	79.8	79.3	78.7	78.4	78.2

Reference: Carey and Lewis (1932).

11.7. In the manufacture of methyl ethyl ketone (MEK) from butanol, the product is separated from unreacted butanol by distillation. The feed to the column consists of a mixture of 0.90 mol fraction MEK, 0.10 mol fraction 2-butanol, with a trace of trichloroethane.

The feed rate to the column is 20 kmol/h and the feed temperature 35°C. The specifications required are: top product 0.99 mol fraction MEK; bottom product 0.99 mol fraction butanol.

Design a column for this separation. The column will operate at essentially atmospheric pressure. Use a reflux ratio 1.5 times the minimum.

(a) determine the minimum reflux ratio
(b) determine the number of theoretical stages
(c) estimate the stage efficiency
(d) determine the number of actual stages needed
(e) design a suitable sieve plate for conditions below the feed point.

Equilibrium data for the system MEK−2-butanol, mol fractions MEK:

liquid phase	0.1	0.2	0.3	0.4	0.5	0.6	0.7	0.8	0.9
vapour phase	0.23	0.41	0.53	0.64	0.73	0.80	0.86	0.91	0.95
boiling point (°C)	97	94	92	90	87	85	84	82	80

11.8. A column is required to recover acetone from an aqueous solution. The feed contains 5 mol% acetone. A product purity of 99.5% w/w is required and the effluent water must contain less than 100 ppm acetone.

The feed temperature will range from 10 to 25°C. The column will operate at atmospheric pressure. For a feed of 7500 kg/h, compare the designs for a sieve plate and packed column, for this duty. Use a reflux ratio of 3. Compare the capital and utility cost for the two designs.

No reboiler is required for this column; live steam can be used.

Equilibrium data for the system acetone–water is given in Example 11.2.

11.9. In the manufacture of MEK, the product is extracted from a solution in water using 1,1,2 trichloroethane as the solvent.

For a feed rate 2000 kg/h of solution, composition 30% w/w MEK, determine the number of stages required to recover 95% of the dissolved MEK; using 700 kg/h trichloroethane (TCE), with counter-current flow.

Tie-line data for the system MEK–water–TCE percentages w/w, from Newman *et al.* (1949).

Water-rich phase		Solvent-rich phase	
MEK	TCE	MEK	TCE
18.15	0.11	75.00	19.92
12.78	0.16	58.62	38.65
9.23	0.23	44.38	54.14
6.00	0.30	31.20	67.80
2.83	0.37	16.90	82.58
1.02	0.41	5.58	94.42

11.10. Chlorine is to be removed from a vent stream by scrubbing with a 5% w/w aqueous solution of sodium hydroxide. The vent stream is essentially nitrogen, with a maximum concentration of 5.5% w/w chlorine. The concentration of chlorine leaving the scrubber must be less than 50 ppm by weight. The maximum flow rate of the vent stream to the scrubber will be 4500 kg/h. Design a suitable packed column for this duty. The column will operate at 1.1 bar and ambient temperature. If necessary, the aqueous stream may be recirculated to maintain a suitable wetting rate.

Note: the reaction of chlorine with the aqueous solution will be rapid and there will be essentially no back-pressure of chlorine from the solution.

12 HEAT-TRANSFER EQUIPMENT

Chapter Contents

Key Learning Objectives

- How to specify and design a shell and tube heat exchanger
- How to design boilers, thermosyphon reboilers and condensers
- How to design a plate heat exchanger
- How to design air coolers and fired heaters
- How to determine whether a reactor can be heated or cooled using a jacket or internal coil

12.1. INTRODUCTION

The transfer of heat to and from process fluids is an essential part of most chemical processes. The most commonly used type of heat-transfer equipment is the ubiquitous shell and tube heat exchanger; the design of which is the main subject of this chapter.

The fundamentals of heat-transfer theory are covered in Coulson *et al.* (1999), and in many other textbooks: Holman (2002), Ozisik (1985), Rohsenow *et al.* (1998), Kreith and Bohn (2000) and Incropera and Dewitt (2001).

Several useful books have been published on the design of heat-exchange equipment. These should be consulted for more details of the construction of equipment and design methods than can be given in this book. A selection of the more useful texts is listed in the bibliography at the end of this chapter. The compilation edited by Schlünder (1983), see also the edition by Hewitt (2002), is probably the most comprehensive work on heat-exchanger design methods available in the open literature. The book by Saunders (1988) is recommended as a good source of information on heat-exchanger design, especially for shell and tube exchangers.

As with distillation, work on the development of reliable design methods for heat exchangers has been dominated in recent years by commercial research organizations: Heat Transfer Research Inc. (HTRI) in the USA and Heat Transfer and Fluid Flow Service (HTFS) in the UK. The HTFS program was developed by the UK Atomic Energy Authority and the National Physical Laboratory, but is now available from Aspen Technology Inc. and as part of the Honeywell UniSim Design Suite; see Chapter 4, Table 4.1. Their proprietary methods are not available in the open literature. They will, however, be available to design engineers in the major operating and contracting companies, whose companies subscribe to these organizations.

The principal types of heat exchanger used in the chemical process and allied industries, which will be discussed in this chapter, are listed below:

1. Double-pipe exchanger: the simplest type, used for cooling and heating
2. Shell and tube exchangers: used for all applications
3. Plate and frame exchangers (plate heat exchangers): used for heating and cooling
4. Plate-fin exchangers
5. Spiral heat exchangers
6. Air cooled: used for coolers and condensers
7. Direct contact: used for cooling and quenching
8. Agitated vessels
9. Fired heaters.

The word 'exchanger' really applies to all types of equipment in which heat is exchanged but is often used specifically to denote equipment in which heat is exchanged between two process streams. Exchangers in which a process fluid is heated or cooled by a plant service stream are referred to as heaters and coolers. If the process stream is vaporized, the exchanger is called a vaporizer if the stream is essentially completely vaporized, a reboiler if associated with a distillation column, and an evaporator if used to concentrate a solution (see Chapter 10). The terms fired exchanger or fired heater are used for exchangers heated by combustion gases, such as boilers; other exchangers are referred to as 'unfired exchangers'.

12.2. BASIC DESIGN PROCEDURE AND THEORY

The general equation for heat transfer across a surface is:

$$Q = UA\Delta T_m \tag{12.1}$$

where

Q = heat transferred per unit time, W
U = the overall heat-transfer coefficient, W/m²°C
A = heat-transfer area, m²
ΔT_m = the mean temperature difference, the temperature driving force, °C.

The prime objective in the design of an exchanger is to determine the surface area required for the specified duty (rate of heat transfer) using the temperature differences available.

The overall coefficient is the reciprocal of the overall resistance to heat transfer, which is the sum of several individual resistances. For heat exchange across a typical heat-exchanger tube the relationship between the overall coefficient and the individual coefficients, which are the reciprocals of the individual resistances, is given by:

$$\frac{1}{U_o} = \frac{1}{h_o} + \frac{1}{h_{od}} + \frac{d_o \ln\left(\dfrac{d_o}{d_i}\right)}{2k_w} + \frac{d_o}{d_i} \times \frac{1}{h_{id}} + \frac{d_o}{d_i} \times \frac{1}{h_i} \tag{12.2}$$

where

U_o = the overall coefficient based on the outside area of the tube, W/m²°C
h_o = outside fluid film coefficient, W/m²°C
h_i = inside fluid film coefficient, W/m²°C
h_{od} = outside dirt coefficient (fouling factor), W/m²°C
h_{id} = inside dirt coefficient, W/m²°C
k_w = thermal conductivity of the tube wall material, W/m°C
d_i = tube inside diameter, m
d_o = tube outside diameter, m.

The magnitude of the individual coefficients will depend on the nature of the heat-transfer process (conduction, convection, condensation, boiling or radiation), on the physical properties of the fluids, on the fluid flow rates, and on the physical arrangement of the heat-transfer surface. As the physical layout of the exchanger cannot be determined until the area is known, the design of an exchanger is of necessity a trial-and-error procedure. The steps in a typical design procedure are given below:

1. Define the duty: heat-transfer rate, fluid flow rates, temperatures.
2. Collect together the fluid physical properties required: density, viscosity, thermal conductivity.

3. Decide on the type of exchanger to be used.
4. Select a trial value for the overall coefficient, U.
5. Calculate the mean temperature difference, ΔT_m.
6. Calculate the area required from equation 12.1.
7. Decide the exchanger layout.
8. Calculate the individual coefficients.
9. Calculate the overall coefficient and compare with the trial value. If the calculated value differs significantly from the estimated value, substitute the calculated for the estimated value and return to step 6.
10. Calculate the exchanger pressure drop; if unsatisfactory return to steps 7 or 4 or 3, in that order of preference.
11. Optimize the design: repeat steps 4 to 10, as necessary, to determine the cheapest exchanger that will satisfy the duty. Usually this will be the one with the smallest area.

Procedures for estimating the individual heat-transfer coefficients and the exchanger pressure drops are given in this chapter.

12.2.1.   Heat-Exchanger Analysis: The Effectiveness–NTU Method

The effectiveness–NTU method is a procedure for evaluating the performance of heat exchangers, which has the advantage that it does not require the evaluation of the mean temperature differences. NTU stands for the Number of Transfer Units, and is analogous with the use of transfer units in mass transfer; see Chapter 11.

The principal use of this method is in the rating of an existing exchanger. It can be used to determine the performance of the exchanger when the heat-transfer area and construction details are known. The method has an advantage over the use of the design procedure outlined above, as an unknown stream outlet temperature can be determined directly, without the need for iterative calculations. It makes use of plots of the exchanger 'effectiveness' versus 'NTU'. The effectiveness is the ratio of the actual rate of heat transfer, to the maximum possible rate.

The effectiveness–NTU method will not be covered in this book, as it is more useful for rating than design. The method is covered in books by Incropera and Dewitt (2001), Ozisik (1985) and Hewitt et al. (1994). The method is also covered by the Engineering Sciences Data Unit in their Design Guides 98003 to 98007 (1998). These guides give large clear plots of effectiveness versus NTU and are recommended for accurate work.

12.3.   OVERALL HEAT-TRANSFER COEFFICIENT

Typical values of the overall heat-transfer coefficient for various types of heat exchanger are given in Table 12.1. More extensive data can be found in the books by Green and Perry (2007), TEMA (Tubular Heat Exchanger Manufacturers Association) (1999) and Ludwig (2001).

TABLE 12.1. Typical Overall Coefficients

Shell and Tube Exchangers		
Hot Fluid	**Cold Fluid**	**U (W/m$^{2\circ}$C)**
Heat exchangers		
Water	Water	800–1500
Organic solvents	Organic solvents	100–300
Light oils	Light oils	100–400
Heavy oils	Heavy oils	50–300
Gases	Gases	10–50
Coolers		
Organic solvents	Water	250–750
Light oils	Water	350–900
Heavy oils	Water	60–300
Gases	Water	20–300
Organic solvents	Brine	150–500
Water	Brine	600–1200
Gases	Brine	15–250
Heaters		
Steam	Water	1500–4000
Steam	Organic solvents	500–1000
Steam	Light oils	300–900
Steam	Heavy oils	60–450
Steam	Gases	30–300
Dowtherm	Heavy oils	50–300
Dowtherm	Gases	20–200
Flue gases	Steam	30–100
Flue	Hydrocarbon vapours	30–100
Condensers		
Aqueous vapours	Water	1000–1500
Organic vapours	Water	700–1000
Organics (some non-condensables)	Water	500–700
Vacuum condensers	Water	200–500
Vaporizers		
Steam	Aqueous solutions	1000–1500
Steam	Light organics	900–1200
Steam	Heavy organics	600–900

Air-cooled Exchangers	
Process Fluid	
Water	300–450
Light organics	300–700
Heavy organics	50–150
Gases, 5–10 bar	50–100
10–30 bar	100–300
Condensing hydrocarbons	300–600

(continued)

Table 12.1. Typical Overall Coefficients—*Cont'd*

Immersed Coils		
Coil	**Pool**	**U (W/m^2 °C)**
Natural circulation		
Steam	Dilute aqueous solutions	500–1000
Steam	Light oils	200–300
Steam	Heavy oils	70–150
Water	Aqueous solutions	200–500
Water	Light oils	100–150
Agitated		
Steam	Dilute aqueous solutions	800–1500
Steam	Light oils	300–500
Steam	Heavy oils	200–400
Water	Aqueous solutions	400–700
Water	Light oils	200–300

Jacketed Vessels		
Jacket	**Vessel**	
Steam	Dilute aqueous solutions	500–700
Steam	Light organics	250–500
Water	Dilute aqueous solutions	200–500
Water	Light organics	200–300

Gasketed-Plate Exchangers		
Hot Fluid	**Cold Fluid**	
Light organic	Light organic	2500–5000
Light organic	Viscous organic	250–500
Viscous organic	Viscous organic	100–200
Light organic	Process water	2500–3500
Viscous organic	Process water	250–500
Light organic	Cooling water	2000–4500
Viscous organic	Cooling water	250–450
Condensing steam	Light organic	2500–3500
Condensing steam	Viscous organic	250–500
Process water	Process water	5000–7500
Process water	Cooling water	5000–7000
Dilute aqueous solutions	Cooling water	5000–7000
Condensing steam	Process water	3500–4500

Figure 12.1, which is adapted from a similar nomograph given by Frank (1974), can be used to estimate the overall coefficient for tubular exchangers (shell and tube). The film coefficients given in Figure 12.1 include an allowance for fouling.

The values given in Table 12.1 and Figure 12.1 can be used for the preliminary sizing of equipment for process evaluation, and as trial values for starting a detailed thermal design.

HEAT-TRANSFER EQUIPMENT

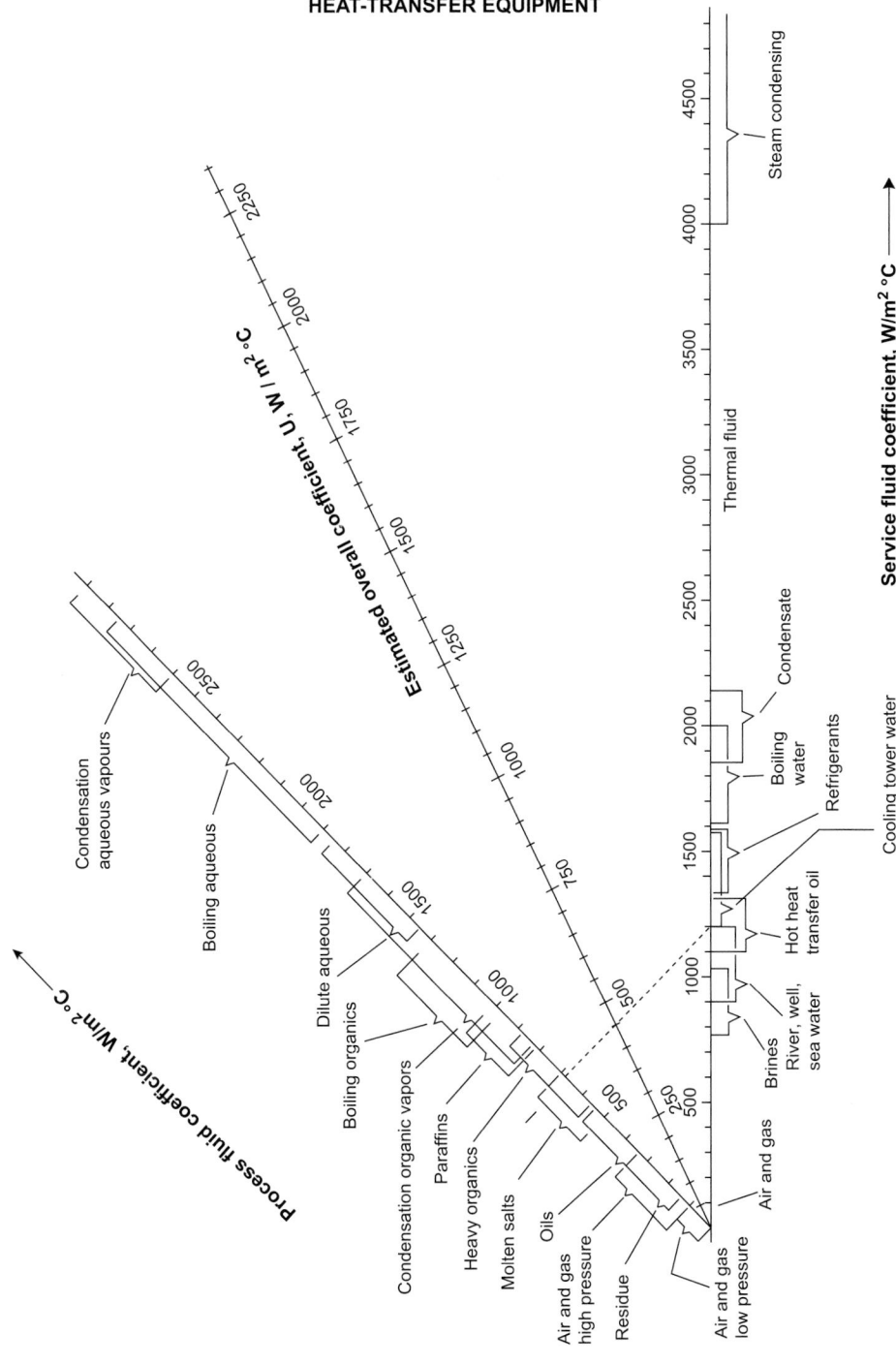

Figure 12.1.   Overall coefficients (join process side duty to service side and read U from centre scale).

12.4. FOULING FACTORS (DIRT FACTORS)

Most process and service fluids will foul the heat-transfer surfaces in an exchanger to a greater or lesser extent. The deposited material will normally have a relatively low thermal conductivity and will reduce the overall coefficient. It is therefore necessary to oversize an exchanger to allow for the reduction in performance during operation. The effect of fouling is allowed for in design by including the inside and outside fouling coefficients in equation 12.2. Fouling factors are usually quoted as heat-transfer resistances, rather than coefficients. They are difficult to predict and are usually based on past experience. Estimating fouling factors introduces a considerable uncertainty into exchanger design; the value assumed for the fouling factor can overwhelm the accuracy of the predicted values of the other coefficients. Fouling factors are often wrongly used as factors of safety in exchanger design. Some work on the prediction of fouling factors has been done by HTRI; see Taborek *et al.* (1972). Fouling is the subject of books by Bott (1990) and Garrett-Price (1985).

Typical values for the fouling coefficients and factors for common process and service fluids are given in Table 12.2. These values are for shell and tube exchangers with plain (not finned) tubes. More extensive data on fouling factors are given in the TEMA standards (1999), and by Ludwig (2001).

The selection of the design fouling coefficient will often be an economic decision. The optimum design will be obtained by balancing the extra capital cost of a larger exchanger against the savings in operating cost obtained from the longer operating

TABLE 12.2. Fouling Factors (Coefficients), Typical Values

Fluid	Coefficient (W/m² °C)	Factor (resistance) (m²°C/W)
River water	3000–12,000	0.0003–0.0001
Sea water	1000–3000	0.001–0.0003
Cooling water (towers)	3000–6000	0.0003–0.00017
Towns water (soft)	3000–5000	0.0003–0.0002
Towns water (hard)	1000–2000	0.001–0.0005
Steam condensate	1500–5000	0.00067–0.0002
Steam (oil free)	4000–10,000	0.0025–0.0001
Steam (oil traces)	2000–5000	0.0005–0.0002
Refrigerated brine	3000–5000	0.0003–0.0002
Air and industrial gases	5000–10,000	0.0002–0.0001
Flue gases	2000–5000	0.0005–0.0002
Organic vapours	5000	0.0002
Organic liquids	5000	0.0002
Light hydrocarbons	5000	0.0002
Heavy hydrocarbons	2000	0.0005
Boiling organics	2500	0.0004
Condensing organics	5000	0.0002
Heat transfer fluids	5000	0.0002
Aqueous salt solutions	3000–5000	0.0003–0.0002

time between cleaning that the larger area will give. Duplicate exchangers should be considered for severely fouling systems so that one exchanger can be taken off line for cleaning while the plant continues to operate using the other exchanger.

When the design engineer adds area to allow for fouling, care must be taken to ensure that the velocity of the fluid is not reduced, otherwise the fouling could be accelerated. For example, if more tubes are added to a shell and tube heat exchanger, then the tube-side flow rate per tube is reduced. Lower tube-side velocity reduces the shear inside the tubes and increases the rate of tube-side fouling. An alternative method of increasing area would be to increase tube length, which comes at the expense of higher pressure drop.

12.5. SHELL AND TUBE EXCHANGERS: CONSTRUCTION DETAILS

The shell and tube exchanger is by far the most common type of heat-transfer equipment used in the chemical and allied industries. The advantages of this type are:

1. The configuration gives a large surface area in a small volume
2. Good mechanical layout: a good shape for pressure operation
3. Uses well-established fabrication techniques
4. Can be constructed from a wide range of materials
5. Easily cleaned
6. Well established design procedures.

Essentially, a shell and tube exchanger consists of a bundle of tubes enclosed in a cylindrical shell. The ends of the tubes are fitted into tube sheets, which separate the shell-side and tube-side fluids. Baffles are provided in the shell to direct the fluid flow and support the tubes. The assembly of baffles and tubes is held together by support rods and spacers (Figure 12.2).

Exchanger Types

The principal types of shell and tube exchanger are shown in Figures 12.3 to 12.8. Diagrams of other types and full details of their construction can be found in the heat exchanger standards (see Section 12.5.1). The standard nomenclature used for shell and tube exchangers is given below; the numbers refer to the features shown in Figures 12.3 to 12.8.

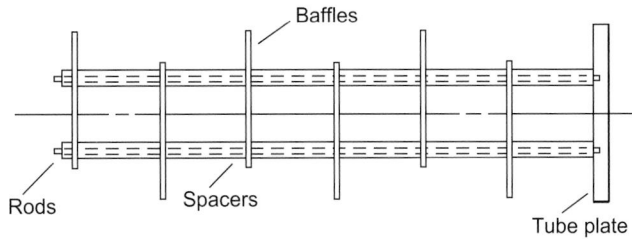

Figure 12.2. Baffle spacers and tie rods.

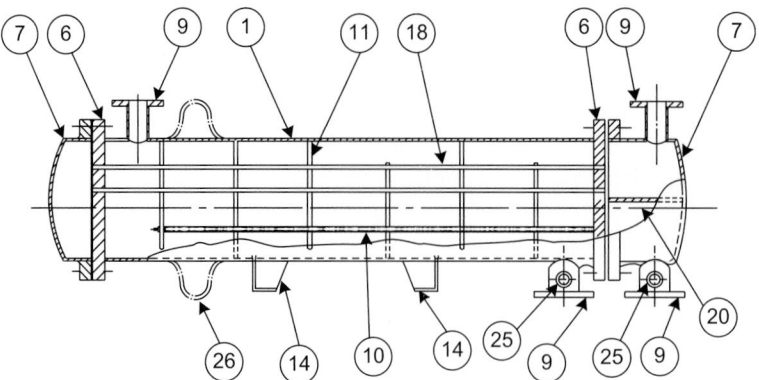

Figure 12.3.   Fixed-tube plate, type BEM (based on figures from BS 3274: 1960).

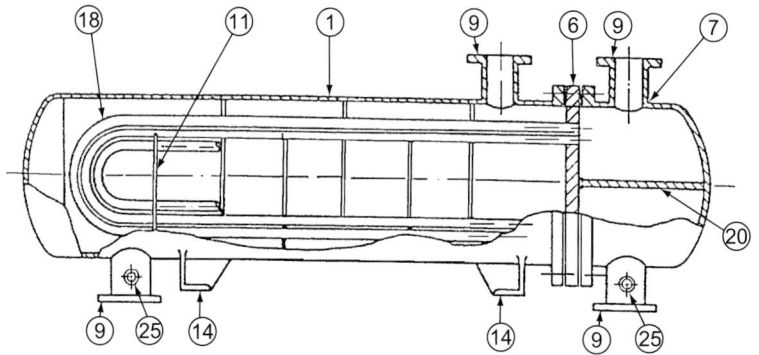

Figure 12.4.   U-tube, type BEU (based on figures from BS 3274: 1960).

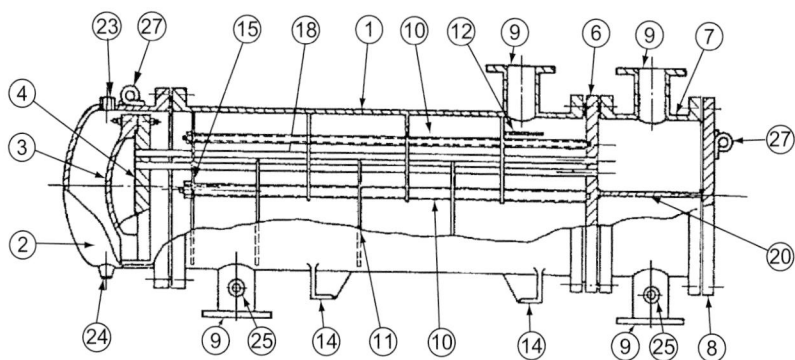

Figure 12.5.   Internal floating head without clamp ring, type AET (based on figures from BS 3274: 1960).

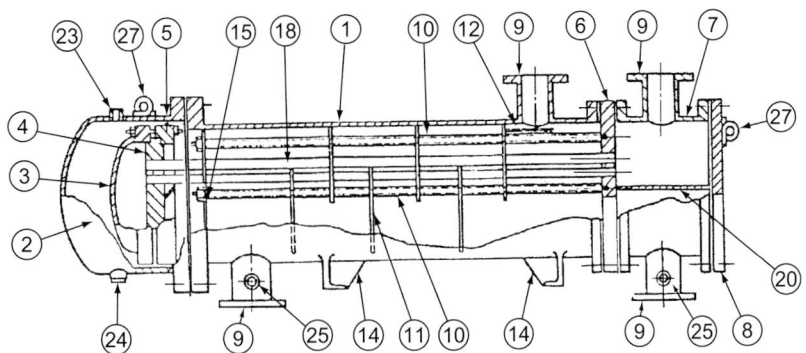

Figure 12.6. Internal floating head with clamp ring, type AES (based on figures from BS 3274: 1960).

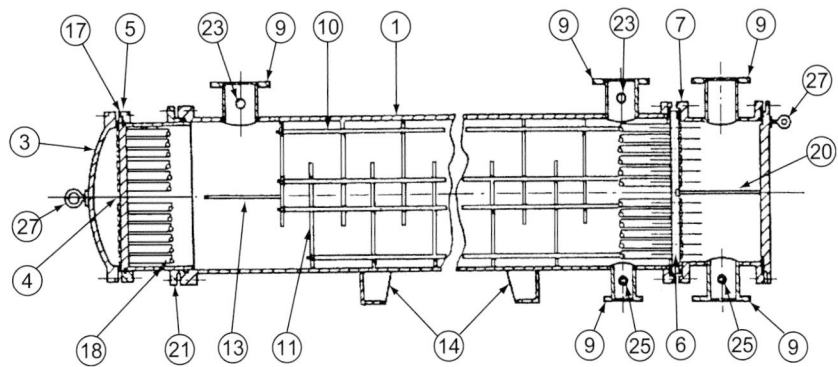

Figure 12.7. External floating head, packed gland, type AEP (based on figures from BS 3274: 1960).

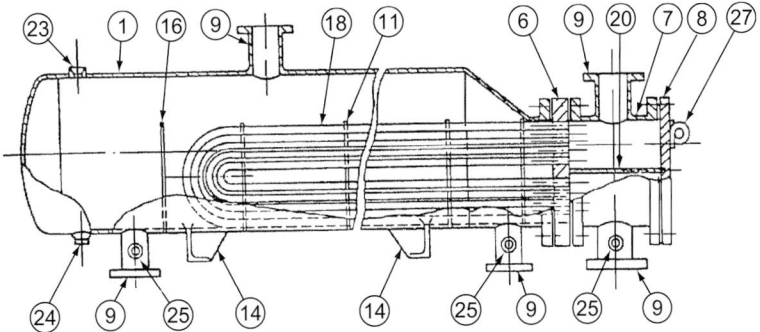

Figure 12.8. Kettle reboiler with U-tube bundle, type AKU (based on figures from BS 3274: 1960).

Nomenclature

Part Number

1. Shell
2. Shell cover
3. Floating-head cover
4. Floating-tube plate
5. Clamp ring
6. Fixed-tube sheet (tube plate)
7. Channel (end-box or header)
8. Channel cover
9. Branch (nozzle)
10. Tie rod and spacer
11. Cross baffle or tube-support plate
12. Impingement baffle
13. Longitudinal baffle
14. Support bracket
15. Floating-head support
16. Weir
17. Split ring
18. Tube
19. Tube bundle
20. Pass partition
21. Floating-head gland (packed gland)
22. Floating-head gland ring
23. Vent connection
24. Drain connection
25. Test connection
26. Expansion bellows
27. Lifting ring

The simplest and cheapest type of shell and tube exchanger is the fixed-tube-sheet design shown in Figure 12.3 (TEMA-type BEM). The main disadvantages of this type are that the tube bundle cannot be removed for cleaning and there is no provision for differential expansion of the shell and tubes. As the shell and tubes will be at different temperatures, and may be of different materials, the differential expansion can be considerable and the use of this type is limited to temperature differences up to about 80°C. Some provision for expansion can be made by including an expansion loop in the shell (shown dotted on Figure 12.3) but their use is limited to low shell pressure; up to about 8 bar. In the other types, only one end of the tubes is fixed and the bundle can expand freely.

The U-tube (U-bundle) type shown in Figure 12.4 requires only one tube sheet and is cheaper than the floating-head types. This is the TEMA-type BEU exchanger, which is widely used but is limited in use to relatively clean fluids as the tubes and bundle are difficult to clean. It is also more difficult to replace a tube in this type.

Exchangers with an internal floating head, Figures 12.5 and 12.6 (TEMA-type AET and AES), are more versatile than fixed head and U-tube exchangers. They are suitable for high temperature differentials and, as the tubes can be rodded from end to end and the bundle removed, are easier to clean and can be used for fouling liquids. A disadvantage of the pull-through design, Figure 12.5, is that the clearance between the outermost tubes in the bundle and the shell must be made greater than in the fixed and U-tube designs to accommodate the floating-head flange, allowing fluid to bypass the tubes. The clamp ring (split flange design), Figure 12.6, is used to reduce the clearance needed. There will always be a danger of leakage occurring from the internal flanges in these floating head designs.

In the external floating head designs, Figure 12.7 (TEMA-type AEP), the floating-head joint is located outside the shell, and the shell sealed with a sliding gland joint employing a stuffing box. Because of the danger of leaks through the gland, the shell-side pressure in this type is usually limited to about 20 bar, and flammable or toxic materials should not be used on the shell side.

The kettle reboiler with U-tubes (TEMA-type AKU) shown in Figure 12.8 is commonly used for reboilers and evaporators that are heated with steam, as steam is a non-fouling service. TEMA type BKU without a removable channel cover is also widely used for kettle reboilers.

12.5.1. Heat-Exchanger Standards and Codes

The mechanical design features, fabrication, materials of construction, and testing of shell and tube exchangers are covered by the standards of the American Tubular Exchanger Manufacturers Association, TEMA. The TEMA standards cover three classes of exchanger: class R covers exchangers for the generally severe duties of the petroleum and related industries, class C covers exchangers for moderate duties in commercial and general process applications, and class B covers exchangers for use in the chemical process industries. The TEMA standards should be consulted for full details of the mechanical design features of shell and tube exchangers; only brief details will be given in this chapter. Internationally, BS 3274 is also sometimes used, although the TEMA codes are most often followed.

The TEMA standards identify heat-exchanger type by a three-letter code. The first letter denotes the tube-side head type, also known as the front end. The second letter identifies the shell type and the third letter defines the rear end. Figure 12.9 illustrates the TEMA nomenclature.

The standards give the preferred shell and tube dimensions, the design and manufacturing tolerances, corrosion allowances, and the recommended design stresses for materials of construction. The shell of an exchanger is a pressure vessel and will be designed in accordance with the appropriate national pressure vessel code or standard; see Chapter 13, Section 13.2. The dimensions of standard flanges for use with heat exchangers are given in the TEMA standards.

In the TEMA and British standards, dimensions are given in feet and inches, so these units have been used in this chapter with the equivalent values in SI units given in brackets.

12.5.2. Tubes

Dimensions

The TEMA design standard allows tube diameters between ¼ in (6.4 mm) and 2 in (50 mm), but tube diameters in the range ⅝ in (16 mm) to 2 in (50 mm) are most often used. The smaller diameters ⅝ to 1 in (16 to 25 mm) are preferred for most duties, as they will give more compact, and therefore cheaper, exchangers. Larger tubes are easier to clean by mechanical methods and are selected for heavily fouling fluids.

The tube thickness (gauge) is selected to withstand the internal and external (shell-side) pressure and give an adequate corrosion allowance. Steel tubes for heat exchangers are covered by BS 3606 (metric sizes); the standards applicable to other materials are given in BS 3274. TEMA standard tubing dimensions are given in table D-7 of the TEMA standards (D7-M in metric units), which is reproduced in Perry's Chemical Engineers Handbook (Green and Perry, 2007). The most commonly used

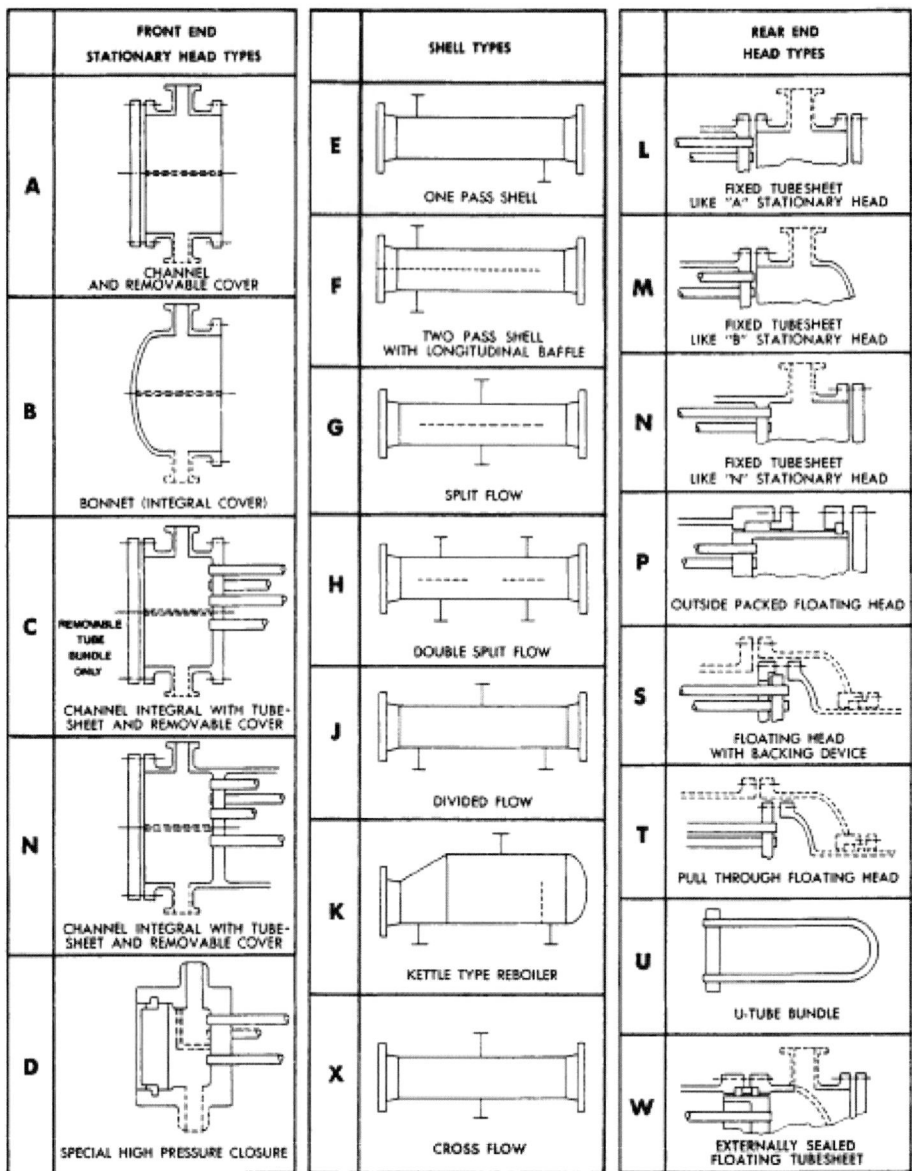

Figure 12.9.   TEMA designations for shell and tube heat exchangers (reproduced with permission from the Tubular Exchanger Manufacturers Association).

thicknesses correspond to even-numbered B.W.G. (Birmingham Wire Gauge) units. Standard diameters and wall thicknesses for steel tubes are given in Table 12.3.

The preferred lengths of tubes for heat exchangers are: 6 ft (1.83 m), 8 ft (2.44 m), 12 ft (3.66 m), 16 ft (4.88 m), 20 ft (6.10 m) and 24 ft (7.32 m). For a given surface area, the use of longer tubes will reduce the shell diameter. This will generally result in a lower cost exchanger, particularly for high shell pressures, but will lead to an

TABLE 12.3. Standard Dimensions for Steel Tubes

Outside Diameter (mm)	Wall Thickness (mm)				
16	1.2	1.7	2.1	—	—
19	—	1.7	2.1	2.8	—
25	—	1.7	2.1	2.8	3.4
32	—	1.7	2.1	2.8	3.4
38	—	—	2.1	2.8	3.4
50	—	—	2.1	2.8	3.4

increase in pressure drop and pump work. The optimum tube length to shell diameter ratio will usually fall within the range of 5 to 10.

If U-tubes are used, the tubes on the outside of the bundle will be longer than those on the inside. The average length needs to be estimated for use in the thermal design. U-tubes will be bent from standard tube lengths and cut to size.

The tube size is often determined by the plant maintenance department standards, as clearly it is an advantage to reduce the number of sizes that have to be held in stores for tube replacement.

As a guide, $\frac{3}{4}$ in (19 mm) is a good trial diameter with which to start design calculations.

Tube Arrangements

The tubes in an exchanger are usually arranged in an equilateral triangular, square, or rotated square pattern; see Figure 12.10.

The triangular and rotated square patterns give higher heat-transfer rates, but at the expense of a higher pressure drop than the square pattern. A square, or rotated square arrangement, is used for heavily fouling fluids, where it is necessary to mechanically clean the outside of the tubes. The recommended tube pitch (distance between tube centres) is 1.25 times the tube outside diameter, and this will normally be used unless process requirements dictate otherwise. Where a square pattern is used for ease of cleaning, the recommended minimum clearance between the tubes is 0.25 in (6.4 mm).

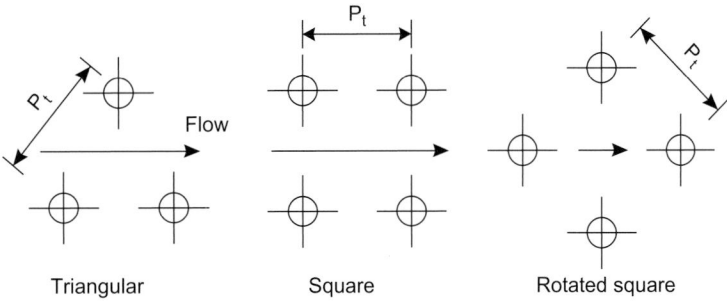

Triangular   Square   Rotated square

Figure 12.10. Tube patterns.

Tube-Side Passes

The fluid in the tube is usually directed to flow back and forth in a number of 'passes' through groups of tubes arranged in parallel, to increase the length of the flow path. The number of passes is selected to give the required tube-side design velocity. Exchangers are built with from one to up to about 16 tube passes. The tubes are arranged into the number of passes required by dividing up the exchanger headers (channels) with partition plates (pass partitions). The arrangement of the pass partitions for two, four and six tube passes are shown in Figure 12.11. The layouts for higher numbers of passes are given by Saunders (1988).

12.5.3. Shells

The British standard BS 3274 covers exchangers from 6 in (150 mm) to 42 in (1067 mm) diameter; and the TEMA standards, exchangers up to 60 in (1520 mm).

Up to about 24 in (610 mm), shells are normally constructed from standard, close tolerance, pipe; above 24 in (610 mm) they are rolled from plate.

For pressure applications the shell thickness would be sized according to the pressure vessel design standards, see Chapter 13. The minimum allowable shell thickness is given in BS 3274 and the TEMA standards. The values, converted to SI units and rounded, are given below:

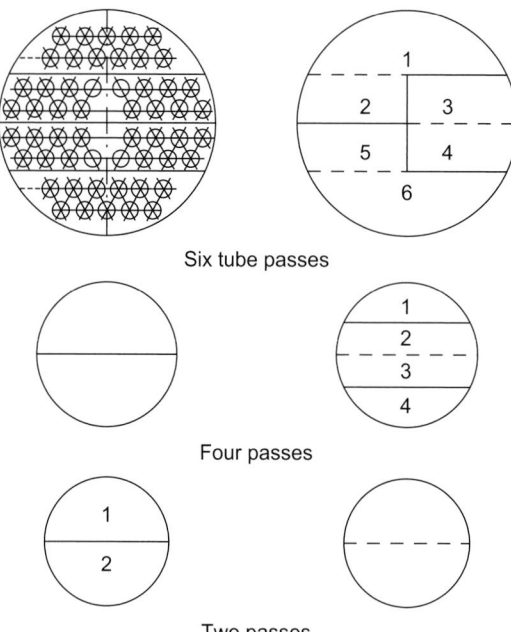

Figure 12.11. Tube arrangements, showing pass-partitions in headers.

Minimum Shell Thickness (mm)

Nominal Shell dia. (mm)	Carbon Steel		Alloy Steel
	Pipe	Plate	
150	7.1	—	3.2
200–300	9.3	—	3.2
330–580	9.5	7.9	3.2
610–740	—	7.9	4.8
760–990	—	9.5	6.4
1010–1520	—	11.1	6.4
1550–2030	—	12.7	7.9
2050–2540	—	12.7	9.5

The shell diameter must be selected to give as close a fit to the tube bundle as is practical; to reduce bypassing round the outside of the bundle; see Section 12.9. The clearance required between the outermost tubes in the bundle and the shell inside diameter will depend on the type of exchanger and the manufacturing tolerances; typical values are given in Figure 12.12.

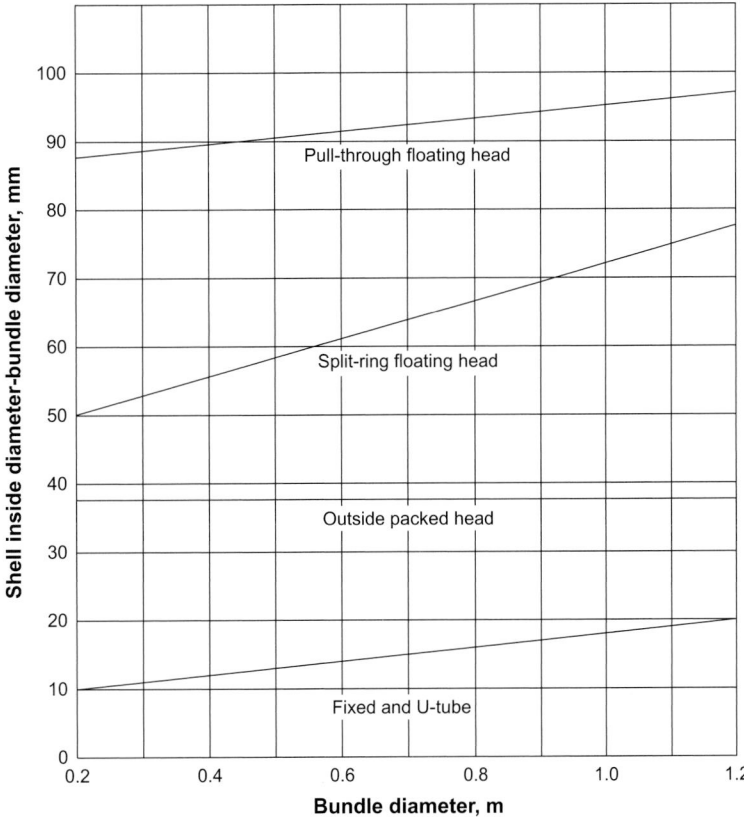

Figure 12.12. Shell-bundle clearance.

12.5.4. Tube-sheet Layout (Tube Count)

The bundle diameter depends not only on the number of tubes but also on the number of tube passes, as spaces must be left in the pattern of tubes on the tube sheet to accommodate the pass partition plates.

An estimate of the bundle diameter D_b can be obtained from equation 12.3b, which is an empirical equation based on standard tube layouts. The constants for use in this equation, for triangular and square patterns, are given in Table 12.4.

$$N_t = K_1 \left(\frac{D_b}{d_o}\right)^{n_1} \tag{12.3a}$$

$$D_b = d_o \left(\frac{N_t}{K_1}\right)^{1/n_1} \tag{12.3b}$$

where

N_t = number of tubes
D_b = bundle diameter, mm
d_o = tube outside diameter, mm.

If U-tubes are used, the number of tubes will be slightly less than that given by equation 12.3a, as the spacing between the two centre rows will be determined by the minimum allowable radius for the U-bend. The minimum bend radius will depend on the tube diameter and wall thickness. It will range from 1.5 to 3.0 times the tube outside diameter. The tighter bend radius will lead to some thinning of the tube wall.

An estimate of the number of tubes in a U-tube exchanger (twice the actual number of U-tubes), can be made by reducing the number given by equation 12.3a by one centre row of tubes.

The number of tubes in the centre row, the row at the shell equator, is given by:

$$\text{Tubes in centre row} = \frac{D_b}{p_t}$$

where p_t = tube pitch, mm.

The tube layout for a particular design will normally be planned with the aid of computer programs. These will allow for the spacing of the pass partition plates and

TABLE 12.4. Constants for Use in Equation 12.3

Triangular pitch, $p_t = 1.25d_o$

No. passes	1	2	4	6	8
K_1	0.319	0.249	0.175	0.0743	0.0365
n_1	2.142	2.207	2.285	2.499	2.675

Square pitch, $p_t = 1.25d_o$

No. passes	1	2	4	6	8
K_1	0.215	0.156	0.158	0.0402	0.0331
n_1	2.207	2.291	2.263	2.617	2.643

the position of the tie rods. Also, one or two rows of tubes may be omitted at the top and bottom of the bundle to increase the clearance and flow area opposite the inlet and outlet nozzles.

Tube count tables that give an estimate of the number of tubes that can be accommodated in standard shell sizes, for commonly used tube sizes, pitches and number of passes, can be found in several books: Kern (1950), Ludwig (2001), Green and Perry (2007), and Saunders (1988).

Some typical tube arrangements are shown in Appendix H.

12.5.5. Shell Types (Passes)

The principal shell arrangements are shown in Figure 12.9. The letters E, F, G, H, J are those used in the TEMA standards to designate the various types. The E shell is the most commonly used arrangement.

Two shell passes (F shell) are occasionally used where the shell and tube side temperature differences are unsuitable for a single pass (see Section 12.6); however, it is difficult to obtain a satisfactory seal with a shell-side baffle and the same flow arrangement can be achieved by using two shells in series.

The divided-flow and split-flow arrangements (G and J shells) are used to reduce the shell-side pressure drop, where pressure drop, rather than heat transfer, is the controlling factor in the design.

12.5.6. Shell and Tube Designation

A common method of describing an exchanger is to designate the number of shell and tube passes: m/n or $m{:}n$; where m is the number of shell passes and n the number of tube passes. So 1/2 or 1:2 describes an exchanger with one shell pass and two tube passes, and 2/4 an exchanger with two shell passes and four tube passes.

12.5.7. Baffles

Baffles are used in the shell to direct the fluid stream across the tubes, to increase the fluid velocity and so improve the rate of transfer. The most commonly used type of baffle is the single segmental baffle shown in Figure 12.13a, other types are shown in Figure 12.13b, c and d.

Only the design of exchangers using single segmental baffles will be considered in this chapter.

If the arrangement shown in Figure 12.13a were used with a horizontal condenser the baffles would restrict the condensate flow. This problem can be overcome either by rotating the baffle arrangement through 90°, or by trimming the base of the baffle (Figure 12.14).

The term 'baffle cut' is used to specify the dimensions of a segmental baffle. The baffle cut is the height of the segment removed to form the baffle, expressed as a percentage of the baffle disc diameter. Baffle cuts from 15% to 45% are used. Generally, a baffle cut of 20% to 25% will be the optimum, giving good heat-transfer rates, without excessive pressure drop. There will be some leakage of fluid round the

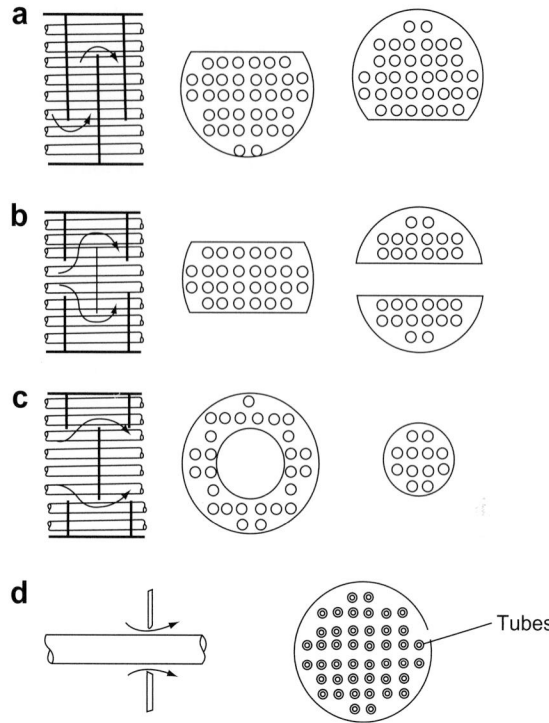

Figure 12.13.   Types of baffle used in shell and tube heat exchangers. (a) Segmental. (b) Segmental and strip. (c) Disc and doughnut. (d) Orifice.

baffle as a clearance must be allowed for assembly. The clearance needed will depend on the shell diameter; typical values, and tolerances, are given in Table 12.5.

Another leakage path occurs through the clearance between the tube holes in the baffle and the tubes. The maximum design clearance will normally be 1/32 in. (0.8 mm).

The minimum thickness to be used for baffles and support plates is given in the standards. The baffle spacings used range from 0.2 to 1.0 shell diameters. A close baffle spacing will give higher heat-transfer coefficients, but at the expense of higher pressure drop. The optimum spacing will usually be between 0.3 to 0.5 times the shell diameter.

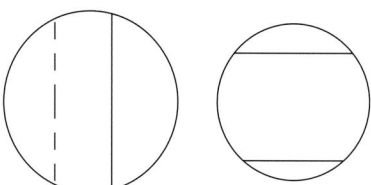

Figure 12.14.   Baffles for condensers.

TABLE 12.5. Typical Baffle Clearances and Tolerances

Shell Diameter, D_s	Baffle Diameter	Tolerance
Pipe shells		
6 to 25 in (152 to 635 mm)	$D_s - \frac{1}{16}$ in (1.6 mm)	$+ \frac{1}{32}$ in (0.8 mm)
Plate shells		
6 to 25 in (152 to 635 mm)	$D_s - \frac{1}{8}$ in (3.2 mm)	$+ 0, - \frac{1}{32}$ in (0.8 mm)
27 to 42 in (686 to 1067 mm)	$D_s - \frac{3}{16}$ in (4.8 mm)	$+0, - \frac{1}{16}$ in (1.6 mm)

12.5.8. Support Plates and Tie Rods

Where segmental baffles are used, some will be fabricated with closer tolerances, $\frac{1}{64}$ in (0.4 mm), to act as support plates. For condensers and vaporizers, where baffles are not needed for heat-transfer purposes, a few will be installed to support the tubes.

The minimum spacings to be used for support plates are given in the standards. The spacing ranges from around 1 m for 16-mm tubes to 2 m for 25-mm tubes.

The baffles and support plate are held together with tie rods and spacers. The number of rods required depends on the shell diameter, and ranges from four 16-mm diameter rods, for exchangers under 380 mm diameter; to eight 12.5-mm rods, for exchangers of 1 m diameter. The recommended number for a particular diameter can be found in the standards.

12.5.9. Tube Sheets (Plates)

In operation, the tube sheets are subjected to the differential pressure between shell and tube sides. The design of tube sheets as pressure-vessel components is covered by the ASME BPV Code and BS 5500 and is discussed in Section 13.11. Design formulae for calculating tube-sheet thicknesses are also given in the TEMA standards.

The joint between the tubes and tube sheet is normally made by expanding the tube by rolling with special tools (Figure 12.15). Tube rolling is a skilled task; the tube must be expanded sufficiently to ensure a sound leak-proof joint, but not over-thinned, weakening the tube. The tube holes are normally grooved (Figure 12.16a), to

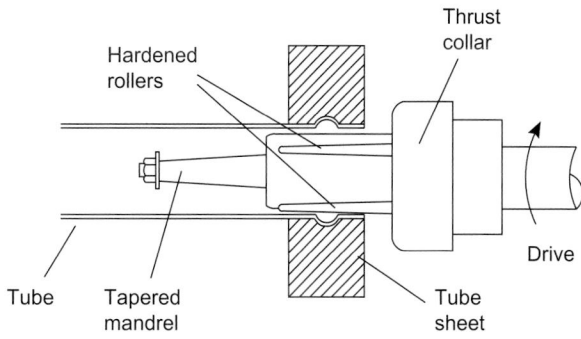

Figure 12.15. Tube rolling.

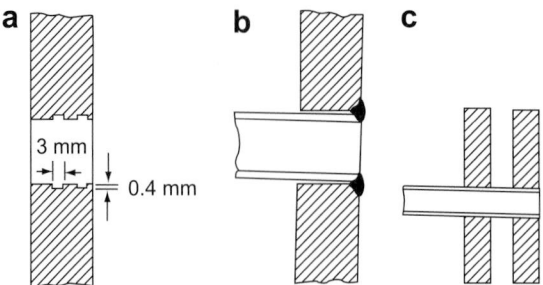

Figure 12.16.   Tube/tube-sheet joints.

lock the tubes more firmly in position and to prevent the joint from being loosened by the differential expansion of the shell and tubes. When it is essential to guarantee a leak-proof joint, the tubes can be welded to the sheet (Figure 12.16b). This adds to the cost of the exchanger, not only due to the cost of welding but also because a wider tube spacing is needed.

The tube sheet forms the barrier between the shell and tube fluids, and where it is essential for safety or process reasons to prevent any possibility of intermixing due to leakage at the tube-sheet joint, double tube sheets can be used, with the space between the sheets vented (Figure 12.16c).

To allow sufficient thickness to seal the tubes, the tube-sheet thickness should not be less than the tube outside diameter, up to about 25 mm diameter. Recommended minimum plate thicknesses are given in the standards.

The thickness of the tube sheet will reduce the effective length of the tube slightly, and this should be allowed for when calculating the area available for heat transfer. As a first approximation, the length of the tubes can be reduced by 25 mm for each tube sheet.

12.5.10.   Shell and Header Nozzles (Branches)

Standard pipe sizes are used for the inlet and outlet nozzles. It is important to avoid flow restrictions at the inlet and outlet nozzles to prevent excessive pressure drop and flow-induced vibration of the tubes. As well as omitting some tube rows (see Section 12.5.4), the baffle spacing is usually increased in the nozzle zone, to increase the flow area. For vapours and gases, where the inlet velocities will be high, the nozzle may be flared, or special designs used, to reduce the inlet velocities; see Figure 12.17a and b. The extended shell design shown in Figure 12.17b also serves as an impingement plate. Impingement plates are used where the shell-side fluid contains liquid drops, or for high-velocity fluids containing abrasive particles.

12.5.11.   Flow-induced Tube Vibrations

Premature failure of exchanger tubes can occur through vibrations induced by the shell-side fluid flow. Care must be taken in the mechanical design of large exchangers where the shell-side velocity is high, say greater than 3 m/s, to ensure that tubes are adequately supported.

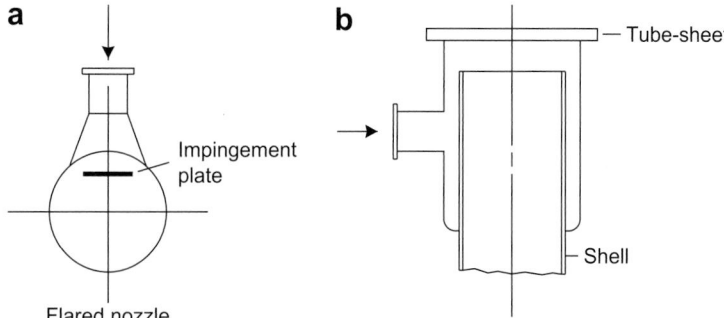

Figure 12.17.   Inlet nozzle designs.

The vibration induced by the fluid flowing over the tube bundle is caused principally by vortex shedding and turbulent buffeting. As fluid flows over a tube, vortices are shed from the down-stream side, causing disturbances in the flow pattern and pressure distribution around the tube. Turbulent buffeting of tubes occurs at high flow rates due to the intense turbulence at high Reynolds numbers.

The buffeting caused by vortex shedding or by turbulent eddies in the flow stream will cause vibration, but large amplitude vibrations will normally only occur above a certain critical flow velocity. Above this velocity, the interaction with the adjacent tubes can provide a feedback path that reinforces the vibrations. Resonance will also occur if the vibrations approach the natural vibration frequency of the unsupported tube length. Under these conditions the magnitude of the vibrations can increase dramatically, leading to tube failure. Failure can occur either through the impact of one tube on another or through wear on the tube where it passes through the baffles.

For most exchanger designs, following the recommendations on support sheet spacing given in the standards will be sufficient to protect against premature tube failure from vibration. For large exchangers with high velocities on the shell side, the design should be analysed to check for possible vibration problems. The computer-aided design programs for shell and tube exchanger design available from commercial organizations, such as HTFS and HTRI (see Section 12.1), include programs for vibration analysis.

Much work has been done on tube vibration over the past 20 years, due to an increase in the failure of exchangers as larger sizes and higher flow rates have been used. Discussion of this work is beyond the scope of this book; for review of the methods used see Saunders (1988) and Singh and Soler (1992). See also the Engineering Science Data Unit Design Guide ESDU 87019, which gives a clear explanation of mechanisms causing tube vibration in shell and tube heat exchangers, and their prediction and prevention.

12.6. MEAN TEMPERATURE DIFFERENCE (TEMPERATURE DRIVING FORCE)

Before equation 12.1 can be used to determine the heat-transfer area required for a given duty, an estimate of the mean temperature difference ΔT_m must be made. This will normally be calculated from the terminal temperature differences: the difference in the fluid temperatures at the inlet and outlet of the exchanger. The well known 'logarithmic mean' temperature difference is only applicable to sensible heat transfer in true co-current or counter-current flow, with linear temperature–enthalpy curves. This situation occurs when the heat capacities of both streams are constant and there is no phase change, or if there is a phase change at constant pressure for a stream that contains a single component. These conditions are only approximated in reality. For counter-current flow (Figure 12.18a), the logarithmic mean temperature difference is given by:

$$\Delta T_{\mathrm{lm}} = \frac{(T_1 - t_2) - (T_2 - t_1)}{\ln \dfrac{(T_1 - t_2)}{(T_2 - t_1)}} \tag{12.4}$$

where

ΔT_{lm} = log mean temperature difference
T_1 = hot fluid temperature, inlet
T_2 = hot fluid temperature, outlet

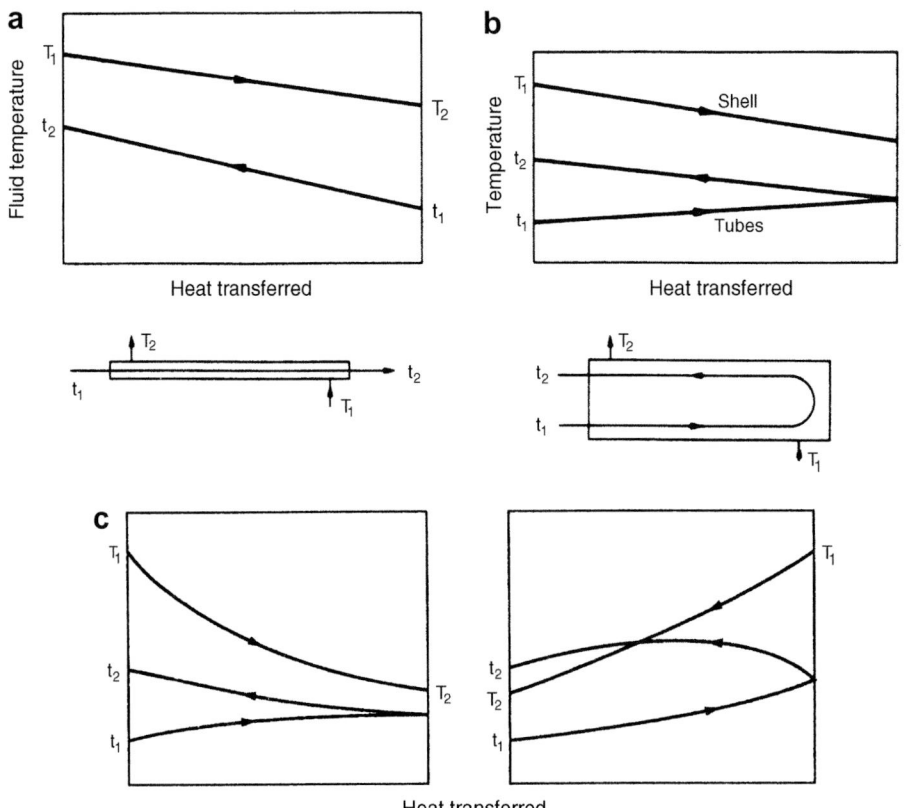

Figure 12.18.   Temperature profiles. (a) Counter-current flow. (b) 1:2 exchanger. (c) Temperature cross.

t_1 = cold fluid temperature, inlet
t_2 = cold fluid temperature, outlet.

The equation is the same for co-current flow, but the terminal temperature differences will be $(T_1 - t_1)$ and $(T_2 - t_2)$. Strictly, equation 12.4 will only apply when there is no change in the specific heats, the overall heat-transfer coefficient is constant, and there are no heat losses. In design, these conditions can be assumed to be satisfied providing the temperature change in each fluid stream is not large.

In most shell and tube exchangers, the flow will be a mixture of co-current, counter-current and cross flow. Figures 12.18b and c show typical temperature profiles for an exchanger with one shell pass and two tube passes (a 1:2 exchanger). Figure 12.18c shows two different cases of temperature cross, where the outlet temperature of the cold stream is above that of the hot stream.

The usual practice in the design of shell and tube exchangers is to estimate the 'true temperature difference' from the logarithmic mean temperature by applying a correction factor to allow for the departure from true counter-current flow:

$$\Delta T_m = F_t \Delta T_{\text{lm}} \tag{12.5}$$

Where ΔT_m = true temperature difference, the mean temperature difference for use in the design equation 12.1

$$F_t = \text{the temperature correction factor}$$

The correction factor is a function of the shell and tube fluid temperatures, and the number of tube and shell passes. It is normally correlated as a function of two dimensionless temperature ratios:

$$R = \frac{(T_1 - T_2)}{(t_2 - t_1)} \tag{12.6}$$

and

$$S = \frac{(t_2 - t_1)}{(T_1 - t_1)} \tag{12.7}$$

R is equal to the shell-side fluid flow rate times the fluid mean specific heat; divided by the tube-side fluid flow rate times the tube-side fluid specific heat.

S is a measure of the temperature efficiency of the exchanger.

For a 1:2 exchanger, the correction factor is given by:

$$F_t = \frac{\sqrt{(R^2 + 1)}\ln\left[\frac{(1 - S)}{(1 - RS)}\right]}{(R - 1)\ln\left[\frac{2 - S[R + 1 - \sqrt{(R^2 + 1)}]}{2 - S[R + 1 + \sqrt{(R^2 + 1)}]}\right]} \tag{12.8}$$

The derivation of equation 12.8 is given by Kern (1950). The equation for a 1:2 exchanger can be used for any exchanger with an even number of tube passes, and is plotted in Figure 12.19. The correction factor for two shell passes and four, or multiples of four, tube passes is shown in Figure 12.20, and that for divided-flow and split-flow shells in Figures 12.21 and 12.22.

Temperature correction factor plots for other arrangements can be found in the TEMA standards and the books by Kern (1950) and Ludwig (2001). Mueller (1973) gives a comprehensive set of figures for calculating the log mean temperature correction factor, which includes figures for cross-flow exchangers.

The following assumptions are made in the derivation of the temperature correction factor F_t, in addition to those made for the calculation of the log mean temperature difference:

1. Equal heat transfer areas in each pass
2. A constant overall heat-transfer coefficient in each pass
3. The temperature of the shell-side fluid in any pass is constant across any cross-section
4. There is no leakage of fluid between shell passes.

Though these conditions will not be strictly satisfied in practical heat exchangers, the F_t values obtained from the curves will give an estimate of the 'true mean temperature difference' that is sufficiently accurate for most designs. Mueller (1973) discusses these assumptions, and gives F_t curves for conditions when all the assumptions are not met; see also Butterworth (1973) and Emerson (1973). Values of

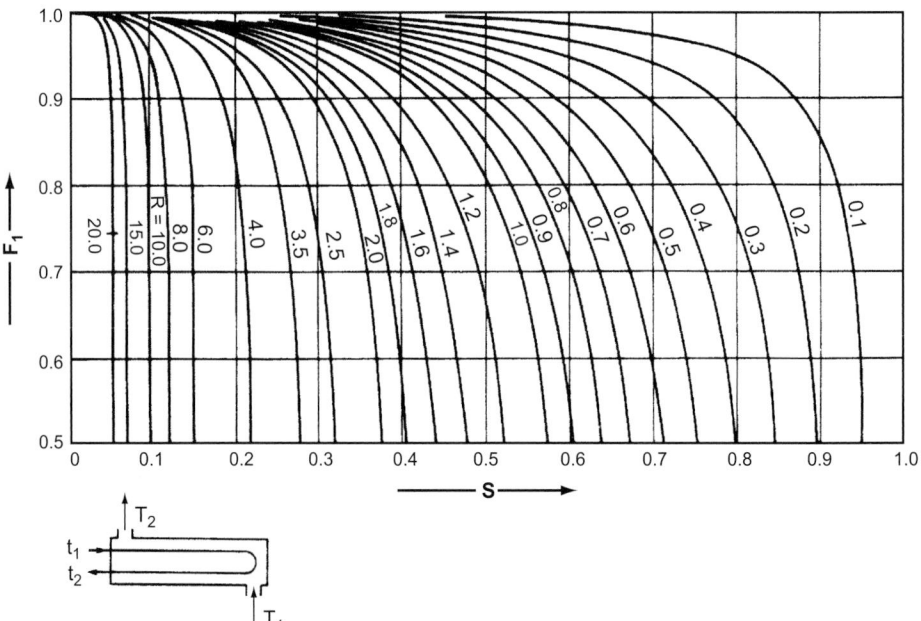

Figure 12.19. Temperature correction factor: one shell pass; two or more even tube passes.

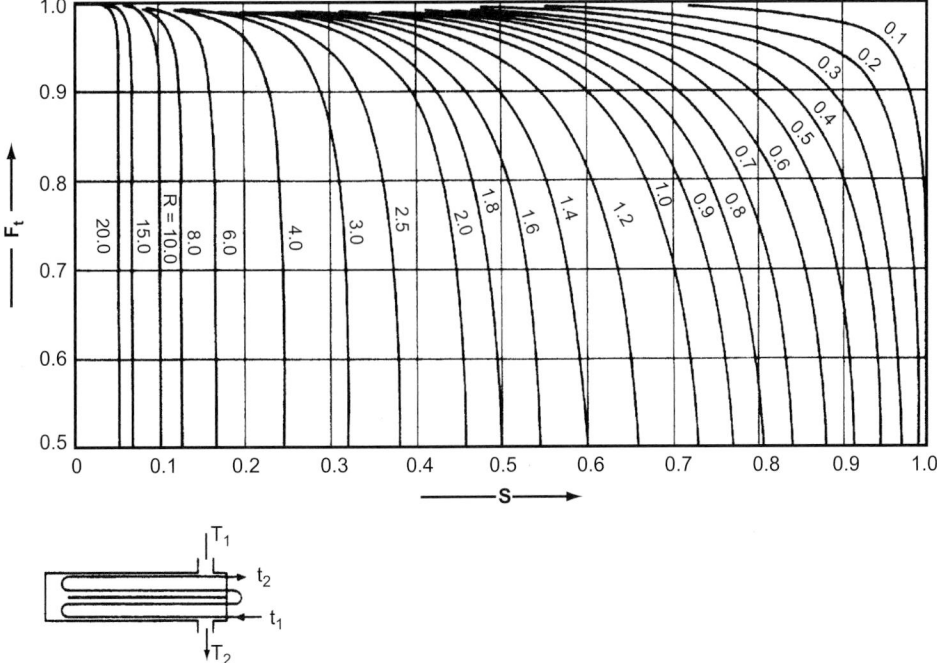

Figure 12.20. Temperature correction factor: two shell passes; four or multiples of four tube passes.

F_t are calculated for heat exchangers in most process simulation programs, as described in Chapter 4.

The shell-side leakage and bypass streams (see Section 12.9) will affect the mean temperature difference, but are not normally taken into account when estimating the correction factor F_t. Fisher and Parker (1969) give curves that show the effect of leakage on the correction factor for a 1:2 exchanger.

The value of F_t will be close to one when the terminal temperature differences are large, but will appreciably reduce the logarithmic mean temperature difference when the temperatures of shell and tube fluids approach each other; it will fall drastically when there is a temperature cross. A temperature cross will occur if the outlet temperature of the cold stream is greater than the outlet temperature of the hot stream, Figure 12.18c.

Where the F_t curve is near vertical, values cannot be read accurately, which will introduce a considerable uncertainty into the design.

An economic exchanger design cannot normally be achieved if the correction factor F_t falls below about 0.75. In these circumstances, an alternative type of exchanger should be considered that gives a closer approach to true counter-current flow. The use of two or more shells in series, or multiple shell-side passes, will give a closer approach to true counter-current flow, and should be considered where a temperature cross is likely to occur.

When both sensible and latent heat is transferred, it will be necessary to divide the temperature profile into sections and calculate the mean temperature difference for each section. The overall heat-transfer coefficient should also be different in each section.

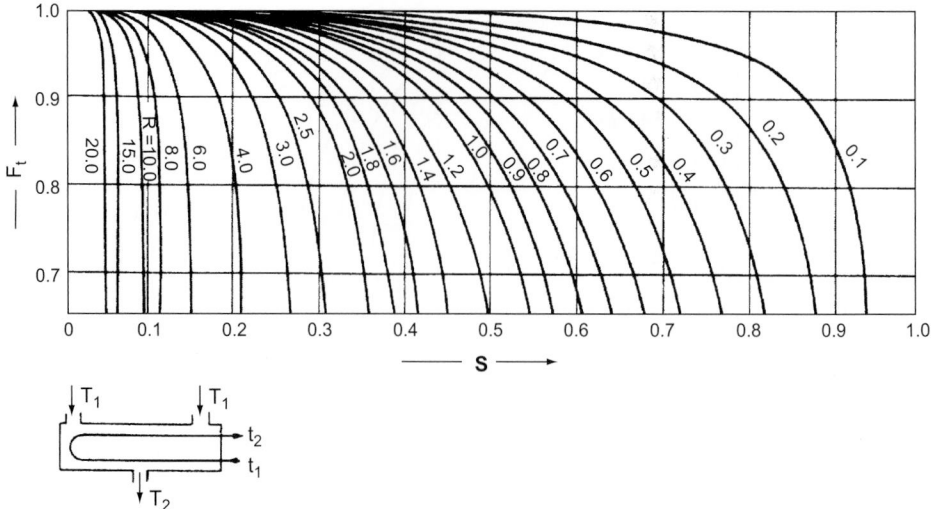

Figure 12.21.   Temperature correction factor: divided-flow shell; two or more even tube passes.

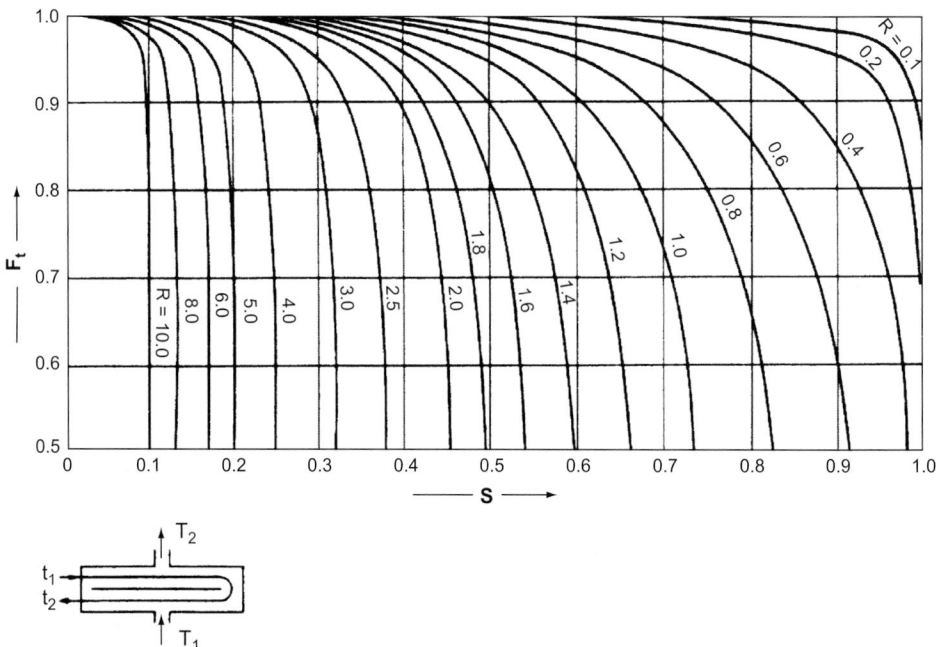

Figure 12.22.   Temperature correction factor: split-flow shell, two tube passes.

12.7. SHELL AND TUBE EXCHANGERS: GENERAL DESIGN CONSIDERATIONS

12.7.1. Fluid Allocation: Shell or Tubes

Where no phase change occurs, the following factors determine the allocation of the fluid streams to the shell or tubes.

Corrosion. The more corrosive fluid should be allocated to the tube side. This will reduce the cost of expensive alloy or clad components.

Fouling. The fluid that has the greatest tendency to foul the heat-transfer surfaces should be placed in the tubes. This gives better control over the design fluid velocity, and the higher allowable velocity in the tubes will reduce fouling. Also, the tubes will be easier to clean.

Fluid temperatures. If the temperatures are high enough to require the use of special alloys, placing the higher temperature fluid in the tubes will reduce the overall cost. At moderate temperatures, placing the hotter fluid in the tubes will reduce the shell surface temperatures, and hence the need for lagging to reduce heat loss, or for safety reasons.

Operating pressures. The higher pressure stream should be allocated to the tube side. High-pressure tubes will be cheaper than a high-pressure shell. The required tube thickness is less for high internal pressure than high external pressure and an expensive high-pressure shell may be avoided.

Pressure drop. For the same pressure drop, higher heat-transfer coefficients will be obtained on the tube side than the shell side, and fluid with the lowest allowable pressure drop should be allocated to the tube side.

Viscosity. Generally, a higher heat-transfer coefficient will be obtained by allocating the more viscous material to the shell side, providing the flow is turbulent. The critical Reynolds number for turbulent flow in the shell is in the region of 200. If turbulent flow cannot be achieved in the shell, it is better to place the fluid in the tubes, as the tube-side heat-transfer coefficient can be predicted with more certainty.

Stream flow rates. Allocating the fluids with the lowest flow rate to the shell side will normally give the most economical design.

12.7.2. Shell and Tube Fluid Velocities

High velocities will give high heat-transfer coefficients but also a high pressure drop. The velocity must be high enough to prevent any suspended solids settling, but not so high as to cause erosion. High velocities will reduce fouling. Plastic inserts are sometimes used to reduce erosion at the tube inlet. Typical design velocities are given below.

Liquids

Tube side, process fluids: 1 to 2 m/s, maximum 4 m/s if required to reduce fouling; water: 1.5 to 2.5 m/s.
Shell side: 0.3 to 1 m/s.

Vapours

For vapours, the velocity used will depend on the operating pressure and fluid density; the lower values in the ranges given below will apply to materials of high molecular weight.

Vacuum	50 to 70 m/s
Atmospheric pressure	10 to 30 m/s
High pressure	5 to 10 m/s

12.7.3. Stream Temperatures

The closer the temperature approach used (the difference between the temperatures of the two streams at a given point, usually calculated at the two ends of the exchanger) the larger will be the heat-transfer area required for a given duty. The optimum value depends on the application, and can only be determined by making an economic analysis of alternative designs. As a general guide, the optimum temperature approach will usually be in the range 10°C to 30°C for heat exchange between process streams. Lower temperature approaches are used for coolers, and a temperature approach of 5°C to 7°C for coolers using cooling water and 3°C to 5°C for those using refrigerated brines is common. The maximum temperature rise in recirculated cooling water is limited to around 30°C. Care should be taken to ensure that cooling media temperatures are kept well above the freezing point of the process materials. Temperature approaches as low as 1°C or 2°C are used in very low temperature sub-ambient processes, such as air separation and natural gas liquefaction. When the heat exchange is between process fluids for heat recovery the optimum approach temperature can be determined by pinch analysis, as described in Chapter 3. The optimum temperature approach for heat recovery depends on the trade-off between capital and energy costs (Figure 3.19). The numerical optimum temperature approach is seldom lower than 20°C, but lower values are often used, as lower temperature approaches lead to more conservative designs with more exchanger area.

12.7.4. Pressure Drop

In many applications, the pressure drop available to drive the fluids through the exchanger will be set by the process conditions, and the available pressure drop will vary from a few millibars in vacuum service to several bars in pressure systems.

When the designer is free to select the pressure drop, an economic analysis can be made to determine the exchanger design that gives the lowest operating costs, taking into consideration both capital and pumping costs; however, a full economic analysis will only be justified for very large, expensive, exchangers. The values suggested below can be used as a general guide, and will normally give designs that are near the optimum.

Liquids

Viscosity	Allowable Pressure Drop
<1 mN s/m^2	35 kN/m^2
1 to 10 mN s/m^2	50–70 kN/m^2

Gas and Vapours

High vacuum	$0.4-0.8 \text{ kN/m}^2$
Medium vacuum	$0.1 \times$ absolute pressure
1 to 2 bar	$0.5 \times$ system gauge pressure
Above 10 bar	$0.1 \times$ system gauge pressure

When a high pressure drop is used, care must be taken to ensure that the resulting high fluid velocity does not cause erosion or flow-induced tube vibration.

12.7.5. Fluid Physical Properties

The fluid physical properties required for heat-exchanger design are: density, viscosity, thermal conductivity and temperature–enthalpy correlations (specific and latent heats). Sources of physical property data are given in Chapter 8. Physical properties are usually obtained from a process simulation model; see Chapter 4. The thermal conductivities of commonly used tube materials are given in Table 12.6.

In the correlations used to predict heat-transfer coefficients, the physical properties are usually evaluated at the mean stream temperature. This is satisfactory when the temperature change is small, but can cause a significant error when the change in temperature is large. In these circumstances, a simple and safe procedure is to evaluate the heat-transfer coefficients at the stream inlet and outlet temperatures and use the lower of the two values. Alternatively, the method suggested by Frank (1978) can be used; in which equations 12.1 and 12.3 are combined:

$$Q = \frac{A[U_2(T_1 - t_2) - U_1(T_2 - t_1)]}{\ln\left[\dfrac{U_2(T_1 - t_2)}{U_1(T_2 - t_1)}\right]} \tag{12.9}$$

where U_1 and U_2 are evaluated at the ends of the exchanger. Equation 12.9 is derived by assuming that the heat-transfer coefficient varies linearly with temperature.

TABLE 12.6. Conductivity of metals

Metal	Temperature (°C)	k_w (W/m°C)
Aluminium	0	202
	100	206
Brass	0	97
(70 Cu, 30 Zn)	100	104
	400	116
Copper	0	388
	100	378
Nickel	0	62
	212	59
Cupro-nickel (10% Ni)	0–100	45
Monel	0–100	30
Stainless steel (18/8)	0–100	16
Carbon steel	40	60
	100	58
	260	51
Titanium	0–100	16

If the variation in the physical properties is too large for these simple methods to be used, it will be necessary to divide the temperature–enthalpy profile into sections and evaluate the heat-transfer coefficients and area required for each section.

12.8. TUBE-SIDE HEAT-TRANSFER COEFFICIENT AND PRESSURE DROP (SINGLE PHASE)

12.8.1. Heat Transfer

Turbulent Flow

Heat-transfer data for turbulent flow inside conduits of uniform cross-section are usually correlated by an equation of the form:

$$\text{Nu} = C\,\text{Re}^a\,\text{Pr}^b \left(\frac{\mu}{\mu_w}\right)^c \tag{12.10}$$

where

$\text{Nu} = \text{Nusselt number} = (h_i d_e / k_f)$
$\text{Re} = \text{Reynolds number} = (\rho\,u_t d_e / \mu) = (G_t d_e / \mu)$
$\text{Pr} = \text{Prandtl number} = (C_p \mu / k_f)$

and

$h_i = $ inside coefficient, W/m$^{2\circ}$C
$d_e = $ equivalent (or hydraulic mean) diameter, m

$$d_e = \frac{4 \times \text{cross-sectional area for flow}}{\text{wetted perimeter}} = d_i \text{ for tubes}$$

$u_t = $ fluid velocity, m/s
$k_f = $ fluid thermal conductivity, W/m°C
$G_t = $ mass velocity, mass flow per unit area, kg/m^{2}s
$\mu = $ fluid viscosity at the bulk fluid temperature, Ns/m^2
$\mu_w = $ fluid viscosity at the wall
$C_p = $ fluid specific heat, heat capacity, J/kg°C.

The index for the Reynolds number, a, is generally taken as 0.8. That for the Prandtl number, b, can range from 0.3 for cooling to 0.4 for heating. The index for the viscosity factor, c, is normally taken as 0.14 for flow in tubes, from the work of Sieder and Tate (1936), but some workers report higher values. A general equation that can be used for exchanger design is:

$$\text{Nu} = C\,\text{Re}^{0.8}\,\text{Pr}^{0.33} \left(\frac{\mu}{\mu_w}\right)^{0.14} \tag{12.11}$$

where C

$= 0.021$ for gases
$= 0.023$ for non-viscous liquids
$= 0.027$ for viscous liquids.

It is not possible to find values for the constant and indices to cover the complete range of process fluids, from gases to viscous liquids, but the values predicted using equation 12.11 should be sufficiently accurate for design purposes. The uncertainty in the prediction of the shell-side coefficient and fouling factors will usually far outweigh any error in the tube-side value. Where a more accurate prediction than that given by equation 12.11 is required, and justified, the data and correlations given in the Engineering Science Data Unit reports are recommended: ESDU 92003 and 93018 (1998).

Butterworth (1977) gives the following equation, which is based on the ESDU work:

$$\text{St} = E\,\text{Re}^{-0.205}\,\text{Pr}^{-0.505} \tag{12.12}$$

where

$\text{St} = \text{Stanton number} = (\text{Nu}/\text{RePr}) = (h_i/\rho u_t C_p)$

and

$E = 0.0225\,\exp(-0.0225(\ln\,\text{Pr})^2).$

Equation 12.12 is applicable at Reynolds numbers greater than 10,000.

Hydraulic Mean Diameter

In some texts the equivalent (hydraulic mean) diameter is defined differently for use in calculating the heat-transfer coefficient in a conduit or channel, than for calculating the pressure drop. The perimeter through which the heat is transferred is used in place of the total wetted perimeter. In practice, the use of d_e calculated either way will make little difference to the value of the estimated overall coefficient, as the film coefficient is only, roughly, proportional to $d_e^{-0.2}$.

It is the full wetted perimeter that determines the flow regime and the velocity gradients in a channel. So, in this book, d_e determined using the full wetted perimeter will be used for both pressure-drop and heat-transfer calculations. The actual area through which the heat is transferred should, of course, be used to determine the rate of heat transfer (equation 12.1).

Laminar Flow

Below a Reynolds number of about 2000 the flow in pipes will be laminar. Providing the natural convection effects are small, which will normally be so in forced convection, the following equation can be used to estimate the film heat-transfer coefficient:

$$\text{Nu} = 1.86\,(\text{RePr})^{0.33}\left(\frac{d_e}{L}\right)^{0.33}\left(\frac{\mu}{\mu_w}\right)^{0.14} \tag{12.13}$$

where L is the length of the tube in metres.

If the Nusselt number given by equation 12.13 is less than 3.5, it should be taken as 3.5.

In laminar flow, the length of the tube can have a marked effect on the heat-transfer rate for length to diameter ratios less than 500.

Transition Region

In the flow region between laminar and fully developed turbulent flow, heat-transfer coefficients cannot be predicted with certainty, as the flow in this region is unstable. The transition region should be avoided in exchanger design. If this is not practicable, the coefficient should be evaluated using both equations 12.11 and 12.13 and the lower value taken.

Heat-transfer Factor, j_h

It is often convenient to correlate heat-transfer data in terms of a heat-transfer 'j' factor, which is similar to the friction factor used for pressure drop. The heat-transfer factor is defined by:

$$j_h = \mathrm{St}\,\mathrm{Pr}^{0.67} \left(\frac{\mu}{\mu_w}\right)^{-0.14} \tag{12.14}$$

The use of the j_h factor allows data for laminar and turbulent flow to be represented on the same graph (Figure 12.23). The j_h values obtained from Figure 12.23 can be used with equation 12.14 to estimate the heat-transfer coefficients for heat-exchanger tubes and commercial pipes. The coefficient estimated for pipes will normally be conservative (on the low side) as pipes are rougher than the tubes used for heat exchangers, which are finished to closer tolerances. Equation 12.14 can be rearranged to a more convenient form:

$$\frac{h_i d_i}{k_f} = j_h \,\mathrm{Re}\,\mathrm{Pr}^{0.33} \left(\frac{\mu}{\mu_w}\right)^{0.14} \tag{12.15}$$

Note: Kern (1950), and others define the heat-transfer factor as

$$j_H = \mathrm{Nu}\,\mathrm{Pr}^{-1/3} \left(\frac{\mu}{\mu_w}\right)^{-0.14}$$

The relationship between j_h and j_H is given by

$$j_H = j_h \,\mathrm{Re}$$

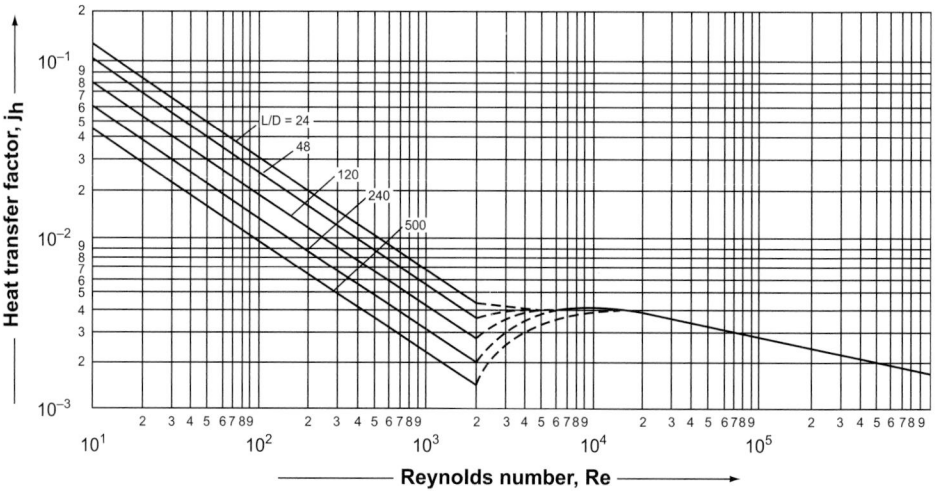

Figure 12.23. Tube-side heat-transfer factor.

Viscosity Correction Factor

The viscosity correction factor will normally only be significant for viscous liquids.

To apply the correction, an estimate of the wall temperature is needed. This can be made by first calculating the coefficient without the correction and using the following relationship to estimate the wall temperature:

$$h_i(t_w - t) = U(T - t) \qquad (12.16)$$

where

t = tube-side bulk temperature (mean)
t_w = estimated wall temperature
T = shell-side bulk temperature (mean).

Usually an approximate estimate of the wall temperature is sufficient, but trial-and-error calculations can be made to obtain a better estimate if the correction is large.

Coefficients for Water

Though equations 12.11 and 12.13 and Figure 12.23 may be used for water, a more accurate estimate can be made by using equations developed specifically for water. The physical properties are conveniently incorporated into the correlation. The equation below has been adapted from data given by Eagle and Ferguson (1930):

$$h_i = \frac{4200(1.35 + 0.02t)u_t^{0.8}}{d_i^{0.2}} \qquad (12.17)$$

where

h_i = inside coefficient, for water, $W/m^2{}^\circ C$
t = water temperature, $^\circ C$
u_t = water velocity, m/s
d_i = tube inside diameter, mm.

12.8.2. Tube-side Pressure Drop

There are two major sources of pressure loss on the tube side of a shell and tube exchanger: the friction loss in the tubes and the losses due to the sudden contraction and expansion and flow reversals that the fluid experiences in flow through the tube arrangement.

The tube friction loss can be calculated using the familiar equations for pressure drop in pipes (see Chapter 5). The basic equation for isothermal flow in pipes (constant temperature) is

$$\Delta P = 8j_f \left(\frac{L'}{d_i}\right)\frac{\rho u_t^2}{2} \qquad (12.18)$$

where j_f is the dimensionless friction factor and L' is the effective pipe length.

The flow in a heat exchanger is clearly not isothermal, and this is allowed for by including an empirical correction factor to account for the change in physical properties with temperature. Normally only the change in viscosity is considered:

$$\Delta P = 8j_f(L'/d_i)\rho \frac{u_t^2}{2}\left(\frac{\mu}{\mu_w}\right)^{-m} \tag{12.19}$$

$$m = 0.25 \text{ for laminar flow, Re} < 2100$$
$$= 0.14 \text{ for turbulent flow, Re} > 2100.$$

Values of j_f for heat-exchanger tubes can be obtained from Figure 12.24; commercial pipes are given in Chapter 5.

The pressure losses due to contraction at the tube inlets, expansion at the exits, and flow reversal in the headers, can be a significant part of the total tube-side pressure drop. There is no entirely satisfactory method for estimating these losses. Kern (1950) suggests adding four velocity heads per pass. Frank (1978) considers this to be too high, and recommends 2.5 velocity heads. Butterworth (1978) suggests 1.8. Lord *et al.* (1970) take the loss per pass as equivalent to a length of tube equal to 300 tube diameters for straight tubes and 200 for U-tubes, whereas Evans (1980) appears to add only 67 tube diameters per pass.

The loss in terms of velocity heads can be estimated by counting the number of flow contractions, expansions and reversals, and using the factors for pipe fittings to estimate the number of velocity heads lost. For two tube passes, there will be two contractions, two expansions and one flow reversal. The head loss for each of these effects is: contraction 0.5, expansion 1.0, 180° bend 1.5; so for two passes the maximum loss will be

$$2 \times 0.5 + 2 \times 1.0 + 1.5 = 4.5 \text{ velocity heads}$$
$$= \underline{\underline{2.25 \text{ per pass}}}$$

From this, it appears that Frank's recommended value of 2.5 velocity heads per pass is the most realistic value to use.

Combining this factor with equation 12.19 gives

$$\Delta P_t = N_p\left[8\,j_f\left(\frac{L}{d_i}\right)\left(\frac{\mu}{\mu_w}\right)^{-m} + 2.5\right]\frac{\rho u_t^2}{2} \tag{12.20}$$

where

$$\Delta P_t = \text{tube-side pressure drop, N/m}^2 \text{ (Pa)}$$
$$N_p = \text{number of tube-side passes}$$
$$u_t = \text{tube-side velocity, m/s}$$
$$L = \text{length of one tube.}$$

Another source of pressure drop is the flow expansion and contraction at the exchanger inlet and outlet nozzles. This can be estimated by adding one velocity head for the inlet and 0.5 for the outlet, based on the nozzle velocities.

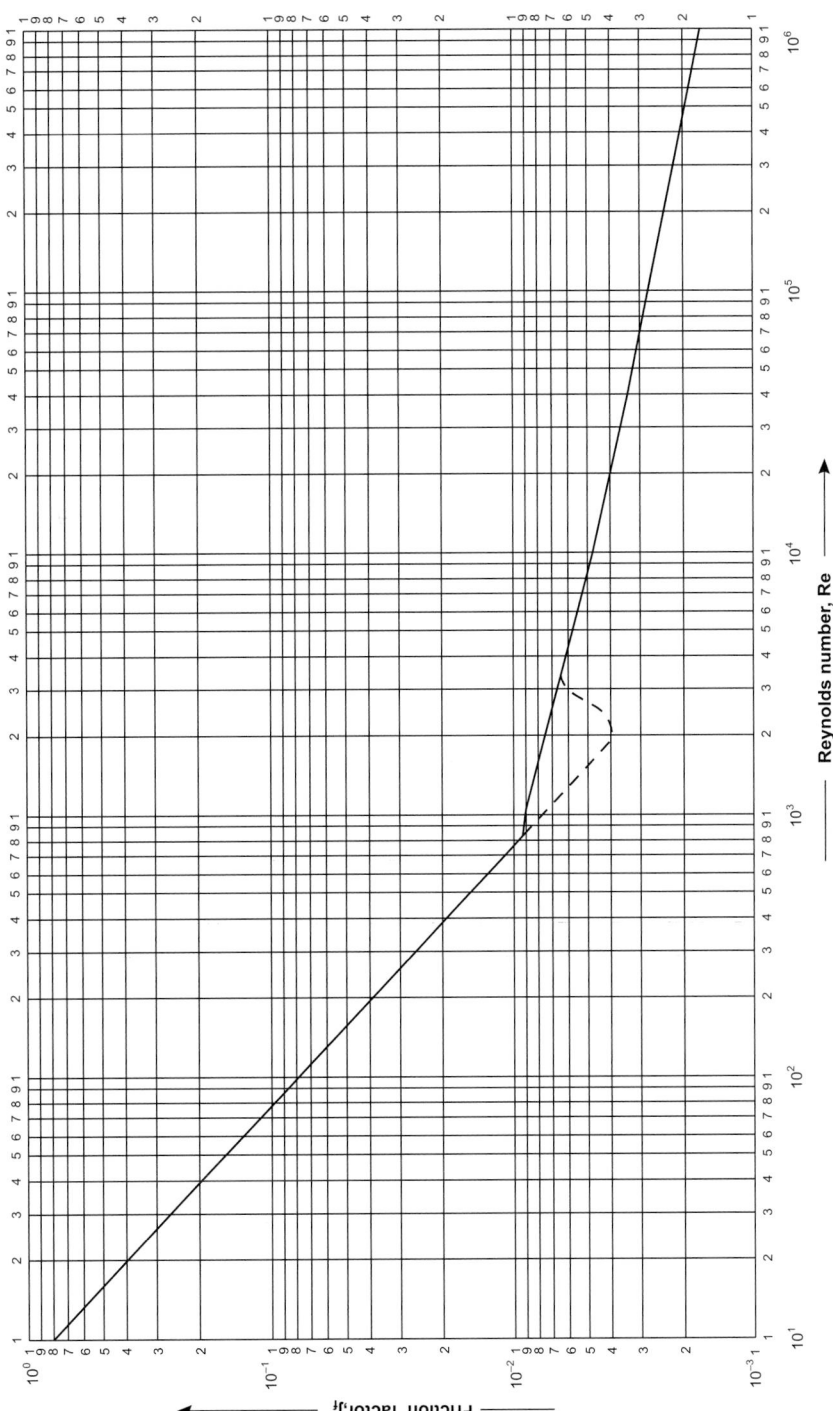

Figure 12.24. Tube-side friction factors. *Note*: The friction factor j_f is the same as the friction factor for pipes ϕ ($= (R/\rho u^2)$).

12.9. SHELL-SIDE HEAT-TRANSFER AND PRESSURE DROP (SINGLE PHASE)

12.9.1. Flow Pattern

The flow pattern in the shell of a segmentally baffled heat exchanger is complex, and this makes the prediction of the shell-side heat-transfer coefficient and pressure drop very much more difficult than for the tube side. Though the baffles are installed to direct the flow across the tubes, the actual flow of the main stream of fluid will be a mixture of cross flow between the baffles, coupled with axial (parallel) flow in the baffle windows; as shown in Figure 12.25. Not all the fluid flow follows the path shown in Figure 12.25; some will leak through gaps formed by the clearances that have to be allowed for fabrication and assembly of the exchanger. These leakage and bypass streams are shown in Figure 12.26, which is based on the flow model proposed by Tinker (1951, 1958). In Figure 12.26, Tinker's nomenclature is used to identify the various streams, as follows:

Stream A: the tube-to-baffle leakage stream. The fluid flowing through the clearance between the tube outside diameter and the tube hole in the baffle.

Stream B: the actual cross-flow stream.

Stream C: the bundle-to-shell bypass stream. The fluid flowing in the clearance area between the outer tubes in the bundle (bundle diameter) and the shell.

Stream E: the baffle-to-shell leakage stream. The fluid flowing through the clearance between the edge of a baffle and the shell wall.

Stream F: the pass-partition stream. The fluid flowing through the gap in the tube arrangement due to the pass partition plates. Where the gap is vertical it will provide a low-pressure drop path for fluid flow.

Note: There is no stream D.

The fluid in streams C, E and F bypasses the tubes, reducing the effective heat-transfer area.

Stream C is the main bypass stream and is particularly significant in pull-through bundle exchangers, where the clearance between the shell and bundle is of necessity large. Stream C can be considerably reduced by using sealing strips; horizontal strips that block the gap between the bundle and the shell (Figure 12.27). Dummy tubes are also sometimes used to block the pass-partition leakage stream F.

The tube-to-baffle leakage stream A does not bypass the tubes, and its main effect is on pressure drop rather than heat transfer.

The clearances will tend to plug as the exchanger becomes fouled and this will increase the pressure drop; see Section 12.9.6.

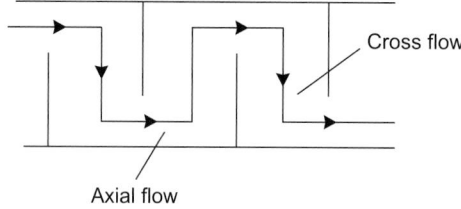

Cross flow

Axial flow

Figure 12.25.   Idealized main stream flow.

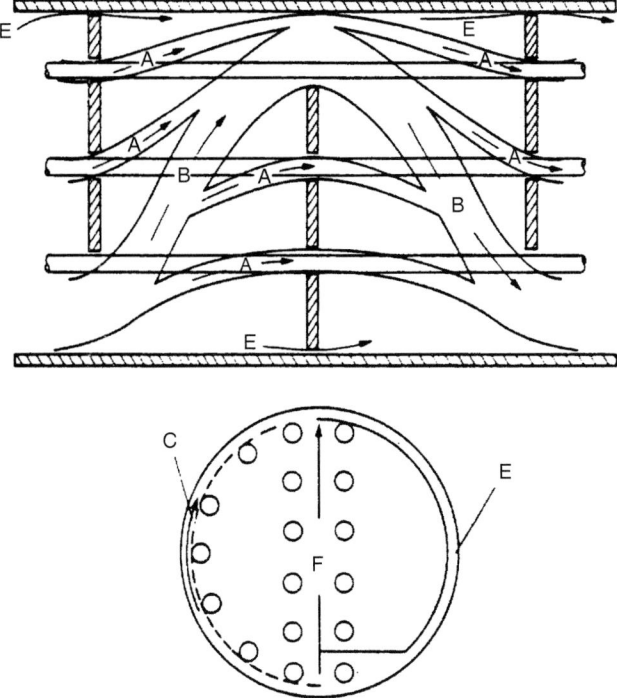

Figure 12.26. Shell-side leakage and by-pass paths.

12.9.2. Design Methods

The complex flow pattern on the shell side, and the great number of variables involved, make it difficult to predict the shell-side heat-transfer coefficient and pressure drop with complete assurance. In methods used for the design of exchangers prior to about 1960 no attempt was made to account for the leakage and bypass streams. Correlations were based on the total stream flow, and empirical methods were used to account for the performance of real exchangers compared with that for cross-flow over ideal tube banks. Typical of these 'bulk-flow' methods are those of Kern (1950) and Donohue (1955). Reliable predictions can only be achieved by comprehensive analysis of the contribution to heat-transfer and pressure drop made

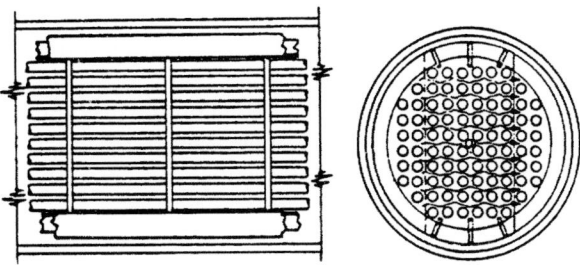

Figure 12.27. Sealing strips.

by the individual streams shown in Figure 12.26. Tinker (1951, 1958) published the first detailed stream-analysis method for predicting shell-side heat-transfer coefficients and pressure drop, and the methods subsequently developed have been based on his model. Tinker's presentation is difficult to follow, and his method is difficult and tedious to apply in manual calculations. It has been simplified by Devore (1961, 1962), using standard tolerances for commercial exchangers and only a limited number of baffle cuts. Devore gives nomographs that facilitate the application of the method in manual calculations. Mueller (1973) has further simplified Devore's method and gives an illustrative example.

Bell (1960, 1963) developed a semi-analytical method based on work done in the cooperative research programme on shell and tube exchangers at the University of Delaware. His method accounts for the major bypass and leakage streams and is suitable for a manual calculation.

The Engineering Sciences Data Unit has also published a method for estimating shell side the pressure drop and heat-transfer coefficient: ESDU Design Guide 83038 (1984). The method is based on a simplification of Tinker's work. It can be used for hand calculations, but as iterative procedures are involved it is best programmed for use with personal computers.

Tinker's model has been used as the basis for the proprietary computer methods developed by Heat Transfer Research Incorporated; see Palen and Taborek (1969); and by Heat Transfer and Fluid Flow Services; see Grant (1973). The HTRI method and software are available from HTRI (www.htri.net). The HTFS programs are available in process simulation programs such as Aspen Technology's Aspen Engineering Suite and Honeywell's UniSim Design Suite; see Chapter 4. The use of the HTFS programs is illustrated in Example 12.4.

Though Kern's method does not take account of the bypass and leakage streams, it is simple to apply and is accurate enough for preliminary design calculations and for designs where uncertainty in other design parameters is such that the use of more elaborate methods is not justified. Kern's method is given in Section 12.9.3 and is illustrated in Examples 12.1 and 12.3.

12.9.3. Kern's Method

This method was based on experimental work on commercial exchangers with standard tolerances and will give a reasonably satisfactory prediction of the heat-transfer coefficient for standard designs. The prediction of pressure drop is less satisfactory, as pressure drop is more affected by leakage and bypassing than heat transfer. The shell-side heat-transfer and friction factors are correlated in a similar manner to those for tube-side flow by using a hypothetical shell velocity and shell diameter. As the cross-sectional area for flow varies across the shell diameter, the linear and mass velocities are based on the maximum area for cross-flow: that at the shell equator. The shell equivalent diameter is calculated using the flow area between the tubes taken in the axial direction (parallel to the tubes) and the wetted perimeter of the tubes; see Figure 12.28.

Shell-side j_h and j_f factors for use in this method are given in Figures 12.29 and 12.30, for various baffle cuts and tube arrangements. These figures are based on data given by Kern (1950) and by Ludwig (2001).

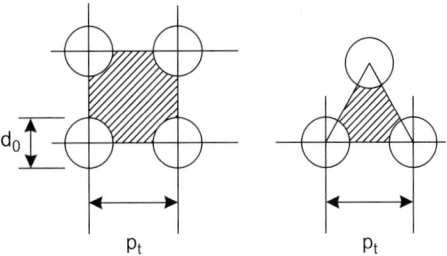

Figure 12.28. Equivalent diameter, cross-sectional areas and wetted perimeters.

The procedure for calculating the shell-side heat-transfer coefficient and pressure drop for a single-shell-pass exchanger is given below.

Procedure

1. Calculate the area for cross-flow A_s for the hypothetical row of tubes at the shell equator, given by:

$$A_s = \frac{(p_t - d_o)D_s l_B}{p_t} \qquad (12.21)$$

where

p_t = tube pitch
d_o = tube outside diameter
D_s = shell inside diameter, m
l_B = baffle spacing, m.

The term $(p_t - d_o)/p_t$ is the ratio of the clearance between tubes and the total distance between tube centres.

2. Calculate the shell-side mass velocity G_s and the linear velocity u_s:

$$G_s = \frac{W_s}{A_s}$$

$$u_s = \frac{G_s}{\rho}$$

where

W_s = fluid flow rate on the shell side, kg/s
p = shell-side fluid density, kg/m^3.

3. Calculate the shell-side equivalent diameter (hydraulic diameter); Figure 12.28. For a square pitch arrangement:

$$d_e = \frac{4\left(\dfrac{p_t^2 - \pi d_o^2}{4}\right)}{\pi d_o} = \frac{1.27}{d_o}(p_t^2 - 0.785 d_o^2) \qquad (12.22)$$

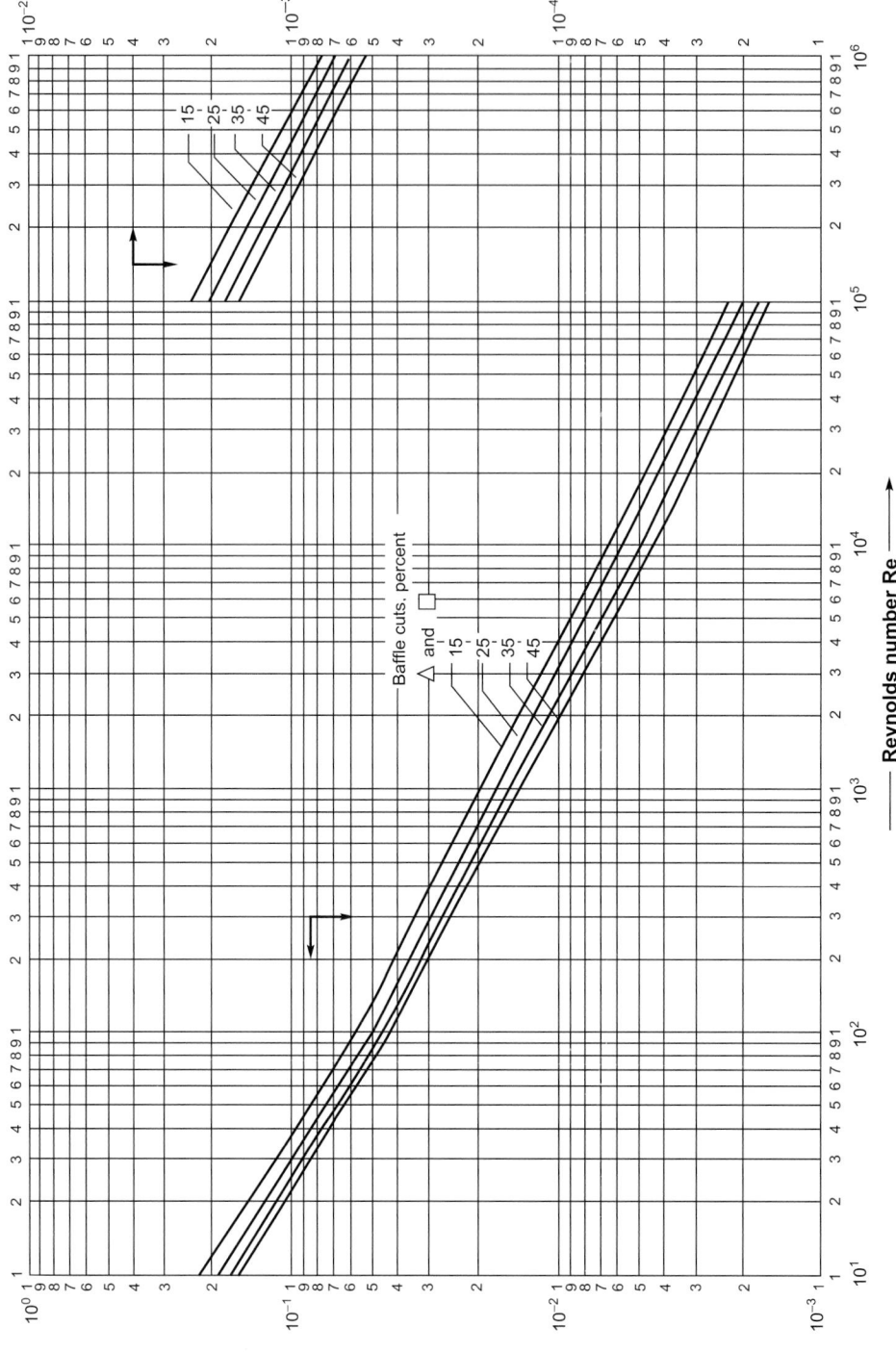

Figure 12.29. Shell-side heat-transfer factors, segmental baffles.

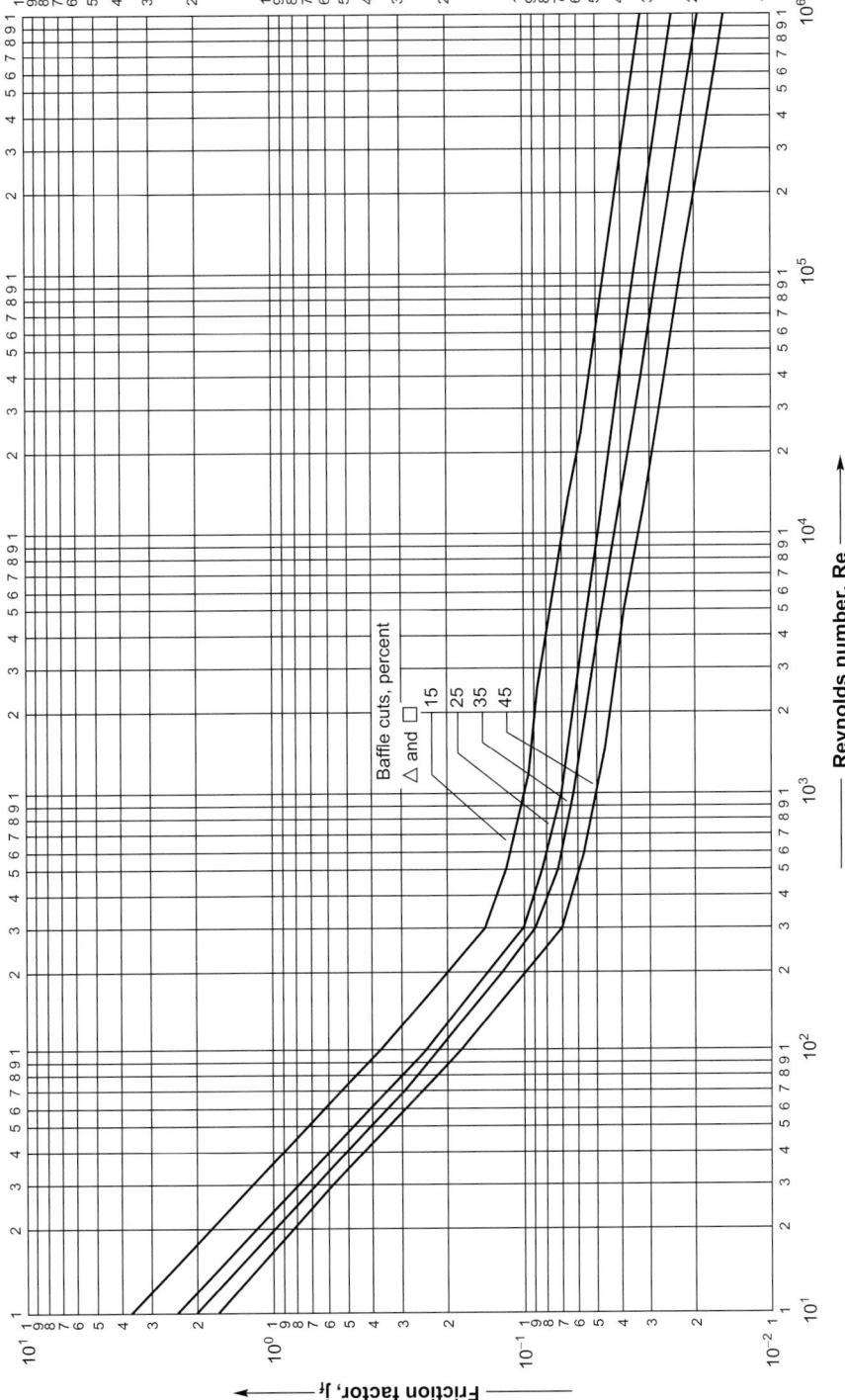

Figure 12.30. Shell-side friction factors, segmental baffles.

For an equilateral triangular pitch arrangement:

$$d_e = \frac{4\left(\frac{p_t}{2} \times 0.87\, p_t - \frac{1}{2}\pi\frac{d_o^2}{4}\right)}{\frac{\pi d_o}{2}} = \frac{1.10}{d_o}(p_t^2 - 0.917d_o^2) \tag{12.23}$$

where d_e = equivalent diameter, m.

4. Calculate the shell-side Reynolds number, given by:

$$\mathrm{Re} = \frac{G_s d_e}{\mu} = \frac{u_s d_e \rho}{\mu} \tag{12.24}$$

5. For the calculated Reynolds number, read the value of j_h from Figure 12.29 for the selected baffle cut and tube arrangement, and calculate the shell-side heat-transfer coefficient h_s from:

$$\mathrm{Nu} = \frac{h_s d_e}{k_f} = j_h\, \mathrm{Re}\,\mathrm{Pr}^{0.33}\left(\frac{\mu}{\mu_w}\right)^{0.14} \tag{12.25}$$

The tube wall temperature can be estimated using the method given for the tube side; Section 12.8.1.

6. For the calculated shell-side Reynolds number, read the friction factor from Figure 12.30 and calculate the shell-side pressure drop from:

$$\Delta P_s = 8\, j_f\left(\frac{D_s}{d_e}\right)\left(\frac{L}{l_B}\right)\frac{\rho u_s^2}{2}\left(\frac{\mu}{\mu_w}\right)^{-0.14} \tag{12.26}$$

where

L = tube length
l_B = baffle spacing.

The term (L/l_B) is the number of times the flow crosses the tube bundle = $(N_b + 1)$, where N_b is the number of baffles.

Shell-Nozzle Pressure Drop

The pressure loss in the shell nozzles will normally only be significant with gases. The nozzle pressure drop can be taken as equivalent to 1½ velocity heads for the inlet and ½ for the outlet, based on the nozzle area or the free area between the tubes in the row immediately adjacent to the nozzle, whichever is the lower.

Example 12.1

Design an exchanger to sub-cool condensate from a methanol condenser from 95°C to 40°C. The flow rate of methanol is 100,000 kg/h. Brackish water will be used as the coolant, with a temperature rise from 25° to 40°C.

Solution

Only the thermal design will be considered.
   This example illustrates Kern's method.
   Coolant is corrosive, so assign to tube side.

$$\text{Heat capacity methanol} = 2.84 \text{ kJ/kg}^\circ\text{C}$$

$$\text{Heat load} = \frac{100,000}{3600} \times 2.84(95 - 40) = 4340 \text{ kW}$$

$$\text{Heat capacity water} = 4.2 \text{ kJ/Kg}^\circ\text{C}$$

$$\text{Cooling water flow} = \frac{4340}{4.2(40 - 25)} = 68.9 \text{ kg/s}$$

$$\Delta T_{lm} = \frac{(95 - 40) - (40 - 25)}{\ln\dfrac{(95 - 40)}{(40 - 25)}} = 31^\circ\text{C} \tag{12.4}$$

Use one shell pass and two tube passes

$$R = \frac{95 - 40}{40 - 25} = 3.67 \tag{12.6}$$

$$S = \frac{40 - 25}{95 - 25} = 0.21 \tag{12.7}$$

From Figure 12.19

$$F_t = 0.85$$
$$\Delta T_m = 0.85 \times 31 = 26^\circ\text{C}$$

From Figure 12.1

$$U = 600 \text{ W/m}^{2\circ}\text{C}$$

Provisional area

$$A = \frac{4340 \times 10^3}{26 \times 600} = 278 \text{ m}^2 \tag{12.1}$$

Choose 20 mm o.d., 16 mm i.d., 4.88-m-long tubes ($\frac{3}{4}$ in $\times$ 16 ft), cupro-nickel. Allowing for tube-sheet thickness, take

$$L = 4.83 \text{ m}$$
$$\text{Area of one tube} = 4.83 \times 20 \times 10^{-3}\pi = 0.303 \text{ m}^2$$
$$\text{Number of tubes} = \frac{278}{0.303} = \underline{\underline{918}}$$

As the shell-side fluid is relatively clean use 1.25 triangular pitch.

$$\text{Bundle diameter } D_b = 20\left(\frac{918}{0.249}\right)^{1/2.207} = 826 \text{ mm} \qquad (12.3b)$$

Use a split-ring floating head type.
From Figure 12.10, bundle diametrical clearance = 68 mm,

$$\text{shell diameter, } D_s = 826 + 68 = 894 \text{ mm}.$$

Note: nearest standard pipe sizes are 863.6 or 914.4 mm.

Shell size could be read from standard tube count tables.

Tube-side coefficient

$$\text{Mean water temperature} = \frac{40 + 25}{2} = 33°C$$

$$\text{Tube cross-sectional area} = \frac{\pi}{4} \times 16^2 = 201 \text{ mm}^2$$

$$\text{Tubes per pass} = \frac{918}{2} = 459$$

$$\text{Total flow area} = 459 \times 201 \times 10^{-6} = 0.092 \text{ m}^2$$

$$\text{Water mass velocity} = \frac{68.9}{0.092} = 749 \text{ kg/s m}^2$$

$$\text{Density water} = 995 \text{ kg/m}^3$$

$$\text{Water linear velocity} = \frac{749}{995} = 0.75 \text{ m/s}$$

$$h_i = \frac{4200(1.35 + 0.02 \times 33)0.75^{0.8}}{16^{0.2}} = 3852 \text{ W/m}^2°C \qquad (12.17)$$

The coefficient can also be calculated using equation 12.15; this is done to illustrate use of this method.

$$\frac{h_i d_i}{k_f} = j_h Re Pr^{0.33}\left(\frac{\mu}{\mu_w}\right)^{0.14}$$

$$\text{Viscocity of water} = 0.8 \text{ mNs/m}^2$$

$$\text{Thermal conductivity} = 0.59 \text{ W/m}°C$$

$$Re = \frac{\rho u d_i}{\mu} = \frac{995 \times 0.75 \times 16 \times 10^{-3}}{0.8 \times 10^{-3}} = 14,925$$

$$Pr = \frac{C_p \mu}{k_f} = \frac{4.2 \times 10^3 \times 0.8 \times 10^{-3}}{0.59} = 5.7$$

$$\text{Neglect } \left(\frac{\mu}{\mu_w}\right)$$

$$\frac{L}{d_i} = \frac{4.83 \times 10^3}{16} = 302$$

From Figure 12.23

$$j_h = 3.9 \times 10^{-3}$$

$$h_i = \frac{0.59}{16 \times 10^{-3}} \times 3.9 \times 10^{-3} \times 14,925 \times 5.7^{0.33} = 3812 \text{ W/m}^2\,°\text{C}$$

Checks reasonably well with value calculated from equation 12.17; use lower figure.

Shell-Side Coefficient

$$\text{Choose baffle spacing} = \frac{D_s}{5} = \frac{894}{5} = 178 \text{ mm}$$

$$\text{Tube pitch} = 1.25 \times 20 = 25 \text{ mm}$$

$$\text{Cross-flow area } A_s = \frac{(25-20)}{25}894 \times 178 \times 10^{-6} = 0.032 \text{ m}^2 \qquad (12.21)$$

$$\text{Mass velocity, } G_s = \frac{100,000}{3600} \times \frac{1}{0.032} = 868 \text{ kg/s m}^2$$

$$\text{Equivalent diameter } d_e = \frac{1.1}{20}(25^2 - 0.917 \times 20^2) = 14.4 \text{ mm} \qquad (12.23)$$

$$\text{Mean shell side temperature} = \frac{95+40}{2} = 68°\text{C}$$

Methanol density $= 750 \text{ kg/m}^3$

Viscosity $= 0.34 \text{ mNs/m}^2$

Heat capacity $= 2.84 \text{ kJ/kg}°\text{C}$

Thermal conductivity $= 0.19 \text{ W/m}°\text{C}$

$$\text{Re} = \frac{G_s d_e}{\mu} = \frac{868 \times 14.4 \times 10^{-3}}{0.34 \times 10^{-3}} = 36,762$$

$$\text{Pr} = \frac{C_p\mu}{k_f} = \frac{2.84 \times 10^3 \times 0.34 \times 10^{-3}}{0.19} = 5.1$$

$$(12.24)$$

Choose 25% baffle cut, from Figure 12.29

$$j_h = 3.3 \times 10^{-3}$$

Without the viscosity correction term

$$h_s = \frac{0.19}{14.4 \times 10^{-3}} \times 3.3 \times 10^{-3} \times 36,762 \times 5.1^{1/3} = 2740 \text{ W/m}^2{}^\circ\text{C}$$

Estimate wall temperature

$$\text{Mean temperature difference} = 68 - 33 = 35^\circ\text{C}$$

$$\text{across all resistances}$$

$$\text{across methanol film} = \frac{U}{h_o} \times \Delta T = \frac{600}{2740} \times 35 = 8^\circ\text{C}$$

$$\text{Mean wall temperature} = 68 - 8 = 60^\circ\text{C}$$

$$\mu_w = 0.37 \text{ mNs/m}^2$$

$$\left(\frac{\mu}{\mu_w}\right)^{0.14} = 0.99$$

which shows that the correction for a low-viscosity fluid is not significant.

Overall Coefficient

Thermal conductivity of cupro-nickel alloys $= 50$ W/m°C.

Take the fouling coefficients from Table 12.2; methanol (light organic) 5000 W/m²°C, brackish water (sea water), take as highest value, 3000 W/m²°C:

$$\frac{1}{U_o} = \frac{1}{2740} + \frac{1}{5000} + \frac{20 \times 10^{-3}\ln\left(\frac{20}{16}\right)}{2 \times 50}$$
$$+ \frac{20}{16} \times \frac{1}{3000} + \frac{20}{16} \times \frac{1}{3812}$$
$$U_o = \underline{\underline{738 \text{ W/m}^2{}^\circ\text{C}}} \tag{12.2}$$

Well above assumed value of 600 W/m²°C.

Pressure Drop

Tube Side
From Figure 12.24, for $Re = 14,925$

$$j_f = 4.3 \times 10^{-3}$$

Neglecting the viscosity correction term

$$\Delta P_t = 2\left(8 \times 4.3 \times 10^{-3}\left(\frac{4.83 \times 10^3}{16}\right) + 2.5\right)\frac{995 \times 0.75^2}{2}$$
$$= 7211 \text{ N/m}^2 = 7.2 \text{ kPa } (1.1 \text{ psi}) \tag{12.20}$$

Low, could consider increasing the number of tube passes.

Shell Side

$$\text{Linear velocity} = \frac{G_s}{\rho} = \frac{868}{750} = 1.16 \text{ m/s}$$

From Figure 12.30, at Re = 36,762

$$j_f = 4 \times 10^{-2}$$

Neglect viscosity correction

$$\Delta P_s = 8 \times 4 \times 10^{-2}\left(\frac{894}{14.4}\right)\left(\frac{4.83 \times 10^3}{178}\right)\frac{750 \times 1.16^2}{2}$$
$$= 272,019 \text{ N/m}^2 \tag{12.26}$$
$$= 272 \text{ kPa } (39 \text{ psi}) \text{ too high},$$

Could be reduced by increasing the baffle pitch. Doubling the pitch halves the shell-side velocity, which reduces the pressure drop by a factor of approximately $(\frac{1}{2})^2$

$$\Delta P_s = \frac{272}{4} = 68 \text{ kPa } (10 \text{ psi}), \text{ acceptable}$$

This will reduce the shell-side heat-transfer coefficient by a factor of $(1/2)^{0.8}$ ($h_o \propto Re^{0.8} \propto u_s^{0.8}$)

$$h_o = 2740 \times \left(\frac{1}{2}\right)^{0.8} = 1573 \text{ W/m}^2{}^{\circ}\text{C}$$

This gives an overall coefficient of 615 W/m²°C – still above assumed value of 600 W/m²°C.

Example 12.2

Gas oil at 200°C is to be cooled to 40°C. The oil flow rate is 22,500 kg/h. Cooling water is available at 30°C and the temperature rise is to be limited to 20°C. The pressure drop allowance for each stream is 100 kN/m².

Design a suitable exchanger for this duty.

Solution

Only the thermal design will be carried out, to illustrate the calculation procedure for an exchanger with a divided shell.

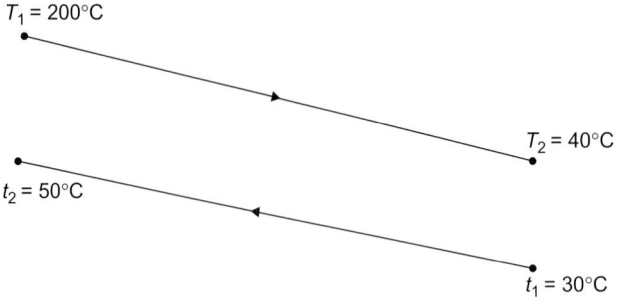

$$\Delta T_{lm} = \frac{(200 - 40) - (40 - 30)}{\text{Ln}\dfrac{(200 - 50)}{(40 - 30)}} = 51.7°C \tag{12.4}$$

$$R = (200 - 50)/(50 - 30) = 8.0 \tag{12.6}$$

$$S = (50 - 30)/(200 - 30) = 0.12 \tag{12.7}$$

These values do not intercept on the figure for a single-shell-pass exchanger (Figure 12.19), so use the figure for a two-pass shell (Figure 12.20), which gives

$$F_t = 0.94$$

so

$$\Delta T_m = 0.94 \times 51.7 = 48.6°C$$

Physical Properties

Water, from steam tables

Temperature, °C	30	40	50
C_p, kJ/kg°C	4.18	4.18	4.18
k, kW/m°C	618×10^{-6}	631×10^{-6}	643×10^{-6}
μ, mNm^{-2}s	797×10^{-3}	671×10^{-3}	544×10^{-3}
ρ, kg m^{-3}	995.2	992.8	990.1

Gas oil, from Kern (1950)

Temperature, °C	200	120	40
C_p, kJ/kg°C	2.59	2.28	1.97
k, W/m°C	0.13	0.125	0.12
μ, mNm^{-2}s	0.06	0.17	0.28
ρ, kg/m^3	830	850	870

Duty

$$\text{Oil flow rate} = 22{,}500/3600 = 6.25 \text{ kg/s}$$

$$Q = 6.25 \times 2.28 \times (200 - 40) = 2280 \text{ kW}$$

$$\text{Water flow rate} = \frac{2280}{4.18(50 - 30)} = 27.27 \text{ kg/h}$$

From Figure 12.1, for cooling tower water and heavy organic liquid, take

$$U = 500 \text{ Wm}^{-2}\text{C}^{-1}$$

$$\text{Area required} = \frac{2280 \times 10^3}{500 \times 48.6} = 94 \text{ m}^2$$

Tube-side Coefficient

Select 20-mm o.d., 16-mm i.d. tubes, 4 m long, triangular pitch $1.25d_o$, carbon steel.

Surface area of one tube $\quad = \pi \times 20 \times 10^{-3} \times 4 = 0.251 \text{ m}^2$

Number of tubes required $\quad = 94/0.251 = 375$, say 376, even number

Cross-sectional area, one tube $\quad = \dfrac{\pi}{4}(16 \times 10^{-3})2 = 2.011 \times 10^{-4} \text{ m}^2$

Total tube area $\quad\quad\quad\quad\quad = 376 \times 2.011 \times 10^{-4} = 0.0756 \text{ m}^2$

Put water through tubes for ease of cleaning.

Tube velocity, one pass $= 27.27/(992.8 \times 0.0756) = 0.363$ m/s

Too low to make effective use of the allowable pressure drop, try four passes.

$$u_t = 4 \times 0.363 = 1.45 \text{ m/s}$$

A floating head will be needed due to the temperature difference. Use a pull-through type.

Tube-side heat-transfer coefficient

$$h_i = \frac{4200(1.35 + 0.02 \times 40)1.45^{0.8}}{16^{0.2}} = 6982 \text{ W/m}^{-2}\,{}^\circ\text{C}^{-1} \tag{12.17}$$

Shell-side Coefficient

From Table 12.4 and equation 12.3b, for four passes, $1.25d_o$ triangular pitch

Bundle diameter, $D_b = 20(376/0.175)^{1/2.285} = 575$ mm

From Figure 12.10, for pull-through head, clearance = 92 mm

Shell diameter, $D_s = 575 + 92 = 667$ mm (26-in pipe)

Use 25% cut baffles, baffle arrangement for divided shell as shown below:

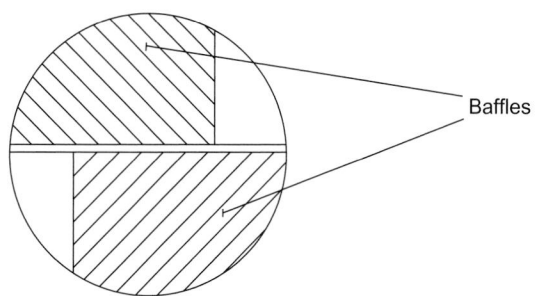

Take baffle spacing as 1/5 shell diameter $= 667/5 = 133$ mm
  Tube pitch, $p_t = 1.25 \times 20 = 25$ mm
  Area for flow, A_s, will be half that given by equation 12.21

$$A_s = 0.5 \times \left(\frac{25 - 20}{25} \times 0.667 \times 0.133 \right) = 0.00887 \text{ m}^2$$

$$G_s = 6.25/0.00887 = 704.6 \text{ kg/s}$$

$$u_s = 704.6/850 = 0.83 \text{ m/s, looks reasonable} \qquad (12.23)$$

$$d_e = \frac{1.10}{20}\left(25^2 - 0.917 \times 20^2 \right) = 14.2 \text{ mm}$$

$$\text{Re} = \frac{0.83 \times 14.2 \times 10^{-3} \times 850}{0.17 \times 10^{-3}} = 58,930$$

From Figure 12.29, $j_h = 2.6 \times 10^{-3}$

$$\text{Pr} = (2.28 \times 10^3 \times 0.17 \times 10^{-3})/0.125 = 3.1$$
$$\text{Nu} = 2.6 \times 10^{-3} \times 58,930 \times 3.1^{1/3} = 223.4 \qquad (12.25)$$

$$h_s = (223.4 \times 0.125)/(14.2 \times 10^{-3}) = 1967 \text{ W/m}^2 \,^\circ\text{C}$$

Overall Coefficient

Take fouling factors as 0.00025 for cooling tower water and 0.0002 for gas oil (light organic). Thermal conductivity for carbon steel tubes 45 W/m°C.

$$1/U_o = 1/1967 + 0.0002 + \frac{20 \times 10^{-3}\ln(20/16)}{2 \times 45}$$
$$+20/16(1/6982 + 0.00025) = 0.00125 \qquad (12.2)$$
$$U_o = 1/0.00125 = 800 \text{ W/m}^{-2}\,^\circ\text{C}^{-1}$$

Well above the initial estimate of 500 W/m²°C, so design has adequate area for the duty required.

Pressure Drops

Tube Side

$$\text{Re} = \frac{1.45 \times 16 \times 10^{-3} \times 992.8}{670 \times 10^{-6}} = 34,378 \qquad (3.4 \times 10^{-4})$$

From Figure 12.24, $j_f = 3.5 \times 10^{-3}$. Neglecting the viscosity correction

$$\Delta P_t = 4\left[8 \times 3.5 \times 10^{-3} \times \left(\frac{4}{16 \times 10^{-3}}\right) + 2.5\right]992.8 \times \frac{1.45^2}{2} = 39,660$$

$$= 40\ \text{kN/m}^2 \qquad\qquad (12.20)$$

Well within the specification, so no need to check the nozzle pressure drop.

Shell Side
From Figure 12.30, for Re $= 58,930$, $j_f = 3.8 \times 10^{-2}$
With a divided shell, the path length $= 2 \times (L/l_b)$
Neglecting the viscosity correction factor

$$\Delta P_s = 8 \times 3.8 \times 10^{-2}\left(\frac{662 \times 10^{-3}}{14.2 \times 10^{-3}}\right) \times \left(\frac{2 \times 4}{132 \times 10^{-3}}\right) \times 850 \times \frac{0.83^2}{2}$$

$$= 251,481 = 252\ \text{kN/m}^2 \qquad\qquad (12.26)$$

Well within the specification, no need to check nozzle pressure drops.
So the proposed thermal design is satisfactory. As the calculated pressure drops are below that allowed, there is some scope for improving the design.

Example 12.3

Design a shell and tube exchanger for the following duty.

Kerosene, 20,000 kg/h (42° API), leaves the base of a kerosene side-stripping column at 200°C and is to be cooled to 90°C by exchange with 70,000 kg/h light crude oil (34° API) coming from storage at 40°C. The kerosene enters the exchanger at a pressure of 5 bar and the crude oil at 6.5 bar. A pressure drop of 0.8 bar is permissible on both streams. Allowance should be made for fouling by including a fouling factor of 0.0003 $(\text{W/m}^{2\circ}\text{C})^{-1}$ on the crude stream and 0.0002 $(\text{W/m}^{2\circ}\text{C})^{-1}$ on the kerosene stream.

Solution

The solution to this example illustrates the iterative nature of heat-exchanger design calculations. An algorithm for the design of shell and tube exchangers is shown in Figure 12.31. The procedure set out in this figure will be followed in the solution.

Step 1: Specification

The specification is given in the problem statement:

20,000 kg/h of kerosene (42° API) at 200°C cooled to 90°C, by exchange with 70,000 kg/h light crude oil (34° API) at 40°C.

The kerosene pressure 5 bar, the crude oil pressure 6.5 bar.

Permissible pressure drop of 0.8 bar on both streams.

Fouling factors: crude stream 0.00035 $(\text{W/m}^{2\circ}\text{C})^{-1}$, kerosene stream 0.0002 $(\text{W/m}^{2\circ}\text{C})^{-1}$.

To complete the specification, the duty (heat-transfer rate) and the outlet temperature of the crude oil needed to be calculated.

$$\text{Mean temperature of kerosene} = (200 + 90)/2 = 145°C$$

At this temperature the specific heat capacity of 42° API kerosene is 2.47 kJ/kg°C (physical properties from Kern, 1950).

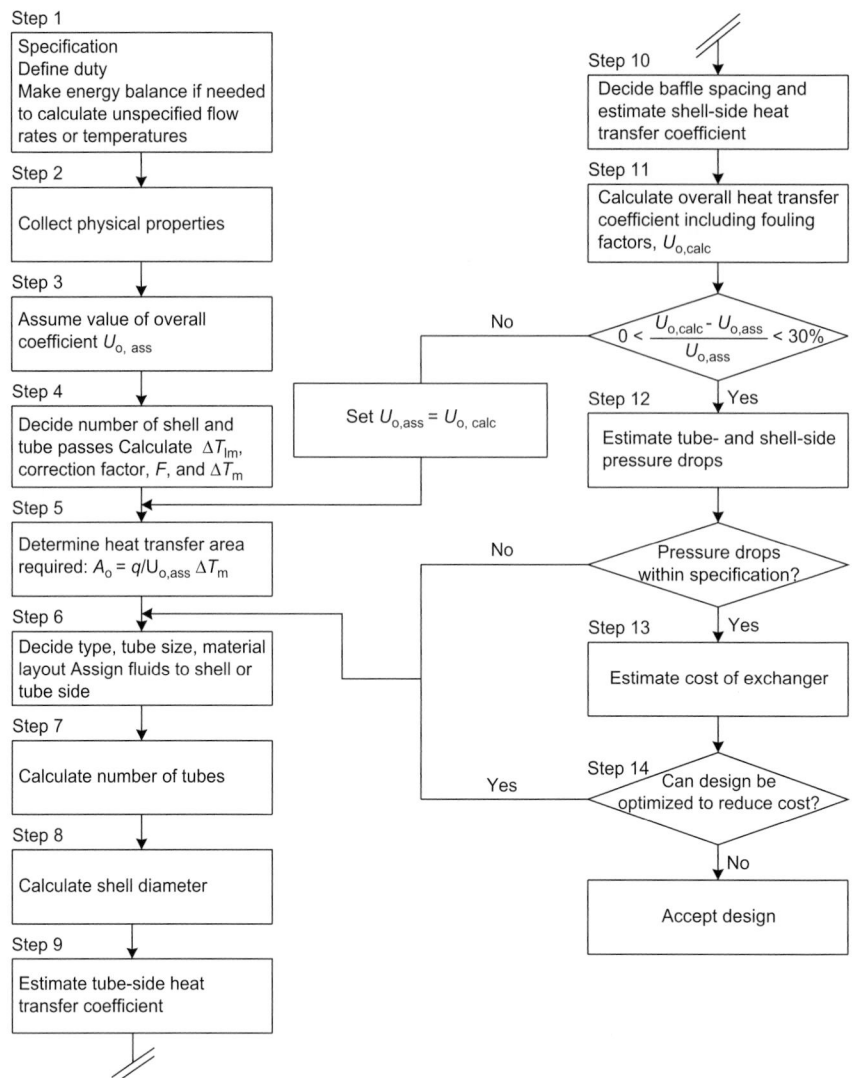

Figure 12.31. Design procedure for shell and tube heat exchangers. Example 12.2 and Figure 12.31 were developed by the author for the Open University Course T333 *Principles and Applications of Heat Transfer*. They are reproduced here by permission of the Open University.

$$\text{Duty} = \frac{20,000}{3600} \times 2.47(200 - 90) = 1509.4 \text{ kW}$$

As a first trial, take the mean temperature of the crude oil as equal to the inlet temperature, 40°C; specific heat capacity at this temperature = 2.01 kJ/kg°C.

An energy balance gives:

$$\frac{70,000}{3600} \times 2.01 \ (t_2 - 40) = 1509.4$$

$t_2 = 78.6°C$ and the stream mean temperature = $(40 + 78.6)/2 = 59.3°C$.

The specific heat at this temperature is 2.05 kJ/kg°C. A second trial calculation using this value gives $t_2 = 77.9°C$ and a new mean temperature of 58.9°C. There is no significant change in the specific heat at this mean temperature from the value used, so take the crude stream outlet temperature to be 77.9°C, say 78°C.

Step 2: Physical Properties

Kerosene	inlet	mean	outlet	
temperature	200	145	90	°C
specific heat	2.72	2.47	2.26	kJ/kg°C
thermal conductivity	0.130	0.132	0.135	W/m°C
density	690	730	770	kg/m^3
viscosity	0.22	0.43	0.80	mN sm^{-2}

Crude oil	outlet	mean	inlet	
temperature	78	59	40	°C
specific heat	2.09	2.05	2.01	kJ/kg°C
thermal conductivity	0.133	0.134	0.135	W/m°C
density	800	820	840	kg/m^3
viscosity	2.4	3.2	4.3	mN sm^{-2}

Step 3: Overall Coefficient

For an exchanger of this type the overall coefficient will be in the range 300 to 500 W/m^2°C (see Figure 12.1 and Table 12.1); so start with 300 W/m^2°C.

Step 4: Exchanger Type and Dimensions

An even number of tube passes is usually the preferred arrangement, as this positions the inlet and outlet nozzles at the same end of the exchanger, which simplifies the pipework.

Start with one shell pass and two tube passes.

$$\Delta T_{lm} = \frac{(200 - 78) - (90 - 40)}{\ln\dfrac{(200 - 78)}{(90 - 40)}} = 80.7°C \tag{12.4}$$

$$R = \frac{(200 - 90)}{(90 - 40)} = 2.9 \tag{12.6}$$

$$S = \frac{(78 - 40)}{(200 - 40)} = 0.24 \tag{12.7}$$

From Figure 12.19, $F_t = 0.88$, which is acceptable.
So

$$\Delta T_m = 0.88 \times 80.7 = 71.0°C$$

Step 5: Heat Transfer Area

$$A_o = \frac{1509.4 \times 10^3}{300 \times 71.0} = 70.86 \text{ m}^2 \tag{12.1}$$

Step 6: Layout and Tube Size

Use a split-ring floating head exchanger for efficiency and ease of cleaning.

Neither fluid is corrosive and the operating pressure is not high, so a plain carbon steel can be used for the shell and tubes.

The crude is dirtier than the kerosene, so put the crude through the tubes and the kerosene in the shell.

Use 19.05-mm (3/4-in) outside diameter, 14.83-mm inside diameter, 5-m-long tubes (a popular size) on a triangular 23.81-mm pitch (pitch/dia. = 1.25).

Step 7: Number of Tubes

Area of one tube (neglecting thickness of tube sheets)

$$= \pi \times 19.05 \times 10^{-3} \times 5 = 0.2992 \text{ m}^2$$

Number of tubes = 70.89/0.2992 = 237, say 240.
So, for two passes, tubes per pass = 120.
Check the tube-side velocity at this stage to see if it looks reasonable.

$$\text{Tube cross-sectional area} = \frac{\pi}{4}(14.83 \times 10^{-3})^2 = 0.0001727 \text{ m}^2$$

$$\text{So area per pass} = 120 \times 0.0001727 = 0.02073 \text{ m}^2$$

$$\text{Volumetric flow} = \frac{70,000}{3600} \times \frac{1}{820} = 0.0237 \text{ m}^3/\text{s}$$

$$\text{Tube-side velocity, } u_t = \frac{0.0237}{0.02073} = 1.14 \text{ m/s}$$

The velocity is satisfactory, between 1 and 2 m/s, but may be a little low. This will show up when the pressure drop is calculated.

Step 8: Bundle and Shell Diameter

From Table 12.4, for two tube passes, $K_1 = 0.249$, $n_1 = 2.207$.
So

$$D_b = 19.05 \left(\frac{240}{0.249}\right)^{1/2.207} = 428 \text{ mm } (0.43 \text{ m}) \tag{12.3b}$$

For a split-ring floating-head exchanger the typical shell clearance from Figure 12.10 is 56 mm, so the shell inside diameter is

$$D_s = 428 + 56 = 484 \text{ mm}$$

Step 9: Tube-side Heat Transfer Coefficient

$$\text{Re} = \frac{\rho u d_i}{\mu} = \frac{820 \times 1.14 \times 14.83 \times 10^{-3}}{3.2 \times 10^{-3}} = 4332$$

$$\text{Pr} = \frac{C_p \mu}{k_f} = \frac{2.05 \times 10^3 \times 3.2 \times 10^{-3}}{0.134} = 48.96 \qquad (12.15)$$

$$\frac{L}{d_i} = \frac{5000}{14.83} = 337$$

From Figure 12.23 $j_h = 3.2 \times 10^{-3}$

$$\text{Nu} = 3.2 \times 10^{-3}(4332)(48.96)^{0.33} = 50.06$$

$$h_i = 50.06\left(\frac{0.134}{14.83 \times 10^{-3}}\right) = 452 \ \text{W/m}^2{}^\circ\text{C}$$

This is clearly too low if U_o is to be 300 W/m²°C. The tube-side velocity did look low, so increase the number of tube passes to four. This will halve the cross-sectional area in each pass and double the velocity.

$$\text{New} \quad u_t = 2 \times 1.14 = 2.3 \ \text{m/s}$$
$$\text{and Re} = 2 \times 4332 = 8664$$
$$j_h = 3.8 \times 10^{-3}$$
$$h_i = \left(\frac{0.134}{14.83 \times 10^{-3}}\right) \times 3.8 \times 10^{-3} (8664)(48.96)^{0.33}$$
$$= 1074 \ \text{W/m}^2{}^\circ\text{C}$$

Step 10: Shell-side Heat-transfer Coefficient

Kern's method will be used.

With four tube passes, the shell diameter will be larger than that calculated for two passes. For four passes, $K_1 = 0.175$ and $n_1 = 2.285$.

$$D_b = 19.05\left(\frac{240}{0.175}\right)^{1/2.285} = 450 \ \text{mm}, \quad (0.45 \ \text{m}) \qquad (12.3b)$$

The bundle to shell clearance is still around 56 mm, giving:
$$Ds = 506 \ \text{mm (about 20 in)}$$

As a first trial take the baffle spacing $= D_s/5$, say 100 mm. This spacing should give good heat transfer without too high a pressure drop.

$$A_s = \frac{(23.81 - 19.05)}{23.81} 506 \times 100 = 10{,}116 \ \text{mm}^2 = 0.01012 \ \text{m}^2 \qquad (12.21)$$

$$d_e = \frac{1.10}{19.05}(23.81^2 - 0.917 \times 19.05^2) = 13.52 \ \text{mm} \qquad (12.23)$$

$$\text{Volumetric flow rate on shell side} = \frac{20,000}{3600} \times \frac{1}{730} = 0.0076 \ \text{m}^3/\text{s}$$

$$\text{Shell side velocity} = \frac{0.0076}{0.01012} = 0.75 \ \text{m/s}$$

$$\text{Re} = \frac{730 \times 0.75 \times 13.52 \times 10^{-3}}{0.43 \times \ 10^{-3}} = 17,214$$

$$\text{Pr} = \frac{2.47 \times 10^3 \times 0.43 \times 10^{-3}}{0.132} = 8.05$$

Use segmental baffles with a 25% cut. This should give a reasonable heat-transfer coefficient without too large a pressure drop.

From Figure 12.29 $j_h = 4.52 \times 10^{-3}$

Neglecting the viscosity correction:

$$h_s = \left(\frac{0.132}{13.52} \times 10^3\right) \times 4.52 \times 10^{-3} \times 17,214 \times 8.05^{0.33} = 1505 \ \text{W/m}^2{}^\circ\text{C}$$

$$(12.25)$$

Step 11: Overall Coefficient

$$\frac{1}{U_o} = \left(\frac{1}{1074} + 0.00035\right)\frac{19.05}{14.83} + \frac{19.05 \times 10^{-3}\text{Ln}\left(\frac{19.05}{14.83}\right)}{2 \times 55} + \frac{1}{1505} + 0.0002$$

$$U_o = 386 \ \text{W/m}^2{}^\circ\text{C} \qquad\qquad (12.2)$$

This is above the initial estimate of 300 W/m$^2{}^\circ$C. The number of tubes could possibly be reduced, but first check the pressure drops.

Step 12: Pressure Drop

Tube side

240 tubes, four passes, tube i.d. 14.83 mm, u_t 2.3 m/s, Re $= 8.7 \times 10^3$.

From Figure 12.24, $j_f = 5 \times 10^{-3}$.

$$\Delta P_t = 4 \left(8 \times 5 \times 10^{-3}\left(\frac{5000}{14.83}\right) + 2.5\right)\frac{(820 \times 2.3^2)}{2}$$

$$= 4(13.5 + 2.5)\frac{(820 \times 2.3^2)}{2} \qquad\qquad (12.20)$$

$$= 138,810 \ \text{N/m}^2, \ 1.4 \ \text{bar}$$

This exceeds the specification. Return to step 6 and modify the design.

Modified design

The tube velocity needs to be reduced. This will reduce the heat-transfer coefficient, so the number of tubes must be increased to compensate. There will also be a pressure drop across the inlet and outlet nozzles. Allow 0.1 bar for this, a typical figure (about 15% of the total); which leaves 0.7 bar across the tubes. Pressure drop is roughly proportional to the square of the velocity and u_t is proportional to the number of tubes per pass. So the pressure drop calculated for 240 tubes can be used to estimate the number of tubes required.

Tubes needed $= 240/(0.6/1.4)^{0.5} = 365$

say, 360 with four passes.

Retain four passes as the heat-transfer coefficient will be too low with two passes.

Second trial design: 360 tubes 19.05 mm o.d., 14.83 mm i.d., 5 m long, triangular pitch 23.81 mm.

$$D_b = 19.05\left(\frac{360}{0.175}\right)^{1/2.285} = 537 \text{ mm, } (0.54 \text{ m}) \tag{12.3b}$$

From Figure 12.10 clearance with this bundle diameter $= 59$ mm

$$D_s = 537 + 59 = 596 \text{ mm}$$

$$\text{Cross-sectional area per pass} = \frac{360}{4}(14.83 \times 10^{-3})^2\frac{\pi}{4} = 0.01555 \text{ m}^2$$

$$\text{Tube velocity, } u_t = \frac{0.02337}{0.01555} = 1.524 \text{ m/s}$$

$$\text{Re} = \frac{820 \times 1.524 \times 14.83 \times 10^{-3}}{3.2 \times 10^{-3}} = 5792$$

L/d is the same as the first trial, 337

$j_h = 3.6 \times 10^{-3}$

$$h_i = \left(\frac{0.134}{14.83} \times 10^{-3}\right)3.6 \times 10^{-3} \times 5792 \times 48.96^{0.33} = 680 \text{ W/m}^2\text{°C} \tag{12.15}$$

This looks satisfactory, but check the pressure drop before doing the shell-side calculation.

$$j_f = 5.5 \times 10^{-3}$$

$$\Delta P_t = 4\left(8 \times 5.5 \times 10^{-3}\left(\frac{5000}{14.83}\right) + 2.5\right)\frac{(820 \times 1.524^2)}{2} \tag{12.20}$$

$$= 66,029 \text{ N/m}^2, 0.66 \text{ bar}$$

Well within specification.

Keep the same baffle cut and spacing.

$$A_s = \frac{(23.81 - 19.05)}{23.81}596 \times 100 = 11,915 \text{ mm}^2, 0.01192 \text{ m}^2 \tag{12.21}$$

$$u_s = \frac{0.0076}{0.01193} = 0.638 \text{ m/s}$$

$$d_e = 13.52 \text{ mm, as before}$$

$$\text{Re} = \frac{730 \times 0.638 \times 13.52 \times 10^{-3}}{0.43 \times 10^{-3}} = 14,644$$

$$\text{Pr} = 8.05$$

$$j_h = 4.8 \times 10^{-3}, j_f = 4.6 \times 10^{-2}$$

$$h_s = \left(\frac{0.132}{13.52 \times 10^{-3}}\right) 4.8 \times 10^{-3} \times 14,644 \times (8.05)^{0.33}$$

$$= 1366 \ \text{W/m}^2\text{°C, looks OK} \tag{12.25}$$

$$\Delta P_s = 8 \times 4.6 \times 10^{-2} \left(\frac{596}{13.52}\right) \left(\frac{5000}{100}\right) \frac{(730 \times 0.638^2)}{2}$$

$$= 120,510 \ \text{N/m}^2, \ 1.2 \ \text{bar} \tag{12.26}$$

Too high; the specification only allowed 0.8 overall, including the loss over the nozzles. Check the overall coefficient to see if there is room to modify the shell-side design.

$$\frac{1}{U_o} = \left(\frac{1}{683} + 0.00035\right) \frac{19.05}{14.83} + \frac{19.05 \times 10^{-3}\ln\left(\frac{19.05}{14.88}\right)}{2 \times 55} + \frac{1}{1366} + 0.0002$$

$$U_o = 302 \ \text{W/m}^2\text{°C}$$

$$U_o \ \text{required} = \frac{Q}{(A_o \Delta T_{\text{lm}})}, \quad A_o = 360 \times 0.2992 = 107.7 \ \text{m}^2,$$

$$\text{so } U_o \ \text{required} = \frac{1509.4 \times 10^3}{(107.7 \times 71)} = 197 \ \text{W/m}^2\text{°C} \tag{12.2}$$

The estimated overall coefficient is well above that required for design, 302 compared to 192 W/m²°C, which gives scope for reducing the shell-side pressure drop.

Allow a drop of 0.1 bar for the shell inlet and outlet nozzles, leaving 0.7 bar for the shell-side flow. So, to keep within the specification, the shell-side velocity will have to be reduced by around $\sqrt{(1/2)} = 0.707$. To achieve this, the baffle spacing will need to be increased to $100/0.707 = 141$, say 140 mm.

$$A_s = \frac{(23.81 - 19.05)}{23.81} 596 \times 140 = 6881 \ \text{mm}^2, \ 0.167 \ \text{m}^2$$

$$u_s = \frac{0.0076}{0.0167} = 0.455 \ \text{m/s}, \tag{12.21}$$

giving: $\text{Re} = 10,443$, $h_s = 1177 \ \text{W/m}^2\text{°C}$, $\Delta P_s = 0.47 \ \text{bar}$, and $U_o = 288 \ \text{W/m}^2\text{°C}$. The pressure drop is now well within the specification.

Step 13: Estimate Cost

The cost of this design can be estimated using the methods given in Chapter 6.

Step 14: Optimization

There is scope for optimizing the design by reducing the number of tubes, as the pressure drops are well within specification and the overall coefficient is well above that needed; however, the method used for estimating the coefficient and pressure drop on the shell side (Kern's method) is not accurate, so keeping to this design will give some margin of safety.

Viscosity Correction Factor

The viscosity correction factor $(\mu/\mu_w)^{0.14}$ was neglected when calculating the heat-transfer coefficients and pressure drops. This is reasonable for the kerosene as it has a relatively low viscosity, but it is not so obviously so for the crude oil. So, before firming up the design, the effect of this factor on the tube-side coefficient and pressure drop will be checked.

First, an estimate of the temperature at the tube wall, t_w is needed.

$$\text{The inside area of the tubes} = \pi \times 14.83 \times 10^{-3} \times 5 \times 360 = 83.86 \text{ m}^2$$

$$\text{Heat flux} = Q/A = 1509.4 \times 10^3/83.86 = 17,999 \text{ W/m}^2$$

As a rough approximation

$$(t_w - t)h_i = 17,999$$

where t is the mean bulk fluid temperature $= 59°C$.

$$\text{So,} \quad t_w = \frac{17,999}{680} + 59 = 86°C.$$

The crude oil viscosity at this temperature $= 2.1 \times 10^{-3} \text{ Ns/m}^2$.

$$\text{Giving} \left(\frac{\mu}{\mu_w}\right)^{0.14} = \left(\frac{3.2 \times 10^{-3}}{2.1 \times 10^{-3}}\right)^{0.14} = 1.06$$

Only a small factor, so the decision to neglect it was justified. Applying the correction would increase the estimated heat-transfer coefficient, which is in the right direction. It would give a slight decrease in the estimated pressure drop.

Summary: The Proposed Design

Split ring, floating head, one shell pass, four tube passes
360 carbon steel tubes, 5 m long, 19.05 mm o.d., 14.83 mm i.d., triangular pitch, pitch 23.18 mm
Heat transfer area 107.7 m^2 (based on outside diameter)
Shell i.d. 597 mm (600 mm), baffle spacing 140 mm, 25% cut
Tube-side coefficient 680 W/m^2°C, clean
Shell-side coefficient 1366 W/m^2°C, clean
Overall coefficient, estimated 288 W/m^2°C, dirty
Overall coefficient required 197 W/m^2°C, dirty
Dirt/Fouling factors:
   tube side (crude oil) 0.00035 (W/m^2°C)$^{-1}$
   shell side (kerosene) 0.0002 (W/m^2°C)$^{-1}$
Pressure drops:
   tube side, estimated 0.40 bar, +0.1 for nozzles; specified 0.8 bar overall
   shell side, estimated 0.45 bar, +0.1 for nozzles; specified 0.8 bar overall.

12.9.4. Commercial Software for Heat-Exchanger Design

Computer methods for detailed design of heat exchangers are available in most of the commercial process simulation programs introduced in Chapter 4; see Table 4.1. For example, Aspen Technology's Aspen Engineering Suite contains the HTFS TASC program and Honeywell's UniSim Design Suite can be linked to Honeywell's Uni-Sim Heat Exchanger program, which is also based on the HTFS methods. The methods developed by Heat Transfer Research Inc. can be licensed from HTRI (www.HTRI.net).

All of the commercial heat-exchanger design programs allow the user to upload process data and stream properties from a process simulation. Some care is needed when uploading data for streams that undergo partial vaporization or have other effects that cause a significant variation in heat capacity or other properties across the exchanger. When there are significant changes in fluid properties between the exchanger inlet and outlet, the designer should break the exchanger into several exchangers in series in the process simulation, so as to obtain several sets of property data at intermediate temperatures for input into the heat exchanger software.

The details of running the commercial heat-exchanger design programs are not addressed here, as each program is slightly different from the others; consult the user manuals and on-line help. The heat-exchanger programs have both rating and design capability. They can be configured to determine a least-cost design for given desired outlet temperatures and allowable pressure drop, or to calculate the outlet stream temperatures and pressures, given details of the exchanger geometry and process fluid inlet conditions. All of the programs allow the user to make quick adjustments to exchanger geometry and then recalculate to see the impact on the stream outlet temperatures and pressure drops.

Example 12.4

Optimize the design of Example 12.3 using commercial heat-exchanger design software.

Solution

This problem was solved using UniSim Heat Exchanger.

Figures 12.32 and 12.33 show the stream data and physical properties. The program was then run with the objective function set to minimum cost, giving the output shown in Figure 12.34 and the setting plan shown in Figure 12.35.

The program selected two tube passes with tubes 6096 mm (20 ft) long and many baffles (72) to obtain good counter-current flow on the shell side. This design could cause problems with plot space, or with supporting the shell, or withdrawing the bundle for cleaning and maintenance. The program was run again with the tube length constrained to be less than 4880 mm (16 ft). This gives a more compact design with four tube passes, 12-ft tubes and only 28 baffles, as shown in Figure 12.36.

The program initially gives a warning 'At some point(s) the cross-flow fraction in the shell-side flow model was less than 30%. This is below the expected range of values and may give poor heat transfer'. When the baffle-to-shell and tube-to-baffle tolerances are specified using the values given in Section 12.5.7 (1.6 mm and 0.8 mm

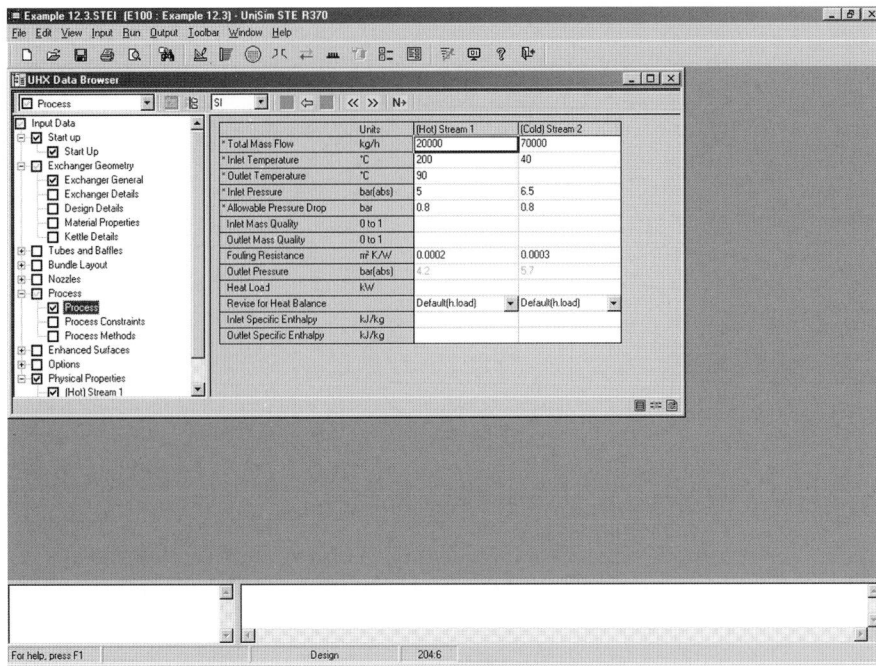

Figure 12.32.   Stream data for Example 12.4.

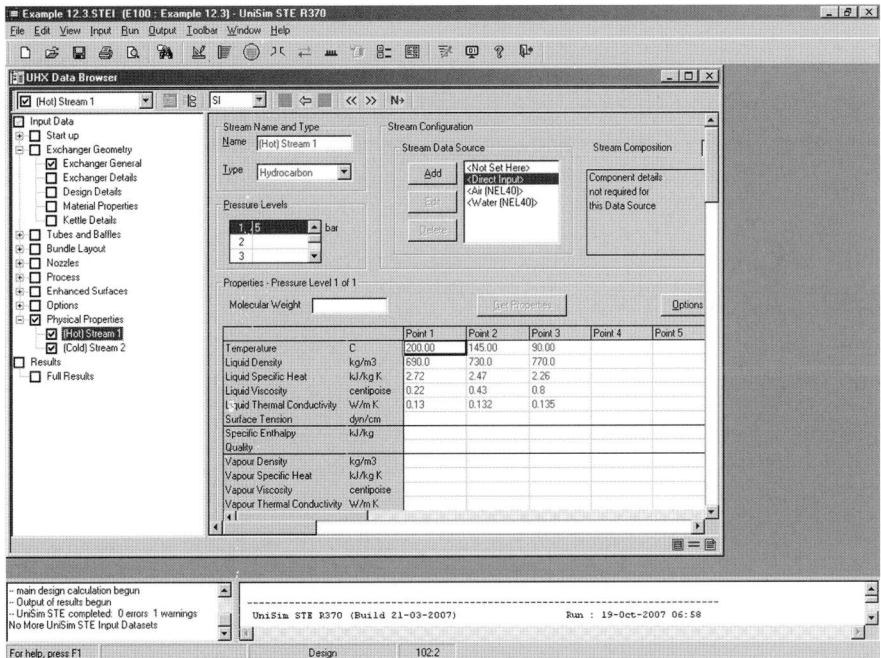

Figure 12.33.   Stream physical properties for Example 12.4.

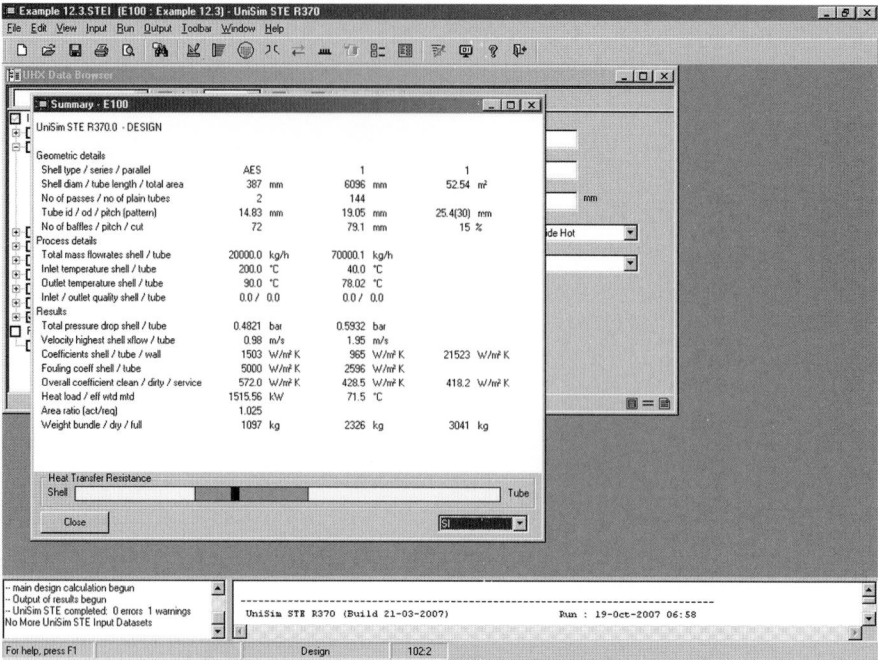

Figure 12.34.   UniSim STE (HTFS) program output.

respectively) then the design converges with no warnings. The TEMA sheet for the resulting design is given in Figure 12.37.

Note that the more compact design did not make as good use of the allowable pressure drop on the shell side and also has a higher bundle weight and area, corresponding to a higher capital cost than the 20-ft-long exchanger initially designed. Note also that both designs developed using the HTFS software need substantially less area than the 107.7 m^2 predicted using Kern's method in Example 12.3.

12.10.   CONDENSERS

This section covers the design of shell and tube exchangers used as condensers. Direct contact condensers are discussed in Section 12.13.

The construction of a condenser is similar to that of other shell and tube exchangers, but with a wider baffle spacing, typically $l_B = D_s$.

Four condenser configurations are possible:

1. Horizontal, with condensation in the shell, and the cooling medium in the tubes
2. Horizontal, with condensation in the tubes
3. Vertical, with condensation in the shell
4. Vertical, with condensation in the tubes.

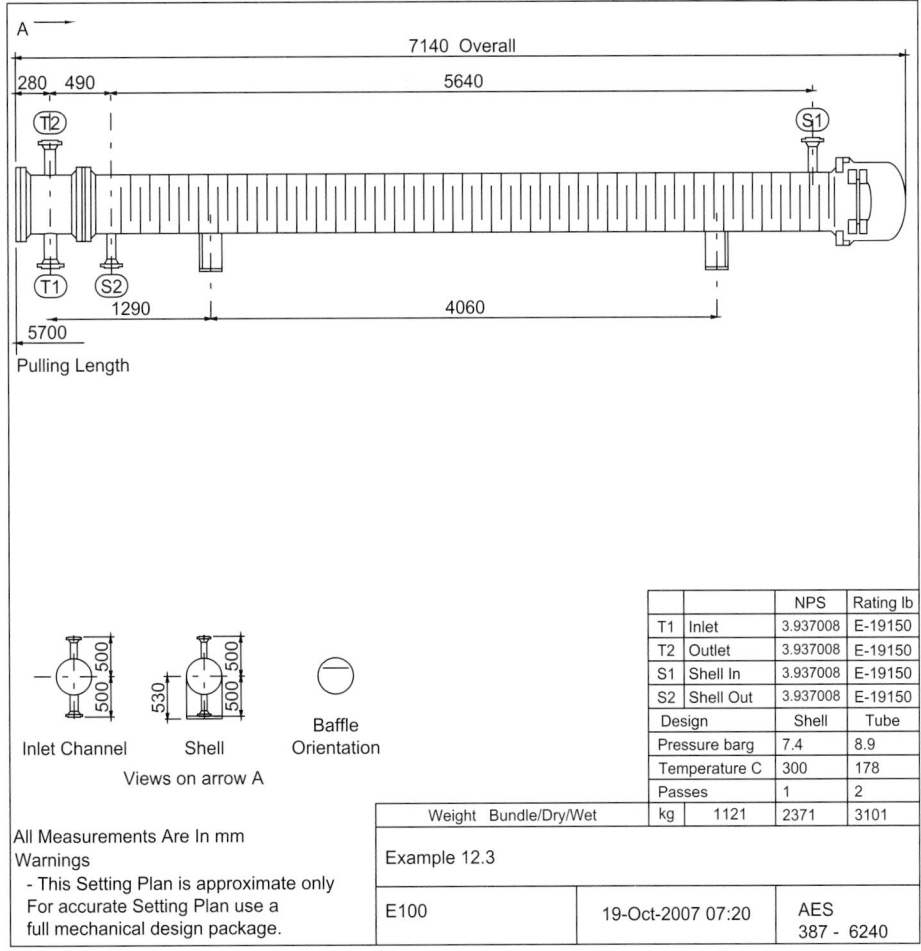

Figure 12.35. HTFS setting plan.

Horizontal shell side and vertical tube side are the most commonly used types of condenser. A horizontal exchanger with condensation in the tubes is rarely used as a process condenser, but is the usual arrangement for heaters and vaporizers using condensing steam as the heating medium.

12.10.1. Heat-Transfer Fundamentals

The fundamentals of condensation heat transfer are covered in Coulson *et al.* (1999).

The normal mechanism for heat transfer in commercial condensers is film-wise condensation. Drop-wise condensation will give higher heat-transfer coefficients but is unpredictable, and is not yet considered a practical proposition for the design of condensers for general purposes.

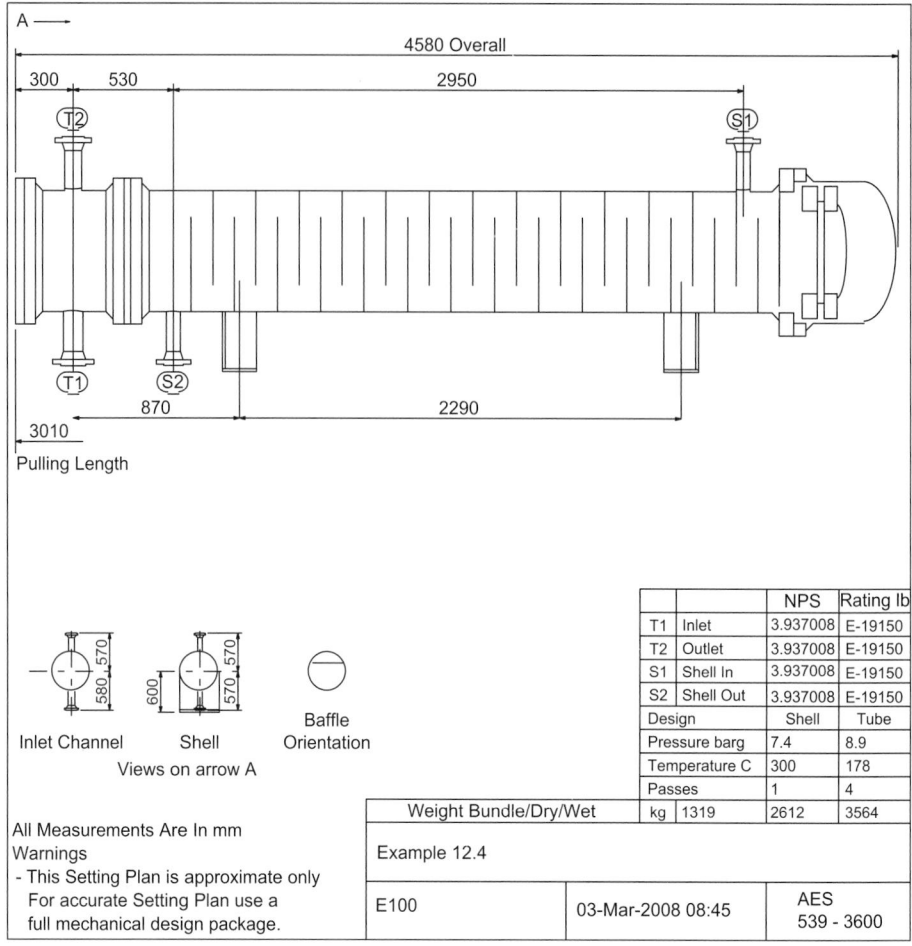

Figure 12.36.   Setting plan with length constrained.

The basic equations for film-wise condensation were derived by Nusselt (1916), and his equations form the basis for practical condenser design. The basic Nusselt equations are derived in Coulson *et al.* (1999). In the Nusselt model of condensation, laminar flow is assumed in the film and heat transfer is assumed to take place entirely by conduction through the film. In practical condensers, the Nusselt model will strictly only apply at low liquid and vapour rates and where the flowing condensate film is undisturbed. Turbulence can be induced in the liquid film at high liquid rates and by shear at high vapour rates. This will generally increase the rate of heat transfer over that predicted using the Nusselt model. The effect of vapour shear and film turbulence are discussed in Coulson *et al.* (1999); see also Butterworth (1978) and Taborek (1974).

Developments in the theory of condensation and their application in condenser design are reviewed by Owen and Lee (1983).

HEAT EXCHANGER SPECIFICATION SHEET

1							Job No.			
2	Customer						Reference No.			
3	Address						Proposal No.			
4	Plant Location						Date		Rev.	
5	Service of Unit	Example 12.4					Item No.	E100		
6	Size 540-3600	Type	AES		Horizontal	Connected	1	parallel	1	series
7	Surf./Unit (Gross)	65.71	m²	Shells/Unit	1	Surface/Shell (Gross)	65.71		m²	

PERFORMANCE OF ONE UNIT

				Shell Side		Tube Side	
9	Fluid Allocation						
10	Fluid Name						
11	Fluid Quantity, Total	kg/h	20000.0		70000.2		
12	Vapour						
13	Liquid		20000.0	20000.0	70000.2	70000.2	
14	Steam						
15	Water						
16	Noncondensable						
17	Temperature (In/Out)	°C	200.0	90.0	40.0	78.0	
18	Density	kg / m³	690.0	770.0	840.0	800.0	
19	Viscosity	centipoise	0.22	0.8	4.3	2.39925	
20	Molecular Weight, Vapor						
21	Molecular Weight, Noncondensable						
22	Specific Heat	kJ/kg K	2.72	2.26	2.01	2.09	
23	Thermal Conductivity	W/m K	0.13	0.135	0.135	0.133	
24	Latent Heat	kJ/kg					
25	Inlet Pressure	bar(abs)	5.0		6.5		
26	Velocity	m/s	0.48		1.84		
27	Pressure Drop , Allow. / Calc.	bar	0.8	0.14338	0.8	0.66002	
28	Fouling Resistance (Min.)	m² K/W	0.0002		0.0003 (0.00039 referred to OD)		
29	Heat Exchanged	1515.56	kW	MTD	71.62	°C	
30	Transfer Rate, Service	345.3	Dirty	345.4	Clean	432.9	W/m² K

CONSTRUCTION OF ONE SHELL

Sketch (Bundle/Nozzle Orientation)

			Shell Side	Tube Side		
33	Design/Test Pressure	bar(g)	7.43	8.93		
34	Design Temperature	°C	300.0	178.0		
35	No. Passes per Shell			4		
36	Corrosion Allowance					
37	Connections In		62.7	90.1		
38	Size & Out		62.7	77.9		
39	Ratings Inter.					
40	Tube No. 305 OD 19.05 mm Thk 2.11 mm	Length 3600 mm Pitch 25.4 mm 30deg				
41	Tube Type plain	Material Carbon Steel				
42	Shell ID 540 OD mm	Shell Cover (Integ.)(Remov.)				
43	Channel or Bonnet Carbon Steel	Channel Cover				
44	Tubesheet - Stationary Carbon Steel	Tubesheet-Floating				
45	Floating Head Cover	Impingement Protection				
46	Baffles-Cross 28 Type Single Segmental	% Cut 25.0 Spacing c/c 113.5 Inlet mm				
47	Baffles-Long	Seal Type				
48	Supports-Tube U-Bend	Type				
49	Bypass Seal Arrangement	Tube-Tubesheet Joint				
50	Expansion Joint	Type				
51	··V²- Inlet Nozzle 4689.0	Bundle Entrance 488.0 Bundle Exit 437.3				
52	Gaskets Shell Side	Tube Side				
53	Floating Head					
54	Code Requirements	Tema Class R				
55	Weight/Shell 2612 Filled with water 3564	Bundle 1319 kg				
56	Remarks					

Figure 12.37. TEMA specification sheet for Example 12.4.

Physical Properties

The physical properties of the condensate for use in the following equations are evaluated at the average condensate film temperature: the mean of the condensing temperature and the tube-wall temperature.

12.10.2. Condensation Outside Horizontal Tubes

$$(h_c)_1 = 0.95 k_L \left[\frac{\rho_L (\rho_L - \rho_v) g}{\mu_L \Gamma} \right]^{1/3} \tag{12.27}$$

where

$(h_c)_1$ = mean condensation film coefficient, for a single tube, $W/m^2°C$
k_L = condensate thermal conductivity, $W/m°C$
ρ_L = condensate density, kg/m^3
ρ_v = vapour density, kg/m^3
μ_L = condensate viscosity, Ns/m^2
g = gravitational acceleration, $9.81\ m/s^2$
Γ = the tube loading, the condensate flow per unit length of tube, $kg/m\ s$.

In a bank of tubes, the condensate from the upper rows of tubes will add to that condensing on the lower tubes. If there are N_r tubes in a vertical row and the condensate is assumed to flow smoothly from row to row (Figure 12.38a), and if the flow remains laminar, the mean coefficient predicted by the Nusselt model is related to that for the top tube by:

$$(h_c)_{N_r} = (h_c)_1 N_r^{-1/4} \tag{12.28}$$

In practice, the condensate will not flow smoothly from tube to tube (Figure 12.38b), and the factor of $(N_r)^{-1/4}$ applied to the single tube coefficient in

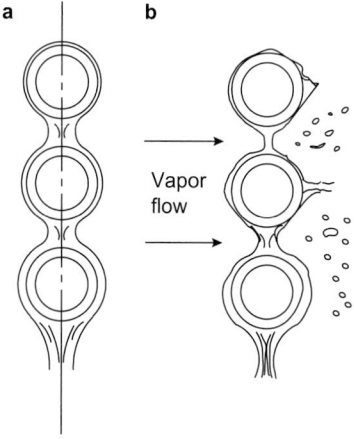

Figure 12.38. Condensate flow over tube banks.

equation 12.28 is considered to be too conservative. Based on results from commercial exchangers, Kern (1950) suggests using an index of 1/6. Frank (1978) suggests multiplying single tube coefficient by a factor of 0.75.

Using Kern's method, the mean coefficient for a tube bundle is given by:

$$(h_c)_b = 0.95 k_L \left[\frac{\rho_L (\rho_L - \rho_v) g}{\mu_L \Gamma_h} \right]^{1/3} N_r^{-1/6} \tag{12.29}$$

where $\Gamma_h = \dfrac{W_c}{L N_t}$

and L = tube length,
W_c = total condensate flow,
N_t = total number of tubes in the bundle,
N_r = average number of tubes in a vertical tube row.

N_r can be taken as two-thirds of the number in the central tube row.

For low-viscosity condensates the correction for the number of tube rows is generally ignored.

A procedure for estimating the shell-side heat transfer in horizontal condensers is given in the Engineering Sciences Data Unit Design Guide, ESDU 84023.

12.10.3. Condensation Inside and Outside Vertical Tubes

For condensation inside and outside vertical tubes the Nusselt model gives:

$$(h_c)_v = 0.926 k_L \left[\frac{\rho_L (\rho_L - \rho_v) g}{\mu_L \Gamma_v} \right]^{1/3} \tag{12.30}$$

where
$(h_c)_v$ = mean condensation coefficient, $W/m^2 {}^\circ C$
Γ_v = vertical tube loading, condensate rate per unit tube perimeter, kg/m s for a tube bundle

$$\Gamma_v = \frac{W_c}{N_t \pi d_o} \text{ or } \frac{W_c}{N_t \pi d_i}$$

Equation 12.30 will apply up to a Reynolds number of 30; above this value waves on the condensate film become important. The Reynolds number for the condensate film is given by:

$$Re_c = \frac{4 \Gamma_v}{\mu_L}$$

The presence of waves will increase the heat-transfer coefficient, so the use of equation 12.30 above a Reynolds number of 30 will give conservative (safe) estimates.

The effect of waves on condensate film on heat transfer is discussed by Kutateladze (1963).

Above a Reynolds number of around 2000, the condensate film becomes turbulent. The effect of turbulence in the condensate film was investigated by Colburn (1934) and Colburn's results are generally used for condenser design (Figure 12.39). Equation 12.30 is also shown on Figure 12.39. The Prandtl number for the condensate film is given by:

$$Pr_c = \frac{C_p \, \mu_L}{k_L}$$

Figure 12.39 can be used to estimate condensate film coefficients in the absence of appreciable vapour shear. Horizontal and downward vertical vapour flow will increase the rate of heat transfer, and the use of Figure 12.39 will give conservative values for most practical condenser designs.

Boyko and Kruzhilin (1967) developed a correlation for shear-controlled condensation in tubes that is simple to use. Their correlation gives the mean coefficient between two points at which the vapour quality is known. The vapour quality x is the mass fraction of the vapour present. It is convenient to represent the Boyko–Kruzhilin correlation as:

$$(h_c)_{BK} = h\prime_t \left[\frac{J_1^{1/2} + J_2^{1/2}}{2} \right] \tag{12.31}$$

where

$$J = 1 + \left[\frac{\rho_L - \rho_v}{\rho_v} \right] x$$

and the suffixes 1 and 2 refer to the inlet and outlet conditions respectively. $h\prime_t$ is the tube-side coefficient evaluated for single-phase flow of the total condensate (the condensate at point 2). That is, the coefficient that would be obtained if the

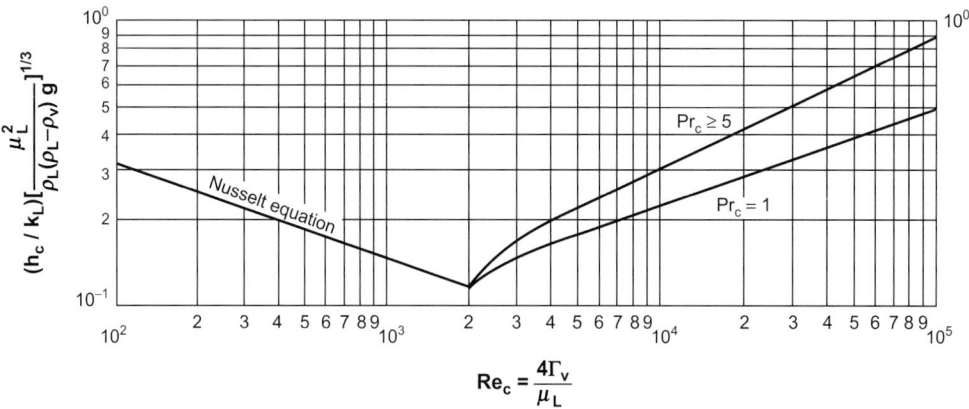

Figure 12.39.   Condensation coefficient for vertical tubes.

condensate filled the tube and was flowing alone; this can be evaluated using any suitable correlation for forced convection in tubes; see Section 12.8.

Boyko and Kruzhilin used the correlation

$$h'_t = 0.021 \left(\frac{k_L}{d_i}\right) \mathrm{Re}^{0.8} \mathrm{Pr}^{0.43} \tag{12.32}$$

In a condenser the inlet stream will normally be saturated vapour and the vapour will be totally condensed.

For these conditions equation 12.31 becomes

$$(h_c)_{BK} = h'_t \left[\frac{1 + \sqrt{\rho_L/\rho_v}}{2}\right] \tag{12.33}$$

For the design of condensers with condensation inside the tubes and downward vapour flow, the coefficient should be evaluated using Figure 12.39 and equation 12.31, and the *higher* value selected.

Flooding in Vertical Tubes

When the vapour flows up the tube, which will be the usual arrangement for a reflux condenser, care must be taken to ensure that the tubes do not flood. Several correlations have been published for the prediction of flooding in vertical tubes (see Green and Perry, 2007). One of the simplest to apply, which is suitable for use in the design of condensers handling low-viscosity condensates, is the criterion given by Hewitt and Hall-Taylor (1970); see also Butterworth (1977). Flooding should not occur if the following condition is satisfied:

$$\left[u_v^{1/2}\rho_v^{1/4} + u_L^{1/2}\rho_L^{1/4}\right] < 0.6\left[gd_i(\rho_L - \rho_v)\right]^{1/4} \tag{12.34}$$

where u_v and u_L are the velocities of the vapour and liquid, based on each phase flowing in the tube alone; and d_i is in metres. The critical condition will occur at the bottom of the tube, so the vapour and liquid velocities should be evaluated at this point.

Example 12.5

Estimate the heat-transfer coefficient for steam condensing on the outside, and on the inside, of a 25-mm o.d., 21-mm i.d. vertical tube 3.66 m long. The steam condensate rate is 0.015 kg/s per tube and condensation takes place at 3 bar. The steam will flow down the tube.

Solution

Physical properties, from steam tables:

Saturation temperature = 133.5°C
$\quad \rho_L = 931\ \mathrm{kg/m}^3$
$\quad \rho_v = 1.65\ \mathrm{kg/m}^3$
$\quad k_L = 0.688\ \mathrm{W/m°C}$
$\quad \mu_L = 0.21\ \mathrm{mNs/m}^2$
$\quad \mathrm{Pr_c} = 1.27.$

Condensation Outside The Tube

$$\Gamma_v = \frac{0.015}{\pi \times 25 \times 10^{-3}} = 0.191 \, \text{kg/sm}$$

$$\text{Re}_c = \frac{4 \times 0.191}{0.21 \times 10^{-3}} = 3638$$

From Figure 12.39:

$$\frac{h_c}{k_L} \left[\frac{\mu_L^2}{\rho_L(\rho_L - \rho_v)g} \right]^{1/3} = 1.65 \times 10^{-1}$$

$$h_c = 1.65 \times 10^{-1} \times 0.688 \left[\frac{\left(0.21 \times 10^{-3}\right)^2}{931(931 - 1.65)9.81} \right]^{-1/3}$$

$$= 6554 \, \text{W/m}^2{}^\circ\text{C}$$

Condensation Inside The Tube

$$\Gamma_v = \frac{0.015}{\pi \times 21 \times 10^{-3}} = 0.227 \, \text{kg/s m}$$

$$\text{Re}_c = \frac{4 \times 0.227}{0.21 \times 10^{-3}} = 4324$$

From Figure 12.39

$$h_c = 1.72 \times 10^{-1} \times 0.688 \left[\frac{(0.21 \times 10^{-3})^2}{931(931 - 1.65)9.81} \right]^{-1/3}$$

$$= 6832 \, \text{W/m}^2{}^\circ\text{C}$$

Boyko–Kruzhilin method:

$$\text{Cross-sectional area of tube} = (21 \times 10^{-3})^2 \frac{\pi}{4} = 3.46 \times 10^{-4} \, \text{m}^2$$

Fluid velocity, total condensation:

$$u_t = \frac{0.015}{931 \times 3.46 \times 10^{-4}} = 0.047 \text{m/s}$$

$$\text{Re} = \frac{\rho u \, d_i}{\mu} = \frac{931 \times 0.047 \times 21 \times 10^{-3}}{0.21 \times 10^{-3}} = 4376$$

$$h_t' = 0.021 \times \frac{0.688}{21 \times 10^{-3}} (4376)^{0.8}(1.27)^{0.43} = 624 \, \text{W/m}^2{}^\circ\text{C} \qquad (12.32)$$

$$h_c = 624 \left[\frac{1 + \sqrt{931/1.65}}{2} \right] = 7723 \, \text{W/m}^2{}^\circ\text{C} \qquad (12.33)$$

$$\text{Take higher value, } h_c = 7723 \, \text{W/m}^2{}^\circ\text{C}$$

Example 12.6

It is proposed to use an existing distillation column, which is fitted with a dephlegmator (reflux condenser) that has 200 vertical, 50-mm i.d., tubes, for separating benzene from a mixture of chlorobenzenes. The top product will be 2500 kg/h benzene and the column will operate with a reflux ratio of 3. Check if the tubes are likely to flood. The condenser pressure will be 1 bar.

Solution

The vapour will flow up and the liquid down the tubes. The maximum flow rates of both will occur at the base of the tube.

$$\text{Vapour flow} = (3 + 1)2500 = 10,000 \text{ kg/h}$$
$$\text{Liquid flow} = 3 \times 2500 = 7500 \text{ kg/h}$$
$$\text{Total area tubes} = \frac{\pi}{4}(50 \times 10^{-3})^2 \times 200 = 0.39 \text{ m}^2$$

Densities at benzene boiling point

$$\rho_L = 840 \text{ kg/m}^3, \quad \rho_v = 2.7 \text{ kg/m}^3$$

Vapour velocity (vapour flowing alone in tube)

$$u_v = \frac{10,000}{3600 \times 0.39 \times 2.7} = 2.64 \text{ m/s}$$

Liquid velocity (liquid alone)

$$u_L = \frac{7500}{3600 \times 0.39 \times 840} = 0.006 \text{ m/s}$$

From equation 12.34 for no flooding

$$[u_v^{1/2}\rho_v^{1/4} + u_L^{1/2}\rho_L^{1/4}] < 0.6[gd_i(\rho_L - \rho_v)]^{1/4}$$

$$[(2.64)^{1/2}(2.7)^{1/4} + (0.006)^{1/2}(840)^{1/4}] < 0.6[9.81 \times 50 \times 10^{-3}(840 - 2.7)]^{1/4}$$
$$[2.50] < [2.70]$$

Tubes should not flood, but there is little margin of safety.

12.10.4. Condensation Inside Horizontal Tubes

Where condensation occurs in a horizontal tube, the heat-transfer coefficient at any point along the tube depends on the flow pattern at that point. The various patterns that can exist in two-phase flow are shown in Figure 12.40, and are discussed in Coulson *et al.* (1999). In condensation, the flow will vary from a single-phase vapour at the inlet to a single-phase liquid at the outlet, with all the possible patterns of flow occurring between these points. Bell *et al.* (1970) give a method for following the change in flow pattern as condensation occurs on a Baker flow-regime map.

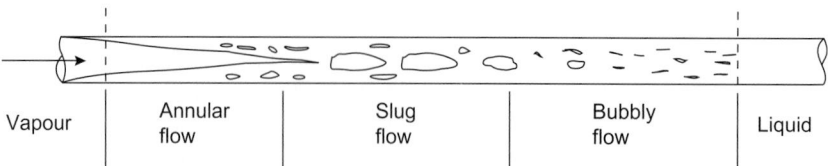

Figure 12.40.   Flow patterns, vapour condensing in a horizontal tube.

Correlations for estimating the average condensation coefficient have been published by several workers, but there is no generally satisfactory method that will give accurate predictions over a wide flow range. A comparison of the published methods is given by Bell *et al.* (1970).

Two flow models are used to estimate the mean condensation coefficient in horizontal tubes: stratified flow (Figure 12.41a) and annular flow (Figure 12.41b). The stratified flow model represents the limiting condition at low condensate and vapour rates, and the annular model the condition at high vapour and low condensate rates. For the stratified flow model, the condensate film coefficient can be estimated from the Nusselt equation, applying a suitable correction for the reduction in the coefficient caused by the accumulation of condensate in the bottom of the tube. The correction factor will typically be around 0.8, so the coefficient for stratified flow can be estimated from:

$$(h_c)_s = 0.76 k_L \left[\frac{\rho_L(\rho_L - \rho_v)g}{\mu_L \Gamma_h} \right]^{1/3}$$

(12.35)

The Boyko–Kruzhilin equation (12.31) can be used to estimate the coefficient for annular flow.

For condenser design, the mean coefficient should be evaluated using the correlations for both annular and stratified flow and the *higher* value selected.

12.10.5.   Condensation of Steam

Steam is frequently used as a heating medium. The film coefficient for condensing steam can be calculated using the methods given in the previous sections; but, as the coefficient will be high and will rarely be the limiting coefficient, it is customary to assume a typical, conservative, value for design purposes. For air-free steam a coefficient of 8000 W/m^2°C (1500 Btu/h ft^2°F) can be used.

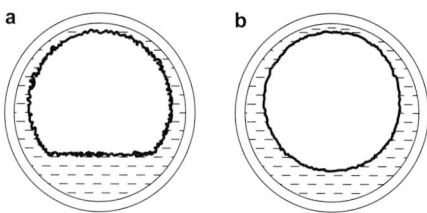

Figure 12.41.   Flow patterns in condensation. (a) Stratified flow. (b) Annular flow.

12.10.6. **Mean Temperature Difference**

A pure, saturated, vapour will condense at a fixed temperature, at constant pressure. For an isothermal process such as this, the simple logarithmic mean temperature difference can be used in equation 12.1; no correction factor for multiple passes is needed. The logarithmic mean temperature difference will be given by:

$$\Delta T_{lm} = \frac{(t_2 - t_1)}{\ln\left[\dfrac{T_{sat} - t_1}{T_{sat} - t_2}\right]} \tag{12.36}$$

where

T_{sat} = saturation temperature of the vapour
t_1 = inlet coolant temperature
t_2 = outlet coolant.

When the condensation process is not exactly isothermal but the temperature change is small, such as where there is a significant change in pressure or where a narrow boiling range multicomponent mixture is being condensed, the logarithmic temperature difference can still be used but the temperature correction factor will be needed for multi-pass condensers. The appropriate terminal temperatures should be used in the calculation.

12.10.7. **Desuperheating and Sub-cooling**

When the vapour entering the condenser is superheated and the condensate leaving the condenser is cooled below its boiling point (sub-cooled), the temperature profile will be as shown in Figure 12.42.

Desuperheating

If the degree of superheat is large, it will be necessary to divide the temperature profile into sections and determine the mean temperature difference and heat-transfer coefficient separately for each section. If the tube-wall temperature is below the dew point of the vapour, liquid will condense directly from the vapour onto the tubes. In

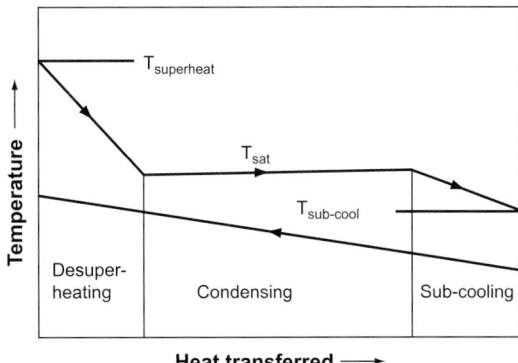

Figure 12.42. Condensation with desuperheating and sub-cooling.

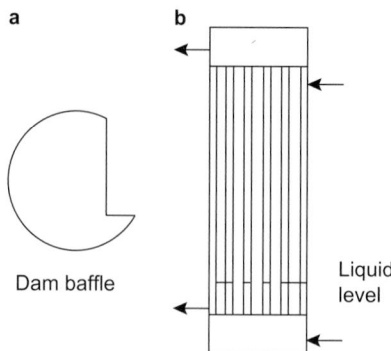

Figure 12.43. Arrangements for sub-cooling.

these circumstances, it has been found that the heat-transfer coefficient in the desu-perheating section is close to the value for condensation and can be taken as the same. So, where the amount of superheat is not excessive, say less than 25% of the latent heat load, and the outlet coolant temperature is well below the vapour dew point, the sensible heat load for desuperheating can be lumped with the latent heat load. The total heat-transfer area required can then be calculated using a mean temperature difference based on the saturation temperature (not the superheat temperature) and the estimated condensate film heat-transfer coefficient.

Sub-cooling of Condensate

Some sub-cooling of the condensate will usually be required to control the net positive suction head at the condensate pump (see Chapter 5), or to cool a product for storage. Where the amount of sub-cooling is large, it is more efficient to sub-cool in a separate exchanger. A small amount of sub-cooling can be obtained in a condenser by controlling the liquid level so that some part of the tube bundle is immersed in the condensate.

In a horizontal shell-side condenser a dam baffle can be used (Figure 12.43a). A vertical condenser can be operated with the liquid level above the bottom tube sheet (Figure 12.43b).

The temperature difference in the sub-cooled region will depend on the degree of mixing in the pool of condensate. The limiting conditions are plug flow and complete mixing. The temperature profile for plug flow is that shown in Figure 12.42. If the pool is perfectly mixed, the condensate temperature will be constant over the sub-cooling region and equal to the condensate outlet temperature. Assuming perfect mixing will give a very conservative (safe) estimate of the mean temperature differ-ence. As the liquid velocity will be low in the sub-cooled region the heat-transfer coefficient should be estimated using correlations for natural convection (see Coulson *et al.*, 1999); a typical value would be 200 $W/m^2 °C$.

12.10.8. Condensation of Mixtures

The correlations given in the previous sections apply to the condensation of a single component, such as an essentially pure overhead product from a distillation column. The design of a condenser for a mixture of vapours is more difficult.

The term 'mixture of vapours' covers three related situations of practical interest:

1. Total condensation of a multicomponent mixture, such as the overheads from a multicomponent distillation.
2. Condensation of only part of a multicomponent vapour mixture, all components of which are theoretically condensable. This situation will occur where the dew point of some of the lighter components is above the coolant temperature. The uncondensed component may be soluble in the condensed liquid, such as in the condensation of some hydrocarbon mixtures containing light 'gaseous' components.
3. Condensation from a non-condensable gas, where the gas is not soluble to any extent in the liquid condensed. These exchangers are often called cooler-condensers.

The following features, common to all these situations, must be considered when developing design methods for mixed vapour condensers:

1. The condensation will not be isothermal. As the heavy component condenses out, the composition of the vapour, and therefore its dew point, changes.
2. Because the condensation is not isothermal, there will be a transfer of sensible heat from the vapour to cool the gas to the dew point. There will also be a transfer of sensible heat from the condensate, as it must be cooled from the temperature at which it condensed to the outlet temperature. The transfer of sensible heat from the vapour can be particularly significant, as the sensible-heat-transfer coefficient will be appreciably lower than the condensation coefficient.
3. As the composition of the vapour and liquid change throughout the condenser their physical properties vary.
4. The heavy component must diffuse through the lighter components to reach the condensing surface. The rate of condensation will be governed by the rate of diffusion, as well as the rate of heat transfer.

Temperature Profile

To evaluate the true temperature difference (driving force) in a mixed vapour condenser, a condensation curve (temperature vs enthalpy diagram) must be calculated, showing the change in vapour temperature versus heat transferred throughout the condenser (Figure 12.44). The temperature profile depends on the liquid flow pattern in the condenser. There are two limiting conditions of condensate–vapour flow:

1. Differential condensation: in which the liquid separates from the vapour from which it has condensed. This process is analogous to differential, or Rayleigh, distillation, and the condensation curve can be calculated using methods similar to those for determining the change in composition in differential distillation; see Richardson *et al.* (2002).
2. Integral condensation: in which the liquid remains in equilibrium with the uncondensed vapour. The condensation curve can be determined using procedures similar to those for multicomponent flash distillation given in Chapter 11. This will be a relatively simple calculation for a binary mixture, but complex and tedious for mixtures of more than two components.

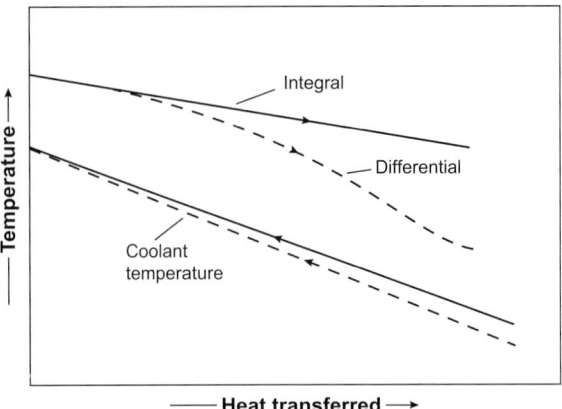

Figure 12.44.   Condensation curves.

It is normal practice to assume that integral condensation occurs. The conditions for integral condensation will be approached if condensation is carried out in one pass, so that the liquid and vapour follow the same path, as in a vertical condenser with condensation inside or outside the tubes. In a horizontal shell-side condenser the condensate will tend to separate from the vapour. The mean temperature difference will be lower for differential condensation, and arrangements where liquid separation is likely to occur should generally be avoided for the condensation of mixed vapours.

Where integral condensation can be considered to occur, the use of a corrected logarithmic mean temperature difference based on the terminal temperatures will generally give a conservative (safe) estimate of the mean temperature difference that can be used in preliminary design calculations.

Estimation of Heat-Transfer Coefficients

Total condensation. For the design of a multicomponent condenser in which the vapour is totally condensed, an estimate of the mean condensing coefficient can be made using the single component correlations with the liquid physical properties evaluated at the average condensate composition. It is the usual practice to apply a factor of safety to allow for the sensible heat transfer and any resistance to mass transfer. Frank (1978) suggests a factor of 0.65, but this is probably too pessimistic. Kern (1950) suggests increasing the area calculated for condensation alone by the ratio of the total heat (condensing + sensible) to the condensing load. Where a more exact estimate of the coefficient is required, and justified by the data, the rigorous methods developed for partial condensation can be used.

Partial condensation. The methods developed for partial condensation and condensation from a non-condensable gas can be divided into two classes:

1. Empirical methods: approximate methods, in which the resistance to heat-transfer is considered to control the rate of condensation, and the mass-transfer resistance is neglected. Design methods have been published by Silver (1947), Bell and Ghaly (1973) and Ward (1960).

2. Analytical methods: more exact procedures, that are based on some model of the heat and mass transfer process, and which take into account the diffusion resistance to mass transfer. The classic method is that of Colburn and Hougen (1934); see also Colburn and Drew (1937) and Porter and Jeffreys (1963). The analytical methods are complex, requiring iterative calculations or graphical procedures. They are suited for computer solution using numerical methods; and proprietary design programs are available. Examples of the application of the Colburn and Drew method are given by Kern (1950) and Jeffreys (1961). The method is discussed briefly in Coulson *et al.* (1999).

An assessment of the methods available for the design of condensers where the condensation is from a non-condensable gas is given by McNaught (1983).

Approximate methods. The local coefficient for heat transfer can be expressed in terms of the local condensate film coefficient h'_c and the local coefficient for sensible heat transfer from the vapour (the gas film coefficient) h'_g, by a relationship first proposed by Silver (1947):

$$\frac{1}{h'_{cg}} = \frac{1}{h'_c} + \frac{Z}{h'_g}$$

(12.37)

where h'_{cg} = the local effective cooling-condensing coefficient

and $$Z = \frac{\Delta H_s}{\Delta H_t} = xC_{pg}\frac{dT}{dH_t},$$

$(\Delta H_s/\Delta H_t)$ = the ratio of the change in sensible heat to the total enthalpy change
(dT/dH_t) = slope of the temperature–enthalpy curve
x = vapour quality, mass fraction of vapour
C_{pg} = vapour (gas) specific heat.

The term dT/dH_t can be evaluated from the condensation curve, h'_c from the single component correlations, and h'_g from correlations for forced convection.

If this is done at several points along the condensation curve the area required can be determined by graphical or numerical integration of the expression:

$$A = \int_0^{Q_t} \frac{dQ}{U\,(T_v - t_c)}$$

(12.38)

where

Q_t = total heat transferred
U = overall heat-transfer coefficient, from equation 12.1, using h'_{cg},
T_v = local vapour (gas) temperature
t_c = local cooling medium temperature.

Gilmore (1963) gives an integrated form of equation 12.37, which can be used for the approximate design of partial condensers

$$\frac{1}{h_{cg}} = \frac{1}{h_c} + \frac{Q_g}{Q_t}\frac{1}{h_g} \qquad\qquad (12.39)$$

where

h_{cg} = mean effective coefficient

h_c = mean condensate film coefficient, evaluated from the single-component correlations, at the average condensate composition, and total condensate loading

h_g = mean gas film coefficient, evaluated using the average vapour flow rate: arithmetic mean of the inlet and outlet vapour (gas) flow rates

Q_g = total sensible heat transfer from vapour (gas)

Q_t = total heat transferred: latent heat of condensation + sensible heat for cooling the vapour (gas) and condensate.

As a rough guide, the following rules of thumb suggested by Frank (1978) can be used to decide the design method to use for a partial condenser (cooler-condenser):

1. Non-condensables <0.5%: use the methods for total condensation; ignore the presence of the uncondensed portion.
2. Non-condensables >70%: assume the heat transfer is by forced convection only. Use the correlations for forced convection to calculate the heat-transfer coefficient, but include the latent heat of condensation in the total heat load transferred.
3. Between 0.5 and 70% non-condensables: use methods that consider both mechanisms of heat transfer.

In partial condensation it is usually better to put the condensing stream on the shell side, and to select a baffle spacing that will maintain high vapour velocities, and therefore high sensible-heat-transfer coefficients.

Fog formation. In the condensation of a vapour from a non-condensable gas, if the bulk temperature of the gas falls below the dew point of the vapour, liquid can condense out directly as a mist or fog. This condition is undesirable, as liquid droplets may be carried out of the condenser. Fog formation in cooler-condensers is discussed by Colburn and Edison (1941) and Lo Pinto (1982). Steinmeyer (1972) gives criteria for the prediction of fog formation. Demisting pads can be used to separate entrained liquid droplets.

12.10.9. Pressure Drop in Condensers

The pressure drop on the condensing side is difficult to predict as two phases are present and the vapour mass velocity is changing throughout the condenser.

A common practice is to calculate the pressure drop using the methods for single-phase flow and apply a factor to allow for the change in vapour velocity. For total condensation, Frank (1978) suggests taking the pressure drop as 40% of the value based on the inlet vapour conditions; Kern (1950) suggests a factor of 50%.

An alternative method, which can also be used to estimate the pressure drop in a partial condenser, is given by Gloyer (1970). The pressure drop is calculated using an average vapour flow rate in the shell (or tubes) estimated as a function of

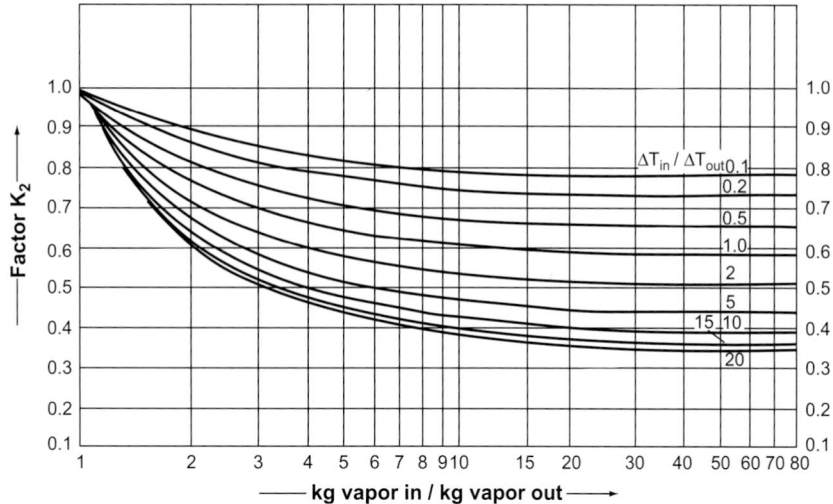

Figure 12.45. Factor for average vapour flow rate for pressure-drop calculation (Gloyer, 1970).

the ratio of the vapour flow rate in and out of the shell (or tubes), and the temperature profile.

$$W_s \text{ (average)} = W_s \text{ (inlet)} \times K_2 \qquad (12.40)$$

K_2 is obtained from Figure 12.45.

$\Delta T_{in}/\Delta T_{out}$ in Figure 12.45 is the ratio of the terminal temperature differences.

These methods can be used to make a crude estimate of the likely pressure drop. A reliable prediction can be obtained by treating the problem as one of two-phase flow. For tube-side condensation the general methods for two-phase flow in pipes can be used; see Collier and Thome (1994) and Coulson *et al.* (1999). As the flow pattern will be changing throughout condensation, some form of step-wise procedure must be used. Two-phase flow on the shell side is discussed by Grant (1973), who gives a method for predicting the pressure drop based on Tinker's shell-side flow model. More sophisticated methods are available in the commercial heat-exchanger design programs from HTFS and HTRI.

A method for estimating the pressure drop on the shell side of horizontal condensers is given in the Engineering Sciences Data Unit Design Guide, ESDU 84023 (1985).

Pressure drop is only likely to be a major consideration in the design of vacuum condensers, and where reflux is returned to a column by gravity flow from the condenser.

Example 12.7

Design a condenser for the following duty: 45,000 kg/h of mixed light hydrocarbon vapours to be condensed. The condenser to operate at 10 bar. The vapour will enter the condenser saturated at 60°C and the condensation will be complete at 45°C.

The average molecular weight of the vapours is 52. The enthalpy of the vapour is 596.5 kJ/kg and the condensate 247.0 kJ/kg. Cooling water is available at 30°C and the temperature rise is to be limited to 10°C. Plant standards require tubes of 20-mm o.d., 16.8-mm i.d., 4.88 m (16 ft) long, of admiralty brass. The vapours are to be totally condensed and no sub-cooling is required.

Solution

Only the thermal design will be done. The physical properties of the mixture will be taken as the mean of those for n-propane (MW = 44) and n-butane (MW = 58), at the average temperature.

$$\text{Heat transferred from vapour} = \frac{45{,}000}{3600}(596.5 - 247.0) = 4368.8 \text{ kW}$$

$$\text{Cooling water flow} = \frac{4368.8}{(40 - 30)4.18} = 104.5 \text{ kg/s}$$

Assumed overall coefficient (Table 12.1) = 900 W/m²°C

Mean temperature difference: the condensation range is small and the change in saturation temperature will be linear, so the corrected logarithmic mean temperature difference can be used.

$$R = \frac{(60 - 45)}{(40 - 30)} = 1.5 \tag{12.6}$$

$$S = \frac{(40 - 30)}{(60 - 30)} = 0.33 \tag{12.7}$$

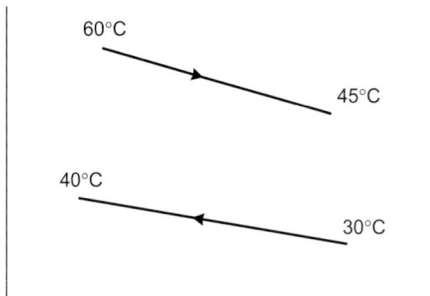

Try a horizontal exchanger, condensation in the shell, four tube passes. For one shell pass, four tube passes, from Figure 12.19, $F_t = 0.92$

$$\Delta T_{\text{lm}} = \frac{(60 - 40) - (45 - 30)}{\ln\dfrac{(60 - 40)}{(45 - 30)}} = 17.4°C$$

$$\Delta T_{\text{lm}} = 0.92 \times 17.4 = 16°C$$

$$\text{Trial area} = \frac{4368.8 \times 10^3}{900 \times 16} = 303 \text{ m}^2$$

Surface area of one tube $= 20 \times 10^{-3} \pi \times 4.88 = 0.305 \text{ m}^2$ (ignore tube-sheet thickness)

$$\text{Number of tubes} = \frac{303}{0.305} = 992$$

Use square pitch, $P_t = 1.25 \times 20 \text{ mm} = 25 \text{ mm}$
Tube bundle diameter

$$D_b = 20 \left(\frac{992}{0.158} \right)^{1/2.263} = 954 \text{ mm} \qquad (12.3b)$$

Number of tubes in centre row $N_r = D_b/P_t = 954/25 = 38$

Shell-Side Coefficient

Estimate tube wall temperature, T_w; assume condensing coefficient of 1500 W/m²°C
Mean temperature

$$\text{Shell-side} = \frac{60 + 45}{2} = 52.5°C$$

$$\text{Tube-side} = \frac{40 + 30}{2} = 35°C$$

$$(52.5 - T_w)1500 = (52.5 - 35)900$$

$$T_w = 42.0°C$$

$$\text{Mean temperature condensate} = \frac{52.5 + 42.0}{2} = 47°C$$

Physical properties at 47°C

$$\mu_L = 0.16 \text{ mNs/m}^2$$

$$\rho_L = 551 \text{ kg/m}^3$$

$$k_L = 0.13 \text{ W/m°C}$$

Vapour density at mean vapour temperature

$$\rho_v = \frac{52}{22.4} \times \frac{273}{(273 + 52.5)} \times \frac{10}{1} = 19.5 \text{ kg/m}^3$$

$$\Gamma_h = \frac{W_c}{L N_t} = \frac{45,000}{3600} \times \frac{1}{4.88 \times 992} = 2.6 \times 10^{-3} \text{ kg/s m}$$

$$N_r = \frac{2}{3} \times 38 = 25 \qquad (12.29)$$

$$h_c = 0.95 \times 0.13 \left[\frac{551(551 - 19.5)9.81}{0.16 \times 10^{-3} \times 2.6 \times 10^{-3}} \right]^{1/3} \times 25^{-1/6}$$

$$= 1375 \text{ W/m}^2°C$$

Close enough to assumed value of 1500 W/m²°C, so no correction to T_w needed.

Tube-side Coefficient

$$\text{Tube cross-sectional area} = \frac{\pi}{4}(16.8 \times 10^{-3})^2 \times \frac{992}{4} = 0.055 \text{ m}^2$$

$$\text{Density of water, at } 35°C = 993 \text{ kg/m}^3$$

$$\text{Tube velocity} = \frac{104.5}{993} \times \frac{1}{0.055} = 1.91 \text{ m/s} \tag{12.17}$$

$$h_i = \frac{4200(1.35 + 0.02 \times 35)1.91^{0.8}}{16.8^{0.2}}$$

$$= 8218 \text{ W/m}^2°C$$

Fouling factors: as neither fluid is heavily fouling, use 6000 W/m^2°C for each side.

$$k_w = 50 \text{ W/m}°C$$

Overall Coefficient

$$\frac{1}{U} = \frac{1}{1375} + \frac{1}{6000} + \frac{20 \times 10^{-3}\ln\left(\frac{20}{16.8}\right)}{2 \times 50} + \frac{20}{16.8} \times \frac{1}{6000} + \frac{20}{16.8} \times \frac{1}{8218}$$

$$U = \underline{\underline{786 \text{ W/m}^2\,°C}} \tag{12.2}$$

Significantly lower than the assumed value of 900 W/m^2°C.
Repeat calculation using new trial value of 750 W/m^2°C.

$$\text{Area} = \frac{4368 \times 10^3}{750 \times 16} = 364 \text{ m}^2$$

$$\text{Number of tubes} = \frac{364}{0.305} = 1194 \tag{12.3b}$$

$$D_b = 20\left(\frac{1194}{0.158}\right)^{1/2.263} = 1035 \text{ mm}$$

$$\text{Number of tubes in centre row} = \frac{1035}{25} = 41$$

$$\Gamma_h = \frac{45,000}{3600} \times \frac{1}{4.88 \times 1194} = 2.15 \times 10^{-3} \text{ kg/m s}$$

$$N_r = \frac{2}{3} \times 41 = 27$$

$$\tag{12.29}$$

$$h_c = 0.95 \times 0.13\left[\frac{551(551 - 19.5)9.81}{0.16 \times 10^{-3} \times 2.15 \times 10^{-3}}\right]^{1/3} \times 27^{-1/6}$$

$$= 1447 \text{ W/m}^2\,°C$$

$$\text{New tube velocity} = 1.91 \times \frac{992}{1194} = 1.59 \text{ m/s}$$

$$h_i = 4200 \left(1.35 + 0.02 \times 35\right) \frac{1.59^{0.8}}{16.8^{0.2}} = 7097 \text{ W/m}^2{}^\circ\text{C} \qquad (12.17)$$

$$\frac{1}{U} = \frac{1}{1447} + \frac{1}{6000} + \frac{20 \times 10^{-3} \ln\left(\frac{20}{16.8}\right)}{2 \times 50}$$

$$+ \frac{20}{16.8} \times \frac{1}{6000} + \frac{20}{16.8} \times \frac{1}{7097} \qquad (12.2)$$

$$U = 773 \text{ W/m}^2{}^\circ\text{C}$$

Close enough to estimate, firm up design.

Shell-Side Pressure Drop

Use pull-through floating head, no need for close clearance.
  Select baffle spacing = shell diameter, 45% cut

From Figure 12.10, clearance = 95 mm

$$\text{Shell i.d.} = 1035 + 95 = 1130 \text{ mm}$$

Use Kern's method to make an approximate estimate.

$$\text{Cross-flow area } A_s = \frac{(25 - 20)}{25} 1130 \times 1130 \times 10^{-6}$$

$$= 0.255 \text{ m}^2 \qquad (12.21)$$

Mass flow rate, based on inlet conditions

$$G_s = \frac{45{,}000}{3600} \times \frac{1}{0.255} = 49.02 \text{ kg/s m}^2$$

$$\text{Equivalent diameter, } d_e = \frac{1.27}{20}(25^2 - 0.785 \times 20^2)$$

$$= 19.8 \text{ mm} \qquad (12.22)$$

$$\text{Vapour viscosity} = 0.008 \text{ mNs/m}^2$$

$$\text{Re} = \frac{49.02 \times 19.8 \times 10^{-3}}{0.008 \times 10^{-3}} = 121{,}325$$

From Figure 12.30, $j_f = 2.2 \times 10^{-2}$

$$u_s = \frac{G_s}{\rho_v} = \frac{49.02}{19.5} = 2.51 \text{ m/s}$$

Take pressure drop as 50% of that calculated using the inlet flow; neglect viscosity correction.

$$\Delta P_s = \frac{1}{2}\left[8 \times 2.2 \times 10^{-2}\left(\frac{1130}{19.8}\right)\left(\frac{4.88}{1.130}\right)\frac{19.5(2.51)^2}{2}\right]$$

$$= 1322 \text{ N/m}^2$$

$$= \underline{1.3 \text{ kPa}}$$

(12.26)

Negligible; more sophisticated method of calculation not justified.

Tube-Side Pressure Drop

Viscosity of water $= 0.6 \text{ mN s/m}^2$

$$\text{Re} = \frac{u_t \rho d_i}{\mu} = \frac{1.59 \times 993 \times 16.8 \times 10^{-3}}{0.6 \times 10^{-3}} = \underline{\underline{44,208}}$$

From Figure 12.24, $j_f = 3.5 \times 10^{-3}$
Neglect viscosity correction

$$\Delta P_t = 4\left[8 \times 3.5 \times 10^{-3}\left(\frac{4.88}{16.8 \times 10^{-3}}\right) + 2.5\right]\frac{993 \times 1.59^2}{2}$$

$$= 53,388 \text{ N/m}^2$$

$$= \underline{53 \text{ kPa}}\,(7.7 \text{ psi}).$$

(12.20)

Acceptable.

12.11.   REBOILERS AND VAPORIZERS

The design methods given in this section can be used for reboilers and vaporizers. Reboilers are used with distillation columns to vaporize a fraction of the bottom product, whereas in a vaporizer essentially all the feed is vaporized.

Three principal types of reboiler are used:

1. Forced circulation, Figure 12.46: in which the fluid is pumped through the exchanger, and the vapour formed is separated in the base of the column. When used as a vaporizer a disengagement vessel will have to be provided.
2. Thermosyphon, natural circulation, Figure 12.47: vertical exchangers with vaporization in the tubes, or horizontal exchangers with vaporization in the shell. The liquid circulation through the exchanger is maintained by the difference in density between the two-phase mixture of vapour and liquid in the exchanger and the single-phase liquid in the base of the column. As with the forced-circulation type, a disengagement vessel will be needed if this type is used as a vaporizer.
3. Kettle type, Figure 12.48: in which boiling takes place on tubes immersed in a pool of liquid; there is no circulation of liquid through the exchanger. This type is also, more correctly, called a submerged-bundle reboiler. In some applications it is possible to accommodate the bundle in the base of the

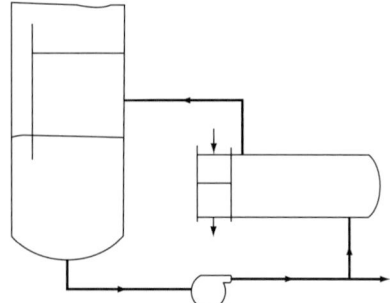

Figure 12.46.   Forced-circulation reboiler.

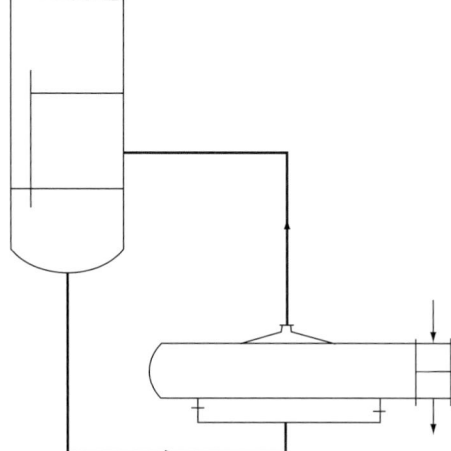

Figure 12.47.   Horizontal thermosyphon reboiler.

column (Figure 12.49), saving the cost of the exchanger shell. This arrangement is commonly known as a 'stab-in' reboiler.

Choice of Type

The choice of the best type of reboiler or vaporizer for a given duty will depend on the following factors:

1. The nature of the process fluid, particularly its viscosity and propensity to fouling
2. The operating pressure: vacuum or pressure
3. The equipment layout, particularly the headroom available.

Forced-circulation reboilers are especially suitable for handling viscous and heavily fouling process fluids; see Chantry and Church (1958). The circulation rate is predictable and high velocities can be used. They are also suitable for low-vacuum operations, and for low rates of vaporization. The major disadvantage of this type is that a pump is required and the pumping cost will be high. There is also the danger that leakage of hot fluid could occur at the pump seal; canned-rotor type pumps can be specified to avoid the possibility of leakage.

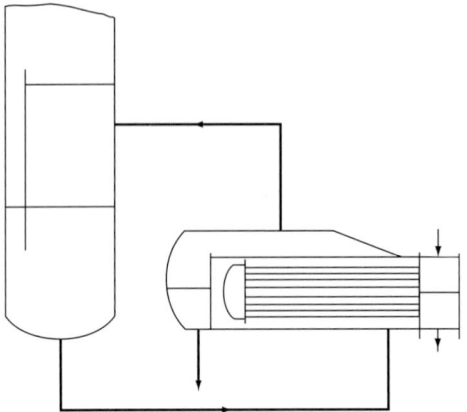

Figure 12.48.   Kettle reboiler.

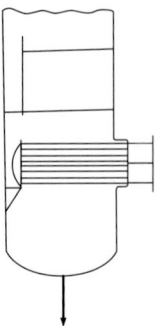

Figure 12.49.   Internal reboiler.

Thermosyphon reboilers are the most economical type for most applications, but are not suitable for high-viscosity fluids or high-vacuum operation. They would not normally be specified for pressures below 0.3 bar. A disadvantage of this type is that the column base must be elevated to provide the hydrostatic head required for the thermosyphon effect. This increases the cost of the column-supporting structure. Horizontal reboilers need less headroom than vertical, but have more complex pipework. Horizontal exchangers are easier to maintain than vertical, as the tube bundle is easier to withdraw.

Kettle reboilers have lower heat-transfer coefficients than the other types, as there is no liquid circulation. They are not suitable for fouling materials, and have a high residence time. They will generally be more expensive than an equivalent thermosyphon type, as a larger shell is needed; but if the duty is such that the bundle can be installed in the column base, the cost will be competitive with the other types. They are often used as vaporizers, as a separate vapour–liquid disengagement vessel is not needed. They are suitable for vacuum operation, and for high rates of vaporization, up to 80% of the feed. Some designs allow a liquid blow-down stream to be withdrawn to prevent accumulation of solids or non-volatile components.

12.11.1. Boiling Heat-Transfer Fundamentals

The complex phenomena involved in heat transfer to a boiling liquid are discussed in Coulson *et al.* (1999). A more detailed account is given by Collier and Thome (1994), Tong and Tang (1997) and Hsu and Graham (1976). Only a brief discussion of the subject will be given in this section: sufficient for the understanding of the design methods given for reboilers and vaporizers.

The mechanism of heat transfer from a submerged surface to a pool of liquid depends on the temperature difference between the heated surface and the liquid (Figure 12.50). At low temperature differences, when the liquid is below its boiling point, heat is transferred by natural convection. As the surface temperature is raised incipient boiling occurs, with vapour bubbles forming and breaking loose from the surface. The agitation caused by the rising bubbles, and other effects caused by bubble generation at the surface, result in a large increase in the rate of heat transfer. This phenomenon is known as nucleate boiling. As the temperature is raised further, the rate of heat transfer increases until the heat flux reaches a critical value. At this point, the rate of vapour generation is such that dry patches occur spontaneously over the surface, and the rate of heat transfer falls off rapidly. At higher temperature differences, the vapour rate is such that the whole surface is blanketed with vapour, and the mechanism of heat transfer is by conduction through the vapour film. Conduction is augmented at very high temperature differences by radiation.

The maximum heat flux achievable with nucleate boiling is known as the critical heat flux. In a system where the surface temperature is not self-limiting, such as a nuclear reactor fuel element or a boiling tube that is heated in a fired heater, operation above the critical flux will result in a rapid increase in the surface temperature, and in the extreme situation the surface will melt. This phenomenon is known as 'burn-out'. The heating media used for process plants are normally self limiting; for example, with a steam-heated

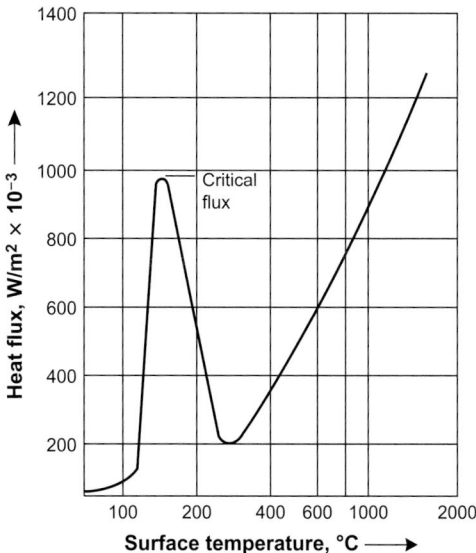

Figure 12.50. Typical pool boiling curve (water at 1 bar).

reboiler the surface temperature can never exceed the saturation temperature of the condensing steam. Care must be taken in the design of electrically heated vaporizers and directly fired vaporizers to ensure that the critical flux can never be exceeded. If the critical flux is exceeded in a directly fired vaporizer then the tube-wall temperature can approach the temperatures of the radiant zone (the bridgewall temperature or even the flame temperature). At these temperatures damage to the tubes will occur.

The critical flux is reached at surprisingly low temperature differences; around 20°C to 30°C for water, and 20°C to 50°C for light organics.

Estimation of Boiling Heat-Transfer Coefficients

In the design of vaporizers and reboilers, the designer will be concerned with two types of boiling: pool boiling and convective boiling. Pool boiling is the name given to nucleate boiling in a pool of liquid, such as in a kettle-type reboiler or a jacketed vessel. Convective boiling occurs where the vaporizing fluid is flowing over the heated surface, and heat transfer takes place both by forced convection and nucleate boiling; as in forced circulation or thermosyphon reboilers.

Boiling is a complex phenomenon, and boiling heat-transfer coefficients are difficult to predict with any certainty. Whenever possible, experimental values obtained for the system being considered should be used, or values for a closely related system.

12.11.2. Pool Boiling

In the nucleate boiling region the heat-transfer coefficient is dependent on the nature and condition of the heat-transfer surface, and it is not possible to present a universal correlation that will give accurate predictions for all systems. Palen and Taborek (1962) have reviewed the published correlations and compared their suitability for use in reboiler design.

The correlation given by Forster and Zuber (1955) can be used to estimate pool boiling coefficients, in the absence of experimental data. Their equation can be written in the form:

$$h_{nb} = 0.00122 \left[\frac{k_L^{0.79} C_{pL}^{0.45} \rho_L^{0.49}}{\sigma^{0.5} \mu_L^{0.29} \lambda^{0.24} \rho_v^{0.24}} \right] (T_w - T_s)^{0.24} (p_w - p_s)^{0.75} \qquad (12.41)$$

where

h_{nb} = nucleate, pool, boiling coefficient, W/m$^{2\circ}$C
k_L = liquid thermal conductivity, W/m°C
C_{pL} = liquid heat capacity, J/kg°C
ρ_L = liquid density, kg/m^3
μ_L = liquid viscosity, Ns/m^2
λ = latent heat, J/kg
ρ_v = vapour density, kg/m^3
T_w = wall, surface temperature, °C
T_s = saturation temperature of boiling liquid, °C
p_w = saturation pressure corresponding to the wall temperature, T_w, N/m^2
p_s = saturation pressure corresponding to T_s, N/m^2
σ = surface tension, N/m.

The reduced pressure correlation given by Mostinski (1963) is simple to use and gives values that are as reliable as those given by more complex equations.

$$h_{nb} = 0.104(P_c)^{0.69}(q)^{0.7}\left[1.8\left(\frac{P}{P_c}\right)^{0.17} + 4\left(\frac{P}{P_c}\right)^{1.2} + 10\left(\frac{P}{P_c}\right)^{10}\right] \qquad (12.42)$$

where

P = operating pressure, bar
P_c = liquid critical pressure, bar
q = heat flux, W/m^2.

Note: $q = h_{nb}(T_w - T_s)$.

Mostinski's equation is convenient to use when data on the fluid physical properties are not available.

Equations 12.41 and 12.42 are for boiling single-component fluids; for mixtures the coefficient will generally be lower than is predicted by these equations. The equations can be used for close boiling range mixtures, say less than 5°C; and for wider boiling ranges with a suitable factor of safety (see Section 12.11.6).

Critical heat flux

It is important to check that the design, and operating, heat flux is well below the critical flux. Several correlations are available for predicting the critical flux. That given by Zuber *et al.* (1961) has been found to give satisfactory predictions for use in reboiler and vaporizer design. In SI units, Zuber's equation can be written as:

$$q_c = 0.131\lambda\left[\sigma g(\rho_L - \rho_v)\rho_v^2\right]^{1/4} \qquad (12.43)$$

where

q_c = maximum, critical, heat flux, W/m^2
g = gravitational acceleration, 9.81 m/s^2.

Mostinski also gives a reduced pressure equation for predicting the maximum critical heat flux:

$$q_c = 3.67 \times 10^4 P_c\left(\frac{P}{P_c}\right)^{0.35}\left[1 - \left(\frac{P}{P_c}\right)\right]^{0.9} \qquad (12.44)$$

Film Boiling

The equation given by Bromley (1950) can be used to estimate the heat-transfer coefficient for film boiling on tubes. Heat transfer in the film-boiling region will be controlled by conduction through the film of vapour, and Bromley's equation is similar to the Nusselt equation for condensation, where conduction is occurring through the film of condensate.

$$h_{fb} = 0.62\left[\frac{k_v^3(\rho_L - \rho_v)\rho_v g\lambda}{\mu_v d_o(T_w - T_s)}\right]^{1/4} \qquad (12.45)$$

where h_{fb} is the film-boiling heat-transfer coefficient; the suffix v refers to the vapour phase and d_o is in metres. It must be emphasized that process reboilers and vaporizers will always be designed to operate in the nucleate boiling region. The heating medium would be selected, and its temperature controlled, to ensure that in operation the temperature difference is well below that at which the critical flux is reached. For instance, if direct heating with steam would give too high a temperature difference, the steam would be used to heat water, and hot water used as the heating medium. Above temperatures where steam can be used, hot oil circuits are often used for reboilers, so as to avoid direct firing of the reboiler.

Example 12.8

Estimate the heat-transfer coefficient for the pool boiling of water at 2.1 bar, from a surface at 125°C. Check that the critical flux is not exceeded.

Solution

Physical properties, from steam tables:

$$\text{Saturation temperature, } T_s = 121.8°C$$
$$\rho_L = 941.6 \text{ kg/m}^3, \quad \rho_v = 1.18 \text{ kg/m}^3$$
$$C_{pL} = 4.25 \times 10^3 \text{ J/kg°C}$$
$$k_L = 687 \times 10^{-3} \text{ Wm°C}$$
$$\mu_L = 230 \times 10^{-6} \text{ Ns/m}^2$$
$$\lambda = 2198 \times 10^3 \text{ J/kg}$$
$$\sigma = 55 \times 10^{-3} \text{ N/m}$$
$$p_w \text{ at } 125°C = 2.321 \times 10^5 \text{ N/m}^2$$
$$p_s = 2.1 \times 10^5 \text{ N/m}^2$$

Use the Foster-Zuber correlation, equation 12.41:

$$h_{nb} = 1.22 \times 10^{-3} \left[\frac{(687 \times 10^{-3})^{0.79}(4.25 \times 10^3)^{0.45}(941.6)^{0.49}}{(55 \times 10^{-3})^{0.5}(230 \times 10^{-6})^{0.29}(2198 \times 10^3)^{0.24}(1.18)^{0.24}} \right]$$
$$\times (125 - 121.8)^{0.24}(2.321 \times 10^5 - 2.1 \times 10^5)^{0.75}$$
$$= \underline{\underline{3738 \text{ W/m}^2°C}}$$

Use the Zuber correlation, equation 12.43:

$$q_c = 1.131 \times 2198 \times 10^3 [55 \times 10^{-3} \times 9.81(941.6 - 1.18)1.18^2]^{1/4}$$
$$= \underline{\underline{1.48 \times 10^6 \text{ W/m}^2}}$$

$$\text{Actual flux} = (125 - 121.8)3738 = \underline{\underline{11,962 \text{ W/m}^2}}$$

Well below critical flux.

12.11.3. Convective Boiling

The mechanism of heat transfer in convective boiling, where the boiling fluid is flowing through a tube or over a tube bundle, differs from that in pool boiling. It will depend on the state of the fluid at any point. Consider the situation of a liquid boiling inside a vertical tube (Figure 12.51). The following conditions occur as the fluid flows up the tube.

1. Single-phase flow region: at the inlet the liquid is below its boiling point (subcooled) and heat is transferred by forced convection. The equations for forced convection can be used to estimate the heat-transfer coefficient in this region.
2. Sub-cooled boiling: in this region the liquid next to the wall has reached boiling point, but not the bulk of the liquid. Local boiling takes place at the wall, which increases the rate of heat-transfer over that given by forced convection alone.
3. Saturated boiling region: in this region bulk boiling of the liquid is occurring in a manner similar to nucleate pool boiling. The volume of vapour is increasing and various flow patterns can form. In a long tube, the flow pattern will eventually become annular, where the liquid phase is spread over the tube wall and the vapour flows up the central core.
4. Dry-wall region: Ultimately, if a large fraction of the feed is vaporized, the wall dries out and any remaining liquid is present as a mist. Heat transfer in this region is by convection and radiation to the vapour. This condition is unlikely to occur in commercial reboilers and vaporizers.

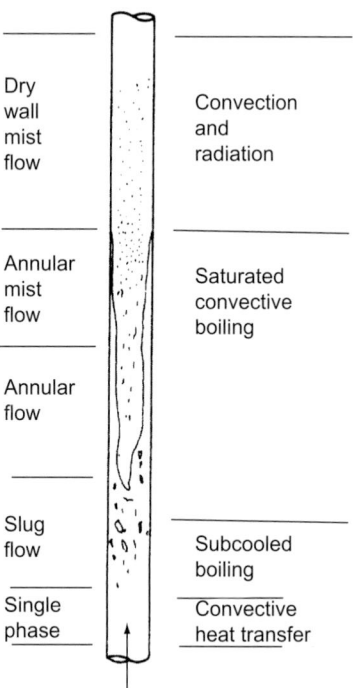

Figure 12.51.   Convective boiling in a vertical tube.

Saturated, bulk, boiling is the principal mechanism of interest in the design of reboilers and vaporizers.

A comprehensive review of the methods available for predicting convective boiling coefficients is given by Webb and Gupte (1992). The methods proposed by Chen (1966) and Shah (1976) are convenient to use in manual calculations and are accurate enough for preliminary design work. Chen's method is outlined below and illustrated in Example 12.9.

Chen's Method

In forced-convective boiling, the effective heat-transfer coefficient h_{cb} can be considered to be made up of convective and nucleate boiling components: h'_{fc} and h'_{nb}.

$$h_{cb} = h'_{fc} + h'_{nb} \qquad (12.46)$$

The convective boiling coefficient h'_{fc} can be estimated using the equations for single-phase forced-convection heat transfer modified by a factor f_c to account for the effects of two-phase flow:

$$h'_{fc} = h_{fc} \times f_c \qquad (12.47)$$

The forced-convection coefficient h_{fc} is calculated assuming that the liquid phase is flowing in the conduit alone.

The two-phase correction factor f_c is obtained from Figure 12.52, in which the term $1/X_{tt}$ is the Lockhart-Martinelli two-phase flow parameter with turbulent flow in both phases (see Coulson *et al.*, 1999). This parameter is given by:

$$\frac{1}{X_{tt}} = \left[\frac{x}{1-x}\right]^{0.9} \left[\frac{\rho_L}{\rho_v}\right]^{0.5} \left[\frac{\mu_v}{\mu_L}\right]^{0.1} \qquad (12.48)$$

where x is the vapour quality, the mass fraction of vapour.

The nucleate boiling coefficient can be calculated using correlations for nucleate pool boiling modified by a factor f_s to account for the fact that nucleate boiling is more difficult in a flowing liquid.

$$h'_{nb} = h_{nb} \times f_s \qquad (12.49)$$

The suppression factor f_s is obtained from Figure 12.53. It is a function of the liquid Reynolds number Re_L and the forced-convection correction factor f_c.

Re_L is evaluated assuming that only the liquid phase is flowing in the conduit, and will be given by:

$$Re_L = \frac{(1 - x)\, G\, d_e}{\mu_L} \qquad (12.50)$$

where G is the total mass flow rate per unit flow area.

Chen's method was developed from experimental data on forced-convective boiling in vertical tubes. It can be applied, with caution, to forced-convective boiling in horizontal tubes, and annular conduits (concentric pipes). Butterworth (1977) suggests that, in the absence of more reliable methods, it may be used to estimate the heat-transfer coefficient for forced-convective boiling in cross-flow over tube bundles, using a suitable cross-flow correlation to predict the forced-convection

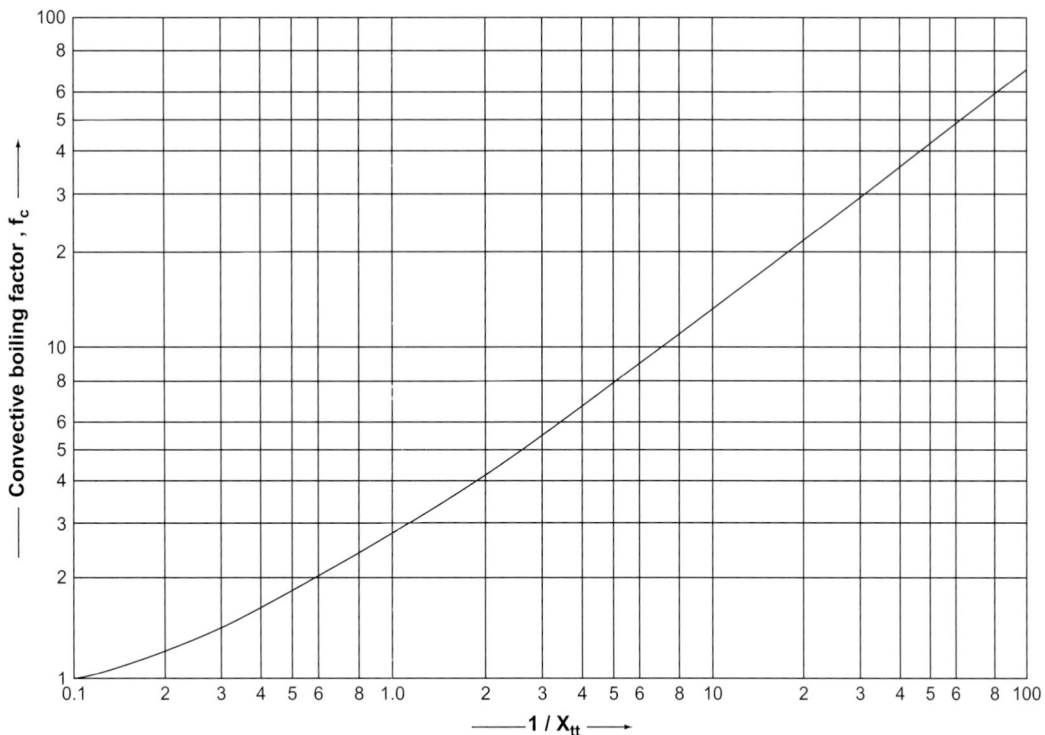

Figure 12.52. Convective boiling enhancement factor.

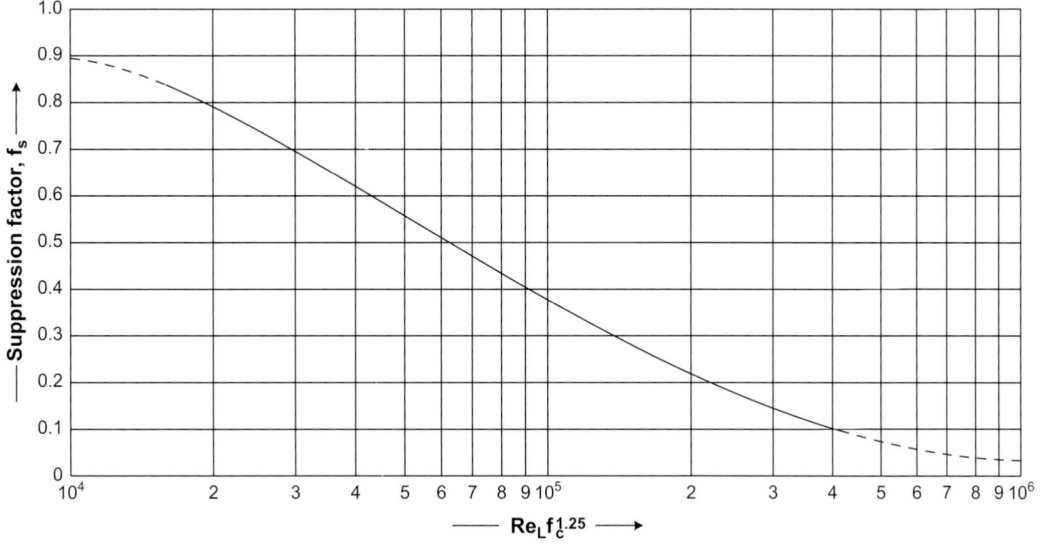

Figure 12.53. Nucleate boiling suppression factor.

coefficient. Shah's method was based on data for flow in horizontal and vertical tubes and annuli.

A major problem that will be encountered when applying convective boiling correlations to the design of reboilers and vaporizers is that, because the vapour quality changes progressively throughout the exchanger, a step-by-step procedure will be needed. The exchanger must be divided into sections and the coefficient and heat-transfer area estimated sequentially for each section.

Example 12.9

A fluid whose properties are essentially those of o-dichlorobenzene is vaporized in the tubes of a forced-convection reboiler. Estimate the local heat-transfer coefficient at a point where 5% of the liquid has been vaporized. The liquid velocity at the tube inlet is 2 m/s and the operating pressure is 0.3 bar. The tube inside diameter is 16 mm and the local wall temperature is estimated to be 120°C.

Solution

Physical properties:

Boiling point $= 136°C$

$$\rho_L = 1170 \text{ kg/m}^3$$
$$\mu_L = 0.45 \text{ mNs/m}^2$$
$$\mu_v = 0.01 \text{ mNs/m}^2$$
$$\rho_v = 1.31 \text{ kg/m}^3$$
$$k_L = 0.11 \text{ W/m}°C$$
$$C_{pL} = 1.25 \text{ kJ/kg}°C$$
$$P_c = 41 \text{ bar}$$

The forced-convective boiling coefficient will be estimated using Chen's method. With 5% vapour, liquid velocity (for liquid flow in tube alone)

$$= 2 \times 0.95 = 1.90 \text{ m/s}$$
$$\text{Re}_L = \frac{1170 \times 1.90 \times 16 \times 10^{-3}}{0.45 \times 10^{-3}} = 79,040$$

From Figure 12.23, $j_h = 3.3 \times 10^{-3}$

$$\text{Pr} = \frac{1.25 \times 10^3 \times 0.45 \times 10^{-3}}{0.11} = 5.1$$

Neglect viscosity correction term.

$$h_{fc} = \frac{0.11}{16 \times 10^{-3}} \times 3.3 \times 10^{-3}(79,040)(5.1)^{0.33}$$
$$= 3070 \text{ W/m}^2°C \qquad\qquad (12.15)$$

$$\frac{1}{X_{tt}} = \left[\frac{0.05}{1-0.05}\right]^{0.9} \left[\frac{1170}{1.31}\right]^{0.5} \left[\frac{0.01 \times 10^{-3}}{0.45 \times 10^{-3}}\right]^{0.1} \quad (12.48)$$

$$= 1.44$$

From Figure 12.52, $f_c = 3.2$

$$h'_{fc} = 3.2 \times 3070 = 9824 \text{ W/m}^2{}^\circ\text{C}$$

Using Mostinski's correlation to estimate the nucleate boiling coefficient

$$h_{nb} = 0.104 \times 41^{0.69}[h_{nb}(136-120)]^{0.7}$$

$$\times \left[1.8\left(\frac{0.3}{41}\right)^{0.17} + 4\left(\frac{0.3}{41}\right)^{1.2} + 10\left(\frac{0.3}{41}\right)^{10}\right]$$

$$h_{nb} = 7.43 \, h_{nb}^{0.7}$$

$$h_{nb} = 800 \text{ W/m}^2{}^\circ\text{C}$$

$$\text{Re}_L f_c^{1.25} = 79,040 \times 3.2^{1.25} = 338,286 \quad (12.42)$$

From Figure 12.53, $f_s = 0.13$

$$h'_{nb} = 0.13 \times 800 = 104 \text{ W/m}^2{}^\circ\text{C}$$

$$h_{cb} = 9824 + 104 = \underline{\underline{9928 \text{ W/m}^2{}^\circ\text{C}}}$$

12.11.4. Design of Forced-circulation Reboilers

The normal practice in the design of forced-convection reboilers is to calculate the heat-transfer coefficient assuming that the heat is transferred by forced convection only. This will give conservative (safe) values, as any boiling that occurs will invariably increase the rate of heat transfer. In many designs, the pressure is controlled to prevent any appreciable vaporization in the exchanger. A throttle value is installed in the exchanger outlet line, and the liquid flashes as the pressure is let down into the vapour–liquid separation vessel.

If a significant amount of vaporization does occur, the heat-transfer coefficient can be evaluated using correlations for convective boiling, such as Chen's method.

Conventional shell and tube exchanger designs are used, with one shell pass and two tube passes, when the process fluid is on the shell side; and one shell and one tube pass when it is in the tubes. High tube velocities are used to reduce fouling (3–9 m/s).

Because the circulation rate is set by the designer, forced-circulation reboilers can be designed with more certainty than natural circulation units.

The critical flux in forced-convection boiling is difficult to predict. Kern (1950) recommends that for commercial reboiler designs the heat flux should not exceed 63,000 W/m² (20,000 Btu/ft²h) for organics and 95,000 W/m² (30,000 Btu/ft²h) for water and dilute aqueous solutions. These values are now generally considered to be too pessimistic.

12.11.5. **Design of Thermosyphon Reboilers**

The design of thermosyphon reboilers is complicated by the fact that, unlike a forced-convection reboiler, the fluid circulation rate cannot be determined explicitly. The circulation rate, heat-transfer rate and pressure drop are all interrelated, and iterative design procedures must be used. The fluid will circulate at a rate at which the pressure losses in the system are just balanced by the available hydrostatic head. The exchanger, column base and piping can be considered as the two legs of a U-tube (Figure 12.54). The driving force for circulation around the system is the difference in density of the liquid in the 'cold' leg (the column base and inlet piping) and the two-phase fluid in the 'hot' leg (the exchanger tubes and outlet piping).

To calculate the circulation rate it is necessary to make a pressure balance around the system.

A typical design procedure will include the following steps:

1. Calculate the vaporization rate required, from the specified duty.
2. Estimate the exchanger area, from an assumed value for the overall heat-transfer coefficient. Decide the exchanger layout and piping dimensions.
3. Assume a value for the circulation rate through the exchanger.
4. Calculate the pressure drop in the inlet piping (single phase).
5. Divide the exchanger tube into sections and calculate the pressure drop section by section up the tube. Use suitable methods for the sections in which the flow is two-phase. Include the pressure loss due to the fluid acceleration as the vapour rate increases. For a horizontal reboiler, calculate the pressure drop in the shell, using a method suitable for two-phase flow.
6. Calculate the pressure drop in the outlet piping (two-phase).
7. Compare the calculated pressure drop with the available differential head, which will depend on the vapour voidage and hence the assumed circulation rate. If a satisfactory balance has been achieved, proceed. If not, return to step 3 and repeat the calculations with a new assumed circulation rate.

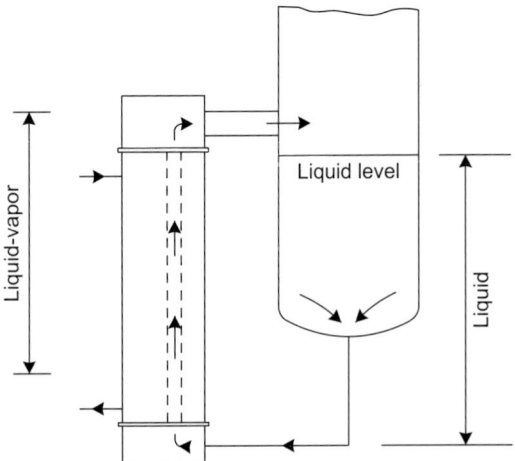

Figure 12.54.   Vertical thermosyphon reboiler, liquid and vapour flows.

8. Calculate the heat-transfer coefficient and heat-transfer rate section by section up the tubes. Use a suitable method for the sections in which the boiling is occurring, such as Chen's method.
9. Calculate the rate of vaporization from the total heat-transfer rate, and compare with the value assumed in step 1. If the values are sufficiently close, proceed. If not, return to step 2 and repeat the calculations for a new design.
10. Check that the critical heat flux is not exceeded at any point up the tubes.
11. Repeat the complete procedure as necessary to optimize the design.

It can be seen that to design a thermosyphon reboiler using hand calculations would be tedious and time-consuming. The iterative nature of the procedure lends itself to solution by computers. Sarma *et al.* (1973) discuss the development of a computer program for vertical thermosyphon reboiler design, and give algorithms and design equations.

Extensive work on the performance and design of thermosyphon reboilers has been carried out by HTFS and HTRI, and proprietary design programs are available from these organizations. The HTFS methods are available in Aspen Technology's Aspen Engineering Suite and in Honeywell's UniSim Design Suite; see Table 4.1.

In the absence of access to a computer program, the rigorous design methods given by Fair (1960, 1963) or Hughmark (1961, 1964, 1969) can be used for thermosyphon vertical reboilers. Collins (1976) and Fair and Klip (1983) give methods for the design of horizontal, shell-side thermosyphon reboilers. The design and performance of this type of reboiler are also reviewed in a paper by Yilmaz (1987).

Approximate methods can be used for preliminary designs. Fair (1960) gives a method in which the heat transfer and pressure drop in the tubes are based on the average of the inlet and outlet conditions. This simplifies step 5 in the design procedure but trial-and-error calculations are still needed to determine the circulation rate. Frank and Prickett (1973) programmed Fair's rigorous design method for computer solution and used it, together with operating data on commercial exchangers, to derive a general correlation of heat-transfer rate with reduced temperature for vertical thermosyphon reboilers. Their correlation, converted to SI units, is shown in Figure 12.55. The basis and limitations of the correlation are listed below:

1. Conventional designs: tube lengths 2.5 to 3.7 m (8 to 12 ft) (standard length 2.44 m), preferred diameter 25 mm (1 in)
2. Liquid in the sump level with the top tube sheet
3. Process-side fouling coefficient 6000 W/m^2°C
4. Heating medium steam, coefficient including fouling, 6000 W/m^2°C
5. Simple inlet and outlet piping
6. For reduced temperatures greater than 0.8, use the limiting curve (that for aqueous solutions)
7. Minimum operating pressure 0.3 bar
8. Inlet fluid should not be appreciably sub-cooled
9. Extrapolation is not recommended.

For heating media other than steam and process-side fouling coefficients different from 6000 W/m^2°C, the design heat flux taken from Figure 12.55 may be adjusted as follows:

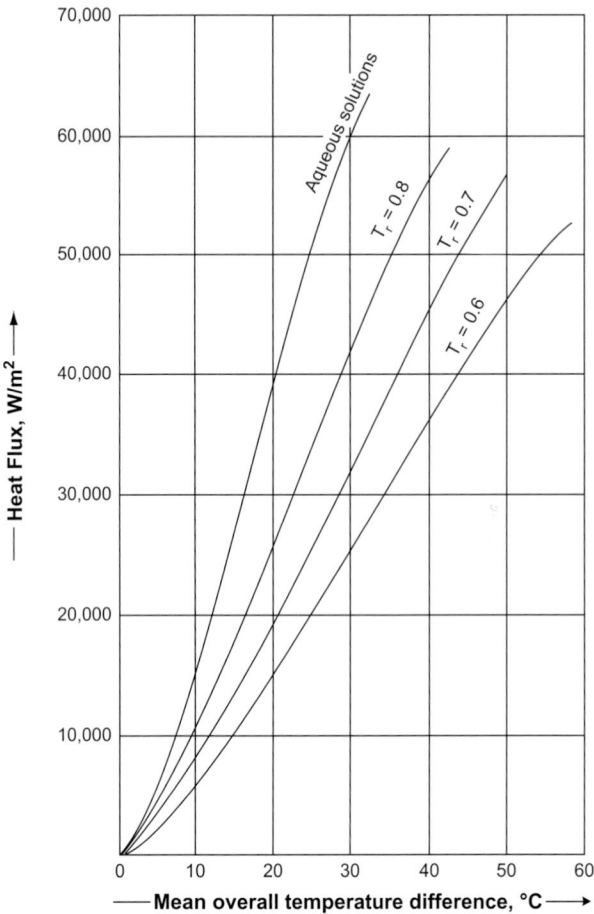

Figure 12.55.   Vertical thermosyphon design correlation.

$$U' = \frac{q'}{\Delta T'} \qquad\qquad (12.51)$$

and

$$\frac{1}{U_c} = \frac{1}{U'} - \frac{1}{6000} + \frac{1}{h_s} - \frac{1}{6000} + \frac{1}{h_{id}}$$

where

q' = flux read from Figure 12.55 at $\Delta T'$
h_s = new shell-side coefficient, W/m^2°C
h_{id} = fouling coefficient on the process (tube) side, W/m^2°C
U_c = corrected overall coefficient.

The use of Frank and Prickett's method is illustrated in Example 12.10.

Limitations on the use of Frank and Prickett's Method

A study by van Edmonds (1994), using the HTFS TREB4 program, found that Frank and Prickett's method gave acceptable predictions for pure components and binary mixtures with water, but that the results were unreliable for other mixtures. Also, van Edmonds' results predicted higher flux values than those obtained by Prickett and Frank.

For preliminary designs for pure components, or near pure components, Prickett and Frank's method should give a conservative estimate of the operating heat flux. It is not recommended for mixtures, other than binary mixtures with water.

Approximate Design Method for Mixtures

For mixtures, the simplified analysis used by Kern (1954) can be used to obtain an approximate estimate of the number of tubes required; see also Aerstin and Street (1978) and Hewitt *et al.* (1994).

This method uses simple, unsophisticated, methods to estimate the two-phase pressure drop through the exchanger and piping, and the convective boiling heat-transfer coefficient. The calculation procedure is set out below and illustrated in Example 12.11.

Procedure

1. Determine the heat duty.
2. Estimate the heat-transfer area, using the maximum allowable heat flux. Take as 39,700 W/m^2 for vertical and 47,300 W/m^2 for horizontal reboilers.
3. Choose the tube diameters and length. Calculate the number of tubes required.
4. Estimate the recirculation ratio, not less than 3.
5. Calculate the vapour flow rate leaving the reboiler for the duty and liquid heat of vaporization.
6. Calculate the liquid flow rate leaving the reboiler for the vapour rate and recirculation ratio.
7. Estimate the two-phase pressure drop though the tubes, due to friction. Use the homogenous model or another simple method, such as the Lochart–Martenelli equation.
8. Estimate the static head in the tubes.
9. Estimate the available head.
10. Compare the total estimated pressure drop and the available head. If the available head is greater by a sufficient amount to allow for the pressure drop through the inlet and outlet piping, proceed. If the available head is not sufficient, return to step 2, and increase the number of tubes.
11. Calculate the convective heat-transfer coefficient using simple methods, such as assuming convection only, or Chen's method; see Section 12.11.3.
12. Calculate the overall heat-transfer coefficient.
13. Calculate the required overall coefficient and compare with that estimated. If satisfactory, accept the design, if unsatisfactory return to step 2 and increase the estimated area.

Maximum Heat Flux

Thermosyphon reboilers can suffer from flow instabilities if too high a heat flux is used. The liquid and vapour flow in the tubes is not smooth but tends to pulsate, and at high heat fluxes the pulsations can become large enough to cause vapour locking. A good practice is to install a flow restriction in the inlet line, a valve or orifice plate, so that the flow resistance can be adjusted should vapour locking occur in operation.

Kern recommends that the heat flux in thermosyphon reboilers, based on the total heat-transfer area, should not exceed 37,900 W/m^2 (12,000 Btu/ft^{2}h). For horizontal thermosyphon reboilers, Collins recommends a maximum flux ranging from 47,300 W/m^2 for 20-mm tubes to 56,800 W/m^2 for 25-mm tubes (15,000 to 18,000 Btu/ft^{2}h). These 'rule of thumb' values are now thought to be too conservative; see Shellene *et al.* (1968) and Furzer (1990). Correlations for determining the maximum heat flux for vertical thermosyphons are given by Lee *et al.* (1956) and Palen *et al.* (1974); and for horizontal thermosyphons by Yilmaz (1987).

General Design Considerations

The tube lengths used for vertical thermosyphon reboilers vary from 1.83 m (6 ft) for vacuum service to 3.66 m (12 ft) for pressure operation. A good size for general applications is 2.44 m (8 ft) by 25 mm internal diameter. Larger tube diameters, up to 50 mm, are used for fouling systems.

The top tube sheet is normally aligned with the liquid level in the base of the column (Figure 12.54). The outlet pipe should be as short as possible, and have a cross-sectional area at least equal to the total cross-sectional area of the tubes.

Example 12.10

Make a preliminary design for a vertical thermosyphon for a column distilling crude aniline. The column will operate at atmospheric pressure and a vaporization rate of 6000 kg/h is required. Steam is available at 22 bar (300 psig). Take the column bottom pressure as 1.2 bar.

Solution

Physical properties, taken as those of aniline:

Boiling point at 1.2 bar = 190°C
Molecular weight = 93.13
T_c = 699 K
Latent heat = 42,000 kJ/kmol
Steam saturation temperature = 217°C.

Mean overall $\Delta T = (217 - 190) = 27°C$.

$$\text{Reduced temperature, } T_r = \frac{(190 + 273)}{699} = 0.66$$

From Figure 12.55, design heat flux = 25,000 W/m^2

$$\text{Heat load} = \frac{6000}{3600} \times \frac{42,000}{93.13} = 751 \text{ kW}$$

$$\text{Area required} = \frac{751 \times 10^3}{25,000} = 30 \text{ m}^2$$

Use 25-mm i.d., 30-mm o.d., 2.44-m-long tubes.

$$\text{Area of one tube} = \pi \times 25 \times 10^{-3} \times 2.44 = 0.192 \text{ m}^2$$

$$\text{Number of tubes} = \frac{30}{0.192} = 157$$

Approximate diameter of bundle, for 1.25 square pitch

$$D_b = 30 \left[\frac{157}{0.215}\right]^{1/2.207} = 595 \text{ mm} \tag{12.3b}$$

A fixed tube sheet will be used for a vertical thermosyphon reboiler. From Figure 12.10, shell diametrical clearance = 14 mm

$$\text{Shell inside dia.} = 595 + 14 = 609 \text{ mm}$$

Outlet pipe diameter; take area as equal to total tube cross-sectional area

$$= 157(25 \times 10^{-3})^2 \frac{\pi}{4} = 0.077 \text{ m}^2$$

$$\text{Pipe diameter} = \sqrt{\frac{0.077 \times 4}{\pi}} = 0.31 \text{ m}$$

Example 12.11

Make a preliminary design for a vertical thermosyphon reboiler for a debutanizer column that has the bottoms composition given below. Take the vapour rate required to be 36 kmol/h.

Bottoms composition: C_3 0.001, iC_4 0.001, nC_4 0.02, iC_5 0.34, nC_5 0.64, kmol. Operating pressure, 8.3 bar. Bubble point of mixture, approximately, 120°C.

Solution

The concentrations of C_3 and iC_4 are small enough to be neglected.
Take the liquid: vapour ratio as 3:1.

Estimate the liquid and vapour compositions leaving the reboiler:
Vapour rate, $V = 36/3600 = 0.1$ kmol/s
$L/V = 3$, so liquid rate, $L = 3\ V = 0.3$ kmol/s and feed, $F = L + V = 0.4$ kmol/s.

The vapour and liquid compositions leaving the reboiler can be estimated using the same procedure as that for a flash calculation; see Section 11.3.3.

	K_i	$A_i = K_i \times L/V$	$V_i = z_i/(1 + A_I)$	$y_i = Vi/V$	$x_i = (Fz_i - V_i)/L$
nC_4	2.03	6.09	0.001	0.010	0.023
iC_5	1.06	3.18	0.033	0.324	0.343
nC_5	0.92	2.76	0.068	0.667	0.627
Totals			0.102	1.001	0.993

(near enough correct)

Enthalpies of vaporization, kJ/mol (taken from Maxwell, 1962):

	x_i	$\Delta H_{vap,i}$	$x_i \, \Delta H_{vap,i}$
nC_4	0.02	16	0.32
iC_5	0.35	17	5.95
nC_5	0.63	19	11.97
Total			18.24

Exchanger duty, feed to reboiler taken as at its boiling point

$$= \text{ vapour flow rate } \times \text{ heat of vaporization}$$
$$= 0.1 \times 10^3 \times 18.24 = \underline{1824 \, \text{kW}}$$

Note: we could also have estimated this using a non-adiabatic flash model in a process
simulation program.

Take the maximum flux as 37,900 W/m²; see Section 12.11.5.
Heat transfer area required $= 1,824,000/37,900 = 48.1 \, \text{m}^2$
Use 25-mm i.d., 2.5-m-long tubes, a popular size for vertical thermosyphon
reboilers.

$$\text{Area of one tube } = 25 \times 10^{-3} \pi \times 2.5 = 0.196 \, \text{m}^2$$
$$\text{Number of tubes required } = 48.1/0.196 = 246$$
$$\text{Liquid density at base of exchanger } = 520 \, \text{kg/m}^3$$

$$\text{Relative molecular mass at tube entry } = 58 \times 0.02 + 72(0.34 + 0.64) = 71.7$$
$$\text{vapour at exit } = 58 \times 0.02 + 72(0.35 + 0.63) = 71.7$$

Two-phase fluid density at tube exit:

$$\text{volume of vapour } = 0.1 \times (22.4/8.3) \times (393/273) = 0.389 \, \text{m}^3$$
$$\text{volume of liquid } = (0.3 \times 71.7)/520 = 0.0413 \, \text{m}^3$$
$$\text{total volume } = 0.389 + 0.0413 = 0.430 \, \text{m}^3$$
$$\text{exit density } = \frac{(0.4 \times 71.7)}{0.430} \times 71.7 = 66.7 \, \text{kg/m}^3$$

Friction Loss

Mass flow rate $= 0.4 \times 71.7 = 28.68 \, \text{kg/s}$

Cross-sectional area of tube $= \dfrac{\pi(25 \times 10^{-3})^2}{4} = 0.00049 \, \text{m}^2$

Total cross-sectional area of bundle $= 246 \times 0.00049 = 0.121 \, \text{m}^2$

Mass flux, $G = \text{mass flow/area} = 28.68/0.121 = 237.0 \, \text{kg m}^{-2} \text{s}^{-1}$
At tube exit, pressure drop per unit length, using the homogeneous model:

homogeneous velocity $= G/\rho_m = 237/66.7 = 3.55 \, \text{m/s}$

Viscosity, taken as that of liquid $= 0.12\,\text{mNsm}^{-2}$

$$\text{Re} = \frac{\rho_m u d}{\mu} = \frac{66.7 \times 3.55 \times 25 \times 10^{-3}}{0.12 \times 10^{-3}} = 49,330$$

Friction factor, from Fig. 12.24 $= 3.2 \times 10^{-3}$

$$\Delta P_f = 8 \times 3.2 \times 10^{-3} \times \frac{1}{25 \times 10^{-3}} \times 66.7 \times \frac{3.55^2}{2} = 430\,\text{N/m}^{-2}\text{per m}$$

$$(12.19)$$

At tube entry, liquid only, pressure drop per unit length:

velocity $= G/\rho_L = 237.0/520 = 0.46$ m/s

$$\text{Re} = \frac{\rho_L\, u\, d}{\mu} = \frac{520 \times 0.46 \times 25 \times 10^{-3}}{0.12 \times 10^{-3}} = 49,833$$

Friction factor, from Fig 12.24 $= 3.2 \times 10^{-3}$

$$\Delta P_f = 8 \times 3.2 \times 10^{-3} \times \frac{1}{25 \times 10^{-3}} \times 520 \times \frac{0.46^2}{2} = 56\,\text{N/m}^{-2}\text{per m}$$

$$(12.19)$$

Taking the pressure drop change as linear along the tube,
Mean pressure drop per unit length $= (430 + 56)/2 = 243\,\text{N/m}^2$
Pressure drop over tube $243 \times 2.5 = 608\,\text{N/m}^2$
The viscosity correction factor is neglected in this rough calculation.

Static Pressure in Tubes

Making the simplifying assumption that the variation in density in the tubes is linear from bottom to top, the static pressure will be given by:

$$\Delta P_s = g \int_0^L \frac{dx}{v_i + x(v_0 - v_i)/L} = \frac{gL}{(v_0 - v_i)} \times \ln(v_0/v_i)$$

where v_i and v_0 are the inlet and outlet specific volumes.

$v_i = 1/520 = 0.00192$ and $v_0 = 1/66.7 = 0.0150\,\text{m}^3/\text{kg}$

$$\Delta P_s = \frac{9.8 \times 2.5}{(0.0150 - 0.00192)} \times \ln(0.0150/0.00192) = 3850\,\text{N/m}^2$$

Total pressure drop over tubes $= 608 + 3850 = 4460\,\text{N/m}^2$

Available Head (Driving Force)

$$\Delta P_s = \rho_{Lg}L = 520 \times 9.8 \times 2.5 = \underline{12,740}\,\text{N/m}^2$$

which is adequate to maintain a circulation ratio of 3:1, including allowances for the pressure drop across the piping.

Heat Transfer

The convective boiling coefficient will be calculated using Chen's method; see Section 12.13.3.

As the heat flux is known and only a rough estimate of the coefficient is required, use Mostinski's equation to estimate the nucleate boiling coefficient; Section 12.11.2.

Take the critical pressure as that for n-pentane, 33.7 bar.

$$h_{nb} = 0.104(33.7)^{0.69}(37,900)^{0.7}[1.8(8.3/33.7)^{0.17}$$
$$+ 4(8.3/33.7)^{1.2} + 10(8.3/33.7)^{10}] \qquad (12.42)$$
$$= 1888.6(1.418 + 0.744 + 0.000) = 4083 \text{ Wm}^{-2}\,^{\circ}\text{C}^{-1}$$

Vapour quality, x = mass vapour/total mass flow = 0.1/0.4 = 0.25

Viscosity of vapour = 0.0084 mNm^{-2}s

Vapour density at tube exit = $(0.1 \times 71.7)/0.389 = 18.43$ kg/m^3

$$1/X_{tt} = [0.25/(1 - 0.25)]^{0.9}[520/18.43]^{0.5}[0.0084/0.12]^{0.1} = 1.51 \quad (12.46)$$

Specific heat of liquid = 2.78 kJkg$^{-1}\,^{\circ}$C^{-1}, thermal conductivity of liquid = 0.12 W/m°C.

$$\text{Pr}_L = (2.78 \times 10^3 \times 0.12 \times 10^{-3})/0.12 = 2.78$$

Mass flux, liquid phase only flowing in tubes = $(0.3 \times 71.7)/0.121$ = 177.8 kg m^{-2}s^{-1}

Velocity = 177.8/520 = 0.34 m/s

$$\text{Re}_L = \frac{520 \times 0.34 \times 25 \times 10^{-3}}{0.12 \times 10^{-3}} = 36,833$$

From Figure 12.23 $j_h = 3.3 \times 10^{-3}$

$$\text{Nu} = 3.3 \times 10^{-3} \times 36,833 \times 2.78^{0.33} = 170.3 \qquad (12.15)$$

$h_i = 170.3 \times (0.12/25 \times 10^{-3}) = 817$ W/m^2°C
again, neglecting the viscosity correction factor.
From Figure 12.52, the convective boiling factor, $f_c = 3.6$

$$\text{Re}_L \times f_c^{1.25} = 36,883 \times 3.6^{1.25} = 182,896 \,(1.8 \times 10^{-5})$$

From Figure 12.53 the nucleate boiling suppression factor, $f_s = 0.23$, so

$$h_{cb} = 3.6 \times 817 + 0.23 \times 4083 = 3880 \text{ W/m}^2\,^{\circ}\text{C}$$

This value has been calculated at the outlet conditions.

Assuming that the coefficient changes linearly for the inlet to outlet, then the average coefficient will be given by:

[inlet coefficient (all liquid) + outlet coefficient (liquid + vapour)]/2

Re$_L$ at inlet = 36,833 $\times$ 0.4/0.3 = 49,111 (4.9 $\times$ 10^{-4})

From Figure 12.23, $j_h = 3.2 \times 10^{-3}$

$\text{Nu} = 3.2 \times 10^{-3} \times 49{,}111 \times 2.78^{0.33} = 220.2$

$h_i = 220.2 \times (0.12/25 \times 10^{-3}) = 1057 \text{ W/m}^2{}^\circ\text{C}$

$$\text{Mean coefficient} = (1057 + 3880)/2 = 2467 \text{ Wm}^{-2}{}^\circ\text{C}^{-1}$$

The overall coefficient, U, neglecting the resistance of the tube wall, and taking the steam coefficient as 8000 W/m^2°C, is given by:

$$1/U = 1/8000 + 1/2467 = 5.30 \times 10^{-4}$$
$$U = \underline{1886} \ \text{Wm}^{-2}{}^\circ\text{C}^{-1}$$

The overall coefficient required for the design $=$ duty/ΔT_{LM}.
$\Delta T_{LM} = 158.8 - 120 = 38.8$°C taking both streams as isothermal.

So, U required $= 37{,}900/38.3 = 990 \ \text{Wm}^{-2}{}^\circ\text{C}^{-1}$

So, the area available in the proposed design is more than adequate and will take care of any fouling.

The analysis could be improved by dividing the tube length into sections, calculating the heat-transfer coefficient and pressure drop over each section, and totalling.

More accurate, but more complex, methods could be used to predict the two-phase pressure drop and heat-transfer coefficients.

The pressure drop over the inlet and outlet pipes could also be estimated, taking into account the bends, and expansions and contractions.

An allowance could also be included for the energy (pressure drop) required to accelerate the liquid–vapour mixture as the liquid is vaporized. This can be taken as two velocity heads, based on the mean density.

12.11.6. Design of Kettle Reboilers

Kettle reboilers, and other submerged bundle equipment, are essentially pool boiling devices, and their design is based on data for nucleate boiling.

In a tube bundle, the vapour rising from the lower rows of tubes passes over the upper rows. This has two opposing effects: there will be a tendency for the rising vapour to blanket the upper tubes, particularly if the tube spacing is close, which will reduce the heat-transfer rate; but this is offset by the increased turbulence caused by the rising vapour bubbles. Palen and Small (1964) give a detailed procedure for kettle reboiler design in which the heat-transfer coefficient calculated using equations for boiling on a single tube is reduced by an empirically derived tube bundle factor, to account for the effects of vapour blanketing. Later work by Heat Transfer Research Inc., reported by Palen et al. (1972), showed that the coefficient for bundles was usually greater than that estimated for a single tube. On balance, it seems reasonable to use the correlations for single tubes to estimate the coefficient for tube bundles without applying any correction (equations 12.41 or 12.42).

The maximum heat flux for stable nucleate boiling will, however, be less for a tube bundle than for a single tube. Palen and Small (1964) suggest modifying the Zuber equation for single tubes (equation 12.43) with a tube density factor. This approach was supported by Palen et al. (1972).

The modified Zuber equation can be written as:

$$q_{cb} = K_b \left(\frac{p_t}{d_o}\right) \left(\frac{\lambda}{\sqrt{N_t}}\right) \left[\sigma g (\rho_L - \rho_v) \rho_v^2\right]^{0.25} \tag{12.52}$$

where

q_{cb} = maximum (critical) heat flux for the tube bundle, W/m^2
K_b = 0.44 for square pitch arrangements
   = 0.41 for equilateral triangular pitch arrangements
p_t = tube pitch
d_o = tube outside diameter
N_t = total number of tubes in the bundle.

Note: For U-tubes, N_t will be equal to twice the number of actual U-tubes.

Palen and Small suggest that a factor of safety of 0.7 be applied to the maximum flux estimated from equation 12.52. This will still give values that are well above those that have traditionally been used for the design of commercial kettle reboilers, such as that of 37,900 W/m^2 (12,000 Btu/ft^2h) recommended by Kern (1950). This has had important implications in the application of submerged-bundle reboilers, as the high heat flux allows a smaller bundle to be used, which can then often be installed in the base of the column, saving the cost of shell and piping.

General Design Considerations

A typical layout is shown in Figure 12.8. The tube arrangement, triangular or square pitch, will not have a significant effect on the heat-transfer coefficient. A tube pitch of between 1.5 and 2.0 times the tube outside diameter should be used to avoid vapour blanketing. Long, thin bundles will be more efficient than short, fat bundles.

The shell should be sized to give adequate space for the disengagement of the vapour and liquid. The shell diameter required will depend on the heat flux. The following values can be used as a guide:

Heat flux (W/m^2)	Shell dia./Bundle dia.
25,000	1.2 to 1.5
25,000 to 40,000	1.4 to 1.8
40,000	1.7 to 2.0

The freeboard between the liquid level and shell should be at least 0.25 m. To avoid excessive entrainment, the maximum vapour velocity $\hat{u}_v$ (m/s) at the liquid surface should be less than that given by the expression:

$$\hat{u}_v < 0.2 \left[\frac{\rho_L - \rho_v}{\rho_v}\right]^{1/2} \tag{12.53}$$

When only a low rate of vaporization is required, a vertical cylindrical vessel with a heating jacket or coils should be considered. The boiling coefficients for internal submerged coils can be estimated using the equations for nucleate pool boiling.

Mean Temperature Differences

When the fluid being vaporized is a single component and the heating medium is steam (or another condensing vapour), both shell-side and tube-side processes are isothermal, and the mean temperature difference will be simply the difference between the saturation temperatures. If one side is not isothermal, the logarithmic mean temperature difference should be used. If the temperature varies on both sides, the logarithmic temperature difference must be corrected for departures from true cross- or counter-current flow (see Section 12.6).

If the feed is sub-cooled, the mean temperature difference should still be based on the boiling point of the liquid, as the feed will rapidly mix with the boiling pool of liquid; the quantity of heat required to bring the feed to its boiling point must be included in the total duty.

Mixtures

The equations for estimating nucleate boiling coefficients given in Section 12.11.1 can be used for close boiling mixtures, say less than 5°C, but will overestimate the coefficient if used for mixtures with a wide boiling range. Palen and Small (1964) give an empirical correction factor for mixtures that can be used to estimate the heat-transfer coefficient in the absence of experimental data:

$$(h_{nb}) \text{ mixture} = f_m(h_{nb}) \text{ single component} \qquad (12.54)$$

where

$$f_m = \exp[-0.0083(T_{bo} - T_{bi})]$$

and T_{bo} = temperature of the vapour mixture leaving the reboiler °C
T_{bi} = temperature of the liquid entering the reboiler °C.

The inlet temperature is the saturation temperature of the liquid at the base of the column and the vapour temperature is the saturation temperature of the vapour returned to the column. The composition of these streams will be fixed by the distillation column design specification.

Example 12.12

Design a vaporizer to vaporize 5000 kg/h n-butane at 5.84 bar. The minimum temperature of the feed (winter conditions) will be 0°C. Steam is available at 1.70 bar (10 psig).

Solution

Only the thermal design and general layout will be done. Select kettle type.
Physical properties of n-butane at 5.84 bar:

Boiling point = 56.1°C
Latent heat = 326 kJ/kg
Mean specific heat, liquid = 2.51 kJ/kg°C
Critical pressure, P_c = 38 bar.

Heat loads:

$$\text{Sensible heat (maximum)} = (56.1 - 0)2.51 = 140.8 \text{ kJ/kg}$$

$$\text{Total heat load} = (140.8 + 326) \times \frac{5000}{3600} = 648.3 \text{ kW}$$

add 5% for heat losses

$$\text{Maximum heat load (duty)} = 1.05 \times 648.3$$
$$= 681 \text{ kW}$$

From Figure 12.1, assume $U = 1000 \text{ W/m}^{2\circ}\text{C}$.

Mean temperature difference; both sides isothermal, steam saturation temperature at 1.7 bar $= 115.2°\text{C}$

$$\Delta T_m = 115.2 - 56.1 = 59.1°\text{C}$$

$$\text{Area (outside) required} = \frac{681 \times 10^3}{1000 \times 59.1} = 11.5 \text{ m}^2$$

Select 25-mm i.d., 30-mm o.d. plain U-tubes,
  Nominal length 4.8 m (one U tube)

$$\text{Number of U tubes} = \frac{11.5}{(30 \times 10^{-3})\pi 4.8} = 25$$

Use square pitch arrangement pitch $= 1.5 \times$ tube o.d
$$= 1.5 \times 30 = 45 \text{ mm}$$

Draw a tube layout diagram, take minimum bend radius
$$1.5 \times \text{tube o.d.} = 45 \text{ mm}$$

Proposed layout gives 26 U-tubes, tube outer limit diameter 420 mm.
Boiling coefficient
  Use Mostinski's equation:
  Heat flux, based on estimated area,

$$q = \frac{681}{11.5} = 59.2 \text{ kW/m}^2$$

$$h_{nb} = 0.104(38)^{0.69}(59.2 \times 10^3)^{0.7}\left[1.8\left(\frac{5.84}{38}\right)^{0.17} + 4\left(\frac{5.84}{38}\right)^{1.2} + 10\left(\frac{5.84}{38}\right)^{10}\right]$$

$$= 4855 \text{ W/m}^{2\circ}\text{C} \tag{12.42}$$

Take steam condensing coefficient as $8000 \text{ W/m}^{2\circ}\text{C}$, fouling coefficient 5000 $\text{W/m}^{2\circ}\text{C}$; butane fouling coefficient, essentially clean, $10,000 \text{ W/m}^{2\circ}\text{C}$.

Tube material will be plain carbon steel, $k_w = 55 \text{ W/m}°\text{C}$

$$\frac{1}{U_o} = \frac{1}{4855} + \frac{1}{10,000} + \frac{30 \times 10^{-3}\ln\frac{30}{25}}{2 \times 55} + \frac{30}{25}\left(\frac{1}{5000} + \frac{1}{8000}\right) \tag{12.2}$$

$$U_o = \underline{1341} \text{ W/m}^{2\circ}\text{C}$$

Close enough to original estimate of $1000 \text{ W/m}^{2\circ}\text{C}$ for the design to stand.

Myers and Katz (1953) give some data on the boiling of n-butane on banks of tubes. To compare the value estimated with their values an estimate of the boiling film temperature difference is required:

$$= \frac{1341}{4855} \times 59.1 = 16.3°\text{C} \ (29°\text{F})$$

Myers data, extrapolated, gives a coefficient of around 3000 Btu/h ft^2°F at a 29°F temperature difference $= 17,100$ W/m^2°C, so the estimated value of 4855 is certainly on the safe side.

Check maximum allowable heat flux. Use modified Zuber equation.

Surface tension (estimated) $= 9.7 \times 10^{-3}$ N/m

$\rho_L = 550$ kg/m^3

$$p_v = \frac{58}{22.4} \times \frac{273}{(273 + 56)} \times 5.84 = 12.6 \text{ kg/m}^3$$

$N_t = 52$

For square arrangement $K_b = 0.44$

$$q_c = 0.44 \times 1.5 \times \frac{326 \times 10^3}{\sqrt{52}} [9.7 \times 10^{-3} \times 9.81(550 - 12.6)12.6^2]^{0.25}$$

$$= 283,224 \text{ W/m}^2$$

$$= 280 \text{ kW/m}^2$$

$\hspace{11cm}$ (12.52)

Applying a factor of 0.7, maximum flux should not exceed $280 \times 0.7 = 196$ kW/m^2. Actual flux of 59.2 kW/m^2 is well below maximum allowable.

Layout

From tube-sheet layout $D_b = 420$ mm

Take shell diameter as twice bundle diameter

$$D_s = 2 \times 420 = 840 \text{ mm}$$

Take liquid level as 500 mm from base,

$$\text{freeboard} = 840 - 500 = 340 \text{ mm, satisfactory.}$$

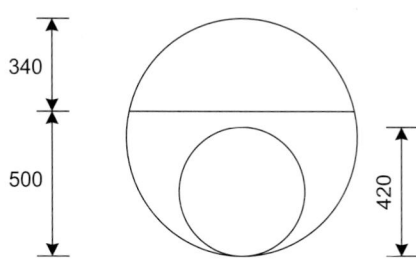

From sketch, width at liquid level $= 0.8$ m.

$$\text{Surface area of liquid} = 0.8 \times 2.4 = 1.9 \text{ m}^2.$$

$$\text{Vapour velocity at surface} = \frac{5000}{3600} \times \frac{1}{12.6} \times \frac{1}{1.9} = \underline{0.06 \text{ m/s}}$$

Maximum allowable velocity

$$\hat{u}_v = 0.2 \left[\frac{550 - 12.6}{12.6} \right]^{1/2} = \underline{1.3 \text{ m/s}}$$

$\hspace{11cm}$ (12.53)

So actual velocity is well below maximum allowable velocity. A smaller shell diameter could be considered.

12.12.   PLATE HEAT EXCHANGERS

12.12.1.   Gasketed-Plate Heat Exchangers

A gasketed-plate heat exchanger consists of a stack of closely spaced thin plates clamped together in a frame. A thin gasket seals the plates around their edges. The plates are normally between 0.5 and 3 mm thick, and the gap between them 1.5 to 5 mm. Plate surface areas range from 0.03 to 1.5 m^2, with a plate width:length ratio from 2.0 to 3.0. The size of plate heat exchangers can vary from very small, 0.03 m^2, to very large, 1500 m^2. The maximum flow rate of fluid is limited to around 2500 m^3/h.

The basic layout and flow arrangement for a gasketed-plate heat exchanger is shown in Figure 12.56. Corner ports in the plates direct the flow from plate to plate. The plates are embossed with a pattern of ridges, which increase the rigidity of the plate and improve the heat-transfer performance.

Plates are available in a wide range of metals and alloys; including stainless steel, aluminium and titanium. A variety of gasket materials can be used; see Table 12.7.

Selection

The advantages and disadvantages of gasketed-plate heat exchangers, compared with conventional shell and tube exchangers are listed below.

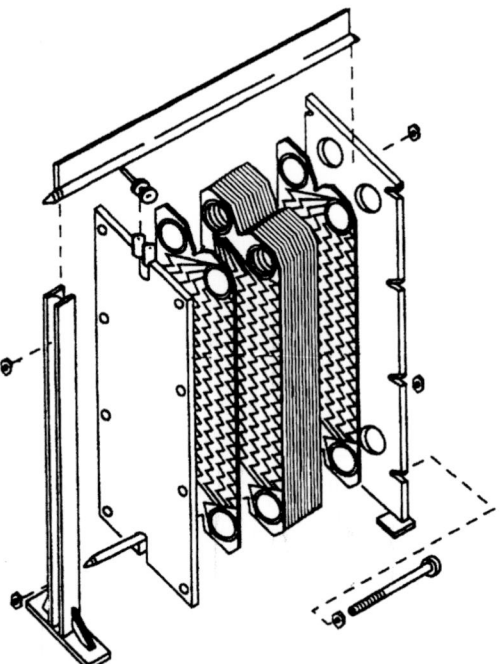

Figure 12.56.   Gasketed-plate heat exchanger.

TABLE 12.7. Typical Gasket Materials for Plated Heat Exchangers

Material	Approximate Temperature Limit (°C)	Fluids
Styrene-butane rubber	85	Aqueous systems
Acrylonitrile-butane rubber	140	Aqueous system, fats, aliphatic hydrocarbons
Ethylene-propylene rubber	150	Wide range of chemicals
Fluorocarbon rubber	175	Oils
Compressed asbestos	250	General resistance to organic chemicals

Advantages
1. Plates are attractive when material costs are high.
2. Plate heat exchangers are easier to maintain.
3. Low approach temps can be used, as low as 1°C, compared with 5 to 10°C for shell and tube exchangers.
4. Plate heat exchangers are more flexible; it is easy to add extra plates.
5. Plate heat exchangers are more suitable for highly viscous materials.
6. The temperature correction factor, F_t, will normally be higher with plate heat exchangers, as the flow is closer to true counter-current flow.
7. Fouling tends to be significantly less in plate heat exchangers; see Table 12.8.

Disadvantages

1. A plate is not a good shape to resist pressure and plate heat exchangers are not suitable for pressures greater than about 30 bar, or for high differential pressures between the two streams transferring heat.
2. The selection of a suitable gasket is critical; see Table 12.7.
3. The maximum operating temperature is limited to about 250°C, due to the performance of the available gasket materials.

Plate heat exchangers are used extensively in the food and beverage industries, as they can be readily taken apart for cleaning and inspection. Their use in the chemical industry depends on the relative cost for the particular application compared with a conventional shell and tube exchanger; see Parker (1964) and Trom (1990).

TABLE 12.8. Fouling Factors (Coefficients), Typical Values for Plate Heat Exchangers

Fluid	Coefficient (W/m²°C)	Factor (m²°C/W)
Process water	30,000	0.00003
Towns, water (soft)	15,000	0.00007
Towns, water (hard)	6000	0.00017
Cooling water (treated)	8000	0.00012
Sea water	6000	0.00017
Lubricating oil	6000	0.00017
Light organics	10,000	0.0001
Process fluids	5000–20,000	0.0002–0.00005

Plate Heat-Exchanger Design

It is not possible to give exact design methods for plate heat exchangers. They are proprietary designs, and are normally specified in consultation with the manufacturers. Information on the performance of the various patterns of plate used is not generally available. Emerson (1967) gives performance data for some proprietary designs, and Kumar (1984) and Bond (1981) have published design data for APV chevron patterned plates.

The approximate method given below can be used to size an exchanger for comparison with a shell and tube exchanger, and to check performance of an existing exchanger for new duties. More detailed design methods are given by Hewitt *et al.* (1994) and Cooper and Usher (1983).

Procedure
The design procedure is similar to that for shell and tube exchangers.

1. Calculate duty, the rate of heat transfer required.
2. If the specification is incomplete, determine the unknown fluid temperature or fluid flow rate from a heat balance.
3. Calculate the log mean temperature difference, ΔT_{lm}.
4. Determine the log mean temperature correction factor, F_t; see method given below.
5. Calculate the corrected mean temperature difference $\Delta T_m = F_t \times \Delta T_{lm}$.
6. Estimate the overall heat-transfer coefficient; see Table 12.1.
7. Calculate the surface area required; equation 12.1.
8. Determine the number of plates required = total surface area/area of one plate.
9. Decide the flow arrangement and number of passes.
10. Calculate the film heat-transfer coefficients for each stream; see method given below.
11. Calculate the overall coefficient, allowing for fouling factors.
12. Compare the calculated with the assumed overall coefficient. If satisfactory, say -0% to $+10\%$ error, proceed. If unsatisfactory, return to step 8 and increase or decrease the number of plates.
13. Check the pressure drop for each stream; see method given below.

This design procedure is illustrated in Example 12.13.

Flow Arrangements

The stream flows can be arranged in series or parallel, or a combination of series and parallel; see Figure 12.57. Each stream can be sub-divided into a number of passes; analogous to the passes used in shell and tube exchangers.

Estimation Of the Temperature Correction Factor

For plate heat exchangers, it is convenient to express the log mean temperature difference correction factor, F_t, as a function of the number of transfer units, *NTU*, and the flow arrangement (number of passes); see Figure 12.58. The correction will normally be higher for a plate heat exchanger than for a shell and tube exchanger operating with the same temperatures. For rough sizing purposes, the factor can be taken as 0.95 for series flow.

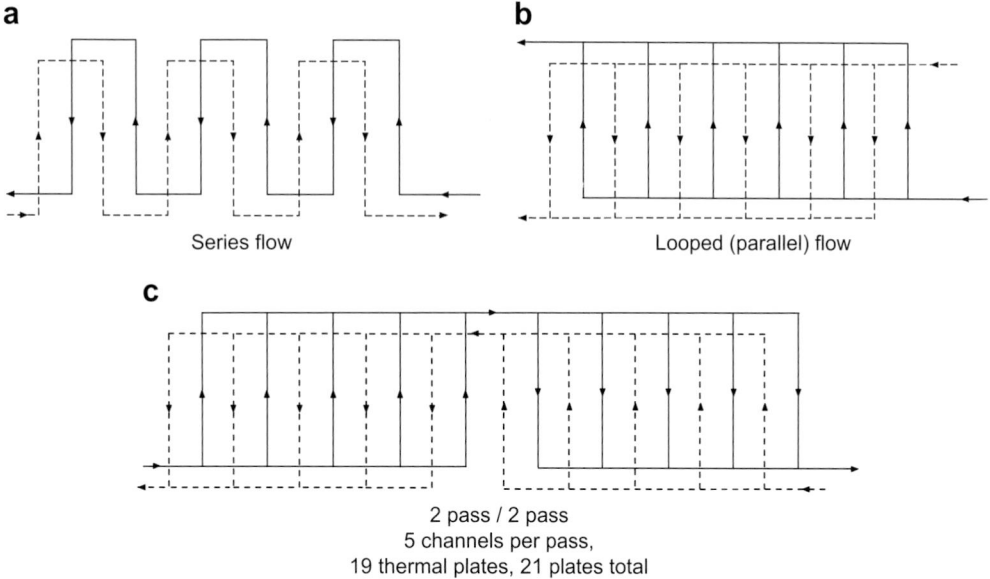

Figure 12.57. Plate heat-exchanger flow arrangements.

The number of transfer units is given by:

$$NTU = (t_o - t_i)/\Delta T_{lm} \qquad (12.55)$$

where

t_i = stream inlet temperature, °C

t_o = stream outlet temperature, °C

ΔT_{lm} = log mean temperature difference, °C.

Typically, the *NTU* will range from 0.5 to 4.0, and for most applications will lie between 2.0 and 3.0.

Heat-Transfer Coefficient

The equation for forced-convective heat transfer in conduits can be used for plate heat exchangers; equation 12.10.

The values for the constant C and the indices *a*, *b* and *c* will depend on the particular type of plate being used. Typical values for turbulent flow are given in the equation below, which can be used to make a preliminary estimate of the area required.

$$\frac{h_p \, d_e}{k_f} = 0.26 \, \mathrm{Re}^{0.65} \, \mathrm{Pr}^{0.4} \left(\frac{\mu}{\mu_w}\right)^{0.14} \qquad (12.56)$$

where

h_p = plate film coefficient

Re = Reynolds number = $\dfrac{G_p d_e}{\mu} = \dfrac{\rho u_p d_e}{\mu}$

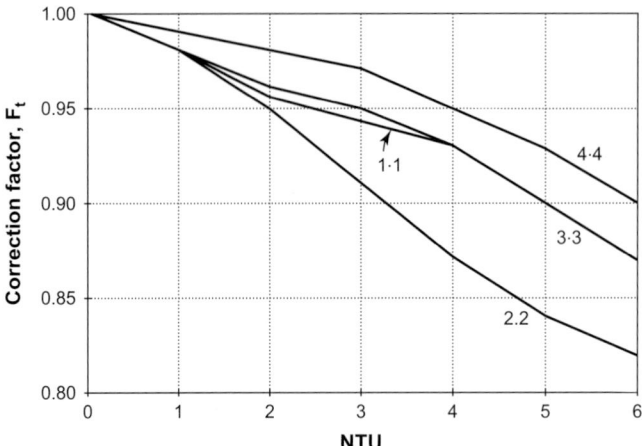

Figure 12.58.   Log mean temperature correction factor for plate heat exchangers (adapted from Raju and Chand, 1980).

G_P = mass flow rate per unit cross-sectional area = w/A_f, $\mathrm{kgm^{-2}s^{-1}}$
w = mass flow rate per channel, kg/s
A_f = cross-sectional area for flow, $\mathrm{m^2}$
u_p = channel velocity, m/s
d_e = equivalent (hydraulic) diameter, taken as twice the gap between the plates, m.

The corrugations on the plates will increase the projected plate area, and reduce the effective gap between the plates. For rough sizing, where the actual plate design is not known, this increase can be neglected. The channel width equals the plate pitch minus the plate thickness.

There is no heat transfer across the end plates, so the number of effective plates will be the total number of plates less two.

Pressure Drop

The plate pressure drop can be estimated using a form of the equation for flow in a conduit; equation 12.18.

$$\Delta P_p = 8j_f(L_p/d_e)\frac{\rho u_p^2}{2} \tag{12.57}$$

where L_p = the path length and $u_p = G_p/\rho$.

The value of the friction factor, j_f, depends on the design of plate used. For preliminary calculations the following relationship can be used for turbulent flow:
$j_f = 0.6\,\mathrm{Re}^{-0.3}$

The transition from laminar to turbulent flow will normally occur at a Reynolds number of 100 to 400, depending on the plate design. With some designs, turbulence can be achieved at very low Reynolds numbers, which makes plate heat exchangers very suitable for use with viscous fluids.

The pressure drop due the contraction and expansion losses through the ports in the plates must be added to the friction loss. Kumar (1984) suggests adding 1.3 velocity heads per pass, based on the velocity through the ports.

$$\Delta P_{pt} = 1.3 \frac{(\rho u_{pt}^2)}{2} N_p \qquad (12.58)$$

where

u_{pt} = the velocity through the ports $w/\rho A_p$, m/s
w = mass flow through the ports, kg/s
A_p = area of the port = $(\pi d_{pt}^2)/4$ m^2
d_{pt} = port diameter, m
N_p = number of passes.

Example 12.13

Investigate the use of a gasketed-plate heat exchanger for the duty set out in Example 12.1: cooling methanol using brackish water as the coolant. Titanium plates are to be specified, to resist corrosion by the saline water.

Summary of Example 12.1

Cool 100,000 kg/h of methanol from 95°C to 40°C, duty 4340 kW. Cooling water inlet temperature 25°C and outlet temperature 40°C. Flow rates: methanol 27.8 kg/s, water 68.9 kg/s.

Physical properties	Methanol	Water
Density, kg/m^3	750	995
Viscosity, mN m^{-2}s	3.4	0.8
Prandtl number	5.1	5.7

Logarithmic mean temperature difference 31°C.

Solution

NTU, based on the maximum temperature difference

$$= \frac{95 - 40}{31} = 1.8$$

Try a 1:1 pass arrangement.
From Figure 12.58, $F_t = 0.96$
From Table 12.2 take the overall coefficient, light organic − water, to be 2000 W/m^2°C.

$$\text{Then, area required} = \frac{4340 \times 10^3}{2000 \times 0.96 \times 31} = 72.92 \text{ m}^2$$

Select an effective plate area of 0.75 m^2, effective length 1.5 m and width 0.5 m; these are typical plate dimensions. The actual plate size will be larger to accommodate the gasket area and ports.

Number of plates = total heat-transfer area / effective area of one plate
$$= 72.92/0.75 = 97$$

No need to adjust this; 97 will give an even number of channels per pass, allowing for an end plate.

Number of channels per pass = $(97 - 1)/2 = 48$

Take plate spacing as 3 mm, a typical value, then:

Channel cross sectional area $= 3 \times 10^{-3} \times 0.5 = 0.0015 \ \text{m}^2$

and hydraulic mean diameter $= 2 \times 3 \times 10^{-3} = 6 \times 10^{-3} \ \text{m}$

Methanol

$$\text{Channel velocity} = \frac{27.8}{750} \times \frac{1}{0.0015} \times \frac{1}{48} = 0.51 \ \text{m/s}$$

$$\text{Re} = \frac{\rho \, u_p d_e}{\mu} = \frac{750 \times 0.51 \times 6 \times 10^{-3}}{0.34 \times 10^{-3}} = 6750$$

$$\text{Nu} = 0.26 \, (6750)^{0.65} \times 5.1^{0.4} = 153.8 \qquad (12.56)$$

$$h_p = 153.8 \left(\frac{0.19}{6 \times 10^{-3}} \right) = 4870 \ \text{Wm}^{-2\circ}\text{C}$$

Brackish Water

$$\text{Channel velocity} = \frac{68.9}{995} \times \frac{1}{0.0015} \times \frac{1}{48} = 0.96 \ \text{m/s}$$

$$\text{Re} = \frac{995 \times 0.96 \times 6 \times 10^{-3}}{0.8 \times 10^{-3}} = 7164$$

$$\text{Nu} = 0.26 \, (7164)^{0.65} \times 5.7^{0.4} = 167.2 \qquad (12.56)$$

$$h_p = 167.2 \left(\frac{0.59}{6 \times 10^{-3}} \right) = \underline{\underline{16,439}} \ \text{Wm}^{-2\circ}\text{C}$$

Overall Coefficient

From Table 12.9, take the fouling factors (coefficients) as: brackish water (seawater) 6000 W/m$^{2\circ}$C and methanol (light organic) 10,000 W/m$^{2\circ}$C.

Take the plate thickness as 0.75 mm. Thermal conductivity of titanium, 21 W/m°C.

$$\frac{1}{U} = \frac{1}{4870} + \frac{1}{10,000} + \frac{0.75 \times 10^{-3}}{21} + \frac{1}{16,439} + \frac{1}{6000}$$

$$U = 1759 \ \text{W/m}^2\text{C, too low}$$

Increase the number of channels per pass to 60, giving $(2 \times 60) + 1 = 121$ plates. Then methanol channel velocity $= 0.51 \times (48/60) = 0.41$ m/s, and Re $= 5400$

cooling water channel velocity $= 0.96 \times (48/60) = 0.77$ m/s, and Re $= 5746$ giving $h_p = 4215$ W/m$^{2\circ}$C for methanol, and 14,244 W/m$^{2\circ}$C for water which gives an overall coefficient of 1640 W/m$^{2\circ}$C.

Overall coefficient required $2000 \times 48/60 = 1600 \, \mathrm{Wm^{-2}{}^\circ C^{-1}}$, so 60 plates per pass should be satisfactory.

Pressure Drops

Methanol

$$j_f = 0.60(5400)^{-0.3} = 0.046$$

Path length = plate length × number of passes = $1.5 \times 1 = 1.5 \, \mathrm{m}$

$$\Delta P_p = 8 \times 0.046 \left(\frac{1.5}{6 \times 10^{-3}} \right) \times 750 \times \frac{0.41^2}{2} = 5799 \, \mathrm{N/m^2} \qquad (12.57)$$

Port pressure loss, take port diameter as 100 mm, area = $0.00785 \, \mathrm{m^2}$
Velocity through port = $(27.8/750)/0.00785 = 4.72 \, \mathrm{m/s}$

$$\Delta P_{pt} = 1.3 \times \frac{750 \times 4.72^2}{2} = 10,860 \, \mathrm{N/m^2} \qquad (12.58)$$

Total pressure drop = $5799 + 10,860 = \underline{\underline{16,659}} \, \mathrm{N/m^2}$, 0.16 bar

Water

$$j_f = 0.6(5501)^{-0.3} = 0.045$$

Path length = plate length × number of passes = $1.5 \times 1 = 1.5 \, \mathrm{m}$

$$\Delta P_p = 8 \times 0.045 \times \left(\frac{1.5}{6 \times 10^{-3}} \right) \times 995 \times \frac{0.77^2}{2} = 26,547 \, \mathrm{N/m^2} \qquad (12.57)$$

Velocity through port = $(68.9/995)/0.0078 = 8.88 \, \mathrm{m/s}$ (rather high)

$$\Delta P_{pt} = 1.3 \times \frac{995 \times 8.88}{2} = 50,999 \, \mathrm{N/m^2} \qquad (12.58)$$

Total pressure drop = $26,547 + 50,999 = \underline{\underline{77,546}} \, \mathrm{N/m^2}$, 0.78 bar

Could increase the port diameter to reduce the pressure drop.

The trial design should be satisfactory, so a plate heat exchanger could be considered for this duty.

12.12.2. Welded Plate

Welded-plate heat exchangers use plates similar to those in gasketed-plate exchangers, but the plate edges are sealed by welding. This increases the pressure and temperature rating to up to 80 bar and temperatures in excess of 500°C. They retain the advantages of plate heat exchangers (compact size and good rates of heat transfer) whilst giving security against leakage. An obvious disadvantage is that the exchangers cannot be dismantled for cleaning, so their use is restricted to applications where fouling is not a problem. The plates are fabricated in a variety of materials.

A welded-plate exchanger can be specified for a high-pressure service if it is contained inside a pressure vessel, and the space between the exchanger and the vessel is pressurized to the same pressure as the fluid inside the exchanger. The differential pressure between the two sides of the exchanger must still be small.

A combination of gasketed-plate and welded-plate construction can also be used: an aggressive process fluid flowing between welded plates and a benign process stream, or service stream, between gasketed plates.

12.12.3. Plate Fin

Plate-fin exchangers consist essentially of plates separated by corrugated sheets, which form the fins. They are made up in a block and are often referred to as matrix exchangers; see Figure 12.59. They are usually constructed of aluminium and joined and sealed by brazing. The main application of plate-fin exchangers has been in the cryogenics industries, such as air separation plants, where large heat-transfer surface areas are needed. They are now finding wider applications in the chemical process industries, where large surface area, compact, exchangers are required. Their compact size and low weight have led to some use in off-shore applications. The brazed aluminium construction is limited to pressures up to around 60 bar and temperatures up to 150°C. The units cannot be mechanically cleaned, so their use is restricted to clean process and service steams. The construction and design of plate-fin exchangers and their applications are discussed by Saunders (1988) and Burley (1991), and their use in cryogenic service by Lowe (1987).

12.12.4. Spiral Heat Exchangers

A spiral heat exchanger can be considered as a plate heat exchanger in which the plates are formed into a spiral. The fluids flow through the channels formed between the plates. The exchanger is made up from long sheets, 150 to 1800 mm wide, formed into a pair of concentric spiral channels. The channels are closed by gasketed end-plates bolted to an outer case. Inlet and outlet nozzles are fitted to the case and connect to the channels; see Figure 12.60. The gap between the sheets varies between 4 and 20 mm, depending on the size of the exchanger and the application. They can be fabricated in any material that can be cold-worked and welded.

Spiral heat exchangers are compact units: a unit with around 250 m^2 area occupying a volume of approximately 10 m^3. The maximum operating pressure is limited to 20 bar and the temperature to 400°C.

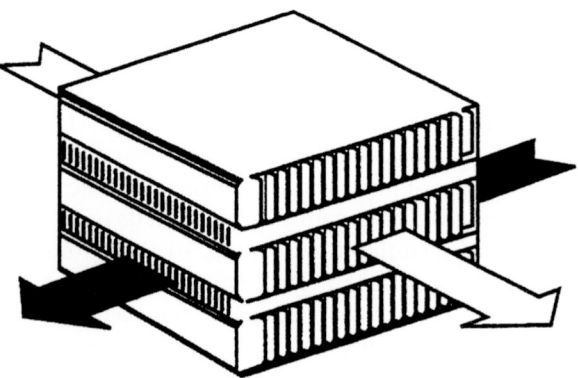

Figure 12.59. Plate-fin exchanger.

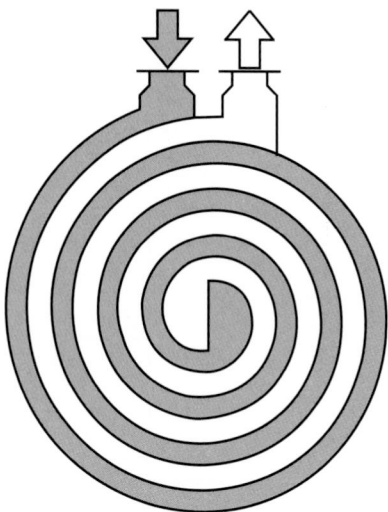

Figure 12.60.   Spiral heat exchanger.

For a given duty, the pressure drop over a spiral heat exchanger will usually be lower than that for the equivalent shell and tube exchanger. Spiral heat exchangers give true counter-current flow and can be used where the temperature correction factor F_t for a shell and tube exchanger would be too low; see Section 12.6. Because they are easily cleaned and the turbulence in the channels is high, spiral heat exchangers can be used for very dirty process fluids and slurries.

The correlations for flow in conduits can be used to estimate the heat-transfer coefficient and pressure drop in the channels, using the hydraulic mean diameter as the characteristic dimension.

The design of spiral heat exchangers is discussed by Minton (1970).

12.13. DIRECT-CONTACT HEAT EXCHANGERS

In direct-contact heat exchange the hot and cold streams are brought into contact without any separating wall, and high rates of heat transfer are achieved.

Applications include: reactor off-gas quenching, vacuum condensers, cooler-condensers, desuperheating and humidification. Water-cooling towers are a particular example of direct-contact heat exchange. In direct-contact cooler-condensers the condensed liquid is frequently used as the coolant (Figure 12.61).

Direct-contact heat exchangers should be considered whenever the process stream and coolant are compatible. The equipment used is simple and cheap, and is suitable for use with heavily fouling fluids and with liquids containing solids; spray chambers, spray columns, and plate and packed columns are used.

There is no general design method for direct-contact exchangers. Most applications involve the transfer of latent heat as well as sensible heat, and the process is one of simultaneous heat and mass transfer. When the approach to thermal equilibrium is rapid, as it will be in many applications, the size of the contacting vessel is not critical and the design can be based on experience with similar processes. For other

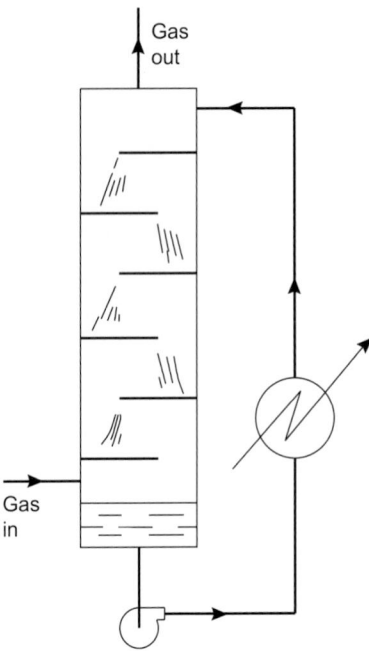

Figure 12.61. Typical direct-contact cooler (baffle plates).

situations, the designer must work from first principles, setting up the differential equations for mass and heat transfer, and using judgment in making the simplifications necessary to achieve a solution. The design procedures used are analogous to those for gas absorption and distillation. The rates of heat transfer will be high, with coefficients for packed columns typically in the range 2000 to 20,000 W/m^3°C (i.e. per cubic metre of packing).

The design and application of direct-contact heat exchangers is discussed by Fair (1961, 1972a, 1972b), and Chen-Chia and Fair (1989); they give practical design methods and data for a range of applications.

The design of water-cooling towers, and humidification, is discussed by Coulson *et al.* (1999). The same principles apply to the design of other direct-contact exchangers.

12.14. FINNED TUBES

Fins are used to increase the effective surface area of heat-exchanger tubing. Many different types of fin have been developed, but the plain transverse fin shown in Figure 12.62 is the most commonly used type for process heat exchangers. Typical fin dimensions are: pitch 2.0 to 4.0 mm, height 12 to 16 mm; ratio of fin area to bare tube area 15:1 to 20:1.

Finned tubes are used when the heat-transfer coefficient on the outside of the tube is appreciably lower than that on the inside, as in heat-transfer from a liquid to a gas, such as in air-cooled heat exchangers.

The fin surface area will not be as effective as the bare tube surface, as the heat has to be conducted along the fin. This is allowed for in design by the use of a fin

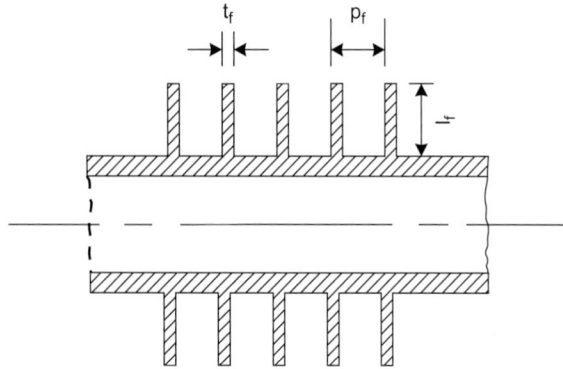

Figure 12.62. Finned tube.

effectiveness, or fin efficiency, factor. The equations describing heat transfer from a fin are derived in Coulson *et al.* (1999); see also Kern (1950). The fin effectiveness is a function of the fin dimensions and the thermal conductivity of the fin material. Fins are therefore usually made from metals with a high thermal conductivity; for copper and aluminium the effectiveness is typically between 0.9 and 0.95.

When using finned tubes, the coefficients for the outside of the tube in equation 12.2 are replaced by a term involving fin area and effectiveness:

$$\frac{1}{h_o} + \frac{1}{h_{od}} = \frac{1}{E_f}\left(\frac{1}{h_f} + \frac{1}{h_{df}}\right)\frac{A_o}{A_f} \tag{12.59}$$

where

h_f = heat-transfer coefficient based on the fin area
h_{df} = fouling coefficient based on the fin area
A_o = outside area of the bare tube
A_f = fin area
E_f = fin effectiveness.

It is not possible to give a general correlation for the coefficient h_f covering all types of fin and fin dimensions. Design data should be obtained from the tube manufacturers for the particular type of tube to be used. For banks of tubes in cross flow, with plain transverse fins, the correlation given by Briggs and Young (1963) can be used to make an approximate estimate of the fin coefficient.

$$\text{Nu} = 0.134\,\text{Re}^{0.681}\,\text{Pr}^{0.33}\left(\frac{p_f - t_f}{l_f}\right)^{0.2}\left(\frac{p_f}{t_f}\right)^{0.1134} \tag{12.60}$$

where

p_f = fin pitch
l_f = fin height
t_f = fin thickness.

The Reynolds number is evaluated for the bare tube (i.e., assuming that no fins exist).

Kern and Kraus (1972) give full details of the use of finned tubes in process heat exchanger design and design methods.

Low Fin Tubes

Tubes with low transverse fins, about 1 mm high, can be used with advantage as replacements for plain tubes in many applications. The fins are formed by rolling, and the tube outside diameters are the same as those for standard plain tubes. Details are given in the manufacturer's data books, Wolverine (1984) and an electronic version of their design manual, www.wlv.com; see also Webber (1960).

12.15. DOUBLE-PIPE HEAT EXCHANGERS

One of the simplest and cheapest types of heat exchanger is the concentric pipe arrangement shown in Figure 12.63. These can be made up from standard fittings, and are useful where only a small heat-transfer area is required. Several units can be connected in series to extend their capacity.

The correlation for forced-convective heat transfer in conduits (equation 12.10) can be used to predict the heat-transfer coefficient in the annulus, using the appropriate equivalent diameter:

$$d_e = \frac{4 \times \text{cross-sectional area}}{\text{wetted perimeter}} = \frac{4(d_2^2 - d_1^2)\frac{\pi}{4}}{\pi(d_2 + d_1)} = d_2 - d_1$$

where d_2 is the inside diameter of the outer pipe and d_1 the outside diameter of the inner pipe.

Some designs of double-pipe exchanger use inner tubes fitted with longitudinal fins.

A variant of the double-pipe heat exchanger is the hairpin exchanger shown in Figure 12.64. A hairpin exchanger is formed by inserting one or more U-tubes into two pipe sections welded to a large flanged end, which is then closed using a removable bonnet. Each straight section of the U-tubes acts as a double-pipe exchanger. True counter-current flow is obtained if a single U-tube is used.

Hairpin exchangers are cheaper than shell and tube exchangers at very small sizes and can be specified for areas from 7 to 150 m^2.

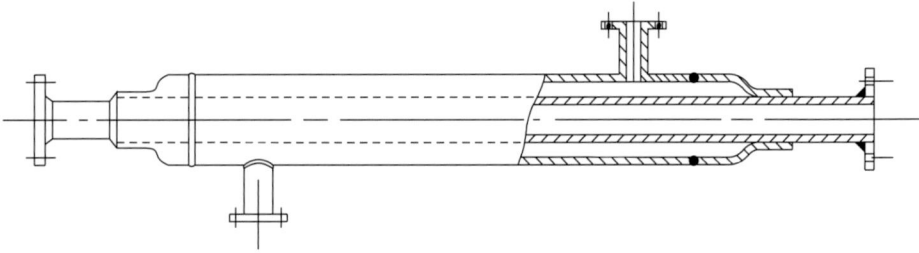

Figure 12.63.   Double-pipe exchanger (constructed for weld fittings).

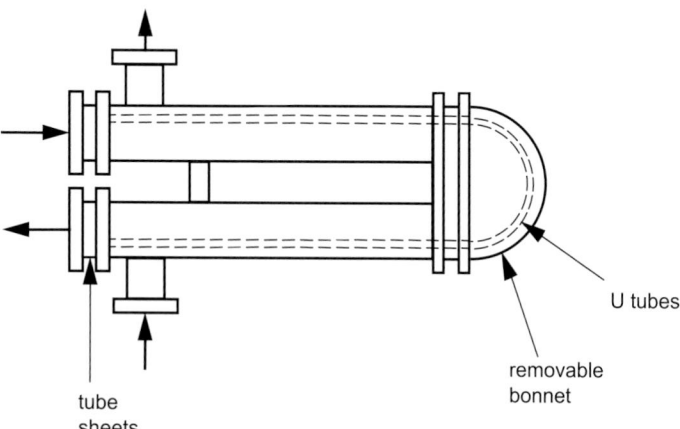

Figure 12.64.   Hairpin exchanger.

12.16. AIR-COOLED EXCHANGERS

Air-cooled exchangers consist of banks of finned tubes over which air is blown or drawn by fans mounted below or above the tubes (forced or induced draft). Typical designs are shown in Figure 12.65.

Air-cooled exchangers should be considered when cooling water is in short supply or expensive. They can be competitive with water-cooled units even when water is plentiful. Frank (1978) suggests that in moderate climates air cooling will usually be the best choice for minimum process temperatures above 65°C, and water cooling for minimum processes temperatures below 50°C. Between these temperatures a detailed economic analysis must be carried out to decide the best coolant. Air-cooled exchangers are used for cooling and condensing.

Cooling water circuits require a humidity driving force to achieve cooling of the water; see Section 14.5 and Coulson *et al.* (1999). In climates that often experience a combination of high temperature and high humidity, air cooling will usually be cheaper than water cooling. Air coolers are also often specified for revamps or additions to existing plants, so as to avoid increasing the cooling tower load and obviate investments in the site utility system.

The design and application of air-cooled exchangers is discussed by Rubin (1960), Lerner (1972), Brown (1978) and Mukherjee (1997). Design procedures are also given in the books by Kern (1950), Kern and Kraus (1972), and Kroger (2004). Lerner and Brown give typical values for the overall coefficient for a range of applications and provide methods for the preliminary sizing of air-cooled heat exchangers.

Details of the construction features of air-cooled exchangers are given by Ludwig (1965). The construction features of air-cooled heat exchangers are covered by the American Petroleum Institute standard, API 661, which has been adopted as the recognized international standard for air coolers, ISO 13706-1:2005.

Air-cooled exchangers are packaged units, and are normally selected and specified in consultation with the manufacturers. Some typical overall coefficients are given in

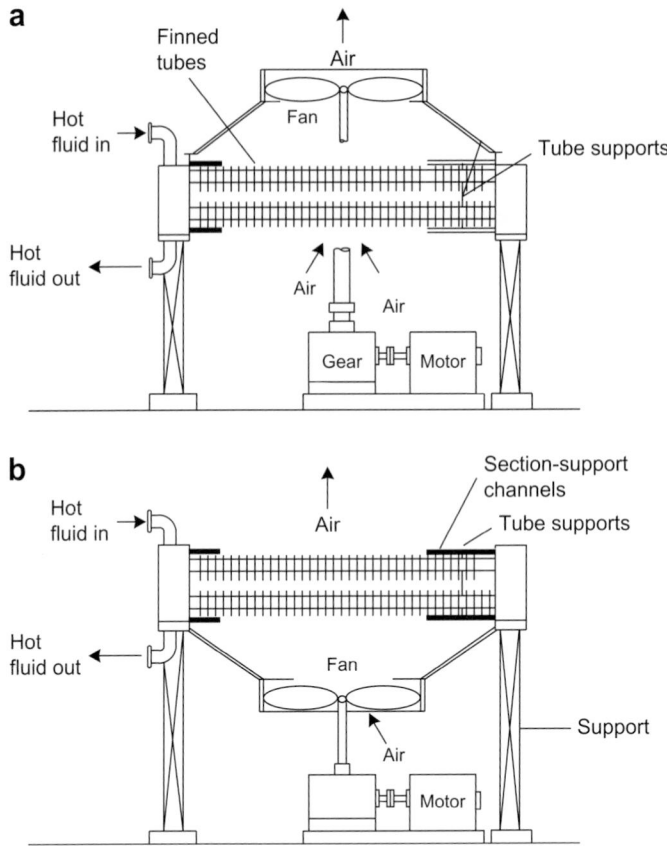

Figure 12.65. Air-cooled exchangers. (*a*) Induced draft. (*b*) Forced draft.

Table 12.1. These can be used to make an approximate estimate of the area required for a given duty. The commercial heat-exchanger design programs developed by HTRI and HTFS include programs for air cooler design; see Section 12.1.

12.16.1. Air Coolers: Construction Details

Air coolers can be designed with the fan mounted either above or below the tube rack, as illustrated in Figure 12.65.

In an induced draft cooler (Figure 12.65a), the fan is mounted above the bank of tubes and air is pulled up over the tubes. The housing around the fan provides for some chimney effect, which can give better cooling when the fan is not running. This design generally has better air distribution across the tube rack and lower chance of air recirculation, but the location of the fan makes maintenance access difficult and the tubes can become damaged during fan maintenance.

Figure 12.65(b) shows a forced draft cooler, in which the fan is mounted below the bank of tubes and air is blown up across the tube bank. The natural draft capability of this design is lower, but the fan is easier to access for maintenance. The fan draws cooler air in a forced draft design, which reduces the power requirement. Forced draft designs can also recirculate air in winter to offset the effect of lower ambient temperature. Care must be taken when designing the cooler and plant layout to ensure that unwanted air recirculation does not occur during normal operation. Forced draft coolers are usually less expensive than induced draft coolers.

The tubes used in air coolers are usually finned, to provide additional surface area to compensate for the poor heat-transfer coefficient on the air side. Because of the use of finned tubes, air coolers are sometimes referred to as 'fin-fan coolers'. The ratio of finned area to bare tube area is typically about 20:1.

The tubes are usually welded into a header at each end of the exchanger and the tube-side flow is usually two pass so that the inlet and outlet are at the same (front) end of the cooler. The opposite (floating) header is mounted to allow for thermal expansion of the tubes. Multiple banks of tubes may be used as long as the air-side pressure drop and fan power are not excessive. Standard tube lengths are typically used, but longer tubes up to 20 m (60 ft) can be accommodated by arranging multiple fans in parallel. The height of the tube bank above the ground must be large enough to give an inlet velocity equal to the face velocity over the tube bundle. A typical height is half of the tube length per fan.

12.16.2. Heat Transfer in Air Coolers

The air-side heat-transfer coefficient in air coolers is usually very low, and dominates the overall heat-transfer coefficient. If the detailed design of the finned tubes is known, then the air-side coefficient can be calculated using equation 12.60 or a similar expression provided by the tube manufacturer. When the finned-tube design is not known, the fin coefficient can be estimated using the expression given by Lohrisch (1966):

$$\mathrm{Nu} = 0.28\,\mathrm{Re}^{0.6}\,\mathrm{Pr}^{0.33} \tag{12.61}$$

Using typical fin dimensions of fin pitch 2.3 mm (12 fins per inch), fin length 15.9 mm and fin thickness 0.48 mm, equation 12.60 reduces to:

$$\mathrm{Nu} = 0.104\,\mathrm{Re}^{0.681}\,\mathrm{Pr}^{0.33} \tag{12.62}$$

Air coolers are typically designed for the highest temperature that will be exceeded for 40 h per year, less 4°C (API 661). This temperature is usually about 40°C ($\sim$ 104°F). At this temperature, dry air has the properties: Pr $\sim$ 0.7, $\rho \sim$ 1.13 kg/m^3, $\mu \sim 1.9 \times 10^{-5}$ Nm^{-2}s and k $\sim$ 0.0272 W/m°C. Using a typical tube diameter of 25.4 mm and face velocity of 2.5 m/s (500 ft/min), we obtain:

From equation 12.61: Nu $\sim$ 34.8, $h_f \sim$ 37.2

From equation 12.62: Nu $\sim$ 25.2, $h_f \sim$ 27.0

The fin efficiency is typically about 0.9 at these values of h_f (Lohrisch, 1966), and the ratio of finned area to bare tube area is typically 20:1. Fouling factors for air are usually quite high and the fouling coefficient can be taken as 5000 W/m^2°C (from

Table 12.2). Taking the average of the values of h_f calculated above, equation 12.59 gives:

$$\frac{1}{h_o} + \frac{1}{h_{od}} = \frac{1}{0.9}\left(\frac{1}{32} + \frac{1}{5000}\right)\frac{1}{20} \tag{12.59}$$

$$\frac{1}{h_o} + \frac{1}{h_{od}} \approx \frac{1}{600} \tag{12.63}$$

where h_o is based on the bare tube area.

Equation 12.63 can be used with equation 12.2 to make an initial estimate of the overall heat transfer coefficient. More accurate correlations are available in the air-cooled exchanger design programs developed by HTRI (www.HTRI.net) and HTFS. The HTFS programs are part of Aspen Technology's Aspen Engineering Suite and Honeywell's UniSim Design Suite; see Table 4.1. Equations for heat transfer to tube banks in natural convection are given by Chu (2005).

12.16.3. Air-cooler design

Air-cooled exchangers are designed to give satisfactory performance under most expected climate conditions. The API 661/ISO 13706 standard recommends that for critical applications the ambient temperature shall be the highest air temperature that is exceeded for 40 h per year. For non-critical applications, the air temperature can be taken as the higher of:

the highest temperature that is exceeded for 400 h/year
the highest temperature that is exceeded for 40 h/year minus 4°C.

A small allowance on ambient temperature is usually made to provide a safety margin to compensate for recirculation; typically 2 to 3°C.

The procedure for designing an air cooler is then as follows.

1. Estimate the cooling duty, including heat of condensation if phase change occurs.
2. Collect physical property data.
3. Estimate the tube inside heat-transfer coefficient and fouling coefficient. Hence, calculate U using equations 12.2 and 12.63.
4. Determine the ambient air temperature for design. Ambient temperature statistics can be found from government meteorological offices and sources such as www.weather.com. Add an allowance for recirculation.
5. Estimate ΔT_{lm}, assuming that the air-side temperature is approximately constant. Hence, estimate ΔT_m, assuming F_t is 0.9. The air-side temperature is obviously not constant, and typically increases by a few °C. The low change in temperature on the air side gives a high value of R and a low value of S unless the process-side outlet temperature is very close to the air temperature. Consequently, F_t is close to 1.0 for air coolers. Assuming $F_t = 0.9$ is a conservative approximation that compensates for the error introduced by using a constant air-side temperature in calculating ΔT_{lm}.
6. Estimate the bare tube area using equation 12.1.

7. Choose a tube diameter (typically 25.4 mm) and length (typically a multiple of 6 m or 20 ft), and hence determine the number of tubes required. Select the tube material of construction.

8. Decide the bundle layout. Determine the number of tubes per bank and number of banks of tubes (generally less than 10 banks per bundle).

9. Determine the bundle area. The tube spacing is typically 50.8or 63.5 mm (2.0 or 2.5 in) on a triangular pitch, but wider spacing may be used if the fin length is large. The bundle area, A_b, is given by:

$$A_b = L' p_t N_{bk} \qquad (12.64)$$

where

L' = effective tube length
p_t = tube pitch
N_{bk} = number of tubes per bank.

10. Estimate the air-side flow rate. A typical face velocity is 2.5 m/s (500 ft/min). The air flow rate is the product of face velocity and bundle area.

11. Estimate the fan power consumption. The fan power, W_f, can be estimated as:

$$W_f = \frac{u_f A_b \Delta P_b}{\eta_f} \qquad (12.65)$$

where

u_f = face velocity
ΔP_b = pressure drop across the bundle
η_f = fan efficiency, typically about 0.7.

In customary units, equation 12.65 is usually written as:

$$\text{Fan power (hp)} = \left(\frac{\text{ACFM } \Delta P_b}{6837 \, \eta_f} \right) \qquad (12.65b)$$

where ACFM = air flow rate in actual cubic feet per minute, and ΔP_b is in inches of water. The pressure drop across the bundle is usually very low, and a value of 150 N/m^2 (0.6 in of water) can be used as an initial estimate.

12. Estimate the overall power consumption by allowing an additional 5% to 10% for motor efficiency.

13. Carry out detailed simulation of the specified geometry to confirm that the heat-transfer coefficients and pressure drops that have been assumed are realistic. This step is usually carried out using commercial design programs. Modify the design if necessary.

14. Determine the air cooler capital and operating costs. Modify the design as necessary, to optimize the total annualized cost.

This procedure is illustrated in Example 12.14.

12.16.4. Air-Cooler Operation and Control

An intrinsic problem of air coolers is that they must be designed to perform under the warmest climate conditions and so are over-sized for the current ambient temperature most of the time. This can be particularly problematic in winter when ambient temperature may be low enough to cause precipitation of components from the process fluid or even freeze the process fluid.

Temperature control of air coolers can be accomplished by varying the air flow or bypassing some hot fluid around the exchanger, as discussed in Chapter 5 and illustrated in Figure 5.24. Several methods are used to vary air flow:

1. Variable-speed drives can be used on the fans. These have higher cost than single-speed fans, but give the best energy efficiency and temperature control.
2. Variable-pitch fans can be specified. Changing the fan pitch varies the air flow and power consumption is reduced at low pitch.
3. Louvers can be used to provide a restriction in the air flow path and reduce the air rate. The louvers may be manually or automatically adjusted. The use of louvers does not reduce power consumption, and louver operation can be impeded by snow or ice, but this is the cheapest method.

In regions that experience cold winters, air-flow control may not be adequate for winterization. Annex C of API 661/ISO 13706 provides guidelines on when winterization is needed and describes other winterization approaches such as internal or external air recirculation.

Example 12.14

Investigate the use of an air-cooled exchanger for the duty set out in Example 12.1. Determine the fan power consumption.

Solution

Summary of Example 12.1

Cool 100,000 kg/h of methanol from 95°C to 40°C, duty 4340 kW. Methanol flow rate 27.8 kg/s, density 750 kg/m^3, viscosity 0.34 mNm^{-2}s, Prandtl number 5.1, thermal conductivity 0.19 W/m°C.

Step 3: Heat-transfer Coefficient

Assuming a typical tube-side velocity of 1.5 m/s and selecting 25.4-mm o.d. (22.1-mm i.d.) tubes:

$$\text{Re} = \frac{\rho u d_i}{\mu} = \frac{750 \times 1.5 \times 22.1 \times 10^{-3}}{0.34 \times 10^{-3}} = 73.1 \times 10^3$$

From equation 12.11, neglecting the viscosity correction:

$$\begin{aligned}
\text{Nu} &= 0.023\,\text{Re}^{0.8}\,\text{Pr}^{0.33} \\
&= 0.023(73.1 \times 10^3)^{0.8}\,(5.1)^{0.33} \\
&= 306.5
\end{aligned}$$

Hence

$$\begin{aligned}
h_i &= 306.5 \times (0.19/(22.1 \times 10^{-3})) \\
&= 2635 \text{ W/m}^{2\circ}\text{C}
\end{aligned}$$

Since the exchanger no longer uses brine as coolant, a cheaper material of construction such as plain carbon steel can be selected. For carbon steel, the thermal conductivity, k_w, is 55 W/m°C over the range of temperature of interest.

Using the same fouling coefficient of 5000 W/m²°C for methanol and substituting equation 12.63 in equation 12.2 gives:

$$\frac{1}{U_o} = \frac{1}{600} + \frac{25.4 \times 10^{-3} \ln\left(\frac{25.4}{22.1}\right)}{2 \times 55} + \frac{25.4}{22.1} \times \frac{1}{2635} + \frac{25.4}{22.1} \times \frac{1}{5000} \quad (12.2)$$

$$\frac{1}{U_o} = \frac{1}{600} + \frac{1}{31120} + \frac{1}{2292} + \frac{1}{4350}$$

From which is it clear that the air-side contribution governs the overall heat-transfer coefficient, and:

$$U_o = 423 \text{ W/m}^2{}^{\circ}C$$

Step 4: Ambient Temperature

If the cooler is located in Northwest Europe then a reasonable initial estimate of the design air temperature is 32°C (90°F). The ambient temperature is obviously highly site specific, and for many regions a product temperature specification of 40°C would immediately rule out the use of air cooling as impractical. Adding 2°C to allow for some recirculation gives a design temperature of 34°C.

Step 5: Estimate ΔT_m

$$\Delta T_{lm} = \frac{(95 - 34) - (40 - 34)}{\ln\left(\frac{95 - 34}{40 - 34}\right)} = 23.7°C$$

Assuming $F_t = 0.9$,

$$\Delta T_m = 0.9 \times 23.7 = 21.3°C$$

Step 6: Estimate Bare Tube Area

$$Q = U A \Delta T_m \quad (12.1)$$

hence

$$A = \frac{4.34 \times 10^6}{423 \times 21.3} = 481.7 \text{ m}^2$$

Steps 7 and 8: Pick Tubes and Determine Bundle Layout

The area of a tube 6 m long and 25.4 mm in o.d. (20 ft, 1 in o.d.) is $\pi \times 25.4 \times 10^{-3} \times 6 = 0.479 \text{ m}^2$, so we require $481.7 / 0.479 = 1006$ tubes. If we increase tube length to 18 m (60 ft) then we need 335 tubes, which can be arranged as five banks of 67 tubes each.

Step 9: Determine Bundle Area

If the tube pitch is 76.2 mm (3 in) then the bundle area is:

$$A_b = 18 \times 76.2 \times 10^{-3} \times 67 = 91.9 \text{ m}^2$$

Step 10: Estimate Air Flow Rate

Assuming a face velocity of 2.5 m/s, the air flow rate is $2.5 \times 91.9 = 229.7$ actual m^3/s.

Steps 11 and 12: Estimate Fan Power And Total Power

From equation 12.65, assuming 150 N/m^2 pressure drop and 70% fan efficiency, the fan power is:

$$W_f = \frac{u_f A_b \, \Delta P_b}{\eta_f} = \frac{2.5 \times 91.9 \times 150}{0.7} = 49.2 \text{ kW} \qquad (12.65)$$

Allowing for motor efficiency of 95%, the total power consumed $= 49.2 / 0.95 = \underline{51.8 \text{ kW}}$.

Note that the fan power consumed is much less than the cooling duty.

Step 13: Confirm Heat-transfer Coefficients and Pressure Drop

Check tube-side heat transfer:

With 335 tubes, we have 167 tubes on one pass and 168 on the other pass. Using 168 tubes, methanol flow rate $= 27.8$ kg/s $= 27.8/750$ m^3/s $= 0.037$ m^3/s.

Tube inside area $= \pi \, (22.1 \times 10^{-3})^2/4 = 3.836 \times 10^{-4}$ m^2 per tube.

So, tube-side velocity $= 0.037/(168 \times 3.836 \times 10^{-4}) = 0.574$ m/s.

This is less than 40% of the assumed 1.5 m/s, so we need to correct the inside heat transfer coefficient. The corrected value of h_i will be:

$$h_i = 2635 \times \left(\frac{0.574}{1.5}\right)^{0.8}$$

$$= 1222 \text{ W/m}^{2\circ} \text{ C}$$

The corrected value of U_o is:

$$\frac{1}{U_o} = \frac{1}{600} + \frac{1}{31120} + \frac{25.4}{22.1} \times \frac{1}{1222} + \frac{1}{4350}$$

$$U_o = 349 \text{ W/m}^{2\circ}\text{C}$$

The area required would now be $481.7 \times (423/349) = 584$ m^2, i.e., 1220 tubes.

Check air-side temperature change and ΔT_m:

Air flow rate $= 229.7$ m^3/s

At 34°C, air has density 1.15 kg/m^3 and $C_p \sim 1$ kJ/kg°C.

So

$$\text{air side temperature change} = \frac{4.34 \times 10^3}{229.7 \times 1.15 \times 1} = 16.4°C$$

Such a large change in temperature on the air side is clearly not acceptable with a temperature approach of 6°C at the cold end of the exchanger. To reach an acceptable air-side temperature change we would need at least three times the air flow calculated, which would cause a factor of three increase in the fan duty. At this point, most experienced design engineers would conclude that because of the low outlet temperature specified for the cooler, an air cooler would not be economically attractive compared to a water cooler in this service.

If water cooling was not possible, then we could consider three air coolers in parallel, each with a single bank of 150 tubes, each 18 m long. Keeping the tube pitch at 76.2 mm, the bundle area is:

$$A_b = 18 \times 76.2 \times 10^{-3} \times 150 = 205.7 \text{ m}^2 (\text{roughly } 18 \text{ m} \times 12 \text{ m})$$

Assuming a face velocity of 2.5 m/s, the air flow rate in each cooler is $2.5 \times 205.7 = 514.3$ actual m^3/s. So the total air flow is 1543 actual m^3/s, roughly three times the initial design.

Hence

$$\text{Air side temperature change} = \frac{4.34 \times 10^3}{514.3 \times 3 \times 1.15 \times 1} = 2.45°C$$

This gives

$$R = \frac{95 - 40}{2.45} = 22.4$$

$$S = \frac{2.45}{95 - 34} = 0.040$$

From which $F_t \sim 0.95$

$$\Delta T_{lm} = \frac{(95 - 36.5) - (40 - 34)}{\ln\left(\dfrac{95 - 36.5}{40 - 34}\right)} = 23.1°C$$

and

$$\Delta T_m = 0.95 \times 23.1 = 21.9°C$$

With 3×150 tubes, we have $3 \times 75 = 225$ tubes per pass, so tube-side velocity $= 0.037/(225 \times 3.836 \times 10^{-4}) = 0.429$ m/s.

$$h_i = 2635 \times \left(\frac{0.429}{1.5}\right)^{0.8}$$

$$= 967 \text{ W/m}^2°C$$

The corrected value of U_o is:

$$\frac{1}{U_o} = \frac{1}{600} + \frac{1}{31120} + \frac{25.4}{22.1} \times \frac{1}{967} + \frac{1}{4350}$$

$$U_o = 321 W/m^2 {}^{\circ}C$$

hence

$$A = \frac{4.34 \times 10^6}{321 \times 21.9} = 617.6 \ m^2$$

The area of a tube 18 m long and 25.4 mm in o.d. (20 ft, 1 in o.d.) is 1.437 m², so we require 617.6 / 1.437 = 430 tubes. This is less than the $3 \times 150 = 450$ tubes assumed in the design, so we now have more than the required area. We could reduce the tube count slightly and iterate towards a converged solution, but the design is probably good enough at this point. The air-side heat-transfer coefficient and pressure drop should now be confirmed and the design finalized using commercial air-cooler design software.

The new fan duty for each bundle is:

$$W_f = \frac{u_f A_b \Delta P_b}{\eta_f} = \frac{2.5 \times 205.7 \times 150}{0.7} = 110.2 \ kW \tag{12.65}$$

Allowing for motor efficiency of 95%, the total power consumed for all three coolers = $3 \times 110.2 / 0.95 = \underline{\underline{348 \ kW}}$.

The design generated in this example is feasible, but has high plot-space requirements as well as higher capital cost than the water cooler. The operating cost of running these air coolers at 0.06 \$/kWh would be $348 \times 0.06 = \$20.9/h$. The cost of providing cooling water to the exchanger of Example 12.1 would be roughly \$6.50/h (based on cooling water cost of \$0.1/1000 gal), so the operating cost of the air cooler is also higher. An air cooler is clearly less economically attractive than a water cooler in this service, and would only be considered if water costs were excessive or if there was flexibility to cool the methanol to a higher outlet temperature.

12.17.   FIRED HEATERS (FURNACES AND BOILERS)

When high temperatures and high flow rates are required, fired heaters are used. Fired heaters are directly heated by the products of combustion of a fuel. The highest temperature at which steam is used for process heating is typically about 250°C. Circulating heating oils are used up to about 330°C, but hot oil loops themselves require a fired heater as the primary heat source. Small vertical cylindrical fired heaters are used for duties up to 45 MW, and larger cabin furnaces are used for higher duties.

Typical applications of fired heaters are:

1. Process feed-stream heaters; such as the feed heaters for high-temperature reactors and refinery crude columns (pipe still furnaces); in which up to 60% of the feed may be vaporized.
2. Reboilers for columns, using relatively small size direct-fired units.

3. Direct-fired reactors; for example, the pyrolysis of dichloroethane to form vinyl chloride.
4. Reformers for hydrogen production, giving outlet temperatures of 800–900°C.
5. Steam boilers.
6. Heaters for hot oil circuits.

12.17.1. Basic Construction

Many different designs and layouts are used, depending on the application; see Berman (1979a) and Trinks *et al.* (2004).

The basic construction consists of a rectangular or cylindrical steel chamber, lined with refractory bricks. Tubes are arranged around the wall, in either horizontal or vertical banks. The fluid to be heated flows through the tubes. Typical layouts are shown in Figure 12.66a, b and c. A more detailed diagram of a pyrolysis furnace is given in Figure 12.67.

Heat transfer to the tubes on the furnace walls is predominantly by radiation. In modern designs, this radiant section is surmounted by a smaller section in which the combustion gases flow over banks of tubes and transfer heat by convection. Extended-surface tubes, with fins or pins, are used in the convection section to improve the heat transfer from the combustion gases. Plain tubes known as 'shock tubes' are used in the bottom rows of the convection section to act as a heat shield from the hot gases in the radiant section. Heat transfer to the shock tubes will be by both radiation and convection. The tube sizes used will normally be between 75 and 150 mm diameter. The tube size and number of passes used depend on the application and the process-fluid flow rate. Typical tube velocities are from 1 to 2 m/s for heaters, with lower rates

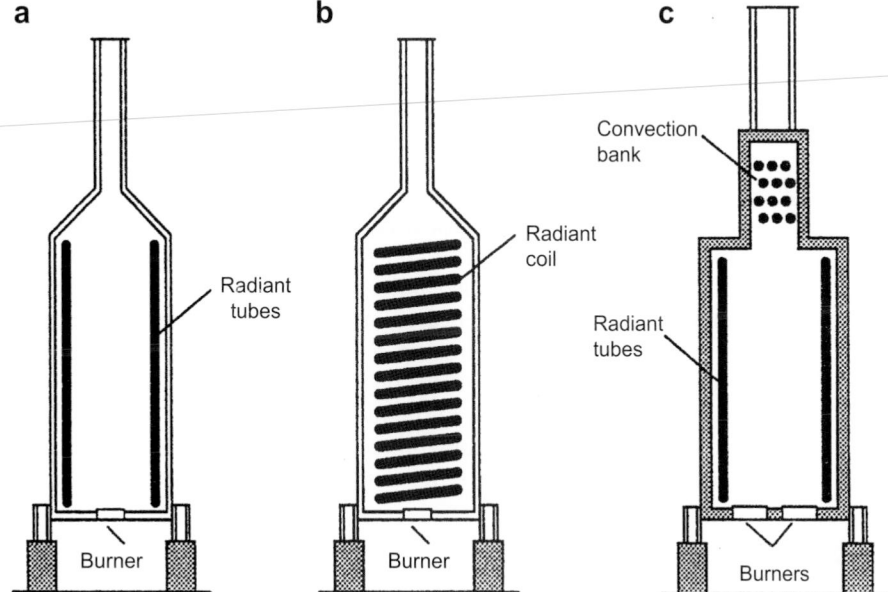

Figure 12.66. Fired heaters. (a) Vertical-cylindrical, all radiant. (b) Vertical-cylindrical, helical coil. (c) Vertical-cylindrical with convection section.

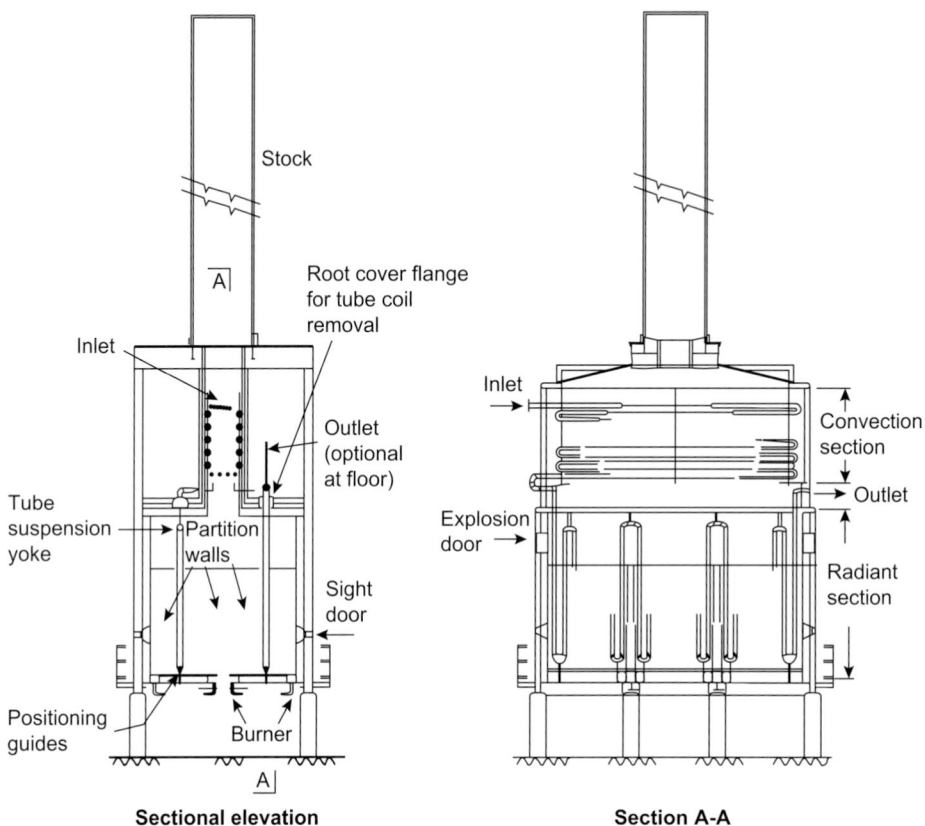

Figure 12.67.   (Foster Wheeler) Multi-zoned pyrolysis furnace.

used for reactors. Carbon steel is used for low-temperature duties; stainless steel and special alloy steels for elevated temperatures. For high temperatures, a material that resists creep must be used. Special metallurgies are also needed if the process fluid undergoes coking or can cause metal dusting corrosion, or if it is a mixed oxidizing–sulphidizing gas that attacks the metal surface; see Chapter 7.

The burners are positioned at the base or along the sides of the radiant section. The combustion air may be preheated in tubes in the convection section.

The fuel is typically natural gas, fuel oil or off-gases from the process. When process off-gases are used, they are often blended with natural gas to allow for start-up and to dampen out any variations in the fuel quality. The selection of fuel is determined by cost and environmental constraints. It will sometimes be necessary to use a more expensive fuel to meet the requirements of environmental emissions permits.

12.17.2.   Design

Commercial computer programs for the design of fired heaters are available from HTFS and HTRI; see Section 12.1. Manual calculation methods, suitable for the preliminary design of fired heaters, are given by Kern (1950), Wimpress (1978) and

Evans (1980). A brief review of the factors to be considered is given in the following sections.

12.17.3. Heat Transfer

Radiant Section

Between 50% and 70% of the total heat is transferred in the radiant section.

The gas temperature depends on the fuel used and the amount of excess air. Around 20% excess air is normally used for gaseous fuels, and 25% for liquid fuels. Excess air is used to prevent the formation of soot and carbon monoxide. The effect of increasing excess air is to reduce the adiabatic flame temperature and increase the stack gas flow rate, hence shifting heat availability from the radiant section to the convective section.

Radiant heat transfer from a surface is governed by the Stefan-Boltzman equation:

$$q_r = \sigma T^4 \tag{12.66}$$

where

q_r = radiant heat flux, W/m^2
σ = Stephen-Boltzman constant, $5.67 \times 10^{-8} \, W/m^2 \, K^4$
T = temperature of the surface, K.

For the exchange of heat between the combustion gases and the hot tubes the equation can be written as:

$$Q_r = \sigma(\alpha A_{cp})F(T_g^4 - T_t^4) \tag{12.67}$$

where

Q_r = radiant heat-transfer rate, W
A_{cp} = the 'cold-plane' area of the tubes
    = number of tubes × the exposed length × tube pitch
α = the absorption efficiency factor
F = the radiation exchange factor
T_g = temperature of the hot gases, K
T_t = tube surface temperature, K.

Part of the radiation from the hot combustion gases will strike the tubes and be absorbed, and part will pass through the spaces between the tubes and be radiated back into the furnace. If the tubes are in front of the wall, some of the radiation from the wall will also be absorbed by the tubes. This complex situation is allowed for by calculating what is known as the cold-plane area of the tubes, A_{cp}; and then applying the absorption efficiency factor α to allow for the fact that the tube area will not be as effective as a plane area. The absorption efficiency factor is a function of the tube arrangement and will vary from around 0.4 for widely spaced tubes, to 1.0 for the theoretical situation when the tubes are touching. It will be around 0.7 to 0.8 when the pitch equals the tube diameter. Values for α are available in handbooks for a range of tube arrangements; see Green and Perry (2007) and Wimpress (1978).

The radiation exchange factor F depends on the arrangement of the surfaces and their emissivity and absorptivity. Combustion gases are poor radiators, because only the carbon dioxide and water vapour, about 20% to 25% of the total, will emit radiation in the thermal spectrum. For a fired heater, the exchange factor will depend on the partial pressure and emissivity of these gases and the layout of the heater. The partial pressure is dependent on the kind of fuel used, liquid or gas, and the amount of excess air. The gas emissivity is a function of temperature. Methods for estimating the exchange factor for typical furnace designs are given in the handbooks; see Green and Perry (2007) and Wimpress (1978).

The heat flux to the tubes in the radiant section will be between 20 and 40 kW/m^2, for most applications. A value of 30 kW/m^2 can be used to make a rough estimate of the tube area needed in this section.

A small amount of heat will be transferred to the tubes by convection in the radiant section, but as the superficial velocity of the gases will be low, the heat-transfer coefficient will be low, around 10 W/m^2°C.

The temperature of the flue gas leaving the radiant section is known as the bridgewall temperature. This temperature can be estimated by assuming that roughly 60% of the heat released by combustion is transferred to the process fluid in the radiant section. The bridgewall temperature is needed for detailed design of the convective section.

Convection Section

The combustion gases flow across the tube banks in the convection section and the correlations for cross-flow in tube banks can be used to estimate the heat-transfer coefficient. The gas-side coefficient will be low, and where extended surfaces are used an allowance must be made for the fin efficiency. Procedures are given in the tube vendors' literature, and in handbooks; see Section 12.14 and Berman (1978b).

The overall coefficient will depend on the gas velocity and temperature, and the tube size. Typical values range from 20 to 50 W/m^2°C.

The shock tubes in the lower rows of the convection section receive heat by radiation from the radiant section. This can be allowed for by including the area of the lower row of tubes with the tubes in the radiant section.

12.17.4. Pressure Drop

Most of the pressure drop on the flue gas side occurs in the convection section. The procedures for estimating the pressure drop across banks of tubes can be used to estimate the pressure drop in this section; see Section 12.9.3.

The pressure drop in the radiant section will be small compared with that across the convection section and can usually be neglected.

12.17.5. Process-Side Heat Transfer And Pressure Drop

The tube inside heat-transfer coefficients and pressure drop can be calculated using the conventional methods for flow inside tubes; see Section 12.8. If the unit is being used as a vaporizer, the existence of two-phase flow in some of the tubes must be taken into account. Berman (1978b) gives a quick method for estimating two-phase pressure drop in the tubes of fired heaters.

Typical approach temperatures, flue gas to inlet process fluid, are around 100°C.

12.17.6. Stack Design

Most fired heaters operate with natural draft, and the stack height must be sufficient to achieve the flow of combustion air required and to remove the combustion products.

It is normal practice to operate with a slight vacuum throughout the heater, so that air will leak in through sight-boxes and dampers, rather than combustion products leak out. Typically, the aim would be to maintain a vacuum of around 2 mm water gauge just below the convection section.

The stack height required depends on the temperature of the combustion gases leaving the convection section and the elevation of the site above sea level. The draft arises from the difference in density of the hot gases and the surrounding air.

The draft in millimetres of water (mmH$_2$O) can be estimated using the equation:

$$P_d = 0.35(L_s)(p')\left[\frac{1}{T_a} - \frac{1}{T_{ga}}\right] \qquad (12.68)$$

where

L_s = stack height, m
p' = atmospheric pressure, millibar (N/m^2 × 10^{-2})
T_a = ambient temperature, K
T_{ga} = average flue-gas temperature, K.

Because of heat losses, the temperature at the top of the stack will be around 80°C below the inlet temperature.

The frictional pressure loss in the stack must be added to the loss in the heater when estimating the stack draft required. This can be calculated using the usual methods for pressure loss in circular conduits; see Section 12.8. The mass velocity in the stack will be around 1.5 to 2 kg/m^2. These values can be used to determine the cross-section needed.

An approximate estimate of the pressure losses in the convection section can be made by multiplying the velocity head ($u^2/2g$) by factors for each restriction; typical values are given below:

0.2–0.5 for each row of plain tubes
1.0–2.0 for each row of finned tubes
0.5 for the stack entrance
1.0 for the stack exit
1.5 for the stack damper.

12.17.7. Thermal Efficiency

Modern fired heaters operate at thermal efficiencies of between 80% and 90%, depending on the fuel and the excess air requirement. In some applications, additional excess air may be used to reduce the flame temperature, to avoid overheating of the tubes.

Where the inlet temperature of the process fluid is such that the outlet temperature from the convection section would be excessive, giving low thermal efficiency, this excess heat can be used to preheat the air to the furnace. Tubes would be installed above the process fluid section in the convection section. Forced draft operation would be needed to drive the air flow through the preheat section.

Heat losses from the heater casing are normally between 1.5 and 2.5% of the heat input.

The largest contribution to heat losses is the sensible heat of the flue gas leaving the stack. There are several practical limits on how much heat can be recovered from the flue gas:

1. For process control reasons, the process heating duty is often carried out in the radiant section only; see Chapter 5.
2. There may not be sufficient process heat requirement in the lower temperature range of the convection section. In this case, the convection section can be used to raise steam or preheat boiler feed water if the heat recovered justifies the cost of the additional piping required.
3. The stack gas should not be cooled to the dew point otherwise condensation can occur, leading to corrosion of the stack. The presence of carbon dioxide and sulphur oxides from the combustion reactions in the flue gas raises the dew point and causes the condensate to be acidic.
4. A large temperature difference is usually specified between the flue gas and the process fluid, so as to decrease the tube cost in the convection section. This leads to high stack temperatures.
5. Many companies avoid cooling the flue gas close to the dew point to prevent the formation of a visible plume from the stack. If the flue gas leaving the stack is near the dew point then when it mixes with cold ambient air a mist will form. This gives the stack the appearance of smoking. If the flue gas is hotter, then the gas can disperse before condensation occurs and the plume is eliminated. The general public prefers not to see plumes coming from chemical plants, so public relations often triumphs over energy efficiency.

12.17.8. Fired-heater Emissions

Fired heaters are a major source of atmospheric emissions and are tightly regulated. Permits are usually required for operation of process heaters, and modifications to the heater or burners often require approval from environmental agencies or re-issuance of the permit.

Fired heater emissions concerns include:

1. Carbon monoxide, unburned hydrocarbons and soot can be formed if combustion is not complete. These emissions are usually minimized by operating with at least 20% excess air.
2. Sulphur oxides and metals can be emitted if sulphur or metals were present in the fuel. These emissions occur mainly when burning heavy fuel oils. Emissions of sulphur oxides can be reduced by switching to a fuel with lower sulphur content, such as natural gas.
3. Nitrogen oxides, NO_x, are formed during combustion. Unfortunately, use of excess air tends to make NO_x formation worse. Formation of NO_x is controlled by using special burner designs such as staged-air or staged-fuel

burners; by using steam injection or flue gas recirculation to reduce the flame temperature; or by catalytic decomposition of the NO_x in the flue gas.

4. Carbon dioxide is formed from combustion of hydrocarbon fuels. The penalties for CO_2 emissions are not yet large enough to drive companies to recover CO_2 from flue gas, but CO_2 can be recovered by scrubbing if necessary. Carbon dioxide capture can be made easier by using novel furnace designs in which the fuel is burnt in oxygen and recirculating carbon dioxide; a system known as 'oxyfuel combustion'.

12.18. HEAT TRANSFER TO VESSELS

The simplest way to transfer heat to a process or storage vessel is to fit an external jacket or an internal coil. If these methods cannot provide sufficient heat-transfer area then a stream is withdrawn from the vessel, pumped through a heat exchanger and returned to the vessel.

12.18.1. Jacketed Vessels

Conventional Jackets

The most commonly used type jacket is that shown in Figure 12.68. It consists of an outer cylinder that surrounds part of the vessel. The heating or cooling medium circulates in the annular space between the jacket and vessel walls and the heat is transferred through the wall of the vessel. Circulation baffles are usually installed in the annular space to increase the velocity of the liquid flowing through the jacket and improve the heat-transfer coefficient, see Figure 12.69a. The same effect can be obtained by introducing the fluid through a series of nozzles spaced down the jacket. The momentum of the jets issuing from the nozzles sets up a swirling motion in the jacket liquid; Figure 12.69d. The spacing between the jacket and vessel wall will depend on the size of the vessel, but will typically range from 50 mm for small vessels to 300 mm for large vessels.

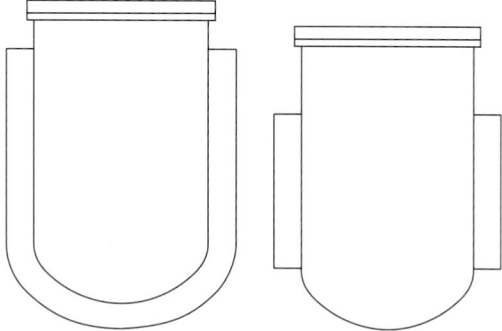

Figure 12.68. Jacketed vessel.

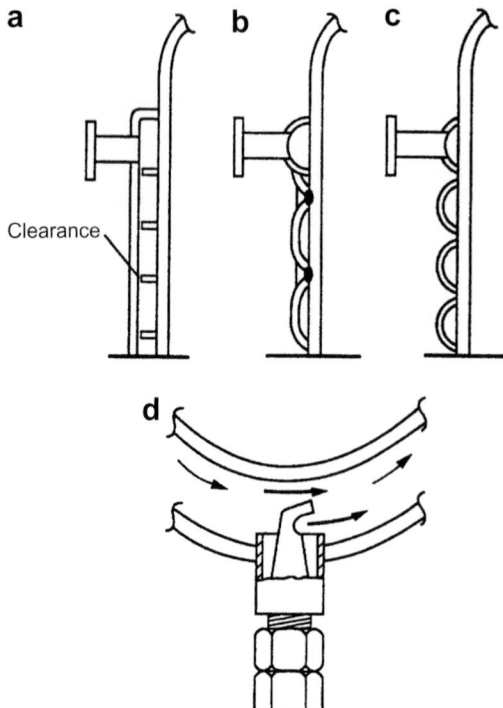

Figure 12.69. Jacketed vessels. (a) Spirally baffled jacket. (b) Dimple jacket. (c) Half-pipe jacket. (d) Agitation nozzle.

Half-pipe Jackets

Half-pipe jackets are formed by welding sections of pipe, cut in half along the longitudinal axis, to the vessel wall. The pipe is usually wound around the vessel in a helix; see Figure 12.69c.

  The pitch of the coils and the area covered can be selected to provide the heat-transfer area required. Standard pipe sizes are used, ranging from 60 to 120 mm outside diameter. The half-pipe construction makes a strong jacket capable of withstanding pressure better than the conventional jacket design.

Dimpled Jackets

Dimpled jackets are similar to the conventional jackets, but are constructed of thinner plates. The jacket is strengthened by a regular pattern of hemispherical dimples pressed into the plate and welded to the vessel wall (Figure 12.69b).

Jacket Selection

Factors to consider when selecting the type of jacket to use are listed below:

  1. Cost: in terms of cost the designs can be ranked, from cheapest to most expensive, as:
    simple, no baffles

agitation nozzles
spiral baffle
dimple jacket
half-pipe jacket.
2. Heat-transfer rate required: select a spirally baffled or half-pipe jacket if high rates are required.
3. Pressure: as a rough guide, the pressure rating of the designs can be taken as:
jackets, up to 10 bar
dimpled jackets, up to 20 bar
half-pipe, up to 70 bar.

So, half-pipe jackets would be used for high pressure.

Jacket Heat-transfer and Pressure Drop

The heat-transfer coefficient to the vessel wall can be estimated using the correlations for forced convection in conduits, such as equation 12.11. The fluid velocity and the path length can be calculated from the geometry of the jacket arrangement. The hydraulic mean diameter (equivalent diameter, d_e) of the channel or half-pipe should be used as the characteristic dimension in the Reynolds and Nusselt numbers; see Section 12.8.1.

In dimpled jackets, a velocity of 0.6 m/s can be used to estimate the heat-transfer coefficient. A method for calculating the heat-transfer coefficient for dimpled jackets is given by Makovitz (1971).

The coefficients for jackets using agitation nozzles are similar to those given by using baffles. A method for calculating the heat-transfer coefficient using agitation nozzles is given by Bolliger (1982).

To increase heat-transfer rates, the velocity through a jacket can be increased by recirculating the cooling or heating liquid.

For simple jackets without baffles, heat transfer will be mainly by natural convection and the heat-transfer coefficient will range from 200 to 400 W/m²°C.

When steam is used in a jacket the heat-transfer coefficient is in the range 4000 to 5000 W/m²°C.

12.18.2.   Internal Coils

The simplest and cheapest form of heat-transfer surface for installation inside a vessel is a helical coil; see Figure 12.70. The pitch and diameter of the coil can be made to suit the application and the area required. The diameter of the pipe used for the coil is typically equal to $D_v/30$, where D_v is the vessel diameter. The coil pitch is usually around twice the pipe diameter. Small coils can be self supporting, but for large coils some form of supporting structure will be necessary. Single or multiple turn coils are used.

Coil Heat-transfer and Pressure Drop

The heat-transfer coefficient at the inside wall and pressure drop through the coil can be estimated using the correlations for flow-through pipes; see Section 12.8. Correlations for forced convection in coiled pipes are also given in the Engineering Sciences Data Unit Design Guide, ESDU 78031 (2001).

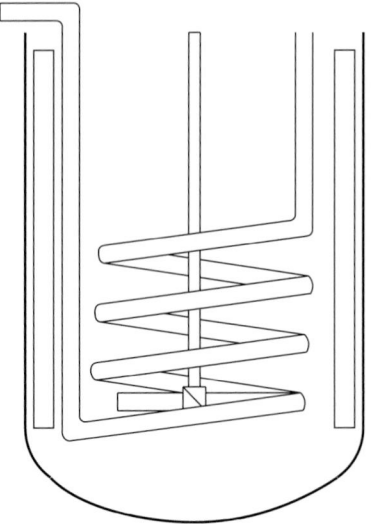

Figure 12.70.   Internal coils.

12.18.3. Agitated Vessels

Unless only small rates of heat transfer are required, as when maintaining the temperature of liquids in storage vessels, some form of agitation will be needed. The various types of agitator used for mixing and blending described in Chapter 10, Section 10.11.2, are also used to promote heat transfer in vessels. The correlations used to estimate the heat-transfer coefficient to the vessel wall, or to the surface of coils, have the same form as those used for forced convection in conduits (equation 12.10). The fluid velocity is replaced by a function of the agitator diameter and rotational speed, $D \times N$, and the characteristic dimension is the agitator diameter.

$$\mathrm{Nu} = C\, \mathrm{Re}^a\, \mathrm{Pr}^b \left(\frac{\mu}{\mu_w}\right)^c \tag{12.10}$$

For agitated vessels

$$\frac{h_v D}{k_f} = C\left(\frac{ND^2\rho}{\mu}\right)^a \left(\frac{C_p\mu}{k_f}\right)^b \left(\frac{\mu}{\mu_w}\right)^c \tag{12.69}$$

where

h_v = heat-transfer coefficient to vessel wall or coil, $\mathrm{W/m^{2}{}^{\circ}C}$
D = agitator diameter, m
N = agitator, speed, rps (revolutions per second)
ρ = liquid density, $\mathrm{kg/m^3}$
k_f = liquid thermal conductivity, $\mathrm{W/m^{\circ}C}$
C_p = liquid specific heat capacity, $\mathrm{J/kg^{\circ}C}$
μ = liquid viscosity, $\mathrm{Nm^{-2}s}$.

The values of the constant C and the indices a, b and c depend on the type of agitator, the use of baffles, and whether the transfer is to the vessel wall or to coils. Some typical correlations are given below.

Baffles will normally be used in most applications.

1. Flat-blade paddle, baffled or unbaffled vessel, transfer to vessel wall, Re < 4000:

$$\text{Nu} = 0.36 \, \text{Re}^{0.67} \, \text{Pr}^{0.33} \left(\frac{\mu}{\mu_w} \right)^{0.14} \tag{12.70a}$$

2. Flat-blade disc turbine, baffled or unbaffled vessel, transfer to vessel wall, Re < 400:

$$\text{Nu} = 0.54 \, \text{Re}^{0.67} \, \text{Pr}^{0.33} \left(\frac{\mu}{\mu_w} \right)^{0.14} \tag{12.70b}$$

3. Flat-blade disc turbine, baffled vessel, transfer to vessel wall, Re > 400:

$$\text{Nu} = 0.74 \, \text{Re}^{0.67} \, \text{Pr}^{0.33} \left(\frac{\mu}{\mu_w} \right)^{0.14} \tag{12.70c}$$

4. Propeller, three blades, transfer to vessel wall, Re > 5000:

$$\text{Nu} = 0.64 \, \text{Re}^{0.67} \, \text{Pr}^{0.33} \left(\frac{\mu}{\mu_w} \right)^{0.14} \tag{12.70d}$$

5. Turbine, flat blades, transfer to coil, baffled, Re = 2000–700,000:

$$\text{Nu} = 1.10 \, \text{Re}^{0.62} \, \text{Pr}^{0.33} \left(\frac{\mu}{\mu_w} \right)^{0.14} \tag{12.70e}$$

6. Paddle, flat blades, transfer to coil, baffled:

$$\text{Nu} = 0.87 \, \text{Re}^{0.62} \, \text{Pr}^{0.33} \left(\frac{\mu}{\mu_w} \right)^{0.14} \tag{12.70f}$$

More comprehensive design data is given by: Uhl and Gray (1967), Wilkinson and Edwards (1972), Penny (1983) and Fletcher (1987).

Example 12.15

A jacketed, agitated reactor consists of a vertical cylinder 1.5 m in diameter, with a hemispherical base and a flat, flanged, top. The jacket is fitted to the cylindrical section only and extends to a height of 1 m. The spacing between the jacket and vessel walls is 75 mm. The jacket is fitted with a spiral baffle. The pitch between the spirals is 200 mm.

The jacket is used to cool the reactor contents. The coolant used is chilled water at 10°C; flow rate 32,500 kg/h, exit temperature 20°C.

Estimate the heat-transfer coefficient at the outside wall of the reactor and the pressure drop through the jacket.

Solution

The baffle forms a continuous spiral channel, section 75 mm $\times$ 200 mm.

Number of spirals = height of jacket/pitch = $1/(200 \times 10^{-3}) = 5$

Length of channel = $5 \times \pi \times 1.5 = 23.6$ m

Cross-sectional area of channel = $(75 \times 200) \times 10^{-6} = 15 \times 10^{-3}$ m^2

$$\text{Hydraulic mean diameter, } d_e = \frac{4 \times \text{cross-sectional area}}{\text{wetted perimeter}}$$

$$= \frac{4 \times (75 \times 200)}{2(75 + 200)} = 109 \text{ mm}$$

Physical properties at mean temperature of 15°C, from steam tables: $\rho = 999$ kg/m^3, $\mu = 1.136$ mNm^{-2}s, Pr = 7.99, $k_f = 595 \times 10^{-3}$ W/m°C.

$$\text{Velocity through channel, } u = \frac{32,000}{3600} \times \frac{1}{999} \times \frac{1}{15 \times 10^{-3}} = 0.602 \text{ m/s}$$

$$\text{Re} = \frac{999 \times 0.602 \times 109 \times 10^{-3}}{1.136 \times 10^{-3}} = 57,705$$

Chilled water is not viscous, so use equation 12.11 with $C = 0.023$, and neglect the viscosity correction term.

$$\text{Nu} = 0.023\text{Re}^{0.8}\text{Pr}^{0.33} \qquad\qquad (12.11)$$

$$\frac{h_j \times 109 \times 10^{-3}}{595 \times 10^{-3}} = 0.023 \, (57705)^{0.8}(7.99)^{0.33}$$

$$\underline{\underline{h_j = 1606 \text{ W/m}^2\text{°C}}}$$

Use equation 12.18 for estimating the pressure drop, taking the friction factor from Figure 12.24. As the hydraulic mean diameter will be large compared to the roughness of the jacket surface, the relative roughness will be comparable with that for heat-exchanger tubes. The relative roughness of pipes and channels and the effect on the friction factor is covered in Chapter 5.

From Figure 12.24, for Re = 5.8×10^4, $j_f = 3.2 \times 10^{-3}$

$$\Delta P = 8 j_f \left(\frac{L}{d_e}\right)\rho\frac{u^2}{2}$$

$$\Delta P = 8 \times 3.2 \times 10^{-3} \left(\frac{23.6}{109} \times 10^{-3}\right)999 \times \frac{0.602^2}{2} \qquad\qquad (12.18)$$

$$= \underline{\underline{1003 \text{ N/m}^2}}$$

Example 12.16

The reactor described in Example 12.15 is fitted with a flat-blade disc turbine agitator, 0.6 m diameter, running at 120 rpm. The vessel is baffled and is constructed of stainless-steel plate 10 mm thick.

The physical properties of the reactor contents are: $\rho = 850\,\text{kg/m}^3$, $\mu = 80\,\text{mNm}^{-2}\text{s}$, $k_f = 400 \times 10^{-3}\,\text{W/m}^\circ\text{C}$, $C_p = 2.65\,\text{kJ/kg}^\circ\text{C}$.

Estimate the heat-transfer coefficient at the vessel wall and the overall coefficient in the clean condition.

Solution

Agitator speed (revs per sec) $= 1200/60 = 2\,\text{s}^{-1}$

$$\text{Re} = \frac{\rho N D^2}{\mu} = \frac{850 \times 2 \times 0.6^2}{80 \times 10^{-3}} = 7650$$

$$\text{Pr} = \frac{C_p \mu}{k_f} = \frac{2.65 \times 10^3 \times 80 \times 10^{-3}}{400 \times 10^{-3}} = 530$$

For a flat-blade turbine use equation 12.70c:

$$\text{Nu} = 0.74\,\text{Re}^{0.67}\,\text{Pr}^{0.33}\left(\frac{\mu}{\mu_w}\right)^{0.14}$$

Neglect the viscosity correction term:

$$\frac{h_v \times 0.6}{400 \times 10^{-3}} = 0.74\,(7650)^{0.67}(530)^{0.33}$$

$$h_v = 1564\,\text{W/m}^2{}^\circ\text{C}$$

Taking the thermal conductivity of stainless steel as 16 W/m°C and the jacket coefficient from Example 12.15:

$$\frac{1}{U} = \frac{1}{1606} + \frac{10 \times 10^{-3}}{16} + \frac{1}{1564}$$
$$U = 530\,\text{W/m}^2{}^\circ\text{C}$$

12.19. REFERENCES

Aerstin, F. and Street, G. (1978) *Applied Chemical Process Design* (Plenum Press).

Bell, K. J. (1960) *Petro/Chem.* **32** (Oct.) C26. Exchanger design: based on the Delaware research report.

Bell, K. J. (1963) *Final Report of the Co-operative Research Program on Shell and Tube Heat Exchangers*, University of Delaware, Eng. Expt. Sta. Bull. 5 (University of Delaware).

Bell, K. J., Taborek, J. and Fenoglio, F. (1970) *Chem. Eng. Prog. Symp. Ser.* No. 102, **66,** 154. Interpretation of horizontal in-tube condensation heat-transfer correlations with a two-phase flow regime map.

Bell, K. J. and Ghaly, M. A. (1973) *Chem. Eng. Prog. Symp. Ser.* No. 131, **69**, 72. An approximate generalized design method for multicomponent/partial condensers.

Berman, H. L. (1978a) *Chem. Eng., NY* 85(14) 99. Fired heaters—Finding the basic design for your application.

Berman, H. L. (1978b) *Chem. Eng., NY* 85(18) 129. Fired heaters—How combustion conditions influence design and operation.

Bolliger, D. H. (1982) *Chem. Eng., NY* **89** (Sept.) 95. Assessing heat-transfer in process-vessel jackets.

Bond, M. P. (1981) *Chem. Engr., London* No. 367 (April) 162. Plate heat exchanger for effective heat-transfer.

Bott, T. R. (1990) *Fouling Notebook* (Institution of Chemical Engineers, London).

Boyko, L. D. and Kruzhilin, G. N. (1967) *Int. J. Heat Mass Transfer* **10**, 361. Heat transfer and hydraulic resistance during condensation of steam in a horizontal tube and in a bundle of tubes.

Briggs, D. E. and Young, E. H. (1963) *Chem. Eng. Prog. Symp. Ser.* No. 59, **61**, 1. Convection heat-transfer and pressure drop of air flowing across triangular pitch banks of finned tubes.

Bromley, L. A. (1950) *Chem. Eng. Prog.* **46**, 221. Heat transfer in stable film boiling.

Brown, R. (1978) *Chem. Eng., NY* **85** (March 27th) 414. Design of air-cooled heat exchangers: a procedure for preliminary estimates.

Burley, J. R. (1991) *Chem. Eng., NY* **98** (Aug.) 90. Don't overlook compact heat exchangers.

Butterworth, D. (1973) *Conference on Advances in Thermal and Mechanical Design of Shell and Tube Heat Exchangers*, NEL Report No. 590. (National Engineering Laboratory, East Kilbride, Glasgow, UK). A calculation method for shell and tube heat exchangers in which the overall coefficient varies along the length.

Butterworth, D. (1977) *Introduction to Heat Transfer*, Engineering Design Guide No. 18 (Oxford University Press).

Butterworth, D. (1978) *Course on the Design of Shell and Tube Heat Exchangers* (National Engineering Laboratory, East Kilbride, Glasgow, UK). Condensation 1 - Heat transfer across the condensed layer.

Chantry, W. A. and Church, D. M. (1958) *Chem. Eng. Prog.* **54** (Oct.) 64. Design of high velocity forced circulation reboilers for fouling service.

Chen, J. C. (1966) *Ind. Eng. Chem. Proc. Des. Dev.* **5**, 322. A correlation for boiling heat-transfer to saturated fluids in convective flow.

Chen-Chia, H. and Fair, J. R. (1989) *Heat Transfer Engineering*, **10** (2) 19. Direct-contact gas-liquid heat-transfer in a packed column.

Chu, C. (2005) *Chem. Eng. Prog.* **101**(11), 46. Improved heat transfer predictions for air-cooled heat exchangers.

Colburn, A. P. (1934) *Trans. Am. Inst. Chem. Eng.* 30, 187. Note on the calculation of condensation when a portion of the condensate layer is in turbulent motion.

Colburn, A. P. and Drew, T. B. (1937) *Trans. Am. Inst. Chem. Eng.* **33**, 197. The condensation of mixed vapours.

Colburn, A. P. and Edison, A. G. (1941) *Ind. Eng. Chem.* **33**, 457. Prevention of fog in condensers.

Colburn, A. P. and Hougen, O. A. (1934) *Ind. Eng. Chem.* **26**, 1178. Design of cooler condensers for mixtures of vapours with non-condensing gases.

Collier, J. G. and Thome, J. R. (1994) *Convective Boiling and Condensation*, 3rd edn (McGraw-Hill).

Collins, G. K. (1976) *Chem. Eng., NY* 83 (July 19th) 149. Horizontal-thermosiphon reboiler design.

Cooper, A. and Usher, J. D. (1983) Plate heat exchangers, in *Heat Exchanger Design Handbook* (Hemisphere Publishing).

Coulson, J. M., Richardson, J. F., Backhurst, J. and Harker, J. H. (1999) *Chemical Engineering: Volume 1*, 6th edn (Butterworth-Heinemann).

Devore, A. (1961) *Pet. Ref.* 40 (May) 221. Try this simplified method for rating baffled exchangers.

Devore, A. (1962) *Hyd. Proc. and Pet. Ref.* 41 (Dec.) 103. Use nomograms to speed exchanger design.

Donohue, D. A. (1955) *Pet. Ref.* 34 (Aug.) 94, (Oct.) 128, (Nov.) 175, and 35 (Jan.) 155, in four parts. Heat exchanger design.

Eagle, A. and Ferguson, R. M. (1930) *Proc. Roy. Soc.* A. 127, 540. On the coefficient of heat-transfer from the internal surfaces of tube walls.

Emerson, W. H. (1967) *Thermal and Hydrodynamic Performance of Plate Heat Exchangers*, NEL. Reports Nos. 283, 284, 285, 286 (National Engineering Laboratories, East Kilbride, Glasgow, UK).

Emerson, W. H. (1973) *Conference on Advances in Thermal and Mechanical Design of Shell and Tube Exchangers*, NEL Report No. 590 (National Engineering Laboratory, East Kilbride, Glasgow, UK). Effective tube-side temperature in multi-pass heat exchangers with non-uniform heat-transfer coefficients and specific heats.

Evans, F. L. (1980) *Equipment Design Handbook*, Vol. 2, 2nd edn (Gulf).

Fair, J. R. (1961) *Petro./Chem. Eng.* 33 (Aug.) 57. Design of direct contact gas coolers.

Fair, J. R. (1960) *Pet. Ref.* 39 (Feb.) 105. What you need to design thermosiphon reboilers.

Fair, J. R. (1963) *Chem. Eng., NY* 70 (July 8th) 119, (Aug. 5th) 101, in two parts. Vaporizer and reboiler design.

Fair, J. R. (1972a) *Chem. Eng. Prog. Sym. Ser.* No. 118, 68, 1. Process heat-transfer by direct fluid-phase contact.

Fair, J. R. (1972b) *Chem. Eng., NY* 79 (June 12th) 91. Designing direct-contact cooler/condensers.

Fair, J. R. and Klip, A. (1983) *Chem. Eng. Prog.* 79 (3) 86. Thermal design of horizontal reboilers.

Fisher, J. and Parker, R. O. (1969) *Hyd. Proc.* 48 (July) 147. New ideas on heat exchanger design.

Fletcher, P. (1987) *Chem. Engr., London* No. 435 (April) 33. Heat transfer coefficients for stirred batch reactor design.

Forster, K. and Zuber, N. (1955) *AIChE Jl* 1, 531. Dynamics of vapour bubbles and boiling heat-transfer.

Frank, O. and Prickett, R. D. (1973) *Chem. Eng., NY* 80 (Sept. 3rd) 103. Designing vertical thermosiphon reboilers.

Frank, O. (1974) *Chem Eng., NY* 81 (May 13th) 126. Estimating overall heat-transfer coefficients.

Frank, O. (1978) Simplified design procedure for tubular exchangers, in *Practical Aspects of Heat Transfer*, Chem. Eng. Prog. Tech. Manual (Am. Inst. Chem. Eng.).

Furzer, I. A. (1990) *Ind. Eng. Chem. Res.* 29, 1396. Vertical thermosyphon reboilers. Maximum heat flux and separation efficiency.

Garrett-Price, B. A. (1985) *Fouling of Heat Exchangers: characteristics, costs, prevention control and removal* (Noyes).

Gilmore, G. H. (1963) Chapter 10, in *Chemical Engineers Handbook*, 4th edn, Perry, R. H., Chilton, C. H. and Kirkpatrick, S. P. (eds) (McGraw-Hill).

Gloyer, W. (1970) *Hydro. Proc.* **49** (July) 107. Thermal design of mixed vapour condensers.

Grant, I. D. R. (1973) *Conference on Advances in Thermal and Mechanical Design of Shell and Tube Exchangers*, NEL Report No. 590 (National Engineering Laboratory, East Kilbride, Glasgow, UK.). Flow and pressure drop with single and two phase flow on the shell-side of segmentally baffled shell-and-tube exchangers.

Green, D. W. and Perry, R. H. (eds) (2007) *Perry's Chemical Engineers' Handbook*, 8th edn (McGraw-Hill).

Hewitt, G. F. and Hall-Taylor, N. S. (1970) *Annular Two-phase Flow* (Pergamon).

Hewitt, G. F. (ed.) (2002) *Heat Exchanger Design Handbook* (Begell House).

Hewitt, G. F., Spires, G. L. and Bott, T. R. (1994) *Process Heat Transfer* (CRC Press).

Holman, J. P. (2002) *Heat transfer*, 9th edn (McGraw-Hill).

Hsu, Y. and Graham, R. W. (1976) *Transport Processes in Boiling and Two-phase Flow* (McGraw-Hill).

Hughmark, G. A. (1961) *Chem. Eng. Prog.* **57** (July) 43. Designing thermosiphon reboilers.

Hughmark, G. A. (1964) *Chem. Eng. Prog.* **60** (July) 59. Designing thermosiphon reboilers.

Hughmark, G. A. (1969) *Chem. Eng. Prog.* **65** (July) 67. Designing thermosiphon reboilers.

Incropera, F. P. and Dewitt, D. P. (2001) *Introduction to Heat Transfer*, 5th edn (Wiley).

Jeffreys, G. V. (1961) *A Problem in Chemical Engineering Design* (Inst. Chem. Eng., London).

Kern, D. Q. (1950) *Process Heat Transfer* (McGraw-Hill).

Kern, D. Q. and Kraus, A. D. (1972) *Extended Surface Heat Transfer* (McGraw-Hill).

Kreith, F. and Bohn, M. S. (2000) *Principles of Heat Transfer*, 6th edn (Thomson-Engineering).

Kroger, D. G. (2004) *Air-cooled Heat Exchangers and Cooling Towers: Thermal-flow Performance Evaluation and Design*, Vol. 1 (PennWell).

Kumar, H. (1984) *Inst. Chem. Eng. Sym. Ser.* No. 86, 1275. The plate heat exchanger: construction and design.

Kutateladze, S. S. (1963) *Fundamentals of Heat Transfer* (Academic Press).

Lee, D. C., Dorsey, J. W., Moore, G. Z. and Mayfield, F. D. (1956) *Chem. Eng. Prog.* **52** (April) 160. Design data for thermosiphon reboilers.

Lerner, J. E. (1972) *Hyd. Proc.* **51**(2). Simplified air cooler estimating.

Lohrisch, F. W. (1966) *Hyd. Proc.* **45**(6), 131. What are optimum conditions for air-cooled exchangers?

LoPinto, L. (1982) *Chem. Eng., NY* **89** (May 17) 111. Fog formation in low temperature condensers.

Lord, R. C., Minton, P. E. and Slusser, R. P. (1970) *Chem. Eng., NY* **77** (June 1st) 153. Guide to trouble free heat exchangers.

Lowe, R. E. (1987) *Chem. Eng., NY* **94** (Aug. 17th) 131. Plate-and-fin heat exchangers for cryogenic service.

Ludwig, E. E. (2001) *Applied Process Design for Chemical and Petroleum Plants*, Vol. 3, 3rd edn (Gulf).

Makovitz, R. E. (1971) *Chem. Eng., NY* **78** (Nov. 15th) 156. Picking the best vessel jacket.

Maxwell, J.B. (1962) *Data Book of Hydrocarbons* (Van Nostrand).

McNaught, J. M. (1983) An assessment of design methods for condensation of vapours from a noncondensing gas, in *Heat Exchangers: Theory and Practice* (McGraw-Hill).

Minton, P. E. (1970) *Chem. Eng., NY* 77 (May 4) 103. Designing spiral plate heat exchangers.

Mostinski, I. L. (1963) *Teploenergetika* 4, 66; English abstract in *Brit. Chem. Eng.* 8, 580 (1963). Calculation of boiling heat-transfer coefficients, based on the law of corresponding states.

Mueller, A. C. (1973) Heat Exchangers, Section 18, in Rosenow, W. M. and Hartnell, H. P. (eds) *Handbook of Heat Transfer* (McGraw-Hill).

Mukherjee, R. (1997) *Chem. Eng. Prog,* 93 (Feb) 26. Effectively design air cooled heat exchangers.

Myers, J.E. and Katz, D.L. (1953) *Chem. Eng. Prog. Symp. Ser.* 49(5), 107. Boiling coefficients outside horizontal tubes.

Nusselt, W. (1916) *Z. Ver. duet. Ing.* 60, 541, 569. Die Oberflächenkondensation des Wasserdampfes.

Ozisik, M. N. (1985) *Heat Transfer: a basic approach* (McGraw-Hill).

Owen, R. G. and Lee, W. C. (1983) *Inst. Chem. Eng. Sym. Ser.* No. 75, 261. A review of recent developments in condenser theory.

Palen, J. W. and Small, W. M. (1964) *Hyd. Proc. 43* (Nov.) 199. A new way to design kettle reboilers.

Palen, J. W., Shih, C. C., Yarden, A. and Taborek, J. (1974) *5th Int. Heat Transfer Conf.,* 204. Performance limitations in a large scale thermosiphon reboiler.

Palen, J. W. and Taborek, J. (1962) *Chem. Eng. Prog.* 58 (July) 39. Refinery kettle reboilers.

Palen, J. W. and Taborek, J. (1969) *Chem. Eng. Prog. Sym. Ser.* No. 92, 65, 53. Solution of shell side flow pressure drop and heat-transfer by stream analysis method.

Palen, J. W., Yarden, A. and Taborek, J. (1972) *Chem. Eng. Symp. Ser.* No. 118, 68, 50. Characteristics of boiling outside large-scale horizontal multitube boilers.

Parker, D. V. (1964) *Brit. Chem. Eng.* 1, 142. Plate heat exchangers.

Penny, W. R. (1983) Agitated vessels, in *Heat Exchanger Design Handbook* (Hemisphere), volume 3.

Porter, K. E. and Jeffreys, G. V. (1963) *Trans. Inst. Chem. Eng.* 41, 126. The design of cooler condensers for the condensation of binary vapours in the presence of a noncondensable gas.

Raju, K. S. N. and Chand J. (1980) *Chem. Eng., NY* 87 (Aug. 11) 133. Consider the plate heat exchanger.

Richardson, J. F., Harker, J. H. and Backhurst, J. (2002) *Chemical Engineering: Volume 2,* 3rd edn (Butterworth-Heinemann).

Rohsenow, W. M., Hartnett, J. P. and Cho, Y. L. (eds) (1998) *Handbook of Heat Transfer,* 3rd edn (McGraw-Hill).

Rubin, F. L. (1960) *Chem. Eng., NY* 67 (Oct. 31st) 91. Design of air cooled heat exchangers.

Sarma, N. V. L. S., Reddy, P. J. and Murti, P. S. (1973) *Ind. Eng. Chem. Proc. Des. Dev.* 12, 278. A computer design method for vertical thermosyphon reboilers.

Saunders, E. A. D. (1988) *Heat Exchangers* (Longmans).

Schlunder, E. U. (ed.) (1983) *Heat Exchanger Design Handbook* (Hemisphere). Five volumes with supplements.

Shah, M. M. (1976) ASHRAE Trans. 82 (Part 2) 66. A new correlation for heat-transfer during boiling flow through tubes.

Sieder, E. N. and Tate, G. E. (1936) *Ind. Eng. Chem.* **28**, 1429. Heat transfer and pressure drop of liquids in tubes.

Silver, L. (1947) *Trans. Inst. Chem. Eng.* **25**, 30. Gas cooling with aqueous condensation.

Singh, K. P. and Soler, A. I. (1992) *Mechanical Design of Heat Exchanger and Pressure Vessel Components* (Springer-Verlag).

Steinmeyer, D. E. (1972) *Chem. Eng. Prog.* **68** (July) 64. Fog formation in partial condensers.

Shellene, K. R., Sternling, C. V., Church, D. M. and Snyder, N. H. (1968) *Chem. Eng. Prog. Symp. Ser.* **64**(82), 102. An experimental study of vertical thermosiphon reboilers.

Taborek, J. (1974) Design methods for heat-transfer equipment: a critical survey of the state of the art, in Afgan, N. and Schlünder, E. V. (eds) *Heat Exchangers: Design and Theory Source Book* (McGraw-Hill).

Taborek, J., Aoki, T., Ritter, R. B. and Palen, J. W. (1972) *Chem. Eng. Prog.* **68** (Feb.) 59, (July) 69, in two parts. Fouling: the major unresolved problem in heat-transfer.

Tinker, T. (1951) *Proceedings of the General Discussion on Heat Transfer*, p. 89, Inst. Mech. Eng., London. Shell-side characteristics of shell and tube heat exchangers.

Tinker, T. (1958) *Trans. Am. Soc. Mech. Eng.* **80** (Jan.) 36. Shell-side characteristics of shell and tube exchangers.

Tong, L. S. and Tang, Y. S. (1997) *Boiling Heat Transfer and Two-Phase Flow*, 2nd edn (CRC Press).

Trinks, W., Mawhinney, M. H., Shannon, R. A., Reed, R. J. and Garvey, J. R. (2004) *Industrial Furnaces*, 6th edn (Wiley).

Trom L. (1990) Hyd. Proc. **69** (10) 75. Consider plate and spiral heat exchangers.

van Edmonds, S. (1994) Masters Thesis, University of Wales Swansea. *A short-cut design procedure for vertical thermosyphon reboilers*.

Uhl, W. W. and Gray, J. B. (eds) (1967) *Mixing Theory and Practice*, 2 vols (Academic Press).

Ward, D. J. (1960) *Petro./Chem. Eng.* **32**, C-42. How to design a multiple component partial condenser.

Webb, R. L. and Gupte, N. S. (1992) *Heat Trans. Eng.*, **13** (3) 58. A critical review of correlations for convective vaporization in tubes and tube banks.

Webber, W. O. (1960) *Chem. Eng.*, NY **53** (Mar. 21st) 149. Under fouling conditions finned tubes can save money.

Wilkinson, W. L. and Edwards, M. F. (1972) *Chem. Engr.*, London No. 264 (Aug) 310, No. 265 (Sept) 328. Heat transfer in agitated vessels.

Wimpress, N. (1978) *Chem. Eng.*, NY **85** (May 22nd) 95. Generalized method predicts fired-heater performance.

Wolverine (1984) *Wolverine Tube Heat Transfer Data Book—Low Fin Tubes* (Wolverine Inc.).

Yilmaz, S. B. (1987) Chem. Eng. Prog. **83** (11) 64. Horizontal shellside thermosiphon reboilers.

Zuber, N., Tribus, M. and Westwater, J. W. (1961) *Second International Heat Transfer Conference*, Paper 27, p. 230, Am. Soc. Mech. Eng. The hydrodynamic crisis in pool boiling of saturated and sub-cooled liquids.

British Standards

BS 3274: 1960 Tubular heat exchangers for general purposes.

BS 3606: 1978 Specification for steel tubes for heat exchangers.

PD 5500 (2003) Unfired fusion welded pressure vessels.

American and International Standards

API 661 / ISO 13706-1:2005 (2006) Air-Cooled Heat Exchangers for General Refinery Service, 6th edn (American Petroleum Institute).

ASME Boiler and Pressure Vessel Code Section VIII (2004) Rules for the construction of pressure vessels. (ASME International).

TEMA (1999) *Standards of the Tubular Exchanger Manufacturers' Association*, 8th edn (Tubular Exchanger Manufacturers' Association, New York).

Engineering Sciences Data Unit Reports

ESDU 78031 (2001) Internal forced convective heat transfer in coiled pipes.

ESDU 83038 (1984) Baffled shell-and-tube heat exchangers: flow distribution, pressure drop and heat-transfer coefficient on the shellside.

ESDU 84023 (1985) Shell-and-tube exchangers: pressure drop and heat-transfer in shellside downflow condensation.

ESDU 87019 (1987) Flow induced vibration in tube bundles with particular reference to shell and tube heat exchangers.

ESDU 92003 (1993) Forced convection heat transfer in straight tubes. Part 1: turbulent flow.

ESDU 93018 (2001) Forced convection heat transfer in straight tubes. Part 2: laminar and transitional flow.

ESDU 98003–98007 (1998) Design and performance evaluation of heat exchangers: the effectiveness-NTU method.

ESDU International plc, 27 Corsham Street, London N1 6UA, UK.

Bibliography

Azbel, D. (1984) *Heat Transfer Application in Process Engineering* (Noyles).

Cheremisinoff, N. P. (ed.)(1986) *Handbook of Heat and Mass Transfer*, 2 vols (Gulf).

Fraas, A. P. (1989) *Heat Exchanger Design*, 2nd edn (Wiley).

Gunn, D. and Horton, R. (1989) *Industrial Boilers* (Longmans).

Gupta, J. P. (1986) *Fundamentals of Heat Exchanger and Pressure Vessel Technology* (Hemisphere).

Kakac, S. (ed.) (1991) *Boilers, Evaporators, and Condensers* (Wiley)

Kakac, S., Bergles, A. E. and Mayinger, F. (eds) (1981) *Heat Exchangers: Thermal-hydraulic Fundamentals and Design* (Hemisphere).

McKetta, J. J. (ed.) (1990) *Heat Transfer Design Methods* (Marcel Dekker).

Palen, J. W, (ed.) (1986) *Heat Exchanger Source Book* (Hemisphere).

Podhorssky, M. and Krips, H. (1998) *Heat Exchangers: A Practical Approach to Mechanical Construction, Design, and Calculations* (Begell House).

Saunders, E. A. D. (1988) *Heat Exchangers* (Longmans).

Schlunder, E. U. (ed.) (1983) *Heat Exchanger Design Handbook*, 5 vols with supplements (Hemisphere).

Shah, R. K. and Sekulic, D. P. (2003) *Fundamentals of Heat Exchanger Design* (Wiley).

Shah, R. K., Subbarao, E. C. and Mashelkar, R. A. (eds) (1988) *Heat Transfer Equipment Design* (Hemisphere).

Singh, K. P. (1989) *Theory and Practice of Heat Exchanger Design* (Hemisphere).

Singh, K. P. and Soler, A. I. (1984) *Mechanical Design of Heat Exchanger and Pressure Vessel Components* (Arcturus).

Smith, R. A. (1986) *Vaporizers: Selection, Design and Operation* (Longmans).

Walker, G. (1982) *Industrial Heat Exchangers* (McGraw-Hill).

Yokell, S. (1990) A Working Guide to Shell and Tube Heat Exchangers (McGraw-Hill).

12.20. NOMENCLATURE

		Dimensions in $\mathbf{MLT}\theta$
A	Heat-transfer area	L^2
A_b	Bundle cross-sectional area of an air-cooled exchanger	L^2
A_{cp}	Cold-plane area of tubes	L^2
A_o	Outside area of bare tube	L^2
A_f	Fin area in equation 12.59	L^2
A_f	Cross-sectional area for flow in equation 12.56	L^2
A_o	Outside area of bare tube	L^2
A_p	Area of a port plate heat exchanger	L^2
A_s	Cross-flow area between tubes	L^2
a	Index in equation 12.10	—
b	Index in equation 12.10	—
C	Constant in equation 12.10	—
C_p	Heat capacity at constant pressure	$L^2T^{-2}\theta^{-1}$
C_{pg}	Heat capacity of gas	$L^2T^{-2}\theta^{-1}$
C_{pL}	Heat capacity of liquid phase	$L^2T^{-2}\theta^{-1}$
c	Index in equation 12.10	—
D	Agitator diameter	L
D_b	Bundle diameter	L
D_s	Shell diameter	L
D_v	Vessel diameter	L
d	Diameter	L
d_e	Equivalent (hydraulic mean) diameter	L
d_i	Tube inside diameter	L
d_{pt}	Diameter of the ports in the plates of a plate heat exchanger	L
d_o	Tube outside diameter	L
d_1	Outside diameter of inner of concentric tubes	L
d_2	Inside diameter of outer of concentric tubes	L
E	Term in equation 12.12	—
E_f	Fin efficiency	—
F	Radiation exchange factor in equation 12.67	—
F	Feed molar flow rate (Example 12.11)	MT^{-1}
F_t	Log mean temperature difference correction factor	—
f_c	Two-phase flow factor	—

f_m	Temperature correction factor for mixtures	—
f_s	Nucleate boiling suppression factor	—
G	Total mass flow rate per unit area	$ML^{-2}T^{-1}$
G_p	Mass flow rate per unit cross-sectional area between plates	$ML^{-2}T^{-1}$
G_s	Shell-side mass flow rate per unit area	$ML^{-2}T^{-1}$
G_t	Tube-side mass flow rate per unit area	$ML^{-2}T^{-1}$
g	Gravitational acceleration	LT^{-2}
H_s	Sensible heat of stream	ML^2T^{-3}
H_t	Total heat of stream (sensible + latent)	ML^2T^{-3}
h_c	Mean heat-transfer coefficient in condensation	$MT^{-3}\theta^{-1}$
$(h_c)_1$	Mean condensation heat-transfer coefficient for a single tube	$MT^{-3}\theta^{-1}$
$(h_c)_b$	Heat-transfer coefficient for condensation on a horizontal tube bundle	$MT^{-3}\theta^{-1}$
$(h_c)_{N_r}$	Mean condensation heat-transfer coefficient for a tube in a row of tubes	$MT^{-3}\theta^{-1}$
$(h_c)_v$	Heat-transfer coefficient for condensation on a vertical tube	$MT^{-3}\theta^{-1}$
$(h_c)_{BK}$	Condensation coefficient from Boyko–Kruzhilin correlation	$MT^{-3}\theta^{-1}$
$(h_c)_s$	Condensation heat-transfer coefficient for stratified flow in tubes	$MT^{-3}\theta^{-1}$
h'_c	Local condensing film coefficient, partial condenser	$MT^{-3}\theta^{-1}$
h_{cb}	Convective boiling heat-transfer coefficient	$MT^{-3}\theta^{-1}$
h_{cg}	Mean effective cooling-condensing heat-transfer coefficient, partial condenser	$MT^{-3}\theta^{-1}$
h'_{cg}	Local effective cooling-condensing heat-transfer coefficient, partial condenser	$MT^{-3}\theta^{-1}$
h_{df}	Fouling coefficient based on fin area	$MT^{-3}\theta^{-1}$
h_f	Heat-transfer coefficient based on fin area	$MT^{-3}\theta^{-1}$

h_{fb}	Film-boiling heat-transfer coefficient	$MT^{-3}\theta^{-1}$
h_{fc}	Forced-convection coefficient calculated assuming liquid is flowing alone (equation 12.47)	$MT^{-3}\theta^{-1}$
h'_{fc}	Forced-convection heat-transfer coefficient in equation 12.46	$MT^{-3}\theta^{-1}$
h_g	Mean gas-film heat-transfer coefficient	$MT^{-3}\theta^{-1}$
h'_g	Local sensible-heat-transfer coefficient, partial condenser	$MT^{-3}\theta^{-1}$
h_i	Film heat-transfer coefficient inside a tube	$MT^{-3}\theta^{-1}$
h_{id}	Fouling coefficient on inside of tube	$MT^{-3}\theta^{-1}$
h_{nb}	Nucleate boiling-heat-transfer coefficient	$MT^{-3}\theta^{-1}$
h'_{nb}	Nucleate boiling coefficient in equation 12.46	$MT^{-3}\theta^{-1}$
h_o	Heat-transfer coefficient outside a tube	$MT^{-3}\theta^{-1}$
h_{od}	Fouling coefficient on outside of tube	$MT^{-3}\theta^{-1}$
h_p	Heat-transfer coefficient in a plate heat exchanger	$MT^{-3}\theta^{-1}$
h_s	Shell-side heat-transfer coefficient	$MT^{-3}\theta^{-1}$
h'_t	Inside film coefficient in Boyko–Kruzhilin correlation	$MT^{-3}\theta^{-1}$
h_v	Heat transfer coefficient to vessel wall or coil	$MT^{-3}\theta^{-1}$
J	Term in Boyko–Kruzhilin correlation, equation 12.31	—
j_h	Heat-transfer factor defined by equation 12.14	—
j_H	Heat-transfer factor defined by equation 12.15	—
j_f	Friction factor	—
K_1	Constant in equation 12.3, from Table 12.4	—
K_2	Constant in equation 12.40	—
K_b	Constant in equation 12.52	—
K_i	Phase equilibrium constant for component I	—
k_f	Thermal conductivity of fluid	$MLT^{-3}\theta^{-1}$
k_L	Thermal conductivity of liquid	$MLT^{-3}\theta^{-1}$
k_v	Thermal conductivity of vapour	$MLT^{-3}\theta^{-1}$

k_w	Thermal conductivity of tube-wall material	$MLT^{-3}\theta^{-1}$
L	Tube length	L
L	Liquid molar flow rate in Example 12.11	MT^{-1}
L'	Effective tube length	L
L_P	Path length in a plate heat exchanger	L
L_s	Stack height	L
l_B	Baffle spacing (pitch)	L
l_f	Fin height	L
m	Index in equation 12.19	—
N	Rotational speed	T^{-1}
N_b	Number of baffles	—
N_{bk}	Number of tubes per bank	—
N_p	Number of passes	—
N_r	Number of tubes in a vertical row	—
N_t	Number of tubes in a tube bundle	—
NTU	Number of transfer units	—
n_1	Index in equation 12.3, from Table 12.4	—
P	Total pressure	$ML^{-1}T^{-2}$
P_c	Critical pressure	$ML^{-1}T^{-2}$
P_d	Stack draft	L
ΔP	Pressure drop	$ML^{-1}T^{-2}$
ΔP_b	Pressure drop across air-cooled exchanger bundle	$ML^{-1}T^{-2}$
ΔP_f	Pressure drop due to friction	$ML^{-1}T^{-2}$
ΔP_p	Pressure drop in a plate heat exchanger	$ML^{-1}T^{-2}$
ΔP_{pt}	Pressure loss through the ports in a plate heat exchanger	$ML^{-1}T^{-2}$
ΔP_s	Shell-side pressure drop	$ML^{-1}T^{-2}$
ΔP_s	Static pressure in thermosiphon tubes	$ML^{-1}T^{-2}$
ΔP_t	Tube-side pressure drop	$ML^{-1}T^{-2}$
p'	Atmospheric pressure	$ML^{-1}T^{-2}$
p_f	Fin pitch	L
p_s	Saturation vapour pressure	$ML^{-1}T^{-2}$
p_t	Tube pitch	L
p_w	Saturation vapour pressure corresponding to wall temperature	$ML^{-1}T^{-2}$
Q	Heat transferred in unit time	ML^2T^{-3}
Q_g	Sensible-heat-transfer rate from gas phase	ML^2T^{-3}

Q_r	Radiant-heat-transfer rate	$\mathbf{ML^2T^{-3}}$
Q_t	Total heat-transfer rate from gas phase	$\mathbf{ML^2T^{-3}}$
q	Heat flux (heat-transfer rate per unit area)	$\mathbf{MT^{-3}}$
q'	Uncorrected value of flux from Figure 12.55	$\mathbf{MT^{-3}}$
q_c	Maximum (critical) flux for a single tube	$\mathbf{MT^{-3}}$
q_{cb}	Maximum flux for a tube bundle	$\mathbf{MT^{-3}}$
q_r	Radiant heat flux	$\mathbf{MT^{-3}}$
R	Dimensionless temperature ratio defined by equation 12.6	—
S	Dimensionless temperature ratio defined by equation 12.7	—
T	Shell-side temperature	θ
T	Temperature of surface	θ
T_a	Ambient temperature	θ
T_{bi}	Temperature of liquid entering reboiler	θ
T_{bo}	Temperature of vapour leaving reboiler	θ
T_c	Critical temperature	θ
T_g	Temperature of combustion gases	θ
T_{ga}	Average flue-gas temperature	θ
T_r	Reduced temperature	—
T_s	Saturation temperature	θ
T_{sat}	Saturation temperature	θ
T_t	Tube surface temperature	θ
T_v	Vapour (gas) temperature	θ
T_w	Wall (surface) temperature	θ
T_1	Hot-side inlet temperature	θ
T_2	Hot-side exit temperature	θ
ΔT	Temperature difference	θ
ΔT_{lm}	Logarithmic mean temperature difference	θ
ΔT_m	Mean temperature difference in equation 12.1	θ
ΔT_s	Temperature change in vapour (gas) stream	θ
t	Tube-side temperature	θ
t	Water temperature in equation 12.17	θ
t_c	Local coolant temperature	θ
t_f	Fin thickness	$\mathbf{L}$
t_i	Stream inlet temperature	θ

t_o	Stream outlet temperature	θ
t_w	Estimated wall temperature	θ
t_1	Cold-side inlet temperature	θ
t_2	Cold-side exit temperature	θ
U	Overall heat-transfer coefficient	$MT^{-3}\theta^{-1}$
U'	Uncorrected overall coefficient, equation 12.51	$MT^{-3}\theta^{-1}$
U_c	Corrected overall coefficient, equation 12.52	$MT^{-3}\theta^{-1}$
U_o	Overall heat-transfer coefficient based on tube outside area	$MT^{-3}\theta^{-1}$
U_1, U_2	Overall heat-transfer coefficients evaluated at the ends of the exchanger	$MT^{-3}\theta^{-1}$
u	Fluid velocity	LT^{-1}
u_L	Liquid velocity, equation 12.34	LT^{-1}
u_p	Fluid velocity in a plate heat exchanger	LT^{-1}
u_{pt}	Velocity through the ports of a plate heat exchanger	LT^{-1}
u_p	Velocity through channels of a plate heat exchanger	LT^{-1}
u_s	Shell-side fluid velocity	LT^{-1}
u_t	Tube-side fluid velocity	LT^{-1}
u_v	Vapour velocity, equation 12.34	LT^{-1}
$\hat{u}_v$	Maximum vapour velocity in kettle reboiler	LT^{-1}
V	Vapour molar flow rate	MT^{-1}
v_i	Specific volume at inlet	L^3M^{-1}
v_o	Specific volume at outlet	L^3M^{-1}
W	Mass flow rate of fluid	MT^{-1}
W_c	Total condensate mass flow rate	MT^{-1}
W_s	Shell-side fluid mass flow rate	MT^{-1}
w	Mass flow through the channels and ports in a plate heat exchanger	MT^{-1}
w_f	Fan power	ML^2T^{-3}
X_{tt}	Lockhart-Martinelli two-phase flow parameter	—
x	Mass fraction of vapour	—
x_i	Mole fraction of component i in liquid phase	—
y_i	Mole fraction of component i in vapour phase	—

Z	Ratio of change in sensible heat to change in total heat of gas stream (sensible + latent)	—
z_i	Mole fraction of component i in feed	—
α	Absorption efficiency factor	—
Γ	Tube loading	$ML^{-1}T^{-1}$
Γ_h	Condensate loading on a horizontal tube	$ML^{-1}T^{-1}$
Γ_v	Condensate loading on a vertical tube	$ML^{-1}T^{-1}$
η_f	Fan efficiency	—
λ	Latent heat	L^2T^{-2}
μ	Viscosity at bulk fluid temperature	$ML^{-1}T^{-1}$
μ_L	Liquid viscosity	$ML^{-1}T^{-1}$
μ_v	Vapour viscosity	$ML^{-1}T^{-1}$
μ_w	Viscosity at wall temperature	$ML^{-1}T^{-1}$
ρ	Fluid density	ML^{-3}
ρ_L	Liquid density	ML^{-3}
ρ_m	Mean density	ML^{-3}
ρ_v	Vapour density	ML^{-3}
σ	Stephan-Boltzmann constant in equation 12.66	$MT^{-3}\theta^{-4}$
σ	Surface tension	MT^{-3}

Dimensional Numbers

Nu	Nusselt number
Pr	Prandtl number
Pr_c	Prandtl number for condensate film
Re	Reynolds number
Re_c	Reynolds number for condensate film
Re_L	Reynolds number for liquid phase
St	Stanton number

12.21. PROBLEMS

12.1. A solution of sodium hydroxide leaves a dissolver at 80°C and is to be cooled to 40°C, using cooling water. The maximum flow rate of the solution will be 8000 kg/h. The maximum inlet temperature of the cooling water will be 20°C and the temperature rise is limited to 20°C.

Design a double-pipe exchanger for this duty, using standard carbon steel pipe and fittings. Use pipe of 50 mm inside diameter, 55 mm outside diameter for the inner pipe, and 75 mm inside diameter pipe for the outer.

Make each section 5 m long. The physical properties of the caustic solution are:

temperature, °C	40	80
specific heat, $kJkg^{-1}{}^{\circ}C^{-1}$	3.84	3.85
density, kg/m^3	992.2	971.8
thermal conductivity, W/m°C	0.63	0.67
viscosity, $mN\ m^{-2}$ s	1.40	0.43

12.2. A double-pipe heat exchanger is to be used to heat 6000 kg/h of 22 mol% hydrochloric acid. The exchanger will be constructed from karbate (impervious carbon) and steel tubing. The acid will flow through the inner, karbate, tube and saturated steam at 100°C will be used for heating. The tube dimensions will be: karbate tube inside diameter 50 mm, outside diameter 60 mm; steel tube inside diameter 100 mm. The exchanger will be constructed in sections, with an effective length of 3 m each.

How many sections will be needed to heat the acid from 15°C to 65°C? Physical properties of 22% HCl at 40°C: specific heat 4.93 $kJkg^{-1}{}^{\circ}C^{-1}$, thermal conductivity 0.39 W/m°C, density 866 kg/m^3.

Viscosity:	temperature	20	30	40	50	60	70°C
	$mN\ m^{-2}$s	0.68	0.55	0.44	0.36	0.33	0.30

Karbate thermal conductivity 480 W/m°C.

12.3. In a food processing plant there is a requirement to heat 50,000 kg/h of town water from 10°C to 70°C. Steam at 2.7 bar is available for heating the water. An existing heat exchanger is available, with the following specification:

Shell inside diameter 337 mm, E type
Baffles 25% cut, set at a spacing of 106 mm
Tubes 15 mm inside diameter, 19 mm outside diameter, 4094 mm long
Tube pitch 24 mm, triangular
Number of tubes 124, arranged in a single pass.
Would this exchanger be suitable for the specified duty?

12.4. Design a shell and tube exchanger to heat 50,000 kg/h of liquid ethanol from 20°C to 80°C. Steam at 1.5 bar is available for heating. Assign the ethanol to the tube side. The total pressure drop must not exceed 0.7 bar for the alcohol stream. Plant practice requires the use of carbon steel tubes, 25 mm inside diameter, 29 mm outside diameter, 4 m long.

Set out your design on a data sheet and make a rough sketch of the heat exchanger. The physical properties of ethanol can be readily found in the literature.

12.5. Ammonia vapour, 4500 kg/h at 6.7 bara pressure, is to be cooled from 120°C to 40°C, using cooling water. The maximum supply temperature of the cooling water available is 30°C, and the outlet temperature is to be restricted to 40°C. The pressure drops over the exchanger must not exceed 0.5 bar for the ammonia stream and 1.5 bar for the cooling water.

A contractor has proposed using a shell and tube exchanger with the following specification for this duty:

Shell: E-type, inside diameter 590 mm
Baffles: 25% cut, 300 mm spacing
Tubes: carbon steel, 15 mm inside diameter, 19 mm outside diameter, 2400 mm long, number 360.
Tube arrangement: eight passes, triangular tube pitch, pitch 23.75 mm
Nozzles: shell 150 mm inside diameter, tube headers 75 mm inside diameter.
It is proposed to put the cooling water though the tubes.
Is the proposed design suitable for the duty?
Physical properties of ammonia at the mean temperature of 80°C: specific heat 2.418 kJkg^{-1}°C^{-1}, thermal conductivity 0.0317 W/m°C, density 4.03 kg/m^3, viscosity 1.21×10^{-5} N m^{-2}s.

12.6. A vaporizer is required to evaporate 10,000 kg/h of a process fluid, at 6 bar. The liquid is fed to the vaporizer at 20°C.

The plant has a spare kettle reboiler available with the following specification. U-tube bundle, 50 tubes, mean length 4.8 m, end to end.

Carbon steel tubes, inside diameter 25 mm, outside diameter 30 mm, square pitch 45 mm.

Steam at 1.7 bara will be used for heating.

Check if this reboiler would be suitable for the duty specified. Only check the thermal design. You may take it that the shell will handle the vapour rate.

Take the physical properties of the process fluid (liquid) as: density 535 kg/m^3, specific heat 2.6 kJkg^{-1}°C^{-1}, thermal conductivity 0.094 W/m°C, viscosity 0.12 mN m^{-2}s, surface tension 0.85 N/m, heat of vaporization 322 kJ/kg.
Vapour density 14.4 kg/m^3
Vapour pressure:

temperature, °C	50	60	70	80	90	100	110	120
pressure, bar		5.0	6.4	8.1	10.1	12.5	15.3	18.5	20.1

12.7. A condenser is required to condense n-propanol vapour leaving the top of a distillation column. The n-propanol is essentially pure, and is a saturated vapour at a pressure of 2.1 bara. The condensate needs to be sub-cooled to 45°C.

Design a horizontal shell and tube condenser capable of handling a vapour rate of 30,000 kg/h. Cooling water is available at 30°C and the temperature rise is to be limited to 30°C. The pressure drop on the vapour stream is to be less than $50 \, kN/m^2$, and on the water stream less than $70 \, kN/m^2$. The preferred tube size is 16 mm inside diameter, 19 mm outside diameter, and 2.5 m long.

Take the saturation temperature of n-propanol at 2.1 bar as 118°C. The other physical properties required can be found in the literature, or estimated.

12.8. Design a vertical shell and tube condenser for the duty given in question 12.7. Use the same preferred tube size.

12.9. In the manufacture of methyl ethyl ketone (MEK) from 2-butanol, the reactor products are precooled and then partially condensed in a shell and tube exchanger. A typical analysis of the stream entering the condenser is, mol fractions: MEK 0.47, unreacted alcohol 0.06, hydrogen 0.47. Only 85% of the MEK and alcohol are condensed. The hydrogen is non-condensable.

The vapours enter the condenser at 125°C and the condensate and uncondensed material leave at 27°C. The condenser pressure is maintained at 1.1 bara. Make a preliminary design of this condenser, for a feed rate of 1500 kg/h. Chilled water will be used as the coolant, at an inlet temperature of 10°C and allowable temperature rise of 30°C.

Any of the physical properties of the components not available in Appendix C, or the general literature, should be estimated.

12.10. A vertical thermosyphon reboiler is required for a column. The liquid at the base of the column is essentially pure n-butane. A vapour rate of 5 kg/s is required.

The pressure at the base of the column is 20.9 bar. Saturated steam at 5 bar will be used for heating.

Estimate the number of 25-mm outside diameter, 22-mm inside diameter, 4-m-long tubes needed.

At 20.9 bar the saturation temperature of n-butane is 117°C and the heat of vaporization 828 kJ/kg.

12.11. An immersed bundle vaporizer is to be used to supply chlorine vapour to a chlorination reactor, at a rate of 10,000 kg/h. The chlorine vapour is required at 5 bar pressure. The minimum temperature of the chlorine feed will be 10°C. Hot water at 50°C is available for heating. The pressure drop on the water side must not exceed 0.8 bar.

Design a vaporizer for this duty. Use stainless-steel U-tubes, 6 m long, 21 mm inside diameter, 25 mm outside diameter, on a square pitch of 40 mm.

The physical properties of chlorine at 5 bar are: saturation temperature 10°C, heat of vaporization 260 kJ/kg, specific heat $0.99 \, kJ \, kg^{-1}{}^{\circ}C^{-1}$, thermal conductivity 0.13 W/m°C, density $1440 \, kg/m^3$, viscosity $0.3 \, mN \, m^{-2} s$, surface tension 0.013 N/m, vapour density $16.3 \, kg/m^3$.

The vapour pressure can be estimated from the equation:

$$\ln(P) = 9.34 - 1978/(T + 246) \qquad Pbar, \, T \, ^{\circ}C$$

12.12. There is a requirement to cool 200,000 kg/h of a dilute solution of potassium carbonate from 70°C to 30°C. Cooling water will be used for cooling, with inlet and outlet temperatures of 20°C and 60°C. A gasketed-plate heat exchanger is available with the following specifications:

Number of plates = 329
Effective plate dimensions: length 1.5 m, width 0.5 m, thickness 0.75 mm
Channel width = 3 mm
Flow arrangement two pass: two pass
Port diameters 150 mm.

Check if this exchanger is likely to be suitable for the thermal duty required, and estimate the pressure drop for each stream.

Take the physical properties of the dilute potassium carbonate solution to be the same as those for water.

12.13. Design an air cooler to cool 30,000 kg/h of diesel oil from 120°C to 50°C. The highest ambient temperature that is exceeded for 40 h/year on average is 40°C.

Physical properties of diesel oil over the temperature range of interest can be taken as: specific heat capacity 2.1 kJ/kg°C, thermal conductivity 0.135 W/m°C, density 800 kg/m^3, viscosity 1.2 mN m^{-2}s.

The temperature dependence of viscosity can be estimated using the method given in Chapter 8.

12.14. The heat duty of the jacketed vessel of Examples 12.15 and 12.16 can be estimated from the coolant flow rate and temperatures. What is the minimum temperature at which the reactor can operate using the jacket as designed? Does the choice of coolant make sense at this temperature? Propose a better design.

12.15. A stirred tank fermentation reactor has height 2 m, diameter 1.5 m and is filled with a fermentation broth that can be assumed to have the physical properties of water. The fermenter must be maintained at a temperature less than 42°C to prevent damage to the cell culture. After allowing for sensible heat losses to the cold feed, the heat that must be removed from the fermenter is 80 kW. Cooling water is available at 20°C and can be returned at any temperature up to 35°C. Recommend a design for cooling the contents of the fermenter.

13 MECHANICAL DESIGN OF PROCESS EQUIPMENT

Chapter Contents

Key Learning Objectives

- What factors a process engineer must consider when setting specifications for a pressure vessel
- How pressure vessels are designed and what determines the vessel wall thickness
- How codes and standards are used in pressure vessel design

13.1. INTRODUCTION

This chapter covers those aspects of the mechanical design of chemical plant that are of particular interest to chemical engineers. The main topic considered is the design of pressure vessels. The design of storage tanks and heat-exchanger tube sheets are also discussed briefly.

The chemical engineer will not usually be called on to undertake the detailed mechanical design of a pressure vessel. Vessel design is a specialized subject, and will be carried out by mechanical engineers who are conversant with the current design codes and practices, and methods of stress analysis. However, the chemical engineer will be responsible for developing and specifying the basic design information for a particular vessel, and needs to have a general appreciation of pressure vessel design to work effectively with the specialist designer.

Another reason why the process engineer must have an appreciation of methods of fabrication, design codes and other constraints on pressure vessel design is that these constraints often dictate limits on the process conditions. Mechanical constraints can cause significant cost thresholds in design, for example, when a costlier grade of alloy is required above a certain temperature.

The basic data needed by the specialist designer will be:

1. Vessel function
2. Process materials and services
3. Operating and design temperature and pressure
4. Materials of construction
5. Vessel dimensions and orientation
6. Type of vessel heads to be used
7. Openings and connections required
8. Specification of heating and cooling jackets or coils
9. Type of agitator
10. Specification of internal fittings.

An elementary understanding of pressure vessel design is needed in the preliminary stages of design, as most correlations for pressure vessel costs are based on the weight of metal required, and hence require an estimate of the vessel wall thickness as well as its volume. In many cases the required wall thickness will be determined by the combination of loads acting on the vessel rather than internal pressure alone.

Pressure vessel information is included in the data sheets for fixed-bed reactors, vapour–liquid contactors and heat exchangers given in Appendix G and available in the on-line material at http://books.Elsevier.com/companions.

There is no strict definition of what constitutes a pressure vessel, and different codes and regulations apply in different countries; however, it is generally accepted that any closed vessel over 150 mm diameter subject to a pressure difference of more than 0.5 bar should be designed as a pressure vessel.

It is not possible to give a completely comprehensive account of vessel design in one chapter. The design methods and data given should be sufficient for the preliminary design of conventional vessels; for the chemical engineer to check the feasibility of a proposed equipment design, to estimate the vessel cost for an economic analysis, and to determine the vessel's general proportions and weight for plant layout purposes. For a more detailed account of pressure vessel design the reader should refer to the books by Singh and Soler (1992), Escoe (1994) and Moss (2003). Other useful books on the mechanical design of process equipment are listed in the bibliography at the end of this chapter.

An elementary understanding of the principles of the 'Strength of Materials' (Mechanics of Solids) will be needed to follow this chapter. Readers who are not familiar with the subject should consult one of the many textbooks available, such as those by Case *et al.* (1999), Mott, R. L. (2001), Seed (2001) and Gere and Timoshenko (2000).

13.1.1. Classification of Pressure Vessels

For the purposes of design and analysis, pressure vessels are sub-divided into two classes depending on the ratio of the wall thickness to vessel diameter: thin-walled vessels, with a thickness ratio of less than 1:10; and thick-walled above this ratio.

The principal stresses (see Section 13.3.1) acting at a point in the wall of a vessel, due to a pressure load, are shown in Figure 13.1. If the wall is thin, the radial stress σ_3 will be small and can be neglected in comparison with the other stresses, and the longitudinal and circumferential stresses σ_1 and σ_2 can be taken as constant over the wall thickness. In a thick wall, the magnitude of the radial stress will be significant, and the circumferential stress will vary across the wall. The majority of the vessels used in the chemical and allied industries are classified as thin-walled vessels. Thick-walled vessels are used for high pressures, and are discussed in Section 13.15.

13.2. PRESSURE VESSEL CODES AND STANDARDS

In all the major industrialized countries the design and fabrication of thin-walled pressure vessels is covered by national standards and codes of practice. In most countries it is a legal requirement that pressure vessels must be designed, constructed and tested in accordance with part or all of the design code. The primary purpose of the design codes is to establish rules of safety relating to the pressure integrity of

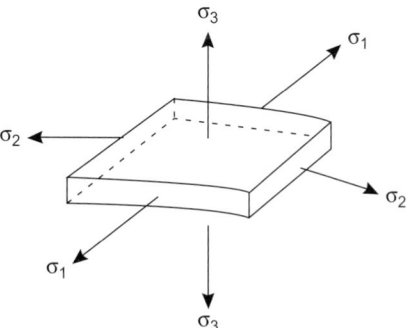

Figure 13.1. Principal stresses in pressure vessel wall.

vessels and provide guidance on design, materials of construction, fabrication, inspection and testing. They form a basis of agreement between the manufacturer and customer, and the customer's insurance company.

There is no international standard for pressure vessel design, but the most widely used standard is the ASME Boiler and Pressure Vessel Code (the ASME BPV Code), which is the required standard in the USA, Canada and many other countries. The 12 sections of the ASME BPV Code are listed in Table 13.1. Most chemical plant and refinery vessels fall within the scope of Section VIII of the ASME BPV Code. Section VIII contains three sub-divisions:

Division 1: contains general rules and is most commonly followed, particularly for low pressure vessels.

Division 2: contains alternative rules that are more restrictive on materials, design temperatures, design details, fabrication methods and inspection, but allow higher design stresses and hence thinner vessel walls. Division 2 rules are usually chosen for large, high-pressure vessels where the savings in metal cost and fabrication complexity offsets the higher engineering and construction costs.

Division 3: contains alternative rules intended for vessels with design pressures greater than 10,000 psig. It does not establish a maximum pressure for vessels designed in accordance with Division 1 or Division 2, but provides alternative rules that can be followed for thicker-walled vessels.

TABLE 13.1.   The 2004 ASME Boiler and Pressure Vessel Code

SECTIONS		
I	Rules for construction of power boilers	
II	Materials	
	Part A	Ferrous metal specifications
	Part B	Non-ferrous metal specifications
	Part C	Specifications for welding rods, electrodes and filler metals
	Part D	Properties (customary or metric versions)
III	Nuclear power plant components	
	NCA	General requirements
	Division 1	
	Division 2	Code for concrete containments
	Division 3	Containments for transport and storage of spent nuclear fuel and high-level radioactive material and waste
IV	Rules for construction of heating boilers	
V	Non-destructive examination	
VI	Recommended rules for the care and operation of heating boilers	
VII	Recommended guidelines for the care of power boilers	
VIII	Rules for the construction of pressure vessels	
	Division 1	
	Division 2	Alternative rules
	Division 3	Alternative rules for the construction of high-pressure vessels
IX	Welding and brazing qualifications	
X	Fibre-reinforced plastic vessels	
XI	Rules for in service inspection of nuclear power plant components	
XII	Rules for construction and continued service of transport tanks	

In the following sections, reference will normally be made to the BPV Code Sec. VIII D.1.The scope of the BPV Code Sec. VIII D.1 covers vessels made from iron, steels and non-ferrous metals. It specifically excludes:

1. Vessels within the scope of other sections of the BPV code. For example, power boilers (Sec. I), fibre-reinforced plastic vessels (Sec. X) and transport tanks (Sec. XII).
2. Fired-process tubular heaters.
3. Pressure containers that are integral parts of rotating or reciprocating devices, such as pumps, compressors, turbines or engines.
4. Piping systems (which are covered by ASME B31.3; see Chapter 5).
5. Piping components and accessories such as valves, strainers, in-line mixers and spargers.
6. Vessels containing water at less than 300 psi (2 MPa) and less than 210°F (99°C).
7. Hot-water storage tanks heated by steam with heat rate less than 0.2 MMBTU/h (58.6 kW), water temperature less than 210°F (99°C) and volume less than 120 gal (450 litres).
8. Vessels having internal pressure less than 15 psi (100 kPa) or greater than 3000 psi (20 MPa).
9. Vessels of internal diameter or height less than 6 in (152 mm).
10. Pressure vessels for human occupancy.

The ASME BPV Code can be ordered from ASME and is also available on-line (for example at www.ihs.com). The most recent edition of the code should always be consulted during detailed design.

In addition to the BPV Code Sec. VIII, the process design engineer will frequently need to consult Section II Part D, which lists maximum allowable stress values under Sec. VIII D.1 and D.2, as well as other materials properties. A comprehensive review of the ASME code is given by Chuse and Carson (1992) and Yokell (1986); see also Green and Perry (2007).

In the European Union, the design, manufacture and use of pressure systems is covered by the Pressure Equipment Directive (Council Directive 97/23/EC). European standard EN 13445, whose use became mandatory in May 2002, provides rules and guidelines similar to those laid out in the ASME BPV Code. The design of fibre-reinforced plastic vessels is covered by European standard EN 13923. The European standards can be obtained from any of the European Union member country national standards agencies; for example, BS EN 13445 can be ordered from www.bsi-global.com. The EN 13445 standard allows for either design by formula (DBF) or design by analysis (DBA), corresponding roughly to Divisions 1 and 2 of the ASME BPV Code. A comparison of both EN 13445 methods and ASME BPV Code Sec. VIII Divisions 1 and 2 was made by Preiss and Zeman (2004), and is available for free download from the European Standards Agency as report CEN/EXPERT/2004/45. They studied nine pressure vessels and concluded that the variance in cost between designs from different manufacturers was generally greater than the difference in cost between designs made under the different national codes. They concluded that the main difference between the codes is that the ASME fatigue results are less conservative than the EN 13445 requirements and are not in conformity with the European Pressure Equipment Directive requirements. The ASME hydraulic test requirements are also less stringent than the European requirements.

In the UK, pressure vessels for use in the chemical and allied industries are designed and fabricated according to the European Standard BS EN 13445, and the former British Standard BS 5500 which is now published by the British Standards Institute as PD 5500. The current (2003) edition of PD 5500 covers vessels fabricated in carbon and alloy steels, and aluminium.

Where national codes are not available, the ASME, British or European codes would normally be used.

Information and guidance on the pressure vessel codes can be found on the internet at www.ihs.com or www.bsi-global.com.

The national codes and standards dictate the minimum requirements, and give general guidance for design and construction; any extension beyond the minimum code requirement will be determined by agreement between the manufacturer and customer. Each of the design codes provides a complete self-consistent set of rules. It is very important that the design should be carried out using only one code, and it is never permissible to mix-and-match rules or allowable stresses from different codes.

The codes and standards are drawn up by committees of engineers experienced in vessel design and manufacturing techniques, and are a blend of theory, experiment and experience. They are periodically reviewed, and revisions are issued to keep abreast of developments in design, stress analysis, fabrication and testing. The latest version of the appropriate national code or standard should always be consulted before undertaking the design of any pressure vessel.

Several commercial computer programs to aid in the design of vessels to the ASME code and other international codes are available. These programs will normally be used by the specialist mechanical engineers who carry out the detailed vessel design. Some examples include:

Pressure Vessel Suite (Computer Engineering Inc.)
PVElite and CodeCalc (COADE Inc.)
TEMA/ASME and COMPRESS (Codeware Inc.).

Some of these programs support several different national codes. Some of them also offer free trial versions of the software, which can be very useful for solving simple problems; see the company web sites for details.

13.3. FUNDAMENTAL PRINCIPLES AND EQUATIONS

This section has been included to provide a basic understanding of the fundamental principles that underlie the design equations given in the sections that follow. The derivation of the equations is given in outline only. A detailed knowledge of the material in this section is not required for preliminary vessel design, but the equations derived here will be referenced and applied in subsequent sections. A full discussion of the topics covered can be found in any text on the strength of materials (mechanics of solids).

13.3.1. **Principal Stresses**

The state of stress at a point in a structural member under a complex system of loading is described by the magnitude and direction of the principal stresses. The principal stresses are the maximum values of the normal stresses at the point, which act on planes on which the shear stress is zero. In a two-dimensional stress system (Figure 13.2), the principal stresses at any point are related to the normal stresses in the x and y directions σ_x and σ_y and the shear stress τ_{xy} at the point by the following equation:

$$\text{Principal stresses, } \sigma_1, \ \sigma_2 = \frac{1}{2}(\sigma_y + \sigma_x) \pm \frac{1}{2}\sqrt{[(\sigma_y - \sigma_x)^2 + 4\tau_{xy}^2]} \qquad (13.1)$$

The maximum shear stress at the point is equal to half the algebraic difference between the principal stresses:

$$\text{Maximum shear stress} = \frac{1}{2}(\sigma_1 - \sigma_2) \qquad (13.2)$$

Compressive stresses are conventionally taken as negative, tensile as positive.

13.3.2. **Theories of Failure**

The failure of a simple structural element under unidirectional stress (tensile or compressive) is easy to relate to the tensile strength of the material, as determined in a standard tensile test, but for components subjected to combined stresses (normal and shear stress) the position is not so simple, and several theories of failure have been proposed. The three theories most commonly used are described below.

Maximum principal stress theory: this postulates that a member will fail when one of the principal stresses reaches the failure value in simple tension, σ_e'. The failure point in a simple tension is taken as the yield-point stress, or the tensile strength of the material, divided by a suitable factor of safety.

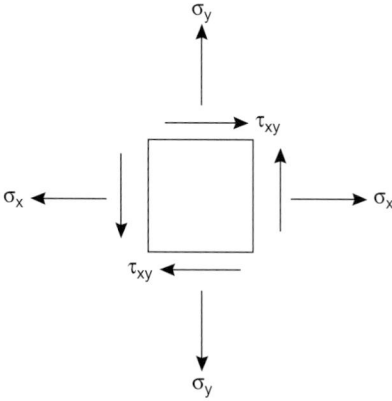

Figure 13.2.   Two-dimensional stress system.

Maximum shear stress theory: this postulates that failure will occur in a complex stress system when the maximum shear stress reaches the value of the shear stress at failure in simple tension.

For a system of combined stresses there are three shear stress maxima:

$$\tau_1 = \frac{\sigma_1 - \sigma_2}{2} \tag{13.3a}$$

$$\tau_2 = \frac{\sigma_2 - \sigma_3}{2} \tag{13.3b}$$

$$\tau_3 = \frac{\sigma_3 - \sigma_1}{2} \tag{13.3c}$$

In the tensile test

$$\tau_e = \frac{\sigma'_e}{2} \tag{13.4}$$

The maximum shear stress will depend on the sign of the principal stresses as well as their magnitude, and in a two-dimensional stress system, such as that in the wall of a thin-walled pressure vessel, the maximum value of the shear stress may be that given by putting $\sigma_3 = 0$ in equations 13.3b and c.

The maximum shear stress theory is often called Tresca's, or Guest's, theory.

Maximum strain energy theory: this postulates that failure will occur in a complex stress system when the total strain energy per unit volume reaches the value at which failure occurs in simple tension.

The maximum shear-stress theory has been found to be suitable for predicting the failure of ductile materials under complex loading and is the criterion normally used in pressure vessel design.

13.3.3. Elastic Stability

Under certain loading conditions, failure of a structure can occur not through gross yielding or plastic failure, but by buckling or wrinkling. Buckling causes a gross and sudden change of shape of the structure; unlike failure by plastic yielding, where the structure retains the same basic shape. This mode of failure will occur when the structure is not elastically stable: when it lacks sufficient stiffness, or rigidity to withstand the load. The stiffness of a structural member is dependent not on the basic strength of the material but on its elastic properties (Young's modulus, E_Y, and Poisson's ratio, v) and the cross-sectional shape of the member. The classic example of failure due to elastic instability is the buckling of tall thin columns (struts), which is described in any elementary text on the strength of materials.

For a structure that is likely to fail by buckling there will be a certain critical value of load below which the structure is stable; if this value is exceeded, catastrophic failure through buckling can occur.

The walls of pressure vessels are usually relatively thin compared with the other dimensions and can fail by buckling under compressive loads. This is particularly important for tall wide vessels such as distillation columns that can experience compressive loads from wind loads.

Elastic buckling is the decisive criterion in the design of thin-walled vessels under external pressure.

13.3.4. Membrane Stresses in Shells of Revolution

A shell of revolution is the form swept out by a line or curve rotated about an axis. (A solid of revolution is formed by rotating an area about an axis.) Most process vessels are made up from shells of revolution: cylindrical and conical sections, and hemispherical, ellipsoidal and torispherical heads; see Figure 13.3.

The walls of thin vessels can be considered to be 'membranes'; supporting loads without significant bending or shear stresses, similar to the walls of a balloon.

The analysis of the membrane stresses induced in shells of revolution by internal pressure gives a basis for determining the minimum wall thickness required for vessel shells. The actual thickness required will also depend on the stresses arising from the other loads to which the vessel is subjected.

Consider the shell of revolution of general shape shown in Figure 13.4, under a loading that is rotationally symmetric; that is, the load per unit area (pressure) on the shell is constant round the circumference, but not necessarily the same from top to bottom.

Let

P = pressure
t = thickness of shell
σ_1 = the meridional (longitudinal) stress, the stress acting along a meridian
σ_2 = the circumferential or tangential stress, the stress acting along parallel circles (often called the hoop stress)
r_1 = the meridional radius of curvature
r_2 = circumferential radius of curvature.

Note: the vessel has a double curvature; the values of r_1 and r_2 are determined by the shape.

Consider the forces acting on the element defined by the points a, b, c, d. Then the normal component (component acting at right angles to the surface) of the pressure force on the element

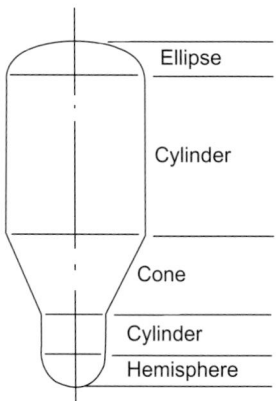

Figure 13.3. Typical vessel shapes.

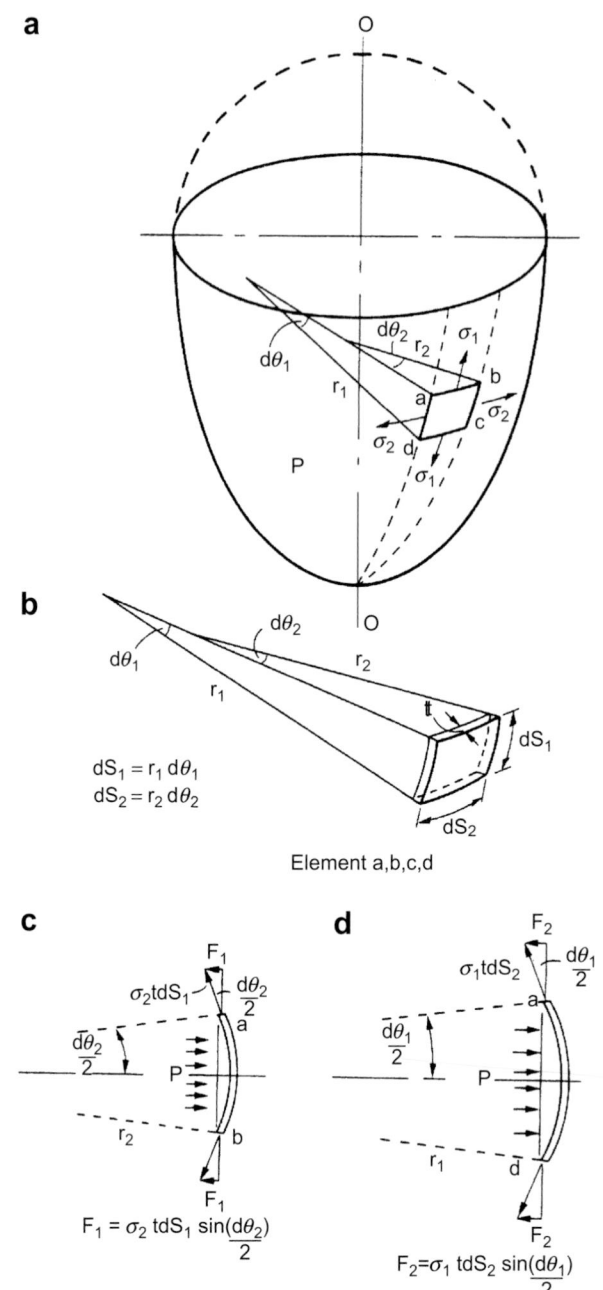

Figure 13.4. a,b. Stress in a shell of revolution. c,d. Forces acting on sides of element *a, b, c, d*.

$$= P\left[2r_1\sin\left(\frac{d\theta_1}{2}\right)\right]\left[2r_2\sin\left(\frac{d\theta_2}{2}\right)\right]$$

This force is resisted by the normal component of the forces associated with the membrane stresses in the walls of the vessel (given by, force = stress × area)

$$= 2\sigma_2 t dS_1 \sin\left(\frac{d\theta_2}{2}\right) + 2\sigma_1 t dS_2 \sin\left(\frac{d\theta_1}{2}\right)$$

Equating these forces and simplifying, and noting that in the limit $d\theta/2 \rightarrow dS/2r$, and $\sin d\theta \rightarrow d\theta$, gives

$$\frac{\sigma_1}{r_1} + \frac{\sigma_2}{r_2} = \frac{P}{t} \tag{13.5}$$

An expression for the meridional stress σ_1 can be obtained by considering the equilibrium of the forces acting about any circumferential line (Figure 13.5). The vertical component of the pressure force is

$$= p\pi(r_2 \sin\theta)^2$$

This is balanced by the vertical component of the force due to the meridional stress acting in the ring of the wall of the vessel:

$$= 2\sigma_1 t \pi (r_2 \sin\theta) \sin\theta$$

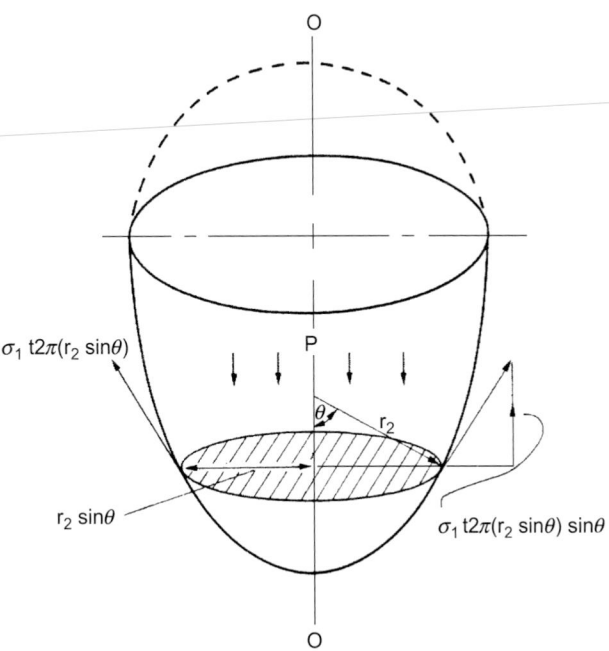

Figure 13.5. Meridional stress, force acting at a horizontal plane.

Equating these forces gives

$$\sigma_1 = \frac{Pr_2}{2t} \tag{13.6}$$

Equations 13.5 and 13.6 are completely general for any shell of revolution.

Cylinder (Figure 13.6a)

A cylinder is swept out by the rotation of a line parallel to the axis of revolution, so

$$r_1 = \infty$$
$$r_2 = \frac{D}{2}$$

where D is the cylinder diameter.

Substitution in equations 13.5 and 13.6 gives

$$\sigma_2 = \frac{PD}{2t} \tag{13.7}$$

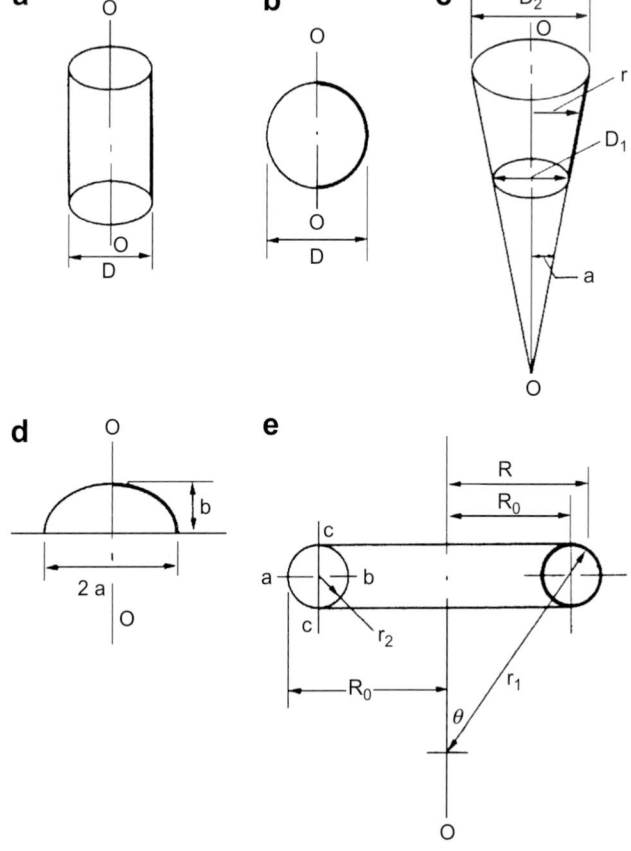

Figure 13.6. Shells of revolution.

$$\sigma_1 = \frac{PD}{4t} \tag{13.8}$$

Sphere (Figure 13.6b)

$$r_1 = r_2 = \frac{D}{2}$$

hence

$$\sigma_1 = \sigma_2 = \frac{PD}{4t} \tag{13.9}$$

Cone (Figure 13.6c)

A cone is swept out by a straight line inclined at an angle α to the axis.

$$r_1 = \infty$$

$$r_2 = \frac{r}{\cos\alpha}$$

substitution in equations 13.5 and 13.6 gives

$$\sigma_2 = \frac{Pr}{t\cos\alpha} \tag{13.10}$$

$$\sigma_1 = \frac{Pr}{2t\cos\alpha} \tag{13.11}$$

The maximum values will occur at $r = D_2/2$.

Ellipsoid (Figure 13.6d)

For an ellipse with major axis $2a$ and minor axis $2b$, it can be shown that (see any standard geometry text)

$$r_1 = \frac{r_2^3 b^2}{a^4}$$

From equations 13.5 and 13.6

$$\sigma_1 = \frac{Pr_2}{2t} \tag{equation 13.6}$$

$$\sigma_2 = \frac{P}{t}\left[r_2 - \frac{r_2^2}{2r_1}\right] \tag{13.12}$$

At the crown (top)

$$r_1 = r_2 = \frac{a^2}{b}$$

$$\sigma_1 = \sigma_2 = \frac{Pa^2}{2tb}$$

At the equator (bottom) $r_2 = a$, so $r_1 = b^2/a$

so

$$\sigma_1 = \frac{Pa}{2t} \tag{13.13}$$

$$\sigma_2 = \frac{P}{t}\left[a - \frac{a^2}{2b^2/a}\right] = \frac{Pa}{t}\left[1 - \frac{1a^2}{2b^2}\right] \tag{13.14}$$

It should be noted that if $\frac{1}{2}(a/b)^2 > 1$, σ_2 will be negative (compressive) and the shell could fail by buckling. This consideration places a limit on the practical proportions of ellipsoidal heads.

Torus (Figure 13.6e)

A torus is formed by rotating a circle, radius r_2, about an axis.

$$\sigma_1 = \frac{Pr_2}{2t}$$

$$r_1 = \frac{R}{\sin\theta} = \frac{R_0 + r_2\sin\theta}{\sin\theta} \qquad \text{(equation 13.6)}$$

and

$$\sigma_2 = \frac{Pr_2}{t}\left[1 - \frac{r_2\sin\theta}{2(R_0 + r_2\sin\theta)}\right] \tag{13.15}$$

On the centre line of the torus, point c, $\theta = 0$ and

$$\sigma_2 = \frac{Pr_2}{t} \tag{13.16}$$

At the outer edge, point a, $\theta = \pi/2$, $\sin\theta = 1$ and

$$\sigma_2 = \frac{Pr_2}{2t}\left[\frac{2R_0 + r_2}{R_0 + r_2}\right] \tag{13.17}$$

the minimum value.
   At the inner edge, point b, $\theta = 3\pi/2$, $\sin\theta = -1$ and

$$\sigma_2 = \frac{Pr_2}{2t}\left[\frac{2R_0 - r_2}{R_0 - r_2}\right] \tag{13.18}$$

the maximum value.
   So σ_2 varies from a maximum at the inner edge to a minimum at the outer edge.

Torispherical Heads

A torispherical shape, which is often used as the end closure of cylindrical vessels, is formed from part of a torus and part of a sphere (Figure 13.7). The shape is close to that of an ellipse but is easier and cheaper to fabricate.

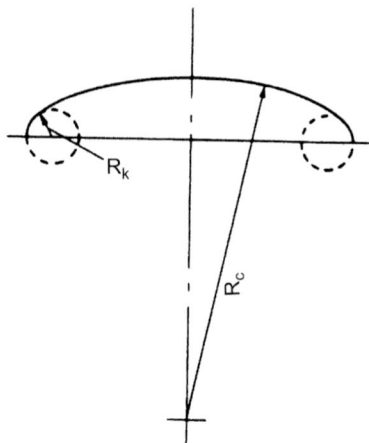

Figure 13.7. Torisphere.

In Figure 13.7, R_k is the knuckle radius (the radius of the torus) and R_c the crown radius (the radius of the sphere). For the spherical portion:

$$\sigma_1 = \sigma_2 = \frac{PR_c}{2t} \tag{13.19}$$

For the torus:

$$\sigma_1 = \frac{PR_k}{2t} \tag{13.20}$$

σ_2 depends on the location, and is a function of R_c and R_k; it can be calculated from equations 13.15 and 13.9.

The ratio of the knuckle radius to crown radius should be made not less than 6/100 to avoid buckling. The stress will be higher in the torus section than the spherical section.

13.3.5. Flat Plates

Flat plates are used as covers for manholes, as blind flanges, and for the ends of small-diameter and low-pressure vessels.

For a uniformly loaded circular plate supported at its edges, the slope ϕ at any radius x is given by

$$\phi = -\frac{dw}{dx} = -\frac{1}{D}\frac{Px^3}{16} + \frac{C_1 x}{2} + \frac{C_2}{x} \tag{13.21}$$

(The derivation of this equation can be found in any text on the strength of materials.)

Integration gives the deflection w:

$$w = \frac{Px^4}{64D} - C_1\frac{x^2}{4} - C_2 \ln x + C_3 \tag{13.22}$$

where

P = intensity of loading (pressure)
x = radial distance to point of interest
D = flexual rigidity of plate = $(E_Y t^3)/(12(1 - v^2))$
t = plate thickness
v = Poisson's ratio for the material
E_Y = modulus of elasticity of the material (Young's modulus).

C_1, C_2, C_3 are constants of integration that can be obtained from the boundary conditions at the edge of the plate.

Two limiting situations are possible:

1. When the edge of the plate is rigidly clamped, not free to rotate; which corresponds to a heavy flange, or a strong joint.
2. When the edge is free to rotate (simply supported); corresponding to a weak joint, or light flange.

1. Clamped Edges (Figure 13.8a)

The edge (boundary) conditions are:

$$\phi = 0 \ at \ x = 0$$

$$\phi = 0 \ at \ x = a$$

$$w = 0 \ at \ x = a$$

where a is the radius of the plate.

Which gives

$$C_2 = 0, \quad C_1 = \frac{Pa^2}{8D}, \quad \text{and} \quad C_3 = \frac{Pa^4}{64D}$$

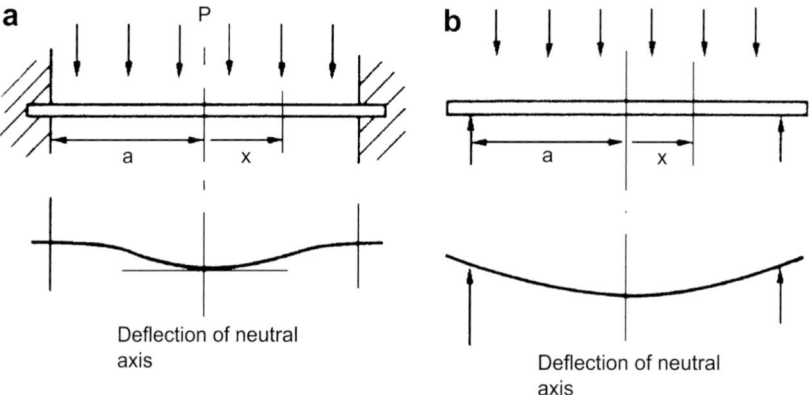

Figure 13.8.   Flat circular plates. (a) Clamped edges. (b) Simply supported.

hence

$$\phi = \frac{Px}{16D}(a^2 - x^2) \tag{13.23}$$

and

$$w = \frac{P}{64D}(x^2 - a^2)^2 \tag{13.24}$$

The maximum deflection will occur at the centre of the plate at $x = 0$:

$$\hat{w} = \frac{Pa^4}{64D} \tag{13.25}$$

The bending moments per unit length due to the pressure load are related to the slope and deflection by

$$M_1 = D\left[\frac{d\phi}{dx} + v\frac{\phi}{x}\right] \tag{13.26}$$

$$M_2 = D\left[\frac{\phi}{x} + v\frac{d\phi}{dx}\right] \tag{13.27}$$

where M_1 is the moment acting along cylindrical sections, and M_2 that acting along diametrical sections.

Substituting for ϕ and $d\phi/dx$ in equations 13.26 and 13.27 gives

$$M_1 = \frac{P}{16}[a^2(1 + v) - x^2(3 + v)] \tag{13.28}$$

$$M_2 = \frac{P}{16}[a^2(1 + v) - x^2(1 + 3v)] \tag{13.29}$$

The maximum values will occur at the edge of the plate, $x = a$:

$$\hat{M}_1 = -\frac{Pa^2}{8}, \quad \hat{M}_2 = -v\frac{Pa^2}{8}$$

The bending stress is given by

$$\sigma_b = \frac{M_1}{I'} \times \frac{t}{2}$$

where $I' =$ second moment of area per unit length $= t^3/12$; hence

$$\hat{\sigma}_b = \frac{6\hat{M}_1}{t^2} = \frac{3}{4}\frac{Pa^2}{t^2} \tag{13.30}$$

2. Simply Supported Plate (Figure 13.8b)

The edge (boundary) conditions are:

$$\phi = 0 \ at \ x = 0$$

$$w = 0 \ at \ x = a$$

$$M_1 = 0 \ at \ x = a(\text{free to rotate})$$

which gives C_2 and $C_3 = 0$.
   Hence

$$\phi = -\frac{1}{D}\frac{Px^3}{16} + \frac{C_1 x}{2}$$

and

$$\frac{d\phi}{dx} = -\frac{1}{D}\left[\frac{3Px^2}{16}\right] + \frac{C_1}{2}$$

Substituting these values in equation 13.26, and equating to zero at $x = a$, gives

$$C_1 = \frac{Pa^2}{8D}\frac{(3+v)}{(1+v)}$$

and hence

$$M_1 = \frac{P}{16}(3+v)(a^2 - x^2) \tag{13.31}$$

The maximum bending moment will occur at the centre, where $M_1 = M_2$:
so

$$\widehat{M}_1 = \widehat{M}_2 = \frac{P(3+v)a^2}{16} \tag{13.32}$$

and

$$\hat{\sigma}_b = \frac{6\widehat{M}_1}{t^2} = \frac{3}{8}(3+v)\frac{Pa^2}{t^2} \tag{13.33}$$

General Equation for Flat Plates

A general equation for the thickness of a flat plate required to resist a given pressure load can be written in the form:

$$t = D\sqrt{\frac{C\,P}{S}} \tag{13.34}$$

where

S = the maximum allowable stress (the design stress)
D = the effective plate diameter
C = a constant, which depends on the edge support.

The limiting value of C can be obtained from equations 13.30 and 13.33. Taking Poisson's ratio as 0.3, a typical value for steels, then if the edge can be taken as completely rigid C = 0.185, and if it is essentially free to rotate C = 0.314.

13.3.6. Dilation of Vessels

Under internal pressure a vessel will expand slightly. The radial growth can be calculated from the elastic strain in the radial direction. The principal strains in a two-dimensional system are related to the principal stresses by

$$\varepsilon_1 = \frac{1}{E_Y}(\sigma_1 - v\,\sigma_2) \tag{13.35}$$

$$\varepsilon_2 = \frac{1}{E_Y}(\sigma_2 - v\,\sigma_1) \tag{13.36}$$

The radial (diametrical strain) will be the same as the circumferential strain ε_2. For any shell of revolution the dilation can be found by substituting the appropriate expressions for the circumferential and meridional stresses in equation 13.36.

The diametrical dilation $\Delta = D\varepsilon_1$.

For a cylinder

$$\sigma_1 = \frac{PD}{4t}$$

$$\sigma_2 = \frac{PD}{2t}$$

Substitution in equation 13.36 gives

$$\Delta_C = \frac{PD^2}{4tE_Y}(2-v) \tag{13.37}$$

For a sphere (or hemisphere)

$$\sigma_1 = \sigma_2 = \frac{PD}{4t}$$

and

$$\Delta_S = \frac{PD^2}{4tE_Y}(1-v) \tag{13.38}$$

So, for a cylinder closed by a hemispherical head of the same thickness the difference in dilation of the two sections, if they were free to expand separately, would be

$$\Delta_C - \Delta_S = \frac{PD^2}{4tE_Y}$$

13.3.7. Secondary Stresses

In the stress analysis of pressure vessels and pressure vessel components, stresses are classified as primary or secondary. Primary stresses can be defined as those stresses that are necessary to satisfy the conditions of static equilibrium. The membrane stresses induced by the applied pressure and the bending stresses due to wind loads are examples of primary stresses. Primary stresses are not self-limiting; if they exceed the yield point of the material, gross distortion, and in the extreme situation, failure of the vessel will occur.

Secondary stresses are those stresses that arise from the constraint of adjacent parts of the vessel. Secondary stresses are self-limiting; local yielding or slight distortion will satisfy the conditions causing the stress, and failure would not be expected to occur in one application of the loading. The 'thermal stress' set up by the differential expansion of parts of the vessel, due to different temperatures or the use of different materials, is an example of a secondary stress. The discontinuity that occurs between the head and the cylindrical section of a vessel is a major source of secondary stress. If free, the dilation of the head would be different from that of the cylindrical section (see Section 13.3.6); they are constrained to the same dilation by the welded joint between the two parts. The induced bending moment and shear force due to the constraint give rise to secondary bending and shear stresses at the junction. The magnitude of these discontinuity stresses can be estimated by analogy with the behaviour of beams on elastic foundations; see Hetenyi (1958) and Harvey (1974). The estimation of the stresses arising from discontinuities is covered in the books by Bednar (1990) and Jawad and Farr (1989).

Other sources of secondary stresses are the constraints arising at flanges, supports, and the change of section due to reinforcement at a nozzle or opening (see Section 13.6).

Though secondary stresses do not affect the 'bursting strength' of the vessel, they are an important consideration when the vessel is subject to repeated pressure loading. If local yielding has occurred, residual stress will remain when the pressure load is removed, and repeated pressure cycling can lead to fatigue failure.

13.4. GENERAL DESIGN CONSIDERATIONS: PRESSURE VESSELS

13.4.1. Design Pressure

A vessel must be designed to withstand the maximum pressure to which it is likely to be subjected in operation.

For vessels under internal pressure, the design pressure (sometimes called maximum allowable working pressure or MAWP) is taken as the pressure at which the relief device is set. This will normally be 5 to 10% above the normal working pressure, to avoid spurious operation of the relief valve during minor process upsets. For example, the API RP 520 recommended practice sets a 10% margin between the normal operating pressure and the design pressure. When deciding the design pressure, the hydrostatic pressure in the base of the column should be added to the operating pressure, if significant.

Vessels subject to external pressure should be designed to resist the maximum differential pressure that is likely to occur in service. Vessels likely to be subjected to vacuum should be designed for a full negative pressure of 1 bar, unless fitted with an effective, and reliable, vacuum breaker.

13.4.2. Design Temperature

The strength of metals decreases with increasing temperature (see Chapter 7), so the maximum allowable stress will depend on the material temperature. Under the ASME BPV Code, the maximum design temperature at which the maximum allowable stress is evaluated should be taken as the maximum working temperature of the material, with due allowance for any uncertainty involved in predicting vessel wall temperatures. Additional rules apply for welded vessels, as described in ASME BPV Code Sec. VIII D.1 part UW. The minimum design metal temperature (MDMT) should be taken as the lowest temperature expected in service. The designer should consider the lowest operating temperature, ambient temperature, auto-refrigeration, process upsets and other sources of cooling in determining the minimum.

Under BS EN 13445, the design temperature of the vessel must not be less than the maximum fluid temperature at the design pressure. This is more conservative than the ASME BPV Code. The BS EN 13445 code is currently limited to temperatures outside the creep range, i.e., less than 370°C for ferritic steels and less than 425°C for austenitic steels (BS EN 13445-3:2002 Clause 5.1). A section on design under creep conditions is in preparation.

13.4.3. Materials

Pressure vessels are constructed from plain carbon steels, low and high alloy steels, other alloys, clad plate, and reinforced plastics.

Selection of a suitable material must take into account the suitability of the material for fabrication (particularly welding) as well as the compatibility of the material with the process environment.

The pressure vessel design codes and standards include lists of acceptable materials, in accordance with the appropriate material standards. The ASME BPV Code Sec. II Part D gives maximum allowable stresses as a function of temperature and maximum temperatures permitted under Sections I, III, VIII and XII of the BPV code for ferrous and nonferrous metals. The design of pressure vessels using reinforced plastics is described in ASME BPV Code Sec. X.

The European standard BS EN 13445-3 Clause 6 gives formulae for calculating the nominal design stresses (maximum allowable stresses) for different grades of steels as a function of the minimum yield strength and minimum tensile strength. Methods for the design of vessels constructed from spheroidal graphitic cast iron are given in BS EN 13445-6 and additional requirements for aluminium vessels are given in BS EN 13445-8. The design of glass-reinforced plastic vessels is described in BS EN 13121:2003.

13.4.4. Maximum Allowable Stress (Nominal Design Strength)

For design purposes it is necessary to decide a value for the maximum allowable stress (nominal design strength) that can be accepted in the material of construction.

This is determined by applying a suitable safety factor to the maximum stress that the material could be expected to withstand without failure under standard test conditions. The safety factor allows for any uncertainty in the design methods, the loading, the quality of the materials and the workmanship.

The basis for establishing the maximum allowable stress values in the ASME BPV Code is given in ASME BPV Code Sec. II Part D, Mandatory Appendix 1. At temperatures where creep and stress rupture strength do not govern the selection of stresses, the maximum allowable stress is the lowest of:

1. the specified minimum tensile strength at room temperature divided by 3.5
2. the tensile strength at temperature divided by 3.5
3. the specified minimum yield strength at room temperature divided by 1.5
4. the yield strength at temperature divided by 1.5.

At temperatures where creep and stress rupture strength govern, the maximum allowable stress is the lowest of:

1. the average stress to produce a creep rate of 0.01%/1000 h
2. F times the average stress to cause rupture at the end of 100,000 h, where $F = 0.67$ for temperatures below 1500°F (815°C); see the code for higher temperatures
3. 0.8 times the minimum stress to cause rupture after 100,000 h.

In some cases where short-time tensile properties govern and slightly greater deformation is acceptable, higher stress values are allowed under ASME BPV Code Sec. VIII D.1. These exceed 67% but do not exceed 90% of the yield strength at temperature. These cases are indicated with a note (G5) in the BPV Code tables. Use of these higher values can result in deformation and changes in the vessel dimensions. They are not recommended for flanges or other applications where changes in dimensions could lead to leaks or vessel malfunction.

The maximum allowable stress values for ASME BPV Code Sec. VIII D.1 are given in ASME BPV Code Sec II Part D Table 1A for ferrous metals and Table 1B for non-ferrous metals. Maximum allowable stress values for Sec. VIII D.2 are given in Sec. II Part D Table 2A for ferrous metals and Table 2B for nonferrous metals. Different values are given for plate, tubes, castings, forgings, bar, pipe and small sections as well as for different grades of each metal.

Typical maximum allowable stress values for some common materials are shown in Table 13.2. These may be used for preliminary designs. The ASME BPV Code should be consulted for the values to be used for detailed vessel design.

The maximum allowed values of the nominal design stress used in BS EN 13445-3 are specified in Clause 6 of the code. Different values of the nominal design stress are used for the normal operating, testing and exceptional load cases. The method used to determine the nominal design stress depends on the elongation of the steel after rupture. For forged austenitic steels with elongation after rupture of between 30% and 35%, the nominal design stress for the normal operating load case is calculated as

TABLE 13.2. Typical Maximum Allowable Stresses for Plate Under ASME BPV Code Sec. VIII D.1 (The Appropriate Material Standards should be Consulted for Particular Grades and Plate Thicknesses)

Material	Grade	Min. Tensile Strength (ksi)	Min. Yield Strength (ksi)	Maximum Temperature (°F)	Maximum Allowable Stress at Temperature °F (ksi = 1000 psi)				
					100	300	500	700	900
Carbon steel	A285 Gr A	45	24	900	12.9	12.9	12.9	11.5	5.9
Killed carbon Steel	A515 Gr 60	60	32	1000	17.1	17.1	17.1	14.3	5.9
Low-alloy steel 1 ¼ Cr, ½ Mo, Si	A387 Gr 22	60	30	1200	17.1	16.6	16.6	16.6	13.6
Stainless steel 13 Cr	410	65	30	1200	18.6	17.8	17.2	16.2	12.3
Stainless steel 18 Cr, 8 Ni	304	75	30	1500	20.0	15.0	12.9	11.7	10.8
Stainless steel 18 Cr, 10 Ni, Cb	347	75	30	1500	20.0	17.1	15.0	13.8	13.4
Stainless steel 18 Cr, 10 Ni, Ti	321	75	30	1500	20.0	16.5	14.3	13.0	12.3
Stainless steel 16 Cr, 12 Ni, 2 Mo	316	75	30	1500	20.0	15.6	13.3	12.1	11.5

Notes:
1. The stress values for type 304 stainless steel are not the same as those given for stainless steel 304L in Table 7.8 of this book.
2. 1 ksi = 1000 psi = 6.8948 N/mm².

the 1% proof strength at the calculation temperature, divided by a safety factor 1.5. This gives a similar value to that calculated in case 4 listed above for the ASME BPV Code. The BS EN 13445-3 code should be consulted for values for other steels.

13.4.5. Welded-Joint Efficiency and Construction Categories

The strength of a welded joint will depend on the type of joint and the quality of the welding. The ASME BPV Code Sec. VIII D.1 defines four categories of weld (Part UW-3):

A. Longitudinal or spiral welds in the main shell, necks or nozzles, or circumferential welds connecting hemispherical heads to the main shell, necks or nozzles.
B. Circumferential welds in the main shell, necks or nozzles or connecting a formed head other than hemispherical.
C. Welds connecting flanges, tube sheets or flat heads to the main shell, a formed head, neck or nozzle.
D. Welds connecting communicating chambers or nozzles to the main shell, to heads or to necks.

Details of the different types of welds used in pressure vessel construction are given in Section 13.12.

The soundness of welds is checked by visual inspection and by non-destructive testing (radiography).

The possible lower strength of a welded joint compared with the virgin plate is usually allowed for in design by multiplying the allowable design stress for the material by a joint efficiency E. The value of the joint efficiency used in design will depend on the type of joint and amount of radiography required by the design code. Typical values are shown in Table 13.3. A joint efficiency of 1.0 is only permitted for butt joints formed by double welding and subjected to full radiographic examination. Taking the factor as 1.0 implies that the joint is equally as strong as the virgin plate; this is achieved by radiographing the complete weld length, and cutting out and remaking any defects. The use of lower joint efficiencies in design, though saving costs on radiography, will result in a thicker, heavier, vessel, and the designer must balance any cost savings on inspection and fabrication against the increased cost of materials.

The ASME BPV Code Sec. VIII D.1 Part UW describes the requirements for pressure vessels fabricated by welding. Limiting plate thicknesses are specified for each type of weld with the exception of double-welded butt joints. Requirements for radiographic examination of welds are also specified. Section UW-13 of the code specifies the types of welds that can be used to attach heads and tube sheets to shells. Section UW-16 gives rules for attachment of nozzles to vessels.

The BPV Code should be consulted to determine the allowed joint types for a particular vessel. Any pressure vessel containing lethal substances will require full radiographic testing of all butt welds.

When designing vessels to other international standards, the standard should always be consulted to determine the rules for joint efficiency, as different codes treat it in different ways. For example, welded-joint efficiency factors are not used, as such, in the design equations given in BS PD 5500; instead limitations are placed on the values of the nominal design strength (maximum allowable design stress) for materials in the lower construction category. The standard specifies three construction categories:

Category 1: the highest class, requires 100% non-destructive testing of the welds; and allows the use of all materials covered by the standard, with no restriction on the plate thickness.

Category 2: requires less non-destructive testing, but places some limitations on the materials that can be used and the maximum plate thickness.

Category 3: the lowest class, requires only visual inspection of the welds, but is restricted to carbon and carbon-manganese steels, and austenitic stainless steel; and

TABLE 13.3. Maximum Allowable Joint Efficiency

Joint Description	Joint Category	Degree of Radiographic Examination		
		Full	**Spot**	**None**
Double-welded butt joint or equivalent	A, B, C, D	1.0	0.85	0.70
Single-welded butt joint with backing strip	A, B, C, D	0.9	0.8	0.65
Single-welded butt joint without backing strip	A, B, C	NA	NA	0.60
Double full fillet lap joint	A, B, C	NA	NA	0.55
Single full fillet lap joint with plug welds	B, C	NA	NA	0.50
Single full fillet lap joint without plug welds	A, B	NA	NA	0.45

limits are placed on the plate thickness and the nominal design stress. For carbon and carbon-manganese steels the plate thickness is restricted to less than 13 mm and the design stress is about half that allowed for categories 1 and 2. For stainless steel the thickness is restricted to less than 25 mm and the allowable design stress is around 80% of that for the other categories.

The European standard BS EN 13445-3 defines four categories of governing welded joints (Clause 5.6):

- Longitudinal or helical welds in a cylindrical shell
- Longitudinal welds in a conical shell
- Main welds in a spherical shell or head
- Main welds in a dished head fabricated from two or more plates

For governing welded joints, the joint efficiency is determined based on the testing group of the weld, as specified in BS EN 13445-5, Clause 6.

13.4.6. Corrosion Allowance

The 'corrosion allowance' is the additional thickness of metal added to allow for material lost by corrosion and erosion, or scaling (see Chapter 7). The ASME BPV Code Sec. VIII D.1 states that the vessel user shall specify corrosion allowances (Part UG-25). Minimum wall thicknesses calculated using the rules given in the code are in the fully corroded condition (Part UG-16). Corrosion is a complex phenomenon, and it is not possible to give specific rules for the estimation of the corrosion allowance required for all circumstances. The allowance should be based on experience with the material of construction under similar service conditions to those for the proposed design. For carbon and low-alloy steels, where severe corrosion is not expected, a minimum allowance of 2.0 mm should be used; where more severe conditions are anticipated this should be increased to 4.0 mm. Most design codes and standards specify a minimum allowance of 1.0 mm, but under the ASME BPV Code Sec. VIII no corrosion allowance is needed when past experience indicates that corrosion is only superficial or does not occur. The European standard BS EN 13445-3 does not specify a minimum corrosion allowance but states that 'an additional thickness sufficient for the design life of the vessel components shall be provided' (Clause 5.2.2).

13.4.7. Design Loads

A structure must be designed to resist gross plastic deformation and collapse under all the conditions of loading. The loads to which a process vessel will be subject in service are listed below. They can be classified as major loads, that must always be considered in vessel design, and subsidiary loads. Formal stress analysis to determine the effect of the subsidiary loads is only required in the codes and standards where it is not possible to demonstrate the adequacy of the proposed design by other means, such as by comparison with the known behaviour of existing vessels.

Major Loads

1. Design pressure: including any significant static head of liquid
2. Maximum weight of the vessel and contents, under operating conditions

3. Maximum weight of the vessel and contents under the hydraulic test conditions
4. Wind loads
5. Earthquake (seismic) loads
6. Loads supported by, or reacting on, the vessel.

Subsidiary Loads

1. Local stresses caused by supports, internal structures and connecting pipes
2. Shock loads caused by water hammer, or by surging of the vessel contents
3. Bending moments caused by eccentricity of the centre of the working pressure relative to the neutral axis of the vessel
4. Stresses due to temperature differences and differences in the coefficient of expansion of materials
5. Loads caused by fluctuations in temperature and pressure.

A vessel will not be subject to all these loads simultaneously. The designer must determine what combination of possible loads gives the worst situation (the 'governing case'), and design for that loading condition.

13.4.8. Minimum Practical Wall Thickness

There will be a minimum wall thickness required to ensure that any vessel is sufficiently rigid to withstand its own weight, and any incidental loads. The ASME BPV Code Sec. VIII D.1 specifies a minimum wall thickness of 1/16 in (1.5 mm) not including corrosion allowance, and regardless of vessel dimensions and material of construction. As a general guide the wall thickness of any vessel should not be less than the values given below, which include a corrosion allowance of 2 mm:

Vessel Diameter (m)	Minimum Thickness (mm)
1	5
1 to 2	7
2 to 2.5	9
2.5 to 3.0	10
3.0 to 3.5	12

13.5. THE DESIGN OF THIN-WALLED VESSELS UNDER INTERNAL PRESSURE

13.5.1. Cylinders and Spherical Shells

For a cylindrical shell the minimum thickness required to resist internal pressure can be determined from equations 13.7 and 13.8.

If D_i is internal diameter and t the minimum thickness required, the mean diameter will be $(D_i + t)$; substituting this for D in equation 13.7 gives

$$t = \frac{P_i(D_i + t)}{2S}$$

where S is the maximum allowable stress and P_i the internal pressure. Rearranging gives

$$t = \frac{P_i D_i}{2S - P_i} \tag{13.39}$$

If we allow for the welded-joint efficiency, E, this becomes

$$t = \frac{P_i D_i}{2SE - P_i} \tag{13.40}$$

The equation specified by the ASME BPV Code (Sec. VIII D.1 Part UG-27) is

$$t = \frac{P_i D_i}{2SE - 1.2 P_i} \tag{13.41}$$

This differs slightly from equation 13.40 as it is derived from the formula for thick-walled vessels.

Similarly, for longitudinal stress the code specifies:

$$t = \frac{P_i D_i}{4SE + 0.8P_i} \tag{13.42}$$

The ASME BPV Code specifies that the minimum thickness shall be the greater value determined from equations 13.41 and 13.42. If these equations are rearranged and used to calculate the maximum allowable working pressure (MAWP) for a vessel of a given thickness, then the maximum allowable working pressure is the lower value predicted by the two equations.

For a spherical shell the code specifies:

$$t = \frac{P_i D_i}{4SE - 0.4P_i} \tag{13.43}$$

Any consistent set of units can be used for equations 13.39 to 13.43.

The European standard BS EN 13445 specifies equation 13.40 or the equivalent equation based on external diameter; consult the design code for details.

13.5.2. Heads and Closures

The ends of a cylindrical vessel are closed by heads of various shapes. The principal types used are:

1. Flat plates and formed flat heads; Figure 13.9
2. Hemispherical heads; Figure 13.10a
3. Ellipsoidal heads; Figure 13.10b
4. Torispherical heads; Figure 13.10c.

Hemispherical, ellipsoidal and torispherical heads are collectively referred to as domed heads. They are formed by pressing or spinning; large diameters are fabricated from formed sections. Torispherical heads are often referred to as dished ends.

The preferred proportions of domed heads are given in the standards and codes.

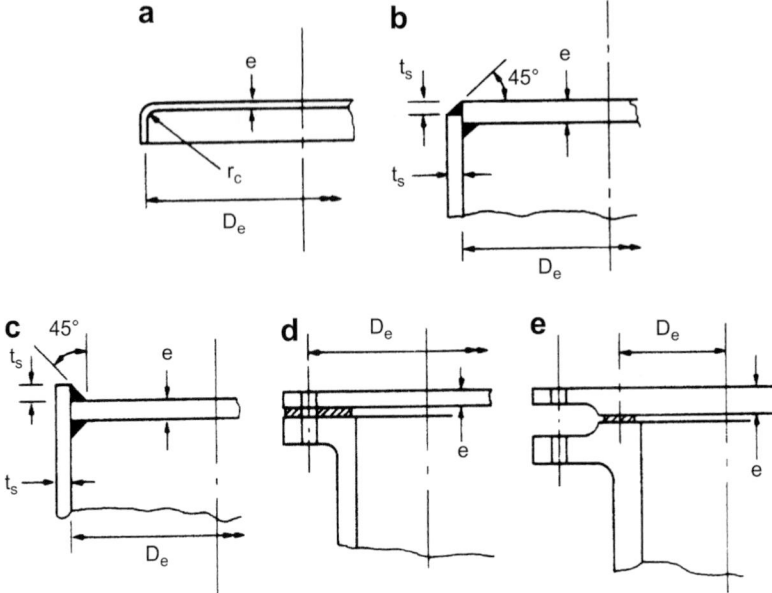

Figure 13.9. Flat-end closures. (a) Flanged plate. (b) Welded plate. (c) Welded plate. (d) Bolted cover. (e) Bolted cover.

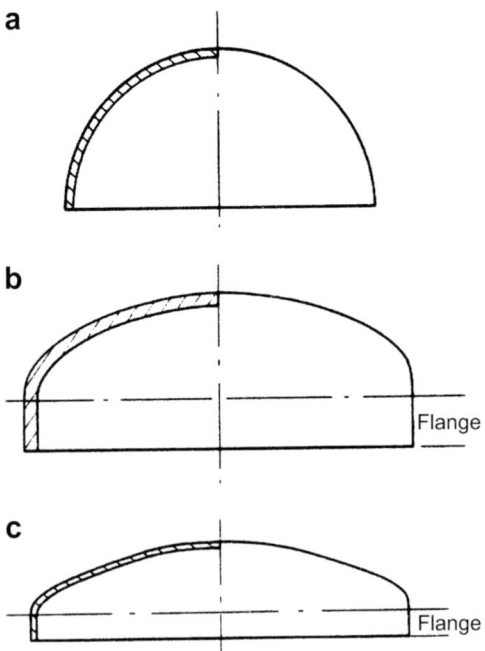

Figure 13.10. Domed heads. (a) Hemispherical. (b) Ellipsoidal. (c) Torispherical.

Choice of Closure

Flat plates are used as covers for manways, and as the channel covers of heat exchangers. Formed flat ends, known as 'flange-only' ends, are manufactured by turning over a flange with a small radius on a flat plate, Figure 13.9a. The corner radius reduces the abrupt change of shape at the junction with the cylindrical section, which reduces the local stresses to some extent: 'flange-only' heads are the cheapest type of formed head to manufacture, but their use is limited to low-pressure and small-diameter vessels.

Standard torispherical heads (dished ends) are the most commonly used end closure for vessels up to operating pressures of 15 bar. They can be used for higher pressures, but above 10 bar their cost should be compared with that of an equivalent ellipsoidal head. Above 15 bar an ellipsoidal head will usually prove to be the most economical closure to use.

A hemispherical head is the strongest shape, capable of resisting about twice the pressure of a torispherical head of the same thickness. The cost of forming a hemispherical head will, however, be higher than that for a shallow torispherical head. Hemispherical heads are used for high pressures.

13.5.3. Design of Flat Ends

Though the fabrication cost is low, flat ends are not a structurally efficient form, and very thick plates would be required for high pressures or large diameters.

The design equations used to determine the thickness of flat ends are based on the analysis of stresses in flat plates; see Section 13.3.5.

The thickness required will depend on the degree of constraint at the plate periphery. The ASME BPV Code specifies the minimum thickness as

$$t = D_e \sqrt{\frac{C P_i}{S E}} \tag{13.44}$$

where

C = a design constant, dependent on the edge constraint
D_e = nominal plate diameter
S = maximum allowable stress
E = joint efficiency.

Any consistent set of units can be used.

Values for the design constant C and the nominal plate diameter D_e are given in the ASME BPV Code for various arrangements of flat-end closures (Sec. VIII D.1 Part UG-34).

Some typical values of the design constant and nominal diameter for the designs shown in Figure 13.9 are given below. For detailed design the ASME BPV Code should be consulted.

(a) Flanged-only end: $C = 0.17$ if corner radius is not more than $3t$, otherwise $C = 0.1$; D_e is equal to D_i.

(b,c) Plates welded to the end of the shell with a fillet weld: angle of fillet 45° and weld depth 70% of the thickness of the shell, $C = 0.33\ t/t_s$, where t_s is the shell thickness; $D_e = D_i$.

(d) Bolted cover with a full-face gasket (see Section 13.10): $C = 0.25$ and D_e is the bolt circle diameter (the diameter of a circle connecting the centres of the bolt holes).

(e) Bolted end cover with a narrow-face gasket: $C = 0.3$ and D_e should be taken as the mean diameter of the gasket.

The European standard EN 13445-3 uses an equation that is similar to equation 13.44, but with a different definition of the constant; consult the standard for details.

13.5.4.   Design of Domed Ends

Design equations and charts for the various types of domed heads are given in the ASME BPV Code and should be used for detailed design. The code covers both unpierced and pierced heads. Pierced heads are those with openings or connections. The head thickness must be increased to compensate for the weakening effect of the holes where the opening or branch is not locally reinforced (see Section 13.6).

For convenience, simplified design equations are given in this section. These are suitable for the preliminary sizing of unpierced heads and for heads with fully compensated openings or branches.

Hemispherical Heads

It can be seen, by examination of equations 13.7 and 13.9, that for equal stress in the cylindrical section and hemispherical head of a vessel, the thickness of the head need only be half that of the cylinder. However, as the dilation of the two parts would then be different, discontinuity stresses would be set up at the head and cylinder junction. For no difference in dilation between the two parts (equal diametrical strain) it can be shown that for steels (Poisson's ratio = 0.3) the ratio of the hemispherical head thickness to cylinder thickness should be 7/17. However, the stress in the head would then be greater than that in the cylindrical section, and the optimum thickness ratio is normally taken as 0.6; see Brownell and Young (1959).

In the ASME BPV Code Sec. VIII D.1, the equation specified is the same as for a spherical shell:

$$t = \frac{P_i D_i}{4SE - 0.4P_i} \tag{13.43}$$

In the European BS EN 13445-3 code, the equation used is also the same as that for a spherical shell, and is the same as equation 13.43, but with denominator $4SE - P_i$.

Ellipsoidal Heads

Most standard ellipsoidal heads are manufactured with a major and minor axis ratio of 2:1. For this ratio, the following equation can be used to calculate the minimum thickness required (ASME BPV Code Sec. VIII D.1 Part UG-32):

$$t = \frac{P_i D_i}{2SE - 0.2P_i} \tag{13.45}$$

The European pressure vessel code treats ellipsoidal heads as a special case of torispherical heads.

Torispherical Heads

There are two junctions in a torispherical end closure: that between the cylindrical section and the head, and that at the junction of the crown and the knuckle radii. The bending and shear stresses caused by the differential dilation that will occur at these points must be taken into account in the design of the heads. The ASME BPV Code gives the design equation (Sec. VIII D.1 Part UG-32):

$$t = \frac{0.885 P_i R_c}{SE - 0.1 P_i} \qquad (13.46)$$

where R_c = crown radius.

The ratio of the knuckle to crown radii should not be less than 0.06, to avoid buckling; and the crown radius should not be greater than the diameter of the cylindrical section. Any consistent set of units can be used with equations 13.43 to 13.46. For formed heads (no welds or joints in the head) the joint efficiency E is taken as 1.0.

The method given in the European standard is more complex; see BS EN 13445-3:2002 Clause 7.5.3.

Flanges (Skirts) on Domed Heads

Formed domed heads are made with a short straight cylindrical section, called a flange or skirt (Figure 13.10). This ensures that the weld line is away from the point of discontinuity between the head and the cylindrical section of the vessel.

13.5.5. Conical Sections and End Closures

Conical sections (reducers) are used to make a gradual reduction in diameter from one cylindrical section to another of smaller diameter.

Conical ends are used to facilitate the smooth flow and removal of solids from process equipment, such as hoppers, spray-dryers and crystallizers.

From equation 13.10 it can be seen that the thickness required at any point on a cone is related to the diameter by the following expression:

$$t = \frac{P_i D_c}{2SE - P_i} \cdot \frac{1}{\cos \alpha} \qquad (13.47)$$

where

D_c is the diameter of the cone at the point
α = half the cone apex angle.

The equation given in the ASME BPV Code is

$$t = \frac{P_i D_i}{2\cos \alpha (SE - 0.6 P_i)} \qquad (13.48)$$

This equation will only apply at points away from the cone-to-cylinder junction. Bending and shear stresses will be caused by the different dilation of the conical and cylindrical sections. A formed section would normally be used for the transition between a cylindrical section and conical section; except for vessels operating at low pressures, or under hydrostatic pressure only. The transition section would be made thicker than the conical or cylindrical section, and formed with a knuckle radius to reduce the stress concentration at the transition (Figure 13.11).

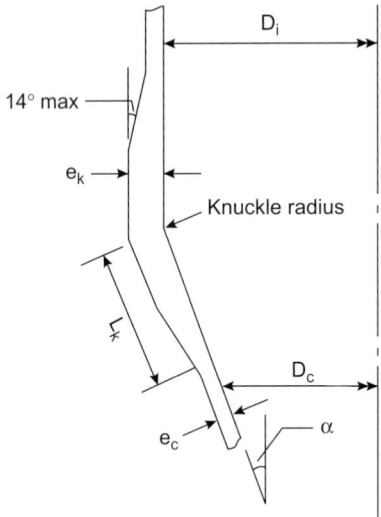

Figure 13.11.   Conical transition section.

The thickness for the conical section away from the transition can be calculated from equation 13.48.

The code should be consulted for details of how to size the knuckle zone.

Example 13.1

Estimate the thickness required for the component parts of the vessel shown in the diagram. The vessel is to operate at a pressure of 14 bar (absolute) and temperature of 260°C. The material of construction will be plain carbon steel. Welds will be fully radiographed. A corrosion allowance of 2 mm should be used.

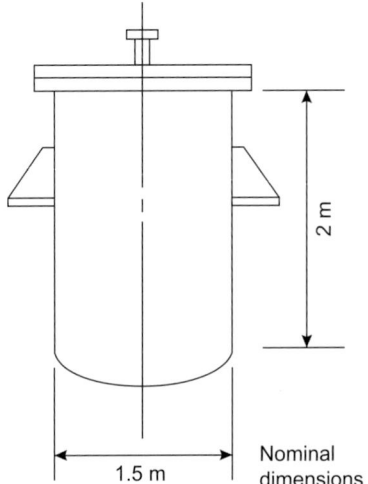

Solution

Design pressure, take as 10% above operating gauge pressure

$= (14 - 1) \times 1.1$
$= 14.3$ bar
$= 1.43$ N/mm^2.

Design temperature 260°C (500°F).
From Table 13.2, maximum allowable stress $= 12.9 \times 10^3$ psi $= 88.9$ N/mm^2.

Cylindrical Section

$$t = \frac{1.43 \times 1.5 \times 10^3}{(2 \times 89 \times 1) - (1.2 \times 1.43)} = 12.2 \text{ mm} \tag{13.41}$$

add corrosion allowance $12.2 + 2 = 14.2$ mm

say <u>15 mm plate</u> or <u>9/16 inch plate</u>

Domed Head

(i) Try a standard dished head (torisphere):

crown radius $R_c = D_i = 1.5$ m
knuckle radius $= 6$ per cent $R_c = 0.09$ m.

A head of this size would be formed by pressing: no joints, so $E = 1$.

$$t = \frac{0.885 \times 1.43 \times 1.5 \times 10^3}{(89 \times 1) - (0.1 \times 1.43)} = 21.4 \text{ mm} \tag{13.46}$$

(ii) Try a 'standard' ellipsoidal head, ratio major: minor axes $= 2:1$

$$t = \frac{1.43 \times 1.5 \times 10^3}{(2 \times 89 \times 1) - (0.2 \times 1.43)} = 12.1 \text{ mm} \tag{13.45}$$

So an ellipsoidal head would probably be the most economical. Take as same thickness as wall, 15 mm or 9/16 in.

Flat Head

Use a full-face gasket $C = 0.25$.
$D_e =$ bolt circle diameter, take as approx. 1.7 m.

$$t = 1.7 \times 10^3 \sqrt{\frac{0.25 \times 1.43}{89 \times 1}} = 107.7 \text{ mm} \tag{13.44}$$

Add corrosion allowance and round-off to <u>111 mm</u> (4 $^3/_8$ in).
This shows the inefficiency of a flat cover. It would be better to use a flanged domed head.

13.6. COMPENSATION FOR OPENINGS AND BRANCHES

All process vessels will have openings for connections, manways and instrument fittings. The presence of an opening weakens the shell, and gives rise to stress concentrations.

The stress at the edge of a hole will be considerably higher than the average stress in the surrounding plate. To compensate for the effect of an opening, the wall thickness is increased in the region adjacent to the opening. Sufficient reinforcement must be provided to compensate for the weakening effect of the opening without significantly altering the general dilation pattern of the vessel at the opening. Over-reinforcement will reduce the flexibility of the wall, causing a 'hard spot', and giving rise to secondary stresses; typical arrangements are shown in Figure 13.12.

The simplest method of providing compensation is to weld a pad or collar around the opening (Figure 13.12a). The outer diameter of the pad is usually between 1½ and 2 times the diameter of the hole or branch. This method, however, does not give the best disposition of the reinforcing material about the opening, and in some circumstances high thermal stress can arise due to the poor thermal conductivity of the pad to shell junction.

At a branch, the reinforcement required can be provided, with or without a pad, by allowing the branch to protrude into the vessel (Figure 13.12b). This arrangement should be used with caution for process vessels, as the protrusion will act as a trap for crud, and crevices are created in which localized corrosion can occur. Forged reinforcing rings (Figure 13.12c) provide the most effective method of compensation, but are expensive. They would be used for any large openings and branches in vessels operating under severe conditions.

The rules for calculating the minimum amount of reinforcement required are complex. For design purposes, consult the ASME BPV Code Sec. VIII D.1 Part UG-37 or BS EN 13445-3:2002 Clause 9.5.

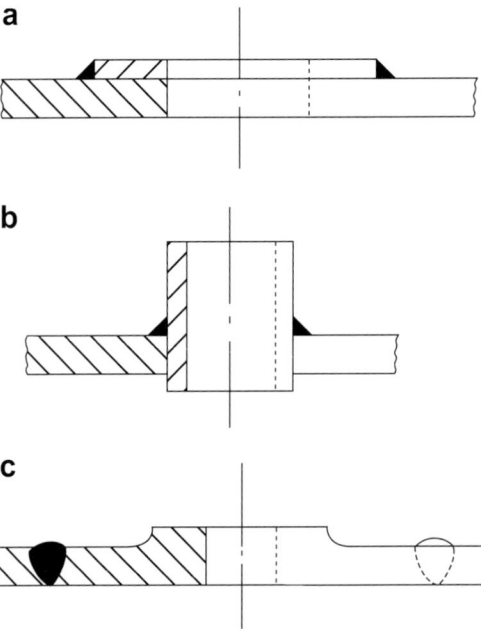

Figure 13.12.   Types of compensation for openings. (a) Welded pad. (b) Inset nozzle. (c) Forged ring.

13.7. DESIGN OF VESSELS SUBJECT TO EXTERNAL PRESSURE

13.7.1. Cylindrical Shells

Two types of process vessel are likely to be subjected to external pressure: those operated under vacuum, where the maximum pressure will be 1 bar (1 atm); and jacketed vessels, where the inner vessel will be under the jacket pressure. For jacketed vessels, the maximum pressure difference should be taken as the full jacket pressure, as a situation may arise in which the pressure in the inner vessel is lost. Thin-walled vessels subject to external pressure are liable to failure through elastic instability (buckling) and it is this mode of failure that determines the wall thickness required.

For an open-ended cylinder, the critical pressure to cause buckling P_c is given by the following expression; see Windenburg and Trilling (1934):

$$P_c = \frac{1}{3}\left[n^2 - 1 + \frac{2n^2 - 1 - v}{n^2\left(\dfrac{2L}{\pi D_0}\right)^2 - 1}\right] \frac{2E_Y}{(1 - v^2)}\left(\frac{t}{D_0}\right)^3$$

$$+ \frac{2E_Y t / D_0}{(n^2 - 1)\left[n^2\left(\dfrac{2L}{\pi D_0}\right)^2 - 1\right]^2} \qquad (13.49)$$

where

L = the unsupported length of the vessel, the effective length
D_0 = external diameter
t = wall thickness
E_Y = Young's modulus
v = Poisson's ratio
n = the number of lobes formed at buckling.

For long tubes and cylindrical vessels this expression can be simplified by neglecting terms with the group $(2L/\pi D0)2$ in the denominator; the equation then becomes:

$$P_c = \frac{1}{3}\frac{2E_Y(n^2 - 1)}{(1 - v^2)}\left(\frac{t}{D_0}\right)^3 \qquad (13.50)$$

The minimum value of the critical pressure will occur when the number of lobes is two, and substituting this value into equation 13.50 gives:

$$P_c = \frac{2E_Y}{(1 - v^2)}\left(\frac{t}{D_0}\right)^3 \qquad (13.51)$$

For most pressure vessel materials, Poisson's ratio can be taken as 0.3; substituting this in equation 13.51 gives:

$$P_c = 2.2E_Y\left(\frac{t}{D_0}\right)^3 \qquad (13.52)$$

For short closed vessels, and long vessels with stiffening rings, the critical buckling pressure will be higher than that predicted by equation 13.52. The effect of stiffening can be taken into account by introducing a 'collapse coefficient', Kc, into equation 13.52.

$$P_c = K_c E_Y \left(\frac{t}{D_0}\right)^3 \tag{13.53}$$

where K_c is a function of the diameter and thickness of the vessel, and the effective length L' between the ends or stiffening rings, and is obtained from Figure 13.13. The effective length for some typical arrangements is shown in Figure 13.14.

It can be shown (see Southwell, 1913) that the critical distance between stiffeners, L_c, beyond which stiffening will not be effective is given by

$$L_c = \frac{4\pi\sqrt{6}D_0}{27}\left[(1 - v^2)^{1/4}\right]\left(\frac{D_0}{t}\right)^{1/2} \tag{13.54}$$

Substituting $v = 0.3$ gives

$$L_c = 1.11 D_0 \left(\frac{D_0}{t}\right)^{1/2} \tag{13.55}$$

Any stiffening rings used must be spaced closer than L_c. Equation 13.53 can be used to estimate the critical buckling pressure and hence the thickness required to resist a given external pressure. A factor of safety of at least 3 should be applied to the values predicted using equation 13.53.

The method outlined above is not recognized by the ASME BPV Code and can only be used for a preliminary estimate of wall thickness required under external pressure. The method recommended by the BPV Code is substantially more complex and takes into account the fact that the maximum allowable stress in compression is different

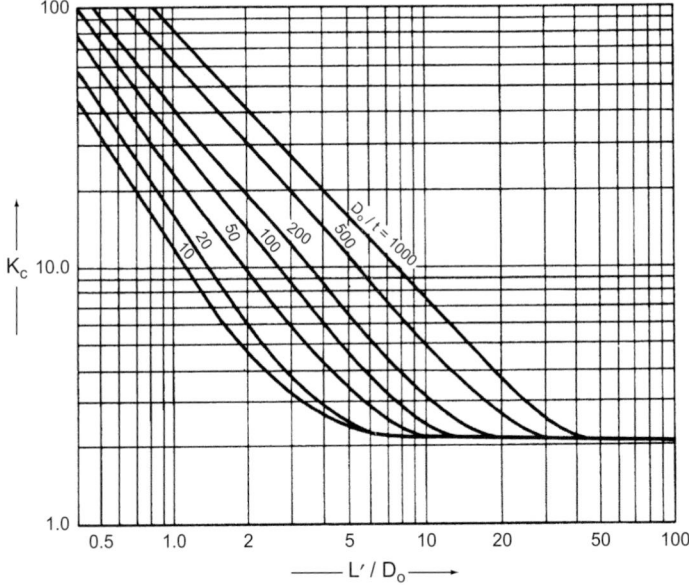

Figure 13.13.   Collapse coefficients for cylindrical shells (after Brownell and Young, 1959).

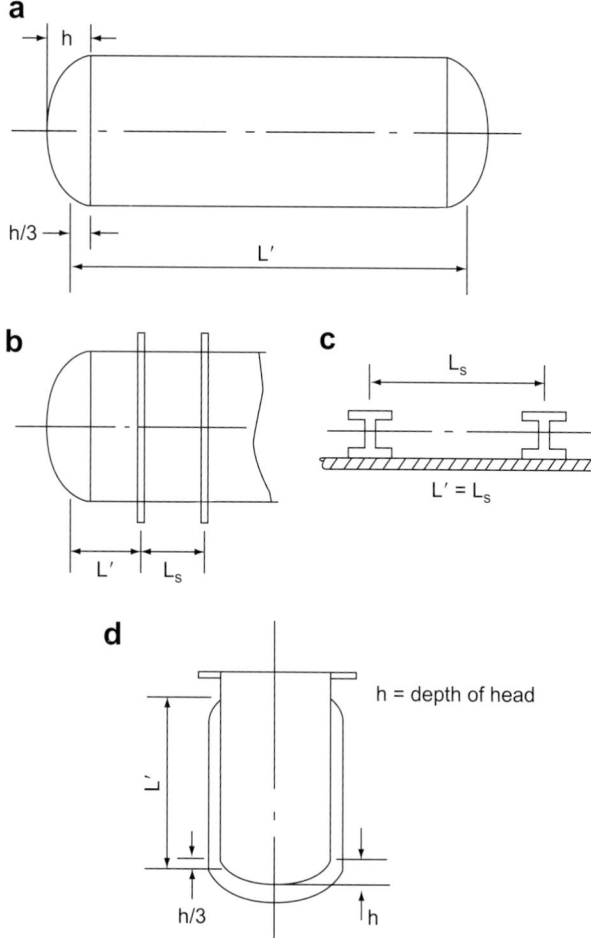

Figure 13.14. Effective length, vessel under external pressure. (a) Plain vessel. (b) With stiffeners (use smaller of L' and L$_s$). (c) I-section stiffening rings. (d) Jacketed vessel.

from that in tension. The ASME BPV Code Sec. VIII D.1 Part UG-28 should be consulted for the approved method for detailed design of vessels subject to external pressure. The EN 13447 code uses the same method as BS PD 5500.

Out of Roundness

Any out of roundness in a shell after fabrication will significantly reduce the ability of the vessel to resist external pressure. A deviation from a true circular cross-section equal to the shell thickness will reduce the critical buckling pressure by about 50%. The ovality (out of roundness) of a cylinder is measured by

$$\text{Ovality} = \frac{2(D_{\max} - D_{\min})}{(D_{\max} + D_{\min})} \times 100, \text{ per cent}$$

For vessels under external pressure this should not normally exceed 1.5%.

13.7.2. Vessel Heads

The critical buckling pressure for a sphere subject to external pressure is given by (see Timoshenko, 1936)

$$P_c = \frac{2E_Y \, t^2}{R_s^2 \sqrt{3(1 - v^2)}} \tag{13.56}$$

where R_s is the outside radius of the sphere. Taking Poisson's ratio as 0.3 gives

$$P_c = 1.21E_Y \left(\frac{t}{R_s}\right)^2 \tag{13.57}$$

This equation gives the critical pressure required to cause general buckling; local buckling can occur at a lower pressure. Karman and Tsien (1939) have shown that the pressure to cause a 'dimple' to form is about one-quarter of that given by equation 13.57, and is given by

$$P_c' = 0.365E_Y \left(\frac{t}{R_s}\right)^2 \tag{13.58}$$

A generous factor of safety is needed when applying equation 13.58 to the design of heads under external pressure. A value of 6 is typically used, which gives the following equation for the minimum thickness:

$$t = 4R_s \sqrt{\frac{P_e}{E_Y}} \tag{13.59}$$

where P_e is the external pressure.

Any consistent system of units can be used with equation 13.59.

Design of vessels using equation 13.59 is not in accordance with the ASME BPV Code, and hence can only be used for initial estimates. For detailed design of heads subject to external pressure the more complex method given in ASME BPV Code Sec. VIII D.1 Part UG-33 must be followed. The method prescribed by the code is derived from the method used for spherical shells subject to external pressure.

Torispherical and ellipsoidal heads can be designed as equivalent hemispheres. For a torispherical head the radius R_s is taken as equivalent to the crown radius R_c. For an ellipsoidal head the radius can be taken as the maximum radius of curvature; that at the top is given by

$$R_s = \frac{a^2}{b} \tag{13.60}$$

where

$2a$ = major axis = D_0 (shell o.d.)
$2b$ = minor axis = $2\,h$
h = height of the head from the tangent line.

Because the radius of curvature of an ellipse is not constant the use of the maximum radius will over-size the thickness required.

Design methods for different shaped heads under external pressure are given in the standards and codes.

13.8. DESIGN OF VESSELS SUBJECT TO COMBINED LOADING

Pressure vessels are subjected to other loads in addition to pressure (see Section 13.4.7) and must be designed to withstand the worst combination of loading without failure. It is not practical to give an explicit relationship for the vessel thickness to resist combined loads. A trial thickness must be assumed (based on that calculated for pressure alone) and the resultant stress from all loads determined to ensure that the maximum allowable stress intensity is not exceeded at any point. When combined loads are analysed, the maximum compressive stress must be considered as well as the maximum tensile stress. The maximum allowable stress in compression is different from the maximum allowable stress in tension, and is determined using the method given in ASME BPV Code Sec. VIII D.1 Part UG-23.

The main sources of load to consider are:

1. Pressure
2. Dead weight of vessel and contents
3. Wind
4. Earthquake (seismic)
5. External loads imposed by piping and attached equipment.

The primary stresses arising from these loads are considered in the following paragraphs, for cylindrical vessels (Figure 13.15).

Primary Stresses

1. The longitudinal and circumferential stresses due to pressure (internal or external), given by

$$\sigma_h = \frac{PD_i}{2t} \tag{13.61}$$

$$\sigma_L = \frac{PD_i}{4t} \tag{13.62}$$

2. The direct stress σ_w due to the weight of the vessel, its contents and any attachments. The stress will be tensile (positive) for points below the plane of the vessel supports, and compressive (negative) for points above the supports; see Figure 13.16. The dead-weight stress will normally only be significant, compared to the magnitude of the other stresses, in tall vessels.

$$\sigma_w = \frac{W_z}{\pi(D_i + t)t} \tag{13.63}$$

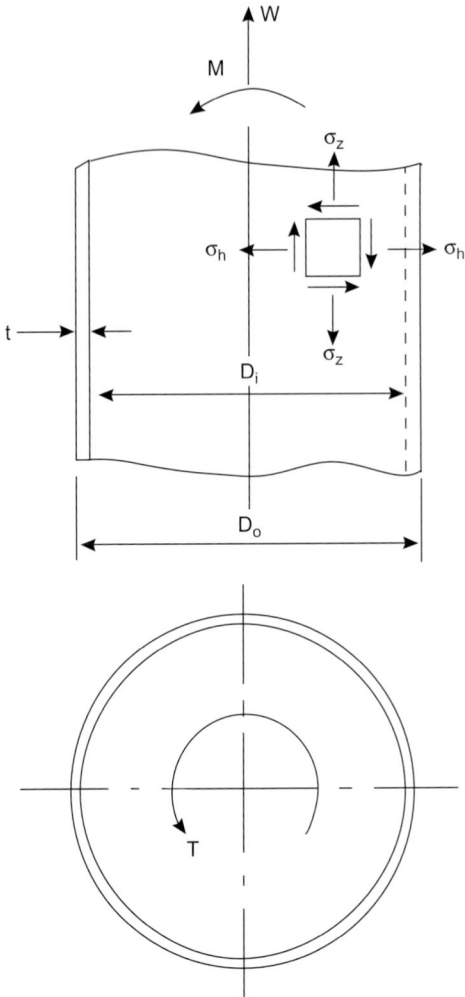

Figure 13.15.   Stresses in a cylindrical shell under combined loading.

where W_z is the total weight that is supported by the vessel wall at the plane considered; see Section 13.8.1.

3. Bending stresses resulting from the bending moments to which the vessel is subjected. Bending moments will be caused by the following loading conditions:
   (a) the wind loads on tall self-supported vessels (Section 13.8.2)
   (b) seismic (earthquake) loads on tall vessels (Section 13.8.3)
   (c) the dead weight and wind loads on piping and equipment that is attached to the vessel, but offset from the vessel centre line (Section 13.8.4)
   (d) for horizontal vessels with saddle supports, from the disposition of dead-weight load (Section 13.9.1)

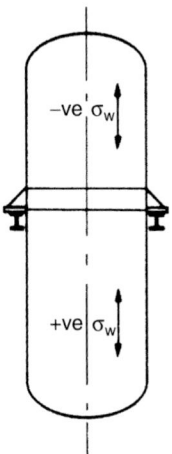

Figure 13.16.   Stresses due to dead-weight loads.

The bending stresses will be compressive or tensile, depending on location, and are given by

$$\sigma_b = \pm \frac{M}{I_v}\left(\frac{D_i}{2} + t\right) \tag{13.64}$$

where M is the total bending moment at the plane being considered and I_v the second moment of area of the vessel about the plane of bending.

$$I_v = \frac{\pi}{64}(D_0^4 - D_i^4) \tag{13.65}$$

4. Torsional shear stresses τ resulting from torque caused by loads offset from the vessel axis. These loads will normally be small, and need not be considered in preliminary vessel designs.

    The torsional shear stress is given by

$$\tau = \frac{T}{I_p}\left(\frac{D_i}{2} + t\right) \tag{13.66}$$

where

$T =$ the applied torque
$I_p =$ polar second moment of area $= (\pi/32)(D_0^4 - D_i^4)$.

Principal Stresses

The principal stresses will be given by

$$\sigma_1 = \frac{1}{2}\left[\sigma_b + \sigma_z + \sqrt{(\sigma_b - \sigma_z)^2 + 4\tau^2}\right] \tag{13.67}$$

$$\sigma_2 = \frac{1}{2}\left[\sigma_b + \sigma_z - \sqrt{(\sigma_b - \sigma_z)^2 + 4\tau^2}\right] \tag{13.68}$$

where

σ_z = total longitudinal stress

$= \sigma_L + \sigma_w \pm \sigma_b$

σ_w should be counted as positive if tension and negative if compressive.

τ is not usually significant.

The third principal stress, that in the radial direction σ_3, will usually be negligible for thin-walled vessels (see Section 13.1.1). As an approximation it can be taken as equal to one-half the pressure loading

$$\sigma_3 = 0.5P \qquad (13.69)$$

σ_3 will be compressive (negative).

Allowable Stress Intensity

The maximum intensity of stress allowed will depend on the particular theory of failure adopted in the design method (see Section 13.3.2). The maximum shear-stress theory is normally used for pressure vessel design.

Using this criterion the maximum stress intensity at any point is taken for design purposes as the numerically greatest value of the following:

$$(\sigma_1 - \sigma_2)$$
$$(\sigma_1 - \sigma_3)$$
$$(\sigma_2 - \sigma_3)$$

The vessel wall thickness must be sufficient to ensure the maximum stress intensity does not exceed the maximum allowable stress (nominal design strength) for the material of construction, at any point. The ASME BPV Code Sec. II Part D or EN 13445-3 Clause 6 should be consulted for the maximum allowable stress values in tension or in compression, depending on the code that is being followed.

Compressive Stresses and Elastic Stability

Under conditions where the resultant axial stress σ_z due to the combined loading is compressive, the vessel may fail by elastic instability (buckling) (see Section 13.3.3). Failure can occur in a thin-walled process column under an axial compressive load by buckling of the complete vessel, as with a strut (Euler buckling); or by local buckling, or wrinkling, of the shell plates. Local buckling will normally occur at a stress lower than that required to buckle the complete vessel. A column design must be checked to ensure that the maximum value of the resultant axial stress does not exceed the critical value at which buckling will occur.

For a curved plate subjected to an axial compressive load, the critical buckling stress σ_c is given by (see Timoshenko, 1936)

$$\sigma_c = \frac{E_Y}{\sqrt{3(1 - \nu^2)}} \left(\frac{t}{R_p}\right) \qquad (13.70)$$

where R_p is the radius of curvature.
Taking Poisson's ratio as 0.3 gives

$$\sigma_c = 0.60 \, E_Y \left(\frac{t}{R_p}\right) \qquad (13.71)$$

By applying a suitable factor of safety, equation 13.71 can be used to predict the maximum allowable compressive stress to avoid failure by buckling. A large factor of safety is required, as experimental work has shown that cylindrical vessels will buckle at values well below that given by equation 13.70. For steels at ambient temperature, $E_Y = 200,000$ N/mm^2, and equation 13.71 with a factor of safety of 12 gives

$$\sigma_c = 2 \times 10^4 \left(\frac{t}{D_0}\right) \text{N/mm}^2 \qquad (13.72)$$

The maximum compressive stress in a vessel wall should not exceed that given by equation 13.72 or the maximum allowable design stress for the material, whichever is the least. For detailed design, the ASME BPV Code Sec. VIII should be consulted and the recommended procedure in the code should be followed.

Stiffening

As with vessels under external pressure, the resistance to failure by buckling can be increased significantly by the use of stiffening rings, or longitudinal strips. Methods for estimating the critical buckling stress for stiffened vessels are given in the standards and codes.

Loading

The loads to which a vessel may be subjected will not all occur at the same time. For example, it is the usual practice to assume that the maximum wind load will not occur simultaneously with a major earthquake.

The vessel must be designed to withstand the worst combination of the loads likely to occur in the following situations:

1. During erection (or dismantling) of the vessel
2. With the vessel erected but not operating
3. During testing (the hydraulic pressure test)
4. During normal operation.

13.8.1.   Weight Loads

The major sources of dead weight loads are:

1. The vessel shell
2. The vessel fittings: manways, nozzles
3. Internal fittings: plates (plus the fluid on the plates); heating and cooling coils
4. External fittings: ladders, platforms, piping
5. Auxiliary equipment that is not self-supported; condensers, agitators
6. Insulation

7. The weight of liquid to fill the vessel. The vessel will be filled with water for the hydraulic pressure test, and may fill with process liquid due to mis-operation.

Note: for vessels on a skirt support (see Section 13.9.2), the weight of the liquid to fill the vessel will be transferred directly to the skirt.

The weight of the vessel and fittings can be calculated from the preliminary design sketches. The weights of standard vessel components: heads, shell plates, manways, branches and nozzles, are given in various handbooks; Megyesy (2001) and Brownell and Young (1959).

For preliminary calculations, the approximate weight of a cylindrical vessel with domed ends and uniform wall thickness can be estimated from the following equation:

$$W_v = C_w \pi \rho_m D_m g (H_v + 0.8 D_m) t \times 10^{-3} \qquad (13.73)$$

where

W_v = total weight of the shell, excluding internal fittings, such as plates, N
C_w = a factor to account for the weight of nozzles, manways, internal supports, etc; which can be taken as
  = 1.08 for vessels with only a few internal fittings
  = 1.15 for distillation columns, or similar vessels, with several manways, and with plate support rings, or equivalent fittings
H_v = height, or length, between tangent lines (the length of the cylindrical section), m
g = gravitational acceleration, 9.81 m/s^2
t = wall thickness, mm
ρ_m = density of vessel material, kg/m^3
D_m = mean diameter of vessel = $(D_i + t \times 10^{-3})$, m.

For a steel vessel, equation 13.73 reduces to:

$$W_v = 240 C_w D_m (H_v + 0.8 D_m) t \qquad (13.74)$$

The following values can be used as a rough guide to the weight of fittings; see Nelson (1963):

(a) caged ladders, steel, 360 N/m length
(b) plain ladders, steel, 150 N/m length
(c) platforms, steel, for vertical columns, 1.7 kN/m^2 area
(d) contacting plates, steel, including typical liquid loading, 1.2 kN/m^2 plate area.

Typical values for the density of insulating materials are (all kg/m^3):

Foam glass	150
Mineral wool	130
Fibreglass	100
Calcium silicate	200

These densities should be doubled to allow for attachment fittings, sealing and moisture absorption.

13.8.2.   **Wind Loads (Tall Vessels)**

Wind loading will only be important on tall columns installed in the open. Columns and chimney-stacks are usually free standing, mounted on skirt supports and not attached to structural steel work. Under these conditions, the vessel under wind loading acts as a cantilever beam; see Figure 13.17. For a uniformly loaded cantilever the bending moment at any plane is given by

$$M_x = \frac{W x^2}{2} \tag{13.75}$$

where x is the distance measured from the free end and W the load per unit length (Newtons per metre run).

So the bending moment, and hence the bending stress, will vary parabolically from zero at the top of the column to a maximum value at the base. For tall columns, the bending stress due to wind loading will often be greater than direct stress due to pressure and will determine the plate thickness required. The most economical design will be one in which the plate thickness is progressively increased from the top to the base of the column; the thickness at the top being sufficient for the pressure load, and that at the base sufficient for the pressure plus the maximum bending moment.

Any local increase in the column area presented to the wind will give rise to a local, concentrated, load; see Figure 13.18. The bending moment at the column base caused by a concentrated load is given by

$$M_p = F_p H_p \tag{13.76}$$

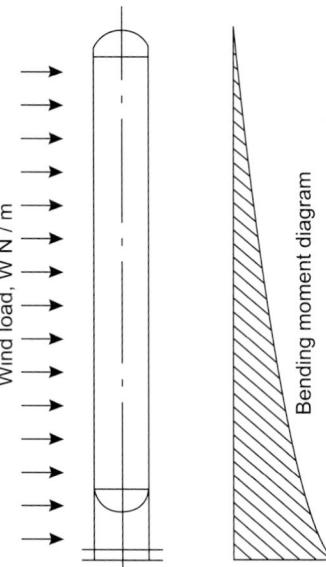

Figure 13.17.   Wind loading on a tall column.

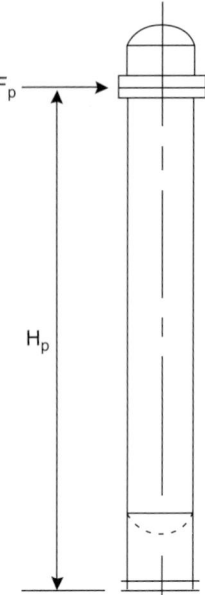

Figure 13.18.   Local wind loading.

where

F_p = local, concentrated, load
H_p = the height of the concentrated load above the column base.

Dynamic Wind Pressure

The load imposed on any structure by the action of the wind will depend on the shape of the structure and the wind velocity.

$$P_w = \frac{1}{2}C_d\rho_a u_w^2 \tag{13.77}$$

where

P_w = wind pressure (load per unit area)
C_d = drag coefficient (shape factor)
ρ_a = density of air
u_w = wind velocity.

The drag coefficient is a function of the shape of the structure and the wind velocity (Reynolds number).

For a smooth cylindrical column or stack, the following semi-empirical equation can be used to estimate the wind pressure:

$$P_w = 0.05u_w^2 \tag{13.78}$$

where

P_w = wind pressure, N/m^2
u_w = wind speed, km/h.

If the column outline is broken up by attachments, such as ladders or pipe work, the factor of 0.05 in equation 13.78 should be increased to 0.07, to allow for the increased drag.

A column must be designed to withstand the highest wind speed that is likely to be encountered at the site during the life of the plant. The probability of a given wind speed occurring can be predicted by studying meteorological records for the site location.

Data and design methods for wind loading are given in the Engineering Sciences Data Unit (ESDU) Wind Engineering Series (www.ihsesdu.com).

Design loadings for locations in the United States are given by Moss (2003), Megyesy (2001) and Escoe (1994).

A wind speed of 160 km/h (100 mph) can be used for preliminary design studies; equivalent to a wind pressure of 1280 N/m^2 (25 lb/ft^2).

At any site, the wind velocity near the ground will be lower than that higher up (due to the boundary layer), and in some design methods a lower wind pressure is used at heights below about 20 m; typically taken as one-half of the pressure above this height.

The loading per unit length of the column can be obtained from the wind pressure by multiplying by the effective column diameter: the outside diameter plus an allowance for the thermal insulation and attachments, such as pipes and ladders.

$$W = P_w \, D_{eff} \qquad\qquad (13.79)$$

An allowance of 0.4 m should be added for a caged ladder. The calculation of the wind load on a tall column, and the induced bending stresses, is illustrated in Example 13.2. Further examples of the design of tall columns are given by Brownell (1963), Henry (1973), Bednar (1990), Escoe (1994) and Jawad and Farr (1989).

Deflection of Tall Columns

Tall columns sway in the wind. The allowable deflection will normally be specified as less than 150 mm per 30 m of height (6 in per 100 ft).

For a column with a uniform cross-section, the deflection can be calculated using the formula for the deflection of a uniformly loaded cantilever. A method for calculating the deflection of a column where the wall thickness is not constant is given by Tang (1968).

Wind-Induced Vibrations

Vortex shedding from tall thin columns and stacks can induce vibrations, which, if the frequency of shedding of eddies matches the natural frequency of the column, can be severe enough to cause premature failure of the vessel by fatigue. The effect of vortex shedding should be investigated for free-standing columns with height-to-diameter ratios greater than 10. Methods for estimating the natural frequency of columns are given by Freese (1959) and DeGhetto and Long (1966).

Helical strakes (strips) are fitted to the tops of tall smooth chimneys to change the pattern of vortex shedding and so prevent resonant oscillation. The same effect will be achieved on a tall column by distributing any attachments (ladders, pipes and platforms) around the column.

13.8.3. Earthquake Loading

The movement of the earth's surface during an earthquake produces horizontal shear forces on tall self-supported vessels, the magnitude of which increases from the base upward. The total shear force on the vessel will be given by:

$$F_s = a_e \left(\frac{W_v}{g} \right) \tag{13.80}$$

where

a_e = the acceleration of the vessel due to the earthquake
g = the acceleration due to gravity
W_v = total weight of the vessel and contents.

The term (a_e/g) is called the seismic constant C_e, and is a function of the natural period of vibration of the vessel and the severity of the earthquake. Values of the seismic constant have been determined empirically from studies of the damage caused by earthquakes, and are available for those geographical locations that are subject to earthquake activity. Values for sites in the US, and procedures for determining the stresses induced in tall columns, are given by Megyesy (2001), Escoe (1994) and Moss (2003).

13.8.4. Eccentric Loads (Tall Vessels)

Ancillary equipment attached to a tall vessel will subject the vessel to a bending moment if the centre of gravity of the equipment does not coincide with the centre line of the vessel (Figure 13.19). The moment produced by small fittings, such as ladders, pipes and manways, will be small and can be neglected. That produced by heavy equipment, such as reflux condensers and side platforms, can be significant and should be considered. The moment is given by

$$M_e = W_e L_o \tag{13.81}$$

where

W_e = dead weight of the equipment
L_o = distance between the centre of gravity of the equipment and the column centre line.

To avoid putting undue stress on the column walls, equipment such as reflux condensers and overhead receiving drums is usually not attached to the top of a column, but is instead located adjacent to the column in the plant structure. Condensers and receiving vessels are often placed above grade level to provide net positive suction head for reflux and overhead pumps sited at grade.

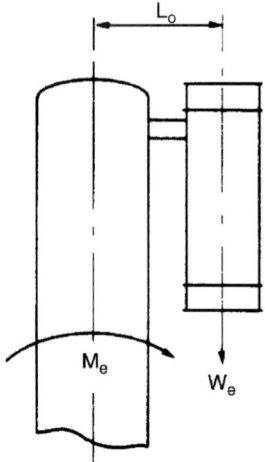

Figure 13.19.   Bending moment due to offset equipment.

13.8.5.   Torque

Any horizontal force imposed on the vessel by ancillary equipment, the line of thrust
of which does not pass through the centre line of the vessel, will produce a torque on
the vessel. Such loads can arise through wind pressure on piping and other attachments. However, the torque will normally be small and usually can be disregarded.
The pipe work and the connections for any ancillary equipment will be designed so as
not to impose a significant load on the vessel.

Example 13.2

Make a preliminary estimate of the plate thickness required for the distillation
column specified below:

Height, between tangent lines	50 m
Diameter	2 m
Hemispherical head	
Skirt support, height	3 m
100 sieve plates, equally spaced	
Insulation, mineral wool	75 mm thick
Material of construction, stainless steel, maximum allowable stress	135 N/mm^2 at design temperature 200°C
Operating pressure	10 bar (absolute)
Vessel to be fully radiographed (joint efficiency 1)	
Process service	Gasoline debutanizer.

Solution

Design pressure; take as 10% above operating pressure

$$= (10 - 1) \times 1.1 = 9.9 \text{ bar, say 10 bar}$$

$$= 1.0 \text{ N/mm}^2$$

Minimum thickness required for pressure loading in accordance with ASME BPV Code Sec. VIII D.1:

$$t = \frac{1 \times 2 \times 10^3}{(2 \times 135 \times 1) - (1.2 \times 1)} = 7.4 \text{ mm} \tag{13.41}$$

A much thicker wall will be needed at the column base to withstand the wind and dead weight loads.

As a first trial, divide the column into five sections (courses), with the thickness increasing by 2 mm per section. Try 10, 12, 14, 16, 18 mm.

Dead Weight of Vessel

Though equation 13.74 only applies strictly to vessels with uniform thickness, it can be used to get a rough estimate of the weight of this vessel by using the average thickness in the equation, 14 mm.

Take

$$C_w = 1.15, \text{ vessel with plates}$$
$$D_m = 2 + 14 \times 10^{-3} = 2.014 \text{ m}$$
$$H_v = 50 \text{ m}$$
$$t = 14 \text{ mm}$$
$$W_v = 240 \times 1.15 \times 2.014(50 + 0.8 \times 2.014)14 \tag{13.74}$$
$$= 401643 \text{ N}$$
$$= 402 \text{ kN}$$

Weight of plates:

plate area $= \pi/4 \times 2^2 = 3.14 \text{ m}^2$

weight of a plate including liquid on it (see section 13.8.1)

$$\approx 1.2 \times 3.14 = 3.8 \text{ kN}$$

100 plates $= 100 \times 3.8 = 380 \text{ kN}$

Weight of insulation:

mineral wool density $= 130 \text{ kg /m}^3$

approximate volume of insulator $= \pi \times 2 \times 50 \times 75 \times 10^{-3} = 23.6 \text{ m}^3$

weight $= 23.6 \times 130 \times 9.81 = 30,049 \text{ N}$

double this to allow for fittings, etc. $= 60 \text{ kN}$

Total weight:

shell	402
plates & contents	380
insulation	60
	842 kN

Note that the weight of the contents of the column would be substantially greater if the column was flooded or entirely filled with liquid. This is the case during hydraulic testing, which should be examined as a different loading scenario.

Wind Loading

Take dynamic wind pressure as 1280 N/m^2, corresponding to 160 kph (100 mph).

Mean diameter, including insulation $= 2 + 2(14 + 75) \times 10^{-3} = 2.18$ m

Loading (per linear metre) $F_w = 1280 \times 2.18 = 2790$ N/m (13.79)

Bending moment at bottom tangent line

$$M_x = \frac{2790}{2} \times 50^2 = 3,487,500 \text{ Nm} \qquad (13.75)$$

Analysis of Stresses

At bottom tangent line
  Pressure stresses:

$$\sigma_L = \frac{1.0 \times 2 \times 10^3}{4 \times 18} = 27.8 \text{ N/mm}^2 \qquad (13.62)$$

$$\sigma_h = \frac{1 \times 2 \times 10^3}{2 \times 18} = 55.6 \text{ N/mm}^2 \qquad (13.61)$$

Dead weight stress:

$$\sigma_w = \frac{W_v}{\pi(D_i + t)t} = \frac{842 \times 10^3}{\pi(2000 + 18)18} \qquad (13.63)$$
$$= 7.4 \text{ N/mm}^2 \text{ (compressive)}$$

Bending stresses:

$$D_o = 2000 + 2 \times 18 = 2036 \text{ mm}$$
$$I_v = \frac{\pi}{64}(2036^4 - 2000^4) = 5.81 \times 10^{10} \text{ mm}^4 \qquad (13.65)$$

$$\sigma_b = \pm\frac{3,487,500 \times 10^3}{5.81 \times 10^{10}}\left(\frac{2000}{2} + 18\right) \qquad (13.64)$$
$$= \pm 61.11 \text{ N/mm}^2$$

The resultant longitudinal stress is:

$$\sigma_z = \sigma_L + \sigma_w \pm \sigma_b$$

σ_w is compressive and therefore negative
σ_z (upwind) $= 27.8 - 7.4 + 61.1 = +81.5$ N/mm^2
σ_z (downwind) $= 27.8 - 7.4 - 61.1 = -40.7$ N/mm^2.

As there is no torsional shear stress, the principal stresses will be σ_z and σ_h. The radial stress is negligible, $(P_i/2) = 0.5 \text{ N/mm}^2$.

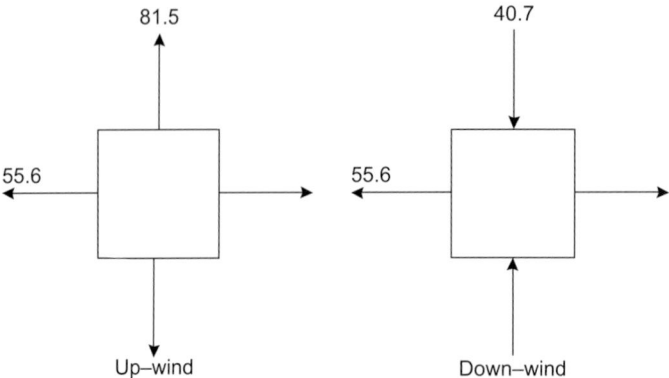

The greatest difference between the principal stresses will be on the down-wind side

$$(55.6 - (-40.7)) = \underline{\underline{96.5 \text{ N/mm}^2}}$$

well below the maximum allowable design stress.

Note that the bending stress due to wind loading is much larger than the dead weight stress. The hydraulic testing case will have a greater dead weight when the vessel is filled with water, but a simple calculation shows that the maximum weight of water in the vessel (neglecting volume of vessel internals) is $\pi/12 \times \rho \times g\,(3D_i^2 L + 2D_i^3) = 1582 \text{ kN}$. If this is added to the total weight calculated above then the dead weight stress will increase by about a factor 3. This is still a lot less than the bending stress due to wind load, so the wind load case is the governing case. The hydraulic test will obviously not be scheduled for a day on which 100 mph winds may occur.

Check Elastic Stability (Buckling)

Critical buckling stress:

$$\sigma_c = 2 \times 10^4 \left(\frac{18}{2036}\right) = \underline{\underline{176.8 \text{ N/mm}^2}} \tag{13.72}$$

The maximum compressive stress will occur when the vessel is not under pressure = $7.4 + 61.1 = 68.5$, well below the critical buckling stress.

So the design is satisfactory. The designer could reduce the plate thickness and recalculate.

13.9. VESSEL SUPPORTS

The method used to support a vessel will depend on the size, shape and weight of the vessel; the design temperature and pressure; the vessel location and arrangement; and the internal and external fittings and attachments. Horizontal vessels are usually mounted on two saddle supports (Figure 13.20). Skirt supports are used for tall,

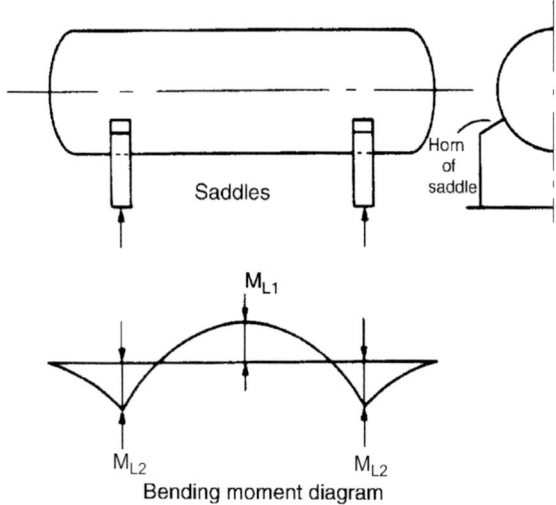

Figure 13.20. Horizontal cylindrical vessel on saddle supports.

vertical columns (Figure 13.21). Brackets, or lugs, are used for all types of vessel (Figure 13.22). The supports must be designed to carry the weight of the vessel and contents, and any superimposed loads, such as wind loads. Supports will impose localized loads on the vessel wall, and the design must be checked to ensure that the resulting stress concentrations are below the maximum allowable design stress. Supports should be designed to allow easy access to the vessel and fittings for inspection and maintenance.

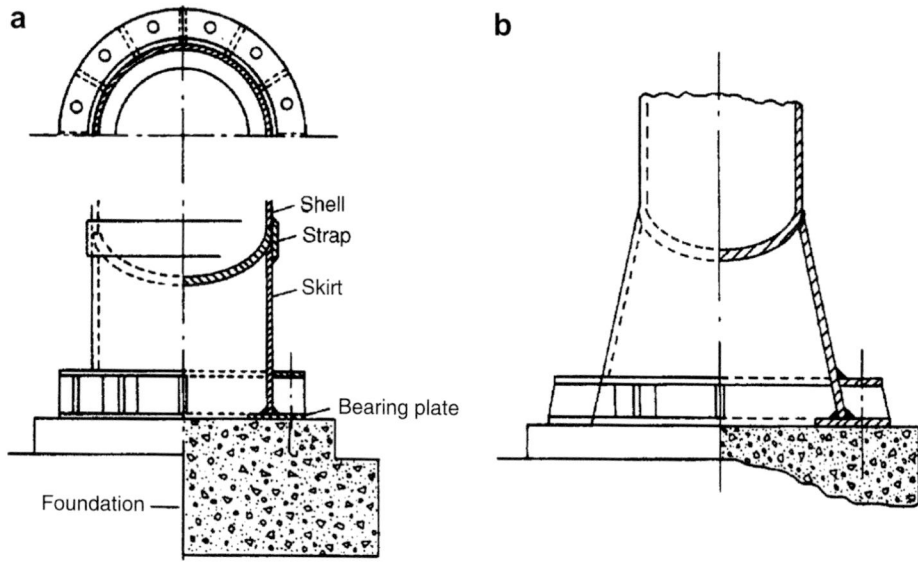

Figure 13.21. Typical skirt-support designs. (a) Straight skirt. (b) Conical skirt.

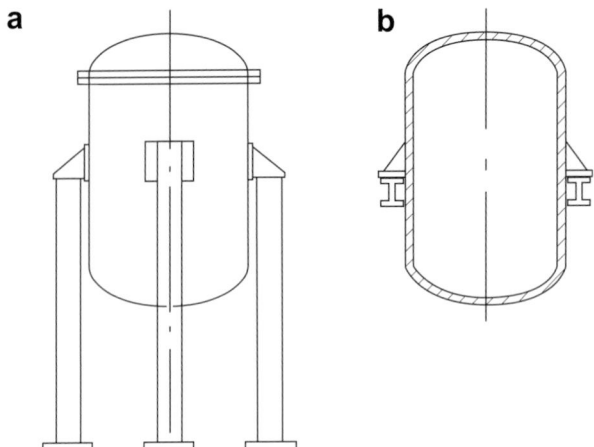

Figure 13.22. Bracket supports. (a) Supported on legs. (b) Supported from steel-work.

13.9.1. Saddle Supports

Though saddles are the most commonly used support for horizontal cylindrical vessels, legs can be used for small vessels. A horizontal vessel will normally be supported at two cross-sections; if more than two saddles are used, the distribution of the loading is uncertain.

A vessel supported on two saddles can be considered as a simply supported beam, with an essentially uniform load, and the distribution of longitudinal axial bending moment will be as shown in Figure 13.20. Maxima occur at the supports and at mid-span. The theoretical optimum position of the supports to give the least maximum bending moment will be the position at which the maxima at the supports and at mid-span are equal in magnitude. For a uniformly loaded beam, the position will be at 21% of the span, in from each end. The saddle supports for a vessel will usually be located nearer the ends than this value, to make use of the stiffening effect of the ends.

In addition to the longitudinal bending stress, a vessel supported on saddles will be subjected to tangential shear stresses, which transfer the load from the unsupported sections of the vessel to the supports, and to circumferential bending stresses. All these stresses need to be considered in the design of large, thin-walled, vessels, to ensure that the resultant stress does not exceed the maximum allowable design stress or the critical buckling stress for the material. A detailed stress analysis is beyond the scope of this book. A complete analysis of the stress induced in the shell by the supports is given by Zick (1951). Zick's method forms the basis of the design methods given in the national codes and standards. The method is also given by Brownell and Young (1959), Escoe (1994) and Megyesy (2001).

Design of Saddles

The saddles must be designed to withstand the load imposed by the weight of the vessel and contents. They are constructed of bricks or concrete, or are fabricated from steel plate. The contact angle should not be less than 120°, and will not normally be greater than 150°. Wear plates are often welded to the shell wall to reinforce the wall over the area of contact with the saddle.

The dimensions of typical 'standard' saddle designs are given in Figure 13.23. To take up any thermal expansion of the vessel, such as that in heat exchangers, the anchor bolt holes in one saddle can be slotted.

Procedures for the design of saddle supports are given by Brownell and Young (1959), Megyesy (2001), Escoe (1994) and Moss (2003).

13.9.2. Skirt Supports

A skirt support consists of a cylindrical or conical shell welded to the base of the vessel. A flange at the bottom of the skirt transmits the load to the foundations. Typical designs are shown in Figure 13.21. Openings must be provided in the skirt for access and for any connecting pipes; the openings are normally reinforced. The skirt may be welded to the bottom head of the vessel (Figure 13.24a), or welded flush with the shell (Figure 13.24b) or welded to the outside of the vessel shell (Figure 13.24c). The arrangement shown in Figure 13.24b is usually preferred.

Skirt supports are recommended for vertical vessels, as they do not impose concentrated loads on the vessel shell; they are particularly suitable for use with tall columns subject to wind loading.

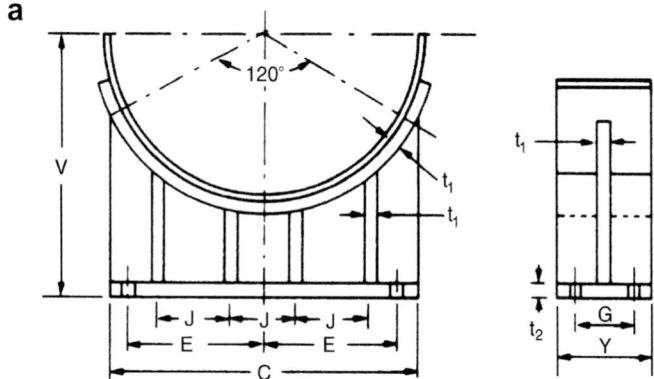

Vessel diam. (m)	Maximum weight (kN)	Dimensions (m)								mm	
		V	Y	C	E	J	G	t_2	t_1	Bolt diam.	Bolt holes
0.6	35	0.48	0.15	0.55	0.24	0.190	0.095	6	5	20	25
0.8	50	0.58	0.15	0.70	0.29	0.225	0.095	8	5	20	25
0.9	65	0.63	0.15	0.81	0.34	0.275	0.095	10	6	20	25
1.0	90	0.68	0.15	0.91	0.39	0.310	0.095	11	8	20	25
1.2	180	0.78	0.20	1.09	0.45	0.360	0.140	12	10	24	30

All contacting edges fillet welded

Figure 13.23. Standard steel saddles (adapted from Bhattacharyya, 1976). (a) For vessels up to 1.2 m.

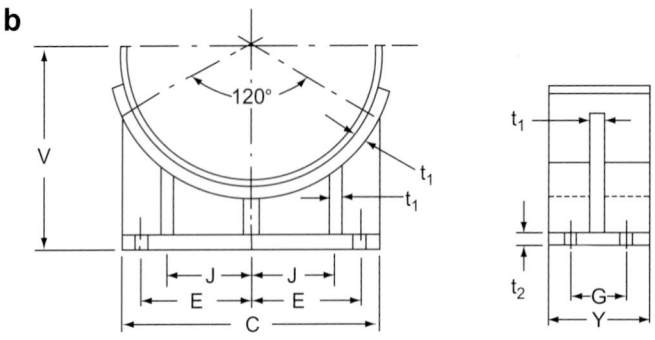

Vessel diam. (m)	Maximum weight (kN)	Dimensions (m)						mm			
		V	Y	C	E	J	G	t_2	t_1	Bolt diam.	Bolt holes
1.4	230	0.88	0.20	1.24	0.53	0.305	0.140	12	10	24	30
1.6	330	0.98	0.20	1.41	0.62	0.350	0.140	12	10	24	30
1.8	380	1.08	0.20	1.59	0.71	0.405	0.140	12	10	24	30
2.0	460	1.18	0.20	1.77	0.80	0.450	0.140	12	10	24	30
2.2	750	1.28	0.225	1.95	0.89	0.520	0.150	16	12	24	30
2.4	900	1.38	0.225	2.13	0.98	0.565	0.150	16	12	27	33
2.6	1000	1.48	0.225	2.30	1.03	0.590	0.150	16	12	27	33
2.8	1350	1.58	0.25	2.50	1.10	0.625	0.150	16	12	27	33
3.0	1750	1.68	0.25	2.64	1.18	0.665	0.150	16	12	27	33
3.2	2000	1.78	0.25	2.82	1.26	0.730	0.150	16	12	27	33
3.6	2500	1.98	0.25	3.20	1.40	0.815	0.150	16	12	27	33

All contacting edges fillet welded.

Figure 13.23 Cont'd.   (b) For vessels greater than 1.2 m.

Skirt Thickness

The skirt thickness must be sufficient to withstand the dead-weight loads and bending moments imposed on it by the vessel; it will not be under the vessel pressure.

The resultant stresses in the skirt will be:

$$\sigma_s \text{ (tensile)} = \sigma_{bs} - \sigma_{ws} \qquad (13.82)$$

and

$$\sigma_s \text{ (compressive)} = \sigma_{bs} + \sigma_{ws} \qquad (13.83)$$

where σ_{bs} = bending stress in the skirt

$$= \frac{4 M_s}{\pi (D_s + t_{sk}) t_{sk} D_s}, \qquad (13.84)$$

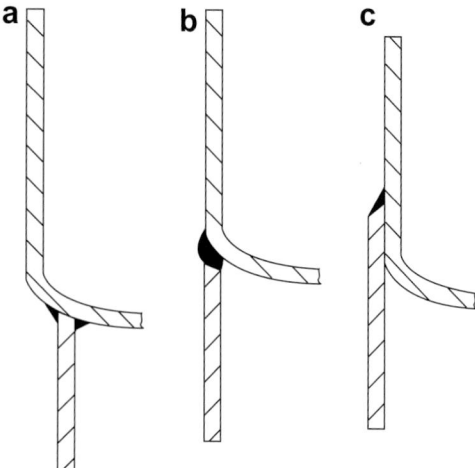

Figure 13.24. Skirt-support welds.

σ_{ws} = the dead weight stress in the skirt

$$= \frac{W_v}{\pi(D_s + t_{sk})\, t_{sk}} \qquad (13.85)$$

where

M_s = maximum bending moment, evaluated at the base of the skirt (due to wind, seismic and eccentric loads; see Section 13.8)

W_v = total weight of the vessel and contents (see Section 13.8)

D_s = inside diameter of the skirt, at the base

t_{sk} = skirt thickness.

The skirt thickness should be such that under the worst combination of wind and dead-weight loading the following design criteria are not exceeded:

$$\sigma_s \,(\text{tensile}) \;<\; S_s\, E \sin \theta_s \qquad (13.86)$$

$$\sigma_s \,(\text{compressive}) \;<\; 0.125\, E_Y \left(\frac{t_{sk}}{D_s}\right) \sin \theta_s \qquad (13.87)$$

where

S_s = maximum allowable design stress for the skirt material, normally taken at ambient temperature, 20°C

E = welded-joint efficiency, if applicable

θ_s = base angle of a conical skirt, normally 80° to 90°.

The minimum thickness should be not less than 6 mm.

Where the vessel wall will be at a significantly higher temperature than the skirt, discontinuity stresses will be set up due to differences in thermal expansion. Methods for calculating the thermal stresses in skirt supports are given by Weil and Murphy (1960) and Bergman (1963).

Base Ring and Anchor Bolt Design

The loads carried by the skirt are transmitted to the foundation slab by the skirt base ring (bearing plate). The moment produced by wind and other lateral loads will tend to overturn the vessel; this will be opposed by the couple set up by the weight of the vessel and the tensile load in the anchor bolts. Various base ring designs are used with skirt supports. The simplest types, suitable for small vessels, are the rolled angle and plain flange rings shown in Figure 13.25a and b. For larger columns, a double ring stiffened by gussets (Figure 13.25c) is used, or chair supports. Design methods for base rings, and methods for sizing the anchor bolts, are given by Brownell and Young (1959). For preliminary design, the short-cut method and nomographs given by Scheiman (1963) can be used. Scheiman's method is based on a more detailed procedure for the design of base rings and foundations for columns and stacks given by Marshall (1958).

Example 13.3

Design a skirt support for the column specified in Example 13.2.

Solution

Try a straight cylindrical skirt ($\theta_s = 90°$) of plain carbon steel, maximum allowable stress 89 N/mm^2 and Young's modulus 200,000 N/mm^2 at ambient temperature.

The maximum dead weight load on the skirt will occur when the vessel is full of water.

$$\text{Approximate weight} = \left(\frac{\pi}{4} \times 2^2 \times 50\right) 1000 \times 9.81 = 1,540,951 \text{ N}$$

$$= 1541 \text{ kN}$$

$$\text{Weight of vessel, from Example 13.2} = 842 \text{ kN}$$
$$\text{Total weight} = 1541 + 842 = 2383 \text{ kN}$$
$$\text{Wind loading, from Example 13.2} = 2.79 \text{ kN/m}$$
$$\text{Bending moment at base of skirt} = 2.79 \times \frac{53^2}{2} \qquad (13.75)$$

$$= 3919 \text{ kNm}$$

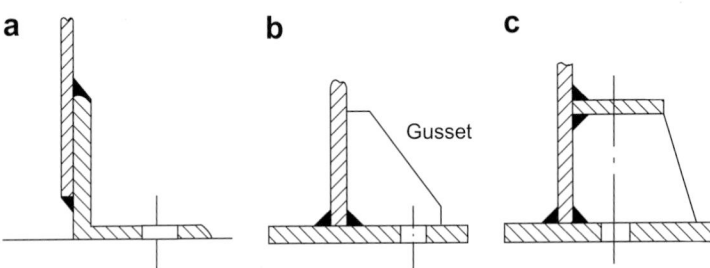

Figure 13.25.   Flange ring designs. (a) Rolled-angle. (b) Single plate with gusset. (c) Double plate with gusset.

As a first trial, take the skirt thickness as the same as that of the bottom section of the vessel, 18 mm.

$$\sigma_{bs} = \frac{4 \times 3919 \times 10^3 \times 10^3}{\pi(2000 + 18)\, 2000 \times 18} \tag{13.84}$$
$$= 68.7 \text{ N/mm}^2$$

$$\sigma_{ws} \text{ (test)} = \frac{2383 \times 10^3}{\pi(2000 + 18)\, 18} = 20.9 \text{ N/mm}^2 \tag{13.85}$$

$$\sigma_{ws} \text{ (operating)} = \frac{842 \times 10^3}{\pi(2000 + 18)\, 18} = 7.4 \text{ N/mm}^2 \tag{13.85}$$

Note: the 'test' condition is with the vessel full of water for the hydraulic test. In estimating total weight, the weight of liquid on the plates has been counted twice. The weight has not been adjusted to allow for this as the error is small, and on the 'safe side'.

$$\text{Maximum } \hat{\sigma}_s \text{ (compressive)} = 68.7 + 20.9 = 89.6 \text{ N/mm}^2 \tag{13.87}$$

$$\text{Maximum } \hat{\sigma}_s \text{ (tensile)} = 68.7 - 7.4 = 61.3 \text{ N/mm}^2 \tag{13.86}$$

Take the joint efficiency E as 0.85.
Criteria for design:

$$\hat{\sigma}_s \text{ (tensile)} < S_s\, E \sin\theta$$
$$61.3 < 0.85 \times 89 \sin 90 \tag{13.86}$$
$$61.3 < 75.6$$

$$\hat{\sigma}_s(\text{compressive}) < 0.125 E_Y \left(\frac{t_{sk}}{D_s}\right)\sin\theta$$
$$89.6 < 0.125 \times 200,000 \left(\frac{18}{2000}\right)\sin 90 \tag{13.87}$$
$$89.6 < 225$$

Both criteria are satisfied; adding 2 mm for corrosion gives a design thickness of <u>20 mm.</u>

13.9.3. Bracket Supports

Brackets, or lugs, can be used to support vertical vessels. The bracket may rest on the building structural steel work, or the vessel may be supported on legs (Figure 13.22).

The main load carried by the brackets will be the weight of the vessel and contents; in addition, the bracket must be designed to resist the load due to any bending moment due to wind, or other loads. If the bending moment is likely to be significant, skirt supports should be considered in preference to bracket supports.

As the reaction on the bracket is eccentric (Figure 13.26), the bracket will impose a bending moment on the vessel wall. The point of support, at which the reaction acts,

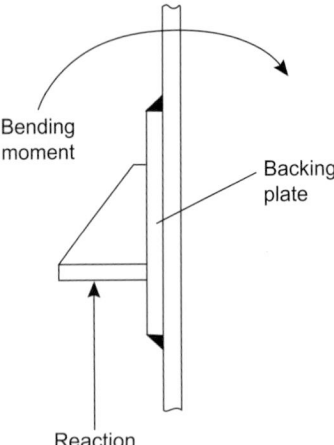

Figure 13.26.   Loads on a bracket support.

should be made as close to the vessel wall as possible, allowing for the thickness of any insulation. Methods for estimating the magnitude of the stresses induced in the vessel wall by bracket supports are given by Brownell and Young (1959) and by Wolosewick (1951). Backing plates are often used to carry the bending loads.

The brackets, and supporting steel work, can be designed using the usual methods for structural steelwork. Suitable methods are given by Bednar (1990) and Moss (2003).

A quick method for sizing vessel reinforcing rings (backing plates) for bracket supports is given by Mahajan (1977).

Typical bracket designs are shown in Figures 13.27a and b. The loads that steel brackets with these proportions will support are given by the following formula.

Single-gusset plate design, Figure 13.27a:

$$F_{bs} = 60L_d\,t_c \qquad\qquad (13.88)$$

Double-gusset plate design, Figure 13.27b:

$$F_{bs} = 120L_d t_c \qquad\qquad (13.89)$$

where

F_{bs} = maximum design load per bracket, N
L_d = the characteristic dimension of bracket (depth), mm
t_c = thickness of plate, mm.

13.10.   BOLTED FLANGED JOINTS

Flanged joints are used for connecting pipes and instruments to vessels, for manhole covers, and for removable vessel heads when ease of access is required. Flanges may also be used on the vessel body, when it is necessary to divide the vessel into sections for transport or maintenance. Flanged joints are also used to connect pipes to other

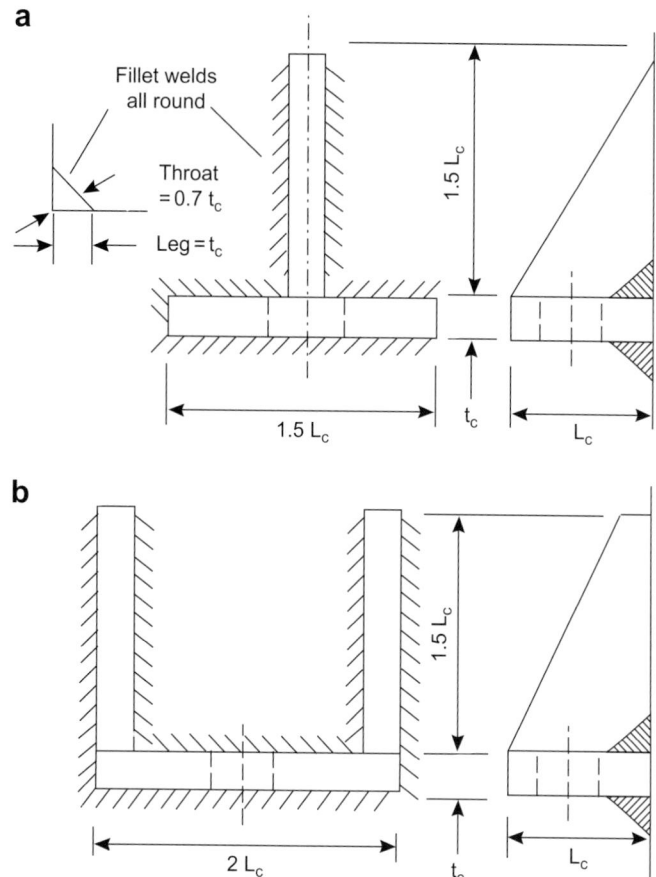

Figure 13.27. Bracket designs. (a) Single-gusset plate. (b) Double-gusset plate.

equipment, such as pumps and valves. Screwed joints are often used for small-diameter pipe connections, below 40 mm. Flanged joints are also used for connecting pipe sections where ease of assembly and dismantling is required for maintenance, but pipework will normally be welded to reduce costs.

Flanges range in size from a few millimetres diameter for small pipes, to several metres diameter for those used as body or head flanges on vessels.

13.10.1. Types of Flange, and Selection

Several different types of flange are used for various applications. The principal types used in the process industries are:

1. Welding-neck flanges
2. Slip-on flanges, hub and plate types
3. Lap-joint flanges
4. Screwed flanges
5. Blank, or blind, flanges.

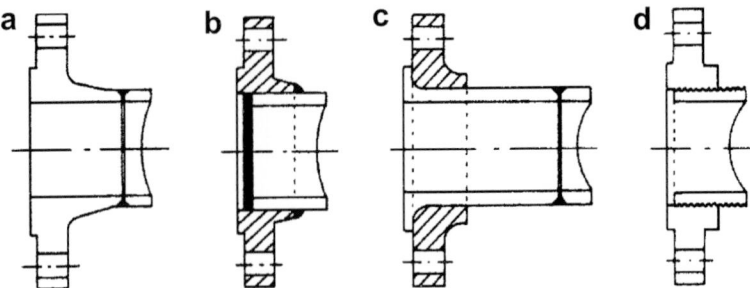

Figure 13.28.　　Flange types. (a) Welding-neck. (b) Slip-on. (c) Lap-joint. (d) Screwed.

Welding-neck flanges, Figure 13.28a: these have a long tapered hub between the flange ring and the welded joint. This gradual transition of the section reduces the discontinuity stresses between the flange and branch, and increases the strength of the flange assembly. Welding-neck flanges are suitable for extreme service conditions where the flange is likely to be subjected to temperature, shear and vibration loads. They will normally be specified for the connections and nozzles on process vessels and process equipment.

Slip-on flanges, Figure 13.28b: these slip over the pipe or nozzle and are welded externally, and usually also internally. The end of the pipe is set back from 0 to 2.0 mm. The strength of a slip-on flange is from one-third to two-thirds that of the corresponding standard welding-neck flange. Slip-on flanges are cheaper than welding-neck flanges and are easier to align, but have poor resistance to shock and vibration loads. Slip-on flanges are generally used for pipe work. Figure 13.28b shows a forged flange with a hub; for light duties slip-on flanges can be cut from plate.

Lap-joint flanges, Figure 13.28c: these are used for pipework. They are economical when used with expensive alloy pipe, such as stainless steel, as the flange can be made from inexpensive carbon steel. Usually a short lapped nozzle is welded to the pipe, but with some schedules of pipe the lap can be formed on the pipe itself, and this will give a cheap method of pipe assembly.

Lap-joint flanges are sometimes known as 'Van-stone flanges'.

Screwed flanges, Figure 13.28d: these are used to connect screwed fittings to flanges. They are also sometimes used for alloy pipe that is difficult to weld satisfactorily.

Blind flanges (blank flanges): these are flat plates, used to blank off flange connections, and as covers for manholes and inspection ports.

13.10.2.　Gaskets

Gaskets are used to make a leak-tight joint between two surfaces. It is impractical to machine flanges to the degree of surface finish that would be required to make a satisfactory seal under pressure without a gasket. Gaskets are made from 'semi-plastic' materials; which will deform and flow under load to fill the surface irregularities between the flange faces, yet retain sufficient elasticity to take up the changes in the flange alignment that occur under load.

A great variety of proprietary gasket materials are used, and reference should be made to the manufacturers' catalogues and technical manuals when selecting gaskets for a particular application. Design data for some of the more commonly used gasket materials are given in Table 13.4. Further data can be found in the ASME BPV Code Sec. VIII D.1 Mandatory Appendix 2, ASME B16.20, BS EN 12560, and in Green and Perry (2007). The minimum seating stress y is the force per unit area (pressure) on the gasket that is required to cause the material to flow and fill the surface irregularities in the gasket face.

The gasket factor m is the ratio of the gasket stress (pressure) under the operating conditions to the internal pressure in the vessel or pipe. The internal pressure will force the flanges' faces apart, so the pressure on the gasket under operating conditions will be lower than the initial tightening-up pressure. The gasket factor gives the minimum pressure that must be maintained on the gasket to ensure a satisfactory seal.

The following factors must be considered when selecting a gasket material:

1. The process conditions: pressure, temperature, corrosive nature of the process fluid
2. Whether repeated assembly and disassembly of the joint are required
3. The type of flange and flange face (see Section 13.10.3).

Up to pressures of 20 bar, the operating temperature and corrosiveness of the process fluid will be the controlling factor in gasket selection. Vegetable fibre and synthetic rubber gaskets can be used at temperatures of up to 100°C. Solid poly-fluorocarbon (Teflon) and compressed asbestos gaskets can be used to a maximum temperature of about 260°C. Metal-reinforced gaskets can be used up to around 450°C. Plain soft-metal gaskets are normally used for higher temperatures.

13.10.3. Flange Faces

Flanges are also classified according to the type of flange face used. There are two basic types:

1. Full-faced flanges, Figure 13.29a: where the face contact area extends outside the circle of bolts, over the full face of the flange.
2. Narrow-faced flanges, Figures 13.29b,c,d: where the face contact area is located within the circle of bolts.

Full face, wide-faced, flanges are simple and inexpensive, but are only suitable for low pressures. The gasket area is large, and an excessively high bolt tension would be needed to achieve sufficient gasket pressure to maintain a good seal at high operating pressures.

The raised face, narrow-faced, flange shown in Figure 13.29b is probably the most commonly used type of flange for process equipment.

Where the flange has a plain face, as in Figure 13.29b, the gasket is held in place by friction between the gasket and flange surface. In the spigot and socket, and tongue and grooved faces (Figure 13.29c), the gasket is confined in a groove, which prevents failure by 'blow-out'. Matched pairs of flanges are required, which increases the cost, but this type is suitable for high-pressure and high-vacuum service. Ring joint flanges (Figure 13.29d) are used for high temperatures and high-pressure services.

TABLE 13.4. Gasket Materials (Based on Table 2-5.1 in ASME BPV Code Sec. VIII D.1 Mandatory Appendix 2, and a Similar Table in BS 5500–2003)

Gasket Material		Gasket Factor m	Min. Design Seating Stress $y(N/mm^2)$	Sketches	Minimum Gasket Width (mm)
Rubber without fabric or a high percentage of asbestos fiber; hardness:					
below 75° IRH		0.50	0		10
75° IRH or higher		1.00	1.4		
Asbestos with a suitable binder for the operating conditions	3.2 mm thick	2.00	11.0		
	1.6 mm thick	2.75	25.5		10
	0.8 mm thick	3.50	44.8		
Rubber with cotton fabric insertion		1.25	2.8		10
	3-ply	2.25	15.2		
Rubber with asbestos fabric insertion, with or without wire reinforcement	2-ply	2.50	20.0		10
	1-ply	2.75	25.5		
Vegetable fiber		1.75	7.6		10
Spiral-wound metal, asbestos filled	Carbon	2.50	20.0		
	Stainless or monel	3.00	31.0		10
Corrugated metal, asbestos inserted or Corrugated metal, jacketed asbestos filled	Soft aluminum	2.50	20.0		
	Soft copper or brass	2.75	25.5		
	Iron or soft steel	3.00	31.0		10
	Monel or 4 to 6% chrome	3.25	37.9		
	Stainless steels	3.50	44.8		
Corrugated metal	Soft aluminum	2.75	25.5		
	Soft copper or brass	3.00	31.0		
	Iron or soft steel	3.25	37.9		10
	Monel or 4 to 6% chrome	3.50	44.8		
	Stainless steels	3.75	52.4		
Flat metal jacketed asbestos filled	Soft aluminum	3.25	37.9		
	Soft copper or brass	3.50	44.8		
	Iron or soft steel	3.75	52.4		
	Monel	3.50	55.1		10
	4 to 6% chrome	3.75	62.0		
	Stainless steels	3.75	62.0		
Grooved metal	Soft aluminum	3.25	37.9		
	Soft copper or brass	3.50	44.8		
	Iron or soft steel	3.75	52.4		10
	Monel or 4 to 6% chrome	3.75	62.0		
	Stainless steels	4.25	69.5		
	Soft aluminum	4.00	60.6		
	Soft copper or brass	4.75	89.5		

TABLE 13.4. Gasket Materials (Based on Table 2-5.1 in ASME BPV Code Sec. VIII D.1 Mandatory Appendix 2, and a Similar Table in BS 5500–2003)—Cont'd

Gasket Material		Gasket Factor m	Min. Design Seating Stress $y(N/mm^2)$	Sketches	Minimum Gasket Width (mm)
Solid flat metal	Iron or soft steel	5.50	124		6
	Monel or 4 to 6%				
	chrome	6.00	150		
	Stainless steels	6.50	179		
Ring joint	Iron or soft steel	5.50	124		
	Monel or 4 to 6%				
	chrome	6.00	150		6
	Stainless steels	6.50	179		

13.10.4. Flange Design

Standard flanges will be specified for most applications (see Section 13.10.5). Special designs would be used only if no suitable standard flange were available; or for large flanges, such as the body flanges of vessels, where it may be cheaper to size a flange specifically for the duty required rather than to accept the nearest standard flange, which of necessity would be over-sized.

Figure 13.30 shows the forces acting on a flanged joint. The bolts hold the faces together, resisting the forces due to the internal pressure and the gasket sealing pressure. As these forces are off-set the flange is subjected to a bending moment. It can be considered as a cantilever beam with a concentrated load. A flange assembly must be sized so as to have sufficient strength and rigidity to resist this bending moment. A flange that lacks sufficient rigidity will rotate slightly, and the joint will leak; see Figure 13.31. The principles of flange design are discussed by Singh and Soler (1992) and Azbel and Cheremisinoff (1982). Singh and Soler give a computer program for flange design.

Design procedures for pressure vessel flanges are given in ASME BPV Code Sec VIII D.1 Mandatory Appendix 2 or in BS EN 1092, 1759 and 1591.

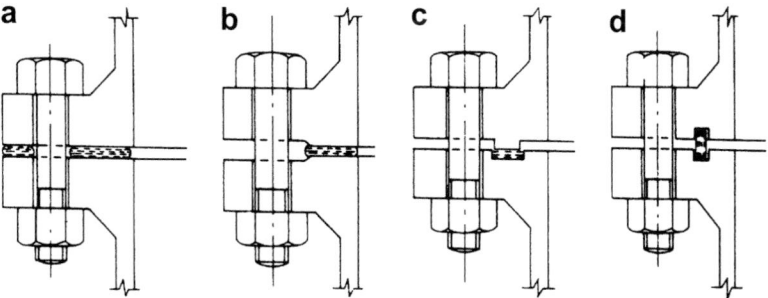

Figure 13.29. Flange types and faces. (a) Full-face. (b) Gasket within bolt circle. (c) Spigot and socket. (d) Ring type joint.

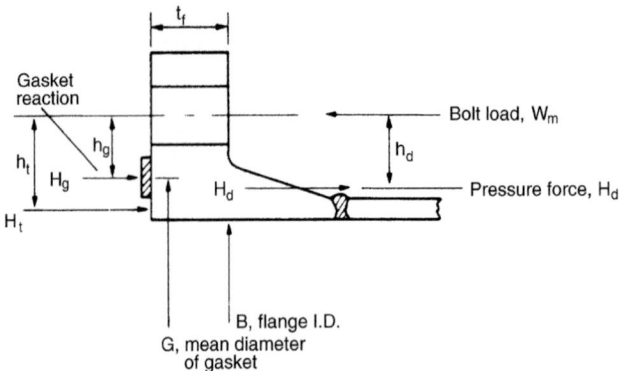

Figure 13.30.   Forces acting on an integral flange.

For design purposes, flanges are classified as integral or loose flanges.

Integral flanges are those in which the construction is such that the flange obtains support from its hub and the connecting nozzle (or pipe). The flange assembly and nozzle neck form an 'integral' structure. A welding-neck flange would be classified as an integral flange.

Loose flanges are attached to the nozzle (or pipe) in such a way that they obtain no significant support from the nozzle neck and cannot be classified as an integral attachment. Screwed and lap-joint flanges are typical examples of loose flanges.

The number of bolts and the bolt size must be chosen such that the bolt load is less than the maximum allowable stress in the bolts. The bolt spacing must be selected to give a uniform compression of the gasket. It will not normally be less than 2.5 times the bolt diameter, to give sufficient clearance for tightening with a wrench or spanner. The following formula can be used to determine the maximum bolt spacing:

$$p_b = 2d_b + \frac{6t_f}{(m + 0.5)} \tag{13.90}$$

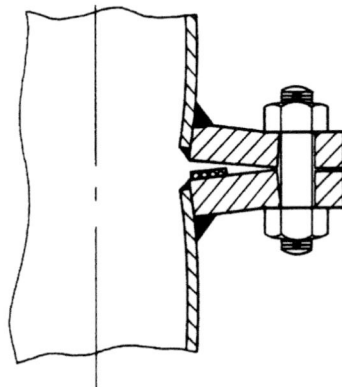

Figure 13.31.   Deflection of a weak flange (exaggerated).

where

p_b = bolt pitch (spacing), mm,
d_b = bolt diameter, mm,
t_f = flange thickness, mm,
m = gasket factor.

Bolting requirements are given in ASME B16.5 or BS EN 1515.

13.10.5. Standard Flanges

Standard flanges are available in a range of types, sizes and materials, and are used extensively for pipes, nozzles and other attachments to pressure vessels.

Standards for flanges and pipe fittings are set by the ASME B16 committee. These include:

ASME B16.5	Pipe flanges and flanged fittings
ASME B16.9	Factory-made wrought buttwelding fittings
ASME B16.11	Forged fittings, socket-welding and threaded
ASME B16.15	Cast bronze threaded fittings
ASME B16.24	Cast copper alloy pipe flanges and flanges fittings
ASME B16.42	Ductile iron pipe flanges and flanges fittings
ASME B16.47	Large diameter steel flanges

An abstract of the American standards is given by Green and Perry (2007).

A typical example of a standard flange design is shown in Figure 13.32. This was based on information in ASME B16.5 Annex F.

Standard flanges are designated by class numbers, or rating numbers, which roughly correspond to the primary service (pressure) rating of a steel flange of those dimensions at room temperature.

The flange class number required for a particular application will depend on the design pressure and temperature, and the material of construction. The reduction in strength at elevated temperatures is allowed for by selecting a flange with a higher rating than the design pressure. For example, for a design pressure of 10 bar (150 psi) a class 150 flange would be selected for a service temperature below 300°C; whereas for a service temperature of, say, 300°C a 300-pound flange would be specified. A typical pressure–temperature relationship for carbon steel flanges is shown in Table 13.5. Pressure–temperature ratings for a full range of materials can be obtained from the design codes.

Designs and dimensions of standard flanges over the full range of pipe sizes are given in ASME B16.5 Annex F. A summary of ASME flange dimensions is given by Green and Perry (2007), which can be used for preliminary designs. The current standards and suppliers' catalogues should be consulted before firming up the design.

The European codes allow both class-designated flanges (BS EN 1759) and PN (*presión nominal* or rated pressure) designated flanges (BS EN 1092). The pressure for PN-designated flanges is the rated pressure in bar. The PN-designated flanges are in metric sizes, and correspond to the former German DIN flanges. Flange dimensions are given in the standards.

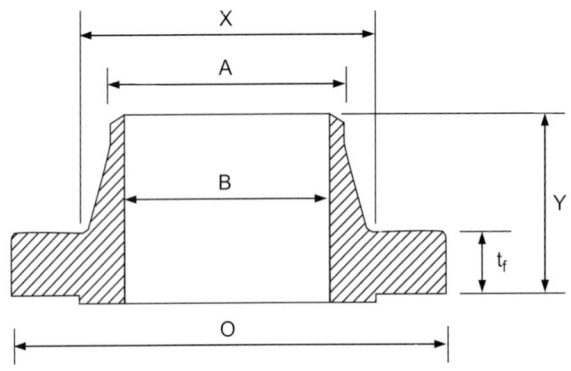

Flange Class	Nominal Pipe Size	Outside Diameter of Flange, O	Thickness of Flange, tf	Diameter of Hub, X	Diameter Beginning of Chamfer, A	Length Through Hub, Y	Bore, B
150	1.00	4.25	0.50	1.94	1.32	2.12	1.05
	2.00	6.00	0.69	3.06	2.38	2.44	2.07
	4.00	9.00	0.88	5.31	4.50	2.94	4.03
	6.00	11.00	0.94	7.56	6.63	3.44	6.07
	8.00	13.50	1.06	9.69	8.63	3.94	7.98
	12.00	19.00	1.19	14.38	12.75	4.44	12.00
	24.00	32.00	1.81	26.12	24.00	5.94	TBS
300	1.00	4.88	0.62	2.12	1.32	2.38	1.05
	2.00	6.50	0.81	3.31	2.38	2.69	2.07
	4.00	10.00	1.19	5.75	4.50	3.32	4.03
	6.00	12.50	1.38	8.12	6.63	3.82	6.07
	8.00	15.00	1.56	10.25	8.63	4.32	7.98
	12.00	20.50	1.94	14.75	12.75	5.06	12.00
	24.00	36.00	2.69	27.62	24.00	6.56	TBS

Note: TBS = To be specified by purchaser.

Figure 13.32.　Standard flange dimensions in inches for welding-neck flanges based on ASME B16.5 Annex F.

TABLE 13.5.　Typical Pressure–Temperature Ratings for Carbon Steel Flanges, A350, A515, A516 (Adapted from ASME B16.5 Annex F Table F2-1.1)

Temperature (°F)	Working Pressure by Flange class (psig)						
	150	300	400	600	900	1500	2500
−20 to 100	285	740	985	1480	2220	3705	6170
200	260	680	905	1360	2035	3395	5655
300	230	655	870	1310	1965	3270	5450
400	200	635	845	1265	1900	3170	5280
500	170	605	805	1205	1810	3015	5025
600	140	570	755	1135	1705	2840	4730
700	110	530	710	1060	1590	2655	4425
800	80	410	550	825	1235	2055	3430

13.11. HEAT-EXCHANGER TUBE PLATES

The tube plates (tube sheets) in shell and tube heat exchangers support the tubes, and separate the shell-side and tube-side fluids (see Chapter 12). One side is subject to the shell-side pressure and the other the tube-side pressure. The plates must be designed to support the maximum differential pressure that is likely to occur. Radial and tangential bending stresses will be induced in the plate by the pressure load and, for fixed-head exchangers, by the load due to the differential expansion of the shell and tubes.

A tube plate is essentially a perforated plate with an unperforated rim, supported at its periphery. The tube holes weaken the plate and reduce its flexual rigidity. The equations developed for the stress analysis of unperforated plates (Section 13.3.5) can be used for perforated plates by substituting 'virtual' (effective) values for the elastic constants E_Y and v, in place of the normal values for the plate material. The virtual elastic constants E_Y' and v' are functions of the plate ligament efficiency (Figure 13.33); see O'Donnell and Langer (1962). The ligament efficiency of a perforated plate is defined as:

$$\lambda = \frac{p_h - d_h}{p_h} \tag{13.91}$$

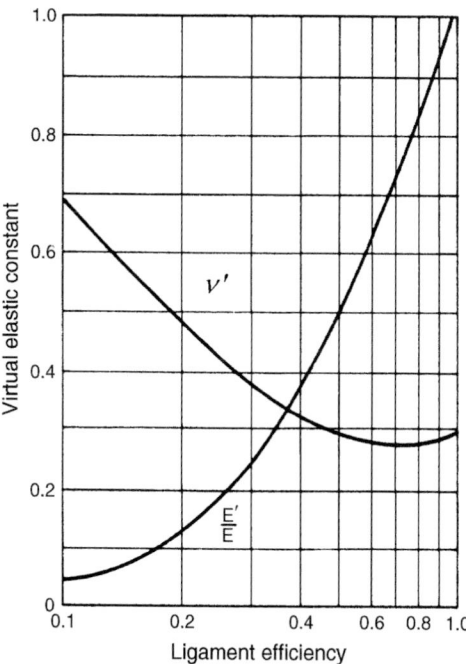

Figure 13.33. Virtual elastic constants.

where

p_h = hole pitch
d_h = hole diameter.

The 'ligament' is the material between the holes (that which holds the holes together). In a tube plate the presence of the tubes strengthens the plate, and this is taken into account when calculating the ligament efficiency by using the inside diameter of the tubes in place of the hole diameter in equation 13.91.

Design procedures for tube plates are given in the ASME BPV Code Sec. VIII D.1 Part UHX, in BS EN 13445-3 Clause 13, and in the TEMA heat-exchanger standards (see Chapter 12). The tube plate must be thick enough to resist the bending and shear stresses caused by the pressure load and any differential expansion of the shell and tubes.

For exchangers with fixed tube plates, the longitudinal stresses in the tubes and shell must be checked to ensure that the maximum allowable design stresses for the materials are not exceeded. Methods for calculating these stresses are given in the standards.

The calculation procedures specified in the ASME BPV Code and BS EN 13445-3 are complex and iterative. Various computer programs are available for mechanical design of heat exchangers, for example:

B-JAC (AspenTech Inc.)
Pressure Vessel Suite (Computer Engineering Inc.)
PVElite and CodeCalc (COADE Inc.)
SnapCAD (Heat Transfer Consultants Inc.)
TEMA/ASME and COMPRESS (Codeware Inc.).

These and other programs can easily be found by searching on-line. Licenses to one or more of these programs will be available in the mechanical engineering design group of most companies. Some University mechanical engineering departments also have licenses to these tools that allow for use in undergraduate design projects.

13.12. WELDED-JOINT DESIGN

Process vessels are built up from preformed parts: cylinders, heads and fittings, joined by fusion welding. Riveted construction was used extensively in the past (prior to the 1940s) but is now rarely seen except on very old plants.

Cylindrical sections are usually made up from plate sections rolled to the required curvature. The sections (strakes) are made as large as is practicable to reduce the number of welds required. The longitudinal welded seams are offset to avoid a conjunction of welds at the corners of the plates.

Many different forms of welded joint are needed in the construction of a pressure vessel. Some typical forms are shown in Figures 13.34 to 13.36.

The design of a welded joint should satisfy the following basic requirements:

1. Give good accessibility for welding and inspection
2. Require the minimum amount of weld metal
3. Give good penetration of the weld metal; from both sides of the joint, if practicable

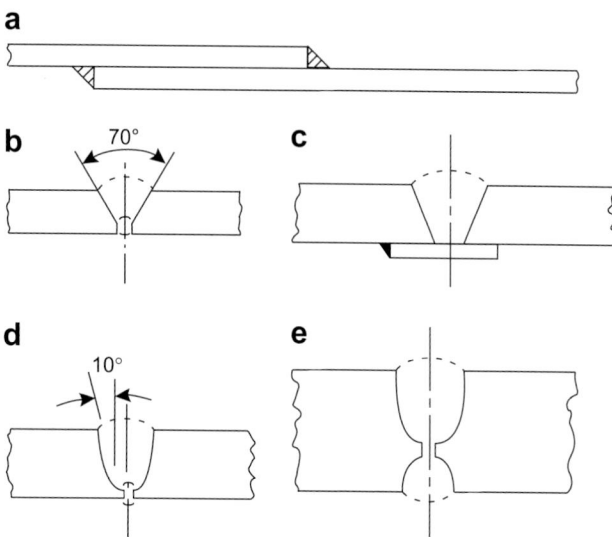

Figure 13.34. Weld profiles: (b to e) butt welds. (a) Lap joint. (b) Single 'V'. (c) Backing strip. (d) Single 'U'. (e) Double 'U'.

4. Incorporate sufficient flexibility to avoid cracking due to differential thermal expansion.

The preferred types of joint, and recommended designs and profiles, are given in the codes and standards. See, for example, ASME BPV Code Sec. VIII D.1, Part UW – Requirements for pressure vessels fabricated by welding, or BS EN

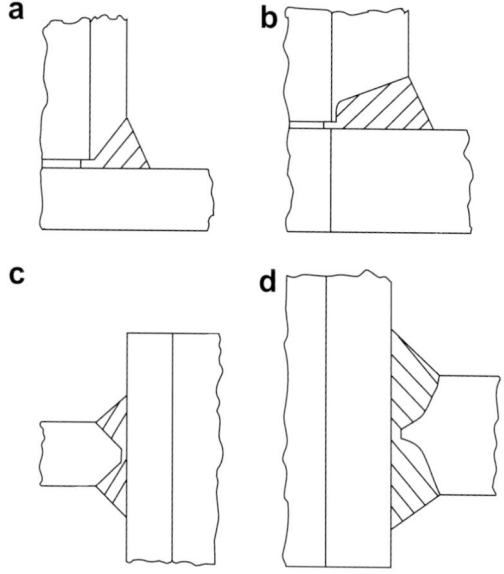

Figure 13.35. Typical weld profiles: branches. (a,b) Set-on branches (c,d) Set-in branches.

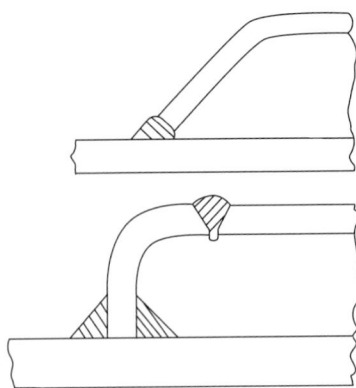

Figure 13.36.   Typical construction methods for welded jackets.

13445-3 Annex P, BS EN 13445-4 and BS EN 1708-1, Welding – Basic weld joint details in steel – Part 1: pressurized components.

The correct form to use for a given joint will depend on the material, the method of welding (machine or hand), the plate thickness and the service conditions. Double-sided V- or U-sections are used for thick plates, and single V- or U-profiles for thin plates. A backing strip is used where it is not possible to weld from both sides. Lap joints are seldom used for pressure vessel construction, but are used for atmospheric pressure storage tanks.

Where butt joints are made between plates of different thickness, the thicker plate is reduced in thickness with a slope of not greater than 1 in 3 (19°) (ASME BPV Code Sec. VIII D.1 Part UW-9, shown in Figure 13.37).

The local heating, and consequent expansion, that occurs during welding can leave the joint in a state of stress. These stresses are relieved by post-weld heat treatment. Not all vessels will be stress relieved. Guidance on the need for post-weld heat treatment is given in ASME BPV Code Sec. VIII D.1 Part UW-40 or BS EN 13445-4, and will depend on the service and conditions, materials of construction, and plate thickness.

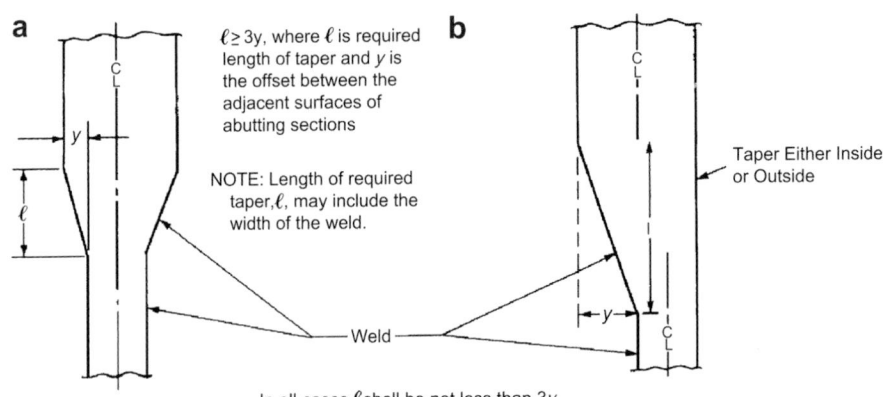

Figure 13.37.   Butt-welding of plates of unequal thickness. Reprinted from ASME 2004 BPVC, Section VIII-Div.1, by permission of the American Society of Mechanical Engineers. All rights reserved.

To ensure that a satisfactory quality of welding is maintained, welding-machine operators and welders working on the pressure parts of vessels are required to pass welder approval tests; which are designed to test their competence to make sound welds. Welding and brazing qualifications are discussed in Section IX of the ASME BPV Code or BS EN 13445-4.

13.13. FATIGUE ASSESSMENT OF VESSELS

During operation, the shell or components of the vessel may be subjected to cyclic stresses.Stress cycling can arise from the following causes:

1. Periodic fluctuations in operating pressure
2. Temperature cycling
3. Vibration
4. 'Water hammer'
5. Fluctuations in the flow of fluids or solids
6. Periodic fluctuation of external loads.

A detailed fatigue analysis is required if any of these conditions is likely to occur to any significant extent. Fatigue failure will occur during the service life of the vessel if the endurance limit (number of cycles for failure) at the particular value of the cyclic stress is exceeded. The codes and standards should be consulted to determine when a detailed fatigue analysis must be undertaken.

13.14. PRESSURE TESTS

The national pressure vessel codes and standards require that all pressure vessels be subjected to a pressure test to prove the integrity of the finished vessel (ASME BPV Code Sec. VIII D.1 Part UG-99, or BS EN 13445-5). A hydraulic test is normally carried out, but a pneumatic test can be substituted under circumstances where the use of a liquid for testing is not practical. Hydraulic tests are safer because only a small amount of energy is stored in the compressed liquid. A standard pressure test is used when the required thickness of the vessel parts can be calculated in accordance with the particular code or standard. The vessel is tested at a pressure 30% above the design pressure. The test pressure is adjusted to allow for the difference in strength of the vessel material at the test temperature compared with the design temperature, and for any corrosion allowance.

Formulae for determining the appropriate test pressure are given in the codes and standards; typically:

$$\text{Test pressure} = 1.30\left[P_d\frac{S_a}{S_n} \times \frac{t}{(t - c)}\right] \tag{13.92}$$

where

P_d = design pressure, N/mm^2
S_a = maximum allowable stress at the test temperature, N/mm^2
S_n = maximum allowable stress at the design temperature, N/mm^2

c = corrosion allowance, mm
t = actual plate thickness, mm.

When the required thickness of the vessel component parts cannot be determined by calculation in accordance with the methods given, the ASME BPV Code requires that a hydraulic proof test be carried out (Sec. VIII D.1 Part UG-101). In a proof test the stresses induced in the vessel during the test are monitored using strain gauges, or similar techniques. In a proof test a duplicate of the vessel or part is tested until the part yields or bursts. The requirements for the proof testing of vessels are set out in ASME BPV Code Sec. VIII D.1 Part UG-101.

13.15. HIGH-PRESSURE VESSELS

High pressures are required for many commercial chemical processes. For example, the synthesis of ammonia is carried out at reactor pressures of up to 1000 bar, and high-density polyethylene processes operate up to 1500 bar.

Although there is no prescribed upper limit on pressure for vessels designed in accordance with ASME BPV Code Section VIII Division 1, the rules given in that section of the code usually cannot be economically satisfied for vessels designed to operate above 3000 psia (200 bar). For pressures greater than about 2000 psia the alternative rules given in Section VIII Division 2 will usually lead to a more economical design. Division 2 restricts the materials that can be used, the allowable operating temperatures (not greater than 900°F) and places stricter requirements on stress analysis and testing. The additional engineering and design costs are usually justified for high-pressure vessels, because the Division 2 rules allow higher maximum allowable stresses and hence lead to thinner-walled vessels.

At the highest operating pressures, typically above 10,000 psia (680 bar) the alternative design rules given in ASME BPV Code Section VIII Division 3 can be followed.

Similarly, the BS EN 13445-3 design by formulae (DBF) method will lead to uneconomic designs for pressures greater than about 2000 psia, compared to designs developed using the design by analysis (DBA) method.

A full discussion of the design and construction of high-pressure vessels and ancillary equipment (pumps, compressors, valves and fittings) is given in the books by Fryer and Harvey (1997) and Jawad and Farr (1989). At high pressures it becomes increasingly difficult to fabricate single-walled vessels with sufficient strength, because of the wall thickness and depth of welds required. Instead, compound vessels with several layers of vessel walls are often used. In a compound vessel, the outer layers can be used to place the inner layers in compression during manufacture, and hence offset the tensile forces that will act most strongly on the inner layers during operation.

13.15.1. Compound Vessels

Shrink-Fitted Cylinders

Compound vessels can be made by shrinking one cylinder over another. The inside diameter of the outer cylinder is made slightly smaller than the outer diameter of the inner cylinder, and the outer cylinder is expanded by heating to fit over the inner. On cooling the outer cylinder contracts and places the inner under compression. The stress distribution in a two-cylinder compound vessel is shown in Figure 13.38; more than two cylinders may be used.

Shrink-fitted compound cylinders are used for small-diameter vessels, such as compressor cylinder barrels. The design of shrink-fitted compound cylinders is discussed by Manning (1947) and Jawad and Farr (1989).

Multilayer Vessels

Multilayer vessels are made by wrapping several layers of relatively thin plate round a central tube. The plates are heated, tightened and welded, and this gives the desired stress distribution in the compound wall. The vessel is closed with forged heads. A typical design is shown in Figure 13.39. This construction technique is discussed by Jasper and Scudder (1941) and Jawad and Farr (1989).

Wound Vessels

Cylindrical vessels can be reinforced by winding on wire or thin ribbons. Winding on the wire under tension places the cylinder under compression. For high-pressure vessels special interlocking strips are used, such as those shown in

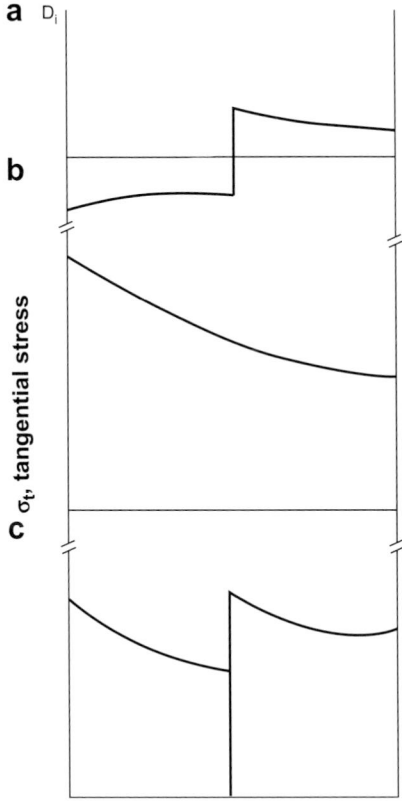

Figure 13.38. Stress distribution in a shrink-fitted compound cylinder. (a) Due to shrinkage. (b) Due to pressure. (c) Combined a+b.

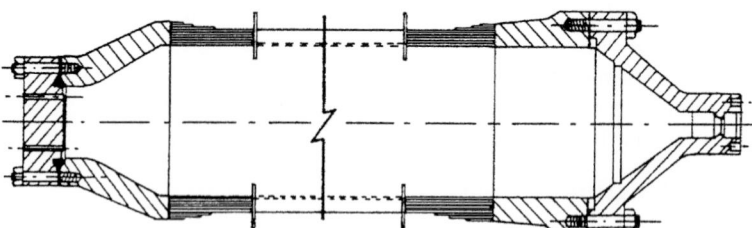

Figure 13.39.   Multilayer construction.

Figure 13.40. The interlocking gives strength in the longitudinal direction and a more uniform stress distribution. The strips may be wound on hot to increase the pre-stressing. This type of construction is described by Birchall and Lake (1947). Wire winding was used extensively for the barrels of large guns.

13.15.2. Autofrettage

Autofrettage is a technique used to pre-stress the inner part of the wall of a monobloc vessel, to give a similar stress distribution to that obtained in a shrink-fitted compound cylinder. The finished vessel is deliberately over-pressurized by hydraulic pressure. During this process the inner part of the wall will be more highly stressed than the outer part and will undergo plastic strain. On release of the 'autofrettage' pressure the inner part, which is now over-size, will be placed under compression by the elastic contraction of the outer part, which gives a residual stress distribution similar to that obtained in a two-layer shrink-fitted compound cylinder. After straining, the vessel is annealed at a relatively low temperature, approximately 300°C. The straining also work-hardens the inner part of the wall. The vessel can be used at pressures up to the 'autofrettage' pressure without further permanent distortion.

The autofrettage technique is discussed by Manning (1950) and Jawad and Farr (1989).

Requirements for pressure vessels fabricated by layered construction are given in ASME BPV Code Sec. VIII D.1 Part ULW and Sec. VIII D.2 Articles D-11 and F-8.

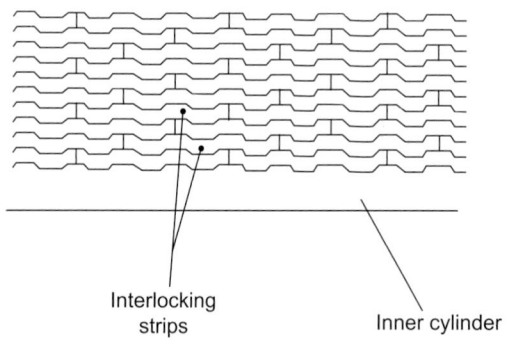

Interlocking
strips

Inner cylinder

Figure 13.40.   Strip-wound vessel.

13.16. LIQUID STORAGE TANKS

Vertical cylindrical tanks, with flat bases and conical roofs, are universally used for the bulk storage of liquids at atmospheric pressure. Tank sizes vary from a few hundred gallons (tens of cubic metres) to several thousand gallons (several hundred cubic metres).

The main load to be considered in the design of these tanks is the hydrostatic pressure of the liquid, but the tanks must also be designed to withstand wind loading and, for some locations, the weight of snow on the tank roof.

The minimum wall thickness required to resist the hydrostatic pressure can be calculated from the equations for the membrane stresses in thin cylinders (Section 13.3.4):

$$t_t = \frac{\rho_L H_L g}{2S_t E} \frac{D_t}{10^3}$$
(13.93)

where

t_t = tank thickness required at depth H_L, mm
H_L = liquid depth, m
ρ_L = liquid density, kg/m^3
E = joint efficiency (if applicable)
g = gravitational acceleration, 9.81 m/s^2
S_t = maximum allowable stress for tank material, N/mm^2
D_t = tank diameter, m.

The liquid density should be taken as that of water (1000 kg/m^3), unless the process liquid has a greater density.

For small tanks a constant wall thickness would normally be used, calculated at the maximum liquid depth.

With large tanks, it is economical to take account of the variation in hydrostatic pressure with depth, by increasing the plate thickness progressively from the top to bottom of the tank. Plate widths of 2 m (6 ft) are typically used in tank construction.

The roofs of large tanks need to be supported by a steel framework; supported on columns in very large-diameter tanks.

The design and construction of atmospheric storage tanks for the petroleum industry are covered by the American Petroleum Industry standards API 650 (2003) and 620 (2002). Other standards are also used internationally, for example European Standards BS EN 14015 for storage at temperatures above ambient, BS EN 14620 for sub-ambient storage and BS EN 13121 for glass-reinforced polymer tanks. The design of storage tanks is covered in the books by Myers (1997), and Jawad and Farr (1989). See also the papers by Debham *et al.* (1968) and Zick and McGrath (1968).

13.17. REFERENCES

Azbel, D. S. and Cheremisinoff, N. P. (1982) *Chemical and Process Equipment Design: Vessel Design and Selection* (Ann Arbor Science).
Bednar, H. H. (1990) *Pressure Vessel Design Handbook*, 2nd edn (Krieger).

Bhattacharyya, B. C. (1976) *Introduction to Chemical Equipment Design, Mechanical Aspects* (Indian Institute of Technology).

Bergman, D. J. (1963) *Trans. Am. Soc. Mech. Eng. (J. Eng. for Ind.)* **85**, 219. Temperature gradients for skirt supports of hot vessels.

Birchall, H. and Lake, G. F. (1947) *Proc. Inst. Mech. Eng* **56**, 349. An alternative form of pressure vessel of novel construction.

Brownell, L. E. (1963) *Hyd. Proc. and Pet. Ref.* **42** (June) 109. Mechanical design of tall towers.

Brownell, L. E. and Young, E. H. (1959) *Process Equipment Design: Vessel Design* (Wiley).

Case, J., Chilver, A. H., and Ross, C. (1999) *Strength of Materials and Structures* (Butterworth-Heinemann).

Chuse, R. and Carson, B. E. (1992) *Pressure Vessels: the ASME Code Simplified*, 7th edn (McGraw-Hill).

Debham, J. B., Russel, J. and Wiils, C. M. R. (1968) Hyd. Proc. **47** (May) 137. How to design a 600,000 b.b.l. tank.

Deghetto, K. and Long, W. (1966) *Hyd. Proc. and Pet. Ref.* **45** (Feb.) 143. Check towers for dynamic stability.

Escoe, A. K. (1994) *Mechanical Design of Process Equipment*, Vol. 1, 2nd edn. Piping and *Pressure Vessels* (Gulf).

Freese, C. E. (1959) *Trans. Am. Soc. Mech. E. (J. Eng. Ind.)* **81**, 77. Vibrations of vertical pressure vessels.

Fryer, D. M. and Harvey, J. F. (1997) *High Pressure Vessels* (Kluwer).

Gere, J. M. and Timoshenko, S. P. (2000) *Mechanics of Materials* (Brooks Cole).

Green, D. W. and Perry, R. H. (eds) (2007) Perry's Chemical Engineers Handbook, 8th edn (McGraw-Hill).

Harvey, J. F. (1974) *Theory and Design of Modern Pressure Vessels*, 2nd edn (Van Nostrand-Reinhold).

Henry, B. D. (1973) *Aust. Chem. Eng.* **14** (Mar.) 13. The design of vertical, free standing process vessels.

Hetenyi, M. (1958) *Beams on Elastic Foundations* (University of Michigan Press).

Jasper, McL, T. and Scudder, C. M. (1941) *Trans. Am. Inst. Chem. Eng.* **37**, 885. Multi-layer construction of thick wall pressure vessels.

Jawad, M. H. and Farr, J. R. (1989) *Structural Design of Process Equipment*, 2nd edn (Wiley).

Karman, von T. and, Tsien, H-S. (1939) J. Aeronautical Sciences 7 (Dec.) 43. The buckling of spherical shells by external pressure.

Mahajan, K. K. (1977) *Hyd. Proc.* **56**(4), 207 Size vessel stiffeners quickly.

Manning, W. R. D. (1947) Engineering **163** (May 2nd) 349. The design of compound cylinders for high pressure service.

Manning, W. R. D. (1950) Engineering **169** (April 28th) 479, (May 5th) 509, (May 15th) 562, in three parts. The design of cylinders by autofrettage.

Marshall, V. O. (1958) *Pet. Ref.* **37** (May) (supplement). Foundation design handbook for stacks and towers.

Megyesy, E. F. (2001) *Pressure Vessel Hand Book*, 12th edn (Pressure Vessel Hand Book Publishers).

Moss, D. R. (2003) *Pressure Vessel Design Manual* (Elsevier/Butterworth-Heinemann).

Mott, R. L. (2001) *Applied Strength of Materials* (Prentice Hall).

Myers, P. E. (1997) *Above Ground Storage Tanks* (McGraw-Hill).

Nelson, J. G. (1963) *Hyd. Proc. and Pet. Ref.* **42** (June) 119. Use calculation form for tower design.

O'Donnell, W. J. and Langer, B. F. (1962) *Trans. Am. Soc. Mech. Eng. (J. Eng. Ind.)* **84**, 307. Design of perforated plates.

Preiss, R. and Zeman, J. L. (2004) ASME PVP Conference, Session 1.4Q (CS-05C) Comparative study EN 13445 / ASME Section VIII Div. 1 & 2. Also available as CEN publication CEN/EXPERT/2004/45 from the European Standards Agency.

Scheiman, A. D. (1963) *Hyd. Proc. and Pet. Ref.* **42** (June) 130. Short cuts to anchor bolting and base ring sizing.

Seed, G. M. (2001) *Strength of Materials: An Undergraduate Text* (Paul & Co. Publishing Consortium).

Singh, K. P. and Soler, A. I. (1992) *Mechanical Design of Heat Exchangers and Pressure Vessel Components* (Springer-Verlag).

Southwell, R. V. (1913) *Phil. Trans.* **213A**, 187. On the general theory of elastic stability.

Tang, S. S. (1968) *Hyd. Proc.* **47** (Nov.) 230. Shortcut methods for calculating tower deflections.

Timoshenko, S. (1936) *Theory of Elastic Stability* (McGraw-Hill).

Weil, N. A. and Murphy, J. J. (1960) *Trans. Am. Soc. Mech. Eng. (J. Eng. Ind.)* **82** (Jan.) 1. Design and analysis of welded pressure vessel skirt supports.

Windenburg, D. F. and Trilling, D. C. (1934) *Trans. Am. Soc. Mech. Eng* **56**, 819. Collapse by instability of thin cylindrical shells under external pressure.

Wolosewick, F. E. (1951) *Pet. Ref.* **30** (July) 137, (Aug.) 101, (Oct.) 143, (Dec.) 151, in four parts. Supports for vertical pressure vessels.

Yokell, S. (1986) *Chem. Eng., NY* **93** (May 12th) 75. Understanding pressure vessel codes.

Zick, L. P. (1951) *Welding J. Research Supplement* **30**, 435. Stresses in large horizontal cylindrical pressure vessels on two saddle supports.

Zick, L. P. and McGrath, R. V. (1968) *Hyd. Proc.* **47** (May) 143. New design approach for large storage tanks.

Bibliography

Useful References on Pressure Vessel Design

Azbel, D. S. and Cheremisinoff, N. P. (1982) *Chemical and Process Equipment Design: Vessel Design and Selection* (Ann Arbor Science).

Bednar, H. H. (1990) *Pressure Vessel Design Handbook*, 2nd edn (Van Nostrand Reinhold).

Chuse, R. and Carson, B. E. (1992) *Pressure Vessels: the ASME Code Simplified*, 7th edn (McGraw-Hill).

Escoe, A. K. (1986) *Mechanical Design of Process Equipment. Vol. 1. Piping and Pressure Vessels; Vol. 2. Shell-and-tube Heat Exchangers, Rotating Equipment, Bins, Silos and Stacks* (Gulf).

Farr, J. R. and Jawad, M. H. (2001) *Guidebook for the Design of ASME Section VIII, Pressure Vessels*, 2nd edn (American Society of Mechanical Engineers).

Gupta, J. P. (1986) *Fundamentals of Heat Exchanger and Pressure Vessel Technology* (Hemisphere).

Jawad, M. H. and Farr, J. R. (1989) *Structural Design of Process Equipment*, 2nd edn (Wiley).

Megyesy, E. F. (2001) *Pressure Vessel Hand Book*, 12th edn (Pressure Vessel Hand Book Publishers).

Moss, D. R. (2003) *Pressure Vessel Design Manual* (Butterworth-Heinemann).

Roake, R. J., Young, W. C., and Budynas, R. G. (2001) *Formulas for Stress and Strain* (McGraw-Hill).

Singh, K. P. and Soler, A. I. (1992) *Mechanical Design of Heat Exchangers and Pressure Vessel Components* (Springer-Verlag).

British and European Standards

BS EN 1092 (2007) Flanges and their joints – Circular flanges for pipes, valves, fittings and accessories, PN designated.
BS EN 1515 (2005) Flanges and their joints – Bolting.
BS EN 1591 (2001) Flanges and their joints – Design rules for gasketed circular flange connections.
BS EN 1708 (1999) Welding. Basic weld joint details in steel. Pressurized components.
BS EN 1759 (2003) Flanges and their joints – Circular flanges for pipes, valves, fittings and accessories, class designated.
BS EN 12560 (2004) Flanges and their joints – Gaskets for class-designated flanges.
BS EN 13121 (2003) GRP tanks and vessels for use above ground.
BS EN 13445-1 (2002) Unfired pressure vessels – Part 1: General.
BS EN 13445-2 (2002) Unfired pressure vessels – Part 2: Materials.
BS EN 13445-3 (2002) Unfired pressure vessels – Part 3: Design.
BS EN 13445-4 (2002) Unfired pressure vessels – Part 4: Fabrication.
BS EN 13445-5 (2002) Unfired pressure vessels – Part 5: Inspection and testing.
BS EN 13445-6 (2006) Unfired pressure vessels – Part 6: Requirements for the design and fabrication of pressure vessels and pressure parts constructed from spheroidal graphitic cast iron.
BS EN 13445-8 (2006) Unfired pressure vessels – Part 8: Additional requirements for pressure vessels of aluminium and aluminium alloys.
BS EN 13923 (2006) Filament-wound FRP pressure vessels. Materials, design, manufacturing and testing.
BS EN 14015 (2004) Specification for the design and manufacture of site-built, vertical, cylindrical, flat-bottomed, above ground, welded steel tanks for the storage of liquids at ambient temperature and above.
BS EN 14620 (2006) Design and manufacture of site-built, vertical, cylindrical, flat-bottomed, steel tanks for the storage of refrigerated, liquefied gases with operating temperature between $0°C$ and $-165°C$.
PD 5500 (2003) Specification for unfired fusion welded pressure vessels (formerly BS 5500).

American and International Standards

API Standard 620 (2002) Design and construction of large, welded, low-pressure storage tanks, 10th edn (American Petroleum Institute).
API Standard 650 (1998) Welded steel tanks for oil storage, 10th edn (American Petroleum Institute).
ASME Boiler and Pressure Vessel Code Section II (2004) Materials (ASME International).
ASME Boiler and Pressure Vessel Code Section VIII (2004) Rules for the construction of pressure vessels (ASME International).
ASME Boiler and Pressure Vessel Code Section IX (2004) Qualification standard for welding and brazing procedures, welders, brazers, and welding and brazing operators (ASME International).
ASME Boiler and Pressure Vessel Code Section X (2004) Fibre-reinforced plastic vessels (ASME International).
ASME B16.5-2003 Pipe flanges and flanged fittings (ASME International).
ASME B16.9-2003 Factory-made wrought buttwelding fittings (ASME International).

ASME B16.11-2001 Forged fittings, socket-welding and threaded (ASME International).

ASME B16.15-1985(R2004) Cast bronze threaded fittings classes 125 and 250 (ASME International).

ASME B16.20-1998(R2004) Metallic gaskets for pipe flanges – ring-joint, spiral-wound, and jacketed (ASME International).

ASME B16.24-2001 Cast copper alloy pipe flanges and flanged fittings (ASME International).

ASME B16.42-1998 Ductile iron pipe flanges and flanged fittings, class 150 and 300 (ASME International).

ASME B16.47-1996 Large diameter steel flanges, NPS 26 through NPS 60 (ASME International).

13.18. NOMENCLATURE

Note that the same nomenclature has been used as in the ASME BPV Code and API recommended practices and standards, as far as possible. This occasionally causes the same symbol to be used with different meanings in different contexts. Where the context is not clear and there is a possibility of confusion, a new symbol has been assigned.

		Dimensions in **MLT**
a	Radius of flat plate	L
$2a$	Major axis of ellipse	L
a_e	Acceleration due to an earthquake	LT^{-2}
$2b$	Minor axis of ellipse	L
C	Constant in equation 13.34 or 13.44	—
C_d	Drag coefficient in equation 13.79	—
C_e	Seismic constant	—
C_w	Weight factor in equation 13.73	—
C_1, C_2, C_3	Constants in equation 13.22	—
c	Corrosion allowance	L
D	Diameter	L
$\mathbf{D}$	Flexual rigidity	ML^2T^{-2}
D_c	Diameter of cone at point of interest	L
D_e	Nominal diameter of flat end	L
D_{eff}	Effective diameter of column for wind loading	L
D_i	Internal diameter	L
D_m	Mean diameter	L
D_{max}	Maximum diameter	L
D_{min}	Minimum diameter	L
D_o	Outside diameter	L

D_s	Skirt internal diameter	L
D_t	Tank diameter	L
d_b	Bolt diameter	L
d_h	Hole diameter	L
E	Joint efficiency, welded joint	—
E_Y	Young's modulus	$ML^{-1}T^{-2}$
E_Y'	Effective Young's modulus for a ligament	$ML^{-1}T^{-2}$
F_{bs}	Load supported by bracket	MLT^{-2}
F_p	Local, concentrated, wind load	MLT^{-2}
F_s	Shear force due an earthquake	MLT^{-2}
g	Gravitational acceleration	LT^{-2}
H_L	Liquid depth	L
H_p	Height of local load above base	L
H_v	Height (length) of cylindrical section between tangent lines	L
h	Height of domed head from tangent line	L
I	Second moment of area (moment of inertia)	L^4
I'	Second moment of area per unit length	L^3
I_p	Polar second moment of area	L^4
I_v	Second moment of area of vessel	L^4
K_c	Collapse coefficient in equation 13.52	—
L	Unsupported length of vessel	L
L'	Effective length between stiffening rings	L
L_c	Critical distance between stiffening rings	L
L_d	Bracket depth	L
L_o	Distance between centre line of equipment and column	L
M	Bending moment	ML^2T^{-2}
M_e	Bending moment due to offset equipment	ML^2T^{-2}
M_p	Bending moment at base due to local load	ML^2T^{-2}
M_s	Bending moment at base of skirt	ML^2T^{-2}
M_x	Bending moment at point x from free end of column	ML^2T^{-2}
M_1	Bending moment acting along cylindrical sections	ML^2T^{-2}
M_2	Bending moment acting along diametrical sections	ML^2T^{-2}

m	Gasket factor	—
n	Number of lobes formed on buckling	—
P	Pressure	$ML^{-1}T^{-2}$
P_c	Critical buckling pressure	$ML^{-1}T^{-2}$
P'_c	Critical pressure to cause local buckling in a spherical shell	$ML^{-1}T^{-2}$
P_d	Design pressure	$ML^{-1}T^{-2}$
P_e	External pressure	$ML^{-1}T^{-2}$
P_i	Internal pressure	$ML^{-1}T^{-2}$
P_{outlet}	Outlet pressure	$ML^{-1}T^{-2}$
P_w	Wind pressure loading	$ML^{-1}T^{-2}$
p_b	Bolt pitch	L
p_h	Hole pitch	L
R_c	Crown radius	L
R_k	Knuckle radius	L
R_o	Major radius of torus	L
R_p	Radius of curvature of plate	L
R_s	Outside radius of sphere	L
r	Radius	L
r_1	Meridional radius of curvature	L
r_2	Circumferential radius of curvature	L
S	Maximum allowable stress (design stress)	$ML^{-1}T^{-2}$
S_a	Maximum allowable stress at test temperature	$ML^{-1}T^{-2}$
S_n	Maximum allowable stress at design temperature	$ML^{-1}T^{-2}$
S_s	Maximum allowable stress for skirt material	$ML^{-1}T^{-2}$
S_t	Maximum allowable stress for tank material	$ML^{-1}T^{-2}$
S_1, S_2	Length elements on surface of revolution in Figure 13.4	L
T	Torque	ML^2T^{-2}
t	Thickness of plate or shell	L
t_c	Thickness of bracket plate	L
t_f	Thickness of flange	L
t_s	Thickness of shell	L
t_{sk}	Skirt thickness	L
t_t	Tank wall thickness	L
u_w	Wind velocity	LT^{-1}
W	Wind load per unit length	MT^{-2}
W_e	Weight of ancillary equipment	MLT^{-2}
W_v	Weight of vessel and contents	MLT^{-2}

W_z	Weight of vessel and contents above a plane at elevation z	MLT^{-2}
w	Deflection of flat plate	L
x	Radius from centre of flat plate to point of interest	L
x	Distance from free end of cantilever beam	L
x_c	Displacement caused by centrifugal force	L
y	Minimum seating pressure for gasket	$ML^{-1}T^{-2}$
α	Cone half cone apex angle	—
Δ	Dilation	L
Δ_c	Dilation of cylinder	L
Δ_s	Dilation of sphere	L
ε	Strain	—
$\varepsilon_1, \varepsilon_2$	Principal strains	—
θ	Angle	—
θ_s	Base angle of conical section	—
λ	Ligament efficiency	—
ν	Poisson's ratio	—
ν'	Effective Poisson's ratio for a ligament	—
ρ_m	Density of vessel material	ML^{-3}
ρ_a	Density of air	ML^{-3}
ρ_L	Liquid density	ML^{-3}
σ	Normal stress	$ML^{-1}T^{-2}$
σ_b	Bending stress	$ML^{-1}T^{-2}$
σ_{bs}	Bending stress in skirt	$ML^{-1}T^{-2}$
σ_c	Critical buckling stress	$ML^{-1}T^{-2}$
σ_e	Stress at elastic limit of material	$ML^{-1}T^{-2}$
σ'_e	Elastic limit stress divided by factor of safety	$ML^{-1}T^{-2}$
σ_h	Circumferential (hoop) stress	$ML^{-1}T^{-2}$
σ_L	Longitudinal stress	$ML^{-1}T^{-2}$
σ_s	Stress in skirt support	$ML^{-1}T^{-2}$
σ_w	Stress due to weight of vessel	$ML^{-1}T^{-2}$
σ_{ws}	Stress in skirt due to weight of vessel	$ML^{-1}T^{-2}$
σ_x	Normal stress in x direction	$ML^{-1}T^{-2}$
σ_y	Normal stress in y direction	$ML^{-1}T^{-2}$
σ_z	Axial stresses in vessel	$ML^{-1}T^{-2}$
$\sigma_1, \sigma_2, \sigma_3$	Principal stresses	$ML^{-1}T^{-2}$
τ	Torsional shear stress	$ML^{-1}T^{-2}$
τ_e	Shear stress at elastic limit of material	$ML^{-1}T^{-2}$

τ_{xy}	Shear stress	$ML^{-1}T^{-2}$
τ_1, τ_2, τ_3	Shear stress maxima	$ML^{-1}T^{-2}$
ϕ	Slope of flat plate	—
ϕ	Angle	—
Superscript $^\wedge$	Maximum	

13.19. PROBLEMS

The problems below do not specify a design code, so any national code can be applied. The code should be applied consistently throughout a given problem and should be stated in the answer.

13.1. Calculate the maximum membrane stress in the wall of shells having the shapes listed below. The vessel walls are 2 mm thick and subject to an internal pressure of 5 bar.
1. An infinitely long cylinder, inside diameter 2 m
2. A sphere, inside diameter 2 m
3. An ellipsoid, major axis 2 m, minor axis 1.6 m
4. A torus, mean diameter 2 m, diameter of cylinder 0.3 m.

13.2. Compare the thickness required for a 2-m-dia. flat plate, designed to resist a uniform distributed load of 10 kN/m², if the plate edge is
(a) completely rigid
(b) free to rotate.
Take the maximum allowable stress for the material as 100 MN/m² and Poisson's ratio for the material as 0.3.

13.3. A horizontal, cylindrical, tank, with hemispherical ends, is used to store liquid chlorine at 10 bar. The vessel is 4 m internal diameter and 20 m long. Estimate the minimum wall thickness required to resist this pressure, for the cylindrical section and the heads. Take the design pressure as 12 bar and the maximum allowable stress for the material as 110 MN/m².

13.4. The thermal design of a heat exchanger to recover heat from a kerosene stream by transfer to a crude-oil stream was carried in Chapter 12, Examples 12.3 and 12.4. Make a preliminary mechanical design for this exchanger. Base your design on the specification obtained from the computer-aided design given in Example 12.4. All material of construction to be carbon steel (semi-killed or silicon killed). Your design should cover:
(a) choice of design pressure and temperature
(b) choice of the required corrosion allowances
(c) choice of the type of end covers
(d) determination of the minimum wall thickness for the shell, headers and ends
(e) a check on the pressure rating of the tubes.

13.5. Make a preliminary mechanical design for the vertical thermosyphon reboiler for which the thermal design was given in Example 12.10 in Chapter 12. The inlet liquid nozzle and the steam connections will be 50 mm inside diameter. Flat-plate end closures will be used on both headers. The reboiler will be hung from four bracket supports, positioned 0.5 m down from the top tube plate. The shell and tubes will be of semi-killed carbon steel.

Your design should cover:

(a) choice of design pressure and temperature
(b) choice of the required corrosion allowances
(c) selection of the header dimensions
(d) determination of the minimum wall thickness for the shell, headers and ends
(e) a check on the pressure rating of the tubes.

13.6. The specification for a sieve-plate column is given below. Make a preliminary mechanical design for the column. Your design should include:

(a) column wall thickness
(b) selection and sizing of vessel heads
(c) the nozzles and flanges (use standard flanges)
(d) column supporting skirt.

You need not design the plates or plate supports.
You should consider the following design loads:

(a) internal pressure
(b) wind loading
(c) dead weight of vessel and contents (vessel full of water).

There will be no significant loading from piping and external equipment. Earthquake loading need not be considered.

Column specification:
Length of cylindrical section 37 m
Internal diameter 1.5 m
Heads, standard ellipsoidal
50 sieve plates
Nozzles: feed, at mid-point, 50 mm inside diameter; vapour out, 0.7 m below top of cylindrical section, 250 mm inside diameter; bottom product, centre of vessel head, 50 mm inside diameter; reflux return, 1.0 m below top of cylindrical section, 50 mm inside diameter
Two 0.6-m-dia. access ports (manholes) situated 1.0 m above the bottom and 1.5 m below the top of the column
Support skirt height 2.5 m
Access ladder with platforms
Insulation, mineral wool, 50 mm thick
Materials of construction: vessel stainless steel, unstabilized (304); nozzles as vessel; skirt carbon steel, silicon killed
Design pressure 1200 kN/m^2
Design temperature 150°C
Corrosion allowance 2 mm.
Make a dimensioned sketch of your design and fill out the column specification sheet given in Appendix G.

13.7. A fixed-bed reactor is to be designed for a hydrocracking process. The reactor will treat 320,000 lb/h of vacuum gas oil (specific gravity 0.85) in the presence hydrogen at 650°F, 2000 psig, 1.0 weight hourly space velocity (WHSV). The catalyst has bulk density of 50 lb/ft³ and void fraction 0.4. The catalyst is to be divided into four beds, to allow a hydrogen quench to be brought in between the beds for temperature control. Make a preliminary mechanical design of the reactor(s). Your design should include:

(a) selection of material of construction
(b) sizing of the vessel(s) including allowance for any internals
(c) determination of the required wall thickness
(d) selection and sizing of vessel heads
(e) the nozzles and flanges (use standard flanges)
(f) a support skirt.

You need not design the vessel internals.

You should consider the following design loads:

(a) internal pressure
(b) wind loading
(c) dead weight of vessel and contents (vessel full of catalyst and gas oil)
(d) ydraulic testing with no catalyst and vessel full of water.

13.8. A jacketed vessel is to be used as a reactor. The vessel has an internal diameter of 2 m and is fitted with a jacket over a straight section 1.5 m long. Both the vessel and jacket walls are 25 mm thick. The spacing between the vessel and jacket is 75 mm.

The vessel and jacket are made of carbon steel. The vessel will operate at atmospheric pressure and the jacket will be supplied with steam at 20 bar. Check if the thickness of the vessel and jacket is adequate for this duty.

Take the allowable design stress as 100 N/mm^2 and the value of Young's modulus at the operating temperature as $180,000 \text{ N/mm}^2$.

13.9. A storage tank for concentrated nitric acid will be constructed from aluminium to resist corrosion. The tank is to have an inside diameter of 6 m and a height of 17 m. The maximum liquid level in the tank will be at 16 m. Estimate the plate thickness required at the base of the tank. Take the allowable design stress for aluminium as 90 N/mm^2.

14 GENERAL SITE CONSIDERATIONS

Chapter Contents

Key Learning Objectives

- Factors that are considered in site selection and plant layout
- Environmental legislation that governs chemical plant operations
- Waste minimization methods that can be used to reduce the environmental impact of a chemical plant

14.1. INTRODUCTION

In the discussion of process and equipment design given in the previous chapters, no reference was made to the plant site. A suitable site must be found for a new project, and the site and equipment layout planned. Provision must be made for the ancillary buildings and services needed for plant operation, and for the environmentally acceptable disposal of effluent. These subjects are discussed briefly in this chapter.

14.2. PLANT LOCATION AND SITE SELECTION

The location of the plant can have a crucial effect on the profitability of a project, and the scope for future expansion. Many factors must be considered when selecting a suitable site, and only a brief review of the principal factors will be given in this section. Site selection for chemical process plants is discussed in more detail by Merims (1966) and Mecklenburgh (1985); see also AIChE (2003). The principal factors to consider are:

1. Location, with respect to the marketing area
2. Raw material supply
3. Transport facilities
4. Availability of labour
5. Availability of utilities: water, fuel, power
6. Availability of suitable land
7. Environmental impact, including effluent disposal
8. Local community considerations
9. Climate
10. Political and strategic considerations.

Marketing Area

For materials that are produced in bulk quantities, such as cement, mineral acids, fuels and fertilizers, where the cost of the product per metric ton is relatively low and the cost of transport is a significant fraction of the sales price, the plant should be located close to the primary market. This consideration is much less important for low volume production, high-priced products, such as pharmaceuticals.

Raw Materials

The availability and price of suitable raw materials will often determine the site location. Plants that produce bulk chemicals are best located close to the source of the major raw material, as long as the costs of shipping product are not greater than the cost of shipping feed. For example, at the time of writing much of the new ethylene capacity that is being added worldwide is being built in the Middle East, close to supplies of cheap ethane from natural gas. Oil refineries, on the other hand, tend to be located close to major population centres, as an oil refinery produces many grades of fuel, which are expensive to ship separately.

Transport

The transport of materials and products to and from the plant can be an overriding consideration in site selection.

If practicable, a site should be selected that is close to at least two major forms of transport: road, rail, waterway (canal or river), or a sea port. Road transport is increasingly used, and is suitable for local distribution from a central warehouse. Rail transport is usually cheaper for the long-distance transport of bulk chemicals.

Air transport is convenient and efficient for the movement of personnel and essential equipment and supplies, and the proximity of the site to a major airport should be considered.

Availability of Labour

Labour will be needed for construction of the plant and its operation. Skilled construction workers are often brought in from outside the site area, but there should be an adequate pool of unskilled labour available locally, and labour suitable for training to operate the plant. Skilled craft workers such as electricians, welders and pipe fitters will be needed for plant maintenance. Local labour laws, trade union customs and restrictive practices must be considered when assessing the availability and suitability of the local labour for recruitment and training.

Utilities (Services)

Chemical processes invariably require large quantities of water for cooling and general process use, and the plant must be located near a source of water of suitable quality. Process water may be drawn from a river, from wells, or purchased from a local authority.

At some sites, the cooling water required can be taken from a river or lake, or from the sea; at other locations cooling towers will be needed.

Electrical power is needed at all sites. Electrochemical processes (for example, chlorine manufacture or aluminium smelting) require large quantities of power and must be located close to a cheap source of power.

A competitively priced fuel must be available on site for steam and power generation.

Environmental Impact and Effluent Disposal

All industrial processes produce waste products, and full consideration must be given to the difficulties and cost of their disposal. The disposal of toxic and harmful effluents will be covered by local regulations, and the appropriate authorities must be consulted during the initial site survey to determine the standards that must be met.

An environmental impact assessment should be made for each new project, or major modification or addition to an existing process; see Section 14.6.6.

Local Community Considerations

The proposed plant must fit in with and be acceptable to the local community. Full consideration must be given to the safe location of the plant so that it does not impose a significant additional risk to the local population. Plants should generally be sited so as not to be upwind of residential areas under the prevailing wind.

On a new site, the local community must be able to provide adequate facilities for the plant personnel: schools, banks, housing, and recreational and cultural facilities.

The local community must also be consulted about plant water consumption and discharge and the effect of the plant on local traffic. Some communities welcome new plant construction as a source of new jobs and economic prosperity. More affluent communities generally do less to encourage the building of new manufacturing plants, and in some cases may actively discourage chemical plant construction.

Land (Site Considerations)

Sufficient suitable land must be available for the proposed plant and for future expansion. The land should ideally be flat, well drained and have suitable load-bearing characteristics. A full site evaluation should be made to determine the need for piling or other special foundations. Particular care must be taken when building plants on reclaimed land near the ocean in earthquake zones because of the poor seismic character of such land.

Climate

Adverse climatic conditions at a site will increase costs. Abnormally low temperatures require the provision of additional insulation and special heating for equipment and pipe runs. Stronger structures are needed at locations subject to high winds (cyclone/ hurricane areas) or earthquakes.

Political and Strategic Considerations

Capital grants, tax concessions, and other inducements are often given by governments to direct new investment to preferred locations, such as areas of high unemployment. The availability of such grants can be the overriding consideration in site selection.

In a globalized economy, there may be an advantage to be gained by locating the plant within an area with preferential tariff agreements, such as the European Union (EU).

14.3.   SITE LAYOUT

The process units and ancillary buildings should be laid out to give the most economical flow of materials and personnel around the site. Hazardous processes must be located at a safe distance from other buildings. Consideration must also be given to the future expansion of the site. The ancillary buildings and services required on a site, in addition to the main processing units (buildings), include:

1. Storage for raw materials and products: tank farms and warehouses
2. Maintenance workshops
3. Stores, for maintenance and operating supplies
4. Laboratories for process quality control
5. Fire stations and other emergency services
6. Utilities: steam boilers, compressed air, power generation, refrigeration, transformer stations

7. Effluent disposal plant: waste water treatment, solid and or liquid waste collection
8. Offices for general administration
9. Canteens and other amenity buildings, such as medical centres
10. Parking lots.

When roughing out the preliminary site layout, the process units are normally sited first and arranged to give a smooth flow of materials through the various processing steps, from raw material to final product storage. Process units are normally spaced at least 30 m apart; greater spacing may be needed for hazardous processes.

The location of the principal ancillary buildings should then be decided. They should be arranged so as to minimize the time spent by personnel in travelling between buildings. Administration offices and laboratories, in which a relatively large number of people will be working, should be located well away from potentially hazardous processes. Control rooms are normally located adjacent to the processing units, but with potentially hazardous processes may have to be sited at a safer distance.

The siting of the main process units determines the layout of the plant roads, pipe alleys and drains. Access roads to each building are needed for construction, and for operation and maintenance.

Utility buildings should be sited to give the most economical run of pipes to and from the process units.

Cooling towers should be sited so that under the prevailing wind the plume of condensate spray drifts away from the plant area and adjacent properties.

The main storage areas should be placed between the loading and unloading facilities and the process units they serve. Storage tanks containing hazardous materials should be sited at least 70 m (200 ft) from the site boundary.

A typical plot plan is shown in Figure 14.1.

A comprehensive discussion of site layout is given by Mecklenburgh (1985); see also House (1969), Kaess (1970) and Meissner and Shelton (1992).

14.4. PLANT LAYOUT

The economic construction and efficient operation of a process unit will depend on how well the plant and equipment specified on the process flow-sheet is laid out.

A detailed account of plant layout techniques cannot be given in this short section. A fuller discussion can be found in the book edited by Mecklenburgh (1985) and in articles by Kern (1977, 1978), Meissner and Shelton (1992), Brandt *et al.* (1992), and Russo and Tortorella (1992).

The principal factors to be considered are:

1. Economic considerations: construction and operating costs
2. The process requirements
3. Convenience of operation
4. Convenience of maintenance

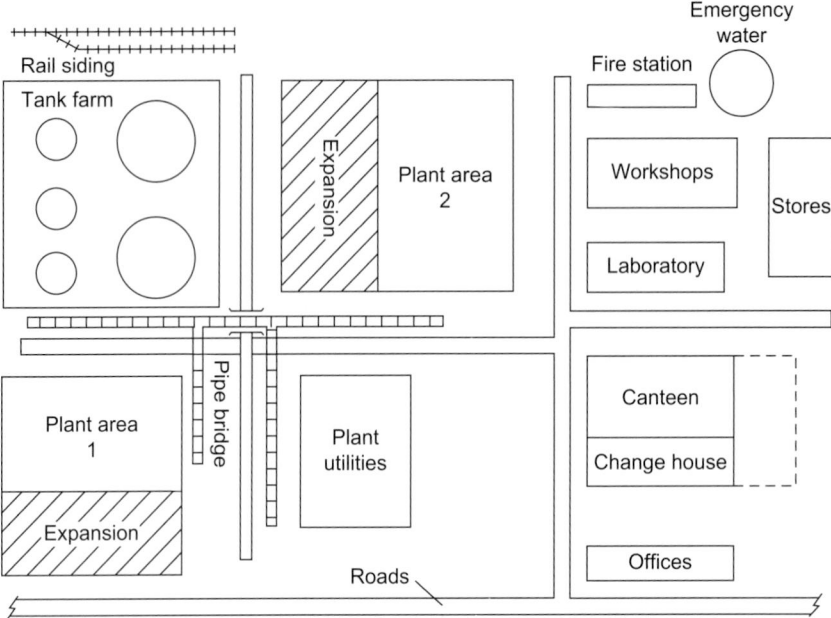

Figure 14.1.   Typical site plan.

5. Safety
6. Future expansion
7. Modular construction.

Costs

The cost of construction can be minimized by adopting a layout that gives the shortest run of connecting pipe between equipment, and the least amount of structural steel work; however, this will not necessarily be the best arrangement for operation and maintenance.

Process Requirements

An example of the need to take into account process considerations is the need to elevate the base of columns to provide the necessary net positive suction head to a pump (see Chapter 5) or the operating head for a thermosyphon reboiler (see Chapter 12).

Operation

Equipment that needs to have frequent operator attention should be located convenient to the control room. Valves, sample points, and instruments should be located at convenient positions and heights. Sufficient working space and headroom must be provided to allow easy access to equipment. If it is anticipated that equipment will need replacement, then sufficient space must be allowed to permit access for lifting equipment.

Maintenance

Heat exchangers need to be sited so that the tube bundles can be easily withdrawn for cleaning and tube replacement. Vessels that require frequent replacement of catalyst or packing should be located on the outside of buildings. Equipment that requires dismantling for maintenance, such as compressors and large pumps, should be placed under cover.

Safety

Blast walls may be needed to isolate potentially hazardous equipment, and confine the effects of an explosion.

At least two escape routes for operators must be provided from each level in process buildings.

Plant Expansion

Equipment should be located so that it can be conveniently tied in with any future expansion of the process.

Space should be left on pipe racks for future needs, and service pipes should be over-sized to allow for future requirements.

Modular Construction

In recent years there has been a move to assemble sections of plant at the plant manufacturer's site. These modules include the equipment, structural steel, piping and instrumentation. The modules are then transported to the plant site, by road or sea.

The advantages of modular construction are:

1. Improved quality control
2. Reduced construction cost
3. Less need for skilled labour on site
4. Less need for skilled personnel on overseas sites.

Some of the disadvantages are:

1. Higher design costs
2. More structural steel work
3. More flanged connections
4. Possible problems with assembly, on site.

A fuller discussion of techniques and applications of modular construction is given by Shelley (1990), Hesler (1990) and Whitaker (1984).

General Considerations

Open, structural steelwork, buildings are normally used for process equipment. Closed buildings are used for process operations that require protection from the weather, for small plants, or for processes that require ventilation with scrubbing of the vent gas.

The arrangement of the major items of equipment often follows the sequence given on the process flow-sheet: with the columns and vessels arranged in rows and the

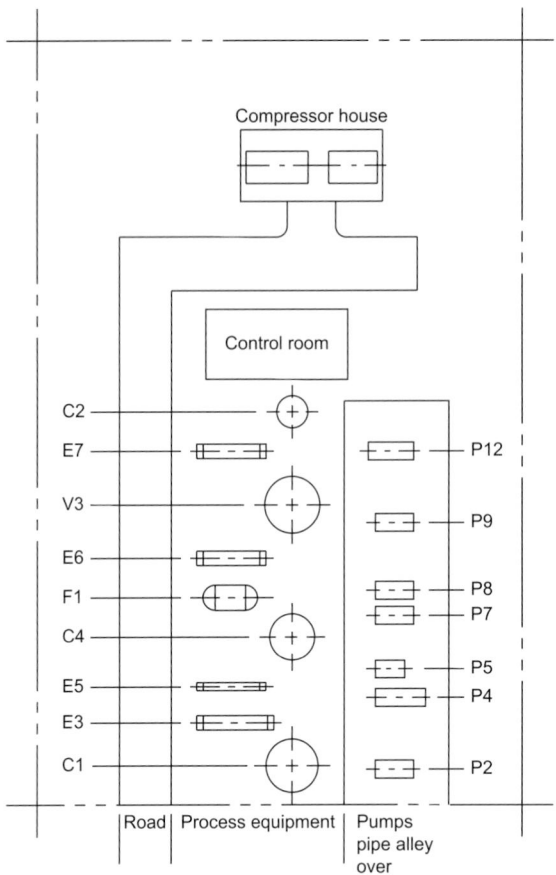

Figure 14.2.　A typical plant layout.

ancillary equipment, such as heat exchangers and pumps, positioned along the outside. A typical preliminary layout is shown in Figure 14.2.

14.4.1.　Techniques Used In Site and Plant Layout

Cardboard cut-outs of the equipment outlines can be used to make trial plant layouts. Simple models, made up from rectangular and cylindrical blocks, can be used to study alternative layouts in plan and elevation. Cut-outs and simple block models can also be used for site layout studies. Once the layout of the major pieces of equipment has been decided, the plan and elevation drawings can be made and the design of the structural steelwork and foundations undertaken.

Large-scale models, to a scale of at least 1:30, are sometimes still made for major projects. These models are used for piping design and to decide the detailed arrangement of small items of equipment, such as valves, instruments and sample points. Piping isometric diagrams can be taken from the finished models. The models are also useful on the construction site, and for operator training. Proprietary kits of parts are available for the construction of plant models.

Computer-aided design (CAD) tools are being increasingly used for plant layout studies and computer models are complementing, if not yet completely replacing, physical models. Several proprietary programs are available for the generation of three-dimensional models of plant layout and piping. Present systems allow designers to zoom in on a section of plant and view it from various angles. Developments of computer technology will soon enable engineers to virtually walk through the plant. A typical computer-generated model is shown in Figure 14.3.

Some of the advantages of computer graphics modelling compared with actual scale models are:

1. The ease of electronic transfer of information. Piping drawings can be generated directly from the layout model. Bills of quantities: materials, valves, instruments, etc. are generated automatically.
2. The computer model can be part of an integrated project information system, covering all aspects of the project from conception to operation.
3. It is easy to detect interference between pipe runs, and pipes and structural steel that occupy the same space.
4. The physical model has to be transported to the plant site for use in the plant construction and operator training. A computer model can be instantly available in the design office, the customer's offices, and at the plant site.
5. Expert systems and optimization programs can be incorporated in the package to assist the designer to find the best practical layout; see Madden *et al.* (1990).

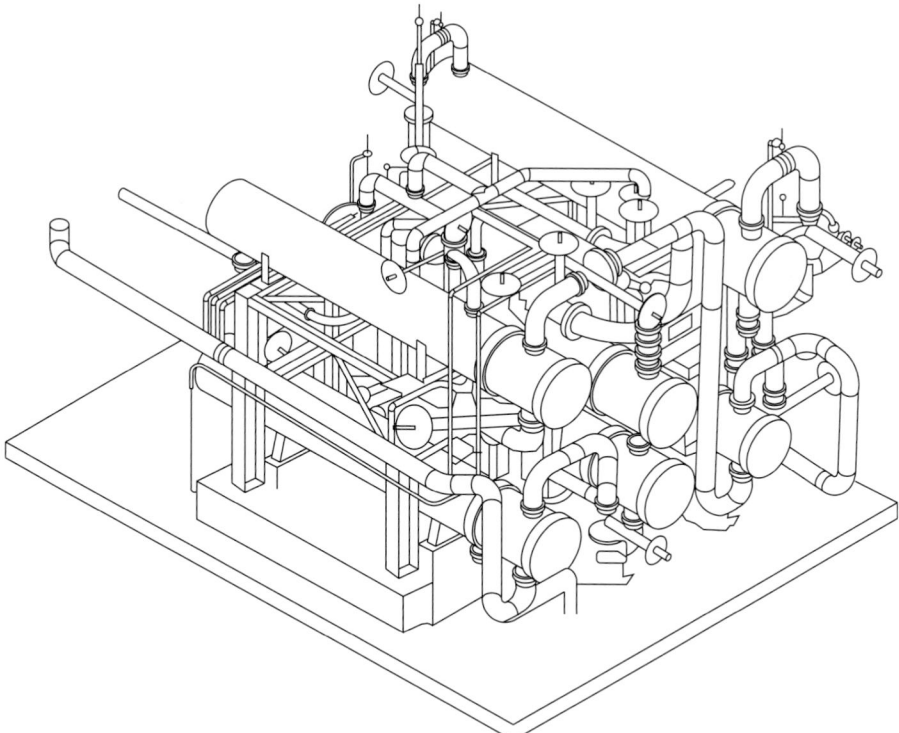

Figure 14.3.   Computer-generated layout 'model'. (Courtesy: Babcock Construction Ltd).

14.5. UTILITIES

The word 'utilities' is used for the ancillary services needed in the operation of any production process. These services are normally supplied from a central site facility, and include:

1. Electricity
2. Steam, for process heating
3. Cooling water
4. Water for general use
5. Demineralized water
6. Compressed air
7. Inert-gas supplies
8. Refrigeration
9. Effluent disposal facilities.

Electricity

The power required for electrochemical processes, motor drives, lighting, and general use, may be generated on site, but will more usually be purchased from the local supply company. The economics of power generation on site are discussed in Section 6.4.3.

The voltage at which the supply is taken or generated will depend on the demand. In the UK, large sites will typically take power at 11,000 or 33,000 V. Transformers at the plant are used to step down the power to the supply voltages used on site. A three-phase 415 V system is used for general industrial purposes and 240 V single-phase is used for lighting and other low-power requirements. In the US, most motors and other process equipment run on 208-V three-phase power, while 120/240-V single-phase power is used for offices, labs and control rooms.

A detailed account of the factors to be considered when designing electrical distribution systems for chemical process plants, and the equipment used (transformers, switch gear and cables), is given by Silverman (1964). Requirements for electrical equipment used in hazardous (classified) locations are given in the Dangerous Substances and Explosive Atmospheres Regulations of 2002 (DSEAR), which are described in Section 9.3.5.

Steam

The steam for process heating is usually generated in water tube boilers, using the most economical fuel available. The design and economics of steam systems are discussed in Section 6.4.4.

Combined Heat and Power (Co-Generation)

The energy costs on a large site can be reduced if the electrical power required can be generated on site and the exhaust steam from the turbines used for process heating. The overall thermal efficiency of such systems can be in the range 70% to 80%, compared with the 30% to 40% obtained from a conventional power station, where the heat in the exhaust steam is wasted in the condenser. Whether a combined heat and power system is worth considering for a particular site depends on the size of the site, the cost of fuel, the balance between the power and heating demands; and particularly on the availability of, and cost of, standby supplies and the price paid for any surplus electric power generated.

On any site it is always worth considering driving large compressors or pumps with steam turbines and using the exhaust steam for local process heating.

Cooling Water

Natural and forced-draft cooling towers are generally used to provide the cooling water required on a site, unless water can be drawn from a convenient river or lake in sufficient quantity. Sea water, or brackish water, can be used at coastal sites, but if used directly will necessitate the use of more expensive materials of construction for heat exchangers (see Chapter 7). The minimum temperature that can be reached with cooling water depends on the local climate. Cooling towers work by evaporating part of the circulating water to ambient air, causing the remaining water to be chilled. If the ambient temperature and humidity are high then a cooling water system will be less effective and air coolers or refrigeration would be used instead.

Water for General Use

The water required for general purposes on a site will usually be taken from the local mains supply, unless a cheaper source of suitable quality water is available from a river, lake or well.

Demineralized Water

Demineralized water, from which all the minerals have been removed by ion exchange, is used where pure water is needed for process use, and as boiler feed-water. Mixed and multiple-bed ion-exchange units are used; one resin converting the cations to hydrogen and the other removing the anions. Water with less than one part per million of dissolved solids can be produced. Reverse osmosis is sometimes used as a pre-treatment step to reduce the size and cost of the ion exchange plant, see Section 10.5.4.

Refrigeration

Refrigeration is needed for processes that require temperatures below those that can be economically obtained with cooling water. For temperatures down to around $10°C$, chilled water can be used. For lower temperatures, down to $-30°C$, salt brines (NaCl and $CaCl_2$) are sometimes used to distribute the 'refrigeration' round the site from a central refrigeration machine. Vapour compression refrigeration machines are normally used.

Compressed Air

Compressed air is needed for general use, and for the pneumatic transmitters and actuators that are sometimes used for process control. Air is normally distributed at a mains pressure of 6 bar (100 psig). Rotary and reciprocating single-stage or two-stage compressors are used. Instrument air must be dry and clean (free from oil).

Inert Gases

Where a large quantity of inert gas is required for the inert blanketing of tanks and for purging (see Chapter 9) this will usually be supplied from a central facility. Nitrogen is normally used, and can be manufactured on site in an air liquefaction plant, or purchased as liquid in tankers. Nitrogen is often supplied through an 'across the fence' contract with one of the air separation companies.

Effluent Disposal

Facilities are required at all sites for the disposal of waste materials without creating a public nuisance; see Section 14.6.3.

14.6. ENVIRONMENTAL CONSIDERATIONS

All individuals and companies have a duty of care to their neighbours, and to the environment in general. In addition to this moral duty, most countries have enacted strict laws to protect the environment and preserve the quality of air, water and land.

Vigilance is required in both the design and operation of process plant to ensure that legal standards are met and that no harm is done to the environment.

Consideration must be given to:

1. All emissions to land, air and water
2. Waste management
3. Smells
4. Noise
5. Visual impact
6. Any other nuisances
7. The environmental friendliness of the products.

14.6.1. Environmental Legislation

It is not feasible to review the entire body of legislation that has been enacted to protect the environment in this chapter. States, provinces and municipalities often pass local legislation that is stricter than the national laws. For example, the South Coast Air Quality Management District (SCAQMD), which sets air quality standards for the Los Angeles basin, has consistently advocated air quality limits that exceed the US national standards. Environmental legislation is also revised frequently, and the design engineer should always check with the local, regional and federal authorities to ensure that the correct standards are being applied in the design and to ensure that the correct information is being generated for permit applications. This section provides a brief overview of some of the main environmental laws in the UK and EU. More information can be found on the web sites of the UK Department of the Environment, Food and Rural Affairs (www.defra.gov.uk/environment), the UK Environment Agency (www.environment-agency.gov.uk), and the EU environment web site (www. ec.europa.eu/environment/index_en.htm). Full text versions of EU directives can be downloaded from http://eur-lex.europa.eu.

Improvement of the quality of environment was one of the founding principles of the European Community, stated in Article 2 of the E.C. treaty. Article 6 of the EC treaty requires environmental protection requirements to be integrated into the implementation of Community policies and activities. In Article 174 of the treaty the requirements of environmental policy are spelled out in more detail:

> *Community policy on the environment shall aim at a high level of protection... It shall be based on the precautionary principle and on the principles that preventive action should be taken, that environmental damage should as a priority be rectified at source and that the polluter should pay.*

These aims have been reinforced through a series of Environment Action Programmes. The Sixth Community Environment Action Programme was passed as Decision 1600/2002/EC of the European Parliament, and covers a 10-year period beginning 22 July 2002. Some specific aims and objectives of this programme include meeting the obligations of Member States under the Kyoto Protocol on greenhouse gas emissions, increasing use of renewable energy to 12% of total energy use by 2010, increasing use of combined heat and power to 18% of total electricity generation, and ensuring that by 2020 chemicals are only produced and used in ways that

> *do not lead to a significant negative impact on health and the environment, recognising that the present gaps of knowledge on the properties, use, disposal and exposure of chemicals need to be overcome ... placing the responsibility on manufacturers, importers and downstream users for generating knowledge about all chemicals (duty of care) and assessing risks of their use.*

This last goal is being implemented through the REACH legislation discussed in Section 9.1.1.

The EU implements environmental policy through Directives of the European Parliament or Commission, which are then passed into laws and enforced by the Member States. Some of the specific Directives and regulations are described below, grouped by topic.

1. Council Decision 2005/370/EC of 17 February 2005 approved the Århus Convention on behalf of the European Community. The Århus Convention was intended to increase public access to information, participation in decision making, and access to justice on environmental matters. Public authorities in the Member States must keep up-to-date, accessible databases of environmental data, plans and policies. The decision to authorize certain industrial activities can only be made after public consultation. This Directive underpins many of the other EU regulations on public access to information and involvement in project approval.

2. In Council Directive 2008/1/EC of 15 January 2008 on integrated pollution prevention and control, the EU defines the obligations of highly polluting industrial and agricultural activities. Industries such as mining, primary metals production, chemicals manufacture, waste management and livestock farming are classified as having high pollution potential and are required to have a permit. A permit can only be issued if the installation uses the best available techniques for pollution prevention, recycles or disposes of waste in the least polluting way possible, prevents large-scale pollution, uses energy efficiently,

ensures accident prevention and damage limitation, and returns the site to its original state once the activity is over. The permit can set emission limit values for pollutant substances, require specific actions for air, soil or water protection, require the operator to monitor for releases, and dictate other actions specific to the site. The permit application must give details on the pollutants formed and the scale of production and this information must be made available to the public as well as the government licensing agency. This Directive replaced the earlier pollution prevention Directive (96/61/EC) and its amendments, and incorporated Regulation (EC) No. 16/2006 of the European Parliament of 18 January 2006, which established the European Pollutant Release and Transfer Register (PRTR). The PRTR is a publicly accessible electronic database covering releases of pollutants to air, water and land from power stations, mining, metalworking, chemical plants, paper plants, and waste and waste-water treatment plants, as well as emissions of listed compounds that exceed threshold values. Member States gather the information and submit it to the European Environment Agency. The integrated pollution prevention and control Directive was implemented in the UK as the Pollution Prevention and Control Act of 1999.

3. Directive 2004/35/EC of 21 April 2004 on environmental liability with regard to the prevention and remedying of environmental damage establishes the legal framework for enforcing the 'polluter pays' principle. The operator of any plant that requires a permit under the Directive on integrated pollution prevention and control can be required to take preventive or remedial actions to prevent environmental damage. If the operator fails to do so, the enforcing agency can take action and recover the cost from the operator. The enforcing agency can also recover the costs of environmental assessments carried out to assess the damage. If there are several operators then they are jointly liable. A proposed Directive under discussion at time of writing would define a set of serious environmental offences that could be punished through criminal law.

4. Council Directive 85/337/EEC of 27 June 1985, and subsequent amendments, requires an Environmental Impact Assessment (EIA) to be carried out for projects that have a significant effect on the environment. Among the projects specified by the Directive are oil refineries, nuclear fuel or waste plants, integrated chemical facilities, power stations above 300 MW, waste and water treatment plants, mining facilities, dams, and large pig or poultry farms. The developer must identify the impact of the project on people, local flora and fauna, the soil, water, air, climate, landscape, material assets and cultural heritage of the site. This information must be made available to the public as well as the authorizing agency of the government before the project can be approved.

5. Directive 2006/12/EC of the European Parliament of 5 April 2006 on waste is a framework act covering all waste products other than gases, waste water, agricultural wastes, animal carcasses, mineral waste, radioactive waste and decommissioned explosives. Dumping and abandoning waste are prohibited and Member States must promote waste prevention, recycling and re-use. The cost of waste disposal must be borne by the producer of the product that gives rise to the waste, in accordance with the 'polluter pays' principle. Additional

rules for hazardous waste are specified in Council Directive 91/689/EEC of 12 December 1991 on hazardous waste, which requires the Member States to identify hazardous wastes, ensure that different categories of hazardous wastes are not mixed with each other or with non-hazardous waste, and specifies requirements for handling and disposing of hazardous waste.

6. Directive 2000/60/EC of 23 October 2000, known as the Water Framework Directive, is a broad legislative framework for water protection and management, including inland surface waters, groundwater, tidal waters and coastal waters. In addition to taking steps to reduce water pollution from discharges or emissions of hazardous substances, Member States must ensure that water-pricing policies provide incentives to use water resources efficiently. This Directive was implemented in England and Wales as the Water Environment (Water Framework Directive) (England and Wales) Regulations of 2003. Additional regulations for groundwater are specified in Directive 2006/118/EC of 12 December 2006.

7. Air quality regulations are not yet covered by a framework Directive, although one is under discussion at time of writing, and it is expected that the current set of laws will be rationalized when this is passed. Air quality limit values have been specified in the following Directives: 1999/30/EC (sulphur dioxide, nitrogen dioxide, nitrogen oxides, lead and suspended particulates); 2000/69/EC (benzene and carbon monoxide); 1999/13/EEC (volatile organic compounds); 2002/3/EC (ozone); and 2004/107/EC (arsenic, cadmium, mercury, nickel and polycyclic aromatic hydrocarbons); while Directive 2001/81/EEC set national emissions ceilings for several other atmospheric pollutants. Atmospheric emission limits can also be set under the integrated pollution prevention and control Directive, discussed above. Most of these Directives were incorporated into UK law under the Air Quality Standards Regulations of 2007. Directive 2003/87/EC of 13 October 2003 established a scheme for greenhouse gas emission allowance trading within the European Community and set up the 'cap and trade' system to reduce European greenhouse gas emissions in line with commitments under the Kyoto Protocol.

8. Council Directive 2002/49/EC of 25 June 2002 requires the Member States to take action to reduce noise pollution, but did not set limit values.

9. Regulation (EC) No 1907/2006 of the European Parliament and of the Council of 18 December 2006 concerns the Registration, Evaluation, Authorisation and Restriction of Chemicals (REACH); see Section 9.1.1.

Much of the recent environmental legislation in the UK has been passed to implement the various Directives of the European Union. Some additional UK laws that have an impact on process industry projects include:

1. The Environment Act of 1995. This broad act specified the overall framework for environmental protection in the UK. It created the Environment Agency and the Scottish Environment Protection Agency (SEPA), and defined their powers and scope, which include enforcement of various environmental laws and regulations. Subsections of the Act also dealt with pollution control, contaminated land, abandoned mines, National Parks, air quality, waste, flood defences, fisheries and the conservation of hedgerows.

2. The Waste Minimisation Act of 1998, which enabled local authorities to take steps to minimize the generation of household, commercial or industrial waste.

Information on United States federal laws can be found on the US Environmental Protection Agency (EPA) web site at www.epa.gov/epahome/laws. Information on Canadian laws is given on the Environment Canada web site at www.ec.gc.ca/envhome. An overview of North American environmental legislation is given in the North American version of this book (Towler and Sinnott, 2008).

Information on environmental regulations in other countries can usually be obtained from government web sites. Countries that have signed the Århus Convention of 1998 are obliged to make environmental regulations and decision-making processes easily accessible to the public. The design engineer should always consult the most recent version of laws and regulations, and must ensure that all the necessary information is developed and provided to the regulatory agencies and public in a timely manner to ensure that the necessary permits can be obtained. Permits are usually obtained before major capital investments are made in the project. Legal counsel should be obtained if there is any question on the interpretation of any regulation, and most large corporations in the process industries employ lawyers who specialize in environmental compliance.

Obtaining permits or modification of existing permits can often be the rate-determining step in a project, particularly if the project faces public opposition. Projects may even need to be suspended or relocated if the permitting process is too lengthy or expensive, as this can have a significant effect on project financial performance.

14.6.2. Waste Minimization

Waste arises mainly as by-products or unused reactants from the process, or as off-specification product produced through mis-operation. There will also be fugitive emissions from leaking seals and flanges, and inadvertent spills and discharges through mis-operation. In emergency situations, material may be discharged to the atmosphere through vents normally protected by bursting discs and relief valves.

Before considering 'end-of-pipe' approaches for treating and managing waste products, the design engineer should always try to minimize production of waste at the source. The hierarchy of waste management approaches is:

1. Source reduction: don't make the waste in the first place. This is the best practice.
2. Recycle: find a use for the waste stream.
3. Treatment: reduce the severity of the environmental impact.
4. Disposal: meet the requirements of the law.

Source reduction is accomplished during process design. Some of the strategies that can be considered include:

1. *Purification of feeds.* Reducing the concentration of impurities in the feed usually leads to reduced side reactions and less waste formation. This approach can also reduce the need for purges and vent streams. Feed impurities also often lead to

degradation of solvents and catalysts. Care must be taken to select a purification approach that does not itself lead to more waste formation.

2. *Protect catalysts and adsorbents.* Deactivated catalysts and adsorbents are solid wastes from the process. In some cases, relatively small amounts of contaminants can cause a load of catalyst or adsorbent to become useless. The catalyst or adsorbent should be protected by using a guard bed of suitable material to adsorb or filter out contaminants before they can damage the catalyst.

3. *Eliminate use of extraneous materials.* When different solvent or mass separating agents are used, this leads to waste formation when the solvents become degraded. If the plant or site uses relatively few solvents then it may be economical to build a solvent-recovery plant. Liquid wastes from spent solvents are very common in fine chemicals and pharmaceuticals manufacture.

4. *Increase recovery from separations.* Higher product recovery leads to less product in the waste streams. Higher purity recycle streams usually lead to less waste formation. These benefits must always be traded off against the extra capital and energy costs involved in driving the separation processes to higher recovery or purity.

5. *Improve fuel quality.* Switching to a cleaner-burning fuel, such as natural gas, reduces the emissions from fired heaters. This must be traded off against the higher cost of natural gas relative to heating oil and coal.

Unused reactants can be recycled and off-specification product reprocessed. Integrated processes can be selected: the waste from one process becoming the raw material for another. For example, the otherwise waste hydrogen chloride produced in a chlorination process can be used for chlorination using a different reaction, as in the balanced, chlorination–oxyhydrochlorination process for vinyl chloride production. It may be possible to sell waste to another company for use as raw material in their manufacturing processes; for example, the use of off-specification and recycled plastics in the production of lower grade products, such as the ubiquitous black plastic bucket.

Processes and equipment should be designed to reduce the chances of misoperation; by providing tight control systems, alarms and interlocks. Sample points, process equipment drains and pumps should be sited so that any leaks flow into the plant effluent collection system, not directly to sewers. Hold-up systems, tanks and ponds, should be provided to retain spills for treatment. Flanged joints should be kept to the minimum needed for the assembly and maintenance of equipment. Fugitive emissions from packings and seals can be reduced by specifying dual seals, dry gas seals or seal-less pumps.

A technique that is sometimes used for waste minimization is the five-step review:

1. Identify waste *components* for regulatory impact.
2. Identify *waste streams* for size and economic impact.
3. List the *root causes* of the waste streams.
4. List and analyse *modifications* to address the root causes.
5. Prioritize and *implement* the best solutions.

The information gathered in the first two steps is often collected in an effluent summary worksheet. An effluent summary lists the regulated pollutants produced by the process and summarizes the quantities produced and where they originate. The effluent summary can be used to focus waste minimization efforts and as a design basis for the design of effluent treatment processes. The information in the effluent summary may also be required for obtaining permits to operate the plant or for preparing more formal environmental impact analyses to convince investors or insurers that environmental impact has been properly addressed. An example effluent summary sheet is given in Appendix G. A template is available in MS Excel format in the on-line material at http://books.elsevier.com/companions.

The American Petroleum Institute Publication 302 (1991) discusses source reduction, recycle, treatment and disposal of wastes. Other source reduction techniques are given by Smith and Petela (1991) and El-Halwagi (1997). The UK Institution of Chemical Engineers has published a guide to waste minimization (IChemE, 1997).

14.6.3. Waste Management

When waste is produced, processes must be incorporated in the design for its treatment and safe disposal. The following techniques can be considered:

1. Dilution and dispersion
2. Discharge to foul-water sewer (with the agreement of the appropriate authority)
3. Physical treatments: scrubbing, settling, absorption and adsorption
4. Chemical treatment: precipitation (for example, of heavy metals), neutralization
5. Biological treatment: activated sludge and other processes
6. Incineration on land, or at sea
7. Landfill at controlled sites
8. Sea dumping – (now subject to tight international control).

Several standards have been written to address waste management systems. The main international standard for waste management systems is ISO 14001 (2004), which has been adopted as the national standard in the countries of the EU. In the US, EPA standard 40 CFR 260 (2006) provides general guidelines, while EPA 40 CFR 264 (2006) gives standards for waste treatment, storage and disposal. Standards for the petroleum industry are given in API publications 300 (1991) and 303 (1992). ASTM standard 11.04 (2006) should also be consulted.

Gaseous Wastes

Gaseous effluents that contain toxic or noxious substances need treatment before discharge into the atmosphere. The practice of relying on dispersion from tall stacks is seldom entirely satisfactory. Gaseous pollutants can be removed by absorption or adsorption. Absorption by scrubbing with water or a suitable solvent or base is probably the most widely used method for high-volume gas streams, while adsorption onto activated carbon or a zeolitic adsorbent is used for smaller gas streams. The design of scrubbing towers is described in Chapter 11. Finely dispersed solids can be removed by scrubbing or using electrostatic precipitators; see Chapter 10. Flammable gases can be burnt. The sources of air pollution and their control are covered in several books: Walk (1997), Heumann (1997), Davies (2000) and Cooper and Ally

(2002). McGowan and Santoleri (2007) discuss methods for reducing emissions of volatile organic compounds (VOCs).

Liquid Wastes

The waste liquids from a chemical process, other than aqueous effluent, will usually be flammable and can be disposed of by burning in suitably designed incinerators. Care must be taken to ensure that the temperatures attained in the incinerator are high enough to completely destroy any harmful compounds that may be formed, such as the possible formation of dioxins when burning chlorinated compounds. The gases leaving an incinerator may be scrubbed, and acid gases neutralized. A typical incinerator for burning gaseous or liquid wastes is shown in Chapter 3 (Figure 3.15). The design of incinerators for hazardous waste and the problems inherent in the disposal of waste by incineration are discussed by Butcher (1990) and Baker-Counsell (1987).

In the past, small quantities of liquid waste, in drums, were disposed of by dumping at sea or in land-fill sites. This is not an environmentally acceptable method and is now subject to stringent controls.

Solid Wastes

Solid waste can be burnt in suitable incinerators or disposed by burial at licensed land-fill sites. As for liquid wastes, the dumping of toxic solid waste at sea is no longer acceptable.

Aqueous Wastes

Aqueous waste streams include process water, utility waste water and site run-off. Water that is used or formed in the process must be sent to effluent treatment. Common process-water effluents include:

- Water contaminated with ammonia or hydrogen sulphide from gas scrubbers
- Salt waters from deionizers, softeners, neutralization steps and washing operations
- Water contaminated with hydrocarbons
- Biologically contaminated water (for example, fermentation broths)
- Spent acid and caustic streams.

The site utility systems produce large waste water flows. A purge known as a 'blowdown' is taken from both the cooling water and the boiler feed water, to prevent the accumulation of solids in either recirculating system. Cooling water blowdown is often the largest contributor to the site waste water. The blowdown streams can be high in minerals, and also contain chemicals such as biocides and corrosion inhibitors that have been added to the boiler feed water or cooling water.

It is also a best practice to collect run-off water from the plant area and treat it in the site waste-water plant before discharging it to the environment. Run-off water can come from rain, fire hydrant flushing and equipment washing. As the water flows over the ground around the plant, it can become contaminated with organic chemicals that have leaked from the plant. Most plants are designed so that all the run-off is collected into local sewers or ditches that are routed to the site waste water treatment plant.

The principal factors that determine the nature of an aqueous industrial effluent and on which strict controls will be placed by the responsible authority are:

1. pH
2. Suspended solids
3. Toxicity
4. Biological oxygen demand.

The pH can be adjusted by the addition of acid or alkali. Spent acid or alkali solutions must usually be neutralized before they can be sent to water treatment plants. Lime (calcium oxide) is frequently used to neutralize acidic effluents. In the case of sulphuric acid, the use of lime leads to formation of calcium sulphate. Calcium sulphate that is potentially contaminated with trace organic material has low value and can be used as road fill. An alternative approach is to neutralize with more expensive ammonia, forming ammonium sulphate, which can be sold as a fertilizer.

Suspended solids can be removed by settling, using clarifiers (see Chapter 10).

For some effluents it is possible to reduce the toxicity to acceptable levels by dilution. Other effluents need chemical treatment.

The oxygen concentration in a water course must be maintained at a level sufficient to support aquatic life. For this reason, the biological oxygen demand of an effluent is of utmost importance. It is measured by a standard test: the BOD5 (5-day biological oxygen demand). This test measures the quantity of oxygen that a given volume of the effluent (when diluted with water containing suitable bacteria, essential inorganic salts, and saturated with oxygen) will absorb in 5 days, at a constant temperature of 20°C. The results are reported as parts of oxygen absorbed per million parts effluent (ppm). The BOD5 test is a rough measure of the strength of the effluent: the organic matter present. It does not measure the total oxygen demand, as any nitrogen compounds present will not be completely oxidized in 5 days. The ultimate oxygen demand (UOD) can be determined by conducting the test over a longer period, up to 90 days. If the chemical composition of the effluent is known, or can be predicted from the process flow-sheet, the UOD can be estimated by assuming complete oxidation of the carbon present to carbon dioxide, and the nitrogen present to nitrate:

$$UOD = 2.67C + 4.57N$$

where C and N are the concentrations of carbon and nitrogen in ppm by weight.

Activated sludge processes are usually used to reduce the biological oxygen demand of an aqueous effluent before discharge. Where waste water is discharged into the sewers with the agreement of the local water authorities, a charge will normally be made according to the BOD value, and any treatment required. Where treated effluent is discharged to water courses, with the agreement of the appropriate regulatory authority, the BOD5 limit will typically be set at 20 ppm. A full discussion of aqueous effluent treatment is given by Eckenfelder *et al.* (1985); see also Eckenfelder (1999).

14.6.4.   Noise

Noise can cause a serious nuisance in the neighbourhood of a process plant. Care must be taken when selecting and specifying equipment such as compressors, air-cooler fans, induced and forced draught fans for furnaces, etc. Excessive noise can also be generated when venting through steam and other relief valves, and from flare stacks. Such equipment should be fitted with silencers. Vendors' specifications should be checked to ensure that equipment complies with statutory noise levels, both for the protection of employees (see Chapter 9) as well as for noise pollution considerations. Noisy equipment should, as far as practicable, be sited well away from the site boundary. Earth banks and screens of trees can be used to reduce the noise level perceived outside the site.

14.6.5. Visual Impact

The appearance of the plant should be considered at the design stage. Few people object to the fairyland appearance of a process plant illuminated at night, but it is a different scene in daylight. There is little that can be done to change the appearance of a modern-style plant, where most of the equipment and piping are outside and in full view, but some steps can be taken to minimize the visual impact. Large equipment, such as storage tanks, can be painted to blend in with, or even contrast with, the surroundings. For example, the Richmond oil refinery in the San Francisco Bay Area has its storage tanks painted to blend in with the surrounding hills. Landscaping and screening by belts of trees can also help improve the overall appearance of the site.

14.6.6. Environmental Auditing

An environmental audit is a systematic examination of how a business operation affects the environment. It will include all emissions to air, land and water; and cover the legal constraints, the effect on community, the landscape and the ecology. Products will be considered, as well as processes.

When applied at the design stage of a new development it is more correctly called an *environmental impact assessment.*

The aims of the audit or assessment are to:

1. Identify environmental problems associated with the manufacturing process and the use of the products, before they become liabilities
2. Develop standards for good working practices
3. Provide a basis for company policy
4. Ensure compliance with environmental legislation
5. Satisfy requirements of insurers
6. Be seen to be concerned with environmental questions: important for public relations
7. Minimize the production of waste: an economic factor.

Environmental auditing is discussed by Grayson (1992). His booklet is a good source of references for commentary on the subject, and to government bulletins.

Life-Cycle Assessment

Life-cycle assessment is a more exhaustive procedure than environmental auditing and is used to compare the long-term sustainability of alternative designs. A life-cycle assessment considers all the environmental costs and impacts of the process, its feed stocks, and the physical plant itself, from initial construction through to final decommissioning and site remediation. The methods for carrying out a life-cycle assessment are given in ISO standards BS EN ISO 14040 and BS EN ISO 14044 (which have superseded the older 14041, 14042 and 14043 standards). A good introduction to life-cycle assessment is given by Clift (2001). Many examples of life-cycle assessments can be found in the journals *Environmental Science and Technology*, *Environmental Progress* and *The International Journal of Life Cycle Assessment*.

14.7. REFERENCES

AIChE (2003) *Guidelines for Facility Siting and Layout* (American Institute of Chemical Engineers).

Baker-Counsell, J. (1987) *Process Eng.* (April) 26. Hazardous wastes: the future for incineration.

Brandt, D., George, W., Hathaway, C. and McClintock, N. (1992) *Chem. Eng.*, NY, **99**(4) 97. Plant layout, Part 2: The impact of codes, standards and regulations.

Butcher, C. (1990) *Chem. Engr., London* No. 471 (April 12th) 27. Incinerating hazardous waste.

Clift, R. (2001) Clean technology and industrial ecology, in *Pollution: Causes, Effects and Control*, 4th edn, Harrison, R.M. (ed.) (Royal Society of Chemistry).

Cooper, C. D. and Ally, F. C. (2002) *Air Pollution Control*, 3rd edn. (Waveland Press).

Davies, W. T. (ed.) (2000) *Air Pollution Engineering Manual* (Wiley – International).

Eckenfelder, W. W., Patoczka, J. and Watkin, A. T. (1985) *Chem. Eng.*, NY, **92**(9) 60. Wastewater treatment.

Eckenfelder, W. W. (1999) *Industrial Water Pollution Control*, 2nd edn (McGraw-Hill).

El-Halwagi, M. M. (1997) *Pollution Prevention Through Process Integration: Systematic Design Tools* (Academic Press).

Grayson, L. (ed.) (1992) *Environmental Auditing* (Technical Communications, UK).

Hesler, W. E. (1990) *Chem. Eng. Prog.*, **86**(10) 76. Modular design: where it fits.

Heumann, W. L. (1997) *Industrial Air Pollution Control Systems* (McGraw-Hill).

House, F. F. (1969) *Chem. Eng.*, NY, **76**(7) 120. Engineers guide to plant layout.

IChemE (1997) *Waste Minimization, a practical guide* (Institution of Chemical Engineers, London).

Kaess, D. (1970) *Chem. Eng.*, NY, **77**(6) 122. Guide to trouble free plant layouts.

Kern, R. (1977) *Chem. Eng.*, NY, **84**:
  (May 23rd) 130. How to manage plant design to obtain minimum costs.
  (July 4th) 123. Specifications are the key to successful plant design.
  (Aug. 15th) 153. Layout arrangements for distillation columns.
  (Sept. 12th) 169. How to find optimum layout for heat exchangers.
  (Nov. 7th) 93. Arrangement of process and storage vessels.
  (Dec. 5th) 131. How to get the best process-plant layouts for pumps and compressors.

Kern, R. (1978) *Chem. Eng.*, NY, **85**:
  (Jan. 30th) 105. Pipework design for process plants.
  (Feb. 27th) 117. Space requirements and layout for process furnaces.
  (April 10th) 127. Instrument arrangements for ease of maintenance and convenient operation.
  (May 8th) 191. How to arrange plot plans for process plants.
  (July 17th) 123. Arranging the housed chemical process plant.
  (Aug. 14th) 141. Controlling the cost factor in plant design.

Madden, J., Pulford, C. and Shadbolt, N. (1990) *Chem. Engr., London*, **474** (May 24th) 32. Plant layout – untouched by human hand?

Mcgowan, T. F. and Santoleri, J. J. (2007) *Chem. Eng., NY*, **114**(2) 34. VOC Emission Controls for the CPI.

Mecklenburgh, J. C. (ed.) (1985) *Process Plant Layout* (Godwin/Longmans).

Meissner, R. E. and Shelton, D. C. (1992) *Chem. Eng., NY*, **99**(4) 97. Plant layout, Part 1: Minimizing problems in plant layout.

Merims, R. (1966) Plant location and site considerations, in *The Chemical Plant*, Landau, R. (ed.) (Reinhold).

Russo, T. J. and Tortorella, A. J. (1992) *Chem. Eng., NY*, **99**(4) 97. Plant layout, Part 3: The contribution of CAD.

Silverman, D. (1964) *Chem. Eng., NY*, **71** (May 25th) 131, (June 22nd) 133, (July 6th) 121, (July 20th), 161, in four parts. Electrical design.

Shelley, S. (1990) *Chem. Eng. NY*, **97**(8) 30. Making inroads with modular construction.

Smith, R. and Petela, E. (1991) *Chem. Engr. London*, **513**, 13. Waste minimization in the process industries: 3. Separation and recycle systems.

Towler, G. and Sinnott, R. (2008) *Chemical Engineering Design: Principles, Practice and Economics of Plant and Process Design* (Butterworth-Heinemann).

Walk, K. (1997) *Air Pollution: Its Origin and Control*, 3rd edn (1997).

Whittaker, R. (1984) *Chem. Eng. NY*, **92**(5) 80. Onshore modular construction.

British and European Standards

BS EN ISO 14001:2004 (2004) Environmental management systems. Requirements with guidance for use, 2nd edn.

BS EN ISO 14040:2006 (2006) Environmental Management – Life Cycle Assessment – Principles and Framework, 2nd edn.

BS EN ISO 14044:2006 (2006) Environmental Management – Life Cycle Assessment – Requirements and Guidelines, 1st edn.

American and International Standards

API Publication 300 (1991) Generation and management of wastes and secondary materials in the petroleum refining industry: 1987-1988 (American Petroleum Institute).

API Publication 302 (1991) Waste Minimization in the Petroleum Industry: Source Reduction, Recycle, Treatment, Disposal: A Compendium of Practices (American Petroleum Institute).

API Publication 303 (1992) Generation and management of wastes and secondary materials (American Petroleum Institute).

ASTM 11.04 (2006) Waste management (American Society for Testing Materials).

EPA 40 CFR 260 (2006) Hazardous waste management systems: General (US Environmental Protection Agency).

EPA 40 CFR 261 (1999) Identification and listing of hazardous waste (US Environmental Protection Agency).

EPA 40 CFR 264 (2006) Standards for owners and operators of hazardous waste treatment, storage and disposal facilities (US Environmental Protection Agency).

NFPA 70 (2006) National electrical code (National Fire Protection Association).

A GRAPHICAL SYMBOLS FOR PIPING SYSTEMS AND PLANT

BASED ON BS 1553: PART 1: 1977

Scope

This part of BS 1553 specifies graphical symbols for use in flow and piping diagrams for process plant.

Symbols (or elements of Symbols) for Use in Conjunction with Other Symbols

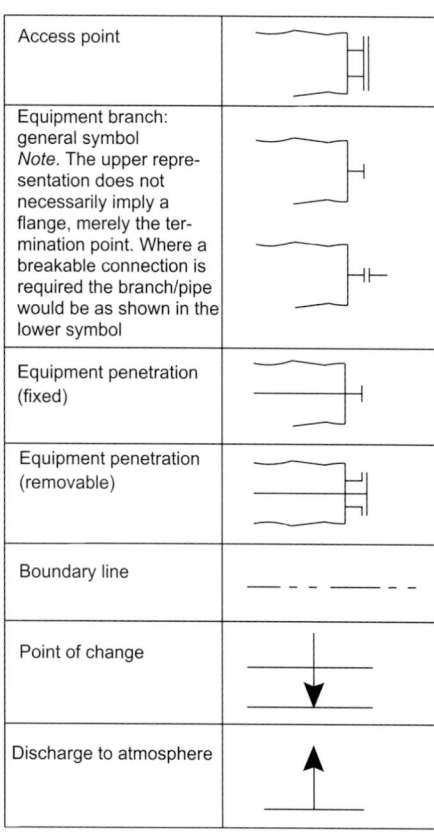

Mechanical linkage	
Weight device	
Electrical device	
Vibratory or loading device (any type)	
Spray device	
Rotary movement	
Stirring device	
Fan	

Access point	
Equipment branch: general symbol *Note*. The upper representation does not necessarily imply a flange, merely the termination point. Where a breakable connection is required the branch/pipe would be as shown in the lower symbol	
Equipment penetration (fixed)	
Equipment penetration (removable)	
Boundary line	
Point of change	
Discharge to atmosphere	

Basic and Developed Symbols for Plant and Equipment

Heat Transfer Equipment

Heat exchanger (basic symbols) Alternative:	
Shell and tube: fixed tube sheet	
Shell and tube: U tube or floating head	
Shell and tube: kettle reboiler	
Air-blown cooler	
Plate type	
Double-pipe type	
Heating/cooling coil (basic symbol)	
Fired heater/boiler (basic symbol)	

Upshot heater	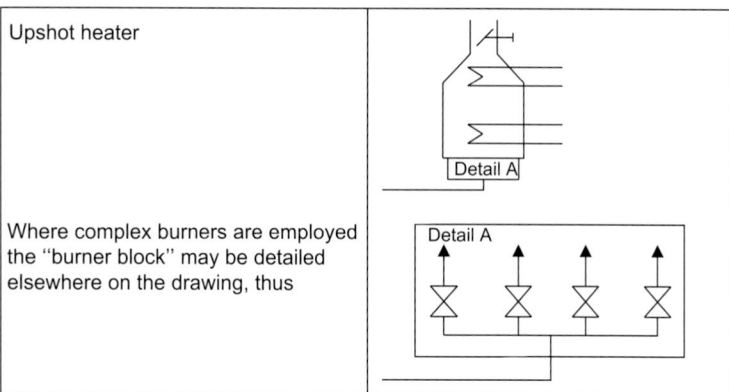
Where complex burners are employed the "burner block" may be detailed elsewhere on the drawing, thus	

Vessels and Tanks

Drum or simple pressure vessel (basic symbol)	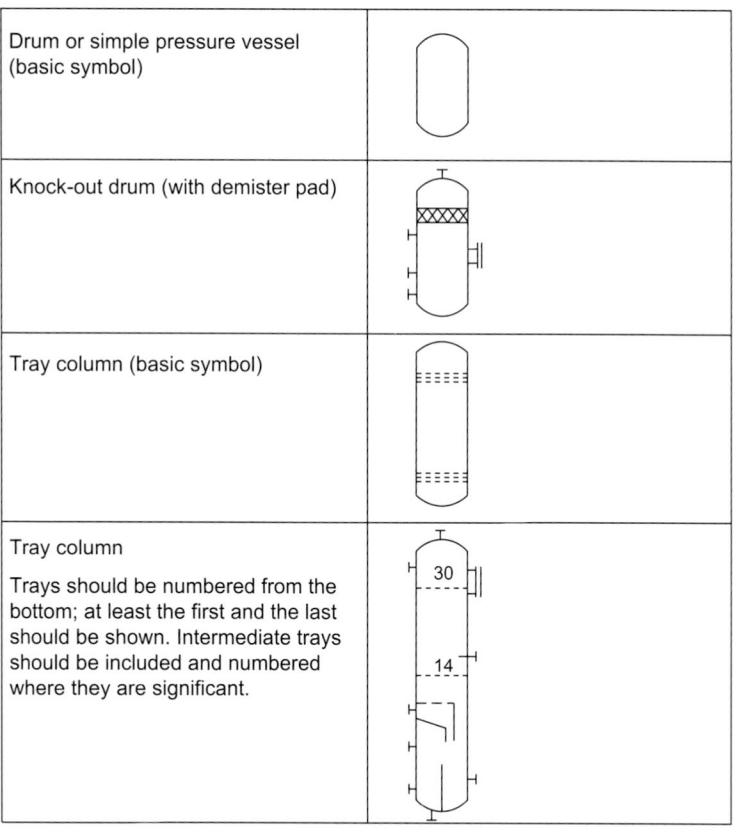
Knock-out drum (with demister pad)	
Tray column (basic symbol)	
Tray column Trays should be numbered from the bottom; at least the first and the last should be shown. Intermediate trays should be included and numbered where they are significant.	

Fluid contacting vessel (basic symbol)	
Fluid contacting vessel Support grids and distribution details may be shown	
Reaction or absorption vessel (basic symbol)	
Reaction or absorption vessel Where it is necessary to show more than one layer of material alternative hatching should be used	
Autoclave (basic symbol)	
Autoclave	

Open tank (basic symbol)	
Open tank	
Clarifier or settling tank	
Sealed tank	
Covered tank	
Tank with fixed roof (with draw-off sump)	
Tank with floating roof (with roof drain)	
Storage sphere	
Gas holder (basic symbol for all types)	

Pumps and Compressors

Rotary pump, fan or simple compressor (basic symbol)	
Centrifugal pump or centrifugal fan	
Centrifugal pump (submerged suction)	
Positive displacement rotary pump or rotary compressor	
Positive displacement pump (reciprocating)	
Axial flow fan	
Compressor: centrifugal/axial flow (basic symbol)	
Compressor: centrifugal/axial flow	
Compressor: reciprocating (basic symbol)	
Ejector/injector (basic symbol)	

Solids Handling

Size reduction	
Breaker gyratory	
Roll crusher	
Pulverizer: ball mill	
Mixing (basic symbol)	
Kneader	
Ribbon blender	
Double cone blender	
Filter (basic symbol, simple batch)	
Filter press (basic symbol)	
Rotary filter, film drier or flaker	

Cyclone and hydroclone (basic symbol)	
Cyclone and hydroclone	
Centrifuge (basic symbol)	
Centrifuge: horizontal peeler type	
Centrifuge: disc bowl type	

Drying

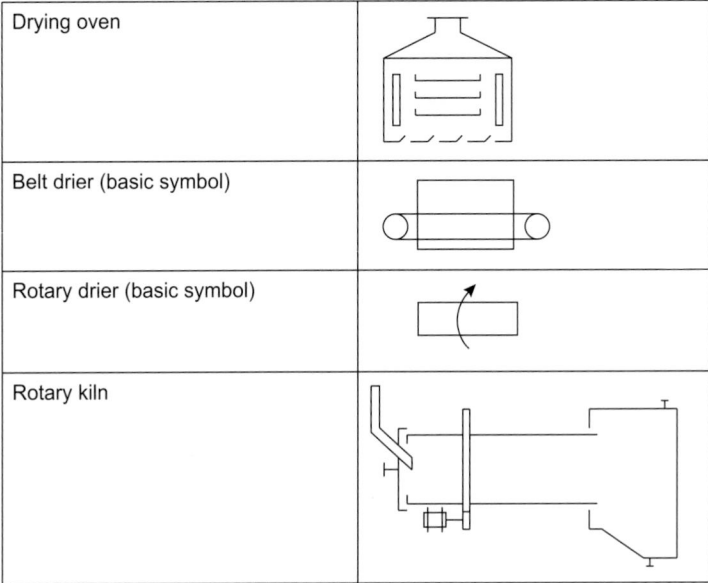

Drying oven	
Belt drier (basic symbol)	
Rotary drier (basic symbol)	
Rotary kiln	

Spray drier	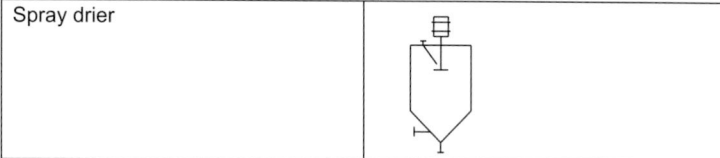

Materials Handling

Belt conveyor	
Screw conveyor	
Elevator (basic symbol)	

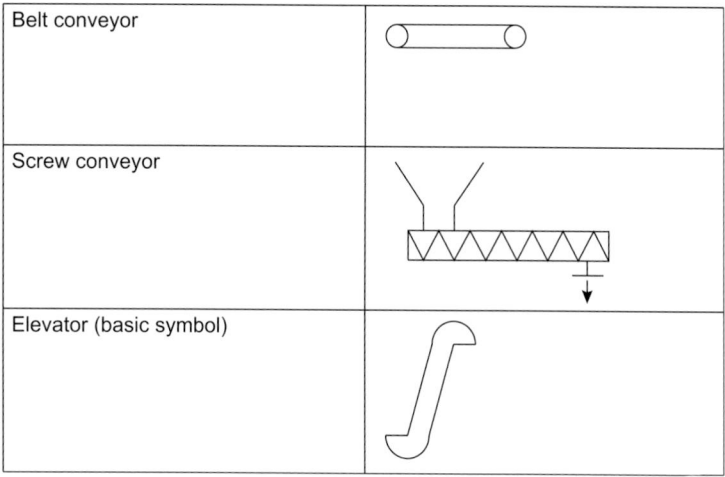

Prime Movers

Electric motor (basic symbol)	
Turbine (basic symbol)	

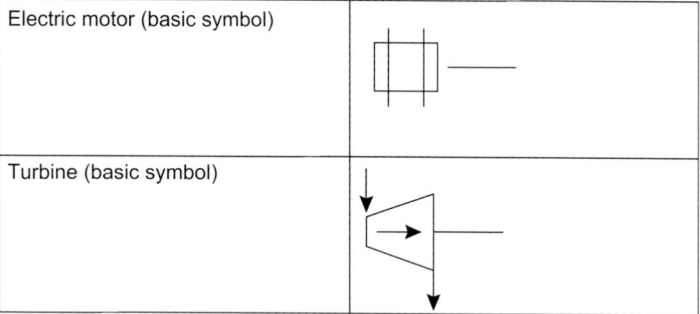

B CORROSION CHART

An R indicates that the material is resistant to the named chemical up to the temperature shown, subject to the limitations given in the notes. The notes are given at the end of the table.

A 'blank' indicates that the material is unsuitable. ND indicates that no data was available for the particular combination of material and chemical.

This chart is reproduced with the permission of IPC Industrial Press Ltd.

> NOTE
> This appendix should be used as a guide only — before a material is used its suitability should be cross-checked with the manufacturer.

METALS

Centigrade	Aluminium (a) 20°	60°	100°	Aluminium Bronze 20°	60°	100°	Brass (b) 20°	60°	100°	Cast Iron (c) 20°	60°	100°	Copper 20°	60°	100°	Gunmetal and Bronze (d) 20°	60°	100°	High Si Iron (14% Si) (c) 20°	60°	100°	Lead 20°	60°	100°	Mild Steel BSS 15 20°	60°	100°	Nickel (cast) 20°	60°	100°
Acetaldehyde	R	R	R	R	R	R	R	R	R	R	ND	ND	R	R	R	R	R	R	R	R	R	R	ND			No data		R	R	R
Acetic acid (10%)	R	R		R	R	R							R	R	R	R	R	R	R	R	R	R	ND					R[20]	R	R
Acetic acid (glac. & anh.)	R[1]	R	R										R	R	R	R	R	R	R	R	R	R	ND					R	R	R
Acetic anhydride	R[1]	R	R	R	R	R				R	R	R	R	R	R	R	R		R	R	R							R	R	R
Acetone	R	R	R	R	R	R	R	R	R	R	R	R	R	R	R	R	R	R	R	R	R	R	R		R[11]			R	R	R
Other ketones	R	R	R	R	R	R	R	R	R	R	R	R	R	R	R	R	R	R	R	R	R	R	R			No data		R	R	R
Acetylene	R	R					R	R	R[82]	R	R	R										R	R	R	R	R	R	R	R	R
Acid fumes	R[2]	R	R	R[2]	R[2]	R[2]																R[2]	R							
Alcohols (most fatty)	R[1]	R	R	R	R	R	R	R	R	R[24]	R	R	R	R	R	R	R	R	R	R	R	R	R		R	R	R	R	R	R
Aliphatic esters	R	R	R	R	R	R	R	R	R	R	R	R	R	R	R	R	R	R	R	R	R	R	R			No data		R	R	R
Alkyl chlorides		No data			No data					R[11]	R	R	R	R	R	R	R	R	R	R	R	R[11]						R	R	R
Alum	R	R	R	R	R	R							R	R	R	R	R	R	R	R	R	R	R[18]	R[10]				R		
Aluminium chloride	R[11]	ND	ND	R[20]	R[20]								R	R	R	R	R	R	R	R		R[4,10]								
Ammonia, anhydrous	R	R	R	R	R	R				R			R	R	R[83]	R	R	R	R	R	R	R	R	R[62]	R	R	R			
Ammonia, aqueous	R	R	R							R	R								R	R	R	R	R		R	R				
Ammonium chloride	R[84]	R	R							R									R	R	R	R	R					R	R	R
Amyl acetate	R	R	R	R	R	R				R[11]	R	R	R	R	R	R	R	R	R	R	R	R[4]	ND	ND		No data		R	R	R
Aniline	R	R	R							R	R	R							R	R	R	R	R			No data		R	R	R
Antimony trichloride					No data					R[11]	R	R		No data		R			R[11]	R	R	R[4]	R					R[11]	R	R
Aqua regia																														
Aromatic solvents	R	R	R	R	R	R	R	R	R	R	R	R	R	R	R	R	R	R	R	R	R	R	R		R[11]					
Beer	R	R	R	R	R	R	R	R	R	R	R	ND	R	R	R	R	R	R	R	R	ND							R	R	R
Benzoic acid	R	R	R	R	R	R	R	R	R				R	R	R	R	R	R	R	R	R	R[4]						R	R	R
Boric acid	R	R		R	R	R	R	R	R				R	R	R	R	R	R	R	R	R	R	R[62]					R	R	R
Brines, saturated	R	R	R	R	R	R				R[84]			R	R	R[20]	R	R		R	R	R	R	R					R	R	R
Bromine	R[11]	R	R	R[20]						R[11]	R								R			R[24]								
Calcium chloride	R	R	R	R						R	R		R	R	R	R	R	R	R	R	R	R[4]						R[20]	R	R
Carbon disulphide	R	R	R	R			R	R	R	R	R	R	R	R	R	R	R	R	R	R	R	R	R	R				R		
Carbonic acid	R	R	R	R	R	R													R	R	R	ND						R		
Carbon tetrachloride	R	R	R	R	R	R	R	R	R	R[11]	R	R	R	R	R	R	R	R	R	R	R	R	R[11]		R[11]	R		R	R	R
Caustic soda & potash				R						R	R		R	R	R	R	R	R							R	R		R	R	R
Chlorates of Na, K, Ba	R[11]	R	R	R	R	R							R	R	R	R	R	R	R	R	R	R	R					R	R	R
Chlorine, dry	R	R	R	R	R	R	R	R	R	R	R	R	R	R	R	R	R	R	R	R	R	R	R	R[4]	R	R	R	R	R	R
Chlorine, wet																			R	R	R	R	R							
Chlorides of Na, K, Mg	R	R	R	R	R	R							R	R	R[20]	R	R	R	R	R	R	R[4]	R[4,22]			No data		R[11]	R	R
Chloroacetic acids					No data									No data			No data											R[11]	R	R
Chlorobenzene	R	ND	ND	R	R	R		No data		R	R	R		No data					R	R	R	R	R		R[11]	R		R	R	R
Chloroform	R[1]	R	R	R	R	R	R	R	R	R	R		R	R	R	R	R	R		No data		R	R		R[11]	R		R	R	R
Chlorosulphonic acid				R[20]	R[20]	R[20]		No data		R[11]	R	R							R	R		R[4]			R					
Chromic acid (80%)																			R	R		R	R							
Citric acid	R	R	R	R	R	R							R	R	R	R	R	R	R	R	R	R	R[25]					R	R	R
Copper salts (most)				R	R	R										R	R	R	R[16]	R	R	R[16]	R					R		
Cresylic acids (50%)	R	R	R	R	R	R							R	R	R	R	R	R	R	R	R	R	R					R	R	R
Cyclohexane	R	R	R	R	R	R	R	R	R	R	R	R	R	R	R	R	R	R	R	R	R	R	R			No data		R	R	R
Detergents, synthetic	R	R	R		No data		R	R	R		No data		R	R	R	R	R	R		No data		R	R			No data		R	R	R
Emulsifiers (all conc.)	R	R	R	R	R	R		No data			No data		R	R			No data			No data		R	R			No data			No data	
Ether	R[1]	R	R	R	R	R	R	R	R	R	R	R	R	R	R	R	R	R	R	R	R	R	R		R	R	R	R	R	R
Fatty acids (> C₆)	R	R	R	R	R	R							R	R	R	R	R	R	R	R	R	R[4]	R	R[58]				R	R	R
Ferric chloride																			R			R[4]								
Ferrous sulphate				R[20]	R[20]	R[20]													R	R	R	R	R	R						
Fluorinated refrigerants, aerosols, e.g. *Freon*	R[11]	ND	ND	R	R	R	R	R	R	R	R	R	R	R	R	R	R	R	R	R	R	R	R		R[11]	ND	ND		No data	
Fluorine, dry	R	R	R	R	R	R[11]							R	R	R	R	R	R				R[4]	R		R	R	R	R	R	R
Fluorine, wet								No data			No data											R	R	ND				R	R	R
Fluosilicic acid																						R	R	R[58]				R[20]	R	R
Formaldehyde (40%)	R			R	R	R	R			R			R	R	R	R	R	R	R	R	R		No data		R			R	R	R
Formic acid	R			R	R	R		No data					R	R	R	R	R	R	R	R	R	R[30]	R[36]					R	R	R

METALS

Nickel-Copper Alloys (e)			Ni Resist (High Ni Iron) (c)			Platinum			Silver			Stainless Steel 18/8 (f)			Molybdenum Stainless Steel 18/8 (f)			Austenitic Ferricr Stainless Steel (x)			Tantalum			Tin (g)			Titanium			Zirconium		
20°	60°	100°	20°	60°	100°	20°	60°	100°	20°	60°	100°	20°	60°	100°	20°	60°	100°	20°	60°	100°	20°	60°	100°	20°	60°	100°	20°	60°	100°	20°	60°	100°
R	R	R	R	ND		R	R	R	R	R	R	R	R	R	R	R	R	R	R	R	R	R	R	R	R		R	R	R	R	R	R
R	R	R	R			R	R	R	R	R	R	R	R	R	R	R	R	R	R	R	R	R	R	R			R	R	R	R	R	R
R	R					R	R	R	R	R	R	R	R		R	R	R[84]	R	R		R	R	R				R	R	R	R	R	R
R	R		R			R	R	R	R	R	R	R[80]			R	R		R	R	R	R	R	R				R	R	R	R	R	R
R	R	R	R	R	R	R	R	R	R	R	R	R	R	R	R	R	R	R	R	R	R	R	R	R	R	R	R	R	R	R	R	R
R	R	R	R	R	R	R	R	R	R	R	R	R	R	R	R	R	R	R	R	R	R	R	R				R	R	R	R	R	R
			R	R	R	R	R	R	R	R	R				R	R	R	R	R	R	R	R	R	R	R		R	ND	ND	R	R	R
R[2]	R	R				R	R	R				R[2]	R	R	R[2]	R	R	R[102]	R[102]	R[102]	R[5]	R	R	R[44]	R	R	R[2]	R[2]	R[2]	R[2]	R	R
R	R	R	R	R	R	R	R	R	R[16]	R	R	R	R	R	R	R	R	R	R	R	R	R	R	R	R	R	R[93]	R[93]	R[93]	R	R	R
R	R	R	R	R	R	R	R	R	R	R	R	R[11]	R	R	R[11]	R	R	R	R	R	R	R	R	R			R			R	R	R
R	R		R	R		R	R	R	R	R	R	R	R[13]					R	R		R	R	R	R			R	R	R	R	R	R
R						R	R	R	R	R	R	R[84]			R[84]	R					R	R	R	R[57]			R	R	R[10]	R	R	R
R	R	R	R	R	R	R	R	R	R	R	R	R	R	R	R	R	R	R	R	R	R	R	R	R	R	R	R	R	R	R	R	R
			R	R	R	R	R	R	R[30]	R	R	R	R	R	R	R	R	R	R	R	R	R	R	R[13]			R	R	R	R	R	R
R			R	R	R	R	R	R	R[73]	R	R	R[84]			R[84]	R		R			R	R	R				R	R	R	R	R	R
R	R	R	R	R	R	R	R	R	R	R	R	R	R	R	R	R	R	R	R	R	R	R	R	R			R	R	R	R	R	R
R			R	R	R	R	R	R	R	R	R	R	R	R	R	R	R	R	R	R	R	R	R	R			R	R	ND	R	R	R
R	ND	ND				R	R	R	R	R	R	R[11]			R[11]	R[11]		R	R		R	R	R				R	R	ND	No data		
																					R	R	R				R					
R	R	R	R	R	R	R	R	R	R	R	R	R	R	R	R	R	R	R	R	R	R	R	R	R	R	R	R	R	ND	R	R	R
R	R	R	R	R	ND	R	R	R	R	R	R	R	R	R	R	R	R	R	R	R	R	R	R	R	R		R	R	R	R	R	R
R	R					R	R	R	R	R	R	R	R	R	R	R	R	No data			R	R	R	R			R	R	R	R	R	R
R	R	R	R	R	R	R	R	R	R	R	R	R[42]			R[42]			R	R	R	R	R	R	R			R	R	R	R	R	R
R	R	R							R	R	R										R	R	R	R[90]						R[90]	R	R
R	R	R	R	R	R	R	R	R	R	R	R				R[42]			R	R	R	R	R	R	R	R		R	R	R	R	R	R
R	R		R	R	R	R	R	R	R	R	R	R	R	ND	R	R	ND	R	R	R	R	R	R	R	R		R	R	R	R	R	R
R	R	R	R	R	R	R	R	R	R	R	R	R	R	R	R	R	R	R	R	R	R	R	R	R	R	R	R	R	R	R	R	R
R	R	R	R	R	R	R	R	R	R	R	R	R[11]	R	R	R[11]	R	R	R	R	R	R	R	R	R[11]	R	R	R	R	R	R	R	R
R	R	R	R	R	R	R	R	R	R	R	R	R	R	R[13]	R	R	R[13]	R[103]	R[103]		R[10]	R	R				R	R[19]	R[15]	R	R	R
R			R	R	R	R	R	R	R	R	R	R[16]	R	R	R[16]	R	R	R	R	R	R	R	R[25]	R	R		R[79]	R[79]	R[79]	R[25]	R[25]	R[25]
R	R	R	R	R	R	R	R	R	R	R	R	R	R	R	R	R	R	R	R	R	R	R	R				R[91]	R	R			
						R	R	R	R	R								R	R	R	R	R	R				R	R	R			
R	R	R	R	R	R	R	R	R	R[70]	R	R	R[84]			R[84]	R		R[56]	R[56]		R	R	R	R[57]	R	R	R	R	R	R	R	R
R			ND	ND	ND	R	R	R	R	R	R							R	R	R	R	R	R	R[2]	R[2]		R	R[2]	R[2]			
R	R		R	R	R	R	R	R	R	R	R	R[11]	R	ND	R[11]	R	R	R	R	R	R	R	R	No data			No data			R	R	R
R	R	R	R	R	R	R	R	R	R	R	R	R[11]	R	R	R[11]	R	R	R	R	R	R	R	R	R[11]	R	R	R	R	ND	R	R	R
R	R					R	R	R	R	R	R	R[84]			No data			R	R	R	R	R	R							R	R	
						R[30]	R	R	R[30]	R	R							R	R	R	R	R	R	R	R	R	R	R	R	R	R	R[19]
R	R					R	R	R	R	R	R	R[13]	R	R	R	R	R[13]	R	R	R	R	R	R	R[20]	R	R	R	R	R[19][27]	R	R	R
R						R[30]	R	R	R[16]	R	R	R[16]	R	R	R[16]	R[16]	R[16]	R	R	R	R	R	R				R	R	R	R[16]	R	R
R			R	ND	ND	R	R	R	R	R	R	R	R	R	R	R	R	R	R	R	R	R	R	R			R	ND	ND	R	R	R
R	R	R	R	R	R	R	R	R	R	R	R	R	R	R	R	R	R	R	R	R	R	R	R	R			R	R	ND	R	R	R
R	R	R	No data			R	R	R	R	R	R	R	R	R	R	R	R	R	R	R	R	R	R	R			R	ND	ND	R	R	R
No data			No data			R	R	R	R	R	R	R	R	R	R	R	R	R	R	R	R	R	R	R			No data			R	R	R
R	R	R	R	R	R	R	R	R	R	R	R	R	R	R	R	R	R	R	R	R	R	R	R				R	R	ND	R	R	R
R	R	R	R	R	R	R	R	R	R	R	R	R	R	R	R	R	R	R	R	R	R	R	R				R	R	R	R	R	R
						R	R											R	R	R	R	R	R				R	R	R			
R	R		R			R	R	R				R	R	R	R	R	R	R	R	R							R	R	R	R	R	R
R	R	R	R	R	R	R	R	R	R	R	R	R[11]	R	R	R[11]	R	R	R	R	R	R	R	R	R	R	R	R	R	R	R	R	R
R	R	R	No data			R	R	R				R	ND	ND	R	ND	ND	R	R								R[5]	R	R			
R	R	R	No data			R	R	R							R	ND	ND															
R	R	R	R[32]	R[32]	ND	R	R	R	R	R	R																					
R	R	R	R	R	R	R	R	R	R	R	R	R	R	R	R	R	R	R	R	R	R	R	R	R	R		R	R	R	R	R	R
R	R					R	R	R	R	R	R	R			R	R		R	R		R	R	R				R[67][69]	R	R[10][20]	R	R	R

METALS

Each metal column gives readings at 20°, 60° and 100° Centigrade (R = resistant, ND = not determined).

Centigrade	Aluminium (a) 20° 60° 100°	Aluminium Bronze 20° 60° 100°	Brass (b) 20° 60° 100°	Cast Iron (c) 20° 60° 100°	Copper 20° 60° 100°	Gunmetal 20° 60° 100°	High Si Iron 20° 60° 100°	Lead (14% Si) (c) 20° 60° 100°	Mild Steel BSS 15 20° 60° 100°	Nickel (cast) 20° 60° 100°
Fruit juices	R R R	R R R			R R R	R R R	R R R		No data	R R R
Gelatine	R R R	R R R	R R R	R R R	R R R	R R R	R R R	R R	No data	R R R
Glycerine	R R R	R R R	R R R	R R R	R R R	R R R	R R R	R R	R R	R R R
Glycols	R R R	R R R	R R R	R R R	R R R	R R R	R R R	R R		R R R
Hexamine							R R R	R R	No data	R R R
Hydrazine	R ND ND			No data			No data	ND	R R R	R ND ND
Hydrobromic acid (50%)				ND ND						
Hydrochloric acid (10%)		R					R R			R
Hydrochloric acid (conc.)		R^{62}						$R^{4,11}$		
Hydrocyanic acid	R R R	R^{20} R^{20} R^{20}					R R R			R R R
Hydrofluoric acid (40%)		R^{62}						R		R^{20}
Hydrofluoric acid (75%)		R^{62}								R
Hydrogen peroxide (30%)	R R R						R			R
(30–90%)	R R R									
Hydrogen sulphide	R R R	R^{11} R R	R^{11} R R	R	R^{11} R R	R^{11} R R	R R R	R R	R^{11} R R	R^{11} R R
Hypochlorites		R					R R R	$R^{4,34,76}$		
Lactic acid (100%)	R R R	R R			R R	R^4 R^4	R R R	ND		R R R
Lead acetate	R^{11} R R	No data		No data						R R R
Lime (CaO)	R^{11}	R R R	R R R	R R R	R R R	R R R	R R R	R^4	R^{11} R R	R R R
Maleic acid	R R R	No data		No data	R R R	No data	R R ND		No data	R R R
Meat juices	R R R	R R R		No data			No data	No data	No data	No data
Mercuric chloride							R			
Mercury				R R		No data	R R R		R R R	R^{27} R R
Milk & its products	R R R	R R R		No data	R R R	R R R	R R R			R R R
Moist air	R R R	R R R		R	R R R	R R R	R R R	R R R		R R R
Molasses	R R R	R^{30} R^{30} R^{30}	R^{30} R	R R R	R^{30} R R	R^{30} R R	R R R			R R R
Naphtha	R R R	R R R	R R R	R R R	R R R	R R R	No data	R R	R	R
Naphthalene	R R R	No data		No data	R R R	No data	R R R	R R	R	R R R
Nickel salts		No data		No data		R R R	R R R	R R		R^{40} R R
Nitrates of Na, K, NH₃	R R R	R^{73} R^{73} R^{73}		R^{11} R R			R R R			R R R
Nitric acid (<25%)							R			
Nitric acid (50%)							R R R			
Nitric acid (95%)	R R R						R R R		R	
Nitric acid, fuming	R ND ND						R R R			
Oils, essential	R R R	R R R	R R R	R R R	R R R	R R R	R R R	No data	R R	R R R
Oils, mineral	R R R	R R R	R R R	R R R	R R R	R R R	R R R	R R	R R	R R R
Oils, vegetable & animal	R R R	R R R	R R R	R R R	R R R	R R R	R R R	R R	R R	R R R
Oxalic acid	R^{50}	R R R	No data		R R R	R R R	R R R	R^4		No data
Ozone	R R R	No data		No data	R ND ND	No data	No data	R R R	R R	R^{11} R R	No data
Paraffin wax	R R R	R R R	R R R	R R R	R R R	R R R	R R R	R R ND		R R R
Perchloric acid		No data					R R R			
Phenol	R R R	R R R	R R R	R R R	R R R	R R R	R R R	R^4 R R^{19}	No data	R R R
Phosphoric acid (25%)	R	R R R			R R R		R R R	R R R		
Phosphoric acid (50%)		R R R					R R R	R R R^4		
Phosphoric acid (95%)		R R R					R R R	R R R		
Phosphorus chlorides		R^{11} R^{11} R^{11}		R^{11} R			R R R^{11}		R^{11} R R	R R R
Phosphorus pentoxide	R^{11} ND ND	No data					R R R		R^{11} R R	No data
Phthalic acid	R R R	R R R		No data		R R R	R R R	R R	No data	R R R
Picric acid	R ND ND						R R R	R^4		R^{11}
Pyridine	R R R	No data		R R R			R R R	R R	No data	R R R
Sea water	R R R	R R R	R^{62} R R	R^{84}	R R R	R R R	R R R	R R		R ND ND
Silicic acid	R R R			No data	R R R	No data	R R ND		No data	R R ND
Silicone fluids	R R R	R R R	R R R	R R R	R R R	R R R	No data		No data	R R R
Silver nitrate				ND ND		R R	R			
Sodium carbonate	R^{42} R	R R R^4	R R R	R^{11} R R	R R R	R R R	R R R	R^4	R R R	R R R
Sodium peroxide				R^{10} R R			R R R	R^{10} R^{10} R^{10}		R R R

METALS

Nickel-Copper Alloys (e)			Ni Resist (High Ni Iron) (c)			Platinum			Silver			Stainless Steel 18/8 (f)			Molybdenum Stainless Steel 18/8 (f)			Austenitic Ferric Stainless Steel (x)			Tantalum			Tin (g)			Titanium			Zirconium		
20°	60°	100°	20°	60°	100°	20°	60°	100°	20°	60°	100°	20°	60°	100°	20°	60°	100°	20°	60°	100°	20°	60°	100°	20°	60°	100°	20°	60°	100°	20°	60°	100°
R			R	R		R	R	R	R	R	R	R[94]	R	R	R[16]	R	R	R	R	R	R	R	R	R	R	R	R[8]	ND	ND	R	R	R
R			R	R	R	R	R	R	R	R	R	R[16]	R	R	R[16]	R	R	R	R	R	R	R	R	R	R	R	R[9]	R	ND	R	R	R
R	R	R	R	R	R	R	R	R	R	R	R	R	R	R	R	R	R	R	R	R	R	R	R	R	R	R	R	R	ND	R	R	R
R	R	R	R	R	R	R	R	R	R	R	R	R	R	R	R	R	R	R	R	R	R	R	R	R	R	R	R	R	ND	R	R	R
R			R	ND	ND	R	R	R	R	R	R	R	R	R	R	R	R	R	R	R	R	R	R	R	R	R	No data			R	R	R
			No data			R[19]	R	R				R	R	R	R	R	R	R	R	R	No data			R[13]			No data			R	ND	ND
R			R			R	R	R	R[45]												R	R	R				R[30]	R	R			
R	R		R			R[70]	R	R	R	R	R							R	R	R				R	R[49]₇₈	R	R	R	R			
R						R[70]	R	R										R	R	R				R[78]	R[78]	ND	R[92]	R[92]	R[92]			
R			R	R	R	R[98]	R	R	R	R		R	R					R	R		R	R					No data			No data		
R	R	R				R	R	R	R	R	R																					
R	R	R				R	R	R	R	R	R																					
R	R					R[87]	R	R				R	R	R	R	R	R	R	R	R	R	R	R	R	R				R	R	R
R[87]						R[87]	R	R				R	R	R[87]	R[63]			R	R		R	R	R	R	R				R	R	R
R	R		R	R		R	R	R				R	R	R	R	R	R	R[11]	R[11]	R[11]	R	R	R	R[11]	R		R	R	R	R	R	R
			R[7]			R	R	R	R[48]									R	R	R							R	R	R	R	R	R
R	R		R			R	R	R	R	R	R	R			R	R		R	R		R	R	R				R	R	R	R	R	R
R	R		No data			R	R	R	R	R	R	R	R	R	R	R	R	R	R	R	No data			R			R	ND	ND	No data		
R	R	R	R	R	R	R	R	R	R	R	R	R	R	R	R	R	R	R	R	R	No data			R			No data			R	R	R
R	R		R	R	R	R	R	R	R	R	R	R[13]	R[13]	R[13]	R[13]	R	R	R	R	R	R	R	R	R[20]	R	R	R	R	R	R	R	R
No data			R	R	R	R	R	R	R	R	R	R	R		R	R		R	R	R	R	R	R	R	R	R	No data			R	R	R
			ND	ND		R	R	R										R	R	R	R	R	R				R	R	R[7]	R	R	R
R	R	R	R	R	R	R	R	R	R	R	R	R	R	R	R	R	R	R	R	R	R	R					R	R	ND	R	R	R
R			R	R	R	R	R	R	R[86]	R	R	R	R	R	R	R	R	R	R	R	R	R	R	R	R	R	No data			R	R	R
R	R	R	R	R	R	R	R	R	R	R	R	R	R	R	R	R	R	R	R	R	R	R	R	R	R	R	R	R	R	R	R	R
R	R	R	R	R	R	R	R	R	R	R	R	R	R	R	R	R	R	R	R	R	R	R	R	R	R	R	No data			R	R	R
R	R	R	R	R	R	R	R	R	R	R	R	R	R	R	R	R	R	R	R	R	R	R	R	R	R	R	No data			R	R	R
R	R	R	R	R	R	R	R	R	R	R	R	R	R	R	R	R	R	R	R	R	R	R	R	R	R	R	R	ND		R	R	R
R	R		R	ND	ND	R	R	R	R	R	R	R[16]	R	R	R[16]	R	R	R	R	R	R	R	R	R	R		R	R	R[31]₃₂	R	R	R
R	R		R	R	R	R	R	R	R[30]	R	R	R	R	R	R	R	R	R	R	R	R	R	R	R	R	R	R	R	R	R	R	R
						R	R	R				R	R	R	R	R	R	R	R	R	R	R	R				R	R	R	R	R	R
						R	R	R	R	R	R	R	R	R	R	R	R	R	R	R	R	R	R				R	R	R	R	R	R
												R			R			R			R	R	R				R	R	R	R	R	R
												R	R[80]					R			R	R	R	R	ND					R	R	R
R	R	R	R	R	R	R	R	R	R	R	R	R	R	R	R	R	R	R	R	R	R	R	R	R	R	R	R	R	ND	R	R	R
R	R	R	R	R	R	R	R	R	R	R	R	R	R	R	R	R	R	R	R	R	R	R	R	R	R	R	R	R	ND	R	R	R
R	R	R	R	R	R	R	R	R	R	R	R	R	R	R	R	R	R	R	R	R	R	R	R	R	R	R	R	R	ND	R	R	R
R	R		R	R	R	R	R	R	R	R	R	R[13]			R[10]	R		R	R	R	R	R	R	R[20]	R	R	R[23]			R	R	R
			R	R	R	R	R	R	R[11]	R	R	R	R	R	R	R	R	R	R	R	R[99]	R	R	R	R	ND	No data			No data		
			R	R	R	R	R	R	R	R	R	R	R	R	R	R	R	R	R	R	R	R	R	R	R		R	R	ND	R[32]	R	R
R			R	R	R	R	R	R	R	R	R	R	R	R	R	R	R	R	R	R	R	R	R	R	R		R	R	ND	R	R	R
R	R					R	R	R	R	R	R	R	R	R	R	R	R	R	R	R	R	R				R	R	R[49]₇₈	R	R	R
R	R					R	R	R	R	R	R	R	R		R	R		R	R	R	R	R				R[49]₇₈	R	R	R	R	R
R						R	R	R	R	R	R	R	R		R	R		R[104]	R[104]	R[104]	R[39]	R	R				R[49]₇₈			R		
R			R[11]	ND	ND	R	R	R	R	R	R										R	R	R				R[11]	R[11]	ND	R		
R			R	R	R	R	R	R	R	R	R	R	R	R[11]	R	R	R[11]	R	R	R	ND	ND	ND	R	R		No data			No data		
R	R	R	R	ND	ND	R	R	R	R	R	R	No data			No data			No data			R	R	R	R			R	ND	ND	R	ND	ND
R				ND	ND	R	R	R	R	R	R	R	R	R	R	R	R	R	R	R	R	R	R				No data			R	R	R
R			R	R	R	R	R	R	R	R	R	R	R	R	R	R	R	R	R	R	R	R	R		R	ND	R	R	ND	R	R	R
R[57]	R	R	R	R	R	R	R	R	R[70]	R	R	R[57]			R[57]			R	R	R	R	R	R	R	R	R	R	R	R	R	R	R
R	R		R	R	R	R	R	R	R	R	R	R	R	R	R	R	R	R	R	R	R	R	R	R	R	R	No data			R	R	R
R	R	R	No data			R	R	R	R	R	R	R	R	R	R	R	R	R	R	R	R	R	R	R	R	R	No data			R	R	R
			ND	ND	ND	R	R	R	R	R	R	R	R	R	R	R	R	R	R	R	R	R	R				R[19]	R[19]	ND	R	R	R
R	R	R	R	R	R	R	R	R	R	R	R	R	R	R	R	R	R	R	R	R	R	R	R	R			R	R	R	R	R	R
No data			R[10]	R[10]	R[10]	R	R	R	R	R	R	R[10]			R[10]	R	R	R	R	R	R	R[10]	R[10]				No data			R	R	R

METALS

Each metal column gives the readings at 20°, 60° and 100° Centigrade.

	Aluminium (a)	Aluminium Bronze	Brass (b)	Cast Iron (c)	Copper	Gunmetal and Bronze (d)	High Si Iron (14% Si) (c)	Lead	Mild Steel BSS 15	Nickel (cast)
	20° 60° 100°	20° 60° 100°	20° 60° 100°	20° 60° 100°	20° 60° 100°	20° 60° 100°	20° 60° 100°	20° 60° 100°	20° 60° 100°	20° 60° 100°
Sodium silicate	R R R	R R R	R R R	R R R	R R R	R R R	R R R	R R	R R R	R R R
Sodium sulphide				R R R			R ND	R[4] R R		R R R
Stannic chloride		R[11]					R R			
Starch	R R R	R R R	No data	R R R	R R R	R R R	R R R	No data	No data	R R R
Sugar, syrups, jams	R R R	R R R	R R R	R R ND	R R R	R R R	R R R		No data	R R R
Sulphamic acid	R[50]	No data					R R R		No data	No data
Sulphates (Na, K, Mg, Ca)	R R R	R R R	R R R	R R R	R R R	R R R	R R R	R R		R R R
Sulphites	R R R	R R R		R[38] R R	R R R	R R R	R[38] R	R R		R R R
Sulphonic acids	No data	No data	No data	R[11]		No data	R R R	R R	No data	R R R
Sulphur	R R R			R R			R R R			R R R
Sulphur dioxide, dry	R R R	R R R	R R R	R R	R R R	R R R		R R R	R R R	R R R
Sulphur dioxide, wet	R[4] R R	R R R						R R R	R R R	R R R
Sulphur trioxide		R[11] R R	R[11] R R		R[11] R R	R[11] R R		R R[4] R	R[11] R R	R R R
Sulphuric acid (< 50%)		R R R			R R R			R R R	R R R	
Sulphuric acid (70%)		R R[62]		R				R R R	R R R	R
Sulphuric acid (95%)		R[62]		R R				R R R	R	
Sulphuric acid, fuming	R[4]			R R R				R R R		R R
Sulphur chlorides				R[11] R[11]			No data	R[4]		No data
Tallow	R R R	R R R	No data	R R R	R R R	No data	R R R	R R	No data	No data
Tannic acid (10%)	R R R	R R R	R R R		R R R	R R R	R R R			R ND ND
Tartaric acid	R R R	R R R	R R R		R R R	R R R	R R R	R[4] R R		R[20] R R
Trichlorethylene	R R R	R R R	R R		R R R	R R R	R R ND	R R	R[11] R	R R R
Vinegar	R R R	R R R					R R R			R R R
Water, distilled	R R R	R[53] R		R R R	R[53] R	R[53] R R	R R R	R[53] R R	R[53] R R	R R R
Water, soft	R[43] R R	R R R	R R R	R	R R R	R R R	R R R			R R R
Water, hard	R[43] R R	R R R	R R R	R R R	R R R	R R R	R R R	R R R	R R R	R R R
Yeast	R R R	No data	No data	R R	R R R	R R R	R R R	No data	No data	R R R
Zinc chloride		R R R						R	R[4] R	R[20] R R

METALS

Nickel-Copper Alloys (e)			Ni Resist (High Ni Iron) (c)			Platinum			Silver			Stainless Steel 18/8 (f)			Molybden Stainless Steel 18/8			Austenitic Stainless Ferric Steel (x)			Tantalum			Tin (g)			Titanium			Zirconium		
20°	60°	100°	20°	60°	100°	20°	60°	100°	20°	60°	100°	20°	60°	100°	20°	60°	100°	20°	60°	100°	20°	60°	100°	20°	60°	100°	20°	60°	100°	20°	60°	100°
R	R	R	R	R	R	R	R	R	R	R	R	R	R	R	R	R	R	R	R	R	R	R	R	R	R	R	R	R	ND	R	R	R
	No data		R	R	R	R	R	R				R	R		R	R		R	R		R	R					R	ND	R[10]	R	R	R
						R	R	R	R[48]	R	R							No data			R	R	R				R[15]	R[15]	R[15]	R[15]	R[15]	R[15]
R	R	R	R	R	R	R	R	R	R	R	R	R	R	R	R	R	R	R	R	R	R	R	R	R	R	R	R	R	ND	R	R	R
R	R	R	R	R	R	R	R	R	R	R	R	R[94]	R	R	R	R	R	R	R	R	R	R	R	R	R	R		No data		R	R	R
R			R[20]			R	R	R	R	R	R	R[44]			R	R[37]		R	R		R	R	R								No data	
R	R	R	R[38]	R	R	R	R	R	R	R	R	R	R	R	R	R	R	R	R	R	R	R	R	R	R		R	ND	R[7,34]	R	R	R
R	R			No data		R	R	R		No data			No data			No data			No data		R	R	R					No data		R	R	R
R	R	R	R	R	R	R	R	R	R[11]	R	R	R	R	R	R	R	R	R	R	R	R	R	R	R	R	R	R	R		R	R	R
R	R		R	R	R	R	R	R	R	R	R	R	R	R	R	R		R	R		R	R	R	R	R	R	R	R	ND	R	R	R
		ND				R	R	R	R	R	R	R			R	R		R	R		R	R					R	R	R	R	R	R
			R	ND	ND	R	R	R	R[11]	R	R				R[11]	R	R	R[11]	R[11]	R[11]	R	R	R							R	R	R
R			R[20]			R	R	R	R	R	R				R[10]			R			R	R	R				R[49,78]	R		R	R	R
R			R			R	R	R	R	R	R							R			R	R	R				R[49,78]	R	R	R	R	R
R						R	R	R				R			R			R	R		R	R	R									
						R	R	R				R	R[80]		R	R[80]		R	R		R	R	R									
R	R		R	R	R	R	R	R											No data		R	R	R					No data				
R	R	R	R	R	R	R	R	R	R	R	R	R	R	R	R	R	R	R	R	R	R	R	R	R	R		R	R	R	R	R	R
R	R	R				R	R	R	R	R	R	R	R	R	R	R	R	R	R	R	R	R	R	R	R		R	R	R	R	R	R
R	R	R	R	R	R	R	R	R	R	R[70]	R	R	R	R	R	R	R	R	R	R	R	R	R	R[20]	R	R	R	R	R[19]	R	R	R
R	R	R	R	R	R	R	R	R	R	R	R	R[11]	R	R	R[11]	R	R	R	R	R	R	R	R	R[11]	R	R	R	R	R	R	R	R
R	R	R				R	R	R	R	R	R	R	R	R	R	R	R	R	R	R	R	R	R	R			R	R	R	R	R	R
R	R	R	R	R	R	R	R	R	R	R	R	R	R	R	R	R	R	R	R	R	R	R	R	R	R	R	R	R	R	R	R	R
R	R	R	R	R	R	R	R	R	R	R	R	R	R	R	R[84]	R	R	R	R	R	R	R	R	R	R	R	R	R	R	R	R
R	R	R	R	R	R	R	R	R	R	R	R	R	R	R	R[84]	R	R	R	R	R	R	R	R	R[57]	R	R	R	R	R	R	R	R
	No data		R	R	R	R	R	R	R	R	R	R	R	R	R	R	R	R	R	R	R	R	R	R	R			No data		R	R	R
R	R	R	R	R	R	R	R	R	R	R	R	R	R	R	R	R	R	R	R	R	R	R	R				R	R	R[52]	R	R	R

THERMOPLASTIC RESINS

Centigrade	Acrylic Sheet (e.g. Perspex) 20°	60°	100°	Acrylonitrile Butadiene Styrene Resins (l) 20°	60°	100°	Nylon 66 Fibre (m) 20°	60°	100°	Nylon 66 Plastics (m) 20°	60°	100°	PCTFE 20°	60°	100°	PTFE 20°	60°	100°	PVDF (y) 20°	60°	100°	Rigid Unplasticised PVC 20°	60°	100°	Plasticised PVC 20°	60°	100°
Acetaldehyde							R	ND	ND	R	R^{50}	ND	R	R	ND	R	R	R	R	R	R	R^6					
Acetic acid (10%)	R	R^{50}		R						R^{50}			R	R	R	R	R	R	R	R	R	R	R	R	R		
Acetic acid (glac. & anh.)													R	R	R^{50}	R	R	R	R	R		R^{50}					
Acetic anhydride	R^{50}						R	R	R		No data		R	R		R	R	R	R	ND	ND						
Acetone							R	R	R	R	R		R	R^{37}		R	R	R	R^{106}	ND	ND						
Other ketones							R	R	R	R	ND	ND	R	R^{37}		R	R	R									
Acetylene		No data			No data			No data			No data			No data		R	R	R	R	ND	ND	R	R			No data	
Acid fumes	R	R^{68}											R	R	R	R	R	R	R	R	R	R	R			No data	
Alcohols (most fatty)							R	R	R	R	R^{50}	R^{50}	R	R	R	R	R	R	R	R	R	R	R	R		No data	
Aliphatic esters							R	R	R	R	ND								R^{50}	R	R	R	R	R		No data	
Alkyl chlorides		No data					R	R	R	R^{46}	ND	ND	R	ND	ND		No data		R	R	R		No data			No data	
Alum	R	R		R	R		R	R	R	R	R	R	R	R	R	R	R	R	R	R	R	R	R	R	R	R	
Aluminium chloride	R	R^{68}		R	R		R^{43}	R	R	R	ND	ND	R	R	R	R	R	R	R^{50}	R	R	R	R	R	R	R	
Ammonia, anhydrous		R						No data		R	ND	ND	R	R	R	R	R	R	R^{107}	R^{107}	R^{107}	R	R				
Ammonia, aqueous	R	R^4		R			R	R	ND	R	ND	ND	R	R	R	R	R	R	R^{107}	R^{107}	R^{107}	R	R			No data	
Ammonium chloride	R	R		R	R		R	R	R	R	ND	ND	R	R	R	R	R	R	R^{50}	R	R	R	R				
Amyl acetate							R	R	R	R	ND	ND	R	R		R	R	R	R	R	R						
Aniline										R^{50}			R	R	ND	R	R	R	R	R	R						
Antimony trichloride	R^{68}	R		R	R					R^{50}	ND	ND		No data			No data		R	R	R	R	R		R	R	
Aqua regia													R	R	R	R	R	R	R	R	R	R	R^{13}			No data	
Aromatic solvents							R	R	R	R	R^{50}	R	R^{14}	R		R^{50}	R	R	R	R	R		No data				
Beer	R	R		R	R		R	R	R	R	R	R	R	R	R	R	R	R	R	R	R	R	ND		R		
Benzoic acid	R	ND		R	R			No data		R^{50}			R	R	ND	R	R	R	R	R	R	R	R^{80}		ND		
Boric acid	R	R^{68}		R	R		R^{43}	R	R	R	R	R	R	R	R	R	R	R	R	R	R	R	R	R	R		
Brines, saturated	R	R		R	R		R	R	R	R	R	R	R	R	R	R	R	R	R	R	R	R	R	R			
Bromine													R	R	R	R^{14}	R	R	R	R	R						
Calcium chloride	R	R		R	R		R	R^{43}	R	R^{50}	ND	ND	R	R	R	R	R	R	R	R	R	R	R	R	R	R	
Carbon disulphide							R	R	ND	R^{50}	ND	ND	R	R	ND	R	R	R	R	R	R	R	R	ND			
Carbonic acid	R	R		R	R			No data		R	R	ND	R	R	R	R	R	R	R	R	R	R	R		R	R	
Carbon tetrachloride							R	R	R	R	ND	ND	R			R^{14}	R	R	R	R	R	R^{14}					
Caustic soda & potash	R	R		R	R		R	R	R	R	R	R	R	R	R	R	R	R	R^{107}	R^{107}	R^{107}	R	R				
Chlorates of Na, K, Ba	R	R^{68}		R	R					R	R	ND	R	R	R	R	R	R	R	R	R	R	R			No data	
Chlorine, dry	ND			R	R								R	R	R	R	R	R	R	R	R	R				No data	
Chlorine, wet	R^4			R	R								R	R	R	R	R	R	R	R	R					No data	
Chlorides of Na, K, Mg	R	R		R	R		R	R	R	R	R	R	R	R	R	R	R	R	R	R	R	R	R		R	R	
Chloroacetic acids		No data											R	R	R	R^2	R^2	R^2	R	ND	ND	R				No data	
Chlorobenzene							R	R	R	R	ND	ND				R^{14}	R	R	R	R	R						
Chloroform							R	R	R				R			R^{14}	R	R	R	R	R						
Chlorosulphonic acid													R	R	ND	R	R	R					ND				
Chromic acid (80%)				R									R	R	R	R	R	R		No data		R^{19}	R^{19}				
Citric acid	R	R		R	R		R	R^{43}	R	R^{50}	ND	ND	R	R	R	R	R	R	R	R	R	R	R	R	R	R	
Copper salts (most)	R^{68}	R		R	R		R	R	R^{31}_{48}	R	R	R	R	R	R	R	R	R	R	R	R	R	R		R	R	
Cresylic acids (50%)													R	R	ND	R	R	R	R	R	R						
Cyclohexane							R	R	R	R	ND	ND	R	R	R	R	R	R	R	R	R					No data	
Detergents, synthetic	R	R		R			R	R	R	R	R	R	R	R	R	R	R	R	R	R	R	R	R		R		
Emulsifiers (all conc.)	R	R			No data		R	R	R	R	R	R	R	R	R	R	R	R	R	R	R	R	R		R	R	
Ether							R	R	R	R	ND	ND				R	R	R	R	R	R					No data	
Fatty acids (>C$_6$)	R	ND		R	R		R	R	ND	R	ND	ND	R	R	R	R	R	R	R	R	R	R	R			No data	
Ferric chloride	R	R		R	R		R	R^{43}		$R^{30,50}$			R	R	R	R	R	R	R^{50}	R	R	R	R	R	R	R	
Ferrous sulphate	R	R		R	R		R	R	R	R	R	R	R	R	R	R	R	R	R	R	R	R	R		R	R	
Fluorinated refrigerants, aerosols, e.g. *Freon*		No data						No data		R	ND	ND	R			R^{14}	R	R	R	R	R	R	R		R	R	
Fluorine, dry		No data											R	R		R^{48}	R	R	R	R	R	R	R				
Fluorine, wet		No data											R	R			No data		R	R	R	R	R				
Fluosilicic acid		No data					R^{43}	R						No data			No data		R	R	R	R^{15}	R			No data	
Formaldehyde (40%)	R	ND		R	R					R^{50}	R^{50}		R	R	ND	R	R	R	R	R	R	R	R^{30}		R		
Formic acid	R^{10}			R^{32}	R^{10}								R			R	R	R	R	R	R	R				No data	

THERMOPLASTIC RESINS															THERMOSETTING RESINS														
Polyethylene Low Density			Polyethylene High Density			Polycarbonate Resins			Polypropylene			Polystyrene			Melamine Resins (o)			Furane Resin			Epoxy Resins (p)			Phenol Form-aldehyde Resins (r)			Polyester Resins		
20°	60°	100°	20°	60°	100°	20°	60°	100°	20°	60°	100°	20°	60°	100°	20°	60°	100°	20°	60°	100°	20°	60°	100°	20°	60°	100°	20°	60°	100°
R[27]			R	R[80]					R	R	ND	No data						No data			R	R		R	ND	ND	No data		
R	R		R[56]	R		R	R	ND	R	R	ND	R			R[4]	ND	ND	R	R	R	R	R		R	R	R	R	R[23]	
R[27]			R[56]	R[50,56]		No data			R	R					No data			R	R	R	R[30]	ND		R	R		R[30]		
ND			R	R[50]					R	R		No data						R	R	R	R[68]	ND		R	ND	ND			
			R	R					R	R	ND				R	R		R	R	ND				R					
	No data		R[56]	R					R	R	ND				R	R		R	R	R	R[4,30]	ND		R					
	No data		R	R		R	ND	ND	No data			No data			R	R	R	No data			No data			No data			No data		
	No data		R	R		No data			R[2]	R		R[2]	R					R	R	R	R[2]	R[30]		R	R	R	No data		
R[27]			R[56]	R		R[46]	ND	ND	R	R	ND	R	R[33]		R	R		R	R	R	R[50]	R[30,71]		R	R		R	R	
			R	R		ND			R	R					R	R		R	R	R	R[50]	R[30,71]		R	R		No data		
	No data		R	R		R			R	R					R	R		R	R	R	R[30,71]	R		No data			No data		
R	R		R	R		R	ND	ND	R	R	R	R	R		R	R		R	R	R	R	R		R	R	R	R	R[30]	R[65]
R	R		R	R		R	ND	ND	R	R	R	R	R	R				ND	ND	ND	R	R		R	R	R	R	R[30]	
R	R		R	R		ND			R	R	R	R	ND	ND	R			ND	ND	ND	No data			No data			R[30]		
R	R		R	R					R	R	R	R	R		R						R	ND					R[30]		
R	R		R	R		R	ND	ND	R	R	R	R	R					R	R	R	R	ND		R	R	R			
			R[50]			ND									R	R		R	R	R	R[30]	ND		R	R		R[30]		
			R[50]						R	R	R				ND			R	R	ND	R[30]	ND		R	ND	ND			
R	R		R	R		R[7]	ND	ND	R	R	ND	No data			ND	ND		No data			R[68]	ND		R	ND	ND			
			R[80]						R[56]			R						R	R		R	R	R	R	R[4,30]		R	R	R
R	R		R	R		R	ND	ND	R	R	ND	R	R		R	ND	ND	No data			R	R		R	R	R	R	R	R
R	R		R[1]			R			R	R	ND	R			R			R	R	R	R	R		R	R	R	R[30]	R	
R	R		R	R		R	ND	ND	R	R	R	R			R	R	R	R	R	R	R	R		R	R	R	R	R[30]	R[65]
R	R		R	R		R	R	R	R	R	R	R	R		ND	ND		R	R	R	R	R		R	R	R	R	R[30]	R[65]
R	R		R	R		R	R	R	R	R	R	R	R		R	ND	ND	R	R	R	R	R		R	R	R	R	R[30]	
			R[50]									R			R			R	ND	ND	No data			R	R				
R	R		R[50]			R	ND	ND	R	R	ND	R	R		R			R	R	R	R[30]	R[30]		No data			R	R	
			R[56]			R	R		R	R	ND	R			R	R		R	R	ND				R	R	ND	R[30]		
R	R		R[19]			R	R	R	R	R	R	R	R		R[10]			R	R	R	R	R[19]					R[13]		
R	R		R	R		R[7]			R	R	ND	No data			ND	ND		No data			R	R		R	ND	ND	R	R[30]	R[65]
			R[80]			ND						No data			No data						R	R[4,30]					R	R[30]	
			R[50]																								R	R[30]	
R	R		R	R		R	ND	ND	R	R	R	R	R		R	R		R	R	R	R	R		R	R	R	R	R[30]	R[65]
	No data		R	R		ND			R	R	ND	No data						R	ND	ND	R[44,50]	ND		No data			R[30]		
			R[80]												R	R		R	R	R	R[30]	ND		R	ND	ND			
															R	R		R	R	R	R[30]	ND		R	R	R			
						ND						No data			No data			No data			No data								
R	R		R[50]			ND			R	ND	ND				R	R		R	R	R	R	R		R	R		R[10]		
	R		R[56]			R	R		R	R	ND	R			R	ND	ND	R	R	R	R	R[4,30]		R	R	R	R	R	R[30]
	R		R	R		R	R		R	R	ND	R			No data			R	R	R	R	R		R	R	R	R	R[30]	
			R[13]	R		R			No data			No data						No data			R	R		R	R		No data		
	No data		R[50]	R[50]		R						R	R		R	R	ND	R[68]	R[68]		R	R	R	No data		
R[56]	R		R[56]	R		R[98]	R[98]		R	R	R	R	R		R	R		R	R	R	R	R		R	ND	ND	R[62]	R[62]	
R[56]	R		R[56]	R		ND			R	R	ND	No data			R	R		R	R	R	R	R		R	R	R	No data		
			R[56]	R[50]					R[50]	R	ND				R	R		R	ND	ND	R	ND		R					
	No data		R[56]	R[56]		ND			R	R	ND	R	R		R	R		R	R	R	R	R		R	R	R	R	R[30]	
	R		R	R		R	R		R	R	R	R	R		No data			R	ND	ND	R	R		R	R	R	R	R[30]	R[65]
	R		R	R		R	ND		R	R	R	R	R		No data			ND	ND	ND	R	R		R	R	R	R	R[30]	R[65]
																		R	ND	ND									
R[50]			R[50]			R			No data						R	R		R	ND	ND	R	R		R	ND	ND	No data		
						ND												No data			R[30]	R					No data		
						ND												No data			R[30]	R[4,30]					No data		
R[3]	R[3]		R[6]	R		ND			R	R	ND	No data						R	R	ND				R	ND	ND	R[15]		
R	R		R[56]	R		R	R		R	R	ND	R			R			R	R	ND	R[4,30]			R	R	R	R[30]	R	
R	R		R[56]	R		R[32]	R[32]		R	ND	ND	R			R						R[4,30]			R			R[15]		

Vertical note (between Polystyrene and Melamine columns): H.D. Polyethylene is suitable for a number of application at 100°C for limited periods, depending on the environment.

THERMOPLASTIC RESINS

	Acrylic Sheet (e.g. Perspex)	Acrylonitrile Butadiene Styrene Resins (l)	Nylon 66 Fibre (m)	Nylon 66 Plastics (m)	PCTFE	PTFE (n)	PVDF (y)	Rigid Unplasticised PVC	Plasticised PVC
Centigrade	20° 60° 100°	20° 60° 100°	20° 60° 100°	20° 60° 100°	20° 60° 100°	20° 60° 100°	20° 60° 100°	20° 60° 100°	20° 60° 100°
Fruit juices	R[68] R	R R	R R ND	R R R	R R R	R R R	R R R	R R	
Gelatine	R R	R R	R R ND	R R R	R R ND	R R R	R R R	R R	No data
Glycerine	R R	R R	R R R	R R[50] ND	R R R	R R R	R R R	R R	R
Glycols	R ND		R R R	R[50] R[50] ND	R R R	R R R	R R R	R R	No data
Hexamine	No data		R[43] R R	No data	R R ND	R R R	R R R		No data
Hydrazine	No data	No data	No data	No data	No data	R R R	No data	No data	No data
Hydrobromic acid (50%)	R R	R			R R R	R R R	R R R	R[32] R	R R
Hydrochloric acid (10%)	R R	R R			R R R	R R R	R R R	R R	R R
Hydrochloric acid (conc.)	R R[50]				R R R	R R R	R R R	R[3] R[3]	R ND
Hydrocyanic acid		R[10] R[10]			R R ND	R R R	R R R	R R	No data
Hydrofluoric acid (40%)		R			R R R	R R R	R R R	R[30]	R
Hydrofluoric acid (75%)					R R R	R R R	R R R	R[19]	
Hydrogen peroxide (30%)		R			R R R	R R R	R R R	R R	R
(30—90%)					R R ND	R R R	R R R	R R[30]	R
Hydrogen sulphide	R ND	No data	No data	R ND ND	R R R	R R R	R R R	R R	R
Hypochlorites	R[34] ND				R R R	R R R	R[107] R[107] R[107]	R[50] R	No data
Lactic acid (100%)	R R[68]	R R	R		R R ND	R R R	R R R	R[13] R	
Lead acetate	R[68] R	R R	R[43] R R	R[50] R[50] ND	No data	R R R	R R R	R R	R R
Lime (CaO)	R R	R R	R R R	R R R	R R R	R R R	R R R	R R	No data
Maleic acid	R[68] R	R R	R[43] R R	R[50] ND ND	R R ND	R R R	R R R	R R[37]	
Meat juices	R R	R R	No data	R R R	R R ND	R R R	R R R	R R	No data
Mercuric chloride	R ND	R	R[43] R R	R[50]	R R R	R R R	R R R	R R	
Mercury	R R	R R	R R ND	R R R	R R R	R R R	R R R	R R	
Milk & its products	R R	R	R R R	R R R	R R R	R R R	R R R	R R	R
Moist air	R R	R R	R R R	R R R	R R R	R R R	R R R	R R	No data
Molasses	R R	R	R R ND	R R R	R R R	R R R	R R R	R R	R R
Naphtha		R	R R R	No data		R R R	R R R	R R	No data
Naphthalene	R[4]	No data	No data	R ND ND		R R R	R R R	R R R	
Nickel salts	R R	R R	R[43] R[31] R	No data	R R R	R R R	R R R	R R	R R
Nitrates of Na, K, NH₃	R R	R R	No data	R R ND	R R R	R R R	R R R	R R	R R
Nitric acid (<25%)	R R[10]				R R R	R R R	R R R	R R[37]	R ND
Nitric acid (50%)					R R R	R R R	R R R	R R[30]	R ND
Nitric acid (95%)					R R R	R R R	R ND ND		
Nitric acid, fuming					R R ND	R R R	R	R[66] R	
Oils, essential	R[62] R[62]	R R	R R R	R R R	No data	R R R	R R R	R[62] R	No data
Oils, mineral	R R	R R	R R R	R R R	R R R	R R R	R R R	R R	No data
Oils, vegetable & animal	R	R R	R R R	R R R	R R ND	R R R	R R R	R R	No data
Oxalic acid	R R	R	R ND ND	R[50] ND ND	R R R	R R R	R R R	R R	R
Ozone	R ND		No data	R[50] ND ND	R R R	R R R	R R R	R R	R
Paraffin wax	R R	R R	R R R	R R R	R R R	R R R	R R R	R R	No data
Perchloric acid	No data	No data			R R R	R R R	R R R	R[52] R[10]	ND
Phenol					R R	R R R	R R R	R R	ND
Phosphoric acid (25%)	R R	R			R R R	R R R	R R R	R R	R R
Phosphoric acid (50%)	R	R			R R R	R R R	R R R	R R	No data
Phosphoric acid (95%)		R			R R R	R R R	R R R	R[55] R	No data
Phosphorus chlorides	No data	No data	No data	No data	R ND ND	R R R	R R R		No data
Phosphorus pentoxide	R[68] R	No data		R R ND	No data	No data	R R R	R R[68]	R ND
Phthalic acid	No data	No data	No data	R	R R ND	R R R	R R R	R R	R R
Picric acid	R[68]	No data	No data	R[50]	No data	R R R	R R R	R[10] R	R[105] ND
Pyridine	No data		R R R	R ND ND	R R ND	R R R	R R	ND	No data
Sea water	R R	R R	R R R	R R R	R R R	R R R	R R R	R R	R R
Silicic acid	No data	R R	R ND ND	No data	R R R	R R R	R R R	R R	No data
Silicone fluids	R[4]₍₃₀₎	No data	R R R	R ND ND	R R R	R R R	R R R	No data	No data
Silver nitrate	R R	R R[64]	R R R	R ND ND	R R R	R R R	R R R	R R	No data
Sodium carbonate	R R	R R	R R R	R ND ND	R R R	R R R	R R R	R R	R ND
Sodium peroxide	R[4]	R ND			R R R	R R R	R R R	R R	R R

THERMOPLASTIC RESINS															THERMOSETTING RESINS														
Polyethylene Low Density			Polyethylene High Density			Polycarbonate Resins			Polypropylene			Polystyrene			Melamine Resins (o)			Furane Resin			Epoxy Resins (p)			Phenol Formaldehyde Resins (r)			Polyester Resins		
20°	60°	100°	20°	60°	100°	20°	60°	100°	20°	60°	100°	20°	60°	100°	20°	60°	100°	20°	60°	100°	20°	60°	100°	20°	60°	100°	20°	60°	100°
R	R		R	R		R			R	R	ND	R^13			R	R			No data		R	R^4,30		R	R	R	R	R	R^65
	No data		R	R		ND			R^30	R	ND		No data		R	R			No data		R	R^4,30			No data			No data	
R	R		R^56	R		R			R	R	R^30	R	R		R	R		R	R	R	R	R		R	R	R	R	R	R^65
R	R		R^56	R		R			R	R	ND	R	R		R	R		R	R	R	R	R		R	R	R	R	ND	ND
	No data		R^56	R		ND				No data			No data		R			ND	ND	ND	R	R^4,30		R	R	R		No data	
	No data		R^56	R		ND			R^13	ND	ND		No data		ND	ND			No data		R	R			No data			No data	
R	R		R	R		ND			R^56	R^27	R^27		No data		R	R		R	ND	ND	R	R^4,30		R	ND	ND		No data	
R	R		R	R^14		R	R		R	R	R^50	R			R			R	R	R	R			R	R	R	R^30	R	R^65
R	R		R^14	R^14					R	R		R									R^4,30			R	ND	ND			
R	R		R	R		ND			R	R	ND		No data			No data			No data		R^4,30				No data			No data	
R	R		R	R^80		R^32			R	ND	ND				R				No data					R	ND	ND	R^10		
R	R		R	R^80		ND			R^50	ND									No data			No data			No data				
R			R	R		R	R					R			R						R^30	ND	ND	R	ND	ND	R		
R			R			ND							No data								R^30	ND	ND						
R			R^13	R		ND			R	R	ND				R			R	R	ND	R	R^30		R	ND	ND	R	R^65	
R	R		R	R		R^7,34			R	R	R	R			R^10						R^4,30						R	ND	
R	R		R^56	R^56					R	R	ND				R				No data		R^4,30			R			R^30	R	
R	R		R	R		ND			R^7	R	ND	R	ND	ND	ND	ND		R	R	R	R	R		R	ND	ND	R	R	R^65
R	R		R	R		R			R^7_11	R	R	R	R		R			R	R	R	R	R			No data		R	R^30	ND
R	R		R^19	R		ND			R	R	R	R	R	ND	R			R	R	R	R			R			R	R^13	R
	No data		R	R		ND			R	R	ND	R	ND	ND	R	R			No data		R^30	R^30			No data			No data	
R	R		R	R		R	R		R^8	R	ND	R	R			No data			No data		R	R		R	R		R	R	R^30
R	R		R	R		R			R	R	R	R	R		R	R		R	ND	ND	R	R			No data		R	R	R
R	R		R	R		R			R	R	R	R	R		R	R		R	R	R	R	R		R	R	R	R	R	R
R	R		R	R		R			R	R	R	R	R		R	R	R	R	R	R	R	R		R	R	R	R	R	R
R	R		R	R		ND			R	R	ND				R	R			No data		R	R		R	R	R	R	R	R
			R	R^80		ND			R^30	ND	ND				R	R		R	R	R	R^30	R		R	R	R	R	R^30	
			R	ND		ND			R	R	R				R	R		R	R	R	R	R		R	R	R	R	R^30	
R	R		R	R		ND			R	R	R^30	R	R			No data			No data		R	R		R	R	R	R	R	R^65
R	R		R	R		R^7,75			R^7	R	ND	R^34	R^34	ND	R^23	R		R	R	R	R			R^75			R	R^30	R^65
R	R		R^80			R_7			R	R	R				R^10			R	R	R	R^10						R		
R^50			R^80																										
R^50			R^50			ND			R	R	R^50				R	R			No data		R	R		R	R	R		No data	
R^50	R		R^50	R		ND			R^30						R	R		R	R	R	R	R		R	R	R	R	R	R
R^50			R^62			R	R		R	R		R			R	R			No data		R	R		R	R	R	R	R	R
R	R		R^50	R		R			R	R		R	R		R	R		R	ND	ND	R^30	ND		R	ND	ND	R	R^30	
			R	R		R	R			No data						No data			No data			No data			No data			No data	
ND	ND		R	R^50		ND	R		R	R	ND	R			R	R		R	R	R	R	R		R	R	R	R	R	R
R^10	R^10		R	R^60		R	ND												No data		R^44,30			R^10					
			R	R					R	R	ND	R^10									R^44,30			R	R				
R	R		R	R		R	R		R	R	R	R	R		R			R	R	R	R	R^30		R	R		R	R	R^65
R	R		R	R		R	R		R	R	ND	R	R		R						R^44,30			R	R		R	R^30	R^65
R^50			R^50			R	R		R	R	ND		No data		R			ND			R^44,30			R	R			No data	
R	R		R	R^50					R	ND	ND		No data					R	R	ND				R	ND	ND		No data	
			R	R		ND			R	R	ND		No data						No data		R^30				No data			No data	
	No data		R	R		ND			R	R	ND		No data		R			R	R	R	R	R^30		R	ND	ND	R	R	R^65
R^13			R	R		ND			R	R	ND		No data		R	ND			No data		R	ND		R	ND	ND	R^10	ND	ND
	No data		R	R^80					R	ND	ND				R	ND	ND		No data		R	R							
R	R		R	R		R			R	R	R	R	R		R	R		R	R	R	R	R		R	R	R	R	R	R
R	R		R	R		R	ND		R	R	ND	R	R		R	R			No data		R	R			No data			No data	
R^50			R^56	R^56		R	R	R	R	R	ND		No data		R	R		R	R	R	R	R			No data			No data	
R	R		R	R		R	ND		R	R	ND	R			R	R			No data		R	R		R	R		R	R^30	R^83
R	R		R	R					R	R	ND	R			R^10	R		R	R	R	R	R		R			R	R	
R	R		R^13	ND		ND				No data			No data		R	R			No data		R	R			No data			No data	

Note (vertical text, Polycarbonate column): These results at 20°C refer to low stress moulded parts made from Makrolon. The chemical resistance of this material may be affected by mechanical stresses and high temperatures. The polymer may however be used at relatively high temperatures for a thermoplastic: it has good heat resistance up to 135°C. Makrolon is the polycarbonic acid ester of 4,4'-dihydroxydipheny-2.2'-propane.

THERMOPLASTIC RESINS

	Acrylic Sheet (e.g. Perspex)			Acrylonitrile Butadiene Styrene Resins (l)			Nylon 66 Fibre (m)			Nylon 66 Plastics (m)			PCTFE			PTFE (n)			PVDF (y)			Rigid Unplasticised PVC			Unplasticised PVC		
Centigrade	20°	60°	100°	20°	60°	100°	20°	60°	100°	20°	60°	100°	20°	60°	100°	20°	60°	100°	20°	60°	100°	20°	60°	100°	20°	60°	100°
Sodium silicate	R	R		R			R	R	R	R	R	R	R	R	R	R	R	R	R	R	R	R	R		R	R	
Sodium sulphide	R	R[68]		R			R	R	ND	R	ND	ND	R	R	R	R	R	R	R	R	R	R	R		R	R	
Stannic chloride	R[68]	R								R[50]	ND	ND	R	R	R	R	R	R	R	R	R	R	R		R	R	
Starch	R	R		R			R	R	R	R	R	R	R	R	R	R	R	R	R	R	R	R	R		R	R	
Sugar, syrups, jams	R	R		R	R		R	R	R	R	R	R	R	R	R	R	R	R	R	R	R	R	R		No data		
Sulphamic acid	No data			No data			R[43]	R	R	No data			No data			No data			R	R	R	No data			No data		
Sulphates (Na, K, Mg, Ca)	R	R		R	R		R	R	R	R	R	R	R	R	R	R	R	R	R	R	R	R	R		R	R	
Sulphites	R	R		R			No data			No data			R	R	R	R	R	R	R	R	R	R	R[50]		No data		
Sulphonic acids	No data			No data			No data									No data			R	R	R	R	R	R			
Sulphur	R	R[68]		R	ND					R	ND	ND	R	R	R	R	R	R	R	R	R	R	R		R	R	
Sulphur dioxide, dry	R	R[68]					No data			R	ND	ND	R	R	R	R	R	R	R	R	R	R	R		R	R	
Sulphur dioxide, wet	R	R[68]		R			No data			R[50]			R	R	R	R	R	R	R	R	R	R	R[50]		ND		
Sulphur trioxide	No data			R	R								No data			R	R	R	R	R	R	R[13]	R		No data		
Sulphuric acid (<50%)	R[25]	R[32]											R	R	R	R	R	R	R	R	R	R	R		R	ND	
Sulphuric acid (70%)													R	R	R	R	R	R	R	R	R	R	R		No data		
Sulphuric acid (95%)													R	R	R	R	R	R	R	R	R	R	R		R[50]		
Sulphuric acid, fuming													R	R	R	R	R	R	R						No data		
Sulphur chlorides	No data			No data						No data			No data			R[30]	R		No data			ND			No data		
Tallow	R[68]	R		R	R		R	R	R	R	R	R	R	R	R	R	R	R	R	R	R	R	R		R	ND	
Tannic acid (10%)	R	ND		No data			R	ND	ND	R	ND	ND	R	R	ND	R	R	R	R	R	R	R	R		R	ND	
Tartaric acid	R	R		R	R		R	ND	ND	R	R[50]	ND	R	R	ND	R	R	R	R	R	R	R	R		R	ND	
Trichlorethylene							R	R	R	R	R	R[50]				R[14]	R	R	R	R	R	R	R	R			
Vinegar	R	R[68]		R	R		R[50]	ND	ND	R	R[50]	R	R	R	R	R	R	R	R	R	R	R	R		R	ND	
Water, distilled	R	R		R	R		R	R	R	R	R[50]	R	R	R	R	R	R	R	R	R	R	R	R		No data		
Water, soft	R	R		R	R		R	R	R	R	R[50]	R	R	R	R	R	R	R	R	R	R	R	R		R	R	
Water, hard	R	R		R	R		R	R	R	R	R[50]	R	R	R	R	R	R	R	R	R	R	R	R		R	R	
Yeast	R	R[68]		R			R	R	R	R	ND	ND	R	R	R	R	R	R	R	R	R	R	ND		R	ND	
Zinc chloride	R	R[68]		R	R		R[43]	R	R				R	R	R	R	R	R	R	R	R	R	R		R	R	

THERMOPLASTIC RESINS															THERMOSETTING RESINS														
Polyethylene Low Density			Polyethylene High Density			Polycarbonate Resins			Polypropylene			Polystyrene			Melamine Resins (o)			Furane Resin			Epoxy Resins (p)			Phenol Form-aldehyde Resins (r)			Polyester Resins		
20°	60°	100°	20°	60°	100°	20°	60°	100°	20°	60°	100°	20°	60°	100°	20°	60°	100°	20°	60°	100°	20°	60°	100°	20°	60°	100°	20°	60°	100°
R	R		R	R		ND			R	R	R	No data			R			R	R	R	R	R		R			R	R	R[65]
R	R		R	R			No data		R	R	R	No data			R			R	R	R	R	R			No data		R	R	R[65]
R	R		R	R			No data		R[7]	R	R	R[1]	R		ND	ND		R	R	ND	R[68]			R	ND	ND		No data	
R	R		R	R		R	ND		R	R	R	R	R		R	R			No data		R	R		R	R	R		No data	
R	R		R	R		R			R	R	ND	R	R		R	R			No data		R	R		R	R	R		No data	
ND	ND			No data			No data		R	R	ND	No data			R	ND	ND	R	ND	ND		No data			No data			No data	
R	R		R	R		R	R	ND	R	R	ND	R	R		R	R		R	R	R	R	R[30]		R	R	R	R	R[30]	R[65]
R[34]			R	R			No data		R	R	ND	R	R		R	R		R	R		R	R		R	R		R	R[30]	R[65]
	No data			No data			No data			No data			No data			No data		R	R	ND	R	R[44]			No data			No data	
R	R		R	R			No data		R	R	ND		No data		R	R		R	R	R	R	R	R		No data			No data	
R	R		R	R			No data		R	R					R			R	R	R	R	R		R	ND	ND	R	ND	ND
R			R	R		ND			R	R	ND				R			R	R	R	R[11]			R	ND	ND	R[30]	ND	ND
						ND				No data														R	ND	ND	R[30]	ND	ND
R	R		R	R		R	R		R	R		R			R[10]			R	R	R	R	R[30]		R	R	R	R[30]	R	
R	R[50]		R	R[50]		R	R	ND	R												R	R[30]					R[30]	R	
R[50]			R[50]	R[80]		R[56]			R[56]																				
			R[60]																										
ND	ND			No data						No data			No data		ND	ND		R	R	R		No data		R	ND	ND		No data	
R			R[50]	ND	ND	R	ND		R	R	ND	R			R	R			No data		R	R		R	R	R			
R	R		R[56]	R		ND			R	R	ND				R	R			No data		R	R[30]		R	R	R	R	R	R
R	R[10]		R	R		R	R		R	R	ND	R	R		R				No data		R	R[30]		R	R	R	R	R	R
			R[50]												R			R	R								R	R[30]	
R	R		R	R			No data		R	R	ND	R			R				No data		R	R[4]		R	ND	ND	R	R[30]	
R	R		R	R		R	R	R	R	R	R	R	R		R	R		R	R	R	R	R		R	R	R	R	R	R[30]
R	R		R	R		R	R	R	R	R	R	R	R		R	R		R	R	R	R	R		R	R	R	R	R	R[30]
R	R		R	R		R	R	R	R	R	R	R	R		R	R		R	R	R	R	R		R	R	R	R	R	R[30]
R	ND		R	R		ND			R	R	ND	No data			R	R			No data		R	R			No data			No data	
R	R		R	R		R	R		R	R	ND	R	R		R			R	R	R	R	ND		R	ND	ND	R	R[30]	R[65]

RUBBERS

	Butyl Rubber and Halo-Butyl Rubber			Ethylene Propylene Rubber (q)			Hard Rubber (Ebonite) (h)			Soft Natural Rubber (h)			Neoprene (i)			Nitrile Rubber			Chlorosulphonated Polyethylene			Polyurethane Rubber (v)			Silicone Rubbers (k)		
	20°	60°	100°	20°	60°	100°	20°	60°	100°	20°	60°	100°	20°	60°	100°	20°	60°	100°	20°	60°	100°	20°	60°	100°	20°	60°	100°
Acetaldehyde	R	R	ND	R	R	ND	R	R	R	R^{80}	R^{80}	ND										ND	ND		R	R	R
Acetic acid (10%)	R^{14}	R	R	R	R^{14}	ND	R	R	R				R	R	R^{14}	R	R		R	R	ND	R^{80}	R^{80}		R	R	R
Acetic acid (glac. & anh.)	R^{14}	R	R	R^{14}	R^{14}	ND	R	R^{14}	R				R^{95}			R^{4}			R^{85}			R^{80}			R^{17}	R	R
Acetic anhydride	R^{80}	R	R	No	data		R		R^{30}				R	R	ND				R^{15}	ND	ND				R^{17}	R	R
Acetone	R	R		R^{60}	R^{60}		R	R	R	R^{60}	R	ND							R^{15}	ND	ND				R^{17}	R	R
Other ketones	R^{13}	R	R	R^{60}	R^{60}		R^{13}_{80}	R	R	R^{30}_{60}	R														R^{17}	R	R
Acetylene	R	R^{80}		No	data		R^{80}	R	R				R^{14}	R	R	R	ND	ND	R^{14}	R	R	ND	ND		No	data	
Acid fumes	R^{2}	R	R	R^{2}	R^{2}	R^{2}	R^{2}	R	R	R^{2}	R	$R^{2.80}$	R^{2}	R	R	R^{2}			R	R	R^{2}	R^{2}	R^{2}		R^{2}	R	R
Alcohols (most fatty)	R	R		R^{60}	R^{60}		R^{30}_{60}	R	R	R^{60}	R		R	R	R^{14}	R	R	R	R	R	R	R^{4}	R^{4}		R^{30}	R	R
Aliphatic esters													No	data											R^{30}	R	R
Alkyl chlorides																									R^{21}	R	R
Alum	R	R	R	R	R	R	R	R	R	R	R	R	R	R	R	R	R	R	R	R	R	R	R		R	R	R
Aluminium chloride	R	R	R	R	R	R	R	R	R	R	R	R	R	R	R	R	R	R	R	R	R	R	R		R	R	R^{4}
Ammonia, anhydrous	R	R	ND	R	R	ND	R	R	R	R^{80}			R	R	R	R	R		R^{10}	ND	ND	R^{80}			R	R	R
Ammonia, aqueous	R	R	R	R	R	R	R	R	R	R	R	R^{80}	R	R	R	R	R		R	R	R	R^{30}	R^{80}		R	R	R
Ammonium chloride	R	R	R	R	R	R	R	R	R	R	R	R	R	R	R	R	R	R	R	R	R	R	R		R	R	R
Amyl acetate	R^{80}																								R^{21}	R	R
Aniline	R	R	ND																						R	R	R
Antimony trichloride	R	R	R	No	data		R	R	R	R	R	R	No	data		No	data		R	R	R / R^{50}	ND	ND		R	R	R
Aqua regia																R^{62}	R										
Aromatic solvents																									R^{21}	R	R
Beer	R	R	R	R	R	R	R	R	R	R	R		R	R	R	R	R	R	R^{86}	R	R	R	R		R	R	R
Benzoic acid	R	R	R	R	R	R	R	R	R	R	R	R	R	R	R	R	R	R	No	data		R	R		R	R	R
Boric acid	R	R	R	R	R	R	R	R	R	R	R	R	R	R	R	R	R	R	R	R	R	R	R		R	R	R
Brines, saturated	R	R	R	R	R	R	R	R	R	R	R	R	R	R	R	R	R	R	R	R	R	R	R		R	R	R
Bromine				No	data																						
Calcium chloride	R	R	R	R	R	R	R	R	R	R	R	R	R	R	R	R	R	R	R	R	R	R	R		R	R	R
Carbon disulphide																R	ND	ND					R		R	R	R
Carbonic acid	R	R	R	R	R	R	R	R	R	R	R	R	R	R	R	R	R	R	R	R	R	R	R		R	R	R
Carbon tetrachloride																									R^{21}	R	R
Caustic soda & potash	R	R	R	R	R	R	R	R	R^{13}	R	R	R	R	R	R	R	R	R	R	R	R	R^{30}	R^{30}		R	R	R^{30}
Chlorates of Na, K, Ba	R	R	R	R	R	R	R	R	R	R	R	R	No	data		No	data		R	R	R	R	R		R	R^{30}	R
Chlorine, dry	R^{50}	R	R	R^{50}	R^{50}	R^{50}	R^{30}	R	R																R	R	R
Chlorine, wet	R^{80}	R	R	R^{50}	R^{50}	R^{50}	R^{13}	R	R										R^{3}	ND	ND				R	R	R
Chlorides of Na, K, Mg	R	R	R	R	R	R	R	R	R	R	R	R^{80}	R	R	R	R	R	R	R	R	R	R	R		R	R	R
Chloroacetic acids	R^{10}						R^{2}	R	R				R	R	ND				R			R^{80}			R	R	R
Chlorobenzene																									R^{21}	R	
Chloroform																											
Chlorosulphonic acid	R^{13}	R^{13}		R^{13}	R^{13}		R^{13}	R	ND										No	data		R^{30}	ND		No	data	
Chromic acid (80%)																			R^{30}	R	ND				R^{19}	R	R
Citric acid	R	R	R	R	R	R	R	R	R	R	R		R	R	R	R	R	R	R	R	R	R	R		R	R	R
Copper salts (most)	R	R	R	R	R	R	R	R	R	R	R	R	R	R	R	R	R	ND	R	R	R	R	R		R	R	R
Cresylic acids (50%)	R^{4}			No	data														R			ND	ND		R^{21}	R	R
Cyclohexane																R	R	R							R^{21}	ND	ND
Detergents, synthetic	R^{13}	R	R	R^{13}	R^{13}	R^{13}	R	R	R	R^{80}	R^{80}	R	R	R	R	R	R	R	R^{30}	R	R	R^{30}	R^{30}		R	R	R
Emulsifiers (all conc.)	R	R	R	No	data		R	R^{4}	R^{4}	No	data		R^{30}	R	R	R	R	R	R^{30}	R	R	ND	ND		R	R	R
Ether																R											
Fatty acids (>C₆)	R^{1}	R^{80}	R	R^{80}	R^{80}		R^{80}	R^{13}_{80}					R	R	R	R	R^{4}		R	R	R	R^{4}_{30}	R^{80}		R	R	R
Ferric chloride	R	R	R	R	R	R	R	R	R	R^{80}	R	R	R	R	R	R	R	R	R	R	R	R	R		R	R	R
Ferrous sulphate	R	R	R	R	R	R	R	R	R	R	R	R	R	R	R	R	R	R	R	R	R	R	R		R	R	R
Fluorinated refrigerants, aerosols, e.g. *Freon*	R^{4}	ND	ND	No	data		No	data		No	data		R^{30}	R	ND	R^{30}	R	R	R^{30}	ND	ND	ND	ND		$R^{4.21}$		
Fluorine, dry	R^{80}	ND	ND	R^{80}	ND	ND	R^{13}	R^{13}	ND				No	data					No	data							
Fluorine, wet	R^{80}	ND	ND	R^{80}	ND	ND	R^{13}	R^{13}	ND				No	data					No	data							
Fluosilicic acid	R	R	R	R	R	R	R	R	R	R	R	R	R	ND	ND	No	data		R	R	R	R	R		R	R	R
Formaldehyde (40%)	R^{80}			R^{80}			R	R^{30}	R	R^{14}			R	ND	ND	R			R			ND	ND		R	R	R
Formic acid	R^{13}	R	R	R^{14}	R^{14}		R	R^{80}		R^{80}			R	R	R	R			R	R	R	R^{80}	R^{80}		R	ND	ND

MISCELLANEOUS

Concrete (s) 20°	60°	100°	Glass (t) 20°	60°	100°	Graphite (u) 20°	60°	100°	Porcelain and Stoneware 20°	60°	100°	Vitreous Enamel (w) 20°	60°	100°	Wood (z) 20°	60°	100°
No data			No data			R	R	R	R	R	R	R	R	R	R	R	R
			R	R	R	R	R	R	R	R	R	R	R	R	R	R	
			R	R	R	R	R	R	R	R	R	R	R	R			
			No data			R	R	R	R	R	R	ND	ND				
R	R	R	R	R	R	R	R	R	R	R	R	R	R	R	R	R	R
R	R	R	No data			R	R	R	R	R	R	R	R	R	R	R	R
R	R	R	R	R	R	R	R	R	R	R	R	R	R	R	No data		
			R[5]	R	R	R	R	R	R	R	R	R[5,11]	R	ND			
R	R	R	R	R	R	R	R	R	R	R	R	R	R	R	R	R	
			R	R	R	R	R	R	R	R	R	R	R	R	R	R	R
			R	R	R	R	R	R	R	R	R	R	R	ND	R	R	
			R	R	R	R	R	R	R	R	R	R	R	R	R	R	
			R[30]			R	R	R	R	R	R	R	R	ND			
No data			No data			R	R	R	R	R	R	R	R	R			
R	R	R	R	R	R[50]	R	R	R	R	R	R	R	ND				
			R	R	R	R	R	R	R	R	R	ND			R	R	
			R	R	R	R	R	R	R	R	R	R	R	R	R	R	R
			R			R	R	R	R	R	R	R	R	R	R	R	
No data			No data			R	R	R	R	R	R	R	R	R			
R	R	R	R	R	R[50]	R	R	R	R	R	R	R	ND				
			R	R	R	R	R	R	R	R	R	ND			R	R	
R	R	R	R	R	R	R	R	R	R	R	R	R	R	R	R	R	R
No data			R			R	R	R	R	R	R	R	R	R	R	R	R
No data			R	R	R	R	R	R	R	R	R	R	R	R	R	R	R
			R	R	R				R	R	R	R	R	ND			
R	R	R	R	R	R	R	R	R	R	R	R	R	R	R	R	R	R
R	R	R[35]	R	R	R	R	R	R	R	R	R	R	R	R	R	R	R
No data			R	R	R	R	R	R	R	R	R	R	R	ND	R	R	R
R	R	R	R	R	R	R	R	R	R	R	R	R	R	ND	R	R	R
R	R	R	R	R	R	R	R	R	R	R	R	R	R	ND	R		
			R	R	R				R	R	R	No data					
R[44]	R	R	R	ND	ND	R	R	R	R	R	R	R	ND		R	R	
R	R	R	R	ND	ND	R	R	R	R	R	R	R	R		R		
R[50]	R	R	R	R	R	R	R	R	R	R	R	R	R	R	R	R	R
R	R	R	R	R	R	R	R	R	R	R	R	R	R	R	R	R	R
R[72]	R	R	ND			R[13]	R	R	R[10]			R[10]					
R	R	R	R	R	R	R	R	R	R	R	R	No data			R	ND	ND
No data			No data			R	R	R	R	R	R	R	R	R			
					ND				R	R	R	No data					
R[12]	R	R	R	R	R	R	R	R	R	R	R	R	R		R	R	
			R	R	R	R	R	R	R	R	R	R	ND	ND			
R	R	R	R	R	R	R	R	R	R	R	R	R	R	R	R	R	
R	R	R	R	R	R	R	R	R	R	R	R	R	R	R	R	R	R
			R	R	R	R	R	R	R	R	R	No data					
			R	R	ND				R	R	R	R	ND	ND	R	R	
			R	R	ND	R	R	R	R	R	R	R	R	R	R	R	
R[51]	R	R	R	R	R	R	R	R	R	R	R	R	R	R	R	R	R
			No data			R	R	R	R	R	R	R	R	ND			
R	R	R	No data			R	R	R	R	R	R	R	R	R	R	R	R
R[81]	R	R	R[13]	R[30]		R	R	R	R	R	R	R	R	R	R	R	
No data						R	R	R	R	R	R	R	R	R	R	R	
R	R	R	R	R	R	R	R	R	R	R	R	R	R	R	R		
			R	R	R	R	R	R	R	R	R	R	R	R	R	R	
			R	R	R	R[13]	R	R	R	R	R	R	ND		R[14]		
			R	R	R	R	R	R	R	R	R	R	ND		R[14]		
R	R	R	No data			R	R	R	R[39]	R	R	No data			R	R	R
No data						R	R	R				R	ND		No data		
No data						R	R	R				No data			No data		
R	R	R				R	R	R							No data		
R[80]			R	R	R	R	R	R	R	R	R	R	R	R	R	R	
			R	R	R	R	R	R	R	R	R	R	R	R			

NOTES

Explanatory notes at lower temperatures may be taken to apply also at higher temperatures unless otherwise shown.

1 Not anhydrous
2 Depending on the acid
3 35%
4 Fair resistance
5 Not HF fumes
6 Up to 40%
7 Saturated solution
8 Pineapple and grapefruit juices 20°C
9 Photographic emulsions up to 20°C
10 10%
11 Anhydrous
12 Not Mg
13 Depending on concentration
14 Discoloration and/or swelling and softening
15 Up to 25%
16 Not chloride/not if chloride ions present
17 Not fluorinated silicone rubbers
18 Up to 60%
19 Up to 50%
20 Not aerated solutions
21 Fluorinated silicone rubbers only
22 ND for Mg
23 5%
24 Pure only
25 Up to 30%
26 If no iron salts or free chlorine
27 May crack under stressed conditions
28 45%
29 55%
30 Depending upon composition
31 Chloride
32 20%
33 Depending on alcohol
34 Data for sodium
35 Fresh
36 Over 85%
37 Some attack at high temperature
38 Neutral
39 Attacked by fluoride ions
40 Sulphate and nitrate
41 Softening point
42 In strong solutions only when inhibited
43 Depending on water conditions
44 Dilute
45 Up to 15%
46 Not methyl
47 Drawn wire
48 Some attack, but protective coating forms
49 Using anodic passivation techniques
50 Some attack/absorption/slow erosion
51 Not sulphate
52 70%
53 In absence of dissolved O_2 and CO_2
54 75%
55 80%
56 May cause stress cracking
57 Pitting possible in stagnant solutions
58 In presence of H_2SO_4
59 Not ethyl
60 May discolour liquid
61 The material can cause decomposition
62 Depending on type
63 95%
64 Slight plating will occur
65 Not recommended under certain conditions of temperature, etc.
66 65%
67 Aerated solution
68 Estimated effect
69 Up to 90%
70 Not oxidising conditions
71 Not lower members of series
72 Not high alumina cement concrete

RUBBERS

	Butyl Rubber and Halo-Butyl Rubber			Ethylene Propylene Rubber (q)			Hard Rubber (Ebonite) (h)			Soft Natural Rubber (h)			Neoprene (i)			Nitrile Rubber			Chlorosulphonated Polyethylene			Polyurethane Rubber (v)			Silicone Rubbers (k)		
	20°	60°	100°	20°	60°	100°	20°	60°	100°	20°	60°	100°	20°	60°	100°	20°	60°	100°	20°	60°	100°	20°	60°	100°	20°	60°	100°
Fruit juices	R[80]	R[80]		R[60]	R[60]		R[65]	R	R	R	R	R	R	R	R	R	R	R	R	R	R	R	R		R	R	R
Gelatine	R	R	R	R	R		R	R	R	R	R	R	R	R	R	R	R	R	R	R	R	R	R		R	R	R
Glycerine	R	R	R	R	R	R	R	R	R	R	R	R	R	R	R	R	R	R	R	R	R	R	R		R	R	R
Glycols	R	R	R	R	R	R	R	R	R	R	R	R	R	R	R	R	R	R[4]	R	R	R	R	R		R	R	R
Hexamine	R	R	ND	No data			R	R	ND	R	R	ND	R	R[37]	R	No data			R	R	R	ND	ND		No data		
Hydrazine	R	R	ND	R	ND	ND	R						R[44]	ND	ND	R			No data			No data			No data		
Hydrobromic acid (50%)	R	R	R	R	R	ND	R	R[37]	R[37]	R[65]	R		R	ND	ND	R			R	R	ND	R[15]	R[15,80]		R	R	R
Hydrochloric acid (10%)	R	R	R	R	R	R	R	R	R	R	R		R	R	R	R	R		R	R		R	R		R	R	R
Hydrochloric acid (conc.)	R[4]	R	R	R	R[4]	R[80]	R[37]	R[37]	R[37]	R	R[80]		R	R[95]	ND				R	R		R[30]			No data		
Hydrocyanic acid	R	R	R	R	R	R	R	R	R	R	R	R	R	R	R	R			R	R	ND	R	R		No data		
Hydrofluoric acid (75%)				R[30]	ND	ND							R						R	R	ND						
Hydrogen peroxide (30%)	R[80]	R	R	R[80]	ND	ND	R[80]						R	R		R			R[87]	R[87]		R[4]	ND		R	R	R
(30–90%)	R[80]	R[87]	ND				R[87]						R	R					R[87]	R[87]		R[30]	R	R	No data		
Hydrogen sulphide	R	R	R	R	R	R	R	R	R	R	R		R	R	ND				R	ND	ND	R	R		No data		
Hypochlorites	R[30]	R[80]	R	R	R	ND	R[30,13]	R		R[80,76,13]			R	R		R			R	R	R	R[30]	ND		R	R	R
Lactic acid (100%)	R	R	ND	R	R	ND	R	R	R	R[14,80]			R	R	R	R	R		R	R	R	R[23]	R[23]		R	R	R
Lead acetate	R	R	R	R	R	R	R	R	R	R[80]	R	R	R	R	R	R	ND	ND	No data			R	R		R	R[30]	R
Lime (CaO)	R	R	R	R	R	R	R	R	R	R	R	R	R	R	R	R	R	R	R	R	R	R	R		R	R	R
Maleic acid	R	R	R	R	R	R	R	R	R	R	R	R	R	R	R	ND			No data			R	R		R	R	R
Meat juices	R	R	R	R	R	R	R[13]	R	R	R[13]	R	R	R	R	R	R	R	R	R	R	R	R	R		R	R	R
Mercuric chloride	R	R	R	R	R	R	R	R	R	R	R	R	R	R	R	R	R		R	R	R	R	R		R	R	R
Mercury	R	R	R	R	R	R	R	R	R	R	R	R	R	R	R	R	R		R	R	R	R	R		R	R	R
Milk & its products	R[80]	R	R	R[80]	R[80]		R	R	R				R	R	R	R	R	R	R	R	R	R	R		R	R	R
Moist air	R	R	R	R	R	R	R	R	R	R	R	R	R	R	R	R	R		R	R	R	R	R		R	R	R
Molasses	R	R	R	R	R	R	R	R	R	R	R	R	R	R	R	R	ND	ND	R[86]	R	R	R	R		R	R	R
Naphtha																R	R		R[38]						R[21]	R	R
Naphthalene																									R[21]	R	R
Nickel salts	R	R	R	R	R	R	R	R	R	R	R	R	R	R	R	R	R	R	R	R	R	R	R		R	R	R
Nitrates of Na, K, NH3	R	R	R	R	R	R	R	R	R	R	R	R	R	R	R	R	R		R	R	R	R	R		R	R	R
Nitric acid (<25%)	R[23]	R[23]	R[23]				R[101]	R		R[101]			R	R		R			R	R					R	R	R
Nitric acid (50%)																			R	R					R[21]		
Nitric acid (95%)																			R						R		
Nitric acid, fuming																			R[89]								
Oils, essential	R[14]	R	ND	R[60]	ND	ND	R[14]			No data			R[14]	ND	ND	R[4]	R[4]		R[30]			R	R		R[30]	R	R
Oils, mineral				R[80]	R[80]	R[80]							R	ND	ND	R	R	R	R[30]			R	R		R[30]	R	R
Oils, vegetable & animal	R	R[14]		R[14]	R[14]		R[80]	R	R	R[14]	R		R[14]	ND	ND	R	R		R[30]			R	R		R	R	R
Oxalic acid	R	R	R	No data			R	R	R	R[80]	R	R	R	R	R	R	R		R	ND	ND	ND	ND		R	R	R
Ozone	R	R		R	R		R	R	R				R	R	R	R[30]			R	R	R	R	R		R	R	R
Paraffin wax	R	R	R	R	R	ND	R	R	R	R	R[14]		R	R	R	R	R	R	R	R	R	R	R		R	R	R
Perchloric acid	R			R															No data			ND	ND		No data		
Phenol	R[80]	ND		R[80]			R[13,80]	R																			
Phosphoric acid (25%)	R	R	R	R	R	R	R	R	R[60]	R	R	R[60]	R	R	R	R	R	ND	R	R	R	R	R		R	R	R
Phosphoric acid (50%)	R	R	R	R	R	R	R	R	R[60]	R	R	R[60]	R	R	R	R			R	R	R	R	R		R[30]	R	R
Phosphoric acid (95%)	R	R	R	R	R	R	R[36]	R	R[60]	R[36]	R	R[60]	R	R	R	R						R	R	R	R	R	R
Phosphorus chlorides	R			No data			No data			No data			No data									ND	ND		No data		
Phosphorus pentoxide	R	R	ND	R	R	ND	R	R		R	R		R	R	R				No data			ND	ND		R	R	R
Phthalic acid	R[13]	R	R	R[13]	R[13]	ND	R[80]	R[80]	R[80]				R	R	R	R	ND	ND	R	R	R	R	R		R	R	R
Picric acid	R[80]	R	R	No data			R	R[30]	R	R	R[30]		R	R	R	R[13]			R	R	R	ND	ND		No data		
Pyridine	R[4]																					ND	ND		No data		
Sea water	R	R	R	R	R	R	R	R	R	R	R	R	R	R	R	R	R	R	R	R	R	R	R		R	R	R
Silicic acid	R	R	R	R	R	R	R	R	R	R	R	R	R	R	R	ND			R	R	R	R	R		R	R	R
Silicone fluids	R	R	ND	No data			R	R	R	R	R	R	R	R	R	R	R	R	R	R	ND	R	ND		R[21]	R	R
Silver nitrate	R	R	R	R[60]	R[60]	R[60]	R[51]			R[80]	R		R	R	R	R	R	ND	R	R	R	R	R		R	R	R
Sodium carbonate	R	R	R	R	R	R	R	R		R	R	R	R	R	R	R	R	R	R	R	R	R	R		R	R	R
Sodium peroxide	R	R	R	R	R	ND	R[13]	R		R[80]	R[80]	ND	R	R[97]	ND	R[13]			R	R	R	ND	ND		R	R	R

MISCELLANEOUS

Concrete (s)			Glass (t)			Graphite (u)			Porcelain and Stoneware			Vitreous Enamel (w)			Wood (z)		
20°	60°	100°	20°	60°	100°	20°	60°	100°	20°	60°	100°	20°	60°	100°	20°	60°	100°
			R	R	R	R	R	R	R	R	R	R	R	R	R	R	R
R	R	R	R	R	R	R	R	R	R	R	R	R	R	R	R	R	R
			R	R	R	R	R	R	R	R	R	R	R	R			
			R	R	R	R	R	R	R	R	R	R	R	R	R		
No data			R	ND	ND	R	R	R	R	R	R	R	R	R	R	R	R
No data			R	R	R	R	R	R	R	R	R	R	R	R	No data		
			R	R	R[50]	R	R	R	R	R	R	No data					
			R	R	R[50]	R	R	R	R	R	R	R	R	R	R		
			R	R	R[50]	R	R	R	R	R	R	R	ND	ND			
			R	R		R	R	R	R	R	R	R	ND	ND	R	R	
						R	R	R									
						R	R	R									
No data			R	R	R	R	R	R	R	R	R	R	R	R			
No data			R	R	R	R	R	R	R	R	R	R	R	R			
						R	R	R	R	R	R	R	R	R	R	R	R
R[72]	R	R	R	R	R	R	R[37]	R[37]	R	R	R	R	R	ND			
			No data			R	R	R	R	R	R	R	R	R	R		
No data			R	R	R	R	R	R	R	R	R	R	R	R	R	R	R
R	R	R	R	R	R	R	R	R	R	R	R	R	R	ND	R[14]		
No data			No data			R	R	R	R	R	R	R	R	R	R	R	R
No data			R	R	R	R	R	R	R	R	R	R	R	R	R	R	R
						R	R	R	R	R	R	No data			R	R	
R	R	R	R	R	R	R	R	R	R	R	R	R	R	ND	R	R	R
R[35]	R	R	R	R	R	R	R	R	R	R	R	R	R	R	R	R	R
R	R	R	R	R	R	R	R	R	R	R	R	R	R	R	R[14]	R[14]	R[14]
R			R	R	R	R	R	R	R	R	R	R	R	R	R	R	
R	R	R	No data			R	R	R	R	R	R	R	R	R	R	R	R
R	R	R	No data			R	R	R	R	R	R	R	R	R	R	R	R
No data			R	R	R	R	R	R	R	R	R	No data			R	R	R
R[73]	R	R	R	ND	ND	R	R	R	R	R	R	R	ND	ND	R	R	
			R	R	R[50]	R	R	R	R	R	R	R	R	R			
			R	R	R[50]				R	R	R	R	R	ND			
			R	R	R[50]				R	R	R	R	R	ND			
			R	R	R[50]				R	R	R	R	ND	ND			
No data			R	R	R	R	R	R	R	R	R	R	R	R	R	R	R
R	R	R	R	R	R	R	R	R	R	R	R	R	R	R	R	R	R
						R	R	R	R	R	R	R	R	R	R	R	R
R	R	R	R	R	R	R	R	R	R	R	R	R	R	R	R	R	R
R	R	R	R	R	R				R	R	R	No data			No data		
R	R	R	R	R	R	R[15]	R[10]		R	R	R	No data					
			R	ND	ND	R	R	R	R	R	R	R	R	R			
			R	R	R	R	R	R	R	R	R	R	R	R			
			R	R	R	R	R	R	R	R	R	R	R	ND			
			R	R	R	R	R	R	R	R	R	R4	R	ND			
			R	ND	ND	R	R	R	R	R	R	No data					
			R	R	R	R	R	R	R	R	R	No data					
			R	R	R	R	R	R	R	R	R	R	R	R	R	R	
No data			R	R	R	R	R	R	R	R	R	R	R	R	No data		
R	R	R	R	R	R	R	R	R	R	R	R	R	R	R	R[14]		
R	R	R	R	R	ND	R	R	R	R	R	R	R	R	ND	R	R	R
R[74]	R	R	R	R	R	R	R	R	R	R	R	R	R	R	R	R	R
No data			R	R	R	R	R	R	R	R	R	R	R	R	R	R	
R[72]	R	R	R			R	R	R	R	R	R	R	R	R			
R[72]	R	R	R	ND	ND	No data			R	R	R	No data					

73 Not ammonium
74 Not chlorsilanes
75 Data for ammonium
76 Data for calcium
77 Data for potassium
78 In presence of heavy metal ions
79 ND for Ba
80 Limited service
81 Except those containing sulphate
82 Provided less than 70% copper
83 Water less than 150 ppm
84 May cause some localised pitting
85 60% in one month
86 Low taste and odour
87 Catalyses decomp. of H_2O_2
88 65%
89 1–2 days
90 Wet gas
91 Less than 0–005% water
92 In absence of heavy metal ions oxidising agents
93 Stress corrosion in MeOH and halides (not in other alcohols)
94 When free of SO_2
95 50% swell in 28 days
96 60% swell in 3 days
97 Could be dangerous in black loaded compounds
98 Not alkaline
99 Ozone 2% Oxygen 98%
100 This is the softening point
101 Nitric acid less than 5% concentration
102 Acid fumes dry. Attack might occur if moisture present and concentrated condensate built up
103 Stainless steels not normally recommended for caustic applications
104 In the absence of impurities
105 10% w/w in alcohol
106 Swelling with some ketones
107 Some stress cracking at high pH

(a) **Aluminium**: In many cases where the chart indicates that aluminium is a suitable material there is some attack, but the corrosion is slight enough to allow aluminium to be used economically.

(b) **Brass**: Some types of brass have less corrosion resistance than is shown on the chart, others have more, e.g. Al brass.

(c) **Cast iron**: This is considered to be resistant if the material corrodes at a rate of less than 0.25 mm per annum. When choosing cast iron, Ni-Resist or high Si iron for a particular application the very different physical properties of these materials must be taken into account.

(d) **Gunmetal**: The data refer only to high tin gunmetals.

(e) **Nickel-copper alloys**: The physical properties are for annealed material. Both the tensile strength and hardness can vary with form and heat treatment condition.

(f) **Stainless steels**: Less expensive 13% chromium steels may be used for some applications instead of 18/8 steels. Under certain conditions the addition of titanium increases the corrosion resistance of 18/8 steels. Also, it produces materials which can be welded without the need for subsequent heat treatment. These steels are, however, inferior in corrosion resistance to the more expensive 18/8/Mo steels.

(g) **Tin**: Data refer to pure or lightly alloyed tin; not to discontinuous tin coatings.

(h) **Soft natural rubber and ebonite**: Performance at higher temperatures depends on method of compounding.

(i) **Neoprene**: Brush or spray applied 1.5 mm thick, and properly cured.

(k) **Silicone rubbers**: Withstand temperatures ranging from −90°C to above 250°C and are resistant to many oils and chemicals. In some cases particularly good resistance is shown by the fluorinated type.

RUBBERS

	Butyl Rubber and Halo-Butyl Rubber			Ethylene Propylene Rubber (q)			Hard Rubber (Ebonite) (h)			Soft Natural Rubber (h)			Neoprene (i)			Nitrile Rubber			Chlorosulphonated Polyethylene			Polyurethane Rubber (v)			Silicone Rubbers (k)		
	20°	60°	100°	20°	60°	100°	20°	60°	100°	20°	60°	100°	20°	60°	100°	20°	60°	100°	20°	60°	100°	20°	60°	100°	20°	60°	100°
Sodium silicate	R	R	R	R	R	R	R	R	R	R	R	R	R	R	R	R	R	ND	R	R	R	R	R		R	R	R
Sodium sulphide	R	R	R	R	R	R	R	R	R	R	R	R	R	R	R	R	R	ND	No data			R	ND		R	R	R
Stannic chloride	R	R	R	R	R	R	R	R	R	R	R	R	R	R	R	R	R	R	R	R	R	R	R		R	R	R
Starch	R	R	R	R	R	R	R	R	R	R	R	R	R	R	R	R	R	R	R	R	R	R	R	R	R	R	R
Sugar, syrups, jams	R[13]	R	R	R[60]	R[60]	R[60]	R[13]	R	R	R[13]	R	R	R	R	R	R	R	R	R	R	R	R	R	R	R	R	R
Sulphamic acid	No data			R	R	ND	R[13]	R		No data			R	ND	ND	No data			R	R	R	R	R		No data		
Sulphates (Na, K, Mg, Ca)	R	R	R	R	R	R	R	R	R	R	R	R	R	R	R	R	R	R	R	R	R	R	R		R	R	R
Sulphites	R	R	R	R	R	R	R	R	R	R[80]	R[80]		R	R	R	R	R	R	R	R	R	R	R		R	R	R
Sulphonic acids	R[13]	R		R[13]	R[13]	R[13]	R[2]	R[2]	R[2]				R	R	R	R	R	R	R	R	R	No data					
Sulphur	R	R	R	R	R	R	R	R	R	R			R	R	R	R[30]	R	R	R	R	R	ND	ND		R	R	R
Sulphur dioxide, dry	R	R	R	R	R	R	R	R	R				R	R	R							ND	ND		R	R	ND
Sulphur dioxide, wet	R	R	R	R	R	R	R	R	R[4]				R	R	R				R[4]	R	R	ND	ND		R	R	ND
Sulphur trioxide																No data									R	R	R
Sulphuric acid (<50%)	R	R	R	R	R	R	R	R	R	R	R		R	R		R			R	R	R	R[25]	R[25]_{80}		R	R	R
Sulphuric acid (70%)				R[80]			R[66]						R						R	R					No data		
Sulphuric acid (95%)																			R								
Sulphuric acid, fuming																											
Sulphur chlorides																No data			No data								
Tallow	R	R	R[4]	R	R[4]	ND	R	R	R	R	R		R	R	R	R	R	R	R	R	R	R	ND		R[30]	R	R
Tannic acid (10%)	R	R	R	R	R	R	R	R	R	R	R	R	R	R	R	R	R		R	R	R	R	R		R	R	R
Tartaric acid	R	R	R	R	R	R	R	R	R	R	R	R	R	R	R	R	R		R	R	R	R	R		R	R	R
Trichlorethylene																R[65]									R[21]	R	R
Vinegar	R	R	R	R	R[14]		R	R	R	R[80]	R		R	R	R	R	R[37]	R	R	R	R	R[80]	R[80]		R	R	R
Water, distilled	R	R	R	R	R	R	R[30]	R	R	R[30]	R	R	R	R	R	R	R	R	R	R	R	R	R		R	R	R
Water, soft	R	R	R	R	R	R	R	R	R	R	R	R	R	R	R	R	R	R	R	R	R	R	R		R	R	R
Water, hard	R	R	R	R	R	R	R	R	R	R	R	R	R	R	R	R	R	R	R	R	R	R	R		R	R	R
Yeast	R	R	R	No data			R	R	R	R	R	R	R	R	R	R	ND	ND	R	R	R	ND	ND		R	R	R
Zinc chloride	R	R	R	R	R	R	R	R	R	R	R	R	R	R	R	R	R	ND	R	R	R	R	R		R	R	R

MISCELLANEOUS																	
Concrete (s)			Glass (t)			Graphite (u)			Porcelain and Stoneware			Vitreous Enamel (w)			Wood (z)		
20°	60°	100°	20°	60°	100°	20°	60°	100°	20°	60°	100°	20°	60°	100°	20°	60°	100°
R	R	R				R	R	R	R	R	R	No data			R		
			R	R	R	R	R	R	R	R	R	No data					
			R	R	R	R	R	R	R	R	R	R	R	R			
R	R	R	R	R	R	R	R	R	R	R	R	R	R	R	R	R	R
			R	R	R	R	R	R	R	R	R	R	R	R	R	R	R
No data			R	R	R	R	R	R	R	R	R	No data			No data		
			R	R	R	R	R	R	R	R	R	R	R	ND			
			R	R	R	R	R	R	R	R	R	R	R	R	R		
			R	R	R	R	R	R	R	R	R	R	R	ND			
R	R	R	R	R	R	R	R	R	R	R	R	R	R	R	R	R	
No data			R	R	ND	R	R	R	R	R	R	R	R	R			
			R	R		R	R	R	R	R	R	R	R	R			
			R	R					R	R	R	R	R	R			
			R	R	R	R	R	R	R	R	R	R	R	R	R^{10}		
			R	R	R	R	R	R	R	R	R	R	R	R			
			R	R	R	R^2	R		R	R	R	R	R	R			
			R	R	R				R	R	R	R	ND	ND			
No data			R	R	R	R	R	R	R	R	R	R	R	ND	No data		
R^{80}			No data			R	R	R	R	R	R	R	R	R	R	R	R
			R	R	R	R	R	R	R	R	R	R	R	R	R	R	R
			R	ND	ND	R	R	R	R	R	R	R	R	R	R	R	R
R	R	R	R	R	R	R	R	R	R	R	R	R	R	R	R	R	R
			R	R	R	R	R	R	R	R	R	R	R	R	R	R	
R^{50}	R	R	R	R	R	R	R	R	R	R	R	R	R	R	R	R	R
R^{50}	R	R	R	R	R	R	R	R	R	R	R	R	R	R	R	R	R
R	R	R	R	R	R	R	R	R	R	R	R	R	R	R	R	R	R
R	R	R	R	R	R	R	R	R	R	R	R	R	R	R	R	R	R
			R	ND	ND	R	R	R	R	R	R	R	R	ND			

(l) **Acrylonitrile butadiene styrene resins**: The information refers to a general purpose moulding grade material.

(m) **Nylon**: Prolonged heating may cause oxidation and embrittle ment. Data on nylon 66 plastics refer to *Maranyl* products. Other nylons, such as types 6 and 610, can behave differently, e.g. towards aqueous solutions of salts.

(n) **P.T.F.E.**: Is attacked by alkali metals (molten or in solution) and by certain rare fluorinated gases at high temperatures and pressures. Some organic and halogenated solvents can cause swelling and slight dimensional changes but the effects are physical and reversible.

(o) **Melamine resins**: The information refers mainly to laminates surfaced with melamine resins, Melamine coating resins are always used in conjunction with alkyd resins and the specifi- cations will depend on the alkyd resin used.

(p) **Epoxy resins**: Data are for cold curing systems.

(q) The information given is based on compounds made from ethylene propylene terpolymer rubber.

(r) **Phenol formaldehyde resins**: These are of several types and care should be taken that the right type is chosen.

(s) **Concrete**: Usually made from Portland cement, but if made from Ciment Fondu or gypsum slag cement might have superior resistance in particular applications.

(t) **Glass**: The information refers to heat-resistant borosilicate glass.

(u) **Graphite**: Data refer to resin-impregnated graphite. Other specially treated graphites have improved corrosion resist- ance to many chemicals.

(v) Chemical resistance of polyurethanes is dependent on the particular structure of the material and is not necessarily applicable to all polyurethanes. Specially designed poly- urethanes can be used at higher temperatures than 60°C but chemical resistance is temperature dependent.

(w) **Vitreous enamel**: Special enamels may be required to withstand particular reagents.

(x) Data is based on Ferralium alloy 255.

(y) Data is based on Solef.

(z) **Wood**: The behaviour of wood depends both on the species used and on the physical conditions of service. Aqueous solutions of some chemicals may cause more rapid degrad- ation. Organic solvents may dissolve out resins, etc. Hydro- gen peroxide (over 50% w/w) produces a fire risk.

C PHYSICAL PROPERTY DATA BANK

Inorganic compounds are listed in alphabetical order of the principal element in the empirical formula.

Organic compounds with the same number of carbon atoms are grouped together, and arranged in order of the number of hydrogen atoms, with other atoms in alphabetical order.

A searchable spreadsheet containing the physical property data and models is available in the on-line material at http://elsevierdirect.com/companions.

$$\begin{aligned}
\text{NO} &= \text{Number in list} \\
\text{MOLWT} &= \text{Molecular weight} \\
\text{TFP} &= \text{Normal freezing point, deg C} \\
\text{TBP} &= \text{Normal boiling point, deg C} \\
\text{TC} &= \text{Critical temperature, deg K} \\
\text{PC} &= \text{Critical pressure, bar} \\
\text{VC} &= \text{Critical volume, cubic metre/mol} \\
\text{LDEN} &= \text{Liquid density, kg/cubic metre} \\
\text{TDEN} &= \text{Reference temperature for liquid density, deg C} \\
\text{HVAP} &= \text{Heat of vaporization at normal boiling point, J/mol} \\
\text{VISA, VISB} &= \text{Constants in the liquid viscosity equation:}
\end{aligned}$$

$$\text{LOG[viscosity]} = [\text{VISA}]\,^{*}\,[(1/T) - (1/\text{VISB})], \text{ viscosity mNs/m}^2, \text{ T deg K}$$

DELHF = Standard enthalpy of formation of vapour at 298 K, kJ/mol
DELGF = Standard Gibbs energy of formation of vapour at 298 K, kJ/mol
CPVAPA, CPVAPB, CPVAPC, CPVAPD = Constants in the ideal gas heat capacity equation:

$$\text{Cp} = \text{CPVAPA} + (\text{CPVAPB})\,^{*}T + (\text{CPVAPC})\,^{*}T\,^{**}2 + (\text{CPVAPD})\,^{*}T\,^{**}3$$
Cp J/mol K, T deg K

ANTA, ANTB, ANTC = Constants in the Antoine equation:

$$\text{Ln (vapour pressure)} = \text{ANTA} - \text{ANTB}/(T + \text{ANTC}), \text{ vapour pressure in mmHg, T in K}$$

To convert mmHg to N/m^2 multiply by 133.32
To convert degrees Celsius to Kelvin add 273.15

TMN = Minimum temperature for Antoine constant, deg C
TMX = Maximum temperature for Antoine constant, deg C

Most of the values in this data bank were taken, with the permission of the publishers, from: The Properties of Gases and Liquids, by Reid, R. C., Sherwood, T. K. and Prausnitz, J. M., 3rd edn, McGraw-Hill.

NO	FORMULA	COMPOUND NAME	MOLWT	TFP	TBP	TC	PC	VC	LDEN	TDEN	HVAP	NO
1	AR	ARGON	39.948	-189.9	-185.9	150.8	48.7	0.075	1373	-183	6531	1
2	BCL3	BORON TRICHLORIDE	117.169	-107.3	12.5	452.0	38.7		1350	11		2
3	BF3	BORON TRIFLUORIDE	67.805	-126.7	-99.9	260.8	49.9					3
4	BR2	BROMINE	159.808	-7.2	58.7	584.0	103.4	0.127	3119	20	30,187	4
5	CLNO	NITROSYL CHLORIDE	65.459	-59.7	-5.5	440.0	91.2	0.139	1420	-12	25,707	5
6	CL2	CHLORINE	70.906	-101.0	-34.5	417.0	77.0	0.124	1563	-34	20,432	6
7	CL3P	PHOSPHORUS TRICHLORIDE	137.333	-112.2	75.8	563.0		0.260	1574	21	27,549	7
8	CL4SI	SILICON TETRACHLORIDE	169.898	-68.9	57.2	507.0	37.5	0.326	1480	20		8
9	D2	DEUTERIUM	4.032	-254.5	-249.5	38.4	16.6	0.060	165	-250	1223	9
10	D2O	DEUTERIUM OXIDE	20.031	3.8	101.4	644.0	216.6	0.056	1105	20	41,366	10
11	F2	FLUORINE	37.997	-219.7	-188.2	144.3	52.2	0.066	1510	-188	6531	11
12	F3N	NITROGEN TRIFLUORIDE	71.002	-206.8	-129.1	234.0	45.3		1537	-129		12
13	F4SI	SILICON TETRAFLUORIDE	104.080	-90.2	-86.2	259.0	37.2		1660	-95		13
14	F6S	SULPHUR HEXAFLUORIDE	146.050	-50.7	-63.9	318.7	37.6	0.198	1830	-50		14
15	HBR	HYDROGEN BROMIDE	80.912	-86.1	-67.1	363.2	85.5	0.100	2160	-57	17,668	15
16	HCL	HYDROGEN CHLORIDE	36.461	-114.2	-85.1	324.6	83.1	0.081	1193	-85	16,161	16
17	HF	HYDROGEN FLUORIDE	20.006	-83.2	19.5	461.0	64.8	0.069	967	20	6699	17
18	H1	HYDROGEN IODIDE	127.912	-50.8	-35.6	424.0	83.1	0.131	2803	-36	19,778	18
19	H2	HYDROGEN	2.016	-259.2	-252.8	33.2	13.0	0.065	71	-253	904	19
20	H2O	WATER	18.015	0.0	100.0	647.3	220.5	0.056	998	20	40,683	20
21	H2S	HYDROGEN SULPHIDE	34.080	-85.6	-60.4	373.2	89.4	0.099	993	-60	18,673	21
22	H3N	AMMONIA	17.031	-77.8	-33.5	405.6	112.8	0.073	639	0	23,362	22
23	H3P	PHOSPHINE	33.998	-133.8	-87.5	324.8	62.7	0.113			14,725	23
24	H4N2	HYDRAZINE	32.045	1.5	113.5	653.0	146.9	0.096	1008	20	44,799	24
25	H4SI	SILANE	32.112	-185.0	-112.2	269.7	48.4		680	88		25
26	HE(4)	HELIUM-4	4.003		-269.0	5.2	2.3	0.057	123	-269	92	26
27	I2	IODINE	253.808	113.6	184.3	819.0	116.5	0.155	3740	180	41,868	27
28	KR	KRYPTON	83.800	-157.4	-153.4	209.4	55.0	0.091	2420	-153	9667	28
29	NO	NITRIC OXIDE	30.006	-163.7	-151.8	180.0	64.8	0.058	1280	-152	13,816	29
30	NO2	NITROGEN DIOXIDE	46.006	-11.3	21.1	431.4	101.3	0.170	1447	20	19,071	30
31	N2	NITROGEN	28.013	-209.9	-195.8	126.2	33.9	0.090	805	-195	5581	31
32	N2O	NITROUS OXIDE	44.013	-90.9	-88.5	309.6	72.4	0.097	1226	-90	16,559	32
33	NE	NEON	20.183	-248.7	-246.2	44.4	27.6	0.042	1204	-246	1842	33
34	O2	OXYGEN	31.999	-218.8	-183.0	154.6	50.5	0.073	1149	-183	6824	34
35	O2S	SULPHUR DIOXIDE	64.063	-75.5	-10.2	430.8	78.8	0.122	1455	-10	24,932	35
36	O3	OZONE	47.998	-192.7	-111.9	261.0	55.7	0.089	1356	-112	11,179	36
37	O3S	SULPHUR TRIOXIDE	80.058	16.8	44.8	491.0	82.1	0.130	1780	45	40,679	37
38	XE	XENON	131.300	-111.9	-108.2	289.7	58.4	0.118	3060	-108	13,013	38
39	CBRF3	TRIFLUOROBROMOMETHANE	148.910	-168.0	-59.2	340.2	39.7	0.200				39

NO	VISA	VISB	DELHF	DELGF	CPVAPA	CPVAPB	CPVAPC	CPVAPD	ANTA	ANTB	ANTC	TMN	TMX	NO
1	107.57	58.76			20.804	-3.211E-05	51.665E-09		15.2330	700.51	-5.84	-192	-179	1
2														2
3														3
4	387.82	292.79			33.859	11.254E-03	-1.192E-05	45.343E-10	15.8441	2582.32	-51.56	-14	81	4
5			52.63	66.99	34.097	44.715E-03	-3.340E-05	10.149E-09	16.9505	2520.70	-23.46	-63	12	5
6	191.96	172.35			26.929	33.838E-03	-3.869E-05	15.470E-09	15.9610	1978.32	-27.01	-101	-9	6
7														7
8	19.67	8.38			30.250	-6.406E-03	11.698E-06	-3.684E-09	15.8019	2634.16	-43.15	-35	91	8
9									13.2954	157.89		-254	-248	9
10	757.92	304.58	-249.41	-234.80					15.6700	714.10	-6.00	-214	-182	10
11	84.20	52.52	-124.68	-127.19	23.216	36.568E-03	-3.462E-05	12.041E-09	15.6107	1155.69	-15.37	-170	-118	11
12														12
13														13
14	251.29	180.75	-1221.71	-1117.88	30.647	-9.462E-03	-17.224E-06	-6.238E-09	19.3785	2524.78	-11.16	-114	-53	14
15	88.08	166.32	-36.26	-53.30	30.291	-7.201E-03	12.460E-06	-3.898E-09	14.4687	1242.53	-47.86	-89	-52	15
16	372.78	277.74	-92.36	-95.33	29.061	66.110E-05	-2.032E-06	25.037E-10	16.5040	1714.25	-14.45	-136	-73	16
17	438.74	199.62	-271.30	-273.40	31.158	-1.428E-02	29.722E-06	-1.353E-08	17.6958	3404.49	15.06	-67	40	17
18	155.15	285.43			27.143	92.738E-04	-1.381E-05	76.451E-10	12.9149	957.96	-85.06	-58	-17	18
19	13.82	5.39	26.38	1.59					13.6333	164.90	3.19	-259	-248	19
20	658.25	283.16	-242.00	-228.77	32.243	19.238E-04	10.555E-06	-3.596E-09	18.3036	3816.44	-46.13	11	168	20
21	342.79	165.54	-20.18	-33.08	31.941	14.365E-04	24.321E-06	-1.176E-08	16.1040	1768.69	-26.06	-83	-43	21
22	349.02	169.63	-45.72	-16.16	27.315	23.831E-03	17.074E-06	-1.185E-08	16.9481	2132.50	-32.98	-94	-12	22
23			229.44	158.64	23.228	44.003E-03	13.029E-06	-1.593E-08						23
24	524.98	290.88	95.25		9.768	18.945E-02	-1.657E-04	60.248E-09	17.9899	3877.65	-45.15	15	70	24
25			32.66	54.93	11.179	12.200E-02	-5.548E-05	68.412E-10	16.3424	1629.99	5.35	-111	-179	25
26									12.2514	33.73	1.79	-269	-269	26
27	559.62	520.55			35.592	65.147E-04	-6.988E-06	28.345E-10	16.1597	3709.23	-68.16	110	214	27
28									15.2677	958.75	-8.71	-160	-144	28
29	406.20	230.21	90.43	86.75	29.345	-9.378E-04	97.469E-07	-4.187E-09	20.1314	1572.52	-4.88	-178	-133	29
30	90.30	46.14	33.87	52.00	24.243	48.358E-03	-2.081E-05	29.308E-11	20.5324	4141.29	3.65	-43	47	30
31			81.60	103.71	31.150	-1.357E-02	26.796E-06	-1.168E-08	14.9542	588.72	-6.60	-219	-183	31
32					21.621	72.808E-03	-5.778E-05	18.301E-09	16.1271	1506.49	25.99	-129	-73	32
33					20.786				14.0099	180.47	-2.61	-249	-244	33
34	85.68	51.50	-297.05	-300.36	28.106	-3.680E-06	17.459E-06	-1.065E-08	15.4075	734.55	-6.45	-210	-173	34
35	397.85	208.42	142.77	162.91	23.852	66.989E-03	-4.961E-05	13.281E-09	16.7680	2302.35	-35.97	-78	7	35
36	313.79	120.34	-395.53	-370.62	20.545	80.093E-03	-6.243E-05	16.973E-09	15.7427	1272.18	-22.16	-164	-99	36
37	1372.80	315.99			16.370	14.591E-02	-1.120E-04	32.423E-09	20.8403	3995.70	-36.66	17	59	37
38														38
39			-649.37	-623.00					15.2958	1303.92	-14.50	-115	-95	39

NO	FORMULA	COMPOUND NAME	MOLWT	TFP	TBP	TC	PC	VC	LDEN	TDEN	HVAP	NO
40	CCLF3	CHLOROTRIFLUOROMETHANE	104.459	-181.2	-81.5	302.0	39.2	0.180			15,516	40
41	CCL2F2	DICHLORODIFLUOROMETHANE	120.914	-157.8	-29.8	385.0	41.2	0.217	1750	-115	19,979	41
42	CCL2O	PHOSGENE	98.916	-128.2	7.6	455.0	56.7	0.190	1361	20	24,409	42
43	CCL3F	TRICHLOROFLUOROMETHANE	137.368	-111.2	23.8	471.2	44.1	0.248			24,786	43
44	CCL4	CARBON TETRACHLORIDE	153.823	-23.2	76.5	556.4	45.6	0.276	1584	25	30,019	44
45	CF4	CARBON TETRAFLUORIDE	88.005	-186.8	-128.0	227.6	37.4	0.140			11,974	45
46	CO	CARBON MONOXIDE	28.010	-205.1	-191.5	132.9	35.0	0.093	803	-192	6046	46
47	COS	CARBONYL SULPHIDE	60.070	-138.9	-50.3	375.0	58.8	0.140	1274	-99		47
48	CO2	CARBON DIOXIDE	44.010	-56.6	-78.5	304.2	73.8	0.094	777	20	17,166	48
49	CS2	CARBON DISULPHIDE	76.131	-111.9	46.2	552.0	79.0	0.170	1293	0	26,754	49
50	CHBR3	BROMOFORM	94.940	-178.3	3.5	464.0	66.1	0.162	1733	0	24,241	50
51	CHCLF2	CHLORODIFLUOROMETHANE	86.469	-160.2	-40.8	369.2	49.8	0.165	1230	16	20,205	51
52	CHCL2F	DICHLOROFLUOROMETHANE	102.923	-135.2	8.8	451.6	51.7	0.197	1380	9	24,953	52
53	CHCL3	CHLOROFORM	119.378	-63.6	61.1	536.4	54.7	0.239	1489	20	29,726	53
54	CHN	HYDROGEN CYANIDE	27.026	-13.3	25.7	456.8	53.9	0.139	688	20	25,234	54
55	CH2BR2	DIBROMOMETHANE	173.835	-52.6	96.8	583.0	71.9		2500	20		55
56	CH2CL2	DICHLOROMETHANE	84.993	-95.1	39.8	510.0	60.8	0.193	1317	25	28,010	56
57	CH2O	FORMALDEHYDE	30.026	-117.2	-19.2	408.0	65.9		815	-20	23,027	57
58	CH2O2	FORMIC ACID	46.025	8.3	100.6	580.0			1226	15	21,939	58
59	CH3BR	METHYL BROMIDE	94.939	-93.7	3.5	464.0	86.1	0.139	1737	-5	23,928	59
60	CH3CL	METHYL CHLORIDE	50.488	-97.8	-24.3	416.3	66.8	0.139	915	20	21,436	60
61	CH3F	METHYL FLUORIDE	34.033	-141.8	-78.4	317.8	58.8	0.124	843	-60		61
62	CH3I	METHYL IODIDE	141.939	-66.5	42.4	528.0	65.9	0.190	2279	20	27,214	62
63	CH3NO2	NITROMETHANE	61.041	-28.6	101.2	588.0	63.1	0.173	1138	20	34,436	63
64	CH4	METHANE	16.043	-182.5	-161.5	190.6	46.0	0.099	425	-161	8185	64
65	CH4O	METHANOL	32.042	-97.7	64.6	512.6	81.0	0.118	791	20	35,278	65
66	CH4S	METHYL MERCAPTAN	48.107	-123.2	5.9	470.0	72.3	0.145	866	20	24,577	66
67	CH5N	METHYL AMINE	31.058	-93.5	-6.4	430.0	74.6	0.140	703	-14	26,000	67
68	CH6N2	METHYL HYDRAZINE	46.072		90.8	567.0	80.4	0.271				68
69	CH6SI	METHYL SILANE	46.145	-156.5	-57.6	352.5						69
70	C2CLF5	CHLOROPENTAFLUOROETHANE	154.467	-106.2	-39.2	353.2	31.6	0.252	1455	25	19,469	70
71	C2CL2F4	1;1-DICHLORO-1;2;2;2-TETRAFLUOROETHANE	170.992	-94.2	3.8	418.6	33.0	0.294	1480	25		71
72	C2CL2F4	1;2-DICHLORO-1;1;2;2-TETRAFLUOROETHANE	170.922	-93.9	3.7	418.9	32.6	0.293	1580	4	23,279	72
73	C2CL3F3	1;2-DICHLORO-1;1;2-TETRAFLUOROETHANE	187.380	-35.0	47.5	487.2	34.1	0.304	1620	16	27,507	73
74	C2CL4	TETRACHLOROETHYLENE	165.834	-22.2	121.1	620.0	44.6	0.290	1645	20	34,750	74
75	C2CL4F2	1;1;2;2-TETRACHLORO-1;2-DIFLUOROETHANE	203.831	24.8	91.5	551.0				25		75
76	C2F4	TETRAFLUOROETHYLENE	100.016	-142.5	-75.7	306.4	39.4	0.175	1519	-76		76
77	C2F6	HEXAFLUOROETHANE	138.012	-100.8	-78.3	292.8		0.224	1590	-78	16,161	77
78	C2N2	CYANOGEN	52.035	-27.9	-20.7	400.0	59.8					78

NO	VISA	VISB	DELHF	DELGF	CPVAPA	CPVAPB	CPVAPC	CPVAPD	ANTA	ANTB	ANTC	TMN	TMX	NO
40			−695.01	−654.40	22.814	19.113E-02	−1.576E-04	44.589E-09						40
41	215.09	165.55	−481.48	−442.54	31.598	17.823E-02	−1.509E-04	43.417E-09						41
42			−221.06	−206.91	28.089	13.607E-02	−1.374E-04	50.702E-09	15.7565	2167.31	−43.15	−60	68	42
43			−284.70	−245.51	40.985	16.308E-02	−1.416E-04	41.462E-09	15.8516	2401.61	−36.30	−33	27	43
44	540.15		−100.48	−58.28	40.717	20.486E-02	−2.270E-04	88.425E-09	15.8742	2808.19	−45.99	−20	101	44
45		290.84	−933.66	−889.03	13.980	20.256E-02	−1.625E-04	45.134E-09	16.0543	1244.55	−13.06	−180	−125	45
46	94.06	48.90	−110.62	−137.37	30.869	−1.285E-02	27.892E-06	−1.272E-08	14.3686	530.22	−13.15	−210	−165	46
47			−138.50	−165.76	23.567	79.842E-03	−7.017E-05	24.535E-09						47
48	578.08	185.24	−393.77	−394.65	19.795	73.436E-03	−5.602E-05	17.153E-09	22.5898	3103.39	−0.16	−119	−69	48
49	274.08	200.22	117.15	66.95	27.444	81.266E-03	−7.666E-05	26.729E-09	15.9844	2690.85	−31.62	−45	69	49
50			−36.34						15.7078	3163.17	−72.18	101	30	50
51			−502.00	−470.89	17.300	16.182E-02	−1.170E-04	30.585E-09	15.5602	1704.80	−41.30	−48	−33	51
52			−298.94	−268.37	23.664	15.814E-02	−1.200E-04	32.636E-09						52
53	394.81	246.50	−101.32	−68.58	24.003	18.933E-02	−1.841E-04	66.570E-09	15.9732	2696.79	−46.16	−13	97	53
54	194.70	145.31	130.63	120.20	21.863	60.625E-03	−4.961E-05	18.154E-09	16.5138	2585.80	−37.15	−39	57	54
55	428.91	294.57	−4.19	−5.61										55
56	359.55	225.13	−95.46	−68.91	12.954	16.232E-02	−1.302E-04	42.077E-09	16.3029	2622.44	−41.70	−44	59	56
57	319.83	171.35	−115.97	−109.99	23.475	31.568E-03	29.852E-06	−2.300E-08	16.4775	2204.13	−30.15	−88	−2	57
58	729.35	325.72	−378.86	−351.23	11.715	13.578E-02	−8.411E-05	20.168E-09	16.9882	3599.58	−26.09	−2	136	58
59	298.15	211.15	−37.68	−28.18	14.428	10.911E-02	−5.401E-05	95.836E-10	16.0252	2271.71	−34.83	−58	53	59
60	426.45	193.56	−86.37	−62.93	13.875	10.140E-02	−3.889E-05	25.665E-10	16.1052	2077.97	−29.55	−93	−7	60
61			−234.04	−210.14	13.825	86.164E-03	−2.071E-05	−1.985E-09	16.3428	1704.41	−19.27	−132	−64	61
62	336.19	229.95	13.98	15.66	10.806	13.892E-02	−1.041E-04	34.855E-09	16.0905	2629.55	−36.50	−13	52	62
63	452.80	261.21	−74.78	−6.95	7.423	19.778E-02	−1.081E-04	20.850E-09	16.2193	2972.64	−64.15	5	136	63
64	114.14	57.60	−74.86	−50.87	19.251	52.126E-03	11.974E-06	−1.132E-08	15.2243	597.84	7.16	−180	−153	64
65	555.30	260.64	−201.30	−162.62	21.152	70.924E-03	25.870E-06	−2.852E-08	18.5875	3626.55	−34.29	−16	91	65
66			−22.99	−9.92	13.268	14.566E-02	−8.545E-05	20.750E-09	16.1909	2338.38	−34.44	−73	27	66
67	311.80	176.30	−23.03	32.28	11.476	14.273E-02	−5.334E-05	47.520E-10	17.2622	2484.83	−32.92	−61	38	67
68			85.41	177.98					15.1424	2319.84	−91.70	−3	127	68
69														69
70					27.834	34.918E-02	−2.891E-04	81.391E-09	15.7343	1848.90	−30.88	−98	−43	70
71					40.453	32.783E-02	−2.752E-04	78.209E-09						71
72			−898.49		38.778	34.399E-02	−2.950E-04	85.076E-09						72
73			−745.67		61.140	28.742E-02	−2.420E-04	69.040E-09						73
74	392.58	281.82	−12.14	22.61	45.971	22.554E-02	−2.294E-04	83.820E-09	15.8424	2532.61	−45.67	−23	87	74
75									16.1642	3259.29	−52.15	34	187	75
76			−659.00	−624.13	29.010	22.772E-02	−2.037E-04	67.784E-09						76
77			−1343.96	−1258.22	26.816	34.579E-02	−2.869E-04	81.350E-09	15.8800	1574.60	−27.22	−133	−63	77
78			309.15	297.39	35.935	92.528E-03	−8.223E-05	29.496E-09	15.6422	1512.94	−26.94	−103	−73	78

NO	FORMULA	COMPOUND NAME	MOLWT	TFP	TBP	TC	PC	VC	LDEN	TDEN	HVAP	NO
79	C2HCL3	TRICHLOROETHYLENE	131.389	-116.4	87.2	571.0	49.1	0.256	1462	20	31,401	79
80	C2HF3O2	TRIFLUROROACETIC ACID	114.024	-15.3	72.4	491.3	32.6		1535	0	16,957	80
81	C2H2	ACETYLENE	26.038	-80.8	-84.0	308.3	61.4	0.113	615	-84		81
82	C2H2F2	1;1-DIFLUOROETHYLENE	64.035			302.8	44.6	0.154				82
83	C2H2O	KETENE	42.038	-135.2	-41.2	380.0	64.8	0.145			20,641	83
84	C2H3CL	VINYL CHLORIDE	62.499	-153.8	-13.4	429.7	56.0	0.169	969	-14	20,641	84
85	C2H3CLF2	1-CHLORO-1;1-DIFLUOROETHANE	100.490	-131.2	-9.8	410.2	41.2	0.231	1100	30		85
86	C2H3CLO	ACETYL CHLORIDE	78.498	-113.0	50.7	508.0	58.8	0.204	1104	20	28,680	86
87	C2H3CL3	1;1;2-TRICHLOROETHANE	133.400	-36.7	113.7	602.0	41.5	0.294	1441	20	33,327	87
88	C2H3F	VINYL FLUORIDE	46.044	-143.2	-37.7	327.8	52.4	0.144				88
89	C2H3F3	1;1;1-TRIFLUOROETHANE	84.041	-111.3	-47.7	346.2	37.6	0.221	782	20	19,176	89
90	C2H3N	ACETONITRILE	41.053	-43.9	81.6	548.0	48.3	0.173	958	20	31,401	90
91	C2H3NO	METHYL ISOCYANATE	57.052		38.8	491.0	55.7				29,601	91
92	C2H4	ETHYLENE	28.054	-169.2	-103.8	282.4	50.4	0.129	577	-110	13,553	92
93	C2H4CL2	1;1-DICHLOROETHANE	98.960	-97.0	57.2	523.0	50.7	0.240	1168	25	28,721	93
94	C2H4CL2	1;2-DICHLOROETHANE	98.960	-35.7	83.4	561.0	53.7	0.220	1250	16	32,029	94
95	C2H4F2	1;1-DIFLUOROETHANE	66.051	-117.0	-24.8	386.6	45.0	0.181			21,353	95
96	C2H4O	ACETALDEHYDE	44.054	-123.0	20.4	461.0	55.7	0.154	778	20	25,749	96
97	C2H4O	ETHYLENE OXIDE	44.054	-112.2	10.3	469.0	71.9	0.140	899	0	25,623	97
98	C2H4O2	ACETIC ACID	60.052	16.6	117.9	594.4	57.9	0.171	1049	20	23,697	98
99	C2H4O2	METHYL FORMATE	60.052	-99.0	31.7	487.2	60.0	0.172	974	20	28,219	99
100	C2H5BR	ETHYL BROMIDE	108.966	-118.6	38.3	503.8	62.3	0.215	1451	25	26,502	100
101	C2H5CL	ETHYL CHLORIDE	64.515	-136.4	12.2	460.4	52.7	0.199	896	20	24,702	101
102	C2H5F	ETHYL FLUORIDE	48.060	-143.3	-37.8	375.3	50.3	0.169				102
103	C2H5N	ETHYLENE IMIDE	43.069	-78.2	56.6				833	25	32,071	103
104	C2H5NO2	NITROETHANE	75.068	-89.2	114.0	595.0	48.5	0.228	1047	20	35,994	104
105	C2H6	ETHANE	30.070	-183.3	-88.7	305.4	48.8	0.148	548	-90	14,717	105
106	C2H6O	DIMETHYL ETHER	46.069	-141.5	-24.9	400.0	53.7	0.178	667	20	21,520	106
107	C2H6O	ETHANOL	46.069	-114.1	78.3	516.2	63.8	0.167	789	20	38,770	107
108	C2H6O2	ETHYLENE GLYCOL	62.069	-13.0	197.2	645.0	77.0	0.186	1114	20	52,544	108
109	C2H6S	ETHYL MERCAPTAN	62.134	-147.9	35.0	499.0	54.9	0.207	839	20	26,796	109
110	C2H6S	DIMETHYL SULPHIDE	62.130	-98.3	37.3	503.0	55.3	0.201	848	20	26,963	110
111	C2H7N	ETHYL AMINE	45.085	-81.2	16.5	456.0	56.2	0.178	683	20	28,052	111
112	C2H7N	DIMETHYL AMIDE	45.085	-92.2	6.8	437.6	53.1	0.187	656	20	26,502	112
113	C2H7NO	MONOETHANOLAMINE	61.084	10.3	170.3	614.0	44.6	0.196	1016	20	50,242	113
114	C2H8N2	ETHYLENEDIAMINE	60.099	10.8	117.2	593.0	62.8	0.206	896	20	41,868	114
115	C3H3N	ACRYLONITRILE	53.064	-83.7	77.3	536.0	35.5	0.210	806	20	32,657	115
116	C3H4	PROPADIENE	40.065	-136.3	-34.5	393.0	54.7	0.162	658	-35	18,631	116
117	C3H4	METHYL ACETYLENE	40.065	-102.7	-23.2	402.4	56.2	0.164	706	-50	22,148	117

NO	VISA	VISB	DELHF	DELGF	CPVAPA	CPVAPB	CPVAPC	CPVAPD	ANTA	ANTB	ANTC	TMN	TMX	NO
79	145.67	196.60	-5.86	19.89	30.174	22.868E-02	-2.229E-04	82.438E-09	16.1827	3028.13	-43.15	-13	127	79
80														80
81			226.88	209.34	26.821	75.781E-03	-5.007E-05	14.122E-09	16.3481	1637.14	-19.77	-79	-71	81
82			-345.41	-321.71	3.073	24.447E-02	-2.099E-04	70.213E-09						82
83			-61.13	-60.33	6.385	16.383E-02	-1.084E-04	26.984E-09	16.0197	1849.21	-35.15	-103	-18	83
84	276.90	167.04	35.17	51.54	5.949	20.193E-02	-1.536E-04	47.730E-09	14.9601	1803.84	-43.15	-88	17	84
85					16.818	27.566E-02	-1.992E-04	53.047E-09						85
86			-244.09	-206.37	25.020	17.107E-02	-9.856E-05	22.190E-09	15.7514	2447.33	-55.53	-36	82	86
87	346.72	304.43	-138.58	-77.54	6.322	34.307E-02	-2.958E-04	97.929E-09	16.0381	3110.79	-56.16	29	155	87
88														88
89			-746.09	-679.22	5.744	31.409E-02	-2.597E-04	84.155E-09	15.8965	1814.91	-29.92	-3	27	89
90	334.91	210.05	87.92	105.67	20.482	10.831E-02	-4.492E-05	32.029E-10	16.2874	2945.47	-49.15	-13	117	90
91	616.78	227.47	-90.02		35.764	10.396E-02	-5.820E-06	-1.687E-08	16.3258	2480.37	-56.31	-43	67	91
92	168.98	93.94	52.33	68.16	3.806	15.659E-02	-8.348E-05	17.551E-09	15.5368	1347.01	-18.15	-153	91	92
93	412.27	239.10	-130.00	-73.14	12.472	26.959E-02	-2.050E-04	63.011E-09	16.0842	2697.29	-45.03	-31	79	93
94	473.95	277.98	-129.79	-73.90	20.486	23.103E-02	-1.438E-04	33.888E-09	16.1764	2927.17	-50.22	-33	100	94
95	319.27	186.56	-494.04	-436.52	8.675	23.957E-02	-1.457E-04	33.942E-09	16.1871	2095.35	-29.16	-35	0	95
96	368.70	192.82	-166.47	-133.39	7.716	18.225E-02	-1.007E-04	23.802E-09	16.2418	2465.15	-37.15	-63	47	96
97	341.88	194.22	-52.67	-13.10	-7.519	22.224E-02	-1.256E-04	25.916E-09	16.7400	2567.61	-29.01	-73	37	97
98	600.94	306.21	-435.13	-376.94	4.840	25.485E-02	-1.753E-04	49.488E-09	16.8080	3405.57	-56.34	17	157	98
99	363.19	212.70	-350.02	-297.39	1.432	27.001E-02	-1.949E-04	57.024E-09	16.5104	2590.87	-42.60	-48	51	99
100	369.80	220.68	-64.06	-26.33	6.657	23.480E-02	-1.472E-04	38.041E-09	15.9338	2511.68	-41.44	-47	60	100
101	320.94	190.83	-111.79	-60.04	-0.553	26.063E-02	-1.840E-04	55.475E-09	15.9800	2332.01	-36.48	-73	37	101
102			-261.67	-209.67	4.346	21.801E-02	-1.166E-04	24.103E-09	16.0686	1966.89	-27.00	-103	-21	102
103			123.51	178.11	-20.771	30.225E-02	-2.063E-04	56.480E-09	16.4227	2610.44	-63.15	-25	86	103
104			-101.32						17.4716	3848.24	-31.96	114	-21	104
105	156.60	95.57	-84.74	-32.95	5.409	17.811E-02	-6.938E-05	87.127E-10	15.6637	1511.42	-17.16	-143	-74	105
106			-184.18	-113.00	17.015	17.907E-02	-5.233E-05	-1.918E-09	16.8467	2361.44	-17.10	-94	-8	106
107	686.64	300.88	-234.96	-168.39	9.014	21.407E-02	-8.390E-05	13.733E-10	18.9119	3803.98	-41.68	-3	96	107
108	1365.00	402.41	-389.58	-304.67	35.697	24.832E-02	-1.497E-04	30.103E-09	20.2501	6022.18	-28.25	91	221	108
109	419.60	206.21	-46.14	-4.69	14.922	23.509E-02	-1.369E-04	31.619E-09	16.0077	2497.23	-41.77	-49	57	109
110	267.34	184.24	-37.56	6.95	24.304	18.748E-02	-6.875E-05	40.989E-10	16.0001	2511.56	-42.35	-47	58	110
111	340.54	192.44	-46.05	37.30	3.693	27.516E-02	-1.583E-04	38.083E-09	17.0073	2616.73	-37.30	-58	43	111
112			-18.84	68.04	-0.172	26.955E-02	-1.329E-04	23.392E-09	16.2653	2358.77	-35.15	-55	37	112
113	1984.10	367.03	-201.72		9.311	30.095E-02	-1.818E-04	46.557E-09	17.8174	3988.33	-86.93	71	204	113
114	839.76	316.41			38.297	24.070E-02	-4.338E-05	-3.948E-08	16.4082	3108.49	-72.15	19	152	114
115	343.31	210.42	185.06	195.44	10.693	22.077E-02	-1.565E-04	46.013E-09	15.9253	2782.21	-51.15	-18	112	115
116			192.26	202.52	9.906	19.774E-02	-1.182E-04	27.821E-09	13.1563	1054.72	-77.08	-99	-16	116
117			185.56	194.56	14.708	18.644E-02	-1.174E-04	32.243E-09	15.6227	1850.66	-44.07	-90	-6	117

NO	FORMULA	COMPOUND NAME	MOLWT	TFP	TBP	TC	PC	VC	LDEN	TDEN	HVAP	NO
118	C3H4O	ACROLEIN	56.064	−87.2	52.8	506.0	51.7		839	20	28,345	118
119	C3H4O2	ACRYLIC ACID	72.064	11.8	140.8	615.0	56.7	0.210	1051	20	46,055	119
120	C3H4O2	VINYL FORMATE	72.064	−57.7	46.4	475.0	57.8	0.210	963	20	32,155	120
121	C3H5CL	ALLYL CHLORIDE	76.526	−134.5	45.1	514.0	47.6	0.234	937	20	27,110	121
122	C3H5CL3	1;2;3-TRICHLOROPROPANE	147.432	−14.7	155.8	651.0	39.5	0.348	1389	20	38,435	122
123	C3H5N	PROPIONITRILE	55.080	−92.7	97.3	564.4	41.8	0.230	782	20	32,280	123
124	C3H6	CYCLOPROPANE	42.081	−127.5	−32.8	397.8	54.9	0.170	563	15	20,055	124
125	C3H6	PROPYLENE	42.081	−185.3	−47.8	365.0	46.2	0.181	612	−50	18,422	125
126	C3H6CL2	1;2-DICHLOROPROPANE	112.987	−100.5	96.3	577.0	44.6	0.226	1150	20	31,401	126
127	C3H6O	ACETONE	58.080	−95.0	56.2	508.1	47.0	0.209	790	20	29,140	127
128	C3H6O	ALLYL ALCOHOL	58.080	−129.2	96.8	545.0	57.1	0.203	855	15	39,984	128
129	C3H6O	PROPIONALDEHYDE	58.080	−80.2	47.8	496.0	47.6	0.223	797	20	28,303	129
130	C3H6O	PROPYLENE OXIDE	58.080	−112.2	34.3	482.2	49.2	0.186	829	20	27,005	130
131	C3H6O	VINYL METHYL ETHER	58.080	−121.7	4.8	436.0	47.6	0.205	750	20	19,050	131
132	C3H6O2	PROPIONIC ACID	74.080	−20.7	140.8	612.0	53.7	0.230	993	20	32,238	132
133	C3H6O2	ETHYL FORMATE	74.080	−79.4	54.2	508.4	47.4	0.229	927	16	30,145	133
134	C3H6O2	METHYL ACETATE	74.080	−98.2	56.9	506.8	46.9	0.228	934	20	30,145	134
135	C3H7CL	PROPYL CHLORIDE	78.542	−122.8	46.4	503.0	45.8	0.254	891	20	27,256	135
136	C3H7CL	ISOPROPYL CHLORIDE	78.452	−117.2	35.7	485.0	47.2	0.230	862	20	26,293	136
137	C3H8	PROPANE	44.097	−187.7	−42.1	369.8	42.5	0.203	582	−42	18,786	137
138	C3H8O	N-PROPYL ALCOHOL	60.096	−126.3	97.2	536.7	51.7	0.219	804	20	41,784	138
139	C3H8O	ISOPROPYL ALCOHOL	60.096	−88.5	82.2	508.3	47.6	0.220	786	20	39,858	139
140	C3H8O	METHYL ETHYL ETHER	60.096	−139.2	7.3	437.8	44.0	0.221	700	20	24,702	140
141	C3H8O2	METHYLAL	76.096	−105.2	41.8	497.0			888	18		141
142	C3H8O2	1;2-PROPANEDIOL	76.096	−60.2	187.3	625.0	60.8	0.237	1036	20	54,177	142
143	C3H8O2	1;3-PROPANEDIOL	76.096	−26.8	214.4	658.0	59.8	0.241	1053	20	56,522	143
144	C3H8O3	GLYCEROL	92.095	17.8	289.8	726.0	66.9	0.255	1261	20	61,127	144
145	C3H8S	METHYL ETHYL SULPHIDE	76.157	−106.0	66.6	533.0	42.6	0.233	837	20	29,517	145
146	C3H9N	N-PROPYL AMINE	59.112	−83.2	48.6	497.0	47.4	0.233	717	20	29,726	146
147	C3H9N	ISOPROPYL AMINE	59.112	−95.3	32.4	476.0	50.7	0.229	688	20	27,214	147
148	C3H9N	TRIMETHYL AMINE	59.112	−117.2	2.9	433.2	40.7	0.254	633	20	24,116	148
149	C4H2O3	MALEIC ANHYDRIDE	98.058	52.8	199.6				1310	60		149
150	C4H4	VINYL ACETYLENE	52.076	−45.6	4.9	455.0	49.6	0.202	710	0	24,493	150
151	C4H4O	FURAN	68.075	−85.7	31.3	490.2	55.0	0.218	938	20	27,105	151
152	C4H4S	THIOPHENE	84.136	−38.3	84.1	579.4	56.9	0.219	1071	16	31,485	152
153	C4H5CL	CHLOROPRENE	88.537		59.4	511.2	42.5	0.266	958	20	29,658	153
154	C4H5CL	CHLOROBUTADIENE	88.537	−60.0	67.8	527.2	39.5	0.265	963	20	29,038	154
155	C4H5N	ALLYL CYANIDE	67.091	−86.5	118.8	585.0	39.5	0.265	835	20	34,332	155
156	C4H5N	PYRROLE	67.091		129.8	640.0			967	21		156

NO	VISA	VISB	DELHF	DELGF	CPVAPA	CPVAPB	CPVAPC	CPVAPD	ANTA	ANTB	ANTC	TMN	TMX	NO
118	388.17	217.14	-70.92	-65.19	11.970	21.055E-02	-1.071E-04	19.058E-09	15.9057	2606.53	-45.15	-38	87	118
119	733.02	307.15	-336.45	-286.25	1.742	31.908E-02	-2.352E-04	69.752E-09	16.5617	3319.18	-80.15	42	177	119
120	428.40	224.83			27.813	18.388E-02	-3.560E-05	-2.335E-07	16.6531	2569.68	-63.15	-33	77	120
121	368.27	210.61	-0.63	43.63	2.529	30.467E-02	-2.278E-04	72.934E-09	15.9772	2531.92	-47.15	-43	77	121
122	818.63	342.88	-185.89	-97.85	26.883	36.220E-02	-2.787E-04	87.881E-09	16.1246	3417.27	-69.15	42	197	122
123	366.77	225.86	50.66	96.21	15.403	22.454E-02	-1.100E-04	19.540E-09	15.9571	2940.86	-55.15	-3	132	123
124			53.34	104.46	-35.240	38.133E-02	-2.881E-04	90.351E-09	15.8599	1971.04	-26.65	-93	-28	124
125	273.84	131.63	20.43	62.76	3.710	23.454E-02	-1.160E-04	22.048E-09	15.7027	1807.53	-26.15	-113	-33	125
126	514.36	281.03	-165.80	-83.15	10.450	36.547E-02	-2.604E-04	77.414E-09	16.0385	2985.07	-52.16	15	135	126
127	367.25	209.68	-217.71	-153.15	6.301	26.059E-02	-1.253E-04	20.377E-09	16.6513	2940.46	-35.93	-32	77	127
128	793.52	307.26	-132.09	-71.30	-1.105	31.464E-02	-2.032E-04	53.214E-09	16.9066	2928.20	-85.15	13	127	128
129	343.44	219.33	-192.17	-130.54	11.723	26.142E-02	-1.300E-04	21.261E-09	16.2315	2659.02	-44.15	-38	77	129
130	377.43	213.36	-92.82	-25.79	-8.457	32.569E-02	-1.989E-04	48.232E-09	15.3227	2107.58	-64.87	-48	67	130
131	318.41	180.98			15.629	23.413E-02	-9.697E-05	10.622E-09	14.4602	1980.22	-25.15	-83	42	131
132	535.04	299.32	-455.44	-369.57	5.669	36.890E-02	-2.865E-04	98.767E-09	17.3789	3723.42	-67.48	42	177	132
133	400.91	226.23	-371.54		24.673	23.161E-02	-2.120E-05	-5.359E-08	16.1611	2603.30	-54.15	-33	87	133
134	408.62	224.03	-409.72		16.550	22.454E-02	-4.342E-05	29.144E-09	16.1295	2601.92	-56.15	-28	87	134
135	374.77	215.00	-130.21	-50.70	-3.345	36.258E-02	-2.508E-04	74.483E-09	15.9594	2581.48	-42.95	-43	77	135
136	306.25	212.24	-146.54	-62.55	1.842	34.876E-02	-2.244E-04	58.615E-09	16.0384	2490.48	-43.15	-48	67	136
137	222.67	133.41	-103.92	-23.49	-4.224	30.626E-02	-1.586E-04	32.146E-09	15.7260	1872.46	-25.16	-109	-24	137
138	951.04	327.83	-256.57	-161.90	2.470	33.252E-02	-1.855E-04	42.957E-09	17.5439	3166.38	-80.15	12	127	138
139	1139.70	323.44	-272.60	-173.50	32.427	18.862E-02	64.058E-06	-9.261E-08	18.6929	3640.20	-53.54	0	111	139
140	303.82	171.66	-216.58	-117.73	18.669	26.854E-02	-1.025E-04	89.514E-10	13.5435	1161.63	-112.40	-68	37	140
141									15.8237	2415.92	-52.58	-3	42	141
142	1404.20	426.74	-424.25		0.632	42.119E-02	-2.981E-04	89.514E-09	20.5324	6091.95	-22.46	84	210	142
143	1813.00	406.96	-409.09		8.269	36.756E-02	-2.162E-04	50.535E-09	17.2917	3888.84	-123.20	107	252	143
144	3337.10	406.00	-585.31		8.424	44.422E-02	-3.159E-04	93.784E-09	17.2392	4487.04	-140.20	167	327	144
145			-59.66	11.43	19.527	28.906E-02	-1.209E-04	12.866E-09	15.9765	2722.95	-48.37	-23	87	145
146			-72.43	39.82	6.691	34.985E-02	-1.822E-04	35.864E-09	15.9957	2551.72	-49.15	-38	77	146
147	433.64	228.46	-83.82		-7.486	41.755E-02	-2.826E-04	83.485E-09	16.3637	2582.35	-40.15	-34	64	147
148			-23.86	98.98	-8.206	39.716E-02	-2.189E-04	46.222E-09	16.0499	2230.51	-39.15	-58	32	148
149	952.48	365.81			-13.075	34.847E-02	-2.184E-04	48.399E-09	16.2747	3765.65	-82.15	79	243	149
150			304.80	306.18	6.757	28.407E-02	-2.265E-04	74.609E-09	16.0100	2203.57	-43.15	-73	32	150
151	389.40	222.70	-34.71	0.88	-35.529	43.208E-02	-3.455E-04	10.743E-08	16.0612	244.70	-45.41	-35	90	151
152	498.60	264.90	115.81	126.86	-30.606	44.799E-02	-3.772E-04	12.527E-08	16.0243	2869.07	-51.80	-13	107	152
153			65.86						14.4844	1938.59	-85.36	27	84	153
154			79.97											154
155	521.30	252.03			21.700	25.715E-02	-1.192E-04	12.292E-09	16.0019	3128.75	-58.15	127	157	155
156			108.35						16.7966	3457.47	-62.73	57	167	156

NO	FORMULA	COMPOUND NAME	MOLWT	TFP	TBP	TC	PC	VC	LDEN	TDEN	HVAP	NO
157	C4H6	ETHYLACETYLENE	54.092	−125.8	8.0	463.7	47.1	0.220	650	16	24,995	157
158	C4H6	DIMETHYL ACETYLENE	54.092	−32.3	27.0	488.6	50.9	0.221	691	20	26,670	158
159	C4H6	1;2-BUTADIENE	54.092	−136.2	10.8	443.7	45.0	0.219	652	20	24,283	159
160	C4H6	1;3-BUTADIENE	54.092	−108.9	−4.5	425.0	43.3	0.221	621	20	22,483	160
161	C4H6O2	VINYL ACETATE	86.091	−100.2	72.8	525.0	43.6	0.265	932	20		161
162	C4H6O3	ACETIC ANHYDRIDE	102.089	−74.2	138.8	569.0	46.8	0.290	1087	20	41,240	162
163	C4H6O4	DIMETHYL OXALATE	118.090	53.8	163.4	628.0	39.8		1150	15		163
164	C4H6O4	SUCCINIC ACID	118.090	182.8	234.8							164
165	C4H7N	BUTYRONITRILE	69.107	−112.2	117.8	582.2	37.9	0.285	792	20	34,415	165
166	C4H7O2	METHYL ACRYLATE	86.091	−76.5	80.3	536.0	42.6	0.265	956	20	32,029	166
167	C4H8	1-BUTENE	56.108	−185.4	−6.3	419.6	37.2	0.240	595	20	21,930	167
168	C4H8	CIS-2-BUTENE	56.108	−138.9	3.7	435.6	42.0	0.234	621	20	23,362	168
169	C4H8	TRANS-2-BUTENE	56.108	−105.6	0.8	428.6	41.0	0.238	604	20	22,772	169
170	C4H8	CYCLOBUTANE	56.108	−90.8	12.5	459.9	49.9	0.210	694	20	24,200	170
171	C4H8	ISOBUTYLENE	56.108	−140.4	−6.9	417.9	40.0	0.239	594	20	22,131	171
172	C4H8O	N-BUTYRALDEHYDE	72.107	−96.4	74.8	524.0	40.5	0.278	802	20	31,527	172
173	C4H8O	ISOBUTYRALDEHYDE	72.107	−65.0	63.8	513.0	41.5	0.274	789	20	31,401	173
174	C4H8O	MERTYL ETHYL KETONE	72.107	−86.7	79.6	535.6	41.5	0.267	805	20	31,234	174
175	C4H8O	TETRAHYDROFURAN	72.107	−108.5	65.9	540.2	51.9	0.224	889	20	29,601	175
176	C4H8O	VINYL ETHYL ETHER	72.107	−115.3	35.6	475.0	40.7	0.260	793	20	26,502	176
177	C4H8O2	N-BUTYRIC ACID	88.107	−5.3	163.2	628.0	52.7	0.292	958	20	42,035	177
178	C4H8O2	1;4-DIOXANE	88.107	11.8	101.3	587.0	52.1	0.238	1033	20	36,383	178
179	C4H8O2	ETHYL ACETATE	88.107	−83.6	77.1	523.2	38.3	0.286	901	20	32,238	179
180	C4H8O2	ISOBUTYRIC ACID	88.107	−46.0	154.7	609.0	40.5	0.292	968	20	41,156	180
181	C4H8O2	METHYL PROPIONATE	88.107	−87.5	79.8	530.6	40.0	0.282	915	20	32,573	181
182	C4H8O2	N-PROPYL FORMATE	88.107	−92.9	80.5	538.0	40.6	0.285	911	16	32,490	182
183	C4H9CL	1-CHLOROBUTANE	92.569	−123.1	78.4	542.0	36.9	0.312	886	20	30,019	183
184	C4H9CL	2-CHLOROBUTANE	92.569	−131.4	68.2	520.6	39.5	0.305	873	20	29,224	184
185	C4H9CL	2-CHLORO-2-METHYL PROPANE	92.569	−25.4	50.8	507.0	39.5	0.295	842	20	27,424	185
186	C4H9N	PYRROLIDINE	71.123		86.5	568.6	56.1	0.249	852	22		186
187	C4H9NO	MORPHOLINE	87.122	−4.8	128.2	618.0	54.7	0.253	1000	20	37,681	187
188	C4H10	N-BUTANE	58.124	−138.4	−0.5	425.2	38.0	0.255	579	20	22,408	188
189	C4H10	ISOBUTANE	58.124	−159.6	−11.9	408.1	36.5	0.263	557	20	21,311	189
190	C4H10O	N-BUTANOL	74.123	−89.3	117.7	562.9	44.2	0.274	810	20	43,124	190
191	C4H10O	2-BUTANOL	74.123	−114.7	99.5	536.0	41.9	0.268	807	20	40,821	191
192	C4H10O	ISOBUTANOL	74.123	−108.0	107.8	547.7	43.0	0.273	802	20	42,077	192
193	C4H10O	2-METHYL-2-PROPANOL	74.123	25.6	82.4	506.2	39.7	0.275	787	20	39,063	193
194	C4H10O	ETHYL ETHER	74.123	−116.3	34.5	466.7	36.4	0.280	713	20	26,712	194
195	C4H10O2	1;2-DIMETHOXYETHANE	90.123	−71.2	85.4	536.0	38.7	0.271	867	20	31,443	195

NO	VISA	VISB	DELHF	DELGF	CPVAPA	CPVAPB	CPVAPC	CPVAPD	ANTA	ANTB	ANTC	TMN	TMX	NO
157			165.29	202.22	12.548	27.436E-02	-1.545E-04	34.499E-09	16.0605	2271.42	-40.30	-73	27	157
158			146.41	185.56	15.927	23.815E-02	-1.070E-04	17.534E-09	16.2821	2536.78	-37.34	-33	47	158
159			162.32	198.58	11.200	27.235E-02	-1.468E-04	30.890E-09	16.1039	2397.26	-30.88	-28	32	159
160	300.59	163.12	110.24	150.77	-1.687	34.185E-02	-2.340E-04	63.346E-09	15.7727	2142.66	-34.30	-58	17	160
161	457.89	235.35	-316.10		15.160	27.951E-02	-8.805E-05	-1.660E-08	16.1003	2744.68	-56.15	-18	106	161
162	502.33	286.04	-576.10	-477.00	-23.128	50.870E-02	-3.580E-04	98.348E-09	16.3982	3287.56	-75.11	35	164	162
163														163
164	0.00					15.072E+00	46.892E-03	-3.143E-04						164
165	438.04	256.84	34.08	108.73	15.211	32.058E-02	-1.638E-04	29.823E-09	16.2092	3202.21	-56.16	34	160	165
166	451.02	245.30			15.165	27.959E-02	-8.805E-05	-1.660E-08	16.1088	2788.43	-59.15	-13	117	166
167	256.30	151.86	-0.13	71.34	-2.994	35.320E-02	-1.982E-04	44.631E-09	15.7564	2132.42	-33.15	-83	22	167
168	268.94	155.34	-6.99	65.90	0.440	29.534E-02	-1.018E-04	-6.155E-10	15.8171	2210.71	-36.15	-73	32	168
169	259.01	153.30	-11.18	63.01	18.317	25.636E-02	-7.013E-05	-8.989E-09	15.8177	2212.32	-33.15	-73	27	169
170			26.67	110.11	-50.254	50.242E-02	-3.558E-04	10.471E-08	15.9254	2359.09	-31.78	-73	17	170
171			-16.91	58.11	16.052	28.043E-02	-1.091E-04	90.979E-10	15.7528	2125.75	-33.15	-83	17	171
172	472.31	233.42	-205.15	-114.84	14.080	34.570E-02	-1.723E-04	28.872E-09	16.1668	2839.09	-50.15	-18	107	172
173	464.06	253.64	-215.87	-121.42	24.463	33.557E-02	-2.057E-04	63.681E-09	15.9888	2676.98	-51.15	-26	97	173
174	423.84	231.67	-238.52	-146.16	10.944	35.592E-02	-1.900E-04	39.197E-09	16.5986	3150.42	-36.65	-16	103	174
175	419.79	244.46	-184.34		-19.104	51.623E-02	-4.132E-04	14.541E-08	16.1069	2768.38	-46.90	-3	97	175
176	349.95	189.02	-140.26		17.279	32.360E-02	-1.471E-04	21.495E-09	15.8911	2449.26	-44.15	-48	67	176
177	640.42	321.13	-476.16		11.740	41.370E-02	-2.430E-04	55.308E-09	17.9240	4130.93	-70.55	62	197	177
178	660.36	308.77	-315.27	-180.91	-53.574	59.871E-02	-4.085E-04	10.622E-08	16.1327	2966.88	-62.15	2	137	178
179	427.38	235.98	-443.21	-327.62	7.235	40.717E-02	-2.092E-04	28.546E-09	16.1516	2790.50	-57.15	-13	112	179
180	588.65	311.24	-484.25		9.814	46.683E-02	-3.720E-04	13.502E-08	16.7792	3385.49	-94.15	57	192	180
181	442.88	238.39			18.204	31.397E-02	-9.353E-05	-1.828E-08	16.1693	2804.06	-58.92	-13	112	181
182	452.97	246.09							15.7671	2593.95	-69.69	7	87	182
183	783.72	260.03	-147.38	-38.81	-2.613	44.966E-02	-2.937E-04	80.805E-09	15.9750	2826.26	-49.05	-18	112	183
184	480.77	237.30	-161.61	-53.51	-3.433	45.594E-02	-2.981E-04	82.564E-09	15.9907	2753.43	-47.15	-23	102	184
185	543.41	253.35	-183.38	-64.14	-3.931	46.515E-02	-2.886E-04	78.712E-09	15.8121	2567.15	-44.15	-38	87	185
186			-3.60	114.76	-51.531	53.382E-02	-3.240E-04	75.279E-09	15.9444	2717.03	-67.90	27	127	186
187	914.14	332.75			-42.802	53.884E-02	-2.666E-04	41.994E-09	16.2364	3171.35	-71.15	27	167	187
188	265.84	160.20	-126.23	-17.17	9.487	33.130E-02	-1.108E-04	-2.822E-09	15.6782	2154.90	-34.42	-78	17	188
189	302.51	170.20	-134.61	20.89	-1.390	38.473E-02	-1.846E-04	28.952E-09	15.5381	2032.73	-33.15	-86	7	189
190	984.54	341.12	-274.86	-150.89	3.266	41.801E-02	-2.242E-04	46.850E-09	17.2160	3137.02	-94.43	15	131	190
191	1441.70	331.50	-292.82	-167.72	5.753	42.454E-02	-2.328E-04	47.730E-09	17.2102	3026.03	-86.65	25	120	191
192	1199.10	343.85	-283.40	-167.43	-7.708	46.892E-02	-2.884E-04	72.306E-09	16.8712	2874.73	-100.30	20	115	192
193	972.10	363.38	-312.63	-177.77	-48.613	71.720E-02	-7.084E-04	29.199E-08	16.8548	2658.29	-95.50	20	103	193
194	353.14	190.58	-252.38	-122.42	21.424	33.587E-02	-1.035E-04	-9.357E-09	16.0828	2511.29	-41.95	-48	67	194
195					32.234	35.672E-02	-1.336E-04	83.987E-10	16.0241	2869.79	-53.15	-11	120	195

NO	FORMULA	COMPOUND NAME	MOLWT	TFP	TBP	TC	PC	VC	LDEN	TDEN	HVAP	NO
196	C4H10O3	DIETHYLENE GLYCOL	106.122	-8.2	245.8	681.0	46.6	0.316	1116	20	57,234	196
197	C4H10S	DIMETHYL SULPHIDE	90.184	-104.0	92.1	557.0	39.6	0.318	837	20	31,778	197
198	C4H10S2	DIETHYL DISULPHIDE	122.244	-101.5	154.0	642.0			998	20	37,723	198
199	C4H11N	N-BUTYL AMINE	73.139	-49.1	77.4	524.0	41.5	0.288	739	20	32,113	199
200	C4H11N	ISOBUTYL AMINE	73.139	-85.2	67.4	516.0	42.6	0.284			30,982	200
201	C4H11N	DIETHYL AMINE	73.139	-49.8	55.4	496.6	37.1	0.301	707	20	27,842	201
202	C4H12SI	TETRAMETHYLSILANE	88.225	-102.2	27.6	448.6	28.2	0.362	646	20	24,685	202
203	C5H4O2	FURFURAL	96.085	-31.0	161.7	657.1	49.2	0.270	1156	25	35,169	203
204	C5H5N	PYRIDINE	79.102	-41.7	115.3	620.0	56.3	0.254	983	20	27,005	204
205	C5H8	CYCLOPENTENE	68.119	-135.1	44.2	506.0			772	20		205
206	C5H8	1;2-PENTADIENE	68.119	-137.3	44.8	503.0	40.7	0.276	693	20	27,591	206
207	C5H8	1-TRANS-3-PENTADIENE	68.119	-87.5	42.0	496.0	39.9	0.275	676	20	27,047	207
208	C5H8	1;4-PENTADIENE	68.119	-148.3	25.9	478.0	37.9	0.276	661	20	25,163	208
209	C5H8	1-PENTYNE	68.119	-105.7	40.1	493.4	40.5	0.278	690	20		209
210	C5H8	2-METHYL-1;3-BUTADIENE	68.119	-146.0	34.0	484.0	38.5	0.276	681	20	26,084	210
211	C5H8	3-METHYL-1;2-BUTADIENE	68.119	-113.7	40.8	496.0	41.1	0.267	686	20	27,256	211
212	C5H8O	CYCLOPENTONE	84.118	-50.7	130.7	626.0	53.7	0.268	950	20	36,593	212
213	C5H8O2	ETHYL ACRYLATE	100.118	-72.2	99.8	552.0	37.5	0.320	921	20	33,285	213
214	C5H10	CYCLOPENTANE	70.135	-93.9	49.2	511.6	45.1	0.260	745	20	27,315	214
215	C5H10	1-PENTENE	70.135	-165.3	29.9	464.7	40.5	0.300	640	20	25,213	215
216	C5H10	CIS-2-PENTENE	70.135	-151.4	36.9	476.0	36.5	0.300	656	20	26,126	216
217	C5H10	TRANS-2-PENTENE	70.135	-140.3	36.3	475.0	36.6	0.300	649	20	26,084	217
218	C5H10	2-METHYL-1-BUTENE	70.135	-137.6	31.1	465.0	34.5	0.294	650	20	25,514	218
219	C5H10	2-METHYL-2-BUTENE	70.135	-133.8	38.5	470.0	34.5	0.318	662	20	26,322	219
220	C5H10	3-METHYL-1-BUTENE	70.135	-168.5	20.1	450.0	35.2	0.300	627	20	24,116	220
221	C5H10O	VALERALDEHYDE	86.134	-91.2	102.8	554.0	35.5	0.333	810	20	33,662	221
222	C5H10O	METHYL N-PROPYL KETONE	86.134	-77.2	102.3	564.0	38.9	0.301	806	20	33,494	222
223	C5H10O	METHYL ISOPROPYL KETONE	86.134	-92.2	94.2	553.4	38.5	0.310	803	20	30,647	223
224	C5H10O	DIETHYL KETONE	86.134	-39.0	101.9	561.0	37.4	0.336	814	20	33,746	224
225	C5H10O2	N-VALERIC ACID	102.134	-34.2	185.5	651.0	38.5	0.340	939	20	49,823	225
226	C5H10O2	ISOBUTYL FORMATE	102.134	-95.2	98.4	551.0	38.8	0.350	885	20	34,206	226
227	C5H10O2	N-PROPYL ACETATE	102.134	-95.2	101.6	549.4	33.3	0.345	887	20	34,206	227
228	C5H10O2	ETHYL PROPIONATE	102.134	-73.9	98.8	546.0	33.6	0.345	895	16	34,248	228
229	C5H10O2	METHYL BUTYRATE	102.134	-84.8	102.6	554.4	34.8	0.340	898	20	34,101	229
230	C5H10O2	METHYL ISOBUTYRATE	102.134	-87.8	92.2	540.8	34.3	0.339	891	20	33,386	230
231	C5H11N	PIPERIDINE	85.150	-10.5	106.5	594.0	47.6	0.289	862	20	34,248	231
232	C5H12	N-PENTANE	72.151	-129.8	36.0	469.6	33.7	0.304	626	20	25,791	232
233	C5H12	2-METHYL BUTANE	72.151	-159.3	27.8	460.4	33.8	0.306	620	20	24,702	233
234	C5H12	2;2-DIMETHYL PROPANE	72.151	-16.6	9.4	433.8	32.0	0.303	591	20	22,768	234

NO	VISA	VISB	DELHF	DELGF	CPVAPA	CPVAPB	CPVAPC	CPVAPD	ANTA	ANTB	ANTC	TMN	TMX	NO
196	1943.00	385.24	-571.50		73.060	34.441E-02	-1.468E-04	18.464E-09	17.0326	4122.52	-122.50	129	287	196
197	407.59	233.32	-83.53	17.79	13.595	39.595E-02	-1.780E-04	26.490E-09	15.9531	2896.27	-54.49	-13	117	197
198			-74.69	22.27	26.896	46.013E-02	-2.710E-04	59.704E-09	16.0607	3421.57	-64.19	39	182	198
199	472.06	243.98	-92.11	49.24	5.079	44.757E-02	-2.407E-04	75.990E-09	16.6085	3012.70	-48.96	-14	100	199
200					9.491	44.296E-02	-2.110E-04	23.329E-09	16.1419	2704.16	-56.15	-22	100	200
201	473.89	229.29	-72.43	72.14	2.039	44.296E-02	-2.183E-04	36.530E-09	16.0545	2595.01	-53.15	-31	77	201
202			-232.41						16.0999	2570.24	-28.73	27	-84	202
203					18.196	28.198E-02	-6.523E-05	-5.476E-08	18.7949	5365.88	5.40	77	277	203
204	618.50	291.58	140.26	190.33	39.791	49.279E-02	-3.558E-04	10.044E-08	16.0910	3095.13	-61.15	12	152	204
205	396.83	218.66	32.95	110.66	-41.512	46.306E-02	-2.579E-04	54.345E-09	15.9356	2583.07	-39.70	-29	105	205
206			145.70	210.55	8.826	38.799E-02	-2.280E-04	52.461E-09	15.9297	2544.34	-44.30	-23	67	206
207			77.87	146.83	30.689	28.110E-02	-6.711E-05	-2.352E-08	15.9182	2541.69	-41.43	-23	67	207
208			105.51	170.36	6.996	39.515E-02	-2.374E-04	55.978E-09	15.7392	2344.02	-41.69	-33	47	208
209			144.44	210.39	18.066	35.035E-02	-1.913E-04	40.976E-09	16.0429	2515.62	-45.97	-43	62	209
210	328.49	182.48	75.78	145.95	-34.122	45.845E-02	-3.337E-04	10.002E-08	15.8548	2467.40	-39.64	-23	57	210
211			129.79	198.75	14.687	35.977E-02	-1.976E-04	42.622E-09	15.9880	2541.83	-42.26	-23	62	211
212	574.71	303.44	-192.76		-40.641	52.251E-02	-3.035E-04	71.301E-09	16.0890	3193.92	-66.15	27	167	212
213	438.08	256.84			16.810	36.898E-01	-1.382E-04	-5.732E-09	16.0890	2974.94	-58.15	1	136	213
214	406.69	231.67	-77.29	38.64	-53.625	54.261E-02	-3.031E-04	64.854E-09	15.8574	2588.48	-41.79	-43	72	214
215	305.25	174.70	-20.93	79.17	-0.134	43.292E-02	-2.317E-04	46.808E-09	15.7646	2405.96	-39.63	-53	52	215
216	305.31	175.72	-28.09	71.89	-13.151	46.013E-02	-2.541E-04	54.554E-09	15.8251	2459.05	-42.56	-53	57	216
217	349.33	176.62	-31.78	69.96	1.947	41.818E-02	-2.178E-04	44.045E-09	15.9011	2495.97	-40.18	-53	57	217
218	369.27	193.39	-36.34	65.65	10.572	39.971E-02	-1.946E-04	33.139E-09	15.8260	2426.42	-40.36	-53	52	218
219	322.47	180.43	-42.58	59.70	11.803	35.090E-02	-1.117E-04	-5.807E-09	15.9238	2521.53	-40.31	-47	62	219
220			-28.97	74.82	21.742	38.895E-02	-2.007E-04	-40.105E-09	15.7179	2333.61	-36.33	-63	42	220
221	521.30	252.03	-227.97	-108.35	14.239	43.292E-02	-2.107E-04	31.623E-09	16.1623	3030.20	-58.15	-46	139	221
222	437.94	243.03	-258.83	-137.16	1.147	48.023E-02	-2.818E-04	66.612E-09	16.0031	2934.87	-62.25	2	137	222
223					-2.914	49.907E-02	-2.935E-04	66.654E-09	14.1779	1993.12	-103.20	-2	133	223
224	409.17	236.65	-258.83	135.36	30.011	39.394E-02	-1.907E-04	33.976E-09	16.8138	3410.51	-40.15	2	127	224
225	729.09	341.13	-490.69	-357.43	13.389	50.325E-02	-2.931E-04	66.193E-09	17.6306	4092.15	-86.55	77	222	225
226					19.850	40.336E-02	-1.436E-04	-7.402E-09	16.2292	2980.47	-64.15	5	136	226
227	489.53	255.83	-466.03		15.420	45.008E-02	-1.686E-04	-1.439E-08	16.2291	2980.47	-64.15	7	137	227
228	463.31	248.72	-470.18	-323.72	19.854	40.344E-02	-1.437E-04	-7.402E-09	16.1620	2935.11	-64.16	3	123	228
229	479.35	254.66												229
230	451.21	246.09												230
231	772.79	313.49	-49.03	-8.37	-53.068	62.886E-02	-3.358E-04	64.267E-09	16.1004	3015.46	-61.15	7	143	231
232	313.66	182.48	-146.54		-3.626	48.734E-02	-2.580E-04	53.047E-09	15.8333	2477.07	-39.94	-53	57	232
233	367.32	191.58	-154.58	-14.82	-9.525	50.660E-02	-2.729E-04	57.234E-09	15.6338	2348.67	-40.05	-57	49	233
234	355.54	196.35	-166.09	-15.24	-16.592	55.517E-02	-3.306E-04	76.325E-09	15.2069	2034.15	-45.37	-13	32	234

NO	FORMULA	COMPOUND NAME	MOLWT	TFP	TBP	TC	PC	VC	LDEN	TDEN	HVAP	NO
235	C5H12O	1-PENTANOL	88.150	-78.2	137.8	586.0	38.5	0.326	815	20	44,380	235
236	C5H12O	2-METHYL-1-BUTANOL	88.150	-70.2	128.7	571.0	38.5	0.322	819	20	45,217	236
237	C5H12O	3-METHYL-1-BUTANOL	88.150	-117.2	131.2	579.5	38.5	0.329	810	20	44,129	237
238	C5H12O	2-METHYL-2-BUTANOL	88.150	-8.8	102.0	545.0	39.5	0.319	809	20	40,612	238
239	C5H12O	2;2-DIMETHYL-1-PROPANOL	88.150	53.8	113.1	549.0	39.5	0.319	783	54	43,124	239
240	C5H12O	ETHYL PROPYL ETHER	88.150	-126.8	63.6	500.6	32.5		733	20	30,522	240
241	C5H12O	METHYL-T-BUTYL ETHER	88.150	-108.2	55.1	407.1	34.3	0.339	741	20	27,646	241
242	C5H12O	BUTYLMETHYL ETHER	88.150	-115.5	70.1	512.8	34.3	0.329				242
243	C6F6	PERFLUOROBENZENE	186.056		80.2	516.7	33.0					243
244	C6F12	PERFLUOROCYCLOHEXANE	300.047		52.5	457.2	24.3					244
245	C6F14	PERFLUORO-N-HEXANE	338.044	-87.2	57.1	451.7	19.0	0.442				245
246	C6H3CL3	1;2,4-TRICHLOROBENZENE	181.449	16.8	213.0	734.9	39.8	0.401				246
247	C6H4CL2	O-DICHLOROBENZENE	147.004	-17.1	180.4	697.3	41.0	0.360	1306	20	39,691	247
248	C6H4CL2	M-DICHLOROBENZENE	147.004	-24.8	172.8	684.0	38.5	0.359	1288	20	38,644	248
249	C6H4CL2	P-DICHLOROBENZENE	147.004	53.1	174.1	685.0	39.5	0.372	1248	55	38,812	249
250	C6H5BR	BROMOBENZENE	157.010	-30.9	156.0	670.0	45.2	0.324	1495	20		250
251	C6H5CL	CHLOROBENZENE	112.559	-45.6	131.7	632.4	45.2	0.308	1106	20	36,572	251
252	C6H5F	FLUOROBENZENE	96.104	-39.2	85.3	560.1	45.5	0.271	1024	20		252
253	C6H5I	IODOBENZENE	204.011	-31.4	188.2	721.0	45.2	0.351	1855	4	39,523	253
254	C6H5NO2	NITROBENZENE	123.112	4.8	210.6	712.0	35.0	0.337	1203	20	44,031	254
255	C6H6	BENZENE	78.114	5.5	80.1	562.1	48.9	0.259	885	16	30,781	255
256	C6H6O	PHENOL	94.113	40.8	181.8	694.2	61.3	0.229	1059	40	45,636	256
257	C6H7N	ANILINE	93.129	-6.2	184.3	699.0	53.1	0.270	1022	20	41,868	257
258	C6H7N	4-METHYL PYRIDINE	93.129	3.7	145.3	646.0	44.6	0.311	955	20	37,472	258
259	C6H10	1;5-HEXADIENE	82.146	-141.2	59.4	507.0	34.5	0.328	692	20	27,470	259
260	C6H10	CYCLOHEXENE	82.146	-103.5	82.9	560.4	43.5	0.292	816	16	30,480	260
261	C6H10O	CYCLOHEXANONE	98.145	-31.2	155.6	629.0	38.5	0.312	951	15	39,775	261
262	C6H12	CYCLOHEXANE	84.162	6.5	80.7	553.4	40.7	0.308	779	20	29,977	262
263	C6H12	METHYLCYCLOPENTANE	84.162	-142.5	71.8	532.7	37.9	0.319	754	16	29,098	263
264	C6H12	1-HEXENE	84.162	-139.9	63.4	504.0	31.7	0.350	673	20	28,303	264
265	C6H12	CIS-2-HEXENE	84.162	-141.2	68.8	518.0	32.8	0.351	687	20	29,140	265
266	C6H12	TRANS-2-HEXENE	84.162	-133.2	67.8	516.0	32.7	0.351	678	20	28,931	266
267	C6H12	CIS-3-HEXENE	84.162	-137.9	66.4	517.0	32.8	0.350	680	20	28,721	267
268	C6H12	TRANS-3-HEXENE	84.162	-113.5	67.1	519.9	32.5	0.350	677	20	28,973	268
269	C6H12	2-METHYL-2-PENTENE	84.162	-135.1	67.3	518.0	32.8	0.351	691	16	29,015	269
270	C6H12	3-METHYL-CIS-2-PENTENE	84.162	-134.9	67.7	518.0	32.8	0.351	694	20	28,847	270
271	C6H12	3-METHYL-TRANS-2-PENTENE	84.162	-138.5	70.4	521.0	32.9	0.350	698	20	29,308	271
272	C6H12	4-METHYL-CIS-2-PENTENE	84.162	-134.2	56.4	490.0	30.4	0.360	669	20	27,591	272
273	C6H12	4-METHYL-TRANS-2-PENTENE	84.162	-141.2	58.5	493.0	30.4	0.360	669	20	27,968	273

NO	VISA	VISB	DELHF	DELGF	CPVAPA	CPVAPB	CPVAPC	CPVAPD	ANTA	ANTB	ANTC	TMN	TMX	NO
235	1151.10	349.62	−298.94	−146.12	3.869	50.451E-02	−2.639E-04	51.205E-09	16.5270	3026.89	−105.00	37	138	235
236	1259.40	349.85	−302.71	−165.71	−9.483	56.773E-02	−3.481E-04	86.374E-09	16.2708	2752.19	−116.30	34	129	236
237	1148.80	349.51	−302.29		−9.542	56.815E-02	−3.485E-04	86.499E-09	16.7127	3026.43	−104.10	25	153	237
238	1502.00	336.75	−329.92	−165.38	−12.087	60.960E-02	−4.204E-04	12.284E-08	15.0113	1988.08	−137.80	25	102	238
239			−293.08	−125.52	12.154	53.968E-02	−3.160E-04	71.217E-09	18.1336	3694.96	−65.00	55	133	239
240	399.87	213.39							15.3549	2423.41	−62.28	−27	87	240
241			−292.99	−125.52	2.533	51.372E-02	−2.596E-04	43.040E-09	16.4174	2913.70	−30.63	−88	88	241
242									15.8830	2666.26	−53.70	69	23	242
243			−957.27	−879.98	36.283	52.670E-02	−4.547E-04	14.558E-08	16.1940	2827.53	−57.66	−3	117	243
244									13.9087	1374.07	−136.80	7	127	244
245									15.8307	2488.59	−59.73	−3	57	245
246	554.35	319.07	29.98	82.73	−14.361	60.876E-02	−5.623E-04	20.725E-08	16.8979	4452.50	−53.00	127	327	246
247	402.20	300.89	26.46	78.63	−14.302	55.056E-02	−4.513E-04	14.294E-08	16.2799	3798.23	−59.84	58	210	247
248	483.82	312.03	23.03	77.20	−13.590	54.931E-02	−4.505E-04	14.269E-08	16.8173	4104.13	−43.15	53	202	248
249	508.18	302.42	105.09	138.62	−14.344	55.349E-02	−4.559E-04	14.478E-08	16.1135	3626.83	−64.64	54	204	249
250	477.76	276.22	51.87	99.23	−28.805	53.507E-02	−4.080E-04	12.117E-08	15.7972	3313.00	−67.71	47	177	250
251	452.06	252.89	−116.64	−69.08	−33.888	56.312E-02	−4.522E-04	14.264E-08	16.0676	3295.12	−55.60	47	147	251
252	565.72	331.21	162.66	187.90	−38.728	56.689E-02	−4.434E-04	13.553E-08	16.5487	3181.78	−37.59	−23	97	252
253					−29.274	55.643E-02	−4.509E-04	14.432E-08	16.1454	3776.53	−64.38	17	197	253
254			−67.49						16.1484	4032.66	−71.81	44	211	254
255	545.64	265.34	82.98	129.75	−33.917	47.436E-02	−3.017E-04	71.301E-09	15.9008	2788.51	−52.36	7	104	255
256	1405.50	370.07	−96.67	−32.91	−35.843	59.829E-02	−4.827E-04	15.269E-08	16.4279	3490.89	−98.59	72	208	256
257	1074.60	357.21	86.92	166.80	−40.516	63.849E-02	−5.133E-04	16.333E-08	16.6748	3857.52	−73.15	67	227	257
258	500.97	285.50	102.28		−17.430	48.818E-02	−2.798E-04	54.512E-09	16.2143	3409.40	−62.65	27	187	258
259			83.74						16.1351	2728.54	−45.45	9	77	259
260	506.92	264.54	−5.36	106.93	−68.651	72.515E-02	−5.414E-04	16.442E-08	15.8243	2813.53	−49.98	27	87	260
261	787.38	336.47	−230.27	−90.81	−37.807	55.391E-02	−1.953E-04	−1.534E-08						261
262	653.62	290.84	−123.22	31.78	−54.541	61.127E-02	−2.523E-04	13.214E-09	15.7527	2766.63	−50.50	7	107	262
263	440.52	243.24	−105.93	35.80	−50.108	63.807E-02	−3.642E-04	80.135E-09	15.8023	2731.00	−47.11	−23	102	263
264	357.43	197.74	−41.70	87.50	−1.746	53.089E-02	−2.903E-04	60.541E-09	15.8089	2654.81	−47.30	−33	87	264
265	344.33	197.95	−52.38	76.28	−9.810	53.089E-02	−2.717E-04	48.274E-09	16.2057	2897.97	−39.30	−28	97	265
266	344.33	197.95	−53.93	76.49	−32.925	69.292E-02	−5.619E-04	20.046E-08	15.8727	2701.72	−48.62	−28	92	266
267	344.33	197.95	−47.65	83.07	−21.729	58.113E-02	−3.362E-04	74.567E-09	15.8384	2680.52	−48.40	−28	92	267
268	344.33	197.95	−54.47	77.67	4.338	55.098E-02	−3.282E-04	80.470E-09	15.9288	2718.68	−47.77	−28	92	268
269			−59.79	71.26	−14.750	56.689E-02	−3.341E-04	79.633E-09	15.9423	2725.89	−47.64	−28	97	269
270			−57.78	73.27	−14.750	56.689E-02	−3.341E-04	79.633E-09	15.9124	2731.79	−46.76	−25	91	270
271			−58.70	71.34	−14.750	56.689E-02	−3.341E-04	79.633E-09	15.9484	2750.50	−48.33	−23	93	271
272			−50.37	82.19	−1.675	53.759E-02	−3.044E-04	67.533E-09	15.7527	2580.52	−46.56	−35	79	272
273			−54.39	79.67	12.627	51.540E-02	−3.007E-04	73.269E-09	15.8425	2631.57	−46.00	−33	81	273

NO	FORMULA	COMPOUND NAME	MOLWT	TFP	TBP	TC	PC	VC	LDEN	TDEN	HVAP	NO
274	C6H12	2;3-DIMETHYL-1-BUTENE	84.162	−157.3	55.6	501.0	32.4	0.343	678	20	27,424	274
275	C6H12	2;3-DIMETHYL-2-BUTENE	84.162	−74.3	73.2	524.0	33.6	0.351	708	20	29,655	275
276	C6H12	3;3-DIMETHYL-1-BUTENE	84.162	−115.2	41.2	490.0	32.5	0.340	653	20	25,665	276
277	C6H12O	CYCLOHEXANOL	100.161	24.8	161.1	625.0	37.5	0.327	942	30	45,511	277
278	C6H12O	METHYL ISOBUTYL KETONE	100.161	−84.2	116.4	571.0	32.7	0.371	801	20	35,588	278
279	C6H12O2	N-BUTYL ACETATE	116.160	−73.5	126.0	579.0	31.4	0.400	898	0	36,006	279
280	C6H12O2	ISOBUTYL ACETATE	116.160	−98.9	116.8	561.0	30.4	0.414	875	20	35,873	280
281	C6H12O2	ETHYL BUTYRATE	116.160	−93.2	120.8	566.0	31.4	0.395	879	20	34,332	281
282	C6H12O2	ETHYL ISOBUTYRATE	116.160	−88.2	111.0	553.0	30.4	0.410	869	20	35,023	282
283	C6H12O2	N-PROPYL PROPIONATE	116.160	−75.9	122.5	578.0			881	20	36,383	283
284	C6H14	N-HEXANE	86.178	−95.4	68.7	507.4	29.7	0.370	659	20	28,872	284
285	C6H14	2-METHYL PENTANE	86.178	−153.7	60.2	497.5	30.1	0.367	653	20	27,800	285
286	C6H14	3-METHYL PENTANE	86.178	−118.2	63.2	504.4	31.2	0.367	664	20	28,093	286
287	C6H14	2;2-DIMETHYL BUTANE	86.178	−99.9	49.7	488.7	30.8	0.359	649	20	26,322	287
288	C6H14	2;3-DIMETHYL BUTANE	86.178	−128.6	58.0	499.9	31.3	0.358	662	20	27,298	288
289	C6H14O	1-HEXANOL	102.177	−44.0	157.0	610.0	40.5	0.381	819	20	48,567	289
290	C6H14O	ETHYL BUTYL ETHER	102.177	−103.2	92.2	531.0	30.4	0.390	749	20	31,820	290
291	C6H14O	DIISOPROPYL ETHER	102.177	−85.5	68.3	500.0	28.8	0.386	724	20	29,349	291
292	C6H15N	DIPROPYLAMINE	101.193	−63.2	109.2	550.0	31.4	0.407	738	20	37,011	292
293	C6H15N	TRIETHYLAMINE	101.193	−114.8	89.5	535.0	30.4	0.390	728	20	31,401	293
294	C7F14	PERFLUOROMETHYLCYCLOHEXANE	350.055		76.3	486.8	23.3					294
295	C7F16	PERFLUORO-N-HEPTANE	388.051	−78.2	82.5	474.8	16.2	0.664	1733	20		295
296	C7H5N	BENZONITRILE	103.124	−13.2	190.8	699.4	42.2		1010	15		296
297	C7H6O	BENZALDEHYDE	106.124	−57.2	178.8	695.0	46.6		1045	20	42,705	297
298	C7H6O2	BENZOIC ACID	122.124	122.4	249.8	752.0	45.6	0.341	1075	130	50,660	298
299	C7H7NO2	O-NITROTOLUENE	137.139	−9.2	222.1	720.0	34.0	0.371	1167	20	45,487	299
300	C7H7NO2	M-NITROTOLUENE	137.139	16.0	233.1	725.0	30.5	0.371	1158	20	46,090	300
301	C7H7NO2	P-NITROTOLUENE	137.139	54.8	238.0	735.0	30.1	0.371	1164	20	46,875	301
302	C7H8	TOLUENE	92.141	−95.2	110.6	591.7	41.1	0.316	867	20	33,201	302
303	C7H8O	METHYL PHENYL ETHER	108.140	−37.5	153.6	641.0	41.7		996	20		303
304	C7H8O	BENZYL ALCOHOL	108.140	−15.4	205.4	677.0	46.6	0.334	1041	25	50,535	304
305	C7H8O	O-CRESOL	108.140	30.9	191.0	697.6	50.1	0.282	1028	40	45,217	305
306	C7H8O	M-CRESOL	108.140	12.2	202.2	705.8	45.6	0.310	1034	20	47,436	306
307	C7H8O	P-CRESOL	108.140	34.7	201.9	704.6	51.5		1019	40	47,478	307
308	C7H9N	2;3-DIMETHYLPYRIDINE	107.156		160.8	655.4			942	25		308
309	C7H9N	2;5-DIMETHYLPYRIDINE	107.156		157.0	644.2			938	0		309
310	C7H9N	3;4-DIMETHYLPYRIDINE	107.156		179.1	683.8			954	25		310
311	C7H9N	3;5-DIMETHYLPYRIDINE	107.156		171.9	667.2			939	25		311
312	C7H9N	METHYLPHENYLAMINE	107.156	−57.2	195.9	701.0	52.0		989	20		312

NO	VISA	VISB	DELHF	DELGF	CPVAPA	CPVAPB	CPVAPC	CPVAPD	ANTA	ANTB	ANTC	TMN	TMX	NO
274			-55.77	79.09	7.025	55.852E-02	-3.696E-04	10.630E-08	15.8012	2612.69	-43.78	-38	87	274
275			-59.24	75.91	2.294	48.274E-02	-2.199E-04	30.417E-09	16.0043	2798.63	-47.71	-23	102	275
276			-43.17	98.22	-12.556	54.847E-02	-2.915E-04	52.084E-09	15.3755	2326.80	-48.24	-48	67	276
277			-294.75	-117.98	-55.534	72.139E-02	-4.086E-04	82.354E-09						277
278	473.65	259.03	-284.03		3.894	56.564E-02	-3.318E-04	82.312E-09	15.7165	2893.66	-70.75	12	152	278
279	537.58	272.30	-486.76		13.620	54.889E-02	-2.278E-04	-7.913E-10	16.1836	3151.09	-69.15	22	162	279
280	533.99	270.49	-495.47		7.310	57.401E-02	-2.576E-04	11.011E-09	16.1714	3092.83	-66.15	16	154	280
281	489.95	264.22			21.508	49.279E-02	-1.938E-04	35.588E-10	15.9987	3127.60	-60.15	15	159	281
282														282
283									16.8641	3558.18	-47.86	19	147	283
284	362.79	207.09	-167.30	-0.25	-4.413	58.197E-02	-3.119E-04	64.937E-09	15.8366	2697.55	-48.78	-28	97	284
285	384.13	208.27	-174.42	-5.02	-10.567	61.839E-02	-3.573E-04	80.847E-09	15.7476	2614.38	-46.58	-33	97	285
286	372.11	207.55	-171.74	-2.14	-2.386	56.899E-02	-2.870E-04	50.325E-09	15.7701	2653.43	-46.02	-33	92	286
287	438.44	226.67	-185.68	-9.63	-16.634	62.928E-02	-3.481E-04	68.496E-09	15.5536	2489.50	-43.81	-43	77	287
288	444.19	228.86	-177.90	-4.10	-14.608	61.504E-02	-3.376E-04	68.203E-09	15.6802	2595.44	-44.25	-38	81	288
289	1179.40	354.94	-317.78	-135.65	4.811	58.908E-02	-3.010E-04	54.261E-09	18.0994	4055.45	-76.49	35	157	289
290	443.32	234.68			23.626	53.675E-02	-2.528E-04	41.567E-09	16.0477	2921.52	-55.15	-8	127	290
291	410.58	219.67	-319.03	-121.96	7.503	58.448E-02	-3.027E-04	58.448E-09	16.3417	2895.73	-43.15	-24	91	291
292	561.11	257.39			6.460	62.928E-02	-3.390E-04	70.715E-09	16.5939	3259.08	-55.15	29	149	292
293	355.52	214.48	-99.65	110.36	-18.430	71.552E-02	-4.392E-04	10.923E-08	15.8853	2882.38	-51.15	-13	127	293
294			-2898.10	-3089.31					15.7130	2610.57	-61.93	17	112	294
295			-3386.70						15.9747	2719.68	-64.50	-3	117	295
296			218.97	261.05	-26.004	57.317E-02	-4.430E-04	13.490E-08						296
297	686.84	314.66	-36.80	22.40	-12.142	49.614E-02	-2.845E-04	51.665E-09	16.3501	3748.62	-66.12	27	187	297
298	2617.60	407.88	-290.40	-210.55	-51.292	62.928E-02	-4.237E-04	10.622E-08	17.1634	4190.70	-125.20	132	287	298
299			-265.86						14.2028	2603.49	-151.52	222	129	299
300			-265.86											300
301									16.0433	3914.07	-90.45	233	129	301
302	467.33	255.24	50.03	122.09	-24.355	51.246E-02	-2.765E-04	49.111E-09	16.0137	3096.52	-53.67	7	137	302
303	388.84	325.85							16.2394	3430.82	-69.58	97	167	303
304	1088.00	367.21	-94.08		-7.398	54.805E-02	-3.357E-04	77.707E-09	17.4582	4384.81	-73.15	112	330	304
305	1533.40	365.61	-128.70	-37.10	-32.276	70.045E-02	-5.924E-04	21.240E-08	15.9148	3305.37	-108.00	97	207	305
306	1785.60	370.75	-132.43	-40.57	-45.008	72.641E-02	-6.029E-04	20.775E-08	17.2878	4274.42	-74.09	97	207	306
307	1826.90	372.68	-125.48	-30.90	-40.633	70.548E-02	-5.757E-04	19.674E-08	16.1989	3479.39	-111.30	97	207	307
308			68.29						17.1492	4219.74	-33.04	147	167	308
309			66.44						16.3046	3545.14	-63.59	77	162	309
310			70.05						16.9517	4237.04	-41.65	127	187	310
311			72.81						16.8850	4106.95	-44.45	127	187	311
312	915.12	332.74	85.41	199.33					16.3066	3756.28	-80.71	47	207	312

NO	FORMULA	COMPOUND NAME	MOLWT	TFP	TBP	TC	PC	VC	LDEN	TDEN	HVAP	NO
313	C7H9N	O-TOLUIDINE	107.156	-14.8	200.1	694.0	37.5	0.343	998	20	45,364	313
314	C7H9N	M-TOLUIDINE	107.156	-30.4	203.3	709.0	41.5	0.343	989	20	45,636	314
315	C7H9N	P-TOLUIDINE	107.156	43.7	200.1	667.0			964	50	44,799	315
316	C7H14	CYCLOHEPTANE	98.189	-8.2	118.7	589.0	37.2	0.390	810	20	33,076	316
317	C7H14	1;1-DIMETHYLCYCLOPENTANE	98.189	-69.8	87.8	547.0	34.5	0.360	759	16	30,312	317
318	C7H14	CIS-1;2-DIMETHYLCYCLOPENTANE	98.189	-53.9	99.5	564.8	34.5	0.368	777	16	31,719	318
319	C7H14	TRANS-1;2-DIMETHYLCYCLOPENTANE	98.189	-117.6	91.8	553.2	34.5	0.362	756	16	30,878	319
320	C7H14	ETHYLCYCLOPENTANE	98.189	-138.5	103.4	569.5	33.9	0.375	771	16	32,301	320
321	C7H14	METHYLCYCLOHEXANE	98.189	-126.6	100.9	572.1	34.8	0.368	774	16	31,150	321
322	C7H14	1-HEPTENE	98.189	-118.9	93.6	537.2	28.4	0.440	679	20	31,108	322
323	C7H14	2;3;3-TRIMETHYL-1-BUTENE	98.189	-109.9	77.8	533.0	29.0	0.400	705	20	28,889	323
324	C7H16	N-HEPTANE	100.205	-90.6	98.4	540.2	27.4	0.432	684	20	31,719	324
325	C7H16	2-METHYLHEXANE	100.205	-118.3	90.0	530.3	27.4	0.421	679	20	30,689	325
326	C7H16	3-METHYLHEXANE	100.205	-173.2	91.8	535.2	28.2	0.404	687	20	30,815	326
327	C7H16	2;2-DIMETHYLPENTANE	100.205	-123.8	79.2	520.4	27.8	0.416	674	20	29,182	327
328	C7H16	2;3-DIMETHYLPENTANE	100.205		89.7	537.3	29.1	0.393	965	20	30,409	328
329	C7H16	2;4-DIMETHYLPENTANE	100.205	-119.2	80.5	519.7	27.4	0.418	673	20	29,517	329
330	C7H16	3;3-DIMETHYLPENTANE	100.205	-134.5	86.0	536.3	29.5	0.414	693	20	29,668	330
331	C7H16	3-ETHYLPENTANE	100.205	-118.6	93.4	540.6	28.9	0.416	698	20	30,978	331
332	C7H16	2;2;3-TRIMETHYLBUTANE	100.205	-24.9	80.8	531.1	29.6	0.398	690	20	28,968	332
333	C7H16O	1-HEPTANOL	116.204	-34.0	176.3	633.0	30.4	0.435	822	20	48,148	333
334	C8H4O3	PHTHALIC ANHYDRIDE	148.118	130.8	286.8	810.0	47.6	0.368			49,614	334
335	C8H8	STYRENE	104.152	-30.7	145.1	647.0	39.9		906	20	36,844	335
336	C8H8O	METHYL PHENYL KETONE	120.151	19.6	201.7	701.0	38.5	0.376	1032	15	43,124	336
337	C8H8O2	METHYL BENZOATE	136.151	-12.4	199.0	692.0	36.5	0.396	1083	20	36,844	337
338	C8H10	O-XYLENE	106.168	-25.2	144.4	630.2	37.3	0.369	880	20	36,383	338
339	C8H10	M-XYLENE	106.168	-47.9	139.1	617.0	35.5	0.376	864	20	36,006	339
340	C8H10	P-XYLENE	106.168	13.2	138.3	616.2	35.2	0.379	861	20	35,588	340
341	C8H10	ETHYL BENZENE	106.168	-95.0	136.1	617.1	36.1	0.374	867	20		341
342	C8H10O	O-ETHYLPHENOL	122.167	-3.4	204.5	703.0			1037	0	48,106	342
343	C8H10O	M-ETHYLPHENOL	122.167	-4.2	218.4	707.6			1025	0	50,828	343
344	C8H10O	P-ETHYLPHENOL	122.167	44.8	217.8	716.4					50,660	344
345	C8H10O	ETHYL PHENYL ETHER	122.167	-30.2	169.8	647.0	34.2		979	4		345
346	C8H10O	2;3-XYLENOL	122.167	74.8	216.9	722.8					47,311	346
347	C8H10O	2;4-XYLENOL	122.167	24.8	210.8	707.6					47,143	347
348	C8H10O	2;5-XYLENOL	122.167	74.8	211.1	723.0					46,892	348
349	C8H10O	2;6-XYLENOL	122.167	48.8	200.9	701.0					44,380	349
350	C8H10O	3;4-XYLENOL	122.167	64.8	226.8	729.8					49,823	350
351	C8H10O	3;5-XYLENOL	122.167	63.8	221.6	715.6					49,404	351

NO	VISA	VISB	DELHF	DELGF	CPVAPA	CPVAPB	CPVAPC	CPVAPD	ANTA	ANTB	ANTC	TMN	TMX	NO
313	1085.10	356.46							16.7834	4072.58	−72.15	102	227	313
314	928.12	354.07					−3.033E-04	46.432E-09	16.7498	4080.32	−73.15	82	227	314
315	738.90	356.02			−15.989	56.815E-02			16.6968	4041.04	−72.15	77	227	315
316			−119.41	63.05	−76.187	78.670E-02	−4.204E-04	75.614E-09	15.7818	3066.05	−56.80	57	162	316
317			−138.37	39.06	−57.891	76.702E-02	−4.501E-04	10.103E-08	15.6973	2807.94	−51.20	−13	117	317
318			−129.62	45.76	−55.643	76.158E-02	−4.484E-04	10.140E-08	15.7729	2922.30	−52.94	−3	127	318
319			−136.78	38.39	−54.521	75.907E-02	−4.480E-04	10.170E-08	15.7594	2861.53	−51.46	−13	117	319
320	433.81	249.72	−127.15	44.59	−55.312	75.111E-02	−4.396E-04	10.040E-08	15.8581	2990.13	−52.47	−3	129	320
321	528.41	271.58	−154.87	27.30	−61.919	78.419E-02	−4.438E-04	93.659E-09	15.7105	2926.04	−51.75	−3	127	321
322	368.69	214.32	−62.34	95.88	−3.303	62.969E-02	−3.512E-04	76.074E-09	15.8894	2895.51	−53.97	−8	127	322
323			−86.54						15.6536	2719.47	−49.56	−20	102	323
324	436.73	232.53	−187.90	8.00	−5.146	67.617E-02	−3.651E-04	76.577E-09	15.8737	2911.32	−56.51	−3	127	324
325	417.46	225.13	−195.06	3.22	−39.389	86.416E-02	−6.289E-04	18.363E-08	15.8261	2845.06	−53.60	−9	117	325
326			−192.43	4.61	−7.046	68.370E-02	−3.734E-04	78.335E-09	15.8133	2855.66	−53.93	−8	117	326
327	417.37	226.19	−206.28	0.08	−50.099	89.556E-02	−6.360E-04	17.358E-08	15.6917	2740.15	−49.85	−19	105	327
328			−199.38	0.67	−7.046	70.476E-02	−3.734E-04	78.335E-09	15.7815	2850.64	−51.33	−11	115	328
329			−202.14	3.10	−7.046	68.370E-02	−3.734E-04	78.335E-09	15.7179	2744.78	−51.52	−17	105	329
330			−201.68	2.64	−7.046	68.370E-02	−3.734E-04	78.335E-09	15.7190	2829.10	−47.83	−13	112	330
331			−189.79	11.01	−7.046	68.370E-02	−3.734E-04	78.335E-09	15.8317	2882.44	−53.26	−13	119	331
332			−204.94	4.27	−22.944	75.195E-02	−4.421E-04	10.048E-08	15.6398	2764.40	−47.10	−7	106	332
333	1287.00	361.83	−332.01	−121.00	4.907	67.784E-02	−3.447E-04	60.457E-09	15.3068	2626.42	−146.60	60	176	333
334			−371.79		−4.455	65.398E-02	−4.283E-04	10.094E-08	15.9984	4467.01	−83.15	136	342	334
335	528.64	276.71	147.46	213.95	−28.248	61.588E-02	−4.023E-04	99.353E-09	16.0193	3328.57	−63.72	32	187	335
336	1316.40	310.82	−86.92	1.84	−29.580	64.100E-02	−4.071E-04	97.217E-09	16.2384	3781.07	−81.15	77	247	336
337	768.94	332.33	−254.06		−21.210	55.015E-02	−1.799E-04	44.254E-09	16.2272	3751.83	−81.15	77	243	337
338	513.54	277.98	19.01	122.17	−15.851	59.620E-02	−3.443E-04	75.279E-09	16.1156	3395.57	−59.46	32	172	338
339	453.42	257.18	17.25	118.95	−29.165	62.969E-02	−3.747E-04	84.783E-09	16.1390	3366.99	−58.04	27	167	339
340	475.16	261.40	17.96	121.21	−25.091	60.416E-02	−3.374E-04	68.203E-09	16.0963	3346.65	−57.84	27	167	340
341	472.82	264.22	29.81	130.67	−43.099	70.715E-02	−4.811E-04	13.008E-08	16.0195	3272.47	−59.95	27	177	341
342			−145.78						17.9610	4928.36	−45.75	77	227	342
343			−146.58						17.1955	4272.77	−86.08	97	227	343
344			−144.65						19.0905	5579.62	−44.15	97	227	344
345	646.88	305.91							16.1673	3473.20	−78.66	112	187	345
346			−157.34						16.2424	3724.58	−102.40	147	227	346
347			−162.78						13.2456	3655.26	−103.80	137	227	347
348			−161.53						16.2328	3667.32	−102.40	137	217	348
349			−161.95						16.2809	3749.35	−85.55	127	207	349
350			−156.50						16.3004	3733.53	−113.90	157	247	350
351			−161.48						16.4192	3775.91	−109.00	137	227	351

NO	FORMULA	COMPOUND NAME	MOLWT	TFP	TBP	TC	PC	VC	LDEN	TDEN	HVAP	NO
352	C8H11N	N;N-DIMETHYLANILINE	121.183	2.4	193.5	687.0	36.3		956	20		352
353	C8H16	1;1-DIMETHYLCYCLOHEXANE	112.216	−33.5	119.5	591.0	29.7	0.416	785	16	32,615	353
354	C8H16	CIS-1;2-DIMETHYLCYCLOHEXANE	112.216	−50.1	129.7	606.0	29.7		796	20	33,662	354
355	C8H16	TRANS-1;2-DIMETHYLCYCLOHEXANE	112.216	−88.2	123.4	596.0	29.7		776	20	32,908	355
356	C8H16	CIS-1;3-DIMETHYLCYCLOHEXANE	112.216	−75.6	120.1	591.0	29.7		766	20	32,825	356
357	C8H16	TRANS-1;3-DIMETHYLCYCLOHEXANE	112.216	−90.2	124.4	598.0	29.7		785	20	33,871	357
358	C8H16	CIS-1;4-DIMETHYLCYCLOHEXANE	112.216	−87.5	124.3	598.0	29.7		783	20	33,787	358
359	C8H16	TRANS-1;4-DIMETHYLCYCLOHEXANE	112.216	−37.0	119.3	590.0	29.7		763	20	32,615	359
360	C8H16	ETHYLCYCLOHEXANE	112.216	−111.4	131.7	609.0	30.3	0.450	788	20	34,332	360
361	C8H16	1;1;2-TRIMETHYLCYCLOPENTANE	112.216		113.7	579.5	29.4				32,615	361
362	C8H16	1;1;3-TRIMETHYLCYCLOPENTANE	112.216		104.8	569.5	28.3				31,694	362
363	C8H16	CIS;CIS;TRANS-1;2;4-TRIMETHYLCYCLOPENTANE	112.216		117.8	579.0	28.8				33,076	363
364	C8H16	CIS;TRANS;CIS-1;2;4-TRIMETHYLCYCLOPENTANE	112.216		109.2	571.0	28.1				33,076	364
365	C8H16	1-METHYL-1-ETHYLCYCLOPENTANE	112.216		121.5	592.0	29.9				33,662	365
366	C8H16	N-PROPYLCYCLOPENTANE	112.216	−117.4	130.9	603.0	30.0	0.425	781	16	34,131	366
367	C8H16	ISOPROPYLCYCLOPENTANE	112.216	−112.7	126.4	601.0	30.0		776	20	34,122	367
368	C8H16	1-OCTENE	112.216	−101.8	121.2	566.6	26.2	0.464	715	20	33,787	368
369	C8H16	TRANS-2-OCTENE	112.216	−87.8	124.9	580.0	27.7		720	20	34,332	369
370	C8H16	N-OCTANE	114.232	−56.8	125.6	568.8	24.8	0.492	703	20	34,436	370
371	C8H18	2-METHYLHEPTANE	114.232	−109.2	117.6	559.6	24.8	0.488	702	16	33,829	371
372	C8H18	3-METHYLHEPTANE	114.232	−120.5	118.9	563.6	25.4	0.464	706	20	33,913	372
373	C8H18	4-METHYLHEPTANE	114.232	−121.0	117.7	561.7	25.4	0.476	705	20	33,913	373
374	C8H18	2;2-DIMETHYLHEXANE	114.232	−121.2	108.8	549.8	25.3	0.478	695	20	32,280	374
375	C8H18	2;3-DIMETHYLHEXANE	114.232		115.6	563.4	26.2	0.468	712	20	33,226	375
376	C8H18	2;4-DIMETHYLHEXANE	114.232		109.4	553.5	25.5	0.472	700	20	32,615	376
377	C8H18	2;5-DIMETHYLHEXANE	114.232	−91.3	109.1	550.0	24.8	0.482	693	20	32,657	377
378	C8H18	3;3-DIMETHYLHEXANE	114.232	−126.2	111.9	562.0	26.5	0.443	710	20	32,490	378
379	C8H18	3;4-DIMETHYLHEXANE	114.232		117.7	568.8	27.0	0.466	719	20	33,298	379
380	C8H18	3-ETHYLHEXANE	114.232		118.5	565.4	26.0	0.455	718	16	33,633	380
381	C8H18	2;2;3-TRIMETHYLPENTANE	114.232	−112.3	109.8	563.4	27.3	0.436	716	20	32,029	381
382	C8H18	2;2;4-TRIMETHYLPENTANE	114.232	−107.4	99.2	543.9	25.6	0.468	692	20	31,028	382
383	C8H18	2;3;3-TRIMETHYLPENTANE	114.232	−100.7	114.7	573.5	28.2	0.455	726	20	32,364	383
384	C8H18	2;3;4-TRIMETHYLPENTANE	114.232	−109.3	113.4	566.3	27.3	0.461	719	20	32,753	384
385	C8H18	2-METHYL-3-ETHYLPENTANE	114.232	−115.0	115.6	567.0	27.1	0.443	719	20	32,988	385
386	C8H18	3-METHYL-3-ETHYLPENTANE	114.232	−90.9	118.2	576.5	28.1	0.455	727	20	32,816	386
387	C8H18O	1-OCTANOL	130.231	−15.5	195.2	658.0	34.5	0.490	826	20	50,660	387
388	C8H18O	2-OCTANOL	130.231	−32.0	179.7	637.0	27.4	0.494	821	20	44,380	388
389	C8H18O	2-ETHYLHEXANOL	130.231	−70.0	184.6	613.0	27.6	0.494	833	20	46,599	389
390	C8H18O	BUTYL ETHER	130.231	−97.9	142.4	580.0	25.3	0.500	768	20	37,263	390

NO	VISA	VISB	DELHF	DELGF	CPVAPA	CPVAPB	CPVAPC	CPVAPD	ANTA	ANTB	ANTC	TMN	TMX	NO
352	553.02	320.03	84.15	231.36					16.9647	4276.08	−52.80	72	207	352
353			−181.12	35.25	−72.105	89.974E-02	−5.020E-04	10.304E-08	15.6535	3043.34	−55.30	10	147	353
354			−172.29	41.24	−68.370	89.723E-02	−5.137E-04	10.986E-08	15.7438	3148.35	−57.31	17	157	354
355			−180.12	34.50	−68.479	91.230E-02	−5.355E-04	11.811E-08	15.7337	3117.43	−54.02	13	151	355
356			−184.89	29.85	−65.163	88.383E-02	−4.932E-04	10.199E-08	15.7470	3081.95	−55.08	11	147	356
357			−176.68	36.34	−64.154	88.258E-02	−5.016E-04	10.685E-08	15.7371	3093.95	−57.76	15	152	357
358			−176.77	37.97	−64.154	88.258E-02	−5.016E-04	10.685E-08	15.7333	3098.39	−57.00	14	152	358
359	506.43		184.72	31.74	−70.363	91.314E-02	−5.309E-04	11.547E-08	15.6984	3063.44	−54.57	10	147	359
360		280.76	−171.87	39.27	−63.891	88.928E-02	−5.108E-04	11.028E-08	15.8125	3183.25	−58.15	20	160	360
361									15.7084	3015.51	−54.59	6	141	361
362									15.6794	2938.09	−53.25	0	131	362
363									15.7543	3073.95	−54.20	10	145	363
364									15.7756	3009.70	−53.23	9	144	364
365									15.8222	3120.66	−55.06	13	149	365
366	454.23	264.22	−148.17	52.63	−55.973	84.490E-02	−4.924E-04	11.175E-08	15.8969	3187.67	−59.99	21	158	366
367									15.8561	3176.22	−55.18	16	154	367
368	418.82	237.63	−82.98	104.29	−4.099	72.390E-02	−4.036E-04	86.750E-09	15.9630	3116.52	−60.39	16	147	368
369	427.64	240.32	−94.58	92.74	−12.820	75.321E-02	−4.442E-04	10.505E-08	15.8554	3134.97	−58.00	16	152	369
370	473.70	251.71	−208.59	16.41	−6.096	77.121E-02	−4.195E-04	88.551E-09	15.9426	3120.29	−63.63	19	152	370
371	643.61	259.51	−215.62	12.77	−89.744	12.422E-01	−1.176E-03	46.180E-08	15.9278	3097.63	−59.46	12	144	371
372			−212.77	13.73	−9.215	78.586E-02	−4.400E-04	96.966E-09	15.8865	3065.96	−60.74	13	145	372
373			−212.23	16.75	−9.215	78.586E-02	−4.400E-04	96.966E-09	15.8893	3057.05	−60.59	12	144	373
374			−224.87	10.72	−9.215	78.586E-02	−4.400E-04	96.966E-09	15.7431	2932.56	−58.08	3	132	374
375			−214.07	17.71	−9.215	78.586E-02	−4.400E-04	96.966E-09	15.8189	3029.06	−58.99	10	142	375
376			−219.56	11.72	−9.215	78.586E-02	−4.400E-04	96.966E-09	15.7797	2965.44	−58.36	5	135	376
377			−222.78	10.47	−9.215	78.586E-02	−4.400E-04	96.966E-09	15.7954	2964.06	−58.74	5	135	377
378	446.20	244.67	−220.27	13.27	−9.215	78.586E-02	−4.400E-04	96.966E-09	15.7755	3011.51	−55.71	6	138	378
379			−213.15	17.33	−9.215	78.586E-02	−4.400E-04	96.966E-09	15.8415	3062.52	−58.29	11	144	379
380	437.60	238.33	−211.01	16.54	−9.215	78.586E-02	−4.400E-04	96.966E-09	15.8671	3057.57	−60.55	13	145	380
381	474.57	257.61	−220.27	17.12	−9.215	78.586E-02	−4.400E-04	96.966E-09	15.7162	2981.56	−54.73	4	136	381
382	467.04	246.43	−224.29	13.69	−7.461	77.791E-02	−4.287E-04	91.733E-09	15.6850	2896.28	−52.41	4	125	382
383			−216.58	18.92	−9.215	78.586E-02	−4.400E-04	96.966E-09	15.7578	3057.94	−52.77	7	142	383
384			−217.59	18.92	−9.215	78.586E-02	−4.400E-04	96.966E-09	15.7818	3028.09	−55.62	7	140	384
385			211.35	21.27	−9.215	78.586E-02	−4.400E-04	96.966E-09	15.8040	3035.08	−57.84	9	142	385
386			−215.12	19.93	−9.215	78.586E-02	−4.400E-04	96.966E-09	15.8126	3102.06	−53.47	10	145	386
387	1312.10	369.97	−360.06	−120.16	6.171	76.074E-02	−3.797E-04	62.635E-09	15.7428	3017.81	−137.10	70	195	387
388					25.879	76.409E-02	−4.224E-04	90.644E-09	14.7108	2441.66	−150.70	72	180	388
389	1798.00	351.17	−365.55		−14.993	86.541E-02	−5.280E-04	12.845E-08	15.3614	2773.46	−140.00	75	185	389
390	473.50	266.56	−334.11	−88.59	6.054	77.288E-02	−4.085E-04	80.847E-09	16.0778	3296.15	−66.15	32	182	390

NO	FORMULA	COMPOUND NAME	MOLWT	TFP	TBP	TC	PC	VC	LDEN	TDEN	HVAP	NO
391	C8H18O5	TETRAETHYLENE GLYCOL	194.229		318.9	795.8	21.0	0.646				391
392	C8H19N	DIBUTYLAMINE	129.247	-62.2	159.6	596.0	25.3	0.517	767	20	39,775	392
393	C8H20SI	TETRAETHYL SILANE	144.333	-82.5	153.4	603.7	26.0	0.582	766	20	36,473	393
394	C9H8	INDENE	116.163		181.9	691.9	38.2	0.377				394
395	C9H10	INDAN	118.179		177.0	681.1	36.3	0.392				395
396	C9H10	ALPHA-METHYL STYRENE	118.179		165.3	654.0	34.0	0.397	911	20	38,309	396
397	C9H10O2	ETHYL BENZOATE	150.178	-34.9	212.7	697.0	32.4	0.451	1046	20	44,799	397
398	C9H12	N-PROPYLBENZENE	120.195	-99.5	159.2	638.3	32.0	0.440	862	20	38,267	398
399	C9H12	ISOPROPYLBENZENE	120.195	-96.1	152.4	631.0	32.1	0.428	862	20	37,556	399
400	C9H12	1-METHYL-2-ETHYLBENZENE	120.195	-80.9	165.1	651.0	30.4	0.460	881	20	38,895	400
401	C9H12	1-METHYL-3-ETHYLBENZENE	120.195	-95.6	161.3	637.0	28.4	0.490	865	20	38,560	401
402	C9H12	1-METHYL-4-ETHYLBENZENE	120.195	-62.4	162.0	640.0	29.4	0.470	861	20	38,435	402
403	C9H12	1;2;3-TRIMETHYLBENZENE	120.195	-25.5	176.0	664.5	34.6	0.430	894	20	40,068	403
404	C9H12	1;2;4-TRIMETHYLBENZENE	120.195	-46.2	169.3	649.1	32.3	0.430	880	16	39,272	404
405	C9H12	1;3;5-TRIMETHYLBENZENE	120.195	-44.8	164.7	637.3	31.3	0.433	865	20	39,063	405
406	C9H18	N-PROPYLCYCLOHEXANE	126.243	-94.5	156.7	639.0	28.1		793	20	36,090	406
407	C9H18	ISOPROPYLCYCLOHEXANE	126.243	-89.8	154.5	640.0	28.4		802	20		407
408	C9H18	1-NONENE	126.243	-81.4	146.5	592.0	23.4	0.580	745	0	36,341	408
409	C9H20	N-NONANE	128.259	-53.5	150.8	594.6	23.1	0.548	718	20	36,940	409
410	C9H20	2;2;3-TRIMETHYLHEXANE	128.259		133.6	588.0	24.9				34,792	410
411	C9H20	2;2;4-TRIMETHYLHEXANE	128.259	-120.2	126.5	573.7	23.7		720	16	34,039	411
412	C9H20	2;2;5-TRIMETHYLHEXANE	128.259	-105.8	124.1	568.0	23.3	0.519	717	16	33,787	412
413	C9H20	3;3-DIMETHYLPENTANE	128.259		146.1	610.0	26.7		752	20	36,006	413
414	C9H20	2;2;3;3-TETRAMETHYLPENTANE	128.259		140.2	607.6	27.4				35,295	414
415	C9H20	2;2;3;4-TETRAMETHYLPENTANE	128.259		133.0	592.7	26.0				34,290	415
416	C9H20	2;2;4;4-TETRAMETHYLPENTANE	128.259	-67.2	122.2	574.7	24.8		719	20	32,866	416
417	C9H20	2;3;3;4-TETRAMETHYLPENTANE	128.259		141.5	607.6	27.2				34,960	417
418	C10H8	NAPHTHALENE	128.174	80.3	217.9	748.4	40.5	0.410	971	90	43,292	418
419	C10H12	1;2;3;4-TETRAHYDRONAPHTHALENE	132.206	-31.2	207.5	719.0	35.2		973	20	39,733	419
420	C10H14	N-BUTYLBENZENE	134.222	-88.0	183.2	660.5	28.9	0.497	860	20	39,272	420
421	C10H14	ISOBUTYLBENZENE	134.222	-51.5	172.7	650.0	31.4	0.480	853	20	37,849	421
422	C10H14	SEC-BUTYLBENZENE	134.222	-75.5	173.3	664.0	29.5		862	20	37,974	422
423	C10H14	TERT-BUTYLBENZENE	134.222	-57.9	169.1	660.0	29.7		867	20	37,639	423
424	C10H14	1-METHYL-2-ISOPROPYLBENZENE	134.222		178.3	670.0	29.0		876	20		424
425	C10H14	1-METHYL-3-ISOPROPYLBENZENE	134.222		175.1	666.0	29.4		861	20	38,142	425
426	C10H14	1-METHYL-4-ISOPROPYLBENZENE	134.222	-73.2	177.1	653.0	28.3		857	20		426
427	C10H14	1;4-DIETHYLBENZENE	134.222	-42.2	183.7	657.9	28.1	0.480	862	20	39,398	427
428	C10H14	1;2;4;5-TETRAMETHYLBENZENE	134.222	78.8	196.8	675.0	29.4	0.480	838	81	45,552	428
429	C10H15N	N-BUTYLANILINE	149.236	-14.2	240.7	721.0	28.4	0.518	932	20	48,944	429

NO	VISA	VISB	DELHF	DELGF	CPVAPA	CPVAPB	CPVAPC	CPVAPD	ANTA	ANTB	ANTC	TMN	TMX	NO
391					7.164	86.164E-02	-2.904E-04	-9.115E-08	20.5564	8215.28	-11.50	227	427	391
392	581.42	286.54			9.764	80.805E-02	-4.392E-04	92.486E-09	16.7307	3721.90	-64.15	49	186	392
393			-314.93						16.6385	3873.18	-39.33	153	-1	393
394					-42.944	68.957E-02	-4.340E-04	91.482E-09	16.4380	3994.97	-49.40	77	277	394
395					-59.639	78.126E-02	-4.841E-04	98.474E-09	16.2601	3789.86	-57.00	77	277	395
396	354.34	270.80			-24.329	69.333E-02	-4.530E-04	11.807E-08	16.3308	3644.30	-67.15	75	220	396
397	746.50	338.47			20.670	68.873E-02	-3.608E-04	50.618E-09	16.2065	3845.09	-84.15	88	258	397
398	527.45	282.65	7.83	137.33	-31.288	74.860E-02	-4.601E-04	10.810E-08	16.0062	3433.84	-66.01	43	188	398
399	517.17	276.22	3.94	137.08	-39.364	78.419E-02	-5.087E-04	12.912E-08	15.9722	3363.60	-63.37	38	181	399
400			1.21	131.17	-16.446	69.961E-02	-4.120E-04	93.282E-09	16.1253	3535.33	-65.85	48	194	400
401			-1.93	126.53	-28.998	72.980E-02	-4.363E-04	99.981E-09	16.1545	3521.08	-64.64	45	190	401
402	463.17	266.08	-2.05	126.78	-27.310	71.762E-02	-4.224E-04	95.417E-09	16.1135	3516.31	-64.23	45	190	402
403			-9.59	124.64	-6.942	63.346E-02	-3.326E-04	66.110E-09	16.2121	3670.22	-66.07	56	206	403
404	872.74	297.75	-13.94	117.02	-4.668	62.383E-02	-3.263E-04	63.765E-09	16.2190	3622.58	-64.59	51	198	404
405	437.52	268.27	-16.08	118.03	-19.590	67.240E-02	-3.692E-04	76.995E-09	16.2893	3614.19	-63.57	48	193	405
406	549.08	293.93	-193.43	47.35	-62.517	98.892E-02	-5.795E-04	12.912E-08	15.8567	3363.62	-65.21	40	186	406
407									15.8260	3346.12	-63.71	57	167	407
408	471.00	258.92	-103.58	112.75	-3.718	81.224E-02	-4.509E-04	97.050E-09	16.0118	3305.03	-67.61	35	175	408
409	525.56	272.12	-229.19	24.83	3.144	67.742E-02	-1.928E-04	-2.981E-08	15.9671	3291.45	-71.33	39	179	409
410			-241.37	24.53	-45.632	10.555E-01	-7.172E-04	19.866E-08	15.8017	3164.17	-61.66	24	163	410
411			-243.38	22.52	-60.311	11.045E-01	-7.712E-04	21.876E-08	15.7639	3084.08	-61.94	18	155	411
412			254.18	13.44	-54.106	10.948E-01	-7.746E-04	22.546E-08	15.7445	3052.17	-62.24	42	147	412
413			231.95	35.09	-67.269	11.262E-01	-7.988E-04	23.061E-08	15.8709	3341.62	-57.57	77	167	413
414			-237.39	34.33	-54.583	10.890E-01	-7.570E-04	21.420E-08	15.7280	3220.55	-59.31	55	167	414
415			-237.22	32.66	-54.583	10.890E-01	-7.570E-04	21.420E-08	15.7363	3167.42	-58.21	45	157	415
416			-242.12	34.04	-67.403	11.681E-01	-8.612E-04	25.736E-08	15.6488	3049.98	-57.13	40	140	416
417			-236.39	34.12	-54.918	10.911E-01	-7.603E-04	21.579E-08	15.8029	3269.07	-58.19	52	152	417
418	873.32	352.57	151.06	223.74	-68.802	84.992E-02	-6.506E-04	19.808E-08	16.1426	3992.01	-71.29	87	252	418
419			27.63	167.05					16.2805	4009.49	-64.98	92	227	419
420	563.84	296.01	-13.82	144.78	-22.990	79.340E-02	-4.396E-04	85.704E-09	16.0793	3633.40	-71.77	62	213	420
421			-21.56						15.9524	3512.47	-69.03	53	203	421
422	582.82	295.82	-17.46		-65.147	98.934E-02	-7.214E-04	21.520E-08	15.9999	3544.19	-68.10	52	203	422
423			-22.69		-86.001	11.020E-01	-8.746E-04	28.265E-08	15.9300	3462.28	-69.87	50	199	423
424									15.9809	3564.52	-70.00	57	208	424
425			-29.31		-48.759	90.644E-02	-6.054E-04	16.274E-08	15.9811	3543.79	-69.22	55	205	425
426									15.9424	3539.21	-70.10	56	207	426
427			-22.27	137.96	-37.417	86.709E-02	-5.560E-04	14.110E-08	16.1140	3657.22	-71.18	62	214	427
428			-45.30	119.53	15.265	65.188E-02	-2.879E-04	32.569E-09	16.3023	3850.91	-71.72	88	227	428
429	1111.10	341.28			-34.068	91.440E-02	-5.560E-04	12.874E-08	16.3994	4079.72	-96.15	112	287	429

NO	FORMULA	COMPOUND NAME	MOLWT	TFP	TBP	TC	PC	VC	LDEN	TDEN	HVAP	NO
430	C10H18	CIS-DECALIN	138.254	−43.2	195.7	702.2	31.4		897	20	39,356	430
431	C10H18	TRANS-DECALIN	138.254	−30.4	187.2	690.0	31.4		870	20	38,519	431
432	C10H19N	CAPRYLONITRILE	153.269	−17.9	242.8	622.0	32.5		820	20		432
433	C10H20	N-BUTYLCYCLOHEXANE	140.270	−74.8	180.9	667.0	31.5		799	20	38,519	433
434	C10H20	ISOBUTYLCYCLOHEXANE	140.270		171.3	659.0	31.2		795	20		434
435	C10H20	SEC-BUTYLCYCLOHEXANE	140.270		179.3	669.0	26.7		813	20		435
436	C10H20	TERT-BUTYLCYCLOHEXANE	140.270	−41.2	171.5	659.0	26.6		813	20		436
437	C10H20	1-DECENE	140.270	−66.3	170.5	615.0	22.1	0.650	741	20	38,686	437
438	C10H20O	MENTHOL	156.269	42.8	216.3	694.0						438
439	C10H22	N-DECANE	142.286	−29.7	174.1	617.6	21.1	0.603	730	20	39,306	439
440	C10H22	3;3;5-TRIMETHYLHEPTANE	142.286		155.6	609.6	23.2				36,676	440
441	C10H22	2;2;3-TETRAMETHYLHEXANE	142.286		160.3	623.1	25.1				36,383	441
442	C10H22	2;2;5-TETRAMETHYLHEXANE	142.286		137.4	581.5	21.9				35,295	442
443	C10H22O	1-DECANOL	158.285	6.9	230.2	700.0	22.3	0.600	830	20	50,242	443
444	C11H10	1-METHYLNAPHTHALENE	142.201	−30.5	244.6	772.0	35.7	0.445	1020	20	46,055	444
445	C11H10	2-METHYLNAPHTHALENE	142.201	34.5	241.0	761.0	35.1	0.462	990	40	46,055	445
446	C11H14O2	BUTYL BENZOATE	178.232	−22.2	249.8	723.0	26.3	0.561	1006	20	48,986	446
447	C11H22	N-HEXYLCYCLOPENTANE	154.297		203.1	660.1	21.4				41,198	447
448	C11H22	1-UNDECENE	154.297	−49.2	192.6	637.0	20.0	0.660	751	20	40,905	448
449	C11H24	N-UNDECANE	156.313	−25.6	195.9	638.8	19.7		740	20	41,533	449
450	C12H8	ACENAPHTHALENE	152.196	95.0	270.0	796.9	32.2	0.487				450
451	C12H10	DIPHENYL	154.212	69.2	255.2	789.0	38.5	0.502	990	74	45,636	451
452	C12H10O	DIPHENYL ETHER	170.211	26.8	258.0	766.0	31.4		1066	30	47,143	452
453	C12H24	N-HEPTYLCYCLOPENTANE	168.324		224.1	679.0	19.5				43,375	453
454	C12H24	1-DODECENE	168.324	−35.2	213.3	657.0	18.5	0.713	758	20	42,998	454
455	C12H26	N-DODECANE	170.340	−9.6	216.3	658.3	18.2	0.720	748	20	43,668	455
456	C12H26O	DIHEXYL ETHER	186.339	−43.2	226.4	657.0	18.2	0.718	794	20	45,636	456
457	C12H26O	DODECANOL	186.339	23.9	259.9	679.0	19.3		835	20		457
458	C12H27N	TRIBUTYLAMINE	185.355		213.4	643.0	18.2		779	20	44,380	458
459	C13H10	FLUORENE	166.223	114.0	297.9	822.3	29.9	0.534				459
460	C13H12	DIPHENYLMETHANE	168.239	26.8	264.3	767.0	29.8		1006	20		460
461	C13H26	N-OCTYLCYCLOPENTANE	182.351		243.7	694.0	17.9				45,427	461
462	C13H26	1-TRIDECENE	182.351	−23.1	232.7	674.0	17.0		766	20	45,008	462
463	C13H28	N-TRIDECANE	184.367	−5.4	235.4	675.8	17.2	0.780	756	20	45,678	463
464	C14H10	ANTHRACENE	178.234	216.5	341.2	883.0					56,522	464
465	C14H10	PHENANTHRENE	178.234	100.5	339.4	878.0					55,684	465
466	C14H28	N-NONYLCYCLOPENTANE	196.378		262.1	710.5	16.5				47,269	466
467	C14H28	1-TETRADECENE	196.378	−12.9	251.1	689.0	15.6		786	0	46,934	467
468	C14H30	N-TETRADECANE	198.394	5.8	253.5	694.0	16.2	0.830	763	20	47,646	468

NO	VISA	VISB	DELHF	DELGF	CPVAPA	CPVAPB	CPVAPC	CPVAPD	ANTA	ANTB	ANTC	TMN	TMX	NO
430			-169.06	85.87	-112.457	11.183E-01	-6.607E-04	14.369E-08	15.8312	3671.61	-69.74	95	222	430
431	702.27	339.66	-182.42	73.48	-97.670	10.446E-01	-5.476E-04	89.807E-09	15.7989	3610.66	-66.49	90	197	431
432														432
433	598.30	311.39	-213.32	56.48	-62.957	10.627E-01	-6.305E-04	14.001E-08	15.9116	3542.57	-72.32	59	212	433
434									15.8141	3437.99	-69.99	82	182	434
435									15.8670	3524.57	-70.78	87	197	435
436									15.7884	3457.85	-67.04	84	177	436
437	518.37		-124.22	121.12	-4.664	90.770E-02	-5.058E-04	10.953E-08	16.0129	3448.18	-76.09	83	187	437
438									19.0161	5539.90	-37.85	212	56	438
439	558.61	288.37	-249.83	33.24	-7.913	96.087E-02	-5.288E-04	11.309E-08	16.0114	3456.80	-78.67	57	203	439
440			-258.74	33.58	-70.372	12.322E-01	-8.646E-04	24.551E-08	15.7848	3305.20	-67.66	40	275	440
441					-58.833	12.313E-01	-8.834E-04	25.849E-08	15.7598	3371.05	-64.09	41	190	441
442					-62.341	12.447E-01	-8.956E-04	26.180E-08	15.8446	3172.92	-66.15	27	165	442
443	1481.80	380.00	-401.93	-104.25	14.570	89.472E-02	-3.921E-04	34.508E-09	15.9395	3389.43	-139.00	103	230	443
444	862.89	361.76	116.94	217.84	-64.820	93.868E-02	-6.942E-04	20.155E-08	16.2008	4206.70	-78.15	107	278	444
445	695.42	351.79	116.18	216.29	-56.518	93.974E-02	-6.469E-04	18.401E-08	16.2758	4237.37	-74.75	104	275	445
446	882.36	350.34			-17.367	86.751E-02	-4.610E-04	72.348E-09	16.3363	4158.47	-94.15	117	297	446
447	617.57	318.65	-209.63	78.25	-58.322	11.279E-01	-6.536E-04	14.729E-08	16.0140	3702.56	-81.55	78	234	447
448	566.26	294.89	-144.86	129.54	-5.585	10.027E-01	-5.602E-04	12.163E-08	16.0412	3597.72	-83.41	72	223	448
449	605.50	305.01	-270.47	41.62	-8.395	10.538E-01	-5.799E-04	12.368E-08	16.0541	3614.07	-85.45	75	225	449
450					-64.623	88.509E-02	-5.853E-04	13.054E-08	16.3091	4470.92	-81.40	177	377	450
451	733.87	369.58	182.21	280.26	-97.067	11.057E-01	-8.855E-04	27.901E-08	16.6832	4602.23	-70.42	70	272	451
452	1146.00	379.29	49.99		-60.730	92.821E-02	-5.870E-04	13.586E-08	16.3459	4310.25	-87.31	145	325	452
453	654.77	333.12	-230.27	86.67	-59.264	12.234E-01	-7.084E-04	15.964E-08	16.0589	3850.38	-88.75	95	256	453
454	615.67	310.07	-165.46	138.00	-6.544	10.978E-01	-6.155E-04	13.410E-08	16.0610	3729.87	-90.88	88	244	454
455	631.63	318.78	-291.07	50.07	-9.328	11.489E-01	-6.347E-04	13.590E-08	16.1134	3774.56	-91.31	91	247	455
456	723.43	323.35			33.536	10.735E-01	-5.535E-04	16.777E-08	16.3372	3982.78	-89.15	100	272	456
457	1417.80	398.89	-443.13	-87.13	9.224	11.032E-01	-5.338E-04	77.791E-09	15.2638	3242.04	-157.10	134	307	457
458	889.06	312.48			7.993	11.978E-01	-6.703E-04	14.486E-08	16.2878	3865.58	-86.15	89	258	458
459					-54.491	90.351E-02	-5.388E-04	92.570E-09	18.2166	6462.60	-13.40	207	407	459
460									14.4856	2902.44	-167.90	200	290	460
461	695.83	346.19	-250.87	95.12	-59.951	13.167E-01	-7.612E-04	17.082E-08	16.0941	3983.01	-95.85	112	276	461
462	658.16	323.71	-186.10	146.37	-7.118	11.911E-01	-6.674E-04	14.511E-08	16.0850	3856.23	-97.94	104	264	462
463	664.10	332.10	-311.71	58.49	-10.463	12.452E-01	-6.912E-04	14.897E-08	16.1355	3892.91	-98.93	107	267	463
464	513.28	405.81	224.83		-58.979	10.057E-01	-6.594E-04	16.056E-08	17.6701	6492.44	-26.13	217	382	464
465			202.64		-58.979	10.057E-01	-6.594E-04	16.056E-08	16.7187	5477.94	-69.39	177	382	465
466	735.19	357.74	-271.51	103.50	-60.809	14.118E-01	-8.156E-04	18.347E-08	16.1089	4096.30	-103.00	127	296	466
467	697.49	336.13	-206.66	154.87	-7.967	12.858E-01	-7.210E-04	15.692E-08	16.1643	4018.01	-102.70	119	284	467
468	689.85	344.21	-332.35	66.86	-10.982	13.377E-01	-7.423E-04	15.981E-08	16.1480	4008.52	-105.40	121	287	468

NO	FORMULA	COMPOUND NAME	MOLWT	TFP	TBP	TC	PC	VC	LDEN	TDEN	HVAP	NO
469	C15H12	1-PHENYLINDENE	192.261		322.0	843.7	27.0	0.598				469
470	C15H14	2-ETHYLFLUORENE	194.277		309.0	811.1	24.6	0.629				470
471	C15H30	N-DECYLCYCLOPENTANE	210.405		279.3	723.8	15.2				49,027	471
472	C15H30	1-PENTADECENE	210.405	-3.8	268.3	704.0	14.6		791	0	48,692	472
473	C15H32	N-PENTADECANE	212.421	9.8	270.6	707.0	15.2	0.880	769	20	49,488	473
474	C16H10	FLUORANTHENE	202.256	110.0	393.0	936.6	26.0	0.660				474
475	C16H10	PYRENE	202.256	151.0	362.0	892.1	26.0	0.637				475
476	C16H12	N-PHENYLNAPHTHALENE	204.272		316.0	840.1	26.3	0.605				476
477	C16H22O4	DIBUTYL-O-PHTHALATE	278.350	-35.2	334.8	750.0	13.6		1047	20	79,131	477
478	C16H32	N-DECYLCYCLOHEXANE	224.432		297.6	717.0	13.4				50,409	478
479	C16H32	1-HEXADECENE	224.432	4.1	284.8	791.0	19.0		788	10	50,451	479
480	C16H32O2	PALMIC ACID	256.431	63.0	348.5	717.0	14.2	0.946	828	102	66,992	480
481	C16H34	N-HEXADECANE	226.448	17.8	286.8	750.0	13.0		773	20	51,246	481
482	C17H34	N-DODECYLCYCLOPENTANE	238.459		310.9	736.0	14.2				52,628	482
483	C17H36O	HEPTADECANOL	256.474	53.8	323.8	733.0	13.2		848	54	60,709	483
484	C17H36	N-HEPTADECANE	240.475	21.8	302.0	733.0	13.2	1.000	778	20	52,921	484
485	C18H12	CHRYSENE	228.294	255.0	448.0	993.6	23.9	0.736				485
486	C18H14	O-TERPHENYL	230.310	56.8	331.8	891.0	39.0	0.769				486
487	C18H14	M-TERPHENYL	230.310	86.8	364.8	924.8	35.1	0.784				487
488	C18H14	P-TERPHENYL	230.310	211.8	375.8	926.0	33.2	0.779				488
489	C18H34O2	OLEIC ACID	282.469	13.3	362.3	797.0	17.0	1.035	893	20	68,131	489
490	C18H36	1-OCTADECENE	252.486	17.6	314.8	739.0	11.3		789	20	54,303	490
491	C18H36	N-TRIDECYLCYCLOPENTANE	252.486		325.4	761.0	12.1				54,345	491
492	C18H36O2	STEARIC ACID	284.485	70.0	371.9	810.0	16.5		844	70	70,049	492
493	C18H38	N-OCTADECANE	254.502	28.1	316.3	745.0	12.1	1.054	777	28	54,512	493
494	C18H38O	1-OCTADECANOL	270.501	57.8	334.8	747.0	14.2		812	59		494
495	C19H38	N-TETRADECYLCYCLOPENTANE	266.513		325.8	772.0	11.2				56,019	495
496	C19H40	N-NONADECANE	268.529	31.8	329.9	756.0	11.1		789	32	56,061	496
497	C20H40	N-PENTADECYLCYCLOPENTANE	280.540		351.8	780.0	10.2				57,694	497
498	C20H42	N-EICOSANE	282.556	36.8	343.8	767.0	11.1		775	40	57,527	498
499	C20H42O	1-EICOSANOL	298.555	65.8	355.8	770.0	12.2				65,314	499
500	C21H42	N-HEXADECYLCYCLOPENTANE	294.567		363.8	791.0	9.7				59,369	500

NO	VISA	VISB	DELHF	DELGF	CPVAPA	CPVAPB	CPVAPC	CPVAPD	ANTA	ANTB	ANTC	TMN	TMX	NO
469					-96.154	11.865E-01	-7.786E-04	17.650E-08	16.4170	4872.90	-97.30	227	427	469
470					-107.036	12.611E-01	-8.156E-04	17.928E-08	16.5199	4789.44	-97.90	207	407	470
471	771.74	368.30	-292.15	111.91	-61.923	15.077E-01	-8.717E-04	19.590E-08	16.1261	4203.94	-109.70	140	313	471
472	739.13	347.46	-227.39	163.16	-9.203	13.825E-01	-7.783E-04	17.028E-08	16.1539	4103.15	-110.60	133	301	472
473	718.51	355.92	-352.99	75.28	-11.916	14.327E-01	-7.972E-04	17.199E-08	16.1724	4121.51	-111.80	135	304	473
474					-80.706	11.715E-01	-7.938E-04	18.600E-08	16.4523	5438.77	-112.40	287	487	474
475					-94.379	11.916E-01	-7.930E-04	17.559E-08	16.4842	5203.08	-107.20	257	477	475
476					-99.516	11.463E-01	-6.113E-04	60.612E-09	16.9691	5351.04	-81.70	227	427	476
477	2588.10	336.24			1.880	12.539E-01	-6.121E-04	69.710E-09	16.9539	4852.47	-138.10	196	384	477
478	925.84	378.69			-69.015	16.542E-01	-9.613E-04	21.428E-08	16.1627	4373.37	-111.80	190	300	478
479	767.48	357.85	-247.98	171.62	-9.705	14.750E-01	-8.298E-04	18.104E-08	16.2203	4245.00	-115.20	147	319	479
480			-723.06						18.9558	7049.18	-55.08	353	153	480
481	738.30	366.11	-373.59	83.74	-13.017	15.290E-01	-8.537E-04	18.497E-08	16.1841	4214.91	-118.70	150	321	481
482	853.53	385.53	-336.12	126.02	-63.263	16.952E-01	-9.768E-04	21.855E-08	16.1915	4395.87	-124.20	168	346	482
483			-546.25	-44.67	-7.792	16.529E-01	-9.345E-04	20.436E-08	15.6161	3672.62	-188.10	191	383	483
484	757.88	375.90	-394.19	92.15	-13.967	16.241E-01	-9.081E-04	19.720E-08	16.1510	4294.55	-124.00	161	337	484
485					-115.757	13.415E-01	-8.311E-04	15.412E-08	16.6038	5915.26	-128.10	377	577	485
486	1094.10	461.27												486
487	940.58	460.94												487
488	911.01	461.10												488
489			-646.02						18.2445	5884.49	-127.26	360	176	489
490	816.19	376.93	-289.22	188.45	-11.329	16.643E-01	-9.374E-04	20.486E-08	16.2221	4416.13	-127.30	171	350	490
491	891.80	392.78	-353.99	137.08	-64.209	17.903E-01	-1.032E-03	23.094E-08	16.2270	4483.13	-131.30	180	361	491
492			-764.51						19.8034	7709.35	-57.83	370	174	492
493	777.40	385.00	-414.83	100.57	-14.470	17.170E-01	-9.592E-04	20.783E-08	16.1232	4361.79	-129.90	172	352	493
494			-566.85	-36.22	-8.704	17.476E-01	-8.524E-04	21.575E-08	15.6898	3757.82	-193.10	201	385	494
495	924.60	399.62	-374.63	145.58	-64.929	18.845E-01	-1.085E-03	24.288E-08	16.2632	4439.38	-138.10	192	375	495
496	793.62	393.54	-435.43	108.98	-15.491	18.125E-01	-1.015E-03	22.052E-08	16.1533	4450.44	-135.60	183	366	496
497	950.57	406.33	-395.28	153.99	-66.093	19.804E-01	-1.140E-03	25.498E-08	16.3092	4642.01	-145.10	203	388	497
498	811.29	401.67	-456.07	117.40	-22.383	19.393E-01	-1.117E-03	25.284E-08	16.4685	4680.46	-141.10	198	379	498
499			-608.13	-19.43	-12.581	19.498E-01	-1.118E-03	25.158E-08	15.8233	3912.10	-203.10	219	406	499
500	977.42	412.29	-415.87	162.41	-66.683	20.741E-01	-1.237E-03	26.682E-08	16.3553	4715.69	-152.10	215	401	500

D CONVERSION FACTORS FOR SOME COMMON SI UNITS

An asterisk (*) denotes an exact relationship.

Length	*1 in	:	25.4 mm
	*1 ft	:	0.3048 m
	*1 yd	:	0.9144 m
	1 mile	:	1.6093 km
	*1 Å (angstrom)	:	10^{-10} m
Time	*1 min	:	60 s
	*1 h	:	3.6 ks
	*1 day	:	86.4 ks
	1 year	:	31.5 Ms
Area	*1 in^2	:	645.16 mm^2
	1 ft^2	:	0.092903 m^2
	1 yd^2	:	0.83613 m^2
	1 acre	:	4046.9 m^2
	1 mile^2	:	2.590 km^2
Volume	1 in^3	:	16.387 cm^3
	1 ft^3	:	0.02832 m^3
	1 yd^3	:	0.76453 m^3
	1 UK gal	:	4546.1 cm^3
	1 US gal	:	3785.4 cm^3
	1 bbl (42 US gal)	:	0.1590 m^3
Mass	1 oz	:	28.352 g
	*1 lb	:	0.45359237 kg
	1 cwt	:	50.8023 kg
	1 long (UK) ton	:	1016.06 kg
	1 short (US) ton	:	907.18 kg
Force	1 pdl	:	0.13826 N
	1 lbf	:	4.4482 N
	1 kgf	:	9.8067 N
	1 tonf	:	9.9640 kN
	*1 dyn	:	10^{-5} N

Temperature difference	*1 deg F (deg R)	:	$\frac{5}{9}$ deg C (deg K)
Energy (work, heat)	1 ft lbf	:	1.3558 J
	1 ft pdl	:	0.04214 J
	*1 cal (internat. table)		4.1868 J
	1 erg	:	10^{-7} J
	1 Btu	:	1.05506 kJ
	1 hp h	:	2.6845 MJ
	*1 kW h	:	3.6 MJ
	1 therm	:	105.51 MJ
	1 thermie	:	4.1855 MJ
Calorific value (volumetric)	1 Btu/ft^3	:	37.259 kJ/m^3
Velocity	1 ft/s	:	0.3048 m/s
	1 mile/h	:	0.44704 m/s
Volumetric flow	1 ft^3/s	:	0.028316 m^3/s
	1 ft^3/h	:	7.8658 cm^3/s
	1 UK gal/h	:	1.2628 cm^3/s
	1 US gpm (gal/min)	:	0.227 m^3/h
	1 US gal/h	:	1.0515 cm^3/s
	1 bpd (bbl/d)	:	6.62 litres/h
Mass flow	1 lb/h	:	0.12600 g/s
	1 ton/h	:	0.28224 kg/s
Mass per unit area	1 lb/in^2	:	703.07 kg/m^2
	1 lb/ft^2	:	4.8824 kg/m^2
	1 ton/sq mile	:	392.30 kg/km^2
Density	1 lb/in^3	:	27.680 g/cm^3
	1 lb/ft^3	:	16.019 kg/m^3
	1 lb/UK gal	:	99.776 kg/m^3
	1 lb/US gal	:	119.83 kg/m^3
Pressure	1 lbf/in^2 (1 psi)	:	6.8948 kN/m^2
	1 ksi (1000 psi)	:	6.8948 MN/m^2
	1 tonf/in^2	:	15.444 MN/m^2
	1 lbf/ft^2	:	47.880 N/m^2
	*1 standard atm	:	101.325 kN/m^2
	*1 atm (1 kgf/cm^2)	:	98.0665 kN/m^2
	*1 bar	:	10^5 N/m^2
	1 ft water	:	2.9891 kN/m^2
	1 in water	:	249.09 N/m^2
	1 in Hg	:	3.3864 kN/m^2
	1 mmHg (1 torr)	:	133.32 N/m^2
Power (heat flow)	1 hp (British)	:	745.70 W
	1 hp (metric)	:	735.50 W
	1 erg/s	:	10^{-7} W
	1 ft lbf/s	:	1.3558 W
	1 Btu/h	:	0.29307 W
	1 ton of refrigeration	:	3516.9 W

Moment of inertia	1 lb ft^2	:	0.042140 kg m^2
Momentum	1 lb ft/s	:	0.13826 kg m/s
Angular momentum	1 lb ft^2/s	:	0.042140 kg m^2/s
Viscosity, dynamic	*1 P (Poise)	:	0.1 N* s/m^2
	1 lb/ft h	:	0.41338 mN s/m^2
	1 lb/ft s	:	1.4882 N s/m^2
Viscosity, kinematic	*1 S (Stokes)	:	10^{-4} m^2/s
	1 ft^2/h	:	0.25806 cm^2/s
Surface energy	1 erg/cm^2	:	10^{-3} J/m^2
(surface tension)	(1 dyn/cm)	:	(10^{-3} N/m)
Mass flux density	1 lb/h ft^2	:	1.3562 g/s m^2
Heat flux density	1 Btu/h ft^2	:	3.1546 W/m^2
	*1 kcal/h m^2	:	1.163 W/m^2
Heat transfer coefficient	1 Btu/h ft^2 F	:	5.6783 W/m^2 K
Specific enthalpy (latent heat, etc.)	*1 Btu/lb	:	2.326 kJ/kg
Specific heat capacity	*1 Btu/lb °F	:	4.1868 kJ/kg K
Thermal conductivity	1 Btu/h ft °F	:	1.7307 W/m K
	1 kcal/h m °C	:	1.163 W/m K

Taken from Mullin, J. W.: *The Chemical Engineer* No. 211 (Sept. 1967), 176. SI units in chemical engineering.
Note: Where temperature difference is involved K = °C.

E   DESIGN PROJECTS I

The problems in this appendix are typical of industrial design problems. They are grouped into sections corresponding to different sectors of the chemical and fuels industries. Most of these are variants on commercially practiced technologies, but many are novel processes that may not yet be commercialized.

The problem statements are intentionally short and little information is given beyond one or two references. Most of the problems are referenced to US patents that give process concepts, chemical paths and yield data, as this is often the starting point for technical and economic analysis in industrial design. There is no copyright on US patents, and all of the referenced patents are available in the on-line material at http://elsevierdirect.com/companions. Patent references are not given for older 'traditional' processes, as flow-sheets and yields for these processes can be found in the encyclopaedias listed in Chapter 8.

An effort has been made to include a range of problems reflecting the broad spectrum of industries in which chemical engineers are employed. It must be recognized, however, that reliable price data for bulk quantities of specialty compounds may be hard to obtain. Many of the problems are therefore based on products and feeds for which the prices are listed in *ICIS Chemical Pricing* or *Oil and Gas Journal*.

Biochemical processes, i.e., processes that use enzymes, cells or micro-organisms to effect chemical transformation or separation, are now prevalent in almost every sector of the chemicals industry. In almost any industry, chemical engineers are faced with process design and evaluation of biological processes. It therefore did not make sense to form a separate category of 'biological processes' or 'biochemicals', as these processes are just alternative routes to making commodity chemicals, polymers, fuels, pharmaceuticals, etc. Fourteen of the 101 design problems in this appendix involve biochemical processing steps. The sectors that do not have at least one biological process are inorganic chemicals, gas processing, electrochemical processes and devices and sensors.

Many of the problems ask for a comparison between two designs and thus require two plants to be designed and costed. Most of the pharmaceutical problems are multi-step processes that also require several plants to be designed. In many cases, the production rate is not given and must be estimated from an analysis of the market.

These problems are intended to be representative of typical problems that a design engineer might face in industry. A shorter selection of more structured problems with more background information is given in the next appendix.

E.1.   COMMODITY CHEMICALS AND POLYMERS

E.1.1.   Acetic acid

Acetic acid is made by carbonylation of methanol. US 5,001,259 (to Hoechst Celanese) describes changes to the reaction medium that improve catalyst stability and productivity. US 3,769,329 (to Monsanto) describes the conventional process. Is it economically attractive to implement the changes proposed by the Hoechst patent in a new world-scale plant?

E.1.2.   Acrolein and acrylic acid

Acrolein and acrylic acid are both made by vapour phase oxidation of propylene. US 6,281,384 (to E. I. du Pont Nemours and Atofina) describes a fluidized-bed process, while US 5,821,390 (to BASF) describes an isothermal reactor cooled by heat transfer to a molten salt. US 6,858,754 and US 6,781,017 (both to BASF) describe alternative processes based on a propane feed. Compare the economics of acrylic acid production from propane with production from propylene. Is the conclusion different if the process is stopped at acrolein?

E.1.3.   Cellulose acetate

Cellulose acetate is used in films, cigarette filters and mouldings. It is made by reacting cellulose with acetic anhydride. US 5,608,050 (to Eastman) describes an acetylation process that improves the thermal stability of the polymer. US 5,962,677 (to Daicel) describes a process for improving the processing properties of the polymer. Estimate the cost of production of the polymer by each route.

E.1.4.   Chloroform and methylene chloride

Chloroform and methylene chloride can be made by chlorination of methyl chloride. US 5,023,387 describes the chlorination process and gives yields. Estimate the cost of production of chloroform for a plant that produces 40,000 metric tons per year. Chloroform is mainly used for making chlorodifluoromethane, which is a precursor for PTFE. Methylene chloride is used as a solvent, but the market for this compound is stagnant because of environmental concerns. How does the cost of production of chloroform change if the plant produces no methylene chloride?

E.1.5.   Dicyclopentadiene

Dicyclopentadiene (DCPD) is usually recovered as a high-value product from the by-product pyrolysis gasoline stream that is generated in steam cracking furnaces (see E.1.7). US 6,258,989 (to Phillips Petroleum) gives a typical pyrolysis gasoline composition and describes a suitable recovery process. Estimate the NPV at 12% interest rate of a plant to recover DCPD from an 800,000 metric ton per year steam naphtha cracker.

E.1.6.   2,6-Dimethylnaphthalene

2,6-Dimethylnaphthalene (2,6-DMN) is used to make 2,6-naphthalenedicarboxylic acid, which can be used to give improved properties to polyester bottle resins. Dimethylnaphthalenes can be made by reaction of butadiene with orthoxylene, yielding principally 1,5-DMN, which can then be isomerized to 2,6-DMN as described in US 6,072,098 (to Mitsubishi Gas Chemical Company). An alternative purification process is described in US 6,737,558 (to ENICHEM). Naphthalenic compounds can also be recovered from the light cycle oil (a diesel-range product) produced in oil refinery catalytic cracking units. Light cycle oil typically contains roughly 2% naphthalene, 4% methyl naphthalenes, 6% dimethyl naphthalenes and 4% trimethyl naphthalenes. The distribution of naphthalene isomers can be approximated as the equilibrium distribution at 900°F. Compare the cost of producing 2,6-DMN from orthoxylene with the cost of recovering it from light cycle oil. Consider using additional processes to enhance the 2,6-DMN yield.

E.1.7.   Ethylene and propylene by steam cracking

Steam cracking of ethane is the most widely used process for making ethylene. US 6,578,378 (to Technip-Coflexip) gives a typical ethane cracker product composition and describes an improved separation process for ethylene recovery. US 5,990,370 (to BP) gives yields for ethane, propane and mixtures. US 5,271,827 (to Stone & Webster) gives details of furnace design and yields for a naphtha feed. Several other separation schemes for ethylene and propylene recovery are described in the literature. Estimate the cost of production for a new steam cracking facility that produces 1 million metric tons per year of ethylene and 600,000 metric tons per year of propylene. What feedstock would you recommend?

E.1.8.   Ethylene by oxidative dehydrogenation

US 6,548,447 and US 6,452,061 (both to Regents of the University of Minnesota) suggest an alternative process for producing olefins such as ethylene from the corresponding paraffin using different catalysts. How does the cost of ethylene produced by this process compare with the cost of ethylene produced by the conventional steam cracking route?

E.1.9.   Ethylene from ethanol

Ethanol is usually made most economically from ethylene, and not vice versa; however, recent high natural gas prices and interest in ethanol from crops as a renewable raw material have prompted interest in the reverse process. US 4,134,926 (to Lummus) describes a process for converting ethanol to ethylene with high yield. What is the cost of ethylene produced by this route based on a fermentation ethanol feed?

E.1.10. Lactic acid by fermentation

US 6,475,759 (to Cargill, Inc.) describes fermentation of corn steep liquor to lactic acid, and US 6,229,046 (also to Cargill, Inc.) describes recovery of lactic acid from the fermentation broth. What is the cost of production of lactic acid by this route?

E.1.11. Linear alkyl benzenes

Linear alkyl benzenes (LAB) are starting compounds for making linear alkyl benzene sulphonates, which are widely used biodegradable surfactants. US 5,012,021 (to UOP) describes a process for making LAB and US 5,196,574 and US 5,344,997 (also to UOP) give yields for several catalysts. Estimate the cost of production of the LAB and determine which catalyst is the best.

E.1.12. 2,6-Naphthalenedicarboxylic acid

2,6-Naphthalenedicarboxylic acid is a precursor to polyethylene naphthalate (PEN), which is used to improve the properties of polyester bottle resins (see also E.1.6). It can be made by the liquid-phase oxidation of 2,6-dimethylnaphthalene as described in US 6,114,575, assigned to BP Amoco. Estimate the cost of production for a plant that produces 250,000 metric tons per year (250 kMTA).

E.1.13. Nitrobenzene

Nitrobenzene is a precursor for aniline and is made by nitration of benzene. US 4,772,757 (to Bayer) describes an improved process with recycle of the nitrating acid. Estimate the cost of production for a plant that produces 150,000 metric tons per year.

E.1.14. Polylactic acid

Polylactic acid is a biodegradable polymer. It can be made from lactic acid, which can be produced by fermentation of glucose. Because it is biodegradable and can be manufactured from agricultural products, polylactic acid is potentially a renewable material. US 6,326,458 assigned to Cargill Inc. describes a process for making poly-lactic acid from lactic acid. Estimate the cost of production of the purified polymer.

E.1.15. Phenol–cyclohexanone

US 6,720,462 describes a new process for phenol with co-production of a ketone. In example 3b, they suggest very high yields of phenol and cyclohexanone byproduct. Estimate the production cost of phenol by this route for a grassroots world-scale plant.

E.1.16. Propylene

Propylene is usually produced as a byproduct of ethylene manufacture. An alternative process is catalytic dehydrogenation of propane, as described in US 4,381,417 (to

UOP). What is the cost of production of propylene by this route for a world-scale plant?

E.1.17. Propylene glycol by fermentation

Propylene glycol (1,2-propanediol) is a commodity chemical. US 6,087,140 (to Wisconsin Alumni Research Foundation) describes a process for fermentation of sugars to propylene glycol using transformed microorganisms. Estimate the maximum price that could be charged for the microorganisms.

E.1.18. Propylene oxide by epoxidation

A novel route for making propylene oxide is by epoxidation of propylene using hydrogen peroxide. US 6,103,915 (to Enichem) and US 5,744,619 (to UOP) give yields for several catalysts. US 5,252,758 describes a process for propylene oxide production. Estimate the cost of propylene oxide production and determine the best catalyst.

E.1.19. Phosgene

Phosgene is an important intermediate in the manufacture of polycarbonate and polyurethane. US 6,500,984 (to General Electric) describes a process for phosgene using carbon and silicon carbide catalysts. US 6,054,104 (to DuPont) describes a process using a silicon carbide catalyst. US 4,231,959 (to Stauffer Chemical) describes a process with recycle of unconverted CO. Estimate the cost of production for a world-scale plant. Which catalyst or combination of catalysts would you recommend?

E.1.20. Pyridine

Pyridine is an important chemical intermediate. US 4,866,179 (to Dairen Chemical) describes a process for forming pyridine from ammonia and a carbonyl compound and gives yields for several aldehydes, ketones and mixtures. US 4,073,787 (to ICI) describes a process based on butadiene, formaldehyde and ammonia. Estimate the cost of production for a world-scale plant and determine which feed is most economical.

E.2. DEVICES AND SENSORS

E.2.1. Fuel processor

A fuel processor is a miniature hydrogen plant that converts a hydrocarbon fuel into hydrogen for use in a fuel cell. US 6,190,623 (to UOP) describes a fuel processor for converting methane. Estimate the volume of a fuel processor unit using this technology for a 3-kW fuel cell system. What would the manufactured cost per unit be at a scale of production of 100,000 units per year?

E.2.2.   Portable oxygen generator

Patients who have difficulty breathing are often given air enriched in oxygen. When the patient is immobile the gas mixture can be supplied from cylinders, but when the patient is mobile this may not be practical, particularly if the patient is weak. An alternative is to supply oxygen or enriched air by means of a portable oxygen generation device. US 6,764,534 (to AirSep Corp.) describes such a device, based on pressure-swing adsorption. Estimate the cost of manufacturing this device based on a production volume of 10,000 units per year.

E.3.   ELECTRONICS AND ELECTROCHEMICAL PROCESSES

E.3.1.   Argon recovery from silicon furnace off-gas

Argon is used as an inert atmosphere in silicon crystallization. US 5,706,674 describes two processes for recovering spent argon. Which process is cheaper? How does the cost of this recovered argon compare with the cost of purchased argon?

E.3.2.   Chlor-alkali manufacture

Chlorine and sodium hydroxide are made by the electrolysis of brine using membrane cells. Conventional and improved membrane cell arrangements are described in US 4,391,693, assigned to Dow Chemical. US 4,470,889 (also to Dow) gives data on membrane materials and performance. What price of electricity is needed for it to be economical to produce chlorine from sea water (3.5 wt% NaCl)?

E.3.3.   Potassium permanganate

Potassium permanganate can be made from potassium hydroxide and manganese dioxide ore using an electrolytic process, as described in US 5,660,712 (unassigned, but clearly owned by Carus Corp.). Estimate the cost of manufacturing potassium permanganate.

E.4.   FOOD PROCESSING AND FORMULATED PRODUCTS

E.4.1.   Aspartame

Aspartame (α-L-aspartyl-L-phenylalanine 1-methyl ester) is a sweetening agent that is roughly 200 times sweeter than sucrose. Routes for preparing this compound are described in US 3,492,131, US 4,440,677 (both to G.D. Searle & Co.) and US 5,476,961 (to the NutraSweet Company). Determine which route gives the lowest cost of production.

E.4.2.   Cocoa processing

Cocoa mass can be separated into cocoa powder and cocoa butter by solvent extraction, as described in US 6,610,343, assigned to Cargill Inc. Cocoa butter is used in various food applications, while cocoa solids provide the flavour for chocolate and chocolate-flavoured foods. The patent also describes several typical recipes for chocolate. Estimate the cost of producing milk chocolate and semisweet chocolate using the recipes given.

E.4.3.   Dicalcium phosphate and phosphoric acid

Dicalcium phosphate is used as a supplement to animal food. Food-grade phosphoric acid is used as an antioxidant and acidulant, for example, giving a sharp taste to soft drinks. US 3,988,420 (to Israel Chemicals Ltd) describes a process for making both products from phosphate rock and hydrochloric or nitric acid. Determine which acid leads to the highest net present value for a plant that produces 5000 metric tons per year of food grade dicalcium phosphate.

E.4.4.   Erythorbic acid

Erythorbic acid (also known as isoascorbic acid) is a preservative. It can be made by fermentation of glucose using various microorganisms, as described in US 3,052,609 assigned to Sankyo Co. Estimate the cost of production and determine which microorganism is preferred.

E.4.5.   Folic acid

Folic acid is a vitamin (sometimes called vitamin M or vitamin Bc) found naturally in mushrooms, spinach and yeast. It is an important dietary supplement during pregnancy, as it reduces the likelihood of spina bifida. US 5,968,788 (to Toray industries) describes conditions for cultivating several strains of yeast or bacteria to increase their yield of folic acid, and gives yields for each species. Determine the optimum strain and recovery process to make a USP product and estimate the cost of production via this route.

E.4.6.   Insect repellent

Insect repellents based on geraniol are described in US 5,521,165 (to International Flavours & Fragrances). Estimate the cost of production of an aerosol dispensed slow-release insect repellent formulation for spraying on skin and clothing.

E.4.7.   Low-fat snacks

The difficulties of making fried snacks using non-digestible fats are described in US 6,436,459 (to Procter and Gamble), which also gives recipes and compositions for potato-based low-fat snacks. Estimate the cost of producing low-fat snacks of the composition and recipe given in example 1, using continuous frying.

E.4.8.   Mannitol

US 6,649,754 (to Merck) describes a process for making mannitol by hydrogenation of a mixture of glucose and fructose. US 3,632,656 (to Atlas Chemical) describes recovery of mannitol from a mixture with sorbitol by crystallization from aqueous solution. US 4,456,774 (to Union Carbide) describes an adsorptive separation of mannitol from sorbitol. US 6,235,947 (to Takeda Chemical Industries) describes a process for recrystallizing mannitol to improve the crystal morphology and hence make a more compressible product that can be used in making tablets. Estimate the cost of production of Mannitol by the Merck route and determine which separation is most economical. What is the additional cost of making the recrystallized product via the Takeda route?

E.4.9.   Margarine

The manufacture of margarine is described in US 4,568,556 (to Procter and Gamble). Estimate the cost of making a stick margarine product of the recipe given in Example II. What is the NPV for a plant that produces 100,000 metric tons per year of margarine?

E.4.10.   Moisturizing lotion

US 5,387,417 describes the formulation of a moisturizing lotion and the preparation of the emulsifying agent. Estimate the cost of production of each of the lotion formulations given in the patent.

E.4.11.   Monosodium glutamate

Monosodium glutamate (MSG) is a flavour enhancer. US 2,877,160 (to Pfizer) and US 2,978,384 (to Koichi Yamada) describe fermentation processes for glutamic acid. US 5,907,059 (to Amylum Belgium & A.E. Staley Manufacturing) describes recovery of the fermentation product and conversion to MSG. Estimate the cost of production via this route.

E.4.12.   Niacinamide (nicotinamide)

Nicotinamide is a vitamin, also known as niacin and vitamin B3. US 4,681,946 (to BASF) describes a process based on amidation of nicotinic acid. US 4,008,241 and US 4,327,219 (both to Lummus) describe a process based on hydrolysis of nicatinonitrile. Which process has the lowest cost of production?

E.4.13.   Riboflavin

Riboflavin (vitamin B2) can be made by fermentation, as described in US 2,876,169 (to Grain Processing Corp.). The fermentation process has undergone many improvements. Newer strains with higher yields are described in US 5,164,303 (to ZeaGen Inc.) and US 4,794,081 (to Daicel Chemical). Alternative chemical routes are described in US 2,807,611 (to Merck) and US 4,687,847 (to BASF). Estimate the cost of producing a USP product via both the chemical and biochemical routes. Which process do you recommend?

E.4.14. α-Tocopherol

α-Tocopherol is the most bioactive form of vitamin E. It can be made by condensation of trimethylhydroquinone with isophytol, as described in US 5,900,494 (to Roche Vitamins) or US 7,153,984 (to DSM B.V.). Determine which process gives the lowest total cost of production.

E.5.   FUELS

E.5.1. Benzene reduction

Gasoline is usually produced as a blend of several petroleum streams that boil in the range of naphtha. A typical gasoline might contain 50% by volume of cracked naphtha with benzene content between 0.5wt% and 2.0 wt% and 25% by volume of catalytically reformed naphtha with benzene content between 1wt% and 3wt%. Estimate the cost per gallon of gasoline of reducing the final benzene content to 0.62% by volume. Compositions of other components in the naphtha streams can be found in the patent literature.

E.5.2. Crude-oil distillation

A typical crude-oil distillation process was described in Chapter 4. Design a crude-oil unit for a refinery that processes a 50:50 mixture (by volume) of Saudi Light and Saudi Heavy crude oils using the cut points given in Chapter 4.

E.5.3. Ethanol by fermentation

Ethanol has a high octane value and is used as a gasoline blending component. It can be manufactured as a renewable fuel by fermentation of sugars using *S. cerevisae*. Compare the costs of producing ethanol from corn in Decatur IL and from sugar cane in Mobile AL.

E.5.4. Hydrocracking

The hydrocracking process is used to crack heavy hydrocarbons to lighter hydrocarbons with addition of hydrogen. It is particularly useful for making distillate fuels such as jet fuel and diesel oil. US 6,190,535 (to UOP) describes a novel hydrocracking process using a hot high-pressure stripping column, and gives an estimate of process yields. Estimate the NPV of a 40-kbd hydrocracker on the US Gulf Coast using this technology.

E.5.5. Isomerization

The catalytic isomerization process is used to convert straight-chain paraffin compounds in light naphtha into branched paraffins that have higher octane numbers and are more valuable as gasoline blending components. US 6,008,427 (to UOP) describes the process flowsheet and US 6,320,089 (also to UOP) gives yields for some new catalysts. Estimate the improvement in octane-barrel yield for the feed of example VI in 6,320,089 using the process of 6,008,427. Estimate the NPV of a 10,000-bpd plant on a USGC basis at 12% interest rate. US 6,472,578 (also to UOP) describes an improved separation scheme. What is the increase in NPV with this new scheme?

E.6.   GAS PROCESSING

E.6.1.   Gas to liquids (Fischer-Tropsch synthesis)

Conversion of natural gas to synthetic crude oil is a possible method for recovering stranded natural gas reserves. US 4,624,968, US 4,477,595 and US 5,118,715 (all to Exxon Corp.) give yields for different catalysts. Determine the cost of producing a liquid product (in $/bbl) if the natural gas is available at $ 0.50 per 1000 scf.

E.6.2.   Hydrogen production

Hydrogen is produced by steam reforming of natural gas. It is used as a raw material for ammonia and methanol production and for various applications in oil refining and chemicals production. Modern steam-reforming plants use pressure-swing adsorption to separate hydrogen from the other reaction products. The pressure-swing adsorption plant can be integrated with the steam-reforming section, as described in US 4,869,894, assigned to Air Products. US 4,985,231 (to ICI) describes a novel reforming reactor and gives examples of typical process yields. Estimate the cost (in $/Mscf) of supplying 100 MMscfd of hydrogen to an oil refinery on the US Gulf Coast.

E.6.3.   Krypton and xenon recovery

Krypton and xenon are valuable gases present in very low concentrations in air. US 6,662,593 assigned to Air Products describes a cryogenic distillation process for air separation with recovery of a stream concentrated in krypton and xenon. What would be the cost of producing purified krypton and xenon by this method? Consider reactive methods for separating krypton and xenon from the concentrated stream as well as the methods suggested in the patent.

E.6.4.   Methanol to olefins

Conversion of natural gas to liquids is currently of great interest, as many large natural gas fields are not close enough to large markets to make construction of a pipeline economically attractive. These 'stranded' reserves can be liquefied, converted into fuels or converted into (higher value) petrochemicals. US 5,714,662 (to UOP) describes a process for converting crude methanol to olefins. What is the cost of production of the ethylene produced by a plant that produces 900,000 t/y of mixed olefins if the cost of producing the natural gas feed is $0.5/MMbtu?

E.6.5.   Natural gas liquefaction

US 6,347,532 (Air Products) describes a process for liquefying natural gas and gives several possible process embodiments. Which of these is the cheapest for the given gas composition? If the gas initially contains 3000 ppmw CO_2, 1250 ppmw H_2S and 28 ppmw COS, and the cost of producing the natural gas feed is $0.5/MMBtu, what is the cost of production of the liquefied natural gas product?

E.6.6.   Natural gas liquids recovery

Natural gas typically contains a range of hydrocarbon compounds, as well as carbon dioxide and hydrogen sulphide. Ethane, propane and butane are often recovered from natural gas for use as petrochemical feed stocks. Typical recovery processes are described in US 4,157,904, assigned to Ortloff Corp. Estimate the cost of recovering ethane and producing a natural gas product that meets pipeline specifications for a plant that processes 150 MMscfd of natural gas with the feed composition given in Example 3 of the patent. Assume the feed also contains 480 ppm of H_2S, 14 ppm of COS and 31 ppm of methyl mercaptan.

E.7.   INORGANIC CHEMICALS

E.7.1.   Ammonia

Ammonia is an important basic chemical and is the starting point for most fertilizer manufacture. A conventional ammonia production process is described in US 4,479,925, assigned to M.W. Kellogg. US 5,032,364 (assigned to ICI) describes a more heat-integrated process, while US 6,216,464 (assigned to Haldor Topsoe A/S) describes a process with power recovery. Estimate the cost of production via each route for a new plant on the US Gulf Coast and for a plant fed with natural gas from a remote gas field in a developing country that is priced at $0.50/MMBtu. Does the price of the natural gas affect the selection of optimum process?

E.7.2.   Bromine

Bromine can be produced by reacting bromide-rich brines with chlorine. The purification of the resulting gas mixture is described in US 3,642,447 (unassigned). If a brine solution containing 0.2 wt% NaBr and 3.4 wt% total salts can be extracted from a well at a cost of $4/metric ton, then would it be economical to produce bromine from this brine?

E.7.3.   Fischer-Tropsch catalyst

The Fischer-Tropsch process is a means of converting synthesis gas into hydrocarbons (see E.6.1). The manufacture of a catalyst for this process is described in US 6,130,184 (to Shell Oil Co.). Estimate the cost of production of this catalyst.

E.7.4. Nitric acid

Nitric acid is made by catalytic oxidation of ammonia. US 5,041,276 (to ICI) gives selectivity data for various catalysts and process conditions. Determine the optimal catalyst and conditions to minimize the cost of production.

E.7.5. Urea

Urea is used as a fertilizer and is made by reacting ammonia with carbon dioxide. The reactions essentially proceed to equilibrium, but the process must be designed to minimize emissions of ammonia. The urea is usually formed into a solid product by prilling. US 6,921,838 (to DSM B.V.) describes a novel process for urea production. Estimate the cost of producing ammonia via this route. Assume that carbon dioxide is available as a byproduct of the ammonia plant (see E.7.1).

E.7.6. Zeolite synthesis

Synthetic zeolites are used in a variety of catalyst and adsorbent applications. Most zeolites are synthesized in batch processes, but US 6,773,694 (to UOP) describes a continuous crystallization process for zeolite formation. The resulting crystals can be dried and formulated into catalysts, adsorbents and other products. Estimate the costs of producing zeolite X and Mordenite by this method.

E.8. PHARMACEUTICALS

These problems are based on some of the highest volume and highest value pharmaceutical compounds at the time of writing. In most cases, the desired product is an active pharmaceutical ingredient (API), although a few of the problems relate to other compounds used in drug formulation.

Many of the high-value API compounds are formed in multi-step syntheses starting from compounds that are themselves specialty chemicals. The patents that are cited give the preparation in the form of a laboratory recipe rather than a process flowsheet, and hence the chemist's recipe must be scaled up to the production recipe. A decision on whether to use batch or continuous production must also be made. These are therefore difficult design problems.

In most cases, the original preparation patent has been cited, as this may be the only route that has received FDA approval. A detailed patent search may reveal alternative routes that can be studied for comparison. Note that several of the products listed under food processing also have pharmaceutical applications, for example as fillers, coatings and sweeteners.

E.8.1. Acetaminophen

Acetaminophen (N-acetyl-p-aminophenol, paracetamol) is an analgesic marketed under a variety of brand names including Tylenol™, Calpol™ and Panadol™. Preparation of the API is described in US 2,998,450 (to Warner Lambert). US 4,474,985 (to Monsanto) describes a process for improving product quality and shelf life. US 5,856,575 (to Council of Scientific Industrial Research) describes an alternative process. Estimate the cost of production by each route.

E.8.2.   Alendronate

Alendronate (4-amino-1-hydroxybutane-1,1-biphosphonic acid) is a biphosphonate drug used to treat osteoporosis and Paget's disease. US 4,621,077 to Istituto Gentili describes the preparation of the API. Estimate the cost of production of the API by the method of example 3.

E.8.3.   Amlodipine besylate

Amlodipine (4-(2-chlorophenyl)-2-[2-(methylamino)ethoxymethyl]-3-ethoxycarbonyl-5-methoxycarbonyl-6-methyl-1,4-dihydropyridine) is an anti-hypertensive, marketed as Norvasc™. US 4,572,909 assigned to Pfizer Inc. describes the preparation of the API and several of the required precursors. Estimate the cost of production of the API.

E.8.4.   Aspirin

Aspirin (acetyl salicylic acid) is a well-known analgesic. Processes for preparing aspirin are described in US 3,235,583, US 3,373,187 (both to Norwich Pharmacal) and US 2,890,240 (to Monsanto). Estimate the cost of production by each route.

E.8.5.   Aspirin (slow release)

Slow-release versions of aspirin for long-term use as an anti-inflammatory drug are described in US 5,855,915. Estimate the cost of production of each of the slow-release formulations given.

E.8.6.   Ciprofloxacin

Ciprofloxacin (1,4-dihydro-1-cyclopropyl-6-fluoro-4-oxo-7-(1-piperazinyl)-3-quino-linecarboxylic acid) is a fluoroquinolone antibiotic drug used mainly to treat respiratory infections and septicaemia. It also enjoyed a brief period of notoriety in 2001 as the preferred antibiotic for treating anthrax. Bayer's patent US 4,670,444 describes the synthesis of the API and several of the required precursors. Estimate the cost of production of the API.

E.8.7.   Citalopram hydrobromide

Citalopram (1-[3-(dimethylamino)-propyl]-1-(4-fluorophenyl)-1,3-dizohydro-5-iso-benzo-furancarbonitrile) is an antidepressant marketed as Celexa™. The original preparation is described in US 4,136,193 (to Kefalas) and an improved route is given in US 4,650,884 (to H. Lundbeck A/S). What is the saving in cost of production of the API by the new route?

E.8.8. Clopidogrel

Clopidogrel is an antithrombotic marketed as Plavix™. Preparation of the API is described in US 4,529,596 (assigned to Sanofi), which also gives several pharmaceutical formulations of the drug. Estimate the cost of production of the API. What composition would you recommend for the tablet formulation, and what is the final cost of production for the tablet form?

E.8.9. Cyclosporin A

Cyclosporins are a group of cyclic non-polar oligopeptides that are immunosuppressants and are produced by *Tolypocladium inflatum Gams* and other fungi. US 4,117,118 gives details of the fermentation and product recovery. Estimate the cost of production and determine which species is preferred.

E.8.10. Doxycyline

Doxycycline (α-6-deoxy-5-oxytetracycline monohydrate) is an antibiotic. US 3,200,149 (to Pfizer) describes the preparation of the API and several formulations. Estimate the cost of producing the tablet formulation given in the patent.

E.8.11. Fexofenadine

Fexofenadine is an antihistamine and is the API for Allegra™ and Telfast™. US 4,254,129 (to Richardson-Merrell) describes the preparation of the API and several formulations of the product, including an aerosol solution (example 11). Estimate the cost of production of the API and the 15-ml aerosol product.

E.8.12. Fluconazole

Fluconazole (2-(2,4-difluorophenyl)-1,3-bis(1H-1,2,4-triazol-1-yl)-propan-2-ol) is an antifungal. US 4,404,216 assigned to Pfizer describes two methods for preparation of the API. Estimate the cost of production of the API by both routes and hence determine which is preferred.

E.8.13. Fluoxetine hydrochloride

Fluoxetine (N-methyl-3-(p-trifluoromethylphenoxy)-3-phenylpropylamine) is an antidepressant. US 4,626,549 (to Eli Lilly & Co.), US 6,028,224 (to Sepracor) and US 6,677,485 (to Ranbaxy) all describe different synthetic routes to this product. Which has the lowest cost of production?

E.8.14. Fluticasone propionate

Fluticasone propionate is an antiallergic drug marketed as Flovent™. US 4,335,121 (assigned to Glaxo) describes the preparation of the API. Estimate the cost of production of the API.

E.8.15.   Granulocyte colony-stimulating factor

Granulocyte colony-stimulating factor (G-CSF) is a hematopoietic stimulant (i.e., encourages formation of new blood cells and is given to patients who have undergone chemotherapy, bone marrow transplants, etc.). It can be produced by expression from genetically modified *E. coli*, as described in US 4,810,643 example 7. Estimate the cost of production of hpG-CSF using the method of this example.

E.8.16.   Guaifenesin

Guaifenesin (guaiacol glyceryl ether, 3-(2-methoxyphenoxy)-1,2-propanediol) is an expectorant that is found in cough medicines such as Actifed™ and Robitussin™. US 4,390,732 (to Degussa) describes preparation of the API and several of its precursors. Estimate the cost of production of the API.

E.8.17.   Ibuprofen

Ibuprofen (2-[4'-isobutylphenyl]propionic acid) is a well-known analgesic marketed as Motrin™, Advil™ and other brands. US 3,385,886 (to Boots Drug Co.) describes the preparation of the API and several formulations. Estimate the cost of production of the tablet form.

E.8.18.   Lansoprazole

Lansoprazole is the API for Prevacid™, a treatment for gastric ulcers. US 4,689,333 (to Takeda Chemical Industries) describes the synthesis of the API and the required precursors. Estimate the cost of production of the API.

E.8.19.   Lisinopril

Lisinopril (N-(1(S)-carboxy-3-phenylpropyl)-L-lysyl-L-proline) is an antihypertensive. The preparation of the API and several product formulations are described in US 4,374,829 (to Merck & Co.). Estimate the cost of production of the API and the tablet formulation.

E.8.20.   Loratadine

Loratadine (11-[N-carboethoxy-4-piperidylidene]-8-chloro-6,11-dihydro-5H-benzo-[5,6]-cyclohepta-[1,2-b]-pyridine) is an antihistamine. US 4,282,233 (to Schering) describes the preparation of the API and also gives recipes for syrup (example 6) and tablet (example 7) formulations. Estimate the cost of the making the API and both formulations.

E.8.21.   S-Ofloxacin

S-Ofloxacin is an optically active fluorinated quinolone with antibacterial properties. US 5,053,407 describes the preparation of the API. Estimate the cost of production of this compound. Which stages of the process would you operate in batch mode and which stages would you operate continuously?

E.8.22.   Omeprazole

Omeprazole (5-methoxy-2-[[(4-methoxy-3,5-dimethyl-2-pyridinyl)methyl]sulfinyl]-1H-benzimidazole) is an antiulcerative marketed as Prilosec™. The prearation of the API in unresolved form is described in US 4,255,431 and US 4,508,905 (both to AB Hässle), while US 5,693,818 (to Astra) describes a route for preparing the optically pure enantiomers. Estimate the cost of producing the S-form by each method.

E.8.23.   Paroxetine

Paroxetine is an antidepressant marketed by GlaxoSmithKline as Paxil™. Example 2 of US 4,007,196 (to A/S Ferrosan) describes preparation on the free base form of paroxetine, while US 4,721,723 (to Beecham Group Plc.) describes the synthesis of the crystalline hydrochloride hemihydrate, which is the preferred form to administer the drug. What is the cost of production of the API in the crystalline hydrochloride hemihydrate form?

E.8.24.   Pseudoephedrine

Pseudoephedrine (2-methylamino-1-phenylpropan-1-ol) is a nasal decongestant. US 4,277,420 (to Monsanto) describes the preparation of the API and several possible precursors. Estimate the cost of production of each route and determine which route is cheapest.

E.8.25.   Risperidone

Risperidone (3-[2-[4-(6-fluoro-1,2-benzisoxazol-3-yl)-1-piperidinyl]ethyl]-6,7,8,9-tetrahydro-2-methyl-4H-pyrido[1,2-a]pyrimidin-4-one) is an antipsychotic marketed by Johnson & Johnson as Risperdal™. The preparation of the API and several formulations are described in US 4,804,663. Estimate the cost of making the API and the tablet and injectable solution formulations.

E.8.26.   Sertraline hydrochloride

Sertraline (cis-(1S)-N-methyl-4-(3,4-dichlorophenyl)-1,2,3,4-tetrahydro-1-naphthaleneamine) in the hydrochloride salt is an antidepressant marketed as Zoloft™. US 4,536,518 (to Pfizer, Inc.) describes the preparation of the API. Estimate the cost of production of the API.

E.8.27.   Simvastatin

Merck & Co. patent US 4,444,784 describes the process for synthesizing simvastatin, which is the API for Zocor™, a cholesterol-lowering drug. Flow-sheet A in the patent gives several possible routes to make the API. Which is the lowest cost?

E.8.28.   Sumatriptan

US 4,816,470 assigned to Glaxo describes the process for making 3-(2-aminoethyl)-N-methyl-1H-indole-5-methanesulponamide, which is the API for Imigran™. Estimate the cost of production for a generic manufacturer to produce this compound.

E.8.29.   Venlafaxine

Venlafaxine (1-[2-(dimethylamino)-1-(4-methoxyphenyl)ethyl]cyclohexanol) is an antidepressant marketed as Effexor™. Synthesis of the API is described in US 4,535,186 assigned to American Home Products. Estimate the cost of production of the API.

E.9.   PULP AND PAPER

E.9.1.   Biopulping

Biological pretreatment has been suggested as a means of improving both mechanical and Kraft pulping. US 6,402,887, assigned to Biopulping International, describes a biological treatment process for wood waste that leads to paper of comparable quality to that produced with virgin wood. Estimate the cost of the biological pretreatment step per pound dry mass of paper product. At what price (recovery cost) of wood waste would this process deliver a 12% internal rate of return?

E.9.2.   Black liquor recovery

Black liquor is a by-product produced in the Kraft process for paper pulping (see E.9.5). Black liquor has a high content of organic compounds and salts and is often incinerated to provide part of the site fuel requirement. US 6,261,411 (unassigned) describes a process for recovery of chemicals from black liquor. Estimate the net present value of a plant that used this technology in a world-scale paper mill.

E.9.3.   Chemimechanical pulping

US 4,900,399 (to Eka AB) and US 5,002,635 (to Scott Paper Co.) both describe chemical pretreatments that claim to improve the properties of mechanical pulp. Estimate the cost of these pretreatment processes for a world-scale mechanical pulping plant. Which method would you recommend?

E.9.4.   Chlorine-free bleaching

Methods of bleaching paper pulp with reduced consumption of chlorine chemicals are of interest to the paper industry, as they reduce the environmental impact of paper manufacture. US 5,004,523 (to US Dept. of Agriculture) and US 5,091,054 (to Degussa) describe a pulp pretreatment using Caro's acid that enhances oxygen delignification and peroxide bleaching. Estimate the cost of this treatment (including the cost of processing any waste streams generated) for a typical Kraft paper mill.

E.9.5.   Kraft pulping

The Kraft process is used for production of pulp for high-quality paper. Improvements to the Kraft process are described in US 5,507,912 (to H.A. Simons Ltd) and US 7,097,739 (to Solutia Inc.). Estimate the annual savings gained by each of these processes relative to conventional Kraft pulping.

E.10.   SPECIALTY CHEMICALS

E.10.1.   Acetophenone

Acetophenone (phenyl methyl ketone) has a wide range of applications in perfumery. It can be recovered from the heavy by-product stream of a phenol process (which otherwise has fuel value) using the process described in US 4,559,110 assigned to Dow Chemical. It can be made by oxidation of ethylbenzene using the process described in US 4,950,794 (to Arco Chemical Technology). It can also be produced as a 'natural' product by fermentation of cinnamic acid using the process described in US 6,482,794 (to International Flavours & Fragrances). Estimate the cost of production via each route.

E.10.2.   Carbon nanotubes

Carbon nanotubes are of current interest as a novel material that can be used in a variety of applications. Large-scale use of these materials has been hindered by the absence of a process that can produce large quantities of high-purity product. US 6,413,487 (to University of Oklahoma) describes a process for large-scale production of single-wall carbon nanotubes. US 6,333,016 (also to University of Oklahoma) gives yield data. US 5,560,898 (to Agency of Industrial Science and Technology, Japan) and US 5,641,466 (to NEC Corp.) describe processes for separating nanotubes from graphitic carbon. US 5,698,175 (to NEC Corp.) describes a process for purifying and uncapping carbon nanotubes. Develop a flow-scheme for production of high-purity single-walled carbon nanotubes and estimate the cost of production.

E.10.3.   3-R Citronellol

3-R citronellol is a fragrance. US 4,962,242 (to Takasago Perfumery Co) describes a preparation from geraniol (example 2). Estimate the cost of production.

E.10.4.   Cleve's acid

1,7-Cleve's acid (1-naphthylamine-7-sulphonic acid) is used in dye manufacture. The preparation is described in US 2,875,243 (to Bayer). Estimate the cost of production.

E.10.5.   Dextrins

Dextrins are used in making pills, bandages, paper, fabrics, glue, matches and a range of other applications that require thickening of pastes. They are made by enzyme hydrolysis of starch, as described in US 6,670,155 (to Grain Processing Corp.). Estimate the cost of producing and purifying dextrin with recovery of the retrograded amylose.

E.10.6. D-Malic acid

Malic acid (hydroxybutanedioic acid) is a chemical intermediate and is also used as a food flavour enhancer. It can be made by several routes. US 5,210,295 (to Monsanto) describes a non-enzymatic process. US 4,772,749 (to Degussa) describes recovery of malic acid from the product of enzymatic conversion of fumaric acid. US 4,912,042 (to Eastman Kodak) describes an enzymatic separation process for separating the *L*- and *D*- isomers. US 5,824,449 (to Ajinomoto Co.) describes a selective fermentation from maleic acid. Estimate the cost of production of *D*-malic acid by each process and determine which is cheapest.

E.10.7. Salicylic acid USP

Salicylic acid is used as a raw material for making aspirin, as well as a starting material for dyes and as a pharmaceutical compound. It is made by heating sodium phenolate with carbon dioxide under pressure, and the process is described in several of the standard reference works listed in Chapter 8. Estimate the cost of production.

E.11. WASTE TREATMENT AND RECOVERY

E.11.1. Nylon recycling

Waste carpet typically contains large quantities of Nylon 6, which can be converted back into caprolactam. Recycling processes are described in US 7,115,671, US 6,111,099 (both to DSM B.V.) and US 5,359,062 (to BASF). Determine the economics of recovering caprolactam from carpet waste if the waste is available at a cost of $-30/$ metric ton (i.e., you are paid $30/ton to accept it). How does this compare to burning the waste carpet in an incinerator with a steam-turbine cogeneration plant?

E.11.2. Sulphur dioxide treatment

Sulphur dioxide is formed whenever sulphur-containing fuels are combusted in air. Sulphur dioxide can lead to the formation of acid rain and is a controlled pollutant in most countries. US 5,196,176, assigned to Paques B.V., describes a biological process for removing sulphur dioxide from a vent gas and converting it to elemental sulphur. Estimate the cost (in $/kWh) of using the Paques process to treat the flue gas from a 1000-MW power station that burns Illinois Number 6 coal in pressurized fluidized-bed combustors.

E.11.3. Sulphur recovery

Many processes release sulphur in the form of H_2S, which is highly toxic, and must be converted to a marketable form such as elemental sulphur or a stable disposable product such as a sulphate salt. US 5,397,556 (to Regents of the University of California) describes a process for converting H_2S to elemental sulphur. How does the cost of sulphur produced by this process compare with the cost of sulphur produced by the conventional modified Claus process?

E.11.4.   Toxic waste disposal

A novel process for toxic waste handling is suggested in US 4,764,282 (to Uniroyal Goodrich Tire Company). A waste liquid is soaked up into ground tyre rubber to form a stable solid that can be transported with reduced risk of spillage. The resulting product can then be incinerated in a fluidized-bed combustor, similar to the fluidized-bed combustors used in coal-fired power stations. Estimate the cost of waste disposal via this route, allowing for a credit for the electricity produced. How does this compare to the cost of toxic waste disposal by conventional incineration?

F   DESIGN PROJECTS II

The design exercises given in this appendix are somewhat more structured than those given in Appendix E. They have been adapted from Design Projects set by the Institution of Chemical Engineers as the final part of the Institution's qualifying examinations for professional Chemical Engineers.

F.1.   ETHYLHEXANOL FROM PROPYLENE AND SYNTHESIS GAS

The Project

Design a plant to produce 40,000 metric tons/year of 2-ethylhexanol from propylene and synthesis gas, assuming an operating period of 8000 h on stream.

The Process

The first stage of the process is a hydroformylation (oxo) reaction from which the main product is n-butyraldehyde. The feeds to this reactor are synthesis gas (CO/H_2 mixture) and propylene in the molar ratio 2:1, and the recycled products of iso-butyraldehyde cracking. The reactor operates at 130°C and 350 bar, using cobalt carbonyl as catalyst in solution. The main reaction products are n- and iso-butyraldehyde in the ratio of 4:1, the former being the required product for subsequent conversion to 2-ethylhexanol. In addition, 3% of the propylene feed is converted to propane whilst some does not react.

Within the reactor, however, 6% of the n-butyraldehyde product is reduced to n-butanol, 4% of the isobutyraldehyde product is reduced to isobutanol, and other reactions occur to a small extent yielding high molecular weight compounds (heavy ends) to the extent of 1% by weight of the butyraldehyde/butanol mixture at the reactor exit.

The reactor is followed by a gas–liquid separator operating at 30 bar from which the liquid phase is heated with steam to decompose the catalyst for recovery of cobalt by filtration. A second gas–liquid separator operating at atmospheric pressure subsequently yields a liquid phase of aldehydes, alcohols, heavy ends and water, which is free from propane, propylene, carbon monoxide and hydrogen.

This mixture then passes to a distillation column which gives a top product of mixed butyraldehydes, followed by a second column which separates the two butyraldehydes into an isobutyraldehyde stream containing 1.3% mol n-butyraldehyde and an n-butyraldehyde stream containing 1.2% mol isobutyraldehyde.

A cracker converts isobutyraldehyde at a per pass yield of 80% back to propylene, carbon monoxide and hydrogen by passage over a catalyst with steam. After separation of the water and unreacted isobutyraldehyde the cracked gas is recycled to the hydroformylation reactor. The isobutyraldehyde is recycled to the cracker inlet. The operating conditions of the cracker are 275°C and 1 bar.

The n-butyraldehyde is treated with a 2% w/w aqueous sodium hydroxide and undergoes an aldol condensation at a conversion of 90%. The product of this reaction, 2-ethylhexanal, is separated and then reduced to 2-ethylhexanol by hydrogen in the presence of a Raney nickel catalyst with a 99% conversion rate. In subsequent stages of the process (details of which are not required), 99.8% of the 2-ethylhexanol is recovered at a purity of 99% by weight.

Feed Specifications

   (i) Propylene feed: 93% propylene, balance propane
  (ii) Synthesis gas: from heavy fuel oil, after removal of sulphur compounds and carbon dioxide: H_2 48.6%; CO 49.5%; CH_4 0.4%; N_2 1.5%.

Utilities

   (i) Dry saturated steam at 35 bar
  (ii) Cooling water at 20°C
 (iii) 2% w/w aqueous sodium hydroxide solution
  (iv) Hydrogen gas: H_2 98.8%; CH_4 1.2%.

Scope of Design Work Required

1. Process Design

   (a) Prepare a material balance for the complete process.
   (b) Prepare a process diagram for the plant showing the major items of equipment. Indicate the materials of construction and the operating temperatures and pressures.
   (c) Prepare energy balances for the hydroformylation reactor and for the isobutyraldehyde cracking reactor.

2. Chemical Engineering Design

Prepare a chemical engineering design of the second distillation unit, i.e., for the separation of n- and isobutyraldehyde. Make dimensioned sketches of the column, the reboiler and the condenser.

3. Mechanical Design

Prepare a mechanical design with sketches suitable for submission to a drawing office of the n- and isobutyraldehyde distillation column.

4. Control System

For the hydroformylation reactor prepare a control scheme to ensure safe operation.

Data

1. Reactions

$CH_3 \cdot CH = CH_2 + H_2$	$\rightarrow CH_3 \cdot CH_2 \cdot CH_3$	$\Delta H^\circ_{298} = -129.5\,kJ/mol$
$CH_3 \cdot CH = CH_2 + H_2 + CO$	$\rightarrow CH_3 \cdot CH_2 \cdot CH_2 \cdot CHO$	$\Delta H^\circ_{298} = -135.5\,kJ/mol$

$$\text{or} \quad \rightarrow CH_3 \cdot \underset{\underset{CHO}{|}}{CH} \cdot CH_3$$

$$\Delta H^\circ_{298} = -141.5\,kJ/mol$$

$C_3H_7CHO + H_2$	$\rightarrow C_4H_9OH$	$\Delta H^\circ_{298} = -64.8\,kJ/mol$
$2CO + 8CO$	$\rightarrow CO_2(CO)_8$	$\Delta H^\circ_{298} = -462.0\,kJ/mol$
$2CH_3 \cdot CH_2 \cdot CH_2 \cdot CHO$	$\rightarrow CH_3 \cdot CH_2 \cdot CH_2 \cdot CH = C - CHO + H_2O$	

$$\underset{\underset{C_2H_5}{|}}{} \qquad \Delta H^\circ_{298} = -262.0\,kJ/mol$$

$$C_4H_8 = \underset{\underset{C_2H_5}{|}}{C} - CHO + 2H_2 \qquad \rightarrow C_4H_9 - \underset{\underset{C_2H_5}{|}}{CH} \cdot CH_2OH \qquad \Delta H^\circ_{298} = -433.0\,kJ/mol$$

2. Boiling Points at 1 Bar

Propylene	$-47.7°C$
Propane	$-42.1°C$
n-Butyraldehyde	$75.5°C$
Isobutyraldehyde	$64.5°C$
n-Butanol	$117.0°C$
Isobutanol	$108.0°C$
2-Ethylhexanol	$184.7°C$

3. Solubilities of Gases at 30 Bar in the Liquid Phase of the First Gas–Liquid Separator

H_2	0.08×10^{-3}	kg dissolved/kg liquid
CO	0.53×10^{-3}	kg dissolved/kg liquid
Propylene	7.5×10^{-3}	kg dissolved/kg liquid
Propane	7.5×10^{-3}	kg dissolved/kg liquid

4.   Vapour–Liquid Equilibrium of the Butyraldehydes at 1 ATM (Ref. 7)

$T°C$	x	y
73.94	0.1	0.138
72.69	0.2	0.264
71.40	0.3	0.381
70.24	0.4	0.490
69.04	0.5	0.589
68.08	0.6	0.686
67.07	0.7	0.773
65.96	0.8	0.846
64.95	0.9	0.927

where x and y are the mol fractions of the more volatile component (iso-butyraldehyde) in the liquid and vapour phases respectively.

REFERENCES

1. Hancock, E. G. (ed.) (1973) *Propylene and its Industrial Derivatives* (New York: John Wiley & Sons) Chapter 9, pp. 333–367.
2. *Carbon Monoxide in Organic Synthesis* (1970) (New York: Falbe-Springer Verlag) pp. 1–75.
3. *Chemical Engineering*, (1974) Sept. 30th, 81, pp. 115–122. Physical and thermodynamic properties of CO and CO_2.
4. *Chemical Engineering* (1975) Jan. 20th, 82, pp. 99–106. Physical and thermodynamic properties of $H_2/N_2/O_2$.
5. *Chemical Engineering* (1975) Mar. 31st, 82, pp. 101–109. Physical and thermodynamic properties of $C_2H_4/C_3H_6/iC_4H_8$.
6. *Chemical Engineering* (1975) May 12th, 82, pp. 89–97. Physical and thermodynamic properties of $CH_4/C_2H_6/C_3H_8$.
7. Wojtasinski, J. G. (1963) *J. Chem. Eng. Data*, July, pp. 381–385. Measurement of total pressures for determining liquid-vapour equilibrium relations of the binary system isobutyraldehyde-n-butyraldehyde.
8. Weber, H. and FALBE, J. (1970) *Ind. Eng. Chem.*, April, pp. 33–7. Oxo Synthesis Technology.
9. *Hydrocarbon Processing* (1971) Nov., p. 166.
10. *Hydrocarbon Processing* (1975) Nov., p. 148.

F.2.   CHLOROBENZENES FROM BENZENE AND CHLORINE

The Project

Design a plant to produce 20,000 metric tons/year of monochlorobenzene together with not less than 2000 metric tons/year of dichlorobenzene, by the direct chlorination of benzene.

The Process

Liquid benzene (which must contain less than 30 ppm by weight of water) is fed into a reactor system consisting of two continuous stirred tanks operating in series at 2.4 bar. Gaseous chlorine is fed in parallel to both tanks. Ferric chloride acts as a catalyst, and is produced *in situ* by the action of hydrogen chloride on mild steel. Cooling is required to maintain the operating temperature at 328 K. The hydrogen chloride gas leaving the reactors is first cooled to condense most of the organic impurities. It then passes to an activated carbon adsorber where the final traces of impurity are removed before it leaves the plant for use elsewhere.

The crude liquid chlorobenzenes stream leaving the second reactor is washed with water and caustic soda solution to remove all dissolved hydrogen chloride. The product recovery system consists of two distillation columns in series. In the first column (the 'benzene column') unreacted benzene is recovered as top product and recycled. In the second column (the 'chlorobenzene column') the mono- and dichlorobenzenes are separated. The recovered benzene from the first column is mixed with the raw benzene feed and this combined stream is fed to a distillation column (the 'drying column') where water is removed as overhead. The benzene stream from the bottom of the drying column is fed to the reaction system.

Feed Specifications

   (i) Chlorine: 293 K, atmospheric pressure, 100% purity
   (ii) Benzene: 293 K, atmospheric pressure, 99.95 wt% benzene, 0.05 wt% water.

Product Specifications

   (i) Monochlorobenzene: 99.7 wt%
   (ii) Dichlorobenzene: 99.6 wt%
   (iii) Hydrogen chloride gas: less than 250 ppm by weight benzene.

Utilities

   (i) Stream: dry saturated at 8 bar and at 28 bar
   (ii) Cooling water: 293 K
   (iii) Process water: 293 K
   (iv) Caustic soda solution: 5 wt% NaOH, 293 K
   (v) Electricity: 440 V, 50 Hz, three phase.

Scope of Design Work Required

1. Process Design

(a) Prepare a materials balance for the process including an analysis of each reactor stage (the kinetics of the chlorination reactions are given below). On-stream time may be taken as 330 days per year.

(b) Prepare energy balances for the first reactor and for the chlorobenzene column (take the reflux ratio for this column as twice the minimum reflux ratio).

(c) Prepare a process flow diagram for the plant. This should show the major items of equipment with an indication of the materials of construction and of the internal layout. Temperatures and pressures should also be indicated.

2. Chemical Engineering Design

Prepare a sieve-plate column design for the chlorobenzene distillation and make dimensioned sketches showing details of the plate layout including the weir and the downcomer.

3. Mechanical Design

Prepare a mechanical design of the chlorobenzene column, estimating the shell thickness, the positions and sizes of all nozzles, and the method of support for the plates and the column shell. Make a dimensioned sketch suitable for submission to a drawing office.

4. Safety

Indicate the safety measures required for this plant bearing in mind the toxic and inflammable materials handled.

Data

1. The Reactions

$$(1) C_6H_6 + Cl_2 \rightarrow C_6H_5Cl + HCl$$
$$(2) C_6H_5Cl + Cl_2 \rightarrow C_6H_4Cl_{12} + HCl$$

The dichlorobenzene may be assumed to consist entirely of the para-isomer and the formation of trichlorobenzenes may be neglected.

The rate equations can be written in first-order form when the concentration of dissolved chlorine remains essentially constant. Thus:

$$r_B = k_1 x_B$$

$$r_M = k_1 x_B - k_2 x_M$$

$$r_D = k_2 x_M$$

where

r is the reaction rate
k_1 is the rate constant for reaction (1) at 328 K $= 1.00 \times 10^{-4}\,\text{s}^{-1}$
k_2 is the rate constant for reaction (2) at 328 K $= 0.15 \times 10^{-4}\,\text{s}^{-1}$
and x denotes mol fraction.

The subscripts B, M and D denote benzene, monochlorobenzene and dichlorobenzene respectively.

Yields for the reactor system should be calculated on the basis of equal liquid residence times in the two reactors, with a negligible amount of unreacted chlorine in the vapour product streams. It may be assumed that the liquid product stream contains 1.5 wt% of hydrogen chloride.

REFERENCE

1. Bodman, S. W. (1968) *The Industrial Practice of Chemical Process Engineering* (The MIT Press).

2. Solubilities

Solubility of the water/benzene system; taken from Seidell, A. S. (1941) *Solubilities of Organic Compounds*, 3rd edn, Vol. II (Van Nostrand):

Temperature (K)	293	303	313	323
g H_2O/100 g C_6H_6	0.050	0.072	0.102	0.147
g C_6H_6/100 g H_2O	0.175	0.190	0.206	0.225

3. Thermodynamic and Physical Properties

	C_6H_6 liquid	C_6H_6 gas	C_6H_5Cl liquid	C_6H_5Cl gas	$C_6H_4Cl_2$ liquid	$C_6H_4Cl_2$ gas
Heat of formation at 298 K (kJ/kmol)	49.0	82.9	7.5	46.1	−42.0	5.0
Heat capacity (kJ/kmol K)						
298 K	136	82	152	92		103
350 K	148	99	161	108	193	118
400 K	163	113	170	121	238	131
450 K	179	126	181	134	296	143
500 K	200	137	192	145	366	155
Density (kg/m^3)						
298 K	872		1100			
350 K	815		1040		1230	

400 K	761	989	1170
450 K	693	932	1100
500 K	612	875	1020
Viscosity (Ns/m^2)			
298 K	0.598×10^{-3}	0.750×10^{-3}	
350 K	0.326×10^{-3}	0.435×10^{-3}	0.697×10^{-3}
400 K	0.207×10^{-3}	0.305×10^{-3}	0.476×10^{-3}
450 K	0.134×10^{-3}	0.228×10^{-3}	0.335×10^{-3}
500 K	0.095×10^{-3}	0.158×10^{-3}	0.236×10^{-3}
Surface tension (N/m)			
298 K	0.0280	0.0314	
350 K	0.0220	0.0276	0.0304
400 K	0.0162	0.0232	0.0259
450 K	0.0104	0.0177	0.0205
500 K	0.0047	0.0115	0.0142

REFERENCES

1. Perry, R. H. and Chilton, C. H. (1973) *Chemical Engineers' Handbook*, 5th edn (McGraw-Hill).
2. Kirk-Othmer (1964) *Encyclopaedia of Chemical Technology*, 2nd edn (John Wiley & Sons).

F.3.   METHYL ETHYL KETONE FROM BUTYL ALCOHOL

The Project

Design a plant to produce 1×10^7 kg/year of methyl ethyl ketone (MEK).
   Feedstock: Secondary butyl alcohol.
   Services available:

Dry saturated steam at 140°C
Cooling water at 24°C
Electricity at 440 V, three phase, 50 Hz
Flue gases at 540°C.

The Process

The butyl alcohol is pumped from storage to a steam-heated preheater and then to a vaporizer heated by the reaction products. The vapour leaving the vaporizer is heated to its reaction temperature by flue gases which have previously been used as reactor heating medium. The superheated butyl alcohol is fed to the reaction system at 400°C to 500°C where 90% is converted on a zinc oxide–brass catalyst to methyl ethyl ketone, hydrogen and other reaction products. The reaction products may be treated in one of the following ways:

(a) Cool and condense the MEK in the reaction products and use the exhaust gases as a furnace fuel.
(b) Cool the reaction products to a suitable temperature and separate the MEK by absorption in aqueous ethanol. The hydrogen off-gas is dried and used as a furnace fuel. The liquors leaving the absorbers are passed to a solvent extraction column, where the MEK is recovered using trichloroethane. The raffinate from this column is returned to the absorber and the extract is passed to a distillation unit where the MEK is recovered. The trichloroethane is recycled to the extraction plant.

Scope of Design Work Required

1. Prepare material balances for the two processes.
2. Size and cost the equipment needed for option b and hence determine which process is preferable.
3. Prepare a material flow diagram of the preferred process.
4. Prepare a heat balance diagram of the preheater–vaporizer–superheater–reactor system.
5. Prepare a chemical engineering design of the preheater–vaporizer–superheater–reactor system and indicate the type of instrumentation required.
6. Prepare a mechanical design of the butyl alcohol vaporizer and make a dimensioned sketch suitable for submission to a drawing office.

Data

Process Data

Outlet condenser temperature $= 32°C$.
   Vapour and liquid are in equilibrium at the condenser outlet.
   Calorific value of MEK $= 41,800\,kJ/kg$.

Reactor Data

The 'short-cut' method proposed in ref. 1 may be used only to obtain a preliminary estimate of the height of catalyst required in the reactor. The reactor should be designed from first principles using the rate equation, below, taken from ref. 1.

$$r_A = \frac{C(P_{A,i} - P_{K,i}P_{H,i}/K)}{P_{Ki}(1 + K_A\,P_{A,i} + K_{AK}P_{A,i}/P_{K,i})}$$

where $P_{A,i}$, $P_{H,i}$ and $P_{K,i}$ are the interfacial partial pressures of the alcohol, hydrogen and ketone in bars, and the remaining quantities are as specified by the semi-empirical equations below:

$$\log_{10} C = \frac{5964}{T_i} + 8.464$$

$$\log_{10} K_A = \frac{3425}{T_i} + 5.231$$

$$\log_{10} K_{AK} = +\frac{486}{T_i} - 0.1968$$

In these equations, the interfacial temperature T_i is in Kelvin, the constant C is in kmol/m²h, K_A is in bar^{-1}, and K_{AK} is dimensionless.

The equilibrium constant, K is given in ref. 1 (although the original source is ref. 2) by the equation:

$$\log_{10} K = -\frac{2790}{T_i} + 1.510 \log_{10} T_i + 1.871$$

where K is in bar.

Useful general information will be found in ref. 3.

REFERENCES

1. Perona, J. J. and Thodos, G. (1957) *AIChE Jl*, 3, 230.
2. Kolb, H. J. and Burwell, R. L. (Jr.) (1945) *J. Am. Chem. Soc.*, 67, 1084.
3. Rudd, D. F. and Watson, C. C. (1968) *Strategy of Process Engineering* (New York: John Wiley & Sons Inc.).

F.4.   ACRYLONITRILE FROM PROPYLENE AND AMMONIA

The Project

Design a plant to produce 1×10^8 kg/year of acrylonitrile ($CH_2{:}CH.CN$) from propylene and ammonia by the ammoxidation process.

Feedstock: Ammonia, 100% NH_3.

Propylene: Commercial grade containing 90% C_3H_6, 10% paraffins, etc., which do not take any part in the reaction.

Services available:

Dry saturated steam at 140°C
Cooling water at 24°C
Other normal services.

The Process

Propylene, ammonia, steam and air are fed to a vapour-phase catalytic reactor (item A). The feed stream composition (mol%) is propylene 7; ammonia 8; steam 20; air 65. A fixed-bed reactor is employed using a molybdenum-based catalyst at a temperature of 450°C, a pressure of 3 bar absolute, and a residence time of 4 s.

Based upon a pure propylene feed, the carbon distribution by weight in the product from the reactor is:

Acrylonitrile	58%
Acetonitrile	2%
Carbon dioxide	16%
Hydrogen cyanide	6%
Acrolein	2%
Unreacted propylene	15%
Other by products	1%

The reactor exit gas is air-cooled to 200°C and then passes to a quench scrubber (B) through which an aqueous solution containing ammonium sulphate 30 wt% and sulphuric acid 1 wt% is circulated. The exit gas temperature is thereby reduced to 90°C.

From the quench scrubber (B) the gas passes to an absorption column (C) in which the acrylonitrile is absorbed in water to produce a 3 wt% solution. The carbon dioxide, unreacted propylene, oxygen, nitrogen and unreacted hydrocarbons are not absorbed and are vented to atmosphere from the top of column (C).

The solution from the absorber (C) passes to a stripping column (D) where acrylonitrile and lower boiling impurities are separated from water. Most of the aqueous bottom product from the stripping column (D), which is essentially free of organics, is returned to the absorber (C), the excess being bled off. The overhead product is condensed and the aqueous lower layer returned to the stripping column (D) as reflux.

The upper layer which contains, in addition to acrylonitrile, hydrogen cyanide, acrolein, acetonitrile, and small quantities of other impurities, passes to a second reactor (E) where, at a suitable pH, all the acrolein is converted to its cyanohydrin. (Cyanohydrins are sometimes known as cyanhydrins.) The product from the reactor (E) is fed to a cyanohydrin separation column (F), operating at reduced temperature and pressure, in which acrolein cyanohydrin is separated as the bottom product and returned to the ammoxidation reactor (A) where it is quantitatively converted to acrylonitrile and hydrogen cyanide.

The top product from column (F) is fed to a stripping column (G) from which hydrogen cyanide is removed overhead.

The bottom product from column (G) passes to the hydroextractive distillation column (H). The water feed rate to column (H) is five times that of the bottom product flow from column (G). It may be assumed that the acetonitrile and other by-products are discharged as bottom product from column (H) and discarded. The overhead product from column (H), consisting of the acrylonitrile water azeotrope, is condensed and passed to a separator. The lower aqueous layer is returned to column (H).

The upper layer from the separator is rectified in a column (I) to give 99.95 wt% pure acrylonitrile.

Scope of Design Work Required

1. Prepare a material balance for the process.
2. Prepare a material flow diagram of the process.
3. Prepare a heat balance for the reactor (A) and quench column (B).
4. Prepare a chemical engineering design of reactor (A) and either column (B) OR column (D).
5. Prepare a mechanical design of the condenser for stripping column (D) and make a dimensioned sketch suitable for submission to a drawing office.
6. Indicate the instrumentation and safety procedure required for this plant bearing in mind the toxic and flammable materials being handled.

REFERENCES

1. Hancock, E. H. (ed.) (1973) *Propylene and its Industrial Derivatives* (London: Ernest Benn Ltd).
2. Sokolov, N. M., Sevryugova, N. N. and Zhavoronkov, N. M. (1969) *Proceedings of the International Symposium on Distillation*, pp. 3, 110–3, 117 (London: I Chem E).

F.5.   UREA FROM AMMONIA AND CARBON DIOXIDE

The Project

A plant is to be designed for the production of 300,000 kg per day of urea by the reaction of ammonia and carbon dioxide at elevated temperature and pressure, using a total recycle process in which the mixture leaving the reactor is stripped by the carbon dioxide feed (DSM process, refs 1 to 4).

Materials Available

(1) Liquid ammonia at 20°C and 9 bar, which may be taken to be 100% pure
(2) Gaseous carbon dioxide at 20°C and atmospheric pressure, also 100% pure.

All normal services are available on site. In particular, electricity, 440 V, three phase, 50 Hz; cooling water at a maximum summer temperature of 22°C; steam at 40 bar with 20°C of superheat.

The on-stream time is to be 330 days/year, and the product specification is fertilizer-grade urea prills containing not more than 1.0% biuret.

The Process

The reaction which produces urea from ammonia and carbon dioxide takes place in two stages; in the first, ammonium carbamate is formed:

$$2NH_3 + CO_2 \rightleftharpoons NH_2COONH_4$$

In the second, the carbamate is dehydrated to give urea:

$$NH_2COONH_4 \rightleftharpoons CO(NH_2)_2 + H_2O$$

Both reactions are reversible, the first being exothermic and going almost to completion, whilst the second is endothermic and goes to 40 to 70% of completion.

Ammonia and carbon dioxide are fed to the reactor, a stainless-steel vessel with a series of trays to assist mixing. The reactor pressure is 125 bar and the temperature is 185°C. The reactor residence time is about 45 min, a 95% approach to equilibrium being achieved in this time. The ammonia is fed directly to the reactor, but the carbon dioxide is fed to the reactor upwardly through a stripper, down which flows the product stream from the reactor. The carbon dioxide decomposes some of the carbamate in the product stream, and takes ammonia and water to a high-pressure condenser. The stripper is steam heated and operates at 180°C, whilst the high-pressure condenser is at 170°C and the heat released in it by recombination of ammonia and carbon dioxide to carbamate is used to raise steam. Additional recycled carbamate solution is added to the stream in the high-pressure condenser, and the combined flow goes to the reactor.

The product stream leaving the stripper goes through an expansion valve to the low pressure section, the operating pressure there being 5 bar. In a steam-heated rectifier, further ammonia and carbon dioxide are removed and, with some water vapour, are condensed to give a weak carbamate solution. This is pumped back to the high-pressure condenser.

A two-stage evaporative concentration under vacuum, with a limited residence time in the evaporator to limit biuret formation, produces a urea stream containing about 0.5% water which can be sprayed into a prilling tower.

Physico-Chemical Data

Heats of reactions:

$$2NH_3 + CO_2 \rightarrow NH_2COONH_4 + 130\ kJ$$
$$NH_2COONH_4 \rightarrow CO(NH_2)_2 + H_2O - 21\ kJ$$

Properties of urea:

Density at 20°C = 1.335 g/cm^3
Heat of solution in water = −250 J/g
Melting point = 133°C
Specific heat = 1.34 J/g at 20°C.

Reactor and Stripper Design

The relationships between temperature, pressure and composition for the urea–CO_2–NH_3–H_2O system are given in refs 5 and 6. These are equilibrium relationships. The reaction velocity may be obtained from the graph in Figure 5 of ref. 5, which is reproduced here for ease of reference (Figure F1). Some stripper design data appear in ref. 7.

Scope of Design Work Required

1. Prepare a mass balance diagram for the process, on a weight per hour basis, through to the production of urea prills.

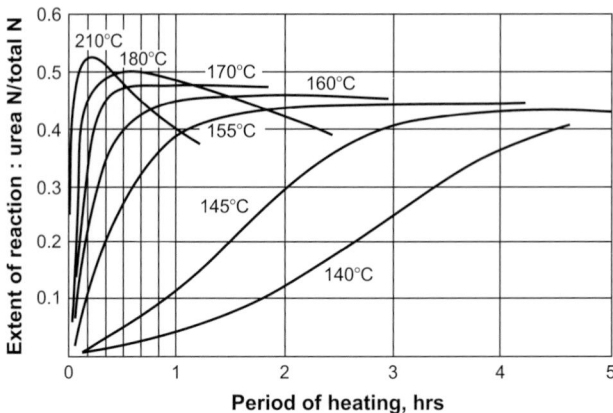

Figure F1.   Rate of dehydration of carbamate.

2. Prepare an energy balance diagram for the reactor–stripper–high-pressure condenser complex.
3. Prepare a process flow diagram, showing the major items of equipment in the correct elevation, with an indication of their internal construction. Show all major pipe lines and give a schematic outline of the probable instrumentation of the reactor and its subsidiaries.
4. Prepare an equipment schedule, listing the main plant items with their size, throughput, operating conditions, materials of construction, and services required.
5. Prepare an outline design of the reactor and carry out the chemical engineering design of the stripper, specifying the interfacial contact area which will need to be provided between the carbon dioxide stream and the product stream to enable the necessary mass transfer to take place.
6. Prepare a mechanical design of the stripper, which is a vertical steam-heated tube bundle rather like a heat exchanger. Show how liquid is to be distributed to the tubes, and how the shell is to be constructed to resist the high pressure and the corrosive process material.
7. Prepare a detailed mechanical design of the reactor in the form of a general arrangement drawing with supplementary detail drawings to show essential constructional features. Include recommendations for the feed of gaseous ammonia, carbon dioxide and carbamate solution, the latter being very corrosive. The design should ensure good gas–liquid contact; suitable instrumentation should be suggested, and provision included for its installation. Access must be possible for maintenance.
8. Specify suitable control systems for the maintenance of constant conditions in the reactor against a 15% change in input rate of ammonia or carbon dioxide, and examine the effect of such a change, if uncorrected, on the steam generation capability of the high-pressure condenser.

REFERENCES

1. Kaasenbrood, P. J. C. and Logemann, J. D. (1969) *Hydrocarbon Processing*, April, pp. 117–121.
2. Payne, A. J. and Canner, J. A. (1969) *Chemical and Process Engineering*, May, pp. 81–88.
3. Cook, L. H. (1966) *Hydrocarbon Processing*, 1966, pp. 129–136.
4. *Process Survey: Urea*. (1969) Booklet published with *European Chemical News*, Jan. 17th, p. 17.
5. Frejacques, M. (1948) *Chimie et Industrie*, July, pp. 22–35.
6. Kucheryavyy, V. I. and Gorlovskiy, D. M. (1969) *Soviet Chemical Industry*, Nov., pp. 44–46.
7. Van Krevelen, D. W. and Hoftyzer, P. J. (1953) *Chemical Engineering Science*, Aug., 2(4) pp. 145–156.

F.6.   HYDROGEN FROM FUEL OIL

The Project

A plant is to be designed to produce 20 million standard cubic feet per day (0.555 × 10^6 standard m^3/day) of hydrogen of at least 95% purity. The process to be employed is the partial oxidation of oil feedstock (refs 1 to 3).

Materials Available

(1) Heavy fuel oil feedstock of viscosity 900 s Redwood One (2.57 × 10^{-4} m^2/s) at 100°F with the following analysis:

Carbon	85% wt
Hydrogen	11% wt
Sulphur	4% wt
Calorific value	18,410 Btu/lb (42.9 MJ/kg)
Specific gravity	0.9435

The oil available is pumped from tankage at a pressure of 30 psig (206.9 kN/m^2 gauge) and at 50°C.

(2) Oxygen at 95% purity (the other component assumed to be wholly nitrogen) and at 20°C and 600 psig (4140 kN/m^2 gauge).

Services Available

(1) Steam at 600 psig (4140 kN/m^2 gauge) saturated
(2) Cooling water at a maximum summer temperature of 25°C
(3) Demineralized boiler feed water at 20 psig (138 kN/m^2 gauge) and 15°C suitable for direct feed to the boilers
(4) Electricity at 440 V, three phase, 50 Hz, with adequate incoming cable capacity for all proposed uses
(5) Waste low-pressure steam from an adjacent process.

On-Stream Time

8050 hours per year.

Product Specification

Gaseous hydrogen with the following limits of impurities:

CO	1.0% vol maximum (dry basis)
CO_2	1.0% vol maximum (dry basis)
N_2	2.0% vol maximum (dry basis)
CH_4	1.0% vol maximum (dry basis)
H_2S	Less than 1 ppm

The gas is to be delivered at 35°C maximum temperature, and at a pressure not less than 300 psig (2060 kN/m² gauge). The gas can be delivered saturated, i.e., no drying plant is required.

The Process

Heavy fuel oil feedstock is delivered into the suction of metering-type ram pumps which feed it via a steam preheater into the combustor of a refractory-lined flame reactor. The feedstock must be heated to 200°C in the preheater to ensure efficient atomization in the combustor. A mixture of oxygen and steam is also fed to the combustor, the oxygen being preheated in a separate steam preheater to 210°C before being mixed with the reactant steam.

The crude gas, which will contain some carbon particles, leaves the reactor at approximately 1300°C and passes immediately into a special waste-heat boiler where steam at 600 psig (4140 kN/m² gauge) is generated. The crude gas leaves the waste heat boiler at 250°C and is further cooled to 50°C by direct quenching with water, which also serves to remove the carbon as a suspension. The analysis of the quenched crude gas is as follows:

H_2	47.6	% vol (dry basis)
CO	42.1	% vol (dry basis)
CO_2	8.3	% vol (dry basis)
CH_4	0.1	% vol (dry basis)
H_2S	0.5	% vol (dry basis)
N_2	1.40	% vol (dry basis)
100.0	% vol (dry basis)	

For the primary flame reaction steam and oxygen are fed to the reactor at the following rates:

Steam	0.75 kg/kg of heavy fuel oil feedstock
Oxygen	1.16 kg/kg of heavy fuel oil feedstock

The carbon produced in the flame reaction, and which is subsequently removed as carbon suspension in water, amounts to 1.5% by weight of the fuel oil feedstock charge. Some H_2S present in the crude gas is removed by contact with the quench water.

The quenched gas passes to an H_2S removal stage where it may be assumed that H_2S is selectively scrubbed down to 15 parts per million with substantially nil removal of CO_2. Solution regeneration in this process is undertaken using the waste low-pressure steam from another process. The scrubbed gas, at 35°C and saturated, has then to undergo CO conversion, final H_2S removal, and CO_2 removal to allow it to meet the product specification.

CO conversion is carried out over chromium-promoted iron oxide catalyst employing two stages of catalytic conversion; the plant also incorporates a saturator and desaturator operating with a hot water circuit.

Incoming gas is introduced into the saturator (a packed column) where it is contacted with hot water pumped from the base of the desaturator; this process serves to preheat the gas and to introduce into it some of the water vapour required as reactant. The gas then passes to two heat exchangers in series. In the first, the unconverted gas is heated against the converted gas from the second stage of catalytic conversion; in the second heat exchanger the unconverted gas is further heated against the converted gas from the first stage of catalytic conversion. The remaining water required as reactant is then introduced into the unconverted gas as steam at 600 psig (4140 kN/m^2 gauge) saturated and the gas/steam mixture passes to the catalyst vessel at a temperature of 370°C. The catalyst vessel is a single shell with a dividing plate separating the two catalyst beds which constitute the two stages of conversion. The converted gas from each stage passes to the heat exchangers previously described and thence to the desaturator, which is a further packed column. In this column the converted gas is contacted countercurrent with hot water pumped from the saturator base; the temperature of the gas is reduced and the deposited water is absorbed in the hot-water circuit. An air-cooled heat exchanger then reduces the temperature of the converted gas to 40°C for final H_2S removal.

Final H_2S removal takes place in four vertical vessels each approximately 60 feet (18.3 m) in height and 8 feet (2.4 m) in diameter and equipped with five trays of iron oxide absorbent. Each vessel is provided with a locking lid of the autoclave type. The total pressure drop across these vessels is 5 psi (35 kN/m^2). Gas leaving this section of the plant contains less than 1 ppm of H_2S and passes to the CO_2 removal stage at a temperature of 35°C.

CO_2 removal is accomplished employing high-pressure potassium carbonate wash with solution regeneration (ref. 4).

Data

1. Basic Data for CO Conversion Section of the Plant

(a) Space Velocity

The space velocity through each catalyst stage should be assumed to be 3500 volumes of gas plus steam measured at NTP per volume of catalyst per hour. It should further be assumed that use of this space velocity will allow a 10°C approach to equilibrium to be attained throughout the possible range of catalyst operating temperatures listed below.

(b) Equilibrium Data for the CO Conversion Reaction

For

$$K_p = \frac{p_{CO} \times p_{H_2O}}{p_{CO_2} \times p_{H_2}}$$

Temp. (K)	K_p
600	3.69×10^{-2}
700	1.11×10^{-1}
800	2.48×10^{-1}

(c) Heat of Reaction

$$CO + H_2O \rightleftharpoons CO_2 + H_2 \; \Delta H = -9.84 \text{ kcal}$$

2. Basic data for CO_2 Removal Using Hot Potassium Carbonate Solutions

The data presented in ref. 4 should be employed in the design of the CO_2 removal section of the plant. A solution concentration of 40% wt equivalent $K_2 CO_3$ should be employed.

Scope of Design Work Required

1. Process Design

(a) Calculate, and prepare a diagram to show, the gas flows, compositions, pressures and temperatures, at each main stage throughout the processes of gasification and purification.

(b) Prepare a mass-balance diagram for the CO conversion section of the plant including the live steam addition to the unconverted gas. Basic data that should be employed for the CO conversion process are presented in the Appendix.

(c) Prepare an energy-balance diagram for the flame reactor and for the associated waste-heat boiler.
(d) Prepare a process flow diagram showing all major items of equipment. This need not be to scale but an indication of the internal construction of each item (with the exception of the flame reactor, waste-heat boiler and quench tower) should be given. The primary H_2S removal stage need not be detailed.
(e) Prepare an equipment schedule for the CO conversion section of the plant, specifying major items of equipment.

2.   Chemical Engineering Design

(a) Prepare a detailed chemical engineering design of the absorber on the CO_2 removal stage.
(b) Prepare a chemical engineering design for the saturator on the CO conversion section.

3.   Mechanical Design

Make recommendations for the mechanical design of the CO_2 removal absorber, estimating the shell and end-plate thickness and showing, by means of sketches suitable for submission to a design office, how:

(a) the beds of tower packing are supported
(b) the liquid is distributed.

Develop a detailed mechanical design of the CO conversion reactor, paying particular attention to the choice of alloy steels versus refractory linings, provisions for thermal expansion, inlet gas distribution, catalyst bed-support design, facilities for charging and discharging catalyst and provisions for instrumentation.

4.   Control

Prepare a full instrumentation flow-sheet of the CO conversion section of the plant, paying particular attention to the methods of controlling liquid levels in the circulating water system and temperatures in the catalyst beds. Derive the unsteady-state equations which would have to be employed in the application of computer control to the CO conversion section of the plant.

REFERENCES

1. Garvie, J. H. (1967) *Chem. Proc. Engng*, Nov., pp. 55–65. Synthesis gas manufacture.
2. *Hydrocarbon Processing—Refining Processes Handbook. Issue A* (1970) Sept., p. 269.
3. Singer, S. C. and Ter Haar, L. W. (1961) *Chem. Eng Prog.*, 57, pp. 68–74. Reducing gases by partial oxidation of hydrocarbons.
4. Benson H. E., Field J. H. and Haynes W. P. (1956) *Chem. Eng Prog.*, 52, pp. 433–438. Improved process for CO_2 absorption uses hot carbonate solutions.

F.7.   CHLORINE RECOVERY FROM HYDROGEN CHLORIDE

The Project

A plant is to be designed for the production of 10,000 tonnes per annum of chlorine by the catalytic oxidation of HCl gas.

Materials Available

(1) HCl gas as by-product from an organic synthesis process. This may be taken to be 100% pure and at 20°C and absolute pressure of 14.7 psi ($100 \, kN/m^2$).
(2) Air. This may be taken to be dry and at 20°C and absolute pressure of 14.7 psi ($100 \, kN/m^2$).

Services Available

(1) Steam at 200 psig ($1400 \, kN/m^2$)
(2) Cooling water at a maximum summer temperature of 24°C.
(3) A limited supply of cooling water at a constant temperature of 13°C is also available.
(4) Electricity at 440 V, three phase, 50 Hz.

On-Stream Time

8000 hours per year.

Product Specification

Gaseous chlorine mixed with permanent gases and HCl. The HCl content not to exceed 5×10^{-5} part by weight of HCl per unit weight of chlorine.

The Process

HCl is mixed with air and fed into a fluidized-bed reactor containing cupric chloride/pumice catalyst and maintained at a suitable temperature in the range 300°C –400°C. The HCl in the feed is oxidized, and the chlorine and water produced in the reaction, together with unchanged HCl and permanent gases, are passed to a packed tower cooler/scrubber, operating somewhat above atmospheric pressure, where they are contacted with aqueous HCl containing 33%–36% by weight of HCl. This acid enters the cooler/scrubber at about 20°C. Most of the water and some of the HCl contained in the gases entering the cooler/scrubber are dissolved in the acid. The liquid effluent from the base of the cooler/scrubber flows to a divider box from which one stream passes to the top of the cooler/scrubber, via a cooler which lowers its temperature to 20°C, and another stream passes to a stripping column ('expeller'). Gas containing 98% by weight of HCl (the other constituents being water and chlorine) leaves the top of the expeller and is recycled to the reactor. A mixture of water and HCl containing 20%–22% by weight of HCl leaves the base of the expeller. This liquid passes, *via* a cooler, to the top of an HCl absorber, which is required to remove almost the whole of the HCl contained in the gases leaving the cooler/scrubber. The liquid leaving the base of the HCl absorber, containing 33%–36% by weight of HCl, is divided into two streams, one of which flows to the expeller while the other is collected as product. The gaseous chlorine leaving the top of the HCl absorber passes to a drier.

Data

Reactor

Catalyst particle size distribution (US Patent 2746 844/1956):

Size range (μm)	Cumulative weight percentage undersize (at upper limit)
50–100	0.39
100–150	15.0
150–200	58.0
200–250	85.0
250–300	96.6
300–350	99.86

Density of catalyst: 40 lb/ft^3 (640 kg/m^3)
Voidage at onset of fluidization: 0.55
Particle shape factor: 0.7
Heat of reaction: 192 kcal/kg of HCl ($\Delta H = -29, 340$ kJ/kmol); see ref. 1
Gas residence time in reactor: 25 s; see ref. 3.

Cooler/Scrubber and Expeller

The overall heat-transfer coefficient between the gas and liquid phases can be taken to be 5.0 Btu/h ft^2 degF (28 W/m^2 °C).

Scope of Design Work Required

1. Prepare a mass-balance diagram for the process, up to but not including the drier, on the basis of weight per hour. Base the calculation on 10,000 long tons/year of chlorine entering the drier together with permanent gases, water and not more than 5×10^{-5} parts by weight of HCl per unit weight of chlorine.
2. Prepare an energy-balance diagram for the reactor and cooler/scrubber system.
3. Prepare a process flow diagram, up to but not including the drier, showing all the major items of equipment, with indications of the type of internal construction, as far as possible in the corrected evaluation. The diagram

should show all major pipe lines and the instrumentation of the reactor and the cooler/scrubber system.

4. Prepare an equipment schedule listing all major items of equipment and giving sizes, capacities, operating pressures and temperatures, materials of construction, etc.

5. Present a specimen pipeline sizing calculation.

6. Work out the full chemical engineering design of the reactor and cooler/scrubber systems.

7. Calculate the height and diameter of the expeller.

8. Prepare a mechanical design of the cooler/scrubber showing by dimensioned sketches suitable for submission to a draughtsman how:

   (a) The tower packing is to be supported

   (b) The liquid is to be distributed in the tower

   (c) The shell is to be constructed so as to withstand the severely corrosive conditions inside it.

9. Discuss the safety precautions involved in the operation of the plant, and the procedure to be followed in starting the plant up and shutting it down.

10. Develop the mechanical design of the reactor and prepare a key arrangement drawing, supplemented by details to make clear the essential constructional features. The study should include recommendations for the design of the bed and means of separation and disposal of dust from the exit gas stream, and should take account of needs connected with thermal expansion, inspection, maintenance, starting and stopping, inlet gas distribution, insertion and removal of catalyst, and the positioning and provision for reception of instruments required for control and operational safety. Written work should be confined, as far as possible, to notes on engineering drawings, except for the design calculations, the general specification and the justification of materials of construction.

11. Assuming that the plant throughout may vary by 10% on either side of its normal design value due to changes in demand, specify control systems for:

    (i) regulation of the necessary recycle flow from the cooler/scrubber base, at the design temperature

    (ii) transfer of the cooler/scrubber make liquor to the expeller.

REFERENCES

1. Arnold, C. W. and Kobe, K. A. (1952) *Chem. Engng Prog.* 48, 293.
2. Fleurke, K. H. (1968) *Chem. Engr., Lond.*, p. CE41.
3. Quant, J., Van Dam, J., Engel, W. F., and Wattimena, F. (1963) *Chem. Engr., Lond.*, p. CE224.
4. Sconce, J. S. (1962) *Chlorine: Its Manufacture, Properties, and Uses* (New York: Rheinhold Publishing Corporation).

F.8.   ANILINE FROM NITROBENZENE

The Project

Design a plant to make 20,000 tonnes per annum of refined aniline by the hydro-genation of nitro-benzene. The total of on-stream operation time plus regeneration periods will be 7500 hours per year.

Materials Available

Nitrobenzene containing <10 ppm thiophene
Hydrogen of 99.5% purity at a pressure of 50 psig ($350 \, kN/m^2$)
Copper on silica gel catalyst.

Services Available

Steam at 200 psig ($1400 \, kN/m^2$), 197°C and 40 psig ($280 \, kN/m^2$), 165°C
Cooling water at a maximum summer temperature of 24°C
Town's water at 15°C
Electricity at 440 V, three phase, 50 Hz.

Product Specification:

Aniline	99.9% w/w min.
Nitrobenzene	2 ppm max.
Cyclohexylamine	100 ppm max.
Water	0.05% w/w max.

The Process

Nitrobenzene is fed to a vaporizer, where it is vaporized in a stream of hydrogen (three times stoichiometric). The mixture is passed into a fluidized-bed reactor containing copper on silica gel catalyst, operated at a pressure, above the bed, of 20 psig ($140 \, kN/m^2$). The contact time, based on superficial velocity at reaction temperature and pressure and based on an unexpanded bed, is 10 s. Excess heat of reaction is removed to maintain the temperature at 270°C by a heat-transfer fluid passing through tubes in the catalyst bed. The exit gases pass through porous stainless-steel candle filters before leaving the reactor.

The reactor gases pass through a condenser/cooler, and the aniline and water are condensed. The excess hydrogen is recycled, except for a purge to maintain the impurity level in the hydrogen to not more than 5% at the reactor inlet. The crude aniline and water are let down to atmospheric pressure and separated in a liquid–liquid separator, and the crude aniline containing 0.4% unreacted nitrobenzene and 0.1% cyclohexylamine as well as water, is distilled to give refined aniline. Two stills are used, the first removing water and lower boiling material, and the second removing the higher boiling material (nitrobenzene) as a mixture with aniline. The vapour from the first column is condensed, and the liquid phases separated to give an aqueous phase and an organic phase. A purge is taken from the organic stream to remove the cyclohexylamine from the system, and the remainder of the organic stream recycled. The cyclohexylamine content of the purge is held to not greater than 3% to avoid difficulty in phase separation. In the second column, 8% of the feed is withdrawn as bottoms product.

The purge and the higher boiling mixture are processed away from the plant, and the recovered aniline returned to the crude aniline storage tank. The aniline recovery efficiency in the purge unit is 87.5%, and a continuous stream of high-purity aniline may be assumed.

The aqueous streams from the separators (amine–water) are combined and steam stripped to recover the aniline, the stripped water, containing not more than 30 ppm aniline or 20 ppm cyclohexylamine, being discharged to drain.

Regeneration of the catalyst is accomplished in place using air at 250°C –350°C to burn off organic deposits. Regeneration takes 24 h, including purging periods.

The overall yield of aniline is 98% theory from nitrobenzene, i.e. from 100 moles of nitrobenzene delivered to the plant, 98 moles of aniline passes to final product storage.

Scope of Design Work Required

1. Prepare a material balance on an hourly basis for the complete process in weight units.
2. Prepare a heat balance for the reactor system, comprising vaporizer, reactor and condenser/cooler.
3. Draw a process flow diagram for the plant. This should show all items of equipment approximately to scale and at the correct elevation. The catalyst regeneration equipment should be shown.
4. Chemical engineering design:
   (a) Vaporizer
   Give the detailed chemical engineering design, and give reasons for using the type chosen. Specify the method of control.
   (b) Reactor
   Give the detailed chemical engineering design for the fluidized-bed and heat-transfer surfaces. Select a suitable heat-transfer fluid and give reasons for your selection. Do not attempt to specify the filters or to design the condenser/cooler in detail.
   (c) Crude aniline separator
   Specify the diameter, height and weir dimensions and sketch the method of interface level control which is proposed.
   (d) Amine water stripper
   Give the detailed chemical engineering design of the column.
5. Prepare a full mechanical design for the reactor. Make a dimensioned sketch suitable for submission to a drawing office, which should include details of the distributor, and show how the heat transfer surfaces will be arranged. An

indication of the method of supporting the candle filters should be shown, but do not design this in detail.

6. Prepare an equipment schedule detailing all major items of equipment, including tanks and pumps. A specimen pipeline sizing calculation for the reactor inlet pipe should be given. All materials of construction should be specified.

7. Describe briefly how the plant would be started up and shut down, and discuss safety aspects of operation.

8. Write a short discussion, dealing particularly with the less firmly based aspects of the design, and indicating the semi-technical work which is desirable.

Data

1. Catalyst properties:
   (a) Grading

0–20 μm	Negligible
20–40 μm	3% w/w
40–60 μm	7% w/w
60–80 μm	12% w/w
80–100 μm	19% w/w
100–120 μm	25% w/w
120–140 μm	24% w/w
140–150 μm	10% w/w
>150 μm	Negligible

   (b) Voidage at minimum fluidization, 0.45
   (c) Shape factor, 0.95
   (d) Bulk density at minimum fluidization, 50 lb/ft^3 (800 kg/m^3)
   (e) Life between regenerations 1500 tonnes of aniline per ton of catalyst, using the feedstock given.

2. Exothermic heat of hydrogenation

$$-\Delta H_{298} = 132,000 \text{ CHU/lb mol } (552,000 \text{ kJ/k mol})$$

3. Mean properties of reactor gases under reactor conditions:

Viscosity	0.02 cP (0.02 mNs/m^2)
Heat capacity at constant pressure	0.66 CHU/lb°C (2.76 kJ/kg°C)
Thermal conductivity	0.086 CHU/h ft^2 (°C/ft)
	(0.15 W/m°C)

4. Pressure drop through candle filters = 5 psi (35 kN/m^2).

5. Density of nitrobenzene:

Temp. (°C)	Density (g/cm^3)
0	1.2230
15	1.2083
30	1.1934
50	1.1740

6. Latent heat of vaporization of nitrobenzene:

Temp. (°C)	Latent heat (CHU/lb)	kJ/kg
100	104	434
125	101	422
150	97	405
175	92.5	387
200	85	355
210	79	330

7. Latent heat of vaporization of aniline:

Temp. (°C)	Latent heat (CHU/lb)	kJ/kg
100	133.5	558
125	127	531
150	120	502
175	110	460
183	103.7	433

8. Specific heat of aniline vapour = 0.43 CHU/lb°C (1.80 kJ/kg°C).
9. Solubility of aniline in water:

Temp. (°C)	% w/w aniline
20	3.1
40	3.3
60	3.8
100	7.2

10. Solubility of water in aniline:

Temp. (°C)	% w/w water
20	5.0
40	5.3
60	5.8
100	8.4

11. Density of aniline/water system:

Temp. (°C)	Density (g/cm³) Water layer	Aniline layer
0	1.003	1.035
10	1.001	1.031
20	0.999	1.023
30	0.997	1.014
40	0.995	1.006
50	0.991	0.998
60	0.987	0.989
70	0.982	0.982

12. Partition of cyclohexylamine between aniline and water at 30°C:

w/w % cyclohexylamine in aniline	w/w % water in aniline	w/w % cyclohexylamine in water	w/w % aniline in water
1.0	5.7	0.12	3.2
3.0	6.6	0.36	3.2
5.0	7.7	0.57	3.2

13. Partition coefficient of nitrobenzene between aniline layer and water layer:

$$C_{a.l.}/C_{w.l} = 300$$

14. Design relative velocity in crude aniline–water separator: 10 ft/h (3 m/h).

15. Equilibrium data for water–aniline system at 760 mmHg abs:

| Temp (°C) | Mole fraction water | |
	Liquid	Vapour
184	0	0
170	0.01	0.31
160	0.02	0.485
150	0.03	0.63
140	0.045	0.74
130	0.07	0.82
120	0.10	0.88
110	0.155	0.92
105	0.20	0.94
100	0.30	0.96
99	0.35–0.95	0.964
0.985	0.9641	
0.9896	0.9642	
0.9941	0.9735	
0.9975	0.9878	
0.9988	0.9932	

16. Equilibrium data for cyclohexylamine–water system at 760 mmHg abs:

| Mole fraction cyclohexylamine | |
Liquid	Vapour
0.005	0.065
0.010	0.113
0.020	0.121
0.030	0.123
0.040	0.124
0.050	0.125
0.100	0.128
0.150	0.131
0.200	0.134
0.250	0.137

17. Temperature coefficient for aniline density: $0.054 \, \text{lb/ft}^3 \, °C (0.86 \, \text{kg/m}^3 \, °C)$ (range 0–100°C).

REFERENCES

1. US Patent 2,891,094 (American Cyanamid Co.).
2. Perry, R. H., Chilton, C. H. and Kirkpatrick, S. D. (eds) (1963) *Chemical Engineers' Handbook*, 4th edn, Section 3 (New York: McGraw-Hill Book Company, Inc.).
3. Leva, M. *Fluidization* (1959) (New York: McGraw-Hill Book Company, Inc.).
4. Rottenburg, P. A. (1957) *Trans. Instn. Chem Engrs*, 35, 21.

As an alternative to ref. 1, any of the following may be read as background information to the process:

5. *Hyd. Proc. and Pet. Ref.* (1961), 40, No. 11, p. 225.
6. Stephenson, R. M. (1966) *Introduction to the Chemical Process Industries* (New York: Reinhold Publishing Corporation).
7. Faith, W. L., Keyes, D. B. and Clark, R. L. (1965) *Industrial Chemicals*, 3rd edn (New York: John Wiley & Sons Inc.).
8. Sittig, M. (1962) *Organic Chemical Processes* (New York: Noyes Press).

G EQUIPMENT SPECIFICATION (DATA) SHEETS

All of the worksheets and specification sheets listed below are available in the on-line material at http://elsevierdirect.com/companions.

(1) Design basis data sheet
(2) Calculation sheet
(3) Problem table algorithm sheet
(4) Pump and line calculation sheet
(5) Cost of production calculation sheet
(6) Economic analysis calculation sheet
(7) Failure mode effect analysis sheet
(8) Fixed-bed reactor data sheet
(9) Fluid-phase splitter data sheet
(10) Fired-heater data sheet
(11) Shell and tube heat-exchanger data sheet
(12) Vapour–liquid contacting column data sheet
(13) Effluent summary data sheet

Company Name		Project Name							
Address		Project Number				Sheet		1	
		REV	DATE	BY	APVD	REV	DATE	BY	APVD
DESIGN BASIS									
Form XXXXX-YY-ZZ									

1 General Information

Owners Name	
Process Unit Name	
Plant Location	
Correspondance Contacts Address Telephone / Fax E-mail	

2 Measurement System

◉ English   ○ Metric

3 Equipment Numbering System

Equipment will be identified by alphabetic prefix as defined here, followed by three-digit serial number unless otherwise indicated

First digit - process section
Second & third digits - equipment count

AC	Air cooler	G	Grinder, mill	PRV	Pressure relief valve
B	Boiler	H	Heater (fired or electric)	R	Reactor
C	Compressor, blower, fan	J	Ejector, jet, turboexpander	SP	Sample point
CT	Cooling tower	M	Motor	T	Storage tank
D	Dryer	ME	Miscellaneous equipment	V	Vessel (including columns)
E	Exchanger	MX	Mixer		
F	Filter, classifier	P	Pump		

4 Primary Products

Product Name Product Grade				
MSDS Form Number				

Production Rate
    Tons per year
    Tons per day
    Other units				

Product Purity (wt%)				
Product shipment mode				
Additional Specifications				

5 Primary Raw Materials
(Attach additional sheets if needed)

Feedstock name Feedstock grade								
MSDS form number								

Feedstock availability
    Tons per year
    Tons per day
    Other units								

Feedstock price ($/lb) (Default: open market price)								
Known feedstock impurities	Name	ppmw	Name	ppmw	Name	ppmw	Name	ppmw
Additional specifications								

6 Site Information

Low ambient temperature (F) High ambient temperature (F) Hight ambient relative humidity (%) Site elevation (ft) Maximum wind loading (mph) Other site design requirements	

Company Name		Project Name						
Address		Project Number				Sheet		2
		REV	DATE	BY	APVD	REV	DATE	BY	APVD
DESIGN BASIS								
Form XXXXX-YY-ZZ								

7 Utility Information

Fuel Gas

Gas source or operation mode	Nat Gas			
Supply header temperature (F)				
Supply header pressure (psia)				
Net calorific value (BTU/lb)				
Marginal availability (lb/h)				
Marginal fuel cost ($/MMBTU)				
Sulfur content (wppm)				
Nitrogen content (wppm)				
Chlorine content (wppm)				
Gas composition (vol%)				
H_2O				
O_2				
N_2				
CO				
CO_2				
H_2S				
H_2				
CH_4				
C_2H_4				
C_2H_6				
C_3H_6				
C_3H_8				
C_4H_8				
iC_4H_{10}				
nC_4H_{10}				
C_5H_{10}				
C5+				

Fuel Oil

Fuel source or operation mode	#2 Heating Oil			
Supply header temperature (F)				
Supply headder pressure (psia)				
Net calorific value (BTU/lb)				
Marginal availability (lb/h)				
Marginal fuel cost ($/MMBTU)				
Fuel viscosity at F				
Fuel viscosity at F				
Flash point (F)				
Pour point (F)				
Sulfur content (wppm)				
Nitrogen content (wppm)				
Ash content (wt %)				

Steam

Steam header classification	VHP	HP	MP	LP
Operating pressure (psia)				
Operating temperature (F)				
Mechanical design pressure (psia)				
Mechanical design temperature (F)				
Marginal availability (lb/h)				
Marginal cost ($/Mlb)				

Coolants

Coolant classification	Cooling Tower Water	Once-Through Water	Chilled Water	
Operating pressure (psia)				
Supply temperature (F)				
Maximum return temperature (F)				
Marginal availability (lb/h)				
Marginal cost ($)				
Marginal cost units				

Process Water Feeds

Water feed stream	Raw Water	Process Water	Boiler Feed Water	Condensate
Supply pressure (psia)				
Supply temperature (F)				
Marginal availability (lb/h)				
Marginal cost ($/1000 gal)				
Total dissolved solids (wt%)				
Hardness as $CaCO_3$ (ppmw)				
Chloride as Cl (ppmw)				
Metallurgy				

Electric power

Power range (kW)				
Voltage (V)				
Phase				
Frequency (Hz)				
Marginal availability (kW)				
Marginal cost ($/kWh)				

Plant air streams

Air stream	Plant Air	Instrument Air	Plant Nitrogen	
Header pressure (psia)				
Header temperature (F)				
Moisture (ppmw)				
Marginal availability (lb/h)				
Marginal cost ($/Mscf)				

	Project Name						
Company Name	Project Number			Sheet		1 of 1	
Address	REV	DATE	BY	APVD	REV	DATE	BY	APVD

CALCULATION SHEET

Form XXXXX-YY-ZZ

Company Name	Project Name							
Address	Project Number				Sheet		1 of 1	
	REV	DATE	BY	APVD	REV	DATE	BY	APVD
PROBLEM TABLE ALGORITHM								
Form XXXXX-YY-ZZ								

1. Minimum temperature approach

ΔT_{min} 0 °C

2. Stream data

Stream No.	Actual temperature (°C)		Interval temperature (°C)		Heat capacity flow rate CP (kW/°C)	Heat load (kW)
	Source	Target	Source	Target		
1	0	0	0	0	0	0
2	0	0	0	0	0	0
3	0	0	0	0	0	0
4	0	0	0	0	0	0
5	0	0	0	0	0	0
6	0	0	0	0	0	0
7	0	0	0	0	0	0
8	0	0	0	0	0	0

3. Problem table

Interval	Interval temp (°C)	Interval ΔT (°C)	Sum CPc - sum CPh (kW/°C)	dH (kW)	Cascade (kW)	(kW)
	0				0	0
1	0	0	0	0	0	0
2	0	0	0	0	0	0
3	0	0	0	0	0	0
4	0	0	0	0	0	0
5	0	0	0	0	0	0
6	0	0	0	0	0	0
7	0	0	0	0	0	0
8						

Company Name Address		Project Name							
		Project Number				Sheet	1	of	1
		REV	DATE	BY	APVD	REV	DATE	BY	APVD
Pump and Line Calculation Sheet									
Form XXXXX-YY-ZZ									

Owner's Name			
Plant Location			
Case Description			
Equipment label		Equipment name	
Plant section			
Process service			
Fluid		Density	kg/m^3
Operating temperature	Normal °C	Viscosity	N.s/m^2
	Min °C	Normal flow rate	kg/s
	Max °C	Design flow rate	kg/s

LINE PRESSURE DROP

SUCTION					DISCHARGE				
Line size			mm		Line size			mm	
Note		Normal	Max.	Units	Note	Flow	Normal	Max.	Units
u_1	Velocity	#DIV/0!	#DIV/0!	m/s	u_2	Velocity	#DIV/0!	#DIV/0!	m/s
Δf_1	Friction loss		0.00	kPa/m	Δf_2	Friction loss		0.00	kPa/m
L_1	Line length			m	L_2	Line length			m
$\Delta f_1 L_1$	Line loss		0.00	kPa	$\Delta f_2 L_2$	Line loss	0.00	0.00	kPa
$\rho u_1^2/2$	Entrance loss	#DIV/0!	#DIV/0!	kPa		Orifice / Flow meter			kPa
(40 kPa)	Strainer			kPa		Control valve			kPa
	(1) Sub-total	#DIV/0!	#DIV/0!	kPa		Equipment			kPa
					S&THX	H 205			kPa
z_1	Static head			m					kPa
$\rho g z_1$		0.0	0.0	kPa					kPa
	Upstream equipment pressure			kPa	Total	(6) Dynamic loss	0	0	kPa
	(2) Sub-total	0.0	0.0	kPa					
					z_2	Static head			m
(2) − (1)	(3) Suction pressure	#DIV/0!		kPa	$\rho g z_2$		0.0	0.0	kPa
	(4) Vapor pressure			kPa		Equip. press (max)			kPa
(3) − (4)	(5) NPSH available	#DIV/0!	0.0	kPa		Contingency	0	0	kPa
(5)/ρg	NPSH available	#DIV/0!	#DIV/0!	m		(7) Sub-total	0.0	0.0	kPa
	NPSH available	#DIV/0!	0.0	m water	(7) + (6)	Discharge pressure	0.0	0.0	kPa
					(3)	Suction pressure	#DIV/0!	0.0	kPa
						(8) Differential pressure	#DIV/0!	0.0	kPa
					(8)/ρg	Pump head	#DIV/0!	#DIV/0!	m
						Control valve			
					Valve/(6)	% Dyn. loss		#DIV/0!	

PUMP DATA

Pump manufacturer				Driver type	
Catalog No.				Power supply	
Pump flow rate	normal	#DIV/0!	m^3/h	Seal type	
	max.	#DIV/0!	m^3/h	Hydraulic power	#DIV/0! kW
Differential pressure		#DIV/0!	kPa	Rated power	kW
		#DIV/0!	m	Efficiency	%
		#DIV/0!	m water	Suction specific speed	
NPSH required			m		
Pump type				Casing design pressure	kPa
No. of stages				Casing design temperature	°C
Impeller type				Casing type	
Mounting				Casing material	

SKETCH

NOTES

1.
2.
3.
4.
5.

Company Name	Project Name								Sheet	1
Address	Project Number									
COST OF PRODUCTION	REV	DATE	BY	APVD	REV	DATE	BY	APVD		
Form XXXXX-YY-ZZ										

Owner's Name		Capital Cost Basis Year	2006	
Plant Location		Units	○ English ● Metric	
Case Description		On Stream	8,000 hr/yr	333.33 day/yr

YIELD ESTIMATE   **CAPITAL COSTS**

	$MM
ISBL Capital Cost	0.000
OSBL Capital Cost	0.000
Engineering Costs	0.000
Contingency	0.000
Total Fixed Capital Cost	0.000
Working Capital	0.000

REVENUES AND RAW MATERIAL COSTS

MASS BALANCE   MB Closure   100%

Key Products	Units	Units/Unit product	Units/yr	Price $/unit	$MM/yr	$/unit main product
Product 1	MT	0	1	0	0.00	0.00
	MT				0.00	0.00
	MT				0.00	0.00
	MT				0.00	0.00
Total Key Product Revenues (REV)	MT	0	1		0.00	0.00

By-products & Waste Streams						
Byproduct 1	MT	0	0	0	0.00	0.00
Byproduct 2	MT	0	0	0	0.00	0.00
Byproduct 3	MT	0	0	0	0.00	0.00
Byproduct 4	MT	0	0	0	0.00	0.00
Off-gas	MT	0	0	0	0.00	0.00
Organic Waste	MT	0	0	0	0.00	0.00
Aqueous Waste	MT	0	0	0	0.00	0.00
	MT				0.00	0.00
	MT				0.00	0.00
Total Byproducts and Wastes (BP)	MT	0	0		0.00	0.00

Raw Materials						
Feed 1	MT	0	1	0	0.00	0.00
Feed 2	MT	0	0	0	0.00	0.00
Feed 3	MT	0	0	0	0.00	0.00
Feed 4	MT	0	0	0	0.00	0.00
	MT				0.00	0.00
	MT				0.00	0.00
	MT				0.00	0.00
Total Raw Materials (RM)		0	1		0.00	0.00

Gross Margin (GM = REV + BP - RM)   0.00   0.00

CONSUMABLES

	Units	Units/Unit product	Units/yr	Price $/unit	$MM/yr	$/unit product
Solvent 1	kg	0	0	0.00	0.00	0.00
Solvent 2	kg	0	0	0.00	0.00	0.00
Solvent 3	kg	0	0	0.00	0.00	0.00
Acid 1	kg	0	0	0.00	0.00	0.00
Acid 2	kg	0	0	0.00	0.00	0.00
Base 1	kg	0	0	0.00	0.00	0.00
Base 2	kg	0	0	0.00	0.00	0.00
Other	kg	0	0	0.00	0.00	0.00
Other	kg	0	0	0.00	0.00	0.00
Other	kg	0	0	0.00	0.00	0.00
Other	kg	0	0	0.00	0.00	0.00
Total Consumables (CONS)					0.00	0.00

UTILITIES

	Units	Units/Unit product	Units/hr	Price $/unit	$MM/yr	$/unit product
Electric	kWh	0.0	0	0.00	0.000	0.00
HP Steam	MT	0.0	0	0.00	0.000	0.00
MP Steam	MT	0.0	0	0.00	0.000	0.00
LP Steam	MT	0.0	0	0.00	0.000	0.00
Boiler Feed	MT	0.0	0	0.00	0.000	0.00
Condensate	MT	0.0	0	0.00	0.000	0.00
Cooling Water	MT	0.0	0	0.00	0.000	0.00
Fuel Fired	GJ	0.0	0	0.00	0.000	0.000
Total Utilities (UTS)					0.00	0.00

Variable Cost of Production (VCOP = RM - BP + CONS + UTS)   0.00   0.00

FIXED OPERATING COSTS

			$MM/yr	$/unit product
Labor				
Number of shift positions	3	4.8 Operators per Shift Position		
		0 $/yr each	0.00	0.00
Supervision		25% of Operating Labor	0.00	0.00
Direct Ovhd.		45% of Labor & Superv.	0.00	0.00
Maintenance		3% of ISBL Investment	0.00	0.00
Overhead Expense				
Plant Overhead		65% of Labor & Maint.	0.00	0.00
Tax & Insurance		2% of Fixed Investment	0.00	0.00
Interest on Debt Financing		0% of Fixed Capital	0.00	0.00
		0% of Working Capital	0.00	0.00
		Fixed Cost of Production (FCOP)	0.00	0.00

ANNUALIZED CAPITAL CHARGES

	$MM	Interest Rate	Life (yr)	ACCR	$MM/yr	$/unit product
Fixed Capital Investment	0.000	15%	15	0.171	0.00	0.00
Royalty Amortization	0.000	15%	10	0.199	0.00	0.00
Inventory Amortization						
Catalyst 1	0.000	15%	3	0.438	0.00	0.00
Catalyst 2	0.000	15%	3	0.438	0.00	0.00
Adsorbent 1	0.000	15%	3	0.438	0.00	0.00
Equipment 1	0.000	15%	5	0.298	0.00	0.00
Equipment 2	0.000	15%	5	0.298	0.00	0.00
Total Annual Capital Charge					0.00	0.00

SUMMARY

	$MM/yr	$/unit product
Variable Cost of Production	0.00	0.00
Fixed Cost of Production	0.00	0.00
Cash Cost of Production	0.00	0.00
Gross Profit	0.00	0.00
Total Cost of Production	0.00	0.00

Company Name		Project Name								
Address		Project Number						Sheet		1
		REV	DATE	BY	APVD	REV	DATE	BY	APVD	
ECONOMIC ANALYSIS										
Form XXXXX-YY-ZZ										

Owner's Name		Capital Cost Basis Year	2006	
Plant Location		Units	◯ English ◉ Metric	
Case Description		On Stream	8,000 hr/yr	333.33 day/yr

REVENUES AND PRODUCTION COSTS		CAPITAL COSTS		CONSTRUCTION SCHEDULE				
	$MM/yr		$MM	Year	% FC	% WC	% FCOP	% VCOP
Main product revenue	0.0	ISBL Capital Cost	0.0	1				
Byproduct revenue		OSBL Capital Cost	0.0	2				
Raw materials cost		Engineering Costs	0.0	3				
Utilities cost		Contingency	0.0	4				
Consumables cost		Total Fixed Capital Cost	0.0	5				
VCOP	0.0			6				
Salary and overheads		Working Capital	0.0	7+				
Maintenance								
Interest								
Royalties								
FCOP	0.0							

ECONOMIC ASSUMPTIONS

Cost of equity		Debt ratio		Tax rate	
Cost of debt				Depreciation method	
Cost of capital				Depreciation period	years

CASH FLOW ANALYSIS

All figures in $MM unless indicated

Project year	Cap Ex	Revenue	CCOP	Gr. Profit	Deprcn	Taxbl Inc	Tax Paid	Cash Flow	PV of CF	NPV
1	0.0	0.0	0.0	0.0	0.0	0.0	0.0	0.0	0.0	0.0
2	0.0	0.0	0.0	0.0	0.0	0.0	0.0	0.0	0.0	0.0
3	0.0	0.0	0.0	0.0	0.0	0.0	0.0	0.0	0.0	0.0
4	0.0	0.0	0.0	0.0	0.0	0.0	0.0	0.0	0.0	0.0
5	0.0	0.0	0.0	0.0	0.0	0.0	0.0	0.0	0.0	0.0
6	0.0	0.0	0.0	0.0	0.0	0.0	0.0	0.0	0.0	0.0
7	0.0	0.0	0.0	0.0	0.0	0.0	0.0	0.0	0.0	0.0
8	0.0	0.0	0.0	0.0	0.0	0.0	0.0	0.0	0.0	0.0
9	0.0	0.0	0.0	0.0	0.0	0.0	0.0	0.0	0.0	0.0
10	0.0	0.0	0.0	0.0	0.0	0.0	0.0	0.0	0.0	0.0
11	0.0	0.0	0.0	0.0	0.0	0.0	0.0	0.0	0.0	0.0
12	0.0	0.0	0.0	0.0	0.0	0.0	0.0	0.0	0.0	0.0
13	0.0	0.0	0.0	0.0	0.0	0.0	0.0	0.0	0.0	0.0
14	0.0	0.0	0.0	0.0	0.0	0.0	0.0	0.0	0.0	0.0
15	0.0	0.0	0.0	0.0	0.0	0.0	0.0	0.0	0.0	0.0
16	0.0	0.0	0.0	0.0	0.0	0.0	0.0	0.0	0.0	0.0
17	0.0	0.0	0.0	0.0	0.0	0.0	0.0	0.0	0.0	0.0
18	0.0	0.0	0.0	0.0	0.0	0.0	0.0	0.0	0.0	0.0
19	0.0	0.0	0.0	0.0	0.0	0.0	0.0	0.0	0.0	0.0
20	0.0	0.0	0.0	0.0	0.0	0.0	0.0	0.0	0.0	0.0

ECONOMIC ANALYSIS

Average cash flow	0.0 $MM/yr	NPV	10 years	0.0 $MM	IRR	10 years	#NUM!
Simple pay-back period	#DIV/0! yrs		15 years	0.0 $MM		15 years	#NUM!
Return on investment (10 yrs)	#DIV/0!		20 years	0.0 $MM		20 years	#NUM!
Return on investment (15 yrs)	#DIV/0!	NPV to yr	1	0.0 $MM			

NOTES

1.
2.
3.

Form 1: Failure Mode Effect Analysis

Company Name		Project Name							
Address		Project Number					Sheet		1
		REV	DATE	BY	APVD	REV	DATE	BY	APVD

FAILURE MODE EFFECT ANALYSIS

Form XXXXX-YY-ZZ

Owner's Name
Plant Location
Case Description

Process Step or Key Input	Failure Mode	Consequence	Impact SEV	Cause	Frequency O	Current Control	Detection DET	Priority RPN	Action
								0	
								0	
								0	
								0	
								0	
								0	
								0	
								0	
								0	
								0	
								0	
								0	
								0	
								0	
								0	
								0	
								0	
								0	

Form 2: Fixed Bed Reactor

Company Name		Project Name								
Address		Project Number				Sheet	1	of		1
		REV	DATE	BY	APVD	REV	DATE	BY	APVD	

FIXED BED REACTOR

Form XXXXX-YY-ZZ

Owner's Name
Plant Location
Case Description   Units   ○ English   ● Metric
Equipment label   Equipment name
Plant section
Process service
Design code   Maximum diameter   m   Total height   m

PROCESS DATA

Bed Number							
Vapor flow direction							
Liquid flow direction							
Top of section	Operating temperature	°C					
	Pressure	bara					
	Vap flow	kg/h					
	Vap density	kg/m3					
	Vap dynamic viscosity	N.s/m2					
	Liq flow	kg/h					
	Liq density	kg/m3					
	Liq dynamic viscosity	N.s/m2					
Bottom of section	Operating temperature	°C					
	Pressure	bara					
	Vap flow	kg/h					
	Vap density	kg/m3					
	Vap dynamic viscosity	N.s/m2					
	Liq flow	kg/h					
	Liq density	kg/m3					
	Liq dynamic viscosity	N.s/m2					
Catalyst type							
Catalyst shape							
Catalyst nominal diameter		mm					
Catalyst bed depth		m					
Catalyst volume		m3					
Catalyst bed bulk density		kg/m3					
Catalyst bed void fraction							
LHSV		/h					
GHSV		/h					
WHSV		/h					

CONSTRUCTION & MATERIALS

Bed Number							
Shell material							
Shell diameter	m						
Shell tangent length	m						
Shell thickness	mm						
Design temperature	°C						
Design pressure	bara						
Test pressure	bara						
Segregation height	overhead	m					
	sump	m					
Pipe branch nominal diameter	mm						
Pipe branch elevation wrt. base of section	m						
Pipe branch nominal diameter	mm						
Pipe branch elevation wrt. base of section	m						
Support grid material							
Support grid elevation wrt. base of section	m						
Distributor type							
Distributor material							
Distributor base elevation wrt. base of section	m						
Demister type							
Demister material							
Demister support grid elevation wrt. base of section	m						
Notes							

Company Name					Project Name							
Address					Project Number			Sheet	1	of	1	
					REV	DATE	BY	APVD	REV	DATE	BY	APVD
FLUID PHASE SPLITTER												
Form XXXXX-YY-ZZ												

Owner's Name				
Plant Location			Units	● English ○ Metric
Case Description				
Equipment label		Equipment name		
Plant section				
Process service				
Design code		Orientation Vertical ▼ Volume		ft3

PROCESS DATA

		IN	VAPOR OUT	ORGANIC OUT	AQUEOUS OUT
Stream No.					
Fluid					
Total fluid flow	lb/h				
Total vapor flow	lb/h				
Total liquid flow	lb/h				
Density	lb/cu ft				
Dynamic viscosity	lbm/ft.s				
Specific heat capacity	Btu/lb.F				
Latent heat	Btu/lb				
Normal temperature	°F				
Max temperature	°F				
Min temperature	°F				
Pressure	psia				
Pressure drop allowed	psi				
Pressure drop calculated	psi				
Pipe branch nominal diameter	in				
Flow velocity (nozzle)	ft/s				
Flow velocity (interface)	ft/s				

Composition	Component	wt%	wt%	wt%	wt%
	Component	mol%	mol%	mol%	mol%

CONSTRUCTION & MATERIALS

Shell
Material
Head type
Length   ft   I.D.   in   Wall thickness   in
Aspect ratio
Design pressure at max temp   psia   Test pressure   psia   Min internal pressure   psia
Baffle material   Baffle type   Baffle pitch   in
Demister material   Demister type   Demister elevatn wrt. base   in
Packing material   Packing type   Packing dimension   in
Interfacial area   ft2   Vapor / Liquid   Liquid / Liquid   ft2

Branches

Inlet 1	Side	Head	Nominal bore	in	CL elevatn wrt. base	in	
Inlet 2	Side	Head	Nominal bore	in	CL elevatn wrt. base	in	
Inlet 3	Side	Head	Nominal bore	in	CL elevatn wrt. base	in	
Outlet 1	Side	Head	Nominal bore	in	CL elevatn wrt. base	in	
Outlet 2	Side	Head	Nominal bore	in	CL elevatn wrt. base	in	
Outlet 3	Side	Head	Nominal bore	in	CL elevatn wrt. base	in	

NOTES

1.
2.
3.
4.
5.

Company Name			Project Name								
Address			Project Number				Sheet	1	of		1
			REV	DATE	BY	APVD	REV	DATE	BY		APVD
FURNACE HEATER											
Form XXXXX-YY-ZZ											

Owner's Name								
Plant Location				Units	◉ English	○ Metric		
Case Description								
Equipment label	Equipment name							
Plant section								
Process service								
Design code	Furnace type							
Furnace Duty	MMBtu/h	No. of Zones						

PROCESS DATA

Zone Name/Number								
Zone Type								
Process Inlet	Stream No.							
	Fluid							
	Total fluid flow	lb/h						
	Total vapor flow	lb/h						
	Total liquid flow	lb/h						
	Fluid vaporized	lb/h						
	Density	lb/cu ft						
	Dynamic viscosity	lbm/ft.s						
	Specific heat capacity	Btu/lb.F						
	Thermal conductivity	Btu.ft/h.ft2.F						
	Latent heat	Btu/lb						
	Normal temperature	°F						
	Max temperature	°F						
	Min temperature	°F						
	Pressure	psia						
	Flow velocity	ft/s						
Process Outlet	Stream No.							
	Fluid							
	Total fluid flow	lb/h						
	Total vapor flow	lb/h						
	Total liquid flow	lb/h						
	Fluid vaporized	lb/h						
	Density	lb/cu ft						
	Dynamic viscosity	lbm/ft.s						
	Specific heat capacity	Btu/lb.F						
	Thermal conductivity	Btu.ft/h.ft2.F						
	Latent heat	Btu/lb						
	Normal temperature	°F						
	Max temperature	°F						
	Min temperature	°F						
	Pressure	psia						
	Flow velocity	ft/s						
Pressure drop allowed		psi						
Pressure drop calculated		psi						
Number of passes								
Process film transfer coefficient		Btu/h.ft2.F						
Fouling coefficient		Btu/h.ft2.F						
Process heat load		Btu/lb						
Heat duty		Btu/h						
Estimated heat loss to surroundings		Btu/h						
Allowed heat flux		Btu/h.ft2						
Fuel	Fuel type							
	Fuel flowrate	lb/h						
	Fuel temperature	°F						
	Fuel pressure	psia						
Air	Air flowrate	lb/h						
	Air temperature	°F						
	Air pressure	psia						
	Air preheat duty	Btu/h						
	Excess air	%						
Flue gas	FG flowrate	lb/h						
	FG temperature in	°F						
	FG temperature out	°F						
	FG moisture	wt%						
	FG dewpoint	°F						
	FG film transfer coeff	Btu/h.ft2.F						
	NOx	ppmw						
	CO	ppmw						
	PM10	ppmw						
Theoretical flame temperature		°F						
Bridgewall temperature		°F						
Max allowable turndown		%						

CONSTRUCTION & MATERIALS

Zone Number								
Zone Type								
Tubes	Material							
	Orientation							
	Count							
	No. of passes							
	Length/pass	ft						
	O.D.	in						
	Total area	ft2						
	Wall thickness	in						
	Design temperature	°F						
	Design pressure	psia						
	Test pressure	psia						
Refractory	Material							
	Sidewall thickness	in						
	Hearth & roof thickness	in						
Casing & Ducting	Material							
	Wall thickness	in						
	Fuel pipe diameter	in						
	Stack diameter	in						
	Stack height	ft						
Burners	Burner type							
	Burner material							
	No. of burners							
	Burner spacing	ft						
	Flame speed	ft/s						
	Flame length	ft						

NOTES

1.
2.
3.
4.
5.

Company Name				Project Name							
Address				Project Number				Sheet	1	of	1
				REV	DATE	BY	APVD	REV	DATE	BY	APVD
SHELL & TUBE HEAT EXCHANGER											
Form XXXXX-YY-ZZ											

Owner's Name						
Plant Location						
Case Description				Units	⦿ English	◯ Metric
Equipment label		Equipment name				
Plant section						
Process service						
Design code	TEMA	Exchanger type				
Shells per unit		Series		Parallel		
Surface per unit	ft2	Surface per shell		ft2		

DATA PER UNIT

		SHELL SIDE		TUBE SIDE	
		IN	OUT	IN	OUT
Stream No.					
Fluid					
Total fluid flow	lb/h				
Total vapor flow	lb/h				
Total liquid flow	lb/h				
Total steam flow	lb/h				
Fluid vaporized / condensed	lb/h				
Density	lb/cu ft				
Dynamic viscosity	lbm/ft.s				
Specific heat capacity	Btu/lb.F				
Thermal conductivity	Btu.ft/h.ft2.F				
Latent heat	Btu/lb				
Normal temperature	°F				
Max temperature	°F				
Min temperature	°F				
Pressure	psia				
Pressure drop allowed	psi				
Pressure drop calculated	psi				
Flow velocity	ft/s				
Number of passes					
Film transfer coefficient	Btu/h.ft2.F				
Fouling coefficient	Btu/h.ft2.F				
Heat duty	Btu/h				
F_T factor					
Effective mean temperature difference	°F				
Minimum surface required	ft2				

CONSTRUCTION & MATERIALS PER SHELL

Tubes
Material

Count		Pitch		in	Square	Triangular	Welded	☑
Length	ft	O.D.		in	Wall thickness		in	
Design pressure at max temp	psia	Test pressure		psia	Max external pressure		psia	
Number of tubes blanked								

Shell
Material

Length	ft	I.D.		in	Wall thickness		in
Design pressure at max temp	psia	Test pressure		psia	Min internal pressure		psia
Baffle material		Baffle type			Baffle pitch		in
Tubesheet material					Tubesheet thickness		in
Bonnet material					Bonnet type		

Branches

Shell side inlet	in N.B.	Shell side outlet		in N.B.
Tube side inlet	in N.B.	Tube side outlet		in N.B.

NOTES

1.
2.
3.
4.
5.

Company Name			Project Name							
Address			Project Number			Sheet	1	of	1	
			REV	DATE	BY	APVD	REV	DATE	BY	APVD
VAPOR LIQUID CONTACTING COLUMN										
Form XXXXX-YY-ZZ										

Owner's Name				
Plant Location		Units	○ English	⦿ Metric
Case Description				
Equipment label	Equipment name			
Plant section				
Process service				
Design code	Maximum diameter	m	Total height	m

PROCESS DATA

Column Section							
Tray or Stage Numbers							
Total vapor flow	kg/h						
Total liquid flow	kg/h						
Top of section	Operating temperature	°C					
	Pressure	bara					
	Vap density	kg/m3					
	Vap dynamic viscosity	N.s/m2					
	Liq density	kg/m3					
	Liq dynamic viscosity	N.s/m2					
	Liq surface tension	dyn/cm					
Bottom of section	Operating temperature	°C					
	Pressure	bara					
	Vap density	kg/m3					
	Vap dynamic viscosity	N.s/m2					
	Liq density	kg/m3					
	Liq dynamic viscosity	N.s/m2					
	Liq surface tension	dyn/cm					
Section pressure drop		bar					

CONSTRUCTION & MATERIALS

Column Section							
Tray or Stage Numbers							
Shell material							
Shell diameter	m						
Shell tangent length	m						
Shell thickness	mm						
Design temperature	°C						
Design pressure	bara						
Test pressure	bara						
Segregation height	overhead	m					
	sump	m					
Pipe branch nominal diameter	mm						
Pipe branch elevation wrt. base of section	m						
Pipe branch nominal diameter	mm						
Pipe branch elevation wrt. base of section	m						
Tray type							
Tray material							
No. trays							
No. liquid passes per tray							
No. holes /tray	total						
	active						
Hole size	mm						
Hole pitch	mm						
Hole area / total area	%						
Tray thickness	mm						
Tray spacing	m						
Weir length	m						
Weir height	mm						
Downcomer clearance	mm						
Pressure drop per plate	bar						
Packing type							
Packing material							
Packing size	mm						
Packing height	m						
Support grid material							
Support grid elevation wrt. base of section	m						
Distributor type							
Distributor material							
Distributor base elevation wrt. base of section	m						
Demister type							
Demister material							
Demister support grid elevation wrt. base of section	m						
Notes							

Company Name			Project Name							
Address			Project Number					Sheet	1	
			REV	DATE	BY	APVD	REV	DATE	BY	APVD
EFFLUENT SUMMARY										
Form: XXXXX-YY-ZZ										

Owner's Name		
Plant Location		Units ○ English ◉ Metric
Case Description		

PROCESS EMISSIONS

Vapor Emissions

Pollutant	Process Source (Stream No. if avail.)	Measurement (estimate) method	Continuous / Intermittent	kg/day	kg/yr	Regulatory Status
Nitrogen Oxides						
Total						
Sulfur Oxides						
Total						
Particulate matter						
Total						
Volatile organic compounds						
Total						
HAPs (list by name)						
Total						

Aqueous Waste Streams

Stream Name	Process Source (Stream No. if avail.)	Water flow kg/day	Contaminant	Contaminent flow kg/day	Contaminent flow metric ton/yr	Concentration (wt%)

Organic Waste Streams

Stream Name	Process Source (Stream No. if avail.)	Measurement (estimate) method	Component	kg/day	kg/yr

Solid Waste Streams

Stream Name	Process Source (Stream No. if avail.)	Measurement (estimate) method	Component	kg/day	kg/yr

UTILITY AND OFFSITES EMISSIONS

Vapor Emissions

Pollutant	Process Source (Stream No. if avail.)	Measurement (estimate) method	Continuous / Intermittent	kg/day	kg/yr	Regulatory Status
Nitrogen Oxides						
Total						
Sulfur Oxides						
Total						
Particulate matter						
Total						
Volatile organic compounds						
Total						

Aqueous Waste Streams

Stream Name	Process Source (Stream No. if avail.)	Water flow kg/day	Contaminant	Contaminent flow kg/day	Contaminent flow metric ton/yr	Concentration (wt%)
Boiler blowdown			TDS			
			Corrosion inhibitor			
Cooling water blowdown			TDS			
			Fouling inhibitor			
			Chloride			

Solid Waste Streams

Stream Name	Process Source (Stream No. if avail.)	Measurement (estimate) method	Component	kg/day	kg/yr

H TYPICAL SHELL AND TUBE HEAT-EXCHANGER TUBE-SHEET LAYOUTS

(a) Fixed tube-sheet exchanger
(b) U-tube exchanger
(c) Floating-head exchanger with split backing ring
(d) Pull through floating-head exchanger

Reproduced with permission from *Heat Exchanger Design*, E. A. D. Saunders (Longman Group).

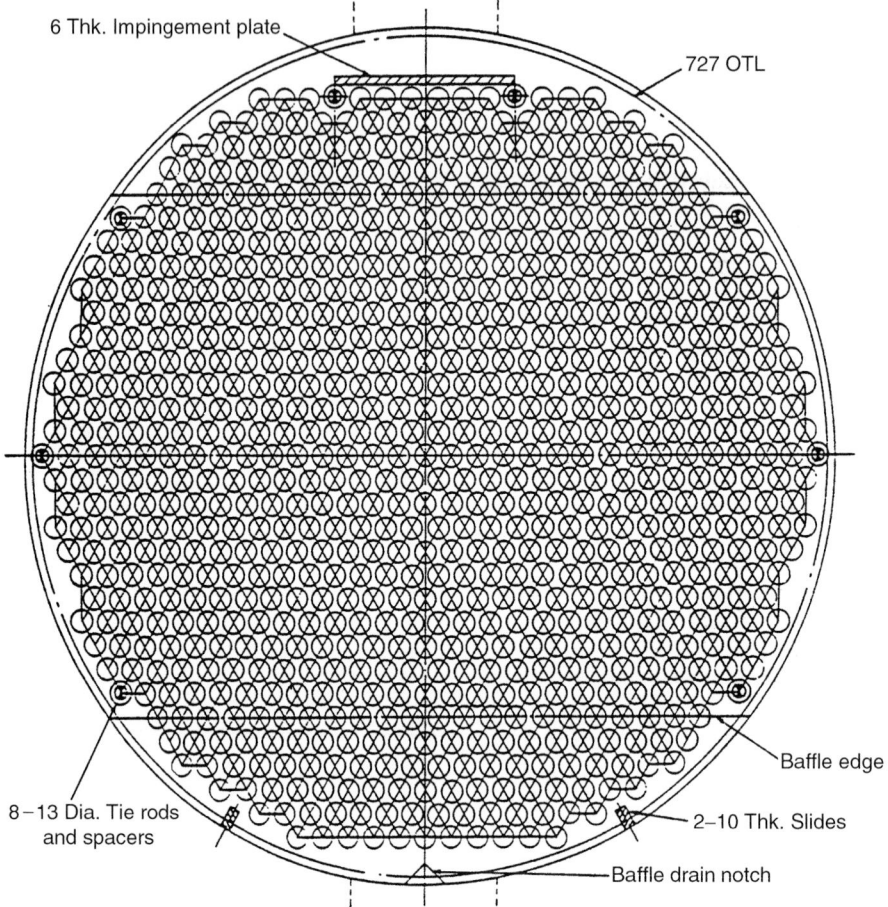

6 Thk. Impingement plate

727 OTL

8–13 Dia. Tie rods and spacers

2–10 Thk. Slides

Baffle edge

Baffle drain notch

(a) Typical tube layout for a fixed tube-sheet exchanger 740 i/Dia. shell, single-pass, 780 tubes, 19.05 o/Dia. on 23.8125 pitch, 30° angle.

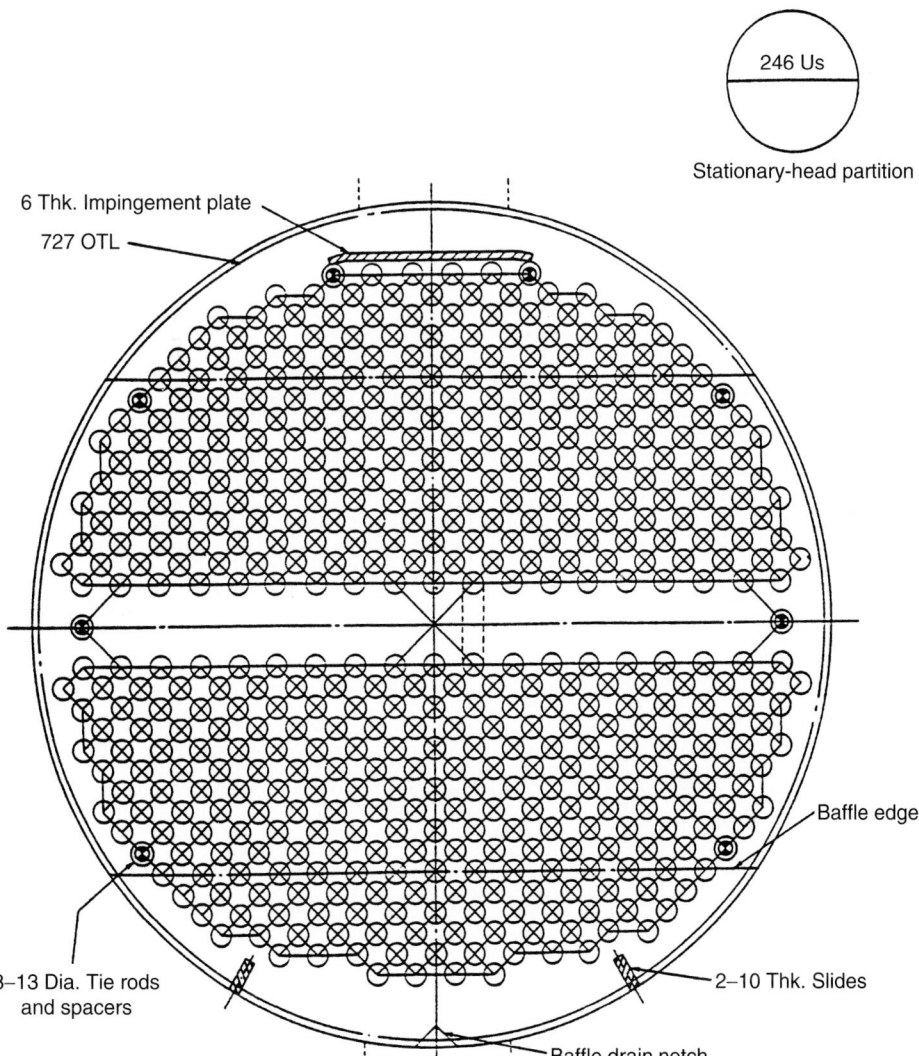

(b) Typical tube layout for a U-tube exchanger 740 i/Dia. shell, two-pass, 246 U-tubes, 19.05 o/Dia. on 25.4 pitch, 45° angle.

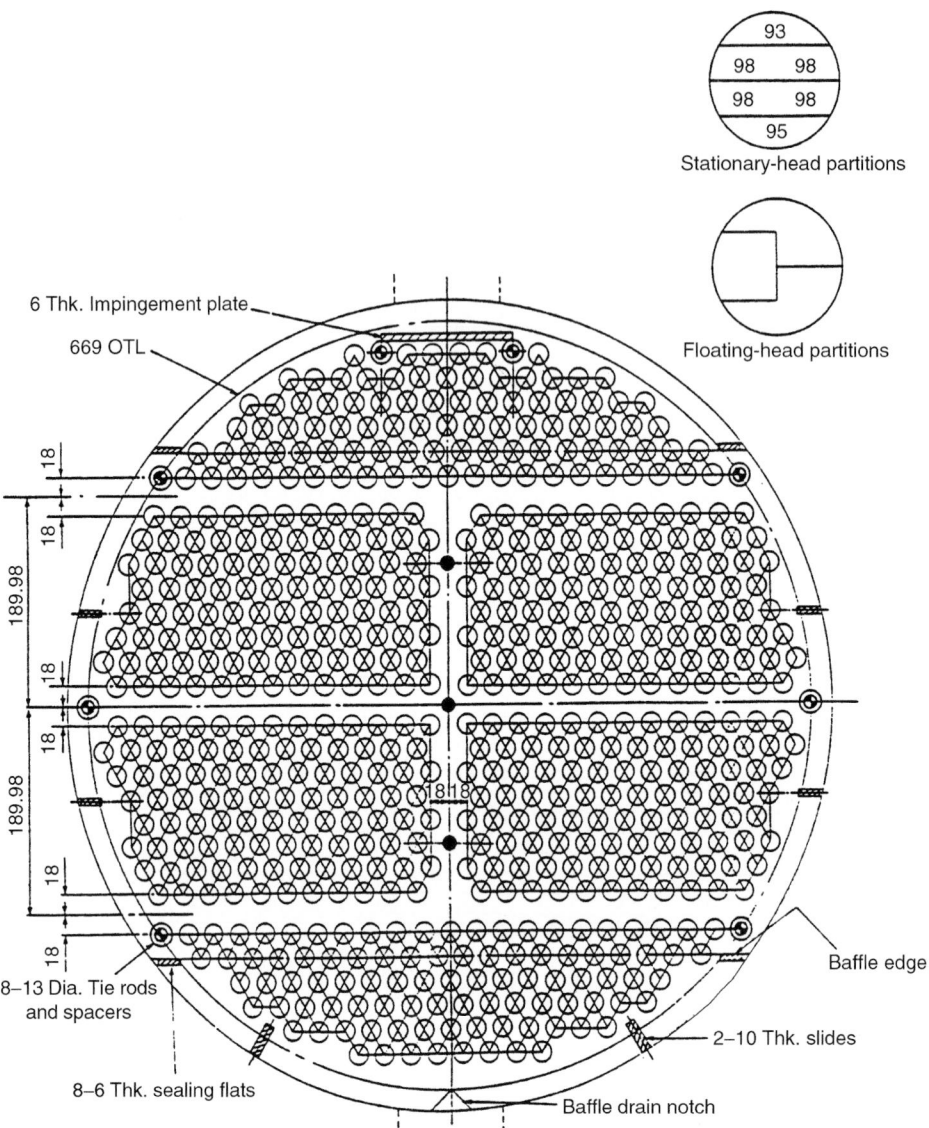

Stationary-head partitions

Floating-head partitions

6 Thk. Impingement plate

669 OTL

189.98

189.98

18

18

18

18

18

8–13 Dia. Tie rods and spacers

8–6 Thk. sealing flats

Baffle edge

2–10 Thk. slides

Baffle drain notch

(c) Typical tube layout for a split backing ring floating-head exchanger 740 i/Dia. shell, six-pass, 580 tubes, 19.05 o/Dia. on 25.4 pitch, 30° angle.

• Denotes 13 Dia. sealing bars.

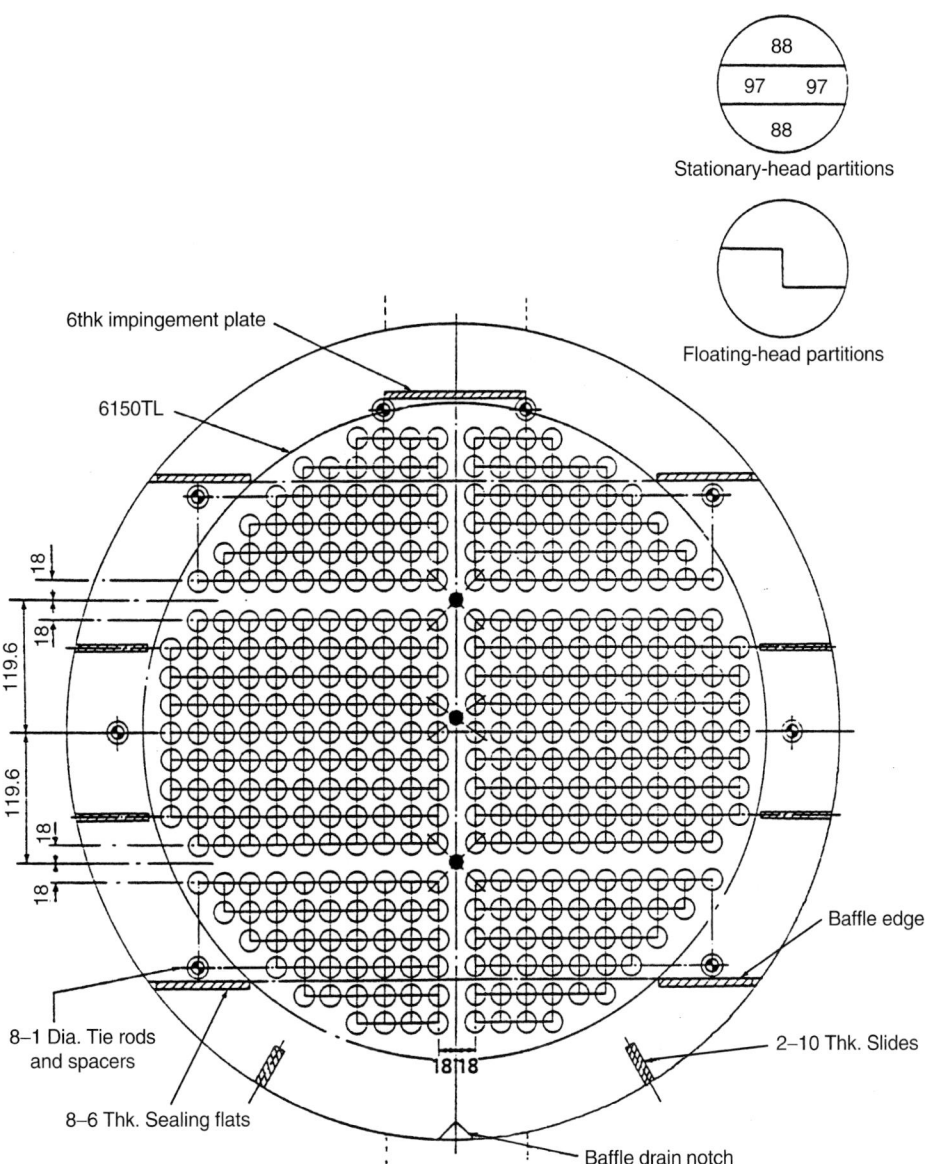

(d) Typical tube layout for a pull-through floating-head exchanger. 740 i/Dia. shell, four-pass, 370 tubes, 19.05 o/Dia. on 25.4 pitch, 90° angle.

• Denotes 13 Dia. sealing bars.

I MATERIAL SAFETY DATA SHEET

Reproduced with permission of Fischer Acros Inc.

1,2-DICHLOROETHANE, EXTRA DRY, WATER <50 PPM

ACC# 00220

SECTION 1 – CHEMICAL PRODUCT AND COMPANY IDENTIFICATION

MSDS Name: 1,2-Dichloroethane, extra dry, water <50 ppm
Catalog Numbers: AC326840000, AC326840010, AC326841000, AC326842500
Synonyms: Ethylene dichloride; 1,2-Ethylene dichloride; Glycol dichloride; EDC; sym-Dichloroethane; 1,2-Dichloroethane; Ethylene chloride
Company Identification:
  Acros Organics N.V.
  One Reagent Lane
  Fair Lawn, NJ 07410
For information in North America, call: 800-ACROS-01
For emergencies in the US, call CHEMTREC: 800-424-9300

SECTION 2 – COMPOSITION, INFORMATION ON INGREDIENTS

CAS#	Chemical Name	Percent	EINECS/ELINCS
107-06-2	1,2-Dichloroethane	>99.9	203-458-1

Hazard Symbols: T F
Risk Phrases: 11 22 36/37/38 45

SECTION 3 – HAZARDS IDENTIFICATION

EMERGENCY OVERVIEW

Appearance: colorless liquid. Flash Point: 56 deg F. **Warning! Flammable liquid and vapor.** May cause central nervous system depression. May cause liver and kidney damage. May cause cancer based on animal studies. Causes eye and skin irritation. Causes respiratory tract irritation. Irritant. May be harmful if swallowed.

Target Organs: Central nervous system, liver, eyes, skin.

Potential Health Effects

Eye: Causes eye irritation. Vapors may cause eye irritation. May cause chemical conjunctivitis and corneal damage.

Skin: Causes skin irritation. May be absorbed through the skin. May cause irritation and dermatitis. May cause cyanosis of the extremities.

Ingestion: May cause central nervous system depression, kidney damage, and liver damage. May cause gastrointestinal irritation with nausea, vomiting, and diarrhea. May cause effects similar to those for inhalation exposure. May be harmful if swallowed.

Inhalation: Inhalation of high concentrations may cause central nervous system effects characterized by nausea, headache, dizziness, unconsciousness, and coma. Causes respiratory tract irritation. May cause liver and kidney damage. Aspiration may lead to pulmonary edema. Vapors may cause dizziness or suffocation. Can produce delayed pulmonary edema. Exposure to high concentrations may produce narcosis, nausea, and loss of consciousness. May cause burning sensation in the chest.

Chronic: Possible cancer hazard based on tests with laboratory animals. Prolonged or repeated skin contact may cause dermatitis. Prolonged or repeated eye contact may cause conjunctivitis. May cause liver and kidney damage. Effects may be delayed.

SECTION 4 – FIRST AID MEASURES

Eyes: Immediately flush eyes with plenty of water for at least 15 minutes, occasionally lifting the upper and lower eyelids. Get medical aid.

Skin: Get medical aid. Flush skin with plenty of water for at least 15 minutes while removing contaminated clothing and shoes. Wash clothing before reuse.

Ingestion: Never give anything by mouth to an unconscious person. Get medical aid. Do NOT induce vomiting. If conscious and alert, rinse mouth and drink 2–4 cupfuls of milk or water.

Inhalation: Remove from exposure and move to fresh air immediately. If not breathing, give artificial respiration. If breathing is difficult, give oxygen. Get medical aid. Do NOT use mouth-to-mouth resuscitation.

Notes to Physician: Treat symptomatically and supportively.

SECTION 5 – FIRE FIGHTING MEASURES

General Information: As in any fire, wear a self-contained breathing apparatus in pressure-demand, MSHA/NIOSH (approved or equivalent), and full protective gear. Vapors may form an explosive mixture with air. During a fire, irritating and highly toxic gases may be generated by thermal decomposition or combustion. Use water spray to keep fire-exposed containers cool. Flammable liquid and vapor. Approach fire from upwind to avoid hazardous vapors and toxic decomposition products. Vapors are heavier than air and may travel to a source of ignition and flash back. Vapors can spread along the ground and collect in low or confined areas.

Extinguishing Media: For small fires, use dry chemical, carbon dioxide, water spray, or alcohol-resistant foam. For large fires, use water spray, fog, or alcohol-resistant foam. Water may be ineffective. Do NOT use straight streams of water.
Flash Point: 56 deg F (13.33 deg C)
Autoignition Temperature: 775 deg F (412.78 deg C)
Explosion Limits, Lower: 6.2%
Upper: 15.9%
NFPA Rating: (estimated) Health: 2; Flammability: 3; Instability: 0

SECTION 6 – ACCIDENTAL RELEASE MEASURES

General Information: Use proper personal protective equipment as indicated in Section 8.
Spills/Leaks: Absorb spill with inert material (e.g. vermiculite, sand or earth), then place in suitable container. Avoid runoff into storm sewers and ditches which lead to waterways. Clean up spills immediately, observing precautions in the Protective Equipment section. Remove all sources of ignition. Use a spark-proof tool. Provide ventilation. A vapor suppressing foam may be used to reduce vapors.

SECTION 7 – HANDLING AND STORAGE

Handling: Wash thoroughly after handling. Remove contaminated clothing and wash before reuse. Ground and bond containers when transferring material. Use spark-proof tools and explosion proof equipment. Avoid contact with eyes, skin, and clothing. Empty containers retain product residue (liquid and/or vapor), and can be dangerous. Keep container tightly closed. Do not pressurize, cut, weld, braze, solder, drill, grind, or expose empty containers to heat, sparks, or open flames. Use only with adequate ventilation. Keep away from heat, sparks, and flame. Avoid breathing vapor or mist.
Storage: Keep away from heat, sparks, and flame. Keep away from sources of ignition. Store in a tightly closed container. Keep from contact with oxidizing materials. Store in a cool, dry, well-ventilated area away from incompatible substances. Flammables-area. Storage under a nitrogen blanket has been recommended.

SECTION 8 – EXPOSURE CONTROLS, PERSONAL PROTECTION

Engineering Controls: Facilities storing or utilizing this material should be equipped with an eyewash facility and a safety shower. Use adequate general or local explosion-proof ventilation to keep airborne levels to acceptable levels.

Exposure Limits

Chemical Name	ACGIH	NIOSH	OSHA - Final PELs
1,2-Dichloroethane	10 ppm TWA	1 ppm TWA; 4 mg/m^3 TWA; 50 ppm IDLH	50 ppm TWA; 100 ppm Ceiling

OSHA Vacated PELs: 1,2-Dichloroethane: 1 ppm TWA; 4 mg/m^3 TWA
Personal Protective Equipment
Eyes: Wear chemical goggles.
Skin: Wear appropriate protective gloves to prevent skin exposure.
Clothing: Wear appropriate protective clothing to prevent skin exposure.
Respirators: A respiratory protection program that meets OSHA's 29 CFR 1910.134 and ANSI Z88.2 requirements or European Standard EN 149 must be followed whenever workplace conditions warrant a respirator's use.

SECTION 9 – PHYSICAL AND CHEMICAL PROPERTIES

Physical State: Liquid
Appearance: colorless
Odor: chloroform-like
pH: Not available
Vapor Pressure: 100 mm Hg @29 deg C
Vapor Density: 3.4 (Air=1)
Evaporation Rate: 6.5 (Butyl acetate=1)
Viscosity: Not available
Boiling Point: 81–85 deg C
Freezing/Melting Point: −35 deg C
Decomposition Temperature: Not available
Solubility: Insoluble
Specific Gravity/Density:1.25 (Water=1)
Molecular Formula: C2H4Cl2
Molecular Weight: 98.96

SECTION 10 – STABILITY AND REACTIVITY

Chemical Stability: Stable at room temperature in closed containers under normal storage and handling conditions.
Conditions to Avoid: Light, ignition sources, excess heat, electrical sparks.
Incompatibilities with Other Materials: Aluminum, bases, alkali metals, ketones, organic peroxides, nitric acid, strong oxidizing agents, strong reducing agents, liquid ammonia, amines.
Hazardous Decomposition Products: Hydrogen chloride, phosgene, carbon monoxide, irritating and toxic fumes and gases, carbon dioxide.
Hazardous Polymerization: Has not been reported.

SECTION 11 – TOXICOLOGICAL INFORMATION

RTECS#:
CAS# 107-06-2: KI0525000
LD50/LC50:

CAS# 107-06-2:
Draize test, rabbit, eye: 63 mg Severe;
Draize test, rabbit, eye: 500 mg/24H Mild;
Draize test, rabbit, skin: 500 mg/24H Mild;
Inhalation, rat: LC50 = 1000 ppm/7H;
Oral, mouse: LD50 = 413 mg/kg;
Oral, rabbit: LD50 = 860 mg/kg;
Oral, rat: LD50 = 670 mg/kg;
Skin, rabbit: LD50 = 2800 mg/kg; <BR.

Carcinogenicity:

CAS# 107-06-2:
ACGIH: A4 – Not Classifiable as a Human Carcinogen.
California: carcinogen; initial date 10/1/87.
NIOSH: potential occupational carcinogen.
NTP: Suspect carcinogen.
OSHA: Possible Select carcinogen.
IARC: Group 2B carcinogen.
Epidemiology: IARC Group 2B: Proven animal carcinogenic substance of potential relevance to humans. IARC Group 2B: No data available on human carcinogenicity, however sufficient evidence of carcinogenicity in animals.
Teratogenicity: See actual entry in RTECS for complete information.
Reproductive Effects: No information found.
Neurotoxicity: No information found.
Mutagenicity: No information found.
Other Studies: See actual entry in RTECS for complete information.

SECTION 12 – ECOLOGICAL INFORMATION

Ecotoxicity: Water flea Daphnia: 218 mg/L; 48H; Bluegill/Sunfish: 430 mg/L; 96H; Static Fathead Minnow: 136 mg/L; 96H; Static – No data available.
Environmental: Terrestrial: Smaller releases on land will evaporate fairly rapidly. Larger releases may leach rapidly through sandy soil into groundwater. Aquatic: If released to surface water, its primary loss will be by evaporation. The half-life for evaporation will depend on wind and mixing conditions and was of the order of hours in the laboratory. However a modeling study using the EXAMS model for a eutrophic lake gave a half-life of 10 days. Atmospheric: Will degrade by reaction with hydroxyl radicals formed photochemically in the atmosphere. Half-life over one month.
Physical: Not expected to biodegrade or bioconcentrate.
Other: For more information, see *HANDBOOK OF ENVIRONMENTAL FATE AND EXPOSURE DATA.*

SECTION 13 – DISPOSAL CONSIDERATIONS

Chemical waste generators must determine whether a discarded chemical is classified as a hazardous waste. US EPA guidelines for the classification determination are listed in 40 CFR Parts 261.3. Additionally, waste generators must consult state and local hazardous waste regulations to ensure complete and accurate classification.

RCRA P-Series: None listed.
RCRA U-Series: CAS# 107-06-2: waste number U077.

SECTION 14 – TRANSPORT INFORMATION

	US DOT	IATA	RID/ADR	IMO	Canada TDG
Shipping Name:	ETHYLENE DICHLORIDE				No information available
Hazard Class:	3				
UN Number:	UN1184				
Packing Group:	II				

SECTION 15 – REGULATORY INFORMATION

US FEDERAL

TSCA
CAS# 107-06-2 is listed on the TSCA inventory.
Health & Safety Reporting List
CAS# 107-06-2: Effective Date: 6/1/87; Sunset Date: 6/1/97.
Chemical Test Rules
None of the chemicals in this product are under a Chemical Test Rule.
Section 12b
None of the chemicals are listed under TSCA Section 12b.
TSCA Significant New Use Rule
None of the chemicals in this material have a SNUR under TSCA.
SARA
CERCLA Hazardous Substances and corresponding RQs
CAS# 107-06-2: 100 lb final RQ; 45.4 kg final RQ.
SARA Section 302 Extremely Hazardous Substances
None of the chemicals in this product have a TPQ.
SARA Codes
CAS # 107-06-2: acute, chronic, flammable.
Section 313
This material contains 1,2-Dichloroethane (CAS# 107-06-2, 99 9%),which is subject to the reporting requirements of Section 313 of SARA Title III and 40 CFR Part 373.
Clean Air Act:
CAS# 107-06-2 is listed as a hazardous air pollutant (HAP). This material does not contain any Class 1 Ozone depletors. This material does not contain any Class 2 Ozone depletors.

Clean Water Act:
CAS# 107-06-2 is listed as a Hazardous Substance under the CWA. CAS# 107-06-2 is listed as a Priority Pollutant under the Clean Water Act. CAS# 107-06-2 is listed as a Toxic Pollutant under the Clean Water Act.

OSHA:
None of the chemicals in this product are considered highly hazardous by OSHA.

STATE
CAS# 107-06-2 can be found on the following state right to know lists: California, New Jersey, Pennsylvania, Minnesota, Massachusetts.

The following statement(s) is(are) made in order to comply with the California Safe Drinking Water Act: WARNING: This product contains 1,2-Dichloroethane, a chemical known to the state of California to cause cancer. California No Significant Risk Level: CAS# 107-06-2: 10 μg/day NSRL.

European/International Regulations
European Labeling in Accordance with EC Directives
Hazard Symbols:
T F

Risk Phrases:
R 11 Highly flammable.
R 22 Harmful if swallowed.
R 36/37/38 Irritating to eyes, respiratory system and skin.
R 45 May cause cancer.

Safety Phrases:
S 45 In case of accident or if you feel unwell, seek medical advice immediately (show the label where possible).
S 53 Avoid exposure – obtain special instructions before use.

WGK (Water Danger/Protection)
CAS# 107-06-2: 3

Canada – DSL/NDSL
CAS# 107-06-2 is listed on Canada's DSL List.

Canada – WHMIS
This product has a WHMIS classification of B2, D2A, D2B.

Canadian Ingredient Disclosure List
CAS# 107-06-2 is listed on the Canadian Ingredient Disclosure List.

Exposure Limits
CAS# 107-06-2: OEL-ARAB Republic of Egypt: TWA 5 ppm (2 mg/m^3) OEL-AUSTRALIA:TWA 10 ppm (40 mg/m^3) OEL-AUSTRIA:TWA 20 ppm (80 mg/m^3) OEL-BELGIUM:TWA 10 ppm (40 mg/m^3) OEL-DENMARK:TWA 1 ppm (4 mg/m^3); Skin OEL-FINLAND:TWA 10 ppm (40 mg/m^3);STEL 20 ppm (80 mg/m^3); CAR OEL-FRANCE:TWA 10 ppm (40 mg/m^3) OEL-GERMANY; Carcinogen OEL-HUNGARY: STEL 4 mg/m^3; Carcinogen OEL-JAPAN: TWA 10 ppm (40 mg/m^3) OEL-THE NETHERLANDS: TWA 50 ppm (200 mg/m^3) OEL-THE PHILIPPINES: TWA 50 ppm (200 mg/m^3) OEL-RUSSIA: TWA 10 ppm OEL-SWEDEN: TWA 1 ppm (4 mg/m^3);STEL 5 ppm (20 mg/m^3); Skin; CAR OEL-SWITZERLAND: TWA 10 ppm (40 mg/m^3);STEL 20 ppm (80 mg/m^3) OEL-TURKEY: TWA 50 ppm (200 mg/m^3) OEL-UNITED KINGDOM: TWA 10 ppm (40 mg/m^3); STEL 15 ppm (60 mg/m^3) OEL IN BULGARIA, COLOMBIA, JORDAN, KOREA check

ACGIH TLV OEL IN NEW ZEALAND, SINGAPORE, VIETNAM check A CGI TLV

SECTION 16 – ADDITIONAL INFORMATION

MSDS Creation Date: 10/19/1998
Revision #6 Date: 4/17/2002

The information above is believed to be accurate and represents the best information currently available to us. However, we make no warranty of merchantability or any other warranty, express or implied, with respect to such information, and we assume no liability resulting from its use. Users should make their own investigations to determine the suitability of the information for their particular purposes. In no event shall Fisher be liable for any claims, losses, or damages of any third party or for lost profits or any special, indirect, incidental, consequential or exemplary damages, howsoever arising, even if Fisher has been advised of the possibility of such damages.

Index

Note: Figures are indicated by *italic page numbers*, Tables by **bold numbers**